SPIN 2000

Related Titles from the AIP Conference Proceedings Subseries on High Energy Physics

564 Quantum Electrodynamics and Physics of the Vacuum: QED 2000, Second Workshop
Edited by Giovanni Cantatore, May 2001, 0-7354-0000-8

562 Particles and Fields: Ninth Mexican School
Edited by Lukas Nellen and Gerardo Herrera Corral, April 2001, 1-56396-998-X

555 Cosmology and Particle Physics: CAPP 2000
Edited by Ruth Durrer, Juan Garcia-Bellido, and Mikhail Shaposhnikov, March 2001, 1-56396-986-6

549 Intersections of Particle and Nuclear Physics: 7th Conference, CIPANP2000
Edited by Zohreh Parsa and William J. Marciano, December 2000, 1-56396-978-5

541 Theoretical High Energy Physics: MRST 2000
Edited by C. R. Hagen, November 2000, 1-56396-966-1

539 Symmetries in Subatomic Physics: 3rd International Symposium
Edited by X.-H. Guo, A. W. Thomas, and A. G. Williams, October 2000, 1-56396-964-5

531 Particles and Fields: Seventh Mexican Workshop
Edited by Alejandro Ayala, Guillermo Contreras, and Gerardo Herrera, July 2000, 1-56396-954-8

512 Nuclear Physics at Storage Rings: Fourth International Conference: STORI99
Edited by Hans-Otto Meyer and Peter Schwandt, June 2000, 1-56396-928-9

508 Hadron Physics: Effective Theories of Low Energy QCD
Edited by A. H. Blin, B. Hiller, M. C. Ruivo, C. A. Sousa, and E. van Beveren, March 2000, 1-56396-927-0

494 New Directions in Quantum Chromodynamics
Edited by Chueng-Ryong Ji and Dong-Pil Min, November 1999, 1-56396-908-4

459 Heavy Quarks at Fixed Target
Edited by Harry W. K. Cheung and Joel N. Butler, February 1999, 1-56396-864-9

412 Intersections Between Particle and Nuclear Physics: 6th Conference
Edited by T. W. Donnelly, December 1997, 1-56396-712-X

To learn more about these titles, or the AIP Conference Proceedings Series, please visit the webpage **http://www.aip.org/catalog/aboutconf.html**

Editors:

Kichiji Hatanaka, Takashi Nakano, and Hiroyasu Ejiri

Research Center for Nuclear Physics
Osaka University
10-1 Mihogaoka, Ibaraki
Osaka 567-0047
JAPAN

E-mail: hatanaka@rcnp.osaka-u.ac.jp
       nakano@rcnp.osaka-u.ac.jp
       ejiri@rcnp.osaka-u.ac.jp

Kenichi Imai
Department of Physics
Kyoto University
Kitashirakawa-Oiwake
Kyoto 606-8502
JAPAN

E-mail: imai@ne.scphys.kyoto-u.ac.jp

The articles on pp. 741–745 and 935–942 were authored by U.S. Government employees and are not covered by the below mentioned copyright.

Authorization to photocopy items for internal or personal use, beyond the free copying permitted under the 1978 U.S. Copyright Law (see statement below), is granted by the American Institute of Physics for users registered with the Copyright Clearance Center (CCC) Transactional Reporting Service, provided that the base fee of $18.00 per copy is paid directly to CCC, 222 Rosewood Drive, Danvers, MA 01923. For those organizations that have been granted a photocopy license by CCC, a separate system of payment has been arranged. The fee code for users of the Transactional Reporting Service is: 0-7354-0008-3/01/$18.00.

© 2001 American Institute of Physics

Individual readers of this volume and nonprofit libraries, acting for them, are permitted to make fair use of the material in it, such as copying an article for use in teaching or research. Permission is granted to quote from this volume in scientific work with the customary acknowledgment of the source. To reprint a figure, table, or other excerpt requires the consent of one of the original authors and notification to AIP. Republication or systematic or multiple reproduction of any material in this volume is permitted only under license from AIP. Address inquiries to Office of Rights and Permissions, Suite 1NO1, 2 Huntington Quadrangle, Melville, N.Y. 11747-4502; phone: 516-576-2268; fax: 516-576-2450; e-mail: rights@aip.org.

L.C. Catalog Card No. 2001089793
ISBN 0-7354-0008-3
ISSN 0094-243X
Printed in the United States of America

SPIN 2000

14th International Spin Physics Symposium

Osaka, Japan   16–21 October 2000

EDITORS
Kichiji Hatanaka
Takashi Nakano
Osaka University, Japan

Kenichi Imai
Kyoto University, Japan

Hiroyasu Ejiri
Osaka University, Japan

Melville, New York, 2001
AIP CONFERENCE PROCEEDINGS ■ VOLUME 570

CONTENTS

Preface xxiii
Committees xxv
Photos xxvii
Opening Address xxxi
    H. Ejiri
Welcome Address xxxiii
    Y. Yamaguchi
Spin Physics 2000 xxxv
    C. Y. Prescott

PLENARY SESSIONS

SPIN: Progress and Prospects 3
    R. L. Jaffe
HERMES at the Turn of the Millennium 24
    E. C. Aschenauer for the HERMES Collaboration
Highlights from the SMC and COMPASS Experiments 34
    F. Bradamante
The Muon g-2 Experiment at Brookhaven 46
    G. Bunce for the Muon g-2 Collaboration
Weak Nucleon Form Factors 59
    P. A. Souder
Polarization in Photodisintegration of the Deuteron 69
    R. Gilman for the Jefferson Lab Hall A Collaboration
Nucleon Electromagnetic Form Factors and Proton Charge Radius 79
    H. Gao
The β-ν Correlation Using a Neutral Atom Trap 89
    P. A. Vetter, S. J. Freedman, B. K. Fujikawa, and N. D. Scielzo
Fundamental Symmetry and Polarized Muons 99
    Y. Kuno
Nuclear Moment Studies with Polarized Radioactive Nuclear Beams 109
    K. Asahi, K. Sakai, H. Ogawa, H. Ueno, Y. Kobayashi, A. Yoshimi, H. Miyoshi,
    K. Yogo, A. Goto, T. Suga, H. Imai, Y. X. Watanabe, K. Yoneda, N. Fukuda,
    N. Aoi, W.-D. Schmidt-Ott, G. Neyens, S. Teughels, A. Yoshida, T. Kubo,
    and M. Ishihara
High Energy Hadron Spin Observables 119
    K.-I. Kubo
Spin Physics in Hypernuclei and Hyperon-Nucleon Interactions 128
    T. Kishimoto
Nucleon Spin Structure Functions 138
    W. Vogelsang
Physics with the First Polarized-Proton Collider RHIC 154
    N. Saito

Spin Dynamics in LEP with 40–100 GeV Beams 169
    R. Assmann, J. Badier, A. Blondel, M. Böge, M. Crozon, B. Dehning, H. Grote, J. P. Koutchouk, M. Placidi, R. Schmidt, F. Sonnemann, F. Tecker, and J. Wenninger

Polarized Ion Sources for High-Energy Accelerators and Colliders 179
    A. N. Zelenski

Polarized Photon Beam Experiments at SPring-8 189
    T. Nakano for the LEPS Collaboration

Experiments with Polarized Photon Beam at GRAAL 198
    J. Ajaka, M. Anghinolfi, Y. Assafiri, O. Bartalini, M. Battaglieri, V. Bellini, J. P. Bocquet, P. Calvat, M. Capogni, M. Castoldi, P. Corvisiero, D. D'Angelo, J. P. Didelez, R. DiSalvo, M. A. Duval, E. Guinault, L. Fichen, C. Gaulard, G. Gervino, F. Ghio, B. Girolami, M. Guidal, E. Hourany, I. Kilvington, V. Kouznetsov, R. Kunne, A. Lapik, P. Levi Sandri, A. Lleres, D. Moricciani, V. Nedorezov, L. Nicoletti, D. Rebreyend, F. Renard, M. Ripani, N. Rudnev, M. Sanzone, C. Schaerf, M. Taiuti, A. Turinge, Q. Zhao, and A. Zucchiatti

Three-Nucleon Spin Observables: Signatures for Three-Nucleon Force Effects 208
    H. Witała, W Glöckle, H. Kamada, A. Nogga, J. Golak, J. Kuroś-Żolnierczuk, and R. Skibiński

Spin-Isospin Responses in Nuclei via Polarization Measurements 218
    H. Sakai

Nuclear Medium Effect Studied by Nucleon Quasifree Scattering 228
    T. Noro

Nonrelativistic Bound States in Quantum Field Theory 238
    A. V. Manohar and I. W. Stewart

Effective Theory for Heavy Quarkonium Production 251
    J. Lee

Recent Developments in the Field of Polarized Solid State Target Materials 261
    S. Goertz, J. Harmsen, J. Heckmann, A. Meier, W. Meyer, E. Radtke, and G. Reicherz

Workshop Report on Spin Polarized Electron Source and Polarimeter 274
    T. Nakanishi

Report on the Workshop "Polarized Sources and Targets" Erlangen 1999 286
    E. Steffens

Prospects of High Energy Polarized ep Colliders 296
    V. W. Hughes

Symposium Highlights 312
    T. Roser

BANQUET SPEECH

Some Milestones in Polarized Proton Beams 323
    E. D. Courant

PARALLEL SESSIONS

1. SYMMETRIES AND SPIN I

Constraints of a Parity-Conserving/Time-Reversal-Non-Conserving Interaction 331
    W. T. H. van Oers

Neutrinos by Double Beta Decays from ^{100}Mo and Nuclear Spin-Isospin Responses 338
    N. Kudomi, H. Ejiri, K. Fushimi, K. Hayashi, T. Kishimoto, M. Komori, K. Kume, H. Kuramoto, K. Matsuoka, H. Ohsumi, K. Takahisa, S. Umehara, and S. Yoshida

Search for Spin Coupled WIMPs with the Large Volume NaI(Tl) Scintillators 343
    S. Yoshida, H. Ejiri, K. Fushimi, K. Hayashi, T. Kishimoto, N. Kudomi, K. Kume, H. Kuramoto, K. Matsuoka, H. Ohsumi, K. Takahisa, Y. Tsujimoto, and S. Umehara

Hyperon Beta-Decay Analysis and the Recent KTeV Data 348
    P. G. Ratcliffe

Novel Spin Maser Mechanism Studied for High-Precision Measurement of Neutron Electric Dipole Moment 353
    A. Yoshimi, K. Asahi, K. Yogo, K. Sakai, H. Ogawa, T. Suzuki, and M. Nagakura

2. SYMMETRIES AND SPIN II

Symmetry Tests in Polarized Z^0 Decays to $b\bar{b}g$ 361
    T. Maruyama for the SLD Collaboration

Polarized Muon Decay: Measurement of the Polarization Vector of the Decay Positrons 366
    K. Bodek, A. Budzanowski, N. Danneberg, W. Fetscher, C. Hilbes, L. Jarczyk, K. Kirch, S. Kistryn, J. Klement, K. Köhler, A. Kozela, J. Lang, G. L. Llácer, M. Markiewicz, X. Morelle, T. Schweizer, J. Smyrski, J. Sromicki, E. Stephan, A. Strzałkowski, and J. Zejma

Measuring the Michel Parameter ξ'' in Polarized Muon Decay 371
    X. Morelle, N. Danneberg, J. Deutsch, J. Egger, W. Fetscher, F. Foroughi, J. Govaerts, C. Hilbes, K. Kirch, P. Knowles, K. Köhler, A. Kozela, J. Lang, Y. W. Liu, R. Medve, O. Naviliat, A. Ninane, R. Prieels, and P. van Hove

Physics Beyond SM at RHIC with Polarized Protons 379
    A. Ogawa, V. L. Rykov, and N. Saito

3. SPIN STRUCTURE OF NUCLEONS

Measurements of the Spin Structure Function g_1 of the Proton and the Deuteron 387
    U. Stösslein for the HERMES Collaboration

Measurement of the Polarised Quark Distributions in the Nucleon
at HERMES 392
    T. Lindemann for the HERMES Collaboration
Polarized Structure Functions of the Deuteron 397
    S. Kumano
Extraction of g_1 in the Resonance Region 402
    R. Fatemi
Preliminary Results for the Spin Structure Function g_2 from
SLAC E155X 407
    R. Prepost for the E155X Collaboration
Neutron Spin Structure and the Extended GDH Sum Rule at Low Q^2 412
    T. Averett for the Jefferson Lab E94-010 Collaboration
Single-Spin Azimuthal Asymmetries in Semi-Inclusive Electro-Production
of Pions at HERMES 417
    P. Di Nezza for the HERMES Collaboration
Estimation of the Proton Transversity from Azimuthal Asymmetries
in DIS 422
    A. V. Efremov
Spin Physics with the HERMES Experiment Using Hadron Identification
by the Ring Imaging Cherenkov Counter 423
    Y. Sakemi for the HERMES Collaboration
Deeply Virtual Compton Scattering at HERMES 428
    M. Amarian for the HERMES Collaboration
Polarized Gluon Distributions from High-p_T Pair Hadron Productions
in Polarized Deep Inelastic Scattering 433
    T. Yamanishi, D. Yu-Bing, and T. Morii
The Shape and Experimental Tests of the Q^2-Invariant Polarized
Gluon Asymmetry 437
    G. P. Ramsey
Measurement of the Gluon Polarization in the Proton at PHENIX 442
    Y. Goto for the PHENIX Collaboration
Extracting $\Delta G(x)$ from the $\vec{p}+\vec{p}\to\gamma+\text{jet}+X$ Reaction with the
STAR Detector at RHIC 447
    S. W. Wissink for the STAR Collaboration
$\vec{p}\vec{p}\to W^{\pm}X$ Asymmetries with STAR at RHIC 452
    S. E. Vigdor for the STAR Collaboration
Future Transversity Measurements at RHIC 457
    M. Grosse Perdekamp and A. Ogawa
Positivity Constraints in Spin Physics 461
    J. Soffer
Twist-2 Polarized Fragmentation Function in the Open Charm Production
in DIS 468
    Y. I. Arestov
Polarized Gluon Distribution Function of Nucleon in Diffractive
Lepto-Production of Charmonium 472
    A. Hayashigaki and K. Suzuki
Determination of Polarized Parton Distribution Functions 477
    M. Hirai, H. Kobayashi, and M. Miyama

Solving the Nucleon Spin Puzzle Based on the Chiral Quark Soliton Model.. 482
    M. Wakamatsu

4. SPIN PHYSICS AND HADRONS

Measurement of Λ^0 Polarization in ν_μ CC Interactions in NOMAD 489
    D. V. Naumov for the NOMAD Collaboration
Hyperon Polarization in Inclusive Hadronic Production.................. 494
    Y. Kanazawa and Y. Koike
Polarization of Hyperons in Photon Induced Reaction at High Energy 499
    K. Suzuki, N. Nakajima, H. Toki, and K.-I. Kubo
Longitudinal and Transverse Lambda Polarization at Hermes 504
    S. Bernreuther for the HERMES Collaboration
Spin Transfer in High Energy Fragmentation Processes.................. 509
    Z-t. Liang and C-x. Liu
Top Quark Physics at the LHC.. 514
    L. Sonnenschein
Total and Differential Cross Sections and Polarization Effects
in pp Elastic Scattering at RHIC...................................... 519
    W. Guryn
New Approaches to the pp Total Cross Section Measurements
at Polarized Colliders 524
    A. A. Bogdanov, S. B. Nurushev, A. Penzo, M. F. Rutzo, O. V. Selyugin,
    M. N. Strikhanov, and A. N. Vasiliev
Measurement of Analyzing Powers and Spin Correlation Coefficients
for Elastic pp Scattering... 529
    F. Bauer for the EDDA Collaboration
A_N for Inclusive $\pi^\pm$ Production at 21.6 GeV/c from C and LH_2......... 534
    H. Spinka, C. Allgower, T. Kasprzyk, K. Krueger, D. Underwood,
    A. Yokosawa, G. Bunce, H. Huang, Y. Makdisi, T. Roser, M. Syphers,
    N. I. Belikov, A. A. Derevschikov, Y. A. Matulenko, L. V. Nogach,
    S. B. Nurushev, A. I. Pavlinov, A. N. Vasiliev, M. Bai, S. Y. Lee, Y. Goto,
    N. Hayashi, T. Ichihara, M. Okamura, N. Saito, H. En'yo, K. Iami, Y. Kondo,
    Y. Nakada, M. Nakamura, H. D. Sato, H. Okamura, H. Sakai, T. Wakasa,
    V. Baturine, A. Ogawa, V. Ghazikhanian, G. Igo, S. Trentalange,
    and C. Whitten

5. SPIN PHYSICS WITH PHOTONS AND ELECTRONS

Spin Effects in Diffractive Hadron Photoproduction 541
    S. V. Goloskokov
Nucleon Resonances in Polarized ω Photoproduction 546
    Y. Oh, A. I. Titov, and T.-S. H. Lee
Polarization Phenomena in Vector Meson Photoproduction on Nucleons
near Threshold... 551
    H. Babacan, T. Babakan, A. Gokalp, and O. Yilmaz

Hard Exclusive Meson Production at HERMES 556
   D. Ryckbosch for the HERMES Collaboration

Spin Structure Function of Virtual Photon and Polarized
Parton Distributions 561
   K. Sasaki and T. Uematsu

Polarization Observables in Wide Angle Compton Scattering 566
   B. B. Wojtsekhowski for the RCS 99-114 Collaboration

Spin Effects in the Fragmentation of Transversely Polarized
and Unpolarized Quarks 571
   M. Anselmino, D. Boer, U. D'Alesio, and F. Murgia

Precise Measurement of the Spin-Dependent Transverse Asymmetry
in Quasielastic $^3\vec{\mathrm{He}}(\vec{e},e')$ and the Neutron Magnetic Form Factor 576
   J.-O. Hansen for the Jefferson Lab E95-001 Collaboration

Single π^0 Electroproduction from CLAS Data at Jefferson Lab 581
   K. Joo for the CLAS Collaboration

Results and Status of Inelastic ed-Scattering Experiments at the Internal
Polarized Deuterium Targets of the VEPP-3 586
   V. N. Stibunov, M. V. Dyug, B. A. Lazarenko, A. Y. Loginov, S. I. Mishnev,
   D. M. Nikolenko, A. V. Osipov, I. A. Rachek, R. S. Sadykov, Y. V. Shestakov,
   A. A. Sidorov, D. K. Toporkov, and S. A. Zevakov

Measurement of Spin Correlation Parameters in the Δ Region
for the $^1\vec{\mathrm{H}}(\vec{e},e')$ Reaction 591
   L. D. van Buuren, D. Szczerba, R. Alarcon, T. S. Bauer, D. Boersma,
   J. F. J van den Brand, H. J. Bulten, M. Ferro-Luzzi, D. W. Higinbotham,
   S. Klous, H. Kolster, J. Lang, F. A. Mul, D. Nikolenko, B. E. Norum,
   I. Passchier, H. R. Poolman, I. Rachek, M. C. Simani, E. Six, H. de Vries,
   K. Wang, and Z.-L. Zhou

6. SPIN PHYSICS IN NUCLEI

Precise Determination of the Spin-Transfer Coefficient $K_{NN'}$
for $\vec{n}p$ Elastic Scattering at 187 MeV 599
   S. W. Wissink, S. Choi, W. A. Franklin, W. W. Jacobs, T. Peterson,
   T. Rinckel, J. Sowinski, E. J. Stephenson, M. Wolanski, and H. Yang

Neutron Densities in ^{120}Sn Observed by Polarized Proton Scattering 604
   H. Sakaguchi, H. Takeda, T. Taki, M. Yosoi, M. Itoh, T. Kawabata,
   T. Ishikawa, M. Uchida, N. Tsukahara, T. Noro, M. Yoshimura, H. Fujimura,
   H. Yoshida, E. Obayashi, A. Tamii, and H. Akimune

Spin-Flip Probability for the ^{26}Mg(^{3}He,t)^{26}Al*(1^+;1.058 MeV) Reaction
at 177 MeV 609
   S. Sakoda, H. Sakai, A. Tamii, T. Ohnishi, K. Yako, M. Hatano, Y. Maeda,
   T. Uesaka, M. B. Greenfield, M. N. Harakeh, A. M. van den Berg,
   V. M. Hannen, B. Krüsemann, R. G. T. Zegers, M. A. de Huu, D. Frekers,
   S. Rakers, H. J. Wörtche, F. Ellingh, M. Hagemann, J. Heyse, N. Blasi,
   and F. Camera

Study of Isospin Structure of 1^+ Spin States in ^{58}Ni and ^{58}Cu by the Comparison of ^{58}Ni(p,p') and ^{58}Ni$(^3\text{He},t)^{58}$Cu Reactions 614
    H. Fujita, Y. Fujita, G. P. A. Berg, Y. Shimbara, A. D. Bacher, C. C. Foster, K. Hara, K. Harada, K. Hatanaka, J. Jänecke, K. Katori, T. Kawabata, T. Noro, D. A. Roberts, H. Sakaguchi, T. Shinada, E. J. Stephenson, T. Taki, H. Ueno, and M. Yosoi

Momentum Transfer Dependence of Spin Isospin Modes in Quasielastic $(\vec{p},\vec{n})$ Reactions 619
    T. Wakasa, H. Sakai, M. Ichimura, K. Hatanaka, H. Okamura, K. Kawahigashi, A. Tamii, H. Otsu, Y. Nakaoka, T. Ohnishi, K. Yako, K. Sekiguchi, T. Yagita, J. Kamiya, S. Sakoda, K. Suda, H. Kato, M. Hatano, and Y. Maeda

Relativistic Calculations of Quasielastic Proton-Nucleus Spin Observables Using a Complete Lorentz Invariant Description of the NN Scattering Matrix 624
    B. I. S. van der Ventel, G. C. Hillhouse, and P. R. De Kock

Calculation of the Complete Set of Spin Transfer Coefficients Including One- and Two-Step Processes in (p,nx) Reaction at 346 MeV 629
    K. Ogata, Y. Watanabe, S. Weili, M. Kohno, and M. Kawai

Relativistic Plane Wave Model for Complete Sets of Spin Transfer Observables for Exclusive Proton-Induced Knockout Reactions 634
    S. M. Wyngaardt, A. A. Cowley, G. C. Hillhouse, B. I. S. van der Ventel, and J. Mano

Spin-Dependent Effective Interaction Studied by the ^{12}C, ^{28}Si$(\vec{p},\vec{p}')$ Reactions at Zero Degrees 639
    A. Tamii, T. Kawabata, H. Akimune, I. Daito, Y. Fujita, M. Fujiwara, K. Hatanaka, K. Hosono, F. Ihara, T. Inomata, T. Ishikawa, M. Itoh, M. Kawabata, M. Nakamura, T. Noro, E. Obayashi, H. Sakaguchi, H. Takeda, T. Taki, H. Toyokawa, H. P. Yoshida, M. Yoshimura, and M. Yosoi

Conventional and Non-Conventional Medium Effects in (p,p') Reactions 644
    E. J. Stephenson and F. Sammarruca

Spin-Dipole Resonances in ^{16}O$(\vec{p},\vec{p}')$ Reaction 649
    T. Kawabata, H. Akimune, H. Fujimura, H. Fujita, Y. Fujita, M. Fujiwara, K. Hara, K. Hatanaka, D. Hirooka, K. Hosono, T. Ishikawa, M. Itoh, J. Kamiya, M. Nakamura, T. Noro, E. Obayashi, H. Sakaguchi, Y. Shimbara, H. Takeda, T. Taki, A. Tamii, H. Toyokawa, N. Tsukahara, H. Ueno, M. Uchida, T. Wakasa, K. Yamasaki, Y. Yasuda, H. P. Yoshida, and M. Yosoi

Measurement of Single and Double Spin-Flip Probabilities in Inelastic Deuteron Scattering on ^{12}C and ^{28}Si at 270 MeV 654
    Y. Satou, S. Ishida, H. Sakai, H. Okamura, H. Otsu, N. Sakamoto, T. Uesaka, T. Wakasa, T. Ohnishi, T. Nonaka, G. Yokoyama, K. Sekiguchi, K. Yako, S. Fukusaka, T. Ichihara, T. Niizeki, K. S. Itoh, and N. Nishimori

DWIA Calculations for Inelastic Scattering of Deuterons at E_d=400 MeV 659
    Y. Hirabayashi, T. Suzuki, and M. Tanifuji

Nuclear Spectroscopy by Means of $(\vec{p},\alpha)$ Reactions on Magic
and Near Magic Nuclei: ^{122}Sn$(\vec{p},\alpha)^{119}$In.................................. 664
    P. Guazzoni, M. Jaskola, V. Y. Ponomarev, L. Zetta, Y. Eisermann, G. Graw,
    R. Hertenberger, A. Vitturi, J. N. Gu, and G. Staudt

Measurement of Analyzing Power for $pp \to pp\pi^0$ Reaction at 392 MeV...... 669
    Y. Maeda, M. Segawa, H. P. Yoshida, M. Nomachi, Y. Shimbara, Y. Sugaya,
    K. Yasuda, K. Tamura, T. Ishida, and T. Yagita

Energy Dependence of ^{12,13}C$(p,\pi^-)^{13,14}$O$_{g.s.}$ Reactions in the Δ_{1232}
Resonance Region 674
    J. Kamiya, K. Hatanaka, T. Noro, T. Wakasa, Y. Maeda, H. P. Yoshida,
    E. Obayashi, D. Hirooka, K. Tamura, H. Sakai, A. Tamii, K. Yako,
    Y. Maeda, and H. Okamura

Effective Charge Anomaly in Neutron-Rich Nuclei Revealed from
Spin-Polarized RI Beam Experiments 679
    H. Ogawa, K. Sakai, H. Ueno, T. Suzuki, K. Asahi, H. Miyoshi, M. Nagakura,
    K. Yogo, A. Goto, T. Suga, T. Honda, N. Imai, Y. X. Watanabe, K. Yoneda,
    A. Yoshimi, N. Fukuda, N. Aoi, Y. Kobayashi, W.-D. Schmidt-Ott,
    G. Neyens, S. Teughels, A. Yoshida, T. Kubo, and M. Ishihara

Role of Deuteron Internal Variables in the ^{3}He$(d,p)^4$He Reaction.......... 684
    T. Uesaka, H. Sakai, H. Okamura, A. Tamii, Y. Satou, N. Sakamoto,
    T. Ohnishi, T. Wakasa, K. Itoh, K. Sekiguchi, K. Yako, K. Suda, S. Sakoda,
    J. Nishikawa, M. Hatano, H. Kato, Y. Maeda, T. Saito, N. Uchigashima,
    T. Wakui, and S. Yamamoto

Tensor Analyzing Power A_{yy} in Deuteron Breakup on Hydrogen and Nuclei
at Large Transverse Momenta of Proton............................... 689
    V. P. Ladygin, L. S. Azhgirey, S. V. Afanasief, V. V. Arkhipov, V. K. Bondarev,
    Y. T. Borzounov, G. Filipov, L. B. Golovanov, A. Y. Isupov, V. I. Ivanov,
    A. A. Kartamyshev, V. A. Kashirin, A. N. Khrenov, V. I. Kolesnikov,
    V. A. Kuznezov, N. B. Ladygina, A. G. Litvinenko, S. G. Reznikov,
    P. A. Rukoyatkin, A. Y. Semenov, I. A. Semenova, G. D. Stoletov,
    A. P. Tzvinev, N. P. Yudin, V. N. Zhmyrov, and L. S. Zolin

Polarization Transfer for the ^{2}H$(d,p)^3$H Reaction at $\theta=0°$ at
a Very Low Energy 694
    T. Katabuchi, K. Kudo, K. Masuno, T. Iizuka, Y. Aoki, and Y. Tagishi

Calculation of Low Energy ^{2}H$(d,p)^3$H Reaction by the Four-body
Faddeev-Yakubovsky Equation 699
    E. Uzu, S. Oryu, and M. Tanifuji

Measurement of Analyzing Power T_{20} in Elastic Electron-Deuteron
Scattering in the Momentum Transfer Range of 0.3–0.8 (GeV/c)^2 704
    H. Arenhövel, L. M. Barkov, S. L. Belostotsky, V. F. Dmitriev, M. V. Dyug,
    R. J. Holt, C. W. de Jager, E. R. Kinney, B. A. Lazarenko, S. I. Mishnev,
    D. M. Nikolenko, V. V. Nelyubin, A. V. Osipov, V. G. Popov, D. H. Potterveld,
    I. A. Rachek, R. S. Sadykov, Y. V. Shestakov, V. N. Stibunov, D. K. Toporkov,
    V. V. Vikhrov, H. de Vries, and S. A. Zevakov

Difference between nd A_y and pd A_y below 16 MeV.................... 709
    T. Fujita, K. Sagara, K. Shigenaga, K. Tsuruta, T. Yagita, T. Nakashima,
    N. Nishimori, H. Nakamura, and H. Akiyoshi

Measurement of Cross Section and Analyzing Powers for *dp* Scattering
at Intermediate Energies and Three-Nucleon Force Effects 714
    K. Sekiguchi, H. Sakai, H. Okamura, N. Sakamoto, A. Tamii, T. Uesaka,
    T. Wakasa, Y. Satou, T. Ohnishi, K. Yako, S. Sakoda, K. Suda, H. Kato,
    Y. Maeda, M. Hatano, and J. Nishikawa

Measurement of Differential Cross Sections and Vector Analyzing Powers for
the $\vec{n}d$ Scattering at 250 MeV 719
    Y. Maeda, H. Sakai, K. Hatanaka, H. Okamura, A. Tamii, T. Wakasa,
    K. Yako, K. Sekiguchi, S. Sakoda, J. Kamiya, K. Suda, H. Kato, M. Hatano,
    D. Hirooka, T. Saito, N. Uchigashima, M. B. Greenfield, and J. Rapaport

Forward Scattering Amplitudes and Contributions of Three-Nucleon
Forces in *nd* Elastic Scattering 724
    S. Ishikawa, M. Tanifuji, and Y. Iseri

7. POLARIZED BEAMS AND POLARIMETERS

Polarized Proton Beam Development at COSY with EDDA as a Fast
Internal Polarimeter 731
    F. Hinterberger for the COSY Team and the EDDA Collaboration

Spin-Flipping with an rf-Dipole and a Full Siberian Snake 736
    A. M. T. Lin, B. B. Blinov, Y. S. Derbenev, T. Kageya, D. Y. Kantsyrev,
    A. D. Krisch, V. S. Morozov, J. R. Murray, D. W. Sivers, V. K. Wong,
    K. Yonehara, V. A. Anferov, C. M. Chu, P. Schwandt, B. von Przewoski,
    V. N. Grishin, V. L. Solovianov, K. Jacobs, and G. T. Zwart

Crossing a Coupling Spin Resonance with an RF Dipole 741
    M. Bai and T. Roser

Beam Polarization Distributions for the Relativistic Heavy Ion Collider 746
    A. Lehrach, A. U. Luccio, W. W. MacKay, and T. Roser

Using the Amplitude Dependent Spin Tune to Study High Order
Spin-Orbit Resonances in Storage Rings 751
    D. P. Barber, G. H. Hoffstätter, and M. Vogt

The Polarized Electron Beam at ELSA 756
    M. Hoffmann, W. v. Drachenfels, F. Frommberger, M. Gowin, K. Helbing,
    W. Hillert, D. Husmann, J. Keil, T. Michel, J. Naumann, T. Speckner, and
    G. Zeitler

Electron Beam Polarization with the Compton Polarimeter at JLab 761
    T. Pussieux for the Compton Polarimeter Collaboration

Instrumentation for the Polarization Transfer Experiment in Proton
Inelastic Scattering at $0°$ 765
    M. Yosoi, H. Akimune, I. Daito, H. Fujimura, Y. Fujita, M. Fujiwara,
    K. Hatanaka, K. Hosono, T. Inomata, T. Ishikawa, M. Itoh, M. Kawabata,
    T. Kawabata, M. Nakamura, T. Noro, E. Obayashi, H. Sakaguchi, H. Takeda,
    T. Taki, A. Tamii, H. Toyokawa, M. Uchida, H. P. Yoshida, and M. Yoshimura

Deuteron Polarimeter DPOL and Calibration of the System 770
    H. Kato, Y. Satou, H. Sakai, A. Tamii, K. Sekiguchi, K. Yako, S. Sakoda,
    M. Hatano, Y. Maeda, N. Sakamoto, K. S. Itoh, T. Ohnishi, H. Okamura,
    T. Uesaka, K. Suda, J. Nishikawa, and T. Wakasa

Radiative Polarization in the BATES South Hall Ring 775
    M. Korostelev and Y. M. Shatunov

Polarised $e^{\pm}$ at HERA: Experience and Expectations after the Luminosity Upgrade... 780
    D. P. Barber, M. Berglund, and E. Gianfelice

Stern-Gerlach Interaction in Fermion Beams............................ 785
    P. Cameron, M. Conte, M. Ferro, G. Gemme, A. U. Luccio, W. W. MacKay,
    M. Palazzi, R. Parodi, and M. Pusterla

Measurement of the Analyzing Power for Proton-Carbon Elastic Scattering in the CNI Region with a 22 GeV/c Polarized Proton Beam 790
    J. Tojo, I. Alekseev, M. Bai, B. Bassalleck, G. Bunce, A. Deshpande, J. Doskow,
    S. Eilerts, D. E. Fields, Y. Goto, H. Huang, V. Hughes, K. Imai, M. Ishihara,
    V. Kanavets, K. Kurita, K. Kwiatkowski, B. Lewis, B. Lozowski, Y. Makdisi,
    H. O. Meyer, B. V. Morozov, M. Nakamura, B. Przewoski, T. Rinkel, T. Roser,
    A. Rusek, N. Saito, B. Smith, D. Svirida, M. Syphers, A. Taketani, T. L. Thomas,
    D. Underwood, D. Wolfe, K. Yamamoto, and L. Zhu

Commissioning of RHIC p-Carbon CNI Polarimeter 795
    H. Huang, I. Alekseev, G. Bunce, A. Deshpande, D. Fields, I. Imai,
    V. Kanavets, K. Kurita, B. Lozowski, W. MacKay, Y. Makdisi, T. Rose,
    N. Saito, H. Spinka, D. Svirida, J. Tojo, and D. Underwood

Deuteron Beam Polarimetry at JINR Accelerator Facility 800
    Y. K. Pilipenko, V. M. Slepnev, and L. S. Zolin

Absolute Calibration of the Deuteron Beam Polarization at Intermediate Energies via the $^{12}C(\vec{d},\alpha)^{10}B^{*}(2^{+})$ Reaction............................ 806
    K. Suda, H. Okamura, N. Sakamoto, A. Tamii, T. Uesaka, Y. Satou,
    T. Ohnishi, K. Sekiguchi, K. Yako, S. Sakoda, J. Nishikawa, H. Kato,
    M. Hatano, Y. Maeda, and H. Sakai

A Spin Polarizer for Low Energy Radioactive Nuclear Beams............. 811
    T. Shimoda, S. Shimizu, E. Doumoto, M. Yagi, M. Asai, M. Nakamura,
    Y. Hirayama, K. Horie, T. Shigematsu, H. Izumi, M. Kawabata,
    and N. Takahashi

8. POLARIZED ION SOURCES AND POLARIZED TARGETS

Improved NMR System with Non-Resonant Cable Arrangement for Target Polarization Measurements 819
    D. G. Crabb, S. Bültmann, G. Court, D. B. Day, M. Houlden, C. Keith,
    S. Penttilä, and Y. Prok

A High Intensity Stern-Gerlach Polarized Hydrogen Source for the Munich MP-Tandem Laboratory Using ECR Ionization and Charge Exchange in Cesium Vapour 825
    R. Hertenberger, Y. Eisermann, A. Metz, P. Schiemenz, and G. Graw

The Polarized Internal Gas Target for ANKE at COSY-Jülich 830
    B. Brüggemann, R. Emmerich, R. Engels, H. Kleines, V. Koptev, P. Kravtsov,
    S. Lemaître, J. Ley, B. Lorentz, S. Lorenz, M. Mikirtytchiants, M. Nekipelov,
    V. Nelyubin, H. P. Schieck, F. Rathmann, J. Sarkadi, H. Seyfarth, E. Steffens,
    H. Ströher, A. Vassiliev, and K. Zwoll

Development of Polarized Negative Hydrogen Ion Source with Resonant
Charge-Exchange Plasma Ionizer 835
    A. S. Belov, S. K. Esin, L. P. Netchaeva, A. V. Turbabin,
    and G. A. Vasil'ev

Development of Polarized ^{3}He Ion Source—from OPPIS
to Spin-Exchange 841
    M. Tanaka, T. Yamagata, K. Yonehara, Y. Arimoto, and N. Shimakura

The New LEGS Highly Polarized Frozen-Spin Solid HD Target Facility 846
    X. Wei, F. Lincoln, M. Lowry, T. Saitoh, and A. M. Sandorfi

The HERMES Polarized Internal Deuterium Gas Target 851
    P. Lenisa for the HERMES Collaboration

Michigan Ultra-Cold Polarized Atomic Hydrogen Jet Target 856
    B. B. Blinov, S. E. Gladycheva, T. Kageya, D. Y. Kantsyrev, A. D. Krisch,
    V. G. Luppov, V. S. Morozov, J. R. Murray, R. S. Raymond, N. S. Borisov,
    V. V. Fimushkin, V. N. Grishin, A. I. Mysnik, and D. Kleppner

Development of a Polarized Proton Target in a Low Magnetic Field
at High Temperature 861
    T. Wakui, M. Hatano, H. Sakai, A. Tamii, and T. Uesaka

Polarized Nuclei in Plastic Scintillators: a New Class of Polarized
Targets 866
    B. van den Brandt, E. I. Bunyatova, P. Hautle, J. A. Konter, S. Mango,
    and I. B. Nemchonok

ABSTRACTS OF POSTERS

Helicity Dependence in Photodisintegration of the Deuteron
(*abstract only*) 877
    B. Wojtsekhowski and W. T. H. van Oers for the DGNP Collaboration

A New Method of UCN Production Using a Spatially Alternating
Magnetic Field with Spin Flips (*abstract only*) 877
    K. Sakai, K. Asahi, H. Ogawa, A. Goto, K. Yogo, T. Suga, H. Miyoshi,
    D. Kameda, M. Utsuro, K. Okumura, M. Hino, and A. Yoshimi

Experimental Sensitivity of Contact Interaction at RHIC (*abstract only*) 878
    J. Murata

The Q^2-Dependence of the Generalised GDH Integral for the Proton
(*abstract only*) 878
    B. Seitz for the HERMES Collaboration

Measurement of the Gluon Polarization with the PHENIX Muon Arms
(*abstract only*) 879
    H. D. Sato for the PHENIX Collaboration

A New Approach to Parton-Density Evolution (*abstract only*) 879
    M. Goshtasbpour and P. G. Ratcliffe

Measurement of Transversal Handedness in 3π Diffractive Production
(*abstract only*) 880
    A. V. Efremov, Y. I. Ivanshin, L. G. Tkatchev, R. Y. Zulkarneev,
    I. Kachaev, and A. Zaitsev

Λ Polarization in Unpolarized Hadron Collisions (*abstract only*) 880
    M. Anselmino, D. Boer, U. D'Alesio, and F. Murgia

Nuclear Responses for Solar-ν, Supernova-ν and $\beta\beta$-ν (abstract only) 881
 H. Ejiri
Search for T-Violation in $K^+ \to \pi^0 \mu^+ \nu$ (abstract only) 881
 Y. Asano for the KEK E246
Study of Heavy Meson Production in NN Collisions with Polarized Beam and Target at COSY (abstract only) 882
 F. Rathmann, M. Düren, P. Jansen, F. Klehr, S. Martin, H. O. Meyer,
 K. Rith, E. Steffens, H. Seyfarth, and H. Ströher
The Observation of Double Strange Stable Dibaryons (abstract only) 882
 P. Z. Aslanyan, V. S. Rikhvitskiy, V. N. Yemelyanenko
TESLA-N: Polarized Electron-Nucleon Scattering at TESLA (abstract only) 883
 F. Ellinghaus and E. C. Aschenauer for the TESLA-N Study Group
A Search for Non-Conventional Medium Effects in (p,p') Reactions (abstract only) 883
 E. J. Stephenson and F. Sammarruca
Isoscalar, Isovector, Spin and Orbital Contributions in $M1$ Transitions (abstract only) 884
 Y. Fujita, T. Adachi, G. P. A. Berg, B. A. Brown, H. Ejiri, H. Fujita,
 H. Fujimura, K. Hara, K. Hatanaka, J. Kamiya, K. Katori, T. Kawabata,
 S. Mizutori, P. von Neumann-Cosel, T. Noro, A. Richter, Y. Shimbara,
 T. Shinada, H. Ueno, A. Weiss, M. Yoshifuku, and M. Yosoi
J^π Decomposition of a Bump at $E_x \simeq 7$ MeV in ^{12}N (abstract only) 884
 S. Fukusaka, H. Sakai, K. Hatanaka, H. Okamura, T. Ohnishi, S. Sakoda,
 K. Sekiguchi, A. Tamii, T. Wakasa, and K. Yako
Polarization Transfer Invariants at $0°$ in Nuclear Reactions (abstract only) 885
 T. Suzuki
Systematic Study of Spin-Isospin Excitations in Neutron Rich Light Nuclei via the $(d,^2\text{He})$ reaction at 270 MeV (abstract only) 885
 T. Ohnishi, H. Sakai, H. Okamura, T. Niizeki, K. Itoh, T. Uesaka, Y. Satou,
 K. Sekiguchi, K. Yakou, S. Fukusaka, N. Sakamoto, and H. Ohnuma
Implications for the πNN Coupling from Spin Transfer Measurements in pp Elastic Scattering at 200 MeV (abstract only) 886
 S. W. Wissink for the IUCF E367 Collaboration
Isospin Identification for $A=25$ Mirror Nuclei by High Resolution (p,p') and $(^3\text{He},t)$ Experiments (abstract only) 886
 Y. Shimbara, H. Fujita, Y. Fujita, T. Adachi, H. Fujimura, K. Harada,
 K. Katori, T. Shinada, H. Ueno, A. D. Bacher, G. P. A. Berg, C. C. Foster,
 K. Hara, K. Hatanaka, J. Jänecke, J. Kamiya, T. Kawabata, S. Mizutori,
 T. Noro, D. A. Roberts, E. J. Stephenson, M. Yoshifuku, and M. Yosoi
Incident Energy Dependence of the Polarization Observables in Deuteron Elastic Scattering at $E_d=50\sim700$ MeV (abstract only) 887
 Y. Iseri and M. Tanifuji
Tensor Analyzing Powers of $^3\text{He}(d,p)^4\text{He}$ Reactions around 430 keV Resonance (abstract only) 887
 M. Tanifuji and H. Kameyama

Measurement of H($\vec{d}$,³He)γ Reaction Using a Large Acceptance
Spectrograph (*abstract only*) 888
    T. Yagita, K. Sagara, M. Kondo, S. Minami, T. Ishida, K. Hatanaka,
    T. Wakasa, J. Kamiya, D. Hirooka, T. Noro, H. P. Yoshida, E. Obayashi,
    K. Takahisa, M. Yoshimura, and H. Akiyoshi

Evidence for the Existence of Supersymmetry in Atomic Nuclei
(*abstract only*) 888
    A. Metz, J. Jolie, G. Graw, R. Hertenberger, J. Gröger, C. Günther,
    N. Warr, and Y. Eisermann

Spin Effects at Fragmentation of Polarized Deuterons into Pions
(*abstract only*) 889
    S. Afanasiev, V. Arkhipov, V. Bondarev, I. Daito, N. Doushita, S. Fukui,
    N. Horikawa, T. Iwata, A. Isupov, V. Kashirin, A. Khrenov, K. Kondo,
    V. Ladygin, A. Litvinenko, A. Malakhov, V. Penev, Y. Pilipenko,
    S. Reznikov, P. Rukoyatkin, I. Rusanov, A. Wakai, and L. Zolin

Quasi-periodicity of Spin Motion in Storage Rings—A New Look
at Spin Tune (*abstract only*) 889
    D. P. Barber, J. Ellison, and K. Heinemann

A Vector and Tensor Polarimeter for Intermediate Energy Deuterons
(*abstract only*) 890
    Y. Satou, S. Ishida, H. Sakai, H. Okamura, H. Otsu, N. Sakamoto, T. Uesaka,
    T. Wakasa, T. Ohnishi, T. Nonaka, G. Yokoyama, K. Sekiguchi, K. Yako,
    S. Fukusaka, T. Ichihara, T. Niizeki, K. S. Itoh, and N. Nishimori

RCNP (n,p) Facility (*abstract only*) 890
    K. Yako, H. Sakai, A. Tamii, K. Sekiguchi, S. Sakoda, Y. Maeda, M. Hatano,
    H. Kato, T. Saito, N. Uchigashima, H. Okamura, K. Suda, M. B. Greenfield,
    T. Wakasa, J. Kamiya, D. Hirooka, and K. Hatanaka

Orientation of Radioactive Nuclei (*abstract only*) 891
    J. Dupák, M. Finger, M. Finger, Jr., A. Janata, T. I. Kracíková,
    N. A. Lebedev, M. Rotter, M. Slunečka, and Y. V. Yushkevich

Suppressing Intrinsic Spin Harmonics in the AGS (*abstract only*) 891
    A. Lehrach, J. W. Glenn, T. Roser, and V. Ranjbar

Application of Internal Gas Target for Beam Polarization Measurement in
the Electron Storage Ring (*abstract only*) 892
    S. I. Mishnev, S. A. Nikitin, D. M. Nikolenko, I. Y. Protopopov, I. A. Rachek,
    Y. M. Shatunov, A. N. Skrinsky, V. N. Stibunov, G. M. Tumaikin,
    D. K. Toporkov, and E. N. Zhilich

On Feasibility of the Experiments with a Polarized Deuteron Beam and
a Polarized Target at Charles University in Relation with Polarized Fusion
(*abstract only*) 892
    Y. A. Plis for the Prague-Dubna-Kharkov-Moscow-Saclay Collaboration

Synchrotron-Sideband Snake Depolarizing Resonances
(*abstract only*) 893
    T. Kageya, V. Anferov, B. Blinov, C. Chu, Y. Derbenev, A. Krisch, S. Lee,
    W. Lorenzon, T. Rinckel, H. Sato, P. Schwandt, D. Sivers, K. Sourkont,
    F. Sperisen, B. von Przewoski, V. Wong, and S. Youssof

Beam-line Polarimeter for Intermediate-Energy Deuteron (*abstract only*).. 893
    T. Uesaka, H. Sakai, H. Okamura, A. Tamii, Y. Satou, N. Sakamoto,
    T. Ohnishi, T. Wakasa, K. Itoh, K. Sekiguchi, K. Yako, K. Suda,
    and S. Sakoda

Development of Deuteron Beam Polarimeter at RCNP (*abstract only*).. 894
    T. Yagita, K. Sagara, M. Kondo, S. Minami, T. Ishida, K. Hatanaka, T. Wakasa,
    J. Kamiya, D. Hirooka, T. Noro, H. P. Yoshida, E. Obayashi, K. Takahisa,
    M. Yoshimura, and H. Akiyoshi

Status of the HERMES Atomic Beam Source and Possible Improvements (*abstract only*).. 894
    A. Nass, N. Koch, M. Raithel, and E. Steffens

Polarization at the Nuclotron (*abstract only*)............................ 895
    V. Angelov, V. P. Ershov, V. V. Fimushkin, G. I. Gai, A. D. Kovalenko,
    L. V. Kutuzova, A. I. Malakhov, V. A. Michailov, Y. K. Pilipenko,
    V. N. Penev, V. M. Slepnev, A. D. Stepanov, V. P. Vadeev, A. I. Valevich,
    V. I. Volkov, L. S. Zolin, and A. S. Belov

Production of Thick CD_2 Targets for Measurements of the $\vec{n}d$ Scattering at 250 MeV (*abstract only*).. 895
    Y. Maeda, H. Sakai, K. Hatanaka, H. Okamura, A. Tamii, T. Wakasa,
    K. Yako, K. Sekiguchi, S. Sakoda, J. Kamiya, K. Suda, H. Kato, M. Hatano,
    D. Hirooka, T. Saito, N. Uchigashima, M. B. Greenfield, and J. Rapaport

The Bochum Polarized Target (*abstract only*)............................ 896
    G. Reicherz, S. Goertz, J. Harmsen, J. Heckmann, A. Meier, W. Meyer,
    and E. Radtke

Polarized Deuteron Target System for Low Energy $\vec{D}(\vec{d},p)T$ Measurement (*abstract only*).. 896
    I. Daito, H. Doushita, S. Hasegawa, N. Horikawa, S. Horikawa, T. Iwata,
    K. Kondo, Y. Miyachi, K. Mori, N. Takabayashi, T. Tojyo, and A. Wakai

WORKSHOP ON POLARIZED ELECTRON SOURCE AND POLARIMETERS

Opening Address 899
    C. Y. Prescott

Atomic and Electronic Engineering of p-GaAs-(Cs,O)-Vacuum Interface 901
    V. E. Andreev, V. V. Bakin, A. N. Litvinov, A. A. Pakhnevich,
    O. E. Tereshchenko, H. E. Scheibler, A. S. Jaroshevich, and
    A. S. Terekhov

Photoemission and STM, STS Study of Cs/p-GaAs(110).................. 908
    T. Yamada, J. Fujii, and T. Mizoguchi

Longitudinal and Transverse Energy Distributions of Electrons Emitted from GaAs(Cs,O).. 912
    D. A. Orlov, M. Hoppe, U. Weigel, D. Schwalm, A. S. Terekhov,
    and A. Wolf

Cesiumoxide-GaAs Interface and Layer Thickness in NEA
Surface Formation... 916
    S. D. Moré, S. Tanaka, S.-i. Tanaka, T. Nishitani, T. Nakanishi,
    and M. Kamada

Temperature Dependence of Electron Spin Dynamics................... 920
    Y. A. Mamaev, Y. P. Yashin, A. V. Subashiev, A. N. Ambrajei,
    and A. V. Rochansky

Latest Results from Time Resolved Intensity and Polarization Measurements
at MAMI... 926
    J. Schuler, K. Aulenbacher, T. Baba, D. v. Harrach, H. Horinaka,
    T. Nakanishi, S. Okumi, E. Reichert, J. Roethgen, and K. Togawa

Photo-Luminescence Study of Superlattice Photocathode................ 930
    T. Matsuyama, M. Mulai, H. Horinaka, K. Wada, T. Nakanishi, S. Okumi,
    K. Togawa, and T. Baba

MeV Mott Polarimetry at Jefferson Lab............................. 935
    M. Steigerwald

Polarized Source Performance and Developments at Jefferson Lab........ 943
    M. Poelker, P. Adderley, J. Clark, A. Day, J. Grames, J. Hansknecht,
    P. Hartmann, R. Kazimi, P. Rutt, C. Sinclair, and M. Steigerwald

New Results from the Mainz Polarized Electron Facilities.............. 949
    K. Aulenbacher, H. Euteneuer, P. Jennewein, K.-H. Kaiser, H. J. Kreidel,
    D. v. Harrach, E. Reichert, J. Schuler, V. Tioukine, M. Wiessner,
    and K. Winkler

New Results from the MIT-Bates Polarized Source and the Test
Beam Setup.. 955
    M. Farkhondeh, E. Tsentalovich, T. Zwart, and E. Ihloff

The 50 kV Inverted Source of Polarized Electrons at ELSA............ 961
    W. Hillert, M. Gowin, and B. Neff

Polarized Electrons in Low Energy Electron Microscopy................ 965
    E. Bauer

Electron Emission from Na/Fe(100) Surfaces by Deexcitation
of Spin-Polarized Helium Metastable Atoms.......................... 972
    Y. Yamauchi, M. Kurahashi, and T. Suzuki

Investigations of the Charge Limit Phenomenon in GaAs
Photocathodes... 976
    T. Maruyama, J. E. Clendenin, E. L. Garwin, R. E. Kirby, G. A. Mulhollan,
    R. Prepost, C. Y. Prescott, and A. V. Subashiev

Polarized Electron Source for Japan Linear Collider.................. 982
    K. Togawa, T. Nakanishi, S. Okumi, C. Suzuki, F. Furuta, K. Wada,
    T. Nishitani, M. Yamamoto, H. Kobayakawa, Y. Takeda, Y. Takashima,
    H. Sugiyama, O. Watanabe, Y. Kurihara, H. Matsumoto, T. Omori,
    Y. Takeuchi, M. Yoshioka, H. Horinaka, K. Wada, T. Matsuyama, T. Saka,
    T. Baba, and T. Kato

Present Status of Experimental S-Band GaAs-Photogun Driven by the
Solid State GaAs Pulse Laser 988
    N. S. Dikansky, R. G. Gromov, E. S. Konstantinov, P. V. Logatchov,
    and A. V. Alexandrov

Fabrication of Photocathode Test-Stand. 992
G. N. Kim, M. W. Lee, D. Son, Y. J. Park, S. J. Park, M. H. Cho, I. S. Ko, and W. Namkung

A Pulsed Electron Source for Atomic Collision Experiments 996
C. D. Schröter, A. Dorn, J. Deipenwisch, C. Höhr, R. Moshammer, and J. Ullrich

Surface Photovoltage Effect on Clean and Negative Electron-Affinity Surfaces of GaAs and Its Superlattice 1000
S. Tanaka, S. D. Moré, T. Nishitani, K. Takahashi, T. Nakanishi, and M. Kamada

Structure and Magnetism of Fe Thin Films Grown on Rh(001) Studied by Spin-Resolved Photoelectron Spectroscopy. 1003
K. Hayashi, M. Sawada, A. Harasawa, A. Kimura, and A. Kakizaki

Electronic Structure and Magnetic Anisotropy of Co/Au(111): a Spin-Resolved Photoelectron Spectroscopy Study 1006
M. Sawada, K. Hayashi, and A. Kakizaki

Reduction of Field Emission Current from Stainless Steel and Copper Surface 1009
C. Suzuki, T. Nakanishi, S. Okumi, T. Gotoh, K. Togawa, F. Furuta, K. Wada, T. Nishitani, M. Yamamoto, J. Watanabe, S. Kurahashi, H. Matsumoto, M. Yoshioka, K. Asano, and H. Kobayakawa

Development of 200 keV Polarized Electron Gun. 1012
K. Wada, M. Yamamoto, T. Nakanishi, S. Okumi, T. Gotoh, K. Togawa, C. Suzuki, F. Furuta, T. Nishitani, J. Watanabe, S. Kurahashi, M. Miyamoto, H. Matsumoto, Y. Takeuchi, and M. Yoshioka

Test of Cesium Telluride Photocathode as a Feasibility Study on Polarized RF-Gun 1015
F. Furuta, H. Sugiyama, T. Nakanishi, S. Okumi, K. Togawa, C. Suzuki, K. Wada, M. Yamamoto, T. Nishitani, J. Watanabe, S. Kurahashi, M. Miyamoto, M. Kuwahara, R. Mizuno, T. Hirose, K. Kimura, H. Kobayakawa, Y. Takashima, M. Yoshioka, and H. Matsumoto

Atomic Hydrogen Cleaning of GaAs Photocathode with a Load-Lock System 1018
M. Yamamoto, K. Wada, T. Nakanishi, S. Okumi, K. Togawa, C. Suzuki, F. Furuta, T. Nishitani, J. Watanabe, S. Kurahashi, and M. Miyamoto

Development of Spin Polarized Electron Photocathodes: GaAs-GaAsP Superlattice and GaAs-AlGaAs Superlattice with DBR. 1021
T. Nishitani, O. Watanabe, T. Nakanishi, S. Okumi, K. Togawa, C. Suzuki, F. Furuta, K. Wada, M. Yamamoto, J. Watanabe, S. Kurahashi, M. Miyamoto, H. Kobayakawa, Y. Takeda, T. Saka, K. Kato, A. K. Bakarov, A. S. Jaroshevich, H. E. Scheibler, A. I. Toropov, and A. S. Terekhov

Fabrication of GaAs/GaAsP Superlattice Photocathode 1024
O. Watanabe, T. Nishitani, K. Togawa, Y. Takashima, T. Nakanishi, Y. Takeda, and H. Kobayakawa

APPENDICES

SPIN2000 Schedule 1029
PES2000 Program 1043
SPIN2000 Partipants List... 1047
PES2000 Participants List 1073
Author Index 1075

Preface

The Fourteenth International Spin Physics Symposium, SPIN2000, was held October 16-21, 2000 at the convention center in the Osaka University campus.

The symposium was supported by the International Union of Pure and Applied Physics, IUPAP. The symposium was hosted by Research Center for Nuclear Physics, RCNP, Osaka University, and was sponsored by the COE (Center of Excellence) conference fund of the Japanese government.

Particle and nuclear spins have played important roles in the progress of particle and nuclear physics. Accordingly the two series of international symposia and conferences have been held. The International Symposia on High Energy Spin Physics have been held every other year since 1974. Recently, the 12th symposium was held at Amsterdam in 1996, and the 13th one at Protvino in 1998. The International Conferences on Polarization Phenomena in Nuclear Physics have been held every five years since 1960, recently at Paris in 1990, and at Bloomington in 1994.

In view of the strong overlap in both the scientific concerns and the technical developments , discussions on combining these high energy spin physics symposia and nuclear polarization conferences have been made lately at both the high energy and nuclear physics communities. Actually the two meetings were held together at Indiana University, Bloomington in 1994.

The present 14th International Spin Physics Symposium is the first combined symposium of the high energy spin physics symposium and the nuclear polarization conference

Nucleons quarks and leptons are fermions with spin 1/2, while gauge bosons of the strong and electro-week interactions associated with these fermions are gluons, photons and weak bosons with spin 1. Consequently the spins are key elements for structures and reactions of the particles and nuclei, and thus spin-polarized particle and nuclear probes are crucial for studies of them.

The present spin physics symposium aimed at discussions on new developments of the particle and nuclear spin physics. The subjects discussed were symmetries and spins, spins in weak interactions, spin structures of nucleons, spin physics in medium-high energy hadron, lepton and photon reactions, spins in nuclear many body systems, polarized ion sources and targets, polarized beams and polarimeters and related subjects.

It is our great pleasure to get many distinguished physicists working in spin physics frontiers from all over the world, and many excellent talks and contributions on spin physics studies.

The participants were nearly 300 from 17 countries. The symposium consisted of 12 plenary sessions with 29 invited talks, 24 parallel sessions with 111 oral presentations of the contributions, and 2 poster sessions with 112 contributed papers. One plenary session was dedicated to the late Prof. Lazarus G. Ratner for his great contributions to spin physics.

Interesting results of experimental and theoretical works were discussed on spin structures and spin dynamics of hadrons and nuclei and on electro-weak and QCD. New technical developments were presented for spin polarization experiments. The first data from RHIC with high energy polarized protons, and those from RCNP/SPring-8 with multi-GeV polarized photons were shown. An impressive speech on the spin physics developments was given at the banquet by Prof. E.D. Courant, the honorary member of the international committee.

The present proceedings of SPIN2000 includes the invited papers presented at the plenary sessions, the papers presented orally at the parallel sessions and the short abstracts of the papers presented at the posted sessions. It includes also a summary of the banquet speech by E.D. Courant, and the proceedings of the satellite workshop on polarized electron source and polarimeter at Nagoya Oct. 12-14, just prior to SPIN2000.

The organizing committee consisted of particle, nuclear and accelerator physicists working currently on spin physics at high energy physics and nuclear physics laboratories and university groups in Japan.

On behalf of the organizing committee, we would like to express our cordial thanks to the International Committee of Spin Physics for the strong encouragements and the valuable suggestions for the first unified symposium in the year of 2000, and to all the participants for wonderful presentations of the nice works and interesting discussions. Many thanks are also due to IUPAP, RCNP, the COE/Japan, and all the supporting staffs and graduate students for their generous supports and great helps.

We hope SPIN2000 makes great contributions to further progresses of spin physics toward the new century. The next spin physics symposium, SPIN2002, will be held at BNL, USA.

Editors:
K. Hatanaka
T. Nakano
K. Imai
H. Ejiri

Organizing Committee

H. Ejiri (Chair)	RCNP
K. Hatanaka (Co-chair)	RCNP
K. Imai (Co-chair)	Kyoto
T. Nakano (Scientific Secretary)	RCNP
N. Horikawa	Nagoya
M. Ishihara	RIKEN
T. Kishimoto	Osaka
K.-I. Kubo	TMU, Tokyo
Y. Kuno	KEK
A. Masaike	Fukui
Y. Mori	KEK
T. Morii	Kobe
Y. Nagashima	Osaka
T. Noro	RCNP
H. Sakai	Tokyo
T.-A. Shibata	TITech, Tokyo
M. Tanifuji	Hosei, Tokyo
H. Toki	RCNP
K. Tokushuku	KEK
S. Yamada	KEK

International Committee

C. Y. Prescott (Chair)	SLAC
A. D. Krisch (Past-chair)	Michigan
T. Roser (Chair-elect)	BNL
D. P. Barber	DESY
J. M. Cameron	IUCF
O. Chamberlain*	Berkeley
E. D. Courant*	BNL
D. G. Crabb	Virginia
A. V. Efremov	Dubna
H. Ejiri	RCNP
G. Fidecaro	CERN
W. Haeberli	Wisconsin
W. Happer	Princeton
V. W. Hughes*	Yale
K. Imai	Kyoto
N. Isgur	TJNAF
Y. M. Shatunov	Novosibirsk
V. Soergel	Munich
L. D. Soloviev	Protvino
E. Steffens	Erlangen
W. T. H. van Oers	Manitoba

* Honorary Member

Opening address

H. Ejiri

Chairperson of SPIN2000, RCNP, Osaka University

Good morning ladies and gentlemen

It is a great pleasure for me to give an opening address for the 14th International Spin Physics Symposium, SPIN2000, at Osaka. which is in the series of the high energy spin physics symposia.

On behalf of the organization committee, I would like to express our hearty welcome to all of you to Osaka for the SPIN2000 symposium.

This symposium is hosted by Research Center for Nuclear Physics, RCNP, Osaka University, and is supported by the COE conference fund of the Ministry of Education, Science, Sports and Culture of Japan and by the International Union of Pure and Applied Physics, IUPAP. We would like to cordially thank them for their generous supports.

Spins have played essential roles in the progress of nuclear and particle physics. Thus the series of the International Symposia on High Energy Spin Physics have been held every other year since 1974, recently at Amsterdam in 1996, and at Protvino in 1998. On the other hand the International Conferences on Polarization Phenomena in Nuclear Physics have been held every five years since 1960, recently at Paris in 1990, and at Bloomington in 1994.

In view of strong overlap in both the scientific concerns and the technical developments , discussions on combining these high energy spin physics symposia and nuclear polarization conferences have been made lately at both the high energy and nuclear physics communities. Actually the two meetings were held together at Bloomington in 1994.

The present 14th International Spin Physics Symposium combines the high energy spin symposium and the nuclear polarization conference. It is our great honor to organize the first unified spin physics symposium at Osaka in the year of 2000. We are very grateful to the International Committee of spin physics for this opportunity and for valuable suggestions and supports .

Spin is a key element in the present and future developments of particle, nuclear and astro physics. Spin plays indeed crucial roles in studies of symmetries and electro-weak interactions, hadron physics and QCD, medium and high energy hadron, lepton and photon reactions, and of structures of hadron-nucleon many body systems. Technical developments of polarized beams, polarimeters, ion sources and targets are also very important. The present symposium aims at discussions of recent new works on these spin physics and related subjects as given in the scientific program.

I would like to convey a hearty welcome from Prof. Nagai, director of RCNP to all of you. In fact RCNP hosted the 6th nuclear polarization conference in 1985. At that time many nuclear polarization studies were being carried out by using the low energy cyclotron with K=140 MeV. Now extensive spin physics programs are going on at RCNP by means of the medium energy ring cyclotron with the proton energy and K = 0.4 GeV, the multi GeV laser electron photons from the 8 GeV Spring-8 electron, and of the Oto underground facilities and detectors. RCNP welcomes you to visit the laboratories.

We are delighted to have here many distinguished physicists working in spin physics frontiers from all over the world, and many excellent talks and contributions on interesting spin physics studies.

We do hope this symposium will be very productive for spin physics and be enjoyable for all of you, and will contribute to the further progress of spin physics in the new century.

Thank you for your attention.

Welcome Address

Y. Yamaguchi

Former president, IUPAP

Ladies and Gentlemen,

I was asked by Professor Ejiri, the chairman of the Organizing Committee of this Symposium, to present the Welcome Address on behalf of IUPAP, since I am the Former President of IUPAP. It is my great honor to serve this job. I wish to express my heartful welcome all of you. SPIN2000 is the 14th International Spin Physics Symposium. Before this Symposium, there used to be held two series of conferences: Nuclear Polarization Conferences and High Energy Spin Physics Symposia, organized by nuclear physics and high energy physics communities, respectively. Now, these two series are unified as the International Spin Physics Symposium common to nuclear and particle physics communities. This SPIN2000 is the first one in the unified series and sponsored by the Monbusho, IUPAP and RCNP. Long ago, particle and nuclear physics are the single field in physics. Then two became gradually separate branches in physics. However, these two branches have recently begun to share again common interest on many important physics issues such as quark-gluon physics, symmetry issues, electro-weak processes and so on. To attack these problems spin is vitally important. Here, at SPIN2000, nuclear physics and high energy physics groups find the excellent opportunity to discuss common aspects relating to spin physics, as you see in the Scientific Program.

Osaka University was not only the birth place of the Meson Theory by Prof. H. Yukawa but also has long history of great contribution to nuclear and particle physics. Especially, RCNP, Osaka University, created 29 years ago, has been one of the most active international research centre on experimental nuclear physics. Moreover, under the leadership of Prof. Ejiri, RCNP has been expanding its activity by adding the Underground Lab and the high energy photon beam facility at SPring-8. The former is suited to study extremely rare phenomena such as double β-decays and neutrino induced reactions. The latter offers one of the central subjects in SPIN2000. These two provide wonderful working sites on many subjects to be discussed here. Thus, RCNP became now excellent working place for nuclear and particle physicists to collaborate. This RCNP is appropriately hosting this Symposium. In addition, RHIC at BNL and the planned JHF offer the active collaboration sites for us. SPIN2000 is just timely organized under such a world-wide trend. I am quite sure that this Symposium shall be most interesting and successful. The 20th Century was the great century in physics. Quantum Physics was born, developed and culminated in the Standard Model. During this remarkable progress, spin often played decisive roles. The same must be true also in the coming century to uncover the ultimate law in nature beyond the Standard Model. Finally, I wish that all participants and accompanied members shall enjoy not only modern Japan but also classic heritage in Osaka and its vicinity.

SPIN PHYSICS 2000

Charles Y. Prescott

Stanford Linear Accelerator Center
Stanford CA 94309

Good Morning. I am very pleased to be here in Osaka today and to participate in the opening of this Symposium. Let me be the first to thank the Organizers and our host, the Research Center for Nuclear Physics, Osaka University, for making this event possible.

Perhaps we should call this the 21st International Spin Physics Symposium. The history of International Spin Physics Symposia goes back to 1960, when the first Symposium, "The International Symposium on Polarization Phenomena of Nucleons", was held in Basel. Symposia for Spin Physics in Nuclear Phenomena were subsequently held every five years. In 1974 the High Energy Spin Physics Symposia started at Argonne. The pace quickened with Spin Physics Symposia for the High Energy community being held in even numbered years. Discussions to unify the Nuclear Spin Physics and High Energy Spin Physics Symposia started at the Nagoya SPIN92 Symposium, leading to a "trial merger" in Bloomington in 1994. That joint Symposium was quite successful, and steps were undertaken to formalize this merger. Today in Osaka we meet in our first fully merged Spin Physics Symposium, a fitting event to start the New Millennium. We look back to 40 years of successes and growth. We look forward to the next 40 years with expectations for great advances in Spin Physics. To our Organizers and hosts, "Thank you for making this event possible", and to our participants, "Welcome to SPIN2000".

High Energy Spin Physics Symposia		Nuclear Spin Physics Symposia	
1974	Argonne	1960	Basel
1976	Argonne	1965	Karlsruhe
1978	Argonne	1970	Madison
1980	Lausanne	1975	Zurich
1982	Brookhaven	1980	Santa Fe
1984	Marseille	1985	Osaka
1986	Protvino	1990	Paris
1988	Minneapolis		
1990	Bonn		
1992	Nagoya		
1994	Bloomington	trial merger	
1996	Amsterdam		
1998	Protvino		
2000	Osaka	fully merged	

PLENARY SESSIONS

SPIN
Progress and Prospects

Robert L. Jaffe

Center for Theoretical Physics[1]
Massachusetts Institute of Technology
Cambridge, Massachusetts 02139

Abstract. I review the progress in fundamental spin physics over the past several years and the prospects for the future. The progress is striking and the prospects are excellent.

INTRODUCTION

I would like to thank the organizers for the honor of delivering the opening talk at this Millennial Conference on Spin in High Energy and Nuclear Physics. Much is new, more will be forthcoming soon. These are exciting times.

Like this conference, my talk will focus on spin in QCD. The organizers asked me to stress progress and prospects, which I will do. The prospects for remarkable advances in the near future – involving spin in one way or another – in electroweak unification and even in gravity compels me to mention those fields as well.

In his welcome to SPIN98 in Protvino, Charlie Prescott, began his talk by pointing out a striking geographical correlation between the historical march of successive spin conferences and the Earth's angular momentum [1]. Our location in Osaka once again displays the "Prescott Effect". I have updated the data in Fig. 1. Word has it that the 2002 conference will move to Brookhaven, once again confirming Prescott's Effect. The rule, of course, displays the remarkably international character of these conferences, which we all hope will continue forever. Parenthetically, as a theorist, I have to note the Anomaly that occurred in 1986–1988, when the conference counter-rotated from the USSR to the USA. Prescott appears to have fudged his data to downplay this intriguing Anomaly – it deserves further study.

Turning to more serious questions: The stuff of the world falls into two categories: (1) Gauge fields, which are bosons and are required by local symmetries of space-time. Gravitons follow from general covariance. W's, Z's, photons and gluons spring from local phase invariances. And (2) Matter, which is composed of spin-$1/2$ particles, quarks, and leptons that carry the quantum numbers of ungauged, global symmetries.

---

[1] This work is supported in part by funds provided by the U.S. Department of Energy (D.O.E.) under cooperative research agreement #DF-FC02-94ER40818.

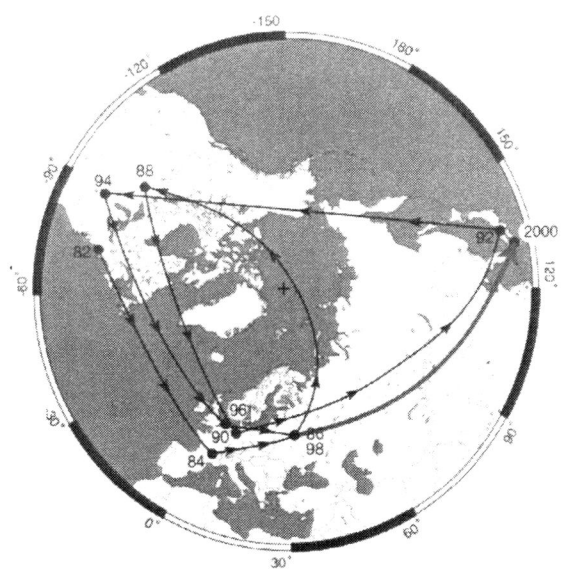

FIGURE 1. The Prescott Effect.

So far there are *no spinless elementary particles at all!* One can argue that this pattern follows in part from the constraints of renormalizability: Massive vector bosons unrelated to gauge symmetries should not appear in our low energy, renormalizable effective Lagrangian. Spinless particles might be propelled to Planck scale masses by quadratically divergent self energies. A couple of questions stand out: Why are there no spinless elementary particles? Why do only the fermions carry the ungauged, global quantum numbers. Why does matter exist at all, since the (Yang-Mills and Einstein) gauge theories are quite consistent and content without them?

Remarkably, we may be close to obtaining new experimental input into these profound questions.

- There is some evidence that a scalar Higgs boson awaits discovery at a mass around 115 GeV. Early indications at LEP will have to wait for FNAL or LHC for confirmation. Of course radiative corrections to the Standard Model now constrain the Higgs mass reasonably well. So unless Nature deals us a major surprise, the discovery of the first spinless elementary particle is imminent.

- A light Higgs suggests (but does not require) that supersymmetry is the natural extension of the Standard Model. If so, we can expect to discover two entirely new forms of matter required by supersymmetry. First, fermions without flavor associated with gauge symmetries – the "ino"s of SUSY like the gluino. Supersymmetric partners of the gauge bosons, they know nothing about the flavor symmetries of ordinary matter. Second, the scalar partners of quarks and leptons, which carry global flavor quantum numbers – the "sparticles" of SUSY – these would be the first bosons carrying the global symmetries which characterize matter.

There are other good reasons to believe in supersymmetry: partial resolution to the hierarchy problem, coupling constant unification, and dark matter candidates are most often mentioned. It is worth remembering, however, that SUSY introduces particles which are qualitatively different from those we have known up to now.

Over the next few years, *spin* will play a central role in testing and looking beyond the Standard Model and in exploring the unresolved mysteries of QCD. Two examples will illustrate the importance of spin in tests of the Standard Model:

- The muon's magnetic moment, $(g-2)_\mu \equiv 2a_\mu$ will be measured with nearly 20 times existing precision by the Brookhaven $g-2$ experiment. At this precision a_μ probes certain extensions of the Standard Model up to energies equivalent to LEP and the Tevatron, and is sensitive to SUSY and other novelties.

- Electric dipole moments (EDM's) fascinate both theorists and experimenters. They probe CP violation, one of the most poorly understood aspects of the Standard Model. If all CP-violation is encoded in the CKM matrix, EDM's are too small to measure. For this reason EDM's are an excellent place to look for CP-violation beyond the Standard Model. The problem of baryogenesis in the early Universe continues to suggest that other sources of CP-violation are waiting to be discovered.

Spin has recently proved itself a very powerful tool to probe the internal structure of hadrons in QCD. We have measured the quark spin contribution to the spin of the nucleon, but we do not understand it. In fact, we know more about the spin of the *graviton*, which has never been observed, than we do about the spin of the nucleon, which composes most of the luminous mass in the Universe. Several new results from the Hermes collaboration at DESY and the SAMPLE collaboration at Bates whet the appetite for future measurements of spin observables in QCD. I will review the outstanding issues in QCD spin physics in the latter 2/3 of this talk.

Finally, I cannot fail to mention the experimental and technical foundations on which our field rests. Spin physics would go nowhere without the extraordinary creativity and devotion of accelerators physicists, who have developed novel methods of accelerating, storing and colliding polarized particles and without experimentalists who have devised high density, high polarization targets, and polarimeters and detectors capable of incredible sensitivity. Were not for this remarkable effort, we theorists might as well do string theory.

BEYOND THE STANDARD MODEL

As an appetizer to QCD, which is the main course at this meeting, here is a quick survey of some issues in spin physics beyond the Standard Model.

The spin of the graviton

We know that the graviton has spin two. Standard tests of general relativity and the measured deceleration of the Hulse-Taylor pulsar assure us of this. Still, it would be nice to have direct observation of gravitational radiation and explicit confirmation of its tensor nature. LIGO, the Laser Interferometry Gravitational Observatory, will do both if Nature is kind enough to provide a strong enough source [2]. LIGO I is scheduled to begin data taking in 2003. The upgrade to LIGO II, with much greater sensitivity, begins in 2005. Perhaps we shall see a direct measurement of the spin of the graviton by the end of this decade.

The anomalous magnetic moment of the muon

This subject is covered in the plenary talk by G. Bunce, so I will be brief. After years of hard work and great patience, the Brookhaven experiment (E821) seems poised to report a value for $a_\mu \equiv (g_\mu - 2)/2$ which will challenge the Standard Model. It is conventional to quote values for a_μ in units of 10^{-10} or 10^{-11} and accuracy in parts per million. Thus the CERN μ^+ value is $a_\mu \times 10^{10} = 116\,591\,00(110)$ has an accuracy of 10 ppm [3]. The E821 number from 1998 running is $a_\mu \times 10^{10} = 116\,591\,91(59)$ (5 ppm), already a twofold improvement in precision over the old CERN experiment [4].

Theory includes electromagnetic, weak and strong corrections. The pure QED terms are known extremely well – they contribute $a_\mu(\text{QED}) \times 10^{11} = 116\,584\,705.7(2)$.[2] Strong (QCD) effects are very significant: one-loop hadronic vacuum polarization gives $a_\mu(\text{QCD-1 loop}) \times 10^{11} = 6924(62)$; two-loop hadronic vacuum polarization effects are important at the level of 1 ppm ($a_\mu(\text{QCD-2 loop}) \times 10^{11} = 101(6)$);[2] and QCD light-by-light scattering enters at the level just below 1 ppm ($a_\mu(\text{QCD–light-by-light}) \times 10^{11} = -85(30)$).[2] The present precision of the theoretical estimate of $(g_\mu - 2)$ is principally limited by the lack of information on QCD light-by-light scattering: $a_\mu(\text{theory}) \times 10^{11} = 116\,591\,62(8)$ (0.66 ppm). Until someone understands how to compute QCD light-by-light scattering more accurately, there is no point carrying experiment beyond 0.5 ppm accuracy. This limit was designed into the BNL experiment: data on tape should allow a precision of 0.5 ppm, and the experiment's ultimate goal is ∼0.35 ppm. At this precision a_μ is sensitive to SUSY radiative corrections from loops involving smuons and neutral and charginos, especially in models with large $\tan\beta$,

$$a_\mu(\text{SUSY}) \approx 140 \times 10^{-11} \left(\frac{100\text{GeV}}{\tilde{m}}\right)^2 \tan\beta \qquad (1)$$

so a_μ probes SUSY masses of order 100 GeV for $\tan\beta \sim 1$. Of course, surprises beyond SUSY may await.

Electric Dipole Moments

The search for an understanding of CP-violation probably commands more resources than any other single issue in high energy physics: ε'/ε, rare K decays, B factories, etc. A classic window into CP-violation is provided by the search for electric dipole moments (EDM's). Khriplovich's PANIC99 talk gives a good summary [5]. I have abstracted Table 1 from his talk[3]. Standard Model (ie. CKM) predictions for

---

[2] See the talk by G. Bunce for the diagrams.
[3] Though the table does not do justice to the complexity of measurements of nuclear EDM's or the sophistication of Khriplovich's talk

TABLE 1. Electric Dipole Moments

Particle	Current Limit	Standard Model	Reasonable Goal
Neutron	$6\text{--}10 \times 10^{-26}$	$10^{-30\text{--}31}$	$10^{-27\text{--}28}$
Electron	4×10^{-27}	10^{-40}	10^{-28}
Nuclei	$\sim 2 \times 10^{-24}$	$\sim 10^{-30}$	—
Muon	10^{-18}	10^{-38}	10^{-24} [a]

[a] Storage ring proposal, Y. Semertzidis [6].

EDM's are far smaller than the reasonable goals of experiments. This means that EDM's provide fertile ground in which to look for sources of CP-violation *beyond* the Standard Model. Of particular interest is Semertzidis's proposal to use the BNL $(g-2)$ ring to improve the limit on the muon EDM by as much as six orders of magnitude [6]. Readers interested in this simple and elegant idea should consult the review by Khriplovich.

SPIN IN QCD

Polarization effects in QCD present a complex landscape. Asymmetries need to be explained. Sometimes we have no explanation but still can use them to probe questions or isolate effects that are perhaps even more interesting. I want to highlight some of the topics I find particularly interesting. This week's program promises many interesting talks, beyond my ability to anticipate – my apologies to all those whose work has been omitted in this brief overview.

I will focus on the overlap between theory and experiment. Many striking asymmetries occur in the low energy or nuclear domain where we have few theoretical insights into QCD [7]. A few dramatic spin-dependent effects occur in the deep inelastic domain, where QCD is transparent. Others occur where deep inelastic and soft domains overlap: the world of parton distribution and fragmentation functions. Here spin effects help elucidate the puzzling nature of hadrons and here I will concentrate.

The topics I will cover include:

- Bjorken's Sum Rule: What it means to "understand" something in QCD.

- Quark and gluon distributions in the nucleon.

- Probing polarized glue in the proton.

- The nucleon's total angular momentum: Progress and frustration.

- Transversity.

- Spin at RHIC.

- Fragmentation and spin: The new HERMES asymmetry and beyond.

- Spin dependent static moments: μ_s, the anapole, etc.

- The Drell-Hearn-Gerasimov-Hosada-Yamamoto Sum Rule

Bjorken's Sum Rule

Occasionally it is worth reminding ourselves what it means to "understand" something in QCD. In the absence of fundamental understanding we often invoke "effective descriptions" based on symmetries and low energy expansions. While they can be extremely useful, we should not forget that a thorough understanding allows us to relate phenomena at very different distance scales to one another. In Bjorken's sum rule, the operator product expansion, renormalization group invariance and isospin conservation combine to relate deep inelastic scattering at high Q^2 to the neutron's β-decay axial charge measured at very low energy. Even target mass and higher twist corrections are relatively well understood. The present state of the sum rule is

$$\int_0^1 dx\, g_1^{ep-en}(x, Q^2) = \frac{1}{6}\frac{g_A}{g_V}\left\{1 - \frac{\alpha_s(Q^2)}{\pi} - \frac{43}{12}\frac{\alpha_s^2(Q^2)}{\pi^2} - 20.215\frac{\alpha_s^3(Q^2)}{\pi^3}\right\}$$
$$+ \frac{M^2}{Q^2}\int_0^1 x^2 dx\left\{\frac{2}{9}g_1^{ep-en}(x, Q^2) + \frac{1}{6}g_2^{ep-en}(x, Q^2)\right\}$$
$$- \frac{1}{Q^2}\frac{4}{27}\mathcal{F}^{u-d}(Q^2) \qquad (2)$$

where the three lines correspond to QCD [8], target mass, and higher twist [9] corrections respectively. g_1 and g_2 are the nucleon's longitudinal and transverse spin dependent structure functions. g_A and g_V are the neutron's β-decay axial and vector charges. $\mathcal{F}$ is a twist-4 operator matrix element with dimensions of [mass]^2, which measures a quark-gluon correlation within the nucleon,

$$\mathcal{F}^u(Q^2)S^\alpha = \tfrac{1}{2}\langle PS|\, g\bar{u}\tilde{F}^{\alpha\lambda}\gamma_\lambda u\big|_{Q^2}|PS\rangle \qquad (3)$$

where g is the QCD coupling, $\tilde{F}$ is the dual gluon field strength, and $|_{Q^2}$ denotes the operator renormalization point.

The most thorough analysis of the Bj Sum Rule I know of is one presented by SMC in 1998 [10]. Their theoretical evaluation gives

$$\int_0^1 dx\, g_1^{ep-en}(x, Q^2)\big|_{\text{theory}} = 0.181 \pm 0.003 \qquad (4)$$

at $Q^2 = 5$ GeV^2. Experiment is not yet able to reach this level of accuracy. The latest data relevant to the Bj Sum Rule is shown in Fig. 2. The value extracted by the SMC is

$$\int_0^1 dx\, g_1^{ep-en}(x, Q^2)\big|_{\text{expt.}} = 0.174 \pm 0.005\, {}^{+0.011}_{-0.009}\, {}^{+0.021}_{-0.006} \qquad (5)$$

at $Q^2 = 5$ GeV^2, and the errors are statistical, systematic, and "theoretical" (eg. generated by running the data to a common Q^2), respectively [10]. Further accuracy is necessary to confirm the target mass corrections and extract the twist-four contribution.

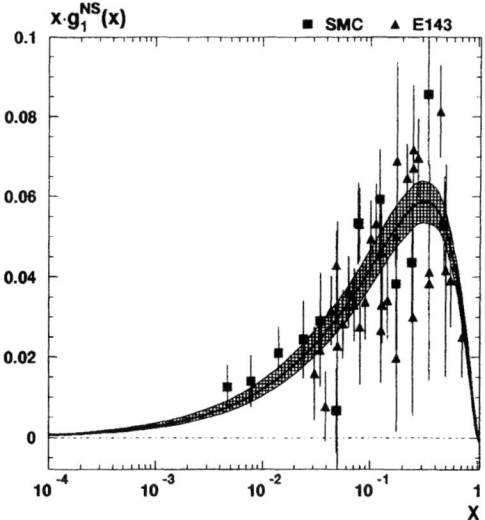

FIGURE 2. SMC analysis of data relevant to the Bjorken sum rule.

Quark and gluon distributions in the nucleon

No summary of recent progress in spin physics is complete without a survey of the polarized quark and gluon distributions in the nucleon. These helicity weighted momentum distributions are the most precise and interpretable information we have about the spin substructure of a hadron. The distributions are usually defined in terms of flavor-SU(3) structure,

$$\begin{aligned} \text{Singlet:} \quad & \Delta\Sigma = \Delta U + \Delta D + \Delta S \\ \text{Nonsinglet, isovector:} \quad & \Delta q_3 = \Delta U - \Delta D \\ \text{Nonsinglet, hypercharge:} \quad & \Delta q_8 = \Delta U + \Delta D - 2\Delta S \end{aligned} \quad (6)$$

where $\Delta Q \equiv q^\uparrow(x,Q^2) + \bar{q}^\uparrow(x,Q^2) - q^\downarrow(x,Q^2) - \bar{q}^\downarrow(x,Q^2)$. Experimenters seem to prefer nonsinglet distributions specialized to the proton and neutron individually,

$$\begin{aligned} \text{Proton nonsinglet:} \quad & \Delta q_{NS}(p) = \Delta U - \tfrac{1}{2}\Delta D - \tfrac{1}{2}\Delta S \\ \text{Neutron nonsinglet:} \quad & \Delta q_{NS}(n) = \Delta D - \tfrac{1}{2}\Delta U - \tfrac{1}{2}\Delta S \end{aligned} \quad (7)$$

so that

$$\begin{aligned} g_1^p &= \tfrac{2}{9}\Delta\Sigma + \tfrac{2}{9}\Delta q_{NS}(p) \\ g_1^n &= \tfrac{2}{9}\Delta\Sigma + \tfrac{2}{9}\Delta q_{NS}(n). \end{aligned} \quad (8)$$

Since the integrated quark spin accounts for only about 30% of the nulceon's spin, it is extremely interesting to know whether the integrated gluon spin in the nucleon is large. Of course the polarized gluon distribution, $\Delta g(x,Q^2)$, cannot be measured directly in deep inelastic scattering because gluons do not couple to the electromagnetic current. Instead Δg is inferred from the QCD evolution of the quark distributions. [See Ref. [10] for details of the process and references to the original literature.] However, evolution of imprecise data only constrains a few low moments of Δg and gives only crude information on global characteristics such as the existence and number of nodes. It is clear that Δg must be measured directly elsewhere.

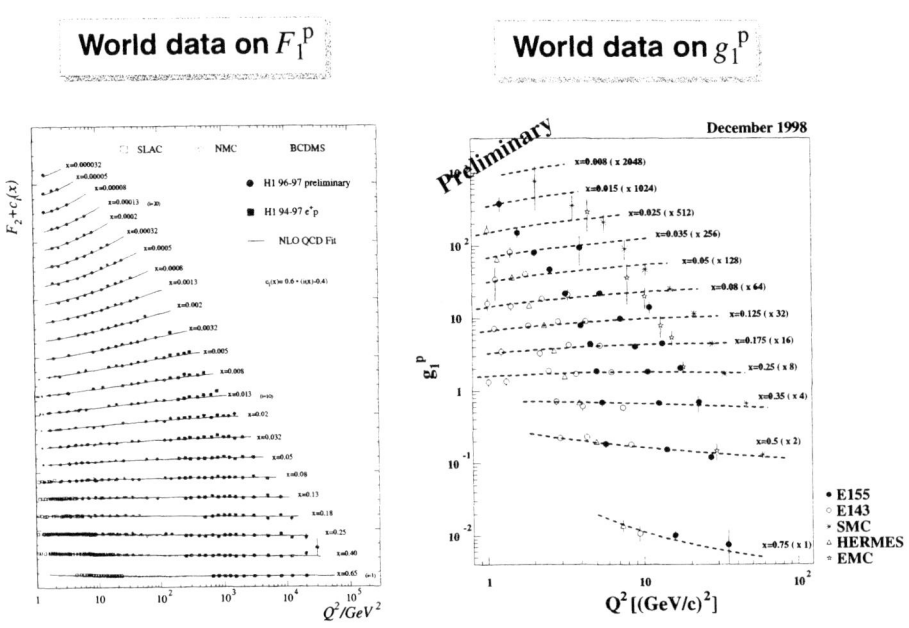

FIGURE 3. World data on spin-average and spin-dependent structure functions [11].

That said, the world's data on polarized structure functions is summarized in Figs. 3 and 4. Fig. 3 is taken from Naomi Makins's talk at DIS2000 and presents the world's data on g_1^p in the same format traditionally used for unpolarized structure function data [11]. The figure highlights the tremendous progress of the past decade as well as the need for much better data if our knowledge of polarized distributions would aspire to the same accuracy as unpolarized distributions. Fig. 4 shows the quark and gluon distributions extracted from the world's data by SMC, together with estimates of systematic and theoretical uncertainties [10]. While the information on quark distributions is fairly precise, it is clear that we know very little about the distribution of polarized gluons in the nucleon.

Probing polarized glue in the nucleon

As must have been clear from the preceeding discussion, a direct measurement of the polarized gluon distribution in the nucleon is probably the highest priority for groups interested in QCD spin physics. Further refinement of the indirect method

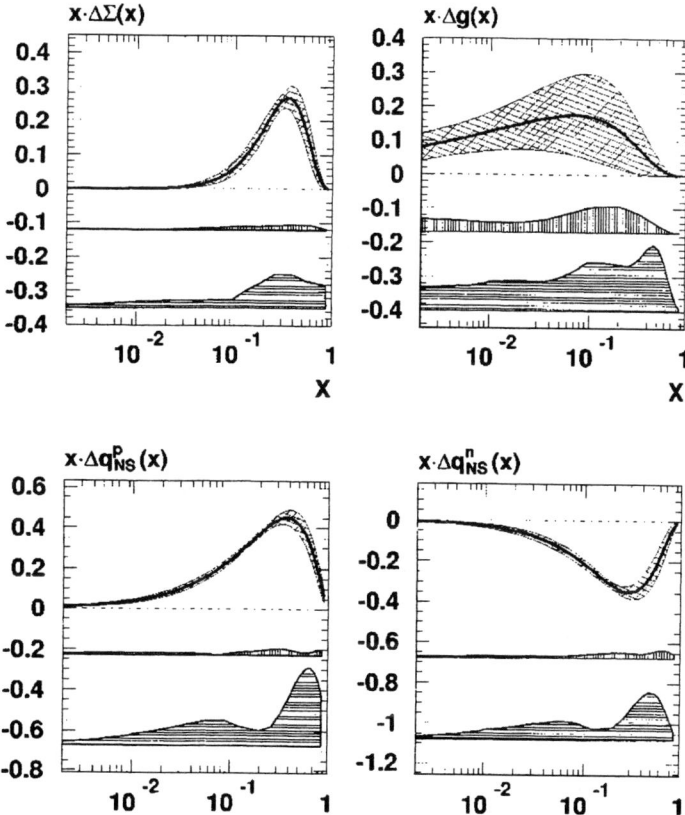

FIGURE 4. Polarized quark and gluon distribution functions. The upper figures show the distribution with a statistical error bound. The lower figures show estimates of systematic and theoretical uncertainties, respectively.

championed by SMC will contribute to this goal, but direct measurement is essential. Several direct methods are being pursued:

- $\bar{c}c$ pair production in $e\vec{p}_{\parallel} \to e'\bar{c}cX$ and related methods.

  The COMPASS Collaboration has proposed to extend this powerful probe of the unpolarized gluon distribution to the polarized case [12]. The basic mechanism is photon-gluon fusion, as shown in Fig. 5. For further discussion see the talk by Bradamante at this meeting.

  Variations on this method include two jet production: $e\vec{p}_{\parallel} \to e'$ jet jet X at large transverse momentum (as originally envisioned by Carlitz, Collins, and Mueller [13]); $\bar{c}c$ photoproduction $\gamma \vec{p}_{\parallel} \to \bar{c}cX$; and pion pair production, $\gamma \vec{p}_{\parallel} \to \pi\pi X$, which Hermes hopes to use a lower center of mass energies where $\bar{c}c$ and two jet production are not available [14].

- Single photon production at high transverse momentum in polarized $\vec{p}_{\parallel}\vec{p}_{\parallel} \to$

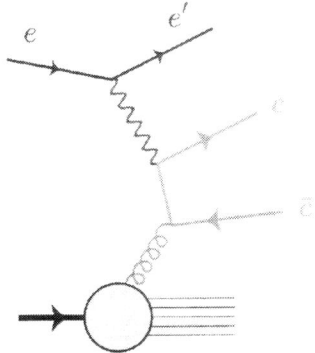

FIGURE 5. Measuring the polarized gluon distribution in $e\vec{p}_\| \to e'c\bar{c}X$.

γ jet X and related methods.

This is a prime goal for the polarized proton program at RHIC [15]. Here the basic mechanism is the QCD Compton process as shown in Fig. 6. This process should be an excellent probe of the polarized gluon distribution. However there is some controversy about higher-order QCD corrections that has yet to be resolved in the unpolarized case. Variations replace the high energy photon with a jet, or in the case of poor jet acceptance, a leading pion at high transverse momentum.

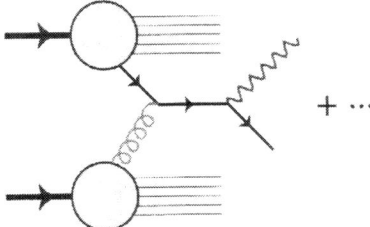

FIGURE 6. Measuring the polarized gluon distribution in $\vec{p}_\|\vec{p}_\| \to \gamma$ jet X.

Estimates of the precision of these methods have become available as better simulations come on line for COMPASS and RHIC. The projections for $\vec{p}_\|\vec{p}_\| \to \gamma$ jet X are shown in Fig. 7. An estimate of the COMPASS sensitivity is shown in Fig. 8. In both cases the experiments are compared with gluon distributions proposed by Gehrmann and Sterling. Clearly, the polarized proton program at RHIC has a major contribution to make in this area.

The nucleon's total angular momentum

In the old days (pre-1988), it was clear that quark and gluon spin distributions could be measured in deep inelastic scattering. In some uncertain sense they were imagined to be part of a relation that gave the nucleon's helicity,

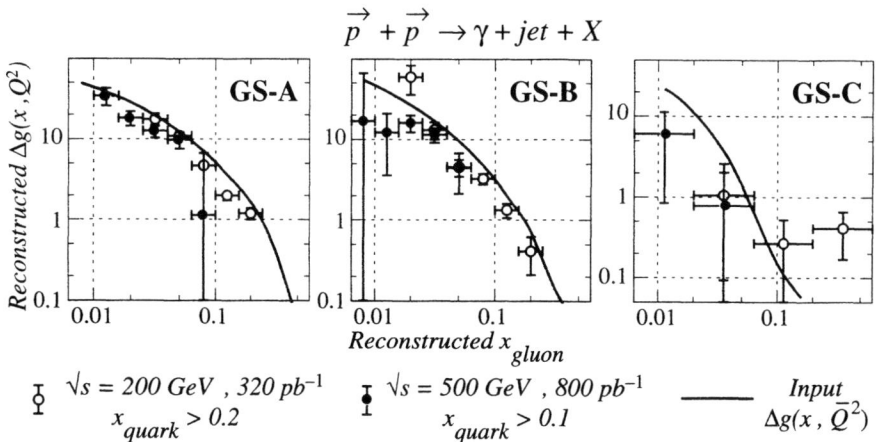

FIGURE 7. Estimates of polarized gluon distribution functions from $\vec{p}_\parallel \vec{p}_\parallel \to \gamma$ jet X at RHIC.

$$\tfrac{1}{2} = \tfrac{1}{2}\Delta\Sigma + \Delta g + \text{the rest} \qquad (9)$$

where "the rest" was not well understood. $\Delta\Sigma$ and Δg were (and are) measurable, gauge invariant, and given by integrals over x, $\Delta\Sigma = \Delta\Sigma(Q^2) = \int_0^1 dx \Delta\Sigma(x, Q^2)$, $\Delta g = \Delta g(Q^2) = \int_0^1 dx \Delta g(x, Q^2)$.

Significant progress occured in the late 80s and 90s as the other pieces of the angular momentum were related to local, gauge invariant operators [16]. This line of work culminated in Ji's decomposition of the nucleon's helicity [17],

$$= \tfrac{1}{2}\Delta\Sigma + \hat{L}_q + \hat{J}_g \qquad (10)$$

where $\hat{L}_q$ is the nucleon matrix element of an operator that rotates quarks' orbital motion about the $\hat{e}_3$-axis in the rest frame. It is one candidate for a definition of the quark orbital angular momentum in the nucleon. $\hat{J}_g$ is the nucleon matrix element of the operator that rotates the gluon about the $\hat{e}_3$ axis. Ji showed that $\hat{J}_g$ cannot be further decomposed into Δg and an orbital contribution given by a local gauge invariant operator. This should not be too surprising because it is well known that Δg itself cannot be expressed in terms of a *local* gauge invariant operator [18]. [In general the operator is non-local, but becomes local in $A^+ = 0$ gauge.] The virtue of eq. 10 is that $\hat{L}_q$ can be measured in deeply virtual Compton scattering (DVCS). Although $\hat{J}_g$ is in principle also measurable in DVCS, it requires a precision study of Q^2 *evolution* and is impossible in practice.

Most recently it has been possible to define gauge invariant *parton distributions* for all the components of the nucleon's angular momentum [19–21],

$$\tfrac{1}{2} = \int_0^1 dx \left\{ \tfrac{1}{2}\Delta\Sigma(x,Q^2) + \Delta g(x,Q^2) + L_q(x,Q^2) + L_g(x,Q^2) \right\} \qquad (11)$$

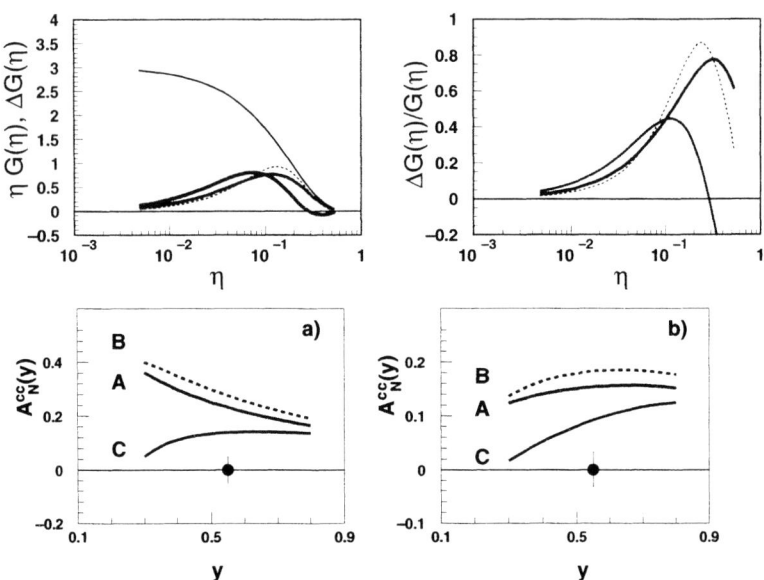

FIGURE 8. Gluon-associated assymetries for $\gamma \vec{p}_\| \to e' c \bar{c} X$ at COMPASS. See Bradamante's talk for details.

where L_q and L_g are Bjorken-x distributions of quark and gluon *orbital* angular momentum in the infinite momentum frame. L_q and L_g are given by the light-cone fourier transforms of bilocal operator products just like other parton distributions. This decomposition has many virtues: the four terms evolve into one another with Q^2 [19,20], each term is the Noether charge associated with the appropriate transformation of quarks or gluons [21]. Thus $L_g(x, Q^2)$ is the observable associated with the orbital rotation of gluons with momentum fraction x, about the infinite momentum axis in an infinite momentum frame. On the other hand, eq. 11 suffers from a significant drawback: unlike Ji's $\hat{L}_q$, we know of no way to measure either $L_q(x, Q^2)$ or $L_g(x, Q^2)$. They do not appear in the description of DVCS.

So the situation with respect to a complete description of the nucleon's angular momentum is frustrating. The theory is under control. Eq. 11 summarizes all we would like to know, but we do not know how to measure what we would like to know.

Transversity

One of the major accomplishments of the recent renaissance in QCD spin physics has been the rediscovery and exploration of the quark *transversity distribution*. First mentioned by Ralston and Soper in 1979 in their treatment of Drell-Yan μ-pair production by transversely polarized protons [22], the transversity was not recognized as a major component in the description of the nucleon's spin until the early 1990's [23–26]. We now know that the transversity, $\delta q(x, Q^2)$, together with the unpolarized distribution, $q(x, Q^2)$, and the helicity distribution, $\Delta q(x, Q^2)$, are required to give a complete

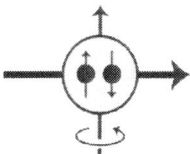

FIGURE 9. Transversity: transversely polarized quarks in a nucleon at infinite momentum.

description of the quark spin in the nucleon at leading twist. An equation tells this story clearly –

$$\mathcal{A}(x,Q^2) = \tfrac{1}{2}q(x,Q^2)\, I \otimes I + \tfrac{1}{2}\Delta q(x,Q^2)\, \sigma_3 \otimes \sigma_3$$
$$+ \tfrac{1}{2}\delta q(x,Q^2)\, (\sigma_+ \otimes \sigma_- + \sigma_- \otimes \sigma_+) \qquad (12)$$

Here, $\mathcal{A}$ is the quark distribution in a nucleon as a density matrix in both the quark and nucleon helicities (hence the external product of two Pauli matrices in each term). q governs spin average physics, Δq governs helicity dependence, and δq governs helicity flip – or transverse polarization – physics.

The transversity can be interpreted in parton language as the probability to find quarks of momentum fraction x, transversely polarized in a transversely polarized nucleon at infinite momentum. This is illustrated in Fig. 9.

The quark momentum distribution is well known and the helicity distribution is becoming better known. In contrast nothing is known about transversity from experiment. This is because it decouples from inclusive DIS. At leading twist helicity and chirality are identical. Transversity corresponds to helicity (and therefore chirality) flip. So transversity decouples from processes with only vector or axial vector couplings. This is shown schematically in Fig. 10a. In order to access transversity it is necessary to flip a quark's helicity in one soft process and compensate with another soft helicity flip process. Two examples where transversity does not decouple are transverse Drell-Yan: $\vec{p}_\perp \vec{p}_\perp \to \mu^+ \mu^- X$ (the original Ralston-Soper process where transversity was discovered) and semi-inclusive DIS where a final state fragmentation function flips helicity: $e\vec{p}_\perp \to e' \vec{h}_\perp X$, of which several examples exist. Both are shown schematically in Fig. 10b and c.

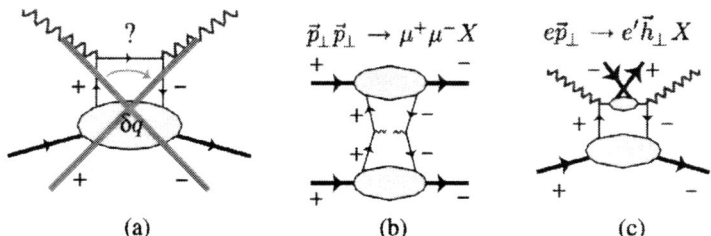

FIGURE 10. Deep inelastic processes relevant to transversity.

Measurements of quark transversity rank high on the agendas of Hermes, COMPASS and RHIC. Interest in this possibility has been piqued by the azimuthal pion asymmetry recently reported by Hermes, as will be discussed below.

Spin commissioning at RHIC

One of the most important landmarks of the past year was the successful spin commissioning run just concluded at RHIC. N. Saito will report in detail in his plenary talk. Polarized protons were accelerated in the AGS, and injected and stored in RHIC. The Coulomb-nuclear interference polarimeter functioned as expected. The Siberian snake in RHIC rotated the polarization as planned. Polarized beam was accelerated in RHIC past depolarizing resonances and the polarization was preserved with the aid of the snake. These milestones mark the beginnings of polarized collider physics at RHIC, and a whole new window on the deep spin structure of hadrons. This new facility would not have been possible without the support of RIKEN and the effort of a team of Japanese experimenters working at BNL and supported by the joint RIKEN/BNL Research Center. Working with both major RHIC detector groups, STAR and PHENIX, the RHIC Spin Collaboration has developed an ambitious program for probing spin structure in QCD.

Fragmentation and spin:
The Hermes asymmetry and beyond

To my mind the single most interesting development in QCD spin physics reported since SPIN98 is the azimuthal asymmetry in pion electroproduction from Hermes [27]. It is interesting in itself and also as an emblem of a new class of spin measurements involving spin-dependent fragmentation processes, which act as filters for exotic parton distribution functions like transversity.

Fragmentation functions allow us to access and explore the spin structure of unstable hadrons, which cannot be used as targets for deep inelastic scattering. Examples include the longitudinal and transverse spin dependent fragmentation functions of the Λ, schematically $\vec{q}_\| \to \vec{\Lambda}_\|$ and $\vec{q}_\perp \to \vec{\Lambda}_\perp$. Since the $\Lambda \to p\pi$ decay is self-analysing it is relatively easy to measure the spin of the Λ. By selecting Λ's produced in the current fragmentation region one can hope to isolate the fragmentation process $q \to \Lambda$. The principal challenge of such measurements is for theorists: we have no theoretical framework for analysing fragmentation functions. Having measured the quark spin structure of the nucleon, we can use flavor-SU(3) to estimate the way quark spins are distributed in the Λ [28]. However we do not know if this information is reflected in the fragmentation process $q \to \Lambda$. Another, perhaps less obvious, example is the tensor fragmentation function of the ρ, denoted schematically by $(q \to \rho_\pm) - (q \to \rho_0)$, where ρ_h are ρ helicity states [29]. ρ decay transmits no spin information, but it distinguishes the longitudinal and transverse helicity states required for this measurement. The data are already available. The challenge to theorists is to make use of it.

Even if we do not know how to interpret fragmentation functions, we can use them as filters, to select parton distribution functions that decouple from completely inclusive DIS. The salient example is the use of a helicity flip fragmentation function to select the quark transversity distribution. As shown in Fig. 10c, by interposing a helicity flip fragmentation function on the struck quark line in DIS, it is possible to access the transversity. There are several candidates for the necessary helicity flip fragmentation function:

- $e\vec{p}_\perp \to e'\vec{\Lambda}_\perp X$

    In this case the helicity flip fragmentation function of the Λ is exactly analogous to the transversity distribution function in the nucleon [30,31]. The only difficulty with this example is the relative rarity of Λ's in the current fragmentation region, and the possibly weak correlation between the Λ polarization and the polarization of the u quarks, which dominate the proton.

- $e\vec{p}_\perp \to e'\pi(\vec{k}_\perp)X$ [32]

    [The "Collins Effect"] In this case the azimuthal angular distribution of the pion relative to the $\vec{q}$ axis can be analyzed to select the interference between pion orbital angular momentum zero and one that correlates with quark helicity flip. In more traditional terms the effect is proportional $\vec{S}_\perp \cdot \vec{q} \times \vec{p}_\pi$. This is multiplied by the quark transversity in the target and an unknown fragmentation function (known as the Collin's function) describing the propensity of the quark to fragment into a pion in a superposition of orbital angular momentum zero and one states. The fact that fragmentation functions depend on z while distribution functions depend on x allows the shape of the transversity distribution to be measured in this manner.

- $e\vec{p}_\perp \to e'\pi\pi X$ [33,34]

    In this case the angular distribution of the two pion final state substitutes for the azimuthal asymmetry.

Last year Hermes announced the observation of an azimuthal asymmetry similar to the Collin's asymmetry described above, but with a longitudinally polarized target: $e\vec{p}_\parallel \to e'\pi(\vec{k}_\perp)X$. Their data are shown in Fig. 11. This asymmetry could be a (suppressed) reflection of the Collins effect because the target spin, while parallel to the electron beam, has a small component, $\mathcal{O}(1/Q)$ perpendicular to the virtual photon. It could also result from competing twist-three helicity flip effects also suppressed by $1/Q$. Unless the Hermes asymmetry is entirely twist three, which seems unlikely, it appears that the prospects for observing a large azimuthal asymmetry from a *transversely* polarized target are very good. Hermes will be running with a transversely polarized target this year and their results will be awaited with considerable excitement.

Spin-dependent static moments

Polarization effects abound in the strong interactions at low energies. Some are quite striking, but most defy theoretical analysis because they occur in two body scattering (or more complex processes), which we do not know how to analyse in QCD. One striking exception, quite similar in many ways to deep inelastic physics, are the spin (and flavor) dependent static moments of the nucleon. In general these are measured in elastic lepton nucleon scattering, $\ell N \to \ell' N$. The general form is

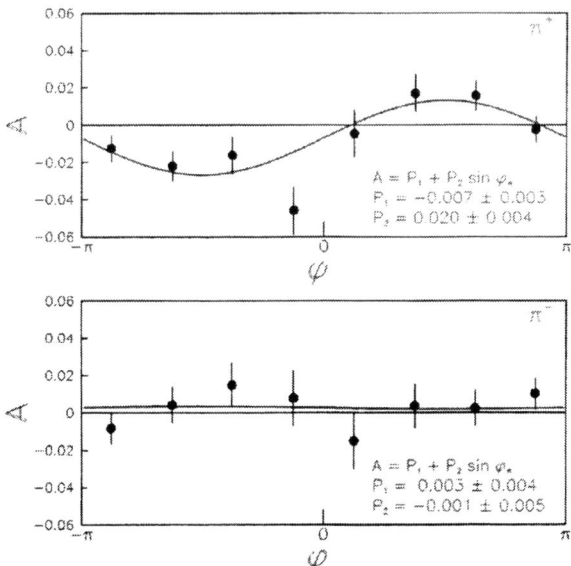

FIGURE 11. The Hermes azimuthal asymmetry.

$$\Gamma \propto \langle PS|\bar{q}\Gamma q|PS\rangle \qquad (13)$$

where Γ is some operator in the spin and/or space coordinates of the quark field. Familiar examples include the axial charges ($\Gamma = \gamma_\mu \gamma_5$), magnetic moments ($\Gamma = \frac{1}{2}\vec{r} \times \vec{\gamma}$), and charge radii ($\Gamma = (\vec{r})^2$). A less familiar example is the "tensor charge" ($\Gamma = \sigma^{0i}\gamma_5$), which though measurable in principle, does not couple to any electroweak current and cannot be measured in $\ell N \to \ell' N$. The isovector, $u - d$, and hypercharge, $u + d - 2s$, flavor combinations are relatively easy to measure given the variety of electroweak currents and baryons related to one another by flavor-SU(3) transformations. However, the third flavor combination, $u+d+s$, does not appear in the electromagnetic or charge-changing weak currents, and cannot be constructed by flavor-SU(3) rotations because it is a flavor-SU(3) singlet and all the other currents are flavor-SU(3) octets.

Much progress has been made in recent years both by theorists, who have learned that the nucleon's tensor charge is related to the lowest moment of its transversity structure function (in analogy to the Bjorken Sum Rule); and by experimenters, who have measured the flavor combination $u + d + s$ (and therefore the strangeness matrix elements) by extracting the Z^0-nucleon coupling via parity violating ep elastic scattering. The Z^0 couples to weak isospin (hence $u - d + s - c \ldots$) that, restricted to light quarks, is a linear combination including the flavor singlet. Table 2 shows a simplified summary of the Dirac and flavor structure of some static matrix elements and how they are measured.

Two flagship measurements in this area are the extraction of μ_s, the nucleon matrix element of $s^\dagger \frac{1}{2}\vec{r} \times \vec{\gamma} s$, from parity violating $ep \to ep'$ (the SAMPLE experiment at Bates), and extraction of $\langle r_s^2\rangle$, the nucleon matrix element of $s^\dagger(\vec{r})^2 s$, from the same process in a different kinematic domain (the HAPPEX experiment at JLab). Of

TABLE 2. Electric Dipole Moments

	Dirac Structure		
Flavor	$\vec{\gamma}\gamma_5$	$\vec{r}\times\vec{\gamma}$	$\sigma^{0i}\gamma_5$
$u-d$	β-decay	Nucleon mag. mom.	Transverse DIS
$u+d-2s$	Hyperon β-decay	Hyperon mag. mom.	Transverse DIS
$u+d+s$	Polarized DIS	Parity odd $\bar{c}p \to ep$	Transverse DIS
	$q+\bar{q}$	$q-\bar{q}$	$q-\bar{q}$

course the latter is not really spin-physics, but it belongs in the same discussion. HAPPEX first run at relatively large momentum transfer saw no sign of a nucleon strange electric form factor [35]. No statement can be made about $\langle r_s^2 \rangle$, however, until data at lower Q^2 become available.

SAMPLE's initial results are quite interesting – for unexpected reasons [36]. The signal of interest, the Z^0-nucleon coupling, is contaminated by a parity violating photon-nucleon interaction, the so called "anapole moment", and by higher-order weak radiative corrections to electron nucleon scattering (see Fig. 12). The anapole and weak radiative corrections are not known a priori and have to be estimated in models [37,38]. They are parameterized by $G_A^e(T=1)$ in Fig. 13. A single measure-

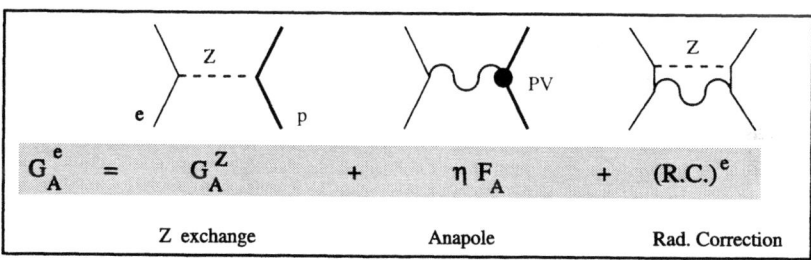

FIGURE 12. Contributions to parity violation in $ep \to ep'$.

ment, say $\vec{A}\, ep \to ep$, gives a line in the $G_M^s - G_A^a(T=1)$ plane. The initial SAMPLE measurement together with the Holstein-Ramsey-Musolf estimate of $G_A^a(T=1)$ gave a large positive estimate for $\mu_s(p)$ (admittedly with large error bars) in contrast to model calculations that typically give negative μ_s [36]. Most recently, SAMPLE has announced measurements off a deuteron target, $\vec{A}\, ed \to ed$, which give an independent line in the $G_M^s - G_A^a(T=1)$ plane [39]. The result, shown in Fig. 13b, suggests the estimate of $G_A^a(T=1)$ may be wrong and that μ_s is closer to zero, though certainly compatible with theoretical models predicting small negative values.

These are only the earliest results in what promises to be a productive study of the strangeness content of the nucleon using very precise measurements of parity violating elastic lepton nucleon scattering.

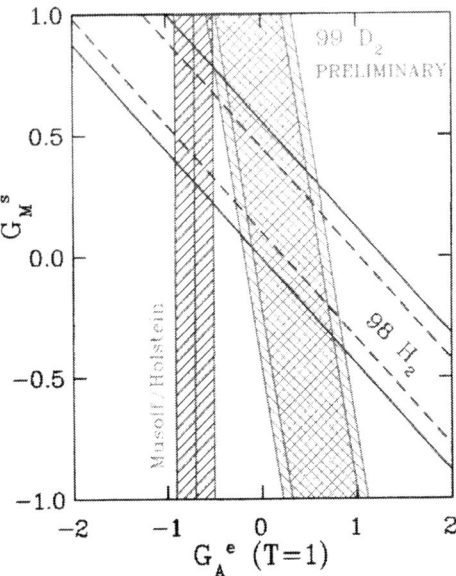

FIGURE 13. Interpretation of SAMPLE measurements.

The Drell-Hearn-Gerasimov-Hosada-Yamamoto Sum Rule

The prospects for a definitive test of this deep and ancient sum rule [40–42] are now excellent. Experiments proposed and/or underway at Mainz and JLab will cover a wide range of energies with high polarization and high statistics. The question I would like to address here is "What does the DHGHY sum rule test?". The sum rule reads,

$$\frac{2\pi^2\alpha}{M^2}\kappa^2 = \int_0^\infty \frac{d\nu}{\nu}\left(\sigma_P(\nu) - \sigma_A(\nu)\right) \quad (14)$$

where κ and M are the anomalous magnetic moment and mass of the target, and $\sigma_{P,A}$ are the total photoabsorption cross sections (as functions of the laboratory photon energy, ν) for target and photon spins parallel and antiparallel.

The sum rule rests on two assumptions:

- Low's low-energy theorem

  Many years ago Low derived an extension of the Thompson limit in Compton scattering [43]. The nucleon's forward Compton amplitude can be written

  $$f(\nu) = f_1(\nu^2)\vec{\varepsilon}'^* \cdot \vec{\varepsilon} + \nu f_2(\nu^2) i\vec{\sigma} \cdot \vec{\varepsilon}'^* \times \vec{\varepsilon} \quad (15)$$

  where f_1 and f_2 are the spin-nonflip and spin-flip amplitudes respectively.

  Using gauge invariance and QED, Low showed

  $$f_2(0) = -\tfrac{1}{2}\frac{\alpha}{M^2}\kappa^2 \;. \quad (16)$$

- An unsubtracted dispersion relation

    Analyticity, crossing and unitarity dictate that the forward Compton amplitudes satisfies dispersion relations, which combine Cauchy's theorem with the optical theorem ($\text{Im} f(\nu) \propto \sigma(\nu)$),

$$\text{Re} f_2(\nu) = \sum_{j=0}^{J_{MAX}} c_j \nu^{2j} + \frac{1}{8\pi^2} \text{P} \int_0^\infty d\nu'^2 \frac{\sigma_A(\nu') - \sigma_P(\nu')}{\nu'^2 - \nu^2} \qquad (17)$$

    The polynomial is usually omitted in writing the dispersion relation, however it is not excluded by analyticity or unitarity. Since it has no imaginary part, it does not affect the measurable cross sections. For the moment let us omit the polynomial.

Then the DHGHY sum rule is obtained by evaluating $f_2(0)$ with the aid of the dispersion relation, and equating it to Low's low energy limit.

What could go wrong with this? Absent any problems with electrodynamics, the only weak point is ignoring the possible polynomial in the dispersion relation. Since we need only $f_2(0)$ only the constant term (c_0) in the polynomial matters. Usually, limits on the growth of amplitudes at high energies are invoked to restrict the order of the polynomial. However, they do not exclude the constant, c_0. In the standard derivations c_0 is simply ignored. This is called the assumption of an "unsubtracted dispersion relation". This is something of a misnomer: If the integral eq. 17 diverged it would be *necessary* to reformulate it by formally subtracting $f_2(0)$ (remember we are assuming $c_1 = 0, c_2 = 0, \ldots$). The resulting integral would be more convergent, but now the constant $f_2(0)$ would appear in the relation,

$$\text{Re} f_2(\nu) = \text{Re} f_2(0) + \frac{\nu^2}{8\pi^2} \text{P} \int_0^\infty d\nu'^2 \frac{\sigma_A(\nu') - \sigma_P(\nu')}{\nu'^2(\nu'^2 - \nu^2)} \qquad (18)$$

This is a "subtracted dispersion relation". Now substitution into the Low's theorem yields nothing useful. Even if the dispersion relation does not *need* subtraction, ie. even if the integral in eq. 17 converges, still the constant c_0 could be non-zero and spoil the sum rule. Therefore, measurement of the high-energy behavior of σ_A and σ_P *does not* determine whether an additive constant is present in the dispersion relation.

Debate about the validity of the DHGHY sum rule usually centers on whether the dispersion integral needs subtraction. Even if it doesn't, the sum rule could be ruined by a non-zero, real constant in $f_2(\nu)$. Such a constant is called, for historical reasons, a "$J=0$ fixed pole". So the question of the validity of the DHGHY sum rule comes down to whether $J=0$ fixed poles occur in QCD. It is known that they do not occur in low orders of perturbation theory. This was first verified when the electroweak anomalous magnetic moment of the muon was first calculated using a generalization of these methods [44]. It has been subsequently studied to higher orders. Brodsky and Primack have argued that it does not occur in ordinary bound states [45] – that the anomalous magnetic moment of hydrogen can be calculated from a generalized DHGHY sum rule with out a $J=0$ fixed pole. Still, the verdict is out in QCD, where bound states are not so simple.

If the DHGHY sum rule is verified experimentally this question will recede to a footnote to history. If, however, experiment fails to confirm it, we will all have a lot to learn about $J=0$ fixed poles!

CONCLUSIONS

My conclusions are brief. We have made striking progress in recent years. The prospects for further progress are excellent. I expect that spin physics will continue to surprise us as it has in the past. The reason is that spin is fundamentally quantum mechanical in its origins so that it beggars our classical intuition. Remember – we don't even know why matter $\equiv$ fermions exists!

REFERENCES

1. C. Prescott, in *SPIN 98 Proceedings of the 13th International Symposium on High Energy Spin Physics Protvino, Russia 8 - 12 September 1998*, N. E. Tyurin, V. L. Solovianov, S. M. Troshin and A. G. Ufimtsev, eds. (World Scientific, Singapore, 1999)
2. B. C. Barish and R. Weiss, Phys. Today **52**, 44 (1999).
3. J. Bailey *et al.* [CERN-Mainz-Daresbury Collaboration], Nucl. Phys. **B150**, 1 (1979).
4. H. N. Brown *et al.* [Muon (g-2) Collaboration], Phys. Rev. D **62**, 091101 (2000) [hep-ex/0009029].
5. I. B. Khriplovich, Nucl. Phys. **A663**, 147 (2000) [hep-ph/9906533].
6. Y. K. Semertzidis [Muon EDM Collaboration], *Prepared for Workshop on Frontier Tests of Quantum Electrodynamics and Physics of the Vacuum, Sandansky, Bulgaria, 9-15 Jun 1998.*
7. A. D. Krisch, in *SPIN 98 Proceedings of the 13th International Symposium on High Energy Spin Physics Protvino, Russia 8 - 12 September 1998*, N. E. Tyurin, V. L. Solovianov, S. M. Troshin and A. G. Ufimtsev, eds. (World Scientific, Singapore, 1999)
8. S. A. Larin and J. A. Vermaseren, Phys. Lett. **B259**, 345 (1991).
9. E. V. Shuryak and A. I. Vainshtein, Nucl. Phys. **B201**, 141 (1982).
10. B. Adeva *et al.* [Spin Muon Collaboration], Phys. Rev. D **58**, 112002 (1998).
11. N. C. R. Makins [for the Hermes Collaboration] Talk presented at DIS2000. To be published in the proceedings.
12. F. Bradamante, Prog. Part. Nucl. Phys. **44**, 339 (2000).
13. R. D. Carlitz, J. C. Collins and A. H. Mueller, Phys. Lett. **B214**, 229 (1988).
14. M. G. Vincter [HERMES Collaboration], RIKEN Rev. **28**, 70 (2000).
15. G. Bunce, N. Saito, J. Soffer and W. Vogelsang, hep-ph/0007218.
16. R. L. Jaffe and A. Manohar, Nucl. Phys. **B337**, 509 (1990).
17. X. Ji, Phys. Rev. Lett. **78**, 610 (1997) [hep-ph/9603249].
18. A. V. Manohar, Phys. Rev. Lett. **66**, 289 (1991).
19. P. Hagler and A. Schafer, Phys. Lett. **B430**, 179 (1998) [hep-ph/9802362].
20. A. Harindranath and R. Kundu, Phys. Rev. **D59**, 116013 (1999) [hep-ph/9802406].
21. S. V. Bashinsky and R. L. Jaffe, Nucl. Phys. **B536**, 303 (1998) [hep-ph/9804397].
22. J. Ralston and D. E. Soper, *Nucl. Phys.* **B152** (1979) 109.
23. X. Artru and M. Mekhfi, *Z. Phys.* **C 45** (1990) 669.

24. R. L. Jaffe and X. Ji, *Phys. Rev. Lett.* **67** (1991) 552.
25. J. L. Cortes, B. Pire, J. P. Ralston, *Z. Phys.* **C 55** (1992) 409.
26. For a review, see R. L. Jaffe, in *Proceedings of the Ettore Majorana International School of Nucleon Structure: 1st Course: The Spin Structure of the Nucleon, Erice, Italy, 1995.*, hep-ph/9602236.
27. A. Airapetian *et al.* [HERMES Collaboration], Phys. Rev. Lett. **84**, 4047 (2000) [hep-ex/9910062].
28. M. Burkardt and R. L. Jaffe, Phys. Rev. Lett. **70**, 2537 (1993) [hep-ph/9302232].
29. A. Schafer, L. Szymanowski and O. V. Teryaev, Phys. Lett. **B464**, 94 (1999) [hep-ph/9906471].
30. R. A. Kunne *et al.*, *Saclay CEN - LNS-Ph-93-01*.
31. R. L. Jaffe, Phys. Rev. D **54**, 6581 (1996) [hep-ph/9605456].
32. J. C. Collins, S. F. Heppelmann and G. A. Ladinsky, Nucl. Phys. **B420**, 565 (1994) [hep-ph/9305309].
33. J. C. Collins and G. A. Ladinsky, hep-ph/9411444.
34. R. L. Jaffe, X. Jin and J. Tang, Phys. Rev. Lett. **80**, 1166 (1998) [hep-ph/9709322].
35. K. A. Aniol *et al.* [HAPPEX Collaboration], Phys. Rev. Lett. **82**, 1096 (1999) [nucl-ex/9810012].
36. D. T. Spayde *et al.* [SAMPLE Collaboration], Phys. Rev. Lett. **84**, 1106 (2000) [nucl-ex/9909010].
37. M. J. Musolf and B. R. Holstein, Phys. Lett. **B242**, 461 (1990).
38. M. J. Musolf, T. W. Donnelly, J. Dubach, S. J. Pollock, S. Kowalski and E. J. Beise, Phys. Rept. **239**, 1 (1994).
39. R. McKeown [for the SAMPLE Collaboration], private communication.
40. S. B. Gerasimov, Sov. J. Nucl. Phys. **2**, 430 (1966).
41. S. D. Drell and A. C. Hearn, Phys. Rev. Lett. **16**, 908 (1966).
42. M. Hosoda and K. Yamamoto Prog. Theor. Phys. Lett. **36**, 425 (1966).
43. F. E. Low, Phys. Rev. **96**, 1428 (1954).
44. G. Altarelli, N. Cabibbo and L. Maiani, Phys. Lett. **B40**, 415 (1972).
45. S. J. Brodsky and J. R. Primack, Annals Phys. **52**, 315 (1969).
46. Smith A., and Doe B., *J. Chem. Phys.* **76**, 4056 (1982).
47. Jones, C., *J. Chem. Phys.* **68**, 5298 (1978).
48. Jones, C., and Smith, A., *Title of Book*, New York: IEEE Press, 1978, ch. 6, pp. 23-26.

HERMES at the Turn of the Millennium

E.C. Aschenauer [a] on behalf of the HERMES Collaboration

[a] *DESY Zeuthen, Platanenallee 6, 15738 Zeuthen, Germany*

Abstract. This report highlight results from the HERMES experiment from the last 5 years on the spin structure of the nucleon. The inclusive spin-structure function $g_1(x,Q^2)$ is described, followed by measurements of the flavor-separated quark polarizations based on semi-inclusive data. Next, first measurements are presented that indicate that both the gluon polarization and the transversity distribution $h_1(x)$ are non-zero Finally, the subject of parton-parton correlations is introduced, along with new data related to higher twist effects and skewed parton distributions.

I INTRODUCTION

Deep-inelastic scattering of polarized leptons from polarized targets has been used successfully for more than a decade to yield information about the spin structure of the nucleon. The central question at issue is the manner in which the partonic components of the nucleon manage to produce its overall spin of $1/2\ \hbar$. The expression

$$\frac{1}{2} = \frac{1}{2}\Delta\Sigma + \Delta g + L_q + L_g \tag{1}$$

illustrates that the total spin of the nucleon must arise from a combination of three sources: the helicity distribution of the quarks ($\Delta\Sigma$), the helicity distribution of the gluons (Δg), and the orbital angular momentum of quarks and gluons (L_q, L_g). The notation $\Delta q(x)$ denotes the *polarized* counterpart of the familiar parton distribution functions (PDF): $\Delta q(x) \equiv q^\uparrow(x) - q^\downarrow(x)$. Here, x is the Bjorken scaling variable, and $q^\uparrow(q^\downarrow)$ indicates quarks whose spins are parallel (anti-parallel) to the spin of the nucleon.

To obtain some feeling for the expected magnitudes of these various sources of the nucleon spin, one may turn to phenomenological models. If one considers the current quarks seen by dynamical QCD processes, relativistic effects must be taken into account because of the light quark masses. Calculations performed in, e.g., the relativistic MIT bag model then suggest that $\Delta\Sigma \simeq 0.60 - 0.75$ [1]; the remainder of the nucleon spin is accounted for by the orbital angular momentum of the moving quarks. To include sea quarks in the picture, one may turn to meson cloud models, for example. In such models, the nucleon is represented as a bare, valence object which may emit a pseudo-scalar meson. It is through the emission of such mesons that the quark sea is generated. One calculation performed in such a framework indicates that the sea quarks should carry a negative polarization, but that the anti-quarks in the sea are unpolarized: all anti-quarks in this picture are part of a spin-0 meson, and therefore carry no polarization [2]. An alternative picture of the nucleon comes from the chiral-quark soliton model, where the nucleon appears as a chiral soliton in a pion field. A recent calculation of this type [3], performed in the large N_c limit, suggests that the $\bar{u}$ and $\bar{d}$ quarks do carry a significant polarization, but of opposite sign: $\Delta \bar{u} \simeq -\Delta \bar{d}$.

This brief sample of phenomenological models makes that progress in our understanding of the spin structure of the nucleon requires precise experimental input.

II INCLUSIVE MEASUREMENTS

The longitudinal spin structure function $g_1(x, Q^2)$ has been measured by a number of experiments, using longitudinally polarized lepton beams and nuclear targets [4]. In the quark-parton model and in leading order QCD, this structure function is given by the charge weighted sum of polarized quark spin distributions Δq of flavor q: $g_1(x) = \frac{1}{2} \sum_q e_q^2 \Delta q(x)$. In the framework of next-to-leading order (NLO) QCD, the relationship between $g_1(x, Q^2)$ and the polarized PDF's is more complex, and includes a contribution from the gluon spin distribution $\Delta g(x)$. The singlet ($\Delta \Sigma$), non-singlet ($\Delta q_3, \Delta q_8$), and gluon (Δg) spin distributions are each prefixed with a coefficient function of different Q^2 dependence. Fits to the data in NLO seek to obtain information about the polarizations of these different flavor combinations by exploiting their distinct Q^2 dependences.

Numerous spin-dependent NLO fits exist in the literature. One example is the global fit performed by the SMC collaboration [5], an analysis distinguished by a careful treatment of uncertainties. The results indicate that the quark spins account for a relatively small fraction of the total nucleon spin, of order 20 - 40% (the value depends on the NLO factorization scheme used in the analysis). Furthermore, the strange sea would seem to have a negative polarization, of order -10%. The gluon polarization Δg, by comparison, is only poorly constrained by the inclusive data but there is some indication that it is positive.

Although the body of data on $g_1(x)$ is still being expanded (see e.g. the new low-x HERMES data presented at this conference [6]) the precision and kinematic coverage of these measurements is still rather limited. Present NLO analyzes also use information from baryon β-decay to constrain the first moments of the non-singlet distributions (Δq_3 and Δq_8). However, the use of the constraint on Δq_8 is based on SU(3)-symmetry among baryons, which is known to be inexact. A recent study [7] explored the dependence of extracted quark polarizations on SU(3)-symmetry breaking effects. The influence of such effects on the quark polarization $\Delta \Sigma$ was found to be small. However, the gluon and strange quark polarizations were found to vary significantly: e.g. values of Δs from -0.02 to -0.15 were obtained.

III QUARK POLARIZATIONS FROM SEMI-INCLUSIVE MEASUREMENTS

To proceed further in our knowledge of the polarized parton distributions, experimenters have turned to *semi-inclusive* asymmetry measurements. In these measurements, a hadron is detected in coincidence with the scattered beam lepton. Through the agency of the fragmentation functions, a probabilistic relation exists between the flavor of the struck quark and the flavor content of the hadrons generated in the final state. By measuring the spin asymmetry A_1^h for hadrons of various types, one may separate the polarized parton distributions by flavor.

While the inclusive measurements only yielded the first moment of the flavor-separated spin distributions, the full x-dependence of the spin distributions can be extracted from semi-inclusive data. Furthermore, inclusive measurements are not able to distinguish between quark and antiquark polarizations, since the inclusive cross-section depends on the square of the charge of the struck quark.

Semi-inclusive asymmetries for positive and negative hadron production have been measured and analyzed by both the HERMES [8,9] and SMC experiments [10]. The SMC measurements were performed on a proton target, while HERMES has used deuterium (1998-2000), hydrogen (1996-1997) and ^{3}He (1995) targets. As in the inclusive case, the precision of the semi-inclusive asymmetries published to date is limited. Consequently, current semi-inclusive analysis impose constraints relating the various flavor

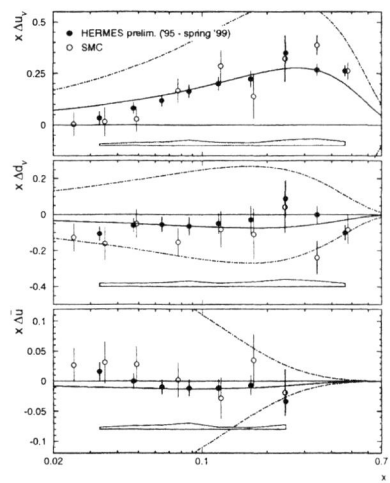

FIGURE 1. The polarized valence quark and the $\bar{u}$ quark distributions as an example for the sea quark distributions.

components of the sea polarization, to improve statistical significance. Two alternatives have been considered: either the polarization $\Delta q_s(x)/q_s(x)$ or the spin distribution $\Delta q_s(x)$ of the sea quarks is assumed to be independent of flavor. The choice between these two hypotheses is found to have a negligible influence on the results.

Figure 1 shows the HERMES and SMC results for the extracted valence ($\Delta u_v, \Delta d_v$) and sea (Δq_s) quark polarizations. All data points are evolved to a common Q^2 of 2.5 GeV/c^2. The dash-dotted lines represent the positivity limit $|\Delta q(x)| \leq q(x)$, while the solid lines show the 'Gluon A' leading order parameterization of Gehrmann and Stirling [11]. As this latter parameterization is based on fits to inclusive measurements, one sees that the results derived from inclusive and semi-inclusive measurements are in good agreement.

IV GLUON POLARIZATION FROM PAIRS OF HIGH-P_T HADRONS

As described above, the gluon polarization is only poorly constrained by existing data. One way to measure $\Delta g(x)$ directly is via the photon gluon fusion process, where a gluon from the nucleon participates in the hard scattering process. A useful experimental signature of this process is the production of jets with high transverse momentum p_T. The transverse momentum produced in the fragmentation process is small, and two back-to-back jets with sufficiently high p_T thus reflect the high p_T of the quark and anti-quark produced in the photon gluon fusion process. At the energies of fixed target experiments, high-p_T hadrons can be used in place of jets [13].

HERMES has recently measured the spin asymmetry in the photoproduction ($Q^2 \simeq 0$) of pairs of high-p_T hadrons, using a polarized hydrogen target [12]. Events were selected that contained at least one positively charged hadron h^+ and at least one negatively charged hadron h^-. In the kinematic region $p_T^{h_1} > 1.5$ GeV/c and $p_T^{h_2} > 1.0$ GeV/c, the spin asymmetry is found to have a negative sign (fig. 2a). The symbols h_1 and h_2 denote respectively the hadrons with the highest and next-to-highest p_T in each event.

The measured asymmetry was interpreted using the PYTHIA Monte Carlo generator at leading order. In this Monte Carlo simulation, several different processes contribute to the two-hadron cross section. However the only presently known process that might contribute to a *negative* asymmetry is the photon-gluon fusion (PGF) process: PGF has an analyzing power of -1 (for massless quarks), and so will produce a negative asymmetry given a positive gluon polarization. In the same region of phase space where a negative asymmetry is observed, the PYTHIA model indicates that the cross section is indeed dominated by photon gluon fusion. The best fit value obtained for the gluon polarization is $\langle \Delta g/g \rangle = 0.41 \pm 0.18$ (stat.) ± 0.03 (syst.), at an average $\langle x_g \rangle = 0.17$. The systematic uncertainty represents the experimental contribution only (fig. 2b).

It must be pointed out that this measurement is subject to theoretical uncertainties of an as-yet undetermined magnitude. At the kinematics of this measurement, no spin-dependent analyzes of higher order QCD processes are available, for example. However, to alter the principal conclusion of this analysis, i.e., that $\langle \Delta g/g \rangle$ at $\langle x_g \rangle = 0.17$ is positive, a significant contribution from a neglected process with a large negative spin asymmetry would be needed.

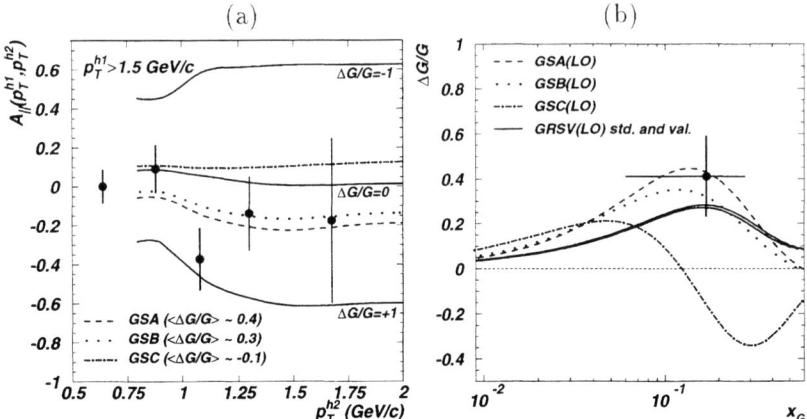

FIGURE 2. (a) $A_\|$ for high-p_T hadron pair production measured at HERMES compared with Monte Carlo predictions for $\Delta G/G = \pm 1$ (lower/upper solid curves), $\Delta G/G = 0$ (middle solid curve), and the phenomenological LO QCD fits of Ref. [11] (dashed, dotted and dot-dashed curves). (b) The extracted value of $\Delta G/G$ compared with phenomenological QCD fits to a subset of the world's data on $g_1^{p,n}(x, Q^2)$. The curves are from Refs. [11,14], evaluated at a scale of 2 $(\text{GeV}/c)^2$.

V HARD EXCLUSIVE PROCESSES

Much recent interest has surrounded the analysis of hard exclusive processes, where a virtual photon at high Q^2 scatters semi-elastically from a nucleon target. The final state contains only the recoiling nucleon along with a single additional particle (such as a meson, or a real photon). Theoretical analysis of such exclusive processes has led to the development of the Skewed Parton Distribution (SAD) formalism (see, for example, Ref. [26]). These SPD's represent a generalization of the familiar structure functions: in the forward limit of the SPD's, one recovers the parton distributions of deep-inelastic scattering; taking the first moments of their x-dependence, one obtains the elastic form factors of the nucleon. The description of hard exclusive processes involves the so-called handbag diagram, where two quarks of different momentum are exchanged with the nucleon in the same event. Such processes thus provide information on parton-parton correlations in the nucleon wave function. Perhaps most intriguing of all, the second moments of the skewed parton distributions are expected to be sensitive to the unknown orbital angular momentum L_q of the quarks.

The cleanest process to study these new parton distributions is deeply virtual compton scattering (DVCS), where a real photon is produced in the exclusive limit. Following Ref. [19] the Φ-dependence of the ep cross section with unpolarized protons and leptons can be written as

$$\frac{d\sigma}{d\Phi dt dQ^2 dx_B} = \frac{1}{32 \cdot (2\pi)^2} \frac{x_B y^2}{Q^4} \frac{1}{\sqrt{1+4x_B^2 m^2/Q^2}} \cdot |T_{BH} + T_{DVCS}|^2 \quad (2)$$

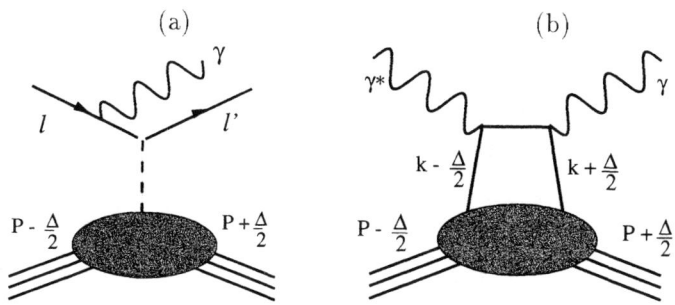

FIGURE 3. Feynman diagram for BH (a) and DVCS (b)

It is unfortunately very difficult to separate experimentally the Bethe-Heitler (BH) process and the DVCS process on the basis of a cross section measurement alone (see figure 3), as the final states are identical. It is easier to study the BH-VCS interference term, equation 3, which can be expressed in terms of helicity amplitudes $M_{h,h'}^{\lambda\lambda'}(Q^2, x_B, \Delta_T)$ where $\lambda(\lambda')$ is the helicity of the initial (final) state photon, $h(h')$ that of the initial (final) state proton and Δ_T is the transverse momentum transfer from the initial proton to the final one.

$$T_{BH}^* T_{DVCS} + T_{DVCS}^* T_{BH} = \frac{e^6}{t} \frac{m}{Q} \cdot \frac{4\sqrt{2}}{x_{bj}} \cdot \frac{1}{\sqrt{1-x_{bj}}} [\cos(\Phi) \frac{1}{\sqrt{\epsilon(1-\epsilon)}} Re\tilde{M}^{1,1}$$

$$-\cos(2\Phi)\sqrt{\frac{1+\epsilon}{1-\epsilon}} Re\tilde{M}^{0,1} - \cos(3\Phi)\sqrt{\frac{\epsilon}{1-\epsilon}} Re\tilde{M}^{-1,1}] + O(\frac{1}{Q^2}) \quad (3)$$

Using lepton beams with longitudinal polarization $h_e = \pm 1/2$ the helicity-dependent terms in the BH-VCS interference, equation 4, probe the absorptive part of VCS, i.e. $Im M_{h,h'}^{\lambda\lambda'}$.

$$T_{BH}^* T_{DVCS} + T_{DVCS}^* T_{BH} \sim \frac{e^6}{t} \frac{m}{Q} \cdot \frac{4\sqrt{2}}{x_{bj}} \cdot \frac{1}{\sqrt{1-x_{bj}}} \cdot 2 \cdot h_e[-\sin(\Phi)\sqrt{\frac{1+\epsilon}{\epsilon}} Im\tilde{M}^{1,1}$$

$$+\sin(2\Phi) Im\tilde{M}^{0,1}] + O(\frac{1}{Q^2}) \quad (4)$$

Hermes [20] has found a significant negative spin-azimuthal asymmetry $A_{LU}^{\sin\phi}$ analyzing $\gamma^* p \longrightarrow \gamma p$ data with different beam helicities in the exclusive limit $M_x \to 0$.

$$A_{LU}^{\sin\phi} = \frac{2\int_0^{2\pi} \sin(\phi) d\phi (d\sigma^+/d\phi - d\sigma^-/d\phi)}{<|P_b|> \int_0^{2\pi} d\phi (d\sigma^+/d\phi + d\sigma^-/d\phi)} \quad (5)$$

Here, ϕ represents the azimuthal angle of the gamma around the virtual photon direction, with respect to the lepton scattering plane. The subscribe LU for the asymmetry refers to the use of a longitudinal polarized beam and an unpolarized target. The single-spin asymmetry (compare Eqn. 5) with respect to missing mass is shown in Figure 4. As expected for hard exclusive processes, the signal peaks at $M_x \sim 1$ GeV.

HERMES has also studied hard exclusive pion production from a polarized proton target. Specifically, the measurement of the spin-azimuthal asymmetry $A_{UL}^{\sin\phi}$ was extended into the exclusive limit $z \to 1$ [21]. A striking behavior is observed: in the semi-inclusive region ($0.2 < z < 0.7$) a small positive value of $A_{UL}^{\sin\phi}$ is seen for pi+ production, while a large negative value is observed at $z \simeq 1$. By comparison, the asymmetry for π^0 production remains positive but becomes larger in magnitude in the exclusive limit. These measurements have yet to be compared with theoretical calculations.

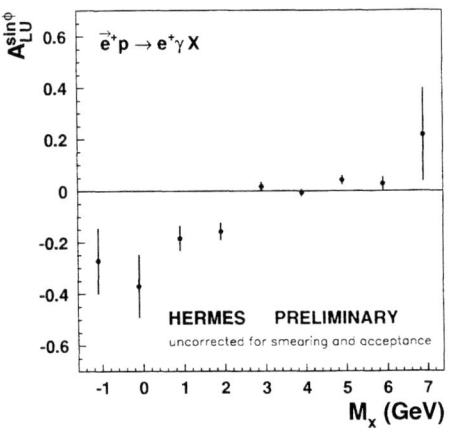

FIGURE 4. The single spin azimuthal asymmetry $A_{LU}^{\sin\phi}$ plotted versus missing mass for the process $\gamma^* p \longrightarrow \gamma p$

VI FUTURE TRANSVERSITY MEASUREMENTS

A complete description of the structure of the nucleon at leading twist requires a third structure function, beyond the already-measured $F_1(x)$ and $g_1(x)$. This structure function is termed *transversity*, and has been assigned the symbol $h_1(x)$. In the same manner as F_1 and g_1 are related respectively to the parton distribution functions $q(x)$ and $\Delta q(x)$, h_1 is related to $\delta q(x)$ – the distribution of *transverse* quark spin in a nucleon polarized transverse to its (infinite) momentum [15]. The function $h_1(x)$ has several interesting properties. In the simplest non-relativistic picture of the nucleon, $\delta q(x)$ is simply equal to the longitudinal helicity distribution $\Delta q(x)$. However, unlike in the case of g_1, the gluon polarization does not mix with quark polarization in h_1, leading to a rather different Q^2 evolution. Further, quarks and anti-quarks contribute to h_1 with opposite sign, making transversity a pure *valence quark* object. The first moment of h_1 thus represents the transverse polarization of valence quarks, an object which offers a promising point of comparison with lattice QCD calculations.

Transversity is as yet unmeasured because it has the unusual property of being chiral-odd: h_1 is not directly observable in inclusive lepton-nucleon scattering experiments. However, it has been suggested that the needed sensitivity can be provided by the semi-inclusive production of pions with modest $p_\perp$ [16]. In such reactions, a chiral-odd fragmentation function known as the Collins function $H_1^\perp(z)$ also appears, thus providing a net chiral-even contribution to the cross-section. If one imagines the plane formed by the virtual photon and the transverse spin direction of the struck quark, the prediction of the Collins function is that leading pions are distributed preferentially above or below this plane, depending on the orientation of the spin of the struck quark.

HERMES has made a first measurement of single-spin asymmetries for semi-inclusive pion production in DIS, using an unpolarized beam and a *longitudinally* polarized proton target [17]. A clear $\sin\phi$ dependence is observed for π^+, indicating that a different out-of-plane direction is indeed preferred in one target orientation compared with the other. No such behavior is seen for π^- production. The $\sin\phi$ moment of the asymmetry (denoted $A_{UL}^{\sin\phi}$) can be interpreted in terms of chiral-odd quark spin-distribution functions closely related to transversity [18].

The obtained non-zero single-spin asymmetries for semi-inclusive pion production with a longitudinal polarized target gives high confidence that HERMES can do a first measurement of h_1 [22] using a transverse polarized proton target in 2001/2002.

To evaluate the level of accuracy that could be obtained when scattering 'unpolarized' leptons off transversely polarized protons at HERMES, simulations were done assuming a target polarization of $P_T = 75\%$ and 7.0 Million reconstructed DIS events. Assuming up-quark dominance for π^+ production, the $\sin\phi$ asymmetry moment is given by

$$A_T^{\pi^+}(x,y,z) = P_T \cdot D_{nn} \cdot \frac{\delta u(x)}{u(x)} \cdot \frac{H_1^{\perp(1)u}(z)}{D_1^u(z)}, \quad (6)$$

where P_T is the target polarization, D_{nn} is the transverse spin transfer (kinematic) coefficient, and $D_1^u(z)$ is the familiar unpolarized fragmentation function. The magnitude of the observed asymmetry depends on the two unknown functions – $\delta u(x)$ and the Collins fragmentation function $H_1^{\perp(1)u}(z)$.

For the simulations [23], $\delta u(x) = \Delta u(x)$ was assumed, which should be suitable for the relatively small values of Q^2 relevant for the HERMES kinematical region. Values for the function $H_1^{\perp(1)u}(z)$ were estimated using the model of Kotzinyan and Mulders [24] that was found to be consistent with the HERMES data. The expected asymmetry $A_T^{\pi^+}(x)$ as it would be measured by HERMES are presented as a function of z in Fig. 5a).

The factorised form of the expression in Eq. (6) with respect to the variables x and z allows in principle the reconstruction of a shape for both of the unknown functions $\delta u(x)$ and $H_1^{\perp(1)u}(z)/D_1^u(z)$, while the individual normalizations can not be fixed without further assumptions. The normalization ambiguity has been resolved by assuming $\delta u(x = 0.25) = \Delta u(x = 0.25)$, as indicated by the asterisk in Fig. 5b).

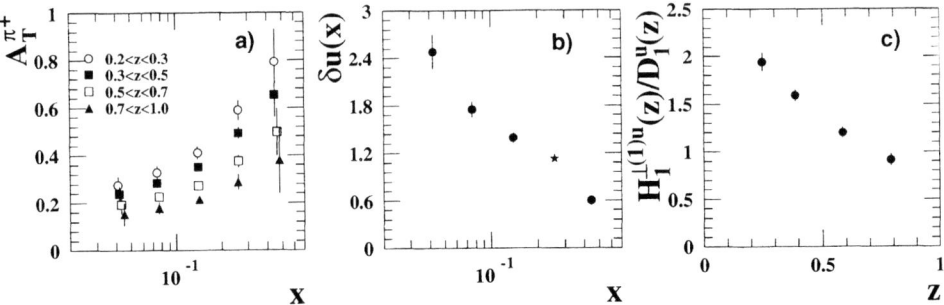

FIGURE 5. *a)* The weighted asymmetry $A_T^{\pi^+}(x)$ as a function of x; *b)* the transversity distribution $\delta u(x)$ as a function of x, and *c)* ratio of the fragmentation functions $H_1^{\perp(1)u}(z)$ to $D_1^u(z)$ as a function of z, as they would be measured by HERMES.

The expected accuracies for the reconstruction of $\delta u(x)$ and $H_1^{\perp(1)u}(z)/D_1^u(z)$ are presented in Figs. 5*b)* and *c)* respectively.

The expected statistical precision is excellent for a first measurement of even the shape of this fundamental distribution $\delta u(x)$. Given the surprising results found for $g_1(x)$ a decade ago, we cannot claim to understand the spin structure of the nucleon unless the single remaining leading-twist structure function h_1 has been measured as well.

VII SUMMARY

The past years have seen significant advances in the experimental exploration of spin-dependent processes, and the coming years promise further exciting developments. Continuing studies of semi-inclusive spin asymmetries are providing new information on the flavor-separated quark spin distributions in the nucleon. HERMES had a record data taking year in 2000. In the years 1998-2000 HERMES collected 8 Million DIS events on a polarized Deuterium target with a RICH detector, which enables the measurement of charged kaon asymmetries, and will hopefully yield a first direct measurement of the strange quark polarization.

Measurements of the gluon polarization are just beginning: HERMES has taken a first look at the asymmetry for production of pairs of high-p_T hadrons, and measurements of this type will be pursued with greater precision by the upcoming COMPASS and RHIC-SPIN experiments. HERMES has investigated the transversity structure function $h_1(x)$ for the first time, and will focus on this subject in coming years with a transversely polarized target program. Finally, spin experiments are beginning to explore nucleon structure at a level of greater complexity, involving parton-parton correlations studied via hard exclusive reactions. The first results from HERMES on DVCS and exclusive pion production have been presented, opening hopefully the possibility to study skewed parton distributions in more detail.

REFERENCES

1. R.L. Jaffe and A. Manohar, *Nucl. Phys.* B **337**, 509 (1990).
2. T.P. Cheng and L.-F. Li, *Phys. Lett.* B **336**, 365 (1996).
3. B. Dressler et al., *Eur. Phys. J.* C **14**, 147 (2000).
4. E142 Coll., P.L. Anthony et al., *Phys. Rev.* D **54**, 6620 (1996);
   E154 Coll., K. Abe et al., *Phys. Rev. Lett.* **79**, 26 (1997);
   HERMES Coll., K. Ackerstaff et al., *Phys. Lett.* B **404**, 383 (1997);
   E143 Coll., K. Abe et al., *Phys. Rev.* D **58**, 112003 (1998);
   HERMES Coll., A. Airapetian et al., *Phys. Lett.* B **442**, 484 (1998);
   SM Coll., B. Adeva et al., *Phys. Rev.* D **58**, 112001 (1998);
   E155 Coll., P.L. Anthony et al., *Phys. Lett.* B **463**, 339 (1999).
5. SMC, B. Adeva et al., *Phys. Rev.* D **58**, 112002 (1998).
6. U. Stösslein, these proceedings.
7. E. Leader, A. Sidorov, D. Stamenov, hep-ph/0004106.
8. HERMES Coll., K. Ackerstaff et al., *Phys. Lett.* B **464**, 123 (1999).
9. Th. Lindemann, these proceedings.
10. SMC, B. Adeva et al., *Phys. Lett.* B **420**, 180 (1998).
11. T. Gehrmann and W.J. Stirling, *Phys. Rev.* D **53**, 6100 (1996).
12. HERMES Coll., A. Airapetian et al., *Phys. Rev. Lett.* **84**, 2584 (2000).
13. A. Bravar, D. von Harrach, A. Kotzinian, *Phys. Lett.* B **421**, 349 (1998).
14. M. Glück, E. Reya, M. Stratmann, W. Vogelsang, *Phys. Rev.* D **53**, 4775 (1996)
15. J.P. Ralston and P.E. Soper, *Nucl. Phys.* B **152**, 109 (1979);
    R. Jaffe and X. Ji, *Nucl. Phys.* B **375**, 527 (1992).
16. J.C. Collins, *Nucl. Phys.* B **396**, 161 (1993).
17. HERMES Coll., A. Airapetian et al., *Phys. Rev. Lett.* **84**, 4047 (2000).
18. E. De Sanctis, W.-D. Nowak, K.A Oganessyan, *Phys. Lett.* B **483**, 69 (2000);
    A.V. Efremov et al., *Phys. Lett.* B **478**, 94 (2000).
19. M. Diehl et al., *Phys. Lett.* B **411**, 193 (1997).
20. M. Amarian, these proceedings.
21. D. Ryckbosch, these proceedings.
22. HERMES Coll.,The HERMES Physics Program and Plans for 2001-2006
23. V.A. Korotkov, W.-D. Nowak and K.A. Oganessyan, DESY-176.
24. A.M. Kotzinian, P.J. Mulders, Phys. Lett. B **406**, 373 (1997).
25. M. Stratmann, *Z. Phys.* C **60**, 763 (1993).
26. M. Vanderhaeghen, P.A.M. Guichon, M. Guidal, *Phys. Rev.* D **60**, 094017 (1999).

Highlights from the SMC and COMPASS experiments

F. Bradamante

University of Trieste and INFN Trieste, Italy

Abstract. The main physics results of the SMC experiment and their impact on spin physics are briefly reviewed. After a short account of the motivations for a new experiment to further investigate the spin structure of the nucleon through semi-inclusive deep inelastic scattering, the physics objectives of the COMPASS experiment and an overall description of the experimental apparatus are given.

INTRODUCTION

The first measurements of the polarized electron-proton scattering were performed at SLAC by the E80 and E130 Collaborations [1,2]. These early data were consistent with the Ellis-Jaffe and the Bjorken sum rules [3,4]. A break-through in this physics occurred when the EMC Collaboration at CERN extended these measurements to a much larger kinematic range by using a polarized muon beam with an energy ten times larger, and found a significant violation of the Ellis-Jaffe sum rule [5,6]. That result implied, in the framework of the Quark-Parton-Model (QPM) that the total contribution of the quark spins to the proton spin is small.

THE SMC EXPERIMENT

The SMC experiment at CERN was proposed in December '88 to perform
- a new measurement of the polarized deep inelastic scattering (DIS) with a polarized proton target
- the first measurement of the polarized DIS with a polarized deuteron target

and thus to determine the spin-dependent structure functions of both the proton and the deuteron, their first moments, and to test the fundamental Bjorken sum rule at the 10% level.

The experiment took data over the period 1992-1996, 3 years with a deuterated butanol target (C_4D_9OD), 1 year with a butanol target (C_4H_9OH), and 1 year with an ammonia target (NH_3). The experiment is completed, all information on the apparatus, the data, the analysis, and the physics results is fully published [7–9].

A major accomplishment of the SMC experiment was the first measurement of the spin structure function of the deuteron, carried out in 1992. The result of that measurement, when combined with the EMC result, was in agreement with the Bjorken sum rule, and implied a violation of the Ellis-Jaffe sum rule also for the neutron. One could safely conclude already in 1993 that the "spin crisis" was a well established phenomenon for both the proton and the neutron, and that it occurred within the boundary given by the Bjorken sum rule.

All the SMC data for the structure functions g_1, are shown, multiplied by x-Bjorken, in Fig. 1 for both the proton and the deuteron. The structure functions g_1

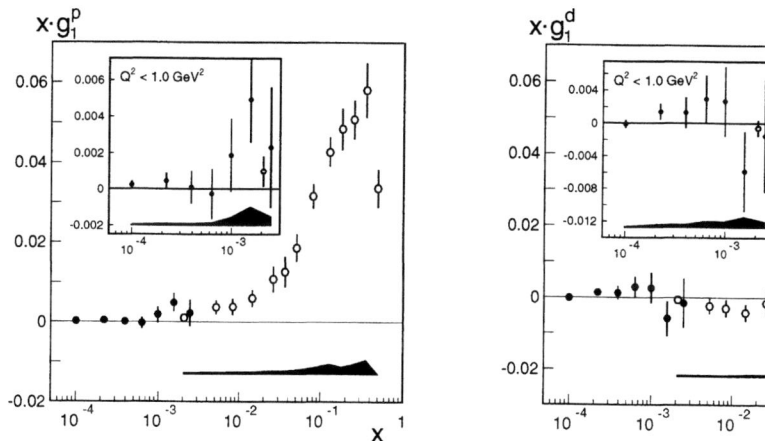

FIGURE 1. The values of xg_1 for the proton (left) and for the deuteron (right) as a function of x. The shaded bands indicate the size of the systematic errors.

depend linearly on the cross-section asymmetry. More precisely from the measured counting rate asymmetries, and knowing the incident muon polarization, the target nucleon polarization, and the polarized nucleon dilution factor of the target, one can extract the cross-section asymmetries for parallel and antiparallel orientations of the beam and target spins,

$$A_\| = \frac{\sigma^{\uparrow\downarrow} - \sigma^{\uparrow\uparrow}}{\sigma^{\uparrow\downarrow} + \sigma^{\uparrow\uparrow}}. \qquad (1)$$

The asymmetry, $A_\|$, and the spin-dependent structure function, g_1, are related to the virtual photon-nucleon asymmetries, A_1 and A_2, by

$$A_\| = D(A_1 + \eta A_2), \quad g_1 = \frac{F_2}{2x(1+R)}(A_1 + \gamma A_2), \qquad (2)$$

in which the factors η and γ depend only on kinematic variables and on the nucleon mass, while the depolarisation factor D depends on kinematic variables and the ratio of total photo-absorption cross sections for longitudinally and transversely polarized virtual photons $R = \sigma_L/\sigma_T$. In the kinematical region of our measurements η and γ are small. The asymmetries A_2^p and A_2^d were measured and found consistent with zero [10–12]. For these two reasons to derive g_1 we neglect the A_2 term in Eq. 2. For more details on the formalism, and on the parametrisation used for F_2 and R the reader is referred to Ref. [8].

After evolving the structure function g_1 to a fixed Q^2, its first moment $\Gamma_1(Q^2) = \int_0^1 g_1(x, Q^2) dx$ was computed: the SMC result, together with the results of the other experiments, is shown in Fig. 2, as a function of Q^2. To extract Γ_1^n from our

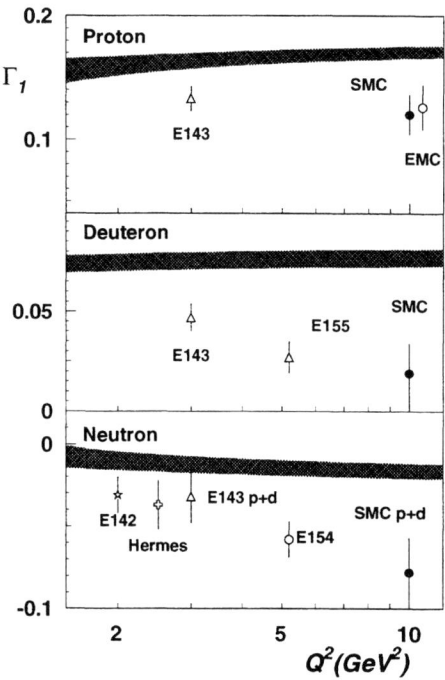

FIGURE 2. First moments of g_1 for the proton, the deuteron and the neutron as a function of the typical Q^2 of the data. The predictions of the Ellis-Jaffe sum rule are shown as hatched bands.

data, we assume that the deuteron structure function g_1^d is related to the proton and neutron structure functions g_1^p and g_1^n by

$$g_1^p + g_1^n = \frac{2g_1^d}{(1 - \frac{3}{2}\omega_D)}, \qquad (3)$$

where $\omega_D = 0.05 \pm 0.01$ is the D-wave state probability in the deuteron. All experimental results are smaller than the Ellis-Jaffe predictions.

The SMC Collaboration has also performed a perturbative QCD (pQCD) analysis in next-to-leading order to determine the polarized parton distribution functions,

Δq_i and ΔG, using all the available data set. The results for the singlet part $\Delta\Sigma$ and the non-singlet part Δq_{NS} for the proton and for the neutron, given by

$$\Delta\Sigma(x, Q^2) = \sum_{i=1}^{n_f} \Delta q_i(x, Q^2),$$

$$\Delta q_{\text{NS}}(x, Q^2) = \sum_{i=1}^{n_f} (e_i^2/\langle e^2 \rangle - 1)\Delta q_i(x, Q^2), \qquad (4)$$

as well as for the gluon polarization ΔG are shown in Fig. 3, together with the

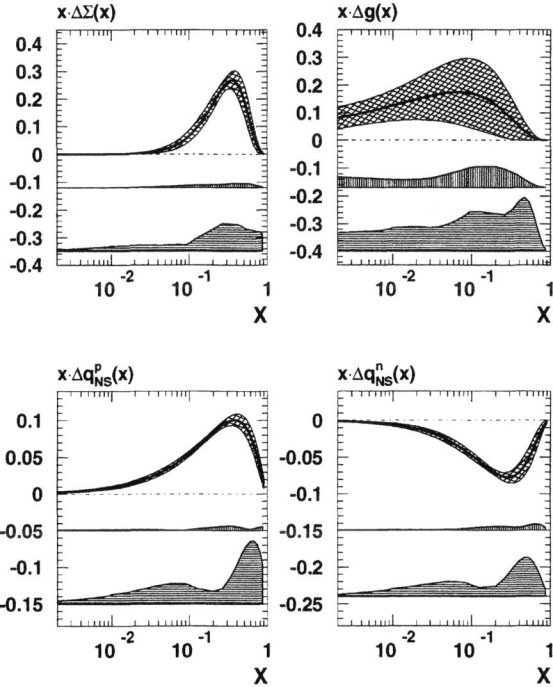

FIGURE 3. Polarized parton distribution functions determined from the pQCD analysis at $Q^2 = 1$ GeV^2. The bands indicate the statistical uncertainty from the QCD fit (crossed hatch), the experimental systematic uncertainty (vertically hatched), and the theoretical uncertainty (horizontally hatched).

uncertainty given by the QCD fit. The fact that the g_1 structure function depends not only on Δq, the polarized quark distribution, but also on ΔG, and that a full analysis of all the data requires a QCD evolution procedure utilising the DGLAP equations is by now well known, and is simply related to the fact that in inclusive DIS the virtual photon is not only absorbed by a quark, but can also fuse with a gluon to give a $q\bar{q}$ pair. It is also well known that, although the uncertainty in the determination of ΔG is very large, this pQCD analysis suggests a solution to the "spin crisis", i.e. a large polarization of the gluon sea.

The SMC experiment has provided another important result, namely the separate valence and sea quark contributions to the nucleon spin could be determined via semi-inclusive measurements of final states. The SMC results [13], together with the more recent data from HERMES Collaboration [14] are shown in Fig. 4.

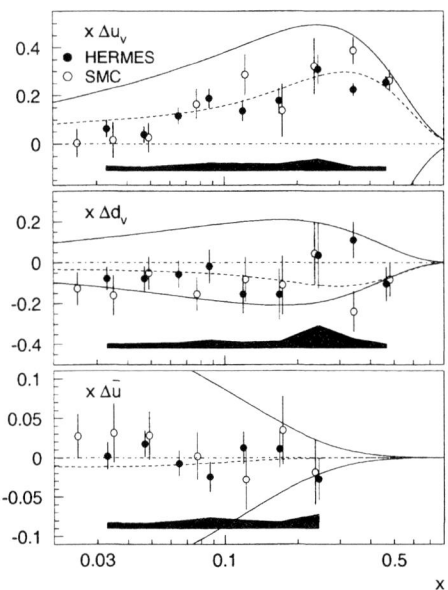

FIGURE 4. The valence and see quark distributions at $Q^2 = 2.5$ GeV^2 from the SMC and Hermes experiments. Also shown are the positivity limit (solid curve) and a parametrisation from Ref. [15].

Although hadrons could not be identified in SMC, by measuring the spin asymmetries for positive and negative hadrons (essentially protons, kaons and pions), and adding the information coming from the inclusive asymmetries, one ends up with six linear equations involving the polarized quark distributions.

The six linear equations are the equations for the inclusive asymmetries, the positive hadron asymmetries, and the negative hadron asymmetries, for both proton and deuteron, which, in the QPM can be written as

$$A_{1p}(x, Q^2) = \frac{\sum_q e_q^2 \left[\Delta q(x, Q^2) + \Delta \bar{q}(x, Q^2)\right]}{\sum_q e_q^2 \left[q(x, Q^2) + \bar{q}(x, Q^2)\right]} \left[1 + R(x, Q^2)\right], \qquad (5)$$

$$A_{1p}^{+(-)}(x,Q^2) = \frac{\sum_{q,h} e_q^2 \left[\Delta q(x,Q^2) D_q^h(Q^2) + \Delta \bar{q}(x,Q^2) D_{\bar{q}}^h(Q^2)\right]}{\sum_{q,h} e_q^2 \left[q(x,Q^2) D_q^h(Q^2) + \bar{q}(x,Q^2) D_{\bar{q}}^h(Q^2)\right]} \left[1 + R(x,Q^2)\right], \quad (6)$$

and corresponding expressions in the case of the deuteron target. The quantities $D_q^h(Q^2)$ represent the probability that a quark q fragments into a hadron h.

In general, it is not possible to solve for all the six unknown quark distributions. The SMC analysis has solved for three distributions, namely the polarized valence quark distributions $\Delta u_v(x) = \Delta u(x) - \Delta \bar{u}(x)$ and $\Delta d_v(x) = \Delta d(x) - \Delta \bar{d}(x)$, and the polarized sea quark distribution $\Delta \bar{q}(x)$. The assumption of a $SU(3)_f$ symmetric sea, i.e. $\Delta \bar{q}(x) = \Delta \bar{u}(x) = \Delta \bar{d}(x) = \Delta s(x) = \Delta \bar{s}(x)$, has allowed to reduce the number of unknown quark distributions to three.

Given the large x range covered by the measurement, one could integrate the polarized quark distributions and obtain the first moments $\Delta q = \int_0^1 x \Delta q_v(x) dx$. The resulting values are

$$\Delta u_v = 0.77 \pm 0.10 \pm 0.08,$$
$$\Delta d_v = -0.52 \pm 0.14 \pm 0.09,$$
$$\Delta \bar{q} = 0.01 \pm 0.04 \pm 0.03.$$

This result is interesting, but one cannot conclude that the quark see is not polarized. Releasing the assumption of a $SU(3)_f$ symmetric sea, and replacing it by only a isospin symmetric sea, $\Delta \bar{u}(x) = \Delta \bar{d}(x) = \Delta \bar{q}(x)$, it is found that the measurement is only sensitive to the polarization of the non-strange sea quarks.

Full particle identification is needed to access the polarization of the strange sea.

FROM SMC TO COMPASS

In 1993, after the first measurements of SMC with a polarized deuteron target, it was already clear to several of us that a new experimental approach was necessary to progress in this field, namely semi-inclusive DIS with a full reconstruction of the hadronic jet. A flavour-tagging procedure already allowed to identify in SMC the struck quark. A suggestion to isolate the photon-gluon fusion process and directly measure ΔG was put forward already in 1988 [16,17], and implied measuring the cross-section asymmetry of open charm in DIS. A new experiment, with full hadron identification and calorimetry, seemed to be necessary.

In addition, a new interesting physics case was rapidly developing, transversity, also demanding for semi-inclusive DIS measurements.

As originally shown by Jaffe and Ji [18], to completely specify the quark state at the twist-two level, to the momentum distribution $q(x)$ and to the helicity distribution $\Delta q(x)$ one has to add the transverse spin distributions $\Delta q_T(x)$, which, summed up over the quark flavours, and weighted by the squared charges, make up a new structure function $h_1(x)$. Due to its odd chirality nature, $h_1(x)$ cannot be measured in inclusive DIS processes. A possible way to access $\Delta q_T(x)$ is via the so-called "Collins asymmetry" [19], namely a possible azimuthal asymmetry of the final hadron with respect to the direction of the transversely polarized quark.

Nowadays transversity has become a big issue, as apparent at this Conference, and is a major part of the programme of many experiments. In 1993 a few enthusiasts met at a workshop in Geneva [1] and laid down the case for a proposal which was submitted to CERN in the fall. The use of the electron beam at LEP was foreseen, in the higher energy range (LEP2 phase), and an internal polarized hydrogen jet-target. The proposal [20] was not recommended for approval by the LEPC committee. Still, I felt the physics case should not be given up, and that we should have tried to pursue our goals using a solid polarized target and the polarized muon beam, in a new experiment, once the SMC experiment was over.

During '94 the physics case for the new experiment and a conceptual design for the future apparatus were agreed upon, and the HMC letter of intent was submitted to SPSC in March '95. A new two-stage spectrometer, with powerful particle identification and calorimetry, and capable of standing a muon beam rate of 10^8/s, was proposed for Hall 888 at CERN, after completion of the SMC experiment. Limited availability of resources suggested a merging of the HMC experimental program with that proposed by the CHEOPS Collaboration, mainly spectroscopy of light quark systems and glueballs, and the investigation of hadronic structure of unstable particles using Primakoff reactions.

The proposal for a COmmon Muon and Proton Apparatus for Structure and Spectroscopy (COMPASS) was submitted in March 1996, and approved on Feb. 6, 1997.

PHYSICS OBJECTIVES OF THE COMPASS EXPERIMENT

Topical to this Conference is the COMPASS programme with the muon beam, therefore I will not mention our programme with the hadronic beams, and refer the reader to the relevant references [21].

The main goal of COMPASS is a direct measurement of ΔG by measuring the cross-section asymmetry $A_{\mu N}^{c\bar{c}}$

$$A_{\mu N}^{c\bar{c}} = \frac{\Delta \sigma^{\mu N \to c\bar{c} X}}{\sigma^{\mu N \to c\bar{c} X}}. \tag{7}$$

At our energies the production of charm goes predominantly via photon-gluon fusion (PGF), according to the diagram shown in Fig. 5, and the quantities $\sigma^{\mu N \to c\bar{c} X}$ and $\Delta \sigma^{\mu N \to c\bar{c} X}$ can be expressed as a convolution of the elementary photon-gluon cross-section with the gluon distributions ΔG and G.

---

[1] The workshop was organised by the late R. Hess in the DPhNC of the University of Geneva in March '93.

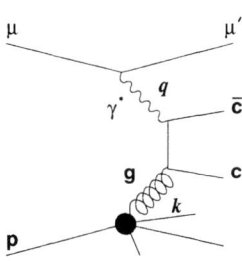
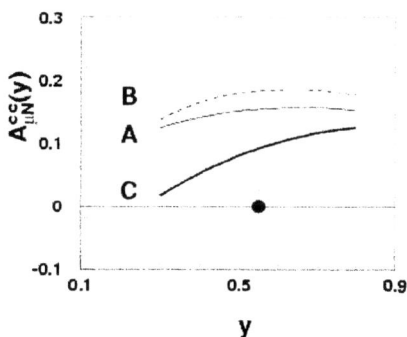

FIGURE 5. The photon-gluon fusion diagram, dominant mechanism for charm production at COMPASS energies.

FIGURE 6. The asymmetry $A^{c\bar{c}}_{\mu N}$ as a function of y, the gluon momentum fraction. The curves refer to three sets of the gluon helicity distribution function.

The open charm asymmetry expected from three possible shapes [22] of the gluon helicity distribution function ΔG are shown in fig. 6. Also shown is the projected precision expected by COMPASS by measuring open charm only by detecting D° and $\overline{D}^\circ$, decaying into the two body ($K+\pi$) channel. The sensitivity of COMPASS to ΔG by this method is expected to be $\delta(\Delta G/G) \simeq 0.11$.

The large acceptance of the COMPASS spectrometer and its capability of identifying all the particles produced in the muon-nucleon scattering open up additional ways to access ΔG.

The most promising additional way to measure ΔG in COMPASS uses the asymmetry of oppositely charged hadron pairs at high p_t [23]. Originally developed for the COMPASS experiment, the method has been recently applied also to the Hermes data [24]. The basic diagram is still the PGF, $\gamma g \to q\bar{q} \to h^+h^-X$, and the hardness of the process is guaranteed by the large p_t. The background from the leading order process $\gamma q \to q$, and the QCD-Compton process, $\gamma q \to \gamma q$, is in general dominating the PGF creation of a light $q\bar{q}$ pair, but suitable kinematic cuts can enhance considerably this process and allow for a statistically precise measurement. As shown in Ref. [23], from a h^+h^- asymmetry measurement for $p_t \geq 1.5$ GeV/c in COMPASS after 1 year of running at 200 GeV/c on a ^{6}LiD target we should get $\delta(\Delta G/G) \simeq 0.05$, dominated by systematic uncertainties.

Apart from ΔG, the COMPASS spectrometer will measure Δq and Δq_T from the relevant identified hadron asymmetries, in semi-inclusive polarized muon - polarized nucleon DIS both in the longitudinal and transversal mode. For the sensitivity, the reader is referred again to the COMPASS proposal [21].

THE COMPASS SPECTROMETER

A common requirement of all the measurements foreseen by COMPASS is the detection and identification of particles over a large angular (±200 mrad) and dynamical (up to ~ 150 GeV) range. To achieve this goal the COMPASS spectrometer comprises two magnetic stages. The first is based on a new large gap (2.3×1.6 m^2) spectrometer magnet (SM1) with 1 Tm bending power. The second uses the existing Forward Spectrometer magnet from the SMC experiment with 4.4 Tm bending power (SM2). Both stages are complemented with charged particle identification with fast RICH detectors, electromagnetic calorimetry, hadronic calorimetry, and muon identification via filtering through thick absorbers.

Due to insufficient funding, the experiment we have started installing this year will have a somewhat reduced configuration (the so-called "initial lay-out", shown in Fig. 7). This configuration is still adequate to start up the muon programme.

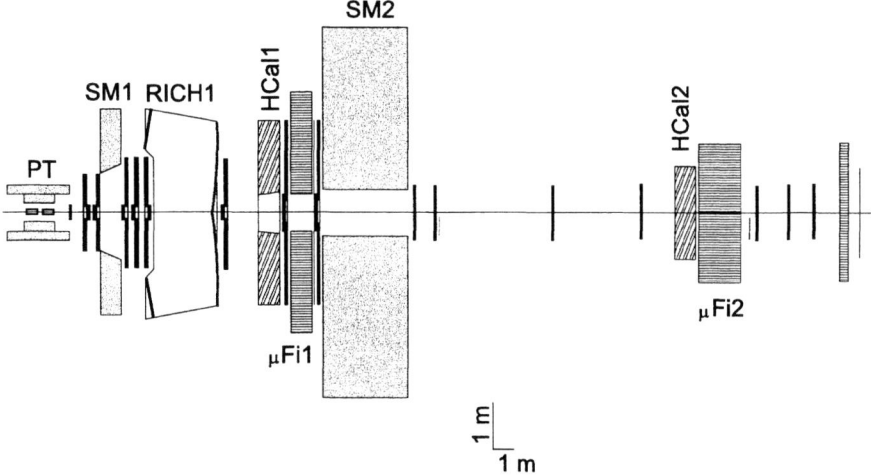

FIGURE 7. Top view of the initial lay-out of the spectrometer for the COMPASS experiment.

A major limitation in the initial lay-out is the absence of large-angle triggering hodoscopes and tracking detectors in the second stage of the spectrometer, which limits the coverage of the apparatus to low Q^2 values (≤ 5 GeV/c). These limitations affect only marginally the measurement of ΔG: most of the events are produced by quasi-real photons, $Q^2 \simeq 0$, and the scattered muon is detected at fairly small angles; the hadrons from D° decay, on the contrary, being emitted around 90° in the D° c.m.s., are produced at fairly large angles in the laboratory, and are detected in the first stage of the spectrometer. Still, the lay-out does not allow the measurement of the quark spin distribution function at large x.

In the following I will mainly quote the detectors which make up the initial lay-out, which are the ones being constructed and tested, and which were on the floor (at least part of them) in the summer of year 2000, to start commissioning the experiment.

The experiment has been planned to run at muon energies from 100 to 200 GeV. The beam is naturally polarised by the π-decay mechanism. The beam polarisation was measured by the SMC Collaboration to be about 80%. The beam intensity is 10^8 muons per spill. The triggering system and the tracking system of COMPASS have been designed to stand the associated rate of secondaries, and use state-of-the-art detectors. Also, fast front-end electronics, multi-buffering, and a large and fast storage of events are essential.

We use the polarised target system of the SMC experiment, which allows for two oppositely polarised target cells, 60 cm long each. A new superconducting magnet with about 60 cm bore diameter will be used, and is presently being built by Oxford Instruments. The magnet can provide both a solenoidal field and a dipole field, for adiabatic spin rotation. Correspondingly, the target polarisation can then be oriented either longitudinally or transversely to the beam direction. We will use two different target materials, NH_3 as proton target, and ^6LiD as deuteron target. Polarisations of 85 % and 50 % can be reached, respectively. The use of ^6LiD with its favourable dilution factor of 0.5 is of the utmost importance for the measurement of ΔG.

To match the expected particle flux in the various locations along the spectrometer, COMPASS will use very different tracking detectors.

The beam region itself is covered by scintillating fiber (SciFi) hodoscopes. For the small area tracking, i.e. for the tracking near the beam, where high location accuracy is needed and the particle flux is large, we use Micromegas's [25] (before SM1) and triple-GEM detectors [26] (downstream of SM1). For large area tracking (LAT), planar drift chambers with small cell size (7×8 mm^2) are used between the polarised target and SM1, and straw-tubes trackers with 6 or 10 mm diameter straws (inner and outer part respectively) are used after SM1. In the initial lay-out only the first straw-tracking station between SM1 and the RICH will be available. In the small angle spectrometer the large area tracking will be based on 11 MWPC's, useful area $\sim 150 \times 80$ cm^2, 2 mm pitch, for a total of 23000 channels.

The charged particle identification relies on the RICH technology. Presently, RICH1 only is being constructed. The length of the radiator (C_4F_{10} gas) vessel is 3 m. The entire downstream surface is covered by mirrors with spherical geometry and a focal length of 3.3 m. We use as UV photon detectors MWPC's with a CsI photocathode [27] (segmented in 8×8 mm^2 pads) which detect photons with wave length shorter than 200 nm, i.e. in the far UV domain. The active area of each of the two photon detectors is 2.8 m^2 and the total number of pads is about 70,000. The front-end elctronics uses a modified version of the Gassiplex chip, and the read-out cards (the so-called BORA PCB's [28]) constitute a major project, utilising hundreds of DSP's. RICH1 has been designed to give a 3 standard deviation π/K separation up to momenta of ~ 60 GeV/c.

The muon filters (μF) serve the identification and detection of the scattered muon. After the first muon wall the tracking is performed by Iarocci tubes and after the second muon wall by drift tubes of 3 cm diameter.

The hadron calorimeters HCAL1 and HCAL2 serve mainly triggering purposes, and a moderate energy resolution is sufficient. The trigger time resolution is 1 ns, which agrees well with experimental requirements. Apart from HCAL, the trigger concept is based on the energy loss of the scattered muons. Two pairs of scintillator counter hodoscopes (one pair, with high granularity, at small angles, very close to the beam, the second pair at larger angles), located at about 35 and 50 m from the target, and a fast matrix coincidence define in the plane scattering angle vs total bending the useful kinematics for $\Delta G/G$ measurement.

A major effort has been dedicated to design front-end electronics with as small dead-time as possible. Originally designed as a zero dead-time experiment up to a trigger rate of 10^5 s^{-1} (fully pipelined read-out), we had to accept a dead-time of about 0.5 μs, mainly due to the RICH and to the calorimetry read-out.

The readout system uses a modern concept, involving highly specialised integrated circuits. The readout chips are placed close to the detectors and the data is concentrated at a very early stage via high speed serial links. At the next level high bandwidth optical links transport the data to a system of readout buffers. The event building system is based on PCs and Gigabit or Fast Ethernet switches and is highly scalable. This high performance network is also used to transfer the assembled data to the computer centre for database formatting, reconstruction, analysis and mass storage. The data are sent via an optical link from the Hall 888 directly to the Computer building for Central Data Recording (CDR).

The estimated power needed to process COMPASS data (the expected raw data size is 300 TB/year) is about 20000 CERN Units. As object-oriented data-base COMPASS uses Objectivity/DB. Objectivity is used to store all the data for the off-line analysis, keeping them under federated data bases. In the off-line farm, the data servers handle the network traffic from the CDR, distribute the raw data to the CPU clients (where they are put in the data base), receive them back from the PCs, and finally send them to a hierarchical storage manager (HSM) system that uses the tape storage complements the disk pool. In parallel, the data servers receive the data to be processed from the HSM, send them to the PCs for processing, collect the output (DST or mDST), and send it to the HSM.

A major effort is also ongoing in writing from scratch the off-line programs (CORAL, the new COmpass Reconstruction and AnaLysis program) using object-oriented technology and C++ language.

CONCLUSIONS

The title of this talk might have been the CERN contribution to the nucleon spin problem. A major physics result was the EMC finding of the smallness of the quark spin contribution to the proton spin. Five years later the SMC experiment has given a decisive contribution in validating this result, extending it to the neutron, and establishing that it occurred within the boundaries of the Bjorken sum rule. Another important contribution of SMC has been the flavour decomposition of the spin structure function in semi-inclusive DIS, which has allowed to determine the separate valence and sea quark contributions to the nucleon spin. Lastly, credits should go to SMC for its contribution to assessing the physics case for COMPASS, which as a Collaboration was largely born within the SMC Collaboration.

I spent some time in recalling how the physics case for transversity, also a major part of the COMPASS programme, grew up at CERN, and mentioned the work done by the HELP Collaboration. And finally I gave an overview of COMPASS, a flexible spectrometer, which in the near future should start addressing a number of physics issues, starting from a direct measurement of the polarization of the gluon sea in a polarized nucleon. A first technical run took place last summer, and the Collaboration is working hard to collect first physics data in the run of year 2001.

REFERENCES

1. M.J. Alguard et al., *Phys. Rev. Lett.* **37**, 1261 (1976).
2. G. Baum et al., *Phys. Rev. Lett.* **51**, 1135 (1983).
3. J. Ellis and R.L. Jaffe, *Phys. Rev.* D **9**, 1444 (1974).
4. J.D. Bjorken, *Phys. Rev.* **148**, 1467 (1966).
5. J. Ashman et al., *Phys. Lett.* B **206**, 364 (1988).
6. J. Ashman et al., *Nucl. Phys.* B **328**, 1 (1989).
7. B. Adeva et al., *Phys. Rev.* D **58**, 112001 (1998).
8. B. Adeva et al., *Phys. Rev.* D **58**, 112002 (1998).
9. B. Adeva et al., *Phys. Rev.* D **60**, 072004 (1999).
10. D. Adams et al., *Phys. Lett.* B **336**, 125 (1994).
11. D. Adams et al., *Phys. Lett.* B **396**, 338 (1997).
12. K. Abe et al., *Phys. Rev. Lett.* **76**, 587 (1996).
13. B. Adeva et al., *Phys. Lett.* B **420**, 180 (1998).
14. K. Ackerstaff et al., *Phys. Lett.* B **464**, 123 (1999).
15. T. Gehrmann and W.J. Stirling, *Phys. Rev.* D **53**, 6100 (1996).
16. G. Altarelli and G.G. Ross, *Phys. Lett.* B **212**, 391 (1988).
17. M. Glück and E. Reya, *Z. Phys.* C **39**, 569 (1988).
18. R.L. Jaffe and X. Ji, *Phys. Rev. Lett.* **67**, 552 (1991).
19. J. Collins, *Nucl. Phys.* B **396**, 161 (1993).
20. The HELP Collaboration, CERN/LEPC 93-14, LEPC/P7 and CERN/LEPC 94-1, LEPC/P7 Add. 1 (Geneva, 1993).
21. COMPASS, A Proposal for a COmmon Muon and Proton Apparatus for Structure and Spectroscopy, CERN/SPSLC 96-14, SPSLC/P297, 1 March 1996.
22. T. Gehrmann and W.J. Stirling, *Z. Phys.* C **65**, 461 (1994).
23. A. Bravar, D. von Harrach, and A. Kotzinian, *Phys. Lett.* B **421**, 349 (1998).
24. A. Airapetian et al., DESY-99-07, July 1999.
25. Y. Giomataris et al., *Nucl. Instr. Meth.* A **376**, 29 (1996).
26. F. Sauli et al., *Nucl. Instr. Meth.* A **386**, 531 (1997).
27. F. Piuz, *Nucl. Instr. Meth.* A **371**, 96 (1996).
28. G. Baum et al., *Nucl. Instr. Meth.* A **433**, 426 (1999).

The Muon g-2 Experiment at Brookhaven

Gerry Bunce for the Muon g-2 Collaboration[1]

Brookhaven National Laboratory, Upton, New York, USA

Abstract. The muon g-2 experiment at Brookhaven is designed to take advantage of very high proton intensity at the AGS to achieve sensitivity well below the contributions to g-2 from W and Z. The experiment will explore the possibility of new physics coupling to the muon at a mass scale of 5 TeV. This talk presents the experiment, results at 5 parts per million based on data collected in 1998, and data analysis of the 1999 data run which will have a statistical error of ±1.3 ppm.

INTRODUCTION

Precise muon g-2 measurements, when compared to precise theoretical predictions of known physics, test at one time whether our understanding of possible contributions to the muon's magnetism is complete, or whether there are new physics contributions present such as Supersymmetry or W boson substructure. Indeed, most proposed new physics will couple to the muon and affect g-2, so that the results of the experiment will greatly constrain or strongly suggest new physics. The goal of the Brookhaven experiment is to reach 1/3 part per million on the g-2 measurement. This level will be 20 times more precise than the beautiful CERN experiment of 20 years ago. At this time the BNL experiment has reported results at ±5 ppm and is completing a new measurement with statistics of ±1.3 ppm. This talk discusses: What is g-2 and why measure it? How do we measure g-2? Results from the 1998 run at BNL, and the 1999 data and analysis.

We had hoped to present a new number here, but we are not ready yet. We have decided to show you the new data and much of the analysis, demonstrating the data quality, and our understanding of it. g-2 is measured by taking the ratio of the difference between the muon precession frequency and cyclotron frequency of the muons in a magnetic field, to the proton NMR resonant frequency in the same average field. Both must be very accurately measured. We analyze the g-2 precession frequency and the average magnetic field separately, with artificial offsets in each, known only to each independent analysis group. When we are satisfied that the data are self-consistent and properly understood, with written reports on each which are evaluated by the collaboration, and with the systematic errors for each determined, then we "open the box" and remove the artificial offsets. We have no idea whether our g-2 precession frequency and magnetic field for the 1999 data, which we present here with offsets, are close or far away from theory.

I WHAT IS G-2 AND WHY MEASURE IT?

The gyromagnetic factor g is the ratio of the magnetic moment to angular momentum. For a point-like spin-1/2 particle, g=2. However, higher order contributions lead to an anomalous magnetic moment, with $a = (g-2)/2$. a is approximately 0.001 for the muon (and electron). Standard Model contributions come from electromagnetic, hadronic, and weak couplings. The hadronic contribution to the theoretical value of muon g-2 can be obtained from measurements of $e^+e^- \to$ hadrons, measured now very accurately at lower center of mass energies in the VEPP2M machine at Novosibirsk. [2] The present theoretical error on g-2 is ±0.7 ppm, essentially all from the hadronic contribution. To make full use of the accuracy of the Brookhaven g-2 measurement, it will be important to reduce the hadronic contribution error still more. The electromagnetic contribution to g-2 is of order $1 \pm (17 \text{ ppb})$, the hadronic contribution is (57.8 ± 0.07) ppm, and the weak (1.3 ± 0.03) ppm. The Standard Model value for g-2 is [3]

$$a_\mu^{SM} = 11659163(8) \times 10^{-10} . \qquad (1)$$

The CERN measurements [4] had ±7.2 ppm accuracy, and the Brookhaven goal is ±0.35 ppm.

II HOW DO WE MEASURE G-2?

We measure the muon anomaly, g-2, directly, as the advance of the muon's spin direction in a storage ring compared to the muon momentum direction. For a≈0.001, and for our muon momentum of 3.1 GeV/c, the muon spin advances about 12° per turn around the storage ring. Thus the experiment will measure g-2 directly to 1/3 parts per million. We store polarized muons in a homogenous magnetic field and measure the difference frequency between spin and momentum precession:

$$\omega_a = \frac{d\theta_a}{dt} = \frac{e}{m_\mu c} a_\mu B , \qquad (2)$$

where $a_\mu = (g_\mu - 2)/2$, and B is the magnetic field.

The polarized muons are obtained from the chain proton+Ni $\to \pi^+ X$, $\pi^+ \to \mu^+ + \nu$, where the pion decay violates parity, and muons near the pion momentum are polarized in the direction of the pion momentum. The Brookhaven AGS provides a total of 6×10^{13} protons in 12 separate rf bunches, 33 msec apart, with each bunch of protons extracted onto a nickel target. The g-2 data are collected for about 1 msec after filling the storage ring with muons from each individual bunch, 12 times for each AGS cycle. The AGS is refilled every 2.5 seconds. A special target station and beam lines were built for the experiment, with nickel chosen as the target material to reduce the possiblity of shattering the target from the sudden heating of the beam. A special beam line collected the muons in a quadrupole channel, and delivered the polarized muons to a new storage ring. It was necessary to build a superconducting injector magnet to introduce the muons into the storage ring, and this device had to both cancel the magnetic field seen by the entering muons (a 1.5 Tesla field over 1.7 meters length along the entering beam), and to contain its own stray magnetic field to a level of about 10^{-6} 2.2 cm away from the injector, at the edge of the storage region. [5]

The storage ring for the experiment was built from one 14 meter diameter magnet, powered by superconducting coils, with iron to shape the field. The diameter was chosen to take advantage of the cancelation of the electric field contribution to the spin motion at a beam momentum of 3.1 GeV/c, so that an electric field could be used to contain the muons in the storage ring without affecting the spin precession to first order, and to take advantage of using iron to shape the magnetic field. The ring was built from a single magnet to avoid ends which would give large field gradients and therefore make precision measurements of the field difficult. Superconductor offers stability and compactness of design. [6] The g-2 storage ring magnet at Brookhaven is the largest diameter superconducting magnet built. A cross section is shown in Figure 1, and the magnet is shown in Figure 2.

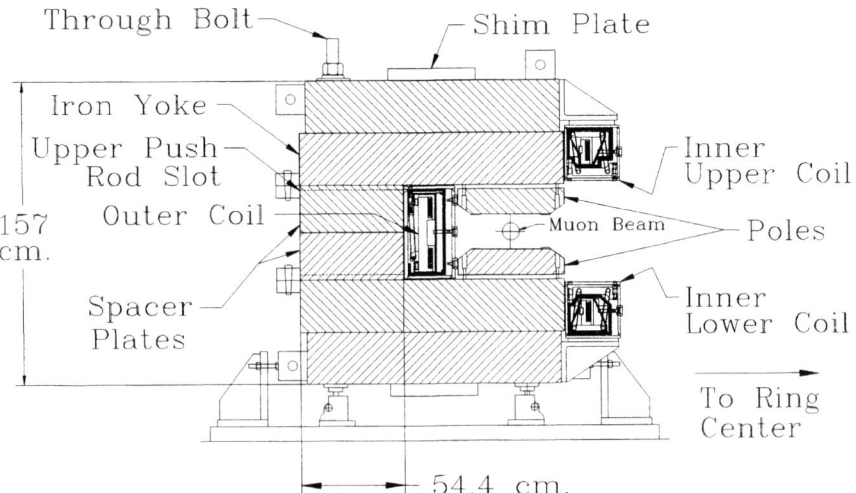

FIGURE 1. A cross section of the muon g-2 storage ring magnet at Brookhaven. The center of the storage ring is at a radius of 7.1 m, and the opening of the C faces the center of the ring.

FIGURE 2. The muon g-2 storage ring magnet.

The g-2 magnet was built with very tight tolerances, but the field quality after construction could be no better than about 1000 ppm uniformity given achievable machining tolerances for large pieces of iron. It was necessary to reach of order 1 ppm uniformity across the storage ring cross section and averaged around the ring. To set the scale of the problem, a human hair is about 50 μm diameter, and a hair difference in the pole gap would change the field by 250 ppm. The magnet had to be extremely stable, and shimable. When we first turned on the magnet, an NMR probe showed that the field fluctuated at a 10^{-8} level per minute, so the required stability was there. Built-in shimming devices allowed the magnetic field to be adjusted to a uniformity of ± 1 ppm. Magnetic field contours in 1/2 ppm steps are shown in Figure 3, for the year 2000 data run. The circle represents the edge of the storage ring. The magnetic field was monitored during the data taking with about 120 NMR probes which surround the vacuum chamber of the storage ring, distributed around the circumference. Every few days, a special NMR trolley is used to bring 17 probes through the storage ring aperture while under vacuum to measure the field every cm around the 44 meter circumference.

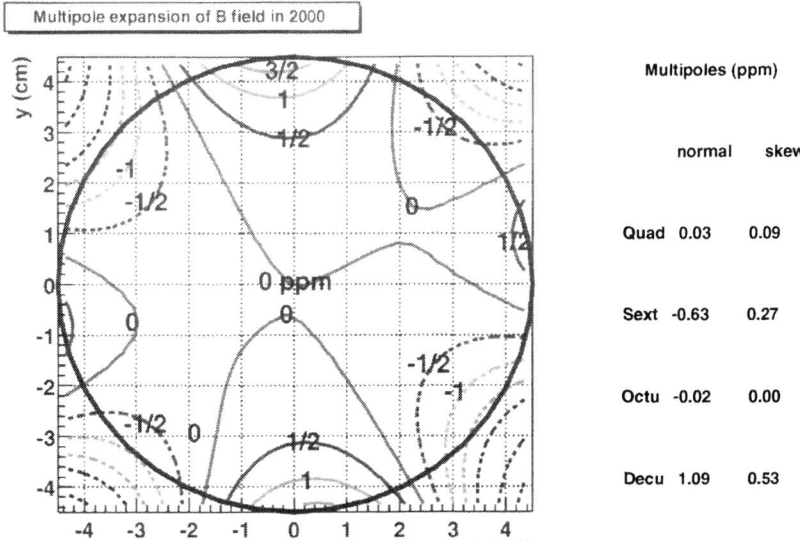

FIGURE 3. Magnetic field contours at 1/2 ppm shown for the year 2000 data run. The circle represents the edge of the storage ring, and the values are averaged over the storage ring circumference. (The contours have meaning only in the measured region, inside the circle.) The data are preliminary.

The Brookhaven experiment uses a new method of collecting and storing the muons. Muons from pion decays in the long beamline are selected in a magnetic spectrometer in the beam line at a momentum about 2% below the pion momentum, and are then injected into the storage ring using first the superconducting injector magnet and then a fast magnetic kicker. The injector allows the muons to be directed tangent and close to the storage region, reducing the required kick (while also traversing a minimum distance in the fringe field of the storage ring). The kicker provides a 10 milliradian kick to the muons, into the storage ring acceptance. This method, as compared to the method used at CERN of injecting pions with their decay to muons providing the kick into orbit, is about 7 times more efficient. However, a kick is required which must be nulled in just a few turns of the muons around the ring. This was successfully done, and was used for 1998 and subsequent data runs. As discussed earlier, electrostatic quadrupoles provide the necessary focussing to keep the muons in the ring. These were built with 4 plates at a radius of 5 cm with 24 KV. There are 4 quads located around the circumference of the storage ring, and the quads cover half of the ring circumference. The edges of the muon beam are scraped during the first 15 μsec after injection to reduce losses during the measurement. This is done by changing the voltage on the quad plates, and using collimators (circles of material at 4.5 cm radius in several locations around the ring) to trim the edges of the beam. In this way, the decays of the muons can be better understood in the data analysis. Also in the storage ring are thin scintillating fiber detectors which can be rotated into the beam to observe the beam storage characteristics for hundreds of turns after injection.

The muons spontaneously decay to positrons and neutrinos. The lifetime of 3.1 GeV/c muons is 64 μsec, and we observe the decay positrons for up to 10 lifetimes. The decay positron energies tag, statistically, the direction of the muon spin. In the muon rest frame, more positrons are emitted in the direction of the muon spin. More positrons are emitted forward in the muon rest frame when the spin points forward, and these are boosted to higher energy in the lab, compared to positrons emitted backward in the muon rest frame. Special calorimeters were built to observe and measure the positron energies. The positrons generally are at lower energy than the parent muons, and spiral inward in the storage ring field. 24 calorimeters, located around the inside circumference of the storage ring, observe the positrons. The calorimeters are built of lead, with arrays of scintillating fibers running radially embedded in the lead. [7] Photomultiplier tubes observe the light, proportional to the positron energy, and the data are recorded in special wave form digitizers, built for the experiment. A very stringent requirement of this system is that the time shift over 3 lifetimes, early to late, be less than 20 picoseconds, and this is achieved.

III G-2 FROM THE 1998 RUN

The "wiggle plot" for 1998 is shown in Figure 4. This shows the number of positrons observed with energy above 1.8 GeV, versus time. The oscillation is the effect of the muon spin precession, there is nothing else oscillating! This is shown on a log scale, the linear falloff of the number of positrons is from the 64 μsec muon lifetime, and the top data are for the first 100 μsec of decays, the second row of data are for the next 100-200 μsec of decays, etc. (Zero time is when the muons are injected into the storage ring.) The value for g-2 is obtained from fitting the frequency of the muon precession, and dividing by the precise average magnetic field seen by the muons. The 1998 result is shown in Figure 5. [8] The result from 1998 has an error of ±5 ppm, and agrees with the Standard Model prediction, as well as being in agreement with previous measurements from CERN [4] and Brookhaven [9].

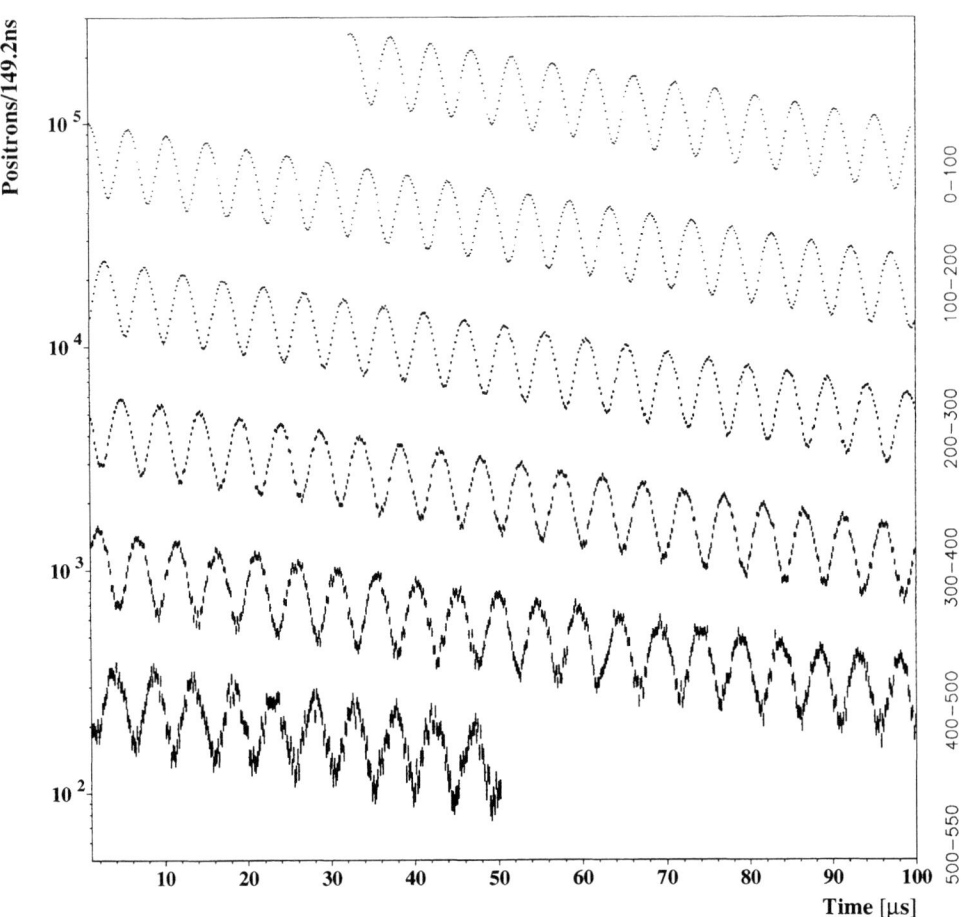

FIGURE 4. Number of positrons observed with energy above 1.8 GeV versus time, for the 1998 data set.

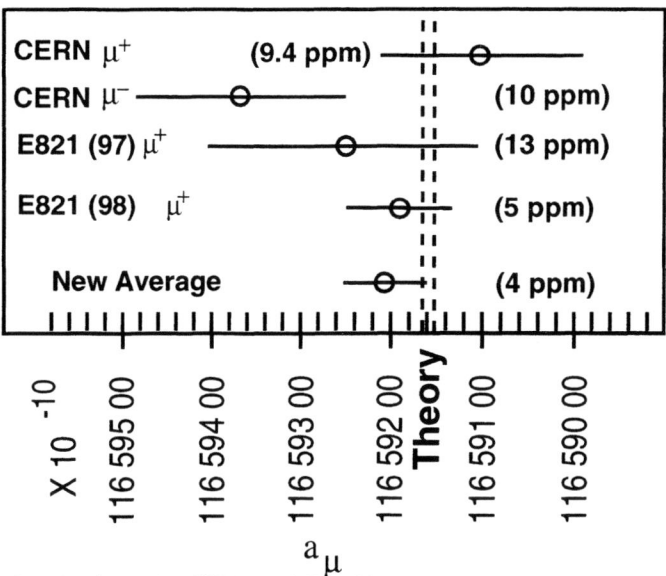

FIGURE 5. Results from the CERN and Brookhaven g-2 experiments, compared with the Standard Model.

IV THE 1999 DATA AND ANALYSIS

As we collect more and more data, the required level of understanding of the data characteristics is ratcheted upward, in order to obtain a good fit of the data set. The 1999 data set has 1 billion positrons, which will give a ±1.3 ppm statistical error on the measurement of g-2. With this data set we are able to see, and must include in our fit, effects from overlapping muon decay events, a coherent betatron oscillation in the data caused by not providing sufficient kick to the muons, and other effects.

When the beam is introduced into the storage ring through the injector magnet window which is necessarily smaller than the transverse aperture of the storage ring, the beam size in the ring will oscillate with the betatron tunes, horizontal and vertical, from the focusing in the ring. For an incomplete kick from the kicker magnet, the injected muons will also oscillate radially around the central orbit with the horizontal betatron frequency. When the scinitllating fiber monitor is rotated into the storage ring, a coherent betatron oscillation is seen for the turns just after injection, as shown in Figure 6. The beam centroid oscillates with the betatron tune, and the tune is different for the two cases shown, without and with scraping. A frequency analysis in Figure 7 shows the peak at the tune and also shows a peak at the proton rotation frequency–since the beam is positive, we are also storing a number of protons. If we fit the "wiggle plot", Figure 4 but for the 1999 data, without including the coherent betatron oscillations (CBO), we obtain a poor fit quality. A frequency analysis of the residuals of the fit shows the CBO effect. Including CBO in the parameterization of the decay spectra greatly improves the probability of the fit. The shift of the g-2 frequency value between the two fits, with and without CBO, is well below the systematic error.

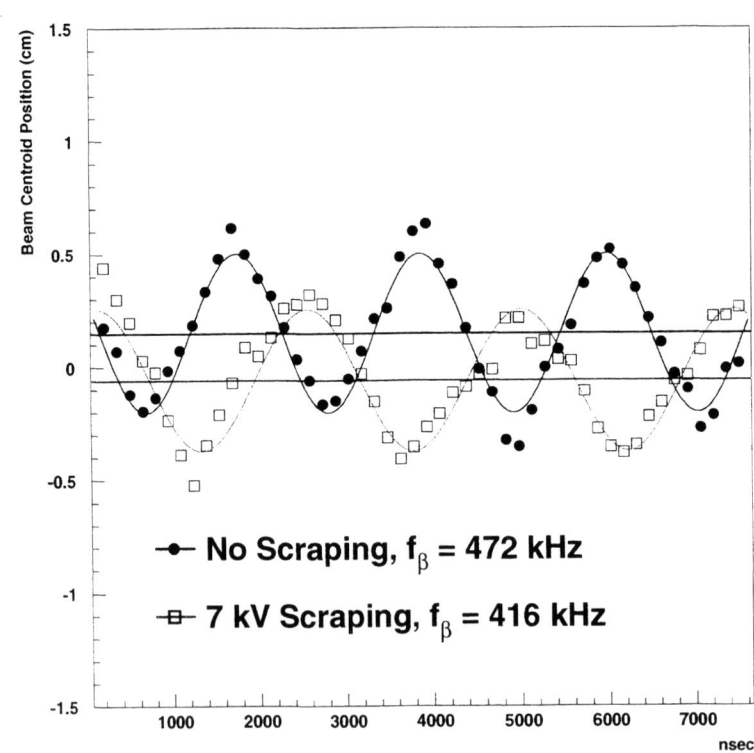

FIGURE 6. The stored beam centroid as measured by the fiber monitor, versus time. To reduce the muon losses during data taking, the stored beam is "scraped" by using imbalanced electrostatic quadrupole voltages in concert with collimators, early in the store. There is a different betatron tune for the scraping condition than for the focussing during data taking ("no scraping").

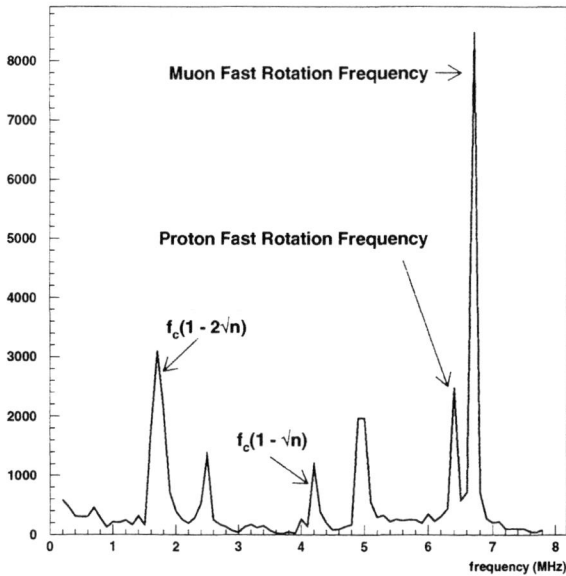

FIGURE 7. A frequency analysis of the fiber monitor data. The coherent betatron oscillation peak is seen at the betatron tune value. A peak is also seen for the proton cyclotron frequency. Since the stored beam is positive, the experiment stores some protons.

The most important "new" effect for the g-2 measurement is to handle overlapping pulses. At early times, the high instantaneous rate gives a small but significant number of overlapped double positron pulses. Generally, due to the snapshots of the pulse height taken by the wave form digitizers every 2.5 nanoseconds, we can distinguish double pulses down to a few nsec pulse separation, but we must correct for overlaps closer than that. When pulses overlap, two lower energy positrons mimic one higher energy positron, at early times, and this will affect the apparent g-2 frequency. That overlapping pulses are present is most obvious comparing the number of unphysically high energy positrons observed at early times versus late times, Figure 8. In the top plot, the positron data at early times are much more frequent than late times, above 3.1 GeV. A method of pile-up subtraction was developed where a pile-up spectrum was obtained from the data itself–from nearby second pulses. The pile-up spectrum is then directly subtracted, giving the results for early and late times shown in the lower plot. The pile-up is reduced by a factor of 10, which then represents a very small contribution to g-2.

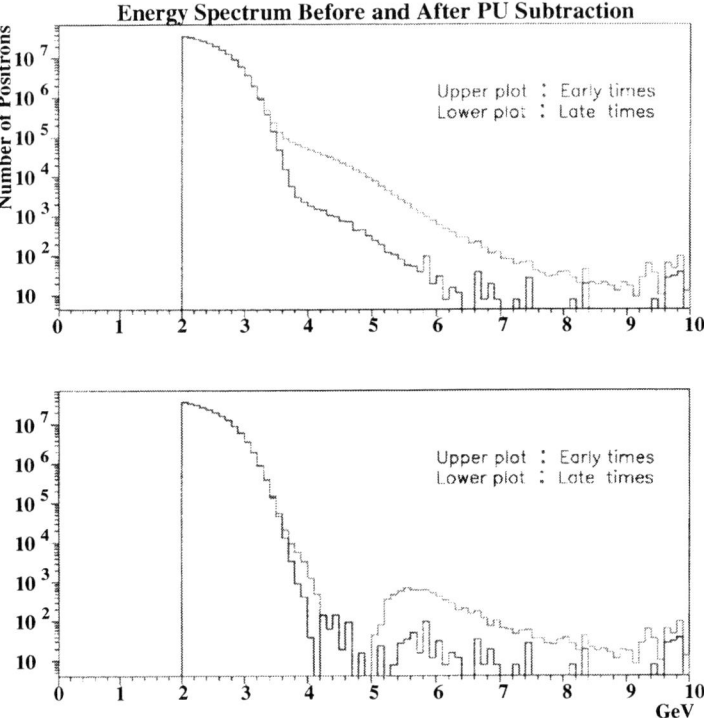

FIGURE 8. The top plot shows the observed positron energy spectrum for early and late times. The data at early times is higher at unphysical energy, above 3.1 GeV. The lower plot shows the spectra for early and late times after directly subtracting the pile-up. We estimate that about 8% of the pile-up remains.

At present, the χ^2/NDF is 0.995 and it does not change significantly from early fit starting times where the calorimeters see very high instantaneous rates, to late starting fit times where the rates are low. The energy spectra seen by the detectors is flat with time, early to late. We have several different methods that we have used to analyze the data, to handle or correct for pile-up, and other data characteristics. We test these methods by detailed simulations, as well as fit quality and the behavior of the fit with starting time, early to late. The magnetic field analysis is done twice independently. As I write this, we expect to "open the box" at our next collaboration meeting in January 2001, and we shall see whether the Standard Model describes muon g-2, or not.

V   FUTURE

In the year 2000, we collected data at a statistical level of ±0.5 ppm, 5 billion positrons. We are preparing for a run in 2001 with negative muons, and we hope to have a comparable statistical level from this run. We expect a final statistical error of ±0.35 ppm from combining positives and negatives. We believe that we will have a comparable systematic error.

We have 3 future experiments planned or proposed. We will search for an electric dipole moment of the muon with the 2001 data and expect a factor of 10 improvement in sensitivity from the CERN measurement which was less than 3.7×10^{-19} e cm. We will use the g-2 ring as a spectrometer to measure the muon neutrino mass from the end point in pion decay, to 8 KeV. [10] We have a letter of intent to search for a muon electric dipole moment to 10^{-24} e cm. [11]

VI   ACKNOWLEDGEMENTS

The members of the experiment, for the 1999 data, are listed in reference [1]. I have relied on an excellent talk given at the 1999 DNP Annual Meeting by Axel Steinmetz. The 1999 muon data analysis used here was done by Cenap Ozben. The other muon analyzers for the 1999 data are Long Duong, Alex Trofimov, Gerco Onderwater, Ivan Logashenko, and Fred Gray (with many other important contributors in the collaboration). I showed the magnetic field analysis of Huaizhang Deng, and Ralf Prigl. I showed the scintillating fiber data analyzed by Ofer Rind.

REFERENCES

1. The g-2 Collaboration for 1999:Boston U.: R.M. Carey, W. Earle, E. Efstathiadis, E.S. Hazen, F. Krienen, I. Logashenko, J.P. Miller, J. Paley, O. Rind, B.L. Roberts, L.R. Sulak, A. Trofimov; BNL: H.N. Brown, G. Bunce, G.T. Danby, R. Larsen, Y.Y. Lee, W. Meng, J. Mi, W.M. Morse, D. Nikas, C. Ozben, C. Pai, R. Prigl, Y.K. Semertzidis, D. Warburton; Cornell U.: Y. Orlov; Fairfield U.: D. Winn; U. Heidelberg: A. Grossmann, K. Jungmann, G. zu Putlitz; U. Illinois: P.T. Debevec, W. Deninger, F. Gray, D.W. Hertzog, C. Onderwater, C. Polly, S. Sedykh, M. Sossong, D. Urner; Max Planck Heidelberg: U. Haeberlen; KEK: A. Yamamoto; U. Minnesota: P. Cushman, L. Duong, S. Giron, J. Kindem, I. Kronkvist, R. McNabb, C. Timmermans, D. Zimmerman; Budker I. Novosibirsk: V.P. Druzhinin, G.V. Fedotovich, B.I. Khazin, N.M. Ryskulov, Yu.M. Shatunov, E. Solodov; Tokyo I. Tech.: M. Iwasaki, M. Kawamura; Yale U.: H. Deng, S.K. Dhawan, F.J.M. Farley, V.W. Hughes, D. Kawall, M. Grosse Perdekamp, J. Pretz, S.I. Redin, E. Sichtermann, A. Steinmetz.
2. R.R. Akhmetshin et al. (CMD2 Collaboration), *Phys. Lett.* **B475**, 1190 (2000).
3. Hughes V.W., Kinoshita T., *Rev. Mod. Phys.* **71**, S133 (1999).
4. Bailey J. et al., Nucl. Phys. **B150**, 1 (1979).

5. Krienen F., Loomba D., Meng W., *Nucl. Instr. Meth.* **A283**, 5 (1989); Meng W., Yamamoto A., et al., Superconducting Inflector for BNL g-2 Experiment, to be submitted to *Nucl. Inst. Meth.* (2001).
6. Danby G.T., et al., *Nucl. Instr. Meth.* **A457/1-2**, 154 (2001).
7. Sedykh S.E., et al., *Nucl. Instr. Meth.*, to be published (2001).
8. H.N. Brown et al., *Phys. Rev.* **D62**, 091101 (2000).
9. R.M. Carey et al., *Phys. Rev. Lett.* **82**, 1632 (1999).
10. Cushman P. et al., Proposal to measure the muon neutrino mass.
11. Semertzidis Y., Letter of intent to measure the muon electric dipole moment.

Weak Nucleon Form Factors

Paul A. Souder[1]

Syracuse University, Syracuse, NY 13244

Abstract.
Recent experiments at JLab and MIT-Bates measuring the parity-violating asymmetry in the scattering of polarized electrons from nucleons have provided new, precise measurements of weak nucleon form factors. The focus of these experiments is the extraction of the contribution of strange quarks to nucleon form factors.

INTRODUCTION

An important issue in the structure of the nucleon is the detailed role of strangeness. About 3% of the momentum of the nucleon is carried by strange quarks [1], and analyses of data on spin structure functions [2–5] suggest that strange quarks also carry some of the spin. Based on early spin structure function data [6], Kaplan and Manohar [7] suggested that strange quarks may also contribute to the elastic nucleon form factors and that this contribution can be isolated by measuring weak form factors.

One way to observe the weak form factors is to measure the parity violating asymmetry in the scattering of polarized electrons from an unpolarized target [8–10]:

$$A^{PV} = \frac{\sigma_R - \sigma_L}{\sigma_R + \sigma_L}, \quad (1)$$

where $\sigma_{L(R)}$ is the cross section for the scattering of left(right) handed electrons. The cross section for unpolarized targets is the sum of three amplitudes, the electromagnetic amplitude f^γ, and the weak amplitudes f_V^Z, and f_A^Z shown in Fig. 1. The parity violating asymmetry isolates the weak amplitudes. In the Standard Model, the axial coupling of the electron to the Z is large, so f_V^Z is also large and provides a useful tool for measuring the weak vector form factors of the target. On the other hand, g^V for the electron and thus f_A^Z is small, so experiments with kinematics chosen to emphasize this term have the potential to measure electroweak radiative corrections.

A large number of experiments, listed in Table 1, have been completed [11–17] or are in progress [18–25] that exploit parity violation in polarized electron scattering.

---

[1] Work supported by the DOE under contract number DE-FG02-84ER40146

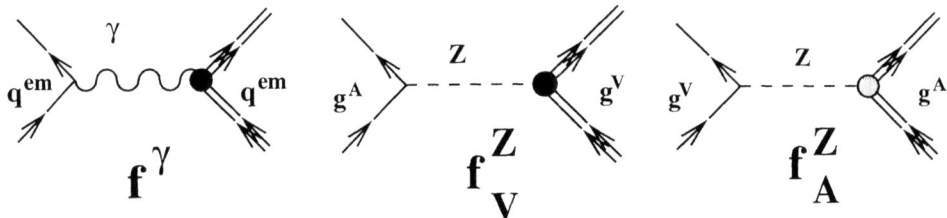

FIGURE 1. Feynman diagrams for electroweak scattering. Since $g^A \gg g^V$ in the Standard Model, f_V^Z is usually the largest parity-violating term.

These experiments have a variety of physics goals as listed in the table. The most common goal recent work is the measurement of weak form factors.

THEORY OF STRANGE QUARKS IN ELASTIC FORM FACTORS

The spin-independent cross section for elastic scattering from the nucleon with four-momentum transfer Q^2 is

$$\frac{d\sigma}{d\Omega} = \left(\frac{d\sigma}{d\Omega}\right)_{Mott} \left[(F_1^{\gamma N})^2 + \tau(F_2^{\gamma N})^2 + 2\tau(F_1^{\gamma N} + F_2^{\gamma N})^2 \tan^2(\theta/2)\right], \quad (2)$$

where θ is the scattering angle, $\tau = Q^2/4M_N^2$, and M_N is the mass of the nucleon. The quantities $F_{1,2}^{\gamma N}$ are the Dirac elastic electromagnetic form factors, which are functions of Q^2. The form factors may be written in terms of contributions from individual quarks as follows:

$$F_{1,2}^{\gamma p} = \frac{2}{3} F_{1,2}^u - \frac{1}{3} F_{1,2}^d - \frac{1}{3} F_{1,2}^s \quad (3)$$

$$F_{1,2}^{\gamma n} = \frac{2}{3} F_{1,2}^d - \frac{1}{3} F_{1,2}^u - \frac{1}{3} F_{1,2}^s \quad (4)$$

where the second relation follows from charge symmetry. Weak scattering from the proton, in which the Z-boson is exchanged, is governed by the weak form factor

$$F_{1,2}^{Zp} = F_{1,2}^u - F_{1,2}^d - F_{1,2}^s - 4\sin^2\theta_W F_{1,2}^{\gamma p}. \quad (5)$$

Once the electromagnetic form factors for both the proton and neutron as well as the weak form factors F_i^{Zp} are measured, the strange form factors $F_{1,2}^s$ may be determined from the above equations.

It is more convenient to use the Sachs form factors: $G_E^i = F_1^i - \tau F_2^i$, and $G_M^i = F_1^i + F_2^i$. The parity-violating asymmetry for the proton is then given by [26]:

$$A^{PV} = \left[\frac{-G_F M_p^2 \tau}{\pi \alpha \sqrt{2}}\right] \left\{(1 - 4\sin^2\theta_W) - \right. \quad (6)$$

$$\frac{[\varepsilon G_E^{p\gamma}(G_E^{n\gamma} + G_E^s) + \tau G_M^{p\gamma}(G_M^{n\gamma} + G_M^s)]}{\varepsilon (G_E^{p\gamma})^2 + \tau (G_M^{p\gamma})^2} -$$

$$\left. \frac{(1 - 4\sin^2\theta_W)\sqrt{\tau(1+\tau)}\sqrt{1-\varepsilon^2}G_M^{p\gamma}(-G_A^{(1)} + \frac{1}{2}F_A^s)}{\varepsilon (G_E^{p\gamma})^2 + \tau (G_M^{p\gamma})^2} \right\}$$

where $\varepsilon = [1 + 2(1+\tau)\tan^2(\theta/2)]^{-1}$ and F_A^s is the strange quark contribution to the axial form factor.

The sensitivity of Eqn. 6 to various form factors depends upon the kinematics. For small values of both τ and θ, the asymmetry is sensitive only to the fundamental weak interaction. A^{PV} is primarily sensitive to G_E^s and G_M^s for small θ and large τ, and primarily to G_M^s and G_A^e for large values of θ. This fact motivates the wide range of kinematics evident in Table 1.

The radiative corrections for the vector form factors terms are known [27]. On the other hand, the radiative correction for the axial term has substantial uncertainties [28]. The SAMPLE collaboration defines a quantity $G_A^e(T = 0)$, which includes all radiative corrections and replaces $G_A^{(1)}$ in the above equation.

TABLE 1. Survey of parity experiments using polarized electrons.

Experiment	Reaction	Physics Goals	A^{PV}
Completed Experiments			
SLAC E122 [11]	$\vec{e}D$ (DIS)	PV of Z	10^{-4}
Mainz [12]	$\vec{e}\,^9$Be QE	New Physics	10^{-5}
Bates [13]	$\vec{e}\,^{12}$C Elastic	New Physics	10^{-6}
SAMPLE-P [14]	$\vec{e}P$ Elastic	$G_M^s(0) = \mu_s$	10^{-5}
SAMPLE-D [15]	$\vec{e}D$ Elastic	G_A^e	10^{-5}
HAPPEX(JLab) [16,17]	$\vec{e}P$ Elastic	$G_M^s + 0.39 G_E^s$	10^{-5}
Approved Experiments			
Mainz [18]	$\vec{e}P$ Elastic	G_M^s, G_E^s	10^{-5}
G^0(JLab) [19]	$\vec{e}P$ Elastic	G_M^s, G_E^s	10^{-5}
^{4}He(JLab) [20]	$\vec{e}\,^4$He Elastic	G_E^s	10^{-5}
Moller(SLAC) [21]	$\vec{e}e$	New Physics	10^{-7}
HAPPEX II(JLab) [22]	$\vec{e}P$ Elastic	$G_M^s + 0.39 G_E^s$	10^{-6}
Pb(JLab) [23]	$\vec{e}$ Pb	Neutron Radius	10^{-6}
SAMPLE-D [24]	$\vec{e}D$ Elastic	G_A^e	10^{-5}
HAPPEX ^{4}He [25]	$\vec{e}\,^4$He Elastic	G_E^s	10^{-5}

A number of papers have made predictions for the sizes of the form factors [29–42]. The predictions may be expressed in terms of the parameters ρ_s and μ_s, which are the low Q^2 limits

$$G_E^s \to \tau \rho_s : \quad G_M^s \to \mu_s. \tag{7}$$

A summary of predictions, given in Table 2, gives a reasonable flavor for the rough size that the strange form factors could be. More reliable calculations are the goal of ongoing work with lattice calculations providing an especially promising approach.

EXPERIMENTAL METHODS

A number of experimental features are common to all polarized electron parity experiments. A typical apparatus is shown in Fig. 2. Polarized electrons are produced by photoemission of circularly polarized laser light from a GaAs cathode. The helicity of the electron beam is determined by the helicity of the light, which is in turn determined by the voltage on a Pockels cell. The electrons are accelerated and then pass through a beam line which is highly instrumented to measure any small correlations between beam parameters and helicity. The beam then strikes the target, and the scattered electrons are detected by a spectrometer. Different types of spectrometers are used to accommodate kinematics for the various experiments. A computer, which monitors all the signals and records the data, also controls the Pockels cell to null any intensity asymmetry and may control coils used to calibrate the sensitivity both of the monitors and of the spectrometers to beam parameters.

Since the measured asymmetries are small, between $10^{-4} - 10^{-7}$ depending on the experiment, it is essential to keep the effects of helicity correlations on the cross section due to any other beam parameter small. Innovative techniques have been developed [43] to maintain the helicity correlations at acceptable levels. The effect of the beam parameters on the number of detected events may be calibrated by dithering steering coils in the beam line.

Recently, impressive progress has been made in the technology required for these challenging experiments. Polarized electron sources with intensities of $100\mu A$ and polarizations of $> 70\%$ have been achieved. Liquid hydrogen and deuterium targets have been developed [44] that can withstand the hundreds of watts that the high intensity beams deposit.

The helicity of the beam is reversed rapidly, typically between 15 and 300 times a second. For pulsed accelerators, including MIT-Bates and SLAC, the signals are integrated to accommodate the high instantaneous rates. For continuous wave accelerators including JLab and Mainz, either integrating or counting methods may be chosen.

TABLE 2. Some published theoretical estimates for ρ_s and μ_s.

Method	ρ_s	μ_s	Author
Pole fits	-2.1 ± 1.0	-0.31 ± 0.09	Jaffe [29]
	-2.9 ± 0.5	-0.24 ± 0.03	Hammer [30]
Kaon Loops	0.2	-0.03	Koepf [31]
	0.5 ± 0.1	-0.35 ± 0.05	Ramsey-Mulolf [32]
	0.3	-0.12	Ito [33]
Unquenched Quarks	0.6	0.04	Geiger [34]
Meson Exchange	0.03	0.002	Meissner [35]
Meson Cloud	–	0.-0.066	Ma [36]
NJL	3.0 ± 0.08	-0.15 ± 0.10	Weigel [37]
Skyrme	1.6	-0.13	Park [38]
	-0.7	-0.05	Park [39]
Chiral Bag Model	–	0.37	Hong [40]
Lattice QCD	1.7 ± 0.7	-0.36 ± 0.20	Dong [41]
Dispersion Relations	0.99	-0.42	Hammer [42]

RECENT RESULTS

The HAPPEX collaboration at JLab used the high-resolution spectrometers in Hall A. Electrons of energy 3.5 GeV scattered elastically at 12.5° were detected with negligible background. The average value of Q^2 was 0.477 $(\text{GeV}/\text{c})^2$.

The HAPPEX experiment took place in two runs. The first was in 1998 [16] and the second in 1999 [17]. The 1998 run used a bulk GaAs crystal which provided a polarization was ~40% and a current on target of ~ 100μA. For the 1999 data, a strained GaAs crystal was used which produced a beam with ~70% polarization and a current of typically 40μA.

For the 1998 run, helicity correlations in the beam parameters were completely negligible. The intensity difference was less than 1 ppm. Position differences at the target were on the order of only a few nm. The helicity-correlated energy difference was also negligible; the position difference was <30 nm at a point on the beam line where the dispersion was ~ 5m. The helicity correlations were larger in the 1999 run, mostly due to the large analyzing power of the strained GaAs crystal. However, the net effect of these correlations on the result was still negligible.

The experimental asymmetry for the combined 1998 and 1999 was [17] $A^{PV} = -14.6 \pm 0.9 \pm 0.5$ ppm where the first error is statistical and the second systematic.

Combining the asymmetry data with Eqn. 6 and with published data on electromagnetic form factors [45-52] gives the result

$$(G_E^s + 0.39 G_M^s)/(G_M^{p\gamma}/\mu_p) = 0.09 \pm 0.05 \pm 0.04 \tag{8}$$

where the first error is the experimental uncertainties added in quadrature, and the second error arises from uncertainties in the electromagnetic form factor data.

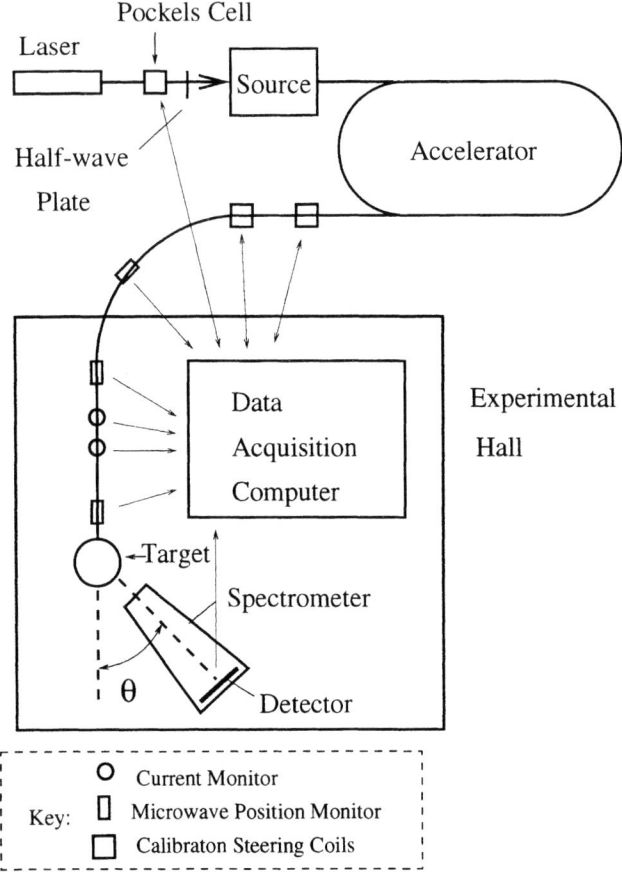

FIGURE 2. Generic Polarized Electron Parity Experiment

This number is consistent with negligible strange form factors. To study the data in the context of the models given above, we assume that the Q^2 dependence of the form factors is

$$G_E^s \sim \tau \rho_s G_M^{\gamma p}/\mu_p : \quad G_E^s \sim \mu_s G_M^{\gamma p}/\mu_p \qquad (9)$$

to obtain the band shown in Figure 3.

We note that a prerequisite for improving the limits on strange form factors, in addition to better parity data, is improved data on electromagnetic form factors. Several such experiments are in progress.

The SAMPLE collaboration at the MIT-Bates Laboratory is in the midst of a program of experiments measuring the asymmetry for electrons scattered from both hydrogen and deuterium by an angle of $\sim 130°$ and $Q^2 \sim 0.1$ (GeV/c)^2. A large solid angle of 1.5 sr is achieved with an array of ellipsoidal mirrors that focus

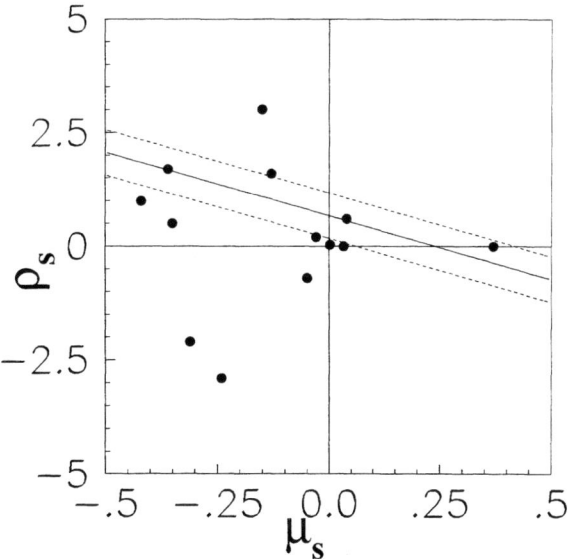

FIGURE 3. Band shows allowed region in the $\rho_s - \mu_s$ parameter space based on the HAPPEX data assuming that the approximation of Eqn. 7 is valid near $Q^2 = 0.48$ (GeV/c)^2.

the Čerenkov light produced by the scattered electrons in the air onto phototubes. Since the beam energy is only 200 MeV, inelastic events are below threshold.

The result of the proton data is [14] $A^{PV} = -4.92 \pm 0.61 \pm 0.73$ ppm where the first error is statistical and the second is systematic. By using Eqn. 6, the results can be interpreted in terms of G^s_M and the radiatively corrected axial form factor G^e_A as shown in Figure 4. Also shown in the figure are the results from the recent deuterium run [15]. The slope of the band from the deuterium data is steeper because of partial cancellation of G^s_M. The region where the bands intersect is consistent with $G^s_M = 0$, but slightly inconsistent with the theoretical prediction for G^e_A [28] shown as the vertical band in the figure. A new SAMPLE run will provide additional deuterium data in the year 2001.

Future Experiments

A number of upcoming experiments will provide more data on strange form factors. An experiment at Mainz by the A4 collaboration [18] will measure elastic scattering from hydrogen at $\theta = 35°$ and $Q^2 = 0.23$ (GeV/c)^2. A unique feature of the experiment is the detector. It will count individual events with an array of 1022

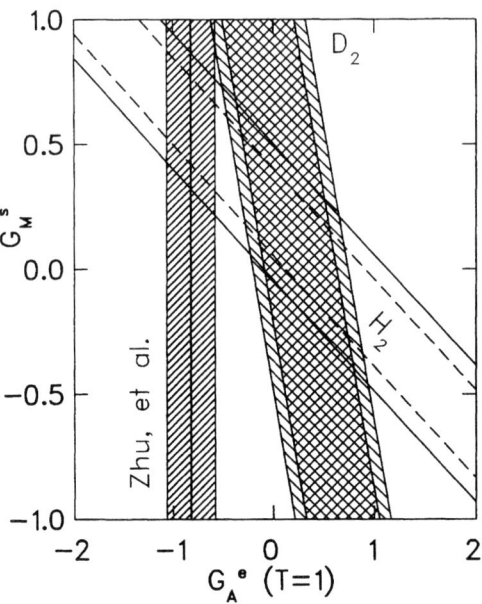

FIGURE 4. Results from the SAMPLE collaboration in terms of G_M^s and the radiatively corrected axial form factor $G_A^e(T=1)$. The bands labeled H$_2$ and D$_2$ are extracted from the data. The vertical band is a theoretical estimate for $G_A^e(T=1)$.

tapered PbF$_2$ crystals. Identification of elastic events is achieved by the excellent resolution of the calorimeter. Initial data-taking is scheduled for the summer of 2000.

The G_0 experiment [19] is a major program at JLab to study strange form factors over a wide kinematic range. The unique feature is a large superconducting toroidal magnet that will serve as the spectrometer. In one configuration, the recoil protons from elastic scattering with $62° < \theta < 78°$ will be detected with plastic scintillators. This kinematics, with a beam energy of 3 GeV, corresponds to electrons with $15° > \theta > 5°$ and $0.16 < Q^2 < 0.95$ (GeV/c)^2. In another configuration, electrons scattered with large angles will be detected. Runs with various beam energies between 0.3 and 0.9 GeV will cover the same Q^2 range as the forward data. The goal is to determine G_E^s and G_M^s separately over a large range of Q^2. Initial running is scheduled for the fall of 2001.

An extension of the HAPPEX experiment, called HAPPEX II [22], will measure elastic scattering from a hydrogen target at $Q^2 \sim 0.1 (\text{GeV}/c)^2$. The new run is motivated by the possibility that the strange form factors are proportional to Q^2 only at low values of Q^2, but fall off much faster at the HAPPEX kinematics. The small angle will be attained with septum magnets that allow the spectrometers in Hall A to reach angles as small as 6°.

Strange form factors may also be measured by elastic scattering from ^{4}He. Since the target is spinless, there is no axial hadronic contribution or uncertain radiative corrections. The asymmetry is given by [10,53]

$$A^{PV} = \frac{G_F Q^2}{\pi \alpha \sqrt{2}} \left[\sin^2 \theta_W + \frac{G_E^s}{2(G_E^p + G_E^n)} \right]. \tag{10}$$

Two experiments have been approved for the Hall A spectrometers at JLab, one at $Q^2 \sim 0.6$ (GeV/c)^2 and the other at $Q^2 \sim 0.1$ (GeV/c)^2.

Currently, experiments measuring parity violation with polarized electrons with goals other than measuring strange form factors are underway. One example is an experiment at JLab which will measure A^{PV} for elastic scattering from ^{208}Pb [23]. The result will determine the radius of the neutron distribution in Pb with the same ease of interpretation as is the case of electromagnetic scattering data for the charge radius [54,55]. The basic idea is that the Z couples mainly to neutrons in the same way that the photon couples to protons.

Another example is experiment E158 [21], which is presently being installed at SLAC. The goal of this project is to perform the most precise measurement of $\sin^2 \theta_W$ at low Q^2 and to determine the running of $\sin^2 \theta_W$ [57]. The experiment is sensitive to physics beyond the Standard Model [56], such as a neutral current mediated by a new Z boson. Another possibility is compositeness of the electron characterized by a new strong interaction with a scale $\Lambda < 15$ TeV.

CONCLUSIONS

Tremendous progress has been made recently in the field of the measurement of parity-violating asymmetries in the scattering of polarized electrons to extract weak form factors. A number of precise results have been published, and more precise data is expected soon. These experiments have already set significant limits on the size of strange elastic form factors of the nucleon.

REFERENCES

1. A. O. Bazarko et al., Z. Phys. C **65**, 189 (1995).
2. K. Abe et al., Phys. Lett. B **405**, 180 (1997).
3. B. Adeva et al., Phys. Rev. D **58**, 112002 (1998).
4. G. Alterelli et al., Acta Phys. Polon. B **29**, 1145 (1998).
5. E. Leader et al., Phys. Lett. B **462**, 189 (1999).
6. J. Ashman et al., Phys. Lett. B **206**, 364 (1988); Nucl. Phys. B **328**, 1 (1989).
7. D. B. Kaplan and A. Manohar, Nucl. Phys. B **310**, 527 (1988).
8. R. D. McKeown, Phys. Lett. B **219**, 140 (1989).
9. E. J. Beise and R. D. McKeown, Comments Nucl. Part. Phys. 20, 105 (1991).
10. D. H. Beck, Phys. Rev. D **39**, 3248 (1989).
11. C. Y. Prescott et al., Phys. Lett. B **84**, 524 (1979).
12. W. Heil et al., Nucl. Phys. B **327**, 1 (1989).
13. P. A. Souder et al., Phys. Rev. Lett. **65**, 694 (1990).

14. D. T. Spayde et al., *Phys. Rev. Lett.* **84**, 1106 (2000).
15. R. Hasty, accepted for publication in *Science*, 2000.
16. K. Aniol et al., *Phys. Rev. Lett.* **82**, 1096 (1999).
17. K. Aniol et al., nucl-ex/0006002.
18. Mainz proposal A4/1-93 (D. von Harrach, spokesperson).
19. JLab experiment 91-017 (D. Beck, spokesperson).
20. JLab experiment 91-004 (E. J. Beise, spokesperson).
21. SLAC experiment E158 (K. S. Kumar, spokesperson, E. W. Hughes and P. A. Souder, deputy spokespersons).
22. JLab experiment 99-115 (K. S. Kumar and D. Lhuillier, spokespersons).
23. JLab experiment 99-012 (R. Michaels and P. A. Souder, spokespersons).
24. MIT-Bates experiment 00-04, T. M. Ito, spokesperson.
25. JLab experiment 00-114 (D. S. Armstrong and R. Michaels, spokespersons).
26. M. J. Musolf et al., *Phys. Rep.* **239**, 1 (1994), and references therein.
27. Particle Data Group, C. Caso et al., *Eur. Phys. J.* C **3**, 1 (1998).
28. S. -L. Zhu et al., *Phys. Rev. D* **62**, 033008 (2000).
29. R. L. Jaffe, *Phys. Lett. B* **229**, 275 (1989).
30. H. -W. Hammer, Ulf-G. Meissner, and D. Drechsel, *Phys. Lett. B* **367**, 323 (1996).
31. W. Koepf, E. M. Henley, and J. S. Pollock, *Phys. Lett. B* **288**, 11 (1992).
32. M. J. Musolf and M. Burkhardt, *Z. Phys. C* **61**, 433 (1994).
33. H. Ito, *Phys. Rev. C* **52**, R1750 (1995).
34. P. Geiger and N. Isgur, *Phys. Rev. D* **55**, 299 (1997).
35. Ulf-G. Meissner et al., *Phys. Lett. B* **408**, 381 (1997).
36. B.-Q Ma, *Phys. Lett. B* **408**, 387 (1997).
37. H. Weigel et al., *Phys. Lett. B* **353**, 20 (1995).
38. N. W. Park, J. Schecter, and H. Weigel, *Phys. Rev. D* **43**, 869 (1991).
39. N. W. Park and H. Weigel, *Nucl. Phys. A* **541**, 453 (1992).
40. S-T. Hong, B-Y. Park, and D-P. Min, *Phys. Lett. B* **414**, 229 (1997).
41. S. J. Dong, K. F. Liu, and A. G. Williams, *Phys. Rev. D* **58**, 074504 (1998).
42. H.-W. Hammer and M. J. Ramsey-Musolf, *Phys. Rev. C* **60**, 045205 (1999).
43. T. Averett et al., *Nucl. Instrum. Methods* **438**, 246 (1999)
44. E. J. Beise et al., *Nucl. Instrum. Methods* **378**, 383 (1996).
45. M. K. Jones et al., *Phys. Rev. Lett.* **84**, 1398 (2000).
46. R. C. Walker et al., *Phys. Rev. D* **49**, 5671 (1994).
47. H. Anklin et al., *Phys. Lett. B* **428**, 248 (1998).
48. E. E. W. Bruins et al., *Phys. Rev. Lett.* **75**, 21 (1995).
49. C. Herberg et al., *Eur. Phys. Jour. A* **5**, 131 (1999).
50. M. Ostrick et al., *Phys. Rev. Lett.* **83**, 276 (1999).
51. I. Passchier et al., *Phys. Rev. Lett.* **82**, 4988 (1999).
52. D. Rohe et al., *Phys. Rev. Lett.* **83**, 4257 (1999).
53. M. J. Musolf, R. Schiavilla, and T. W. Donnelley *Phys. Rev. C* **50**, 2173 (1994).
54. C. J. Horowitz, *Phys. Rev. C* **57**, 3430 (1998).
55. C. J. Horowitz, S. J. Pollock, P. A. Souder, and R. Michaels, nucl-th/9912038.
56. K. S. Kumar et al., *Mod. Phys. Lett.* **A10**, 2979 (1995).
57. A. Czarnecki and W. J. Marciano, *Phys. Rev. D* **53**, 1066 (1996).

Polarization in Photodisintegration of the Deuteron

R. Gilman*
for the Jefferson Lab Hall A Collaboration

*Rutgers University [1]
Piscataway, New Jersey 08855 USA, and
Thomas Jefferson National Accelerator Laboratory [2]
Newport News, Virginia 23606 USA

Abstract. Recoil proton polarization was measured in deuteron photodisintegration at $\theta_{cm} = 90°$ for photon energies up to 2.4 GeV. The induced polarization p_y is consistent with vanishing above 1 GeV, in surprising disagreement with meson-baryon calculations. From pQCD it is generally expected that hadron helicity is conserved, but, the polarization transfer observables do not vanish. As the data are inconsistent with existing meson baryon model predictions, and with expectations from pQCD, we suggest that nonperturbative quark models will be needed to understand the underlying reaction dynamics.

INTRODUCTION

Whether a meson baryon or a quark gluon picture is more appropriate to describe high momentum transfer exclusive reactions is a fundamental issue in nuclear physics. It is often suggested that, at high momentum transfers / short distances, conventional meson baryon models will break down, and quark models will be needed. While self-consistent perturbative quantum chromodynamics (pQCD) calculations are believed to not generally describe exclusive reactions at currently accessible kinematics, various nonperturbative quark model approaches have been suggested.

The issue has been studied in several electromagnetic exclusive reactions at large momentum transfer, including deuteron photodisintegration [1], elastic electron deuteron scattering [2,3], and threshold deuteron electrodisintegration [4]. To date, the evidence is that meson baryon models work well, although sometimes the underlying reaction dynamics are ambiguous, as meson baryon and quark models give

---

[1] Supported by U.S. National Science Foundation grant PHY 9803860
[2] Southeastern Universities Research Association (SURA) manages the Thomas Jefferson National Accelerator Facility under DOE contract DE-AC05-84ER40150.

similar results. For example, in elastic ed scattering, polarization measurements [3] show that only meson baryon models can describe the data well up to $Q^2 = 1.7$ GeV^2, while cross section measurements at higher momentum transfer [2] are consistent with both meson baryon calculations and quark models.

Similarly, in deuteron photodisintegration, the highest energy cross sections show an s^{-11} scaling behavior at $\theta_{cm} = 90°$ that is consistent with a meson baryon calculation [5], a nonperturbative QCD model [6], and pQCD [7]. The onset of the scaling is at $-t \approx 1$ GeV^2, which corresponds to a spatial resolution of ≈ 0.2 fm, and to a momentum transfer Q^2 of nearly 4 GeV^2 in ed elastic scattering [8].

Thus, we have measured reoil proton polarizations, with circularly polarized beam, to provide data that may be better able to discriminate among the various models that appear to explain the cross section measurements. We extend the range of induced polarization measurements from a maximum energy of near 1 GeV to about 2.4 GeV. We also present the first results for the double-polarization observables with circularly polarized beam and recoil proton polarization. Predictions for these observables are available from pQCD, and several meson baryon calculations of the induced polarization exist [9-13] - though only [12] extends the polarization predictions to energies above 1 GeV. Before the experiment ran, the general expectation was that pQCD predicts that the induced polarization vanishes, while the meson baryon calculations predict that resonance contributions lead to an induced polarization that is generally large and strongly energy dependent. Calculations are also underway in nonperturbative quark models [14,15].

In pQCD, vector interactions conserve quark helicity; to the extent that orbital angular momentum effects can be neglected, quark helicity conservation then leads to hadron helicity conservation (HHC) [16]. (In contrast, there are recent arguments for hadronic reactions that quark exchange diagrams will lead to violation of HHC [17].) HHC requires that the sum of the components of the hadronic spins along their respective momentum directions is the same for both initial and final states, and gives predictions for some spin observables. In particular, the recoil polarization induced component p_y and polarization transfer component C_x vanish. This can be seen from the expressions for these polarization components. There are 12 independent complex amplitudes. The induced polarization p_y and the polarization transfer C_x are proportional to the imaginary and real parts, respectively, of the same combination of amplitudes [18]. For example, p_y is given by

$$f(\theta)p_y = 2\Im \sum_{i=1}^{3}[F_{i,+}^* F_{i+3,-} + F_{i,-}F_{i+3,+}^*], \qquad (1)$$

with $f(\theta) = \sum_{i=1}^{6}\sum_{\pm}|F_{i,\pm}|^2$. The hadron helicity conserving amplitudes are $F_{1,+} = <\lambda_p\lambda_n|T|\lambda_\gamma\lambda_d> = <\frac{1}{2}\frac{1}{2}|T|11>$, $F_{3,-} = <-\frac{1}{2}-\frac{1}{2}|T|1-1>$, and $F_{5,\pm} = <\pm\frac{1}{2}\mp\frac{1}{2}|T|10>$. Thus, each helicity-conserving amplitude multiplies a helicity non-conserving amplitude, and both observables vanish in case of HHC. The polarization transfer C_z is given by

$$f(\theta)C_z = \sum_{i=1}^{6}\sum_{\pm} \pm|F_{i\pm}|^2. \qquad (2)$$

With HHC, $f(\theta)C_z \to F_{1,+}^2 - F_{3,-}^2 + F_{5,+}^2 - F_{5,-}^2$, which is generally nonzero. However, at $\theta_{cm} = 90°$, relations are expected between the helicity conserving amplitudes [19]. Both $F_{1,+}$ and $F_{3,-}$ should be small, since they couple an initial state of $S = 0$ or 2 to a final state of $S = 1$, thus requiring orbital angular momentum effects. $F_{5,+}$ and $F_{5,-}$ should be nearly equal, since the difference between the two amplitudes is only that the proton and neutron helicities are reversed. Thus, C_z should also approximately vanish with HHC.

We consider now the most recently published cross section data, from Hall C at the Thomas Jefferson National Accelerator Facility (JLab) [1], to illustrate the variety of physical models in which explanations of photodisintegration have been attempted. Fig. 1 presents existing high energy cross section data, and partially illustrates the ambiguity in interpretation mentioned above.

- The cross sections at 90° and 69° follow the pQCD scaling law, although pQCD is generally not expected to describe exclusive reactions at energies of a few GeV.

- At 90° and for energies up to 1.6 GeV, the meson baryon model calculations of the Bonn group [12] (not shown) also is in very good agreement with the data.

- The asymptotic meson exchange model [5] (not shown) also generally agrees with the cross sections up to ≈3 GeV. The model will eventually break down, becuase it does not quite scale as s^{-11}. This model includes a conventional description of the deuteron plus an additional hard component, normalized to fit the data at 1 GeV. The hard component presumably results from the underlying quark degrees of freedom, and thus is perhaps better considered to be a hybrid model.

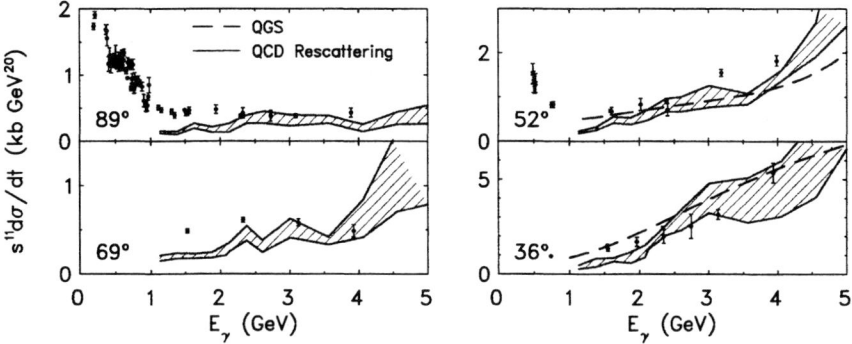

FIGURE 1. $s^{11}d\sigma/dt$ vs E_γ for different center of mass angles. Data and curves are described in the text.

- The quark gluon string (QGS) [20] model takes the reaction to be dominated by three quark exchange, which can be calculated with Regge trajectory approaches. This nonperturbative QCD model works well at small $-t$.

- The QCD rescattering model [6] is a large $-t$, large E_γ nonperturbative QCD model. It calculates absolute cross sections for photodisintegration from the deuteron wave function, the hard photo-quark coupling, and the nucleon-nucleon scattering amplitude, taken from nucleon nucleon data. Thus, this approach includes, through the nucleon nucleon data, soft nonperturbative physics. The resulting calculation (hatched region) describes the data well, even beyond the limits of the models validity.

- Miller [22] and Radyushkin [23] have independently shown how, in a meson baryon and in a nonperturbative quark exchange model, respectively, the cross section should more or less follow phase space factors multiplied by nucleon form factors. This behavior is not too different from pQCD, at least over a limited kinematic range. This helps explain why the data and calculations in different frameworks are similar.

EXPERIMENT

Experiment E89-019 [21] was done at JLab Hall A. Longitudinally polarized electrons were incident on a copper radiator, producing an untagged Bremsstrahlung beam of circularly polarized photons. The photons and electrons then struck a liquid deuterium target. Eight beam energies from 0.5 to 2.5 GeV were used. Protons from the target were detected at a center of mass angles near 90° in one of the Hall A high resolution spectrometers, equipped with a focal plane polarimeter (FPP). The photon energy was reconstructed from the measured proton momentum and scattering angle. Background contributions from the target windows and from electrodisintegration were subtracted. Events with final-state pions were excluded by requiring the reconstructed photon energy to be near the electron beam energy.

Polarized protons scatter in the FPP carbon analyzer with an azimuthal asymmetry. The distributions were analyzed by means of a maximum likelihood method to obtain the induced and transferred polarization components. Spin precession in the spectrometer magnetic fields and the helicity of the electron beam were taken into account.

The polarimeter calibration was checked with ep elastic scattering data at each of the eight momentum settings. In ep elastic scattering, the ratio of the proton electromagnetic form factors G_E and G_M determines the polarization transfer components, usually called P_l and P_t. The ratio of these two components also directly determines the ratio of the form factors G_E/G_M, and the magnitudes determine the product of beam helicity and polarimeter analyzing power, hA_C. Here, A_C indicates a carbon analyzer, as has been used in most previous intermediate energy polarimeters. With Møller polarimeter measurements of the beam helicity, A_C is

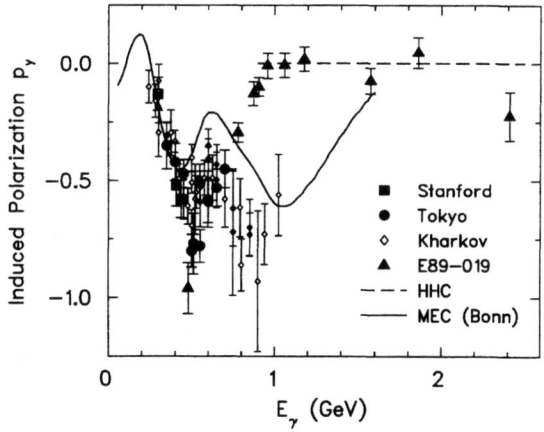

FIGURE 2. Induced polarization p_y in the $\vec{\gamma}d \to \vec{p}n$ reaction at $\theta_{cm} = 90°$.

determined, and agrees well with previous work [24]. Note that, while A_C does not affect the extracted form factor ratio G_E/G_M, it does affect the size of the uncertainty of the ratio.

Neglecting two photon exchange, the ep scattering has no induced polarization; thus, it also directly measures the small instrumental asymmetries, providing small corrections to the induced polarization p_y. Uncertainty in the instrumental asymmetry model dominates the systematic uncertainties for the induced polarization, ±0.04. For the polarization transfer data, the false asymmetries largely cancel with the helicity difference, leaving a smaller systematic uncertainty of about ±0.02. The systematic uncertainties tend to be small, smaller than the statistical uncertainties, because the polarizations measured are generally small, and uncertainties related to the carbon analyzing power and beam polarization are multiplicative factors.

RESULTS

The preliminary result for the normal component p_y of the induced polarization is shown as triangles in Fig. 2. Our data agree with earlier lower energy measurements from Stanford [25], Tokyo [26], and Kharkov [18,27], but disagree with the higher energy Kharkov results [28]. The higher energy Kharkov data were taken under much more difficult conditions than our measurement, which featured much more extensive calibrations, and lower backgrounds. We find that the induced polarization is small, consistent with vanishing above $E_\gamma \approx 1$ GeV. The six highest energy data points average to -0.02 ± 0.02.

The solid curve shows the result of the meson-baryon calculation of the Bonn group [12]. There is fair qualitative agreement between the calculation and the data in the Δ-resonance region, but we find no evidence in our data for the predicted structure near 1 GeV. Instead, our p_y data appear to support HHC (dashed curve), and thus pQCD. In the Bonn calculation the structure results largely from the D_{13} and D_{33} resonances interfering with the Born amplitude. Since there are only a few resonances with large photo-nucleon couplings in this energy region, the disagreement is an unexpected result, and suggests that the meson-baryon picture, at least as calculated, is not applicable.

The question is to what extent the Bonn calculation can be considered reliable and representative of meson baryon theory in general, and thus whether one can conclude anything about the general applicability of meson baryon theory. The answer is probably that no general conclusions can be made at this time. In the Bonn calculation, and others [9–11,13], final state interactions appear to have small effects on p_y. It should be noted that these models generally start from older potentials, derived largely from nucleon nucleon scattering up to 350 MeV. To match s, one considers nucleon nucleon scattering at a beam energy twice the photon energy; nucleon nucleon scattering at 2 - 5 GeV is highly inelastic, and thus probably the final state interactions in these calculations are not under good control.

If the final state interactions are indeed small, the Born amplitude is largely real, and the resonance – Born interference generates the induced polarization. The induced polarization results from the imaginary part of an interference of amplitudes, and one expects that isolated resonances lead to an energy dependent polarization that is maximal at the peak of the resonance. Especially since polarizations are more sensitive to the resonances than are the cross sections, overlapping resonances including those with small photo-couplings will change the detailed predictions from this simple picture.

The Bonn calculation includes π, ρ, η, and ω exchange, plus 17 well-established nucleon and Δ resonances with mass less than 2 GeV and $J \leq 5/2$, for which resonance parameters were taken from the Particle Data Group. Earlier polarization calculations [9–11], which included only the Δ resonance, have similar predictions to [12] at low energies. The inclusion of the Roper, S_{11}, and D_{13} was studied in [13], also with similar predictions to [12]. Thus, there is qualitative agreement between several calculations and the intuitive expectation.

Given approximations in the calculations, which are not discussed, uncertainties in resonance parameters, and possible problems in final state interactions, it is clear that one cannot conclude anything generally about whether meson baryon models are in principle capable of representing the induced polarization data shown above. We speculate, however, that hybrid models such as [5] are likely to succeed, since it appears that the cross section at large energies comes largely from the hard component of the deuteron, *if* final state interactions generate small polarizations in this framework.

Both the scaling of the cross section and the vanishing of p_y start near 1 GeV, and are consistent with pQCD predictions. Because of the failure of existing meson-baryon calculations, and the surprising agreement with HHC in a region in which pQCD is not expected to explain the data, further tests are required. More tests

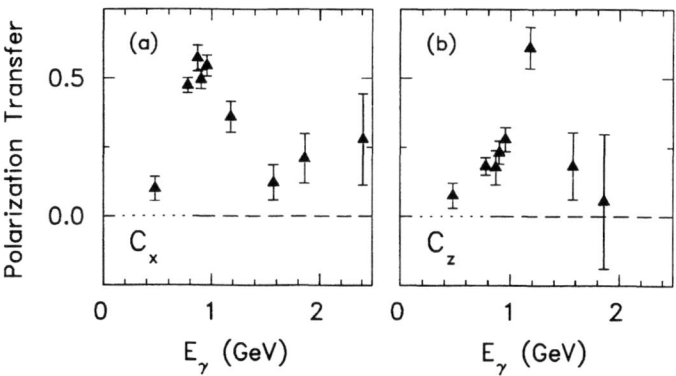

FIGURE 3. Polarization transfer in the $\vec{\gamma}d \to \vec{p}n$ reaction at $\theta_{cm} = 90°$, C_x (a) and C_z (b).

are possible with the polarization transfer observables C_x and C_z.

Our preliminary results are shown in Fig. 3; neither other data nor calculations exist. While C_z appears to approach zero near 2 GeV, the C_x data are non-zero and indicate that hadron helicity is *not* conserved in this energy range. The generally accepted expectation is that pQCD leads to HHC; thus we conclude that one should not try to explain the data as arising from pQCD. While this conclusion agrees with theoretical bias [29] that perturbative processes will lead to only a small fraction of the observed strength, we note here that there is a weakness to this argument. There are both nonperturbative and perturbative corrections to the pQCD polarization prediction of HHC that we have shown. For example, even neglecting possible orbital angular momentum contributions [17], one expects that there are perturbative quark helicity flip contributions [16]. Including these contributions will lead to a violation of HHC, but with a smooth approach to vanishing polarizations, which we cannot rule out.

We construct here a toy model to indicate how these observables could approach the HHC limit. Each helicity-flip should cause the amplitude to decrease faster by one power of t. This result is well known for the nucleon form factors, for which it is expected that $-tF_2/F_1$ approaches a constant at large $-t$. We assume the helicity conserving amplitudes have the form $F_i = A_i f(s,t)$, where A is a complex magnitude and $f(s,t)$ contains the kinematic variations. The leading corrections come from single helicity flip amplitudes, that we assume to have the form $F_j = A_j f(s,t)/t$. At sufficiently large $-t$ - possibly much larger than the range in this experiment - the cross section is dominated by non-helicity-flip amplitudes, and the expression given above for C_x can be seen to lead to a $1/t$ behavior. Presumably p_y vanishes in the data due to a lack of an imaginary part of the interference. With the additional assumptions of [19], C_z should approach zero like $1/t^2$. Figure 4 shows that these t dependences, *normalized* to the data, are in good, but not compelling agreement. Exploring this possibility further requires more precise data.

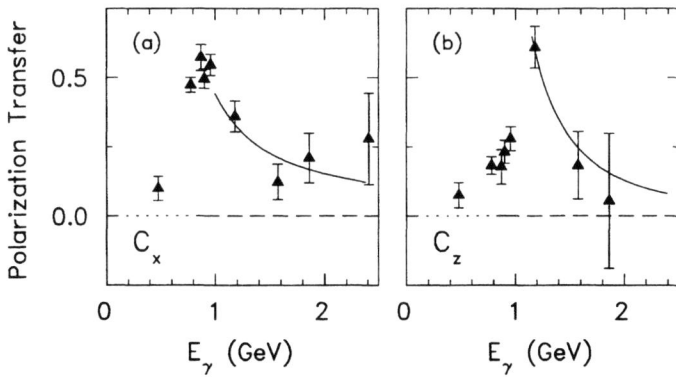

FIGURE 4. Polarization transfer in the $\vec{\gamma}d \to \vec{p}n$ reaction at $\theta_{cm} = 90°$, C_x (a) and C_z (b), compared to a toy model "prediction" of the approach to helicity conservation, normalized to the data.

We have shown above that existing MEC and pQCD do not predict the polarization data. We now consider several nonperturbative quark models, which have been applied to deuteron photodisintegration cross sections. The QCD rescattering model [6] provides absolute cross section calculations that agree with the data. The nonperturbative contributions are effectively included in the hard scattering pn amplitude, which is taken from the pn data. Preliminary induced polarization calculations [14] from this model indicate that p_y is small, for energies above about 2.5 GeV, as it is for E_γ from 1 to 2.4 GeV in the present data. There are also cross section calculations in the quark-gluon string model [30], and polarization calculations are underway [15]. The quark exchange model [31] suggests that the amplitudes are nearly real; this can explain p_y, but calculations are needed for C_x and C_z. The reduced nuclear amplitudes model [32] does not reproduce the cross sections well, and assumes helicity conservation, which would incorrectly predict that C_x vanishes.

CONCLUSIONS

The induced polarization above about 1 GeV is small, consistent with vanishing. The polarization transfers C_x and C_z are consistent with a steady decrease in magnitude starting slightly above 1 GeV, though the precision of the data is such that this result is not compelling.

These results have not been predicted in any theoretical approach. The induced polarization is clearly inconsistent with the only available high energy meson baryon polarization calculation. The results of this calculation are typical of what is seen in other meson baryon calculations, but one cannot rule out that an improved future calculation in this framework will agree with the data. No calculations of polarization transfer exist for comparison.

The vanishing of the induced polarization is naturally explained by HHC within pQCD, but the polarization transfer observable C_x does not vanish. This clearly shows, with no assumptions, that HHC is not valid for deuteron photodisintegration in this energy region [33]. Thus, the data disagree with the usually expected behavior from pQCD. But the data do not rule out a possible perturbative approach to HHC, as demonstrated in a toy model.

If meson baryon theory and pQCD are both not applicable, then one is left by elimination with nonperturbative quark and hybrid models. These new benchmark data provide clear tests for the polarization calculations that will soon be available with these models.

REFERENCES

1. C. Bochna et al., Phys. Rev. Lett. **81**, 4576 (1998); J.E. Belz *et al.*, Phys. Rev. Lett. **74**, 646 (1995); S.J. Freedman *et al.*, Phys. Rev. C **48**, 1864 (1993); J. Napolitano *et al.*, Phys. Rev. Lett. **61**, 2530 (1988).
2. L.C. Alexa *et al.*, Phys. Rev. Lett. **82**, 1374 (1999).
3. D. Abbott *et al.*, Phys. Rev. Lett. **84**, 5053 (2000).
4. R.G. Arnold *et al.*, Phys. Rev. C **42**, R1 (1990).
5. A.E.L. Dieperink and S.I. Nagorny, Phys. Lett. B **456**, 9 (1999).
6. L.L. Frankfurt *et al.*, Phys. Rev. Lett. **84**, 3045 (2000); L.L. Frankfurt *et al.*, Nucl. Phys. A **663**, 349 (2000).
7. S.J. Brodsky and G.R. Farrar, Phys. Rev. Lett. **31**, 1153 (1973).
8. R. Holt, Phys. Rev. C **41**, 2400 (1990).
9. J.M. Laget, Nucl. Phys. A **312**, 265 (1978).
10. W. Liedemann and H. Arenhövel, Nucl. Phys. A **465**, 573 (1987).
11. H. Tanabe and K. Ohta, Phys. Rev. C **40**, 1905 (1989).
12. Y. Kang et al., *Abstracts of the Particle and Nuclear Intersections Conference*, (MIT, Cambridge, MA 1990); Y. Kang, Ph.D. dissertation, Bonn (1993).
13. M. Schwamb, H. Arenhövel, P. Wilhelm, Few Body Syst. **19**, 121 (1995).
14. M.M. Sargsian, private communication.
15. E. De Sanctis, L.A. Kondratyuk, private communications.
16. See S.J. Brodsky and G.P. Lepage, Phys. Rev. D **24**, 2848 (1981), and references therein.
17. T. Gousset, B. Pire, and J.P. Ralston, Phys. Rev. D **53**, 1202 (1996).
18. V.P. Barranik *et al.*, Nucl. Phys. A **451**, 751 (1986).
19. D. Sivers, private communication; S.I. Nagornyĭ, Yu.A. Kasatkin, and I.K. Kirichenko, Yad. Fiz. **55**, 345 (1992) [Sov. J. Nucl. Phys. **55**, 189 (1992)].
20. L.A. Kondratyuk *et al.*, Phys. Rev. C **48**, 2491 (1993).
21. JLab Experiment E89-019, R. Gilman, R.J. Holt, and Z.-E. Meziani, spokespersons; K. Wijesooriya *et al.*, submitted to Phys. Rev. Lett.
22. G.A. Miller, private communication.
23. A. Radyushkin, private communication.

24. B. Bonin et al., Nucl. Instrum. Methods A **288**, 379 (1990); M.W. McNaughton et al., Nucl. Instrum. Methods A **241**, 435 (1985).
25. F.F. Liu et al., Phys. Rev. **165**, 1478 (1968).
26. T. Kamae et al., Phys. Rev. Lett. **38**, 468 (1977); T. Kamae et al., Nucl. Phys. B **139**, 394 (1978); H. Ikeda et al., Phys. Rev. Lett. **42**, 1321 (1979); H. Ikeda et al., Nucl. Phys. B **172**, 509 (1980).
27. A.S. Bratashevskij et al., Nucl. Phys. B **166**, 525 (1980); A.S. Bratashevskij et al., Yad. Fiz. **31**, 860 (1980) [Sov. J. Nucl. Phys. **31**, 444 (1980)]; A.S. Bratashevskij et al., Pis'ma Zh. Eksp. Teor. Fiz. **35**, 489 (1982) [JETP Lett. **35**, 605 (1982)]; A.S. Bratashevskij et al., Yad. Fiz. **43**, 785 (1986) [Sov. J. Nucl. Phys. **43**, 499 (1986)]; A.A. Zybalov et al., Nucl. Phys. A **533**, 642 (1991); V.B. Ganenko et al., Z. Phys. A **341**, 205 (1992).
28. A.S. Bratashevskij et al., Pis'ma Zh. Eksp. Teor. Fiz. **34**, 410 (1981); A.S. Bratashevskij et al., Pis'ma Zh. Eksp. Teor. Fiz. **36**, 174 (1982) [JETP Lett. **36**, 216 (1982)]; A.S. Bratashevskij et al., Yad. Fiz. **44**, 960 (1986) [Sov. J. Nucl. Phys. **44**, 619 (1986)].
29. N. Isgur and C. Llewellyn Smith, Nucl. Phys. B **317**, 526 (1989); A. V. Radyushkin, Nucl. Phys. A **523**, 141c (1991).
30. E. De Sanctis et al., Few Body Syst. Suppl. **6**, 229 (1992); L. A. Kondratyuk et al. Phys. Rev. C **48**, 2491 (1993).
31. A. Radyushkin, private communication.
32. S.J. Brodsky and J.R. Hiller, Phys. Rev. C **28**, 475 (1983).
33. Note also recent Σ asymmetry results from Yerevan: F. Adamian et al., Eur. Phys. J. A **8**, 423 (2000).

Nucleon Electromagnetic Form Factors and Proton charge radius

Haiyan Gao

Laboratory for Nuclear Science and Department of Physics
Massachusetts Institute of Technology
Cambridge, MA 02139
U.S.A.

Abstract. Nucleon electromagnetic form factors are fundamental quantities related to the charge and current distribution inside the nucleon. They have been studied in the past using unpolarized electron scattering. With the development in polarized beam and polarized target technologies, polarization experiments have provided more precise data on these quantities. In this talk, I review the recent progress on this subject from polarized electron scattering experiments and provide some future outlook. I also discuss the planned high precision measurement of the proton charge radius at the MIT-Bates Laboratory.

INTRODUCTION

The electromagnetic form factors of the nucleon have been a longstanding subject of interest in nuclear and particle physics. They are fundamental quantities describing the distribution of charge and magnetization within nucleons and allow sensitive tests of nucleon models based on Quantum Chromodynamics (QCD), as well as provide a basis for calculations of processes involving the electromagnetic interaction with complex nuclei. This advances our knowledge of nucleon structure and provides a basis for the understanding of more complex strongly interacting matter in terms of quark and gluon degrees of freedom.

The proton electromagnetic form factors have been determined with good precision at low values of the squared four-momentum transfer, Q^2, using Rosenbluth separation [1] of elastic electron-proton cross sections, and more recently at higher Q^2 using a polarization transfer technique [2]. The neutron form factors are known with much poorer precision than proton form factors because of the lack of free neutron targets. Over the past decade, with the advent of polarized beams and targets, the precise determination of both the neutron electric form factor, G_E^n, and the magnetic form factor, G_M^n, has become a focus of experimental activities. Considerable attention has been devoted to the precise measurement of G_M^n. While knowledge of G_M^n is interesting in itself, it is also required for the determination of

G_E^n, which is usually measured via the ratio G_E^n/G_M^n. Further, precise data on the nucleon electromagnetic form factors are essential for the analysis of parity violation experiments [3,4] designed to probe the strangeness content of the nucleon.

PROTON ELECTROMAGNETIC FORM FACTOR

The proton electric (G_E^p) and magnetic (G_M^p) form factors have been studied extensively in the past from unpolarized electron-proton elastic scattering using the Rosenbluth separation technique. The unpolarized ep cross section in the one-photon exchange picture is

$$\frac{d\sigma}{d\Omega} = \frac{\alpha^2 E' cos^2(\theta/2)}{4E^3 sin^4(\theta/2)}[G_E^{p\,2} + \frac{\tau}{\epsilon}G_M^{p\,2}](\frac{1}{1+\tau}), \tag{1}$$

where ϵ is the virtual photon longitudinal polarization, $\tau = \frac{Q^2}{4M}$, Q^2 is the squared of the four-momentum transfer, and M is the proton mass. In the Rosenbluth method [1], the separation of $G_E^{p\,2}$ and $G_M^{p\,2}$ is achieved by measuring the cross section at a given Q^2 by varying the incident electron beam energy and the scattered electron angle. While the $G_E^{p\,2}$ term dominates the cross section at low Q^2, the $G_M^{p\,2}$ term dominates at large Q^2. Thus, the extraction of G_E^p at large Q^2 becomes difficult using the Rosenbluth technique.

While precise information on G_E^p and G_M^p is important for understanding the underlying electromagnetic structure of the nucleon, it is also very interesting to study the ratio of these two form factors, $\frac{G_E^p}{G_M^p}$ as a function of Q^2. Although the standard dipole parametrization, $G_D = (1 + \frac{Q^2}{0.71})^{-2}$, seems to describe the Q^2 dependence of both form factors well up to $Q^2 \sim 0.5$ (GeV/c)^2, new data from a polarization transfer experiment [2], which directly measures this ratio shows very intriguing behavior at higher Q^2.

For one-photon exchange, the scattering of longitudinally polarized electrons results in a transfer of polarization to the recoil proton with only two nonzero components, P_t perpendicular to, and P_l parallel to the proton momentum in the scattering plane. The form factor ratio can be determined from a simultaneous measurement of the two recoil polarization components in the scattering plane.

$$\frac{G_E^p}{G_M^p} = -\frac{P_t}{P_l}\frac{E+E'}{2M}tan(\theta/2), \tag{2}$$

where E and E' are the incident and scattered electron energy, respectively, and θ is the electron scattering angle. Fig. 1 shows the recent result from the Jefferson Lab Hall A experiment [2] on the proton form factor ratio together with the existing data from unpolarized ep elastic scattering measurements. The new data from the polarization measurement shows unprecedented precision compared to the data from unpolarized measurements and $\frac{\mu G_E^p}{G_M^p}$ shows intriguing behavior, dropping to 0.5 at Q^2 above 3 (GeV/c)^2.

NEUTRON ELECTROMAGNETIC FORM FACTOR

Unlike the proton electromagnetic form factors, data on the neutron form factors are of inferior quality due to the lack of free neutron targets. The most precise information on G_E^n at low Q^2 prior to any polarization experiment is from the elastic electron-deuteron scattering experiment by Platchkov et al. [5]. However, the extracted G_E^n values are extremely sensitive to the deuteron structure. Fig. 2 shows the G_E^n values extracted with the Paris potential together with the fit of the data (dash-dotted curve). Results from fitting the G_E^n data extracted with the Nijmegen potential, the Argonne V14 (AV14) and the Reid-Soft Core (RSC) NN potentials are shown as solid, dashed and dotted curves, respectively. The large spread represents the uncertainty of G_E^n due to the deuteron structure, and the absolute scale of G_E^n contains a systematic uncertainty of about 50% from the measurement by Platchkov et al. [5].

The development of polarized targets and beams has allowed more complete studies of electromagnetic structure than has been possible with unpolarized reactions. In quasielastic scattering, the spin degrees of freedom introduce new response functions into the differential cross section, thus providing additional information on nuclear structure [6]. Experiments with longitudinally polarized electron beams

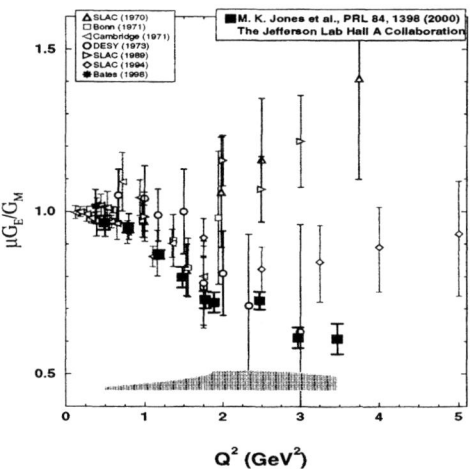

FIGURE 1. New Jefferson Lab data on the proton form factor ratio, $\frac{\mu G_E^p}{G_M^p}$ as a function of Q^2 together with the existing data from unpolarized measurements.

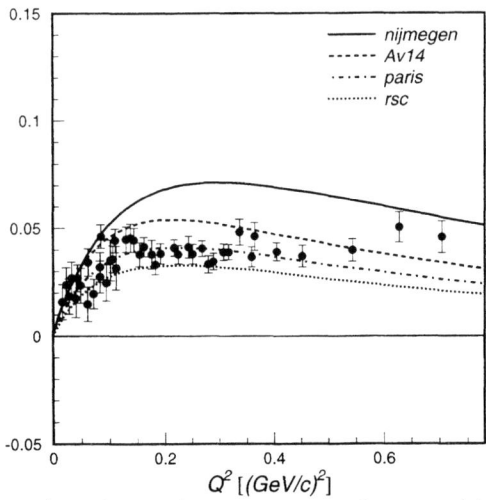

FIGURE 2. The electric form factor of the neutron as a function of four-momentum transfer from Platchkov et al. [5].

and recoil neutron polarimeters have been carried out at MIT-Bates [7] and Mainz [8,9] and G_E^n has been extracted from the $d(\vec{e}, e'\vec{n})$ process. Recently, the neutron electric form factor was extracted for the first time [10] from $\vec{d}(\vec{e}, e'n)$ reaction in which a vector polarized deuteron target from an atomic beam source was employed. Using the polarization degrees of freedom, the proton contribution to the scattering process is suppressed and more precise information on the neutron charge form factor can be extracted.

Polarized ^{3}He is a good candidate for an effective neutron target because its ground state wave function is dominated by the S-state in which the proton spins cancel and the nuclear spin is entirely due to the neutron. Therefore, inelastic scattering of polarized electrons from polarized ^{3}He in the vicinity of the quasielastic peak should be useful for studying the neutron electromagnetic form factors.

The idea of using polarized ^{3}He nuclear target as an effective neutron target was first investigated by Blankleider and Woloshyn in closure approximation [11]. Friar et al. [12] have studied the model dependence in the spin structure of the ^{3}He wave function and its effect on the quasielastic asymmetry. The plane wave impulse approximation (PWIA) calculations performed independently by two groups [13,14] using spin-dependent spectral functions show that the spin-dependent asymmetries are very sensitive to the neutron electric or magnetic form factors at certain kinematics near the top of the quasielastic peak. Recently, Fadeev calculations have been carried out which include the final state interaction (FSI) [15], FSI and meson exchange current (MEC) [16]. These state-of-the-art three-body calculations are very important for extracting the neutron form factors from double polarization electron-^{3}He scattering experiments.

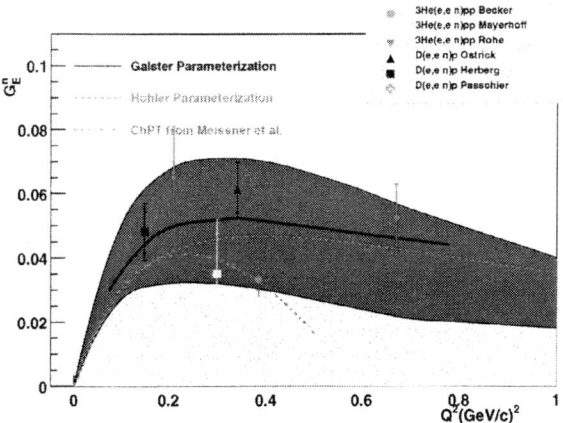

FIGURE 3. Recent data on G_E^n from polarization experiments.

Recently, the $^3\vec{He}(\vec{e}, e'n)$ reaction, in which the proton contribution to the asymmetry is minimized, has been used at MAINZ and NIKHEF in studying the neutron electric form factor. The first measurement on G_E^n from $^3\vec{He}(\vec{e}, e'n)$ was reported by Meyerhoff et al. [17] in which a high pressure polarized ^{3}He target achieved by the metastability-exchange optical pumping technique and the compression method was employed. More recent MAINZ measurement by Becker et al. [18] and Rohe et al. [19] using the same technique show much improved statistical accuracy. No FSI corrections have been applied in the extracted G_E^n values from these experiments shown in Fig. 3. With corrections for FSI applied [20], the extracted values of G_E^n are in good agreement with those from polarized deuterium measurements. To summarize, at present our best knowledge of G_E^n from these polarization measurements is $\sim 30\%$ for $Q^2 \le 0.6 (\text{GeV}/c)^2$.

Until recently, most data on G_M^n had been deduced from elastic and quasielastic electron-deuteron scattering experiments. For inclusive measurements, this procedure requires the subtraction of a large proton contribution and suffers from large theoretical uncertainties due to the deuteron model employed and corrections for final-state interactions (FSI) and meson-exchange currents (MEC). The proton subtraction is avoided in coincidence $d(e, e'n)$ experiments [21], and the sensitivity to nuclear structure can be greatly reduced by measuring the cross section ratio $d(e, e'n)/d(e, e'p)$ at quasielastic kinematics. Several recent experiments [22–24] have employed the latter technique to extract G_M^n with uncertainties of <2% in the low Q^2 region. While this precision is excellent, the results of these experiments [21–24] are not fully consistent. Recently, this ratio technique has been employed in Hall B at Jefferson Lab over a large Q^2 region (from 0.2 to 4.8 $(\text{GeV}/c)^2$) [25] and preliminary results on G_M^n from this experiment is anticipated in the very near future.

An alternative approach to a precision measurement of G_M^n is the inclusive quasielastic $^3\vec{\text{He}}(\vec{e}, e')$ process. In comparison to deuterium experiments, this technique employs a different target and relies on polarization degrees of freedom. It is thus subject to completely different systematics. A pilot experiment using this technique was carried out at MIT-Bates and a result for G_M^n was extracted [26]. The first precision measurement of G_M^n using a polarized ^{3}He target was carried out recently in Hall A at Jefferson Lab [27].

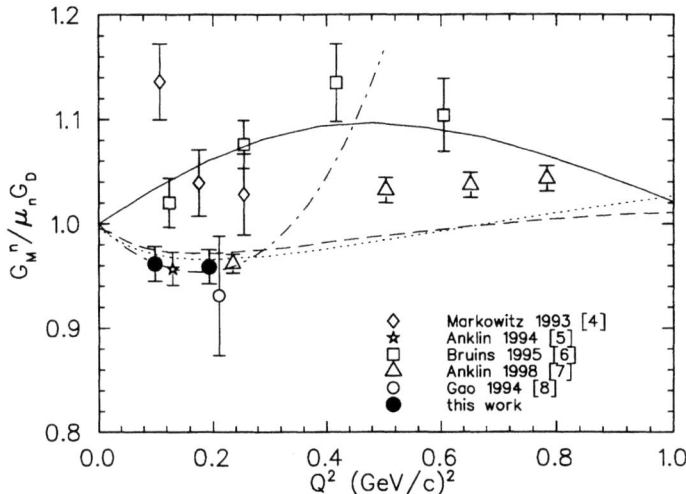

FIGURE 4. The neutron magnetic form factor G_M^n in units of the standard dipole form factor $(1 + Q^2/0.71)^{-2}$, as a function of Q^2, and a few selected theoretical models. The Q^2 points of Anklin 94 [22] and Gao 94 [26] have been shifted slightly for clarity. The solid curve is a recent cloudy bag model calculation [29], the long dashed curve is a recent calculation based on a fit of the proton data using dispersion theoretical arguments [30], and the dotted curve is from the Höhler [31] parametrization. The dash-dotted curve is an analysis based on the relativistic baryon chiral perturbation theory [32].

The experiment was carried out in Hall A at the Thomas Jefferson National Accelerator Facility (JLab), using a longitudinally polarized continuous wave electron beam of 10 μA current incident on a high-pressure polarized ^{3}He gas target [28]. The target was polarized by spin-exchange optical pumping at a density of 2.5×10^{20} nuclei/cm^3. The beam and target polarizations were approximately 70% and 30%, respectively, and the beam helicity was flipped at a rate of 1 Hz (30 Hz for part of the experiment). To improve the optical pumping efficiency, the target contained a small admixture of nitrogen ($\sim 10^{18}$ cm^{-3}). Backgrounds from the target cell walls and the nitrogen admixture were determined to be a few percent of the full target yield in calibration measurements using a reference cell with the same dimensions as those of the ^{3}He target cell.

Six kinematic points were measured corresponding to $Q^2 = 0.1$ to 0.6 (GeV/c)^2 in steps of 0.1 (GeV/c)^2. An incident electron beam energy, E_i of 0.778 GeV was employed for the two lowest Q^2 values of the experiment and the remaining points were completed at $E_i = 1.727$ GeV. Electrons scattered from the target were observed in the two Hall A High Resolution Spectrometers, HRSe and HRSh. Both spectrometers were configured to detect electrons in single-arm mode. The HRSe was set for quasi-elastic kinematics while the HRSh detected elastically scattered electrons. Since the elastic asymmetry can be calculated very well at low Q^2 using the well-known elastic form factors of ^{3}He [33], the elastic measurement allows precise monitoring of the product of the beam and target polarizations, $P_t P_b$.

The state-of-the-art three-body calculation treats the ^{3}He target state and the 3N scattering states in the nuclear matrix element in a consistent way by solving the corresponding 3N Faddeev equations [34]. The MEC effects were calculated using the prescription of Riska [35], which includes π- and ρ-like exchange terms. While the agreement between the data and full calculations is very good at $Q^2 = 0.1$ and 0.2 (GeV/c)^2, the full calculation is not expected to be applicable at higher Q^2 because of its fully non-relativistic framework. A full calculation within the framework of relativity is highly desirable.

To extract G_M^n for the two lowest Q^2 kinematics, the transverse asymmetry data were averaged over a 30 MeV bin around the quasi-elastic peak. The full Faddeev calculation including MEC [16] was employed to generate $A_{T'}$ as a function of G_M^n in the same ω region. By comparing the measured asymmetries with the predictions, the G_M^n values at $Q^2 = 0.1$ and 0.2 (GeV/c)^2 were extracted. The extracted values of G_M^n are shown in Fig. 4 along with results from previous measurements and several theoretical calculations. The uncertainties shown are the quadrature sum of the statistical and experimental systematic uncertainties.

PROTON CHARGE RADIUS

Recent results from lattice QCD calculations suggest that the proton root-mean-square (rms) charge radius can be calculated from first principles with an uncertainty of only a few percent, and this field is rapidly evolving due to both improvements in computer architecture and new algorithms. Thus, precise information on this fundamental quantity is essential in terms of testing the QCD prediction from the lattice calculation.

Accurate information about this fundamental quantity is also essential in conducting high-precision tests of Quantum Electrodynamics (QED) from hydrogen Lamb shift measurements. The standard Lamb shift measurement probes the 1057 MHz fine structure transition between the $2S_{1/2}$ and $2P_{1/2}$ states in hydrogen. The hydrogen Lamb shift can be calculated to high precision from QED using higher order corrections. The proton rms charge radius is an important input in calculating the hadronic contribution to the hydrogen Lamb shift.

The two most precise and widely cited determinations of the proton charge radius in the literature give $r_p = 0.805(11)$ fm [36] and $r_p = 0.862(12)$ fm [37], respectively, differing from each other by more than 7%. While the recent precision hydrogen Lamb shift measurements [38–42] are in better agreement with the QED predictions using the smaller value of the proton charge radius without the two-loop binding effects, they are consistent with the larger value of the proton charge radius when two-loop binding effects are included in the QED calculations. The past analyses also rely on an assumption of the ratio of the electric and magnetic form factors of the proton which has been shown to be incorrect by recent experiments at Jefferson Laboratory [2]. Before accurate comparisons between theory and experiment can be made, both in QCD and QED, a new, precise measurement of the proton charge radius is urgently needed.

Recently, a precision measurement of r_p [43], which will employ the novel technique of combining the determination of $\frac{G_E^p}{G_M^p}$ from the ep elastic asymmetry measurement and the relative differential cross section measurement, is planned at Bates with BLAST. For a symmetric detector configuration, two independent asymmetries corresponding to the same value of Q^2 can be measured simultaneously with respect to the electron beam helicity in the left and right sectors of the detector. The ratio of these two asymmetries, the so-called super ratio, allows the determination of $\frac{G_M^p}{G_E^p}$ to be independent of the knowledge of the beam and target polarizations. The large acceptance detector allows the simultaneous measurement of the proton form factor ratio as a function of Q^2. This is a unique advantage provided by the combination of a polarized internal gas target in an electron storage ring, and a symmetric large acceptance detector system. With the ratio $\frac{G_E^p}{G_M^p}$ determined from the asymmetry measurement, only relative cross section measurements over the same $Q^2 \to 0$ region are needed in order to determine r_p precisely: because it is determined by the slope of G_E^p in the $Q^2 = 0$ limit. The planned measurement will improve the precision of this fundamental quantity by a factor of **three** compared with the single most precise measurement of this quantity from electron scattering experiments. This information will be sufficiently precise to permit high precision tests of QED from hydrogen Lamb shift measurements and to provide reliable tests of Lattice QCD calculations.

FUTURE OUTLOOK

The intriguing result on $\frac{G_E^p}{G_M^p}$ at high Q^2 from Jefferson Lab [2] motivated a lot of interests on this subject both experimentally and theoretically. An extension of this measurement to higher Q^2 ($Q^2 = 5.5$ (GeV/c)^2) is currently in progress [44] at Jefferson Lab. Recent polarization experiments have demonstrated that G_E^n can be determined with much better precision using polarization degrees of freedom. Experiments are in progress and planned at Jefferson Lab on measurement of G_E^n at high Q^2 [45,46]. In the low Q^2 region, Bates with its BLAST detector and high intensity stored electron beam can deliver the definite data on G_E^n at long distances. BLAST will measure $G_E^n(Q^2)$ at $Q^2 = 0.1$ to 0.8 (GeV/c)^2 with high precision both on deuterium and ^{3}He within the same apparatus. With overall uncertainties $\leq 5\%$ it should be possible to carry out a high precision comparison of G_E^n in the deuteron and ^{3}He. This will allow the most precision search for the predicted modification of the neutron pion cloud in the nuclear medium. With future new precision data on G_E^n from Jefferson Lab and Bates, our knowledge of the neutron charge distribution will be improved to a level comparable to that of the proton. With possible future energy upgrade of CEBAF to 12 GeV [47] at Jefferson Lab, nucleon electromagnetic form factor measurements can be extended to much higher Q^2 values, making the connection between non-perturbative QCD to perturbative QCD.

ACKNOWLEDGEMENT

I thank M. Jones and D. Higinbotham for providing Figure 1 and Figure 2. This work is supported by the U.S. Department of Energy under contract number DE-FC02-94ER40818.

REFERENCES

1. M.N. Rosenbluth, Phys. Rev. **79**, 615 (1950).
2. M. Jones et al., Phys. Rev. Lett. **84**, 1398 (2000).
3. B. Mueller et al., Phys. Rev. Lett. **78**, 3824 (1997).
4. K.A. Aniol et al., Phys. Rev. Lett. **82**, 1096 (1999).
5. S. Platchkov et al., Nucl. Phys. **A510**, 740 (1990).
6. T.W. Donnelly and A.S. Raskin, Ann. Phys. (N.Y.) **169** (1986) 247.
7. T. Eden et al., Phys. Rev. C **50**, R1749 (1994).
8. M. Ostrick et al., Phys. Rev. Lett. **83**, 276 (1999).
9. C. Herberg et al., Eur. Phys. Jour. **A5**, 131 (1999).
10. I. Passchier et al., Phys. Rev. Lett. **82**, 4988 (1999).
11. B. Blankleider and R.M. Woloshyn, Phys. Rev. **C29**, 538 (1984).
12. J.L. Friar et al., Phys. Rev. **C42** (1990) 2310.
13. R.-W. Schulze and P.U. Sauer, Phys. Rev. **C48** (1993) 38.
14. C. Ciofi degli Atti, E. Pace and G. Salmè, Phys. Rev. **C51** (1995) 1108; C. Ciofi degli Atti, E. Pace and G. Salmè, in *Proceedings of the 6th Workshop on Perspectives in Nuclear Physics at Intermediate Energies*, ICTP, Trieste May 1993, (World Scientific); C. Ciofi degli Atti, E. Pace and G. Salmè, Phys. Rev. **C51**, 1108 (1995); G. Salmè, private communication.
15. S. Ishikawa et al., Phys. Rev. **C57** (1998) 39; and private communication.
16. V.V. Kotlyer, H. Kamada, W. Glöckle, J. Golak, Few-Body Syst. **28**, 35 (2000).

17. M. Meyerhoff et al., Phys. Lett. **B327**, 201 (1994).
18. J. Becker et al., Eur. Phys. J. A **6**, 329 (1999).
19. D. Rohe et al., Phys. Rev. Lett. **83**, 4257 (1999).
20. J. Golak, G. Ziemer, H. Kamada, H. Witała, W.Glöckle, nucl-th/0008008, submitted to Phys. Rev. **C**.
21. P. Markowitz et al., Phys. Rev. C **48** (1993) R5.
22. H. Anklin et al., Phys. Lett. **B336** (1994) 313.
23. E.E.W. Bruins et al., Phys. Rev. Lett. **75** (1995) 21.
24. H. Anklin et al., Phys. Lett. **B428** (1998) 248.
25. W. Brook, private communication; Jefferson Lab experiment E94-017, spokespersons: W. Brooks, M. Vineyard.
26. H. Gao et al., Phys. Rev. C **50**, R546 (1994); H. Gao, Nucl. Phys. **A631**, 170c (1998).
27. W. Xu et al., Phys. Rev. Lett. **85**, 2900 (2000).
28. J.S. Jensen, Ph.D. Thesis, California Institute of Technology, 2000 (unpublished); P.L. Anthony et al., Phys. Rev. D, **54** 6620 (1996); http://www.jlab.org/e94010/tech_notes.html.
29. D.H. Lu, A.W. Thomas, A.G. Williams, Phys. Rev. C **57**, 2628 (1998).
30. P. Mergell, U.-G. Meißner, D. Drechsel, Nucl. Phys. **A596**, 367 (1996).
31. G. Höhler et al., Nucl. Phys. **B114**, 505 (1976).
32. B. Kubis, U.-G. Meißner, hep-ph/0007056.
33. A. Amroun et al., Nucl. Phys. **A579**, 596 (1994); C.R. Otterman et al., Nucl. Phys. **A435**, 688 (1985); P.C. Dunn et al., Phys. Rev. C **27**, 71 (1983).
34. J. Golak et al., Phys. Rev. C **51**, 1638 (1995).
35. D.O. Riska, Phys. Scr. **31**, 471 (1985).
36. L.N. Hand, D.G. Miller, and R. Wilson, Rev. Mod. Phys. **35**, 335 (1963).
37. G.G. Simon et al., Nucl. Phys. **A333**, 381 (1980).
38. M. Weitz et al., Phys. Rev. Lett. **72**, 328 (1994).
39. E.W. Hagley and F.M. Pipkin, Phys. Rev. Lett. **72**, 1172 (1994).
40. D.J. Berkeland, E.A. Hinds, and M.G. Boshier, Phys. Rev. Lett. **75**, 2470 (1995).
41. S. Bourzeix et al., Phys. Rev. Lett. **76**, 384 (1996).
42. A. Van Wijngaarden et al., Can. Journal of Phys. **76**, 95 (1998).
43. MIT-Bates proposal PR-00-02, Spokespersons: H. Gao, J.R. Calarco.
44. Jefferson Lab experiment E99-007, spokespersons: C. Perdrisat, E. Brash, M. Jones, V. Punjabi.
45. Jefferson Lab experiment E93-038, spokespersons: R. Madey, B. Anderson, and S. Kowalski.
46. Jefferson Lab experiment E93-026, spokesperson: D. Day.
47. The white paper on *The Science Driving the 12 GeV Upgrade of CEBAF*.

The β-ν Correlation Using a Neutral Atom Trap

P.A. Vetter*, S.J. Freedman*,[†], B.K. Fujikawa*, and N.D. Scielzo[†]

*Lawrence Berkeley National Laboratory[1], Berkeley, California 94720
[†]Physics Department, University of California, Berkeley, California 94720

Abstract. By confining up to 500,000 radioactive ^{21}Na atoms in a magneto-optic trap, we have studied the beta-neutrino correlation. The ^{21}Na is produced on-line at the Lawrence Berkeley National Laboratory 88" Cyclotron and trapped using common laser cooling techniques. We detect the low-energy recoil ^{21}Ne daughter nucleus in coincidence with the decay positron. Using the recoil's time-of-flight, we determine the initial momentum of the recoil nucleus and infer the β-ν correlation. Our current data set has a statistical uncertainty to determine the correlation parameter a to better than 2%, and we are studying potential systematic errors at the few percent level. Other beta decay correlations can be studied using magneto-optically trapped radioactive atoms. Precise measurements of correlation parameters can provide limits on non-Standard Model scalar and tensor interactions, or anomalous induced form factors.

INTRODUCTION

The motivation for applying neutral atom laser trapping techniques to the study of beta decay is intimately related to the observables in beta decay. The differential decay probability can be written [1]:

$$d\Gamma = d^3\mathbf{p}_e d^3\mathbf{p}_\nu F(Z,E)\xi \left\{ 1 + a\frac{\mathbf{p}_e \cdot \mathbf{p}_\nu}{E_e E_\nu} + \frac{\langle \mathbf{J}_i \rangle}{J_i} \cdot \left[A\frac{\mathbf{p}_e}{E_e} + B\frac{\mathbf{p}_\nu}{E_\nu} + D\frac{\mathbf{p}_e \times \mathbf{p}_\nu}{E_e E_\nu} \right] \right\} \quad (1)$$

In the allowed approximation, the correlation parameters a, A, B, and D depend on fundamental coupling constants for electroweak currents, C_i, ($i = S, T, V, A$). For example, the β-ν correlation parameter is

$$a = \frac{\left(|C_V|^2 + |C_V'|^2 - |C_S|^2 - |C_S'|^2\right) + \frac{|\langle GT \rangle|^2}{3|\langle F \rangle|^2}\left(|C_T|^2 + |C_T'|^2 - |C_A|^2 - |C_A'|^2\right)}{\left(|C_V|^2 + |C_V'|^2 + |C_S|^2 + |C_S'|^2\right) + \frac{|\langle GT \rangle|^2}{|\langle F \rangle|^2}\left(|C_T|^2 + |C_T'|^2 + |C_A|^2 + |C_A'|^2\right)} \quad (2)$$

[1)] This work was supported by the Director, Office of Energy Research, Office of Basic Energy Sciences, of the U.S. Department of Energy under Contract No. DE-AC03-76SF00098.

The term ξ in Eq. 1 depends on the matrix elements connecting the initial and final nuclear states:

$$\xi = |C_V|^2 |\langle F \rangle|^2 + |C_A|^2 |\langle GT \rangle|^2 \quad (3)$$

where $\langle F \rangle$ and $\langle GT \rangle$ are the Fermi and Gamow-Teller matrix elements of the β transition, respectively. The Standard Model allowed couplings are $C_V = C'_V = 1$ and $C_A = C'_A \approx 1.25$. In the electroweak Standard Model at the fermion level, only vector and axial vector couplings in a maximally parity-violating combination (V-A only) are allowed. Measuring the correlation parameters a, A, and B tests for non-Standard Model physics: (V+A) currents from left-right symmetric physics, scalar and tensor effective couplings (induced for example by massive leptoquarks) or second class currents.

Measuring the correlation terms in Eq. 1 requires precise knowledge of the momenta and spins. Magneto-optic traps (MOT's) provide ideal sources for precise measurements of decay correlations: a pointlike sources of activity suspended in vacuum, negligible source scattering, isotopic selectivity, and polarizability using hyperfine optical pumping. These benefits were suggested in the first demonstration of the technique on short-lived radioactive atoms in [2] and have been emphasized in reviews of the field [3]. Preliminary beta-neutrino correlation results from our laboratory and the TRINAT group at TRIUMF [4], along with results on the beta asymmetry parameter from a Los Alamos group [5] have demonstrated the advantages of a MOT for decay correlation measurements.

The current limits on non-Standard Model (V+A) currents expected from left-right symmetric models (reviewed in reference [6]) are restrictive on the mass and mixing angle of any additional W bosons with right-handed couplings. However, theoretical complications such as a different CKM matrix for right-handed couplings, or right handed neutrino mixing justifies continued searches for (V+A) currents. In contrast to the situation with left-right symmetric theories, the limits on the existence of scalar or tensor currents are much less stringent, more complex, and model-dependent. Despite complicated interpretation of potential violations of the Standard Model, the β-ν correlation directly searches for non-Standard Model scalar and tensor couplings. Scalar and tensor currents are only limited to be less than $10 - 15\%$ of the vector and axial vector couplings [7]. From Eq. 2, we see that different beta decay transitions have different sensitivities to the presence of scalar and tensor couplings in the β-ν correlation, according to their allowed-order matrix elements. Of particular interest are the pure Fermi and pure Gamow-Teller transitions, (where either $\langle GT \rangle$ or $\langle F \rangle = 0$) which exclusively limit S and T terms, respectively. Measurements on different transitions will provide a stringent search for the existence of S and T currents. Figure 1 shows the dependence of the β-ν correlation on the Fermi fraction $\mathcal{F}$ of the transition, where

$$\mathcal{F} = \left(1 + \left|\frac{C_A \langle GT \rangle}{C_V \langle F \rangle}\right|^2\right)^{-1} \quad (4)$$

Clearly, the data (ordered numbers in Figure 2 refer to Table 1) agree with the (V-A) form of the Standard Model. Figure 2 shows the fractional difference (residuals) between the data and predicted (V-A) line. A nonzero slope to the residuals in Fig. 2 would suggest new scalar currents, while a nonzero offset would indicate tensor currents. Besides the presence of scalar or tensor currents, the β-ν correlation is sensitive to second class weak currents – as classified by opposite behavior under G parity transformation [14]. The β-ν correlation is directly sensitive to the induced tensor form factor, a recoil-order term in beta decay [15]. The induced tensor term is a non-Standard Model interaction. Its presence in the β-ν correlation would appear as nucleus specific differences in Figure 2. The variety of desired limits on all possible coupling constants C_i and C_i' and the search for second class currents makes the β-ν correlation an interesting observable.

EXPERIMENT

The ^{21}Na activity is produced using an intense proton beam from the 88" Cyclotron at E.O. Lawrence Berkeley National Laboratory. The apparatus is shown schematically in Figure 3, and has been previously described in references [2] and [16].

The cyclotron provides a 2 μA beam of protons at 25 MeV for the reaction ^{24}Mg(^{1}H,^{4}He)^{21}Na. The target consists of nine disks of compressed MgO powder

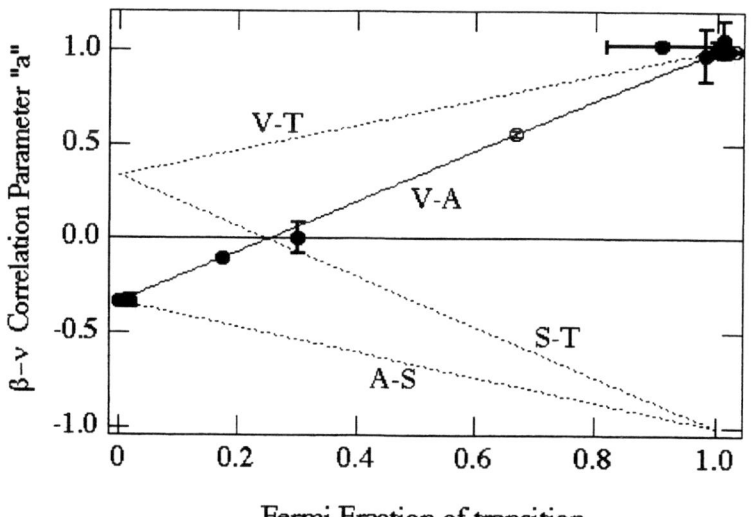

FIGURE 1. Measurements of the β-ν correlation as a function of the Fermi fraction of the transition. References for the points are the same (reading points left to right) as in Fig. 2.

(total target thickness ≈ 350 mg/cm², held at normal incidence to the beam by a "comb" of 0.025 mm Ta foil. The target disks are contained in a sealed ceramic crucible which is heated to roughly 1200 C to enhance the diffusion of the ^{21}Na from the refractory MgO. Using the MgO target, the holdup time of the ^{21}Na in the source oven has been measured to be about one second, in contrast to the 40 seconds estimated in reference [2], with a metallic Mg target. We have tested a variety of different target materials in an effort to discover a system with greater diffusion of the ^{21}Na out of oven, but have yet to improve on MgO.

The ^{21}Na evolves from the target as a neutral atomic beam through four narrow collimator tubes in the side of the target crucible. The forward flux of the atomic beam is increased roughly a factor of ten by the use of counterpropagating cooling lasers in a two-dimensional "optical molasses" configuration immediately after leaving the oven nozzles. These laser beams reduce the transverse velocity of the atomic beam. The atomic beam is slowed from its initial thermal velocity ($v_{rms} \approx 1200$ m/s) by a counterpropagating (slowdown) laser beam. The tapered magnetic field of our Zeeman slower [17] contiuously adjusts the atomic resonance frequency to compensate for the changing Doppler shift of the atoms as they are slowed. Since the publication of references [2] and [16], we have improved the performance of the Zeeman slower by careful tuning of the magnetic field profiles to ensure adiabatic passage of atoms through the slower to a lower final velocity [18]. The trap loading efficiency and lifetime of atoms in the trap have also been im-

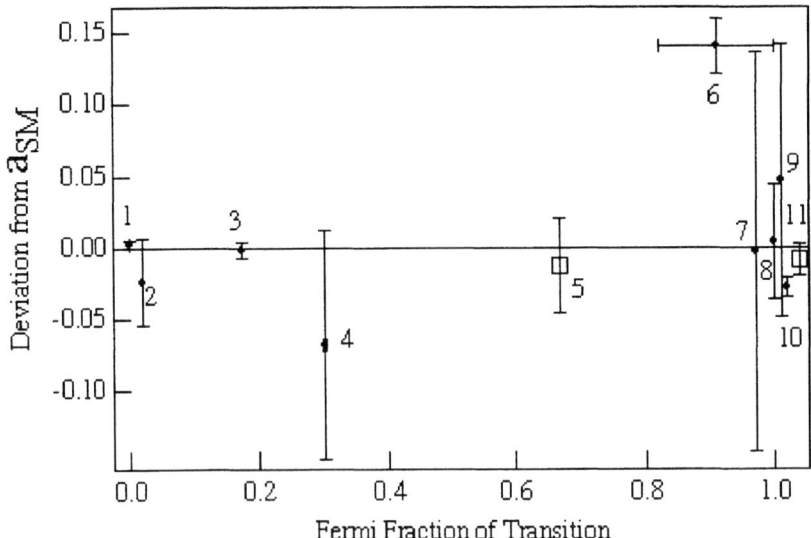

FIGURE 2. Deviations of measurements of the β-ν correlation from the Standard Model (V-A) calculation. References for the points are in Table 1.

TABLE 1. Measurements of a as in Figure 2

Point	Nucleus	$a_{\beta\nu}$	reference
1	^{6}He	-.3308(30)	[8]
2	^{23}Ne	-.33(33)	[8]
3	n	-.1017(51)	[9]
4	^{19}Ne	0.00(8)	[10]
5	^{21}Na	0.552(33)*	this work, (prelim.)
6	^{33}Ar	1.02(2)	[11]
7	^{35}Ar	0.97(14)	[10]
8	^{32}Ar	1.0036(40)	[11]
9	^{18}Ne	1.060(95)	[12]
10	^{32}Ar	0.9989(65)	[13]
11	^{38m}K	0.992(11)*	[4], (prelim.)

proved by using a more spatially homogeneous slowdown laser beam with a "dark spot" [19]. A shadow is cast in the slowdown laser before it enters the trapping chamber, with the spot positioned so that the MOT is in the umbra in the slowdown beam. Atoms in the MOT are unaffected by the strong unbalanced force of the slowing laser, which would otherwise reduce the capture velocity of the trap and degrade the trap lifetime.

The atomic beam loaded magneto-optic trap is a convenient technique for beta decay correlation studies since it combines an efficient capture of the atoms with

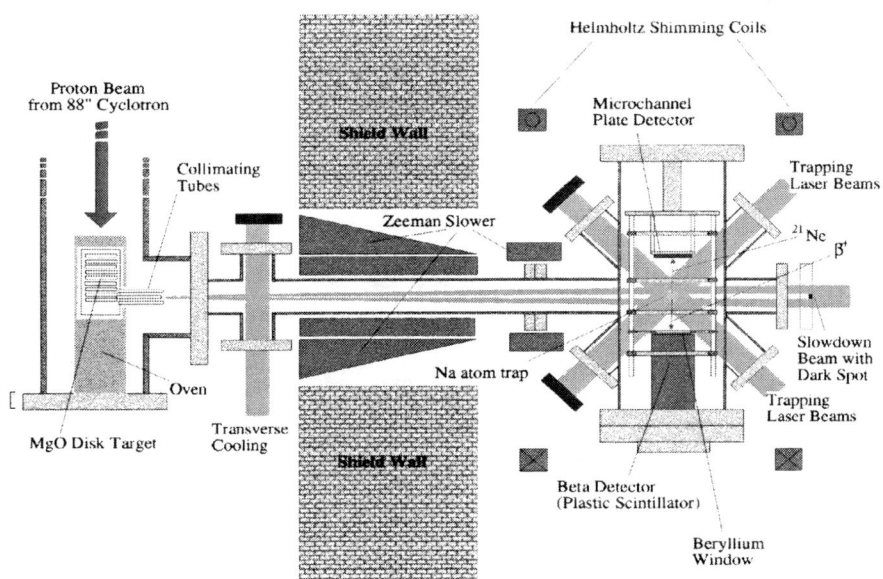

FIGURE 3. Schematic view of the apparatus for production, trapping, and detection of ^{21}Na.

a relatively low-background environment for counting the activity. The other commonly used loading technique for magneto-optic trapping is vapor-cell loading [20], in which a vapor of the desired species suffuses the capture trapping chamber. The MOT loads atoms with velocities less than its capture velocity (typically tens of m/s), and the low-velocity portion of the vapor is replenished by re-thermalization in the cell. The untrapped vapor population can produce a large background decay rate for decay correlation experiments. This load technique requires transferring atoms into a second MOT, located in a differentially pumped ultra-high-vacuum chamber. This additional step concentrates the activity of interest in a relatively background-free volume. Because the lifetime of atoms in a MOT is limited (usually by collisions with residual gas atoms which eject atoms from the trap), and because the capture efficiency of a MOT, regardless of load technique, is less than unity (and often less than 50%), the trapping chamber will contain untrapped radioactive atoms. This can create problematic backgrounds in beta decay experiments.

The atom trap in this work is located between a beta and an ion detector. The beta detector is a 1 cm thick cylinder of plastic scintillator, 54 mm diameter, 6.7 cm from the trapped atoms. Following beta decay, the recoil nuclei emerge in one of many possible charge states. The charged and neutral recoil ^{21}Ne are detected by a microsphere plate (MSP) detector (active area diameter 23 mm) located 5.45 cm from the trap. The MOT is surrounded by several ring and plate electrodes at potentials up to 10 kV in order to accelerate the ions toward the MSP. The potentials are chosen so that the electric field focuses all charged recoil ^{21}Ne daughter ions onto the active area of the MSP, regardless of their initial momentum. A hit in the beta detector provides a start signal for a time-amplitude converter to measure the time of flight of the recoil nucleus. If a coincident recoil is recorded within the 3 μs time window, the beta detector pulse amplitude is also recorded.

The initial momentum of the ^{21}Ne determines its total time of flight in the electric fields: ^{21}Ne initially moving towards the beta detector have a longer flight time than those moving towards the MSP. The shape of the time-of-flight histogram for the ion events encodes the initial kinematics of the recoil nuclei. The kinematics of the decay given by integrating Eq. 1 produce a known time-of-flight distribution for recoil ions. If the nuclei have no net polarization or alignment, then the beta decay rate is simply

$$\Gamma = K \left(1 + a \frac{v_\beta}{c} \cos \theta_{\beta\nu} \right). \tag{5}$$

The detected ion TOF spectrum can be calculated by Monte-Carlo simulations of the beta decay and ion trajectories in the electric field of the trapping chamber. These simulations generate template timing curves. The data are fit to a sum of contributions from the isotropic and beta-neutrino correlated time-of-flight curves. This fit generates the measured value of a.

Each charge state of ^{21}Ne in Fig. 4 is observed to have a sharp rise time. The leading edge of the arrival time of each charge state depends on the trap position and the net electric field. The sharpness of the leading edge is determined by the geometric size of the atom cloud: the more spatially diffuse the cloud, the softer the leading time edge. Using CCD cameras, we image the fluorescence of the trapped atoms along two axes every two minutes during data acquisition. From these images, we can measure the position and radial density of the trap cloud. These parameters are used as input to the Monte-Carlo time-of-flight spectra calculations. Our uncertainty in absolute trap position and radius currently contributes a roughly 3% systematic uncertainty in a, but this uncertainty can easily be reduced.

The electric field is known from the direct measurement of plate potentials and spacings, and can also be extracted directly from time-of-flight data using the time separation between charge states. We find that the electric field deduced from the data is consistent with the calculated fields to about 1%.

DATA

We currently have a data set from several radioactive trapping runs which consists of $\approx 175,000$ coincident beta-ion events. Approximately one-third of the available data set is shown in Figure 4 along with the best fit to Monte-Carlo generated curves from our simulations. Several charge state peaks are detected, ranging from neutral neon to positive ions with charge up to +4. The detection efficiency of the MSP for the different charge states should be uniform to within a few percent, based on studies of microchannel plate efficiencies [21]. The neutrals are unguided by the electric field, so a relatively small number are detected, and the detection efficiency of the MSP for such low energy (0-220 eV) events is unknown and probably energy dependent.

Data were taken both with the atom trap present and without the trap. Coincidence events taken without the trap suggest that the background from untrapped ^{21}Na is quite small and consistent with estimated accidental rates. However, the background TAC spectrum does show a broad peak at 1200 - 1600 ns, which probably corresponds to singly ionized ^{21}Ne originating from ^{21}Na decays on the thin aluminum plate directly in front of the beta detector. This background peak can be subtracted from the data to improve the resolution of the neutral peak, but uncertain detection efficiency makes interpretation of the neutral peak data difficult. The neutral ^{21}Ne events are not included in our analysis for a.

ANALYSIS

Analyzing the data to extract the electron-neutrino correlation from the ion time-of-flight spectrum requires a complete calculation of the kinematics of the ions, from the initial distribution of momentum of these ions through the fields. The electric field and ion flight trajectories are calculated using a commercial ion optics software package, SIMION. The program accepts as input the geometry of the trap chamber and electrodes, the potential at all surfaces, and the initial kinematics of the beta decay (generated by a Monte-Carlo routine to randomize the initial beta direction and account for the acceptance of the beta detector). The full simulation of the ion trajectories produces a time-of-flight spectrum for ions consisting of two parts: one curve proportional to the uncorrelated term in the decay rate, and one curve proportional to the β-ν correlation coefficient a. The time-of-flight of the correlated curve produces an enhancement of the count rate at short times for each charge state as the recoil ions are preferentially emitted opposite to the $\beta+$ and neutrino, and it has a depleted count rate at comparatively long times.

The largest systematic uncertainty in the experiment comes from the sensitivity of the beta detector to positron annihilation radiation and to scattered positrons. We have made Monte-Carlo studies of recoil ion time-of-flight distributions caused by these types of "backscatter" events, and these simulations can qualitatively account for the presence of an extended "tail" in the TOF histogram at 800-900 ns, but the scattering simulations are not sophisticated enough to completely remove such events from the data. Our uncertainty in modelling the backscattering and gamma-ray event components contributes at least 5% uncertainty to a. We are constructing an improved beta detector to greatly reduce the acceptance of gammas and betas scattered from the chamber.

Another source of systematic uncertainty arises from the polarization state of the nuclei of the trapped atoms. If the nuclei are polarized (or aligned), then the presence of the terms with A and B (or c, the alignment term, not shown for

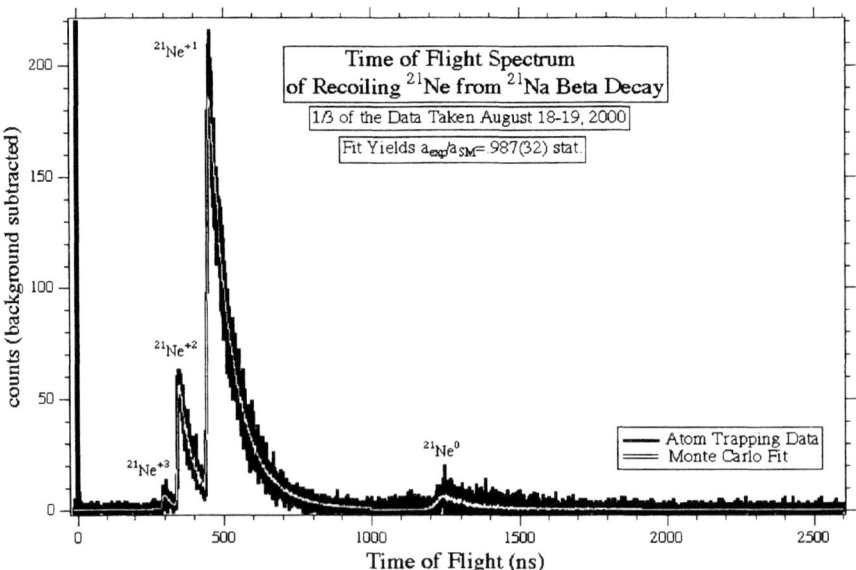

FIGURE 4. A sample of the background subtracted TOF spectrum. The best fit curve is a sum of Monte-Carlo generated templates. The fit contains 7 parameters.

the sake of brevity) in Eq. 1 cause kinematic preferences in the recoil-ion TOF spectrum which would project onto our measurement of a. We were able to probe the polarization state of the atoms using a circular dichroism measurement. Using a large trap of ^{23}Na with the same optical setup as for ^{21}Na, we pass a polarized probe laser through the atom cloud and then onto a photodiode. The absorptivity of the atoms is measured as the laser frequency is scanned, and the lineshape is digitized. The polarization state of the laser is varied to left and right circularly polarized and horizontally and vertically polarized. The lineshapes for the different polarization states are compared to one another and to lineshapes calculated assuming different nuclear polarization and/or alignments. The data for the different polarization states of the probe laser agreed to within measurement uncertainty (dominated by noise and laser frequency drift). This suggests that any residual nuclear polarization must be smaller than 3-4%, and the resulting systematic effect on a less than 1.5%. This technique can be improved easily by the use of faster polarization modulation of the probe beams. Another promising technique for precise polarimetry is the use of polarized phase-contrast imaging of small, dense atomic samples [22], but this is more applicable to colder, denser, polarized samples of atoms confined in purely magnetic traps or optical dipole traps. These techniques seem very promising for studying beta decay correlations with nuclear spin, namely the A, B, and D coefficients.

In total, our data set has a statistical uncertainty for a of about 1.7%, and a combined systematic uncertainty estimated at 6-9%. The systematic uncertainty is dominated by our inability to account for backscattered events in the tail of the ^{21}Ne$^+$ peak.

Based on our current data set, we believe that laser-trapped radioactive atoms are capable of providing new and interesting measurements of beta decay correlation parameters. Our study of the β-ν correlation provides another datum in the search for exotic scalar, tensor, or second class currents. We believe that there are no insurmountable obstacles for $< 1\%$ uncertainty measurments. Measurements by the TRIUMF group and the Los Alamos group with similar MOT techniques should also contribute significantly to "beyond Standard Model" searches. Trapping other atomic species of interest for electroweak and nuclear physics is also possible – the MOT technique has been applied to alkalis, alkaline earth, noble gases, and a few select transition elements, and radioactive atoms have been trapped by several other groups for such work.

ACKNOWLEDGMENTS

We would like to thank Michelle Cyrier for assistance during data acquisition. This experiment is supported by the Director, Office of Energy Research, Office of High Energy Physics, and Nuclear Physics Division of the United States Department of Energy under contract number DE-AC03-76SFF00098.

REFERENCES

1. Jackson, J.D., Treiman, S.B., and Wyld, H.W., *Phys. Rev.* **106**, 517-521 (1957).
2. Lu, Z-T., Bowers, C.J., Freedman, S.J., Fujikawa, B.K. Mortara, Shang, S-Q., Coulter, K.P., and Young, L., *Phys. Rev. Lett.* **72**, 3791-3794 (1994).
3. Voytas, P.A., Behr, J.A., Ghosh, A., Gwinner, G., Orozco, L.A., Simsarian, J.E., Sprouse, G.D., and Xu, F., *Hyperfine Interactions* **97/98**, 529-534 (1996). Sprouse, G.D., and Orozco, L.A., *Ann. Rev. Nucl. Part. Sci.* **47**, 429-461 (1997).
4. Behr, J.A. et al. , *AIP Conf. Proc.* **457**, 148-154 (1999), and private communication.
5. Crane, S., Brice, S., Hime, A., Vieira, D., and Zhao, X., private communication.
6. Severijns, N. et al. , *Nucl. Phys. A* **629**, 423c-432c (1998).
7. Yerozolimsky, B.G., *Nucl. Inst. and Methods A* **440**, 491 (2000). Gaponov, Yu. V., *Physics of Atomic Nuclei* **62**, 1206 (1999).
8. Johnson, C.H., Pleasonton, F., and Carlson, T.A., *Phys. Rev.* **132**, 1149 (1963).
9. Stratowa, C., Dobrozemsky, R., and Weinzierl, P., *Phys. Rev. D* **18**, 3970 (1978).
10. Allen, J.R., Burman, R.L., Hermannsfeldt, W.B., Stähelin, P., and Baird, T.H., *Phys. Rev.* **116**, 134 (1959).
11. Schardt, D., and Riisager, K., *Z. Phys. A* **345**, 265 (1993). c.f. Adelberger, E.G., *Phs. Rev. Lett.* **70**, 2856 (1993).
12. Egorov, V. et al., *Nucl. Phys.* **A621**, 745 (1997).
13. Adelberger, E.G. et al., *Phys. Rev. Lett.* **83**, 1299 (1999).
14. Wilkinson, D.H., *Eur. Phys. J. A* **7**, 307 (2000).
15. Grenacs, L., *Ann. Rev. Nucl. Part. Sci.* **35**, 455-499 (1985).
16. Rowe, M.A., Freedman, S.J., Fujikawa, B.K., Gwinner, G., Shang, S.Q., and Vetter, P.A., *Phys. Rev. A* **59**, 1869-1873 (1999).
17. Phillips, W.D., and Metcalf, H., *Phys. Rev. Lett.* **48**, 596 (1982).
18. Napolitano, R.J., Zilio, S.C., and Bagnato, V.S., *Opt. Commun.* **80**, 110 (1990).
19. Miranda, S.G., Muniz, S.R., Telles, G.D., Marcassa, L.G., Helmerson, K., and Bagnato, V.S., *Phys. Rev. A* **59**, 882-885 (1999).
20. Stephens, M., and Wieman, C., *Phys. Rev. Lett.* **72**, 3787-3790 (1994).
21. Straub, H.C. et al., *Rev. Sci. Inst.* **70**, 4238 (1999).
22. Bradley, C.C., Sackett, C.A., and Hulet, R.G., *Phys. Rev. Lett.* **78**, 985 (1997).

Fundamental Symmetry and Polarized Muons

Yoshitaka Kuno

Department of Physics, Osaka University[1]
Toyonaka, Osaka 506

Abstract. A list of various physics topics to study the fundamental symmetries in the processes involving the muon is presented. A future plan on high-intensity muon sources which is needed to explore muon physics further is also presented as well as a high-intensity neutrino source based on muon decay.

INTRODUCTION

The muon was discovered in 1937 by S. Neddermeyer and C. Anderson in a cloud chamber. The mass of the muon is about 200 times that of the electron. Since then, we have found a total of six leptons, together with six quarks. Among them, however, owing to its unique properties, the muon is still playing major roles in studying particle and nuclear physics [1]. In particular, the use of the muon polarization is known to be quite useful to study the fundamental symmetries.

In this talk, we like to summarize the physics programs with muons and muon (spin) polarization. In the following section, the studies of the normal muon decay ($\mu \to e\nu\bar{\nu}$) are presented. In the subsequent section, searches for the muon lepton flavor violating processes are described with special emphasis of the use of muon polarization. Then, a plan to construct a high-intensity muon source (called "PRISM") in Japan is shown. It would give future progress in the muon physics. Then, a high-intensity neutrino source (based on a muon storage ring) which is called a "neutrino factory" is outlined.

NORMAL MUON DECAY

The muon decays, with almost 100 % branching ratio, into $\mu^- \to e^-\nu_e\bar{\nu}_e$ (and $\mu^+ \to e^+\bar{\nu}_\mu\nu_e$). In the SM, the normal muon decay is described by the V-A interaction. In its extensions to the SM, the energy spectrum of a decay electron

---
[1] Email: kuno@phys.sci.osaka-u.ac.jp

(positron), its angular distribution if muons are polarized, and its spin polarization, are sensitive to the type of interaction on muon decays, including new possible interactions besides the V-A interaction. If they are included, the muon differential decay rate is given in a few parameters by [2]

$$\frac{d^2\Gamma}{dx d\cos\theta_e} = \frac{m_\mu}{4\pi^3} W_{e\mu}^4 G_F^2 \sqrt{x^2 - x_0^2} \left(F_{IS}(x) \pm P_\mu \cos\theta_e F_{AS}(x)\right)\left(1 + \vec{P}_e(x,\theta) \cdot \hat{\zeta}\right), \quad (1)$$

where $W_{e\mu} = (m_\mu^2 + m_e^2)/(2m_\mu)$, $x = E_e/W_{e\mu}$ and $x_0 = m_e/W_{e\mu}(= 9.7 \times 10^{-3}) \leq x \leq 1$. E_e is an energy of the $e^\pm$. m_e and m_μ are masses of the $e^\pm$ and the muon, respectively. The plus (minus) sign corresponds to $\mu^+(\mu^-)$ decay. θ_e is an angle between the muon polarization $\vec{P}_\mu$ and the electron (or positron) momentum, and $\hat{\zeta}$ is a directional vector of the measurement of $e^\pm$ spin polarization. $\vec{P}_e(x,\theta_e)$ is the polarization vector of the $e^\pm$. The functions of $F_{IS}(x)$ and $F_{AS}(x)$ are the isotropic part, the anisotropic part of $e^\pm$ energy spectrum, respectively. They are given by

$$F_{IS}(x) = x(1-x) + \frac{2}{9}\rho(4x^2 - 3x - x_0^2) + \eta x_0(1-x), \quad (2)$$

$$F_{AS}(x) = \frac{1}{3}\xi\sqrt{x^2 - x_0^2}\left[1 - x + \frac{2}{3}\delta(4x - 3 + (\sqrt{1 - x_0^2} - 1))\right]. \quad (3)$$

where ρ, η, ξ, and δ are called Michel parameters [3].

When the spin polarization of decay $e^+(e^-)$ is detected in the $\mu^+ \to e^+ \nu_e \bar{\nu}_\mu$ ($\mu^- \to e^- \nu_\mu \bar{\nu}_e$) decay, $\vec{P}_e(x,\theta_e)$ in Eq.(1) can be measured. It is given by

$$\vec{P}_e(x,\theta_e) = P_{T1} \cdot \frac{(\vec{z} \times \vec{P}_\mu) \times \vec{z}}{|(\vec{z} \times \vec{P}_\mu) \times \vec{z}|} + P_{T2} \cdot \frac{\vec{z} \times \vec{P}_\mu}{|\vec{z} \times \vec{P}_\mu|} + P_L \cdot \frac{\vec{z}}{|\vec{z}|}, \quad (4)$$

where $\vec{z}$ is the direction of $e^\pm$ momentum, and $\vec{P}_\mu$ is the muon spin polarization. P_{T1}, P_{T2}, and P_L are the polarization component transverse to the $e^\pm$ momentum and in the decay plane, that transverse to the $e^\pm$ momentum and normal to the decay plane, and that along the $e^\pm$ momentum direction, respectively. Here, a non-zero value of the triple T-odd correction, P_{T2}, would imply violation of time-reversal invariance. To include new interaction for those observables, the other Michel parameters of ξ, ξ', and ξ'', are introduced.

Table 1 summarizes the present knowledge of the Michel decay parameters [4]. The precise measurements of the Michel decay parameters would place constraints on various new physics which would induce a small deviation from the V-A couplings. The current constraints on the general four-fermion couplings are summarized in Ref. [2].

Transverse $e^\pm$ Polarization in Muon Decay

P_{T2}, which is a triple-vector correlation as shown in Eq.(4), is odd under the time-reversal operation. If the time reversal invariance is held, P_{T2} should vanish. The

TABLE 1. Experimental values of some of the Michel decay parameters.

Michel parameter	SM value	Experimental value	Sensitive observables
ρ	3/4	0.7518 ± 0.0026	F_{IS}
η	0	-0.007 ± 0.013	F_{IS} and P_{T1}
δ	3/4	0.7486 ± 0.0038	F_{AS} and P_L
ξ	1	1.0027 ± 0.0084	$F_{AS}^{\dagger}$ and P_L
ξ'	1	1.00 ± 0.04	P_L
ξ''	0	0.65 ± 0.36	P_L

† Only the product of ξP_μ is measured.

measurement of P_{T2} would provide a sensitive search for T-violation. In particular, it is unique since it is a pure leptonic process. The current limit is $P_{T2} = (7 \pm 23) \times 10^{-3}$. The other observable, P_{T1}, in polarized muon decay is sensitive to the Michel Parameter η. It is the better observable since the Michel spectrum shape is not useful to determine the parameter η. There is a new experiment going on at PSI (R-94-10) to improve the both limits of P_{T2} and P_{T1} with a precision of 3×10^{-3}. [5]

Longitudinal $e^{\pm}$ Polarization in Muon Decay

The longitudinal polarization of $e^{\pm}$ from the normal muon decay is sensitive to the Michel parameters, ξ, ξ' and ξ''. If the muon is not polarized, P_L is given by

$$P_L(x, \cos\theta_e) = \xi' \tag{5}$$

and sensitive to ξ', where x is the normalized $e^{\pm}$ energy described before and θ_e is an angle between the $e^{\pm}$ momentum direction and the muon spin direction. If the muon is polarized, P_L is shown by

$$P_L(x, \cos\theta_e) = \xi + \frac{\xi P_\mu \cos\theta_e(2x - 10)}{(3 - 2x) + \xi P_\mu \cos\theta_e(2x - 1)} \cdot \frac{(\xi'' - \xi\xi')}{\xi} \tag{6}$$

Therefore, it is sensitive to the Michel parameter $(\xi'' - \xi\xi')/\xi$. In particular when $x \sim 1$ and $\cos\theta_e \sim -1$,

$$P_L(x = 1, \cos\theta_e = -1) \sim \xi + \frac{-\xi P_{mu}}{1 - \xi P_\mu} \cdot \frac{(\xi'' - \xi\xi')}{\xi} \tag{7}$$

It is seen that the effect is enhanced by a factor of $-\xi P_\mu/(1 - \xi P_\mu)$ which is quite a large number if $P_\mu = 1$ (and $\xi = 1$ in the SM). The new experiment (R-97-06) is being undertaken at PSI [6].

POLARIZATION OF MUONS FROM HADRON DECAYS

The hadron decays including the muon in the final state is also a useful process to study the fundamental symmetries. One example concerning K decay is the transverse muon spin polarization $P_\mu^\perp$ in $K^+ \to \pi^0 \mu^+ \nu$ ($K_{\mu3}^+$) decay [7], where $P_\mu^\perp$ is defined as the component of muon spin polarization normal to the decay plane, determined by the μ^+ and π^0 momentum vectors. It is given by

$$P_\mu^\perp \equiv \frac{\vec{s}_{\mu^+} \cdot (\vec{p}_{\pi^0} \times \vec{p}_{\mu^+})}{|\vec{p}_{\pi^0} \times \vec{p}_{\mu^+}|}, \tag{8}$$

where $\vec{s}_{\mu^+}$ is the muon spin vector and $\vec{p}_{\mu^+}$ and $\vec{p}_{\pi^0}$ are the momentum vectors of the muon and neutral pion, respectively. Since the T-reversal operation changes the sign of $P_\mu^\perp$, a non-zero value of $P_\mu^\perp$ would signal T-violation. If CPT invariance is obeyed, T-violation implies CP-violation.

A search for $P_\mu^\perp$ in $K_{\mu3}^+$ decay is sensitive to new mechanisms of CP violation beyond the Standard Model. The physics motivation of new CP violation arises from the observed baryon asymmetry in the universe, which cannot be explained by the CP violation in the Standard Model alone [8]. Therefore there must be new additional sources of CP violation. Furthermore, recent theoretical progress of electroweak baryogenesis suggests that new CP violation sources might exist at the electroweak scale which can be accessible experimentally. In general, neither an effective vector (V) nor axial-vector (A) interaction introduce $P_\mu^\perp$, but only effective scalar (S) or pseudo-scalar (P) interactions give a non-zero $P_\mu^\perp$ [9]. Some examples of theoretical models are the three Higgs doublet models [10,11], supersymmetric models with squark family mixing [12], leptoquark models or supersymmetric model with R-parity breaking [13].

As a clean search for T-violating phenomena, the measurement of $P_\mu^\perp$ in $K_{\mu3}^+$ decay has several striking advantages. First, the final-state electromagnetic interaction (FSI), which would otherwise mimic a fake T-odd effect, is negligible, of the order of 10^{-6}, in $K_{\mu3}^+$ decay [14]. This is due to the fact that only one charged particle exists in the final state. However, this small FSI is not always the case; for instance, in triple correlations in nuclear β decays and $P_\mu^\perp$ in $K_L^0 \to \pi^- \mu^+ \nu$ ($K_{\mu3}^0$) decay, the FSI is predicted [15] to be as large as 10^{-3}. This advantage in $K_{\mu3}^+$ decay allows a wider window to search for T-violation mechanisms, free of FSI-induced background. Second, $P_\mu^\perp$ in $K_{\mu3}^+$ decay has no contribution from the CKM phase in the minimal Standard Model at the tree level, and higher-order contributions are extremely small ($\sim 10^{-6}$). This implies that the observation of a non-zero $P_\mu^\perp$ value would be a definite signature of new physics beyond the minimal Standard Model.

The present experimental upper limit on $P_\mu^\perp$ in $K_{\mu3}^+$ decay, which was obtained from the previous experiment using in-flight K^+ decays at Brookhaven National Laboratory (BNL) [16] is $P_\mu^\perp = (-4.2 \pm 6.7) \times 10^{-3}$ in the K^+ rest frame. This yielded the measure of CP violation, $\text{Im}\xi = \text{Im}(f^+/f^-) = -0.016 \pm 0.025$, or

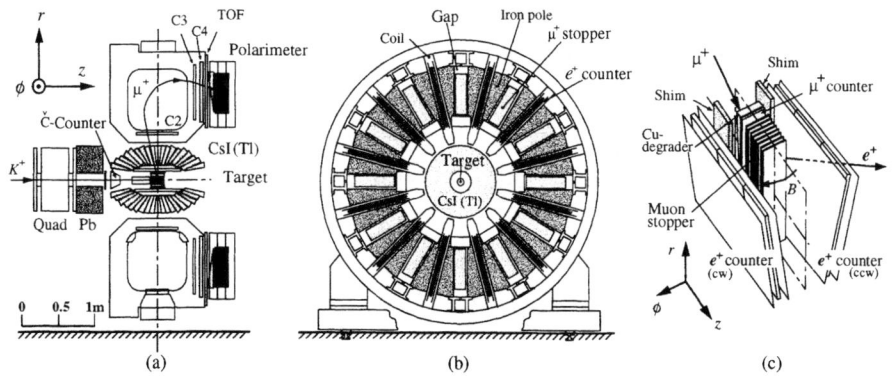

FIGURE 1. E246 detector; (a) side view, (b) end view, and (c) one sector of muon polarimeter.

$|\text{Im}\xi| \leq 0.049$ at 90 % confidence level[2], where f^+ and f^- are the form factors in the $K_{\mu 3}$ matrix element.

KEK Experiment E246 Detector

A new experiment, KEK-PS E246, employs K decays at rest, in contrast to the previous experiments which used in-flight K decays. There are several significant advantages of the use of the K decays at rest. For instance, it allows clean and precise determination of K-decay kinematics. This, together with the carefully-designed detector, will reduce systematic errors.

Schematic side and end views of the E246 detector are shown in Fig.1. In E246, the $K_{\mu 3}^+$ events which have the π^0 moving along the detector axis, either forward or backward from the target, will be accepted as "gold-plated events" For those events, the decay plane can be set radially from the detector axis. $P_\mu^\perp$ is directed azimuthally in a *screw-sense* around the detector axis. $P_\mu^\perp$ would then manifest itself as a difference in the e^+ counts between the cw and ccw counters from the muon stopper.

There are two important techniques for suppressing bias asymmetries in E246. First, summing over all magnet sectors would cancel any non-screw type biases. The second is the comparison of the $K_{\mu 3}^+$ samples with forward π^0 and backward π^0 directions. Since $P_\mu^\perp$ is of the same magnitude but opposite in sign between these two samples and any biased asymmetries are likely to be independent of π^0 directions, their comparison would enable us to reduce the systematic errors significantly.

---

[2]) Since the $P_\mu^\perp$ measurements in $K_{\mu 3}^0$ decay might have FSI effects, only the $K_{\mu 3}^+$ data is quoted instead of the combined result.

The result presented here is based on the data taken in 1996 and 1997. A total number of gold-plated $K^+_{\mu3}$ samples of 2.1 M events is obtained. The analyzing power α was experimentally obtained from the $K^+_{\mu3}$ samples with the π^0 moving perpendicular to the detector axis. Extensive studies of the possible systematic errors was done before examining $P^\perp_\mu$, based on real data as much as possible. As a result of all the examinations, a combined systematic error of less than 1.0×10^{-3} is obtained.

The E246 initial result of $P^\perp_\mu$ in $K^+ \to \pi^0 \mu^+ \nu$ decay from the data collected in 1996 and 1997 [18] is

$$P^\perp_\mu = -0.0042 \pm 0.0049(stat) \pm 0.0009(sys). \qquad (9)$$

which yields the ratio of the two form factors in the $K_{\mu3}$ matrix elements,

$$\mathrm{Im}\xi = -0.013 \pm 0.016(stat) \pm 0.003(syst). \qquad (10)$$

The 90% confidence limits are given as $|P^\perp_\mu| < 0.011$ and $|\mathrm{Im}\xi| < 0.033$. At this moment, the statistical error dominates the systematic error. E246 has taken more data since then and they will be analyzed soon.

MUON LEPTON FLAVOR VIOLATION

In particle physics, the field of low-energy muon physics, in particular searches for muon lepton flavor violation (LFV), has received growing attention from both theorists and experimentalists [1]. This surge in interest can be attributed to the fact that many supersymmetric (SUSY) models (in particular SUSY-GUT) predict large branching ratios for such LFV processes, in some cases as large as one or two orders of magnitude below the present experimental limits [19]. Figure 2 shows predicted branching ratios of the muon LFV process, $\mu^- - e^-$ conversion in a muonic atom, in the SUSY $SU(5)$ model.

Furthermore, the recent experimental hints of non-vanishing neutrino masses and mixing suggested by the atmospheric neutrino anomaly might invoke additional LFV contributions in SU(5) SUSY-GUT models [20]. This model includes a heavy right-handed Majorana neutrino of $10^{14} - 10^{15}$ GeV/c^2 with the ν_μ-ν_τ mixing of $\sin^2(2\theta_{\nu_\mu \nu_\tau}) \sim 1$ (as suggested by the Super Kamiokande). This contribution enhances LFV by a factor of about $(m_\tau/m_\mu)^2$ over the minimal SU(5) SUSY-GUT model.

LFV processes of major interest are those with muons, which are for example $\mu^+ \to e^+ \gamma$, $\mu^- + (A,Z) \to e^- + (A,Z)$ conversion in a muonic atom, $\mu^+ \to e^+ e^+ e^-$, and muonium-antimuonium conversion.

$\mu^+ \to e^+ \gamma$ with Polarized Muons

This angular distribution is known to be useful to discriminate different models since the felicity of e^+ in $\mu^+ \to e^+ \gamma$ is sensitive to the mechanism of lepton flavor

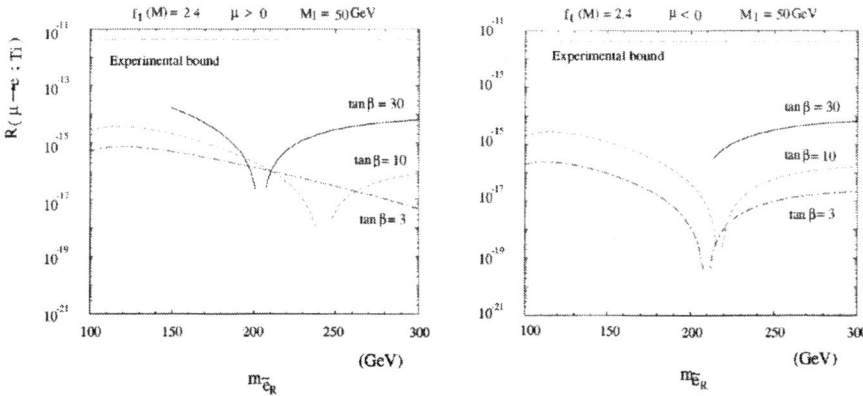

FIGURE 2. Prediction of $\mu - e$ conversion branching ratio from SUSY-GUT models.

mixing. For instance, the minimal SU(5) SUSY-GUT model predicts a non-zero A_L and a vanishing A_R, which yields a $(1 + P_\mu \cos\theta)$ distribution. On the other hand, the simplest version of the SO(10) SUSY-GUT models predict approximately-equal felicity amplitudes for right-handed and left-handed e^+s ($A_L \approx A_R$), which result in an almost uniform angular distribution. For some non-unified supersymmetric models, A_L is vanishing but A_R is non-zero, which gives a $(1 - P_\mu \cos\theta)$ distribution.

It is known that the experimental backgrounds could be potentially further reduced by polarized muons [21]. One of the major backgrounds to the search for $\mu^+ \to e^+\gamma$ is an accidental background which comes from accidental coincidence of a e^+ in a normal muon decay ($\mu^+ \to e^+\nu\bar{\nu}$) accompanied by a high energy photon. The sources of a high energy photon might be either that in $\mu^+ \to e^+\nu\bar{\nu}\gamma$ decay, or external bremsstrahlung or annihilation-in-flight of e^+s in the normal muon decay.

It turns out that e^+s in the normal Michel muon decay follows a $(1 + P_\mu \cos\theta_e)$ distribution. We have studied the inclusive angular distribution of a high energy photon (*e.g.* ≥ 50 MeV) from $\mu^+ \to e^+\nu\bar{\nu}\gamma$. It is found to be emitted preferentially along the muon spin direction; namely follows a $(1 + P_\mu \cos\theta_\gamma)$ distribution, where θ_γ is an angle of the photon direction with respect to the muon spin direction.

This inclusive angular distribution of a high energy photon in $\mu^+ \to e^+\nu\bar{\nu}\gamma$ implies that the accidental background could be suppressed for $\mu^+ \to e_L^+\gamma$ where high energy photons must be detected at the opposite direction to the muon polarization. A similar suppression mechanism of accidental background can be seen for $\mu^+ \to e_R^+\gamma$ when high energy e^+s are detected at the opposite direction to the muon polarization. As a result, the selective measurements of either e^+s or photons anti-parallel to the muon spin direction would give the same accidental background suppression for $\mu^+ \to e_R^+\gamma$ and $\mu^+ \to e_L^+\gamma$ decays respectively.

HIGH INTENSITY MUON SOURCE – PRISM –

To make substantial improvements in sensitivity, a high-intensity muon beam is necessary. Most of the muon LFV experiments use a stopped muon beam. Therefore, the necessary requirements to aim for significant progress are (1) a narrow energy spread in a muon beam (to increase stopped efficiency and improve detector resolution for daughter particles from rare muon decay), and (2) low contamination of a muon beam (to reduce backgrounds, in particular pions in a beam).

PRISM is a project in Japan to make a high intensity muon source with a narrow energy-spread and less contamination [22]. PRISM stands for "Phase-Rotation Intense Slow Muons". The aimed intensity is $10^{12}\mu^{\pm}$/sec, four orders of magnitude higher than that available at present. Its schematic layout is shown in Fig.3. PRISM would combine high-field pion capture of about 10 T, the $\pi - \mu$ decay section of a 10-m long superconducting solenoid magnet, and the phase rotation section.

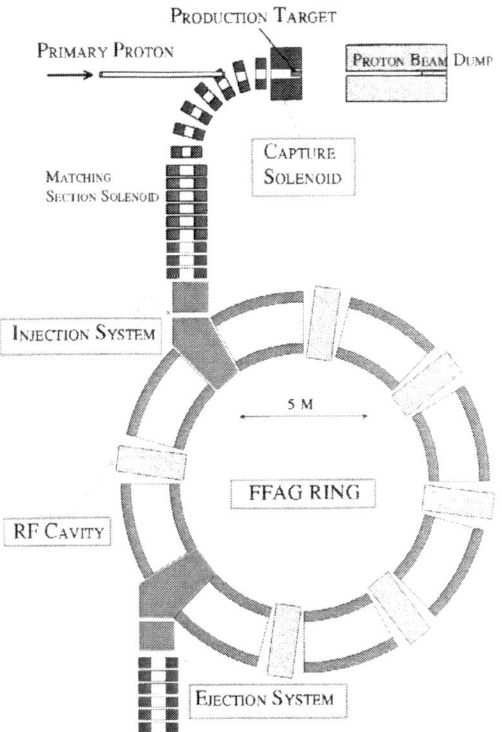

FIGURE 3. Schematic layout of PRISM

The phase rotation is to accelerate slow muons and to decelerate fast muons by a strong radio-frequency (RF) electric field, yielding a narrow longitudinal momentum spread. To identify fast and slow muons by their time of flight from the production time, a very narrow pulsed proton beam must be used.

One of the features of PRISM is to do phase rotation at a Fixed-Field Alternating Gradient synchrotron (FFAG), which has several advantages such as a large momentum acceptance and large emittance. The present design of the FFAG ring has a diameter of about 10 m. About five turns of muons in the FFAG ring will complete phase rotation. Since PRISM is focused on experiments with stopped muons, the central muon momentum of the FFAG ring is set to 68 MeV/c (corresponding to a kinetic energy of 20 MeV). From the simulation studies of phase rotation at FFAG, the original momentum spread of ± 20 % is reduced down to ± 2 % after phase rotation. If it is combined with the planned 50-GeV PS at the KEK-JAERI Joint Project [23], PRISM would provide about $10^{12} \mu^{\pm}$/sec.

The ideas of high-field pion capture and phase rotation have emerged in studies of a $\mu^+\mu^-$ collider at the high-energy frontier. Although there are many common R&D items between a low-energy muon source and a $\mu^+\mu^-$ collider, there are discussions on whether the front-end muon collider (FMC) could be directly used in experiments with muons. The FMC will run with a pulsed beam of slow repetition (at typically 15 Hz). However, most experiments with muons require a beam with a high duty factor, because of the reduction of the instantaneous rate. Thus, independent R&D items exist in a low-energy muon source, such as in PRISM.

FROM PRISM TO A NEUTRINO FACTORY

In future, PRISM would be extended to a muon accumulator ring for neutrino sources (called a neutrino factory) [24,25]. In a neutrino factory, neutrinos from muon decays in the straight section of the muon storage ring of a several 10 GeV are used for physics experiments. Possible experiments of long-baseline neutrino oscillation are, for instance, to determine θ_{13} in the 3-generation lepton mixing matrix (MNS matrix), the matter effect, the sign of Δm_{31}^2, and to search for CP violation in the neutrino sector.

The neutrino factory project, which would be connected to the KEK/JAERI Joint Project, in Japan would proceed in a staging approach [26]. Its first stage is to aim for $10^{19}\nu$s/year without muon ionization cooling with muon energy rage of 10 – 20 GeV. The second stage with muon ionization cooling would aim for $10^{20} - 10^{21}\nu$s/year, covering muon energy of up to 50 GeV. A detector for a neutrino factory needs charge identification to separate ν and $\overline{\nu}$. A possible detector distance is from 300 km to 2000 km. Optimization of possible physics cases will be under consideration.

REFERENCES

1. Kuno Y., and Okada Y., Review of Modern Physics, **73** 151 (2001).

2. Fetcher W. and Gerber H.J., in *Particle Data Group : Review of Particle Properties*, Euro. Phys. Journal **3**, p.282 (1988).
3. Michel L., Proc. Phys. Soc. A **63**, 514 (1950).
4. Particle Data Group, Euro. Phys. Journal, **3** 1, (1998).
5. Fetcher W., in this proceedings.
6. Morelle X.,in this proceedings.
7. Sakurai J.J., Phys. Rev. **109**, 980 (1980).
8. Mclerran L., Shaposhnikov M., Turok N., and Voloshin M., Phys. Lett. **B 25 6**, 451 (1991); Turok N., and Voloshin M., Phys. Lett. **B 256**, 451 (1991); Turok N., and Zadrozny J., Nucl. Phys. **B 358**, 471 (1991); Dine M., Huet P., Singleton R., and Susskind L., Phys. Lett. **B 257**, 351 (1991).
9. Leurer M., Phys. Rev. Lett. **62**, 1967 (1989); Castoldi P., Frère J.-M., and Kane G.L., Phys. Rev. D **39**, 2633 (1989).
10. Weinberg S., Phys. Rev. Lett. **37**, 657 (1976), and as a recent review, Cheng H.-Y., Int. J. Mod. Phys. A **7**, 1059 (1992) and references therein.
11. Bélanger G., and Geng C.Q., Phys. Rev. D **44**, 2789 (1991).
12. Wu G.-H. and Ng J.N., Phys. Lett. B **392**, 93 (1997).
13. Fabbrichesi M. and Vissani F., Phys. Rev. D **55**, 5334 (1997).
14. Zhitnitskii A.R., *Yad. Fiz.* **31**, 1014 (1980) [*Sov. J. Nucl. Phys*, **31**, 529 (1980)].
15. Adkins G.S., Phys. Rev. D **28**, 2885 (1983) and references therein.
16. Campbell M.K., *et al.*, Phys. Rev. Lett. **47**, 1032 (1981); Blatt, S.R., *et al.*, Phys. Rev. D **27**, 1056 (1983).
17. Garisto R., and Kane G., Phys. Rev. D **44**, 2038 (1991).
18. Abe M., *et al.*, Phys. Rev. Lett. **83**, 4253 (1999).
19. Barbieri R., Hall L.J., and Strumia A., Nucl. Phys. B **445**, 219 (1995).
20. Hisano J. and Nomura D., Phys. Rev. D **59**, 115005 (1999).
21. Kuno Y. and Okada Y., Phys. Rev. Lett. **77**, 434 (1996).
22. Kuno Y., *et al.*, *Proceedings of Workshop on High Intensity Secondary Beam with Phase Rotation* edited by Y. Kuno and N. Sasao (1998) p.71.
23. The Joint Project Team of JAERI and KEK, "The Joint Project for High-Intensity Proton Accelerators", KEK Report 99-4, JAERI-Tech 99-056, JHF-99-3, (1999).
24. Geer S., *Phys. Rev.* D **57** (1998) 6989 and Erratum: *ibid.* D **59** (1999) 039903.
25. CERN Report CERN 99-02 and ECFA 99-197 "Prospective Study of Muon Storage Rings at CERN", edited by Autin B., Blondel A., and Ellis J. (1999).
26. Kuno Y. and Mori Y., *Proceedings of "High Intensity Muon Sources* (World Scientific) edited by Y. Kuno and T. Yokoi (1999), P.119.

Nuclear Moment Studies with Polarized Radioactive Nuclear Beams

K. Asahi[1,2], K. Sakai[1], H. Ogawa[1,2], H. Ueno[2], Y. Kobayashi[2],
A. Yoshimi[2], H. Miyoshi[1], K. Yogo[1], A. Goto[1], T. Suga[1], H. Imai[3],
Y.X. Watanabe[2], K. Yoneda[3], N. Fukuda[3], N. Aoi[3],
W.-D. Schmidt-Ott[4], G. Neyens[5], S. Teughels[5], A. Yoshida[2],
T. Kubo[2], and M. Ishihara[6]

[1] *Dept. of Physics, Tokyo Institute of Technology, Meguro-ku, Tokyo 152-8551, Japan*
[2] *Inst. of Physical and Chemical Research (RIKEN), Wako-shi, Saitama 351-0198, Japan*
[3] *Dept. of Physics, Univ. of Tokyo, Bunkyo-ku, Tokyo 113-0033, Japan*
[4] *II. Physikalisches Inst., Univ. Göttingen, Bunsenstrasse 7-9, D-37073 Göttingen, Germany*
[5] *Inst. voor Kern- en stralingsfysica, Univ. of Leuven, B-3001 Leuven, Belgium*
[6] *RIKEN-BNL Research Center, Brookhaven National Laboratory, Upton, NY 11973, USA*

Abstract. Beams of spin-polarized radioactive nuclei are produced in the projectile fragmentation reaction induced by intermediate-energy heavy ion beams, and are used for the measurement of magnetic dipole and electric quadrupole moments of nuclei in the regions far from stability. In this paper, after remarking the mechanism of fragment polarization and the experimental techniques to make use of it, we discuss the recent results obtained for the electromagnetic moments of light neutron-rich nuclei at RIKEN. Analyses of the observed magnetic moments in terms of shell models show intriguing results which imply that some of the basic nuclear properties, such as the order in energy of the single-particle orbits, the pairing energy between the excess neutrons, and the effective g_s-factor for a halo nucleon, tend to alter as the degree of neutron excess increases. Also, the measured electric quadrupole moments reveal remarkable quenching of the E2 effective charge in nuclei with large N/Z ratios.

INTRODUCTION

The domain of nuclei that can be produced in laboratory at considerable rates have been expanded remarkably, thanks to the rapid progress made in the radioactive nuclear beam (RNB) techniques in this decade at a number of nuclear physics facilities in the world. Exploiting the new opportunities thus opened, experimental and theoretical studies have been extensively made, and a number of interesting new findings have emerged [1–6]: Some of nuclei on the neutron dripline show a huge extended tail of the neutron spatial distribution (neutron halo), representing a new form of nuclear matter. The behavior of the observed binding energies reveals that the shell closure occurs at different neutron numbers than the traditional ones when the neutron excess is radically increased [7]. Present experimental technology in physics with RNBs allows us to investigate not only rather simple nuclear observables such as radii, masses, and lifetimes, but also a number of more elaborate properties and dynamics of nuclei, by subjecting the RNB nuclei to *e.g.* the Coulomb excitation and dissociation, proton elastic/inelastic scattering in the inversed kinematics, in-beam gamma-ray spectroscopy via secondary fragmentation reactions, *etc.* In this context the RNB physics may be considered to be now at its second stage of evolution. The work presented in this paper is intended to take a step forward in the RNB physics by introducing the spin polarization of RNB as a key experimental grip. The availability of spin-polarized nuclei potentially offers very important experimental opportunities in studies of nuclear structures, fundamental interactions, and symmetries [8]. At present, however, we at RIKEN mainly concentrate on nuclear structure studies via electromagnetic moments measurements.

FRAGMENT POLARIZATION MECHANISM

There are two major approaches to obtain beams of radioactive nuclei. The first one is based on the production of radioactive nuclei via the target fragmentation (TF) reaction induced by a high-energy proton beam, taking advantage of high available proton-beam intensities and thick adoptable targets. The activities produced come immediately to rest in the target, diffuse out, get ionized, and then are accelerated to form a beam which is then mass separated in a magnet of an isotope separator on line (ISOL). The other approach for the RNB production, the projectile fragmentation (PF) method, relies on the fragmentation of beam particles. Heavy ion beams at several tens of MeV/u or higher are used. The activities, projectile fragments, are produced almost at rest in the projectile frame, or as a beam well foucused in angle and velocity in the laboratory frame. The fragment beam thus obtained is immediately delivered to an in-flight isotope separation by means of combined magnetic-rigidity and momentum-loss analyses. The latter method using the PF reaction is rather new as compared to the former (TF-ISOL method) which have been well developed at ISOLDE/CERN. Because of the fast, chemistry-independent nature of the in-flight isotope separation mechanism, the PF method applies to quite wide regions of nuclei without need for elaborate technical developments. For this reason, it may be considered that the RNB physics largely owes its recent progress to the advent of the PF method.

The momentum distribution of projectile fragments in the PF reaction is characterized by a sharp, Gaussian-like peak located around the beam velocity, and is fairly well explained in terms of a simple participant-spectator model [9]. In this model, the collision between a projectile and a target nucleus is considered as a simple ensemble of independent collisions between individual nucleons in the projectile

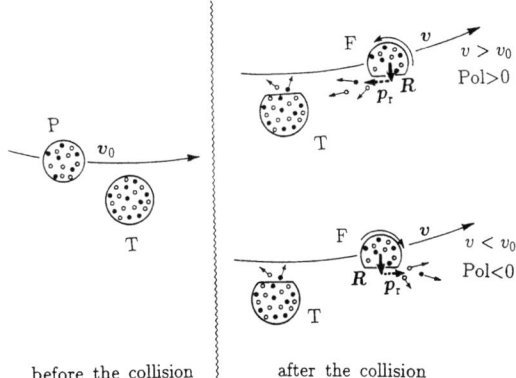

FIGURE 1. Predicted correlation between the fragment spin and linear momentum in a model of projectile fragmentation. A projectile P incident at a velocity v_0 is transmuted to a fragment F through removal of nucleons at the position $\vec{R}$. See text for further explanation.

and those in the target. Nucleons in a part of projectile volume which overlaps with the target at collision is removed from the projectile as illustrated in Fig. 1, leaving the remaining portion (projectile fragment) in its original motion. The fragment momentum then is given as $\vec{p} = M\vec{v}_0 - \vec{p}_r$, where M and $\vec{v}_0$ denote the fragment mass and the projectile velocity. $\vec{p}_r$ is the internal momentum of the nucleon(s) removed from the projectile. Noting that the angular momemtum, $\vec{L}_r = \vec{R} \times \vec{p}_r$ (with $\vec{R}$ representing the c.m. position of the removed nucleons measured from the projectile center), accompanies the removed nucleons, one may be lead to an expectation that the fragment spin, given as $\vec{L} = -\vec{L}_r$, is correlated with the linear momentum $\vec{p}$, and thus the fragment would be spin polarized by simply selecting its linear momentum. An experiment to examine this expectation was performed by us for the fragmentation of ^{14}N projectile at 40 MeV/u on a Au target [10]. The data indeed revealed a substantial polarization of ^{12}B fragments with the momentum dependence compatible with expectation from the above simple model. Subsequent experiments since then [11–13] have shown that the polarization is a general phenomenon common to fragmentation reactions. The size of polarization obtained in this method typically ranges from 2 to 10 %.

EXPERIMENTAL TECHNIQUES

Typical arrangement for the nuclear moment measurements with polarized RNBs at RIKEN is shown in Fig. 2. A primary beam of energies around $E/A = 100$ MeV/u from the RIKEN Ring Cyclotron is introduced into a target chamber through a beam swinger, so that fragments emitted from the target at finite angles are accepted by the fragment separator RIPS [14]. Fragments of the objective nuclide are isotope-separated and momentum-analyzed in RIPS, and finally focused

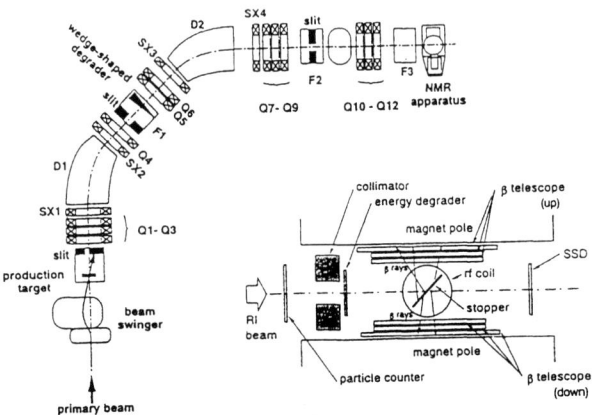

FIGURE 2. Arrangement for the moment measurements with polarized RNBs.

on a collection point F3, at which a stopping material and a device for the β-ray detected nuclear magnetic resonance (β-NMR) experiments are located. The polarization of fragments in the direction perpendicular to the reaction plane (referred to as the z-direction below) is preserved during the transportation through RIPS to the stopper, and held in the stopper under an applied static magnetic field $\boldsymbol{B}_0$ in the z-direction until the fragments decay by the β-ray emission. The β-ray angular distribution from polarized nuclei is given as $W(\theta) = 1 + \frac{v}{c} A_\beta P \cos\theta$, where A_β and P denote the asymmetry parameter and the degree of polarization, respectively, and $v/c \approx 1$ is the ratio of the β particle velocity to the speed of light. A change in the up/down ratio $U/D \equiv W(0°)/W(180°) \approx 1 + 2A_\beta P$ of the β-ray intensities provides a sensitive index for the spin flip resulting from the nuclear magnetic resonace (NMR), which is incorporated in the measurements as described shortly. An rf field $\boldsymbol{B}_1$ perpendicular to $\boldsymbol{B}_0$ is applied, with its frequency ν being swept over the range from $\nu_1 - \frac{1}{2}\Delta\nu$ to $\nu_1 + \frac{1}{2}\Delta\nu$. If the range includes the Larmor frequency $\nu_{\rm L} = \mu B_0/hI$, the nuclear spins are inverted (the adiabatic fast passage inversion; AFP) and the U/D ratio changes as $1 + 2A_\beta P \to 1 - 2A_\beta P$. Thus, the NMR is searched for by scanning the ν_1 value while monitoring the β-ray U/D ratio.

Once the NMR is successfully found, the degree of polarization P for the fragment beam can be deduced from the observed size of change in asymmetry. In order to pursue the NMR search, however, it is highly desirable to know P *before* the NMR effect is found, since the sensitivity of the search depends on the size of P. This situation represents a sort of dilemma inherent in the present 'β-NMR on fragmentation-induced polarization' experiment. Recently, we have developed a means to overcome this dilemma: The degree of polarization for fragments is measured prior to the NMR search by means of the adiabatic field rotation tech-

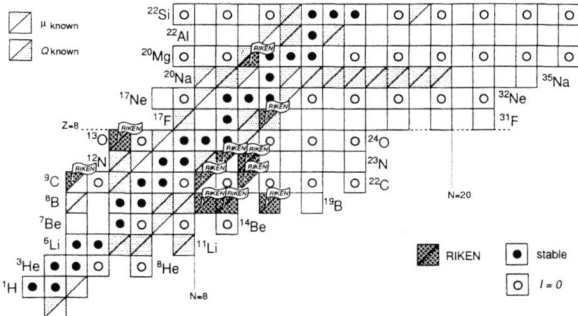

FIGURE 3. The nuclides for which the magnetic moments or the elecric quadrupole moments have been determined.

nique [15]. Using this technique, the best experimental conditions are searched for by changing the settings for the emission angle θ_L and momentum p, in order to optimize the figure of merit given by $F = P^2 Y$ where Y is the yield for the objective fragments.

In a measurement of electric quadrupole moment Q, a single crystal stopper is employed, in which an electric field gradient eq is known to act on an implanted radioactive atom. In the spin $I = 1$ case, the quadrupole coupling between eq and Q causes the NMR line to split into two frequencies $\nu_\pm = \nu_L \pm \frac{3}{8}\nu_Q$, where $\nu_Q \equiv eqQ/h$. The measurement is made by applying the B_1 field having both of the two frequency components $\nu_\pm$ with the ν_Q value being scanned.

STRUCTURE OF NEUTRON-RICH NUCLEI STUDIED FROM ELECTROMAGNETIC MOMENTS

The above method of 'β-NMR on fragmentation-induced polarization' have been applied to the measurement of magnetic dipole and electric quadrupole moments for numbers of nuclei situated in the light-mass, neutron-rich region. In Fig. 3 we show nuclides for which the magnetic dipole or the electric quadrupole moments are determined at RIKEN. Until present, ten new magnetic dipole moments and six new electric quadrupole moments have been determined.

Below in this report, we discuss physics implications derived from the results of these experiments.

Weakening of the singlet pairing energy for excess neutrons

Basically, the magnetic moment is expressed as $\mu = \langle \psi_{I,M=I} | \hat{\mu}_z | \psi_{I,M=I} \rangle$ in terms of the M1 operator $\hat{\boldsymbol{\mu}} = \sum_{k=1}^{A} (g_s \boldsymbol{s}_k + g_\ell \boldsymbol{\ell}_k) \mu_N$ and the nuclear magneton μ_N. Suppose that a nucleus is known to have a spin value I. Since the expression for μ contains much different weighting factors g_s and g_ℓ between the spin s and orbital angular momentum ℓ (for example, if the bare values are taken, $g_s = +5.59$, $g_\ell = +1$ for proton and $g_s = -3.83$, $g_\ell = 0$ for neutron),[1] the predicted μ should differ distinctively for different (s_k, ℓ_k) compositions, even the total spin $I = \sum_{k=1}^{A}(s_k + \ell_k)$ is the same. The μ moment thus reflects sensitively what single particle orbits contribute to the nuclear wave function. In the following, we investigate from the measured μ values what importances certain configurations share, and from such information the nuclear structures and interactions are discussed, most cases with the help of shell model calculations but in some particular cases quite model independently. The ^{17}N moment below as well as the ^{9}C-^{9}Li isoscaler moment [19,20] would be the latter cases.

The magnetic moment for the ^{17}N ground state ($I^\pi = 1/2^-$) has been measured to be $|\mu(^{17}\mathrm{N})| = 0.352 \pm 0.002\mu_N$ [21]. The $^{17}\mathrm{N}_{\mathrm{g.s.}}$ nucleus, with 7 protons and 10 neutrons, is described in a $0\hbar\omega$ p-sd space by a proton hole in the p shell and two neutrons in the sd shell. We note that the configurations which can enter here are quite restricted: In a configuration where the sd neutrons are coupled to $J_n^\pi = 0^+$ the proton hole should be in the $p_{1/2}$ orbit to conform with the empirical spin-parity $I^\pi = 1/2^-$. Thus the major component of $^{17}\mathrm{N}_{\mathrm{g.s.}}$ wave function should be expressed by $\psi_0 = |\pi(p_{1/2})_{-1}, \nu[(sd)_2]^{0+}\rangle^{J^\pi=1/2^-}$. For other configurations the sd neutrons may couple to $J_n^\pi = 2^+$,[2] but in this case the proton hole has to be promoted to the $p_{3/2}$ orbit, as $\psi_A = |\pi(p_{3/2})_{-1}, \nu[(d_{5/2})_2]^{2+}\rangle^{J^\pi=1/2^-}$ or $\psi_B = |\pi(p_{3/2})_{-1}, \nu[d_{5/2}, s_{1/2}]^{2+}\rangle^{J^\pi=1/2^-}$. Thus the $^{17}\mathrm{N}_{\mathrm{g.s.}}$ wave function is written as $\psi = c_0\psi_0 + c_A\psi_A + c_B\psi_B$, and since there appear no cross term between ψ_0 and $\psi_{A,B}$ the magnetic moment is expressed as

$$\mu(^{17}\mathrm{N}) = c_0^2\langle\psi_0|\hat{\mu}_z|\psi_0\rangle + c_A^2\langle\psi_A|\hat{\mu}_z|\psi_A\rangle + c_B^2\langle\psi_B|\hat{\mu}_z|\psi_B\rangle \tag{1}$$

The moments $\langle\psi_A|\hat{\mu}_z|\psi_A\rangle = -2.01\mu_N$ and $\langle\psi_B|\hat{\mu}_z|\psi_B\rangle = -1.54\mu_N$ are distinctively large compared to the part $\langle\psi_0|\hat{\mu}_z|\psi_0\rangle = -0.276\mu_N$ from the major configuration. Consequently the $^{17}\mathrm{N}_{\mathrm{g.s.}}$ moment is very sensitive to the small admixing components $\psi_{A,B}$ of $J_n = 2$ configurations. Thus, from the experimental μ, the amount of the $J_n = 2$ type admixture was determined without recourse to theory. The result, (5.6 ± 0.8) %, for the probability of the 2^+ configurations of the two sd neutrons is much larger compared to calculations [22] with the PSDMK [23] and PSDWB [24] effective interactions. The enhancement of 2^+ neutron configurations has been also

---

[1]) In actual calculations referred to in the discussions below, an effective operator taking the effective g-factors of Refs. [16–18] is employed in order to include effects of the meson exchange currents and second-order configuration mixing.

[2]) The $J_n^\pi = 1^+$ configuration is forbidden for two neutrons in the same orbit, that is, $\nu[(d_{5/2})_2]^{1+}$ and $\nu[(s_{1/2})_2]^{1+}$ are not allowed; $\nu[s_{1/2}, d_{5/2}]$ cannot make $J_n = 1$; although the configurations $\nu[s_{1/2}, d_{3/2}]$ and $\nu[d_{5/2}, d_{3/2}]$ may enter, their strengths should be small because of a much higher energy of the $d_{3/2}$ single-particle orbit.

indicated in the μ measurements for ^{15}B [25] and ^{17}B [21]: All the μ results for ^{17}N, ^{15}B and ^{17}B are well accounted for by assuming that the 0^+ pairing energy for the sd neutrons is diminished by 30 % from the one usually employed in the region close to stability.

Lowering of the $2s_{1/2}$ orbit

A tendency of the $\nu s_{1/2}$ single particle orbit to lower in energy in neutron-rich $N = 7$ isotones has been early pointed out [26]. Whether the same phenomenon occurs in other isotones would be interesting in the study of neutron-rich nuclei. The magnetic moment of the ^{14}B ground state ($I^\pi = 2^-$) has been determined to be $|\mu(^{14}\text{B})| = 1.185 \pm 0.005\mu_N$ [25]. The ^{14}B$_\text{g.s.}$ state is dominated by two configurations $[\pi p_{3/2}, \nu s_{1/2}]^{J^\pi=2^-}$ and $[\pi p_{3/2}, \nu d_{5/2}]^{J^\pi=2^-}$, for which the $\hat{\mu}_z$ expectation values are much different: $\langle \pi p_{3/2}, \nu s_{1/2} | \hat{\mu}_z | \pi p_{3/2}, \nu s_{1/2} \rangle^{2^-} = +1.880 \mu_N$ and $\langle \pi p_{3/2}, \nu d_{5/2} | \hat{\mu}_z | \pi p_{3/2}, \nu d_{5/2} \rangle^{2^-} = -0.98 \mu_N$. The magnetic moment therefore should provide a good measure of the relative importance between these configurations. The experimental μ turns out to be substantially larger than standard shell model calculations [22], suggesting an underestimation of the $[\pi p_{3/2}, \nu s_{1/2}]^{J^\pi=2^-}$ component. Agreement is obtained when ε is lowered by about 1 MeV from the value which is normally employed in calculations in the region of nuclei close to stability. Following the ^{14}B moment experiment, Aoi et al. [27] confirmed the lowered ε from the position of a newly assigned 1^+ state in ^{14}B in their β-neutron-γ spectroscopic work on ^{14}Be.

Does the magnetic moment of halo nucleon tend to recover its free space value?

The ground state of ^{11}Be, with its anomalous spin-parity $I^\pi = 1/2^+$ and small neutron separation energy $S_n = 503$ keV, is known to have a one-neutron halo structure. Then, an interesting question arises whether the spin g-factor for a neutron in a halo, $g_s^\text{eff}(\text{halo})$, tends to recover the free nucleon value, since a halo neutron is considered to spend most of time in space outside the nucleus. The $g_s^\text{eff}(\text{halo})$ may be extracted from an observed ^{11}Be$_\text{g.s.}$ magnetic moment. More quantitatively, in a weak coupling model the ^{11}Be$_\text{g.s.}$ wave function is expressed as $|^{11}\text{Be}_\text{g.s.}(1/2^+)\rangle = \alpha|^{10}\text{Be}(0^+) \otimes \nu s_{1/2}\rangle^{J^\pi=1/2^+} + \beta|^{10}\text{Be}(2^+) \otimes \nu d_{5/2}\rangle^{J^\pi=1/2^+}$, where $\alpha^2 \approx 0.55$ is predicted by the Variational Shell Model [28]. The ^{11}Be$_\text{g.s.}$ magnetic moment is then given as [28,29]

$$\mu(^{11}\text{Be}_\text{g.s.}) = \alpha^2 \mu(\nu s_{1/2}) + (1-\alpha^2)\left[\frac{7}{15}\mu(\nu d_{5/2}) - \frac{1}{3}\mu(2^+)\right]. \quad (2)$$

The $\mu(^{11}\text{Be}_\text{g.s.})$ is experimentally known [31] to be $\mu(^{11}\text{Be}_\text{g.s.}) = -1.6816 \pm 0.0008\mu_N$, while the ^{10}Be$(2^+)$ core moment $\mu(2^+)$ is evaluated as $\mu(2^+) = 1.786 \mu_N$ in Ref. [30]

from the shell-model wave function. The $d_{5/2}$ neutron moment $\mu(\nu d_{5/2})$ is expressed as $\mu(\nu d_{5/2}) = \frac{1}{2}g_s^{\text{eff}}\mu_N$, in terms of the effective spin g-factor g_s^{eff} for (non-halo) neutron inside a nucleus. Thus, if the (non-halo) neutron g_s^{eff} factor is known, we obtain the g_s-factor for the halo $s_{1/2}$ state neutron through Eq. (2).

The β-NMR measurement of the ^{15}C$_{\text{g.s.}}$ magnetic moment was carried out. A preliminary analysis of the data gives $|\mu(^{15}\text{C})| = (1.720\pm0.009)\mu_N$ [35]. Shell-model calculations indicate that the ^{15}C$_{\text{g.s.}}$ wave function is dominated by configurations of $\pi(0^+)\otimes\nu s_{1/2}$ to 97-98 % in probability. After correction for the remaining 2-3 % contribution from other types of configurations, we obtain the "experimental" $2s_{1/2}$ neutron single particle moment, $\mu(\nu s_{1/2})^{\text{exp}} = -(1.77 \pm 0.05)\mu_N$, or the effective g_s-factor for well-bound (i.e. non-halo) neutron, $g_s^{\text{eff}} = (0.92 \pm 0.02)g_s^{\text{bare}}$, where $g_s^{\text{bare}} = -3.83$ is the free-neutron g factor.

By inserting g_s^{eff} in Eq. (2), the g_s value for the halo neutron is obtained. Interestingly, $g_s^{\text{eff}}(\text{halo}) \approx 0.99 g_s^{\text{bare}}$ is obtained if we employ $\alpha^2 \approx 0.55$ predicted by VSM. On the other hand, if a larger value of $\alpha^2 \approx 0.84$ derived from $p(^{11}\text{Be},^{10}\text{Be})d$ reaction [32] an almost normal value of $g_s^{\text{eff}}(\text{halo}) \approx 0.90 g_s^{\text{bare}}$ is obtained. Thus, further studies are needed to attain a conclusive result on the present problem.

Quenching of the effective charges in neutron-rich nuclei

We now turn to the quadrupole moment results. The nuclear electric quadrupole moment is written as $Q = \langle\psi_{I,M=I}|\sum_{k=1}^{A}e_k\hat{Q}_k|\psi_{I,M=I}\rangle$, where $\hat{Q}_k = \sqrt{\frac{16\pi}{5}}r_k^2 Y_{20}(\hat{r}_k)$ and e_k is the electric charge for the k-th nucleon. Since $e_k = 0$ for neutron, there should appear no direct contribution from neutrons, but this point affords further attention: The addition of a valence neutron would induce a polarization of the core into configurations outside the adopted model space. Such an effect is included by introducing effective charges $e_p^{\text{eff}} = e + \delta e_p$ and $e_n^{\text{eff}} = \delta e_n$ for proton and neutron. Empirically, the polarization charges are known to take values $\delta e_p \approx 0.29e$ and $\delta e_n \approx 0.49e$ for nuclei in the region close to stability with a mass range $A = 17\text{--}39$. It is interesting to examine how such a property of nucleons in a nucleus changes when going farther from stability.

The electric quadrupole moment for the ^{18}N ground state has been measured [33] to be $|Q(^{18}\text{N})| = 12.3 \pm 1.2$ emb. Because of the $p_{1/2}$ character of a ^{18}N proton hole, the proton contribution to the Q-moment is very small: Rewriting as $Q = e_p^{\text{eff}}Q_p + e_n^{\text{eff}}Q_n$ where $Q_p = \langle\psi_{I,I}|\sum_{\text{proton}}\hat{Q}_k|\psi_{I,I}\rangle$ and $Q_n = \langle\psi_{I,I}|\sum_{\text{neutron}}\hat{Q}_k|\psi_{I,I}\rangle$, shell model calculations with the PSDMK and PSDWBT interactions predict Q_p = 1.7-2.0 mb while Q_n = 31.8-33.4 mb. Note that the calculated $Q_{p,n}$ are quite stable against the choice of the effective interaction. Inserting the usually taken values of effective charges, $e_p^{\text{eff}} \approx 1.3e$ and $e_n^{\text{eff}} \approx 0.5e$, the shell model values for Q substantially overestimate the experimental Q. Because Q_p is very small, the small $Q(^{18}\text{N})$ implies that e_n^{eff} should be smaller. The agreement is obtained if $e_n^{\text{eff}} = 0.3e$ is taken.

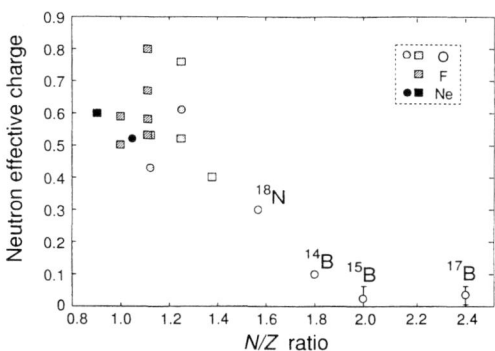

FIGURE 4. Empirical effective charge for neutrons in the sd orbits, extracted from the experimental static Q-moments (circles) and E2 transition probabilities (boxes).

The quenching of e_n^{eff} turns out to be even more striking in the measured Q moments for ^{15}B [34] and ^{17}B [35]. The ^{15}B and ^{17}B isotopes have, respectively, two and four extra neutrons on top of the closed-shell $N=8$ isotope ^{13}B. Shell model calculations predict a rapid increase of the matter Q moment for neutrons Q_n from 0.0 mb through 45.3 mb to 60.5 mb^3 for isotopes ^{13}B, ^{15}B, and ^{17}B, respectively, while that for proton stays almost constant, $Q_p = 31.2$–25.1 mb. Thus, the shell model values for Q increases by about 40 % and 55 % as going from ^{13}B to ^{15}B and ^{17}B isotopes, if the standard value of $e_n^{\text{eff}} \approx 0.5e$ is assumed for the neutron effective charge. To the contrary, the experimental Q, $|Q(^{15}\text{B})| = 38.01 \pm 1.08$ emb [34] and $|Q(^{17}\text{B})| = 38.8 \pm 1.5$ emb (preliminary) [35], are surprisingly close to that of ^{13}B, $|Q(^{13}\text{B})| = 36.9 \pm 1.0$ emb [36,37].[4] In a weak coupling model in which ^{15}B and ^{17}B are described by two and four neutrons coupled to the ^{13}B core, the similarities of $Q(^{15}\text{B})$ and $Q(^{17}\text{B})$ to $Q(^{13}\text{B})$ would imply smallness of neutron part of Q in ^{15}B and ^{17}B. Therefore it may be interesting to assume that the effective charge for neutron is quenched while that for proton remains unaltered, although other possibilities cannot be excluded. To obtain agreement between theory and experiment under this assumption, anomalously small values for the neutron effective charge, $e_n^{\text{eff}} \leq 0.1e$, are needed. The experimental Q of ^{14}B, $|Q(^{14}\text{B})| = 29.84 \pm 0.75$ emb [34] is found well reproduced by calculations with $e_n^{\text{eff}} \approx 0.1e$. The e_n^{eff} values extracted from the present experiments are plotted in Fig. 4 as a function of the N/Z ratio, together with those for nuclei close to stability taken from literature. One may find an apparent tendency that the quenching of e_n^{eff} develops gradually as N/Z increases.

---

[3] Values from calculations with the PSDMK interaction. We stress again that the calculated Q proved very stable against the choice of the effective interactions, and therefore the discrepancy observed here is quite substantial.

[4] Re-evaluated from $Q(^{13}\text{B})$ reported in Ref. [36] (which takes the old data for $Q(^{12}\text{B})$ as a reference) by taking the new $Q(^{12}\text{B})$ value from Ref. [37] as a reference.

REFERENCES

1. I. Tanihata et al., *Phys. Lett.* B **160**, 380 (1985).
2. T. Motobayashi et al., *Phys. Lett.* B **346**, 9 (1995).
3. H. Sakurai et al., *Phys. Lett.* B **448**, 180 (1999).
4. T. Nakamura et al., *Phys. Rev. Lett.* **83**, 1112 (1999).
5. H. Iwasaki et al., *Phys. Lett.* B **481**, 7 (2000).
6. K. Yoneda et al., *Phys. Lett.* B, in press (2001).
7. A. Ozawa et al., *Phys. Rev. Lett.* **84**, 5493 (2000).
8. K. Sugimoto, M. Ishihara, and N. Takahashi, *Treatise on Heavy-Ion Science* **3**, ed. A. Bromley, Plenum Publ. Co., 1985, ch. 5, pp. 397-536.
9. A.S. Goldhaber, *Phys. Lett.* B **53**, 306 (1974).
10. K. Asahi et al., *Phys. Lett.* B **251**, 488 (1990).
11. K. Asahi et al., *Hyperfine Int.* **75**, 101 (1992).
12. H. Okuno et al., *Hyperfine Int.* **78**, 97 (1993).
13. H. Okuno et al., *Phys. Lett.* B **335**, 29 (1994).
14. T. Kubo, *Nucl. Instrum. Methods* B **70**, 309 (1992).
15. H. Ogawa et al., in preparation.
16. A. Arima, K. Shimizu, W. Bentz, and H. Hyuga, *Adv. Nucl. Phys.* **18**, 1 (1987).
17. B.A. Brown and B.H. Wildenthal, *Nucl. Phys.* A **474**, 290 (1987).
18. I.S. Towner, *Phys. Rep.* **155**, 263 (1987).
19. K. Matsuta et al., *Nucl. Phys.* A **588**, 153c (1995).
20. M. Huhta et al., *Phys. Rev.* C **57**, R2790 (1998).
21. H. Ueno et al., *Phys. Rev.* C **53**, 2142 (1996).
22. B.A. Brown, A. Etchegoyen, and B.D.M. Rae, Computer code OXBASH, *MSU Cyclotron Laboratory Report* No. 524 (1986).
23. D.J. Millener and D. Kurath, *Nucl. Phys.* A **255**, 315 (1975).
24. E.K. Warburton and B.A. Brown, *Phys. Rev.* C **46**, 923 (1992).
25. H. Okuno et al., *Phys. Lett.* B **354**, 41 (1995).
26. I. Talmi and I. Unna, *Phys. Rev. Lett.* **4**, 469 (1960).
27. N. Aoi et al., *Nucl. Phys.* A **616**, 181c (1997).
28. T. Otsuka et al., *Phys. Rev. Lett.* **70**, 1385 (1993).
29. T. Suzuki and T. Otsuka, Proc. of RIKEN Symposium Shell Model 2000, Wako-shi, Japan, March 2000, to appear in *Nucl. Phys.* A (2001).
30. Toshio Suzuki et al., *Phys. Lett.* B **364**, 69 (1995).
31. W. Geithner et al., *Phys. Rev. Lett.* **83**, 3792 (1999).
32. S. Fortier et al., *Phys. Lett.* B **461**, 22 (1999).
33. H. Ogawa et al., *Phys. Lett.* B **451**, 11 (1999).
34. H. Izumi et al., *Phys. Lett.* B **366**, 51 (1996).
35. H. Ogawa et al., Proc. of RIKEN Symposium Shell Model 2000, Wako-shi, Japan, March 2000, to appear in *Nucl. Phys.* A (2001).
36. R.C. Haskell and L. Madansky, *J. Phys. Soc. Jpn.* **34** Suppl., 167 (1973).
37. T. Minamisono et al., Phys. Rev. Lett. **69**, 2058 (1992).

High Energy Hadron Spin Observables
– Microscopic QRC model and spin asymmetry –

Ken-ichi Kubo

Physics, School of Science, Tokyo Metropolitan University, 1-1 Minami-osawa, Hachioji, Tokyo 192-0397, Japan
and
Tokyo Metropolitan College of Aeronautical Engineering, 8-52-1 Minami-senju, Arakawa-ku, Tokyo 116-0003, Japan

Abstract. We briefly review concept of the quark recombination (QRC) model and a general success of the model. To solve the existing problem, so called anomalous spin observables, in the high energy hyperon spin phenomena, we propose a mechanism; the primarily produced quarks, which are predominantly u and d quarks, act as the leading partons to form the hyperons. Extension of the quark recombination concept with this mechanism is successful in providing a good account of the anomalous spin observables. Another kind of anomaly, the non-zero analyzing power and spin depolarization in the Λ hyperon productions, are also discussed and well understood by the presently proposed mechanism. Recently, a further difficulty was observed in an exclusive $\Lambda K^+ p$ production and we will indicate a possible diagram for resolving it.

INTRODUCTION

It has been generally established that, in contrast to the perturbative QCD predictions, spin effects in the hadron collisions are large over a wide incident energy range [1] even at extremely high energies such as 2000GeV [2]. Among them, the hyperon spin polarization mechanisms have been well studied with the various theoretical models assuming a soft scattering process [3–5]. The left-right asymmetries of $\pi^{\pm,0}$ from the inclusive $\vec{p}p$ collisions [6] have also attracted attention and the observed characteristics have been well described by those theoretical calculations [3–5].

Essential results of the spin polarizations in the inclusive hadron reactions are well interpreted by the parton recombination picture proposed by DeGrand and Miettinen (DM) [3]. However, the DM model predicts only order of magnitudes of the hadron spin observables, and does not give spin observables as functions of x_F nor P_T, longitudinal (Feynman) and transverse momentum variables, respectively. Hence we have developed the so-called microscopic quark recombination (QRC) model in the relativistic framework [7]. We have demonstrated the general success of the model for most of the existing high-energy inclusive experimental data [7]. Then we have concluded that the QRC formalism itself contains the dynamics producing spin asymmetry, which was assumed as an empirical rule in the DM model.

In the following, we summarize first the concept of the DM model and the QRC model to show what the normal cases are in order to define 'anomalous' or 'puzzling' spin observables. In these models, the projectile makes a hard collision with the target and one of the fragments (leading parton which can be a quark or a diquark), which moves fast with the original velocity and has a color, makes a color flux behind with its colored counterpart. As the color flux is prolonged by the initial parton momentum, then the pair creation takes place somewhere in the color flux and the leading parton together with the one of the created pair form a pre-hadron. At the time the pre-hadron is formed, the leading (valence) parton is moving fast and the picked up (sea) parton which is one of the created pair is moving slow. They are pulled each other towards the averaged velocity to form the hadron eventually to be observed experimentally.

The confining force takes place in this momentum averaging process, which causes the spin dependent effect (Thomas precession) and produces the spin polarization of the ejectile hadron. DeGrand and Miettinen brought the effect into an empirical rule [3] that parton(s) moving fast (slow) carries up (down) spin mostly, respectively. We shortly call it as "Fast-spin-up and Slow-spin-down" in our work. Finally the sign of the spin polarization is determined by two ingredients; one is the dynamics of the spin dependent effect and the other is the wave functions of the projectile and ejectile hadrons, for which the DM model and the QRC model use the SU(6) valence quark model.

Thus the DM model consists of a leading parton (projectile valence quark(s)) with high velocity and an associated parton (sea quark(s)) with low velocity. The model then predicts correct signs and rough magnitudes of hadron spin polarizations for valence quark reactions and no spin polarization for no-valence quark reactions such as $p \to \bar{\Lambda}$, $\Sigma^- \to \bar{\Lambda}$ and $p \to \Omega^-$ (completely strange baryons), where we use the notation e. g. $p \to \bar{\Lambda}$ for inclusive reactions such as pp and pA collisions leading to $\bar{\Lambda}X$. The cases to be described by the mechanism explained above are the normal ones.

In fact the measurements [8, 9] give results essentially consistent with zero for these no-valence quark cases, as shown in Figs.1(a) and (b) for Λ. We then had a good understanding that existence of a leading parton plays an essential role in the spin polarizations until two anomalous cases were discovered experimentally. One is the $\bar{\Xi}^+$ polarization [10] shown in Fig.1(c) which is found to have almost the same polarization as that of Ξ^- (not shown here) and the other is $\bar{\Sigma}^-$ [11] shown in Fig. 1(d) with the same sign but less magnitude compared to Σ^+. We call this case as the *first puzzle* and we have reported our successful understanding for the puzzle [12]. Here we will shortly show essence of the result.

There exists another anomalous spin observable. The measurements [13] of analyzing power A_N and spin depolarization D_{NN} in $\vec{p}p \to \Lambda X$ reactions, where

$\vec{p}$ presents incident protons with spin perpendicularly polarized in the scattering plane, show apparent non-zero values in the kinematical region above $x_F = 0.5$, contrary to the prediction of zero by the DM model with SU(6) baryon wave functions. Namely in the model, ud diparton with spin $S = 0$ from the projectile, and s parton from a created pair form Λ, but the initial spin carrier, u parton in the proton, does not participate in the Λ formation. Therefore Λ has no way to know whether or not the incident proton is spin polarized, and consequently A_N and D_{NN} become zero. Contrary to this prediction, the experiments show finite values to these spin observables [13], therefore we call it the *second puzzle*. To solve the puzzle, here we propose a new spin transfer mechanism and show the process is quite promising to provide finite values of the A_N and D_{NN} in hyperon productions.

Recently, a measurement of the spin depolarization D_{NN} of an exclusive collision at 3.67 GeV/c has been reported [14]. Natural extension of the diagram proposed for interpretation of the second puzzle predicts D_{NN} with positive sign in contrast to observation with negative sign. This new difficulty is referred to as the *third puzzle* in this article and we will discuss a possible mechanism by noting requirement of parity conservation.

THE FIRST PUZZLE;
EXTENSION OF THE STANDARD QRC MODEL

Multiplicity distributions of the products in pp, pA and AA collisions at high energies are well-known to be dominated by π and other light mesons made of ud quarks and antiquarks [15] produced primarily or secondarily. This is understood as due to the Schwinger mechanism for the quark pair creation, which is extremely sensitive to the quark mass. This fact indicates that many of the mesons with velocity as high as the incident hadrons are produced and hence may suggest a mechanism according to which constituent ud quarks and antiquarks of the mesons, hereafter referred to as *primary ud-quark*, participate as leading partons. They recombine with a successively produced parton of a created pair to make final hyperons. More generally, it is not necessary for π and other light mesons to be produced, but any mechanisms like the tube model and the string model are conceivable, in which light quarks and antiquarks are produced in the primary interaction [16]. The essential requirement is that those primary ud and anti-ud partons move with such high enough velocity compared with that of the picked up partons and that those particles are regarded as valence partons. This requirement is a consequence of the DM model as explained in the Introduction; the model requires valence parton(s) with large velocity (carrying up-spin) and sea parton(s) with small velocity (down-spin).

Accepting this extended mechanism we calculate the spin polarizations of outgoing hadrons by assuming the pion as incident particle. For the momentum distribution functions of the valence and the created partons, we use those of the pion [17].

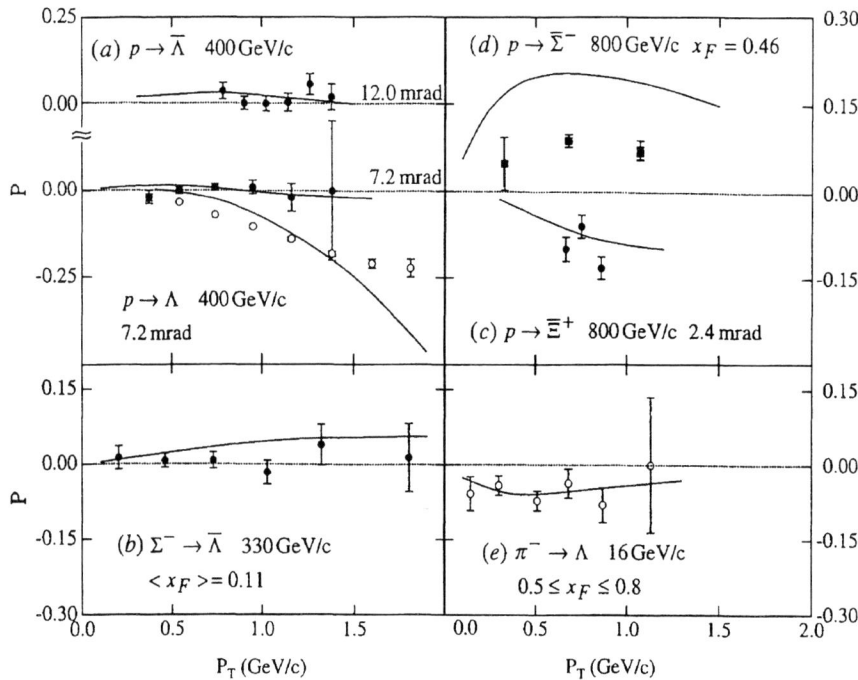

FIGURE 1. The hyperon spin polarizations as a function of P_T in the kinematical regions indicated. The observed data are taken from the works quoted in [12] and shown by the open symbols for the hyperons and by the solid symbols for the anti-hyperons. The solid curves for the anti-hyperons show the results obtained by the QRC process with the present primary ud-quark mechanism, and for the Λ polarizations in (a) and (e) are obtained by the standard QRC process.

The results of the calculation are shown by the solid curves and compared with the observed data in Fig.1. The calculated $p \to \bar{\Lambda}$ polarizations for the two different kinematical regions are shown in (a). They are as small as the observations and much smaller than the $\pi^- \to \Lambda$ result shown in (e) which predicts quite well the observed data. We are careful to say 'consistent with zero' for the $\Sigma^- \to \bar{\Lambda}$ as seen in (b), so that altogether the further confusion pointed out above for the p and $\Sigma^- \to \bar{\Lambda}$ and the $\pi^- \to \Lambda$ is cleared up. The Λ polarization is shown in (a) for comparison. These small $\bar{\Lambda}$ polarizations are found to be caused in those kinematical regions by a large cancellation between the scalar and vector diquark contributions in association with the Λ wave function in SU(6) symmetry. The $\bar{\Xi}^+$ calculation (c) reasonably well reproduces the measurements and the $\bar{\Sigma}^-$ (d) gives the observed shape of the P_T distributions with the correct sign but too large magnitude by a factor two.

The present mechanism predicts various relations among the hyperon polarizations. First, there should be the relations; $P(p \to \bar{\Lambda}) = P(\pi^- \to \bar{\Lambda}) = P(\pi^- \to \Lambda)$ as noted already. This relation indicates large polarizations for $p \to \bar{\Lambda}$ in the kinematical x_F region indicated in (e); $x_F = 0.5 - 0.8$. Note that the kinematics in the $\bar{\Lambda}$ in (a) and (b) corresponds roughly to $x_F = 0.1 - 0.3$. Second, the other prediction is $P(p \to \bar{\Xi}^0) = P(p \to \bar{\Xi}^+)$, therefore, $= P(p \to \Xi^-)$ as discussed in (c). A further prediction is the same polarizations of $\bar{\Sigma}^+$ and $\bar{\Sigma}^0$ as that of $\bar{\Sigma}^-$ which is shown in (d). It is very interesting to test these relationships by measurements in order to confirm further the present mechanism.

THE SECOND PUZZLE;
HOW TO TRANSFER SPIN INFORMATION

As for the finite values of A_N and D_{NN} in the $\vec{p}p \to \Lambda X$ reaction [13], the question is to find a mechanism to transfer the initial spin direction of incident proton to the outgoing Λ. This may be possible in the framework of the QRC model by considering an annihilation and creation mechanism, as shown in Fig.2, where the initial valence u quark, which carries the proton's spin information, annihilates with $\bar{u}$ in the target proton and then $s\bar{s}$ pair is created through the gluon propagation, and the s quark recombines with ud scalar diquark to form Λ. From the Clebsh-Gordon coefficients, we can estimate that the s quark has spin up with probability of 75% and spin down with that of 25%. Suppose the u quark with spin up annihilates with the $\bar{u}$ quark with spin up. Then intermediate gluon carries spin up and hence the created s and $\bar{s}$ quarks carry spin up, therefore the Λ particle with spin up is produced. On the other hand, if the u quark has spin down, the intermediate gluon carries zero spin direction and the s and $\bar{s}$ quarks do not have preferred spin directions, therefore for this case the initial spin information cannot be transferred to the produced Λ particle. In this way we can transfer spin information to s quark to form Λ particle.

However, this process should not be all. The standard QRC process also participates to Λ production. Since we do not know amount of the two contributions in kinematical region of interest, we write the probability of creating spin up s quark as $(1+\gamma)/2$ and spin down s quark as $(1-\gamma)/2$, for the case of the incident proton with spin up. Then using the DM empirical rule and the parameter ϵ, $\vec{p} \to \Lambda$ production cross-sections can be expressed as,

$$\sigma_{\uparrow\uparrow(\uparrow\downarrow)} = \frac{1 \pm \gamma}{2}(1 \mp \epsilon), \quad \sigma_{\downarrow\uparrow(\downarrow\downarrow)} = \frac{1 \mp \gamma}{2}(1 \mp \epsilon), \tag{1}$$

where e.g. $\uparrow\downarrow$ means proton(u) spin up and $\Lambda(s)$ spin down, respectively. Using these cross sections, we obtain the relations,

$$P = -\epsilon, \quad D_{NN} = \gamma, \quad A_N = -\gamma\epsilon = D_{NN}P. \tag{2}$$

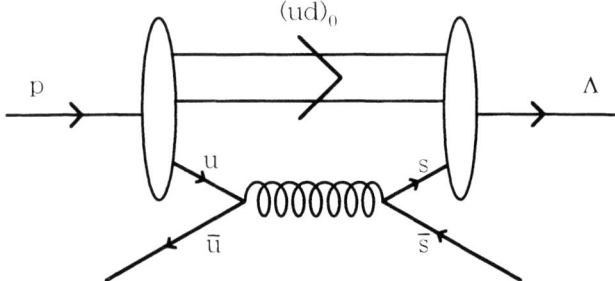

FIGURE 2. The diagram for a production of Λ from pp collision through $u\bar{u}$ annihilation and $s\bar{s}$ creation.

We are not able to predict the spin depolarizing parameter γ which needs information on the dynamics of the quark-antiquark annihilation and creation through gluon exchange, but we can calculate the spin polarization $P(x_F, P_T)$ from our QRC model. Therefore in this article, we test the consistency of the last relation in eq.(2), since we find γ to be just the spin depolarization itself for which measurements exist [13]. Using our calculated polarization $P(x_F, P_T)$ and accepting the observed values for $D_{NN}(x_F, P_T)$, we obtain A_N and compare it with the measurements. In Fig.3, the results are shown by the solid curves. The kinematical parameters $P_T = 1$ GeV/c for (a) and $x_F = 0.2 - 1.0$ for (b) were used in the $P(x_F, P_T)$ calculation, therefore the obtained results should be rather compared with the solid circles in the both Figs.3 (a) and (b); the signs and slopes obtained (in a) and estimated (in b) by the present calculation are consistent with the measurements and this fact indicates the present u annihilation and s creation mechanism being quite promising to solve the second puzzle.

THE THIRD PUZZLE;
THE DIAGRAM FOR THE EXCLUSIVE $\Lambda K^+ P$ PRODUCTION

Recently, D_{NN} of an exclusive $\vec{p}p \to \Lambda K^+ p$ collision at 3.67 GeV/c has been observed [13]. The result shows negative sign, that is different from sign of the previous inclusive Λ production case. Our annihilation and creation diagram Fig. 2 can be extended to the exclusive production to form K^+ by merging $\bar{s}$ having spin-up with the counter parton u of the created $u\bar{u}$ pair having spin-down. This, however, produces Λ with spin-up, therefore the calculated spin depolarization D_{NN} could have positive sign in contrast to the negative sign observed if the spin flip

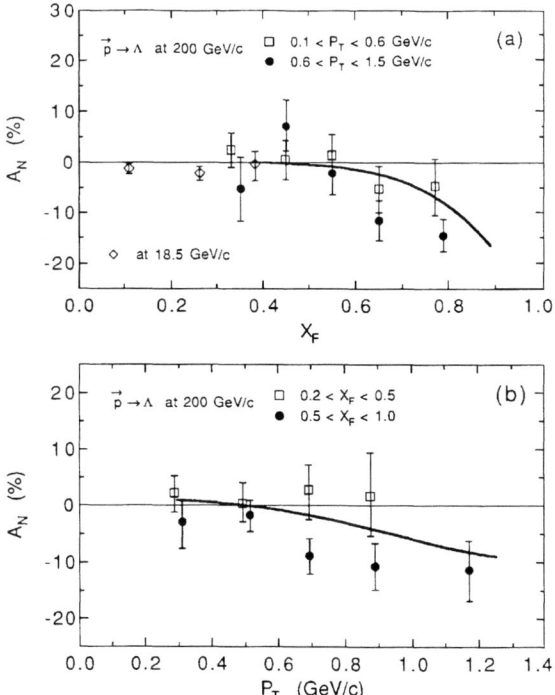

FIGURE 3. The A_N of $\vec{p}p \to \Lambda X$ reaction. The solid curves are obtained using the calculated P and the observed D_{NN}, according to $A_N = D_{NN}P$ found in the present spin transfer process, for the kinematical parameters $P_T = 1$ GeV/c in (a) and $x_F = 0.2 - 1.0$ in (b). The experimental data are taken from the works quoted in the text.

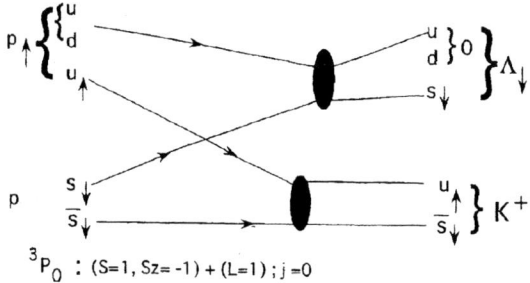

FIGURE 4. The leading order diagram based on the originally used Λ spin polarization in the inclusive $\vec{p}p \to \Lambda X$ collision and extended to the exclusive $\vec{p}p \to \Lambda K^+ p$ collision.

cross section is small.

Therefore something may be lack in our interpretation of the exclusive production mechanism mentioned above. Perhaps, lack is consideration of parity. In fact, it is apparent that the simply extended diagram from Fig.2 does not conserve parity. Hence we consider P-wave contribution, namely we create $u\bar{u}$ pair with $3P_0$ configuration. Then we can easily show that the new diagram is possible to produce Λ with spin-down and possibly D_{NN} with negative sign.

However, there exists a more fundamental (leading order) graph which was originally used in the Λ inclusive production. We extend this diagram to the exclusive $pp \to K^+ \Lambda p$ production with the parity conservation condition. One of the possible four new graphs is shown in Fig. 4 and we find that spin direction of the produced Λ is down, so that we may obtain negative value for the spin depolarization for this graph.

We similarly estimated the spin dependent cross sections as described in Sec. 3 but by introducing the two parameters, λ and κ with positive sign, characterizing the DM empirical rule since now we concern the two hadrons, Λ and K^+, production case. Using these cross sections, we obtained the following relations corresponding to Eq. 2,.

$$P = -\lambda, \quad A_N = \kappa, \quad D_{NN} = -\lambda\kappa = PA_N. \tag{3}$$

This predicts a negative sign for D_{NN} of the present exclusive production, which is consistent with the observed result.

SUMMARY

To resolve the anomalous spin observables found in the anti-hyperon spin polarizations, we have proposed new mechanism; the primary ud quark mechanism. We have applied our microscopic quark recombination model to the spin polarizations and shown that the calculations with the present mechanism predict large polarizations where large polarizations are observed and vice versa, so that the first puzzle clearly disappears.

For the second puzzle of the unexpected analyzing power and spin depolarization in the $\vec{p}p \to \Lambda X$ reaction, we have indicated a possible spin transfer mechanism (u annihilation and s creation) in the intermediate stage, which provides finite value for these observables. A simple relationship among the three spin observables (P, A_N, D_{NN}) has been found and we have confirmed that the relation provides the result consistent with the measurements. Thus the second problem has also been clearly solved. The spin depolarizing parameter γ, and therefore the sea quark polarizing dynamics through the u annihilation and s creation is left for future study. It is a useful finding that spin depolarization provides a direct information for this q-q interaction study.

For the third difficulty of difference in sign between the two D_{NN}'s of Λ observed in the inclusive $\vec{p}p \to \Lambda X$ and the exclusive $\vec{p}p \to \Lambda K^+ p$ collisions, we have proposed a possible mechanism with satisfying parity conservation. This mechanism is an extension of the the diagram originally used and established for the spin polarization calculation of $pp \to \Lambda X$ inclusive collision. We are now calculating D_{NN} for the exclusive $\vec{p}p \to \Lambda K^+ p$ by using this mechanism.

Through all these systematic studies of spin transfer mechanisms both in the normal and anomalous hadron production processes with our basic microscopic QRC model, we like to contribute to precise understanding of nucleon structure, the QCD spin physics and the confinement interactions.

The collaborators: Y. Yamamoto, Y. Kitsukawa from Tokyo Metropolitan University, H. Toki, N. Nakajima from RCNP Osaka University, and K. Suzuki from University of Tokyo.

REFERENCES

1. G. Bunce et al., Phys. Rev. Lett.**36**, 1113(1976).
2. S. Erhan et al., Phys. Lett. **82B**, 301(1979).
3. T. M. DeGrand and H. I. Miettinen, Phys. Rev. **D24**, 2419(1981); 31, 661E(1985).
4. J. Szwed, Phys. Lett. 105B, 40(1981). B. Andersson, G. Gustafson and G. Ingelman, Phys. Lett. **85B**, 417(1979), Phys. Reports **97**, 31(1983). J. Soffer and N. A. Toernqvist, Phys. Rev. Lett. **68**, 907(1992). X. Artru, J. Czyzewski and H. Yabuki, LYCEN/9423(1994). See also ref.7.
5. C. Boros, Z.-T. Liang and T.-C. Meng, Phys. Rev. Lett. **70**, 1751(1993), Phys. Rev. **D51**, 4867(1995), **D53**,R2279(1996).
6. V. D. Apokin et al., Phys. Lett. **243B**, 461(1990), D. L. Adams et al., **261B**, 201(1991), **264B**, 462(1991).
7. Yuichi Yamamoto, K.-I. Kubo and H. Toki, Prog. Theor. Phys. **98**, 95(1997).
8. P. Skubic et al., Phys. Rev. **D18**, 3115(1978). K. Heller et al., Phys. Rev. Lett. **41**, 607(1978). E. J. Ramberg et al., Phys. Lett. **338B**, 403(1994). M. I. Adamovich et al., Z. Phys. **A 350**, 379(1995).
9. K. B. Luk et al., Phys. Rev. Lett. **70**, 900(1993).
10. P. M. Ho et al., Phys. Rev. Lett. **65**, 1713(1990), Phys. Rev. **D 44**, 3402(1991).
11. A. Morelos et al., Phys. Rev. Lett. **71**, 2172(1993).
12. K.-I. Kubo, Yuichi Yamamoto and H. Toki, Prog. Theor. Phys. **101**, 615(1999).
13. A. Bravar et al., Phys. Rev. Lett. **75**, 3073(1995), **78**, 4003(1997).
14. F.Balestra et al., Phys. Rev. Lett. **83**, 1534(1999).
15. See following Proceedings of the serial Intern. Confs, published in Nucl.Phys. **A** ; Ultra-Relativistic Nucleus-Nucleus Collisions (Quark Matter, every year from 1983), Nucleus-Nucleus Collisions (every three years from 1988), T. S. Biro et al., Phys. Lett. **347B**, 6(1995).
16. K. Abe et al., Phys. Rev. Lett. **78**, 3442(1997).
17. P. J. Sutton et al., Phys. Rev. **D45**, 2349(1992).

Spin Physics in Hypernuclei and Hyperon-Nucleon Interactions

Tadafumi Kishimoto

Department of Physics, Osaka University, Toyonaka, Osaka, 560-0043, Japan

Abstract. Role of spin in the recent study of hyperon nucleon interaction is reviewed. I discuss two phenomena. One is the spin-orbit splitting of Λ single particle states in hypernuclei. Recently it confirmed to be very small for which gamma ray spectroscopy played an essential role. Another is the weak nonmesonic decay of Λ hypernuclei. Spin polarization and asymmetric weak decay with respect to the polarization axis gave a new information on the study of strangeness changing weak hyperon nucleon interaction.

SPIN DEPENDENT HYPERON NUCLEON INTERACTION

Spin-orbit Splitting of single Λ states in hypernuclei

Hyperon nucleon interaction has been given by the study of hypernuclei. Observation of Λ single-particle states, first by the (K^-, π^-) reaction [1] and then later (π^+, K^+) reaction [2,3], clarified the gross structure of the Λ-nucleus interaction. The dominant central part is found to be roughly 2/3 of that of nucleon. This is quantitatively reproduced by one boson exchange (OBE) models of the hyperon-nucleon (YN) interaction and qualitatively understood in quark models in that the Λ is composed of u, d, and s quarks and the s (strange) quark contributes little to the nuclear force. On the other hand, little has been known about the spin-dependent interaction. Especially, spin-orbit (ℓs) splitting was found to be smaller than that for the nucleon [4–6] although no experiment has given a definite value so far. The ℓs-splitting in the ΛN interaction has been a major goal in the study of hypernuclei.

The smallness of the ℓs-splitting of single-Λ states was first suggested by the (K^-, π^-) reaction on ^{16}O [4]. The difference of energy between the $((p_{1/2})_n^{-1}, (p_{1/2})_\Lambda)0^+$ and $((p_{3/2})_n^{-1}, (p_{3/2})_\Lambda)0^+$ states in $^{16}_{\Lambda}O$ is almost the same as that of the $(p_{1/2})_n^{-1}$ and $(p_{3/2})_n^{-1}$ states in ^{15}O, setting an upper limit of 0.3 MeV for the ℓs-splitting of single-p_Λ states. A small splitting was also indicated by the ^{13}C$(K^-,\pi^-)^{13}_\Lambda$C reaction and 0.36 ± 0.3MeV was obtained [5,7]. Hypernuclear γ rays observed in $^9_\Lambda$Be indicated that the 3/2$^+$ and 5/2$^+$ states, which have $((s_{1/2})_\Lambda, 2^+(^8\text{Be}))$ configuration, are too close to be separated by NaI detectors suggesting small ℓs-splitting [6]. However, the possibility that the 5/2$^+$ state was not populated in the (K^-,π^-) reaction was not completely ruled out.

$^{13}_\Lambda$C is an ideal hypernucleus with which to extract the ℓs-splitting. The 1/2$^-$ and 3/2$^-$ states at around 11 MeV are dominantly represented as $(p_{1/2})_\Lambda$ and $(p_{3/2})_\Lambda$ coupled to the 0^+ (^{12}C) core, respectively. Therefore the energy difference between the 1/2$^-$ and 3/2$^-$ states directly gives the ℓs-splitting of single-p_Λ states. So far, the ℓs-splitting has been measured mostly by magnetic spectrometers. The best energy resolution of magnetic spectrometers achieved for the study of hypernuclei is around 2 MeV. Since the ℓs-splitting is predicted to be 0~1 MeV, one wishes to measure it with precision better than 0.1 MeV. Since the p_Λ states in $^{13}_\Lambda$C are just below the particle emission threshold, γ rays can be observed by which the energy resolution is greatly improved [8].

The AGS-E929 experiment

The experiment (AGS-E929) was carried out at the D6 beam line of the alternating-gradient synchrotron (AGS) of BNL. The ^{13}C(K^-,π^-) reaction at P_K=0.93 GeV/c was used to produce $^{13}_\Lambda$C. The incident K^- momentum was chosen so as to maximize the production rate of the states. The K^- beam intensity was typically $\sim 8 \times 10^4$/spill for 5×10^{12} proton/spill. A spill consisted of 1.4 seconds of continuous beam every 4 seconds. The beam-line spectrometer measured the incident K^- momentum. The momentum of the outgoing π^- was measured by the 48D48 spectrometer, which has a momentum resolution of 15 MeV/c (FWHM). Its large angular acceptance (0~ 16°) [9] made simultaneous measurement of forward and backward scattering angles possible. This is vital for the small ℓs-splitting.

Figure 1 shows the detectors around the target region. BS is a plastic scintillator which defines the incident beam. K^-'s were tagged by TOF with a counter further upstream in the beam line. BQC is the differential-type quartz Čerenkov counter which vetoes π^-'s in the beam. FAC is the aerogel Čerenkov counter which tags π^-'s right after the target. The (K^-,π^-) trigger was very clean with these conditions. A benzene liquid scintillator, whose carbon was 99% enriched ^{13}C, was used as an active target. It was contained in four quartz containers of 30(thickness) × 15(height) × 60(width) mm^3, giving a total thickness of 120 mm. The biggest background is the K^- decay in-flight mostly from $K^- \to \pi^-\pi^0$ and $K^- \to \mu^-\nu$. These events gave (K^-,π^-) trigger signals identical to hypernuclear production. The active target was quite effective to suppress them. Every (K^-,π^-) event deposits energy in the active target due to energy loss of the K^- and π^-. Λ hypernuclei give additional energy deposit in the target due to their weak decay products. Optimization of container shape gave good separation of the weak-decay signal to kaon decay signal [10,11]. Plastic scintillators above and below the active target (DEC) gave supplementary signals to tag the weak decay. The excitation energy (E_{ex}) spectrum of the $^{13}C(K^-,\pi^-)^{13}_\Lambda C$ reaction was obtained from the measured K^- and π^- momenta. The p_Λ states appear at around E_{ex} =11 MeV. The overwhelming K^- decay and limited momentum resolution obscure the E_{ex} spectrum. However, tagging by the active target and DEC made it almost background free. A relatively wide cut ($0 < E_{ex} < 25$ MeV) for the p_Λ state was chosen to maximize 11 MeV γ rays where clean γ ray spectra were obtained as shown later.

γ rays from the $^{13}_\Lambda C$ were measured by two detectors located below and above the target as shown in figure 1. Each detector consisted primarily of an array of 36 NaI crystals, each of which had a dimension of $6.5 \times 6.5 \times 30$ cm^3. Four plastic scintillators were placed in front of each NaI array to veto charged particles. High beam-intensity can be tolerated by this segmented NaI detector. It is rare that an 11 MeV γ ray is fully contained in one NaI crystal. Signals from up-to three adjacent crystals were added to obtain the full energy peak.

Energy spectra of γ rays in coincidence with π^-'s scattered at $0° < \theta_\pi < 7°$, $7° < \theta_\pi < 10°$ and $10° < \theta_\pi < 16°$ are shown in figure 2(upper). We can clearly see a peak corresponding to p_Λ-to-GS transitions, although the two transitions of interest are not resolved. Doppler shift due to recoil of $^{13}_\Lambda C$ was corrected on an event-by-event basis. The recoil momentum and γ ray direction were obtained from the reaction vertex, given by the drift chamber information, and the position of the NaI crystal which had maximum energy deposited. The correction is typically less

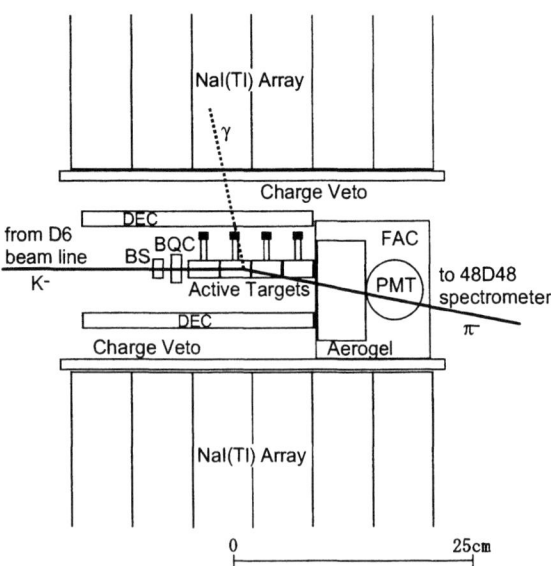

FIGURE 1. Detector system at the target region is shown schematically. See text for the description of each detector element.

than 100 keV.

The response function of the NaI detector for 11 MeV γ rays was obtained by a GEANT simulation which included the detector geometry and the procedure to sum energy of NaI crystals. With these conditions it reproduces well a peak in each spectrum as shown in figure 2(upper). A shoulder approximately 0.5 MeV lower than the 11 MeV peak of interest is due mainly to single escape and a tail extends to the low energy region. Since the peak width is well fit as a single transition, we conclude that the ℓs-splitting is small compared to our resolution. We derived peak positions by fitting a single γ ray peak for each of the three spectra.

Angular distributions of π^- for the $(p_{1/2})_\Lambda$ and $(p_{3/2})_\Lambda$ states, calculated by the distorted wave impulse approximation (DWIA), are shown in figure 2 (lower) [12]. The $(p_{1/2})_\Lambda$ state is dominant at $0° < \theta_\pi < 7°$, on the other hand the $(p_{3/2})_\Lambda$ state is dominant at $10° < \theta_\pi < 16°$. Both states are almost equally excited at $7° < \theta_\pi < 10°$. In order to obtain the ℓs-splitting, the relative yields of $(p_{1/2})_\Lambda$ and $(p_{3/2})_\Lambda$ states must be estimated for each spectrum. Since two states cannot be separated experimentally, relative yield in each spectrum is calculated by using the theoretical differential cross section (figure 2(lower)) and the acceptance of the 48D48 spectrometer for scattered π^-'s. The acceptance was estimated by a GEANT simulation which included the magnetic field distribution and relevant information of all counters. The simulation reproduces the profile of K^-'s and π^-'s and thus

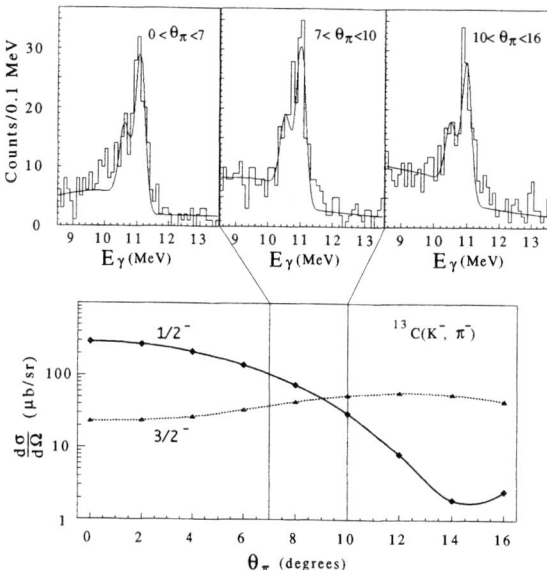

FIGURE 2. γ ray spectra taken in coincidence with scattered π^-'s (upper) and calculated differential cross section of $1/2^-$ and $3/2^-$ states (lower) are shown.

reproduces the acceptance.

Spin-orbit splitting of p_Λ state in $^{13}_\Lambda$C

Peak positions are plotted as a function of predicted yield ratio (R) in figure 3 where $R = (N(1/2^-) - N(3/2^-))/(N(1/2^-) + N(3/2^-))$. Here $N(1/2^-)$ and $N(3/2^-)$ stand for γ ray yield from the $1/2^-$ and $3/2^-$ states, respectively. The error bars on the peak positions include only statistical errors in the fitting. The ℓs-splitting is obtained by linear fitting of the three points. It is $152 \pm 54(statistics) \pm 36(systematic)$ keV.

The observed ℓs-splitting is quite small. The ℓs-splitting for single nucleon states in p orbit around this mass region is $3 \sim 5$ MeV thus the ℓs-splitting for the single-Λ state in p orbit is about 20-30 times smaller than that for the nucleon. Furthermore it is determined first that the $p_{1/2}(\Lambda)$ state appears higher than the $p_{3/2}(\Lambda)$ state, as is the case for nucleon.

Recently YN interactions have been refined in both an OBE model [13] and a quark model [14]. Hypernuclear structure calculation with a cluster model shows that the ℓs-splitting in $^{13}_\Lambda$C is 0.39-0.96 MeV for YN interactions based on an OBE model and it is ~ 0.2 MeV for a YN interaction based on a quark model [15]. Systematic study of light Λ-hypernuclei shows that the YN interactions based on an OBE model need to be modified so that a smaller ℓs-splitting, as indicated by the present experiment, can be accomodated [16]. A new mechanism will be required for the unified understanding of the baryon-baryon (NN, YN and YY)

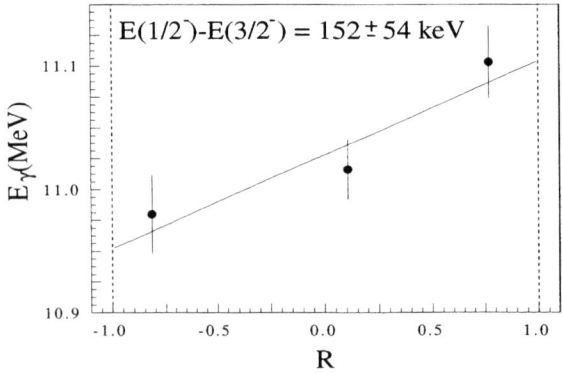

FIGURE 3. Peak positions obtained by fitting the γ ray spectra are shown as a function of R.

interaction. In summary the ℓs-splitting of single-Λ states has been observed for the first time. It is much smaller than that for the nucleon. The state-of-the-art calculation of the YN interaction based on OBE models is unable to reproduce the present result.

Other studies on the spin dependent hyperon nucleon interaction

Recently gamma ray spectroscopy in the study of hypernuclei had appreciable progress with high resolution gamma ray spectrometers. The small but finite value of the spin-orbit splitting was measured by the NaI detector as shown above. Germanium(Ge) detector has the best energy resolution among gamma ray detectors. Thus use of the Ge detector opens new field in the hypernuclear physics though it was achieved only recently since low primary beam intensity required large efficiency for which vast amout of investiment had to be given. The splitting corresponding to spin-spin interaction in $^7_\Lambda$Li was measured [17] by observing spin-flip M1 transition. Recently a splitting that directly gives the two body ΛN spin-orbit force is observed in the $^9_\Lambda$Be [18]. These measurements gave precise energy for states of the present interest which allow detailed comparison with the theoretical calculation. However, NaI detector is sometimes relevant for measurement of energetic gamma rays ($\sim$ 10MeV) like $^{13}_\Lambda$C. Here order was determeined by high statistics measurement with angular distribution pions.

Another important issue in the spin dependent hyperon nucleon interaction is Σ-nucleon spin-orbit interaction. The Σ hypernuclei have a long history. It was originally predicted to have spin-orbit interaction even larger than NN force. In the early stage of study many thought that the large ΣN spin-orbit splitting was observed [19]. However, existence of the Σ hypernuclei is even challenged recently. Study of ΣN spin-orbit interaction is one of the most important issues in the near future since we now know that ΛN spin-orbit splitting is very small.

WEAK HYPERON NUCLEON INTERACTION

Weak nonmesonic decay decay of Λ hypernuclei

Weak hyperon nucleon (YN) interaction can be studied by the weak nonmesonic decay (NM-decay) of Λ hypernuclei. The NM-decay ($\Lambda n \to nn$ (Γ_n), $\Lambda p \to pn$ (Γ_p)) is strangeness-changing YN weak process. One can study both parity-conserving and parity-non-conserving part of the weak YN interaction since no strong interaction can change Flavor(Strangeness). Recently, polarized $^{12}_\Lambda$C hypernuclei were produced by the ^{12}C(π^+, K^+) reaction, and their asymmetric NM-decay was observed [20]. The asymmetry parameter was found to be quite large ($\alpha^{NM} = -1.0 \pm 0.4$) which first demonstrated that the large parity violation is involved in the process.

Further study was carried out for the $^5_\Lambda$He which was excited by the ^{6}Li(π^+, K^+p) reaction at $P_\pi=1.05$ GeV/c where pions were provided by the K6 beam line of KEK-PS (PS-E278). It has the least ambiguity from nuclear physics. The large asymmetry parameter of the $^5_\Lambda$He M-decay (essentially equal to that of free Λ) can give its polarization by the experiment which was not possible in the $^{12}_\Lambda$C. Details of the experiment have been presented elsewhere [21–23]. Measured polarization of $^5_\Lambda$He (P) was consistent with the calculation [21] as shown in table 1. The asymmetry parameter of the NM-decay was derived to be $\alpha^{NM} = 0.22 \pm 0.20$ by $A_p = P\alpha^{NM}\varepsilon_p$. Its sign is opposite to theoretical calculations based on the meson-exchange model [24]. On the other hand, the model gave consistent result for the $^{12}_\Lambda$C [25,24].

TABLE 1. Measured polarization (P), calculated one (P_{cal}), proton asymmetry (A_p) and the obtained asymmetry parameter (α^{NM}) are shown.

Reaction		$\theta = 2 \sim 7°$	$\theta = 7 \sim 15°$
(π^+, K^+p)	P	-0.247 ± 0.082	-0.393 ± 0.094
$-4 < E_X < 4$	P_{cal}	-0.181	-0.368
GS region	A_p	0.076 ± 0.058	0.027 ± 0.077
	α^{NM}		0.22 ± 0.20

Experimental results might simply represent a statistical fluctuation. However, one would think that something is missing in the short-range part. Then the meson-exchange model reproduces well the $^{12}_\Lambda$C due to long-range part (ΛN relative p) but not the $^5_\Lambda$He. An effort to describe the short-range part of the interaction has been made by including a direct quark-exchange mechanism [26,27]. The model improves a longstanding discrepancy that the theory predicts too much Γ_p over Γ_n [27]. The asymmetry parameter requires relative phase of relevant mechanisms thus further study is needed. I wish to stress that this is a clear puzzle in nuclear physics and clear puzzles triggered progress in physics.

Weak ΛN interaction studied by the $pn \to p\Lambda$ reaction

The weak NM-decay of Λ hypernuclei may not directly correspond to the two body YN interaction since it takes place in a nucleus. Study of the inverse reaction ($pn \to p\Lambda$) can clarify the situation. The Q value of the reaction is $m_\Lambda - m_n = 176$MeV for which proton beam energy has to be larger than 370 MeV for free neutron target. Currently ^{9}Be is under consideration as a target. A 400 MeV proton beam is available at RCNP (Research Center for Nucler Physics) at Osaka University. The cross section is related to the weak NM-decay rate by

$$\frac{1}{\tau_{\Lambda p \to pn}} = <v\sigma(\Lambda p \to pn)>_{av} \int \rho_N |u_\Lambda|^2 d^3r \qquad (1)$$

where $\tau_{\Lambda p \to pn}$ is partial life of hypernuclear NM-decay, ρ is the density distribution of a nucleon, u_Λ is the wave function of a Λ. The estimated cross section by this equation is $\sim 10^{-39} cm^2$ [28]. Similar values were obtained by recent theoretical calculations [29–31]. No Λ is produced by the strong interaction which inevitably accompany kaon production up to 1.6 GeV for a neutron target and 0.7 GeV for infinitely heavy nuclear target. Study of energy dependence will give us relevant information to understand the process.

It has been recently pointed out that there are many spin dependent observable one can measure in the reaction since we can control initial proton spin and measure final Λ spin [32]. For example, the parity violation can also be studied by the analyzing power of the inverse reaction with longitudinally polarized proton beam. It is represented by

$$A = \frac{\sigma(h=1) - \sigma(h=-1)}{\sigma(h=1) + \sigma(h=-1)} \quad (2)$$

where $h = J_p \cdot P_p / |J_p||P_p|$. The polarization of the proton beam can be as large as 0.8. One would expect almost unity effect in the $pn \to p\Lambda$ reaction though it is very small ($10^{-7} \sim 10^{-8}$) in the proton-proton scattering where only parity violating part can be studied. However, one has to overcome very small cross section. In this reaction one can search for the T odd correlation $(\vec{J}_p \times \vec{k}_p) \cdot \vec{J}_\Lambda$. This process gives interesting new testing ground for the T violation. One always has to investigate the accuracy achievable by the final state interaction since it was proven that there is no null test for the T violation experiment [33]. Currently the precision we expect is inferior to other studies though it is always worth testing for the new reaction until limited by the final state interaction.

Experiment at RCNP

The signature of the reaction is the detection of π^- and proton from Λ decay. The $pn \to p\Lambda$ reaction takes place on a neutron which is moving in a nucleus with Fermi momentum. Outgoing particles (π^- and p) from Λ decay and from background process are studied by the GEANT. A detector is designed to reduce backgrounds keeping detection efficiency of π^- and p from Λ high [34]. The lifetime of Λ is 2.6×10^{-10} sec. The Λ has momentum typically 0.6 GeV/c thus the decay length of Λ is $\sim$4 cm. Since there is no other physical process that produces π^-'s and protons several centimeters from the target, the detection of decay vertex should be the genuine signal of the Λ production. Figure 4 shows the conceptual design of our detector system. It primarily consists of a collimater, silicon microstrip detectors (SSD) at the inner most region, a CDC and trigger counters in a solenoid magnet.

A collimater at the target protect our detector from vast amount of particles from the target. The particles from the target cannot be completely shielded out in reality. The leakage have yet to be reduced for which further study is underway. Relevant vertex information is obtained by two layers of SSD. Strip width is 0.2

mm in z direction and 0.1 mm in azimuthal direction. We will have to read almost 10 k channels of the signal. A preamplifier with multiplexer and flash ADC's with synchronized read-out, we developed recently, enable us to handle such huge data. The CDC at the outer region of the SSD is capable to give radial trajectory and longitudinal trajectory. It was recently completed and we are making final check.

Trigger counters at the outer most layer are designed sensitive only to π^-. We will use the solenoid magnet to give the momentum and charge of particles. Currently a solenoid magnet with 3 k gauss field is available. A few MeV/c momentum resolution will be obtained for 100 MeV/c particles. Pions have large transverse momentum ($\sim$ 100 MeV/c). Invariant mass of Λ will be reconstructed.

The whole system has an efficiency of $\sim$5% for Λ produced by the quasifree process on a neutron in the ^{9}Be. The efficiency is dominantly given by Λ decay length and detection efficiency of protons from Λ decay. The expected yield is $\sim$4 events/day assuming realistic beam intensity. The 30 days running times gives the $\sim$100 events. Here we should stress that we are studying a new quantity and order-of-magnitude constraint on the cross section gives us great progress. Detailed design study is under progress.

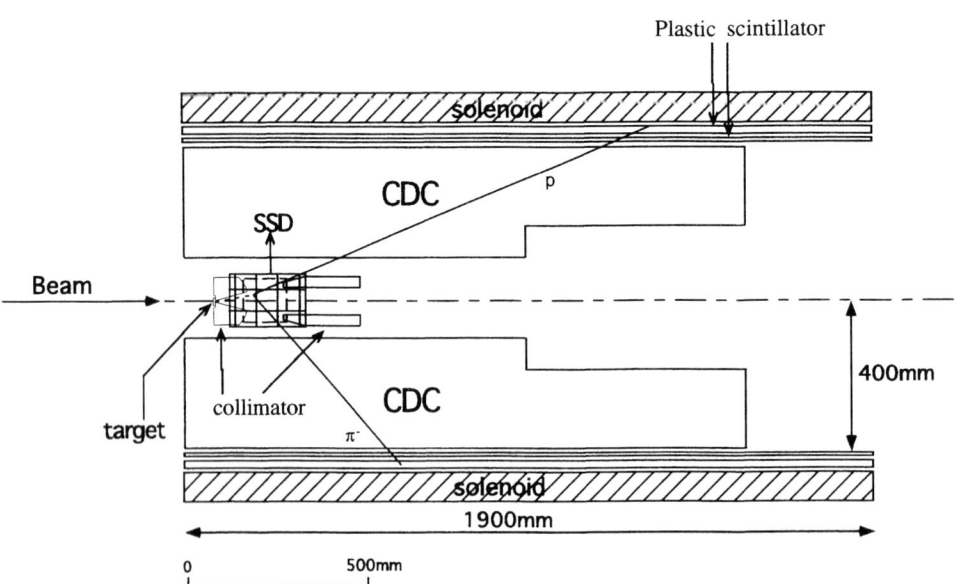

FIGURE 4. Schematic layout of the proposed decay counter system for the detection of $pn \to p\Lambda$ reaction.

REFERENCES

1. R. Chrien and C. Dover, Ann. Rev. Nucl. Science 39 (1989), 113 and references therein.
2. P. H. Pile et al., Phys. Rev. Lett. 66, 2585 (1991).
3. T. Hasegawa et al., Phys. Rev. Lett. 74, 224 (1995); T. Hasegawa et al., Phys.Rev. C 53, 1210 (1996).
4. W. Brückner et al., Phys. Lett. 79B, 157 (1978).
5. M. May et al., Phys. Rev. Lett. 47, 1106 (1981).
6. M. May et al., Phys. Rev. Lett. 51, 2085 (1983).
7. E. H. Auerbach et al., Phys. Rev. Lett. 47 (1981) 1110
8. T. Kishimoto et al., AGS research proposal E929 (1996).
9. R.W. Stotzer, et al., Phys. Rev. Lett.78, 3646 (1997).
10. H. Kohri, PhD thesis, Osaka univ., 2000, unpublished.
11. S. Ajimura et al., Nucl. instr. Meth., to be published.
12. T. Motoba, private communication (1998).
13. T. A. Rijken, V. G. Stoks, Y. Yamamoto, Phys. Rev. C59, 21 (1999).
14. Y. Fujiwara, C. Nakamoto and Y. Suzuki, Phys. Rev. Lett.76 (1996) 2242.
15. E. Hiyama, M. Kamimura, T. Motoba, T. Yamada, and Y. Yamamoto, Phys. Rev. Lett. 85, 270 (2000).
16. D. J. Millener, private communication (2000).
17. H. Tamura et al., Phys. Rev. Lett. 84, 5963 (2000).
18. H. Tamura and T. Akikawa private communication (2000).
19. H. Yamazaki et al., Phys. Rev. Lett. 54, 102 (1985).
20. S. Ajimura et al., Phys. Lett. B282 (1992) 293.
21. S. Ajimura et al., Phys. Rev. Lett. 80, 3471 (1998).
22. S. Ajimura et al.,Phys. Rev. Lett. 84, 4052 (2000).
23. S. Ajimura et al., Nucl. Phys. A663, 493 (2000).
24. A. Parreño, A. Ramos, and C. Bennhold, Phys. Rev. C **56**, 339 (1997).
25. A. Ramos, E. van Meijgaard, C. Bennhold, and B. K. Jannings, Nucl. Phys. A544, 703 (1992).
26. C.-Y. Cheung, D.P. Heddle, and L.S. Kisslinger, Phys. Rev. C **27**, 335 (1983); D.P. Heddle and L.S. Kisslinger, *ibid.* **33**, 608 (1986).
27. T. Inoue, S. Takeuchi, and M. Oka, Nucl. Phys. A **597**, 563 (1996); T. Inoue *et al.*, *ibid.* **633**, 312 (1998).
28. T. Kishimoto, Proc. on Weak and Electromagnetic Interactions in Nuclei (WEIN'95), 1995, Osaka, 514.
29. J. Haidenbauer et al., Phys. Rev. C52 3496 (1995).
30. M. Oka, Nucl. Phys. **A639**, 317c (1998).
31. A. Parreño, A. Ramos, N.G. Kelkar, C. Bennhold Phys. Rev. Cbf 59, 2122 (1999).
32. H. Nabetani, T. Ogaito, T. Sato, T. Kishimoto, Phys. Rev. C60, 017001, (1999).
33. H. E. Conzett, Proc. IV Int Symp. on Weak and Electromagnetic Inteactions in Nuclei"WEIN'95, June 1995, Osaka, World Scientific, 68 (1995).
34. T. Kishimoto et al., RCNP proposal E-122.

Nucleon Spin Structure Functions

Werner Vogelsang

*RIKEN-BNL Research Center, Bldg. 510a, Brookhaven National Laboratory,
Upton, New York 11973 – 5000, U.S.A.*

Abstract. We briefly review and discuss some aspects of nucleon spin structure functions, focusing on experimental results and their implications.

INTRODUCTION

High-energy spin physics has been going through a period of great popularity and rapid developments ever since the measurement of the proton's spin structure function g_1^p by the EMC [1] more than a decade ago. As a result of combined experimental and theoretical efforts, we have gained some fairly precise information concerning, for example, the quark spin contribution to the nucleon's spin. Yet, many other interesting and important questions, most of which came up in the wake of the EMC measurement, remain unanswered so far, the most prominent 'unknown' being the nucleon's spin-dependent gluon density. Future dedicated spin experiments are expected to provide answers to these questions.

DEEP-INELASTIC SCATTERING WITH SPIN

So far, information on the spin structure of the nucleon mainly comes from inclusive deep-inelastic scattering (DIS) of polarized leptons off polarized nucleon targets, depicted in Fig. 1. The cross section for DIS is given as a product of a calculable leptonic tensor $\mathcal{L}_{\mu\nu}$ with the hadronic tensor $\mathcal{W}^{\mu\nu}$ that is parameterized in terms of structure functions:

$$\frac{d^2\sigma}{dxdy} \propto \mathcal{L}_{\mu\nu}(k,q,s)\, \mathcal{W}^{\mu\nu}(P,q,S)\,, \tag{1}$$

where $x = -q^2/2Pq$, $y = Pq/Pk$, and

$$\mathcal{W}^{\mu\nu}(P,q,S) = \frac{1}{4\pi}\int d^4z\, e^{iq\cdot z} \langle P,S| [\mathcal{J}_\mu(z),\mathcal{J}_\nu(0)] |P,S\rangle = -g^{\mu\nu} F_1(x,Q^2)$$

$$+\frac{P^\mu P^\nu}{P\cdot q} F_2(x,Q^2) - i\varepsilon^{\mu\nu\rho\sigma}\frac{q_\rho P_\sigma}{2\, P\cdot q} F_3(x,Q^2) + i\varepsilon^{\mu\nu\rho\sigma} q_\rho \left[\frac{S_\sigma}{P\cdot q}\, g_1(x,Q^2)\right.$$

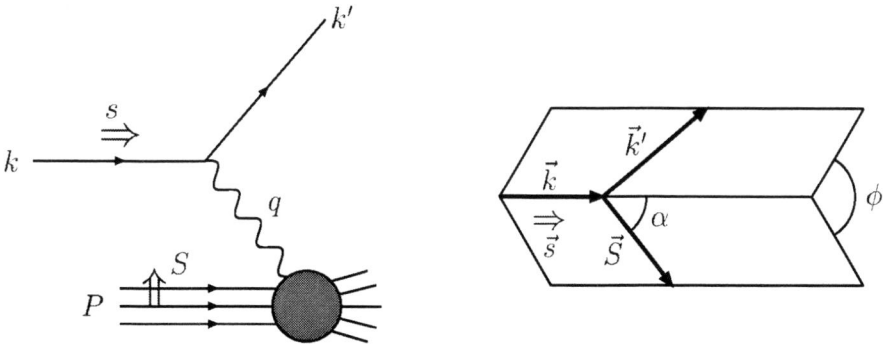

FIGURE 1. Left: Polarized DIS. The double arrows denote the spin vectors of the lepton and the nucleon. Right: Definition of angles in polarized DIS in the fixed-target laboratory frame.

$$+ \frac{S_\sigma(P \cdot q) - P_\sigma(S \cdot q)}{(P \cdot q)^2} g_2(x, Q^2) \Bigg] + \left[\frac{P^\mu S^\nu + S^\mu P^\nu}{2P \cdot q} - \frac{S \cdot q}{(P \cdot q)^2} P^\mu P^\nu \right] g_3(x, Q^2)$$
$$+ \frac{S \cdot q}{(P \cdot q)^2} P^\mu P^\nu g_4(x, Q^2) - \frac{S \cdot q}{P \cdot q} g^{\mu\nu} g_5(x, Q^2) \,. \quad (2)$$

The F_i are referred to as the 'unpolarized' structure functions, whereas the g_i are the 'polarized' ones, since their associated tensors depend on the nucleon spin vector S^μ. Note that parity-violating interactions are required for F_3, g_3, g_4, g_5 to be nonzero. Differences of cross sections with opposite settings of the nucleon spin vector give access to the g_i. For example, for longitudinally polarized leptons of helicity λ that scatter off nucleons polarized at an angle α (see Fig. 1), one has (see, e.g., [2]):

$$\frac{d^2\sigma^{(\alpha)}}{dxdyd\phi} - \frac{d^2\sigma^{(\alpha+\pi)}}{dxdyd\phi} \propto \cos\alpha \left[\lambda xy(2-y) g_1 + (1-y) g_4 + xy^2 g_5 \right]$$
$$- \frac{2Mxy}{Q} \sqrt{1-y} \sin\alpha \cos\phi \left[\lambda yx g_1 + 2\lambda x g_2 + \frac{2-y}{2y} g_3 - \frac{1-y}{y} g_4 - xy g_5 \right], \quad (3)$$

where non-leading effects associated with the nucleon mass M have been neglected. In the case of pure-photon interactions, scattering off a longitudinally polarized target ($\alpha = 0$) determines g_1, whereas for transverse polarization of the target ($\alpha = \pi/2$), one measures a linear combination of g_1 and g_2, the entire cross section being suppressed as M/Q in this case.

PARTON MODEL AND QCD

In the parton model, the DIS process is expressed as incoherent scattering of the lepton off individual quarks. As a result, one finds, for example,

$$g_1(x) = \frac{1}{2}\sum_q e_q^2 [\Delta q(x) + \Delta \bar{q}(x)] , \qquad (4)$$

in case of photon exchange. Here,

$$\Delta q(x) \equiv q_{\Rightarrow}^{\rightarrow}(x) - q_{\Rightarrow}^{\leftarrow}(x) \qquad (q = u, d, s, \ldots) , \qquad (5)$$

where $q_{\Rightarrow}^{\rightarrow}(x)$ ($q_{\Rightarrow}^{\leftarrow}(x)$) measures for a longitudinally polarized nucleon the number density of quarks with momentum fraction x and the same (opposite) helicity as the nucleon. As a further example, for charged-current interactions via W^- exchange one has:

$$g_1^{(W^-)}(x) = \Delta u(x) + \Delta \bar{d}(x) + \Delta s(x) , \quad g_5^{(W^-)}(x) = \Delta u(x) - \Delta \bar{d}(x) - \Delta s(x) . \qquad (6)$$

It turns out that the partonic picture fails for the structure functions g_2 and g_3 [2,3].

QCD predicts Q^2 dependence ('scaling violations') of the structure functions. For structure functions with partonic interpretation, the leading effect is simply that their parton densities start to depend on Q^2, in a way governed by evolution equations that resum large logarithms resulting from collinear parton emission [4,5]:

$$\frac{d}{d\ln Q^2}\begin{pmatrix}\Delta q \\ \Delta g\end{pmatrix}(x,Q^2) = \frac{\alpha_s(Q^2)}{2\pi}\int_x^1 \frac{dz}{z}\begin{pmatrix}\Delta P_{qq}(z) & \Delta P_{qg}(z) \\ \Delta P_{gq}(z) & \Delta P_{gg}(z)\end{pmatrix}\begin{pmatrix}\Delta q \\ \Delta g\end{pmatrix}\left(\frac{x}{z},Q^2\right) . \qquad (7)$$

Here, the ΔP_{ij} describe the collinear splitting of longitudinally polarized partons at lowest order in QCD perturbation theory [4]. The evolution of Δq, $\Delta \bar{q}$ involves the spin-dependent gluon density Δg, which is defined in analogy with Eq. (5).

Corrections to this picture arise from yet higher-order (in practice: next-to-leading order (NLO)) QCD corrections, and from contributions suppressed by powers of $1/Q^2$. These change the structure of the expression for g_1 in Eq. (4), which becomes:

$$g_1(x,Q^2) = \frac{1}{2}\sum_q e_q^2 \left\{\left[1 + \frac{\alpha_s(\mu^2)}{\pi}C_q^{(1)}\left(x,\frac{Q^2}{\mu^2}\right)\right] \otimes \left[\Delta q + \Delta \bar{q}\right](x,\mu^2) \right.$$
$$\left. + \left[\frac{\alpha_s(\mu^2)}{\pi}C_g^{(1)}\left(x,\frac{Q^2}{\mu^2}\right)\right] \otimes \Delta g(x,\mu^2)\right\} + \mathcal{O}\left(\frac{1}{Q^2}\right) , \qquad (8)$$

where $\otimes$ denotes a convolution. Eq. (8) is a typical example of factorization in high-energy scattering: the explicit $\mathcal{O}(\alpha_s)$ corrections in (8) result from the partonic channels $\gamma^* q \to qg$ and $\gamma^* g \to q\bar{q}$. The cross sections for these are singular when an emitted parton becomes collinear to the initial one. These collinear singularities, associated with long-distance physics, are factorized into the polarized quark and antiquark densities, and the hard-scattering cross sections $C_q^{(1)}$ and $C_g^{(1)}$ are the finite remainders of this procedure. Since singularities can be subtracted in different ways, the $C_i^{(1)}$, and hence the parton densities, are not unique at NLO, but depend on the convention ('scheme') adopted in factorization. Factorization inevitably introduces dependence on a scale that separates long-distance and hard short-distance physics: the factorization scale μ. As the μ dependence of the $C_i^{(1)}$ is calculable, and since g_1 is a physical quantity and thus independent of μ, one can perturbatively determine the μ-dependence of the (otherwise non-perturbative) parton densities, which is exactly expressed by the evolution equations in Eq. (7). We note that consistency requires that corrections to evolution itself be taken into account at NLO, achieved by $\alpha_s/2\pi \, \Delta P_{ij} \longrightarrow \alpha_s/2\pi \, \Delta P_{ij} + (\alpha_s/2\pi)^2 \, \Delta P_{ij}^{(1)}$ in (7), with the NLO splitting functions $\Delta P_{ij}^{(1)}$ given in [6]. The scale μ is arbitrary; $\mu = Q$ is a common choice for DIS. Finally, the factorization formula in (8) is correct up to $1/Q^2$ 'higher twist' corrections arising from target-mass corrections and more dynamical origins [7].

INFORMATION FROM DIS DATA

Analyses of the DIS data for g_1 aim at extracting information on the nucleon's polarized parton densities, and at tests of perturbative QCD evolution. Generally, the NLO framework as sketched above is used [8–10]; attempts to take into account higher-twist contributions (on the basis of fits or estimates) are usually not made.

Fig. 2 shows the results of a recent NLO analysis [8] of the world data [1] on the DIS spin asymmetries $A_1^N \approx g_1^N/F_1^N$ ($N = p, n, d$). There is a very good overall agreement of the NLO fit with the data, which is a non-trivial finding as the data have been taken at various values of Q^2. Fig. 3 displays the results for the Q^2 dependence of g_1 in various x bins for the same NLO fit, again compared to data. The data indeed line up well with the predictions of QCD evolution of polarized parton densities, even though the precision, the number, and in particular the Q^2 leverage of the data points do not allow for tests nearly as precise as those delivered by unpolarized DIS measurements of F_2.

We emphasize that the curves in Figs. 2 and 3 were obtained from a fit to the data on the asymmetry A_1^N, rather than to those on g_1^N. In particular, the unpolarized F_1^N in the denominator of A_1^N was calculated in NLO QCD, whereas the published data on g_1^N have been derived from A_1^N by means of $F_1^N = F_2^N/[2x(1+R^N(x,Q^2))]$, with *measured* F_2^N and R^N. To the extent that NLO QCD correctly describes data for F_2^N and R^N, the two approaches are entirely equivalent; however, at low Q^2 and/or high x, higher-twist effects do seem to play a non-negligible role in the unpolarized case, making an NLO QCD description somewhat inadequate. In analyses of F_2 [11], one therefore usually applies cuts like $Q^2 \gtrsim 4$ GeV^2, $W^2 =$

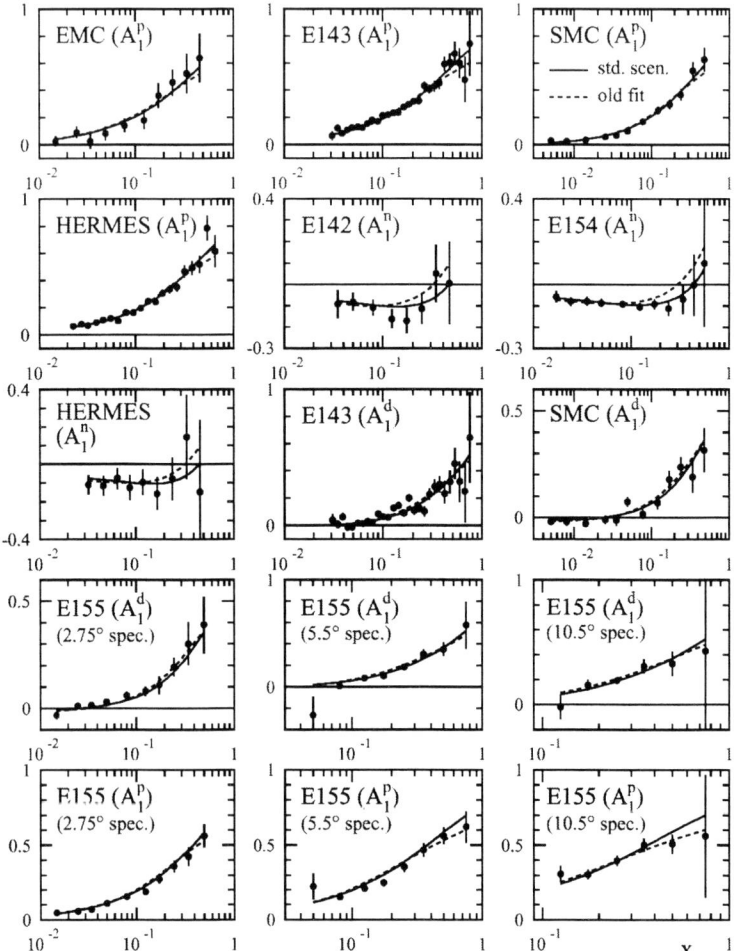

FIGURE 2. NLO QCD analysis [8] of world data [1] on A_1^N.

$Q^2(1-x)/x \gtrsim 10$ GeV^2, in order to exclude regions that are susceptible to higher-twist effects. Ideally, one would want to apply similar constraints in the polarized case; however, cuts of this form would dramatically reduce the number of data points. A *leading-twist* analysis based on measured A_1^N, F_2^N and R^N thus runs a risk of yielding a distorted picture due to higher-twist contributions in the data. An analysis of A_1^N, with F_1^N taken from theory, does not a priori warrant a better situation; however, if one admits higher-twist terms to the analysis, these come out small in the fitting procedure, hinting at a cancelation of higher-twist contributions in A_1^N [8,12]. Parton distributions obtained in a leading-twist analysis of A_1^N –

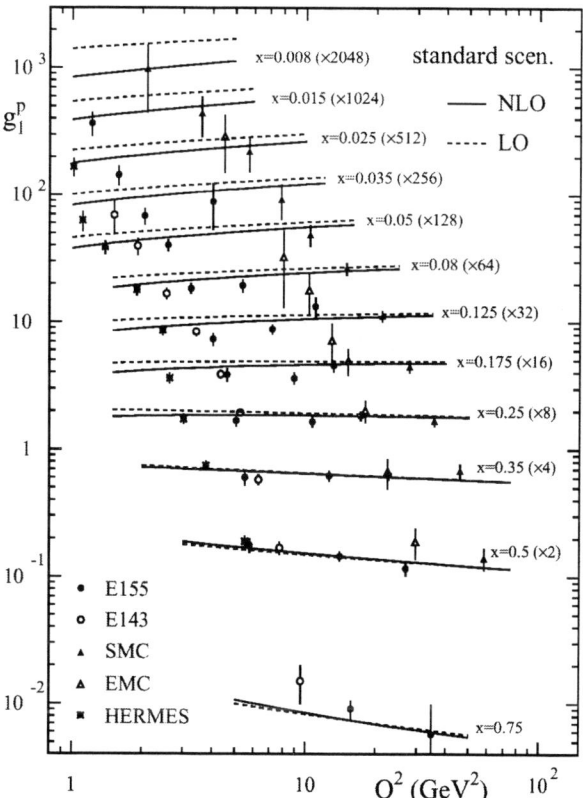

FIGURE 3. g_1^p as a function of Q^2 for fixed x, along with the results of the NLO QCD fit [8].

even without stringent kinematical cuts – therefore appear to be more reliable. We emphasize strongly that much further insight into this issue would be given by data at similar x but larger Q^2, such as from a polarized ep collider.

Beyond the fact that the NLO analysis works well, the key finding concerning the nucleon's spin structure emerging from the fits is that quarks carry only a small fraction of the proton spin. Experiments [1] have been able to determine the integral $\Gamma_1^N(Q^2) \equiv \int_0^1 g_1^N(x, Q^2) dx$. From Eq. (8), one has to NLO

$$\Gamma_1^p(Q^2) = \left[1 + \frac{\alpha_s(Q^2)}{\pi} C_q^{(1)}\right] \left\{ \frac{\Delta \mathcal{A}_3}{12} + \frac{\Delta \mathcal{A}_8}{36} + \frac{\Delta \Sigma(Q^2)}{9} \right\} + \frac{\alpha_s(Q^2)}{3\pi} C_g^{(1)} \Delta \mathcal{G}(Q^2) , \tag{9}$$

where the $C_i^{(1)}$ are the first moments (x-integrals) of the NLO coefficient functions, with $C_q^{(1)} = -1$, $C_g^{(1)} = 0$ in the $\overline{\text{MS}}$ scheme [6,13], and where

$$\Delta\Sigma(Q^2) = \int_0^1 dx \left[\Delta u + \Delta\bar{u} + \Delta d + \Delta\bar{d} + \Delta s + \Delta\bar{s}\right](x, Q^2),$$

$$\Delta\mathcal{G}(Q^2) = \int_0^1 dx\, \Delta g(x, Q^2). \tag{10}$$

Likewise, the $\mathcal{A}_i$ are the first moments of the flavor non-singlet combinations of quark densities. They are proportional to the nucleon matrix elements of the respective quark non-singlet axial currents, $\langle P, S | \bar{q}\gamma^\mu\gamma^5\lambda_i q | P, S\rangle$, which in turn can be related by SU(3) rotations to the β-decay parameters F, D of the baryon octet [14]. Here, the (all-order) Q^2 independence of the $\mathcal{A}_i$, which arises thanks to the fact that the $\bar{q}\gamma^\mu\gamma^5\lambda_i q$ are conserved currents, is crucial. A measurement of $\Gamma_1^p(Q^2)$ therefore gives access to $\Delta\Sigma(Q^2)$, and the analysis [1,8–10] results in a small value $\Delta\Sigma(Q_0^2 = 5\text{ GeV}^2) \approx 0.2$, often referred to as the 'spin crisis', since $\Delta\Sigma(Q^2)/2$ is by definition the contribution of quark spins to the proton spin.

A measurement of $\Gamma_1(Q^2)$ obviously relies on an estimate of the contribution to the integral from x outside the measured region. The extrapolation to small x certainly yields one main uncertainty in the value of $\Delta\Sigma(Q^2)$. It turns out that the size of the polarized gluon density very much drives the behavior of g_1 toward low x, through DGLAP evolution [8,15]. Roughly speaking, the larger Δg is at the initial scale, the more negative does g_1 become at small x, when Q^2 increases. We also note that for the present data sets, the points at the smallest x also have Q^2 values that are small by DIS standards and could possibly be still sensitive to higher-twist contributions. The uncertainties related to these points can be diminished by performing measurements of polarized DIS at smaller x, higher Q^2, and of Δg directly, over a sizable x range. Another source of uncertainty is associated with flavor-SU(3) breaking effects in the determinations of the F, D values, and in the rotations required to relate the latter to the $\mathcal{A}_i$ [16].

As a perturbative explanation for the smallness of $\Delta\Sigma(Q^2)$, it was proposed [17] that gluons could play an important direct role in this quantity. The starting point was that $\Delta\Sigma$ is related to the nucleon matrix element of the flavor-*singlet* axial quark current $j_5^\mu = \bar{q}\gamma^\mu\gamma^5 q$, which is non-conserved due to the axial anomaly, $\partial_\mu j_5^\mu = f\frac{\alpha_s}{2\pi}\epsilon^{\mu\nu\rho\sigma}\text{Tr}\left[G_{\mu\nu}\tilde{G}^{\mu\nu}\right]$, with the gluon field strength tensor $G_{\mu\nu}$ and its dual $\tilde{G}^{\mu\nu}$. As a result, $\Delta\Sigma$ which is Q^2-independent at lowest order, starts to depend on Q^2 at $\mathcal{O}(\alpha_s^2)$ [18]; the corresponding two-loop anomalous dimension was first calculated in [19]. The evolution of $\Delta\Sigma(Q^2)$ decreases its value, but is very mild and cannot explain why $\Delta\Sigma$ is small at experimentally relevant Q^2. However, it was recognized [17] that $j_5^\mu - fK^\mu$ is conserved, where $K^\mu \equiv \frac{\alpha_s}{4\pi}\epsilon^{\mu\nu\rho\sigma} A_\nu^a \left(G_{\rho\sigma}^a - \frac{g}{3}f_{abc}A_\rho^b A_\sigma^c\right)$ is the gluonic Chern-Simons current. This makes it appealing to identify the matrix element of $j_5^\mu - fK^\mu$ with a 'true' (then Q^2-independent) $\Delta\Sigma'$, and that of fK^μ with $\Delta\mathcal{G}(Q^2)$, implying in essence that the $\overline{\text{MS}}$-scheme $\Delta\Sigma(Q^2)$ in Eq. (9) ought to be regarded as $\Delta\Sigma' - f\alpha_s(Q^2)\Delta\mathcal{G}(Q^2)/2\pi$. In addition, it turns out that [4,17] $\alpha_s(Q^2)\Delta\mathcal{G}(Q^2)/2\pi$ does not decrease at large Q^2, as a proper $\mathcal{O}(\alpha_s)$ correction would, but tends to a constant and thus could indeed potentially be a significant contribution, allowing in turn for $\Delta\Sigma'$ to be large. The problem with all this, however, is that the forward matrix element of fK^μ is *not* gauge invariant [20]. This makes the precise meaning of the above split of $\Delta\Sigma(Q^2)$ unclear, except that its perturbative part corresponds to a factorization scheme transformation (expressed by $\mathcal{C}_g^{(1)} = -1/2$, instead of $\mathcal{C}_g^{(1)} = 0$, in Eq. (9)), that is, to a mere redefinition of NLO parton densities.

We emphasize at this point that beyond all these considerations, $\Delta g(x, Q^2)$ clearly is interesting by its own right. Its anticipated precise measurement over a wide x-range in polarized-proton collisions at RHIC [21], along with future information from polarized lepton-nucleon scattering [22], will mean an important advance in our understanding of the nucleon spin structure.

LESS INCLUSIVE MEASUREMENTS

It is clear from Eq. (4) that inclusive DIS (via photon exchange) can only give access to the combinations $\Delta q + \Delta\bar{q}$, but not to quark and antiquark densities separately. This 'lock' can be broken by DIS measurements in which a specific leading hadron $h = \pi^\pm, K^\pm, \ldots$ is identified in the final state [23], usually referred to as semi-inclusive DIS (SIDIS). Instead of Eq. (4), one then has in the parton model:

$$g_1^h(x,z) = \frac{1}{2} \sum_q e_q^2 \left[\Delta q(x) D_q^h(z) + \Delta\bar{q}(x) D_{\bar{q}}^h(z) \right], \qquad (11)$$

where D_q^h denotes a fragmentation function for the transition $q \to hX$. Measurements of this type have been performed by SMC [24] and HERMES [25]. Fig. 4 shows a comparison of the HERMES data for $A_1^{h^+}$ off a proton target with two theoretical predictions [8] that both fit the inclusive data, but differ in their assumptions concerning the decomposition and the flavor symmetry breaking of the polarized sea (see also [9] for a recent joint analysis of DIS and SIDIS data). The potential of SIDIS to disentangle between models is apparent. On the other hand, the fragmentation process involved in SIDIS introduces a significant uncertainty as the fragmentation functions are not known precisely in the kinematical regions covered in present SIDIS experiments. In addition, a *full* flavor separation of the polarized nucleon sea seems hard to achieve even with more precise future data, due to the fact that with photon exchange u quarks are strongly favored. It is also worth mentioning that in the unpolarized case, SIDIS has not really served as a tool to determine unpolarized parton densities. Measurements of $W^\pm$ and Drell-Yan dimuon production at RHIC [21,26] will be very powerful tools for determining $\Delta u, \Delta\bar{u}, \Delta d, \Delta\bar{d}$. Inclusive DIS via Z or W exchange, possible at a polarized ep collider, would test various other combinations of parton densities, see Eq. (6).

Recently, the spin asymmetry for *photo*production of pairs of oppositely charged hadrons was measured by HERMES [27]. Since the photon is almost real in this

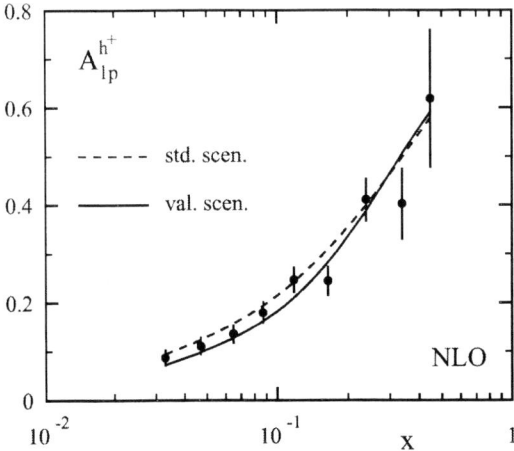

FIGURE 4. SIDIS spin asymmetry [25] for positively charged hadrons produced off proton target. The curves correspond to the two scenarios presented in [8].

case, $Q^2 \approx 0$, the large momentum scale required for a partonic and perturbative interpretation of the reaction needs to be set by the transverse momenta, p_T^h, of the hadrons. For large p_T^h, the spin asymmetry for $\gamma p \to h^+ h^- X$ could constrain the spin gluon density $\Delta g(x, Q^2)$ through the dominance of the partonic reaction $\gamma g \to q \bar{q}$. In practice, for the conditions of the HERMES experiment [27], one has $p_T^{h_1}, p_T^{h_2} \sim \mathcal{O}(1 \div 1.5)$ GeV, which makes it unlikely that perturbative QCD is in the position to describe a process relying on a threefold factorization. Indeed, within a LO framework, one finds [28] that the unpolarized cross section for $\gamma p \to h^+ h^- X$ shows a dramatic dependence on the renormalization/factorization scale and is not at all under control; see Fig. 5. Conclusions on Δg from the data [27] therefore appear impossible; the same statement applies to results on the spin asymmetry for the reaction $\gamma p \to hX$ presented in [29]. As visible in Fig. 5, the situation quickly improves if larger energies are available, such as at hadron colliders.

THE "OTHER" STRUCTURE FUNCTION: G_2

The structure function g_2 has also attracted a lot of attention because of its direct sensitivity to twist-three effects. The operator-product expansion for polarized DIS shows that matrix elements of twist-two and twist-three operators contribute to g_2 on equal footing [2,3], and that the twist-two part of g_2 is uniquely determined by that of g_1:

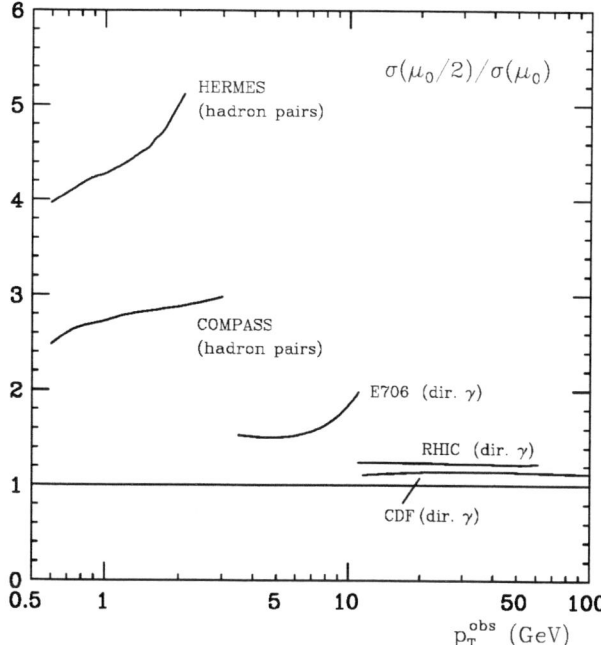

FIGURE 5. Scale dependence of perturbative QCD descriptions of unpolarized cross sections relevant for a determination of the gluon density. The prompt photon calculations are NLO, and $p_T^{obs} = p_T^\gamma$, $\mu_0 = p_T^\gamma$. In the case of the hadron pairs [28], only leading order is available. Here p_T^{obs} refers to the transverse momentum, $p_T^{h^+}$, of the positively charged hadron, and we have integrated over $p_T^{h^-} > 1.5$ GeV as in the HERMES experiment [27]. The default scale was here chosen to be $\mu_0 = p_T^{h^+} + p_T^{h^-}$ [28].

$$g_2^{tw-2}(x, Q^2) = -g_1^{tw-2}(x, Q^2) + \int_x^1 \frac{dy}{y} g_1^{tw-2}(y, Q^2) . \qquad (12)$$

This relation is due to Wandzura and Wilczek [30]. It shows that, indeed, a precise measurement of g_2 offers the unique possibility to access twist-three effects.

Bag and chiral soliton models have been used [3,31] to predict g_2, including its twist-three part, g_2^{tw-3}. In order to evolve the predictions from the low model scales to experimentally relevant Q^2, one needs to know the evolution of g_2^{tw-3}. This question is also relevant for evolving data to a common Q^2. It turns out, unfortunately, that the evolution of g_2^{tw-3} is very complicated and not of a typical DGLAP type. For the nth moment of g_2^{tw-3}, $n-2$ operators contribute and mix under evolution [3,32,33]. This means that, for example, $\int_0^1 dx x^4 g_2^{tw-3}(x, Q^2)$ evolves via a 3×3 matrix and requires three boundary conditions at the input scale. The reason for this behavior is that g_2^{tw-3} is only one projection of a more general higher-twist function $G(x, y, Q^2)$, which itself does have a DGLAP-type evolution, but does not pass on this property to $g_2^{\text{tw}-3}$. $G(x,y,Q^2)$ is related to quark-gluon correlations; radiative corrections have been calculated based on this picture [33–35].

The evolution of $g_2^{\text{tw}-3}$ simplifies dramatically in the limit $N_C \to \infty$, where it takes a DGLAP form. This was worked out in [36] for the flavor non-singlet case; very recently first similar steps for the singlet sector were taken [35]. The simplified non-singlet evolution for $N_C \to \infty$ has been used in the model calculations [31]. Figs. 6 and 7 show the latest, most precise data on $g_2(x,Q^2)$ [37], along with the Wandzura-Wilczek result and with the model predictions. The data are close to a level of precision at which an extraction of $g_2^{\text{tw}-3}$ becomes possible. We emphasize that for the Wandzura-Wilczek curves shown in Figs. 6 and 7, the experimental results for g_1 have been used, while in Eq. (12) only the *twist-two* part of g_1 appears. Future attempts of analyzing twist-three effects in g_2 will need to address this issue.

TRANSVERSITY

The transversity distributions [38–41] measure differences of probabilities for finding quarks with transverse spin aligned and anti-aligned with the transverse

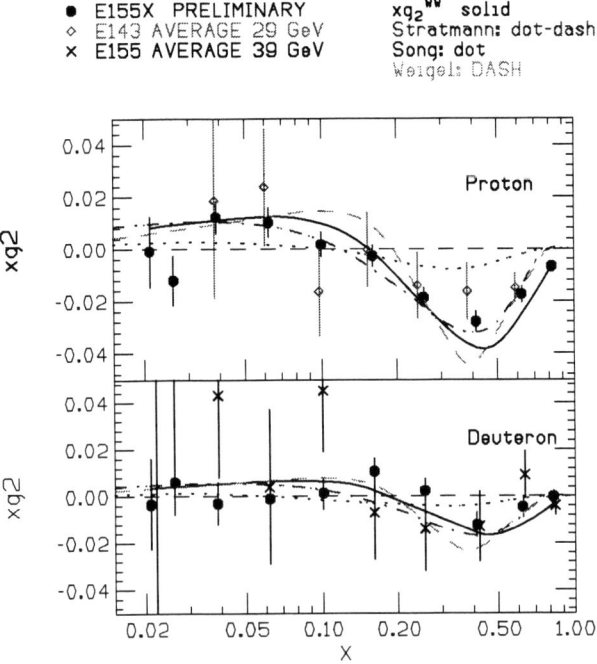

FIGURE 6. x-dependence (left) of g_2 for proton and deuteron targets [37], compared to the Wandzura-Wilczek expectation [30] and to model calculations [31].

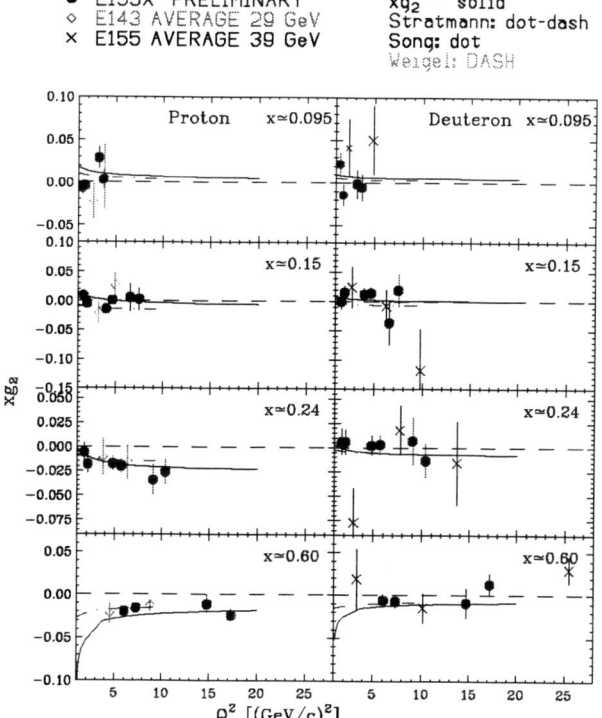

FIGURE 7. Q^2-dependence of g_2 for proton and deuteron targets [37], compared to the Wandzura-Wilczek expectation [30] and to model calculations [31].

nucleon spin. They are twist-two and hence equally interesting and fundamental as the Δq, Δg; they just have evaded measurement so far since they decouple from inclusive DIS [39,41]. This property can be seen [42] if we work in a helicity basis and view the quark densities as u-channel discontinuities of polarized quark-hadron forward scattering, denoted by $\mathcal{A}(H, h; H', h')$, where H, h (H', h') refer to the helicities of the incoming (outgoing) hadron and quark, respectively. One then has $q = \mathcal{A}(++;++) + \mathcal{A}(+-;+-)$, $\Delta q = \mathcal{A}(++;++) - \mathcal{A}(+-;+-)$, but $\delta q = \mathcal{A}(++;--)$. Thus, for transversity to contribute, the quark has to undergo a helicity flip in the hard scattering, which is not allowed (for massless quarks) at the DIS quark-photon vertex due to helicity conservation.

Another important consequence is that, unlike the situation for unpolarized and longitudinally polarized densities, there is *no* transversity gluon distribution [39–41]. This is due to angular momentum conservation: a gluonic helicity-flip amplitude would require the hadron to absorb two units of helicity, which a spin-1/2 target cannot do. Because of the absence of a gluon distribution, the transversity densities all obey rather simple 'non-singlet-type' evolution equations, e.g., at lowest order,

$$\frac{d\delta q(x,Q^2)}{d\ln Q^2} = \frac{\alpha_s(Q^2)}{2\pi} \int_x^1 \frac{dz}{z} \delta P_{qq}(z)\, \delta q\left(\frac{x}{z}, Q^2\right). \tag{13}$$

The transversity splitting functions δP_{qq} [41,43,44] are such that the transversity densities evolve to zero at all x, if $Q^2 \to \infty$. More precisely, again at LO, one finds

$$x^{-0.34}\delta q(x,Q^2) \stackrel{Q^2\to\infty}{\longrightarrow} \delta(x) \int_0^1 dx\, x^{-0.34}\delta q(x,Q_0^2) = \text{const} \times \delta(x). \tag{14}$$

From the joint description of the quark distributions in terms of the $\mathcal{A}(H,h;H',h')$, one finds Soffer's inequality [45],

$$q(x) + \Delta q(x) \geq 2|\delta q(x)|, \tag{15}$$

which holds for all quark flavors, and separately for their antiquarks. As was demonstrated in [43,46–48], the inequality is preserved under QCD evolution.

The helicity flip required for transversity to contribute to hard scattering can occur if there are two soft hadronic vertices in the process. In this case, transverse spin can be carried from one hadron to the other, along a quark line. One possibility is to have two transversely polarized hadrons in the initial state, as will be realized at RHIC [21]. Here, Drell-Yan dimuon production [38,48] and possibly jet production will give access to the δq, even though it will be very challenging experimentally.

Alternatively, one can have one transversely polarized initial hadron and take advantage of a final-state fragmentation process that is sensitive to transverse polarization. Here, the other initial particle could be a lepton, as in DIS, or another proton, as again at RHIC. It was first shown in [49] that in SIDIS the azimuthal distribution of the outgoing hadron about the jet axis can be used as a measure of the transverse polarization of the quark initiating the jet. The same is true [49–51] for the azimuthal distribution of pion pairs about the final-state jet axis. Time-reversal invariance, however, precludes non-zero effects, unless phases are generated by final-state interactions in the fragmentation process that do not average to zero upon summation over unobserved hadrons. It is a priori not clear whether such a net phase will exist, even though there are now early and exciting experimental results [52] hinting at a non-zero Collins effect and at its viability for measurements of the δq. In order to ensure that the final-state interaction phase does not average to zero, Ref. [53] proposed to focus on the interference between S and P waves of two-pion systems with invariant mass around the ρ. This S-P wave interference in the $q \to \pi\pi$ formation is described by the conceptually new *interference* fragmentation functions [53]. The price to be paid for obtaining sensitivity to transversity in all of the ways suggested in [49,51,53] is thus the introduction of another unknown component; however, one may hope that the involved fragmentation functions can be determined independently in e^+e^- annihilation. Studies of the experimental situation at RHIC concerning the proposal of [53] look very encouraging [54].

ACKNOWLEDGMENTS

I am grateful to the organizers of SPIN2000 for their invitation and for a good conference. I thank D. de Florian, M. Glück, E. Reya and M. Stratmann for collaboration on some of the topics presented here. I am also grateful to E. Aschenauer, A. Belitsky, D. Boer, A. Brüll, G. Bunce, A. Deshpande, M. Grosse-Perdekamp, R.L. Jaffe, P. Ratcliffe, N. Saito and D. Stamenov for useful discussions, and to R. Prepost for the plots of the SLAC g_2 data. I thank RIKEN, Brookhaven National Laboratory and the U.S. Department of Energy (contract number DE-AC02-98CH10886) for providing the facilities essential for the completion of this work.

REFERENCES

1. For a recent compilation of the data on polarized DIS, see: Hughes E., Voss R., *Ann. Rev. Nucl. Part. Sci.* **49**, 303 (1999); more recently appeared: Anthony P., et al., E155 Collab., *Phys. Lett.* **B463**, 339 (1999); hep-ph/0007248.
2. Anselmino M., Efremov A., Leader E., *Phys. Rep.* **261**, 1 (1995); Blümlein J., Kochelev N., *Nucl. Phys.* **B498**, 285 (1997).
3. Jaffe R., Ji X., *Phys. Rev.* **D43**, 724 (1991).
4. Altarelli G., Parisi G., *Nucl. Phys.* **B126**, 298 (1977).
5. Dokshitser Yu., *Sov. Phys. JETP* **46**, 641 (1977); Lipatov L., *Sov. J. Nucl. Phys.* **20**, 95 (1975); Gribov V., Lipatov L., *Sov. J. Nucl. Phys.* **15**, 438 (1972).
6. Mertig R., van Neerven W., *Z. Phys.* **C70**, 637 (1996); Vogelsang W., *Phys. Rev.* **D54**, 2023 (1996); *Nucl. Phys.* **B475**, 47 (1996).
7. Matsuda A., Uematsu T., *Nucl. Phys.* **B168**, 181 (1980); Piccione A., Ridolfi G., *Nucl. Phys.* **B513**, 301 (1998); Blümlein J., Tkabladze A., *Nucl.Phys.* **B553**, 427 (1999); Gornicki P., et al., *Phys.Lett.* **B343**, 369 (1995); Mankiewicz L., et al., *Phys.Lett.* **B383**, 463 (1996); Erratum *ibid.* **B393**, 487 (1997).
8. Glück M., Reya E., Stratmann M., Vogelsang W., hep-ph/0011215.
9. de Florian D., Sassot R., *Phys. Rev.* **D62**, 094025 (2000).
10. Other recent NLO analyses include: Altarelli G., Ball R., Forte S., Ridolfi G., *Acta Phys. Polon.* **B29**, 1145 (1998); Leader E., Sidorov A., Stamenov D., *Phys. Lett.* **B488**, 283 (2000); Bartelski J., Tatur S., hep-ph/0004251; Ghosh D., Gupta S., Indumathi D., *Phys. Rev.* **D62**, 094012 (2000); Goto Y., et al., *Phys. Rev.* **D62**, 034017 (2000).
11. Martin A., Roberts R., Stirling J., Thorne R., *Eur. Phys. J.* **C4**, 463 (1998); Lai H.L., et al., CTEQ Collab., *Eur. Phys. J.* **C12**, 375 (2000); Glück M., Reya E., Vogt A., *Eur. Phys. J.* **C5**, 461 (1998).
12. Leader E., Sidorov A., Stamenov D., in: Proceedings of the 9th Lomonosov Conference on Elementary Particle Physics, Moscow, September 1999, ed. A. Studenikin.
13. Kodaira J., Matsuda S., Sasaki K., Uematsu T., *Nucl. Phys.* **B159**, 99 (1979); Ratcliffe P., *Nucl. Phys.* **B223**, 45 (1983); Bodwin G., Qiu J., *Phys. Rev.* **D41**, 2755 (1990); Cheng H.-Y., *Chin. J. Phys.* **35**, 25 (1997).

14. Bjorken J., *Phys. Rev.* **148**, 1467 (1966); Ellis J., Jaffe R., *Phys. Rev.* **D9**, 1444 (1974).
15. Altarelli G., Ball R., Forte S., Ridolfi G., *Nucl. Phys.* **B496**, 337 (1997).
16. Lipkin H., *Phys. Lett.* **B256**, 284 (1991); **B337**, 157 (1994); Karliner M., Lipkin H., *Phys. Lett.* **B461**, 280 (1999); Ratcliffe P., *Phys. Lett.* **B365**, 383 (1996).
17. Altarelli G., Ross G., *Phys. Lett.* **B212**, 391 (1988); Altarelli G., Stirling J., *Part. World* **1**, 40 (1989); Carlitz R., Collins J., Mueller A., *Phys. Lett.* **B214**, 229 (1988).
18. Jaffe R., *Phys. Lett.* **B193**, 101 (1987).
19. Kodaira J., *Nucl. Phys.* **B165**, 129 (1980).
20. Manohar A., Jaffe R., *Nucl. Phys.* **B337**, 509 (1990); see also: Bass S., *Eur. Phys. J.* **A5**, 17 (1999).
21. Saito N., PHENIX Collab., these proceedings. For an overview of the RHIC spin program, see also: Bunce G., Saito N., Soffer J., Vogelsang W., hep-ph/0007218.
22. Bradamante F., COMPASS Collab., these proceedings; Aschenauer E., HERMES Collab., these proceedings; Bosted P., et al., E161 Collab., SLAC Proposal E161; Deshpande A., et al., proceedings of '2nd eRHIC workshop', Yale, April 2000, BNL-52592; Anselmino M., et al., TESLA-N study group, hep-ph/0011299.
23. Close F., Milner R., *Phys. Rev.* **D44**, 3691 (1991).
24. Adeva B., et al., SMC, *Phys. Lett.* **B420**, 180 (1998).
25. Ackerstaff K., et al., HERMES Collab., *Phys. Lett.* **B464**, 123 (1999).
26. Bourrely C., Soffer J., *Phys. Lett.* **B314**, 132 (1993); *Nucl. Phys.* **B423**, 329 (1994).
27. Airapetian A., et al., HERMES Collab., *Phys. Rev. Lett.* **84**, 2584 (2000).
28. de Florian D., Stratmann M., Vogelsang W., work in preparation.
29. Anthony P., et al., E155 Collab., *Phys. Lett.* **B458**, 536 (1999).
30. Wandzura W., Wilczek F., *Phys. Lett.* **B172**, 195 (1977).
31. Stratmann M., *Z. Phys.* **C60**, 763 (1993); Song X., McCarthy J., *Phys. Rev.* **D49**, 3169 (1994), Erratum *ibid.* **D50**, 4718 (1994); Weigel H., Gamberg L., Reinhardt H., *Phys. Rev.* **D55**, 6910.
32. Shuryak E., Vainshtein A., *Nucl. Phys.* **201**, 141 (1982).
33. Ji X., Chou C., Phys. Rev. **D42**, 3637 (1990).
34. Ji X., Lu W., Osborne J., Song X., *Phys. Rev.* **D62**, 094016 (2000).
35. Belitsky A., Ji X., Lu W., Osborne J., hep-ph/0007305; Braun V., Korchemsky G., Manashov A., hep-ph/0010128.
36. Ali A., Braun V., Hiller G., *Phys. Lett.* **B266**, 117 (1991).
37. Abe K., et al., E143 Collab., *Phys. Rev.* **D58**, 112003 (1998); Anthony P., et al., E155 Collab., *Phys. Lett.* **B458**, 529 (1999); Bosted P., et al., E155x Collab., *Nucl. Phys.* **A663**, 297 (2000); Prepost R., these proceedings.
38. Ralston J., Soper D., *Nucl. Phys.* **B152**, 109 (1979).
39. Jaffe R., Ji X., *Phys. Rev. Lett.* **67**, 552 (1991); *Nucl. Phys.* **B375**, 527 (1992).
40. Ji X., *Phys. Lett.* **B289**, 137 (1992).
41. Artru X., Mekhfi M., *Z. Phys.* **C45**, 669 (1990).
42. Goldstein G., Jaffe R., Ji X., *Phys. Rev.* **D52**, 5006 (1995).
43. Vogelsang W., *Phys. Rev.*, **D57**, 1886 (1998).
44. Hayashigaki A., Kanazawa Y., Koike Y., Phys. Rev. **D56**, 7350 (1997); Kumano S., Miyama M., Phys. Rev. **D56**, 2504 (1997).

45. Soffer J., *Phys. Rev. Lett.* **74**, 1292 (1995); Sivers D., *Phys. Rev.* **D51**, 4880 (1995).
46. Barone V., *Phys. Lett.* **B409**, 499 (1997).
47. Bourrely C., Soffer J., Teryaev O., *Phys. Lett.* **B420**, 375 (1998).
48. Martin O., Schäfer A., Stratmann M., Vogelsang W., *Phys. Rev.* **D57**, 3084 (1998); *ibid.* **D60**, 117502 (1999).
49. Collins J., *Nucl. Phys.* **B396**, 161 (1993).
50. Ji X., *Phys. Rev.* **D49**, 114 (1994).
51. Collins J., Heppelmann S., Ladinsky G., *Nucl. Phys.* **B420**, 565 (1994).
52. Airapetian A., et al., HERMES Collab., *Phys. Rev. Lett.* **84**, 4047 (2000); Bravar A., SMC, *Nucl. Phys. (Proc. Suppl.)* **B79**, 520 (1999).
53. Jaffe R., Jin X., Tang J., *Phys. Rev. Lett.* **80**, 1166 (1998); *Phys. Rev.* **D57**, 5920 (1998).
54. Grosse-Perdekamp M., PHENIX Collab., these proceedings.

Physics with the First Polarized-Proton Collider RHIC

Naohito Saito*,†

*RIKEN (The Institute of Physical and Chemical Research), Wako, Saitama 351-0198, Japan
and
†RIKEN BNL Research Center, Brookhaven National Laboratory, Upton, NY 11973, USA

Abstract. Polarized pp collisions at high energy will provide a unique opportunity to study the spin structure of the nucleon and the symmetries in the nature. The physics program with the polarized beams at RHIC is overviewed with the expected sensitivities and the successful commissioning of the first polarized proton beam at RHIC is reported.

INTRODUCTION

Spin is one of the most fundamental properties of an elementary particle such as an electric charge, a mass, and intrinsic symmetries. Therefore, it is important to understand the spin of hadrons in terms of more fundamental degree of freedom, the spin of quark and gluon and their orbital motion. In addition, the *axial vector* nature of spin has been useful in testing symmetries in the fundamental process such as parity and time-reversal invariance. In this context, importance of the physics with the polarized hadrons can be understood in two ways. One is elucidation of the spin structure of the hadrons, and the other is utilizing known spin structure to reveal violation/conservation of symmetries in reactions.

The spin structure of the nucleon has been studied for over two decades using deep-inelastic scattering of longitudinally polarized leptons off longitudinally polarized-nucleon target (polarized-DIS) [1–6]. A goal is to obtain a complete picture of the nucleon spin in terms of quark and gluon degree of freedom, which are summarized in Table 1. The first moments of the spin structure functions g_1^p and g_1^n whose parton model interpretation at leading order are

$$g_1^p(x, Q^2) = \frac{1}{2}\left[\frac{4}{9}\Delta\mathcal{U}(x,Q^2) + \frac{1}{9}\Delta\mathcal{D}(x,Q^2) + \frac{1}{9}\Delta\mathcal{S}(x,Q^2)\right], \quad (1)$$

where $\Delta\mathcal{U} \equiv \Delta u + \Delta \bar{u}$ etc., have been measured to a precision of $\sim 20\%$. The error is largely dominated by uncertainties in extrapolation to the unmeasured x-

TABLE 1. *Unpolarized* and *polarized* quark and gluon distributions. Gluon *transversity* does not exist for the nucleon which has spin $\frac{1}{2}$. Q^2 dependence is dropped for simplicity.

	quark	gluon
spin averaged distribution	$q(x) \equiv q^+(x) + q^-(x) = q^\uparrow(x) + q^\downarrow(x)$	$g(x) \equiv g^+(x) + g^-(x)$
helicity distribution	$\Delta q(x) \equiv q^+(x) - q^-(x)$	$\Delta g(x) \equiv g^+(x) - g^-(x)$
transversity distribution	$\delta q(x) \equiv q^\uparrow(x) - q^\downarrow(x)$	

region[1]. The fraction of the proton spin carried by quark spin, or the first moment of flavor-singlet quark distribution, $\Delta\Sigma(x) \equiv \Delta\mathcal{U}(x) + \Delta\mathcal{D}(x) + \Delta\mathcal{S}(x)$ has been determined in various global analyses and the typical value ranges 0.1-0.2[2]. It is the only measured piece of the proton-spin sum rule;

$$\frac{1}{2}^{\text{proton}} = \int_0^1 dx \left[\frac{1}{2}\Delta\Sigma(x) + \Delta g(x)\right] + L_z, \qquad (2)$$

where L_z represents contribution from orbital angular momenta. The experimental information on other pieces is indispensable to complete the picture of the spin structure of the nucleon. Especially direct measurements of gluon and anti-quark polarization are essential and missing, since polarized-DIS is primarily sensitive to only sum of the quark and anti-quark contributions.

Once these spin structure functions are measured to a reasonable precision, the polarized proton beams can be regarded as *polarized quark and gluon beams* with known luminosities and energies, which can be used to explore searches of new physics. Especially parity violating effects can be extracted by using longitudinally polarized quark beams, which can potentially reveal the substructure of a quark [7]. Furthermore, there is a proposal to utilize polarized gluon beam to determine the CP state of Higgs particle [8].[3]

These measurements will be done using the first polarized proton collider, Relativistic Heavy Ion Collider (RHIC) at Brookhaven National Laboratory (BNL), which is illustrated in Figure 1. The machine parameters are listed in Table 2. RHIC is the first collider for both polarized protons and heavy ions. There are five experiments, BRAHMS, STAR (Figure 2(a)), PHENIX(Figure 2(b)) PHOBOS, and PP2PP. STAR, PHENIX, and PP2PP plan spin physics measurements, and a potential of spin physics with BRAHMS and PHOBOS is also high.

To realize the polarized pp collisions at RHIC, helical dipoles for *Siberian Snake* and *Spin Rotators* and polarimeter are developed in addition to an optically-pumped polarized ion source [10]. The *Snake* will be used to maintain polarization

---
[1]) For the measured region, the precision for the proton (neutron) is 5% (18%) [4].
[2]) Here we refer to next-to-leading order fit in $\overline{MS}$ scheme
[3]) Unfortunately maximum energy of RHIC seems too low even with light MSSM Higgs and large $\tan\beta$ assumption [9]. However, the polarized parton distributions to be measured in RHIC SPIN will provide the basis of future collider with polarization options.

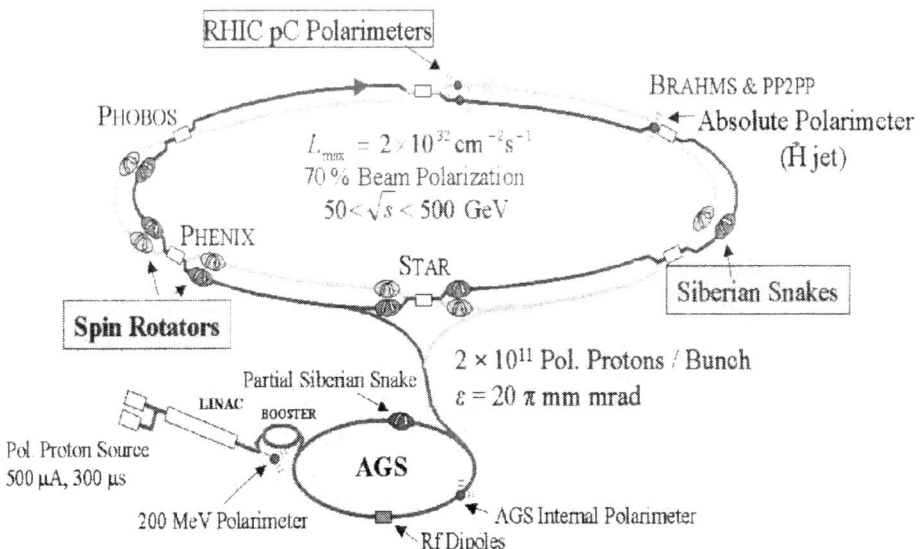

FIGURE 1. Layout of the first polarized collider, RHIC.

in accelerating and storing beams by flipping spin direction by 180° every half or full revolution to cancel the spin precession accumulated. The *Spin Rotators* will be used to obtain given spin direction at collision points and to return it to the original direction. The first *Siberian Snake*, which is a set of four helical dipoles, was installed in one of the RHIC rings with the newly developed polarimeter. Commissioning of the polarimeter and *Siberian Snake* was successful as will be described later.

In this contribution, we review the physics with the polarized proton beams at RHIC assuming the integrated luminosities of 800 pb^{-1} for $\sqrt{s} = 500$ GeV and 320 pb^{-1} for $\sqrt{s} = 200$ GeV, which corresponds to about four months with 40% efficiency with full luminosities. More comprehensive review can be found in Ref. [11]. In the next section, current knowledge on the spin structure of the nucleon is briefly summarized. RHIC spin measurements are explained in the following section. Commissioning of the RHIC polarized collider is described finally, followed by summary.

THE SPIN STRUCTURE OF THE NUCLEON

Efforts to elucidate the spin structure of the nucleon can be divided into three generations of experiments. The first generation of the experiments at SLAC showed an agreement with the quark parton models which suggested that the proton spin is fully carried by the quark spin [1]. The second generation of experiment

TABLE 2. RHIC accelerator specification.

Configuration	Two Concentric Super-conducting magnet Rings	
Circumference	3.8 km	
Interaction point	6	
Number of Bunch	60; 120 in enhanced mode	
Ion Species	ranges from p to Gold, $p+A$ is possible	
	$p + p$	Au+Au
Injection	Linac→Booster→ AGS	Tandem →Booster→ AGS
E_{cm}(Maximum)	500 GeV	200 GeV
Luminosity	$2\times 10^{32} \text{cm}^{-2}\text{sec}^{-1}$ [a]	$2\times 10^{26} \text{cm}^{-2}\text{sec}^{-1}$
Polarization	70%	–

[a] The luminosity for pp is for enhanced mode.

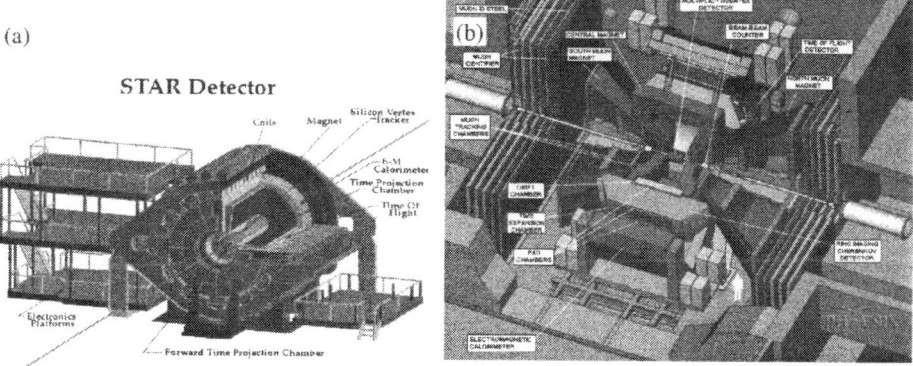

FIGURE 2. Two large scale experiments at RHIC, (a) STAR and (b)PHENIX.

has extended its kinematical coverage of the measurement into small-x region and revealed the discrepancy from the naive quark model, and more improved calculations such as Ellis-Jaffe sum rule [2]. The third generation of measurements improved the experimental precision greatly and confirmed the Bjorken sum rule to ∼10%. These measurements include both polarized proton and neutron targets to reveal isospin dependence, direct separation of two spin structure functions g_1 and g_2 [5], and the first measurement with the pure polarized-proton target [6]. In parallel to these efforts, there was an attempt to obtain gluon polarization in the proton utilizing polarized proton and anti-proton beams [12]. Most of the world data on the cross section asymmetry $\mathcal{A}_1$ ($\approx g_1/F_1$) for the proton are plotted in Figure 3(a). Results from one of global analyses [13] in leading order (LO) and next-to-leading order (NLO) are also shown in the plot.

Now we are turning into the forth generation of the experiments employing open geometry spectrometer in polarized-DIS (HERMES and COMPASS) or polarized proton beams (RHIC SPIN) to access gluon and anti-quark density directly.

It should be also noted that it is very important to measure the small-x region in both unpolarized and polarized cases, since any quantum number carried by the proton can be obtained only by integrating the measured distribution function in $0 \leq x \leq 1$. Figure 3 shows how $\Delta\Sigma$ depends on the extrapolation in small-x down to 0. $\Delta\Sigma(x_{min}) = \int_{x_{min}}^{1} \Delta\Sigma(x)dx$ is shown as a function of x_{min} for various values of $\alpha_{\bar{q}}$, which specifies the small-x behavior of anti-quark polarization. It is clear that the value of $\Delta\Sigma$ depends on $\alpha_{\bar{q}}$, which addresses the importance of the small-x measurements.

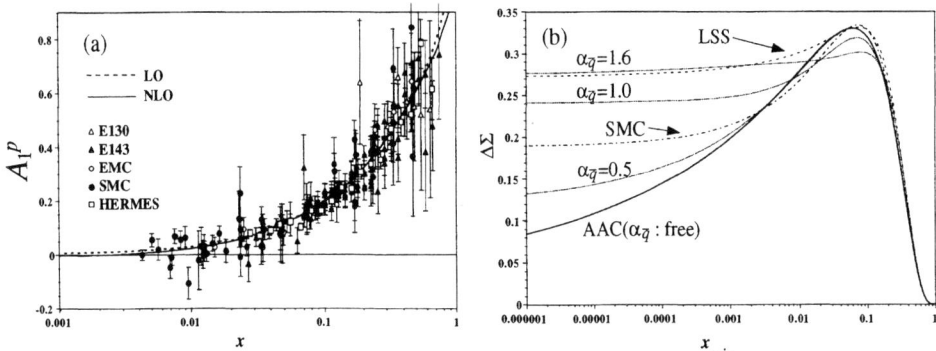

FIGURE 3. Selected plots from precision measurements of polarized DIS: (a) asymmetry $\mathcal{A}_1$ for the proton and (b) total fraction of quark spin contribution to the proton spin, as a function of lower limit of the integration [13].

In obtaining the function $\Delta\Sigma(x)$ from global analyses of $\mathcal{A}_1^p(x, Q^2)$ and $\mathcal{A}_1^n(x, Q^2)$, two more quark distributions e.g. $a_3(x) = \Delta\mathcal{U}(x) - \Delta\mathcal{D}(x)$ and $a_8(x) = \Delta\mathcal{U}(x) + \Delta\mathcal{D}(x) - 2\Delta\mathcal{S}(x)$ are extracted. Separation of the last two functions from $\Delta\Sigma(x)$ relys on their different coupling to $\Delta g(x)$. Therefore, the determination of $\Delta g(x)$ will benefit to obtain $\Delta\Sigma(x)$ more precisely. More details are fully discussed in Vogelsang's contribution [14].

MEASUREMENTS IN RHIC SPIN PROGRAM

To obtain complete picture of the spin structure of the nucleon, it is essential to combine spin-dependent hadron-induced reactions with different probes. Needless to say, photon, the primary probe in DIS is only sensitive to electric charge, therefore it is difficult either to separate flavors or to probe gluons. In Drell-Yan production of lepton pairs in hadron collisions, intermediate colorless neutral object, photon guarantees main contribution to be a combination of quark and anti-quark distributions. Prompt photon production is dominated by gluon Compton scattering, which measures a combination of gluon and quark distributions. Open heavy flavor production is dominated by gluon-fusion process, thus sensitive to the product of gluon distributions. The main goal of the spin structure studies of RHIC SPIN program is to complete the picture by providing qualitatively new information from *polarized hadron collisions*.

TABLE 3. Initial state spin asymmetries in pp collisions. $(+)$ and $(-)$ refers to the helicity states of the beams and $\uparrow$ and $\downarrow$ represent vertically *Up* and *Down* polarization.

Asymmetries	definition	remarks
A_{LL}	$\frac{\sigma(++)+\sigma(--)-\sigma(+-)-\sigma(-+)}{\sigma(++)+\sigma(--)+\sigma(+-)+\sigma(-+)}$	often used to extract Δf
A_{TT}	$\frac{\sigma(\uparrow\uparrow)+\sigma(\downarrow\downarrow)-\sigma(\uparrow\downarrow)-\sigma(\downarrow\uparrow)}{\sigma(\uparrow\uparrow)+\sigma(\downarrow\downarrow)+\sigma(\uparrow\downarrow)+\sigma(\downarrow\uparrow)}$	often used to extract δq
A_L	$\frac{\sigma(+)-\sigma(-)}{\sigma(+)+\sigma(-)}$	sensitive to parity violation
A_N	$\frac{\sigma(\uparrow)-\sigma(\downarrow)}{\sigma(\uparrow)+\sigma(\downarrow)}$	sensitive to higher twist effects
A_{LT}	$\frac{\sigma(+\uparrow)+\sigma(-\uparrow)-\sigma(+\downarrow)-\sigma(-\downarrow)}{\sigma(+\uparrow)+\sigma(-\uparrow)+\sigma(+\downarrow)+\sigma(-\downarrow)}$	sensitive to Δq, δq and higher-twist

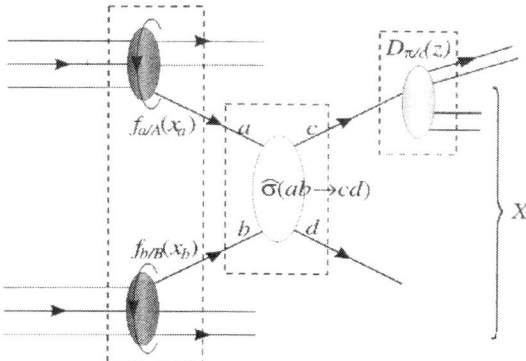

FIGURE 4. Hadron-hadron interaction factorized into distribution function, hard scattering cross section, and fragmentation function.

A hadron-hadron reaction, e.g. $pp \to \pi X$ can be factorized into three steps [31]:
i) emission of partons (with fractional longitudinal momentum of x) from the initial hadrons, which is characterized by parton distribution functions, $f_{a/A}(x,\mu)$,
ii) hard scattering of partons, whose cross section is described by $\frac{d\hat{\sigma}}{dt}(ab \to cd)$, and
iii) partons fragment into hadrons (with fractional longitudinal momentum z), whose distribution is represented by $D_{\pi/c}(z,\mu)$.

Therefore the cross section can be written as

$$E\frac{d^3\sigma}{dp^3} = \Sigma_{a,b,c} \int f_{a/A}(x_a,\mu) f_{b/B}(x_b,\mu) \frac{d\hat{\sigma}}{dt}(ab \to cX) D_{\pi/c}(z,\mu) d\Omega. \quad (3)$$

The factorization also holds for the polarized initial states [31]. The important point here is that structure functions and fragmentation functions are universal and once they are measured in one factorized process, they can be used to describe other processes. We discuss two examples to show how this works.

- **Gluon polarization from prompt photon production**
The leading process to produce prompt photon with large transverse momentum in pp collision is gluon-Compton. The longitudinal double-spin asymmetry, $\mathcal{A}_{LL}$ can be expressed at leading order as

$$\mathcal{A}_{LL} \approx \frac{\Delta g(x_1)}{g(x_1)} \cdot \frac{\Sigma_q e_q^2 [\Delta q(x_2) + \Delta \bar{q}(x_2)]}{\Sigma_q e_q^2 [q(x_2) + \bar{q}(x_2)]} \cdot \hat{a}_{LL}(qg \to \gamma q) + (1 \leftrightarrow 2). \quad (4)$$

In this expression, $\mathcal{A}_{LL}$ is the experimental observable, the second factor in the right hand side coincides with the asymmetry $\mathcal{A}_1^p$ measured in polarized-DIS to leading order, and the asymmetry in hard scattering, $\hat{a}_{LL}(qg \to \gamma q)$ is calculable in perturbative QCD. Therefore $\Delta g(x)/g(x)$ is the only unknown and it can be extracted from the measurement.

- **New physics searches with parity violation in Jet production**
Jet production in pp collision is dominated by quark-quark scattering in high E_T region. Thus single longitudinal-spin asymmetry $\mathcal{A}_L$ can be expressed as

$$\mathcal{A}_L \approx \sum_{i,j} \frac{\Delta q_i(x_1)}{q_i(x_1)} \cdot \frac{q_j(x_2)}{q_j(x_2)} \cdot \frac{\Delta \sigma}{\sigma}(q_i q_j \to q_i q_j) + (i \leftrightarrow j). \quad (5)$$

The quark flavor here is dominated by u and d due to their abundance in large x region. These distributions will be measured in W production precisely, as will be described later. The ratio of spin dependent cross section to spin averaged cross section will be the only unknown in the expression, thus it will be determined from the measurement to compare with the standard model expectation. Any deviation from the SM will immediately indicate the presence of new physics.

Asymmetries for various reactions in polarized pp collisions are listed with their primary goals of measurements in Table 4.

Gluon Helicity Distribution

Historically gluon distribution has been studied using the cross sections for prompt photon production in hadron collisions, both in fixed target and in collider experiments (Figure5(b)). In addition, jet production, which is dominated by gg-scattering in low-p_T ($x_T \equiv p_T/(\sqrt{s}/2) \leq 0.1$) and qg-scattering in mid-p_T ($0.1 \leq x_T \leq 0.3$) (Figure 5(c)), $t\bar{t}$ production in $p\bar{p}$ collisions (Figure 5(d)), and Q^2-evolution of the structure function $F_2(x,Q^2)$ has been used to constrain the

TABLE 4. Spin asymmetries for various pp reactions with major goals of measurements.[a]. $\mathcal{A}_{LT}$ is not listed here but the asymmetry for $pp \to \gamma^* X$ can be found in Ref. [15].

process	$\mathcal{A}_{LL}$	$\mathcal{A}_L$	$\mathcal{A}_{TT}$	$\mathcal{A}_N$
$pp \to \gamma$ (jet) X	$\Delta g \otimes \mathcal{A}_1^p$ [16]	–	~ 0 [17]	twist-3 [18]
$pp \to jet$ X	$\Delta g \otimes (\Delta g + \Delta \Sigma)$ [19]	W/Z/contact interaction [20]	~ 0 [17]	–
$pp \to Q\bar{Q}X$	$\Delta g \otimes \Delta g$ [21]	Z/MSSM Higgs [22]	–	–
$pp \to J/\psi X$	$\Delta g \otimes \Delta g$ [23]	–	–	–
$pp \to \chi_2 X$	$\Delta g \otimes \Delta g$ [24]	–	–	–
$pp \to W^+ X$	$\Delta u \otimes \Delta \bar{d}$ [25]	$\Delta u, \Delta \bar{d}$ [25]	~ 0 [26]	–
$pp \to W^- X$	$\Delta d \otimes \Delta \bar{u}$ [25]	$\Delta d, \Delta \bar{u}$ [25]	~ 0 [26]	–
$pp \to \gamma^* X$	$\Delta q \otimes \Delta \bar{q}$ [27,15]	γ^*/Z mixing [28]	$\delta q \otimes \delta \bar{q}$ [29]	twist-3 [30]

[a] References shown not necessarily represent initial work.

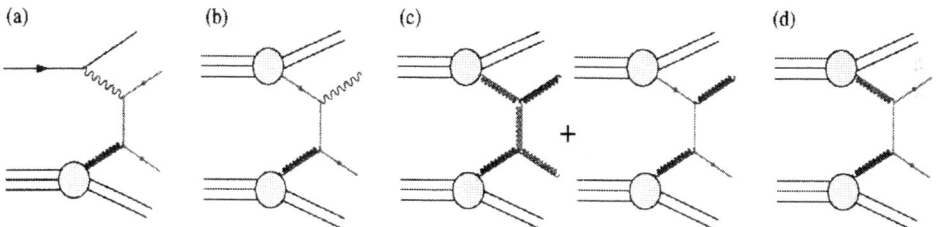

FIGURE 5. Hadron induced reactions relevant to gluon distribution: (a) photon-gluon fusion (b) gluon-Compton (c) gg and gq scattering, and (d) gluon fusion processes.

gluon distribution in smaller-x. These efforts have led to a gluon distribution with a reasonable precision, although more improvements are required.

As for the polarized case, the efforts have been limited to fixed target experiments, since there has been no polarized collider. Furthermore, one of the major source of information, prompt photon experiment with polarized beam and target was not possible due to the limitation of polarized beam intensity and both beam and target polarization.

In the Fermilab-E704/581 experiments, asymmetries ($\mathcal{A}_{LL}$) for high-p_T π^0 and high-mass multi-γ pair productions in polarized pp collisions have been measured [12]. Both processes were used as a jet surrogate, which is a mixture of gg and qg scattering. The data were compared with the model calculation of $\mathcal{A}_{LL}$ and suggested that models with smaller Δg are preferred.

In lepton scattering experiment, $\Delta g(x)$ has been determined through Q^2-evolution of spin dependent structure function $g_1(x, Q^2)$ but the uncertainty is still large enough to leave a room for any model [4].

Recently a proposal to use pairs of oppositely charged hadrons to probe photon-gluon fusion process (Figure 5(a)) in lepton scattering has been made [32]. HERMES experiment at DESY has measured the asymmetry [33], but the interpretation is still controversial due to a large scale dependence of perturbative QCD calculation [14].

At RHIC, its high polarization and luminosities allow us to measure $\mathcal{A}_{LL}$ for prompt photon precisely even in $p_T > 10$ GeV/c, where perturbative QCD calculation can be safely applied. As shown in Figure 6(a), statistical errors will be small enough to separate the models on the gluon polarization. When we detect the away-side jet, we may go one step further by reconstructing x_{gluon}, the momentum fraction carried by gluon, although there remain ambiguities due to intrinsic transverse momenta, assignment of x for either quark or gluon, and contribution from annihilation process, $\bar{q}q \to \gamma g$. Such attempts on simulation data seem to be successful as shown in Figure 6(b). Usually these studies utilize a QCD event generator, and the validity of such a generator has to be checked experimentally.

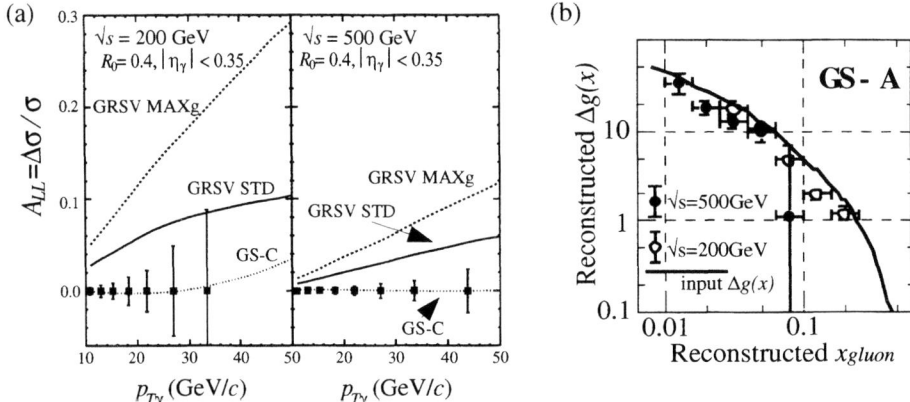

FIGURE 6. (a) Predicted asymmetries for prompt photon production in pp collisions at $\sqrt{s} = 200$ and 500 GeV. The error bars indicate the projected statistical precision. (b) Reconstructed $\Delta g(x)$ is compared with the input model [34].

One of the advantage of the RHIC SPIN measurements of $\Delta g(x)$ is that we can study it in many channels as shown in Table 4. Jet production, open and bound heavy quark production are sensitive to the gluon polarization as show in Table 4. Sensitivities of jet production to Δg for the STAR detector are shown in Figure 7(a) for $\sqrt{s} = 500$ GeV. In addition, it will become possible to cover a wide range of x and Q^2 by using all channels available. The covered x-Q^2 region from selected measurements at $\sqrt{s} = 200$ GeV is shown together with the x-Q^2 region covered in F_2 measurements at HERA and g_1 measurements at CERN, SLAC and DESY in Figure 7(b). By using open charm and bottom production, we can cover x-Q^2 region which overlaps well with the area covered by g_1 measurements. Since separation of flavor-singlet quark distribution, $\Delta\Sigma(x)$ from non-singlet contributions in g_1 is done based on the different coupling to $\Delta g(x)$, the measurement of $\Delta g(x)$ in the same x-Q^2 region as g_1 measurements should maximally benefit the precision determination of $\Delta\Sigma(x)$.

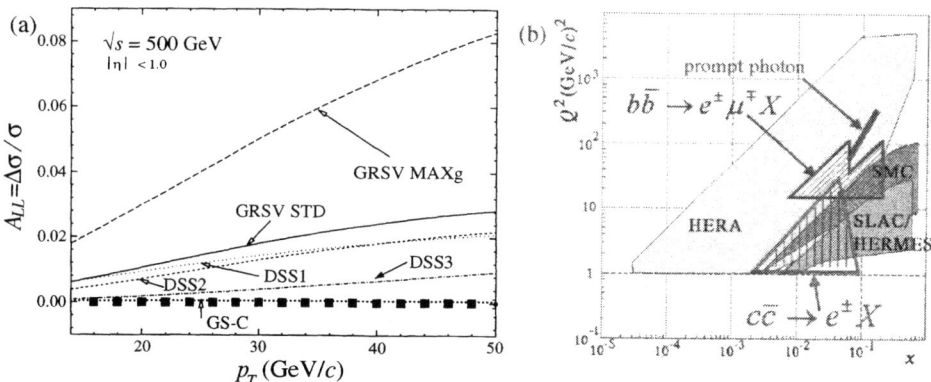

FIGURE 7. (a) Asymmetry versus jet transverse momentum [19] with the projected statistical precision for STAR (which is smaller than the size of box, everywhere), and (b) x-Q^2 region covered by selected gluon measurements by PHENIX at RHIC overlayed on the region covered by F_2 measurements at DESY-HERA, and g_1 measurements at CERN, SLAC, and DESY.

Flavor Decomposition of Quark and Anti-Quark Polarization, Transversity, and New Physics Search

The dominant source of information on polarized parton distributions, polarized DIS is only sensitive to the electric charge squared, and it is rather difficult to separate the contributions of quark and anti-quark. At RHIC, the parity violating asymmetry for W production, will directly measure the polarization of u, $\bar{d}$, d, and $\bar{u}$ quarks in the polarized proton:

$$A_L^{W^+} = \frac{\Delta u(x_1)\bar{d}(x_2) - \Delta \bar{d}(x_1)u(x_2)}{u(x_1)\bar{d}(x_2) + \bar{d}(x_1)u(x_2)}. \quad (6)$$

To obtain W^- asymmetry, u and d should be interchanged. These measurements will be done by STAR in the decay mode of $W \to e\nu$ and PHENIX in $W \to e\nu$ and $W \to \mu\nu$ at $\sqrt{s}$=500 GeV. Another possibility is to utilize semi-inclusive DIS (SIDIS) of polarized lepton off the polarized nucleon. The detection of the final state hadron will enhance the contribution of a specific flavor, thus useful in the flavor decomposition of the quark polarization.

Sensitivity of RHIC SPIN W measurements with PHENIX Muon Arms and HERMES SIDIS measurements are compared on the models of polarization of quarks [35,36] as functions of x. Error bars associated with closed circles represent projected statistical precision from W measurements at RHIC (PHENIX only). Projected errors from all HERMES data (as of November 2000) are represented by the error bars with squares. HERMES cannot separate the flavor dependence of $\bar{u}$ and $\bar{d}$ without additional model assumptions, currently $\Delta \bar{u}/\bar{u} = \Delta \bar{d}/\bar{d}$, while W production explicitly separates them.

In addition to these helicity distributions of quark and anti-quark, RHIC SPIN is also sensitive to the transversity distributions. One of the cleanest measurements will be provided by the Drell-Yan production of lepton pairs, whose sensitivity at $\sqrt{s}=200$ GeV is shown in Figure 8(b). Although the sensitivity seems marginal in this plot, the measurement will serve as the first and cleanest measurement and will become most powerful with the proposed upgrade in luminosity ($\times 25$) and energy (< 650 GeV) [37]. In addition, a new class of measurements utilizing spin transfer asymmetry and interference fragmentation function are being studied [38].

Basing on the measured polarized quark distributions, we can perform a search for new physics beyond the standard model. Sensitivities with jet production is illustrated in Figure 8(c). The measurements will provide significant constraints on the size and/or the chiral structure of possible new physics signals. Further studies on possible searches for new physics are underway [9,39].

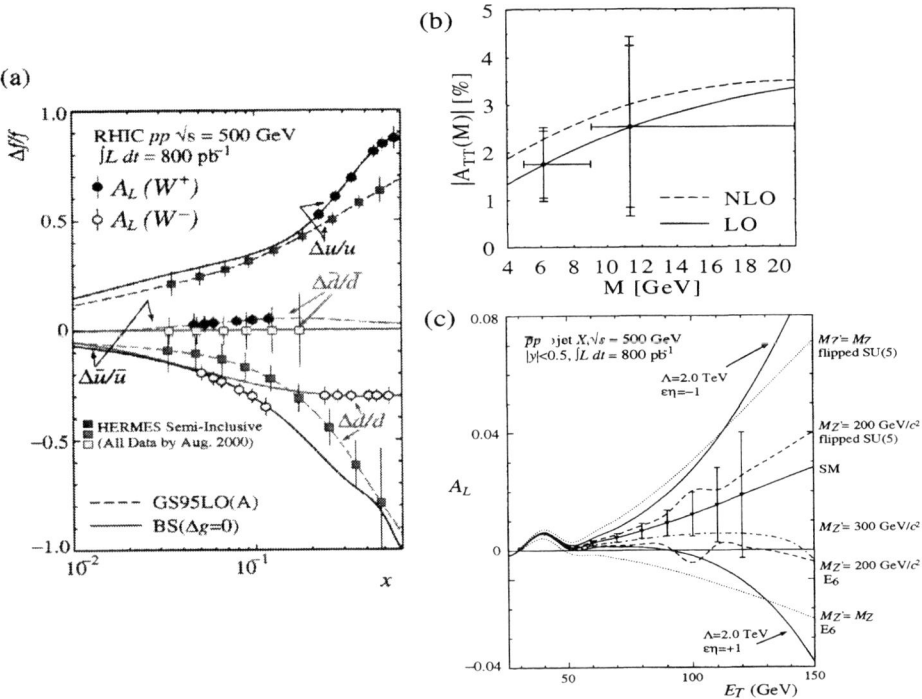

FIGURE 8. (a) Polarization of u, d, $\bar{u}$, $\bar{d}$ as functions of x modeled by Bourreley-Soffer, and Gehrmann-Stirling. Sensitivities of HERMES-SIDIS measurements and RHIC SPIN W measurements are shown. (b) double transverse-spin asymmetry for Drell-Yan dimuon production at $\sqrt{s} = 200$ GeV. (c) Parity violating asymmetry $\mathcal{A}_L$ for jet production compared with the SM, contact interaction, and leptophobic Z' [7].

RHIC SPIN COMMISSIONING

The first year of the RHIC operation was very successful. The first collision of the gold beams was recorded on June 13, 2000, at STAR experiment, followed by other three experiments. The beam energy was increased up to 65 GeV per nucleon, and millions of collision events are recorded to tape at each experiment, and three physics papers have been generated to date [40–42].

Following this successful physics run with gold ion beams, we commissioned one RHIC ring with the polarized proton beam. The preparation of the polarized proton beam in the AGS was done in parallel to the luminosity heavy-ion run at RHIC.

In the commissioning, the polarimeter is the most critical device to tune the machine for the maximal polarization. It should be fast and reliable to provide a polarization measurement with a 10% relative error in few minutes. We have developed the polarimeter basing on the proton-Carbon elastic scattering in the Coulomb-Nuclear Interference (CNI) region [43] for this purpose. The analyzing power $\mathcal{A}_N$ is expected to be as large as 4% at $-t = 3 \times 10^{-3}$ $(\text{GeV}/c)^2$. In prior to this spin commissioning we have measured the asymmetry in Experiment 950 in AGS, using the polarized proton beam of 21.7 GeV/c which is in the range of RHIC transfer energy [44]. This measurement provides the basis of absolute polarization measurement at RHIC.

The recoiled Carbon was detected with the Si-strip detector which is 15 cm distant from the carbon target and distributed in azimuth, $\phi = 45°$, $135°$, $225°$, and $315°$. This configuration allows us to measure the polarization vector in the vertical plane. These four detector can be grouped into *Left* and *Right* to measure vertical asymmetry $\mathcal{A}_y$, thus vertical component of the spin vector in the plane, and *Up* and *Down* for side-way asymmetry ($\mathcal{A}_x$) for a side-way component.

A typical correlation of ADC and TDC is shown in the top left plot in Figure 9. A band of carbon is clearly seen. In the mass distribution obtained from time-of-flight and kinetic energy measured by Si-strip, peaks for carbon and helium are clearly seen as shown in the bottom left plot in Figure 9, which ensured the superb identification of carbon in the measurement.

The proton beam with vertical polarization was injected to RHIC ring at 24.3 GeV ($G\gamma = 46.5$, where G is anomalous magnetic moment of the proton and γ is a Lorentz factor [4]) with *Siberian Snake* off. In this case, the stable spin direction is vertical, and the measured asymmetry (Figure 9(a)) clearly showed the vertical polarization. The polarization is estimated to be 19±1% (statistical error only), basing on the analyzing power measured in E950. Then we turned the Snake on adiabatically to rotate the spin to horizontal direction (Figure 9(b)). The beam was accelerated to 25.1 GeV ($G\gamma = 48$), where we expected the spin direction to be rotated by 180°, and we observed the expected asymmetry (Figure 9(c)). It is also demonstrated that no polarization was maintained with snake off. Polarized beam

---

[4] $G\gamma$ is often used rather than beam energy, because it represents the number of spin precession per revolution.

was further accelerated up to 29.2 GeV ($G\gamma = 55.7$) with significant polarization, but we observed no polarization at 31.6 GeV ($G\gamma = 60.3$), which is just above strong depolarization resonance.

The absolute values of asymmetries ($= \sqrt{\mathcal{A}_x^2 + \mathcal{A}_y^2}$) measured are summarized in Figure 10 as a function of $G\gamma$. The commissioning of the polarimeter and the first ring towards the polarized collider was successful, although the estimated polarization was clearly needed to be improved for physics running. More details of commissioning can be found in Huang's contribution [43].

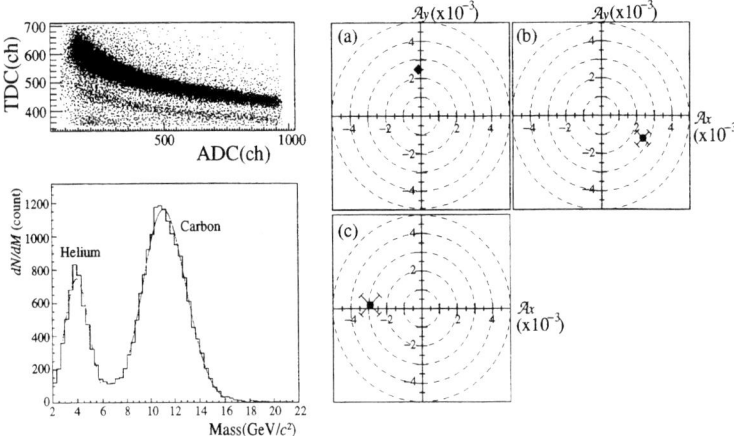

FIGURE 9. Plots to show performance of the Si detectors. Top-left: a typical TDC-ADC correlation in Si detectors. Bottom-left: invariant mass distribution calculated from the time-of-flight and the kinetic energy measured by Si. Measured asymmetries ($\mathcal{A}_x, \mathcal{A}_y$) at (a) $G\gamma=46.5$ with *Snake* off, and (b) *Snake* on, and (c) $G\gamma=48.0$ with *Snake* on.

SUMMARY

We reviewed the physics with the first polarized proton collider, RHIC and its status towards the first spin physics run in 2001. Clearly RHIC SPIN will provide a unique opportunity for high-energy hadron physics. The beam polarization has to be improved at RHIC for the physics run, and we have a plan to achieve it including the installation of additional three *Snakes* to overcome depolarization resonances, and dedicated polarization studies in the AGS.

In the end, we would like to urge theorists to post *your* predictions as soon as possible. If you calculate after the data, those will become *postdictions*.

ACKNOWLEDGMENTS

I am grateful to the organizers of Spin2000 for their invitation. A part of the presentation is based on the article written with G. Bunce, W. Vogelsang, and

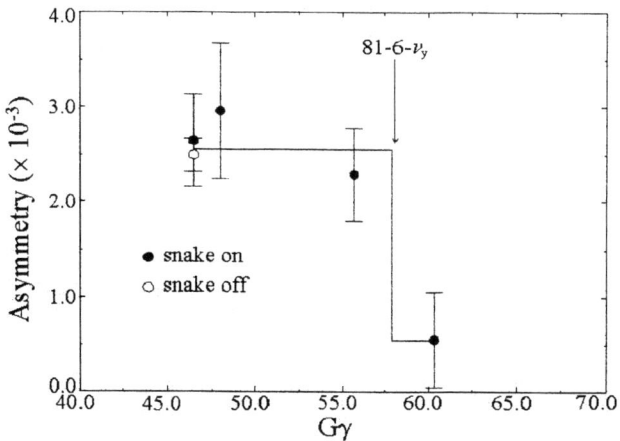

FIGURE 10. Asymmetries measured at RHIC with the first polarized beam are summarized as a function of $G\gamma$.

J. Soffer. I would like to thank E.C. Aschenauer, C. Balazs, L.C. Bland, D. Boer, G. Bunce, M. Grosse Perdekamp, Y. Goto, H. Huang, V.W. Hughes, R.L. Jaffe, X. Ji, B.Z. Kopeliovich, A. Masaike, J. Murata, J.F. Owens, T. Roser, H.D. Sato, A. Schäfer, J. Soffer, H. Spinka, M. Stratmann, T.L. Trueman, J. Tojo, W. Tung, S. Vigdor, W. Vogelsang, K. Yazaki, and all members of AAC for useful discussions. This work is supported by RIKEN BNL Collaboration for RHIC Spin Physics and by the US Department of Energy.

REFERENCES

1. Alguard, M.J., et al., SLAC-E80, *Phys. Rev. Lett.*, **37**,1261 (1976); Alguard, M.J., et al., SLAC-E80, *Phys. Rev. Lett.*, **41**,70 (1978); Baum, G., et al., SLAC-E130, *Phys. Rev. Lett.*, **51**, 1135 (1983).
2. Ashman, J, et al., CERN-EMC Collaboration, *Phys. Lett.* **B206**, 364 (1988); *Nucl. Phys*, **B328**, 1 (1989).
3. Adeva, B., et al., CERN-SMC Collaboration, *Phys. Lett.* **B412**, 414 (1997); *Phys. Rev.* **D58**, 112001 (1998).
4. Adams, D., et al, *Phys. Rev.*, **D58**, 112002 (1998).
5. Anthony, P.L.,et al., SLAC-E142, *Phys. Rev.* **D54** 6620 (1996); Abe, K., et al., SLAC-E143, *Phys. Rev.* **D58** 112003 (1998); Abe, K., et al., SLAC-E154, *Phys. Rev. Lett.* **79**, 26 (1997); Anthony, P.L., et al., SLAC-E155, *Phys. Lett.* **B463**, 339 (1999).
6. Ackerstaff, K., et al., HERMES Collaboration, *Phys. Lett.* **B404** 383 (1997); Airapetian, A., et al., *Phys. Lett.* **B442** 484 (1998).
7. Taxil, P., and Virey, J.M., *Phys. Lett.* **B364** 181 (1995); *Phys. Rev.* **D55** 4480 (1997); *Phys. Lett.* **B383** 355 (1996); *Phys. Lett.* **B441** 376 (1998).

8. Gunion, J.F., Yuan, T.C., and Grzadkowski, B., *Phys. Rev. Lett.* **71** 488 (1993); Erratum-ibid **71** 2681 (1993).
9. Rykov, V., Ogawa, A., and Saito, N., these proceedings.
10. Zelenski, A., these proceedings.
11. Bunce, G., Saito, N., Soffer, J., and Vogelsang, W., *Ann. Rev. Nucl. Part. Sci.* **50** 525 (2000).
12. Adams, D.L., *et al.*, FNAL-E704, *Phys. Lett.* **B261**, 197 (1991); *Phys. Lett.* **B336**, 269 (1994).
13. Goto., Y., *et al.*, Aymmetry Analysis Collaboration, *Phys. Rev.* **D62**, 034017 (2000).
14. Vogelsang, W., these proceedings.
15. Jaffe, R.L., and Ji, X., *Phys. Rev. Lett.* **67**, 552 (1991).
16. Papavassilou, C., Mobed, N., and Svec, M., *Phys. Rev.* **D26**, 3284 (1982); Berger, E., and Qiu, J., *Phys. Rev.* **D40**, 778 (1989) and references therein.
17. Jaffe, R.L., and Saito, N., *Phys. Lett* **B382**, 165 (1996).
18. Qiu, J., and Sterman, G., *Phys. Rev. Lett.* **67**, 2264 (1991).
19. de Florian, D., Frixione, S., Signer, A., and Vogelsang, W., *Nucl. Phys.* **B539**, 455 (1999).
20. Bourrely, C., Soffer, J., Renard, F.M., and Taxil, P., *Phys. Rept.* **177** 319 (1989).
21. Karliner, M., and Robinett, R.W., *Phys. Lett.*, **B324**, 209 (1994).
22. Kao, C., Atwood, D., and Soni, A., *Phys Lett.* **B395** 327 (1997).
23. Tkabladze, A., and Teryaev, O., *Phys. Rev.* **D56**, 7331 (1997).
24. Jaffe, R.L., and Kharzeev, D., *Phys. Lett.* **B455**, 306 (1999).
25. Bourrely, C., and Soffer, J., *Phys. Lett.* **B314**, 132 (1993).
26. Boer, D., *Phys. Rev.* **D62**, 094029 (2000).
27. Ratcliffe, P., *Nucl. Phys.* **B223**, 45 (1983).
28. Leader, E., and Sridhar, K, *Phys. Lett.* **B311**, 324 (1993)
29. Vogelsang, W., and Weber, A., *Phys. Rev.*, **D48**, 2073 (1993)
30. Hammon, N., Teryaev, O., and Schäfer, A., *Phys. Lett.* **B390** 409 (1997).
31. Collins, J.C., Soper, D.E., and Sterman, G., *Nucl. Phys.*, **B261**, 104 (1985).
32. Bravar, A., von Harrach, and D., Kotzinian, A., *Phys. Lett.*, **B421**, 349 (1998).
33. Airapetian, A., *et al.*, *Phys. Rev. Lett.* **84**, 2584 (2000).
34. Bland, L.C., for STAR Collaboration, *RIKEN Rev.* **28** 8 (2000).
35. Bourrely, C., Soffer, J., *Nucl. Phys.* **B445** 341 (1995).
36. Gehrmann, T., Stirling, W.J., *Phys. Rev.* **D53** 6100 (1996).
37. Roser, T., private communication.
38. Grosse Perdekamp, M., these proceedings.
39. Murata, J., poster presentation at this symposium.
40. Back, B.B. *et al.*, PHOBOS Collaboration, *Phys. Rev. Lett.* **85** 3100 (2000).
41. Ackermann, K.H., *et al.*, STAR Collaboration, nucl-ex/0009011.
42. Adcox, K., *et al.*, PHENIX Collaboration, nucl-ex/0012008.
43. Huang, H., *et al*, these proceedings.
44. Tojo, J., *et al.*, BNL-AGS E950 Collaboration, these proceedings.

Spin Dynamics In LEP With 40-100 GeV Beams

R. Assmann[*], J. Badier[#], A. Blondel[#], M. Böge[*+], M. Crozon[&],
B. Dehning[*], H. Grote[*], J.P. Koutchouk[*], M. Placidi[*], R. Schmidt[*],
F. Sonnemann[*], F. Tecker[*], J. Wenninger[*]

[*]*European Organization for Particle Physics (CERN), CH-1211 Geneva 23, Switzerland*
[#]*Laboratoire de Physique Nucléaire et des Hautes Energies, Ecole Polytechnique, IN^2P^3-CRNS, F-91128 Palaiseau Cedex, France*
[&]*College de France, Lab. De Physique Corpusculaire, IN^2P^3-CNRS, F-75231 Paris Cedex 05, France*
[+]*Present address: PSI – Paul Scherrer Institut, Villingen, Switzerland*

Abstract. Radiative spin polarization has been studied in the Large Electron-Positron Collider (LEP) at CERN for beam energies from 40 GeV to 100 GeV. The data cover a unique range of spin dynamics, not previously accessible with other storage rings. After optimization of machine parameters and the successful application of new Harmonic Spin Matching techniques, a transverse beam polarization of 57 % was obtained at 44.7 GeV. At 60.6 GeV the maximum level reached 8 %. The observed energy dependence of radiative spin polarization at LEP is in excellent agreement with the theoretically expected behavior. The LEP data provide the first experimental confirmation for a theory of depolarization at very high energies, first developed in the 1970s by Derbenev and Kontratenko. The results will help to guide the design of any future high energy electron-positron storage ring requiring polarized beams.

INTRODUCTION

The Large Electron-Positron collider (LEP) was operated at CERN between 1989 and 2000 [1]. As the largest storage ring to date LEP accelerated electron and positron beams from 22 GeV to more than 104 GeV [2]. The compensation of synchrotron radiation losses and the necessary beam stability were achieved with an accelerating radio-frequency voltage of up to 3.65 GV.

The lepton beams in LEP polarized spontaneously due to the Sokolov-Ternov effect [3]. We shortly review the main features of radiative spin-polarization for LEP, without going into any detail. For a planar storage ring the polarization builds up in the vertical transverse direction and (without imperfections) reaches an asymptotic degree of 92.4 %. The classical "spin vector" processes around the vertical axis with a frequency f_{spin} that is a multiple ν of the particle revolution frequency f_{rev}. The number ν is called the spin tune and can be expressed as a simple function of the beam energy E:

$$\nu = \frac{E}{440.6486 \text{ MeV}} \qquad (1)$$

This relationship was used for a precise determination of the LEP beam energy [4]. The exponential build-up time τ_p of radiative polarization is a function of the storage

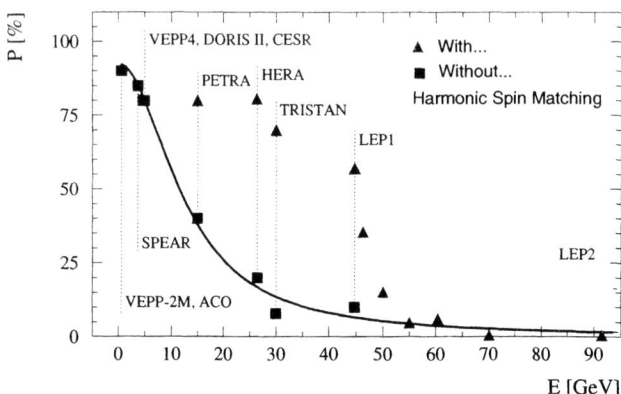

FIGURE 1. Overview of highest measured polarization degrees in electron-positron storage rings. Measurements with (triangle) and without (square) Harmonic Spin Matching are shown. The grey area indicates the energy range of the LEP collider.

ring bending radius and beam energy. Using LEP parameters we can write $\tau_p = 1/\lambda \approx (3.8 \cdot 10^{12}\,\text{s}) \cdot (E/\text{GeV})^{-5}$. λ is the polarizing rate. The polarizing time decreases rapidly with beam energy. For LEP at 100 GeV it is as small as 6 minutes, to be compared to 5.7 hours at 45 GeV.

Depolarization is caused by unavoidable imperfections in the vertical orbit of planar storage rings and is enhanced by synchrotron radiation. It is characterized by a depolarization time τ_d. The asymptotic degree P of polarization is reduced to

$$P = \frac{92.4\%}{1 + \tau_p/\tau_d} \quad (2)$$

and the "effective" build-up time is shorter ($1/\tau_p^{\text{eff}} = 1/\tau_p + 1/\tau_d$). Polarization theories aim at estimating the term τ_p/τ_d. At LEP the behavior of polarization was studied in a unique range of high beam energies. Measurements at LEP and other lepton storage rings are summarized in Fig. 1. The LEP data cover a range from about 40 to 100 GeV that was not accessible with other storage rings. In this paper we describe and analyze the observed behavior of spin dynamics over this large range of beam energy.

OPTIMIZATION OF TRANSVERSE POLARIZATION

Careful optimization of the vertical orbit is required in order to maximize polarization at the beam energies of LEP. A good starting point for polarization was established with some basic machine optimization:

Precise control of the vertical orbit. Residual offsets after orbit correction were minimized with a yearly vertical realignment of all quadrupoles in LEP (~ 150 μm rms residual error after realignment). The knowledge of the beam offsets in the quadrupoles was improved by determining the alignment of the beam position monitors

(BPMs) with respect to the magnetic centers of the quadrupoles. To that purpose a beam-based method was used ("k-modulation") with about 70 µm rms accuracy [5].

Accurate setting of spin tune (or beam energy). Depolarization is a resonant phenomenon. Spin resonances occur at spin tunes v_{dep} that are the sum of an integer plus multiples of the machine tunes Q_x, Q_y and Q_s:

$$v_{depol} = k \pm k_x \cdot Q_x \pm k_y \cdot Q_y \pm k_s \cdot Q_s \quad , \quad k, k_x, k_y, k_s \in N \qquad (3)$$

The machine tunes and the spin tune ν are set to values that maximize the distance of the spin tune to the most significant depolarizing resonances. Typically, ν is set close to a half-integer. Precise calibration of the beam energy at lower energies and its extrapolation with magnetic measurements [6] improves the accuracy in the setting of the optimized working point.

Optimization of the accelerator optics. The strength of magnets, the effect of imperfections on the beam and the efficiency of the orbit correction are functions of the accelerator optics. Polarization measurements at LEP were made with a variety of optics. Cell phase advances in the horizontal and vertical plane of 90/60, 90/90, 60/60 and 101/45 degrees were used, the later optics being the most favorable [7].

Even, if the spin tune is set to maximize the distance to the most important depolarizing resonances, residual depolarization will occur. It has been shown that the achievable polarization is mainly given by the strengths of the Fourier orbit harmonics (in spin precession frame) close to the spin tune [8]. For a spin tune ν=k+0.5, i=k,k+1 and some numerical factors γ_i the equilibrium polarization P can be written as:

$$P \approx \frac{92.4 \ \%}{1 + \sum_i \gamma_i a_i^2} \qquad (4)$$

where the term a_i is the complex amplitude of the Fourier harmonic i of the vertical orbit. Relationship (4) opens the way to an efficient optimization of polarization. With vertical orbit bumps the amplitude of the orbit harmonics i can be reduced: $a_i \rightarrow a_i - a_i^{bump}$. This procedure is called "Harmonic Spin Matching" [8,9,10]. The method was improved at LEP by calculating the harmonics directly from the measured orbit. Such a deterministic correction is much less time consuming than empirical optimization. Fig. 2 shows an example of deterministic Harmonic Spin Matching. The method requires an accurate measurement of the vertical orbit, as discussed above.

A maximum polarization level of 57 % was achieved at 44.7 GeV with this method and in combination with empirical Harmonic Spin Matching. The measurement is shown in Fig. 3. During the experiment a rise-time measurement was performed on a separate bunch. This allows an accurate determination of the absolute polarization scale. The beneficial effect of Harmonic Spin Matching is shown in Fig. 1, where LEP measurements with and without Harmonic Spin Matching are shown.

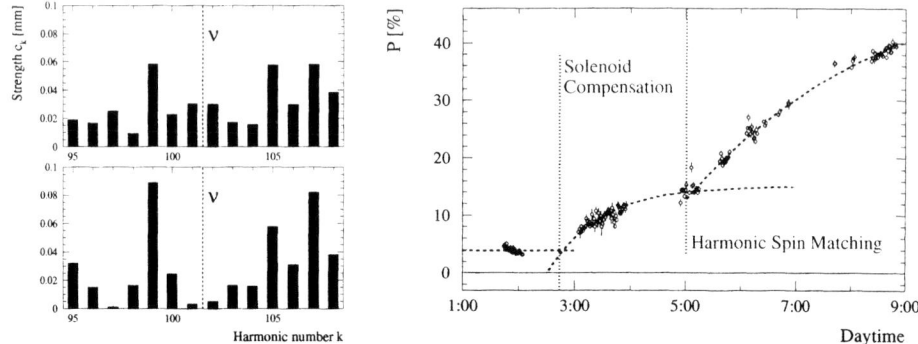

FIGURE 2. Measured strengths of harmonics before (left top) and after (left bottom) deterministic Harmonic Spin Matching. Measured beam polarization at 44.7 GeV on a selected bunch versus time (right). Solenoid spin compensation and deterministic Harmonic Spin Matching are indicated.

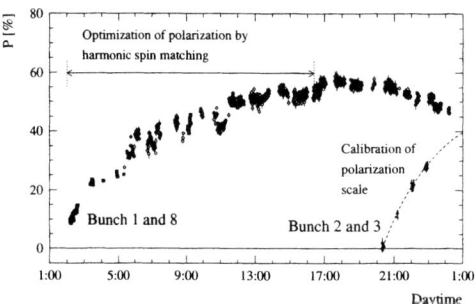

FIGURE 3. Measured LEP beam polarization on selected bunches versus time (at 44.7 GeV). The absolute scale of polarization was calibrated with an accurate measurement of the effective polarization build-up time.

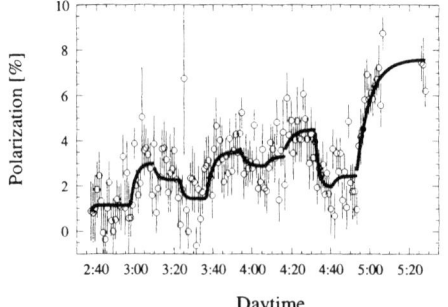

Time	HSM bumps settings				Fit results
hr:min	137 (cos)	137 (sin)	138 (cos)	138 (sin)	P_1 (%)
02:35	0.0	0.0	0.0	0.0	1.15 ± 0.23
02:58	2.0	0.0	0.0	0.0	3.03 ± 0.37
03:10	2.0	0.0	2.0	0.0	2.28 ± 0.30
03:23	2.0	0.0	2.0	2.0	1.45 ± 0.27
03:36	2.0	2.0	2.0	0.0	3.51 ± 0.28
03:54	2.0	2.0	2.0	-2.0	2.88 ± 0.25
04:06	2.0	2.0	2.0	0.0	3.33 ± 0.40
04:16	4.0	2.0	2.0	0.0	4.53 ± 0.34
04:33	6.0	2.0	2.0	0.0	1.86 ± 0.27
04:41	3.0	4.0	2.0	0.0	2.66 ± 0.35
04:53	3.0	2.0	0.6	-0.6	7.69 ± 0.36

FIGURE 4. Example of educated Harmonic Spin Matching at 60.6 GeV. The closest orbit harmonics were changed in known steps and the polarization level was measured versus time (left). The data was used to fit the asymptotic polarization levels (right). The final measurement shows the polarization after compensating the fitted orbit harmonics.

Another type of Harmonic Spin Matching was developed at LEP. Vertical orbit bumps (inducing orthogonal amplitudes a_i^{bump}) are changed in known steps for different harmonics i. For each step the asymptotic polarization level is fitted. With a minimum of five polarization measurements (for different a_i^{bump}) the unknown harmonics a_i of the vertical orbit and some residual depolarization $(\tau_p/\tau_d)_0$ can be determined:

$$P_{asym} = \frac{92.4\ \%}{1+\left(\tau_p/\tau_d\right)_0 + \sum_i \gamma_i \left(a_i - a_i^{bump}\right)^2} \quad (5)$$

An experiment at 60.6 GeV is shown in Fig. 4. The polarization level was measured versus time. Overlaid are polarization fits that were used to determine the asymptotic degree of polarization for each setting of harmonic spin bumps. The results from the first 10 measurements were included in a fit to determine the orbit harmonics. The last measurement in Fig. 4 shows the polarization with compensation of the fitted harmonics. The best polarization level at 60.6 GeV reached 7.7 %.

ENERGY DEPENDENCE OF POLARIZATION

The expected energy dependence of polarization is included in the theory for depolarization at ultra-high energies by Derbenev and Kondratenko [11]. We follow their approach. With $v^2 \lambda / Q_s^3 \ll 1$ subsequent passings of spin resonances are correlated. Polarization can then be described with Equation 2 and:

$$\frac{\tau_p}{\tau_d} = \frac{11}{18} v^2 \sum_{k,m} \frac{|w_k|^2 \langle T_m^2 \rangle}{\left[\left(k - v - mQ_s\right)^2 - Q_s^2\right]^2} \quad (6)$$

Here, w_k is the complex strength of the spin resonance at integer k, v is the spin tune averaged over the particle ensemble and m an integer giving the order of the synchrotron sideband resonance. Betatron spin resonances with the transverse tunes Q_x and Q_y do not appear. For high energy lepton storage rings they are much weaker than synchrotron resonances and can be neglected. For a given rms strength of imperfections the statistical average value of $|w_k|^2$ is proportional to the square of the beam energy (or spin tune): $|w_k|^2 \propto v^2$. Equation 6 contains a term T_m:

$$\langle T_m^2 \rangle = I_m \left(\frac{\sigma_v^2}{2 Q_s^2}\right) \cdot \exp\left(-\frac{\sigma_v^2}{2 Q_s^2}\right) \quad (7)$$

The I_m are the modified Bessel functions. As representative LEP parameters we use $|w_k|^2 = 2 \times 10^{-10} \cdot v^2$, $Q_s = 0.077$ and a spin tune spread $\sigma_v = 6.67 \times 10^{-6} \cdot v^2$. The actual values w_k depend on the distribution of vertical orbit offsets. The Q_s values used at LEP varied from 0.0625 to 0.11.

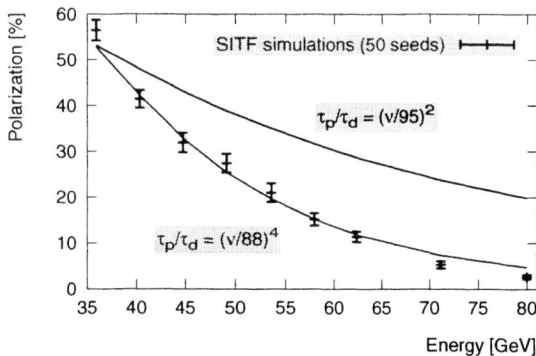

FIGURE 5. The simulated energy dependence of polarization is compared with curves that represent an increase of depolarization τ_p/τ_d with the second or fourth power of energy.

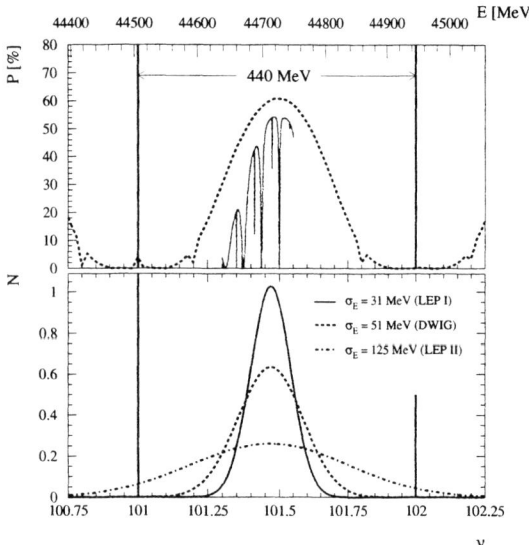

FIGURE 6. Illustration of depolarization enhancement at high beam energies. The upper graph shows a numerical simulation of LEP polarization versus beam energy around 44.7 GeV, both in linear (dashed) and higher order (solid) approximation. The spin resonances are clearly visible with the strongest contributions from 101+Q_s and 102-Q_s. The lower graph shows the energy distribution in the beam. The narrow curve is for 44.7 GeV and corresponds to the simulation in the upper graph. The energy distribution becomes much wider if the damping wigglers (DWIG) are excited or the beam energy is increased (LEP II). As a result spin resonances are excited more strongly and polarization is suppressed.

The magnitude of the spin tune spread σ_v determines the size of the T_m term. Higher order spin resonances are not important and $\langle T_m^2 \rangle \cong 1$, if the spin tune spread is much smaller than the synchrotron tune. The achievable polarization level is then only affected by linear spin resonances ($v_{dep} = k \pm Q_s$). In the following this is called the "linear" theory. Polarization is expected to drop with the fourth power of beam energy (for

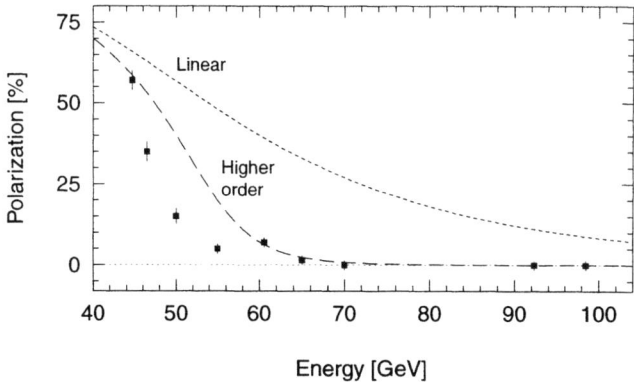

FIGURE 7. Maximum polarization levels measured for different energies in LEP. Note that the measurements at 44.7 GeV and 60.6 GeV were fully optimized. Measurements at other energies below 60.6 GeV were used for energy calibration purposes and are only partially optimized. The theoretically expected energy dependence of polarization is shown with $|w_k|^2 = 2 \times 10^{-10} \cdot v^2$ for both linear and higher order theory.

the same rms imperfections). This basic result from linear theory is compared in Fig. 5 with a numerical simulation, finding good agreement.

If the spin tune spread becomes larger than the synchrotron tune, the linear and higher order synchrotron resonances ($v_{dep} = k \pm k_s \cdot Q_s$, $k_s \geq 1$) limit the achievable polarization. This is referred to as "higher-order theory". Polarization drops even faster with beam energy than in the linear regime. An intuitive picture why depolarization is so strongly enhanced is shown in Fig. 6. As the spin tune spread becomes much larger at the high beam energies of LEP2, it does not only overlap higher order resonances, but also the strong linear and integer spin resonances.

The maximum levels of transverse polarization observed at LEP are shown in Fig. 7 for different beam energies. The measurement at 44.7 GeV is extrapolated to higher beam energies with Equations 6 and 7, assuming the same residual imperfections in the vertical orbit. Because we assume the same resonance strength $|w_k|^2 = 2 \times 10^{-10} \cdot v^2$ for the calculation of both the linear and higher order theory, the linear value is always above the higher order prediction. Experimentally a sharp drop in radiative spin polarization is observed at LEP, in good agreement with the expectation from higher order theory. In particular the measurement at 60.6 GeV is in excellent agreement with the theory. Measurements between 44.7 GeV and 60.6 GeV are below the expectation because they were not fully optimized.

The decrease in polarization is mainly due to the enhancement of depolarization from higher-order synchrotron resonances. It is much steeper than with the fourth power of energy that one would expect in linear theory. The LEP measurements are the first experimental confirmation of the theory that Derbenev and Kontratenko developed in the 1970s.

Above 70 GeV the condition $v^2 \lambda / Q_s^3 \ll 1$ is violated for LEP parameters and the spin dynamics enters into the regime of uncorrelated passings of spin resonances [11].

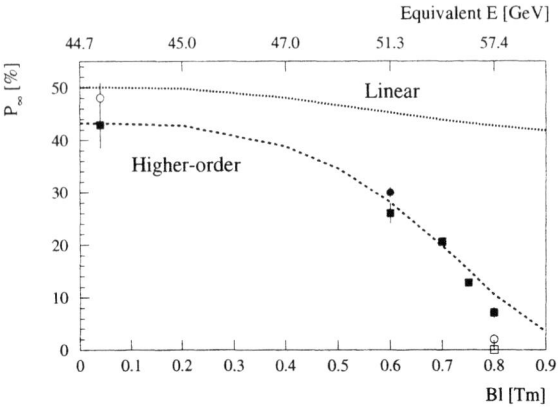

FIGURE 8. Observed polarization level at 44.7 GeV for different excitations Bl of the LEP damping wigglers. The upper scale indicates the beam energy that would produce the same spin tune spread. The polarization measurements are compared to the expectations from linear and higher-order theory.

Equations 6 and 7 no longer apply. Derbenev and Kontratenko have also studied this regime of spin dynamics [11] but it is beyond the scope of this paper to review the details of this theory. LEP is the first storage ring that operated in this special regime of spin dynamics. Measurements of polarization were performed at LEP up to 98.5 GeV. The data is shown in Fig. 7. No polarization was observed above 65 GeV. A hypothetical increase of polarization with beam energy for very intense synchrotron radiation [11,12] was not seen and would only be expected for LEP with a beam energy around 200 GeV.

The theory of Derbenev and Kontratenko was confirmed independently with the asymmetric damping wigglers in LEP. Those wigglers decrease the polarization build-up time but at the same time increase the spin tune spread. From Equation 7 we do then expect an increase of depolarization. The asymmetric wigglers therefore allow "simulating" the increased energy and spin tune spread at higher beam energies. The vertical orbit and other parameters can be kept conveniently stable. In Fig. 8 the measured polarization and the theoretical prediction are compared for different settings of the damping wigglers. Strong depolarization was observed for large excitations of the wigglers, in excellent agreement with the theoretical expectation.

POLARIZATION WITH COLLIDING BEAMS AT 45 GEV

Polarized beams in LEP were used for accurate energy calibration by resonant depolarization. The description of this method and its results for LEP are published in [4]. Originally it was foreseen to directly exploit the particle physics potential of polarized beams in LEP. In order to propose and implement such an option, it had to be shown that the transverse polarization can be rotated into the longitudinal direction in the interaction points of LEP and that polarization is preserved during collision. An appropriate spin rotator for LEP was designed and its performance was demonstrated in simulations [13].

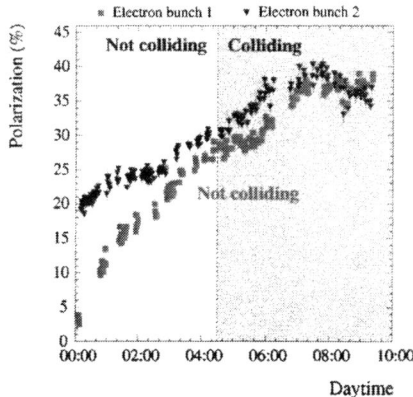

FIGURE 9. Polarization measurements during a beam-beam experiment. Electron bunch 1 (square) was not colliding during the whole experiment. Electron bunch 2 (triangle) was initially not colliding and was put into collision at 4:30h with a vertical beam-beam parameter of 0.04.

The strength of depolarization induced by beam-beam collisions was measured at 45 GeV. The measurements are shown in Fig. 9. The polarization of two electron bunches was monitored simultaneously. Polarization was slowly building up on both bunches while the vertical orbit was optimized for maximum asymptotic polarization. For the first part of the experiment both bunches were not colliding with the positron beam. At a time when both bunches had a similar level of polarization, one bunch was put into collision with a positron bunch. The measured vertical beam-beam parameter was 0.04. The polarization on both electron bunches kept increasing until it reached an asymptotic value of 38 %. As seen from Fig. 9, the difference in absolute polarization level between the colliding and the non-colliding electron bunch was smaller than 4 %. The beam-beam induced depolarization τ_p/τ_d was smaller than 0.3, imposing a "beam-beam" polarization limit of above 70 % for LEP with one collision point. Depolarization due to beam-beam effects is not a serious problem in a 45 GeV storage ring with LEP parameters. Collisions of polarized beams were, however, incompatible with the existing LEP experiments, and this option was finally abandoned.

CONCLUSION

Radiative spin polarization has been studied at LEP for beam energies from 40 GeV to 100 GeV. The data covers a unique range of spin dynamics, not previously accessible with other storage rings. After careful optimization of machine parameters and the successful use of new implementations of Harmonic Spin Matching, transverse beam polarization of 57 % was observed at 44.7 GeV. At 60.6 GeV the highest level reached 8 %. The observed energy dependence of radiative spin polarization in LEP is in excellent agreement with the theoretically expected behavior. The LEP data provide the first experimental confirmation for a theory of depolarization at very high energies, first developed in the 1970s by Derbenev and Kontratenko. With colliding beams at one interaction point, polarization levels of 38 % were observed. The experiment showed that beam-beam depolarization is weak for a 45 GeV storage ring with LEP parameters. The LEP results on polarization verify long standing theories and will help to guide the design of any future high energy electron-positron storage ring requiring polarized beams.

ACKNOWLEDGMENTS

The following persons contributed to different aspects of the LEP polarization project over the last eleven years: L. Arnaudon, F. Bordry, B. Bouchet, E. Bravin, J. Camas, W. Coosemans, A. Drees, J.P. Ferry, the late G.E. Fisher, P. Grosse-Wiesmann, K.N. Henrichsen, M. Hildreth, A. Hofmann, R. Jacobsen, M. Jonker, L. Knudsen, L. Lawson-Choroco, J. Mann, J. Matheson, M. Mayoud, G. Mugnai, J. Miles, R. Olivier, R. Olsen, C. Pan, G. Ramseier, F. Roncarolo, E. Torrence, J. Uythoven and A. Verdier. They provided important help in making polarization at LEP a success. Their contributions are gratefully acknowledged. We thank W. Blum, F. Dydak, L. Evans, K. Hübner, S. Myers, and the late J. Thresher for their encouragement and support. Our colleagues in the LEP energy calibration group were our only customers for polarized beams and always showed a keen interest and support for improvements in transverse beam polarization. Last not least, we thank D.P. Barber and R. Brinkmann from DESY, A. Chao from SLAC, K. Yokoya from KEK, Y.S. Derbenev, Y. Eidelman, A.M. Kontratenko, Yu.M. Shatunov, and A.N. Skrinsky from Novosibirsk, S. Mane from BNL and J. Buon from the University Paris-Orsay for their constructive criticism and their support in polarization theory and simulation.

REFERENCES

1. S. Myers. CERN Yellow Report 91-08.
2. G. Arduini et al. "LEP Operation and Performance with 100-GeV Colliding Beams". CERN-SL-2000-045-OP, Jun 2000. EPAC2000.
3. A.A. Sokolov and I.M. Ternov, "On Polarization and Spin Effects in the Theory of Synchrotron Radiation", *Sov. Phys. Dokl.* 8(1964)1203.
4. L. Arnaudon et al, "Accurate Determination of the LEP Beam Energy by Resonant Depolarization". *Z. Phys.* C66(1995)45-62.
5. B. Dehning et al., "Dynamic Beam Based Calibration of Beam Position Monitors". CERN SL/98-38-BI. EPAC98.
6. R. Assmann et al., "Calibration of centre-of-mass energies at LEP1 for precise measurements of Z properties". *Eur. Phys. J.* C 6(1999), 187-223.
7. R. Assmann and A. Verdier, "An Optics with 101/45 Degree Phase Advance for LEP". SL-Note-2000-013 OP.
8. V.N. Baier and Yu.F. Orlov, "Quantum Depolarization of Electrons in a Magnetic Field". *Sov. Phys. Dokl.* 10(1966)1145.
9. R. Rossmanith and R. Schmidt, "Compensation of Depolarizing Effects in Electron Positron Storage Rings". *Nucl. Instr. Meth.* A236(1985)231.
10. D.P. Barber et al, "High Spin Polarization at the HERA Storage Ring". *Nucl. Instr.Meth.* A338(1994)166-184.
11. Ya.S. Derbenev, A.M. Kondratenko and A.N. Skrinsky, "Radiative Polarization at Ultra-High Energies". *Part. Acc.* 1979, Vol. 9, pp. 247-266.
12. R. Assmann. "The regimes of polarization in a high energy e+e- storage ring". PAC99.
13. H. Grote, "Spin rotator". 2nd Workshop on LEP performance, Chamonix, France, 19-25 Jan. 1992. Ed. by J. Poole. CERN, Geneva, 1992. CERN SL 92-29 DI.

Polarized Ion Sources For High-Energy Accelerators And Colliders

A.N.Zelenski

Brookhaven National Laboratory, Upton, N.Y., 11973, USA

Abstract. The recent progress in polarized ion source development is reviewed. In dc operation a 1.0 mA polarized H⁻ ion current is now available from the Optically–Pumped Polarized Ion Source (OPPIS) . In pulsed operation a 10 mA polarized H⁻ ion current was demonstrated at the TRIUMF pulsed OPPIS test bench and a 2.5 mA peak current was obtained from an Atomic Beam Source (ABS) at the INR Moscow test bench. The possibilities for future improvements with both techniques are discussed. A new OPPIS for RHIC spin physics is described. The OPPIS reliably delivered polarized beam for the polarized run at RHIC. The results obtained with a new pulsed ABS injector for the IUCF Cooler Ring are also discussed.

INTRODUCTION

Collider experiments with polarized beams at RHIC /1/ and HERA /2/ , will provide essential tests of QCD and the electroweak interaction. Polarization asymmetries and parity violation are the strong signatures for identification of the fundamental processes, which are otherwise inaccessible. Such experiments will require the maximum available luminosity and therefore polarization must be obtained as an extra beam quality without sacrificing intensity. This is already implemented at the electron accelerators . In a storage ring an electron beam is self polarized by the Sokolov-Ternov effect and for linac accelerators a great effort in polarized electron source development was finally rewarded by achievement of 80% polarization and high beam intensity which will be sufficient to run high-current accelerators like CEBAF at full polarized intensity.

There are a number of proposals to polarize the high-energy proton beam in the storage ring by the Stern-Gerlach effect or spin-filter techniques. But so far the only feasible option is to accelerate the polarized beam produced in the source and make sure that polarization will survive during acceleration and storage. High intensity unpolarized H⁻ ion sources are presently a common choice for high-energy accelerators due to the advantage of stripping injection into the accelerator ring. Typical currents for such injectors are in the 20-50 mA range. The present status of high-current polarized ion sources and future prospects for intensity increases to the 20-50 mA range will be discussed.

POLARIZATION TECHNIQUES

In any type of polarized ion source the first step is the generation of an electron-spin polarized atomic beam. The polarization is then transferred to the protons by hyperfine interaction and finally the beam is ionized (a nuclear spin polarized hydrogen beam can be used as a polarized internal target in ring accelerators or colliders). The difference is in the velocity of the atomic beam. It is comparatively easy to polarize a "slow" (thermal energy) beam by using separating magnets. The advantages of using "fast" (a few keV energy) beams are higher intensity and simple, more efficient ionization (W.Haeberli /3/).

a). **Polarization of "slow" atomic beams:** An atomic hydrogen beam is produced in an RF dissociator. A typical beam velocity is about $(1-2)310^5$ cm/s, which is achieved by cooling of the dissociator nozzle to a temperature of 30-80 K. The acceptance of separating sextupole magnets is proportional to $\sim \mu B/kT$. Field strength is limited to about 1.5 T by available magnetic materials. The atomic beam of a selected spin-state is directed into the small acceptance of the ionizer (or the storage cell). The initial beam velocity spread and beam scattering in the residual gas along the way (in particular, inside the low vacuum conductance separating magnets) reduce the beam intensity. It appears that everything has already been optimized and a beam intensity of $(6-7)310^{16}$ atoms/s can not be significantly improved upon. In pulsed operation, the higher atomic beam intensity of up to 2310^{17} atoms/s within a diameter of 15 mm ionized acceptance was reported. Perhaps, in a transition mode at the front of the beam pulse the scattering is less and the pulsed dissociator works better. The higher field of up to 4-5 T can be obtained in syperconducting sextupole magnets and a large cold aperture can be used for cryogenics pumping. D.Toporkov has recently built a new polarized deuterium source with high field (up to 4.8 T) superconducting sextupoles and cryogenic pumping. The D beam intensity is about 50% higher than in conventional sources /4/.

There is a possibility of polarization of hydrogen (deuterium) atoms in spin-exchange collisions with optically-pumped alkali-metal atoms in a vapor cell /5/. Polarization losses are caused by wall collisions and recombination. A flux of atomic H in excess of 10^{18} atoms/s and an electron polarization of 60 % has been obtained by using a dry-film wall coating to reduce depolarization. Selective ionization must be used to obtain a high nuclear polarization.

In dc atomic beam sources the beam is usually ionized in an ECR-type ionizer and then converted to an H⁻ ion beam in the cesium charge-exchange cell. The total efficiency of the atomic H beam conversion to the H⁻ ion beam is about 0.1% and a typical dc polarized H⁻/D⁻ ion current is about 10 uA /6/. In a pulsed mode the ionization by a 30-50 keV atomic cesium beam is more efficient and up to 40 uA polarized H⁻ ion current was obtained at BNL /7/. The ionization efficiency is about 0.5%. In a D⁻ enriched plasma ionizer an order of magnitude higher efficiency was obtained (see A.Belov presentation at this conference).

b). **Polarization of "fast" (0.5-3.0 keV energy) atomic beams:** A "fast" atomic H beam can be easily converted to an H⁻ ion beam just by passing the beam through an alkali-metal vapor cell. The H⁻ ion yield is about 9% in a sodium vapor cell for 0.5-3.0 keV beam energies and about 16-22% in a Rb vapor cell for 0.5-1.0 keV atomic beam energies. The electron –spin polarization of the "fast" H beam is produced either in a charge-exchange process, when primary protons capture polarized electrons from optically-pumped alkali-metal atoms in a vapor cell, or in spin-exchange collisions /8/. This technique is called an "Optically-Pumped Polarized Ion Source" (OPPIS), although polarized electrons can be captured from aferromagnetic foil (as in the original Zavoiski proposal), or from hydrogen, or an alkali-metal atomic beam polarized by separating magnets. The "fast" atomic beam intensity in dc operation is about 10^{17} atoms/s (within the ionizer cell's acceptance). This intensity is similar or higher than the "slow" atomic beam intensity in the dc ABS but the ionization efficiency is two orders of magnitude higher. In excess of 1.0 mA polarized H⁻ ion beam current was obtained in the dc OPPIS /9/. In pulsed operation the "fast" beam intensity of 2310^{18} atoms/s was obtained within the ionizer acceptance and 10 mA pulsed polarized H⁻ ion current and 40 mA of a pulsed polarized proton current was achieved. The feasibility of a 30 mA polarized H⁻ ion current in a pulsed OPPIS was demonstrated /10/.

PULSED ATOMIC BEAM SOURCES AT INR MOSCOW AND IUCF

Conventional dc atomic beam sources have been reviewed in detail everywhere /3,6/. A new ABS for the Muenchen tandem accelerator was just completed and a polarized H⁻ ion current of about 10 µA was obtained (see R. Hertenberger presentation at this conference). Significant progress since the 1999 spin conference in Protvin has been achieved in pulsed ABS development. A second ABS with a resonant charge-exchange ionizer, based on original INR Moscow design, was built and put into operation at IUCF in 1999. A new source delivers polarized H⁻/D⁻ ion beams for injection to the Cooler Injection Synchrotron, which has replaced the IUCF cyclotron as an injector for the Cooler Ring (see A.Belov, V.Derenchuk presentation at this conference). Similar to the INR ABS, a small volume dissociator, fast gas pulsing and pulsed RF power (2 kW) were used. In a short pulse, an atomic beam intensity of about 2310^{17} was obtained within a 15 mm diameter ionizer acceptance. It is 3 times higher than for similar geometry ABS's which operate in the dc mode.

The plasma injector and extraction system were built at INR Moscow. Simultaneously with the polarized H⁻ beam, an unpolarized D⁻ ion beam of an order of magnitude higher intensity was produced in this source. A very high D⁻ ion beam current of 40 mA was obtained recently at INR Moscow as a result of the plasma ionizer D^+ to D⁻ converter optimization (at IUCF a 20-30 mA unpolarized D⁻ current is used routinely for the unpolarized beam physics program). So far the peak polarized beam intensity scales linearly with the unpolarized ion beam current and an H⁻ ion peak intensity of 2.5 mA was obtained at INR (see A.Belov presentation at this conference). Polarization of about 80% is expected. The IUCF source produces about 1.5 mA (peak current) H⁻ and D⁻ ion current of 80% polarization, which meets the requirements of the polarized Cooler injector.

The estimation of the plasma ionizer efficiency in a short pulse mode is ambiguous because in a short 50-100 μs pulse the supply of "fresh" atoms is less than the number of atoms which is already "stored" in the volume of a 30 cm long ionizer cell. The polarized H⁻ ion current drops rather sharply, therefore the efficiency estimation depends strongly on the averaging time. If it is averaged over a 150 μs pulse duration, which is approximately the polarized H atoms replacement time in the volume of the ionizer cell, then an efficiency of about 6-8 % can be deduced. This result exceeds the initial expectation, which was based on an assumption that the limiting factor would be H⁻ ion stripping by plasma electrons. Precautions were taken to reduce the electron density and there are speculations that a pure D^+D^- plasma is actually produced in the ionizer. At the estimated plasma temperature of about 10 eV, the charge-exchange production cross-section ($H^0 + D^-$ to $H^- + D^0$) is about $43 \cdot 10^{-15}$ cm², and neutralization cross-section ($H^- + D^+$ to $H^0 + D^0$) is about $153 \cdot 10^{-15}$ cm², therefore, up to 20% of the polarized atoms in the cell can be converted to H⁻ ions. Taking into account that a polarized ion beam is affected strongly by the space-charge of the higher current unpolarized D⁻ ion beam, it might be that the achieved efficiency is already close to its maximum. The use of superconducting separating sextupoles and further dissociator and plasma ionizer optimization might produce a polarized H⁻ ion beam with a peak current of 10 mA.

OPPIS FOR RHIC SPIN PHYSICS

The polarization facilities at RHIC will provide 70% polarized proton-proton collisions at energies up to sqrt(s) = 500 GeV and a luminosity of $2 \cdot 10^{32}$/cm² s /1/. This luminosity will be obtained with 57 bunches of polarized proton beam having $2 \cdot 10^{11}$ particles/bunch intensity in each ring. The polarized source must produce in excess of 0.5 mA H⁻ ion current during a 300 μs pulse, or current×duration >150 mA μs, within a normalized emittance of less than 2 pi mm mrad. This current corresponds to $9 \cdot 10^{11}$ particles/pulse. Assuming 50% beam losses in the LEBT, RFQ, LINAC, and injection to the AGS Booster, that gives $4.5 \cdot 10^{11}$ polarized protons per booster pulse and finally $2 \cdot 10^{11}$ particles in the RHIC bunch. A 1.6 mA DC polarized H⁻ ion current was obtained at the TRIUMF OPPIS during the feasibility studies for the polarized beam in the FNAL Tevatron collider /9/. The ECR-type primary proton source used in the TRIUMF OPPIS is similar to the OPPIS which was first constucted at KEK /11/. Polarized beam is not presently required at KEK, and Y.Mori suggested using this source as an injector for RHIC. The source has been upgraded at TRIUMF to meet the RHIC requirements . After upgrade completion, the OPPIS was moved to BNL for installation at the RHIC injector in October of 1999 and has been used for the RHIC spin comissioning in September of 2000.

a). **ECR proton source upgrade:** A 28 GHz ECR source is used at TRIUMF vs. 18 GHz at KEK. In the KEK OPPIS the protons are produced in a 6.4 kG field and extracted at a 25 kG field which is nessesary to obtain high polarization. With the 28 GHz frequency at TRIUMF the resonance field is 10 kG . This gives a factor of 2-3 current gain, other conditions being similar, for the TRIUMF OPPIS. In DC operation, the extraction grids are hot, which prevents Rb metal deposition and provides reliable long-term practically spark-free operation. After modification to 28 GHz DC operation, a 1.0 mA H⁻ ion current was obtained with a 199 hole extraction system within the specified emittance (the conventional oven-type sodium ionizer cell was used in initial tests). The same extraction system produced twice as much current in the TRIUMF OPPIS. The difference is partly due to the longer distance between the ECR source and the ionizer (the KEK superconducting solenoid has a room temperature yoke and the TRIUMF solenoid has a cold yoke). In addition, the large hole in the KEK solenoid yoke disturbs the magnetic field symmetry and might be responsible for transverse field components which mis-steer the proton beam. This displacement was observed by direct measurements of the polarized atomic beam profile at the entrance of the ionizer cell. The biggest problem was the degradation of the source performance, within 12 hrs, to 50% or less of the initial current obtained with fresh grids and cavity assembly. Similar effects were observed in the past for the TRIUMF OPPIS operation, but due to an ample excess of current for routine beam production, the initial decrease was not a problem. After systematic tests the explanation has been found in the ECR gas composition. With a fresh source assembly there is water vapor contamination to the hydrogen in the discharge tube. Water is desorbed from the boron-nitride cups which isolate the plasma from the copper cavity walls. As the cavity dries out the ECR current goes through a maximum (in a few hours) and then drops. A controlled water vapor supply was set up, comprising a water reservoir at 0 deg. C, a needle valve and bypass pumping. When properly tuned, the H⁻ ion current recovered to its best value and remained stable for hundreds of hours of operation. Another remarkable feature was very quiet ECR operation. Similar behavior was observed with an oxygen admixture to the hydrogen supply. It was speculated that an oxygen admixture helps to activate the wall surface for better electron emission to the ECR plasma, the same as in ECR sources of multiply-charged ions.

b). **Pulsed laser development:** DC Ti:sapphire lasers were used for optical pumping in both the KEK and TRIUMF sources. Higher laser power is required for the high-current OPPIS which is easier to obtain in a pulsed operation. Optical pumping of the high density Rb vapor in the RHIC OPPIS is produced by a pulsed flashlamp pumped Cr 3+ doped LiSrAl F (Cr:LiSAF) crystal operating at 795 nm wavelength ($5S_{1/2}$ to $5P_{1/2}$ transition D1 line for Rb atoms). It is an excellent replacement for the initially used Ti:sapphire crystal, due to much longer upper level lifetime (a 67 μs for Cr:LiSAF vs. 3.4 μs for Ti:sapphire). A 400 μs laser pulse duration was easily obtained with the LiSAF crystal and a pulsed power of up to 1.0 kW. The maximum duration of the Ti:sapphire laser pulse doesn't exceed 200 μs. The wavelength is tuned by a 3-plate birefringent filter and a 0.5 mm thick uncoated etalon. A Burleigh WTA-2000 pulsed wavemeter is used for wavelength measurements and on-line monitoring. The laser linewidth is about 8 GHz. The Faraday rotation technique is used for polarization measurements. Nearly 100% Rb electron polarization was measured at a Rb cell thickness of 10^{14} atoms/cm² in a two cm diameter optically-pumped neutralizer cell.

At BNL the LiSAF laser is installed in a clean room and the laser beam is delivered to the OPPIS entrance window in a 20 m long beam transport line. Two focusing lenses are used to match the laser beam to the Rb cell diameter. About half of the initial laser power was measured through the 15 mm diameter collimator at the entrance of the Rb cell. The LiSAF crystal is cooled by deionized water (about 10 MOhm/cm) to reduce the crystal dissolution rate. The water circulation and temperature stabilization is produced by a NESLAB HX-75 chiller-circulator. The laser pulse repetition rate was 1 Hz and flashlamp input energy is about 60 J. The simmer arc current is 0.5 A. Flashlamp life-time (with a 4.0" arc gap) at these conditions is more than a million pulses. Laser operation was reliable during several months of OPPIS tests and polarized run in AGS and RHIC.

c). **Sodium-jet ionizer cell:** The polarized H⁻ ion beam emittance is mostly determined by the ionizer cell aperture diameter and solenoid magnetic field, due to the well known effect of emittance growth during ionizatioin in a magnetic field. The contribution of the primary neutral H beam emittance is neglegable. A field of 0.15 T is required to reduce polarization losses to below 2.5 % during ionization. Therefore, the specification for beam emittance of 2.0 πmm mrad gives the limit for the sodium cell aperture diameter of 2.0 cm. The sodium vapor flow and corresponding sodium consumption, deposition and penetration into the low field region is proportional to the cube of the cell diameter in the oven-type cell. The laser beam diameter in the optically-pumped cell has to be larger than the proton beam diameter to ensure high electron polarization of Rb atoms . The laser beam must pass through the ionizer cell, therefore the larger cell aperture results in a higher polarized ion current.

The neutral atomic beam of 3.0 keV energy enters the ionizer, and H⁻ ions produced in the cell can then be accelerated to 35 keV energy (which is required for injection to the RFQ) by ionizer biasing to – 32.0 kV. A large cell aperture is essential for this purpose because the neutral beam, collimated to 2.0 cm in diameter before the cell, should not touch the biased cell parts, otherwise secondary electron emission will cause sparking. A new jet-type ionizer cell with transverse sodium flow was developed to allow large apertures (see Fig.1) .

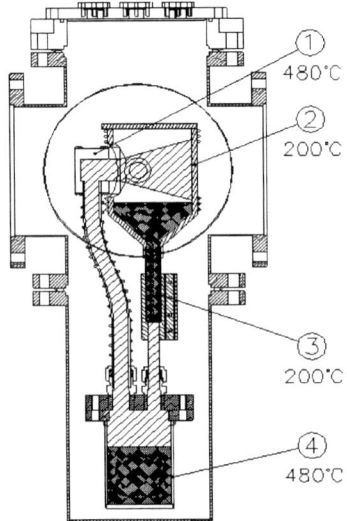

FIGURE 1. Sodium jet ionizer cell: 1- nozzle; 2- collector; 3- return line 4- sodium reservoir.

The reservoir is loaded with about 150 g of sodium metal and heated to 480 deg. C. At this temperature the sodium vapor density is about 10 atoms/cm^3. The vapor is delivered through a hot transport tube to the nozzle assembly, which produces a horizontal vapor jet having an effective thickness of about 5310^{15} atoms/cm^2, sufficient for H⁻ yield saturation. A nozzle slit is 2.0 mm wide and 20 mm tall. The transport tube and nozzle temperatures are mainted at 480 deg.C. The sodium vapor condenses at the collector walls, which are air-cooled to about 180 deg.C. At this temperature the sodium vapor density is 2310^{12} and the sodium viscosity is low. Liquid sodium flows easily down the return tube and back to the reservoir. The return tube temperature is kept at about 200-250 deg.C by an attached cooling line. The backstream vapor flow through the return tube is negligible due to the low conductance at 200 deg.C. Sodium in the jet-cell circulates along the path reservoir-nozzle-collector-return line-reservoir and the system provides continual, stable operation for hundreds of hours with 150 g of sodium. Without the circulation the cell works for only 3 hours, measured in a test with the collector water cooled to 25 C. The frozen sodium in the collector had a volcano shape perfectly confined within the 10 cm collector length. The sodium flow outside the cell was much less than with an oven-type cell. The whole ionizer assembly, including the solenoid magnet is electrically isolated from the rest of the OPPIS by 50 mm thick Delrin flanges.

An immediate acceleration reduces polarized current losses and improves polarization, since the beam energy during spin-transfer collisions can be kept optimal for efficient polarization transfer and ionization (below 3.0 keV). Sodium losses are greatly reduced in the jet-cell, which is essential because a large diameter is necessary to exclude direct exposure of the biased cell parts to the 3.0 keV energy atomic beam. The atomic beam is reduced to a diameter of 20 mm by a grounded collimator. The jet-cell apertures are 30 mm in diameter. The acceleration to 35 keV is produced in a two gap extraction system of 25 mm and 75 mm lengths (see Fig. 2). The extraction voltage applied to the first gap is 4.0 kV and to the second gap is 28.0 kV.

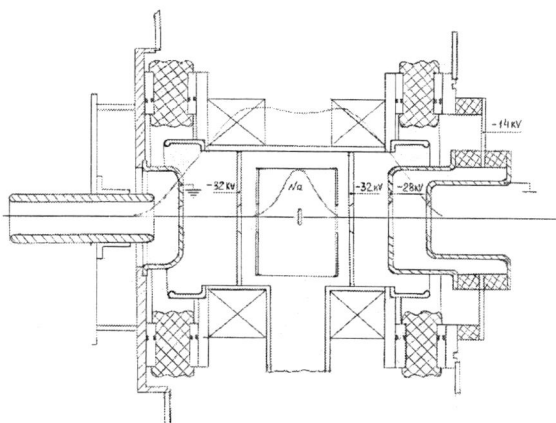

FIGURE 2. H- beam acceleration after Na-jet ionizer cell.

e).**Experimental results:** For final tests the BNL OPPIS was installed at the operational TRIUMF OPPIS bench. The TRIUMF Oxford made superconducting solenoid and polarimeters were used in these measurements. A focusing effect of the acceleration gap and improved transport efficiency of the higher energy beam allowed a very high DC polarized H$^-$ ion current of 1.6 mA to be obtained. This current is similar to the previous record of 1.64 mA /9/, but, at this time it was obtained with a 120 hole ECR extraction system instead of a 199 hole, therefore higher polarization was expected. The proton polarization was optimized by use of a Lamb-shift polarimeter and measured after acceleration to 300 keV with a nuclear scattering polarimeter based on the Li-6 (p , He-3) He-4 reaction (see Fig.3).

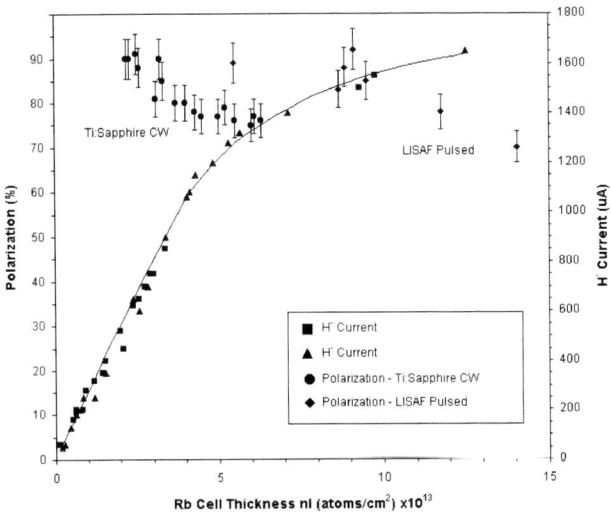

Figure 3: H- current and polarization vs Rb vapor thickness.

At low Rb thickness a polarization of 85% was achieved in DC mode with a single 4.5 W Ti:sapphire laser. At higher Rb density the polarization drops, but the use of a pulsed LiSAF laser restores the polarization to 85 % at 1.6 mA beam intensity. The polarized source was shipped to BNL and installed at the 200 MeV linac injector complex in September of 1999. The OPPIS was used for the RHIC SPIN comissioning in August-September 2000. The OPPIS worked continuously and reliably during a three week final run. So far, a 800 uA pulsed polarized H$^-$ ion current ha been obtained out of the source and 350 uA was accelerated to 200 MeV for injection to the booster. This gives $63 10^{11}$ H$^-$ions in a 300 µs long pulse, which is sufficient to produce the required $23 10^{11}$ particles/bunch in RHIC. A 65-72 % polarization was measured at 200 MeV beam energy after the linac by using a proton-carbon scattering polarimeter . Some polarization losses were obseved in the 35 keV LEBT line due to large spin precession in the bending dipoles and focusing solenoid in front of the RFQ. The Lamb-shift polarimeter is presently completed and will be used for optimization of the OPPIS polarization at a beam energy of a few keV.

PULSED OPPIS DEVELOPMENT

An order of magnitude higher polarized beam intensity will be required for the future polarized RHIC luminosity upgrade and also for the proposed polarized proton facility at the HERA electron-proton collider. The BNL OPPIS is basically a DC source, the pulsed beam is produced only at acceleration after the ionizer cell, the laser for optical pumping is also pulsed. A much higher current of 20-30 mA was obtained in a pulsed INR, Moscow -type OPPIS with an atomic hydrogen injector in experiments at TRIUMF /10,12/. For polarization measurements, the atomic H injector was installed at the extended TRIUMF OPPIS test bench, the ECR proton source was replaced with a pulsed He ionizer cell, and a new 45 cm long Rb cell was installed. The proton polarization was measured by the Lamb-shift polarimeter, which was tested and calibrated with the dc polarized beam from the TRIUMF OPPIS. The source was operated at a 1 Hz repetition rate and about 100 µs pulse duration. The optical pumping was produced by the pulsed Cr:LiSAF laser as described above. The TRIUMF superconducting solenoid was designed for ECR source applications. While the total length is 105 cm , the flat field region of a 2.45 T produced by the main coil is only 30 cm long. The fields of the other two coils are limited, and to produce a more or less flat field 60 cm long, only 10.0 kG is achievable. The low field in the optically-pumped Rb cell causes about 40 % polarization losses. The results of polarization measurements (see Fig.4) are to be compared with the dashed curve which is calculated for polarization in the 10 kG field.

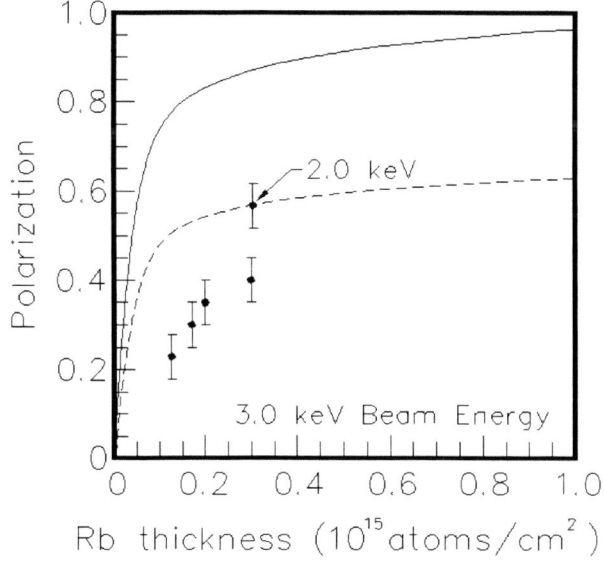

FIGURE 4. Solid line – expected polarization in the 25kG field.

The polarized H⁻ ion current is reduced to 8.0 mA with the He ionizer in operation because of nonhomogeneity of the magnetic field in the cell. The polarized proton current of 50 mA was obtained by replacement of the sodium ionizer cell by a pulsed He cell. The results of current and polarization measurements are presented in Table.I.

TABLE 1. Results of Current and Polarization Measurements.			
Beam energy (keV)	2.0	3.0	4.0
Peak Current I (mA)	0.1	0.5-1.0	0.5-1.5
H⁻ Current (mA)	5.0	8.0	14.0
H⁺ Current (mA)	16.0	50.0	-
Polarization (%)	5565	4265	3065

A new superconducting solenoid is being designed at BNL which will produce a 30 kG field with a flat top 80 cm long. The calculated proton polarization with such a solenoid is 80-85 %. The use of the jet-type sodium ionizer cell and immideate beam acceleration should further increase the achivable polarized H⁻ ion current as it was demonstrated in the BNL OPPIS. The expected polarized H⁻ ion current from the pulsed OPPIS is 30-50 mA, which is close to the present operational unpolarized H⁻ ion sources.

ACKNOWLEDGEMENTS

I am thankful to A.Belov and V.Derenchuk for information about the status of their developments and to D.Toporkov and E.Steffen for useful discussions on the limitation of the atomic beam intensity. I acknowledge Y.Mori's (KEK) original proposal and further numerous contributions to the BNL OPPIS development. I would like to thank D.Dutto, P.Levy, G.Wight (TRIUMF), J.Alessi, T.Roser, A.Lechrach (BNL), M.Okamura (RIKEN), and V.Klenov, S. Kokhanovski, V.Zoubets (INR, Moscow) for their contributions to the OPPIS development. The pulsed OPPIS development has been partly funded by DESY through the SPIN Collaboration (spokesperson A.Krish) and by INR, Moscow.

REFERENCES

1. J.Bunce et al., "Polarized protons at RHIC", Particle World, 3, p.1, (1992).
2. "Prospects of the Spin Physics at HERA", DESY Report 200-95, (1995).
3. W.Haeberli, "Sources of polarized negative ions", Lausanne, p.199,(1990).
4. D.Toporkov, talk given at this conference.
5. J.Fedchak et al., "The Argonne laser-driven polarized D target", AIP Conf.Proc.421, p.129, (1997).
6. T.B. Clegg , "Review of high intensity polarized ion sources", Ref.5, p.336.
7. J.Alessi et al., "Polarized ion sources for AGS", Helvetica Physica Acta, v.59, p.563, (1986).
8. A.Zelenski, "Optically-pumped polarized ion sources", Int.Workshop on Polarized beams and targets", Cologne 1995, World Scientific, Singapore, p.111, (1995).
9. A.Zelenski et al., The TRIUMF high-current DC OPPIS", Proc.1995 IEEE PAC, Dallas, p.864, (1995)
10. A.Zelenski et al., "OPPIS development for precision experiments and high-energy colliders", Ref.5, p.372
11. Y.Mori et al., AIP Conf.Proc.117, p.123, (1983)
12. A.Zelenski et al., "OPPIS for RHIC and HERA colliders", DESY Report , 1999.

Polarized Photon Beam Experiments at SPring-8

T. Nakano for the LEPS collaboration

RCNP, Osaka University,
10-1 Mihogaoka, Ibaraki, Osaka 567-0047, JAPAN

Abstract. The GeV photon beam at SPring-8 is produced by backward-Compton scattering of laser photons from 8 GeV electrons. Polarization of the photon beam will be ∼100 % at the maximum energy with fully polarized laser photons. We report the status of the new facility and the prospect of ϕ photo-production study with this high-quality beam. Preliminary results from the first physics run are presented.

INTRODUCTION

Hadron-hadron total cross sections (including a γp total cross section) in a wide energy range are reproduced very well in terms of two s-dependent terms with $s^{-0.5}$ and $s^{0.08}$ dependences, where s is the Mandelstam s variable [1]. The Reggeon exchange model clearly suggests that the $s^{-0.5}$ term originates from the ρ meson ($T = 1$, $J^\pi = 1^-$.) trajectory. On the other hand, the $s^{0.08}$ term requires the introduction of an unobserved Regge trajectory known as the pomeron trajectory [2], whose $\alpha(t = 0)$ has to be 1.08. One can identify the Pomeron trajectory with a glueball trajectory since the first particle state on the trajectory appears at $m^2 \sim 4$ GeV^2 with $J = 2$ (2^{++} glueball).

At high energies, diffractive photo-production of a ϕ meson from a proton target is well described as a pomeron-exchange process in the framework of the Regge theory and of the Vector Dominance Model (VDM) [3]; a high energy photon converts into a ϕ meson and then it is scattered from a proton by an exchange of the pomeron [4–6]. However, at low energies other contributions arising from meson(π, η)-exchange [6], a scaler(0^{++} glueball)-exchange [7], and $s\bar{s}$ knock-out [8] may be detectable.

In contrast to ω photo-production, the meson-exchange contribution in the ϕ photo-production is highly suppressed by the OZI rule. The ϕ production measurements near the threshold may reveal the existence of another gluon-exchange trajectory (a 0^{++} glueball trajectory) whose contribution falls off rapidly as the incident gamma energy increases [7,9]. If the contribution from the 0^{++} glueball trajectory is small, it is difficult to observe because the meson-exchange contribution becomes comparable. However, precise measurements of spin observables with linearly polarized photons will be useful to decompose these two contributions [9]. For natural-parity exchange such as pomeron and 0^{++} glueball exchanges, the decay plane of K^+K^- is concentrated in the direction of the photon polarization vector. For unnatural-parity exchange processes like π and η exchange processes, it is perpendicular to the polarization vector.

In the measurement of ϕ production, one can also address the question concerning the $s\bar{s}$ components in nucleon. Spin observables magnify small amplitudes hidden in a dominant amplitude by interference effects. The interference between a pomeron-exchange amplitude and a small amplitude due to a direct knockout of $s\bar{s}$ in the nucleon may cause a large asymmetry of the production cross sections between the spin parallel and anti-parallel configurations of the polarized photons and polarized nucleons [9].

In addition, a Compton-like direct ϕ radiation contribution arising from a ϕNN vertex can be studied by detecting extremely forward protons from u-channel ϕ photo-production. It is interesting to compare the cross-section with the existing data for u-channel ω photo-production [10] in order to understand the nature of large OZI violation in BR($p\bar{p} \to \phi\gamma$)/BR($p\bar{p} \to \omega\gamma$) [11].

LEPS FACILITY

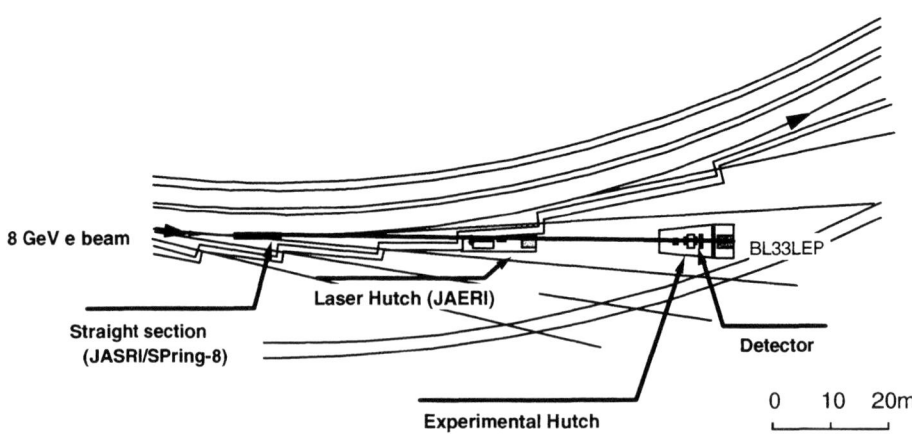

FIGURE 1. Plan view of the Laser-Electron Photon facility at SPring–8 (LEPS).

The Spring-8 facility is the most powerful third-generation synchrotron radiation facility with 61 beamlines. We use a beamline, BL33LEP (Fig. 1), for the quark nuclear physics studies. The beamline has a 7.8-m long straight section between two bending magnets. Polarized laser photons are injected from a laser hutch toward the straight section where Backward-Compton scattering (BCS) [12] of the laser photons from the 8 GeV electron beam takes place (Fig. 2). The BCS photon beam is transfered to the experimental hutch, 60 m downstream of the straight section. The maximum energy of the BCS photon is expressed by

$$k_2 = \frac{4k_1 E_e^2}{m_e^2 + 4k_1 E_e}, \qquad (1)$$

where k_1 is the energy of the laser photon, E_e is the energy of the electron, and m_e is the electron mass. For a 351-nm Ar laser and a 8-GeV electron beam, the maximum energy is 2.5 GeV well above the threshold of ϕ-phtoproduction from a nucleon (1.57 GeV).

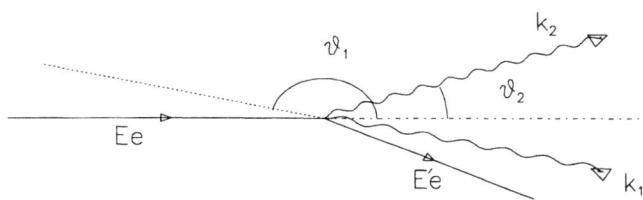

FIGURE 2. Backward-Compton scattering process.

If laser lights are 100 % polarized, a backward-Compton-scattered photon is highly polarized at the maximum energy. The polarization drops as the photon energy decreases as shown in Figure 3 and 4. However, an energy of laser photons is easily changed so that the polarization remains reasonably high in the energy region of interest (Figure 5). The intensity, position, and polarization of the laser lights which do not interact with the electron beam are monitored at the end of the beamline.

The incident phton energy is determined by measuring the energy of a recoil electron with a tagging counter. The tagging counter is located at the exit of the bending magnet after the straight section. It consists of multi-layers of a 0.1 mm pitch silicon strip detector (SSD) and plastic scintillator hodoscopes. Electrons in

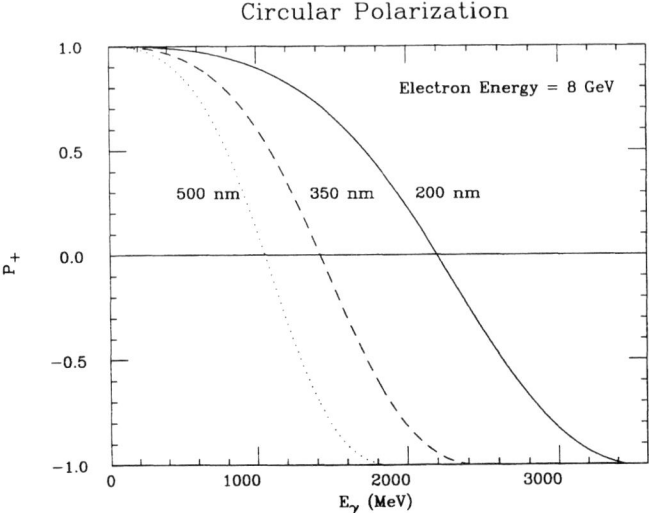

FIGURE 3. Circular polarization as a function of photon energy. Laser photons at various wavelengths are assumed to have 100 % circular polarization.

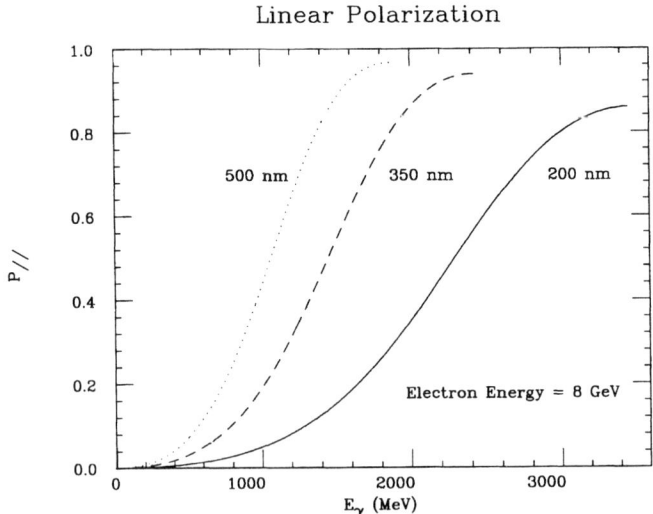

FIGURE 4. Linear polarization as a function of photon energy. Laser photons at various wavelengths are assumed to have 100 % linear polarization.

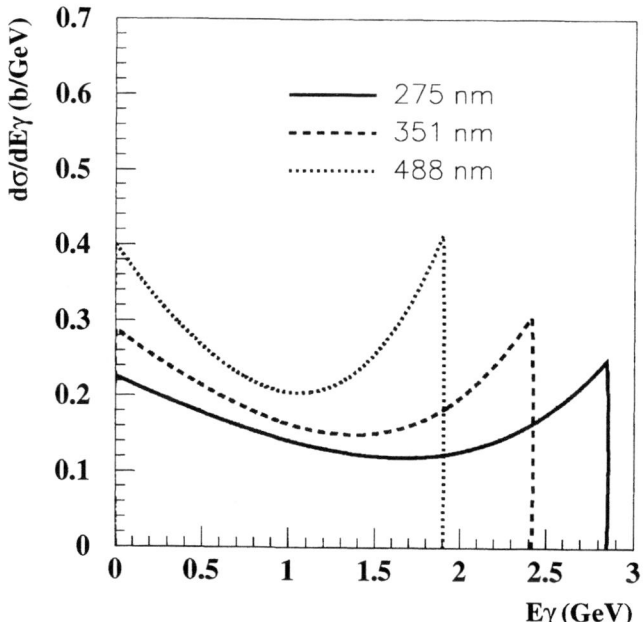

FIGURE 5. Differential cross-section of the Compton scattering at three typical wavelengths of Ar laser.

the energy region of 4.5 – 6.5 GeV are detected by the counter. The corresponding photon energy is 1.5 – 3.5 GeV. The position resolution of the system is much better than a required resolution. The energy resolution (RMS) of 15 MeV for the photon beam is limited by the energy spread of the electron beam and an uncertainty of a photon-electron interaction point.

The operation of the BCS beam at SPring-8 started in July, 1999. Ar laser at 351-nm wave length is used to produce 2.4-GeV BCS photon beam. The intensity of the beam is about 2.5×10^6 photons/sec for 5-W laser output.

DETECTOR

The LEPS detector (Fig. 6) consists of a plastic scintillator to detect charged products after a target, an aerogel cerenkov counter with a refractive index of 1.008, charged-particle tracking counters, a dipole magnet, and a time-of-flight TOF wall. The design of the detector is optimized for a ϕ photo-production in forward angles.

The opening of the dipole magnet is 135-cm wide and 55-cm height. The length of the pole is 60 cm, and the field strength at the center is 1 T. The vertex detector consists of 2 planes (x- and y-) of single-sided SSDs (SVTX) and 5 planes (x,x',y,y',u) multi-wire drift chamber (DC1), which are located upstream of the magnet. The stereo ambiguity (pairing ambiguity) for two-track events are solved

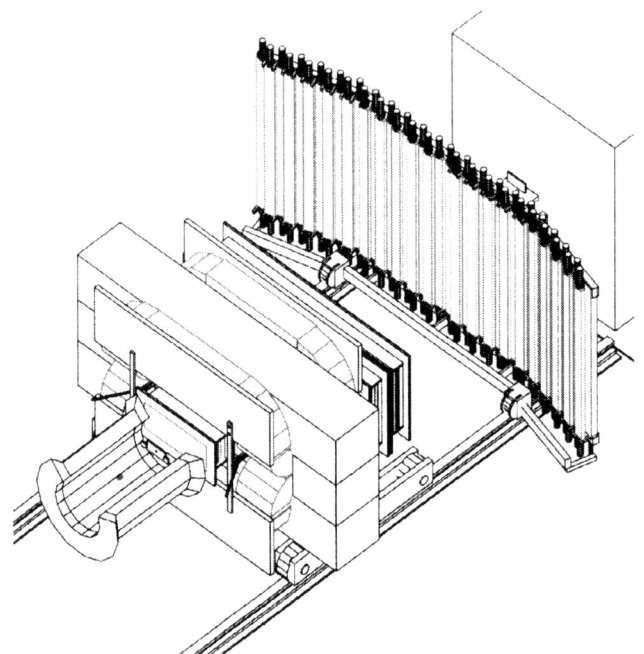

FIGURE 6. The LEPS detector setup.

with DC1. Two sets of MWDCs (DC2 and DC3) are located downstream of the magnet. The active area size of DC2 and DC3 is 200cm(W) × 80cm(H). Each set has 5 planes; x,x',u,u', and v.

The identification of momentum analyzed particles is performed by measuring a time of flight from the target to the TOF wall. The start signal for the TOF measurement is provided by a RF signal from the 8-GeV ring, where electrons are bunched at every 2 nsec (508MHz) with a width (σ) of 12 psec. Since the speeds of the both electron beam and a laser-electron photon are same, the arrival time of the laser-electron photon at the target is synchronized with the RF signal. A stop signal is provided by the TOF wall consisting of 40 2m-long plastic scintillation bar with a cross section of 4cm (t) × 12 cm (w). The resolution of the TOF counter is about 100 psec.

The first physics run with a CH_2 target started in May, 2000. The trigger required a tagging counter hit, no charged particle before the target, charged particles after the target, no signal in the aerogel cerenkov counter, at least one hit on the TOF wall. A typical trigger rate was about 100 counts per second. Figure 7 shows a preliminary mass distribution of charged particles reconstructed from momentum and TOF information.

A ϕ meson is identified through the reconstruction of the KK invariant mass. The estimated mass resolution is 600 keV, which is much smaller than the ϕ width. The measurement error of a momentum transfer is about 0.01 GeV^2, and it mainly

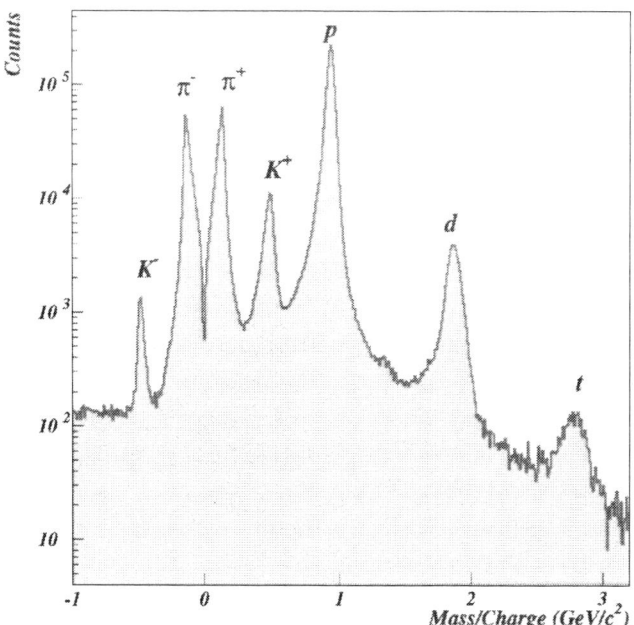

FIGURE 7. A preliminary mass distribution of charged particles reconstructed from momentum and TOF information.

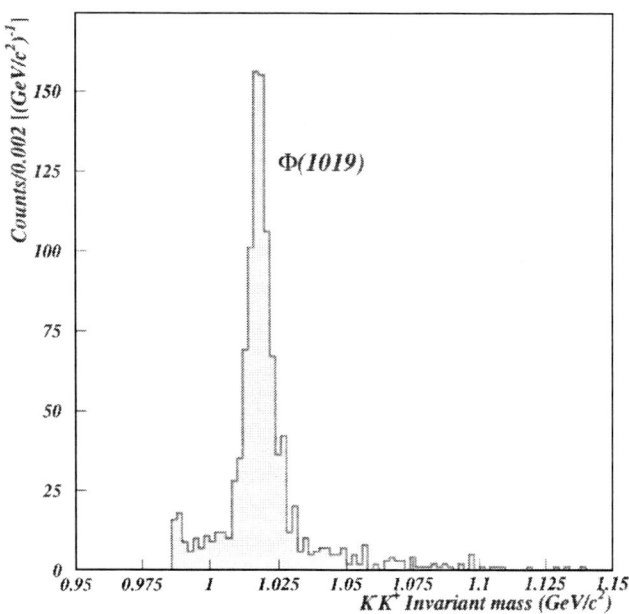

FIGURE 8. A preliminary two-kaon invariant mass distribution. A ϕ peak is clearly identified.

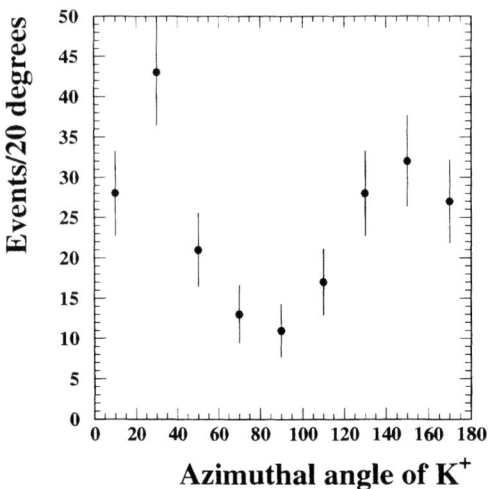

FIGURE 9. The azimuthal angular distribution of K^+K^- decay plane (very preliminary). The photon polarization vector was in parrallel with the y axis.

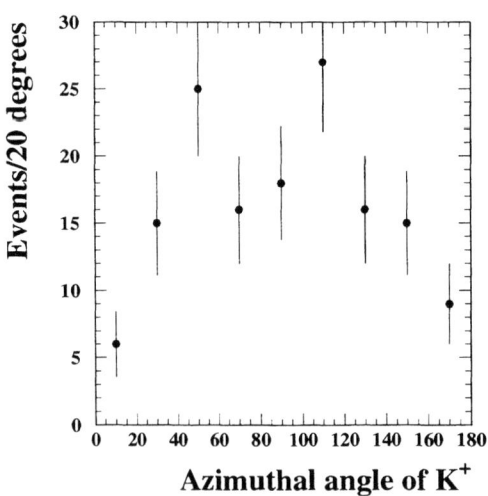

FIGURE 10. The azimuthal angular distribution of K^+K^- decay plane (very preliminary). The photon polarization vector was perpendicular to the y axis.

comes from the measurement error of the incident photon energy. Figure 8 shows a preliminary two-kaon invariant mass distribution. Figure 9 and Figure 10 show the azimuthal angular distribution of K^+K^- decay plane when the photon polarization vector was in parallel with or perpendicular to the y axis. The event are more concentrated in the direction which is along to the polarization vector. And it indicates the ϕ photoproductiion is dominated by natural-parity exchange processes. Experiments with a LH_2 target will start in December, 2000. Differential crossections and spin observables will be measured in the photon energy from the threshold to 2.4 GeV.

REFERENCES

1. A. Donnachie and P. V. Landshoff, Phys. Lett. B 296 (1992) 227.
2. I.Y. Pomeranchuk, Sov. Phys. 7, (1958) 499.
3. J.J. Sakurai, Ann. Phys. 11 (1960) 1; J.J. Sakurai, Phys. Rev. Lett. 22 (1969) 981.
4. T.H. Bauer *et al.*, Rev. Mod. Phys. 50 (1978) 261.
5. A. Donnachie and P.V. Landshoff, Nucl. Phys. B267 (1986) 690.
6. M.A. Pichowsky and T.-S. H. Lee, Phys. Rev. D56 (1997) 1644.
7. T. Nakano and H. Toki, in Proc. of Intern. Workshop on Exciting Physics with New Accelerator Facilities, SPring-8, Hyogo, 1997, World Scientific Publishing Co. Pte. Ltd., 1998, p.48.
8. A.I. Titov, Y. Oh, and S.N. Yang, Phys. Rev. Lett. 79 (1997) 1634;
   A.I. Titov, Y. Oh, and S.N. Yang, Phys. Rev. C58 (1998) 2429.
9. A.I. Titov, T.-S. H. Lee, and H. Toki, Phys. Rev. C59 (1999) 2993.
10. R.W. Clifft *et al.*, Phys. Lett. B 72 (1977) 144.
11. C. Amsler *et al.*, Phys. Lett. B 346 (1995) 363.
12. R.H. Milburn, Phys. Rev. Lett. 10 (1963) 75; E. Hourany, in these proceedings.

Experiments with polarized photon beam at GRAAL

J. Ajaka[1], M. Anghinolfi[2], Y. Assafiri[1], O. Bartalini[3], M. Battaglieri[2], V. Bellini[4], J.P. Bocquet[5], P. Calvat[5], M. Capogni[1,3], M. Castoldi[2], P. Corvisiero[2], A. D'Angelo[3], J.P. Didelez[1], R. Di Salvo[1,3], M.A. Duval[1], E. Guinault[1], L. Fichen[1], C. Gaulard[9], G. Gervino[6], F. Ghio[7], B. Girolami[7], M. Guidal[1], E. Hourany[1], I. Kilvington[11], V. Kouznetsov[8], R. Kunne[1], A. Lapik[8], P. Levi Sandri[9], A. Lleres[5], D. Moricciani[3], V. Nedorezov[8], L. Nicoletti[5], D. Rebreyend[5], F. Renard[5], M. Ripani[2], N. Rudnev[8], M. Sanzone[2], C. Schaerf[3], M. Taiuti[2], A. Turinge[10], Q. Zhao[1], A. Zucchiatti[2]

[1] *IN2P3, Institut de Physique Nucléaire, 91406 Orsay, France.*
[2] *INFN Genova and Dipartimento di Fisica, 16146 Genova, Italy*
[3] *INFN sezione di Roma II and Università di Tor Vergata, Roma, Italy*
[4] *INFN Laboratori Nazionali del Sud and Università di Catania, Catania, Italy*
[5] *IN2P3, Institut des Sciences Nucléaires, 38026 Grenoble, France*
[6] *INFN sezione di Torino and Università di Torino, Torino, Italy*
[7] *INFN sezione Sanità and Instituto Superiore di Sanità, Roma, Italy*
[8] *Institute for Nuclear Research, Moscow, Russia*
[9] *Laboratori Nazionali di Frascati, Frascati, Italy*
[10] *I. Kurchacov Institute of Atomic Energy, Moscow, Russia*
[11] *European Synchrotron Radiation Facility, 38026 Grenoble, France.*

Abstract. Since 1996 the GRAAL experiment is running at the ESRF, using a linearly polarized photon beam produced by backscattering a laser beam on the electron beam in the storage ring. A liquid hydrogen target and a large acceptance detector are used. All sizeable photoproduction reactions are measured and analysed. We present the results of η meson, one and two pions, and ω photoproduction, which are partly published or still preliminary. Predictions or fits of these data by theoretical models are given.

INTRODUCTION

GRAAL is an experiment with a polarized photon beam at the European Synchrotron Radiation Facility (ESRF) in Grenoble, France. The experiment studies the resonances of the nucleon in photoproduction reactions on a hydrogen target with a linearly polarized photon beam of energy ranging from 500 to 1500 MeV.

At one line of the ring of the ESRF, which is filled with an electron beam of 6 GeV and 200 mA, a tagged and polarized photon beam was produced by Compton backscattering an argon laser beam on the electron beam in the ring. The produced photon beam first hits a liquid hydrogen target surrounded by a 4π detector, then crosses a thin monitor and finally is stopped into a thick calorimeter. The energy of the photons is measured by detecting the correlated electrons in the Compton backscattering, near the intersection region in the ring.

For each selected line of the laser spectrum, the produced photon beam has a flat energy spectrum of Compton shape and a corresponding maximum energy which is 1100 MeV for the green line and 1500 MeV for the UV line. The polarization of the laser beam is transmitted to the produced photon beam with a high rate at the upper part of the energy spectrum and close to 100% at the maximum energy. GRAAL has operated by selecting either the green line of the argon laser to cover from 600 to 1100 MeV or the UV line of the laser to cover from 900 to 1500 MeV. These provide for a given reaction two complementary energy ranges with high degrees of polarization.

The detection system is composed of three distinct parts. A central part covering from $\theta=25°$ to $155°$ and consisting of three layers, two cylindrical wire chambers, a barrel of scintillators and a ball of 480 cristals of BGO. The BGO ball is a calorimeter of high resolution for the neutral mesons ($\pi^°$ and η) through their decays into 2γ or 6γ. The forward part consists also of three layers, two plane wire chambers, a double wall of scintillator bars and a shower wall. These walls placed at 3 m from the the target provide time of flight measurements for charged and neutral particles. The backward detector is composed of two disks of scintillators sensitive to charged particles.

Despite the absence of a magnetic field, this detection setup has several particle identification capabilities. For instance, the invariant mass spectrum reconstructed for 2 γ detected in the BGO ball displays two sharp peaks corresponding to $\pi^°$ and η mesons. Also, in each of the scintillator walls, the plot of the energy loss ΔE versus the time of flight shows clearly the characteristic line of the proton and the time of flight spectrum of the shower wall presents a narrow peak for the γ and a wide spectrum for the neutrons which are slower. On the other hand, the granularity of each of the three layers is good, so allowing to measure the θ and ϕ angles of charged and neutral particles.

As to the reaction identification, it is achieved with overdetermination to most channels by applying the energy and momentum conservation laws. With higher overdetermination the identification is easier and the cleaning from the background is better.

GRAAL has a high performance in the measurement of the beam asymmetry Σ corresponding to a linear and transversal polarization of the photon beam. For that purpose, the whole setup has by construction an azimuthal symmetry around the beam direction. In addition, the methods of measurement and analysis were optimized to correct from azimuthal variations of various electronic origins (thresholds, gains, pedestals) : (i) during the measurements the direction of the polarization was alternated between vertical and horizontal and (ii) in the off-line analysis either the data were added up for the two directions of polarization to obtain results for unpolarized beam or the data were sorted according to the polarization direction.

Definitely, the data obtained for unpolarized beam contain azimuthal variations originating only from operating conditions and therefore are used to control them.

As to the data corresponding to one chosen direction of the polarization, they are used to extract the beam asymmetry Σ. For this, the events corresponding to a given photoproduction channel are binned in incident energy E_γ and polar angle θ_M of the produced meson. Then, for those events falling into a cell (E_γ, θ_M) the spectrum of the ϕ-angle of the produced meson is plotted and fitted with an expression $1+P\Sigma\cos(2\phi)$. The degree of polarization P is known, one deduces Σ.

With the described setup, three rounds of experiments are performed using a liquid hydrogen target of 6 cm, a beam intensity of 2×10^6 γ/sec and a duty cycle of the electron beam of 2/3 :

1) in 1996-1997, experiments with an energy range of 600-1100 MeV:

The sizeable channels in this energy range are the single pion photoproduction, $\gamma p \to p\pi^0$ and $\gamma p \to n\pi^+$, the double pion photoproduction $\gamma p \to p\pi^0\pi^0$, $\gamma p \to n\pi^+\pi^0$ and $\gamma p \to p\pi^+\pi^-$ and the η photoproduction $\gamma p \to p\eta$.

2) in 1998-2001, experiments with an energy range of 900-1500 MeV:

In this energy range the thresholds of kaon and ω meson photoproduction are crossed. So, we are measuring and analysing the following channels: (i) $\gamma p \to K^+\Lambda$ and $\gamma p \to K^+\Sigma^0$, (ii) $\gamma p \to p\omega$, and (iii) the extension to higher energy of the channels measured in the 600-1100 MeV energy range.

3) in 2002, experiments with a double polarization of the beam and the target.

A polarized target of new type, pure HD, will be installed at the end of 2001 and experiments with circular polarizations of the beam and longitudinal polarizations of the target will be performed to measure the DHG sum rule.

In the following sections, the results of GRAAL for the beam asymmetry Σ will be presented and an interpretation given. The degree of impact on the data bases will be outlined. The results in η and single pion photoproduction were partially published. The results of two pions and ω photoproduction are preliminary together with the accompagnying interpretation.

PHOTOPRODUCTION OF η MESON

The η meson is well detected with GRAAL detector through its various decay modes and in particular its decay into 2 γ. The photoproduction of η is identified over a negligeable background and is measured with good statistics, consequently to its high cross section of about 10 μb. The results of GRAAL complete recent data from Mainz and Bonn. All these recent data are obtained with tagged photon beams and large acceptance detectors.

There is a strong contrast between the poor quality of the scarce old data and the richness of the presently measured ones, so that the old data can be dropped out.

The interpretation of the old data all over the three last decades has shown the dominance of S11(1535) resonance excitation with respect to other resonances and the smallness of Born term and meson exchange term in t-channel, below 1 GeV.

During the last years several groups of theoreticians emerged and worked on the recent data of η photoproduction using effective Lagrangian with hadronic degrees of freedom or with direct quark-meson interaction in a quark model approach. A significant effort was spent on the simultaneous prediction or fit of complementary observables comprising the cross section and the single or double polarization observables.

The η photoproduction might be the simplest channel to be completely studied: because η is pseudoscalar there are required 7 independent observables to completely determine its transition amplitudes and because η has an isospin zero only the resonances of isospin I=1/2 could be excited. The Σ results of GRAAL in η photoproduction were completely new below 1 GeV, so that their interpretation was performed by the various existing models either separately or integrated in the new data base of the recent results. In both cases the GRAAL results singled out the quantitave contribution of D13(1520) in the photoproduction process and pushed the field towards a complete measurement.

Interpretation of Σ results

In figure 1, there are the Σ results of GRAAL in 6 bins of energy ranging from the threshold to 1100 MeV. Two sets of experimental points are displayed which correspond to two configurations of events measured at GRAAL [1].

The Σ values plotted versus the angle θ of η meson in the center of mass are positive and large and are very high at the highest energies and forward angles. The dashed and dotted curves are predictions by effective Lagrangien calculations published in 1994 [2] and 1995 [3]. The positive and large values of asymmetry obtained in these models come from the interference between the dominant S11 and the weakly excited D13. The continuous line is a result of a fit with an isobaric model performed by Bijan Saghai. The author tried to reproduce the high values at high energies and forward angles by including several resonances, S11(1535), D13(1520), P13(1720), D15(1675) and a "missing" resonance P13(1880). This gave a limited improvement. These high values were subsequently confirmed by GRAAL results obtained in the 900 - 1500 MeV experiment.

Impact of GRAAL data on the data base

A simultaneous interpretation was performed for the three sets of observables, $d\sigma/d\Omega$ measured at Mainz, the target asymmetry T measured at Bonn and the beam asymmetry Σ of GRAAL. Predictions of effective Lagrangien models exist and recently an almost model-independent calculation was carried out successively by two groups, based on an expansion into multipoles near the threshold of the transition amplitudes [4], [5]. Under the assumption of the dominance of s-wave and the additional contribution of p- and d- waves only by their interference with s-wave, the expansion could be limited to a few terms. Simple expressions of the three observables $d\sigma/d\Omega$, T and Σ are obtained. The fit of the 3 sets of data by these expressions determine the coefficients into the expansion and hence the multipoles. In figure 2, the dashed and dotted lines are predictions of an effective Lagrangien model corresponding to calculations with and without D13 respectively. While the two calculations give similar results for $d\sigma/d\Omega$ and T, they give a dramatic change for Σ: without D13, values close to zero and even negative are found, and, with D13, values compatible with the experimental ones are seen. In addition, not only the agreement is poor in T, but also the theoretical result does not show a nodal angular structure at low energy.

The continuous line is the result of the fit by the expansion into multipoles. This fit reproduces the 3 sets of observables, nevertheless with two drawbacks: the high values of Σ at high energies and forward angles are not reproduced even after allowing the contribution of higher waves through an additional term in the expression of Σ, and the fit of the nodal structure is possible only at the expenses of the wrong sign in the phase between S11 and D13. The authors claimed the confirmation of the experimental results in T and when extracting numerical values for various basic quantities they carried out results in two cases, case 1 for the 3 observables fitted together and case 2 for $d\sigma/d\Omega$ and Σ.

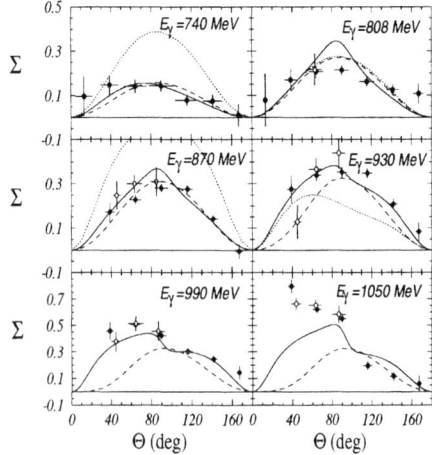

FIGURE 1. Beam asymmetry Σ of $\gamma p \to \gamma \eta$. The experimental data are GRAAL data. The theoretical curves are explained in the text.

The extracted values of the branching ratio of D13 into η are 0.08±0.01 and 0.05±0.02 in cases 1 and 2 respectively, close to the prediction of the quark model, which is of 0.09. As to the helicity amplitude $A_{1/2}$ and the ratio $A_{3/2}/A_{1/2}$ for the excitation of D13, they were found equal, in the two cases, to the following values:

$A_{1/2}$=-79±9 (to be compared to -24±9 of the PDG tables)

$A_{3/2}/A_{1/2}$=-2.1±0.2 (to be compared to -6.9±2.6 of the PDG tables).

This striking disagreement between these quantities extracted in η photoproduction of D13 and those of PDG tables based on pion photoproduction results has been previously found in the extraction of $A_{1/2}$ for S11 from η photoproduction using an expansion into multipoles near the threshold.

Higher energy results of GRAAL in $d\sigma/d\Omega$ and Σ which are under completion would help to explain the disagreement. But the better is to go further into measuring a complete set of observables, with the possibilities of simultaneous polarization of the beam and the target which is becoming available in the new modern facilities.

PHOTOPRODUCTION OF ONE PION

At GRAAL, the two channels of π^0 and π^+ photoproduction on the proton target were measured, analysed and partly published [6]. The cross sections and the beam

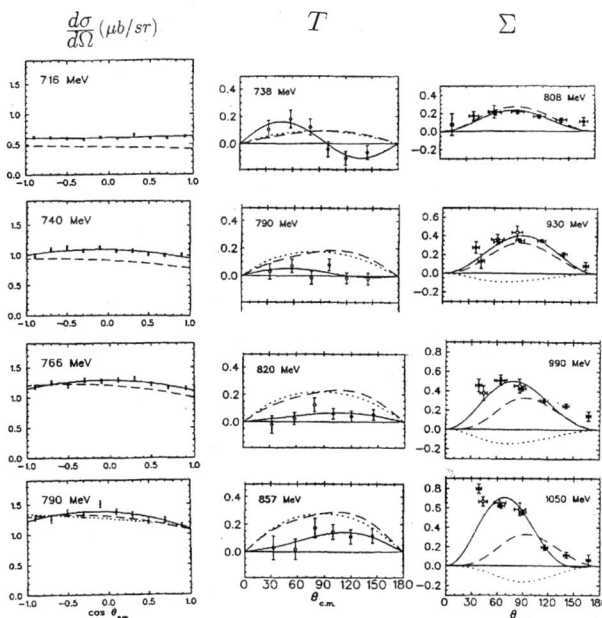

FIGURE 2. Cross section $d\sigma/d\Omega$ of Mainz, target asymmetry T of Bonn and beam asymmetry Σ of GRAAL of the reaction $\gamma p \to \eta p$ and their interpretation given in reference [5].

asymmetries are being extracted in the whole energy range from 500 to 1500 MeV.

For these channels, the existing data base is rich but with much more cross section data than polarization observables data, and with a poor coverage for energies higher than 1 GeV and forward or backward angles.

For the data of these channels, a compilation and a refined partial waves analysis was performed by VPI group with a code called SAID [7].

In figure 3, are shown the GRAAL data of Σ observable for the reaction $\gamma p \to n\pi^+$ [6]. The GRAAL data in solid circles or triangles agree with previous data and are of higher precision and extend them to backward angles. The VPI calculation which is in continuous line reproduces the experimental data except at some energies and backward angles. As to the predictions of Drecksel et al. [8], in dashed lines, they reproduce better the data at backward angles but deviate from them at forward angles.

Very precise data were obtained at GRAAL for the differential and total cross sections and the beam asymmetry Σ of the reaction $\gamma p \to p\pi^0$. The VPI calculation reproduces these results with small localized deviations.

PHOTOPRODUCTION OF TWO PIONS

We analyzed the following three channels:
(1) $\gamma p \to n\pi^+\pi^0$, (2) $\gamma p \to p\pi^0\pi^0$ and (3) $\gamma p \to p\pi^+\pi^-$. The π^0 was detected in the BGO calorimeter and only for channel (3) the detection of the

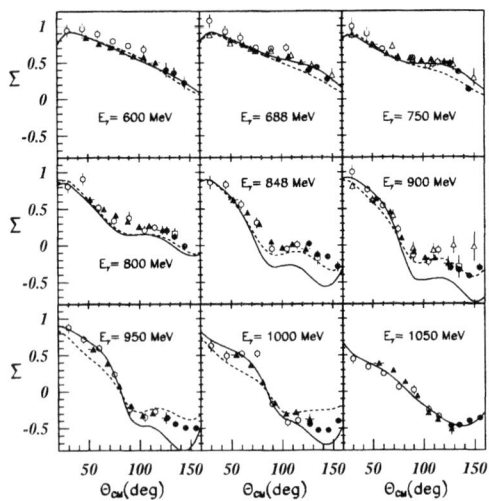

FIGURE 3. Beam asymmetry Σ of $\gamma p \to n\pi^+$. Black triangles and circles are GRAAL results for two configurations of events [6]. Open triangles, squares and circles are previous data. Continuous line is the VIP calculation [7] and dashed line is the calculation of reference [8].

proton was restricted to the forward scintillator detector.

Here, to apply the single meson photoproduction procedure we have to group one meson (M_1) to the nucleon N and to measure the beam asymmetry of the second meson (M_2): $\gamma\, p \rightarrow (N\, M_1)\, M_2$. To simplify we have applied the procedure for two extreme cases of invariant masses of the system (NM_1): by (i) taking the whole invariant mass spectrum and by (ii) taking a narrow band of invariant masses selected by a window around the peak in the invariant mass spectrum located at the Δ mass value (1232 MeV).

Also, the single meson photoproduction procedure was applied to the system of two pions considered as a single entity.

The efficiency correction was applied to the data before extracting the observable Σ. It turned out that the values of Σ were not significantly sensitive to this correction.

Our preliminary results were compared to a two pions calculation performed by Valencia group. Their model studies together all the two pions photoproduction channels on the proton and the neutron [9]. It uses an effective Lagrangien where the components are pions, nucleons and nucleonic resonances. Their calculation was performed in the energy range of 600 to 800 MeV, where they considered the contributions from N*(1440) which is excited on shell at these energies and the N*(1520) which has a large coupling to the photons.

This model was applied to calculate the total cross section of $\gamma p \rightarrow n\pi^+\pi^0$

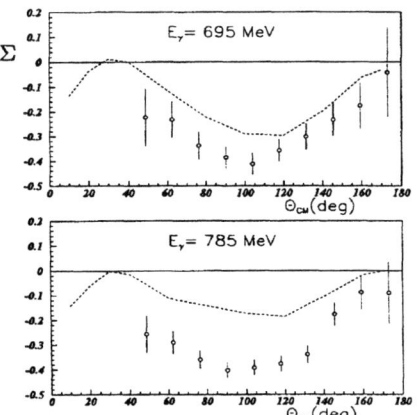 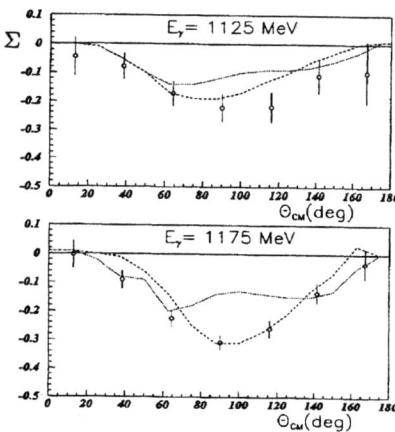

FIGURE 4. On the left side, the beam asymmetry Σ of $\gamma p \rightarrow (n\pi^+)\pi^0$. The experimental data are GRAAL data. The dashed lines are the results of Valencia team model. On the right side, the beam asymmetry Σ of ω photoproduction. The experimental data are GRAAL data. The dashed and dotted lines are the quark model results with all resonances and with all except P13(1670) respectively.

and $\gamma p \to n\pi^0\pi^0$ measured at Mainz. The model was applied to calculate the beam asymmetry measured at GRAAL for the various two pions channels presented above. We give in figure 4 (left side) the predictions of the model [10] for reaction (1) where the general behaviour together with the sign of the experimental asymmetries are reproduced.

PHOTOPRODUCTION OF ω MESON

In ω meson photoproduction, the identification and selection of the events were carried out for the case of the ω-decay into three pions (branching ratio of 89%):

$\gamma p \to p \omega \qquad \omega \to \pi^+ \pi^- \pi^0$

where the π^0 was detected in the BGO ball and the proton in the forward scintillator detector.

With a linear polarization of the photons, the distribution of the angles θ and ϕ of the normal to the decay plane of ω in the rest frame of ω is given by :

$$W(cos\theta, \phi, \Phi) = W^0(cos\theta, \phi) - P_\gamma . cos(2\Phi) . W^1(cos\theta, \phi) - P_\gamma . sin(2\Phi) . W^2(cos\theta, \phi)$$

where, Φ is the angle between the production plane of ω and the polarization of the beam, P_γ is the degree of polarization of the beam, and the expressions W^0, W^1, W^2 are functions of spin density matrix elements ρ_{ik}^α which are to be determined by fitting the experimental data and compared to values extracted from theoretical models. These matrix elements are in fact double polarization beam-ω meson observables.

Here, we present preliminary results of the beam asymmetry in the production of omega with integration over its decay products. Integrating $W(cos\theta, \phi, \Phi)$ over θ and ϕ yields :

$$W(\Phi) = 1 + P_\gamma . \frac{2\rho_{11}^1 + \rho_{00}^1}{2\rho_{11}^0 + \rho_{00}^0} . cos(2\Phi)$$

We measured the beam asymmetry Σ by fitting at each couple of incident energy and θ_ω angle bins the Φ distribution by an expression $K.(1 + P_\gamma.\Sigma.cos(2\Phi))$, without applying the efficiency correction which was not ready yet.

We compare the Σ values to the analog theoretical ones $\frac{2\rho_{11}^1 + \rho_{00}^1}{2\rho_{11}^0 + \rho_{00}^0}$ obtained in a quark model [11].

In this model an effective Lagrangian is used to study the s- and u- channel resonance contributions.

$$L_{eff} = -\overline{\psi}\gamma_\mu p^\mu \psi + \overline{\psi}\gamma_\mu e_q A^\mu \psi + \overline{\psi}(a\gamma_\mu + \frac{ib\sigma_{\mu\nu}q^\nu}{2m_q})\phi_m^\mu \psi$$

where ψ and $\overline{\psi}$ represent the quark and antiquark fields and ϕ_m^μ denotes the vector meson field. The two parameters a and b represent the vector and tensor couplings of the quark to the vector meson and m_q=330 MeV is the constituent quark mass.

The t-channel natural parity exchange is taken into account through the Pomeron exchange, while the unnatural parity exchange is described by the π° exchange.

The results of this model are consistent with the known characteristics of ω meson photoproduction at high energies, i.e., a dominance of Pomeron and π° exchanges in the total cross section with a forward diffractive peaking in the differential cross sections. Nevertheless, the contribution of the resonances to the total cross section near the threshohld, within the GRAAL energy range, is sizeable. In addition, the contributions of the resonances in the differential cross sections are dominant at medium and large θ angles. Furthermore, in the beam asymmetry spectra Σ measured at GRAAL in ω photoproduction the contribution of the diffractive phenomenon is close to zero at all angles.

The high sensitivity, in this model, of Σ observable to the resonance excitation allowed to search among the isospin I=1/2 resonances selected here by the zero isospin of ω, which ones decay through ω mesons. Two resonances P13(1720) and F15(1680) have been found to play a dominant role. In figure 4 (right side), the dashed line gives the contribution of all resonances taken together and the dotted lines show the contributions of all except the P13(1720) resonance. The good agreement between the experimental results and the predictions of the model is very encouraging and the high sensitivity of the theoretical result to F13(1720) and F15(1680) resonance make these two resonances as good candidate to be decaying into ω meson.

REFERENCES

1. Ajaka J. et al., *Phys. Rev. Lett.*, **81**, 1797 (1998).
2. Bennhold C. et al., *Nucl. Phys.* **A 530**, 625 (1991); Tiator L. et al., *Nucl. Phys.* **A 580**, 455 (1994)
3. Knochlein G., Drechsel D., and Tiator L., *Z Phys.*, **A 352**, 327 (1995).
4. Mukhopadhyay N. C. and Mathur N., *Phys. Lett.* **B444**, 7 (1998).
5. Tiator L. et al., *Phys. Rev.* **C 60**, 035210.
6. Ajaka J. et al., *Phys. Lett.* **B 475**, 372(2000)
7. Arndt R. A., Strakovski I. I.,and Workman R. L., *Phys. Rev.* **C53**, 430 (1996)
8. Drechsel D. et al. *Nucl. Phys.* **A 645**, 145 (1999).
9. Gomez Tejedor J.A., and Oset E., *Nucl. Phys.* **A 571**, 667-693 (1994), and *Nucl. Phys.* **A 600**, 413-435 (1996)
10. Nacher J. C. and Oset E., private communication.
11. Zhao Q., *Nucl.Phys.* **A 675**, 217-221 (2000)

Three-nucleon spin observables: signatures for three-nucleon force effects

H.Witała*, W.Glöckle†, H.Kamada†, A.Nogga†, J.Golak*†,
J.Kuroś-Żołnierczuk*, R.Skibiński*

*Institute of Physics, Jagellonian University, PL-30059 Cracow, Poland
†Institut für Theoretische Physik II, Ruhr Universität Bochum, D-44780, Bochum, Germany

Abstract.
The numerical solutions of three-nucleon (3N) Faddeev equations with modern, high precision nucleon-nucleon (NN) interactions are compared to new nucleon-deuteron (Nd) data at nucleon laboratory energies between 100 and 200 MeV. The large discrepancies between theory and data clearly point to the action of three-nucleon forces (3NF). Successes and failures in the description of those data using in addition different, present day 3NF models are described. This indicates flaws in the present day 3NF models. However the large 3NF effects found theoretically for different 3N spin observables are promising to pin down the proper spin structure of 3NF's.

INTRODUCTION

A major goal in nuclear physics is to establish the form of nuclear forces to be used in the nuclear hamiltonian. Presently the construction of nuclear forces is mostly guided by meson theory and the most advanced potential in that meson-exchange picture is represented in the so-called CD Bonn potential [1] which describes very precisely all existing two-nucleon (2N) data below the π threshold. In addition more phenomenological NN potentials have been constructed which together with the CD Bonn potential are called a new generation of realistic NN forces: AV18 [2], Nijm I,II and 93 [3]. They all describe the NN data set with an unprecedented precision of χ^2 per data point very close to one.

In recent years it became possible to solve exactly three- and four-nucleon bound states using standard integration and differentiation methods [4,5]. Stochastic techniques allow to go beyond A=4 and now low energy states of nuclei up to A=8 can be calculated [6,7]. In all cases realistic NN forces lead to clear cut underbinding. For A=3 and A=4 nuclei this is examplified in Table 1.

There is an interesting correlation between 3- and 4-nucleon binding energies, the so called Tjon line (Fig.1). It gives the hope, that curing the binding energy for one nucleus will also cure the other one. A natural idea to explain this underbinding

TABLE 1. 3N and 4N binding energies for various NN potentials together with expectation values T of the kinetic energy.

Potential	^{3}H		^{3}He		^{4}He	
	E_B [MeV]	T [MeV]	E_B [MeV]	T [MeV]	E_B [MeV]	T [MeV]
CD Bonn	-8.012	37.42	-7.272	36.55	-26.26	77.15
AV18	-7.623	46.73	-6.924	45.68	-24.28	97.83
Nijm I	-7.736	40.73	-7.085	39.97	-24.98	84.19
Nijm II	-7.654	47.51	-7.012	46.62	-24.56	100.31
Exp.	-8.48	—	-7.72	—	-28.30	—

is the consideration of 3N forces which appear if one restricts the Hilbert space to three nucleons but considers processes leading to excited states of them. A natural process which leads to a 3NF is the $\pi - \pi$ exchange between three nucleons with an intermediate Δ excitation of one nucleon state [8]. This process is incorporated into the Urbana IX 3NF, where it is supplemented by a phenomenological short range spin- and isospin-independent part [9].

A more general 2π-exchange mechanism underlies the Tuscon Melbourne (TM) model which is around since quite some time [10]. In addition to its underlying basic building block, the πN amplitude, it contains a strong form factor parametrization

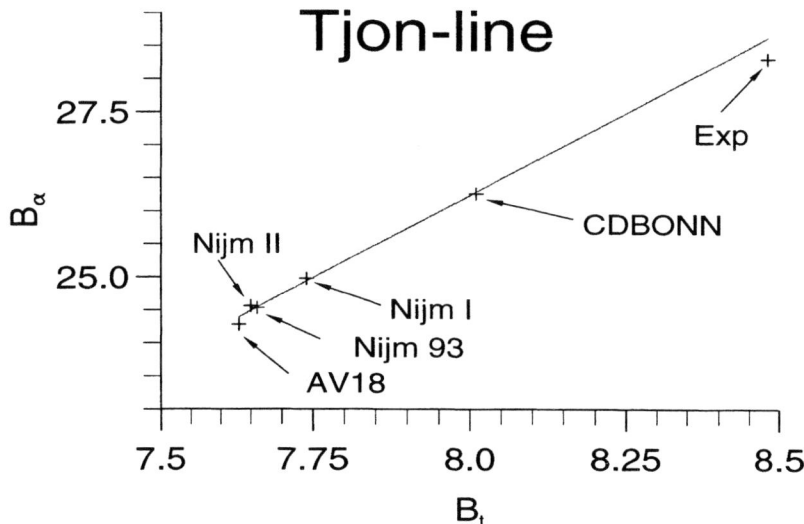

FIGURE 1. Correlation between binding energies of ^{3}H(B_t) and ^{4}He(B_α).

TABLE 2. Cut-off parameters Λ, adjusted 3N binding energies and resulting α particle binding energies for various force combinations. Bold faced values have been adjusted to the experiment by choosing the given Λ's. Expectation values of the kinetic energy are also shown.

Potential	$\Lambda\,[m_\pi]$	^{3}H		^{3}He		^{4}He	
		E_B[MeV]	T[MeV]	E_B[MeV]	T[MeV]	E_B[MeV]	T[MeV]
CDBonn+TM	4.784	**-8.480**	39.10	-7.734	38.24	-29.15	83.92
CDBonn+TM	4.767	-8.464	39.03	**-7.720**	38.18	-29.06	83.71
AV18+TM	5.156	**-8.476**	50.76	-7.756	49.69	-28.84	111.84
AV18+TM	5.109	-8.426	50.51	**-7.709**	49.47	-28.56	110.92
AV18+TM'	4.756	-8.444	50.55	**-7.728**	49.54	-28.36	110.14
NijmI+TM	5.035	-8.392	43.35	**-7.720**	42.59	-28.60	93.58
NijmII+TM	4.975	-8.386	51.02	**-7.720**	50.13	-28.54	113.09
AV18+UrbIX	—	-8.478	51.28	-7.760	50.23	-28.50	113.21
Exp.		-8.48	—	-7.72	—	-28.30	—

with a cut-off parameter Λ which will be specified below. In a meson exchange picture additional processes could be thought of containing other meson exchanges like $\pi - \rho, \rho - \rho$, and also different intermediate excited states might play a role. Some 3NF models with respect to those extensions have already been developed and applied [11–14]. Recently also chiral pertubation theory was used to develop consistent NN and 3N force models. This might help to find the most important spin-isospin structures [15].

By properly adjusting the Λ parameter in the TM 3NF to the ^{3}H or ^{3}He binding energy, one predicts a small overbinding in the 4N system as given in Table 2. Taking the AV18 potential with the Urbana IX 3NF one can reach even a reasonable description of low energy bound states of up to A=8 [16]. However in the latter case there seems to be among other defects an insufficient spin-orbit splitting of nuclear levels what may be caused by a wrong spin structure of the present day 3NF's. It is clear that todays nuclear hamiltonians are not able to describe the $A \geq 4$ systems accurately.

It is evident that both Nd elastic scattering and breakup processes, with their rich set of spin observables is a source of valuable information on 3NF's (their proper spin and momentum structure). In the next section we review briefly the 3N scattering formalism and compare in some examples the predictions of various NN potentials alone and combined with different 3N forces to new data at energies between 100 and 200 MeV where one can expect pronounced 3NF effects. Finally we give a summary and an outlook.

THREE NUCLEON SCATTERING RESULTS

We solve the following type of Faddeev equations for an amplitude $T|\phi\rangle$,

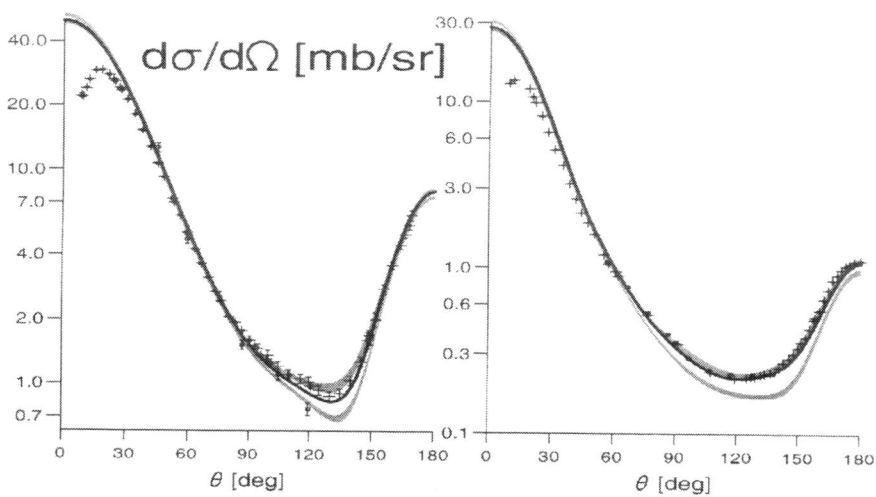

FIGURE 2. The differential cross section in elastic Nd scattering at nucleon energy E=65 MeV (left) and E=135 MeV (right). The light shaded band contains NN force predictions (AV18, CD-Bonn, Nijm I,II and 93), the darker shaded band the NN force predictions + TM 3NF individually adjusted according to Table 2. The solid curves are the AV18+Urbana IX predictions. Data at 65 MeV are from [20] (pd - crosses) and [21] (nd - circles), and at 135 MeV from [22] (pd - circles) and [23] (pd - crosses).

$$T|\phi\rangle = tP|\phi\rangle + (1+tG_0)V_4^{(1)}(1+P)|\phi\rangle \\ + tPG_0T|\phi\rangle + (1+tG_0)V_4^{(1)}(1+P)G_0T|\phi\rangle. \quad (1)$$

from which all physical observables for elastic Nd scattering and breakup process can be obtained [17]

The ingredients of Eq.(1) are the off-the-energy shell NN t-matrix t, the sum P of a cyclic and anticyclic permutation of 3 objects, the free 3N propagator G_0, and the initial channel state $|\phi\rangle$, composed of a deuteron and a momentum eigenstate of the projectile nucleon. On top of two-nucleon forces also a 3NF is included, where $V_4^{(1)}$ is that part of it, which is symmetrical under exchange of nucleons 2 and 3. The iteration of Eq.(1) yields an infinite sequence of terms containing consecutive t-matrices and the 3NF with free propagations in between. In [18] the connection to the cross sections and the spin observables are given. Using the modern NN forces: AV18, CD Bonn, Nijm I,II and 93 one gets in general predictions for 3N scattering observables which agree quite well with the 3N data at lower energies (below $\approx$ 30 MeV). A fairly complete overview of those theoretical predictions in comparison to very many data is presented in [18]. At higher energies discrepancies develop. We examplify this situation for the elastic scattering cross section in Fig.2. The strong discrepancy in the minimum of the cross section using NN forces only is removed

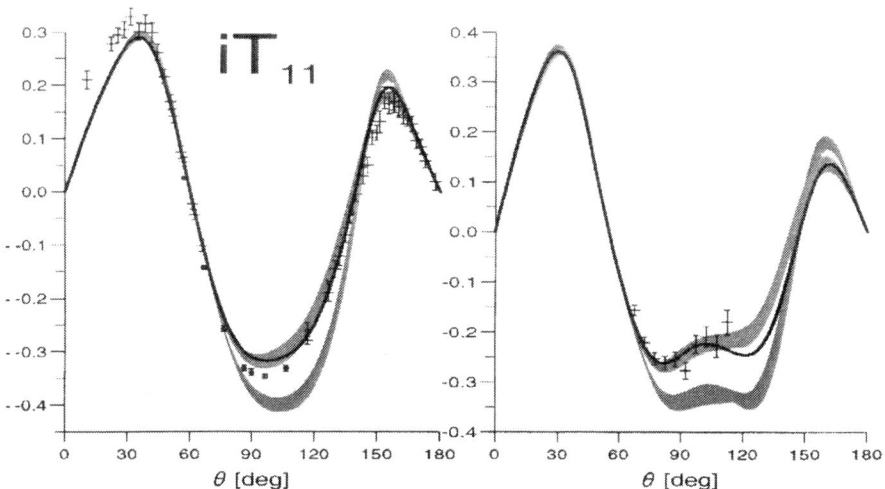

FIGURE 3. The deuteron vector analyzing power iT_{11} for elastic Nd scattering at 135 MeV (left) and 190 MeV (right). Curves as in Fig.2, pd data at 135 MeV are from [22] (circles), [23] (crosses), and at 190 MeV from [24].

when 3NF's reproducing the experimental triton binding energy are included [19]. A similar situation exists for the deuteron vector analyzing power iT_{11} (Fig.3). However, in the case of the nucleon analyzing power A_y there are quite different effects given by TM and Urbana IX 3NF's and in addition they do not reproduce the experimental data (Fig.4). Also for the tensor analyzing powers the situation is very challenging. New high precision pd data at 135 MeV [23] can neither be reproduced by pure 2N force predictions nor by adding the TM or the Urbana IX 3NF. Also here the predictions of these two 3NF's are quite different (Fig.5). There are many spin transfer and spin correlation coefficients for which drastic 3NF effects exist and for which the TM and the Urbana IX 3N forces give very different predictions. As examples we show in Fig.6 the spin correlation coefficient $C_{xy,x}$ and in Fig.7 the spin transfer coefficient $K_{xz}^{y'}$ from the deuteron to the nucleon. Also spin observables in the breakup process are quite promising and indicate very large 3NF effects in specific geometrical configurations at higher energies. As examples we show in Fig.8 A_y and tensor analyzing powers A_{xx}, A_{yy} and A_{xz} at 200 MeV incoming proton energy for specific breakup configurations. Not only large 3NF effects are clearly seen but also the predictions given by the TM and the Urbana IX 3NF's are totally different. This gives hope that with more precise data on spin observables it will be possible to nail down the proper spin structure of the 3NF.

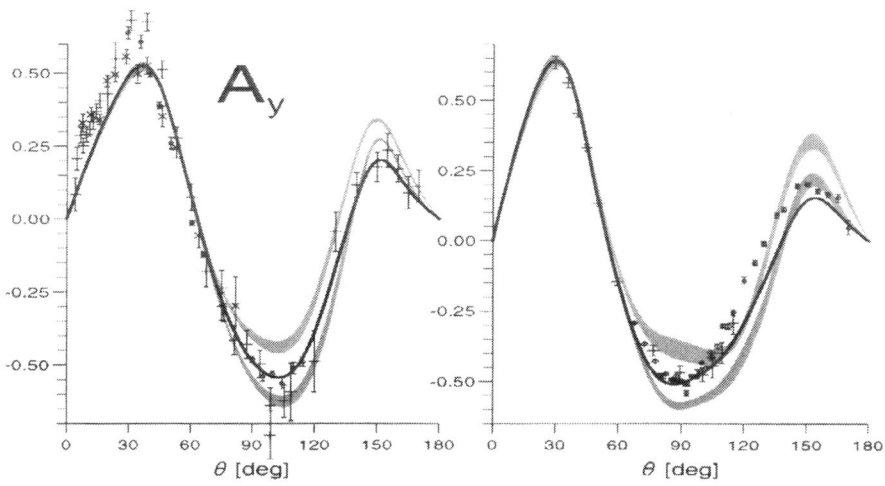

FIGURE 4. The analyzing power A_y for elastic Nd scattering at 135 MeV (left) and 190 MeV (right). Curves as in Fig.2, pd data at 135 MeV are from [25] (circles 150 MeV), [26] (crosses 146 MeV), [27] (x's 155 MeV), and at 190 MeV from [25] (crosses), [28] (circles 198 MeV) and [29] (squares 197 MeV).

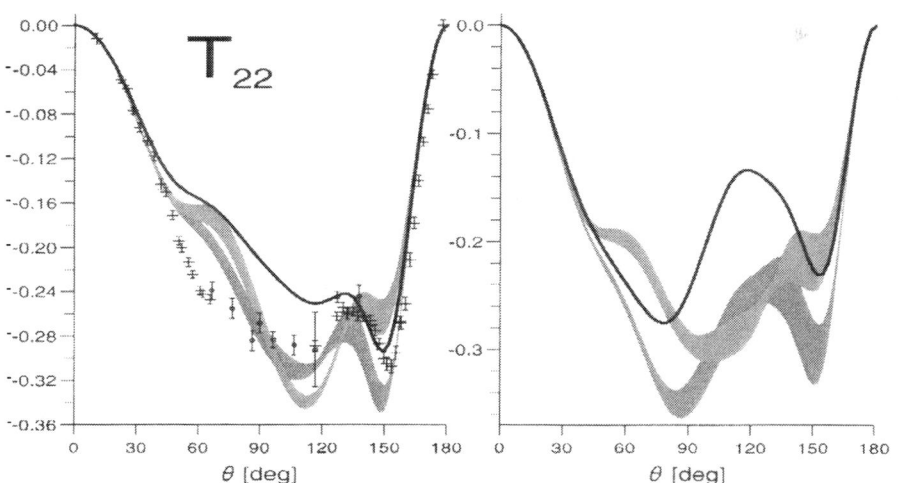

FIGURE 5. The tensor analyzing power T_{22} for elastic Nd scattering at 135 MeV (left) and 190 MeV (right). Curves as in Fig.2, pd data at 135 MeV are from [22] (circles) and [23] (crosses).

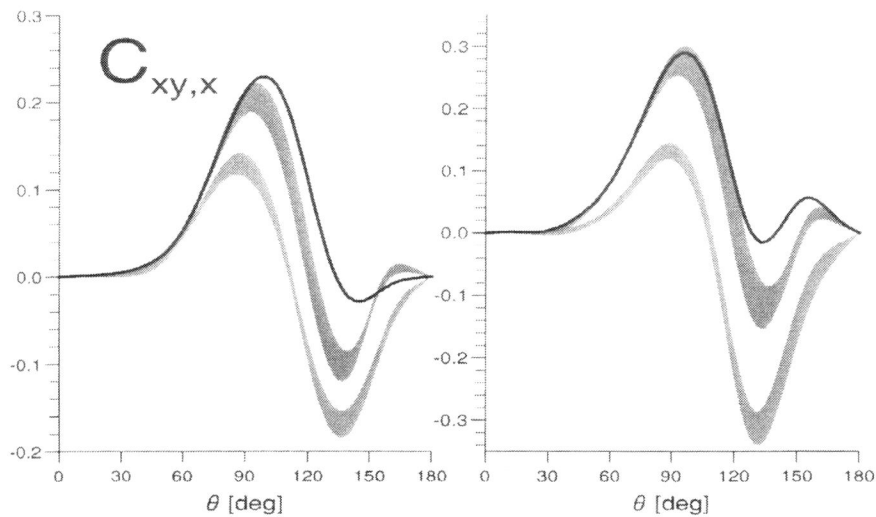

FIGURE 6. The spin correlation coefficient $C_{xy,x}$ for elastic Nd scattering at 135 MeV (left) and 190 MeV (right). Curves as in Fig.2.

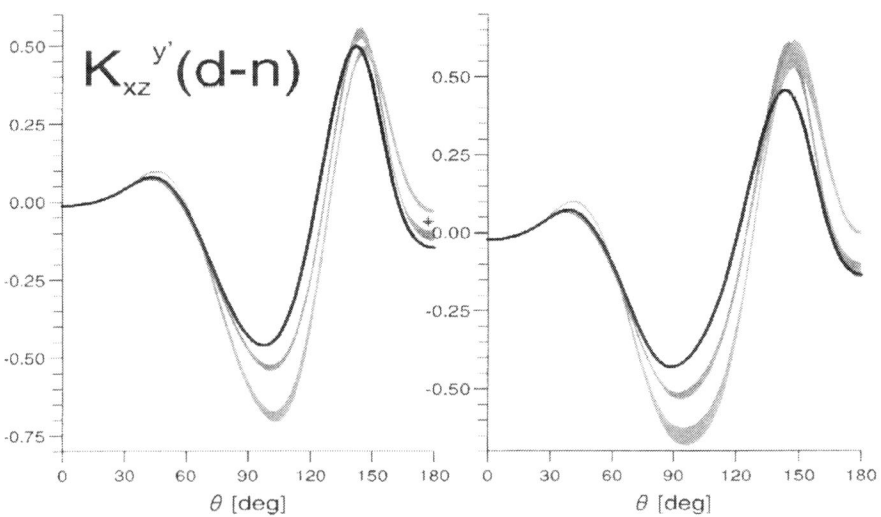

FIGURE 7. The spin transfer coefficient $K_{xz}^{y'}$ from deuteron to nucleon for elastic Nd scattering at 135 MeV (left) and 190 MeV (right). Curves as in Fig.2, pd datum at 135 MeV from [23].

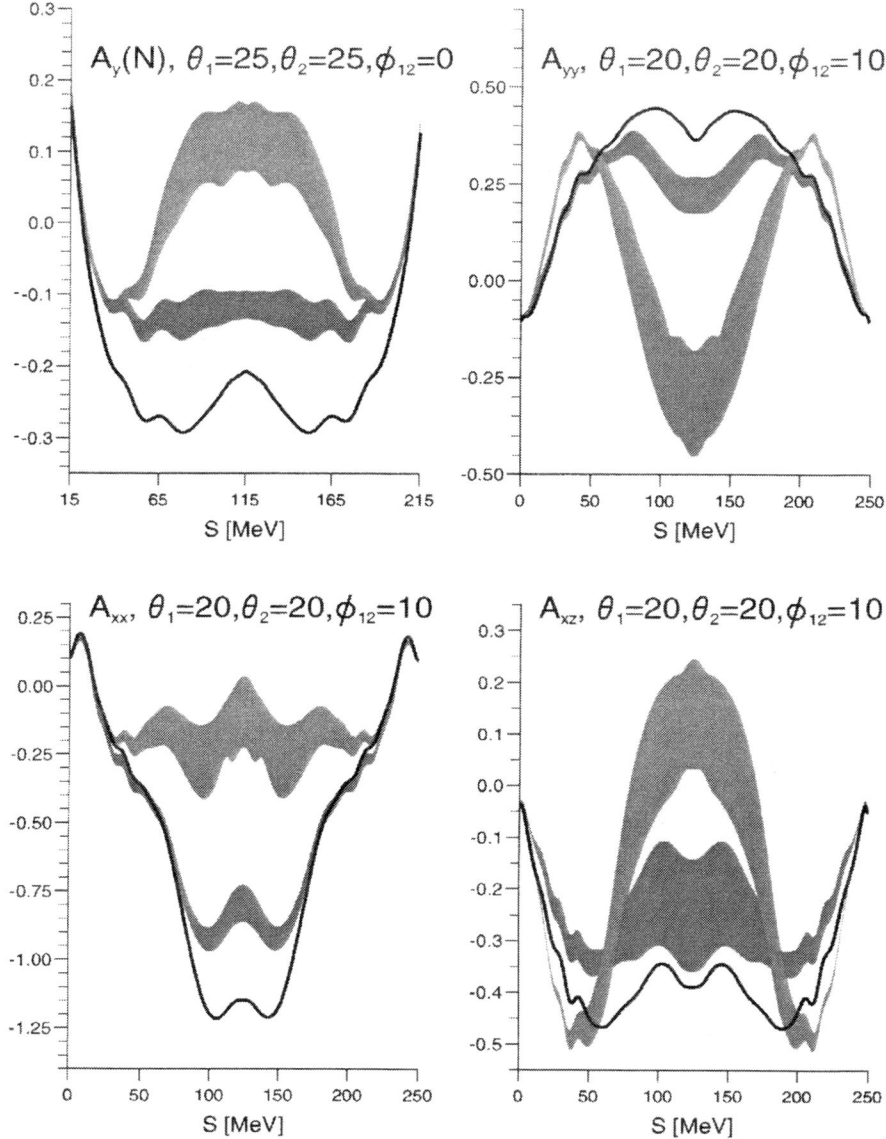

FIGURE 8. The analyzing power $A_y(N)$ and tensor analyzing powers A_{xx}, A_{yy} and A_{xz} for some specific break-up configurations at 200 MeV. Curves as in Fig.2.

SUMMARY AND OUTLOOK

In many spin observables one sees very drastic effects of 3N forces at higher energies in both the elastic Nd scattering and the deuteron breakup process. Different 3NF models such as TM and Urbana IX lead for some spin observables to very distinct predictions. The effects are angular and energy dependent and become large at higher energies. It seems that with a sufficiently rich and precise data basis one should be able to nail down the proper spin structure of 3N forces. This study of looking for the proper spin structure of 3NF's can possibly be guided by the chiral effective field theory approach [15]. A first step in this direction, where nuclear forces constructed in a systematic manner from chiral perurbation theory have been applied to three- and four-nucleon systems, appear be promising [30].

ACKNOWLEDGMENTS

This work was supported by the Deutsche Forschungsgemeinschaft and by the Polish Committee for Scientific Research under Grant No. 2P03B02818. The numerical calculations have been performed on the Cray T90 and T3E of the NIC in Jülich, Germany.

REFERENCES

1. R.Machleidt, nucl-th/0006014, *and references therein*.
2. R.B.Wiringa, V.G.J.Stoks, R.Schiavilla, *Phys. Rev.* **C51**, 38(1995).
3. V.G.J.Stoks, R.A.M.Klomp, C.P.F.Terheggen, J.J. de Swart, *Phys. Rev.* **C49**, 2950(1994).
4. A.Nogga, H.Kamada, W.Glöckle, *Phys. Rev. Lett.* **85**, 944(2000).
5. M.Viviani, *Nucl. Phys.* **A 631**, 111c(1998).
6. J.Carlson, R.Schiavilla, *Rev. of Mod. Phys.* **70**, 743(1998).
7. R.B.Wiringa, S.C.Pieper, J.Carlson, V.R.Pandharipande, *Phys. Rev.* **C62**, 014001 (2000).
8. J.Fujita, H.Miyazawa, *Prog. Theor. Phys.* **17**, 360(1957).
9. B.S.Pudliner, V.R.Pandharipande, J.Carlson, S.C.Pieper, R.B.Wiringa, *Phys. Rev.* **C56**, 1720(1997).
10. S.A. Coon et al., *Nucl. Phys.* **A317**, 242(1979)
    S.A.Coon, W.Glöckle, *Phys. Rev.* **C23**, 1970(1981).
11. S.A.Coon, M.T.Peña, *Phys. Rev.* **C48**, 2559(1993).
12. S.A.Coon, M.T.Peña, D.O.Riska, *Phys. Rev.* **C52**, 2925(1995).
13. B.D.Keister, R.B.Wiringa, *Phys. Lett.* **B173**, 5(1986).
14. T.-Y.Saito, J.Haidenbauer, nucl-th/0003064.
15. U. van Kolck, *Phys. Rev.* **C 49**, 2932(1994)
    E.Epelbaum et al., *Nucl. Phys.* **A637**, 107(1998); **A671**, 295(2000).
16. R.B.Wiringa, *private communication*.

17. D.Hüber, H.Kamada, H.Witała, W.Glöckle, *Acta Phys. Polonica* **B** 28, 167(1997).
18. W.Glöckle, H.Witała, D.Hüber, H.Kamada and J.Golak, *Phys. Rep.* **274**, 107(1996).
19. H.Witała, W.Glöckle, D.Hüber, J.Golak, H.Kamada *Phys. Rev. Lett.* **81**, 1183(1998).
20. S.Shimizu et al., *Phys. Rev.* **C52**, 1193(1995).
21. H.Rühl et al., *Nucl. Phys.* **A524**, 377(1991).
22. N.Sakamoto et al., *Phys. Lett.* **B367**, 60(1996).
23. H.Sakai, K.Sekiguchi, H.Witała, W.Glöckle, M.Hatano, H.Kamada, H.Kato, Y.Maeda, A.Nogga, T.Ohnishi, H.Okamura, N.Sakamoto, S.Sakoda, Y.Satou, K.Suda, A.Tamii, T.Uesaka, T.Wakasa, K.Yako, *Phys. Rev. Lett.* **84**, 5288(2000); H.Sakai et al., *to be published in the proceedings of 16th International Conference on "Few-body Problems in Physics", March 6-10, 2000, Taipei.*
24. R.V.Cadman et al., nucl-ex/0010006.
25. R.Bieber et al., *Phys. Rev. Lett.* **84** 606(2000).
26. H.Postma, R.Wilson, *Phys. Rev.* **121**, 1129(1961).
27. K.Kuroda, A.Michałowicz, M.Poulet, *Nucl. Phys.* **88**, 33(1966).
28. R.E.Adelberger, C.N.Brown, *Phys. Rev.* **D5**, 2139(1972).
29. S.P.Wells et al., *Nucl. Instrum. Methods Phys. Res.*, **Sect.A325**, 205(1993).
30. E.Epelbaum, H.Kamada, A.Nogga, H.Witała, W.Glöckle, U-G.Meisner, nucl-th/0007057.

Spin-Isospin Responses in Nuclei via Polarization Measurements

Hide Sakai*†

*Department of Physics/CNS[1], University of Tokyo, Bunkyo, Tokyo 113-0033, Japan
†RIKEN, Wako, Saitama 351-0198, Japan

Abstract. High quality (p,n) data obtained by NTOF+NPOL2 facility at RCNP were presented. From the measurement of ^{90}Zr(p,n) reaction at 295 MeV, the quenching value for the Gamow-Teller transition in terms of the Ikeda's sum rule of $S_{\beta-} - S_{\beta+} = 3(N-Z)$ is derived as 0.90 ± 0.05. By using this quenching value, the Landau-Migdal parameters representing short-range correlation in isospin-spin interactions are deduced as $(g'_{NN}, g'_{N\Delta}) = (0.6, 0.2)$. This small $g'_{N\Delta}$ value favors the pion condensation. The complete set of the polarization transfer coefficients, $D_{LL'}$, $D_{SS'}$, $D_{NN'}$, $D_{LS'}$ and $D_{SL'}$, for the (p,n) quasi-elastic scattering has been measured at 350 MeV. The spin-longitudinal cross section $ID_q (\approx R_L)$ and the spin-transverse cross section $ID_p (\approx R_T)$ are deduced. ID_q is found to be consistent with the DWIA+RPA calculation with $(g'_{NN}, g'_{N\Delta}) = (0.6, 0.3)$. This result supports strongly the existence of the pionic enhancement in nuclei.

I INTRODUCTION

Since Hideki Yukawa proposed the pion exchange mechanism as a dominant carrier of nuclear force in 1935 [1], pionic nuclear collectivity(enhancement) is the heart of nuclear physics. The strong attraction produced by pions leads to various interesting phenomena such as pion condensation or its precursor phenomena.

Recently G.F. Bertsch, L. Frankfurt and M. Strikman argued in the article [2] that the pionic field at short distances is greatly suppressed based on the three experimental results;

1. EMC experiment at CERN,
2. Drell-Yan process in high-energy proton-nucleus collision experiment, and
3. R_L(longitudinal response) of QES(Quasi-Elastic Scattering).

The title of the article is "Where Are the Nuclear Pions?". This puts into serious question the conventional meson-exchange picture of the nucleon-nucleon interaction. Subsequently this question is answered theoretically by G.E. Brown *et al.* in Ref. [3] with the title "Where the Nuclear Pions Are". They proposed a theoretical

---

[1] Center for Nuclear Study

resolution based on arguments of partial restoration of chiral symmetry with nuclear density. In this talk I would like to try to answer experimentally the question raised by Bertsch et al.

The simplest effective interaction for nuclear medium is so-called $\pi+\rho+g'$ model [4] and expressed as

$$V(q,\omega) = [V_L(q,\omega)(\boldsymbol{\sigma}_1 \cdot \hat{\boldsymbol{q}})(\boldsymbol{\sigma}_2 \cdot \hat{\boldsymbol{q}}) + V_T(q,\omega)(\boldsymbol{\sigma}_1 \times \hat{\boldsymbol{q}})(\boldsymbol{\sigma}_2 \times \hat{\boldsymbol{q}})](\boldsymbol{\tau}_1 \cdot \boldsymbol{\tau}_2). \quad (1)$$

Here $V_L(q,\omega)$ and $V_T(q,\omega)$ are the longitudinal and transverse spin-isospin parts of interactions, respectively. They are expressed as

$$V_L(q,\omega) = C_0 g' + V_\pi(q,\omega), \quad (2)$$
$$V_T(q,\omega) = C_0 g' + V_\rho(q,\omega). \quad (3)$$

$V_\pi(V_\rho)$ is the one-pion(rho) exchange interaction. g' is the Landau-Migdal parameter which represents short-range correlations. C_0 is the unit of Landau-Migdal parameter and it is $C_0 = (\frac{f}{m_\pi})^2 \approx 392$ MeVfm^3.

When the Δ degree of freedom is taken into account besides the nucleon(N) degree of freedom, there are three different g's relevant to the interactions between particl-hole(p-h) and p-h, p-h and Δ-h and Δ-h and Δ-h denoted as g'_{NN}, $g'_{N\Delta}$, and $g'_{\Delta\Delta}$, respectively. It is frequently assumed so-called *universality ansatz*, $g'_{NN}=g'_{N\Delta}=g'_{\Delta\Delta}(=g')$ for convenience of calculations.

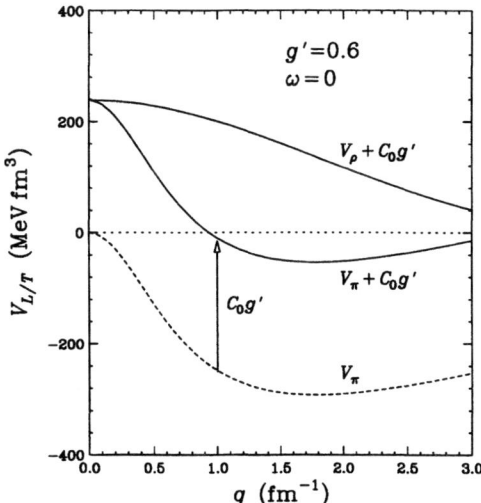

FIGURE 1. Momentum dependence of the longitudinal and transverse of the spin-isospin interaction. The short range correlation parameter is taken to be $g' = 0.6$. The one-pion-exchange interaction V_π is given for comparison.

Figure I shows the momentum dependence of V_L and V_T with $\omega = 0$ MeV and $g' = 0.6$. Bare V_π is plotted by dashed line. Since g' shifts V_π to repulsive direction, V_L becomes attractive in the region where $q > 1.0$ fm^{-1}. Physical conditions such as pion condensation in neutron star or its precursor phenomena in ordinary nuclei critically depend on the size of this attraction. The behavior of V_L at large q is essentially determined by the value of g'. Therefore it is extremely important to get information on g' as well as to confirm the attraction of V_L at high q region.

In this presentation, recent measurements of the (p,n) reaction which solved the quenching problem of the Gamow-Teller(GT) strength in nuclei are firstly shown(Sec. II). Since this is the phenomenon at $q \sim 0$, the quenching value provides the information on the Landau-Migdal parameters g'_{NN} and $g'_{N\Delta}$ as discussed by Ericson et al. [5]. To show the consistency of thus obtained g'_{NN} and $g'_{N\Delta}$ values, the longitudinal response function $R_L(ID_q)$ at $q \approx 1.7$ fm^{-1} in which the pionic correlation(enhancement) should appear is secondly introduced. The R_L is deduced from the complete set of the polarization transfer coefficients for the $(\vec{p}, \vec{n})$ QES (Sec. IV).

II QUENCHING OF THE GAMOW-TELLER STRENGTH

The Gamow-Teller(GT) quenching (missing strength in terms of the Ikeda's sum rule [6]) is one of the most interesting phenomena found in the early 1980. The quenching values derived from the (p,n) reaction at IUCF are plotted in Fig. 2 by open circles taken from Ref. [7]. Note that the systematic uncertainties are not included in this plot.

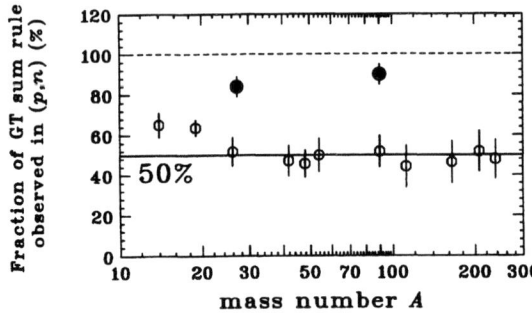

FIGURE 2. Fraction of the GT sum rule strength taken from Gaarde [7] (open circle). Solid circles are taken from Wakasa et al. [14,15]

The importance of the GT quenching lies in its large size. This large quenching of the strength by about 40–50% of the sum-rule value is rather well explained using the value, $g'(= g'_{NN} = g'_{N\Delta} = g'_{\Delta\Delta}) = 0.6 - 0.8$ assuming the universality ansatz of the Landau-Migdal parameters [8,9]. This large $g'_{N\Delta}$ value made not only the pion condensation [9] but also its precursor phenomena [10] very unlikely.

Experimentally, however, it has been questioned that how much is contained in small amplitudes at excitation energies higher in the continuum as a tail of the GT giant resonance. Such small amplitudes after integrating over the continuum might fill the gap of quenching. However, it is a difficult experimental challenge to seek such small amplitudes out.

A GT strength search in the continuum by the (p,n) reaction at 300 MeV

The experiments were carried out at the neutron time-of-flight facility [11] at the Research Center for Nuclear Physics (RCNP). Polarized protons from the ring cyclotron with $T_p = 295$ MeV bombarded self supporting ^{90}Zr and ^{27}Al targets. Neutrons from the (p,n) reaction were detected by the neutron detector/polarimeter NPOL2 [11–13]. Since the experimental results on ^{90}Zr and ^{27}Al targets are published elsewhere [14,15], in the following we show mainly the results on ^{90}Zr.

The double differential cross sections as a function of scattering angles and excitation energies are plotted in Fig. 3 for $\theta_{lab} = 0.0° - 12.3°$.

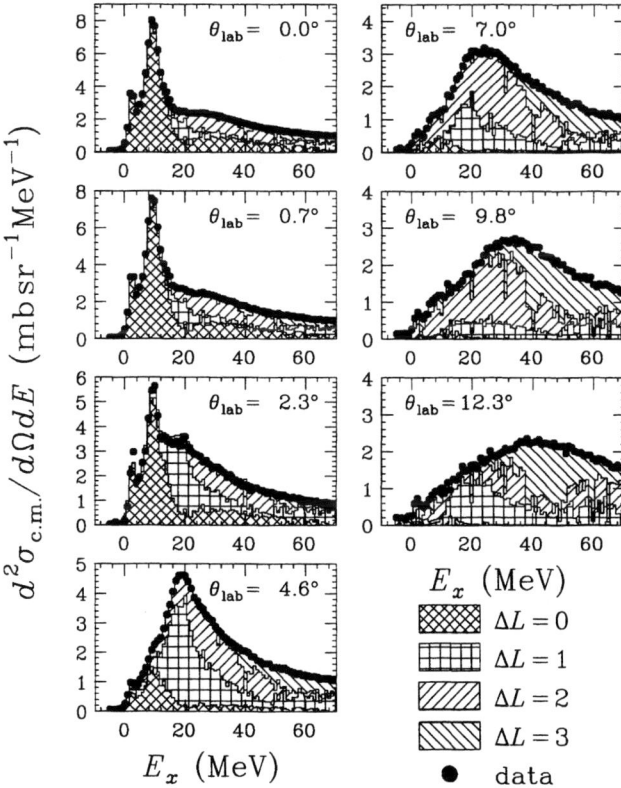

FIGURE 3. Differential cross sections (filled circles) for the ^{90}Zr(p,n) reaction at $T_p = 295$ MeV. Histograms represent the result of MDA. See text for detail.

A multipole decomposition analysis technique(MDA) is applied to the cross sections to extract $\Delta L = 0, 1, 2$ and 3. For each 1.0 MeV excitation energy interval, the experimentally obtained angular distribution $\sigma^{\text{exp}}(q, E_x)$ was fitted by means of the least square method with the sum of calculated angular distributions $\sigma^{\text{calc}}_{\Delta L}(q, E_x)$

weighted with fitting coefficients $a_{\Delta L}$ as

$$\sigma^{\text{exp}}(q, E_x) = \sum_{\Delta L} a_{\Delta L} \sigma_{\Delta L}^{\text{calc}}(q, E_x) \ . \tag{4}$$

The fitting procedure has been performed for several possible p-h combinations (80640 combinations) of the calculated angular distributions. The combination of calculated angular distributions giving the minimum chi-square value was chosen. The result of the MDA is shown in Fig. 3. The present MDA clearly shows fairly large contribution of the $\sigma_{\Delta L=0}$ component up to 50 MeV excitation.

The (p,n) cross section for the $\Delta L = 0$ transfer may be related to the corresponding $B(\text{GT})$ values by using the relation

$$\sigma_{\Delta L=0}(q, \omega) = \hat{\sigma}_{\text{GT}} \cdot F(q, \omega) \cdot B(\text{GT}) \ , \tag{5}$$

where $\hat{\sigma}_{\text{GT}}$ is the GT unit cross section and $F(q, \omega)$ describes the dependence on momentum transfer q and energy loss ω. The $\hat{\sigma}_{\text{GT}}$ value of 3.6 ± 0.6 mb/sr is used in the present analysis. The $F(q, \omega)$ is defined to be unity at $(q, \omega) = (0, 0)$, and can be obtained from the the distorted-wave impulse approximation (DWIA) calculation for the $\Delta L = 0$ transfer. By using Eq.(5), $\sigma_{\Delta L=0}(q, \omega)$ can be converted into $B(\text{GT})$.

The total GT strength summed over the region up to 50 MeV excitation becomes $S_{\beta-} = \sum B(\text{GT}) = 28.0 \pm 1.6$ after subtracting the contribution of the Iso-Vector Spin Monopole (IVSM) strength.

The $S_{\beta+}$ strength for ^{90}Y has been derived by MDA for the $^{90}\text{Zr}(n,p)^{90}\text{Y}$ reaction at 198 MeV by Raywood et al. [16]. It is $S_{\beta+} = 1.0 \pm 0.3$. Thus the total $S_{\beta-}$ strength minus the total $S_{\beta+}$ strength becomes

$$S_{\beta-} - S_{\beta+} = (28.0 \pm 1.6) - (1.0 \pm 0.3) = 27.0 \pm 1.6. \tag{6}$$

Note that errors are statistical ones only. The value of 27.0 is 90 % of the Ikeda's sum-rule value $3(N - Z) = 30$. A similar quenching value of 0.84 ± 0.05 has been obtained for ^{27}Al [15]. These new quenching values are plotted in Fig. 2 by solid circles.

Thus we can conclude that the quenching of the GT strength is mainly due to the configuration mixing mechanism and consequently the Δ-h admixture into $1p$-$1h$ GT state is small.

III LANDAU-MIGDAL PARAMETERS, g'_{NN} AND $g'_{N\Delta}$

The Landau-Migdal parameters are estimated by Suzuki and Sakai [17] by using these experimental results. The random-phase approximation (RPA) in p-h and Δ-h space is used and in the RPA, the couplings are assumed to be due to the zero-range forces whose strengths are given by the three Landau-Migdal parameters, g'_{NN}, $g'_{N\Delta}$ and $g'_{\Delta\Delta}$.

Estimated results are

$$0.18 < g'_{N\Delta} < 0.23, \qquad 0.58 < g'_{NN} < 0.59,$$

for $0 < g'_{\Delta\Delta} < 1.0$ [18]. The most important conclusions drawn from the small GT quenching value are :

1. the universality assumption does not hold.
2. g'_{NN} and $g'_{N\Delta}$ values are around 0.6 and 0.2, respectively.

It is important to refine the Landau-Migdal parameters theoretically. Recently Arima et al. have derived $g'_{N\Delta}$ by taking into account the effect of the finite-range $\pi + \rho$ exchange interactions [19]. According to them, the value of $g'_{N\Delta}$ increases by about 0.1 thus $g'_{N\Delta} \approx 0.3$.

A Critical density of pion condensation in neutron star

Various spin physics discussed by assuming the universality ansatz should be re-analyzed in the light of new result. One of them is the pion condensation which might be realized in neutron star.

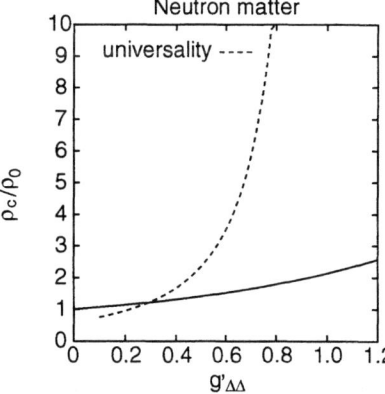

FIGURE 4. Critical density for π^0 condensation in pure neutron matter as a function of $g'_{\Delta\Delta}$ with universality ansatz (dashed) and with $g'_{N\Delta} = 0.2$ and $m^* = 0.8 m_N$ (solid). See text for detail.

In Fig. 4, the critical density ρ_c for π^0 condensation in neutron matter is shown as a function of $g'_{\Delta\Delta}$ by solid curve [20] for $g'_{N\Delta} = 0.2$ with the effective mass $m^* = 0.8 m_N$. The dashed curve is that with the universality ansatz. The ρ_c goes rapidly to infinite at $g' \approx 0.8$ in the universality ansatz, and thus the pion condensation becomes unlikely. On the other hand, when $g'_{N\Delta} = 0.2$ is used, the dependence of ρ_c is very much softened $1.4\rho_0 < \rho_c < 2.2\rho_0$ Thus the pion condensation becomes likely. This result is consistent with the prediction by Akmal-Pandharipande [21] who employed the data equivalent modern NN potential AV18 with the three-nucleon force UIX. Their result is $\rho_c \sim 1.3\rho_0$.

IV (p, n) QES MEASUREMENT AND LONGITUDINAL RESPONSE FUNCTION

The longitudinal response function R_L and transverse response function R_T are defined as

$$R_L(q,\omega) = \frac{1}{A}\sum_n |\langle n|\sum_i \tau_i^-(\boldsymbol{\sigma}_i\cdot\hat{\boldsymbol{q}})e^{i\boldsymbol{q}\boldsymbol{r}_i}|0\rangle|^2 \delta(\omega-(E_n-E_0)), \qquad (7)$$

$$R_T(q,\omega) = \frac{1}{2A}\sum_n \sum_\mu |\langle n|\sum_i \tau_i^-(\boldsymbol{\sigma}_i\times\hat{\boldsymbol{q}})_\mu e^{i\boldsymbol{q}\boldsymbol{r}_i}|0\rangle|^2 \delta(\omega-(E_n-E_0)). \qquad (8)$$

In 1982, Alberico et al. [22] theoretically predicted based on the simple $\pi+\rho+g'$ nuclear correlations that R_L is enhanced and softened while R_T is quenched and hardened in the quasi-elastic region so that

$$R_L/R_T \gg 1.$$

However, the ratio R_L/R_T extracted from the experiments [23] is

$$R_L/R_T \leq 1,$$

which is against the predicted pionic nuclear collectivity.

The enhancement of R_L may be related with the excess of the pion distribution in nuclei [24] as

$$\delta n_A(q) = \frac{C_0 F^2(q)}{2c_q} A q^2 \int_0^\infty d\omega \frac{\delta R_L(q,\omega)}{[\epsilon_q+\omega]^2}, \qquad (9)$$

with form factor $F^2(q)$, pion energy $\epsilon_q = \sqrt{q^2+m_\pi^2}$ and $\delta R_L = R_L - R_L^N$, where R_L^N is the response function for the *nucleon* (isospin average).

A How to extract R_L and R_T from experiment

R_L and R_T can be deduced from the complete polarization transfer measurement of the $(\vec{p},\vec{n})$ QES. Observables are $D_{LL'}$, $D_{SS'}$, $D_{NN'}$, $D_{LS'}$ and $D_{SL'}$ besides the unpolarized double differential cross section I.

These D_{ij}'s in the laboratory frame are transformed into the spin-longitudinal and spin-transverse cross sections, ID_q and ID_p, in the c.m. (q,p,n) frame as

$$ID_q = \frac{I}{4}[1-D_{NN'}+(D_{SS'}-D_{LL'})\cos\alpha_2-(D_{SL'}+D_{LS'})\sin\alpha_2], \qquad (10)$$

$$ID_p = \frac{I}{4}[1-D_{NN'}-(D_{SS'}-D_{LL'})\cos\alpha_2+(D_{SL'}+D_{LS'})\sin\alpha_2], \qquad (11)$$

where α_2 is defined in Ref. [25].

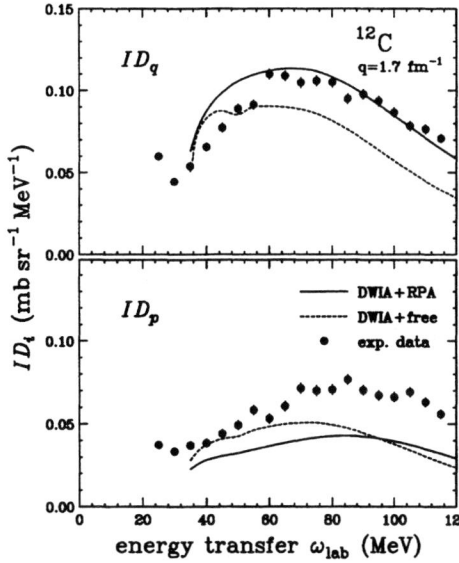

FIGURE 5. Results of ID_q and ID_p at $q = 1.7$ fm^{-1} for ^{12}C$(\vec{p}, \vec{n})$ (solid dot) with DWIA predictions with (solid) and without (dashed) RPA correlations with $(g'_{NN}, g'_{N\Delta}) = (0.6, 0.3)$.

In the frame work of the plane-wave impulse approximation (PWIA), ID_q and ID_p are related with R_L and R_T, respectively, as

$$ID_q = CN_{eff}|E|^2 R_L,$$
$$ID_p = CN_{eff}|F|^2 R_T,$$

where C is the kinematical factor, $|E|(|F|)$ is the NN scattering amplitude and N_{eff} is the effective neutron number [25]. R_L and R_T obtained in this manner are compared to theoretical predictions. However, more direct and accurate comparison is possible, if ID_q and ID_p can be predicted by theory. Such elaborate calculations are available in terms of DWIA by Kawahigashi et al. [26].

B Experiment

The complete D_{ij} measurements for the ^{2}He, ^{6}Li, ^{12}C, ^{40}Ca, ^{208}Pb $(\vec{p}, \vec{n})$ reactions at $T_p = 346$ MeV were carried out at $\theta_{lab} = 22°$ ($q = 1.7$ fm^{-1}) [27]. ID_q and ID_p are obtained from D_{ij} values and plotted in Fig. 5 for ^{12}C [28].

C Comparison of ID_q and ID_p with DWIA calculations

Ichimura group carried out DWIA calculations of ID_q and ID_p with the following sophistication:

1. Continuum RPA (nuclear finite size effects are taken into account),

2. Inclusion of Δ excitation and removal of the universality ansatz,
3. Spreading widths in particle and hole states,
4. Effective mass (m*).

Predicted results with $(g'_{NN}, g'_{N\Delta}, g'_{\Delta\Delta})= (0.6, 0.3, 0.5)$ and $m^* = 0.7 m_N$ are plotted with(without) RPA correlations by solid(dashed) curves.

Experimental ID_q is well reproduced by the calculation which is consistent with the Landau-Migdal parameters deduced from the analysis of the GT strength. Present result clearly shows the expected enhancement of the pionic correlations in nuclei and thus supports the precursor phenomena of pion condensation. However, the experimental ID_p is much larger than the theoretical one. This is the main reason of the experimentally observed $R_L/R_T \leq 1$. In this context, recent calculation of the two-step contribution by Nakaoka-Ichimura [29] is interesting. The two-step contribution in ID_p is significantly larger than that in ID_q.

V SUMMARY

Based on the high quality (p, n) data measured at 295 MeV, we searched the GT strength in the highly excited continuum via the MDA technique and found significant strengths there. The quenching value of the GT strength in terms of Ikeda's sum rule $S_{\beta-} - S_{\beta+} = 3(N - Z)$ is derived as 0.90 ± 0.05 for the ^{90}Zr(p, n) reaction [14].

From this quenching value, the Landau-Migdal parameters g'_{NN} and $g'_{N\Delta}$ are deduced to be around 0.6 and 0.2, respectively [17]. It is also shown that the universality ansatz does not hold. Recent calculation including the finite-range $\pi + \rho$ exchange shows $g'_{N\Delta} \approx 0.3$ [19]. Such a small $g'_{N\Delta}$ value favors the pion condensation [20].

The spin-longitudinal cross section $ID_q (\approx R_L)$ and the spin-transverse cross section $ID_p (\approx R_T)$ are deduced from the $(\vec{p}, \vec{n})$ QES at 350 MeV [27]. ID_q is found to be consistent with the predictions [26] with $(g'_{NN}, g'_{N\Delta})= (0.6, 0.3)$ which are close to values deduced from the GT quenching. These results strongly support the enhancement of pionic correlations in nuclei.

In conclusion, I would like to answer to G.F. Bertsch et al. [2],

"The Nuclear Pions Are Here."

VI ACKNOWLEDGMENTS

I am very much indebted to my many collaborators, M.B. Greenfield, K. Hatanaka, M. Hatano, D. Hirooka, J. Kamiya, H. Kato, Y. Maeda, C. Morris, H. Okamura, J. Rapaport, S. Sakoda, K. Sekiguchi, K. Suda, A. Tamii, T. Tatsumi, T. Wakasa, and K. Yako. In particular, T. Wakasa has not only contributed to the physics which I presented but also helped me in preparing the figures of this manuscript. I am grateful to M. Ichimura and T. Suzuki for illuminating and valuable discussions. I would like to thank A. Arima for his encouragements. This work is supported financially in part by the Grant-in-Aid for Scientific Research Nos. 6342007, 04402004 and 10304018 of Ministry of Education, Science, Culture and Sports of Japan.

REFERENCES

1. H. Yukawa, *Proc. Phys. Math. Soc. Japan* **17** (1935) 48.
2. G.F. Bertsch, L. Frankfurt and M. Strikman, *Science* **259** (1993) 773.
3. G.E. Brown, M. Buballa, Z. Bang Li and J. Wambach, *Nucl. Phys.* **A593** (1995) 295.
4. for example, F. Osterfeld, *Rev. Mod. Phys.* **64** (1992) 491.
5. M. Ericson et al., *Phys. Lett.* **B45** (1973) 19; M. Rho, *Nucl. Phys.* **A231** (1974) 493; K. Ohta and M. Wakamatsu, *Nucl. Phys.* **A234** (1974) 445.
6. K. Ikeda, S. Fujii and J.I. Fujita, *Phys. Rev. Lett.* **3** (1963) 271.
7. C. Gaarde, *Nucl. Phys.* **A396** (1983) 127c.
8. E. Oset and M. Rho, *Phys. Rev. Lett.* **42** (1979) 47; A. Bohr and B. R. Mottelson, *Phys. Lett.* **B100** (1981) 10.
9. J. Meyer-ter-Vehn, *Phys. Rep.* **74** (1981) 323.
10. E. Oset, H. Toki and W. Weise, *Phys. Rep.* **83** (1982) 281, and references therein.
11. H. Sakai et al., *Nucl. Instrum. Methods*, A **369** (1996) 120.
12. T. Wakasa et al., *Nucl. Instrum. Methods*, A **404** (1998) 355.
13. H. Sakai et al. *Nucl. Instrum. Methods*, A **320** (1992) 479.
14. T. Wakasa et al., *Phys. Rev.* C **55** (1997) 2909.
15. T. Wakasa et al., *Phys. Lett.* B **426** (1998) 257.
16. K.J. Raywood et al., *Phys. Rev.* **C41** (1990) 2836.
17. T. Suzuki and H. Sakai, *Phys. Lett.* B **455** (1999) 25.
18. W. H. Dickhoff et al., *Phys. Rev.* **C23** (1981) 1154.
19. A. Arima, W. Bentz, T. Suzuki and T. Suzuki, to be published in *Phys. Lett.*
20. T. Suzuki, H. Sakai and T. Tatsumi, Proc. of Int. Conf. on Nuclear Responses and Medium Effects, Osaka, 1998, Universal Academy Press, Tokyo, 1999, p77.
21. A. Akmal and V.R. Pandharipande, *Phys. Rev.* C **56** (1997) 2261.
22. W. M. Alberico, M. Ericson, and A. Molinari, *Nucl. Phys.* A **379** (1982) 429.
23. T.M. Taddeucci et al., *Phys. Rev. Lett.* **73** (1994) 3516 and references therein.
24. D.S. Koltun, *Phys. Rev.* C **57** (1998) 1210 and references therein.
25. M. Ichimura and K. Kawahigashi, *Phys. Rev.* C **45** (1992) 1822.
26. K. Kawahigashi, K. Nishida, A. Itabashi and M. Ichimura, submitted to *Phys. Rev.* C.
27. T. Wakasa et al., *Phys. Rev.* C **59** (1999) 3177.
28. Recently a normalization error was found in data of Ref. [27] and therefore it is corrected for in Fig. 5.
29. Y. Nakaoka and M. Ichimura, *Prog. Theoret. Phys.* **102** (1999) 599.

Nuclear Medium Effect Studied by Nucleon Quasifree Scattering

Tetsuo Noro

Research Center for Nuclear Physics, Osaka University, Ibaraki 567-0047, Japan

Abstract. Spin observables in proton quasifree scattering have been measured aiming at investigating the effect of hadron-mass modifications in nuclear field. In addition to the analyzing power A_y at an incident energy of 392MeV, a density dependent reduction, which is extracted from target dependence, of induced polarization P was newly observed at 1 GeV. Those data are consistent with a model calculation where the hadron-mass modifications are taken into account. The spin-transfer coefficients D_{ij}, at 392 MeV, do not show distinct density dependence while the calculation predicts strong dependence for some of the coefficients. But it is shown that these coefficients are quite sensitive to each meson-mass reduction. A discussion is given also for finite $P-A_y$ and $D_{pq}+D_{qp}$ values caused by a relativistic effect, which are eliminated by the time reversal invariance for the proton-proton scattering in free space.

INTRODUCTION

An atomic nucleus is a system where nucleons and mesons are packed governed by the strong interaction. Since these hadrons are also composite particles, it is not surprising that their structure and properties are modified in nuclear field to some extent.

In a framework of QCD, it is predicted theoretically that partial restoration of the chiral symmetry in nuclear field causes modification of hadron masses even at the nuclear saturation density [1-3]. Modification of nucleon mass is also predicted in quantum hadrodynamics, where nuclear system is described semi-phenomenologically by using hadron degrees of freedom [4]. In this framework, effect of strong scaler-meson field is treated as reduction of effective nucleon mass and the lower component of the nucleon Dirac spinor is predicted to be enhanced. It is one of challenges in current nuclear physics to find experimental evidence of such medium effects.

At several hundred MeV of incident energies, the interactions between nucleons are well described by theoretical models based on meson exchange forces. Thus it is expected that above mass modifications cause modification of nucleon-nucleon (NN) interactions and are detectable in some nuclear reactions.

The nucleon quasifree scattering is a reaction where an incident nucleon is scattered with and knocks out one of bound nucleons in a target nucleus. It is, therefore, regarded as an NN scattering in nuclear field and has a possibility to be a tool to study NN interactions in the field. In particular, spin observables are expected to play important roles in such studies because they are less distorted by strong absorption effect by surrounding nuclear potential and also because they are sensitive to the lower component, which includes a spin operator, of nucleons and spin of exchanged mesons.

Here we report on our recent study of nuclear medium effect in nucleon quasifree scattering, especially exclusive $(p, 2p)$ reactions.

A_Y DATA AT 392 MEV AND NON-RELATIVISTIC AND RELATIVISTIC MEDIUM EFFECTS

It was first pointed out clearly by Horowitz and collaborators [6,7], based on Los Alamos data [5], that the analyzing power A_y for quasifree scattering of protons are significantly reduced from the value for free nucleon-nucleon scattering and they claimed that this evidence is one of the most distinct relativistic signature. Such reduction of A_y was also found for a proton knockout reaction, exclusive measurement of the quasifree scattering or $(p, 2p)$ reaction, leading to the $1s_{1/2}$ state of residual ^{15}N nucleus [8]. They has concluded in a recent paper that the reduction found is much stronger than those predicted by non-relativistic DWIA calculations with density dependent t matrix or a relativistic DWIA calculation [9].

At RCNP Osaka, we have measured $1s_{1/2}$ or $2s_{1/2}$ knockout reactions for several kinds of target nuclei at a zero-recoil condition and found that such reduction is monotonically decreasing function of nuclear density, which is estimated by using a distorted wave impulse approximation (DWIA) [10]. The data correspond to the forward detection angle of 25.5° are plotted in Fig. 1. The figure also shows that PWIA (dashed line) and DWIA (solid line) calculations with free t matrix fail to explain this reduction. At the same time, it is also shown that the distortion effect, difference of those calculations, does not play essential roles for this kinematical condition.

One of possibilities which causes such reduction is contribution of multistep processes. On this point, we have examined from three kinds of view point, namely:

- Separation-energy spectra of ^{12}C$(p, 2p)$ reaction leading to deep hole states of residual ^{11}B nuclei [11]
- Comparison of experimental data with a DWIA calculation on recoil-momentum dependence of differential cross section for $1s_{1/2}$ knockout from ^{12}C target [11]
- Comparison of experimental cross section of ^{40}Ca$(p, p'p'')$ reaction for a wide range of angle pairs and outgoing energies with a model calculation of pre-equilibrium processes [12].

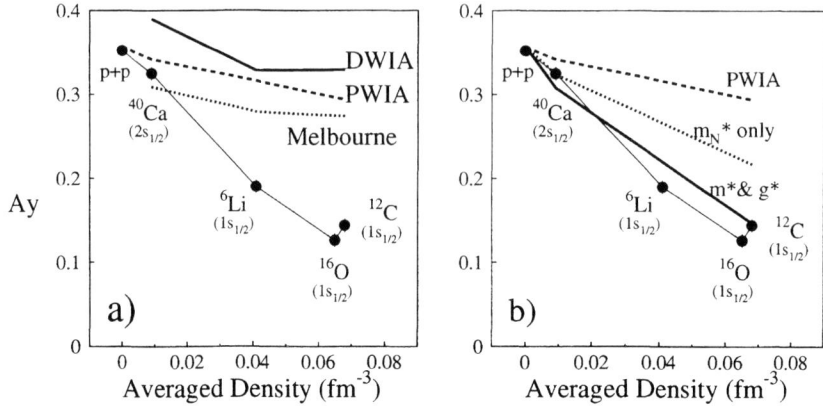

FIGURE 1. Experimental A_y data as a function of mean nuclear density which contributes to DWIA cross sections. The data are connected by thin lines for eye guide. The left pannel shows comparison with calculations by using free NN interaction (DWIA and PWIA) or g matrix made by Amos and collaborators (Melbourne: DW calculation). The right pannel is comparison with PW calculations where effect of hadron-mass modificaitons are taken into account. See the text for deteil.

In addition to these, we also confirmed that DWIA calculations well reproduce both of differential cross sections and analyzing powers for $(p, 2p)$ reactions leading to low-lying discrete states of several nuclei [13].

From those examinations, we conclude that concentration of multistep processes does not cause observed distinct A_y reduction, which is about 50% of the value estimated by DWIA with free t matrix in the case of ^{12}C target. All of those results strongly suggest existence of some medium effect in the NN interactions.

Next, these data have been compared with a calculation where non-relativistic medium effect is taken into account. Amos and collaborators have succeeded to reproduce proton-nucleus elastic data in a wide region of incident energies using g-matrices which were obtained by solving the Bethe-Goldstone equation [14]. The dotted line in Fig. 1a) is a DWIA calculation by using their g-matrix in on-shell approximation. A density-dependent version of the computer code THREEDEE [15] was used for this calculation. The result shows some reduction from the DWIA calculation with free t matrix but it completely fail to reproduce the density dependence found in the experimental data. Since the calculation shows a significant reduction even for ^{40}Ca target case, where the averaged density is only several percent of the saturation density, the result suggests that the difference of two calculations is caused by difference of free NN interactions used. This comparison supports a claim given by Horowitz's group that non-relativistic calculations does not reproduce the A_y reduction in nucleon quasifree scattering.

Effects of hadron-mass modifications are estimated by a model calculation which follows a procedure proposed by Horowitz and Iqbal [6]. A T matrix of the $(p,2p)$ reaction is written in the relativistic framework with zero-range approximation as

$$T = F_k \int \bar{\Psi}_1(\mathbf{r})\bar{\Psi}_2(\mathbf{r})\hat{F}\Phi(\mathbf{r})\Psi_0(\mathbf{r})d\mathbf{r},$$

where $\Psi_i(\mathbf{r})$ is four component wave function of an incident ($i=0$) or outgoing ($i=1,2$) proton. The wave function of the bound nucleon is described as $\Phi(\mathbf{r})$, $\hat{F}$ is a Lorenz invariant NN amplitude, and F_k is a kinematical factor.

In their procedure, this T matrix is expressed in a Schrödinger equivalent form,

$$T = F_k \int \chi_1^*(\mathbf{r})\chi_2^*(\mathbf{r})\langle \bar{U}_1\bar{U}_2|\hat{F}|U_3U_0\rangle \phi(\mathbf{r})\chi_0(\mathbf{r})d\mathbf{r},$$

and $\langle \bar{U}_1\bar{U}_2|\hat{F}|U_3U_0\rangle$ is equated to the t matrix in the Schrödinger framework. Here $\chi_i(\mathbf{r})$'s and $\phi(\mathbf{r})$ are two component wave functions and U_i's are defined by $\Psi_i(\mathbf{r}) = U_i\chi_i(\mathbf{r})$ and $\Phi(\mathbf{r}) = U\phi(\mathbf{r})$.

In the present calculations, non-relativistic distortion effects are neglected. Namely, all of χ_i's are replaced by plane waves and momentum operator in the lower components of U_i's are replaced by asymptotic values of momentum vectors. Effect of nucleon mass modification is estimated by replacing the mass parameters in the U_i's. For the Lorents invariant amplitudes $\hat{F}$, we use the relativistic Love Franey interactions [16]. Effect of meson-mass and coupling-constant modifications is estimated by simply changing corresponding parameters in the interactions.

Values of these effective masses and coupling constants at saturation density used are taken Ref. [17]. In actual calculation, linear dependences of these parameters to nuclear density are assumed and values corresponding to the averaged densities, given in the horizontal axes, are used for each target case.

The result is shown in Fig. 1b by a thick solid line. The dotted line is a result when only nucleon mass is modified and dashed line is a PWIA calculation with free NN interaction. The observed reduction is almost completely reproduced by the solid line and a half of the reduction is caused by the nucleon-mass modification.

POLARIZATION MEASUREMENT AT 1 GEV

Even though a clear density dependence of A_y has been observed, there exist serious limitations in studying medium effect at 392 MeV, which is expected to be coped with at higher energies, typically at 1 GeV. Those limitations, and advantages at 1 GeV, are:

- In the case of Fig. 1 data, contribution from the exchange terms of the NN amplitudes amount to 20–30%, typically, of the total amplitudes. At 1 GeV, the exchange contribution reduces and simpler relations between each meson exchange term and spin observables are expected.

- Measurements in a wider momentum-transfer (q) region are possible in the case of 1 GeV. This q dependence is expected to give information to distinguish effects of meson-mass, nucleon-mass, and coupling-constant modifications.
- At 400 MeV, clear $1s_{1/2}$-knockout bumps are observed only for $1p$-shell target nuclei such as ^{12}C and ^{16}O and not observed in the case of heavier target as ^{40}Ca. On the other hand at 1 GeV, a clear $1s_{1/2}$ bump have been reported even for ^{208}Pb target [18].

In order to extend our work to 1 GeV region, an experimental program by using a synchro-cyclotron at Petersburg Nuclear Physics Institute has been started [19]. Since no polarized beams are available, induced polarizations P_1 and P_2 of forward and backward outgoing protons, respectively, have been measured. A pair spectrometer system which consists of two QQD type spectrometers has been used for detection of two outgoing protons in $(p,2p)$ reactions. Polarization of outgoing particles has been measured by using a polarimeter system, consisting of a carbon block and multi-wire proportional chambers on each focal plane. As the data in Fig. 1, the measurement was performed for a zero-recoil condition. Detection angle of backward outgoing protons is fixed at 53.05 degree and other angle and energy values are set so as to fulfill this condition at the center of $1s_{1/2}$-knockout bump. The forward detection angles and energies of forward and backward protons range within 22–26 degrees, 730–745 MeV, and 220–270 MeV, respectively.

Figure 2 shows preliminary data, plotted to averaged densities as Fig. 1. Even though the data themselves show only a moderate decrease with the density, a PWIA result, shown as a solid line, increases and data are significantly reduced from this line.

In the same figure, a PWIA estimation with nucleon mass modification, calculated in the Shrödinger equivalent framework as the dotted line in Fig. 1b) is plotted by dashed curve. As the case of A_y at 392 MeV, this estimation shows clear density dependence and explains about a half of experimentally found reduction.

The facts that the energies of incoming and outgoing protons are quite different between 392 MeV and 1 GeV case, and nevertheless that the reduction rate compared with relativistic estimations is almost the same for both cases again suggest that this reduction is structure or interaction originated.

SPIN-TRANSFER MEASUREMENT

Another extension of the study is to measure spin-transfer coefficients on the same reaction. The final goal of this extension is to determine in-medium NN amplitudes in spin space experimentally and to study medium modification of each meson exchange separately. For such purpose, all of spin-transfer coefficients to both of forward and backward outgoing protons are required in addition to differential cross section and analyzing power. So far, however, spin-transfers to backward outgoing protons have not been measured and only a limited result has been obtained.

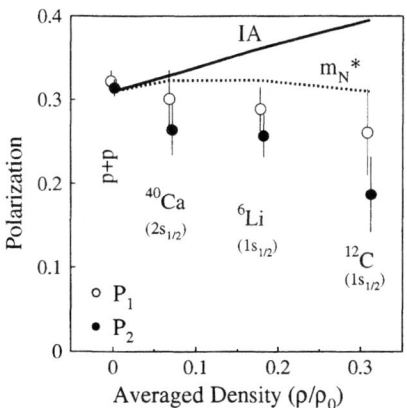

FIGURE 2. Induced polarization data for $(p, 2p)$ reactions at the incident energy of 1 GeV. P_1 and P_2 correspond to polarizations of forward and backward outgoing protons, respectively. The solid line is a PWIA calculation with free NN interaction and dotted line is a calculation when the nucleon mass is reduced.

Figure 3 shows experimental data of induced polarization P, analyzing power A_y and five spin transfer coefficients $D_{i'j}$'s to forward outgoing protons for the same kinematical condition as the data in Fig. 1. Calculations with DWIA and PWIA in the final energy description are also plotted in the same figure as well as a DWIA calculation in the initial energy prescription. There are significant differences between two prescriptions except for P and A_y and the experimental data are reasonably reproduced by the final energy prescription. It is also found from this figure that distortion effects, differences between PWIA and DWIA calculations, are almost negligible for all of spin transfer coefficients for this kinematical condition.

Effects of mass modifications are also estimated by the same procedure as described above for A_y and the results for a few observables are given in Fig. 4. The solid and dashed lines are calculations with the same condition as corresponding lines in Fig. 1b. The mass-modification effect, which reproduce the A_y reduction, gives significant effect also for $D_{S'S}$ and it results worse fit to the data. It is mentioned here that the PWIA results in this figure is not the same as those in Fig. 3 since the relativistic Love Franey interaction is used here while a t matrix from a phase shift analysis is used in the case of Fig. 3.

It is well known that the effects of the scaler and vector fields are mostly cancel each other. Even though a large effect of mass modification is given for $D_{i'j}$'s in Fig. 4, there still exists the same kind of cancellation. The dotted (dot-dashed) line in the same figure is a result when only omega-(sigma-)meson mass is modified by a small amount, 0.9 of free mass at the saturation density, which corresponds to 0.98

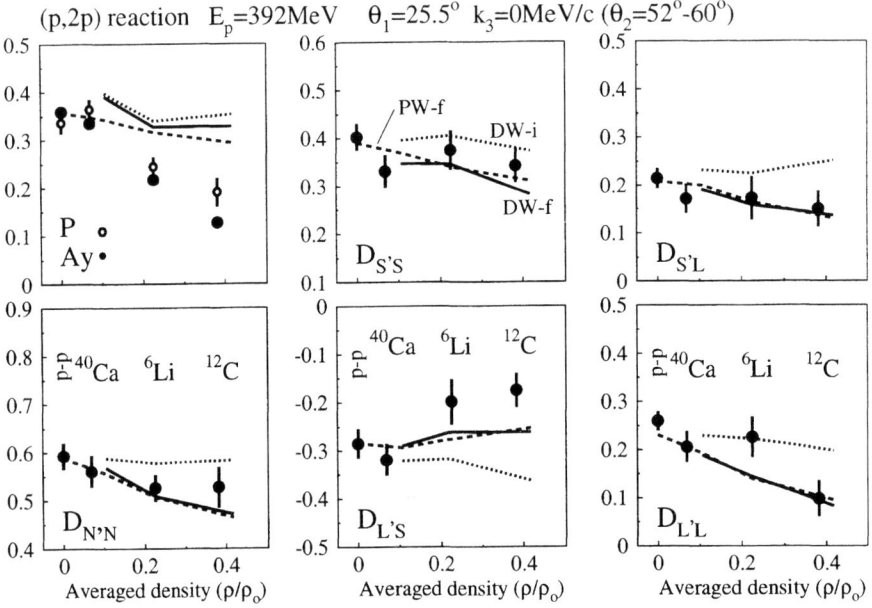

FIGURE 3. Experimental $D_{i'j}$ data. The lines are PWIA and DWIA calculations with free NN interactions in the initial- and final-energy prescription.

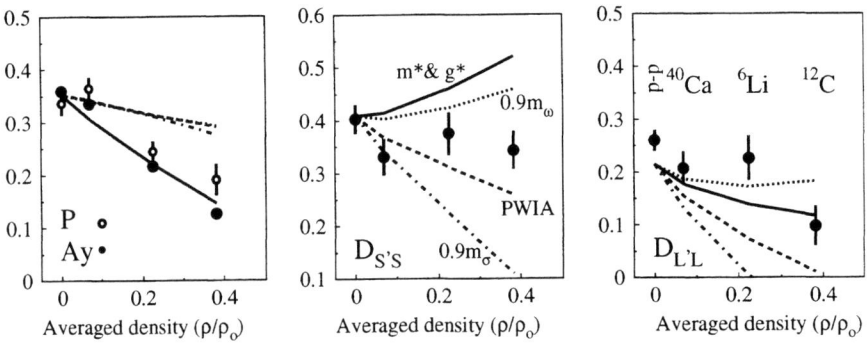

FIGURE 4. PWIA calculations with nucleon and meson mass modifications. See the text for calculation lines.

for ^{6}Li target case. It is found that such small modifications cause large deviation of $D_{i'j}$'s. Namely, spin transfer coefficients are quite sensitive to each meson mass and results of calculations fluctuate significantly with minor modifications of mass parameters. This means, on the other hand, that $D_{i'j}$ has a possibility to offer constraints to reduction rates of meson masses, such as the Brown-Rho scaling low [1]. It is worth mentioned that A_y is almost free from such critical cancellation effects.

INEQUALITIES BETWEEN SPIN-OBSERVABLES CAUSED BY RELATIVISTIC EFFECT

In NN elastic scattering, the polarization (P) and analyzing power (A_y) are equal to the extent that time reversal symmetry holds. There is also an equality between two spin transfer coefficients, $D_{qp}=-D_{pq}$, where q and p are directions of the transfered momentum and an average of initial and final momentum, respectively, in the center of mass system.

In the case of quasifree scattering, these equalities are broken in general, even if its reaction process is simply one step, because of energy off-shell effect and also because of distortion effect. Here we examine these effects in Schrödinger equivalent expression of the relativistic DWIA.

A trivial effect is given by distorting potentials which are the same as those in non-relativistic framework. But for the kinematical condition of the measurements in Fig. 1 and Fig. 3, these distortion effects themselves are not so large. As a result, it was examined for the case of ^{12}C($p, 2p$) that the absolute value of $P-A_y$ and $D_{qp}+D_{pq}$ caused by this distortion effect are 0.005 and 0.015, respectively, which are much smaller than the experimental errors of present data.

More significant contribution is derived from a lower-component contribution combined with an off-shell condition. The non-relativistic t-matrix of the NN scattering are expressed with five terms in spin space, under assumptions of parity and time reversal invariance, as

$$t = A + B\sigma_{1n}\sigma_{2n} + C(\sigma_{1n} + \sigma_{2n}) + E\sigma_{1p}\sigma_{2p} + F\sigma_{1q}\sigma_{2q},$$

where n is a direction normal to the reaction plane and p and q are the same as above. In the Shrödinger equivalent expression, this t-matrix is equated to $\langle \bar{U}_1 \bar{U}_2 | \hat{F} | U_3 U_0 \rangle$ as described in previous section.

Since U_i, two components of Dirac spinor, is a function of the momentum as well as the effective nucleon mass, difference of the momentum values between the initial and final channels causes additional terms in above t matrix. Actually it is easily derived that an additional term

$$D(k^2 - k'^2)(\sigma_{1p}\sigma_{2q} + \sigma_{1q}\sigma_{2p})$$

appears if $k \neq k'$ where k and k' are initial and final momentum values in cm system.

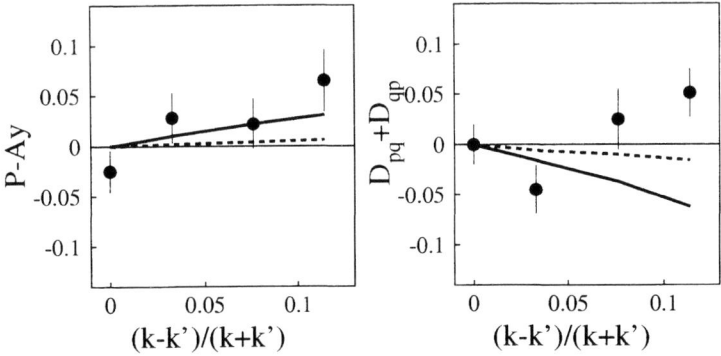

FIGURE 5. Experimental and calculated $P-A_y$ and $D_{pq}+D_{qp}$. The solid lines are the relativistic effect and the dashed lines are non-relativistic ones, which are described in the text.

This term is equivalent to the *D-term* proposed by KMT [20] and causes finite values for both of $P\text{-}A_y$ and $D_{qp}+D_{pq}$.

These values are calculated and plotted in Fig. 5 as well as the experimental data and estimation of the non-relativistic distortion effects. The figure shows that this relativistic off-shell effect causes significant deviations from zero for both of those observables. Since these deviations are the comparable size with present error bars it is difficult to extract a definite result at present, but improved data and calculations will give new information on reliability of relativistic treatments for nuclear reactions.

SUMMARY

The nucleon quasifree scattering, especially exclusive measurement of it, at the incident energy of several hundred MeV is potentially ideal reaction to investigate the NN interaction in nuclear field and thus it is expected to be informative in studying medium effects on hadrons in nuclei. Actually, it is concluded from DWIA calculations that such effect is likely observed in this reaction if hadron properties are modified as functions of nuclear local density and they cause significant change of NN interactions even at the saturation density, which model calculations suggest.

Through measurements of spin observables for this reaction, it was found that the reduction of analyzing power and induced polarization for this reaction is monotonically density dependent. It was also found from the measurement at 392 MeV and 1 GeV that this feature is general one beeing independent of incident and outgoing energies. On the other hand, spin transfer data at 392 MeV show no significant density dependent deviations from PWIA and DWIA calculations in the final energy prescription. A model calculation with hadron-mass modifications exclusively reproduce the A_y reductions, but it fail to reproduce the spin-transfer data and further theoretical effort is required to explain these density dependences. Since D_{ij}'s are quite sensitive to each each meson mass, there is possibility to get better fit by fine tuning of parameters and, at the same time, these data give critical constraint to them.

Present experimental data also suggest finite values, several percent at most, for both of $P-A_y$ and $D_{pq}+D_{qp}$, which are zero in the case of free NN scattering under parity and time-reversal invariances. The same size of non-zero values are predicted in a PWIA calculation where a relativistic off-shell effect is taken into account. This is a new evidence of relativistic effect and will offer additional criterion in testing a long-standing problem, reliability of the relativistic model where nucleons are treated as Dirac particles.

Nuclear quasifree scattering gives an effective means in studying deep inside of nucleus and nuclear medium which characteristic is still not well known.

REFERENCES

1. G. E. Brown and M. Rho, *Phys. Rev. Lett.* **66**, 2720 (1991).
2. R. J. Furnstahl, D. K. Griegel, and T. D. Cohen, *Phys. Rev.* **C46**, 1507 (1992).
3. T. Hatsuda, *Nucl. Phys.* **A544**, 27c (1992).
4. B. Serot and J. D. Walecka, *Int. J. Mod. Phys.* **E6**, 515 (1997) and references therein.
5. J. A. McGill et al., *Phys. Lett.* **134B**, 157 (1984).
6. C. J. Horowitz and M. J. Iqbal, *Phys. Rev.* **C33**, 2059 (1986).
7. C. J. Horowitz and D. P. Murdock, *Phys. Rev.* **C37**, 2032 (1988).
8. C. A. Miller et al., in *Proc. of the 7th Int. Conf. on Polarization Phenomena in Nuclear Physics*, (Paris, 1990) C6-595.
9. C. A. Miller et al., *Phys. Rev.* **C57**, 1756 (1998).
10. K. Hatanaka et al., *Phys. Lev. Lett.* **78**, 1014 (1997).
11. T. Noro et al., *Nucl. Phys.* **A629**, 324c (1998).
12. A. A. Cowley et al., *Phys. Rev.* **C57**, 1749 (1998).
13. Obayashi et al., RCNP Annual Report 1998, p.11.
14. P. J. Dortmans and K. Amos, *J. Phys.* G **17**, 901 (1991).
15. N. S. Chant and P. G. Roos, it Phys. Rev. **C27**, 1060 (1983).
16. C. J. Horowitz, it Phys. Rev. **C31**, 1340 (1985).
17. G. Krein et al., *Phys. Rev.* **C51**, 2646 (1995).
18. A. A. Vorob'ev et al., Yad. Fiz. **58**, 1923 (1995); Phys. Atom. Nucl. **58**, 1817 (1995).
19. SC-150/2 collaboration at PNPI, private communications.
20. A. K. Kerman, H. McManus and R. M. Thaler, *Ann. Phys.* **8**, 551 (1959).

Nonrelativistic Bound States in Quantum Field Theory

Aneesh V. Manohar and Iain W. Stewart

*Physics Department 0319, University of California at San Diego,
9500 Gilman Drive, La Jolla, CA 92093*

Abstract. Nonrelativistic bound states are studied using an effective field theory. Large logarithms in the effective theory can be summed using the velocity renormalization group. For QED, one can determine the structure of the leading and next-to-leading order series for the energy, and compute corrections up to order $\alpha^8 \ln^3 \alpha$, which are relevant for the present comparison between theory and experiment. For QCD, one can compute the velocity renormalization group improved quark potentials. Using these to compute the renormalization group improved $\bar{t}t$ production cross-section near threshold gives a result with scale uncertainties of 2%, a factor of 10 smaller than existing fixed order calculations.

INTRODUCTION

Nonrelativistic bound states in QED and QCD provide an interesting and highly nontrivial problem to which effective field theory methods can be applied [1,2]. The QCD bound states we will consider are heavy $\bar{Q}Q$ states such as $\bar{t}t$ bound states or the Υ system. In QED, the classic examples are Hydrogen, muonium ($\mu^+ e^-$), and positronium. Each of these systems has three important scales, m the fermion mass, mv the fermion momentum, and mv^2, the fermion energy. (For Hydrogen and muonium, m is the electron mass or the reduced mass of the two particles.) The velocity v is of order the coupling constant (α_s or α), and we will only consider the case $v \ll 1$, $mv^2 \gg \Lambda_{\rm QCD}$ so that nonperturbative effects are small.

The goal is to correctly separate the scale m, mv and mv^2 for nonrelativistic bound state problems using an effective field theory, and to sum large logarithms using the renormalization group. The large logarithms in this case are $\ln p/m$, $\ln E/m$ and $\ln p/E$ which are proportional to $\ln v$, and lead to $\ln \alpha$ contributions to bound state energies. Furthermore, for QCD, the effective theory also determines the scale of the strong coupling constant, i.e. whether one should use $\alpha_s(m)$, $\alpha_s(mv)$ or $\alpha_s(mv^2)$. The nonrelativistic effective theory, NRQCD/NRQED, has

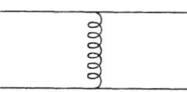

FIGURE 1. A potential gauge boson exchange. The typical momentum and energy transfered are mv and mv^2.

been studied extensively in the past [1–17]. What is new is the precise formulation of the effective theory, and the way in which the renormalization group is scaling is implemented.

NEW RESULTS

There are many interesting new results that have been obtained for QED and QCD [11–15,25–28]. For QED, one finds a universal description of $\ln \alpha$ terms. A single renormalization group equation gives the Lamb shift, hyperfine splitting and decay widths for Hydrogen, muonium and positronium. The renormalization group method allows us to compute for the first time the $\alpha^8 \ln^3 \alpha$ Lamb shift in positronium and the $\alpha^8 \ln^3 \alpha$ Lamb shift in Hydrogen and muonium including recoil corrections. It also resolves a controversy in the literature about the $\alpha^8 \ln^3 \alpha$ Hydrogen Lamb shift in the limit $m_p \to \infty$.

The renormalization group method allows one to understand the structure of the QED perturbation series, and why the $\ln \alpha$ corrections terminate. The leading order series has a single term that contributes at order $\alpha^5 \ln \alpha$ to the energy, and the next-to-leading order series terminates after three terms, $\alpha^6 \ln \alpha$, $\alpha^7 \ln^2 \alpha$, and $\alpha^8 \ln^3 \alpha$. One also finds some infinite series of terms in QED, but they have the form $(\alpha^3 \ln^2 \alpha)^n$, rather than $(\alpha \ln \alpha)^n$.

In QCD, one is able to sort out the scales for α_s, and decide whether the strong coupling is $\alpha_s(m)$, $\alpha_s(mv)$, or $\alpha_s(mv^2)$. We also obtain the renormalization group improved computations of the bound states potentials in QCD. There are numerous applications of these results, and I will show an example of the dramatic improvement one obtains for the $\bar{t}t$ production cross-section near threshold [27,28].

THE PROBLEM

The basic problem can be seen by drawing a few Feynman diagrams. A typical gauge boson exchange in the t channel such as Fig. 1 has momentum transfer of order $p \sim mv$. A wavefunction graph or radiated gauge boson graph such as Figs. 2 have gauge boson momenta of order $E \sim mv^2$. More interesting diagrams such as those in Figs. 3 involve gauge bosons with momenta of order p and order E. In a graph such as Fig. 4, the vacuum polarization insertions make the effective coupling of the two gluons $\alpha_s(mv)$ and $\alpha_s(mv^2)$ respectively. One result which should be clear from Fig. 4 is that graphs can involve $\alpha_s(mv)$ and $\alpha_s(mv^2)$ *simultaneously*. We will return to this important point later on.

FIGURE 2. Graphs containing ultrasoft photons, with energy and momentum of order mv^2.

FIGURE 3. Graphs containing gauge bosons carrying momentum of order mv and mv^2.

MOMENTUM REGIONS AND DEGREES OF FREEDOM

The Feynman integrals in the full theory can be evaluated using the threshold expansion [29]. The important momentum regions (in Feynman gauge) are referred to in the literature as hard ($E \sim m$, $p \sim m$), potential ($E \sim mv^2$, $p \sim mv$), ultrasoft ($E \sim mv^2$, $p \sim mv^2$) and soft ($E \sim mv$, $p \sim mv$). The threshold expansion momentum regions are often used to describe bound state computations; however it is important to note that *the threshold expansion is not an effective field theory*. To construct an effective field theory, one needs to include only modes that can be on-shell. The effective theory therefore has nonrelativistic fermions (which are potential modes), and soft and ultrasoft gauge boson modes. The hard fermion and gauge boson momentum regions, the soft fermion momentum region, and the potential gauge boson momentum region do not require modes in the effective theory.

The desired effective theory is valid for energies and momenta much smaller than the fermion mass m. One can try expanding in powers of E/m and p/m as in heavy quark effective theory, so that the expansion parameter is $1/m$. For example, the dispersion relation $E = \sqrt{\mathbf{p}^2 + m^2}$ gives terms in the Lagrangian of the form

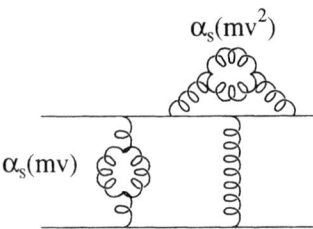

FIGURE 4. An example of a graph involving both $\alpha(mv)$ and $\alpha(mv^2)$.

$$L = \psi^\dagger \left(E - \frac{\mathbf{p}^2}{2m} + \frac{\mathbf{p}^4}{8m^3} + \ldots \right) \psi. \tag{1}$$

The lowest order propagator is $1/(E+i\epsilon)$, which gives $\theta(t)$ in position space. This is the static propagator of HQET: fermions propagate forward in time, but do not move in space. This propagator is acceptable for some calculations involving heavy quarks. For example, one can compute the static potential between fixed sources using this propagator. However, for $t\bar{t}$ production, the quarks are produced at the same point, and they remain at the same point for all time if the static propagator is used. This is too singular, and the HQET expansion breaks down. In general, it is essential for treating nonrelativistic bound states that the heavy fermions move. For this to occur, the lowest order propagator should be $1/(E-\mathbf{p}^2/2m+i\epsilon)$, so that E and $\mathbf{p}^2/2m$ are of the same order in the effective theory power counting. This implies that the $1/m$ expansion cannot be used; instead one must use an expansion in powers of v, where E and $\mathbf{p}^2/2m$ are both of order v^2 [1,2].

The effective theory expansion parameter is the velocity v, and formally, α must also be treated as order v. Thus order α^2 radiative corrections to the leading term are just as important as order v^2 relativistic corrections. The effective theory below the scale m has:

- Nonrelativistic fermions with propagator

$$\frac{1}{E - \mathbf{p}^2/2m + i\epsilon}$$

- Ultrasoft gauge bosons coupled via interactions that are multipole expanded [6].

- Potentials $V(\mathbf{p},\mathbf{p}')$ for the scattering of an incoming Q and $\bar{Q}$ with momenta $\mathbf{p}$ and $-\mathbf{p}$ to outgoing Q and $\bar{Q}$ with momenta $\mathbf{p}'$ and $-\mathbf{p}'$.

- Soft gauge bosons. The importance of introducing soft fields in the effective theory was first pointed out by Griesshammer [30].

The effective theory has two different gauge boson fields, soft bosons and ultrasoft bosons. This does not lead to any double counting if graphs are evaluated in dimensional regularization.

The static theory is not the $m \to \infty$ limit or the $v \to 0$ limit of the effective theory. For this reason, the static potential and the effective theory potential are not equal.

RENORMALIZATION GROUP EVOLUTION

The nonrelativistic bound state system has three important mass scales, m, mv and mv^2.

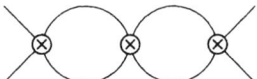

FIGURE 5. An ultraviolet divergent two-loop graph involving the iteration of three potentials.

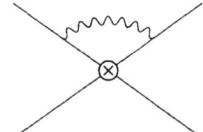

FIGURE 6. One-loop ultrasoft photon renormalization of the potential.

Two-stage running

The conventional method of implementing the renormalization group is as follows

- Start at $\mu = m$
- Scale μ from m to mv
- Integrate out the soft modes at mv
- Scale μ from mv to mv^2

This is referred to as the two-stage method, because there are two-stages of renormalization group evolution. Consider a loop graph involving time-ordered products of potentials, such as Fig. 5. This graph contains a logarithm of the form $\ln\sqrt{mE}/\mu$. When μ is set to mv, this logarithm has the from $\ln\sqrt{E/mv^2}$, and is small. Thus the logarithms in the graph are summed by renormalization group evolution of μ from m to mv.

The graph in Fig. 6 involving an ultrasoft photon exchange contains a logarithm of the form $\ln E/\mu$. The μ in this ultrasoft graph is scaled all the way down (in two stages) to mv^2, at which point the logarithm is $\ln E/mv^2$, and also small.

However, this two-stage method of implementing the renormalization group turns out to be incorrect for nonrelativistic bound states. The reason is that the scales mv and mv^2 are correlated—one cannot be varied independently of the other. Instead one needs to use an alternative one-stage scaling procedure.

One-stage running

In one stage running, one introduces two different μ parameters, μ_S and μ_U [11]. In dimensional regularization in $4-2\epsilon$ dimensions, the soft photon coupling is multiplied by μ_S^ϵ, the ultrasoft photon coupling by μ_U^ϵ, and the potentials by $\mu_S^{2\epsilon}$. Note that this is only possible because we have two different photon fields to represent the soft and ultrasoft photons in the effective theory. Then

- Set $\mu_S = m\nu$, $\mu_U = m\nu^2$

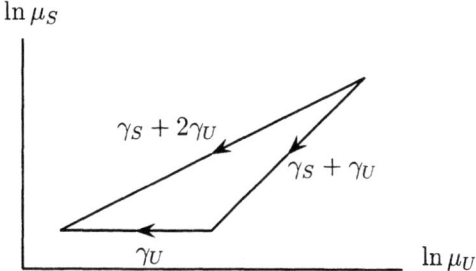

FIGURE 7. Paths in the (μ_U, μ_S) plane for one-stage and two-stage running.

- Start at $\nu = 1$ and scale to $\nu = v$.

This procedure is referred to as the velocity renormalization group, because one runs in velocity ν rather than momentum [11]. The logarithms in Figs. 5 and 6 are now $\ln \sqrt{mE}/m\nu$ and $\ln E/m\nu^2$, which are minimized when $\nu = v$. Thus this method also minimizes logarithms in the diagrams, and sums them by renormalization group evolution.

The difference between the two renormalization group methods can be seen in Fig. 7 [25]. In two-stage running, there is only a single μ, so that $\mu_S = \mu_U = \mu$, and they are lowered together from m to mv. At this point, the soft modes are integrated out, and μ_U for the ultrasoft modes is lowered to mv^2. The integration path in Fig. 7 is along the lower edges of the triangle. In one-stage running, the integration path is along the diagonal. It is convenient to define two anomalous dimensions, γ_S and γ_U by taking the derivatives of Green's functions with respect to $\ln \mu_S$ and $\ln \mu_U$, respectively. One can show by explicit calculation that

- The two paths give different answers. The integration is path dependent because $\nabla \times \gamma \neq 0$.

- One-stage running using the velocity renormalization group agrees with explicit QED calculations at order $\alpha^3 \ln^2 \alpha$, $\alpha^7 \ln^2 \alpha$ and $\alpha^8 \ln^3 \alpha$.

The moral is that for nonrelativistic bound states, one should run in velocity rather than momentum.

RUNNING POTENTIALS

The running potential $V(\mathbf{p}, \mathbf{p}')$ has an expansion

$$V(\mathbf{p}, \mathbf{p}') = \mathbf{V}^{(-1)} + \mathbf{V}^{(0)} + \mathbf{V}^{(1)} + \mathbf{V}^{(2)} + \ldots \qquad (2)$$

where $V^{(n)}$ is of order v^n in the velocity power counting. The first three terms in the expansion have the form

TABLE 1. Numerical values for the $\bar{t}t$ singlet potentials. The values at $\nu = 1$ are the matching values at $\mu = m_t$. The values at $\nu = v$ are the velocity renormalization group improved values, where $v = 0.14$ has been used.

Coefficient	$U_c^{(s)}$	$mU_k^{(s)}$	$m^2U_r^{(s)}$	$m^2U_2^{(s)}$	$m^2U_s^{(s)}$	$m^2U_\Lambda^{(s)}$	$m^2U_t^{(s)}$
$\nu = 1$	-1.81	-0.36	-1.81	0	0.60	0.15	2.71
$\nu = v$	-2.47	-0.03	-1.49	0.63	0.53	0.16	3.11

$$V^{(-1)} = \frac{U_c}{\mathbf{k}^2},$$
$$V^{(0)} = \frac{U_k}{|\mathbf{k}|}, \quad (3)$$
$$V^{(1)} = U_2 + U_s \mathbf{S}^2 + \frac{U_r(\mathbf{p}^2 + \mathbf{p}'^2)}{2\mathbf{k}^2} - \frac{i\mathbf{U}_\Lambda \cdot (\mathbf{p}' \times \mathbf{p})}{\mathbf{k}^2} + U_t\left(\boldsymbol{\sigma_1} \cdot \boldsymbol{\sigma_2} - \frac{3\mathbf{k} \cdot \boldsymbol{\sigma_1}\mathbf{k} \cdot \boldsymbol{\sigma_2}}{\mathbf{k}^2}\right),$$

where $V^{(0)} \sim 1/m$, and $V^{(1)} \sim 1/m^2$. In QCD, each of the coefficients can be written as $U \to U^{(1)}1 \otimes 1 + U^{(T)}T^A \otimes \bar{T}^A$, where 1 and $T^A/\bar{T}^A$ are color matrices acting on the quark/antiquark lines. The anomalous dimensions for the coefficients U_c–U_t have been computed, and the details are given in Refs. [12–14]. The renormalization group improved static potential was computed in Ref. [31]. An important point to note is that graphs can involve both soft and ultrasoft gluons, so that the anomalous dimensions involve *both* $\alpha_s(mv)$ and $\alpha_s(mv^2)$. As an example, the running of $U_2^{(1)}$ is given by

$$m^2U_2^{(1)}(\nu) = \frac{14C_1}{3}\alpha_s(m\nu)\alpha_s(m)\ln\left(\frac{m\nu}{m}\right) - \frac{32\pi C_1}{3\beta_0}\alpha_s(m)\ln\left[\frac{\alpha_s(m\nu)}{\alpha_s(m\nu^2)}\right] \quad (4)$$

where $C_1 = 2/9$ for QCD. Note that Eq. (4) depends on $\alpha_s(m)$, $\alpha_s(m\nu)$, and $\alpha_s(m\nu^2)$. The running coefficients in the singlet channel ($U^{(s)} = U^{(1)} - C_F U^{(T)}$) for $\bar{t}t$ production are presented in Table 1. The renormalization group improved coefficients U_r and U_2 differ significantly from their matching values, because they depend on the ultrasoft scale through $\alpha_s(mv^2)$. The other coefficients only have a soft anomalous dimension, and do not run as much.

The renormalization group improved potentials can be used to calculate the renormalization group improved cross-section for $\bar{t}t$ production in the threshold region. Fig. 8 shows a sample fixed order calculation of R, the ratio of $\sigma(e^+e^- \to \bar{t}t)/\sigma(e^+e^- \to \mu^+\mu^-)$ [32]. The scale uncertainty is of order 20%. The renormalization group improved version of the results is shown in Fig. 9. There is a dramatic reduction in the scale uncertainty, which is now around 2%, as well as an improvement in convergence for the normalization. The small theoretical uncertainty means that an accurate measurement of the cross-section can be used

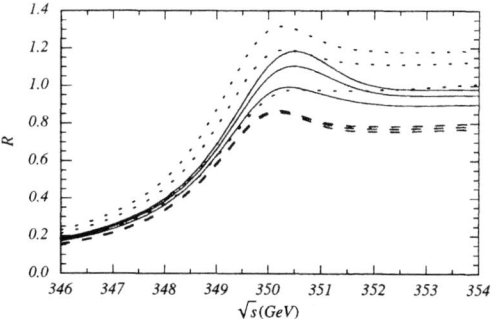

FIGURE 8. Fixed order computation of $\bar{t}t$ production near threshold. The curves are LO (dotted), NLO (dashed) and NNLO (solid.) Uses the $1S$ mass-scheme [33–35]

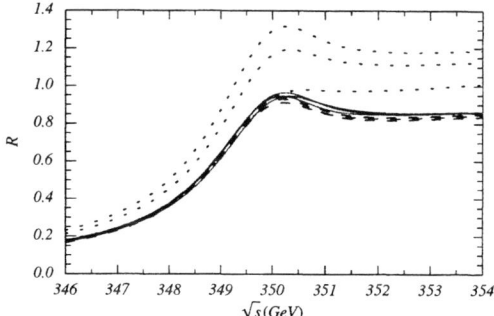

FIGURE 9. Renormalization group improved computation of $\bar{t}t$ production near threshold. The curves are LL (dotted), NLL (dashed) and NNLL (solid.). Uses the $1S$ mass-scheme [33–35]

to study new physics. For example, a standard model Higgs boson of mass around 115 GeV changes the cross-section by $\sim 5\%$, and is measurable.

QED

The velocity renormalization group method gives very interesting and important results when applied to QED [15]. The basic potentials we will need for QED are summarized in Table 2.

The bound state energy levels can be determined to order α^4 by computing the matrix elements of $V^{(0)}$ and $V^{(1)}$ between Coulomb wavefunctions. Time-ordered products of two potentials, such as $T\left[V^{(0)}V^{(0)}\right]$, $T\left[V^{(0)}V^{(1)}\right]$ and $T\left[V^{(1)}V^{(1)}\right]$ first contribute at order α^6.

Integrating the renormalization group equations for $V^{(0)}$ and $V^{(1)}$ using the leading order anomalous dimension gives a series of the form

$$\alpha\left(1 + \alpha\ln\alpha + \alpha^2\ln^2\alpha + \alpha^3\ln^3\alpha + \ldots\right),$$

TABLE 2. Table of potentials for QED. The first column is the potential, the second gives typical terms in the potential, the third gives the power counting in α and v, the fourth gives the order in the v counting scheme when $v \sim \alpha$, and the fifth gives the contribution of the potential to the bound state energy.

		Power Counting	Order	E
$V^{(-1)}$	$\frac{\alpha}{\mathbf{k}^2}$	$\frac{\alpha}{v}$	1	α^2
$V^{(0)}$	$\frac{\alpha}{m|\mathbf{k}|}$	α^2	α^2	α^4
$V^{(1)}$	$\frac{\alpha}{m^2}, \frac{\alpha \mathbf{S}^2}{m^2}$	αv	α^2	α^4
$V^{(2)}$	$\frac{\alpha|\mathbf{k}|}{m^3}$	$\alpha^2 v^2$	α^4	α^6
$V^{(3)}$	$\frac{\alpha \mathbf{k}^2}{m^4}$	αv^3	α^4	α^6
$\vdots$	$\vdots$	$\vdots$	$\vdots$	$\vdots$

which contributes

$$\alpha^4 \left(1 + \alpha \ln \alpha + \alpha^2 \ln^2 \alpha + \alpha^3 \ln^3 \alpha + \ldots \right)$$

to the energy. Integrating the next-to-leading order anomalous dimensions gives

$$\alpha^4 \alpha \left(1 + \alpha \ln \alpha + \alpha^2 \ln^2 \alpha + \alpha^3 \ln^3 \alpha + \ldots \right)$$

terms in the energy. The next-to-next-to-leading anomalous dimension gives

$$\alpha^4 \alpha^2 \left(1 + \alpha \ln \alpha + \alpha^2 \ln^2 \alpha + \alpha^3 \ln^3 \alpha + \ldots \right),$$

terms in the energy, which are the same order as those obtained by using the leading order anomalous dimension for the $V^{(2)}$ and $V^{(3)}$ potentials which first contribute at order α^6. Thus one can compute the

$$\begin{array}{cccc} \alpha^5 \ln \alpha & \alpha^6 \ln^2 \alpha & \alpha^7 \ln^3 \alpha & \ldots \\ \alpha^6 \ln \alpha & \alpha^7 \ln^2 \alpha & \alpha^8 \ln^3 \alpha & \ldots \end{array}$$

series in the energy using $\gamma_{\rm LO}$, $\gamma_{\rm NLO}$ for $V^{(0,1)}$.

LEADING ORDER

The Coulomb potential and $V^{(0)}$ do not run in QED at leading and next-to-leading order, so one is left with the running of $V^{(1)}$. The anomalous dimensions are evaluated for a particle of mass m_1 and charge $-e$ interacting with a second particle of mass m_2 and charge Ze. Evaluating the graphs in Fig. 10 gives

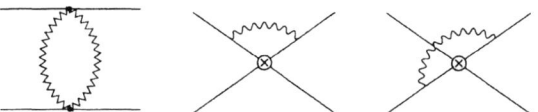

FIGURE 10. One-loop running of $V^{(1)}$ in QED. The first graph has a soft photon, and the other graphs have ultrasoft photons and a potential.

$$\nu \frac{dU_2}{d\nu} = \frac{14Z^2\alpha^2}{3m_1m_2} + \frac{2\alpha}{3\pi}\left(\frac{1}{m_1} + \frac{Z}{m_2}\right)^2 U_c \quad (5)$$

where the first term is the soft contribution from Fig. 10a and the second is the ultrasoft contribution from Fig. 10b,c. Note that the ultrasoft contribution has been multiplied by two, since the anomalous dimension for the velocity renormalization group is $\gamma_S + 2\gamma_U$. The other coefficients in $V^{(1)}$ (U_r, U_s, U_Λ, U_t) have zero anomalous dimension at this order.

Since the Coulomb potential and α do not run in QED, one can combine the two terms,

$$\nu \frac{dU_2}{d\nu} = \gamma_0 U_c \quad (6)$$

which defines

$$\gamma_0 = \frac{2\alpha}{3\pi}\left(\frac{1}{m_1^2} + \frac{Z}{4m_1m_2} + \frac{Z^2}{m_2^2}\right). \quad (7)$$

γ_0 is a constant in QED since α does not run. Integrating Eq. (6) gives

$$U_2(\nu) = U_2(1) + \gamma_0 U_c \ln \nu, \quad (8)$$

where U_2 is evaluated at $\nu = v = \alpha$. Since γ_0 is a constant, $U_2(\nu)$ only has a $\ln \nu$ term, and terms of the form $\ln^n \nu$, with $n > 1$ vanish. As a result, the leading order energy series Eq. () terminates after a single term, so one has an $\alpha^5 \ln \alpha$ contribution to the energy, but the $\alpha^6 \ln^2 \alpha$, etc. terms vanish. At low orders, the absence of terms other than $\alpha^5 \ln \alpha$ in the leading order series has been noticed before by an explicit examination of Feynman graphs. This is the first general proof that all the terms beyond $\alpha^5 \ln \alpha$ in the leading order series vanish

The matrix element of U_2 gives the energy shift

$$\Delta E = \langle U_2(\nu) \rangle = \gamma_0 U_c \ln \nu |\psi(\mathbf{0})|^2 = -\frac{8Z^4\alpha^5 m_R^3}{3\pi n^3}\left(\frac{1}{m_1^2} + \frac{Z}{4m_1m_2} + \frac{Z^2}{m_2^2}\right)\ln Z\alpha, \quad (9)$$

where we have used

$$|\psi(\mathbf{0})|^2 = \frac{(m_R Z\alpha)^3}{\pi n^3} \quad (10)$$

for the nS state, and m_R is the reduced mass. This is the famous $\alpha^5 \ln \alpha$ correction to the Lamb shift first computed by Bethe, including all recoil corrections.

NEXT-TO-LEADING ORDER

At next-to-leading order, the anomalous dimension for $V^{(1)}$ is

$$\nu \frac{dU_{2+s}}{d\nu}\bigg|_{NLO} = \rho_{ccc}\, U_c^3 + \rho_{cc2}\, U_c^2\,(U_{2+s} + U_r) + \rho_{c22}\, U_c\left(U_{2+s}^2 + 2U_{2+s}U_r + \frac{3}{4}U_r^2 - 5U_t^2\mathbf{S}^2\right)$$

$$+ \rho_{ck}\, U_c U_k + \rho_{k2}\, U_k\,(U_{2+s} + U_r/2) + \rho_{c3}\, U_c\left(U_3 + U_{3s}\mathbf{S}^2 + \frac{1}{2}U_{rk}\right), \quad (11)$$

where $U_{2+s} = U_2 + U_s \mathbf{S}^2$, and the coefficients are

$$\rho_{ccc} = -\frac{m_R^4}{64\pi^2}\left(\frac{1}{m_1^3} + \frac{1}{m_2^3}\right)^2, \quad \rho_{c22} = -\frac{m_R^2}{4\pi^2},$$
$$\rho_{cc2} = -\frac{m_R^3}{8\pi^2}\left(\frac{1}{m_1^3} + \frac{1}{m_2^3}\right), \quad \rho_{c3} = \frac{2m_R}{\pi^2}, \quad (12)$$
$$\rho_{ck} = \frac{m_R^2}{2\pi^2}\left(\frac{1}{m_1^3} + \frac{1}{m_2^3}\right), \quad \rho_{k2} = \frac{2m_R}{\pi^2}.$$

The anomalous dimension Eq. (11) can be integrated by substituting the leading order running, Eq. (8) for the coefficients on the right-hand side. Since only U_2 runs at leading order, the right hand side has at most a $\ln^2\nu$, so that the integral has at most a $\ln^3\nu$ term. This implies that the next-to-leading order series Eq. () terminates after the first three terms, $\alpha^6 \ln\alpha$, $\alpha^7 \ln^2\alpha$, and $\alpha^8 \ln^3\alpha$.

The only term that contributes to the $\ln^3\alpha$ correction is the U_2^2 term of Eq. (11). Integrating gives a contribution to $U_2(\nu)$ of the form

$$\frac{1}{3}\gamma_0^2\, \rho_{c22}\, U_c^3(1) \ln^3\nu, \quad (13)$$

which is spin-independent, and has no imaginary part. There is no contribution to the decay width or hyperfine splitting at this order. The Lamb shift at this order is obtained by multiplying Eq. (13) by the matrix element of the unit operator, $|\psi(0)|^2$, to give

$$\Delta E = \frac{64 m_R^5 \alpha^8 Z^6}{27\pi^2 n^3}\left(\frac{1}{m_1^2} + \frac{Z}{4m_1 m_2} + \frac{Z^2}{m_2^2}\right)^2 \ln^3(Z\alpha) \quad (14)$$

which is approximately 8 KHz for the $2P$–$2S$ Lamb shift in Hydrogen. Substituting $Z = 1$ and $m_1 = m_2 = m_e$ gives the $\alpha^8 \ln^3\alpha$ Lamb shift for positronium

$$\Delta E = \frac{3 m_e \alpha^8 \ln^3\alpha}{8\pi^2 n^3}. \quad (15)$$

The positronium Lamb shift is a new result, as are the recoil terms in the Hydrogen Lamb shift. In the limit $m_1/m_2 \to 0$, the Hydrogen Lamb shift has been computed previously by several groups. There is an analytic computation by Karshenboim [36] and a numerical computation by Goidenko et al. [37] that agree with our result. There are also numerical computations by Malampalli and Sapirstein [38], and by

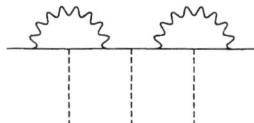

FIGURE 11. Four-loop diagram that contributes to the $\alpha^8 \ln^2 \alpha$ Lamb shift

Yerokhin [39] which agree with each other, but disagree with the other results. Recently, there has been a computation by Pachucki [40] that agrees with our result. Yerokhin [41] has emphasized that the complete $\alpha^8 \ln^3 \alpha$ Lamb shift might not be contained in the loop-after-loop calculations of Refs. [38,39].

The other calculations rely on extracting the logarithm from four-loop diagrams such as Fig. 11. The velocity renormalization group factors the graph into the product of a two-loop anomalous dimension ρ_{c22}, and the square of a one-loop anomalous dimension γ_0^2.

The $\alpha^7 \ln^2 \alpha$ hyperfine splitting, and the $\alpha^3 \ln^2 \alpha$ and $\alpha^2 \ln \alpha$ contributions to $\Delta\Gamma/\Gamma$ for o- and p-positronium can be computed using Eq. (11), and agree with known results.

CONCLUSIONS

The methods presented here give a systematic way of separating scales in nonrelativistic bound state problems. All large logarithms are summed using the velocity renormalization group. The method provides a universal description of QED logarithms. The agreement with known results at order $\alpha^5 \ln \alpha$, $\alpha^6 \ln \alpha$, $\alpha^7 \ln^2 \alpha$, and $\alpha^8 \ln^3 \alpha$ is a highly non-trivial check of the formalism. In QED, one finds that the leading order series terminates after one term, and the next-to-leading order series terminates after three terms. In addition, the method resolves a controversy about the $\alpha^8 \ln^3 \alpha$ Lamb shift for Hydrogen, and gives the first calculation of the $\alpha^8 \ln^3 \alpha$ energy shift for positronium.

In QCD, one can distinguish $\alpha_s(mv)$ and $\alpha_s(mv^2)$, and both can appear simultaneously in the same anomalous dimension. The renormalization group improved potentials can be used to compute $\bar{t}t$ production, and reduce the scale uncertainties by a factor of ten.

The velocity renormalization group should also be applicable to other problems with correlated scales. In the bound state problem, one can generate the scale mv in loop graphs from the scale m and mv^2, $mv = \sqrt{m \times mv^2}$. Similar effects can occur at finite temperature, where one has the scales T, gT and g^2T, and some of the ideas described here might be applicable to that problem as well.

This work was supported in part by the Department of Energy under grant DOE-FG03-97ER40546 and by NSERC of Canada.

REFERENCES

1. W.E. Caswell and G.P. Lepage, Phys. Lett. **167B**, 437 (1986) 437.
2. G.T. Bodwin, E. Braaten and G.P. Lepage, Phys. Rev. **D51**, 1125 (1995), Erratum ibid. **D55**, 5853 (1997).
3. P. Labelle, Phys. Rev. **D58**, 093013 (1998).
4. M. Luke and A.V. Manohar, Phys. Rev. **D55**, 4129 (1997).
5. A. V. Manohar, Phys. Rev. **D56**, 230 (1997).
6. B. Grinstein and I.Z. Rothstein, Phys. Rev. **D57**, 78 (1998).
7. M. Luke and M.J. Savage, Phys. Rev. **D57**, 413 (1998).
8. A. Pineda and J. Soto, Nucl. Phys. Proc. Suppl. **64**, 428 (1998).
9. A. Pineda and J. Soto, Phys. Rev. **D58**, 114011 (1998).
10. A. Pineda and J. Soto, Phys. Rev. **D59**, 016005 (1999).
11. M.E. Luke, A.V. Manohar, and I.Z. Rothstein, Phys. Rev. **D61**, 074025 (2000).
12. A.V. Manohar and I.W. Stewart, Phys. Rev. **D62**, 014033 (2000).
13. A.V. Manohar and I.W. Stewart, Phys. Rev. **D62**, 074015 (2000).
14. A.V. Manohar and I.W. Stewart, hep-ph/0003107.
15. A.V. Manohar and I.W. Stewart, Phys. Rev. Lett. **85**, 2248 (2000).
16. N. Brambilla, A. Pineda, J. Soto and A. Vairo, Nucl. Phys. **B566**, 275 (2000).
17. B. A. Kniehl and A. A. Penin, Nucl. Phys. **B563**, 200 (1999).
18. S.R. Lundeen and F.M. Pipkin, Phys. Rev. Lett. **46**, 232 (1981).
19. H. Hellwig et al. IEEE Trans. **IM-19**, 200 (1970).
20. W. Liu et al., Phys. Rev. Lett. **82**, 711 (1999).
21. A. Manohar and H. Georgi, Nucl. Phys. **B234**, 189 (1984).
22. Quantum Electrodynamics, ed. T. Kinoshita, (World Scientific, Singapore, 1990).
23. K. Pachucki, Hyp. Int. **114**, 55 (1998).
24. M.I. Eides, H. Grotch, and V.A. Shelyuto, hep-ph/0002158.
25. A.V. Manohar, J. Soto, and I.W. Stewart, Phys. Lett. **B486**, 400 (2000).
26. A.V. Manohar and I.W. Stewart, UCSD/PTH 00-24.
27. A.H. Hoang, A.V. Manohar, I.W. Stewart, and T. Tebuner, UCSD/PTH 00-25.
28. A.H. Hoang, A.V. Manohar, I.W. Stewart, and T. Tebuner, UCSD/PTH 00-26.
29. M. Beneke and V.A. Smirnov, Nucl. Phys. **B522**, 321 (1998).
30. H. W. Griesshammer, Nucl. Phys. **B579**, 313 (2000).
31. A. Pineda and J. Soto, hep-ph/0007197.
32. A. H. Hoang et al., Eur. Phys. J. direct **C3**, 1 (2000).
33. A. H. Hoang, Z. Ligeti and A. V. Manohar, Phys. Rev. Lett. **82**, 277 (1999).
34. A. H. Hoang, Z. Ligeti and A. V. Manohar, Phys. Rev. **D59**, 074017 (1999).
35. A. H. Hoang and T. Teubner, Phys. Rev. **D60**, 114027 (1999).
36. S.G. Karshenboim, Sov. Phys. JETP **76**, 541 (1993).
37. I. Goidenko et al., Phys. Rev. Lett. **83**. 2312 (1999).
38. S. Mallampalli and J. Sapirstein, Phys. Rev. Lett. **80**, 5297 (1998).
39. V. A. Yerokhin, Phys. Rev. **A62**, 012508 (2000).
40. K. Pachucki, unpublished.
41. V. A. Yerokhin, hep-ph/0010134.

Effective Theory for Heavy Quarkonium Production

Jungil Lee

Deutsches Elektronen-Synchrotron DESY, D-22603 Hamburg, Germany

Abstract. As a stringent test of the nonrelativistic QCD factorization approach, which has been successful in explaining the larger-than-expected production rates of heavy quarkonia, we discuss the recent measurements of polarization of $J^{PC} = 1^{--}$ heavy quarkonia by the CDF Collaboration at the Fermilab Tevatron compared with the predictions based on the nonrelativistic QCD factorization approach.

INTRODUCTION

Heavy quarkonium is a very nice probe for our understanding of the production mechanisms for heavy quarks and the nonperturbative QCD effects that bind the heavy quark-antiquark pair into quarkonium. The purely leptonic decays of the $J^{PC} = 1^{--}$ quarkonium states allow very clean measurements of their cross sections. The nonrelativistic QCD (NRQCD) factorization approach provides a systematic framework for calculating the inclusive production rates of heavy quarkonium [1]. In the NRQCD factorization approach, the larger-than-expected cross sections observed at hadron colliders are explained by introducing phenomenological parameters that describe the probabilities for the formation of the quarkonium state from color-octet heavy quark-antiquark pairs [2,3].

In high-energy $p\bar{p}$ collisions, the dominant contribution to the charmonium production rate at large p_T comes from gluon *fragmentation* [3] into a color-octet $Q\bar{Q}$ pair with $^3S_1$ state. Due to the approximate heavy quark spin symmetry of the nonrelativistic QCD [1], 1^{--} quarkonium states from the pair should have a large transverse polarization at sufficiently large p_T [4,5]. A convenient measure of the polarization is $\alpha = (1 - 3\sigma_L/\sigma)/(1 + \sigma_L/\sigma)$, where σ_L/σ is the longitudinal polarization fraction in the hadron CM frame. It describes the angular distribution of leptons from the decay of the 1^{--} quarkonium state with respect to the quarkonium momentum. The polarization variable α for direct J/ψ and direct ψ' at the Tevatron have been predicted by Beneke and Rothstein [5], by Beneke and Krämer [6] and by Leibovich [7]. They predicted that α should be small for p_T below about 5 GeV, but then should begin to rise dramatically. The first measurements of the polarization of J/ψ and $\psi(2S)$ by the CDF collaboration [8] have shown no evidence for this predicted increase. However, the discrepancies with the theoretical predictions are significant only in the highest p_T bin, so a definitive conclusion must await the higher statistics measurements that will be possible in Run II of the Tevatron. The CDF Collaboration has also measured the polarization of inclusive $\Upsilon(1S)$ in Run IB of the Tevatron [9]. The results for the p_T bins from 2 to 20 GeV and from 8 to 20 GeV are both consistent with no polarization. Since the cross section falls rapidly with p_T, this indicates that there is little if any polarization for p_T below about 10 GeV. To determine whether this result is compatible with the NRQCD prediction, we need a quantitative calculation of the polarization for *inclusive* $\Upsilon(1S)$ mesons. A recent prediction of inclusive $\Upsilon(1S)$ polarization by Braaten and Lee agrees with the CDF data in the p_T bin from 8 to 20 GeV [10].

In this talk, we review the recent progress of the heavy quarkonium phenomenology based on the NRQCD factorization approach. As the most important experimental results, we especially concentrate on the polarization measurement of $J^{PC} = 1^{--}$ heavy quarkonia by the CDF Collaboration.

PROMPT J/ψ POLARIZATION AT THE TEVATRON

The theoretical ingredients needed to calculate the polarization of *directly* produced $J^{PC} = 1^{--}$ mesons have been available for several years [7,11]. They were used by Beneke and Krämer and by Leibovich to predict the polarization of prompt $\psi(2S)$ at the Tevatron [6,7]. The calculation of the polarization of *prompt J/ψ*, *i.e.*

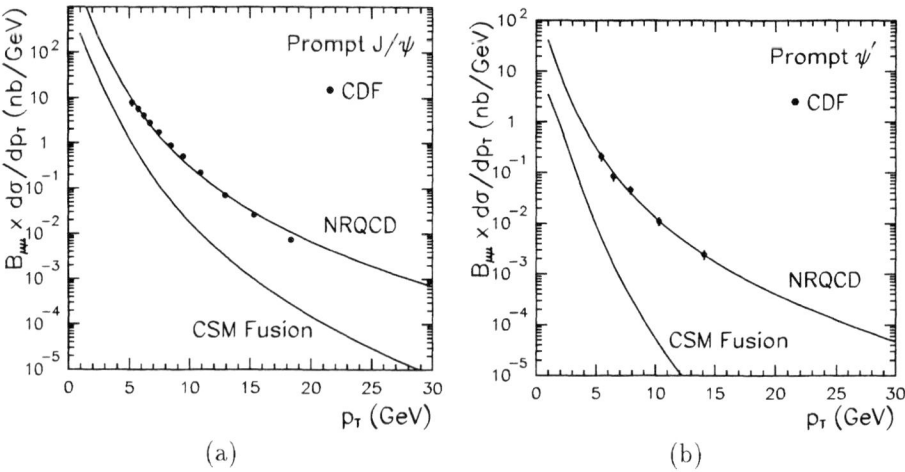

FIGURE 1. Prompt cross sections of (a) J/ψ and (b) ψ' at Run I. The lower lines represent the color-singlet model contribution via gluon fusion. The NRQCD matrix elements are from Ref. [14].

J/ψ not coming from B decay but including the feeddown from higher resonances, or *inclusive* $\Upsilon(nS)$ is complicated by the contribution from the higher resonances. Prompt J/ψ signal includes J/ψ mesons that come from decays of the higher charmonium states χ_{c1}, χ_{c2}, and ψ'. They account for about 15%, 15%, and 10% of the prompt J/ψ signal, respectively [2]. And decays of $\chi_b(1P)$, $\Upsilon(2S)$, and $\chi_b(2P)$ account for about 27%, 11%, and 11% of the inclusive $\Upsilon(1S)$ signal, respectively [12]. The missing ingredients in the calculation of the polarizations of prompt J/ψ and inclusive $\Upsilon(1S)$ were the cross sections for polarized χ_{cJ} and χ_{bJ}. The necessary parton cross sections were recently calculated by Kniehl and Lee [13] and used to predict the polarization of prompt J/ψ at the Tevatron [14].

In order to predict the polarization of prompt J/ψ at the Tevatron, we need values for the scalar matrix elements. The color-singlet matrix elements $\langle O_1^{\psi(nS)}(^3S_1)\rangle$ and $\langle O_1^{\chi_{c0}}(^3P_0)\rangle$ can be determined phenomenologically from the decay rates for $\psi(nS) \to \ell^+\ell^-$ and $\chi_{c2} \to \gamma\gamma$ [15]. The color-octet matrix elements are phenomenological parameters that must be determined from production data. To predict the polarization at the Tevatron, it is preferable to use the matrix elements extracted directly from Tevatron data in order to cancel theoretical errors associated with soft gluon radiation. There have been several previous extractions of the color-octet matrix elements [6,14,16–18] from the CDF data on the p_T distributions of J/ψ, χ_c, and ψ' [2]. We use the numerical values obtained by an updated analysis [14]. In the fusion cross section, we include the parton processes $ij \to c\bar{c} + k$, with $i,j = g,q,\bar{q}$ and $q = u,d,s$. In the fragmentation cross section, we include only the $g \to c\bar{c}_8(^3S_1)$ term, since this is the only fragmentation process for which the fragmentation function is of order α_s. The NRQCD predictions of differential cross sections of prompt J/ψ and ψ' vs. the transverse momentum are shown in Fig. 1.

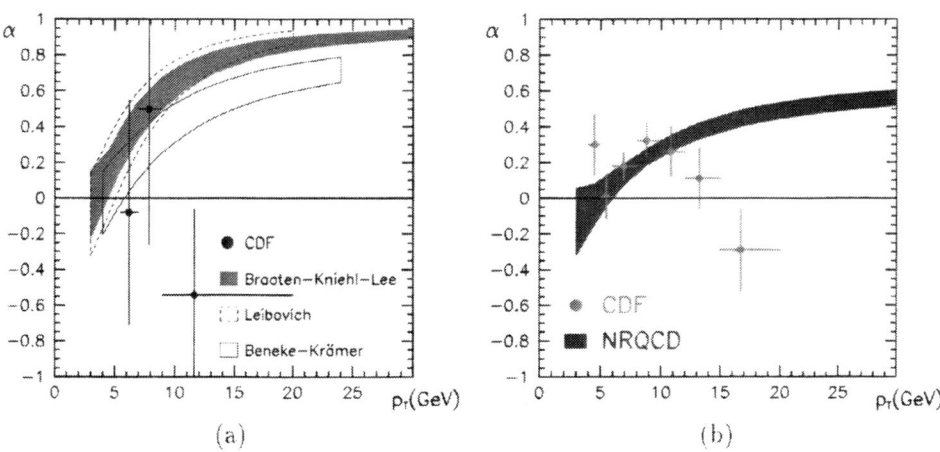

FIGURE 2. Plarization variable α for prompt (a) J/ψ and (b) ψ'. From Ref. [14].

By using the NRQCD matrix elements obtained by fitting the production rate, we can proceed to calculate the polarization variable α [14]. We present currently available predictions in the form of an error band obtained by combining the theoretical errors described in Ref. [14]. In Fig. 2(a), we compare the three theoretical predictions for direct ψ' as a function of p_T with the CDF data [8]. The differences among the predictions are due to the theoretical input [14]. The predictions agree with the CDF data in the first two p_T bins, while they disagree with the data in the highest p_T bin. But the error bars in the CDF data are too large to draw any definitive conclusions. We next consider the polarization variable α for prompt J/ψ. It has been recently predicted by Braaten, Kniehl and Lee [14]. In Fig. 2(b), we compare the prediction for α as a function of p_T with the CDF data [8]. The result for α is small around $p_T = 5$ GeV, but it increases with p_T. It is in good agreement with the CDF measurement at intermediate values of p_T, but it disagrees in the highest p_T bin, where the CDF measurement is consistent with 0. In the moderate p_T region, the contributions from ψ' and from χ_c add to give an increase in the transverse polarization of prompt J/ψ compared to direct J/ψ. In the high p_T region, the contributions from ψ' and χ_c tend to cancel [14].

The CDF measurement of the polarization of prompt J/ψ presented a serious challenge to the NRQCD factorization formalism for inclusive quarkonium production. If the result continues to disagree with the predictions of the NRQCD factorization approach, it would indicate a serious flaw in our understanding of inclusive charmonium production. The study on the NLO contribution to the gluon fragmentation function has been started [19,20]. But the prediction of the transverse polarizstion at large p_T may not be escapable.

Υ POLARIZATION AT THE TEVATRON

In this section, we present quantitative predictions for the polarization of inclusive $\Upsilon(1S)$, $\Upsilon(2S)$, and $\Upsilon(3S)$ at the Tevatron using the NRQCD factorization formalism [10]. Since the expansion parameters $\alpha_s(m_Q)$ and v_Q are smaller than those of charmonium, the NRQCD prediction of inclusive Υ polarization should therefore be more reliable than that of prompt J/ψ. The studies on the bottomonium states have been accelerated very recently. The CDF collaboration has studied the bottomonium data from Run IB at the Tevatron [9]. Based on the new CDF data, various NRQCD matrix elements for the bottomonium states were determined by Braaten, Fleming and Leibovich [21]. We use their NRQCD matrix elements for direct bottomonium production in our calculation.

The cross section σ for inclusive $\Upsilon(nS)$ is the sum of the direct cross section and the direct cross sections for the higher bottomonium states $\Upsilon(mS)$ and $\chi_b(mP)$ weighted by the inclusive branching fractions $B_{H\to\Upsilon(nS)}$ for $H \to \Upsilon(nS) + X$:

$$\sigma[\Upsilon(nS)]_{\rm inc} = \sigma[\Upsilon(nS)] + \sum_H B_{H\to\Upsilon(nS)}\,\sigma[H]\,. \tag{1}$$

For $\Upsilon(3S)$, we consider only the direct channel, neglecting any possible feeddown from higher states, such as $\chi_b(3P)$. For $\Upsilon(2S)$, we take into account the direct channel and the feeddown from $\chi_b(2P)$ and $\Upsilon(3S)$. For $\Upsilon(1S)$, we include the direct channel and the feeddown from $\chi_b(1P)$, $\Upsilon(2S)$, $\chi_b(2P)$, and $\Upsilon(3S)$.

The cross section σ_L for inclusive $\Upsilon_L(nS)$ [1] is the sum of the direct cross section for $\Upsilon_L(nS)$ and the direct cross sections for the higher spin states $\Upsilon(mS)_\lambda$ and $\chi_{bJ}(mP)_\lambda$ weighted by $B_{H\to\Upsilon(nS)}$ and by the conditional probability $P_{H_\lambda\to\Upsilon_L(nS)}$ for H_λ to decay into $\Upsilon_L(nS)$ given that it decays into $\Upsilon(nS)$ [10]:

$$\sigma_L[\Upsilon(nS)]_{\text{inc}} = \sigma[\Upsilon_L(nS)] + \sum_{H,\lambda} B_{H\to\Upsilon(nS)} P_{H_\lambda\to\Upsilon_L(nS)} \sigma[H_\lambda], \qquad (2)$$

For $\chi_{bJ}(nP)_\lambda \to \Upsilon_L(nS)$, $n=1,2$, they are given by simple angular-momentum factors for the radiative transition. For $\chi_{bJ}(2P)_\lambda \to \Upsilon_L(1S)$ and $\Upsilon(mS)_\lambda \to \Upsilon_L(nS)$, we must average over the various decay paths weighted by their branching fractions. The important steps in the decay paths are of 3 kinds. The observed hadronic transitions $\Upsilon(mS) \to \Upsilon(nS) + \pi\pi$ preserve the spin λ. For the radiative transitions $\chi_{bJ}(2P)_\lambda \to \Upsilon_L(1S)+\gamma$ and $\Upsilon(mS)_\lambda \to \chi_{bJ}(nP)_{\lambda'}+\gamma$, the probabilities for each spin state are given by simple angular-momentum factors [22].

The polarization variable α for $\Upsilon(nS)$ has been expressed as a ratio of linear combinations of the direct cross sections for $\Upsilon(nS)$ and higher bottomonium states. The *NRQCD factorization formula* for the direct cross section for a bottomonium state H of momentum P and spin quantum number λ has the schematic form

$$d\sigma[p\bar{p}\to H_\lambda(P)+X] = \sum_n d\sigma[p\bar{p}\to b\bar{b}_n(P)+X]\,\langle O_n^{H_\lambda(P)}\rangle, \qquad (3)$$

where the summation index n extends over all the color and angular momentum states of the $b\bar{b}$ pair. The $b\bar{b}$ cross section can be expressed as

$$d\sigma[p\bar{p}\to b\bar{b}_n(P)+X] = f_{i/p} \otimes f_{j/\bar{p}} \otimes d\hat{\sigma}[ij \to b\bar{b}_n(P)+X], \qquad (4)$$

where $f_{i/p}(x,\mu)$ and $f_{j/\bar{p}}(x,\mu)$ are parton distribution functions (PDF's) and a sum over the partons i,j is implied. The parton cross sections $d\hat{\sigma}$ can be calculated using perturbative QCD. All dependence on the state H is contained within the nonperturbative matrix elements $\langle O_n^{H_\lambda(P)}\rangle$. In general, they are Lorentz tensors that depend on the momentum P and the polarization tensor of H_λ. The Lorentz indices are contracted with those of $d\sigma$ to give a scalar cross section. The symmetries of NRQCD can be used to reduce the tensor matrix elements $\langle O_n^{H_\lambda(P)}\rangle$ to scalar matrix elements $\langle O_n^H\rangle$ that are independent of P and λ. This reduces the variable α to a ratio of linear combinations of the NRQCD matrix elements.

A nonperturbative analysis of NRQCD reveals how the various matrix elements scale with the typical relative velocity v of the heavy quarks. The most important matrix elements for the production of the S-wave states $\Upsilon(nS)$ and $\eta_b(nS)$ can

---
[1] We denote the longitudinally polarized $\Upsilon(nS)$ state by $\Upsilon_L(nS)$.

be reduced to one color-singlet parameter $\langle O_1^{\Upsilon(nS)}(^3S_1)\rangle$, which scales like v^3, and three color-octet parameters $\langle O_8^{\Upsilon(nS)}(^3S_1)\rangle$, $\langle O_8^{\Upsilon(nS)}(^1S_0)\rangle$, and $\langle O_8^{\Upsilon(nS)}(^3P_0)\rangle$, all of which scale like v^7. The most important matrix elements for the production of the P-wave states $\chi_{bJ}(nP)$ and $h(nP)$ can be reduced to a color-singlet parameter $\langle O_1^{\chi_{b0}}(^3P_0)\rangle$ and a single color-octet parameter $\langle O_8^{\chi_{b0}}(^3S_1)\rangle$, both of which scale like v^5. At higher orders in v, so many new matrix elements enter that the predictive power of the NRQCD approach is lost. We therefore assume the matrix elements enumerated above are sufficient to describe the bottomonium cross sections.

The first determination of the color-octet matrix elements for bottomonium production was a pioneering analysis by Cho and Leibovich [16] of the data on bottomonium production from Run IA of the Tevatron [23]. Due to the limited statistics, they had to use educated guesses for some of the matrix elements. An updated theoretical analysis based on the new CDF data from Run IB [9] has been made by Braaten, Fleming, and Leibovich [21]. Their color-singlet matrix elements are given in Table II of Ref. [21]. Those for $\Upsilon(nS)$ were determined phenomenologically from the leptonic decay rates of $\Upsilon(nS)$, while those for $\chi_{bJ}(nP)$ were estimated from potential models. Their color-octet matrix elements are given in Table V of Ref. [21]. They were determined by fitting the CDF data on the differential cross sections for $\Upsilon(1S)$, $\Upsilon(2S)$, and $\Upsilon(3S)$ at $p_T > 8$ GeV and from the fractions of $\Upsilon(1S)$ from the decays of $\chi_b(1P)$ and $\chi_b(2P)$ [12].

The leading terms in the parton cross sections in (4) depend on the region of the transverse momentum p_T of the $b\bar{b}$ pair. For p_T in the range 8 GeV $< p_T <$ 30 GeV, the leading terms are *fusion* contributions from the parton processes $ij \to b\bar{b} + k$. For p_T much greater than $2m_b$, *fragmentation* contributions from parton processes such as $ij \to g + k$, followed by $g \to b\bar{b}$, become important. For charmonium, fragmentation effects change the differential cross sections by less than 3% at $p_T = 5$ and less than 11% at $p_T = 10$ GeV. Since m_b is 3 times larger than m_c, we expect fragmentation effects to change the differential cross sections for bottomonium by less than 11% for p_T less than 30 GeV. We will therefore avoid the complications of fragmentation by restricting our predictions to $p_T < 30$ GeV.

For p_T much smaller than $2m_b$, parton processes such as $ij \to b\bar{b} + ggg...$ involving the multiple emission of soft gluons become important and it is necessary to resum their effects. In the analysis of Ref. [21], this problem was avoided by using only the CDF data for $p_T > 8$ GeV to fit the color-octet matrix elements. Since we will be using the matrix elements from that analysis, we will also restrict our predictions to $p_T > 8$ GeV.

Having restricting our attention to the region 8 GeV $< p_T <$ 30 GeV, it should be safe to use only the fusion cross sections for $d\hat{\sigma}$ in (4). We include the parton processes $ij \to b\bar{b} + k$, with $i,j,k = g,q,\bar{q}$ and $q = u,d,s,c$. We treat the c quark as a massless parton. The leading-order parton cross sections $d\hat{\sigma}$ are proportional to $\alpha_s^3(\mu)$. They are given in Refs. [7] and [11] for all the relevant $b\bar{b}$ color and spin states with the exception of color-singlet ^3P_J states, for which they are given in Ref. [13].

We follow the analysis of Ref. [21] as closely as possible. We use a common renormalization and factorization scale μ for $f_{i/p}$, $f_{j/\bar{p}}$, and α_s, taking its central value to be $\mu_T = (m_b^2 + p_T^2)^{1/2}$ and allowing it to vary within the range $\frac{1}{2}\mu_T$ to $2\mu_T$. We set $m_b = 4.77 \pm 0.11$ GeV. We consider two choices for the PDF's for comparison: CTEQ5L and MRST98LO [24]. We evaluate $\alpha_s(\mu)$ from the one-loop formula with $n_f = 5$ using the boundary conditions $\alpha_s(M_Z) = 0.127$ for CTEQ5L and $\alpha_s(M_Z) = 0.125$ for MRST98LO. We consider two extreme cases: $\langle O_8(^1S_0)\rangle = 0$ and $\langle O_8(^3P_0)\rangle = 0$. The NRQCD matrix elements and their statistical errors are given in Table II and V of Ref. [21]. The color-octet matrix elements in Table V also have an additional upper and lower error associated with changing μ by a factor of 2. This allows the correlation between the errors in μ and the color-octet matrix elements to be taken into account.

The polarization variable α can be expressed as a ratio of linear combinations of NRQCD matrix elements. The errors in the matrix elements from Ref. [21] give large uncertainties in the numerator and the denominator, but they tend to cancel in the ratio. As our central value for α, we take the average value from the 4 combinations corresponding to the two choices $\langle O_8(^1S_0)\rangle = 0$ or $\langle O_8(^3P_0)\rangle = 0$ and the two choices of PDF's. The detailed description on the error estimation can be found in Ref. [10].

Our results for the polarization variable α for $\Upsilon(1S)$, $\Upsilon(2S)$, and $\Upsilon(3S)$ for the

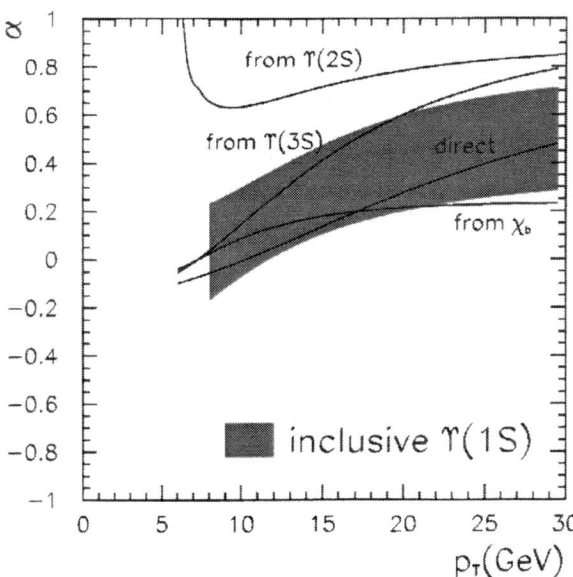

FIGURE 3. Polarization variable α vs. p_T at $\sqrt{s} = 2.0$ TeV for inclusive $\Upsilon(1S)$ (shaded band) [10]. The curves are the central values for direct $\Upsilon(1S)$, $\Upsilon(1S)$ from $\Upsilon(2S) + \pi\pi$, and $\Upsilon(1S)$ from $\chi_b(1P) + \gamma$ or $\chi_b(2P) + \gamma$.

Tevatron at $\sqrt{s} = 2.0$ TeV are shown as shaded bands in Figs. 3, 4(a), and 4(b), respectively. The curves in Figs. 3–4 are the central values of α for direct $\Upsilon(nS)$, for $\Upsilon(nS)$ from $\chi_b(mP) + \gamma$, and for $\Upsilon(nS)$ from $\Upsilon(mS) + \pi\pi$. These channels together provide a complete decomposition of the inclusive rate. The fractions of the inclusive rate from each of these channels vary slowly with p_T and add up to 1.

The predictions for α for $\sqrt{s} = 1.8$ TeV are essentially identical to the predictions for $\sqrt{s} = 2.0$ TeV in Figs. 3–4. The cross sections σ_L and σ are both smaller by about 16%, but the change cancels in the ratio. Integrating over p_T in the range 8 GeV $< p_T <$ 20 GeV, we obtain $\alpha = 0.13 \pm 0.18$ [10]. This is in good agreement with the value measured by the CDF collaboration: $\alpha = 0.03 \pm 0.28$ [9].

In Figs. 3–4, the curves for α for direct $\Upsilon(nS)$ and for $\Upsilon(nS)$ from $\Upsilon(mS) + \pi\pi$ increase steadily with p_T. The curves for $\Upsilon(nS)$ from $\chi_b(mP) + \gamma$ increase at first, but then flatten out at a value of α just above 0.2. Thus the feeddown from $\chi_b(mP)$ tends to wash out the polarization. We therefore expect α to increase more rapidly with p_T for $\Upsilon(2S)$ and $\Upsilon(3S)$ than for $\Upsilon(1S)$. A much larger transverse polarization for $\Upsilon(2S + 3S)$ than for $\Upsilon(1S)$ has also recently been observed in bottomonium production in p-Cu collisions at $\sqrt{s} = 38.8$ GeV [25].

The predictions in Figs. 3–4 are based on NRQCD matrix elements extracted from CDF data in the range 8 GeV $< p_T <$ 20 GeV. The extrapolation of these results to lower values of p_T should be considered unreliable. This is evident in the curve for direct $\Upsilon(1S)$ from $\Upsilon(2S)$ in Fig. 3 and direct $\Upsilon(2S)$ in Fig. 4. The

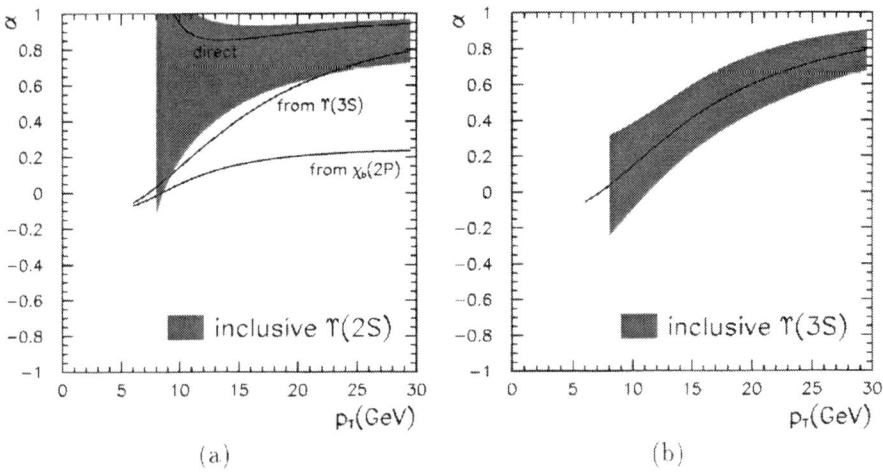

FIGURE 4. (a) Polarization variable α vs. p_T at $\sqrt{s} = 2.0$ TeV for inclusive $\Upsilon(2S)$ (shaded band) [10]. The curves are the central values for direct $\Upsilon(2S)$, $\Upsilon(2S)$ from $\Upsilon(3S) + \pi\pi$, and $\Upsilon(2S)$ from $\chi_b(2P) + \gamma$. (b) Polarization variable α vs. p_T at $\sqrt{s} = 2.0$ TeV for $\Upsilon(3S)$ (shaded band) [10]. The curve is the central value.

dramatic changes in these curves at the smaller values of p_T are artifacts of the fit of Ref. [21] having given negative central values for the color-octet matrix elements $\langle O_8(^1S_0) \rangle$ or $\langle O_8(^3P_0) \rangle$ for $\Upsilon(2S)$.

In Run II of the Tevatron, the much higher statistics will allow more accurate measurements of the polarization of $\Upsilon(1S)$ in several bins of p_T. It may also allow measurements of the polarizations of $\Upsilon(2S)$ and $\Upsilon(3S)$. Significant improvements in the theoretical predictions are also possible. The most important improvement is taking into account the effects of multiple soft-gluon emission at low p_T. This would allow more accurate determinations of the color-octet matrix elements, since the entire p_T range of the CDF data from Run I could then be used in the fits. It is also important to include fragmentation effects, so that the predictions can be extrapolated with confidence to large p_T.

DISCUSSION

The CDF measurement of the polarization of prompt J/ψ presents a serious challenge to the NRQCD factorization formalism for inclusive quarkonium production. The qualitative prediction that α should increase at large p_T seems inescapable. The studies on the bottomonium states have been accelerated very recently. The CDF collaboration has studied the bottomonium data from Run IB at the Tevatron [9]. Based on the new CDF data, various NRQCD matrix elements for the bottomonium states were determined by Braaten, Fleming and Leibovich [21]. Based on their analysis, we have obtained the NRQCD predictions of the polarization of $\Upsilon(1S)$, $\Upsilon(2S)$ and $\Upsilon(3S)$. Our result is compatible with the recent CDF measurement for $\Upsilon(1S)$ in the p_T bin from 8 to 20 GeV, which is consistent with no polarization. Our results also indicate that a nonzero transverse polarization should be observable for all three $\Upsilon(nS)$ states

In Run II of the Tevatron, the data samples for both charmonium and bottomonium states should be more than an order of magnitude larger than in Run I, which will allow both the production rate and the polarization to be measured with higher precision and out to larger values of p_T. The results which will be obtained from the Run II of the Tevatron will definitely give a great impact to the quarkonium phenomenology.

The author thanks Eric Braaten for the enjoyable collaboration on the subject discussed here. This work was supported in part by and by the KOSEF and the DFG through the German-Korean scientific exchange program DFG-446-KOR-113/137/0-1.

REFERENCES

1. G. T. Bodwin, E. Braaten, and G. P. Lepage, *Phys. Rev. D* **51**, 1125 (1995); **55**, 5855(E) (1997).

2. CDF Collaboration, F. Abe *et al.*, *Phys. Rev. Lett.* **79**, 572 (1997); *ibid.* **79**, 578 (1997).
3. E. Braaten and S. Fleming, *Phys. Rev. Lett.* **74**, 3327 (1995).
4. P. Cho and M. B. Wise, *Phys. Lett.* B **346**, 129 (1995).
5. M. Beneke and I. Z. Rothstein, *Phys. Lett.* B **372**, 157 (1996); **389**, 769(E) (1996).
6. M. Beneke and M. Krämer, *Phys. Rev.* D **55**, 5269 (1997).
7. A. K. Leibovich, *Phys. Rev.* D **56**, 4412 (1997).
8. CDF Collaboration, T. Affolder *et al.*, *Phys. Rev. Lett.* **85**, 2886 (2000).
9. R. Cropp (CDF Collaboration), *hep-ex/9910003*;
   Vaia Papadimitriou, *FERMILAB-CONF-00-308-E*.
10. E. Braaten and J. Lee, *DESY 00-185, hep-ph/0012244*.
11. M. Beneke, M. Krämer, and M. Vänttinen, *Phys. Rev.* D **57**, 4258 (1998).
12. CDF Collaboration, T. Affolder *et al.*, *Phys. Rev. Lett.* **84**, 2094 (2000).
13. B. Kniehl and J. Lee, *Phys. Rev.* D **62**, 114027 (2000).
14. E. Braaten, B. Kniehl and J. Lee, *Phys. Rev.* D **62**, 094005 (2000).
15. Particle Data Group, D. E. Groom *et al.*, *Eur. Phys. J.* C **15**, 1 (2000).
16. P. Cho and A. K. Leibovich, *Phys. Rev.* D **53**, 150 (1996); **53**, 6203 (1996).
17. B. A. Kniehl and G. Kramer, *Eur. Phys. J.* C **6**, 493 (1999); *Phys. Rev.* D **60**, 014006 (1999).
18. M. Cacciari, M. Greco, M. L. Mangano, and A. Petrelli, *Phys. Lett.* B **356**, 553 (1995).
19. E. Braaten and J. Lee, *Nucl. Phys.* B **586**, 427 (2000).
20. J. Lee, *DESY 00-092, hep-ph/0009111*.
21. E. Braaten, A. K. Leibovich, and S. Fleming, *hep-ph/0008091*.
22. P. Cho, M. B. Wise, and S. P. Trivedi, *Phys. Rev.* D **51**, 2039 (1995).
23. CDF Collaboration, *Phys. Rev. Lett.* **75**, 4358 (1997).
24. A. D. Martin, R. G. Roberts, W. J. Stirling, and R. S. Thorne, *Eur. Phys. J.* C **4**, 463 (1998); CTEQ Collaboration, H. L. Lai *et al.*, *Eur. Phys. J.* C **12**, 375 (2000).
25. FNAL E866/NuSea Collaboration, *hep-ex/0011030*.

Recent Developments in the Field of Polarized Solid State Target Materials

St. Goertz, J. Harmsen, J. Heckmann, A. Meier, W. Meyer, E. Radtke, G. Reicherz

Institut für Experimentalphysik I, Ruhr-Universität Bochum, Bochum, Germany

Abstract. With the help of systematic EPR and NMR studies under various conditions the polarization properties of some established target materials could be considerably improved. In ^{6}LiD preliminary polarization results of the still ongoing investigations are 52 %[1] at $150\,mK$ and 18 % at $1\,K$ both at a magnetic field of $2.5\,T$. Radiation doped d-butanol was polarized up to 48 %[1] and 13 % under the same conditions. Additionally first investigations were started into a group of materials, the alkanes, which have not been tested for polarization purposes so far. Difficulties specific to these compounds could be overcome with the result of promising $+16.2\,\%/-13.5\,\%$ at $1\,K/2.5\,T$ in a hybrid sample of chemically doped n-pentane/butanol. In this context new insights could be also gained into the polarization physics of radiation doped polyethylene.

INTRODUCTION

Since almost 40 years experiments with polarized solid state targets are one of the major tools in order to examine the spin structure of nucleons and nuclei as a consequence of the fundamental interactions. During that time a lot of improvements could be achieved on both the polarized target technique and the polarized target material side, i.e. advances in magnet and cryogenic technologies on the one hand and the invention of target materials with better and better dilution factors and radiation hardness on the other hand. Even so, the polarized solid target has to keep pace with the new generation of experiments looking for very rare phenomena in the dynamics of elementary particles, which often require the use of high intense beams delivered by the modern particle accelerators.

It is not only the heat load caused by the energy loss of the projectiles the target cryostat has to withstand, also the limited resistance of the target material against radiation damage restricts the usable beam flux to some $100\,nA$. For these high fluxes a ^{4}He-evaporation cryostat, which delivers a cooling power of about

---

[1] These results were obtained shortly after the conference was held.

$1\,W$ at a temperature of $1\,K$, must be used. In connection with a target material of high radiation hardness (ammonia and the lithium hydrides or their deuterated equivalents) luminosities of some $10^{35}\,cm^{-2}sec^{-1}$ can be achieved. Doses of some $10^{15}\,e^-/cm^2$ may be accumulated before the radiation induced defects reach a critical concentration in the material, beyond which the maximum achievable polarization decreases.

For experiments running with particle energies not really high enough to completely neglect the binding energies of nucleons in nuclei, the radiation hard target materials suffer from the non vanishing spin of the nuclei of those atoms, which form the molecules of the solid together with the hydrogen and deuterium atoms. Whereas for instance the use of ^{6}LiD in a DIS experiment leads to an effective dilution factor of almost 50 %, it would be unclear how to correct the data of an experiment at intermediate energies for the highly polarized ^{6}Li nuclei. The same holds not only for the nitrogen nucleus in ammonia, but strictly speaking also for the deuterium nucleus, which is used as a carrier of polarized neutrons because of its low binding energy. Besides comparing measurements with a ^{3}He gas target whenever possible, the theoretical knowledge of the deuteron and the influence of the nuclear forces on the measured observable in a given experiment are of extreme importance.

In contrast to the latter problem, which is inherent in a polarized solid target, *it would be very desirable to have a target material available with a high resistance against radiation damage but free of polarized background, i.e. a material consisting (besides hydrogen or deuterium) only of spinless nuclei.*

In fact, one of these criteria is fullfilled by another class of target materials, the alcohols. But the radiation sensitivity of the alcohols does not allow their use as materials for intense particle beams. The reason for that may lie either in a high probability of breaking the carbon hydride bindings by means of the experimental beam or in the usual way of implanting the necessary paramagnetic centers into these compounds. Here the unpaired electron spins, from which the polarization is transferred to the nucleon spins via the dynamic nuclear polarization (DNP) process, are provided by dissolving a small amount of a chemical radical into the liquid substances. The particle beam may not only be able to damage the solvent but also the large (molecular weight more than 150) radical molecules leading to the disappearance of the unpaired spin. Thus, for instance in the case of the commonly used butanol, the critical dose is more than one order of magnitude lower than in the radiation hard materials mentioned above, in which an intense particle beam is even used to create the necessary paramagnetic impurities.

The comparatively small radiation hardness together with the absence of polarized background favour the alcohols for experiments with low particle fluxes. In particular they are used for the so-called 'frozen spin experiments', in which the target polarization – once created using a strong magnetic field of some Teslas – can be maintained by means of a weak magnetic field over long periods of time (i.e. several hundred hours) on condition that the temperature is diminished to less than about $70\,mK$. These weak fields of some kG are either supplied by the fringe field of an external magnet or by a thin superconducting coil, which can be implemented in the target cryostat itself [1]. Both techniques offer enough space to place a detector of large angular acceptance around the end of the target cryostat with the material inside. The advantages of a further development of the frozen spin target materials may be seen by the following arguments:

In contrast to the deuterated material ^{6}LiD, in which a deuterium (and ^{6}Li) polarization of more than 50 % can be obtained already at the relatively low magnetic field of $2.5\,T$ [2,3], *the usual deuteron polarizations of the chemically doped alcohols do not exceed* 30 % − 35 % *under the same conditions.*[2] *This situation can be improved by a further development of the preparation and doping procedures:*

Considerably higher polarizations have been observed in d-butanol doped by electron irradiation at the temperature of liquid argon. This contribution reports about the status of characterizing the corresponding radical as well as about the first polarization measurements.

The second point capable of improvement is the relatively low content of free hydrogen or deuterium nuclei compared to all nuclei in the frozen spin materials used so far. The alkanes, which are oxygen free carbon hydrides, may be an alternative to the alcohols with a considerably higher dilution factor. First polarization studies of a special alkane, the n-pentane, have been successfully performed.

All materials mentioned up to now require special care concerning the temperatures they are exposed to. In 'radiation doped' substances recombination of the paramagnetic defects sets in at temperatures noticeable lower than the materials melting points, whereas in 'chemically doped' substances it is the glassy state of the solid, which can not be recovered once the temperature in the material exceeded the so-called 'devitrification point', a temperature, at which the solid transforms to a partially crystalline state. This transformation is accompanied by the loss of the homogeneous radical distribution in the material, which drastically reduces the achieveable polarization. The devitrification temperature can be increased by adding a certain amount of water to the alcohols, but in any case the temperature stability is restricted to the materials melting point, which is around $-90°C$ for butanol. *On the other hand a target material being stable at higher temperatures at least for some minutes would greatly simplify the loading process of a target cryostat.* In particular in horizontal arrangements like the cryostat used for the measurement of the GDH sumrule [4] at Mainz and Bonn, which itself is part of the beamline, the loading and sealing procedures make great demands on the temperature stability of the target material.

A further property, which is common to all materials mentioned so far, is the difficulty of producing a solid block with precise thickness out of them. The exact knowledge of the targets areal density is an indispensable condition when measuring spin dependent differences of cross sections instead of asymmetries, which are normalized to the unpolarized cross sections.

---

[2] With the exception of the Cr(V) doped d-butanol used 1994 in the SMC experiment at CERN [5], which relaxation time in the frozen spin mode has never been tested.

Both of the demands mentioned can be fullfilled by the 'infinitly long' carbon hydride polyethylene available as $[-CH_2-]_n$ and $[-CD_2-]_n$. From a certain chain length on, these materials are solid at room temperature such that it is possible to form almost arbitrary shapes and sizes out of them. The usual way of doping polyethylene makes use of the fact that molecules of a stable radical like TEMPO are able to diffuse into thin foils or very small grains of this material at elevated temperatures. But the diffusion depth is limited to about a hundred microns such that the material has to be pressed afterwards, in order to get thicker targets. An alternative method is the dissolution of the material into a toluene-TEMPO mixture followed by an evaporation of the solvent. Both methods have been invented by the polarized target group of the Paul Scherrer Institute in Switzerland [6].

Again alternatively, due to the solid state at room temperature, it is possible not only to irradiate the material at low temperature[3] (i.e. in liquid argon or liquid nitrogen) but also near room temperature. This method circumvents the problem of shaping the already doped material. First results of these investigations may be also found in this contribution.

THE EST PICTURE OF THE DNP PROCESS

The first model, which was able to describe the DNP process in non metallic substances, the so-called solid state effect (SSE) has been proposed by Abragam in 1955 [8]. The most prominent example for the validity of this model and at the same time the first material, which has been used as a real polarized target, is lanthanum magnesium nitrate (LMN) doped with neodynium ions. In this material the microwave frequencies for optimum positive and negative polarization differ by two times the Larmor frequency of the protons at the magnetic field used, an observation, which is in perfect agreement with the prediction of the SSE model [9]. One assumption for the validity of the SSE model is that the width of the electron paramagnetic resonance (EPR) must not exceed the nucleon Larmor frequency, a condition, which is violated in all target materials used today since the EPR lines may be broadened by both an anisotropic g-factor and a magnetic interaction with the surrounding nuclei (HFS). Figure 1 shows the so-called frequency curve in ^{6}LiD for two different temperatures. Instead of exhibiting a sharply defined frequency for each of the polarization directions, the data are best fitted with a convolution of a gaussian and a lorentzian line shape derivative with the distances of the extrema even different for different temperatures. This behaviour can be understood in the framework of the equal spin temperature (EST) theory [10,11], in which it is the electronic spin system, which recieves a certain temperature due to irradiation with saturating microwaves:

If the concentration of the unpaired electron spins is high enough, the dipol-dipol interaction is able to mediate between those spins, which resonance frequencies would otherwise be strongly separated by inhomogeneously broadening mechanisms like

---

[3] A method, which was tried by the Nagoya polarized target group for the first time [7].

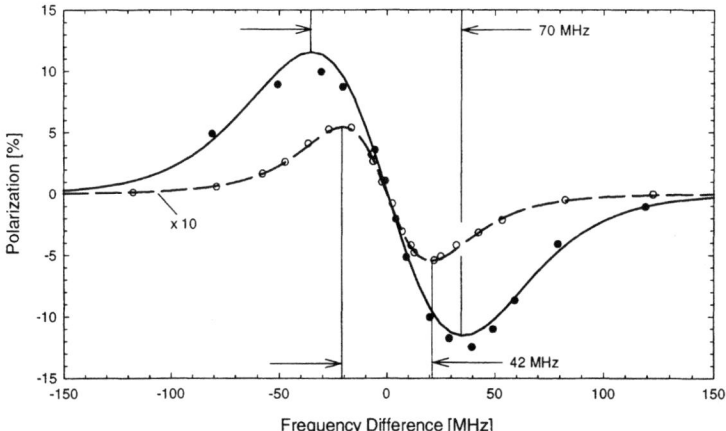

FIGURE 1. Deuteron polarization as a function of the microwave frequency in ^{6}LiD. Solid curve: $T = 1\,K$, dashed curve: $T = 77\,K$

g-factor anisotropies or HFS interactions. When irradiating the system slightly off the resonance center with microwaves intense enough to equalize the two Zeeman levels, not only those electrons are involved, which have the correct Larmor frequency, but the whole resonance line is participating in the transitions. The energy mismatch is compensated by 'cooling' or 'heating' the electron dipolar reservoir. Due to the already mentioned similarity of the electron resonance width and the energy difference of the nucleon Zeeman levels, heat can be transferred from the nucleon Zeeman reservoir to the electron dipolar reservoir. The common spin temperature T_S of both of the systems can get several hundred times smaller than the actual temperature of the lattice T_L.

Here the most important prediction is that the dynamic polarization can be calculated by means of the *Brillouin function* in the same way as in the thermal equilibrium case $(T_S = T_L)$, i.e. as an increasing function of the ratio of the magnetic (μB) and the thermal (kT) energy:

$$P(B/T_S) = \mathcal{B}\left(\frac{g_n\,\mu_n\,B}{2\,k\,T_S}\right) = \begin{cases} \text{spin-1/2} & \tanh\left(\dfrac{g_n\,\mu_n\,B}{2\,k\,T_S}\right) \\ \text{spin-1} & \dfrac{4\tanh\left(\frac{g_n\,\mu_n\,B}{2\,k\,T_S}\right)}{3 + \tanh^2\left(\frac{g_n\,\mu_n\,B}{2\,k\,T_S}\right)} \end{cases} \quad (1)$$

For spin-1/2 and spin-1 systems the Brillouin function reduces to simple functions of the hyperbolic tangens. Although in general the spin temperature is not known, in a substance containing at least two different spin species equation (1) can be simply checked by comparing the polarizations of the different nuleons or nuclei, respectively. An additional, although much more sophisticated way to examine the spin temperature behaviour of a particular material is the observation of the electron paramagnetic resonance lineshapes for different microwave power levels. This method opens up the possibility to directly observe the consequences of saturating a spin system, which is subjected to strong dipolar interactions. Under certain conditions it is even possible to extract both the Zeeman and the dipolar longitudinal relaxation times as well as the dipolar energy of the electrons from these 'saturation curves' [12]. In particular the data obtained from the electron irradiated lithium hydrides can be nicely fitted to the corresponding theoretical expressions making these compounds the best understood target materials by far.

NEW RESULTS ON ESTABLISHED MATERIALS

This section presents preliminary results on alternative preparation methods for the 'state of the art' target materials ^{6}LiD and (d-)butanol. Although polarized polyethylene is rather a recent development, it will be included in this section, because the investigations into this material turned out to be a natural extension of the work done in the irradiated alcohols.

Lithium Deuteride

For the first running period of the COMPASS[4] polarized muon program with the objective of examining the gluon contribution to the nucleon spin the 'isoscalar' target material ^{6}LiD will be used, since it offers a Figure of Merit second to none. In comparison to ammonia, the second best choice, the gain is roughly

$$\frac{F(LiD)}{F(NH_3)} = \left(\frac{P_{LiD}}{P_{NH3}} \cdot \frac{f_{LiD}}{f_{NH3}}\right)^2 \simeq \left(\frac{0.5}{1} \cdot \frac{0.5}{0.17}\right)^2 \simeq \left(\frac{3}{2}\right)^2 \simeq 2 \quad , \qquad (2)$$

which is essentially a consequence of the much higher dilution factor f of ^{6}LiD. In 1997 a new research program has been started in Bochum in order to optimize this compound for the COMPASS target, where advantage can be taken of earlier results on this material from the Saclay and the Bonn polarized target groups [13,14]. For the moment the highest polarization obtained from a ^{6}LiD sample of high degree of hydration and isotopical purity (95 % ^{6}Li), which has been synthesized at the Bochum University [3], is 52 % at $2.5\,T$ and $150\,mK$.

Also an alternative irradiation procedure motivated by EPR measurements is currently under further investigation. This 'annealing method' consists of an electron irradiation at a temperature considerably lower ($140\,K$) than the one known to be the optimum ($185\,K$), followed by an exposition of the material to room temperature for some minutes. A series of samples irradiated with either the 'annealing

---

[4] Common Muon and Proton Apparatus for Structure and Spectroscopy, NA58 at CERN

method' or the 'traditional method' were also tested in an evaporation cryostat of high cooling power at a temperature of $1\,K$ and a magnetic field of $2.5\,T$. In the course of these tests a special sample prepared with the annealing method and irradiated with the relatively small dose of $1 \cdot 10^{17}\,e^-/cm^2$ could be polarized up to $18\,\%$, which is an improvement by a factor of 1.5 compared to earlier results [14,15].

In particular with the prospect of the TESLA-N project, a planned fixed target experiment at the upcoming TESLA collider at DESY, this is an important step forward. The TESLA-N polarized target will be equipped with a high cooling power evaporator and a magnet of at least $5\,T$. In order to compare this new result with those already obtained in a running high current target, e.g. the E155(X) target at SLAC, one can make use of the agreement between the EST theory and the experimentally found magnetic field dependence of the lithium hydrides. Figure 2

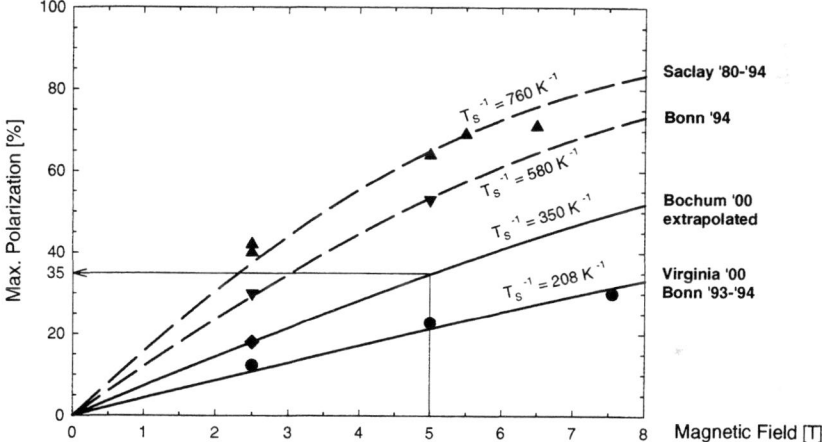

FIGURE 2. Inverse spin temperatures of the ^{6}LiD world data, obtained under dilution cryostat conditions (dashed) and under evaporation cryostat conditions (solid).

shows three sets of polarization data. Within one set the samples are similarly or at least comparably treated, whereas there are more or less big differences between the different sets corresponding to the individual preparation method and the kind of polarization apparatus used. The fit of the data sets to the EST theory according to equation (1) allows the extraction of the corresponding inverse spin temperatures. From the inverse temperature of $350\,K^{-1}$ of the new Bochum point it seems to be justified to extrapolate a polarization of $35\,\%$ for a $1\,K/5\,T$-environment. This has to be contrasted to the $208\,K^{-1}$ being the inverse spin temperature obtained from the material used in the SLAC experiments mentioned above. Together with an additional gain of about $30\,\%$ due to the 'in-situ irradiation' by the experimental beam [16] one ends up with an estimated maximum polarization around $45\,\%$ for the new material.

Irradiated Butanol and Polyethylene

The initial motivation to look for an alternative doping method in butanol was the ambition of clarifying the 'mysterious' magnetic field dependence of the polarization in chemically doped d-butanol, which has been demonstrated in EDBA (Cr(V)) doped samples of this substance[5] [17]. The question to answer was, whether this effect would be also present when using paramagnetic centers of a completely different nature. From the Abingdon workshop 1979 [18] it is known that irradiated butanol would polarize up to some percent, high enough, in order to permit a study of this phenomenon.

But in contrast to the expectations already the first samples of normal and deuterated butanol, which were electron irradiated with $1 \cdot 10^{16}\, e^-/cm^2$ at $90\,K$ gave a proton and deuteron polarization at $1\,K/2.5\,T$ of $\pm 15\,\%$ and $\pm 10\,\%$, respectively. EPR measurements exhibited a radical concentration one order of magnitude higher than in the chemically doped standard samples. Therefore a d-butanol sample with only a tenth of the radiation dose was prepared, which even polarized up to $\pm 13\,\%$. Meanwhile the same sample was also polarized at the low temperatures of a dilution cryostat up to $48\,\%^1$ and $56\,\%^1$ at $2.5\,T$ and $5\,T$, respectively [19]. The $2.5\,T$-value surpasses earlier results on chemically doped d-butanol by about $50\,\%$ relative, while the $5\,T$-value is the highest one ever seen in d-butanol at all.

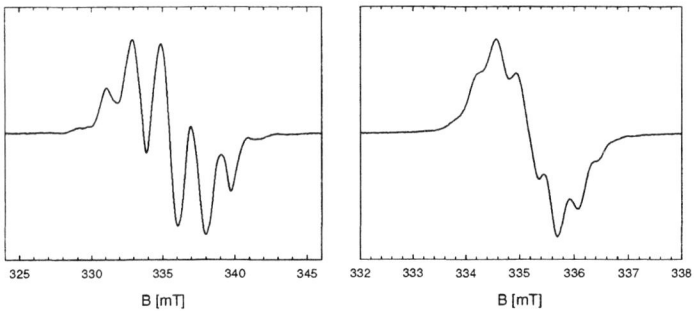

FIGURE 3. EPR signals of electron irradiated h-butanol (left) and d-butanol (right)

In order to optimize the polarization behaviour of a certain material, it is important to understand the nature of the 'DNP pumping system'. Therefore a lot of work has been done in order to characterize the radiation induced paramagnetic defects in the alcohols by the observation of their EPR resonance signals. The interaction of the unpaired electrons with the surrounding nuclear magnetic moments causes a hyperfine structure in the resonance line, which can be often attributed to a certain structure of the paramagnetic center (e.g. the F-center in the lithium hydrides or the $N\dot{H}_2$ radical in ammonia.). Figure 3 shows the EPR

---

[5] In striking contradiction with the prediction of the EST theory (equation (1)) the polarization was shown to be a decreasing function of the magnetic field and even vanishing at $5\,T$.

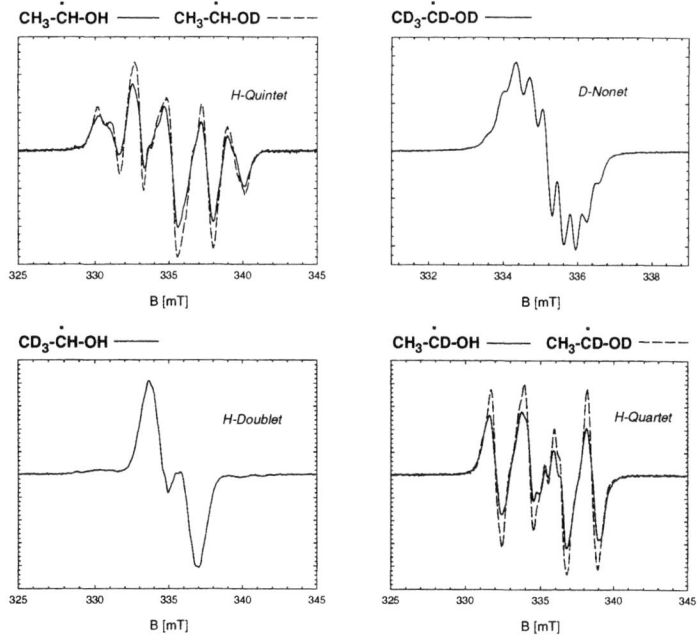

FIGURE 4. EPR signals of irradiated ethanol with different isotopical compositions

signals obtained from irradiated h- and d-butanol with partially resolved hyperfine structure corresponding to the intramolecular interactions with the adjacent proton and deuteron magnetic moments. According to the existence of four butanol isomeres there is a high transition probability of the primary n-butanol radical into radicals of an isomere of higher stability, which complicates the identification of the butanol radical structure. In contrast to that no such transition can occur in ethanol leading to a rather simple interpretation of the spectra, which all are in agreement with a loss of a hydrogen or deuterium atom at the carbinol position (Figure 4). The unpaired electron interacts with the protons or deuterons bound to the so-called α- and β-carbons[6] giving $(n + 1)$ or $(2n + 1)$ components for n equivalent[7] protons or deuterons, respectively.

In order to understand the defect structures in longer carbon hydride chains the question arose, how the corresponding defect would look like in an 'infinitely long' molecule, the polyethylene. For that purpose CH_2 pellets with a molecular weight of 125,000 were treated in the same way as the alcohols. The EPR signal taken at $77\,K$ from a CH_2 sample irradiated at $90\,K$ is composed of six lines, which may

---

[6] The α-carbon is the one, which lost the hydrogen, the β-carbons are the next neighbours.
[7] Strictly speaking the α- and β-protons/deuterons are not equivalent, but the corresponding HFS coupling constants are too similar as to be resolved.

be easily explained by the interaction of an unpaired electron originating from a hydrogen vacancy somewhere in the chain (See Figure 5(a)). After an exposition

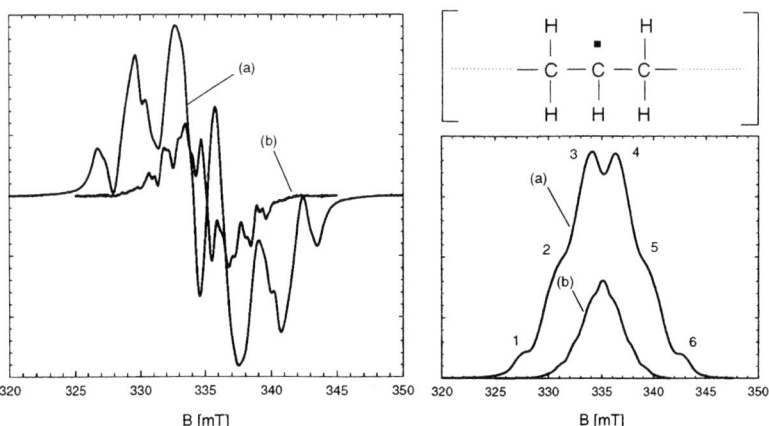

FIGURE 5. EPR of irradiated CH_2 - left: resonance lines from material irradiated at 90 K (a) and at $0°C$ (b) - right: corresponding integrated signals and suggested radical structure for (a)

to room temperature the resonance had almost completely disappeared with the exception of a tiny remnant of a different structure. Fortunately, this structure with an obviously higher temperature stability can be created selectively, too. The signals (b) correspond to the same kind of material, but irradiated at $0°C$ with a dose 20 times higher than in the former case. Even so, the radical density is only about 20 % of the one measured in the low temperature irradiated samples. Polarization tests of samples prepared in both ways were performed at $1\,K/2.5\,T$. The low temperature irradiated CH_2 sample rapidly (build up time = $3.5\,min$) polarized up to 10 %, whereas in the high temperature irradiated one the maximum polarization was extrapolated to about 15 %. Using the high temperature stability of the radicals, the latter sample was handled at room temperature for about $20\,min$ before loading into the cryostat.

Although exhibiting a relatively high polarization as well as a remarkable temperature stability (radical decay constant less than 1 %/h at room temperature), with more than $5\,h$ the build up time is still too long for using the high temperature irradiated CH_2 as a real target material. Usually the build up time is very sensitive on the radical concentration. But due to its square root dependence on the irradiation dose at high temperatures increasing the concentration is a time consuming purpose.

This kind of investigations will be also extended to the deuterated material CD_2, in which even at the comparitively high temperature of about $500\,mK$ already 18 % deuteron polarization had been achieved at $2.5\,T$ [7].

THE ALKANES, A HIGH DILUTION FACTOR ALTERNATIVE TO THE ALCOHOLS

Starting from the shortest alcohol, methanol, and increasing the chain length, the (proton) dilution factor runs from 1/8 up to the asymptotical value 1/7 for an 'infinitly long' alcohol. In practice one is restricted to pentanol being the longest alcohol, which readily forms a glass when droped into liquid nitrogen. On the other hand, in an oxygen free carbon hydride, the highest dilution factor is given by the shortest molecule possible, which is methane with 1/4. The dilution factor decreases with increasing chain lengths down to again 1/7, the value for polyethylene. This behaviour is summarized for the alcohols and the alkanes, respectively by

$$f_{alcohols}(N) = \frac{N+1}{7N+9} \quad \text{and} \quad f_{alkanes}(N) = \frac{N+1}{7N+1} , \qquad (3)$$

where N represents the number of carbon atoms in the molecule. For the first polarization tests on this group of substances n-pentane was chosen because of its liquid state at room temperature and its comparatively high freezing point of $143\,K$.

In the alcohols the presence of the OH-group causes a certain polarity, which is strongest in methanol and decreases towards longer chains. A fact, which not only makes the shorter alcohols simply solidified in a glassy state, but which is also responsible for the formation of at least one 'shielding layer' of alcohol molecules around a dissolved radical molecule. This layer prevents two radical molecules from getting close to each other. In contrast to that the alkane molecules are mirror symmetric and thus non polar, which favour the formation of a crystalline state even when cooled down rapidly. Whereas this problem can be circumvented by mixing different alkanes, such a mixture would still allow the formation of clusters of radical molecules with the possibility of overlapping wavefunctions of the unpaired spins. These 'exchange narrowed' spins give a clear signature in their EPR resonances in the form of a single narrow line overlaying the broad and highly structured resonance of the single electrons. Due to their very fast relaxation it is not possible to decrease the spin temperature of the paired electrons below the lattice temperature or in other words no dynamic polarization can be obtained. Instead of putting together different alkanes, both problems can be avoided at the same time by mixing the specific alkane with about 20 % of an alcohol. The corresponding EPR signal of such a 'hybrid' looks similar to those obtained from pure alcohols with no sign of exchange interaction. So far a sample containing 0.5 weight percent TEMPO dissolved in 20 weight percent butanol and 80 weight percent n-pentane was polarized up to $+16.2/-13.5\,\%$ at $1\,K/2.5\,T$ in a time comparable to the usual build up times in standard chemically doped butanol. With an effective dilution factor of 16 % its Figure of Merit is still 1.5 times higher than the one of butanol. Examinations of alternative material compositions as well as tests at lower temperatures are the plans for the near future.

SUMMARY AND OUTLOOK

In order to keep pace with the requirements of modern polarized experiments it is not only the technique of the polarized target, which has to try new avenues, but also the polarized solid state target materials have to be further developed. The most crucial properties to improve are their maximum polarization, their content of polarizable nucleons and their resistance against radiation damage. With the prospect of more and more sophisticated cryogenic set-ups, which have to fit into state of the art '4π-detectors', target materials, which are less demanding with regard to the compliance of the cold chain would be very desirable, too. In order to initiate and to pursue new developments in this field, a more fundamental understanding of the physics of the DNP process is of great importance. Although on the market for a pretty long time, the concept of spin temperature still seems to be the most promising basis in order to push forward our knowledge about dynamic nuclear polarization. Besides polarization measurements probably the most important tool, in order to compare theory with experiment, is the observation of the electron spin system under the conditions of a real polarization experiment. With the help of EPR measurements of undistorted resonances as well as in the saturation regime some ansatzes for new preparation methods and materials have been found. The future of this kind of investigations will lie in the combination of EPR and NMR experiments and eventually in their simultaneous observation in order to breed systems of paramagnetic centers most suitable to polarize the nuclear spin systems.

REFERENCES

1. Gehring R. et al., *NIM A* **418**, 233 (1998)
2. van den Brandt et al. *Proc. 9th Int. Symp. on High Energy Spin Physics* ed. W. Meyer, W. Thiel, E. Steffens, 320 (Bonn 1990)
3. Meier A., PhD thesis in preparation
4. Bradtke Ch. et al., *NIM A* **436**, 430 (1999)
5. Spin Muon Collaboration, Adeva, H. et al., *NIM A* **372**, 339 (1996)
6. van den Brandt et al. *NIM A* **356**, 36 (1995)
7. Doushita N. et al. *Proc. Int. Workshop on Polarized Sources and Targets* ed. A. Gute, St. Lorenz, E. Steffens, 344 (Erlangen 1999)
8. Abragam A. *Phys. Rev.* **98**, 1729 (1955)
9. Schmugge T.J. and Jeffries C.D. *Phys. Rev. A* **138**, 1785 (1965)
10. Abragam A. and Goldman M., *Rep. Prog. Phys.* **41**, 395 (1978)
11. Borghini M. *Proc. Int. Conf. on Polarized Targets* ed. G. Shapiro, 1 (Berkeley 1971)
12. Goertz St. et al. *Proc. Int. Workshop on Polarized Sources and Targets* ed. A. Gute, St. Lorenz, E. Steffens, 360 (Erlangen 1999)
13. Durand G. and Ball J. *Proc. 10th Int. Symp. on High Energy Spin Physics* ed. T. Hasegawa, N. Horikawa, A. Masaike and S. Sawada, 355 (Nagoya 1992)
14. Goertz St. et al., *NIM A* **356**, 20 (1995)

15. Bültmann St. et al. Internal Report, University of Virginia, Charlottesville (2000)
16. Bültmann St. et al. *NIM A* **425**, 23 (1999)
17. Trentange S. et al.,*Proc. 9th Int. Symp. on High Energy Spin Physics*
    ed. W. Meyer, W. Thiel, E. Steffens, 325 (Bonn 1990)
18. Crabb D.G., *Proc. Second Workshop on Polarized Target Materials*
    ed. G.R. Court, S.F.J. Cox, D.A. Cragg and T.O. Niinikoski, 33 (Abingdon 1979)
19. Harmsen J., PhD thesis in preparation

Workshop Report on Spin Polarized Electron Source and Polarimeter

T. Nakanishi

Department of Physics, Nagoya University, Nagoya 464-8602, Japan

Abstract. Three days satellite-workshop on polarized electron source (PES) and polarimeter was held at Iris-Aichi hotel in Nagoya during Oct.12 to 14, just prior to the Spin-2000 symposium. This is the seventh in a row of similar satellite-workshop, which was first held at Stanford in 1983.

It is a good tradition of this workshop to invite not only scientists working in high-energy physics but also those in atomic, surface and semiconductor physics. It is inevitable since polarized electrons are produced in GaAs-type semiconductors by optical pumping and emitted through a potential barrier by using the NEA (Negative Electron Affinity) technique. Therefore the exchange of information between both communities are very important for advance of this research field. In fact, there were totally 58 participants and a half of them, 29 (12) persons came from 11 (2) institutions of high-energy physics, and another half, 29 (24) persons came from 16 (11) institutions of material science. (the numbers in parentheses shows domestic one) Fortunately, we could share so called "gemütlich" atmosphere with each other during two-day working and one-day excursion.

The workshop contributions consist of 23 oral and 13 poster presentations. There were many interesting and important reports, but it is impossible for me to make a complete review of all contributions due to the page limit. Therefore, I will try here to give a rough sketch of present achievements and further problems in this field. I ask you for more details to see each report included in this proceeding.

1. Semiconductor photoemission source

Recently, the polarized electron beam has been used in most of high-energy electron accelerators as an important tool to observe the spin dependent interactions. In material science field, it also gathers attention in relation with the new "spintronics" technology, where electrons in the nano-structure crystals will be used as "moving spins" instead of "moving charges" in traditional electronics technology. In answer to these needs, the frontier of PES technologies has been still expanded and the interesting efforts to this direction are reported in this workshop.

In order to provide a basis for understanding on physics and technology behind the recent developments, a brief review of principle of GaAs-type PES is given at first. It is based on a combination of two fundamental processes; (1) production of spin polarized electrons in conduction band and (2) emission of polarized electrons into vacuum.

(1) Polarization mechanism

The band gap excitation with circular polarized photons is used for creation of highly polarized electrons, since the maximum electron polarization (P) is determined by the fine structure at the top of valence band (Γ8). For example, heavy-hole (hh) and light-hole (lh) bands are degenerate for GaAs crystal and the highest P is limited to be less than 50%. It was reported that P slightly higher than 50% can be obtained if the valence electrons near Γ8 are excited and only hot electrons are extracted by St. Petersburg group. Anyway for the thin layer of strained GaAs (or GaAsP, InGaAsP) or the InGaAs-AlGaAs (or GaAs-GaAsP) superlattice structure, this degeneracy is removed by several tens of meV and much higher P can be obtained, as shown in Fig. 1.

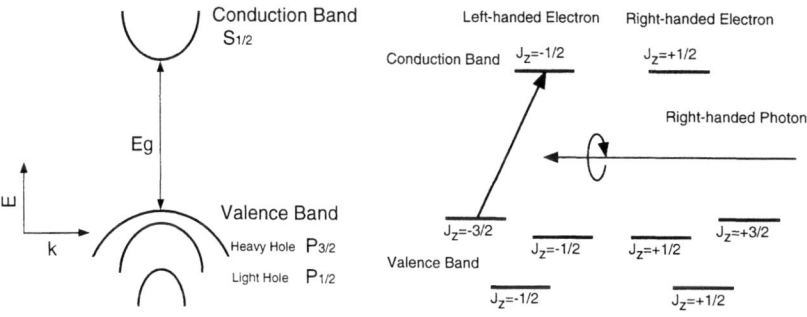

Fig. 1 : Polarization mechanism using the optical pumping method

(2) Emission mechanism

For emission of polarized electrons staying at the bottom of conduction band (Γ6), it is required to pull down the vacuum level by amount of more than electron affinity of GaAs (~4 eV). Therefore the NEA (Negative Electron Affinity) surface is indispensable for this GaAs-type photoemitter. The heavily P doping can bend the band level by (0.4-0.6) eV at the surface. Additional formation of the thin (GaAs-Cs-O) interface which works like an electric dipole layer at the surface is required to change the sign of electron affinity from positive (+) to negative (−) by a few hundreds of meV. Then polarized electrons can tunnel through the interface layer into vacuum, as shown in Fig. 2.

This NEA surface brings two big advantages to GaAs-PES, those are much higher quantum efficiency (QE) and polarization than those obtained by other metal and PEA-semiconductor cathodes. On contrary to the advantage of NEA surface, it brings several difficulties (or limitations) for the PES operations. Those are problems of (1) NEA cathode lifetime and (2) NEA surface charge limitation (SCL), which are discussed in following Section 2.

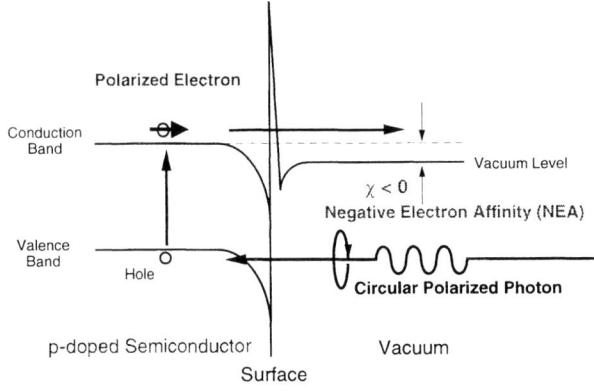

Fig. 2 : Polarized electron emission mechanism using the NEA surface

In addition, the formation mechanism of NEA surface of GaAs was not yet understood completely in atomic level, since it was too difficult to control and observe such a surface. Remarkable advances in this subject were reported at the workshop as described in Section 3.

2. Problems related to NEA surface

2-1. Lifetime problem

The decay rate of QE, defined as the cathode lifetime, is one of the most important parameters for operation and maintenance of PES. To preserve the active NEA surface, it is most important to reduce the positive ion back-bombardments to the cathode as shown in Fig. 3.

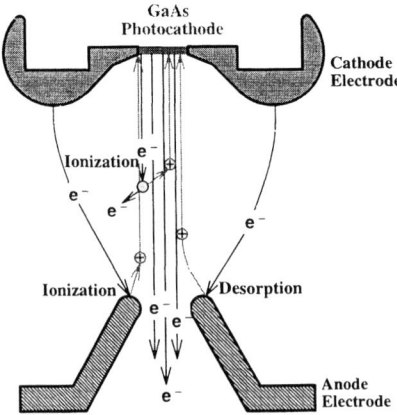

Fig. 3 : Various mechanisms bringing the degradation of NEA surface

Those ions are produced by the beam itself which collides with residual gas or additional out-gas produced at the anode surface by field emission dark current from the cathode electrode. Therefore the NEG pumping and the load-lock system are employed to achieve the extremely good UHV ($p \leq 10^{-11}$ torr) and reduction of dark current to the very small level ($I \leq 10$ nA), respectively, as the standard technologies of PES gun.

It is reported from Mainz that soft X-rays produced by even very small beam loss ($\sim 10^{-4}$) also limited the beam-current lifetime product below 10 Coulomb. This limit could be relaxed to be ~100 Coulomb by the anodized cathode from which parasitic photoemission from the outer ring can be eliminated. The same method has been also employed successfully at Jefferson Lab.

2-2. Surface charge limit problem

The maximum current, extracted from the NEA activated cathode, is limited by so called "surface charge limit (SCL)" effect, rather than space charge limit effect. Some fraction of the electrons excited from valence to conduction band can be trapped at the surface and thus induce a drop of the magnitude of band-bending (called as photovoltage effect) and thus a rise of electron affinity. This causes the lower tunneling probability through a potential barrier in the later portion of electrons in the pulse. Obviously this effect must be overcome for the multi-bunch beam generation in linear colliders. For example at JLC, about 100 bunches having 2×10^{10} electrons per bunch (0.7 ns width, separated by 2.8 ns) for the C- or X-band scheme are required.

The solutions to relax the SCL problem should exist in two directions; (1) reduction of electrons trapped in the band-bending region (BBR), or (2) reduction of the trapped electrons by faster recombination with holes in valence band, as shown in Fig. 4.

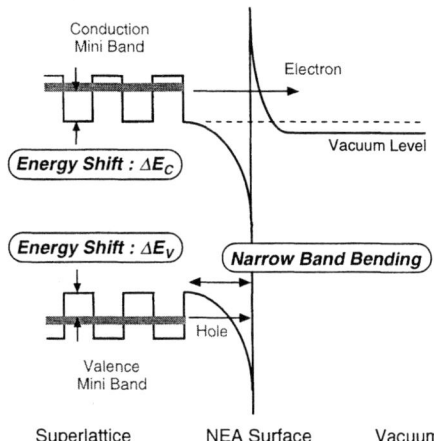

Fig. 4 : Two possible mechanisms to relax the SCL effect

The latter method has been pursued at SLAC by using the heavily p-doping GaAs at the surface, which is expected to decrease the BBR thickness so that the holes can tunnel more easily through the BBR barrier to the surface. By the laser pump-probe (pulse-width: 2 ns, separation time: variable) method, they found that no SCL (or photovoltage) effect is observed for an unstrained GaAs (100 nm) sample with p-doping of 5×10^{19} cm^{-3}, while it is slightly remained for the 2×10^{19} cm^{-3} sample (recovery time ~8 ns). The polarization dependence on the highly p-doped layer thickness was also measured using the strained GaAs layer (100 nm) samples with surface p-doping of 4×10^{19} cm^{-3}. The initial thickness of high-doped layer was 40 nm and it was reduced gradually by anodization procedure. They found that the large spin relaxation happens in such highly doped BBR, but the polarization of ~80% was preserved for the thickness less than 5 nm as shown in Fig. 5. As a next step, it is planed to observe the SCL effect for such highly polarized photocathode by the same pump-probe method.

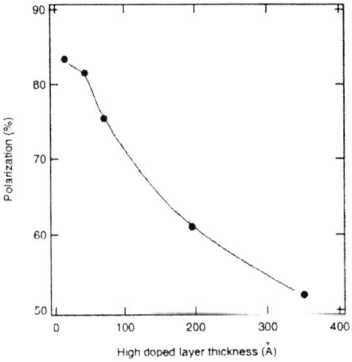

Fig. 5 : The max. polarization as a function of the high-doped layer thickness

The former method to relax the SCL effect has been pursued by Nagoya group, using the photocathode with a superlattice (SL) structure. For example in the GaAs-GaAsP SL structure, due to confinement effect, the lowest level of conduction band is shifted upward from the bottom of GaAs layer and the highest level of heavy-hole mini-band is also shifted downward from the top of GaAs layer. Obviously these shifts can simultaneously increase the escape probability for conduction electrons and the recombination probability for valence holes, respectively. The typical amount of such shift is estimated to be (70-150) meV (as large as amount of NEA). A test experiment to produce the sub-nanosecond double bunches with a separation time of 2.8 ns was done at Nagoya, using a 0.7 ns double bunch laser. For example, Fig. 6 shows the double bunch generation data taken with the GaAs-GaAsP SL (SLSP#9) photocathode having Pol~80% and QE~0.3%, where no SCL effect is observed and the saturated charge (0.6×10^{10} electrons)/ bunch is limited only by the space charge effect.

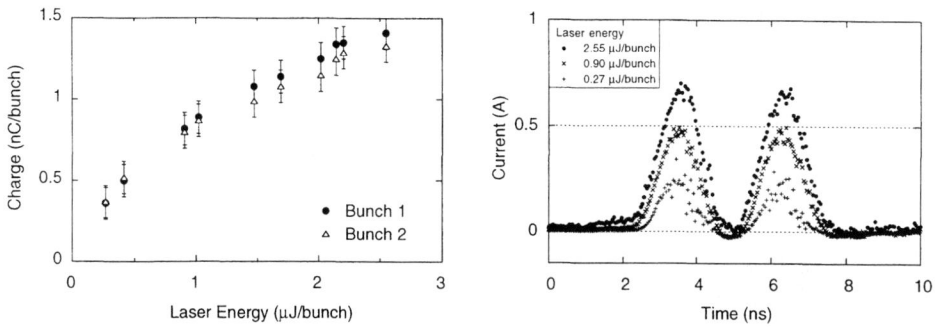

Fig. 6 : Double bunch beam produced from the GaAs-GaAsP superlattice

The usefulness of the InGaAs-AlGaAs SL photocathode with Pol~80% and QE~0.5% was also demonstrated at Bonn, where 100 mA peak-current beam could be produced without the SCL effect. Therefore, at this moment, the SL structure seems best suitable to overcome the SCL effect.

3. Microscopic observation of NEA surface

In spite of a long research history about the NEA surface, there are still no definite conclusions or understandings on the NEA formation mechanism of p-GaAs/Cs-O interface, since it is so difficult to make the clean GaAs surface and to observe such mono-layer interface in atomic scale. However, recent progresses of surface-analysis instrumentation are remarkable, and the goal of this study seems to be near hand. In facts, there are three interesting reports for the microscopic NEA studies by 1) STM (Scanning Tunneling Microscope) for the Cs/p-GaAs(110) surface, 2) photoemission spectroscopy with UV synchrotron radiation for clean and NEA surface, and 3) preparation of the highest quality NEA surface.

3-1. Study by STM

The Cs/p-GaAs(110) interface was studied by STM at Gakushuin Univ. (Tokyo) to find the relationship between photoemission and various surface configurations of Cs (1D lines, polygons and coherently c(4×4) ordered polygons) grown on the cleaved p-GaAs(110) with Zn doping of 10^{19} cm^{-3}. They observed not only the topographic images by STM but also measured the band edges by tunneling current (I) vs. bias voltage (V) and the local work-functions (ϕ) by tunneling current (I) vs. separation (S) between the sample and the tip. It was confirmed that only the coherently c(4×4) ordered Cs polygons surface (Cs: (0.6-0.7)ML) can emit the photoelectrons, in which the work-function is reduced down to 1.3 eV and the NEA state (~0.1 eV) can be made, as shown in Fig. 7.

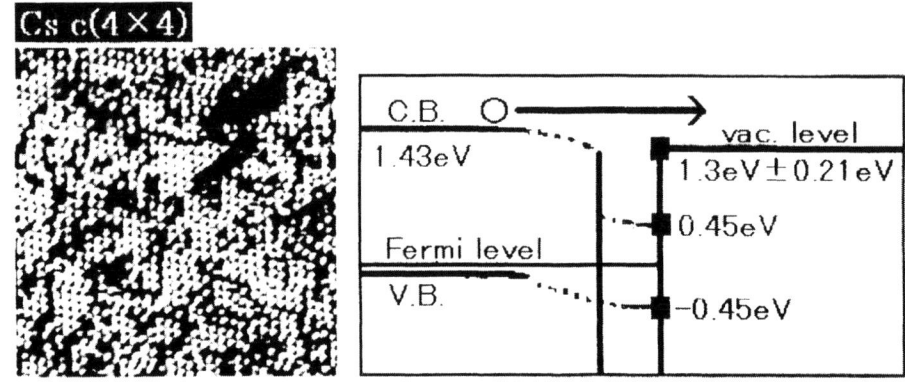

Fig. 7 : The STM image and the band structure diagram for the coherently c(4×4) ordered Cs polygons surface

The QE for this state was measured to be ~0.04% which would correspond to the first QE peak by Cs deposition in the yo-yo method, and should be enhanced by two orders of magnitude by oxygen deposition. As far as I know, this is the first experiment to clarify the role of (p-GaAs)-Cs dipole layer in atomic scale without any doubts. For approaching the final goal of microscopic study of NEA surface, it is desirable to clarify also the role of oxygen deposition using the similar method.

3-2. Study by ESCA

The GaAs surface and GaAs/Cs-O interface can be also studied by ESCA (Electron Spectroscopy for Chemical Analysis) which is well known as a powerful tool to search the chemical bonding status at surface by analyzing the chemical shifts of photoelectron spectrum. The difference of chemical interactions of Cs and Oxygen with GaAs in various activation treatments (Cs-rich, standard yo-yo, and over-layers) was studied at UVSOR facility in Okazaki by analyzing the chemical shifts of Ga-3d and As-3d peaks. The photovoltage effect was also observed not only as a peak-shift in ESCA, but also as a sudden drop and slow recovery in the temporal intensity profile of photoelectrons observed by the pump (laser) probe (SOR) method. From those data the decrease of band-bending due to photovoltage effect could be estimated. This work will be continued to study more details about NEA surface.

3-3. Study to prepare the highest quality NEA surface

Systematic studies have been also continued at Novosibirsk to fabricate the NEA surface which can provide the beam with the highest QE ($\geq$30% at 700 nm) and the lowest angular divergence. They insisted that the GaAs(100) surface prepared for NEA activation should be (1) atomically clean, (2) atomically flat, (3) low density defect and (4) Ga-stabilized. From this point of view, several surface-treatment methods for the GaAs (Cs-O) activation are also discussed, and the HCl-treatment with subsequent UHV annealing or Atomic hydrogen treatment are recommended as the best one.

A rough sketch of the desirable micro-structure of NEA surface might be drawn as shown in Fig. 8, summarizing the reports of this workshop.

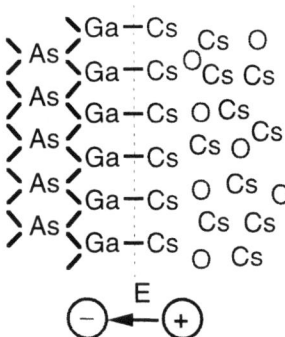

Fig. 8 : A rough sketch of NEA surface drawn in atomic scale

4. Special studies for the beam production using NEA surface

4-1. Production of an intense ultra-cold electron beam

In order to study the photoelectron escape process at the NEA surface in more detail, both of longitudinal and transverse energy distribution curves have been measured at MPI (Heidelberg). The adiabatic transverse expansion method was used to measure the mean transverse energy (MTE) which corresponds to transverse temperature of photoelectrons at the source. A new data of differential transverse energy distributions (electron distribution as a function of their transverse energy $E\perp$ with fixed longitudinal energy $E_{\parallel}^{f}$) was taken at 295K and 90K. For example, the data of complete energy distribution, $N(E\perp, E_{\parallel})$ taken at 90K is shown in Fig. 9.

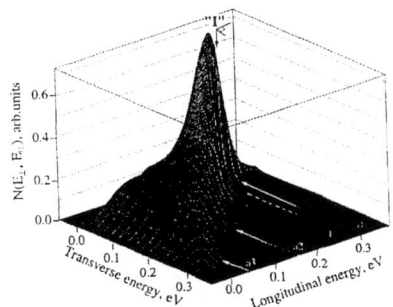

Fig. 9 : The complete energy distribution, $N(E\perp, E_{\parallel})$ from GaAs(Cs,O) at 90K

It shows a possibility to produce an electron beam with very narrow transverse and longitudinal energy spreads (≤10 meV) by using both of cooling the cathode and blocking the low E∥ parts of the distribution. It was also shown that main parts of transverse distribution corresponds to the electrons which underwent elastic and inelastic scattering (suffered energy loss) in the BBR as well.

A pulsed electron source (≤500 ps with ~3 MHz repetition) for "reaction microscope study of atomic collision" has been also developed at Freiburg Univ. This gun has a special spectrometer for simultaneous measurement of the energy- and angular- distributions of photoemitted electrons from NEA surface, in which the TOF (Time of Flight) technique and the position-sensitive MCP (Micro-Channel Plate) detector are employed.

4-2. Production of ultra-short (picosecond) polarized electron beam

An interesting measurement of the time resolved intensity and polarization for the picosecond electron bunch have been continued at Mainz using the RF-streak method. The time resolution of their system has been improved less than 2 ps including a laser pulse duration of 0.8 ps. The pulse-tail of photoelectrons extracted from a bulk GaAs cathode through NEA surface is well fitted by so called diffusion model. While those from the thin GaAs (200 nm thickness) layer is rather best fitted by exponential decay curve with time constant of ~1.4 ps that is as small as the laser pulse duration. It suggests that some ballistic electrons travel through crystal to surface with no interactions and thus the observed spin relaxation should take place not in bulk but in BBR for such thin active layer cathode. Temporal profiles of beam-intensity and polarization were also measured for the InGaAs-AlGaAs superlattice cathode (SLSA#4 made in Nagoya). Its initial and average beam polarization were ~86% and ~80%, respectively with a polarization decay time of ~55 ps.

The photoluminescence study for the same SLSA#4 photocathode using the sub-picosecond laser was done at Osaka-Pref. Univ. The measured lifetime and spin-relaxation time for conduction band electrons were ~270 ps and ~120 ps, respectively at room temperature. The measured luminescence polarization for band edge recombination was ~65%. From those data the initial polarization of conduction band electrons was estimated to be ~94% which seems to be roughly consistent with that (~86%) of Mainz data.

5. Polarimeter

There is a nice summary report on the 5 MeV Mott polarimeter developed at Jefferson-Lab, which contains (1) accurate determination of theoretical Sherman-function ($\Delta S \leq 1\%$), (2) Correct measurement of left-right asymmetries with pure energy spectra (background subtraction is not necessary), and (3) Au foil-thickness extrapolation to target thickness zero (uncertainty ≤0.5%). A systematic uncertainty of polarization measurement obtained by their system is concluded to be less than 1.1%

A very low-energy spin detector based on spin dependent transmission of electrons through ultra-thin ferromagnetic metal layers has been developed at Ecole Polytechnique. The low-energy electrons (a few eV above vacuum-level) is injected to the sand-witch structures of Au-Co-Au, or Au-Co-Au-Co-Au multi-layers, and the transmitted current is observed by a retarding potential detector. As the intrinsic Sherman function is high (66% at vacuum level), the good figure of merit of $\geq 2\times 10^{-4}$ should be obtained if the transmission efficiency in the target can be further improved. In near future, this work might have the intimate connection with so called "solid-state spin detector" which is required for development of "spintronics" technology.

6. Polarized electron beam in material science

Recently the polarized electron beam has been used widely for spin physics in the field of magnetic films and semiconductors science, and there are several reports in this workshop.

6-1. SPLEEM

At Arizona State Univ., a SPLEEM (Spin Polarized Low Energy Electron Microscopy) system was developed to observe the spin-dependent reflectivity (R↑, R↓) images of domain structure of thin ferromagnetic (Co) film systems. The subtracted (R↑−R↓) image contains only the spin dependent contrast ($\propto$P•M, P: beam polarization, M: magnetization of specimen) and is used to evaluate the reorientation or the transition process of magnetic domain structures during the film growth. Each image can be taken within a few seconds with good resolutions (depth: atomic, lateral: ≥ 10 nm). If the present polarization (~25%) is improved, it will make the important contribution to obtain the larger asymmetry and the better signal to noise ratio.

6-2. Spin-resolved photoelectron spectroscopy

Electronic-band-structure and magnetism of ferromagnetic materials has been also studied at KEK-Photon Factory, using SARPES (Spin and Angle Resolved Photo-Electron Spectroscopy) which consists of a hemispherical electron energy analyzer and a compact retarding type Mott detector. Another apparatus for SARIPES (Spin and Angle Resolved Inverse Photoemission Spectroscopy) has been developed at Hiroshima Univ. to study "spin-dependent unoccupied band structure" of magnetic materials.

6-3. Spin-STM

The idea to develop SP-STM (Spin-sensitive Tunneling Microscope) was already proposed about 10 years ago to observe the atomic scale imaging of the magnetic moments at the surface. However, it is not yet achieved in spite of many efforts. In this workshop, the SP-STM result using optically pumped GaAs tips was reported by Univ. of Electro-communications (Tokyo) group. They succeeded in fabricating the tip with top radius of ≤50 nm by means of photo-lithography and an-isotropic etching of GaAs(100) wafer, as shown in Fig. 10.

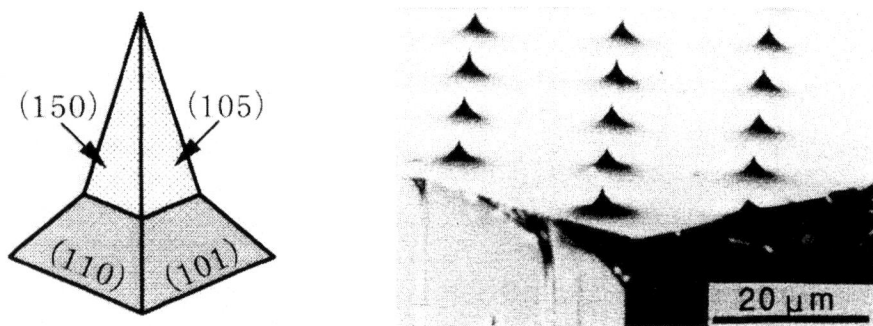

Fig. 10 : Micro-fabricated GaAs tips for the SP-STM

They could observe the images of domain like structure (in the Co thin layers grown on Au(111) surface) with typical width of (150-500) nm, which coincides with magnetic domain size obtained by MFM observation. The necessity of further basic studies (tip surface state etc.) is still pointed out to have more clear evidence of the spin-sensitive tunneling currents.

7. Present status and future plan of polarized electron beam at high-energy accelerators

The status reports are made from the existing facilities of polarized electrons of high-energy and nuclear physics accelerator laboratories, such as Bonn, Jefferson-Lab., Mainz, MIT/Bates, Novosibirsk and SLAC.

At Jefferson-Lab., the significant progresses and performances of PES system have been achieved in recent three years. For example, the up-graded Ti:sapphire laser can now deliver 500 mW (2 W maximum) at 499 MHz to the high polarization cathodes and it was demonstrated by an injector test that high average current more than 430 µA can be extracted from the gun. Further efforts are continued for near-future experiments to achieve the higher polarization (≥80%) and to reduce the helicity correlated systematic effects of polarized beam system.

At Bonn, a 50 keV gun with an InGaAs-AlGaAs SL photocathode came into operation. Together with the successful study to reduce the spin-flip depolarization during acceleration in ELSA, it made possible already to start the GDH experiment.

At MIT/Bates, totally over 900 Coulombs (peak~10 mA, average~120 μA) of polarized beam from the bulk-GaAs cathode could be delivered successfully to the SAMPLE experiment that measured the extremely small helicity-asymmetry of ~10^{-6}. Further effort to use the high polarization cathode is continued.

At SLAC, the preparation of polarized beam (6×10^{11} electrons in ~400 ns pulse) is in progress for the first observation of parity violation effect in Møller scattering (E-158) at 45.0 and 48.3 GeV to determine the precise W-S angle.

As future plans of polarized electron beam projects, there are several reports.

A polarized electron-proton (or ion) collider is planed at U.S., and two schemes are discussed, i.e. a ring-ring and a linac-ring options. In the latter option, it demands a CW source with very high average current of (100-200) mA, and a substantial level of R/D is needed.

The R/D works have been continued for the PES required by the future electron-positron linear colliders at SLAC and KEK-Nagoya. As described in section 2-2, the prospect to produce the multi-bunch beam by the polarized DC-gun with superlattice-cathodes becomes now clear for the C- and X-band linac schemes.

The idea of polarized RF-gun was proposed more than 10 years ago to produce the picosecond polarized beam with the smaller emittance than that of DC-gun. However the feasibility of this gun is not yet demonstrated, since the lifetime of NEA surface of GaAs-type photocathode under high field gradient in the RF-cavity is too short as experienced at Novosibirsk. Therefore this subject is not only attractive but also challenging. It was announced that the first workshop of polarized RF gun is scheduled on April 18-20, 2001 at FNAL to make a road-map and a list of R/D objectives to overcome this problem.

Acknowledgements

The PES-2000 Workshop Organizing Committee (T. Nakanishi, H. Horinaka, A. Kakizaki, H. Kobayakawa, T. Saka, M. Yoshioka) would like to thank all participants for the valuable contributions to this workshop. Special thanks for Professors, C. Prescott, H. Ejiri, A. Masaike, K. Imai (Spin-2000 Symposia Committee) and Professors, H. Sugawara, Y. Kimura, S. Iwata, M. Kihara (KEK) for encouragement to held this satellite-workshop. We would like to thank also for financial supports from Nagoya Univ. and Foundation of Accelerator Science. Finally, thanks for secretaries and students of Nagoya Univ., who supported the workshop and offered the pleasant times in the welcome party and the excursion in Naka-sendo trail.

Report on the Workshop "Polarized Sources and Targets" Erlangen 1999

Erhard Steffens

Physikalisches Institut, University of Erlangen-Nürnberg,
Erwin-Rommel-Str.1, D-91058 Erlangen, Germany
steffens@physik.uni-erlangen.de

Abstract. The International Workshop PST99, devoted to Rudolf Fleischmann, was conducted at Erlangen from September 29 to October 2, 1999, and attended by about 120 scientists from nine countries. The program consisted of three parallel sessions on polarized ion sources and gas targets, polarized electron sources and polarized solid targets. In addition, the plenary sessions included reports on spin experiments and applications, and summary talks of all three topics. The workshop provided an overview of existing sources and targets at operating facilities and highlighted the developments necessary for a new generation of spin experiments.

INTRODUCTION

The series of workshops dedicated to instrumental tools for the study of spin-dependent quantities in nuclear and particle physics has a long tradition, probably going back to the International Conference on Polarized Targets and Ion Sources at Saclay in 1966 [1]. In 1981, a series of topical workshops was initiated by Alan Krisch [2,3] which led to workshops on Polarized Ion Sources at regular intervals [4], later supplemented by Targets and Electron Sources. These workshops, organized under the auspices of the International Committee for High Energy Spin Physics, were conducted in the years between Spin Conferences.

The workshop PST99 on which I am reporting was a follower of Cologne (1995) [5] and Urbana (1997) [6]. In 1999 for the first time all the traditional topics for spin physics tools were discussed in a single meeting. The workshop was approved by the International Committee for Spin Physics.

PST99 was devoted to Professor Emeritus Rudolf Fleischmann (Erlangen), one of the pioneers of polarized ion sources and author of the first paper on Atomic Beam Sources (ABS) [7], published in 1956 with G. Clausnitzer and H. Schopper. We were pleased that R. Fleischmann was able to attend the opening session (see

Figure 1) with University Rector Prof. G. Jasper and Dean of the Faculty of Natural Science I, Prof. K. Rith.

The workshop was sponsored generously by the Deutsche Forschungsgemeinschaft (DFG), by DESY (Hamburg), Pfeiffer Vacuum, the Physikalisches Institut and the University of Erlangen-Nürnberg and several other Institutions. PST99 was attended by about 120 scientists from nine countries, including Central Europe, South Africa, Japan, USA, Canada and Russia. The program was set up by an Advisory Committee consisting of: K. Aulenbacher (Mainz), N. Horikawa (Nagoya), W. Meyer (Bochum), H. Paetz gen. Schieck (Cologne), G. Anton, K. Rith and E. Steffens (Erlangen). The Local Organizing Committee consisted of N. Koch, W. Kretschmer, K. Rith, F. Schmidt, D. Seyboth and E. Steffens.

The program of the three parallel sessions was organized by the following conveners:

- Topic A – *Polarized Ion Sources and Gas Targets*
  H. Paetz gen. Schieck (Cologne) and E. Steffens (Erlangen)

- Topic B – *Polarized Electron Sources*
  K. Aulenbacher (Mainz)

- Topic C – *Polarized Solid Targets*
  N. Horikawa (Nagoya) and W. Meyer (Bochum)

The different topics of the workshop were linked by plenary sessions with reports on spin experiments and applications, and summary talks on the three parallel sessions. In addition, a social program was conducted, consisting of a reception at the Faculty Club "Unicum" and an excursion to the historical City of Bamberg with guided tour and Workshop Dinner at the "Altenburg".

The proceedings of PST99 were published in December 1999, following the style of the Riken-BNL workshops. Every talk gets one or two type-set summary pages plus 6 - 12 copies of transparencies, reduced to half size. The full reference is [8]:

FIGURE 1. Willy Haeberli and Rudolf Fleischmann at the opening session.

Proceedings of the Int. Workshop on Polarized Sources and Targets,
Erlangen 1999, A. Gute, S. Lorenz and E. Steffens (Edts.)
University of Erlangen-Nürnberg (1999), **ISBN 3-00-005510-X**

In the following, a summary of the three parallel and the plenary sessions is presented. Excellent reviews of polarized target material by S. Goertz [9] and of polarized electron sources by T. Nakanishi [10] have been presented at this symposium. Therefore I will be very brief on these topics. Whenever possible conclusions on the material presented at PST99 will be given.

I POLARIZED ELECTRON SOURCES (PES)

There were some 21 talks and one poster presented at PST99 on this subject. A comprehensive summary was given by the convener of this session, K. Aulenbacher (Mainz). Modern PES are nearly exclusively based on strained GaAs photocathodes, coated with Cs and operated under UHV conditions for long lifetime. The laser light can be modulated according to the pulse structure of the linac. Extraction potentials of more than 100kV are employed. These sources have major advantages over previous designs. Electron polarization is produced within the cathode. Polarization and time structure of the extracted beam is defined by the laser beam. GaAs sources are or were in routine operation at Jefferson Lab, Mainz, MIT-Bates, SLAC and at the recently dismantled AmPS ring at NIKHEF. – Topics discussed at PST99 were:

1. *Photocathode lifetime and availability* – This is strongly dependent on the vacuum quality. Attempts are in progress to employ XUV techniques. Integrated beam charges of 100C were reported by the MIT-Bates group.

2. *Higher bunch charges and current densities* – These are the most demanding requirements for future linear colliders. Promising results from super-lattice photocathodes were reported by the St.Petersburg group.

3. *Switching asymmetries* – Experience from Mainz and MIT-Bates shows that in CW sources these asymmetries are sufficiently small in order to enable for high precision experiments.

4. *Increasing polarization beyond 80%* – Here reports were given on a better understanding of the escape process and the band structure, which may lead to a further increase in polarization.

5. *Surface studies and the generation of cold beams* – Reports from Heidelberg and Novosibirsk showed the status of these very bright and monocromatic beams. The effect of a detoriated Cs surface layer was clearly visible using LEED diagnostic techniques.

6. *Challenges* – Work is in progress on PES at future linear colliders, like TESLA, NLC/JLC and CLIC. The most demanding case is the generation of polarized positrons from Bremsstrahlung, generated by polarized electrons.

II POLARIZED SOLID TARGETS

The session on solid targets included 22 talks, summarized at PST99 by Don Crabb (Virginia). The main conclusions were the following:

1. *Target material for high-intensity beams* – The best choice is radiation doped NH_3 (17%) for proton targets and ND_3 (31%) or Li^6D (50%). In brackets the ideal fraction of polarizable nucleons is given, which is important for high-energy experiments in which the target nucleons are indistinguishable. These targets have been operated at up to 10^{11}e/s. After degradation, the material can be annealed *in situ* and repolarized several times up to a total dose of about 10^{17} electrons for NH_3 has been reached. For ND_3 the dose effect is less pronounced. Lithium deuteride has a superior radiation resistance, but the polarizing dynamics is less understood.

2. *Targets for specific experiments* – Targets for the following experiments have been discussed at Erlangen:

   - **EG-1** for CLAS (JLab) – 5T, 1K (operational)
   - **GDH** (MAMI-Mainz, ELSA-Bonn) – frozen spin (operational)
   - **COMPASS** (CERN) – large volume (under construction)
   - **TESLA-N** (DESY) – 5T, 1K (proposed)

3. *High-statistics experiments* – Suchlike experiments are meaningful only if the systematic error can be reduced below the statistical error, thus calling for NMR systems precise to 1-2 % and keeping other systematics under control. The precision of calibration becomes extremely important for high fields, very small targets and for deuterium targets due to their small signal height. In addition a high radiation resistance is required.

4. *New materials*

   - *Chemically-doped material* – Materials discussed at PST99 were butanol, doped with *Tempo* (Bochum), CD_2-foil, doped with *Tempo* (Nagoya), and plastic scintillator, doped with *Tempo* (PSI).
   - *Brute-force polarized "pure" HD targets* – After many years of development at Syracuse, a prototype target has been built for the LEGS facility at BNL. A similar project us underway for GRAAL at ESF (Grenoble). No figures on measured performance were available at the time of PST99.

III POLARIZED ION SOURCES AND GAS TARGETS

There is a close relation between these two topics due to common techniques like Stern-Gerlach separation and optical pumping which are used to produce polarized atoms. These atoms are then ionized in an ion source or employed as polarized target. In addition, polarized atoms may be applied in other fields like medicine or surface physics. There were 34 talks and 15 posters presented at Erlangen on these topics. The summary was given by T.B. Clegg (Chapel Hill).

A Polarized Ion Sources

1. *Status reports on existing ion source facilities* – Reports were given from KVI Groningen, NAC Faure, South Africa, COSY Jülich, the Nucletron Dubna, and FSU Tallahassee, illustrating the performance of sources for protons, deuterons and even lithium ions, obtained in routine operation. These sources are all of the atomic beam type, differing in the type of ionizer being employed, e.g. via electron impact, colliding Cs beam or surface ionization. The outstanding performance of the TRIUMF H^- optically pumped OPPIS source was covered in A. Zelenski's talk about their parity violation experiment in pp scattering at 221 MeV. Systematic errors on A_z in the range of a few 10^{-8} have been obtained.

2. *New developments in polarized ion sources*

    - *Commissioning of new ion sources* – The new CIPIOS pulsed H^- ion source for the pulsed injector synchrotron CIS at the IUCF cooler ring has been successfully run in, as reported by V. Derenchuk. The source was jointly developed with INR Moscow (A. Belov). Ionization of the thermal atoms takes place by interaction with an intense D^- plasma beam. Peak currents of 1.2mA with $400\mu s$ pulse length have been obtained.
      A new source for H^- and D^- ions with ECR ionizer and double charge exchange in Cs is beeing commissioned at the Munich MP tandem. Initial D^- currents of $13\mu A$ at 83% polarization were reported by G. Graw at the workshop.

    - *OPPIS H^- sources for RHIC and HERA* – The source for RHIC is based on the former KEK source and has been upgraded at TRIUMF. More than 80% polarization at 1.6mA were reported by A. Zelenski. As shown in the talk by T. Takeuchi, careful studies on ion and spin motion were conducted in order to ensure a high performance. – The injector project for HERA aims at pulse currents of more than 20mA. The present studies indicate that with higher solenoid fields and laser power this goal is within reach.

    - *Study of ionization inside a storage cell* – The concept of using a plasma jet which has been studied at INR Moscow for pulsed mode [6], is being

investigated at TUNL/Chapel Hill for DC operation. According to S. Lemaitre, H^+ and D^+ currents in the order of 1mA are expected.
Ionization by a fast Cs^0 beam inside a storage cell is under study at Cologne in order to upgrade the intensity of the COSY source.

- *Spin exchange ^3He source* – The prototype source at RCNP employs "collisional pumping" to achieve nuclear polarization of ^3He ions by multiple charge exchange on a dense optically pumped Rb target. Recent measurements have led to a rather low spin-exchange cross section as compared to theoretical expectations which can partly be compensated for by an increase in Rb thickness. The group of M. Tanaka is going ahead with a proposal for a polarized $^3He^{++}$ ion source for injection into the cyclotron.

B Polarized Gas Targets for Storage Rings

1. *Targets based on Atomic Beam Sources* (ABS) – Targets for ^1H, ^2H usually employ an ABS in order to generate a polarized atomic beam of high vector or tensor polarization. As the beam itself represents a rather low areal density of a few $10^{11} atoms/cm^2$ storage cells are used in order to increase the thickness by a factor of up to several hundred, depending on the geometry of the cell. Several targets of this kind were discussed at PST99.

   - *PINTEX target at IUCF* – The status was presented by W. Haeberli. The target cell consists of thin Teflon foil at 300K, surrounded by Si strip de-

FIGURE 2. Participants of the workshop PST99

tectors, thus enabling the detection of low-energy recoils. A versatile spin handling system for the stored beam in conjunction with fast switching through all cartesian vector polarization components of the hydrogen target has led to impressive efficiency and precision.

A very interesting study on the residual polarization of molecules after recombination from polarized atoms is underway at IUCF, presented by T. Wise. The molecules recombine in a strong longitudinal guide field similar to the HERMES conditions. Using a longitudinal polarized proton beam the average polarization of the target gas can be measured. Its molecular fraction can be varied in a controlled way by means of a "recombiner". Results were presented at Spin2000.

- *HERMES target at HERA* – This target runs in the HERA electron ring with polarized e^+ or e^- of up to 50mA since 1996. Developments on the ABS like the microwave dissociator or attenuation studies were presented by N. Koch. The target density has been pushed to about $1 \cdot 10^{14} atoms/cm^2$. Due to the low rates in deep-inelastic scattering the precise determination of the target polarization by a target polarimeter itself is mandatory. The HERMES sampling polarimeter and the various methods to correct for the molecular fraction and the inhomogenous distribution along the axis of the target cell were discussed by C. Baumgarten and H. Kolster. For hydrogen, the error is about 4%, dominated by the error on the polarization carried by molecules (see PINTEX target). For deuterium which is runnig since 1998, the error might be considerably smaller due to the very low recombination rate.

- *VEPP-3 target* – D. Toporkov reported about the cryogenic ABS delivering for deuterium $6.4 \cdot 10^{16}$ atoms/s in 3 substates into the 350mm long, 20mm diameter feed tube of the storage cell. The source is equipped with MF and SF transitions. By using the polarimeter, efficiencies close to 100% were found. The Al storage cell with drifilm coating is cooled with cold nitrogen gas. The central 16cm out of the 40cm long storage cell are viewed by the detector. From the asymmetry in $\vec{ed}$ elastic scattering an average fraction of $(40 \pm 3)\%$ of the maximum polarization has been deduced, indicating damage of the cell coating by synchrotron radiation.

- *AmPS target* – This target has been switched off and is partly re-used in the BLAST detector (MIT-Bates). Summaries of the running and experimental results were given by J. v.d. Brand and M. Ferro-Luzzi. The storage cell target for ^1H and ^2H was supplied by an ABS. A simple target polarimeter served to optimize the settting of the HFTs. The *extracted ion* polarimeter was mainly used to correct for the molecular fraction of the target. The target polarization was deduced from elastic ed scattering. Another set-up enabled to run with polarized ^3He produced by optical pumping. In total, it has been a successful facility leading to a considerable amount of new data.

- *Targets for COSY* – Two experiments at COSY are preparing for polarized target running:
  EDDA - The status of the target was presented by D. Eversheim. The polarized *jet* provides an areal density of about $2 \cdot 10^{11}/cm^2$ at 85% polarization. A storage cell is in preparation in order to increase the density.
  ANKE - The design and status of the storage cell target for the ANKE spectrometer was presented by H. Seyfarth. A number of related developments like the design of high-field permanent 6-poles and a profile monitor for the dissociated beam were presented in separate contributions. First operation of the new ABS was expected for the beginning of 2000.

- *RHIC polarimeter* – A polarized hydrogen jet plus detector has been proposed as polarimeter for the RHIC proton beam. The basic idea was presented by A. Penzo. By comparing single spin asymmetries with polarized target **or** polarized beam one is able to calibrate the pp analyzing power. This should allow for a measurement of the polarization of the high-energy proton beam at the 2% level.

2. *Targets based on optical pumping (OP)* – Optically pumped targets for storage rings include ^3He targets by direct OP and metastability exchange [5,6], and ^1H targets by spin exchange OP of a small alkali fraction. As the pumping is done by lasers, the latter technique is called *Laser Driven Source* (LDS). At PST99, only the LDS technique has been discussed.

   - LDS test stand at Erlangen – In the report by F. Schmidt the diagnostic technique was discussed which is necessary to understand the complexity of the processes taking place within the hot pumping cell in the presence alkali vapor and atomic hydrogen. A study of spin relaxation of Rb within the pumping cell was presented by W. Nagengast.

   - The LDS prototype target at the IUCF cooler ring – The status of the IUCF ^1H and ^2H LDS target, a joint project of Illinois, Argonne and Erlangen [6], was presented by C. Jones. The stored vertically polarized proton beam was employed to determine the effective target polarization, averaged over atoms and the significant molecular background. For hydrogen, the average polarization reached 10%, for deuterium about 15%, indicating strong relaxation and recombination in the target.

3. *Ultracold target source for hydrogen* – This technique is under study at Michigan. The status was presented by V. Luppov. At T=0.3K separation of the two electron spin states $m = \pm 1/2$ takes place in the fringe field of a 12T solenoid. The "+"-component is accelerated in the field gradient and leads to a cold polarized beam with narrow velocity distribution, which is focussed by means of a parabolic mirror and a sextupole magnet. The initial results of the Mark-II source were reported. The areal density was an order of magnitude below the goal of $5 \cdot 10^{12}/cm^2$.

C High Pressure Targets and Applications of Polarized Atoms

1. *Polarized* ^3He – At Mainz, high density polarized ^3He gas is produced by OP of metastable atoms and mechanical compression to several bars with moderate loss in polarization. The gas can be stored in glass bulbs with relaxation times of about 100h and produced in remote Filling Stations. These targets are used as Spin Filters for thermal neutrons from reactors or spallation sources. D. Hofmann described how they are applied as polarizers and analyzers for the primary and scattered neutrons, respectivley. M. Ebert showed the application of polarized ^3He gas in lung imaging. By measuring the relaxation rate one is sensitive to the local oxygen concentration and its evolution in time which may lead to a new diagnostic tool.
The alternative method for high pressure ^3He targets by spin exchange OP is being studied by the Michigan group. These cells are applied as targets in high energy scattering experiments or as spin filters. K. Coulter reported how by using refillable cells the requirements of different experiments can be met in an optimum way. In the same way, polarized ^{129}Xe is produced for NMR and MR-Imaging. – Spin exchange is also applied at RIKEN in ^3He targets for nuclear physics experiments, as described by T. Uesaka.

2. *Polarized* ^8Li – These radioactive spin-two atoms are "self-analyzing", i.e. their polarization can be detected via the β ray asymmetry with respect to the direction of the guide field. D. Fick reported how they are produced by 24MeV ^7Li ions, thermalized and polarized via OP, followed by adsorption on a surface to be studied. For the surface studies UHV conditions are required. Spin relaxation is used as a tool to measure e.g. electric field gradients at surfaces.

3. *Polarized* ^{129}Xe – These atoms are of interest for medical applications [6] studied by the Michigan group. They may also be used for NMR on surfaces. In contrast to ^8Li, these are stable atoms. H. Jänsch demonstrated that one can obtain NMR from adsorbed layers of polarized Xe atoms with reasonable signal-to-noise down to coverages below one monolayer. This may lead to a new class of spin relaxation experiments on surfaces.

4. Magnetic Resonance Imaging Based on Equilibrium Polarization – In his plenary talk, A. Oppelt (Siemens) presented the status of modern MR Imaging devices and its impressive development and application in medicine. Due to the flexibility in rf and gradient pulse sequences special requirements like increased tissue contrast, visualization of flow and diffusion patterns can be met. MRI is a standard method for image-guided minimal-invasive techniques and surgical procedures. Via relaxation MRI is sensitive to the concentration of Oxygen, enabling e.g. the study of blood supply of the brain. By means of

contrast agents the applicability of MRI is extended further, e.g. to direct observation of the blood stream or accumulation processes within the tissue.

IV ACKNOWLEDGEMENT

I gratefully acknowledge the material for preparing my workshop summary at Spin2000 provided by T.B. Clegg, W. Haeberli and P. Lenisa, and the assistance by M. Henoch and A. Gute in preparing this manuscript. I thank all the individuals and institutions mentioned in the introduction and in the proceedings for making PST99 a success.

REFERENCES

1. Proceed. of the Int. Conf. on *Polarized Targets and Ion Sources*, Saclay 1966. Ed. by Physique Centre d'Etudes Nucleaires de Saclay.
2. Proceed. of the Workshop on *Polarized Proton Ion Sources*, Ann Arbor 1981. A.D. Krisch, A.T.M. Lin (Edts.), AIP Conf. Proceed.**80**, AIP (1982).
3. Proceed. of the Workshop on *Polarized Antiproton Sources*, Bodega Bay 1985. A.D. Krisch, A.T.M. Lin, O. Chamberlain (Edts.), *AIP Conf. Proceed.***145**, AIP (1986).
4. Proceed. of the Int. Workshop on *Polarized Proton Ion Sources*, Vanvcouver 1983. G. Roy, P. Schmor (Edts.), AIP Conf. Proceed. **117**, AIP (1984).
5. Proceed. of the Int. Workshop on *Polarized Beams and Polarized Gas Targets*, Cologne 1995. H. Paetz gen. Schieck, L. Sydow (Edts.), World Scientific (1996).
6. Proceed. of the 7th Int. Workshop on *Polarized Gas Targets and Polarized Beams*, Urbana, IL, 1997. R.J. Holt, M.A. Miller (Edts.), *AIP Conf. Proceed.***421**, AIP (1998).
7. G. Clausnitzer, R. Fleischmann, H. Schopper, *Z. Physik* **144**, 336 (1956).
8. Proceedings of the Int. Workshop on *Polarized Sources and Targets*, Erlangen 1999, A. Gute, S. Lorenz and E. Steffens (Edts.), University of Erlangen-Nürnberg (1999), ISBN 3-00-005510-X.
9. S. Goertz: *Recent Developments in the Field of Solid Polarized Target Materials*, these proceedings.
10. T. Nakanishi: Workshop Report on *Spin Polarized Electron Sources and Polarimeters*, these proceedings.

Prospects of High Energy Polarized ep Colliders

V.W. Hughes

Department of Physics, Yale University
P.O. Box 208121, New Haven, CT 06520-8121, USA

I INTRODUCTION

Studies of spin effects in a new regime have often led to important discoveries. A list of some outstanding examples follows in Table 1.

Table 1 : Some Surprises With Spin

1. Space quantization associated with quantized spin directions.
   Stern, Gerlach, 1921.

2. Atomic fine structure and electron spin magnetic moment.
   Goudsmit, Uhlenbeck, 1926.

3. Proton anomalous magnetic moment; $\mu_p = 2.79\ nm$.
   Stern, 1933.

4. Electron spin anomalous magnetic moment. $\mu_e = \mu_0(1.00119)$; QED.
   Kusch, 1947.

5. Electroweak interference from $\vec{e}_1 d$ DIS parity nonconservation.
   Prescott & SLAC-Yale Collaboration, 1978.

6. Proton spin structure; puzzle or crisis.
   EMC, 1989.

Hence from an historical viewpoint the spin variable is a promising one for discovery.
  Much has been learned about the spin structure of the nucleon in the past 20 years and excellent survey talks have been given by R. Jaffe, F. Bradamante and W. Vogelsang at our Symposium. Still important uncertainties in our knowledge remain which could be resolved with an extension in the kinematic range of the data. There are two high energy polarized ep colliders for which serious studies have been and are now in progress: polarized HERA at DESY and eRHIC at BNL. The physics opportunities and accelerator problems have been discussed, and indeed the additional possibilities of polarized ed and e ^{3}He colliders are considered.

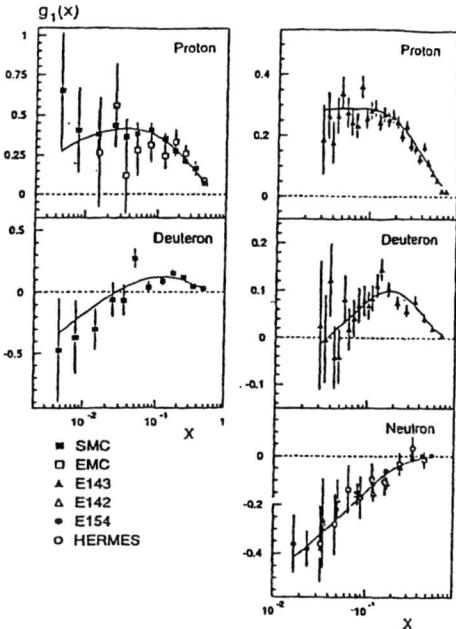

FIGURE 1. The structure function g_1 of the proton, the deuteron, and the neutron, as a function of x, from the CERN muon-scattering experiments (*left*) and the SLAC and DESY electron-scattering experiments (*right*). Only statistical errors are shown. Solid lines show a next-to-leading order QCD fit.

II CURRENT INFORMATION ON NUCLEON SPIN STRUCTURE AND OUTSTANDING PROBLEMS

The present world data on the spin structure functions for the proton, deuteron and neutron are shown in Figure 1 as a function of x at the measured Q^2 values [1]. The data extend from $x = 3 \times 10^{-3}$ to $x = 0.7$ with Q^2 between 1 and 60 GeV^2. A fit based on pQCD with α_s taken from the Particle Data Tables adequately represents the data and provides a confirmation of the successful application of pQCD to spin structure functions. However, the accuracy of the pQCD calculations is limited importantly by uncertainties associated with the choices of factorization and renormalization scales and initial parton distributions. The limited extent of the kinematic range with $x > 0.003$ is an important contributor to the uncertainties and to the extension of the pQCD calculations to small x where data are absent. Since evaluation of the important sum rules of QCD require values of the spin structure functions g_1 from $x = 0$ to $x = 1$, measured values of g_1 are needed to as low an x value as possible.

It is well known that present world data confirm the basic Bjorken sum rule to about 10% and are in substantial disagreement (3 to 5 standard deviations) with

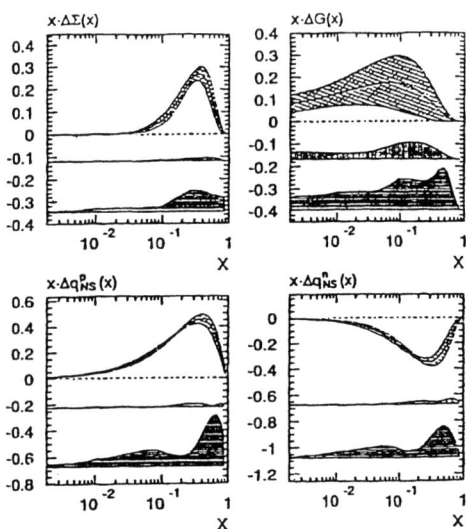

FIGURE 2. Polarized parton distribution function at $Q^2 = 1\ GeV^2$, obtained from a perturbative QCD analysis. The statistical uncertainties are indicated by the cross-hatched bands. The systematic experimental and theoretical uncertainties are indicated by the vertically and horizontally hatched bands, respectively.

the Ellis–Jaffe sum rule for the proton and neutron [2]. In addition, present data using pQCD provide information on the polarized parton distributions(Figure 2) [1]. Note that the polarized gluon appears to be positive but is determined only with very large errors. Lack of information on the first moment of $\Delta g(x)$ leads to large uncertainties in the first moment $\Delta\Sigma = \Delta u + \Delta d + \Delta s$ and also in those of Δu, Δd and Δs individually.

Present data from semi-inclusive DIS provides a determination of the valence and sea quark spin distributions of Δu_v, Δd_v and $\Delta \bar{q}$ (Figure 3) [3,4].

In addition the spin dependent structure function $g_2(x)$ has been measured (Figure 4) [5].

Despite the impressive amount of information now available on nucleon spin structure functions (still much less than that on unpolarized structure functions), important outstanding problems remain. Among these are the following:

1. The polarized gluon distribution $\Delta g(x, Q^2)$ and the contribution of polarized gluons to the nucleon spin.

2. Orbital angular momentum in the nucleon and its contribution to nucleon spin.

3. Behaviour of $g_1(x)$ at low x. Accurate determination of the first moment of g_1^p, g_1^d and g_1^n and a precise test of the fundamental Bjorken sum rule.

4. Extension of tests of pQCD and accurate determination of α_s.

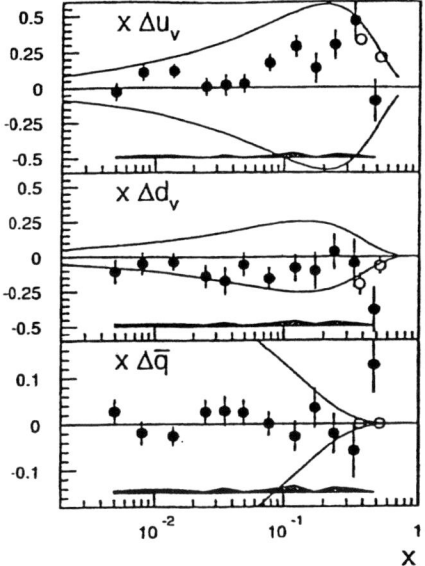

FIGURE 3. The polarized quark distributions $x\Delta u_v(x)$, $x\Delta d_v(x)$ and $x\Delta \bar{q}(x)$ measured by SMC, assuming $\Delta \bar{u}(x) = \Delta \bar{d}(x)$. The open circles are obtained when the sea polarization is set to zero; the solid circles are obtained without this assumption. The error bars are statistical and the shaded areas represent the systematic uncertainty. The solid lines show the limit $\pm xq(s)$ from unpolarized quark distribution at $Q^2 = 10$ GeV. Bottom: Curves correspond to $\pm x(\bar{u}(x) + \bar{d}(x))/2$.

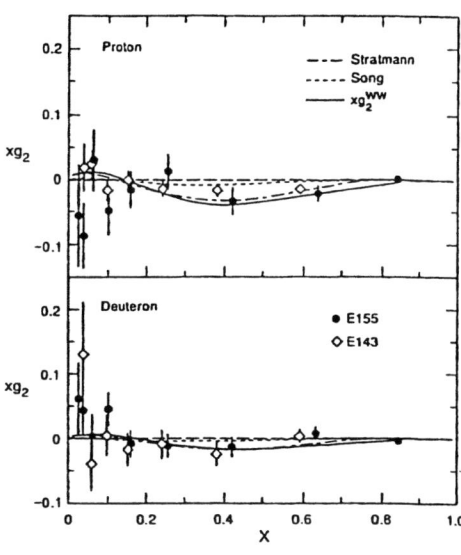

FIGURE 4. The spin-dependent structure function xg_2 as a function of x, measured by SLAC experiments E143 and E155, for the proton and deuteron. Only statistical errors are shown; the systematic errors are much smaller. The solid curve shows a twist-2 calculation of g_2^{ww}. Also shown are bag model calculations at $Q^2 = 5.0$ GeV^2 by Stratmann and Song.

5. Other spin structure functions: the parity violating g_s and transversity h_1.

6. Hadronic spin structure of the proton.

Data in an extended kinematic range are required or will contribute to all of these problems. The importance of obtaining data in an extended kinematic range was exemplified dramatically by the EMC experiment at CERN [6] which followed earlier SLAC experiments and measured g_1^p down to $x = 0.01$, whereas the SLAC measurements extended down only to $x = 0.1$. The results are shown in Figure 5a. The SLAC data are in agreement with the theoretical curve which incorporated the Ellis-Jaffe and Bjorken sum rules, as well as the standard quark model which ascribed most of the proton spin to the quark spins. However, the EMC data below $x = 0.1$ disagree with the theoretical curve. The implication as shown in Figure 5b is that the quark spins contribute very little to the proton spin. This unexpected result constitutes the proton spin puzzle or crisis and has stimulated the enormous experimental and theoretical effort in this field for the past 15 years.

III NEW HIGH ENERGY POLARIZED EP COLLIDERS

The energies of HERA and eRHIC colliders are listed and the available kinematic ranges in x and Q^2 are shown in Figure 6 for a HERA collider and for eRHIC, as well as the ranges used thus far in fixed target experiments. The HERA collider would extend the kinematic range of fixed targets to lower x by a factor of 100 and to higher Q^2 and achieve a center of mass energy $\sqrt{s} = 300\ GeV$. eRHIC will increase these kinematic ranges by a factor of 10 and achieve $\sqrt{s} = 100\ GeV$. For both colliders the possibilities for polarized ed and e ^{3}He collisions are being investigated.

The center of mass energies and luminosities for various collisions are listed in Table 2.

Table 2 : Center of Mass Energies and Luminosities For Various Collisions

Facility	Particles and Energies(GeV)	$\sqrt{s}(GeV)$	L $(cm^{-2}sec^{-1})$
RHIC	$p(250) \times p(250)$	500	2×10^{32}
HERA	$e(30) \times p(900)$	300	7×10^{31}
HERA(A)	$e(30) \times A(450)$	200	$(7 \times 10^{31})/A$
eRHIC	$e(10) \times p(250)$	100	1.5×10^{33}
eRHIC(Au)	$e(10) \times$ Au(100)	60	5×10^{30}
FNAL	$\mu(600) \times p(1)$	35	$\sim 10^{31}$
EPIC	$e(3) \times p(50)$	25	$> 10^{33}$
SMC	$\mu(200) \times p(1)$	20	$\sim 10^{32}$
SLAC	$e(50) \times p(1)$	10	1.3×10^{33}
HERMES	$e(30) \times p(1)$	7	$\sim 10^{31}$

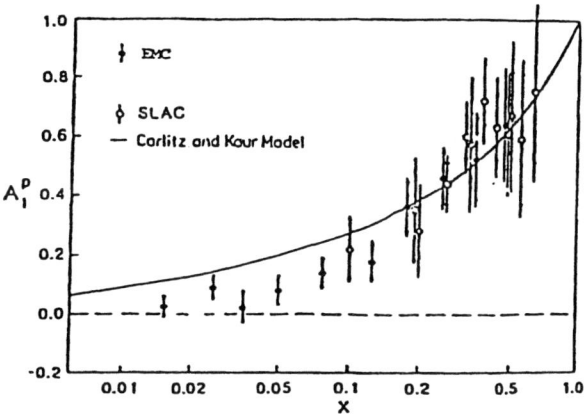

Figure 5 : (a) The asymmetry A_1 for the proton as a function of x.

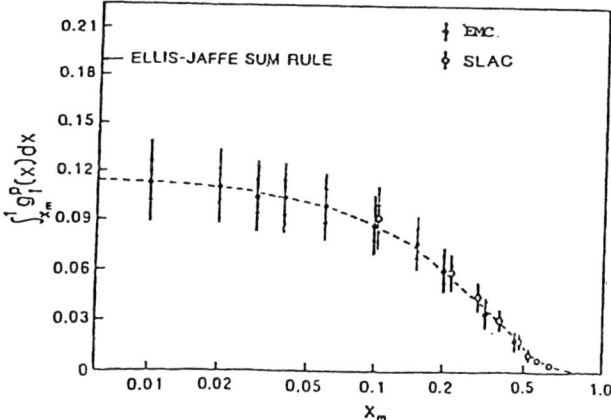

FIGURE 5. (b) The convergence of the integral $\int_{x_m}^{1} g_1^p dx$ as a function of x_m, where x_m is the value of x at the low edge of each bin.

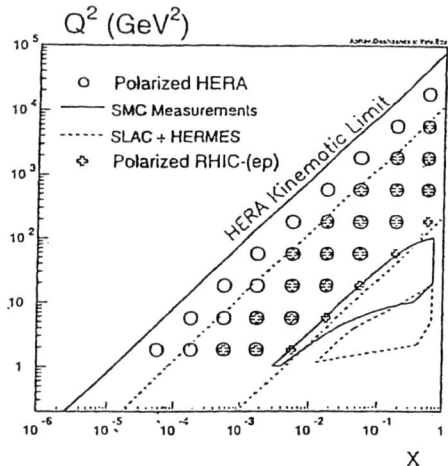

FIGURE 6. The $x - Q^2$ ranges with polarized colliders and fixed targets

- Polarized HERA: $E_p \sim 820 - 925~GeV$, $E_e \sim 27.6~GeV$, $\sqrt{s} \sim 300 - 320~GeV$
- Polarized eRHIC: $E_p \sim 50 - 250~GeV$, $E_e \sim 10~GeV$, $\sqrt{s} \sim 45 - 100~GeV$

IV THE POLARIZED HERA COLLIDER

HERA now has a storage ring with polarized $e^{\pm}$ of 27.5 GeV and a storage ring with unpolarized protons of 800-900 GeV. Two major collider detectors - ZEUS and H1 - have been operating for many years. Hence obtaining polarized protons in the high energy ring and providing spin rotators for longitudinal proton polarization at the collision points are the only requirements to study polarized ep collisions. Over the past 6 years much work and many workshops have been devoted to a polarized HERA collider.

Table 3 : A Polarized HERA Collider

	Electron Beam	Proton Beam
Beam Energy	26-30 GeV	800-930 GeV
Polarization Status	Polarized Sokolov Ternov Effect(STE) $P_e \sim 60\%$ in 1/2 hour	Negligible polarization due to STE Need polarized source
Expected Polarization	70%	70%
Uncertainty $\delta P/P$	$\leq 2\%$	$\leq 3\%$
Integrated Luminosity	$\sim 500~pb^{-1}$, 3 year running with $150 - 170~pb^{-1}/year$	

The most recent and complete information is given in the Proceedings of the Workshop "Polarized Protons at High Energies-Accelerator Challenges and Physics Opportunities", 1999 [7].

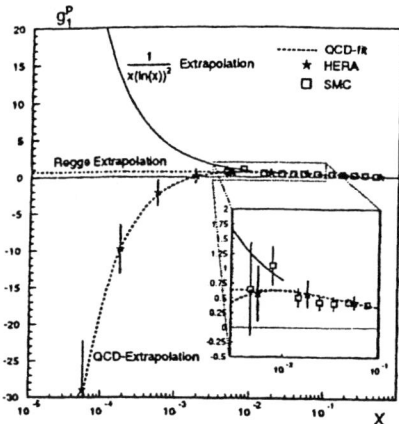

FIGURE 7. The statistical uncertainty on g_1^p from possible measurements at HERA with 500 pb^{-1} is shown along with different theoretical predictions for the low x in the kinematic region $x < 0.003$.

A Particle Physics Topics

A polarized HERA collider could measure the spin structure function $g_1^p(x)$ to very low $x \sim 10^{-4}$ and to very high $Q^2 \sim 10^4\ GeV^2$. The behaviour of $g_1(x)$ at low x is of great interest to test pQCD and other theoretical predictions. Figure 7 shows the theoretical prediction of pQCD, Regge theory and a strong powerlike positive rise. The perturbative QCD curve is an extrapolation from a pQCD fit to SMC data which goes down only to $x = 3 \times 10^{-3}$. The insert in the figure shows an enlargement of this higher x region to which the SMC data extend. The very dramatic decrease in $g_1(x)$ at low x for the pQCD curve corresponds to the rapid rise in $F_2(x)$ found at HERA, which is in agreement with the pQCD prediction for $F_2(x)$.

The points indicated by HERA show the expected statistical errors from future HERA measurements and are plotted on the extrapolated pQCD fit. The depolarization factor D is small in this low x region, so the measured asymmetries will be small. Systematic errors associated with electron energy losses, multiple scattering and event migration have been considered and are small. Uncertainties in electron polarization P_e and in proton polarization P_p produce a normalization type error of about 5%. False asymmetries associated with detector and beam instabilities can be kept small due to the rapid bunch to bunch modulation of the proton polarization direction.

In addition to the importance of studying $g_1(x)$ itself at low x, measurement of $g_1(x)$ at low x will improve our knowledge of the first moment of $g_1(x)$,

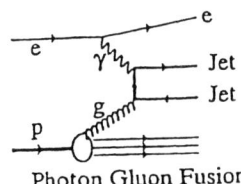

Figure 8 : (a) Feynman diagram for the process contributing to the di-jet cross section for Photon Gluon Fusion at LO.

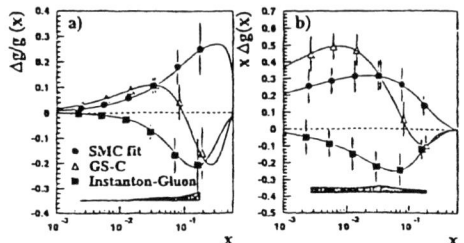

FIGURE 8. (b) Statistical accuracy possible for the measurement of $\Delta g(x)$ using 2 jets from the PGF process shown with different predictions for $\Delta g(x, Q^2) = 20\ GeV^2$

$\Gamma_1 = \int_0^1 dx g_1(x)$ both because of the known measured values at lower x and because of the reduction in the theoretical errors in the pQCD prediction. Better knowledge of Γ_1^p will improve the test of the Bjorken sum rule. If polarized d or ^{3}He is obtained in the HERA ring, further accuracy in determining the Bjorken sum could be achieved.

The polarized gluon distribution $\Delta g(x, Q^2)$ as well as polarized quark distributions can be determined by fitting the distributions to the $g_1(x, Q^2)$ data using pQCD. Our present information on Δg is obtained in this way. The first moment $\Gamma_1^g = \int_0^1 \Delta g(x) dx$, which is the contribution of polarized gluons to the proton spin. is now known to be $\Gamma_1^g = 1.0 \pm 0.3(stat) \pm 0.3(syst) \pm 1.0(theor)$. The additional low x data on $g_1(x)$ from HERA together with the associated reduction in theoretical error for determining $g_1(x, Q^2)$ should reduce the statistical plus theoretical uncertainty in the first moment $\delta \Gamma_1^g$ to ± 0.4.

The most reliable current information on the unpolarized gluon distribution $g(x, Q^2)$ in the proton comes from data on $F_2(x, Q^2)$, to which low x data from HERA contributes importantly. However, additional valuable information on $g(x, Q^2)$ comes from HERA studies of dijets from the proton gluon fusion (PGF) process [8]. Similarly for $\Delta g(x, Q^2)$ dijets from the PGF process for polarized ep collisions at HERA will provide useful information on $\Delta g(x, Q^2)$, particularly on its shape in the measured kinematic range. The PGF process [8] in leading order is shown in Figure 8a. The principal background comes from QCD-Compton process. The results of a study of the determination of Δg from dijets at HERA is shown in Figure 8b [9].

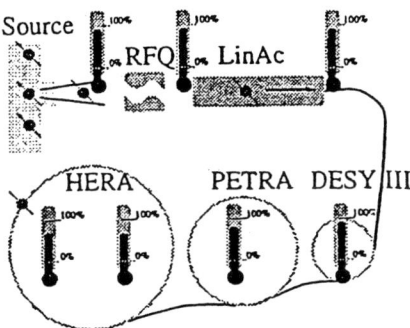

FIGURE 9. The various stages of the proton acceleration from the polarized source to HERA. [DESY III, 7.5 GeV; PETRA, 40 GeV; HERA, 800 GeV]

A combination of existing g_1 data together with projected data from HERA on $g_1(x)$ and on dijets should determine Γ_1^g with the statistical plus theoretical error less than 0.3.

Other promising particle physics to which a polarized HERA collider would contribute are listed in Table 4. Generally, for whatever observables are studied in unpolarized ep scattering, complementary studies can be made in polarized ep scattering. Notable are determination of the parity violating structure function g_5 and the polarized parton distribution in the photon.

B Accelerator Physics Topics

The system of accelerators to produce the 800-900 GeV proton in the HERA storage ring is shown in Figure 9 and the characteristics of a polarized HERA collider were listed in Table 3.

An electron ring at 27.5 GeV with $e^\pm$ polarized by the Sokolov-Ternov effect is operational. HERA is now in the midst of a luminosity upgrade to produce the luminosities indicated in the Table 2. The increased luminosity will be achieved principally by increasing the phase space density at the interaction point with superconductiong combined-function (focusing and bending) magnets in the interaction region.

Table 4 : Physics Topics and their Potential

Polarized Structure Functions g_1, g_5, g_2	Very Good
Polarized Gluon Distribution Δg:	Very Good
• NLO-pQCD fits of g_1	Very Good
• Di-jet events in DIS	Very Good
• 2-Track events in DIS	Very Good
• Combined fits: g_1 + Di-jets	Very Good
• Photoproduction	Very Good
Polarized semi-inclusive measurements	Good
Polarized parton distribution in photon Δq^γ	Very Good
Diffraction/Vector Meson	Not Good
DHG Sum rule: $(\sigma \uparrow\downarrow - \sigma \uparrow\uparrow)$ at $Q^2 = 0$	Very Good
$(W^\pm, Z^0)$ Production	Not Good
High Q^2 anomaly $\rightarrow$ polarized HERA	Very Good
Target fragmentation	Very Good
Λ Polarization	Good
Deeply virtual polarized Compton scattering	Good
$\vec{p}\vec{p}$ scattering with HERA-$\vec{N}$. Internal polarized p jet target	Very Good

Two major high energy detectors ZEUS and H1 are in operation in HERA. These are general purpose magnetic detectors with nearly hermetic calorimetric coverage. ZEUS has tracking chambers inside a superconducting solenoidal magnet, surrounded by calorimeters and muon chambers. There are electron detectors downstream in the direction of the electron beam and a proton spectrometer and neutron calorimeter in the proton beam. In addition, there are luminosity detectors.

An early study of a system to produce polarized protons in the HERA ring was done jointly by Krisch and the SPIN Collaboration and by Barber and the DESY team [10]. Two critical problems were identified: (1) Development of a high intensity (20 mA) and high polarization ($P_p > 0.7$) H^- source, (2) Retention of high polarization in the acceleration and storage of the protons.

Considerable progress has been made on both these problems in the past several years. At BNL in Fall 2000 a polarized H^- atomic beam source with an intensity of 1 mA and a polarization of 0.8, based on an optically pumped polarized ion source(OPPIS), was used to inject into the system of accelerators for RHIC.

The general scheme of OPPIS is indicated in Figure 10. Stages 1 through 4 serve to form an H^+ beam of a few keV energy. Then H^+ are neutralized to H by electron capture in optically pumped polarized Rb vapor to produce electronically polarized H. The fast electronically polarized H beam is converted into a fast nuclear spin polarized H atomic beam by the electron-proton hyperfine interaction during the passage of the fast atomic beam through a region where the magnetic field is reversed (region 6 in Figure 10). Finally the fast nuclear spin polarized atomic beam is ionized in a Na vapor cell(region 7). Further recent developments

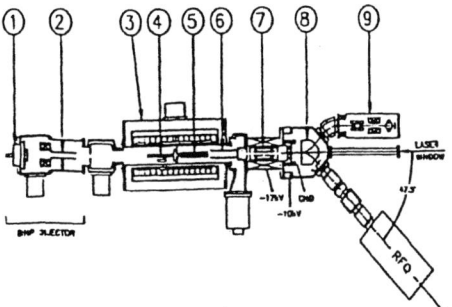

FIGURE 10. Proposed Optically Pumped Polarized Ion Source: 1. plasmatron proton source, 2. H_2 neutralizer, 3. superconducting solenoid, 4. He ionizer, 5. optically pumped Rb vapor cell, 6. deflection plates, 7. Na vapor ionizer, 8. bending magnet, 9. RFG.

in H^+ sources, in the laser for optically pumping Rb and in the Na ionizer promise that a 20 mA, high polarization H^- source can be achieved [11].

Major advances have also been made on the topic of proton spin retention during acceleration and storage in a high energy storage ring. At RHIC for the RHIC-SPIN project high energy polarized protons have been produced in a RHIC storage ring. This involved a system of accelerators at BNL similar to that at HERA(Figure 9). Polarized protons were injected from the AGS at 24 GeV into a RHIC ring with a single Siberian snake and accelerated to 40 GeV. The proton polarization was measured from the asymmetry in pC elastic scattering via Coulomb Nuclear Interference. Although the data have not been fully analyzed, it appears that the expected polarization was achieved [12].

In addition to the dramatic experimental achievement at RHIC, major advances in the understanding of spin motion in HERA have been make at DESY by Barber and his colleagues through extensive theoretical and numerical spin-orbit studies based on a rigorous definition of the proton spin vector. The HERA proton ring was not initially designed for polarized protons. Furthermore the multitude of depolarizing resonances at high energy and the non-flat ring make obtaining polarized protons at 800 GeV a challenging problem. Simulations have shown that with a careful choice of orbital tunes and Siberian snakes polarization can be preserved during acceleration up to 800 GeV out to about 2 sigma in horizontal and vertical phase space(Figure 11). It has been concluded that electron cooling of the proton beam in the PETRA accelerator is needed to reduce its transverse emittance. Further studies of beam dynamics are in progress at DESY.

Director Albrecht Wagner of DESY has indicated that a full study presenting the physics motivations and practicality from the accelerator physics viewpoint of a polarized ep collider would be welcomed by 2003, when the laboratory would be able to make a decision about providing a polarized proton beam in the HERA

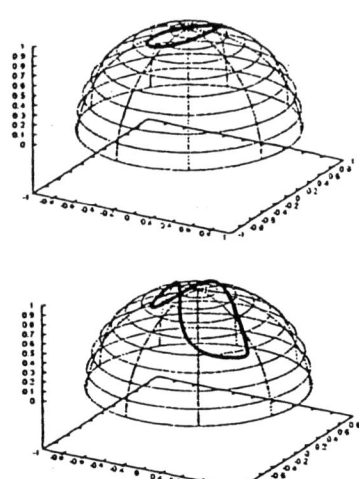

FIGURE 11. SPRINT simulation of the resultant spin vector at 800(802) GeV proton beam above (below). Shown is the effect of 4π mm mr deviation from the nominal spin direction.

ring.

V ERHIC

For eRHIC two proton rings each with 250 GeV polarized protons will be available. Last fall polarized protons were successfully injected and accelerated to about 40 GeV in one ring containing a single Siberian snake. About one year ago BNL became interested in the possibility of adding an electron beam of about 10 GeV either from a linac or a ring to the RHIC tunnel so that polarized ep or eA collisions could be studied. Several workshops have taken place. The most complete information on a polarized ep collider is given in the eRHIC Workshop at Yale [13]. A new general purpose magnetic detector similar to ZEUS and H1 will be needed for the eRHIC facility.

A Particle Physics Topics

Most of the topics discussed for a polarized HERA collider in section 4.1 are relevant to eRHIC as well. The kinematic range for eRHIC is less (Figure 6) and $\sqrt{s} = 100\ GeV$, but the luminosity is $1.5 \times 10^{33}\ cm^{-2}s^{-1}$ and thus about a factor of 10 greater than for HERA.

The spin structure function g_1^p could be determined with good precision down to $x = 5 \times 10^{-4}$ (Figure 12) due to the high luminosity. For given x and Q^2 the dilution factor D(y) is larger than for HERA. The data on g_1 at low x will be of

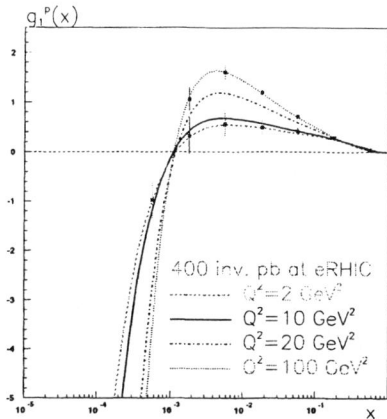

FIGURE 12. The statistical uncertainty on g_1^p from possible measurements at eRHIC with 400 pb^{-1} is shown.

great interest and together with eRHIC dijet data should determine the gluon first moment Γ_1^g to about 0.3. In addition, the accuracy of the data in the region of $x = 10^{-3}$ over a wide Q^2 range will provide a determination of α_s comparable in precision to its presently known value.

As for HERA dijets of high transverse momentum from PGF will determine the polarized gluon distribution as indicated in Figure 13.

Another potential research is on transversity or the structure function h_1. For observation in semi-inclusive DIS the azimuthal distribution of the final pion will provide information on h_1 down to $x = 5 \times 10^{-4}$ and hence extend results already reported by HERMES on transversity. Other topics mentioned for HERA in section 4.1 can also be considered for eRHIC.

B Accelerator Physics Topics

The eRHIC accelerator complex will consist of a 10 GeV electron accelerator together with a RHIC ring with 250 GeV polarized protons. The electron accelerator might be either a superconducting linac with polarized electrons or an electron ring. A schematic diagram of eRHIC with a linac is shown in Figure 14 together with a list of its characteristics. Several notable features of eRHIC are the following. One intersection region is envisaged for polarized ep collisions where a new general purpose collider detector will be located. Energy recovery from the electron beam is planned by decelerating the return beam in the linac. The high intensity high polarization electron beam to be delivered as a bunched beam will be produced by laser photoemission from a strained GaAs photocathode as is currently done at SLAC and at JLab. Electron cooling of the proton beam is planned for the RHIC rings independent of the eRHIC project.

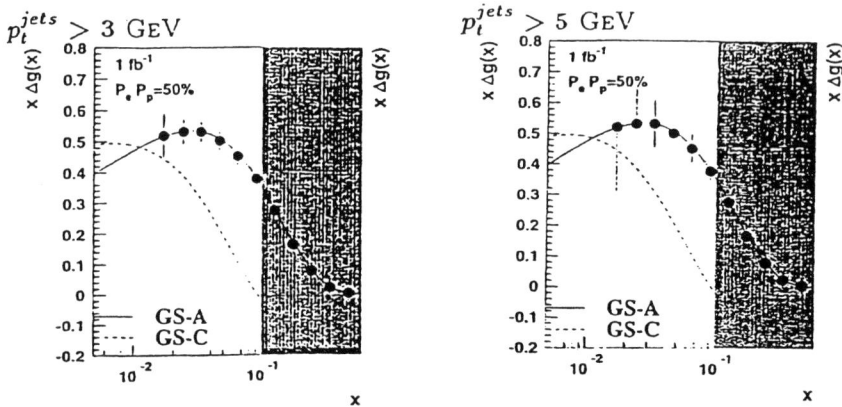

FIGURE 13. Statistical accuracy possible for the measurement of $\Delta g(x)$ using 2 jets from the PGF process for eRHIC shown

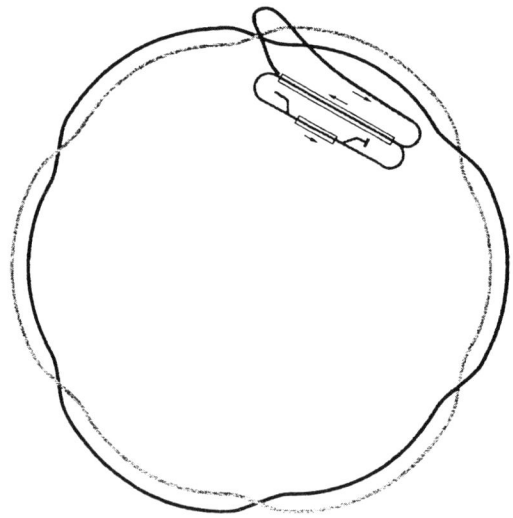

FIGURE 14. eRHIC showing proton ring and electron linac with energy recovery.

Beginning about 1 year ago serious study has gone into the eRHIC accelerator system, but the design is still at an early stage.

VI CONCLUSION

A large part of what we know about the proton and its spin structure has come from studying the relatively simple scattering of the understood lepton probe(e or μ) from the proton. A polarized ep collider at HERA and/or at eRHIC with large kinematic range would provide an opportunity for important new information on proton spin structure.

VII ACKNOWLEDGMENTS

This work was supported in part by the US Department of Energy. The author thanks I. Ben-Zvi and A. Deshpande for useful discussions and H. Deng for help in preparation of the manuscript.

REFERENCES

1. SMC Collaboration, B. Adeva et al. *Phys. Rev.* **D58**, 112002 (1998)
2. E.W. Hughes and R. Voss, *Ann. Rev. Nucl. Part. Sci.* **49**, 303 (1999)
3. SMC Collaboration, B. Adeva et al. *Phys. Lett.* **B369**, 93 (1996)
4. E.C. Ashenauer (HERMES) in SPIN 2000 Symposium Proc.
5. SLAC E143, E155 and W. Vogelsang SPIN 2000 Symposium Proc.
6. EMC Collaboration, J. Ashman et al. *Nucl. Phys.* **B328**, 1 (1989)
7. DESY-PROC 1999-03 "Polarized Protos at High Energies - Accelerator Challenges and Physics Opportunities", Ed. A. DeRoeck, D. Barber and G. Rädel
8. H. Abramowicz and A.C. Caldwell, *Rev. Mod. Phys.* **71**, 1275 (1999)
9. A. DeRoeck et al. *Eur. Phys. J*, **C6**, 121 (1999)
10. SPIN Collaboration and DESY Polarization Team. "Acceleration of Polarized Protons to 820 GeV at HERA", UM HE 96-20, Nov. 1996
11. A.N. Zelenski in SPIN 2000 Symposium Proc.
12. N. Saito in SPIN 2000 Symposium Proc.
13. The 2nd eRHIC Workshop, Yale, April 2000, Ed. A. Deshpande, V.W. Hughes and R. Venugopalan

Symposium Highlights*

Thomas Roser

*Brookhaven National Laboratory, Upton
N.Y. 11973, USA*

Abstract. Some of the highlights of the 14th International Symposium on Spin Physics are presented with emphasis on recent and planned progress in experimental tools and facilities.

INTRODUCTION

It is a great pleasure to summarize the 14th International Symposium on Spin Physics which was in fact the first symposium to combine the series of Symposia on High Energy Spin Physics and on Polarization Phenomena in Nuclear Physics. The Symposium was opened by R. Jaffe who reminded everybody that 'All elementary particles and their interactions have spin'. In fact, we have yet to discover a spin zero particle with the elusive Higgs particle maybe being the first. Although spin is ubiquitous within particle physics the effects on experimental observations are often subtle and need very sophisticated tools, and it has been a sign of a mature field to focus on spin effects. This has been the case for nuclear and medium energy physics for some time now and more recently also included electro-weak processes and nucleon structure. The latter, in particular, marked the beginning of precision measurements of quantities that ultimately will test our understanding of strong (non-perturbative) Quantum-Chromo-Dynamics.

For these highlights I will focus on recent progress in experimental tools and facilities available to the investigation of spin effects. This is mainly due to fact that there were several excellent talks that summarized the status of spin physics theory as well as to my own background in experimental and accelerator physics. Many new facilities and tools were in fact presented at this conference which lets us anticipate exciting progress over the next few years. In anticipation of what is to come I organized this talk in terms of news headlines that were generated during this symposium.

---

*) Work performed under the auspices of the U.S. Department of Energy

POLARIZED RADIOACTIVE BEAMS AT RIKEN

The RIKEN radioactive beam facility has produced proton and neutron rich radioactive nuclei that acquire polarization during the producing fragmentation process. K. Asahi reported on a beautiful experimental set-up to measure the few percent beam polarization by stopping the radioactive nuclei and detecting the asymmetry in the weak β-decay. Finally the magnetic moment of these short-lived nuclei was measured to high precision using NMR techniques. The present beam energy of 130 MeV/amu is planned to be upgraded to about 400 MeV/amu.

HIGH ENERGY POLARIZED PHOTONS AT SPRING8

A new high energy polarized photon facility started operation at SPRING8. The photons are produced by compton back scattering from the 100 mA, 8 GeV electron beam using a high power 3.5 eV laser. T. Nakano layed out the plans to use the 2.4 GeV polarized photons with an intensity of about $2.5 \times 10^6 \gamma/s$ to explore photo production of exotic hadrons above the kaon threshold and to test the DHG sum rule using a polarized proton target in the future.

RECORD INTENSITY POLARIZED ELECTRON BEAM AT JLAB
UNPOLARIZED OPERATION TO END SOON

Responding to the demand for more polarized electron beam at JLab the new polarized electron source has delivered record intensity and polarization. Using the now standard method of producing polarized electrons with a circularly polarized laser from a strained GaAs photo-cathode 75% polarization and 50 - 75 microamps of electron beam was reached during the last run. This is expected to increase to about 100 microamps soon. At this intensity all operations at JLab can be satisfied with polarized beam although, in fact, most experiments already require polarized beam.

To monitor the beam polarization on-line and non-destructively a new compton polarimeter was developed as reported by T. Pussieux in the parallel sessions. To achieve the required laser intensity a Fabry-Perot laser cavity was successfully commissioned. It will now be possible to measure the beam polarization with an accuracy of about 2% for a 40 microamps, 4 GeV electron beam.

Several experiments are using the intense polarized electron beam. H. Gao described measurements of the spin transfer coefficient of the reaction $^1H(e,e'p)$, which allows the determination of the ratio of the electric to magnetic form factor. The new data covers a Q^2 range up to 3.5 GeV^2 and is much more precise than previous data.

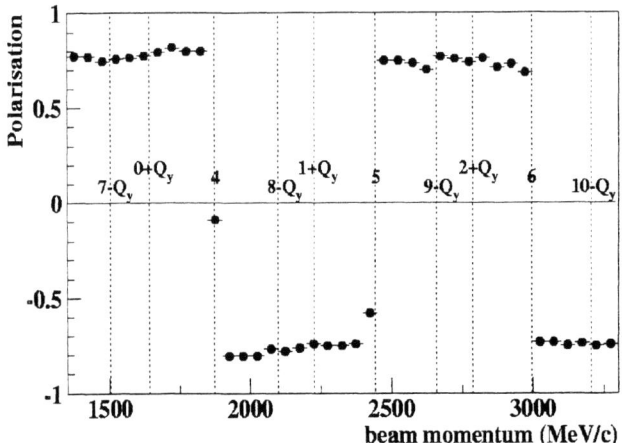

FIGURE 1. Beam polarization achieved at COSY using tune jumps to cross the spin resonances

P. Souder discussed the measurement of the weak nucleon form factor using parity violation in electron proton scattering. Through the coupling of the Z boson to the sea quarks this gives access to the strange radius and magnetic moment of the proton. Two experiments are presently under way: HAPPEX at JLAB and SAMPLE at Bates. The results indicate a small and positive strange magnetic moment and already rule out several of the proposed theoretical models.

HIGHLY POLARIZED BEAMS AT COSY

F. Hinterberger provided an update in the polarized proton acceleration at COSY. As shown in Fig. 1 using a tune jump system to overcome the intrinsic resonances a proton beam momentum of 3.5 GeV/c was reached with about 75% polarization. After the closure of SATURNE medium energy polarized proton beams are now again available. The high quality polarized proton beam at COSY offers the opportunity to continue, at higher energy, the highly successful program of high precision experiments conducted at IUCF.

PHYSICISTS ARE EAGERLY AWAITING NEW RESULTS FROM "G-2"

Precise measurement of the anomalous magnetic moment of the muon is a very sensitive probe of corrections to QED. Electro-weak corrections are calculable and the hadronic contribution can be determined from electron-positron scattering. Any remaining difference constitutes a very sensitive probe of physics beyond the standard model. The experiment at BNL is measuring the difference between the spin and momentum precession in a highly homogenous magnetic field. G. Bunce showed beautiful precession data from the 1999 run with much more data from the 2000 run available but not yet analyzed. The so far collected data should give an accuracy of about 0.5 ppm which is about 20 times more accurate than the result from the previous CERN experiment.

POLARIZED PROTON SOURCES REACH RECORD INTENSITIES

Polarized proton sources have recently made tremendous progress. In recent installations at IUCF and BNL intensities of about 1 mA have been achieved. A. Belov described the new pulsed Atomic Beam Source at IUCF that reached a peak intensity of 1.5 mA and about 80% polarization. During his overview of the status of polarized sources A. Zelenski showed results from the new Optically Pumped Polarized Ion Source developed at TRIUMF and installed at BNL. It has already met the design intensity of 1×10^{12} polarized protons in a $300\mu s$ pulse. After many years of development both source techniques have now reached a very high level of performance and for many applications such as the RHIC collider operation the polarized source is not the intensity limiting component anymore.

POLARIZED PROTONS ACCELERATED TO HIGHEST ENERGY AT RHIC
PLANS FOR UNPOLARIZED OPERATION CANCELLED

The Relativistic Heavy Ion Collider (RHIC) at BNL has started operation this year with data collected at all four initial detectors. Although this first operation period was dedicated to colliding gold beams, preparations to collide polarized proton beams in RHIC, sponsored by RIKEN, have started and a first commissioning period with polarized protons was completed. Fig. 2 shows a lay-out of the hardware needed for polarized proton acceleration at RHIC.

Of particular interest is the design of the Siberian snakes (two for each ring) and the spin rotators (four for each collider experiment) for RHIC. Each snake or spin rotator consists of four $2.4\,m$ long, $4\,T$ super-conducting helical dipole magnet modules each having a full 360 degree helical twist. Using helical magnets minimizes orbit excursions within the extend of the snake or spin rotator which is most important at injection energy. Nevertheless the bore of the helical magnets has to be 10 cm in diameter to accommodate the 3 cm orbit excursions.

Only a single Siberian snake in one of the two RHIC rings was available this year. This meant that vertically polarized protons needed to be injected without the snake powered, the snake was then turned on, which rotates the spin into the

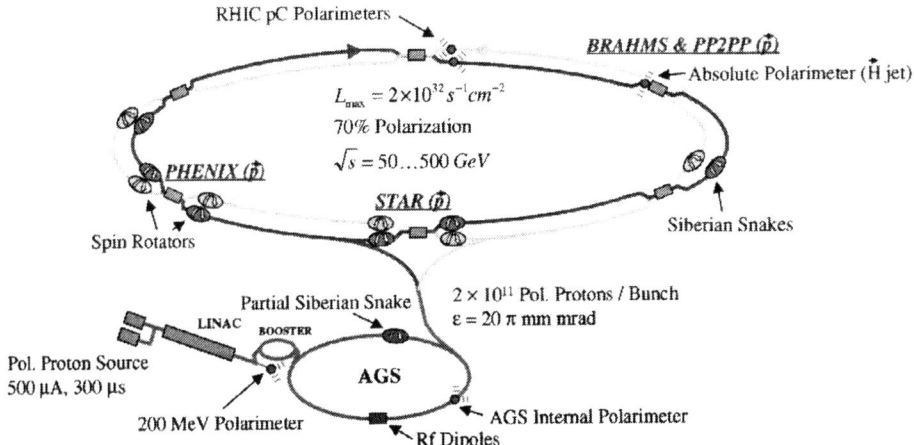

FIGURE 2. The Brookhaven hadron facility complex, which includes the AGS Booster, the AGS, and RHIC. The hardware for polarized proton operation in RHIC includes two snakes per ring, four spin rotators per detector for achieving helicity-spin experiments, a proton-carbon polarimeter per ring for beam polarization monitoring, and a polarized hydrogen gas jet for absolute polarization calibration.

horizontal plane, and then finally the beam was accelerated. Even under this more complicated scenario polarized protons were successfully accelerated to about 30 GeV, the highest energy accelerated polarized proton beam ever achieved. Fig. 3 shows the beam polarization measured in RHIC during this first commissioning run. More than 20 years after Y. Derbenev and A. Kodratenko made their proposal to use local spin rotators to stabilize polarized beams in high energy rings it has now been demonstrated that their concept is working flawlessly even in the presence of strong spin resonances at high energy. It also confirms the initial tests of the Siberian snake concept at low energy that were performed at IUCF.

In addition to maintaining polarization the fast, accurate, and reliable measurement of the beam polarization is of great importance. Very small angle elastic scattering in the Coulomb-Nuclear interference region offers the possibility for an analyzing reaction with a high figure-of-merit which is not expected to be strongly energy dependent. For polarized beam commissioning in RHIC an ultra thin carbon ribbon was used as an internal target and the recoil carbon were detected to measure both vertical and radial polarization component. H. Huang showed data from the operation of this high energy polarimeter that showed excellent particle identification. It was demonstrated that this polarimeter can be used to monitor polarization of high energy proton beams in a almost non-destructive manner.

For the upcoming polarized proton run it is planned to install all four Snakes

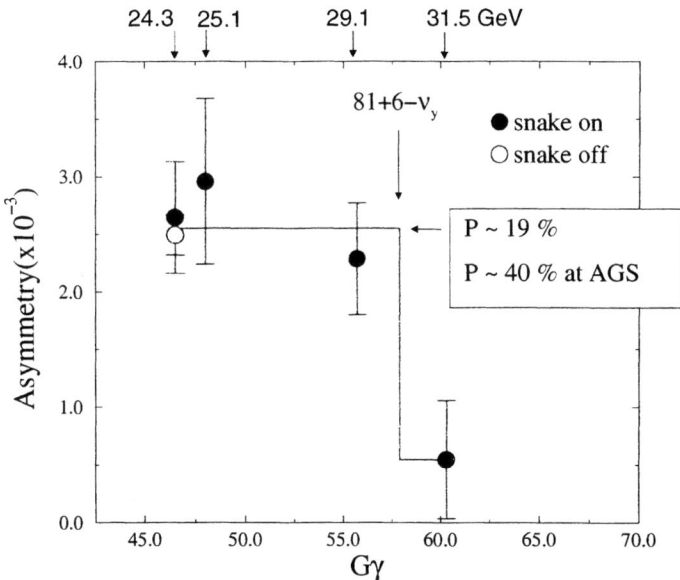

FIGURE 3. Measured polarization of the first polarized proton beam in RHIC at injection without Siberian snake (open circle) and accelerated with snake (filled circles). The graph shows the asymmetry measured with the RHIC proton-carbon polarimeter. The polarization values given were obtained from calibration measurements performed previously in the AGS. Significant polarization was lost at injection when the snake was off and at the first very strong spin resonance. Spin tracking indicates that this latter loss is due to an improper betatron tune setting and the large uncorrected closed orbit distortions.

as well as the proton-carbon polarimeter in the second ring. This should allow for acceleration to the full RHIC energy of 250 GeV. Collisions are planned at 100 GeV on 100 GeV at which energy one of the two snakes per ring can be turned off to produce longitudinal polarization at the detetors. The luminosity is expected to be about $5 \times 10^{30}\ cm^{-2}\ s^{-1}$.

RECORD HERMES DATA RUN
6 MILLION DIS EVENTS IN 2000

The HERMES experiment uses a polarized internal gas target in the HERA electron ring to collect deep inelastic scattering (DIS) data. E. Aschenauer reported that this past year HERMES was able to break all records with increased target thickness and excellent HERA machine performance. Beam and target polarizations were typically about 55 % and 85 %, respectively.

Compared to the CERN SMC experiment, the lower beam energy of about 28 GeV limits the range in x to about 0.01 and higher for the determination of the spin structure function. However, in this range the data is of very high quality and is in very good agreement with previous data from CERN and SLAC measurements. The recently collected deuteron data is expected to give very precise information on the g_1 spin structure function of the neutron.

The very high data rate and recent detector upgrades also allowed HERMES to expand the scope of DIS measurements. Measuring semi-inclusive production of hadrons, presumably coming from the struck parton, in principle allows the flavor decomposition of g_1. Also, pion production showed an azimuthal asymmetry around the direction of the struck parton, indicating that the fragmentation function is spin dependent and can be used to analyze the parton polarization.

NEXT PHASE OF PROTON SPIN STRUCTURE STUDIES TO BEGIN SOON

F. Bradamante discussed the plans for the new CERN experiment COMPASS. It is a major upgrade of the previous SMC experiment to include the detection of the hadrons produced in semi-inclusive polarized DIS. As described above for HERMES, semi-inclusive DIS in principle allows for the measurement of flavor separated quark spin structure functions. Open charm production from photon-gluon fusion will give access to gluon polarization. Also, using the spin-dependence of the fragmentation functions the quark transversity function h_1 should be accessible. The higher energy of the polarized CERN muon beam of about 200 GeV will make all of these measurements easier to analyze and also allows the measurement to be extended to lower x values. The experiment is scheduled to start in 2001.

Around the same time the first polarized proton collision at RHIC are scheduled to occur. N. Saito described the plans of the spin measurements of the two main RHIC detector collaborations STAR and PHENIX. At RHIC center-of-mass collision energies of 200 to 500 GeV the hard scattering events can be factorized similarly to DIS into a product of the spin structure functions of the two partons in the initial state, and the analyzing power of the hard interaction. If a leading hadron instead of a direct photon or jet is observed an additional fragmentation function for the production of this hadron has to be included.

Direct photon production from gluonic compton scattering in pp collisions will give direct access to the gluon polarization. This is the "gold plated" mode for both PHENIX and STAR and will give high precision data on the gluon polarization distribution from 0.01 to 0.2 for x.

The high energy will also allow the observation of W production from the interaction of a valence quark with a sea anti-quark. Since W production is a weak process it has a very large parity violating asymmetry which can be observed at RHIC for the first time. But it will also provide the direct measurement of the sea anti-quark polarization distribution of the proton.

It is planned to start with collisions at a center-of-mass energy of 200 GeV for the first two years and continue in 2003 with collisions at 500 GeV. Just like the semi-inclusive DIS measurements it is also possible at RHIC to use spin-dependent fragmentation functions to access transversity distributions. In this case the colliding proton beams will be vertically polarized.

CONCLUSIONS

There has been stunning progress in the performance of polarized sources and the acceleration of polarized beams and several long standing major issues are being addressed by ongoing experiments or experiments that are scheduled to start next year. These are truly exciting times in spin physics.

BANQUET SPEECH

Some Milestones in Polarized Proton Beams

Ernest D. Courant

Brookhaven National Laboratory

My connection with polarized protons began in 1962, when **Vernon Hughes** suggested I look into the possibility of getting polarized beams into high-energy accelerators. So I must thank Vernon for getting me started in this field, which I have found stimulating and fascinating ever since. Tonight I want to talk about the development of polarized proton beams.

I learned about the BMT equation and, via **Froissart** and **Stora**, how it leads to depolarizing resonances because the anomalous magnetic moment transforms differently from the regular moment. I found that depolarization could best be avoided and polarized beams could best be accelerated in a machine with many identical periods, such as the Princeton PPA – but that machine was soon euthanized, before anyone could seriously look into its potential as a polarized accelerator.

Another good candidate was the ZGS at Argonne. We all know how **Alan Krisch, Larry Ratner** et al made the ZGS into the pioneering polarized proton accelerator – and found that, contrary to the prevailing conventional wisdom at the time, polarization asymmetries did not disappear at high energies but in fact became more pronounced up to the limiting energy of the ZGS.

The fact that the ZGS came to be a unique facility for this work was not enough to keep it from being shut down at the height of its powers – though its life was extended for a couple of months for the final polarized proton run.

At a workshop in Ann Arbor in 1977, organized by **Alan Krisch**, we speculated on whether the ZGS success in getting polarized protons to 12 GeV could be extended to higher energies – the Brookhaven AGS, CERN PS and the next generation of accelerators and colliders in the works – up to the SPS, Fermilab, or the ISABELLE collider then contemplated at Brookhaven. Conclusion: At the AGS, the "brute force" methods used at the ZGS might just work: Pass through intrinsic resonances (of which there are relatively few because of the 12-fold symmetry of the AGS) rapidly with fast pulsed quadrupoles; and correct orbit error harmonics at each integer resonance with an elaborate correction system. But it was clear that going to even higher energies would be prohibitively difficult – the resonances would just be too dense, and get stronger at higher energies.

A new idea, which we learned about at this workshop, promised to come to the rescue. We had some preprints from **Derbenev, Kondratenko** and **Skrinsky** in Novosibirsk, which proposed a scheme where the beam would be successively deflected by three to eight magnets, alternately vertical and horizontal, to produce a wiggly orbit, and this could be arranged so that the spin precession, instead of resonating with the orbit at many energies, would maintain a constant frequency, so that the troublesome resonances would just disappear! We named this scheme the "Siberian Snake".

This idea just cried out to be tried in practice. But unfortunately it was not practical for the AGS or any accelerator existing at the time, because (a) the snake, to produce the necessary 180° rotation, has to be at least about 10 meters long, much longer than the AGS straight sections; and (b) the excursions of the snaky orbit would be large at moderate energies – around 10 cm at 15 GeV, but decreasing in inverse proportion to the energy. But it was seen at once that the idea was just right for the very high energy machines then contemplated, such as ISABELLE. Almost immediately a lot of work was done on the concept – many varieties of snakes were devised, especially by **Klaus Steffen** at DESY. It was seen that in large machines two or more Siberian snakes would be preferable to just one. Plans were made to incorporate a pair of snakes in ISABELLE. Later, when the Superconducting Supercollider (SSC) was designed, provisions were made to include 26 "snake pits" in it to enable it to handle polarized protons.

But, as we all know, first ISABELLE and then the SSC were aborted. Therefore the wiggly Siberian snake had to wait until this year at RHIC (the reincarnation of ISABELLE) – twenty-three years later!

An alternate way to get the resonance suppression of the snake is to use a solenoid that rotates the spin by 180°; at low energy this can be done with just a few Tesla-meters. **Alan Krisch** (again) noted that the IUCF cooler ring in Indiana, in the 100-200 MeV range, has long straight sections, and a superconducting solenoid happened to be available. He persuaded IUCF to let him install this solenoid in the cooler ring as a snake-like spin rotator (one should not really call it a snake, since it doesn't wiggle the orbit, but it is generally called by that name anyway). Over the past ten years or so he and his group have conducted a whole series of studies – still ongoing – showing that such a snake does indeed permit passage through energies where there would be depolarizing resonances without it.

Lots of other things have happened since that seminal 1977 workshop. CERN decided against trying polarization at the PS and its successors; their management felt it wouldn't be worth the trouble. But at Brookhaven – largely thanks to Alan's initiative, and with leadership by **Larry Ratner, Kent Terwilliger** and others, – polarized protons were injected and accelerated to 22 GeV in 1980, with the help of 10 pulsed quadrupoles and 94 individually tuned small orbit correctors. Needless to say, setting these up for a polarization run was excruciatingly laborious, but it did work.

Larry Ratner, Thomas Roser, Mei Bai and others have simplified this operation. A modest solenoid can rotate the spin by 5% of the full 180° of a real snake. This enhances all the integral resonances to make them so strong that the spin reverses completely on passage through the resonant energy, so polarization is preserved. An intrinsic resonance can be handled by using an rf dipole to excite coherent betatron oscillations; this again makes the resonance strong enough to produce complete spin flip. This has to be carefully tailored so as to preserve the emittance, and turn off the oscillation once the resonance is passed – we heard about further developments of this at this meeting.

Finally the new RHIC heavy-ion collider has been designed to handle polarized protons (heavy ions of mass 1, charge 1 and spin _). Each of the two rings contains two Siberian snakes. These snakes are made with helical superconducting dipoles (as pointed out by **Yuri Shatunov** and **Vadim Ptitsyn**, helices are a bit more efficient and have smaller orbit excursions than discrete deflecting magnets). I should mention that the Siberian snake project at Brookhaven has been largely supported and funded by RIKEN in Japan. So far one snake has been installed in one of the rings, and last month polarized protons from the AGS were injected at 25 GeV and accelerated to about 30 GeV without losing polarization – the first time snakes have demonstrated their worth for high energy polarization (and, incidentally, the current world's record for energy of accelerated polarized protons). Next year we expect to have snakes in both RHIC rings and to begin work with polarized colliding beams – a goal we have been waiting for for a long time.

And maybe polarized ^{3}He and deuteron beams can be tried too, giving a handle on the spin behavior of the neutron as well.

PARALLEL SESSIONS

1. SYMMETRIES AND SPIN I

Constraints of a Parity-Conserving/ Time-Reversal-Non-conserving Interaction[1]

Willem T.H. van Oers

Department of Physics, University of Manitoba, Winnipeg, MB, Canada R3T 2N2
and
TRIUMF, 4004 Wesbrook Mall, Vancouver, BC Canada V6T 2A3

Abstract. Time-Reversal-Invariance non-conservation has for the first time been unequivocally demonstrated in a direct measurement, one of the results of the CPLEAR experiment. What is the situation then with regard to time-reversal-invariance non-conservation in systems other than the neutral kaon system? Two classes of tests of time-reversal-invariance need to be distinguished: the first one deals with parity violating (P-odd)/time-reversal-invariance non-conserving (T-odd) interactions, while the second one deals with P-even/T-odd interactions (assuming CPT conservation this implies C-conjugation non-conservation). Limits on a P-odd/T-odd interaction follow from measurements of the electric dipole moment of the neutron. This in turn provides a limit on a P-odd/T-odd pion-nucleon coupling constant which is less than 10^{-4} times the weak interaction strength. Limits on a P-even/T-odd interaction are much less stringent. The better constraint stems also from the measurement of the electric dipole moment of the neutron. Of all the other tests, measurements of charge-symmetry breaking in neutron-proton elastic scattering provide the next better constraint. The latter experiments were performed at TRIUMF (at 477 and 347 MeV) and at IUCF (at 183 MeV). Weak decay experiments (e.g., the transverse polarization of the muon in $K^+ \to \pi^0 \mu^+ \nu_\mu$) have the potential to provide comparable or possibly better constraints.

INTRODUCTION

Time-reversal-invariance non-conservation has for the first time been unequivocally demonstrated in a direct measurement in the CPLEAR experiment [1]. The experiment measured the difference in the transition probabilities $P(\overline{K}^0 \to K^0)$ and $P(K^0 \to \overline{K}^0)$. Assuming CPT conservation but allowing for a possible breaking of the $\Delta S = \Delta Q$ rule, the result obtained for A_T

$$A_T = \frac{R(\overline{K}^0 \to K^0) - R(K^0 \to \overline{K}^0)}{R(\overline{K}^0 \to K^0) + R(K^0 \to \overline{K}^0)} = [6.6 \pm 1.3(\text{stat.}) \pm 1.0(\text{syst.})] \times 10^{-3} \quad (1)$$

---

[1] Work supported in part by the Natural Sciences and Engineering Research Council of Canada.

is in good agreement with the measure of CP violation in neutral kaon decay. A more recent reported result on CP violation is a large asymmetry in the distribution of $K_L \to \pi^+\pi^- e^+ e^-$ events in the CP-odd/T-odd angle ϕ between the decay planes of the $\pi^+\pi^-$ and e^+e^- pairs in the K_L centre of mass system. The overall asymmetry found was $[13.6 \pm 2.5(\text{stat.}) \pm 1.2(\text{syst.})]\%$ [2]. The question then arises: what is the situation with regard to time-reversal-invariance in systems other than the kaon system?

Tests of time-reversal-invariance can be distinguished as belonging to two classes: the first one deals with P-odd/T-odd interactions, while the second one deals with P-even/T-odd interactions (assuming CPT conservation this implies C-conjugation non-conservation). But it should be noted that constraints on these two classes of interactions are not independent since the effects due to P-odd/T-odd interactions may also be produced by P-even/T-odd interactions in conjunction with Standard Model parity violating radiative corrections. The latter can occur at the 10^{-7} level and consequently could present a limit on the constraint of a P-even/T-odd interaction, derived from experiment. Limits on a P-odd/T-odd interaction follow from measurements of the electric dipole moment (edm) of the neutron (which currently stands at $< 6 \times 10^{-26}$ e.cm [95% C.L.]). This provides a limit on a P-odd/T-odd pion-nucleon coupling constant which is less than 10^{-4} times the weak interaction strength. Measurements of ^{129}Xe and ^{199}Hg edm's ($< 8 \times 10^{-28}$ e.cm [95% C.L.]) give similar constraints. [see Ref. 3]

Experimental limits on a P-even/T-odd interaction are much less stringent. Following the standard approach of describing the nucleon-nucleon interaction in terms of meson exchanges, it can be shown that only charged rho-meson exchange and A_1-meson exchange can lead to a P-even/T-odd interaction [4]. The better constraints stem from measurements of the edm of the neutron and next from measurements of charge symmetry breaking in neutron-proton (n-p) elastic scattering. All other experiments, like gamma decay experiments [5], detailed balance experiments [6], polarization - analyzing power difference measurements [7], and five-fold correlation experiments with polarized incident nucleons and aligned nuclear targets [8], have been shown to be at least an order of magnitude less sensitive. Haxton, Hoering, and Musolf [3] have deduced constraints on a P-even/T-odd interaction from nucleon, nuclear, and atomic edm's with the better constraint coming from the edm of the neutron. In terms of a ratio to the strong rho-meson nucleon coupling constant, they deduced for the P-even/T-odd rho-meson nucleon coupling: $|\overline{g_\rho}| < 0.53 \times 10^{-3} \times |f_\pi^{\text{DDH}}/f_\pi^{\text{meas.}}|$. But the ratio of the theoretical to the experimental value of f_π may be as large as 15! [9]. It has been pointed out that constraints derived from one-loop contributions to the edm of the neutron exceed the two-loop limits by more than an order of magnitude and are much more stringent [10]. However, measurements of edm's yield constraints only under the assumption that parity conservation is restored at the mass scale associated with physics beyond the Standard Model. If parity violation persists at short distances, it are the direct searches which will provide the least ambiguous bounds. Consequently, it can be argued that searches for permanent edm's and direct P-even/T-odd effects provide complementary information [11].

As corollary it is very difficult to accommodate a P-even/T-odd interaction in the Standard Model. It requires C-conjugation non-conservation, which cannot be introduced at the first generation quark level. It can neither be introduced into the gluon self-interaction. Consequently, one needs to consider C-conjugation non-conservation between quarks of different generations and/or between interacting fields [12].

CHARGE SYMMETRY BREAKING IN NEUTRON PROTON ELASTIC SCATTERING

Charge symmetry breaking (CSB) in n-p elastic scattering manifests itself as a non-zero difference of the neutron (A_n) and proton (A_p) analyzing powers, $\Delta A \equiv A_n - A_p = 2 \times [\text{Re}(b^*f) + \text{Im}(c^*h)]/\sigma_0$. Here the complex amplitude f is charge symmetry breaking, while the complex amplitude h is both charge symmetry breaking and time-reversal-invariance non-conserving. The complex amplitudes b and c belong to the usual five n-p scattering amplitudes and σ_0 is the unpolarized differential cross section. The three precision experiments performed (at TRIUMF at 477 MeV [13] and at 347 MeV [14], and at IUCF at 183 MeV [15]) have unambiguously shown that charge symmetry is broken and that the results for ΔA at the zero-crossing angle of the average analyzing power are very well reproduced by meson exchange model calculations (see Fig. 1). A P-even/T-odd interaction produces a term in the scattering amplitude which is simultaneously charge symmetry breaking (the complex amplitude h in the expression above). Thus, Simonius [16] deduced an upper limit on a P-even/T-odd CSB interaction from a comparison of the experimental results with the theoretical predictions for the three n-p CSB experiments. The upper limit so derived is $|\bar{g}_\rho| < 6.7 \times 10^{-3}$ [95% C.L.]. This is therefore comparable to the upper limit deduced from the electric dipole moment of the neutron, taking present experimental limits of f_π, and is considerably lower than the limits inferred from direct tests of a P-even/T-odd interaction. For instance the detailed balance experiments give a limit of $|\bar{g}_\rho| < 2.5 \times 10^{-1}$ [see Ref. 8], while measurements of the five-fold correlation parameter $A_{y,xz}$ in polarized neutron transmission through nuclear spin-aligned ^{165}Ho give a limit of $|\bar{g}_\rho| < 5.9 \times 10^{-2}$ even though the measured value of $A_{y,xz}$ was $(8.6 \pm 7.7) \times 10^{-6}$ [8]. It is effectively only the valence proton in ^{165}Ho which contributes to $A_{y,xz}$. Even though it is inconceivable in the Standard Model to account for a P-even/T-odd interaction, as remarked above, there is at least a need to clarify the experimental situation by providing a better experimental result.

Such an experimental constraint may be provided by an improved upper limit on the electric dipole moment of the neutron. In fact a new measurement with a sensitivity of 4×10^{-28} e.cm has been proposed at the Los Alamos Neutron Science Center [17]. Performing an improved n-p elastic scattering CSB experiment also appears to be a very attractive possibility. One can calculate with a great deal of confidence the contributions to CSB due to one-photon exchange and due to the n-

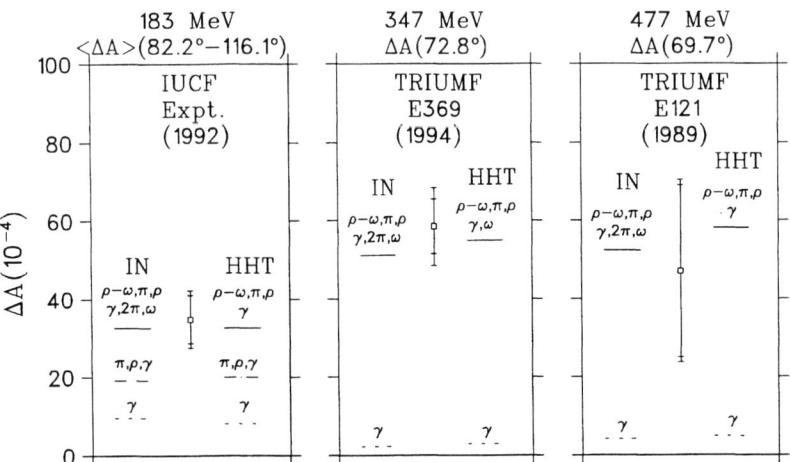

FIGURE 1. Experimental results of ΔA at the zero-crossing angle at incident neutron energies of 183, 347, and 477 MeV compared with theoretical predictions of Iqbal and Niskanen, and Holzenkamp, Holinde, and Thomas. The inner error bars present the statistical uncertainties; the outer error bars have the systematic uncertainties included (added in quadrature). For further details see Ref. 14.

p mass difference affecting charge one-pion and rho-meson exchange. Furthermore, one can select an energy where the $\rho^0 - \omega$ meson mixing contribution changes sign at the same angle where the average of the analyzing powers A_n and A_p changes sign and therefore does not contribute to ΔA. This occurs at a neutron energy of 320 MeV and is caused by the particular interplay of the n-p phase shifts and the form of the spin/isospin operator connected with the $\rho^0 - \omega$ mixing term. But also the one-photon exchange term changes sign at about the same angle at 320 MeV. The contribution due to two-pion exchange with an intermediate Δ is expected to be less than one tenth of the overall CSB effect, essentially determining an upper limit on the theoretical uncertainty (see Fig. 2) [18]. It has been shown that simultaneous γ-π exchanges can only contribute to ΔA through second order processes and can therefore be neglected [19]. Also the effects of inelasticity are negligibly small at 320 MeV. It appears therefore well within reach to reduce the theoretical uncertainty in the comparison of experiment and theory. Subtracting the calculated ΔA from the measured ΔA permits establishing an upper limit on a P-even/T-odd/CSB interaction.

In the TRIUMF CSB experiments polarized neutrons were scattered from unpolarized protons and vice versa. The polarized (or unpolarized) neutron beam was obtained using the (p,n) reaction with a 369 (and 497) MeV polarized (or unpolarized) proton beam incident on a 0.20 m long LD_2 target. At these energies one makes use of the large sideways-to-sideways polarization transfer coefficient r_t at 9° in the lab. The only difference in obtaining the polarized and unpolarized 347 MeV neutron beams was turning off the pumping laser light in the optically pumped po-

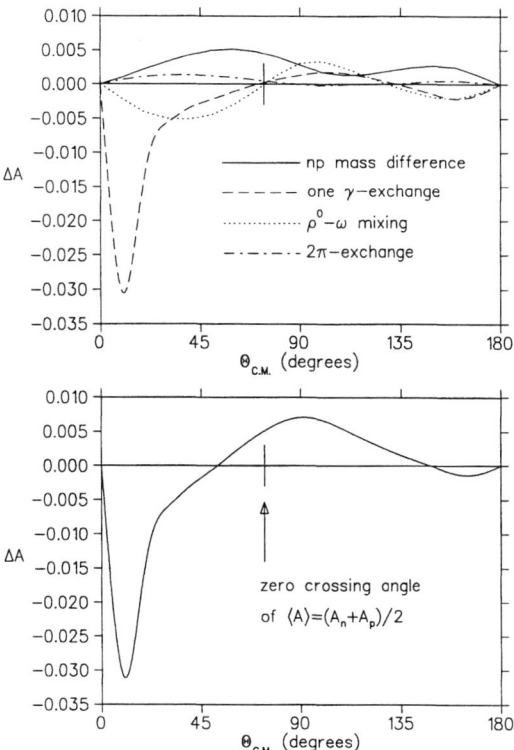

FIGURE 2. Angular distributions of the different contributions to ΔA at an incident neutron energy of 320 MeV. (Ref. 18) Note that the ρ^0 - ω mixing contribution passes through zero at the same angle as the average of A_n and A_p (vertical bar). The lower part of the figure gives the total ΔA angular distribution.

larized ion source (OPPIS). The polarized proton target was of the frozen spin type with butanol beads as target material. The same target after depolarization was used as the unpolarized proton target. Great care was taken that the two interleaved phases of the experiments were performed with identical beam and target parameters except for the polarization states. At 347 MeV scattered neutrons and recoiling protons were detected in coincidence in the c.m. angular range 53.4 to 86.9 degrees in two left-right symmetric detector systems. Rather than measuring A_n and A_p directly (which would limit the accuracy attainable by not having polarization calibration standards of the required precision), the zero-crossings of A_n and A_p were determined by fitting the partial angular distributions with polynomials, deduced from $n-p$ phase shift analyses. The difference ΔA followed then by multiplying the difference in the zero-crossing angles by the average slope of the analyzing powers (the experiment measured the slope of A_p at the zero-crossing angle, which is a good approximation for the average slope at the zero-crossing angle and introduces a negligible error). The execution of the experiments depended on a great deal of simultaneous monitoring and online control measurements. Both the statistical and systematic errors, obtained in the 347 MeV experiment, can be considerably improved upon (by a factor three to four). With the OPPIS developments which have taken place in the intervening years and with a biased Na-ionizer cell it will be possible to obtain up to 50 μA of 342 MeV 80% polarized proton beam incident on the neutron production target (a factor of 50 increase in neutron beam intensity at 320 MeV over the previous 347 MeV CSB experiment). In addition various systematic error reducing improvements can be introduced in the experimental arrangements and procedures. Such an experiment would constitute a measurement of CSB in n-p elastic scattering of unprecedented precision of great value on its own and would simultaneously provide a greatly improved upper limit on a P-even/T-odd interaction.

PARTICLE DECAYS

Searches for P-even/T-odd interactions are also made in particle decays, e.g., in the decay $K^+ \to \mu^+ \pi^0 \nu_\mu$. In this reaction, a non-zero value of the muon polarization transverse to the decay plane would be an indication of time-reversal-invariance non-conservation. Several experiments have been performed to measure the transverse muon polarization in both neutral and charged kaon decay. There is a unique feature to the transverse muon polarization in that it does not have contributions from the Standard Model at tree level and that higher order effects are of order 10^{-6}. When only one charged particle is present in the final state, a final state interaction, which can mimic a time-reversal-invariance breaking effect, is greatly reduced and is estimated to occur only at the same level of 10^{-6}. The more recent effort of measuring the time-reversal-invariance non-conserving transverse muon polarization is at KEK using a stopped K^+ beam. The experiment reported a result for P_T = -0.0042 $\pm$ 0.0049(stat.) $\pm$ 0.0009(syst.), based on the data taken in 1996 and 1997, which translates into a value of Imξ = -0.013 $\pm$ 0.016(stat.) $\pm$ 0.003(syst.) [20]. The quantity ξ is defined as the ratio of two form factors, $f_+(q^2)$ and $f_-(q^2)$, in the $K_{\mu 3}$ decay. Imξ must be equal to zero for time-reversal-invariance to hold. With the data already in hand and with the approved data taking time, it is anticipated to arrive at a statistical error of $\pm$0.007 in Imξ. The best previous experimental limits were obtained with both neutral and charged kaons at the BNL-AGS [21]. A combination of both experimental results provided a limit on the imaginary part of the hadron form factor, Imξ = -0.010 $\pm$ 0.019. A new search for the time-reversal-invariance non-conserving transverse muon polarization with in-flight decays of $K^+ \to \mu^+ \pi^0 \nu_\mu$ was proposed at the BNL-AGS [22]. It was intended to obtain a sensitivity to the transverse muon polarization of $\pm$0.00013, corresponding to a sensitivity to Imξ of $\pm$0.0007. Similar searches for the time-reversal-invariance non-conserving transverse τ polarization in B semileptonic decays, $B \to M \tau \nu_\tau$ are under consideration. Significant transverse τ lepton polarizations have been predicted. Clearly, a non-zero value of the transverse muon polarization in $K_{\mu 3}$ decay, or of the τ lepton in semi-leptonic B decays would constitute evidence for new physics.

SUMMARY

The searches made so far for a P-even/T-odd interaction have resulted in only very modest constraints on such an interaction. Most promising are the continued efforts to measure the electric dipole moment of the neutron on the one hand and charge symmetry breaking in neutron-proton elastic scattering at around 320 MeV on the other hand. But also measurements of transverse lepton polarizations in K, and B decays have the potential to set better experimental limits on a P-even/T-odd interaction.

REFERENCES

1. Angelopoulos, A. et al., (CPLEAR Collaboration), Phys. Lett. B444, 43 (1998).
2. Alavi-Harati, A. et al., (KTeV Collaboration), Phys. Rev. Lett. 84, 408 (2000).
3. Haxton, W.C., Hoering, A. and Musolf, M.J., Phys. Rev. D50, 3422 (1994).
4. Simonius, M., Phys. Lett. 58B, 147 (1975).
5. Boehm, F. in Symmetries and Fundamental Interactions in Nuclei, ed. Haxton, W.C. and Henley, E.M. (World Scientific, Singapore, 1997), p. 67.
6. Blanke, E. et al., Phys. Rev. Lett. 51, 355 (1983).
7. Conzett, H.E. in Polarization Phenomena in Nuclear Physics - 1980, ed. Ohlsen, G.G., Brown, R.E., Jarmie, N., McNaughton, W.W., and Hale, G.M. AIP Conference Proceedings No 69, p. 1452.
8. Huffman, P.R. et al., Phys. Rev. C55, 2684 (1997).
9. Page, S.A. et al., Phys. Rev. C35, 1119 (1987); Bini, M. et al., Phys. Rev. C38, 1195 (1988).
10. Ramsey-Musolf, M.J., Phys. Rev. Lett. 83, 3997 (1999).
11. Ramsey-Musolf, M.J., Proceedings of the Workshop on Fundamental Physics with Pulsed Neutron Beams, in press.
12. Simonius, M. in Intersections between Particle and Nuclear Physics - 1984, ed. Mischke, R.E. AIP Conference Proceedings No 123, p. 1115.
13. Abegg, R. et al., Phys. Rev. D39, 2464 (1989).
14. Zhao, J. et al., Phys. Rev. C57, 2126 (1998).
15. Vigdor, S.E. et al., Phys. Rev. C46, 410 (1992).
16. Simonius, M. Phys. Rev. Lett. 78, 4161 (1997).
17. LANSCE Proposal, spokespersons Cooper, M.J. and Lamoreaux, S.K.
18. Iqbal, J. private communication (1999).
19. Friar, J.L. and Coon, S.A. Phys. Rev. C53, 588 (1996).
20. Abe, M. et al., Phys. Rev. Lett. 83, 4253 (1999).
21. Morse, W.M. et al., Phys. Rev. D21, 1750 (1980); Blatt, S.R. et al., Phys. Rev. D27, 1056 (1983).
22. BNL-AGS proposal, spokespersons Divan, M.V., Ma, Hong and Adair, R.

Neutrinos by Double Beta Decays from ^{100}Mo and Nuclear Spin-Isospin Responses

N. Kudomi[a], H. Ejiri[a], K. Fushimi[c], K. Hayashi[a],
T. Kishimoto[c], M. Komori[c], K. Kume[a], H. Kuramoto[a],
K. Matsuoka[c], H. Ohsumi[d], K. Takahisa[a], S. Umehara[b],
and S. Yoshida[a]

a) RCNP, Osaka Univ., Ibaraki, Osaka 567, Japan
b) Facul. of Integ. Arts and Sci., The Univ. of Tokushima, Tokushima, 770, Japan
c) Dept. of Phys., Osaka Univ., Toyonaka, Osaka 560, Japan

Abstract. Spectroscopic studies of neutrino-less double beta decays($0\nu\beta\beta$) of ^{100}Mo were made by means of ELEGANT V. The data at Oto lab., being combined with the data at Kamioka, gives stringent limits on the half-life of $T_{1/2}$ for the $0\nu\beta\beta$ and the effective Majorana neutrino mass of $\langle m_\nu \rangle <$ 2.1 eV(90%C.L.). Spin-isospin responses for neutrinos associated with $\beta\beta$ of ^{100}Mo are discussed. A perspective of double beta decay of ^{100}Mo and a possible proposal of MOON are discussed.

I INTRODUCTION

Double beta decays are of current interest from both astroparticle and nuclear physics view points [1-3]. The neutrino-less double beta decays($0\nu\beta\beta$) provides one with very sensitive tests for the Majorana neutrino mass, the right handed weak current and so on. Two neutrino double beta decay ($2\nu\beta\beta$) gives directly the nuclear matrix element [5]. This is used to verify nuclear structure calculations and to select appropriate spin-isospin interactions $H_{\tau\sigma}$ to be used for evaluating the nuclear matrix element $M^{0\nu}$ for $0\nu\beta\beta$. The $0\nu\beta\beta$ and $2\nu\beta\beta$ transition rates are given $[T^{0\nu}_{1/2}(\nu)]^{-1} = G^{0\nu}|M^{0\nu}|^2(\langle m_\nu \rangle + C_{RHC}\langle RHC \rangle)^2$ and $[T^{2\nu}_{1/2}]^{-1} = G^{2\nu}|M^{2\nu}|^2$, where $G^{0\nu}$ and $G^{2\nu}$ are the phase space factor and $M^{0\nu}$ and $M^{2\nu}$ are the nuclear matrix element for the mass term. C_{ij}, with i, j being m, λ and η, are the nuclear responses in units of $M^{0\nu}$. Since $\beta\beta$ transition rates depend largely on the nuclear matrix elements, it is important to study $\beta\beta$ decays on several nuclei to extract universal values of (limits on) physics quantities beyond the standard theory.

The nuclear matrix element $M^{2\nu}$ for $2\nu\beta\beta$ is derived experimentally from the observed $2\nu\beta\beta$ rate. The transition rate for $0\nu\beta\beta$ associated with the neutrino exchange is written in terms of the neutrino mass term $\langle m_\nu \rangle$ and the RHC terms of $\langle \lambda \rangle$ and $\langle \eta \rangle$ [1-3]. These $0\nu\beta\beta$ transition rates include both the nuclear matrix elements (form factors) relevant to the nuclear structures and the physics quantities relevant to the particle physics.

II DOUBLE BETA DECAYS BY ELEGANT V

So far, $0\nu\beta\beta$ decay rates have been studied on several nuclei, unique features of the present work of ^{100}Mo with ELEGANT V [7] are as follows. (i) Energy and angular correlations of two β rays and γ rays are measured. Thus limits on the $0\nu\beta\beta$ processes for individual terms can be obtained. (ii) ^{100}Mo has large phase space factors for $\beta\beta$ because of the large Q values. Furthermore, the $2\nu\beta\beta$ matrix element is known to be large from the previous work [6]. (iii) Origins of the background events are well investigated by the $\beta(e)$ and γ correlation studies. Thus corrections for them are possible.

The obtained limits on $\langle m_\nu \rangle$, $\langle \lambda \rangle$, $\langle \eta \rangle$, $\langle g_B \rangle$ and others, depending somewhat on the matrix elements used, are most stringent for ^{100}Mo. They are same orders of magnitudes as the values derived from other nuclei. Thus the present data with the unique points as given above, together with other data, may give stringent limits on the relevant values beyond the standard theory.

ELEGANT V consists of three drift chambers for measuring two β trajectories, plastic scintillator arrays(PL) for β ray energies and arrival times, and NaI scintillator(NaI) arrays for γ and X rays [7]. They are schematically shown in ref. [7,9]. The total weight of ^{100}Mo(94.5% enrichment) is 171gr. PL's and NaI's are calibrated by γ-rays from ^{22}Na [8]. The detection efficiencies of DC's are checked by the β-ray from ^{90}Sr passing through DC's.

III ANALYSES AND RESULTS

The $0\nu\beta\beta$ and $2\nu\beta\beta$ of ^{100}Mo are measured at Oto Cosmo Observatory with 1400m.w.e.. The sum energy spectrum is obtained by selecting events with several conditions for live time of 7582hrs(Fig. 1). The major part of the spectrum in 2.0~3.0MeV is the $2\nu\beta\beta$ component.

Background contributions from the natural radioactive contaminations were estimated as follows. The ^{214}Bi and ^{208}Tl are two major isotopes, which may give background events. These isotopes are decay products of ^{238}U- and ^{232}Th-chain isotopes contained in the ^{100}Mo source film and the detector components. ^{214}Bi is produced also from Rn contained in the air around the source and the detector element. The total amounts were evaluated from the single β event rates in coincidence with γ-rays characteristics of the decays. The obtained contents are shown in Table 1. On the basis of the estimated ^{214}Bi and ^{208}Tl contents, Monte Carlo calculations were carried out to evaluate the BG(fake) rates caused by these isotopes, which may survive after the $\beta\beta$ selections. The major BG $\beta\beta$ events come from the

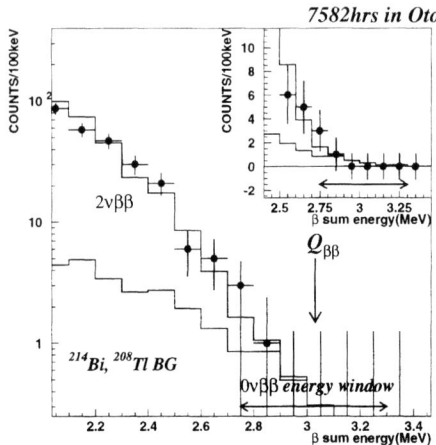

FIGURE 1. The sum energy spectrum($E_1 + E_2$) for the two electron events. Here, the obtained spectrum in 2.0~3.0MeV was reproduced with $\chi^2 = 1.4$ by the sum of evaluated ^{208}Tl and ^{214}Bi energy spectra and the $2\nu\beta\beta$ spectrum.

β's and conversion-electrons, and the single β's or the Compton-electrons passing through the source film. The observed spectrum is reproduced by the sum of the estimated BG spectra and the $2\nu\beta\beta$ spectrum with the half-life of 1.15×10^{19}y [6].

There are no excess of $0\nu\beta\beta$ counts beyond the statistical fluctuations of the $2\nu\beta\beta$+BG events. The limits on the half-lives are obtained from the number of the observed counts and the number of the estimated $2\nu\beta\beta$+BG counts. The standard method of the likelihood analysis was used [11,12]. The results of the measurement at Oto, being combined with the previous results at Kamioka are summarized in Table 2. Using the nuclear matrix elements given in ref. [10], the half-life limits lead to the upper limits on physics quantities(Table 2).

The present work measures the energy and angular correlations of two β rays, which are important for the study of individual double beta processes. It is also

TABLE 1. The ^{214}Bi and ^{208}Tl contents, which contribute to the $0\nu\beta\beta$ energy window.

Origin	Location	Amount	
^{214}Bi	Source	$(8.3 \pm 1.7) \times 10^{-3}$	Bq/kg
	DC-gas	$(2.2 \pm 0.5) \times 10^{-2}$	Bq/m^3
	Open Space	$(6.5 \pm 1.5) \times 10^{-2}$	Bq/m^3
	PL	$< 3.9 \times 10^{-3}$	Bq/kg
	NaI	$(1.08 \pm 0.06) \times 10^{-4}$	Bq/kg
^{208}Tl	Source	$(1.9 \pm 0.7) \times 10^{-4}$	Bq/kg
	PL	$(1.05 \pm 0.18) \times 10^{-4}$	Bq/kg
	NaI	$(6.2 \pm 0.9) \times 10^{-6}$	Bq/kg

important to study the double beta decay of several nuclei, because the values of $\langle m_\nu \rangle$, $\langle \lambda \rangle$, $\langle \eta \rangle$, $\langle g_B \rangle$ and others depend somewhat on the calculated matrix elements with some approximations.

The nuclear responses for $\beta\beta$ are given by $|M^{0\nu}|^2$ and $|M^{2\nu}|^2$, respectively. $M^{0\nu}$ is crucial for extracting $\langle m_\nu \rangle$ from the observed rate of $T^{0\nu}$. The $M^{0\nu}$ and $M^{2\nu}$ in medium nuclei, which are mostly spin-isospin matrix elements, are very small because of the large spin-isospin core polarization [4,5] Thus theoretical calculations for them are hard. It has been shown that the nuclear matrix element $M^{2\nu}$ is expressed by a separable form as $M^{2\nu} = M_s(GT)M_{s'}(GT)/\Delta_s$, where $M_s(GT)$ and $M_{s'}(GT)$ are the single β GT($J^+ = 1^+$) matrix element from the initial and final nuclei to the intermediate single particle hole state, respectively [3–5]. The single β matrix elements are obtained by charge exchange spin flip interactions. The $0\nu\beta\beta$ is mainly due to virtual ν exchange between two nucleons in a nucleus. Neutrino momenta involved are 1~50MeV/c, and accordingly intermediate state with E_{ex} up to 50MeV, and L up to $5\hbar$ are involved in $0\nu\beta\beta$ process. Thus multi-pole spin isospin giant resonances with $J = 0 \sim 5$ play important roles for $\beta\beta - \nu$ responses.

The nuclear matrix element may be expressed as a separable form as $M^{0\nu} = \sum_J M_S(TSLJ)M_{S'}(TSLJ)/\Delta_S$. Then $M_S(TSLJ)$ and $M_{S'}(TSLJ)$ for single particle-hole states are obtained from charge-exchange hadron reactions and/or single β decay rates. Double charge-exchange reactions are interesting for obtaining directly double spin isospin responses relevant to $\beta\beta$ decays.

IV PERSPECTIVE OF ν STUDIES WITH ^{100}MO

The neutrino mass limit to be studied is given by using the relation of $Y^{0\nu} \sim \sqrt{(Y^{BG})}$, where $Y^{0\nu}$ is yields of the $0\nu\beta\beta$ signal, and Y^{BG} that of the BG. It is expressed as $\langle m_\nu \rangle^{-1} = N_\beta^{1/2} G^{0\nu\ 1/2} M^{0\nu} t^{1/4} N_{BG}^{-1/4}$ where N_β and N_{BG} are number of counts for $0\nu\beta\beta$ signal and BG.

The limit with the present ELEGANT V for the running time of t~4yr is around 1.3eV. In order to study the $\langle m_\nu \rangle$ in the 0.03eV region of the current particle and astrophysics interest, one needs to enlarge the number N_β of ^{100}Mo isotopes.

It is shown by Ejiri, et al., [13] that the spectroscopic method for two β-rays from ^{100}Mo are used for two types of the low energy neutrino studies. One is for

TABLE 2. Limits with 90(68)%CL on the half-lives for the $0\nu\beta\beta$ decay of ^{100}Mo by combining the results at Oto lab. and Kamioka Lab., and those on particle physics quantity.

Decay Mode		$T_{1/2}$ (10^{23}yr)	Physics Quantity 90(68)%CL
$0^+ \to 0^+$	$\langle m_\nu \rangle$ term	>0.55(1.0)	$\langle m_\nu \rangle$ <2.1(1.5)eV
	$\langle \lambda \rangle$ term	>0.34(0.64)	$\langle \lambda \rangle$ < 3.1(2.5) × 10^{-6}
	$\langle \lambda \rangle$ term*	>0.42(0.85)	$\langle \lambda \rangle$ < 2.7(2.0) × 10^{-6}
	$\langle \eta \rangle$ term	>0.49(0.93)	$\langle \eta \rangle$ < 2.1(1.5) × 10^{-8}
Majoron emission		>0.046(0.077)	$\langle g_B \rangle$ < 7.9(6.1) × 10^{-5}

the exclusive measurement of the Majorana neutrino mass to the level of 0.03eV. The other is for the real-time studies of the low energy solar neutrino with specified neutrino sources of pp, ^{7}Be and ^{8}B-neutrinos.

^{100}Mo has just the adequate level structure with the large $Q_{\beta\beta}$=3.034MeV for the $\beta\beta$ decay and small negative Q_β=0.168MeV for the inverse β-decay to make these studies feasible. The unique features of ^{100}Mo are as the followings [13]. 1)The β_1 and β_2 with the large energy sum of $E_1 + E_2$ are measured in coincidence for the $0\nu\beta\beta$ studies, while the inverse β-decay induced by the solar ν and the successive β-decay are measured sequentially in an adequate time window for the low energy solar-ν studies. The isotope ^{100}Mo is just the one that satisfies the conditions for the $\beta\beta - \nu$ and solar-ν studies. 2)The large Q value of $Q_{\beta\beta}$=3.034 MeV gives a large phase-space factor $G^{0\nu}$ to enhance the $0\nu\beta\beta$ rate and a large energy sum of $E_1 + E_2 = Q_{\beta\beta}$ to place the $0\nu\beta\beta$ energy signal well above most BG except ^{208}Tl and ^{214}Bi. The energy and angular correlations for the two β-rays can be used to identify the ν-mass term. 3)The low threshold energy of 0.168 MeV for the solar-ν absorption allows observation of low energy sources such as pp and ^{7}Be. The capture rates are 640 SNU and 207SNU for ^{7}Be ν and pp ν, respectively. The solar-ν sources are identified by measuring the inverse-β energies. Only the ^{100}Tc ground state can absorb ^{7}Be ν and pp ν. 4)The measurement of two β-rays (charged particles) enables one to localize in space and in time the decay-vertex points for both the $0\nu\beta\beta$ and solar-ν studies.

REFERENCES

1. W.C. Haxton and G.J. Stephenson, Jr. *Prog. in Part. Nucl. Phys.* **12**, 409(1984).
2. M. Doi, T. Kotani and E. Takasugi, *Prog. Theor. Phys. Suppl.* **83**, 1(1985).
3. H. Ejiri, Int. J. Mod. Phys. **E 6**, 1(1997)
4. H. Ejiri, Phys. Rep. (2000), to be published.
5. H. Ejiri and H. Toki, *Joun. of Phys. Soc. of Japan,* **65**, 7(1996).
6. H. Ejiri, et al., *Phys. Lett.* **B 258**, 17(1991); H. Ejiri, et al., *J. Phys. G. Nucl. Part. Phys.* **17**, S155(1991).
7. H. Ejiri, et al., *Nucl. Instr. and Meth.* **A 302**, 304(1991). K. Nagata, et al. *Nucl. Instr. and Meth.* **A 362**, 261(1995).
8. N. Kudomi, *Nucl. Instr. and Meth.* **A430**, 96(1999)
9. H. Ejiri, et al., Nucl. Phys. A611, 85(1996).
10. T. Tomoda, *Rep. Prog. Part. Phys.* **54**, 53(1991).
11. Particle Data Group, Phys. Rev. **D 50**, (1994)1173.
12. O. Helene, Nucl. Instr. and Meth in Phys. Res. **A390**, 383(1997)
13. H. Ejiri, Paper presented at the Workshop on Low Energy Neutrino Physics, July 23, 1999. H. Ejiri, et al., Phys. Rev. Lett. **85**, 2917(2000).
14. H. Akimune et al., Phys. Lett **B 394**, 23(1997).
15. A. Garcia et al., Phys. Rev. **C 47**, 2910(1993).

Search for Spin Coupled WIMPs with the Large Volume NaI(Tl) Scintillators

S. Yoshida[a], H. Ejiri[a], K. Fushimi[b], K. Hayashi[a], T. Kishimoto[c],
N. Kudomi[a], K. Kume[a], H. Kuramoto[a], K. Matsuoka[c],
H. Ohsumi[c], K. Takahisa[a], Y. Tsujimoto[a], and S. Umehara[b]

[a] *RCNP, Osaka University,Ibaraki, Osaka 567-0047, Japan.*
[b] *Department of Physics, Osaka University, Toyonaka, Osaka 560-0043, Japan.*
[c] *Faculty of Integrated Arts and Sciences, The University of Tokushima, Tokushima 770-0814, Japan.*

Abstract. The cold dark matter search has been carried out at Oto Cosmo Observatory with the large volume NaI(Tl) scintillators of ELEGANT V. The new limits on WIMPs could be obtained by the analysis of the annual modulation.

INTRODUCTION

Experimental and theoretical studies for the universe indicate that at least 90% of the mass in the universe is composed mostly of invisible matters(DM) [1]. The cold DM such as Weakly Interacting Massive Particles(WIMPs) hypothetically forms a large fraction of our galaxy.

WIMPs could be directly detected through the nuclear recoil induced by their elastic and inelastic scattering with detector target [2]. The expected recoil energy spectrum has a exponential shape with a characteristic energy less than 100keV. The event rate of the WIMPs signals, which depends on the model of WIMPs, is the order of 1event/kg/day. Thus the measurement with the low energy threshold and with extremely low background condition is required for WIMPs search.

In order to satisfy requirements mentioned above, the large volume low-background NaI detectors, which primarily have been developed as one component of the detector system ELEGANT V for studying $\beta\beta$ [4], are applied for this study. It should be noted that a large volume detector(730kg) has a statistically great advantage for this purpose, and the 100% isotopic contents on odd A isotopes of NaI detector (^{23}Na and ^{127}I respectively) make sensitive to not only spin independent interactions with WIMPs but spin dependent ones.

In addition, it is difficult to identify the WIMPs because the expected raw spectrum has no distinctive feature. A convincing proof of the detection of WIMPs would be to find the unique signature in the data, which cannot be faked by the systematic background. One of such distinctive features would be the annual modulation of the WIMPs signal, caused by the motion of the earth around the sun [3]. The purpose of this work is to look for the annual modulation in the data collected at Oto Cosmo Observatory.

EXPERIMENTAL DATA AND ANALYSIS

ELEGANT V [5] consists of three drift chambers and 16 modules of plastic scintillators for the study of $\beta\beta$ decay, 20 modules of NaI(Tl) scintillators for the measuring of the recoil energy. The whole detector system is covered with 10cm thick OFHC copper and 15cm thick low activity lead. In order to purge the Rn gas in the vicinity of the detectors, which is one of the most serious seasonal modulated background origin, the pure nitrogen gas ($\sim$7.0 l/min) is introduced into air tight. The Rn concentration in the extracted gas is continuously monitored on real time with high sensitive Rn monitor system [6]. The measured Rn concentration was about 150mBq/m^3, and the seasonal modulation was suppressed less than 60mBq/m^3. This value was satisfied with the requirement from WIMPs search with the annual modulation.

The data considered here has been collected at Oto Cosmo Observatory (1400m.w.e) from the middle of May,1999 to the beginning of July,2000. The energy spectrum at the low energy region was obtained as in Fig.1. The peak around 50keV comes from the ^{210}Pb contamination inside the NaI scintillators, and suffers the serious background at low energy region of interest for WIMPs search. Both of the position and the resolution of this peak were used to check the stability of the NaI(Tl) scintillators [7]. The modules, to be taken into account the analysis, were selected as followings, (i)the energy threshold is lower than 5keV, (ii)the amount of ^{210}Pb contamination is less than 10mBq/kg, and (iii)the event rate in the energy window 10 - 20keV, where no modulation is expected, is stable in time. After these selections, 9 modules were selected for the analysis at the energy bin of 4 - 5keV, and 11 modules at the energy higher than 5keV.

The procedure used to extract the periodic component is following [8]. The observed event rate is expressed, in first order, as $S_{obs}(t) = \langle B \rangle + \langle S_0 \rangle + S_m \cos(\omega t)$, where brackets mean the average of Poisson distributed variables, ω is inverse modulation period, t is the day from June 2nd when the maximum is expected. S_0 and S_m are the unmodulated and modulated components of WIMPs signal, respectively. The modulation amplitude and its dispersion can be calculated, including of the collection for the measurement period, as followings,

$$S_m = \frac{\sum_j [\cos(\omega t_j) - \beta] S_j}{N(\alpha - \beta^2)}, \quad \sigma(S_m) = \frac{\sqrt{\sum_j [\cos(\omega t_j) - \beta]^2 S_j}}{N(\alpha - \beta^2)}. \quad (1)$$

with $\beta = \frac{1}{N}\sum_j \cos(\omega t_j) = 0.924$, $\alpha = \frac{1}{N}\sum_j \cos^2(\omega t_j) = 0.121$, N is the total live time (340.5 days), and S_j is the observed event rate at j-th day.

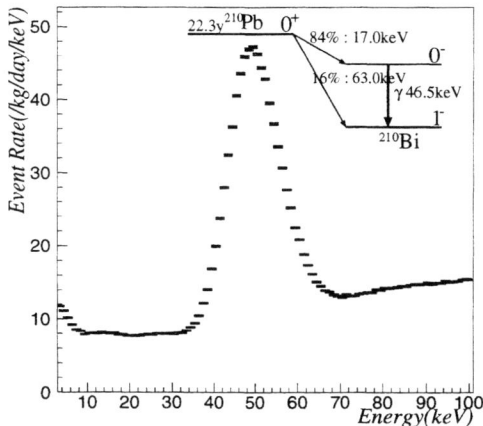

FIGURE 1. The energy spectrum of the NaI(Tl) scintillators with the exposure of 2123kg×day and the decay scheme of ^{210}Pb.

The analyses were carried out for each energy bin between 4 to 10keV, and also for higher energy region to check the stability of systematics. The obtained values are given in Table 1.

No significant deviation above the statistical fluctuation was found in the data. Consequently, the exclusion plots can be obtained by comparing the experimental upper limits on the modulation components with expected ones. The expected values were calculated with the assumptions of astrophysical parameters [9](local density, velocities of WIMPs model), nuclear physics parameters(spin matrix elements [10], nuclear form factors) and experimental conditions(energy threshold, resolution and light output responses [11]). The parameters used for the calculation are summarized in Table 2. The form factors used here for the spin independent interaction are Helm form factors [12]. For the spin dependent case, form factors are more complex ones [10].

For the comparison of the results obtained with different nuclei, the exclusion plots were represented by the cross section for proton. They are shown in Fig.2 for the interactions of both spin independent and dependent.

DISCUSSION AND FUTURE PERSPECTIVES

In case of the spin independent interaction, the cross section is scaled according to A^2. Nuclear recoils occur mostly on iodine, the expected differential energy spectrum forms steeply decreasing and most of the signals below 4keV(threshold). The sensitivity comes from the event rate at the lowest energy. The present work

TABLE 1. The modulation amplitudes and their dispersions for the cosine modulation analysis.

Energy bin (keV)	Exposure (kg × day)	S_m	$\sigma(S_m)$
		(/day/kg/keV)	
4-5	328.5 × 340.5	0.009	0.019
5-6		-0.007	0.014
6-7		0.009	0.011
7-8		0.003	0.011
8-9	401.5 × 340.5	-0.005	0.010
9-10		0.004	0.010
(keV)		(/day/kg)	
10-15		-0.011	0.021
15-20		0.014	0.020

TABLE 2. List of parameters used for the calculation of WIMPs spectra.

Parameter	Value
WIMPs local density	0.3GeV/cc
WIMPs velocity distribution	220km/sec
Escape velocity	600km/sec
Earth velocity	230±15km/sec
Spin factor ^{23}Na	0.089
^{127}I	0.126
Quenching factor ^{23}Na	0.40
^{127}I	0.05

was not sensitive enough to neither to exclude nor prove the claimed candidate of DAMA/NaI experiment [13]. The detector with much low background and/or the lower energy threshold are required to improve the sensitivity for the spin independent interaction.

On the other hand, the most stringent limit was obtained for the spin dependent interaction with the low WIMPs mass. The quenching factor of Na nucleus in the NaI scintillator is larger than that of Iodine nucleus. In addition, the spin form factor of ^{23}Na is the two orders of magnitude larger than that of ^{127}I at the characteristic recoil energy of 100keV. Thus, the sensitivity for spin dependent interaction is efficient to improve the published limits.

REFERENCES

1. J.R. Primack, D. Seckel and B. Sadoulet, Ann.Rev.Nucl.Sci. 38 (1988) 751.

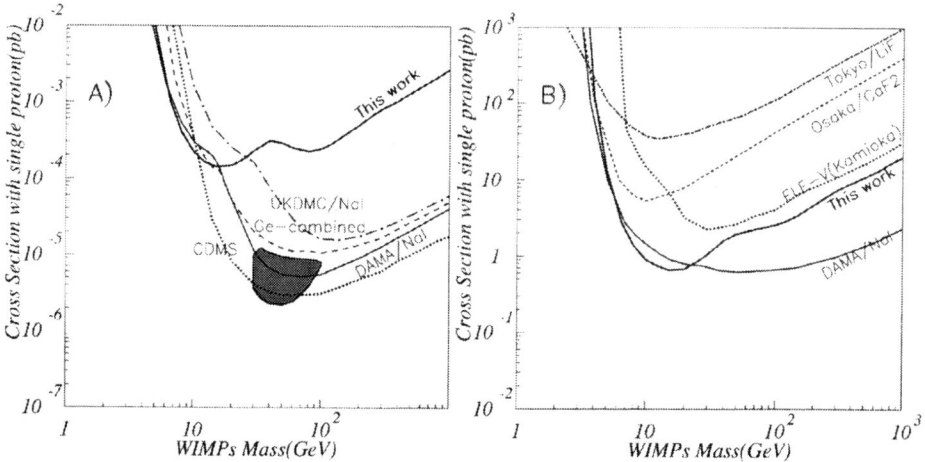

FIGURE 2. The exclusion plots(90% C.L.) on $\sigma_{p-\chi}$ for the interaction of A) spin independent and B) dependent from this work(preliminary) and other experiments [13–19], including the positive result of DAMA/NaI experiment [13].

2. A. Drukier et al., Phys.Rev.D33 (1986) 3495.
3. K. Freese et al., Phys.Rev.D37 (1988) 3388.
4. H. Ejiri et al., Nucl.Phys.A611 (1996) 85, K. Kume et al., Nucl.Phys.A577 (1994) 405c.
5. H. Ejiri et al., Nucl.Instr.Meth.A302 (1991) 304.
6. E. Choi et al., to be published in Nucl.Instr.Meth.A
7. S. Yoshida et al., Nucl.Phys.B87 (2000) 58.
8. D. Abriola et al., Astropart.Phys. 10 (1999) 133.
9. J.D. Lewin and P.F. Smith, Astropart.Phys. 6 (1996) 87.
10. M.T. Ressell et al., Phys.Rev.C56 (1997) 534.
11. K. Fushimi et al., Phys.Rev.C47 (1993) R425.
12. R.H. Helm, Phys.Rev. 104 (1956) 1466.
13. R. Bernabei et al.,Phys.Lett.B389 (1996) 757, Phys.Lett.B436 (1998) 379, Phys.Lett.B450 (1999) 448.
14. P.F. Smith et al., Phys.Lett.B379 (1996) 299,
15. CDMS Collaboration, Phys.Rev.Lett. 84 (2000) 5699.
16. L. Baudis et al., Phys.Rev.D59 (1998) 022001, D. Reusser et al., Phys.Lett.B255 (1991) 143, E. García et al., Phys.Rev.D51 (1995) 1458.
17. K. Fushimi et al., Astropart.Phys. 12 (1999) 185.
18. T. Kishimoto and I. Ogawa, private communication(2000).
19. W. Ootani et al., Phys.Lett.B461 (1999) 371.

Hyperon Beta-Decay Analysis and the Recent KTeV Data

Philip G. Ratcliffe

Dipartimento di Scienze CC.FF.MM.,
Università degli Studi dell'Insubria—sede di Como
via Valleggio 11, 22100 Como, Italy
and
Istituto Nazionale di Fisica Nucleare—sezione di Milano
via Celoria 15, 20133 Milano, Italy
pgr@fis.unico.it

Abstract. The analysis of hyperon semi-leptonic decay data is addressed with reference to SU(3) breaking and isospin mixing between Λ^0 and Σ^0. Various approaches to SU(3) breaking are discussed and compared. The phenomenological implications of $\Lambda^0 - \Sigma^0$ mixing are not to be underestimated: it can induce vector couplings in decays otherwise purely axial and may also modify rates. In regard of the KTeV data on $\Xi^0 \to \Sigma^+ e \bar{\nu}$, predictions are presented and the impact of present a future data on the extraction of F and D is also examined. In addition, the implications of the new data for the use of octet baryon beta decays in determining V_{us} are considered.

I INTRODUCTION

Hyperon semi-leptonic decay (HSD) data are the sole present source of information on the F and D parameters, vital for the analysis of polarised deep-inelastic scattering experiments. In addition, they may provide access to the Cabibbo-Kobayashi-Maskawa (CKM) matrix element, V_{us}. Recent data for $\Xi^0 \to \Sigma^+ e\bar{\nu}$ from the KTeV collaboration [1] at Fermilab have added new interest to this area.

Beside the accurately measured neutron β-decay rate and angular asymmetries, there is a body of data on the rest of the baryon octet [2]. In SU(3) such decays are described via two parameters, F and D, relating to strong-interaction effects and two further parameters, V_{ud} and V_{us}, the CKM matrix elements (heavy-flavour contributions may be neglected). The F and D parameters are important in connection with in the Ellis-Jaffe sum rule [3]; a 15% reduction in the ratio F/D from its accepted value (~ 0.6) would remove the discrepancy with polarised DIS data and alleviate the "proton-spin puzzle" [4].

As SU(3) is violated at about the 10% level, a reliable description of the breaking is important. A test of any scheme lies in the predictions made for new decays: such as, the process $\Xi^0 \to \Sigma^+ e\bar{\nu}$, which has now been measured by the KTeV collaboration [1,5]. In this talk, after outlining the data and an approach to SU(3) breaking, I shall present a prediction for this decay [6] and discuss future developments.

II THE HSD DATA

A number of hyperon semi-leptonic decays have been measured with varying degrees of accuracy and depth of information; fig. 1 depicts the measured baryon

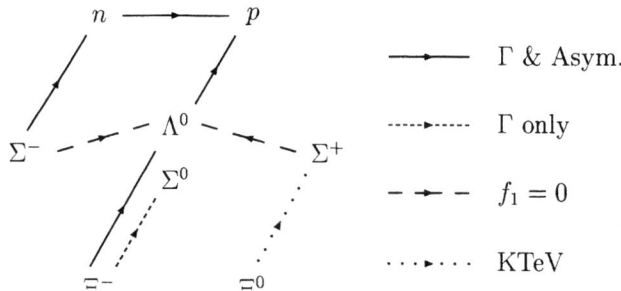

FIGURE 1. The SU(3) scheme of the measured baryon octet β-decays: the solid lines represent decays for which both rates and asymmetry measurements are available; the short dash, only rates; the long dash, $f_1 = 0$ decays; and the dotted line, the KTeV measurement.

octet β-decays, indicating the type of data available, while the present world HSD data are collected in table 1. Note that several of the rates and asymmetries have now been measured to better than 5%. Moreover, I should add that a past discrepancy in the neutron β-decay data has now been resolved.

TABLE 1. Present world HSD rate and angular-correlation data [2]. Numerical values marked g_1/f_1 are extracted from angular and spin correlations.

Decay $A \to B\ell\nu$	Rate (10^6 s^{-1}) $\ell = e^\pm$	Rate (10^6 s^{-1}) $\ell = \mu^-$	g_1/f_1 $\ell = e^-$	g_1/f_1 SU(3)
$n \to p$	1.1278 ± 0.0024 [a]		1.2670 ± 0.0035 [b]	$F + D$
$\Lambda^0 \to p$	3.161 ± 0.058	0.60 ± 0.13	0.718 ± 0.015	$F + \frac{1}{3}D$
$\Sigma^- \to n$	6.88 ± 0.23	3.04 ± 0.27	-0.340 ± 0.017	$F - D$
$\Sigma^- \to \Lambda^0$	0.387 ± 0.018			$-\sqrt{\frac{2}{3}}D$ [c]
$\Sigma^+ \to \Lambda^0$	0.250 ± 0.063			$-\sqrt{\frac{2}{3}}D$ [c]
$\Xi^- \to \Lambda^0$	3.35 ± 0.37 [d]	2.1 ± 2.1 [e]	0.25 ± 0.05	$F - \frac{1}{3}D$
$\Xi^- \to \Sigma^0$	0.53 ± 0.10			$F + D$

[a] Rate given in units of 10^{-3} s^{-1}.
[b] Scale factor 1.9 included in error (PDG practice for discrepant data).
[c] Absolute expression for g_1 given (as $f_1 = 0$).
[d] Scale factor 2 included in error (as above).
[e] Data not used in these fits.

III SU(3) BREAKING AND FIT RESULTS

SU(3) breaking is well described using centre-of-mass (CoM), or recoil, corrections [7–9]. One approach (A here) is to account for the extended nature of the baryon by applying momentum smearing to its wave function. For the decay $A \to B\ell\nu$, CoM corrections to g_1 lead to

$$g_1 = g_1^{SU(3)} \left[1 - \frac{\langle p^2 \rangle}{3 m_A m_B} \left(\frac{1}{4} + \frac{3 m_B}{8 m_A} + \frac{3 m_A}{8 m_B}\right)\right].$$

A similar approach (B) relates the breaking to mass-splitting effects in the interaction Hamiltonian via first-order perturbation theory [10]. The correction here takes on the following form:

$$g_1 = g_1^{SU(3)} \left[1 - \epsilon(m_A + m_B)\right].$$

Both approaches normalise the corrections to the reference-point correction for $g_1^{n \to p}$ and depend a single new parameter ($\langle p^2 \rangle$ or ϵ). Corrections to f_1 are negligible in A and assumed so in B, in accordance with the Ademollo-Gatto theorem. Any further global normalisation correction to the $|\Delta S=1|$ rates is disfavoured; in [7] a calculated value of $\sim 8\%$, was used, this is excluded by present-day fits.

Table 2 displays the results of three fits: S (symmetric), A and B. Note that the value of V_{ud} (and hence V_{us}, fixed here via CKM unitarity) is determined mainly by the super-allowed nuclear *ft* values. However, when V_{ud} and V_{us} (with or *without* imposition of unitarity) are extracted from HSD data *alone*, all parameter values remain essentially unchanged. Thus, unitarity appears to be well respected.

TABLE 2. SU(3) symmetric and breaking fits, including V_{ud} from nuclear *ft*.

Fit	V_{ud}	F	D	χ^2/DoF
S	0.9748 (4)	0.466 (6)	0.800 (6)	2.3
A	0.9740 (4)	0.460 (6)	0.808 (6)	0.8
B	0.9740 (4)	0.459 (6)	0.809 (6)	0.8

IV Λ^0 AND Σ^0 MIXING

While at the level of isospin violation itself the effects are obviously small, their influence in HSD may, in fact, be significant. It has been pointed out [11] that isospin violation can induce mixing between Λ^0 and Σ^0, described via

$$\Lambda^0 = \cos\phi \Lambda_8 + \sin\phi \Sigma_8,$$
$$\Sigma^0 = -\sin\phi \Lambda_8 + \cos\phi \Sigma_8.$$

The suggested phenomenological mixing angle is $\phi = -0.86°$ [12]. Now, consider, e.g., the $\Sigma^{\pm} \to \Lambda^0$ decay: in exact SU(2) f_1 is zero, thus angular or spin correlations vanish here. If, however, the Λ^0 contains a small admixture of Σ^0, this is no longer true. While there is no strong signal in the fits for such mixing, intriguingly, the values returned are around $-0.8° \pm 0.8°$ in both SU(3) symmetric and broken fits.

V A PREDICTION

Rates and parameters may now be predicted for any other decay in the octet; table 3 compares the predictions obtained for $\Xi^0 \to \Sigma^+ e\bar{\nu}$ from the above three fits. Recall that $g_1/f_1 = F + D$ for this decay, thus allowing for important cross

TABLE 3. The axial coupling, rate and branching ratio (BR) for $\Xi^0 \to \Sigma^+ e\bar{\nu}$. The errors are those returned by the fitting routine.

Fit	g_1/f_1	Γ (10^6 s^{-1})	BR (10^{-4})
S	1.267 (0) [a]	0.901 (42)	2.61 (09)
A	1.151 (27)	0.796 (44)	2.31 (10)
B	1.136 (30)	0.781 (46)	2.26 (12)

[a] Zero error is assigned to g_1/f_1 in the symmetric fit as it would be essentially that of neutron β-decay.

checks. The variation between the two SU(3) breaking fits is well within statistical errors, I therefore combine the two, obtaining the following mean values:

$$g_1/f_1 = 1.14 \pm 0.03 \pm 0.01 \quad \text{and} \quad \Gamma = (0.79 \pm 0.05 \pm 0.01) \cdot 10^6 \, \text{s}^{-1},$$

where the second error estimates the systematic uncertainty due to the difference between fits, the corresponding branching ratio is $BR = (2.29 \pm 0.12) \cdot 10^{-4}$.

Let me now briefly compare with a $1/N_c$ approach [13]: the quoted fit there results in a very low $F/D = 0.46$ and for $\Xi^0 \to \Sigma^+ e\bar{\nu}$ predicts

$$g_1/f_1 = 0.91 \quad \text{and} \quad \Gamma = 0.68 \cdot 10^6 \, \text{s}^{-1} \quad \text{(fit B of ref. [13])},$$

Both values are smaller than those presented here, which are in turn smaller than naïve SU(3). To comprehend the difference between the predictions, note that the analysis of ref. [13] includes baryon-decuplet non-leptonic decay data, which dominate; and the overall fit (i.e., χ^2) is poor. However, applied to the HSD data alone, the results are similar to those reported here [14]. These differences should be distinguishable by an experiment with good statistics, such as KTeV [1].

The preliminary KTeV results are as follows [1,5]:

$$g_1/f_1 = 1.24^{+0.20}_{-0.17} \pm 0.07 \quad \text{and} \quad BR = (2.54 \pm 0.11 \pm 0.16) \cdot 10^{-4},$$

the errors are expected to be at least halved when the full data set is analysed. Although not crucial to the F and D determination (owing to low sensitivity to the breaking scheme), depending upon the final central values, the KTeV data should have strong impact on the determination of V_{us}.

VI CONCLUSIONS

Before concluding, let me call attention to an all too often overlooked point: although easier to analyse (no extra corrections are necessary), the present data for angular correlations alone show *no* evidence of SU(3) breaking. Furthermore, compared to the full data set, they lack statistical power. *Only full analyses can display the true picture* [9]. Here I would comment that the observed consistency with unitarity (*i.e.*, between V_{ud} and V_{us}) and the fact that there is no disagreement between *complete* analyses (with the possible exception of [13]) suggest that the PDG [2] might reconsider this sector as a contending source for estimating V_{us}. Indeed, unitarity appears better satisfied here than in the $K_{\ell 3}$ data.

A complete comprehension of SU(3) breaking is still wanting: witness the octet-decuplet discrepancy. Moreover, the system is as yet not truly over-constrained; in this context, I might also mention another decay (already measured but not accurately) for which large corrections are expected: namely, $\Xi^- \to \Sigma^0 e \bar{\nu}$. There too, $g_1/f_1 = F + D$, permitting additional sensitive cross checks, especially in combination with the KTeV results.

Concluding then, I would stress that, while the data do manifest significant departures from SU(3), and there is even modest evidence for SU(2) mixing, the mass-splitting driven schemes discussed here provide an adequate description. That said, there remains much to be understood, *e.g.*, the long-standing question of second-class currents. Thus, any new *precise* data are more than welcome and the contribution of the KTeV collaboration will be invaluable.

REFERENCES

1. Affolder, A. *et al.* (KTeV E832/E799 collab.), Phys. Rev. Lett. **82**, 3751 (1999).
2. Particle Data Group, Groom, D.E. *et al.*, Eur. Phys. J. C **15**, 1 (2000).
3. Ellis, J. and Jaffe, R.L., Phys. Rev. D **9**, 1444 (1974); *erratum ibid.* D **10**, 1669 (1974).
4. Close, F.E. and Roberts, R.G., Phys. Lett. B **316**, 165 (1993).
5. Alavi-Harati, A. (KTeV collab.), e-print hep-ex/9903031.
6. Ratcliffe, P.G., Phys. Rev. D **59**, 014038 (1999).
7. Donoghue, J.F., Holstein, B.R. and Klimt, S.W., Phys. Rev. D **35**, 934 (1987).
8. Ratcliffe, P.G., Phys. Lett. B **242**, 271 (1990).
9. Ratcliffe, P.G., Phys. Lett. B **365**, 383 (1996).
10. Ratcliffe, P.G., invited plenary talk in proc. of *Deep Inelastic Scattering off Polarized Targets: Theory Meets Experiment* (DESY-Zeuthen, Sept. 1997), eds. Blümlein, J., De Roeck, A., Gehrmann, T. and Nowak, W.-D. (DESY 97-200, 1997), p. 128.
11. Karl, G., Phys. Lett. B **328**, 149 (1994); *erratum ibid.* B **341**, 449 (1995).
12. Karl, G., in proc. of *Hyperon 99* (Fermilab, Sept. 1999), eds. Jensen, D.A. and Monnier, E. (Fermilab-Conf-00/059-E, 2000), p. 41.
13. Flores-Mendieta, R., Jenkins, E. and Manohar, A.V., Phys. Rev. D **58**, 094028 (1998).
14. Manohar, A., private communication.

Novel spin maser mechanism studied for high-precision measurement of neutron electric dipole moment

A. Yoshimi*, K. Asahi*[†], K. Yogo[†], K. Sakai[†], H. Ogawa[†], T. Suzuki[†], and M. Nagakura[†]

*The Institute of Physical and Chemical Research (RIKEN), Wako-shi, Saitama, Japan
[†]Department of Physics, Tokyo Institute of Technology, Meguro-ku, Tokyo, Japan

Abstract. We propose a new scheme of nuclear spin maser and have performed the experimental and theoretical analyses for its operation, aiming at the significant improvement in sensitivity of neutron electric dipole moment (EDM). The proposed spin-masing scheme incorporates an artificial feedback mechanism, and is applicable to extremely low density systems of spins such as a bottled ensemble of UCNs. Its operation was experimentally demonstrated by using a spin-polarized ^{129}Xe gas as a model substance. The frequency analysis suggests that a frequency precision allowing a neutron EDM search with a sensitivity of 10^{-28} ecm level is attainable by this maser scheme.

INTRODUCTION

The EDM of neutron is considered to be an important site where T- and hence the CP-violation is looked for, since it suffers neither from a theoretical uncertainty of nuclear structure origin which in the case of nuclear EDMs obscured the comparison between theory and experiment, nor from the Schiff shielding which in the case of atoms strongly hindered the EDM detection. The upper limit on the size of neutron EDM, d_n, set by recent experiments is $|d_n| < 6.3 \times 10^{-26}$ ecm [1], whereas some of the currently most promising theories predict $|d_n|$ in a range $10^{-25}-10^{-28}$ ecm [2,3]. Therefore an improvement in sensitivity of neutron EDM measurements by 2–3 orders of magnitude would already be of great significance. Experimentally, the effect of an EDM manifests itself in the precession frequency ν of spin s subjected to a magnetic field B and an electric field E parallel or anti-parallel to B. In the case of neutron whose spin is $s = 1/2$, ν should change by $\delta\nu = 4dE/h$ when the direction of E is reversed. Here, d denotes the electric dipole moments. The major difficulty of an EDM search in neutron is the extreme smallness of $\delta\nu$. For example, if we take a value of $E = 10$ kV/cm, the EDM of a size $|d_n| = 10^{-26}$ ecm gives a frequency change as small as $\delta\nu = 10^{-7}$ Hz. Thus, a high precision in frequency, or a long precession time, is essential for EDM searches.

A SPIN MASER WITH ACTIVE FEEDBACK SYSTEM

The spin precession times τ_{pre} employed in recent neutron EDM experiments [1] are as long as 150 s. An inevitable limitation of τ_{pre} would be given by the neutron lifetime $\tau_n = 1280$ s. Such the limitation in τ_{pre} may be overcome by introducing a spin maser technique. Unfortunately, this technique can only be applied to a system of high spin density such as ^{3}He [4,5] and ^{129}Xe [6] gases. As will be discussed shortly, however, this limitation of a spin maser may be overcome. The principle of a spin maser is illustrated in Fig. 1 (a) where the polarized nuclear spins precess about the static field B_0. A pickup coil (inductance L) around the cell is connected to a capacitor C to compose a tank circuit whose resonance frequency is tuned to the Larmor frequency $\omega_0 \equiv \gamma B_0$. Here, γ denotes the gyro-magnetic ratio. Thus the energy of a current induced in the coil by the spin precession is stored in C and then fed again to the coil to produce a feedback field $B_{\text{FB}}(t)$ on the spins. The action of the $B_{\text{FB}}(t)$ results in a non-decaying spin precession in spite of a transverse relaxation [7]. Because the amplitude of the B_{FB} field is determined by a combined

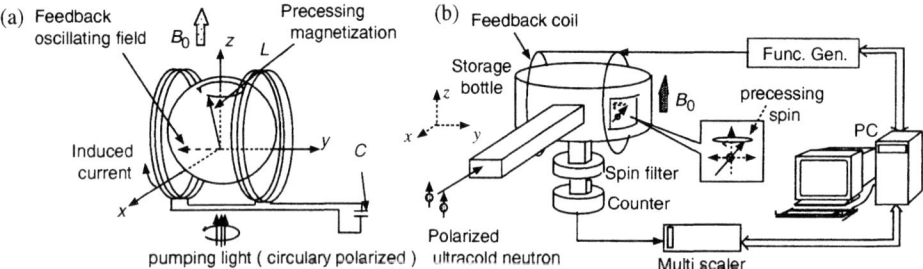

FIGURE 1. Schematic view of the spin maser. (a) Conventional spin maser. (b) Spin maser of neutron operating in a new masing scheme.

factor of the spin magnetization and the coil Q-factor, which is represented by a radiation damping time τ_{RD} [4], the maser oscillation requires a large number of polarized nuclear spins (typically $\approx 10^{18}$ cm^{-3}) as well as a large Q-factor. Thus conventional spin maser technique is not applicable to neutrons because the number density attainable for stored ultracold neutrons (UCN) is far below the level required for maser oscillation. Looking into processes undertaking the maser operation, however, one notices that the large density required is for the detection of the phase of precession from the induction current, according to which the $B_{\text{FB}}(t)$ is to be invoked. In the case of neutrons, the phase can be detected better in another way taking advantages of neutron. The conceptual setup for the new masing scheme based on this method is illustrated in Fig. 1 (b). Neutrons which are spin polarized along the z-axis with a degree of polarization P_0 are continuously introduced into a storage bottle in which a static field B_0 is applied. The number of UCNs in the bottle reaches its equilibrium value N^{eq} as a result of competition between the continuous introduction of UCNs and the disappearance due to absorption at the wall or β decays. A fraction of the stored neutrons are continuously extracted through a sampling hole on the bottle, and are led to a magnetized iron foil where spin analysis is made. Then the count rate at a counter placed behind the iron foil varies as $N_{\text{cnt}}(t) = N_0[1 + A_{\text{filter}}P_x(t)]$, where A_{filter} denotes the analyzing power of the iron foil and N_0 a constant. The instantaneous value of the x-component polarization $P_x(t)$ is deduced through a computer analysis of $N_{\text{cnt}}(t)$, and is sent to a function generator which feeds a time-varying current into a feedback coil wound around the neutron bottle to produce the field $B_{\text{FB}}(t)$. If the coil produces a rotating field $\boldsymbol{B}_{\text{FB}}(t) = (B_x, B_y, 0)$ whose amplitude is proportional to that of the transverse polarization with the phase shifted by 90°, as $B_x = (\alpha/\gamma)P_y$ and $B_y = -(\alpha/\gamma)P_x$, the transverse and longitudinal components of the polarization vector $\boldsymbol{P}$ in the equilibrium are given as [8] $|P_T^{\text{eq}}| = \sqrt{(P_0 - (\alpha\tau_{\text{sto}})^{-1})/(\alpha\tau_{\text{sto}})}$ and $P_z^{\text{eq}} = (\alpha\tau_{\text{sto}})^{-1}$, provided that $\alpha > (P_0\tau_{\text{sto}})^{-1}$ is realized. Here, τ_{sto} represents the disappearance time of individual neutron from the bottle. This represents a situation that the polarization vector keeps precessing with a constant transverse component P_T^{eq} when the gain α is set large so as to satisfy the condition $\alpha > (P_0\tau_{\text{sto}})^{-1}$. In the proposed spin maser with "active" feedback, the sensitivity for the precession is radically higher (by factors of $\approx 10^{15}$ in typical), thanks to the particle-by-particle polarization analysis, and furthermore the feedback gain α can be set at any favored value, thus offering a possibility of applying to UCN spins.

ACTIVE SPIN MASER OPERATION WITH SPIN POLARIZED ^{129}Xe NUCLEI AND DISCUSSION

In what follows, we demonstrate the operation of an "active" spin maser by taking a polarized ^{129}Xe gas as a model substance. The spin polarized ^{129}Xe ($I = 1/2$) gas with a typical polarization of $P_0 \approx 10\%$ was prepared through spin exchange with optically pumped Rb vapor [9,10]. The experimental setup is shown in Fig. 2. An enriched ^{129}Xe gas of 230 torr was confined in a spherical glass cell, and the cell is located in a three-layer cylindrical magnetic shield made of Permalloy. A static field $B_0 = 3.0$ G is produced by a solenoid coil wound inside the shield. The ^{129}Xe cell is continuously irradiated by a circularly polarized light from an 18 W diode laser. The precession signal $V_s(t)$ from a pickup coil was phase sensitively detected in a lock-in amplifier, as shown in Fig 2 (b), and the resulting outputs $V_X(t) = \frac{1}{2}V_sV_r \cos[(\omega - \omega_r)t + (\phi_0 - \phi_{0r})]$ and $V_Y(t) = \frac{1}{2}V_sV_r \sin[(\omega - \omega_r)t + (\phi_0 - \phi_{0r})]$ were processed by analog operational circuits to yield a signal $V_{\text{fb}}(t) \equiv V_X(t)V_{r1}(t) - V_Y(t)V_{r2}(t) = \frac{1}{2}V_sV_r^2 \sin(\omega_s t + \delta_s)$, where the reference signals $V_{r1}(t) = V_r \sin(\omega_r t + \phi_{0r})$ and $V_{r2}(t) = -V_r \cos(\omega_r t + \phi_{0r})$ were provided by a two-output function generator. The $V_{\text{fb}}(t)$ signal was fed to a drive coil to produce a feedback field $B_{\text{FB}}(t)$. The operation of thus constructed "active" maser was monitored by recording an output signal $V_{\text{mon}}(t)$ from another lock-in amplifier. The obtained result is shown in Fig. 3 (a). The field homogeneity was deliberately worsen and the laser power weakened, in order to let the maser condi-

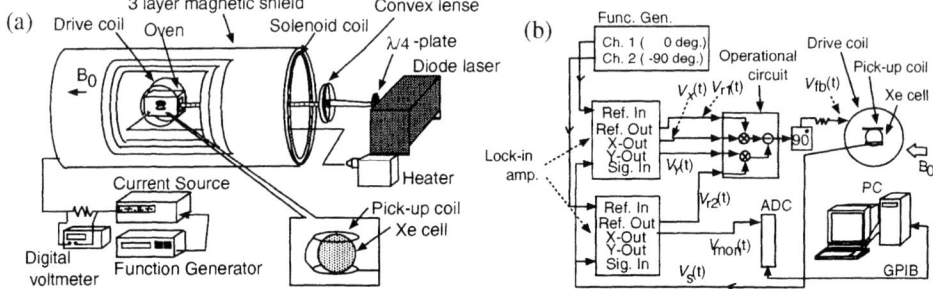

FIGURE 2. (a) Experimental setup of spin maser. (b) Block diagram of active feedback system.

tion be unsatisfied with a shortened $T_2(=62.3$ s) and elongate $\tau_{\mathrm{RD}}(=93.9$ s). Thus we certainly confirmed that the spin maser in this condition did not oscillate. Then we proceeded to the active maser operation by switching on the $V_{\mathrm{fb}}(t)$ at the drive coil. The precession signal immediately appeared, as shown in Fig. 3 (a).

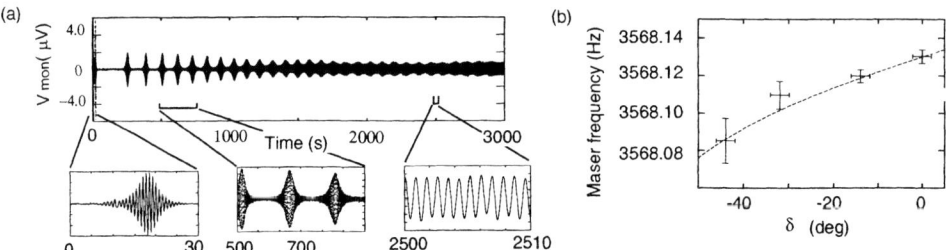

FIGURE 3. (a) Maser signal of ^{129}Xe in active maser operation. (b) Maser frequency measured as a function of the phase shift δ. Dashed line represented the theoretical frequency shift from Eq.(1).

We now investigate the frequency ω of the signals in Fig. 3, since the error in ω should be the most essential in applications to EDM experiments. Theoretically, the response of a spin maser to a small deviation δ in phase of the feedback signal may lead to a frequency shift from the Larmor frequency $\omega_0 = \gamma B_z$ as [8]

$$\omega = \omega_0 + \frac{\tan \delta}{T_2}. \tag{1}$$

The adequacy of the expression (1) for an active maser was tested experimentally by inserting a phase shifter after the feedback circuit. The measured frequency shifts are well reproduced by dashed line representing the relation (1) as shown in Fig. 3 (b).

In the application of the active maser mechanism to neutrons, as depicted in Fig. 1, the spin precession is sensed through change in the count rate $N(t)$ at a neutron counter behind a spin filter foil. When neutrons are sampled at a rate N_{smp}, the uncertainty of the precession frequency estimate from time duration τ is given by $\sigma_\omega = 1/(A_{filter}P_T)\sqrt{24/N_{smp}}\,\tau^{-3/2}$ with the precision rapidly improving with increasing measurement time τ [8,11]. We should also take into account that the maser frequency itself is modified through Eq.(1) by the phase deviation δ between the feedback signal and the actual spin precession. By considering the statistical fluctuation of δ due to the counting fluctuation of neutrons, the fluctuation of the maser frequency ω around ω_0 is given as $\sigma_\omega^{maser} = \sqrt{8}/(A_{filter}P_0\tau_{sto}\sqrt{N_{smp}\tau})$. Thus, for longer measurement times ($\tau > \sqrt{12}\tau_{sto}$), σ_ω^{maser} dominates over σ_ω. Setting e.g., $\tau_{sto} = 500$ s, $A_{filter} = 1$, $P_0 = 1$, and $N_{smp} = 10^4$ s^{-1} the measurement for 30 days will lead to the frequency precision of 10^{-8}–10^{-9} Hz which corresponds to the EDM sensitivity of 10^{-27}–10^{-28} ecm, improving the experimental sensitivity by 2–3 orders from the present experimental limit.

ACKNOWLEDGMENTS

This work was supported by Matsuo Foundation, Monbusho Grant-in-Aid for Scientific Research on Priority Areas (A), and President's Special Research Grant of RIKEN. One of the Authors (A.Y.) is grateful to the STA Special Postdoctoral Researchers Program.

REFERENCES

1. P.G. Harris et al., *Phys. Rev. Lett.* **82**, 904 (1999).
2. A.I. Sanda, *Phys. Rev. D* **32**, 2992 (1985).
3. J. Ellis, *Nucl. Instrum. Meth. Phys. Res. A* **284** (1989) 33.
4. M.G. Richards, B.P. Cowan, M.F. Secca, and K. Machin, *J. Phys. B* **21**, 665 (1988).
5. K. Sakai et al., *Nucl. Instrum. Meth. Phys. Res. A* **402**, 244 (1998).
6. T.E. Chupp et al., *Phys. Rev. Lett.* **72**, 15 (1994) 2363.
7. K. Asahi et al., *Czechoslovak J. Phys.* **50** Suppl. S1, 179 (2000).
8. A. Yoshimi et al., to be published.
9. W. Happer et al., *Phys. Rev. A* **29**, 3092 (1984).
10. H. Sato et al., *Nucl. Instrum. Meth. Phys. Res. A* **402**, 241 (1998).
11. Y. Chibane et al., *Meas. Sci. Technol.* **6**, 1671 (1995).

2. SYMMETRIES AND SPIN II

Symmetry Tests in Polarized Z^0 Decays to $b\bar{b}g$ [1]

Takashi Maruyama

representing the SLD Collaboration

Stanford Linear Accelerator Center, Stanford, CA 94309, USA

Abstract. We have made the first direct symmetry tests in the decays of polarized Z^0 bosons into fully-identified $b\bar{b}g$ states, collected in the SLD experiment at SLAC. We searched for evidence of parity violation at the $b\bar{b}g$ vertex by studying the asymmetries in the b-quark polar- and azimuthal-angle distributions, and for evidence of T-odd, CP-even or odd, final-state interactions by measuring angular correlations between the three-jet plane and the Z^0 polarization. We found results consistent with Standard Model expectations and set 95% C.L. limits on anomalous contributions.

The unique sample of polarized Z^0 bosons produced in annihilations of longitudinally-polarized electrons with unpolarized positrons at the SLAC Linear Collider (SLC) can be employed for fundamental symmetry tests. Here we use $e^+e^- \to Z^0 \to$ three-jet events to test symmetry properties of the Standard Model (SM). The $b\bar{b}g$ final state provides a particularly interesting probe for possible beyond-SM processes that couple to massive particles.

The tree-level differential cross section for $e^+e^- \to q\bar{q}g$ can be expressed [1]

$$2\pi \frac{d^4\sigma}{d(\cos\theta)d\chi dx d\bar{x}} = [\frac{3}{8}(1+\cos^2\theta)\,\sigma_U + \frac{3}{4}\sin^2\theta\,\sigma_L$$
$$+\frac{3}{4}\sin^2\theta\cos 2\chi\,\sigma_T + \frac{3}{2\sqrt{2}}\sin 2\theta\cos\chi\,\sigma_I]\,h_f^{(1)} + [\frac{3}{4}\cos\theta\,\sigma_P - \frac{3}{\sqrt{2}}\sin\theta\cos\chi\,\sigma_A\,]\,h_f^{(2)},$$

(1)

where θ is the polar angle of the thrust axis w.r.t. the electron beam, and χ is the azimuthal angle of the event plane w.r.t. the quark-electron plane. The thrust axis is defined to be along the most energetic jet and to point into (opposite) the hemisphere containing the quark (antiquark) if the quark (antiquark) has the higher energy. The sign of $\cos\chi$ is defined in terms of momentum vectors, $\text{sign}(\cos\chi) =$

---

[1] This work was supported in part by Department of Energy Contract No. DE-AC03-76SF00515

sign$(({\vec q}\times {\vec g}) \cdot ({\vec q}\times {\vec e^-}))$. The functions $h_f^{(1)}$ and $h_f^{(2)}$ contain the dependence on the beam polarization and the electroweak couplings [1]; QCD contributions are expressed in terms of $\sigma_i \equiv d^2\sigma_i/dxd\bar{x}$, with $i = U, L, T, I, A, P$, where x and $\bar{x}$ are the scaled momenta of the quark and anti-quark, respectively, While the first four terms are even under P reversal, the last two terms are P-odd, and are sensitive to any parity-violating interactions at the $Z^0 q\bar{q}$ or $q\bar{q}g$ vertices. More generally there can be, in addition, three terms that are odd under T reversal [2]. However, these terms vanish at tree level in a theory that respects CPT invariance.

Integrating Eq. 1 over x, $\bar{x}$ and χ [3],

$$\frac{d\sigma}{d\cos\theta} \propto 1 + \alpha\cos^2\theta + 2P_Z A_P \cos\theta, \qquad (2)$$

where $P_Z = (P_{e^-} - A_e)/(1 - P_{e^-} \cdot A_e)$, P_{e^-} is the signed electron-beam polarization, the parity-violation parameter $A_P = A_b \cdot \hat{\sigma}_P/(\hat{\sigma}_U + \hat{\sigma}_L)$, A_e (A_b) is the electroweak coupling of the Z^0 to the initial (final) state, and $A_j = 2v_j a_j/(v_j^2 + a_j^2)$ in terms of the vector, v_j, and axial-vector, a_j, couplings of fermion j to the Z^0; $\alpha = (\hat{\sigma}_U - 2\hat{\sigma}_L)/(\hat{\sigma}_U + 2\hat{\sigma}_L)$, and $\hat{\sigma}_i \equiv \int d^2\sigma_i$. Similarly,

$$\frac{d\sigma}{d\chi} \propto 1 + \beta\cos 2\chi - \frac{3\pi}{2\sqrt{2}} P_Z A'_P \cos\chi, \qquad (3)$$

where $A'_P = A_b \cdot \hat{\sigma}_A/(\hat{\sigma}_U + \hat{\sigma}_L)$, and $\beta = \hat{\sigma}_T/(\hat{\sigma}_U + \hat{\sigma}_L)$. Given the value of A_b, measurement of A_P and A'_P allows one to test the QCD predictions for $\hat{\sigma}_P/(\hat{\sigma}_U+\hat{\sigma}_L)$ and $\hat{\sigma}_A/(\hat{\sigma}_U + \hat{\sigma}_L)$. Furthermore, the ratio A'_P/A_P yields $\hat{\sigma}_A/\hat{\sigma}_P$ independently of A_b.

In terms of the polar angle ω, w.r.t. the electron-beam direction, of the vector $\vec{n}$ normal to the event plane,

$$\frac{d\sigma}{d\cos\omega} \propto 1 + \gamma\cos^2\omega + \frac{16}{9} P_Z A_T \cos\omega, \qquad (4)$$

where $\gamma = (2\hat{\sigma}_L - \hat{\sigma}_U - 6\hat{\sigma}_T)/(3\hat{\sigma}_U + 2\hat{\sigma}_L + 2\hat{\sigma}_T)$; the asymmetry term is one of the three T-odd terms mentioned above. The vector $\vec{n}$ can be defined in several ways; for example: 1) using the two highest-energy jets, $\vec{n} = \vec{p}_1 \times \vec{p}_2$; or 2) using the quark and anti-quark momenta, $\vec{n} = \vec{p}_b \times \vec{p}_{\bar{b}}$. The asymmetry is CP-even in the first definition, and CP-odd in the second; in both cases in the SM $A_T = 0$ at tree level. Higher-order corrections to $e^+e^- \to b\bar{b}g$ yield $|A_T| < 10^{-5}$ [4]. The asymmetry in $\cos\omega$ is hence potentially sensitive to beyond-SM processes [5]. In an earlier paper [6] we studied only the CP-even case using a lower-statistics sample of flavor-inclusive Z^0 decays.

The data were recorded in the SLC Large Detector (SLD). This analysis used charged tracks measured in the central drift chamber and in the CCD-based vertex detectors (VXD) [7]. About 70% of the data were taken with the new VXD installed in 1996 (VXD3), and the rest with the previous detector (VXD2). Only well-reconstructed tracks were used for the b-jet tagging. In each event jets were reconstructed using the "Durham" algorithm [8]. To select planar three-jet events we required exactly three reconstructed jets to be found with a jet-resolution parameter value of $y_c=0.005$, the sum of the angles between the three jets to be greater than 358°, and that each jet contain at least two charged tracks; 74,886 events satisfied these criteria. The jet energies were calculated by using the measured jet directions and solving the three-body kinematics assuming massless jets. The jets were then labeled such that $E_1 > E_2 > E_3$.

To select $b\bar{b}g$ events the long lifetime and large invariant mass of B-hadrons were exploited. An algorithm [9,10] was applied to the set of well reconstructed tracks in each jet in an attempt to reconstruct a decay vertex. Vertices were required to contain at least two tracks, and to be separated from the interaction point (IP) by at least 1 mm. We calculated (described below) that the probability for reconstructing at least one such vertex was ∼ 91% (77%) in $b\bar{b}g$ events, ∼ 45% (26%) in $c\bar{c}g$ events, and ∼ 2% (2%) in light-quark events recorded in VXD3 (VXD2). An event was selected as $b\bar{b}g$ if at least one jet contained a vertex with invariant mass [10] > 1.5 GeV/c². A total of 14,658 events satisfied this requirement and were subjected to further analysis. We calculated that this selection is 84% (69%) efficient for identifying a sample of $b\bar{b}g$ events with 84% (87%) purity, and containing 14% (11%) $c\bar{c}g$ and 2% (2%) light-flavor backgrounds.

Figs. 1(a) and (b) show the observed $\cos\theta$ distributions for event samples produced by left- and right-handed electron beams, respectively. The histograms show the estimated backgrounds, which are mostly $c\bar{c}g$ events. A maximum-likelihood fit yields an asymmetry parameter of $A_P = 0.855 \pm 0.050$ (stat.). Figs. 2(a) and (b) show the observed χ distributions for events produced with left- and right-handed electron beams, respectively. A maximum-likelihood fit yields an asymmetry parameter of $A'_P = -0.013 \pm 0.033$ (stat.).

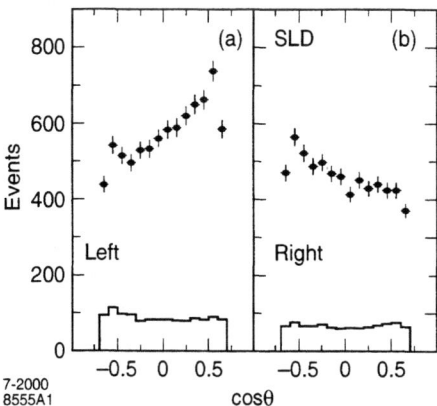

FIGURE 1. Polar-angle distribution of the signed-thrust axis direction for (a) left- and (b) right-handed electron beam. The histograms represent the simulated backgrounds.

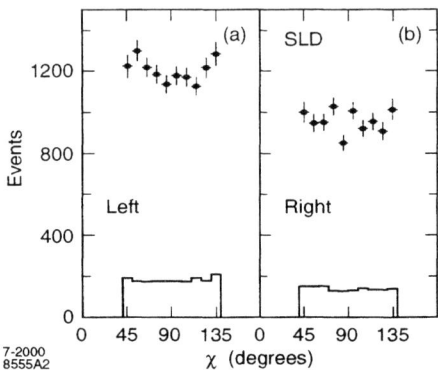

FIGURE 2. As Figure 1, for the azimuthal-angle distribution.

Figs. 3(a) and (b) show the left-right-forward-backward asymmetry in $|\cos\omega| \equiv z$,

$$\tilde{A}_{FB}(z) \equiv \frac{\sigma_L(z) - \sigma_L(-z) + \sigma_R(-z) - \sigma_R(z)}{\sigma_L(z) + \sigma_L(-z) + \sigma_R(-z) + \sigma_R(z)}$$

for the two definitions of the direction of $\vec{n}$: (1) $\vec{p_1} \times \vec{p_2}$, and (2) $\vec{p_b} \times \vec{p_{\bar{b}}}$. No asymmetry is apparent. Maximum-likelihood fits (Fig. 3) yielded $A_T^+ = -0.014 \pm 0.016$ (stat.) and $A_T^- = -0.035 \pm 0.024$ (stat.).

From the measured values of A_P and A'_P, and assuming the SM expectation of $A_b \simeq 0.935$ for $\sin^2\theta_w = 0.232$, we find, respectively

$$\frac{\hat{\sigma}_P}{\hat{\sigma}_U + \hat{\sigma}_L} = 0.914 \pm 0.053 \, (stat.) \pm 0.063 \, (syst.),$$

$$\frac{\hat{\sigma}_A}{\hat{\sigma}_U + \hat{\sigma}_L} = -0.014 \pm 0.035 \, (stat.) \pm 0.002 \, (syst.).$$

These values are consistent with the $\mathcal{O}(\alpha_s^2)$ QCD expectation (for massless quarks) of $\hat{\sigma}_P/(\hat{\sigma}_U + \hat{\sigma}_L) = 0.93$ and $\hat{\sigma}_A/(\hat{\sigma}_U + \hat{\sigma}_L) = -0.06$, respectively, calculated using JETSET 7.4. These yield $\hat{\sigma}_A/\hat{\sigma}_P = -0.015 \pm 0.038$ independent of the assumed A_b, and consistent with the expected value -0.065. We also used the A_P and A'_P values to set a 95% C.L. limit on an anomalous axial-vector coupling of the gluon to the b-quark, parametrized by a factor $(1 + \epsilon\gamma_5)\gamma_\nu$ in the $b\bar{b}g$ coupling, of $\epsilon < 0.34$. The measured values of A_T^+ and A_T^- correspond to 95% C.L. limits of $-0.045 < A_T^+ < 0.016$ and $-0.082 < A_T^- < 0.012$ (Fig. 3).

In conclusion, we have made the first symmetry tests in polarized Z^0 decays to $b\bar{b}g$. From the forward-backward polar-angle asymmetry of the signed-thrust axis and the azimuthal-angle asymmetry we found the parity-violation parameters A_P and A'_P, respectively, to be consistent with $\mathcal{O}(\alpha_s^2)$ QCD expectations. We set a corresponding 95% C.L. limit on parity violation at the $b\bar{b}g$ vertex. Using the event-plane normal polar-angle distributions we set 95% C.L. limits on the T-odd, CP-even and odd asymmetry parameters A_T^+ and A_T^-, respectively.

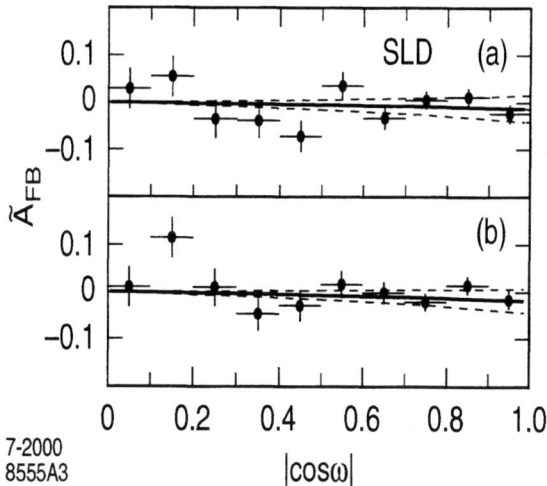

FIGURE 3. Left-right forward-backward asymmetry in $\cos\omega$ for (a) CP-even case, and (b) CP-odd case. The solid curve is the best fit to the data sample, and the dashed curves correspond to the 95% C.L. limits.

REFERENCES

1. H. A. Olsen et al., Nucl. Phys. **B171** 209 (1980).
2. K. Hagiwara et al., Nucl. Phys. **B358** 80 (1991).
3. P. N. Burrows, P. Osland, Phys. Lett. **B400** 385 (1997).
4. A. Brandenburg, L. Dixon, Y. Shadmi, Phys. Rev. **D53** 1264 (1996).
5. See, eg. C. D. Carone, H. Murayama, Phys. Rev. Lett. **74** 3122 (1995).
6. K. Abe et al., Phys. Rev. Lett. **75** (1995) 4173.
7. C. J. S. Damerell et al., Nucl. Inst. Meth. **A288** 236 (1990); K. Abe et al., Nucl. Inst. Meth. **A400** 287 (1997).
8. S. Catani et al., Phys. Lett. **B263** 491 (1991).
9. D. J. Jackson, Nucl. Inst. Meth. **A388** 247 (1997).
10. K. Abe et al., Phys. Rev. Lett. **84** 4300 (2000).

Polarized Muon Decay: Measurement of the Polarization Vector of the Decay Positrons

K. Bodek[1,2], A. Budzanowski[3], N. Danneberg[1], W. Fetscher[1], C. Hilbes[1], L. Jarczyk[2], K. Kirch[1], S. Kistryn[2], J. Klement[1], K. Köhler[1], A. Kozela[1,3], J. Lang[1], G. Llosá Llácer[1], M. Markiewicz[1], X. Morelle[4], T. Schweizer[1], J. Smyrski[2], J. Sromicki[1], E. Stephan[5], A. Strzałkowski[2], J. Zejma[1,2]

[1] *Institut für Teilchenphysik, ETH Zürich, CH 8093 Zürich, Switzerland*[1]
[2] *Institute of Physics, Jagellonian University, Kraków, Poland*
[3] *H. Niewodniczanski Institute of Nuclear Physics, Kraków, Poland*
[4] *Paul Scherrer Institut, CH-5232 Villigen-PSI, Switzerland*
[5] *Institute of Physics, University of Silesia, Katowice, Poland*

Abstract.
In the standard model (SM) of electroweak interactions the positron from the decay of polarized positive muons is mainly longitudinally polarized. The measurement of the two transverse polarization components, therefore, is a sensitive tool for contributions from additional, exotic, interactions.

The energy dependence of the transverse polarization component P_{T_1}, which lies in the plane spanned by muon-spin and positron momentum, yields the low energy parameter η and thus an improved model-independent value of the Fermi coupling constant. A non-zero value of the transverse component P_{T_2}, which is perpendicular to the above mentioned plane, would be the first observation of time reversal violation in a purely leptonic decay.

The μ_{P_T} experiment at the Paul Scherrer Institute determines the three polarization components simultaneously with the same apparatus by making use of three different reactions (spatial and temporal dependence of annihilation-in-flight with polarized electrons as well as muon decay asymmetry). The use of a stroboscopic method greatly reduces systematic errors. The measurement of the longitudinal polarization serves mainly as a test of the sensitivity of the apparatus, while the measurement of the two transverse components will improve the current experimental limits. The preliminary results are $P_{T_1} = (7 \pm 13) \times 10^{-3}$, $P_{T_2} = (19 \pm 13) \times 10^{-3}$.

---

[1] This project is supported in part by the Swiss National Science Foundation and by the Polish Committee for Scientific Research under Grant No. 2P03B05111.

INTRODUCTION

Measurements of muon decay are low energy tests of the standard model. In fact, only a few years ago it has been shown that $V - A$, as one of the basic assumptions of the standard model, *follows* from the results of a selected set of muon decay experiments (including inverse muon decay) [1]. The experimental limits obtained up to now, however, still allow for substantial contributions from non-standard couplings which differ in their spin structure from the V - A interaction. The limits on these couplings can be efficiently improved by performing experiments with polarized muons and positrons. The measurement of the transverse positron polarization P_{T_1} as a function of the positron energy, in particular, offers the possibility to obtain the low energy parameter η without the suppression factor m_e/m_μ, which makes the determination of η from the electron energy spectrum extremely difficult. The simultaneous measurement of the polarization component P_{T_2} allows one to test time reversal invariance.

I OBSERVABLES AND INTERACTIONS

Fig. 1 shows the kinematic variables for muon decay. While the e^+ from μ^+ decay is mainly longitudinally polarized (polarization P_L), there also is a transverse polarization component P_{T_1} lying in the plane of muon polarization $\mathbf{P}_\mu$ and positron momentum $\mathbf{k}_e$. Within the standard model P_{T_1} is negligibly small at large positron

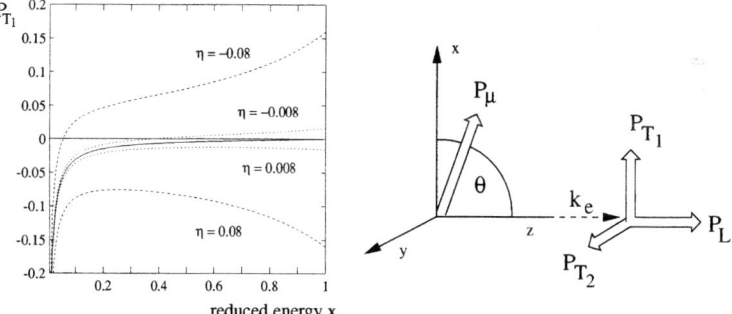

FIGURE 1. Transverse positron polarization P_{T_1} as a function of the reduced positron energy. The standard model predicts $\eta = 0$ (solid curve).

energies, but substantial at lower energies and reaches the value $-1/3$ in the limiting case of a positron at rest (see Fig. 1, $\eta = 0$). Due to the low rate at small positron energies the energy averaged transverse polarization predicted by the standard model is $< P_{T_1} > = -0.003$ and therefore at present cannot be detected.

One can, however, obtain large transverse polarizations by including additional interactions. In the representation of ref. [1] the matrix element for muon decay is given by

$$\mathcal{M} = \frac{4G_{\mathrm{F}}}{\sqrt{2}} \sum_{\substack{\gamma=\mathrm{S,V,T} \\ \varepsilon,\mu=\mathrm{R,L}}} g^{\gamma}_{\varepsilon\mu} \langle \overline{e}_{\varepsilon} | \Gamma^{\gamma} | (\nu_e)_n \rangle \langle \overline{\nu}_m | \Gamma_{\gamma} | (\mu)_{\mu} \rangle \qquad (1)$$

The index γ labels the type of interaction (S = 4-scalar, V = 4-vector, T = 4-tensor). The indices ε, μ indicate the chiralities of the spinors of the observed (charged) leptons. The chiralities n, m of the neutrinos are uniquely determined for given γ, ε and μ.

The transverse polarization component P_{T_1} yields the low energy parameter η *without* the suppression factor m_e/m_μ of η in the energy spectrum of the decay positron. With the experimental knowledge that V - A is dominant [1], and neglecting exotic contributions in second order, one obtains

$$\eta = \frac{1}{2} Re\{g^{\mathrm{S}}_{\mathrm{RR}}\} \qquad (2)$$

In the general case there will be a phase between V - A and an additional interaction which leads to a transverse component P_{T_2} *perpendicular* to the plane of muon polarization and positron momentum, and which violates time reversal invariance. Correspondingly one derives a value for $Im\{g^{\mathrm{S}}_{\mathrm{RR}}\}$ from the energy dependence of P_{T_2}. Here $g^{\mathrm{S}}_{\mathrm{RR}}$ represents a scalar, charge-changing interaction with right-handed charged leptons [3].

A more precise value of η is urgently needed for a model-independent determination of the Fermi coupling constant G_{F}: The influence of the uncertainty in the experimental value of η on the value of G_{F} is at present 20 times larger than the one of the more precisely known muon life time [3].

II EXPERIMENTAL SETUP

The experimental setup [4] is shown in Fig. 2. A beam of highly polarized muons ($P_\mu \approx 91\%$) enters the beryllium stop target with bunches every 20 ns. The polarization of the stopped muons precesses in a homogeneous magnetic field with the same frequency as the accelerator RF. Thus every new muon bunch is added coherently with the same direction as the polarization vector. Decay e^+ emitted parallel to the B-field are tracked by drift chambers and can annihilate with polarized e^- in a magnetized foil. The two annihilation quanta are then detected by a hexagonal array of 127 BGO crystals. A valid annihilation event requires a coincidence of two plastic scintillator counters before the magnetized foil with two separated clusters of BGO detectors and an anticoincidence with a plastic counter array in front of the BGO wall. A possible transverse polarization would be detected as a harmonic time dependence of the annihilation rate for a given detector pair.

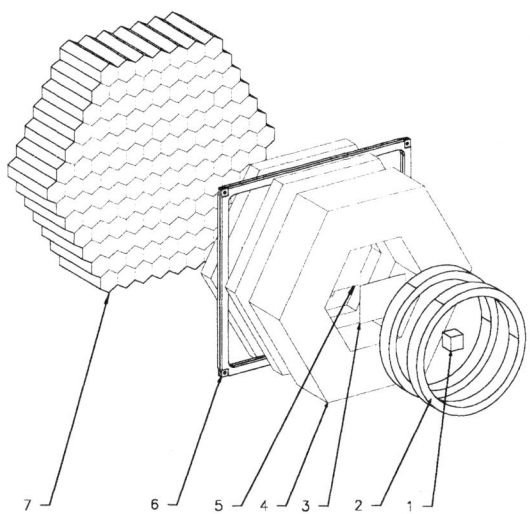

FIGURE 2. Experimental setup: 1 - Be target, 2 - spin precession magnet, 3 and 5 - plastic trigger counters, 4 - drift chamber (10 planes), 6- iron yoke of the magnetized Vacoflux foil, 7 - BGO calorimeter. Two additional drift chambers (2 planes each) sandwich the magnetized foil. An array of plastic veto counters (ANTI) in front of and cosmic trigger scintillators on top and below the BGO wall are not shown.

III EXPERIMENTAL RESULTS

In fall of 1999 we had the first data taking run. The data have been analysed and preliminary results are given.
The time distribution of the annihilation events contains two effects:

1. Since the accepted decay positrons are emitted into a cone whose axis coincides with the symmetry axis of the apparatus and is perpendicular to the precession plane of the muon polarization, there is a small remnant μSR effect (i.e., a time-dependent rate variation due to the decay asymmetry with respect to the precessing muon spin). This effect depends on the azimuthal angle of emission φ of the positron and yields time zero, i.e. the position of the precessing polarization vector $\boldsymbol{P}_\mu$ of the muon.

2. The effect due to a possible transverse polarization $\boldsymbol{P}_T$, in contrast, does not depend on φ, but only on the relative orientation of $\boldsymbol{P}_T$ and the electron polarization in the magnetized foil. This measurement yields the absolute value of $\boldsymbol{P}_T$,

Both P_{T_1} and P_{T_2} are consistent with zero at the present precision (see Fig. 3).

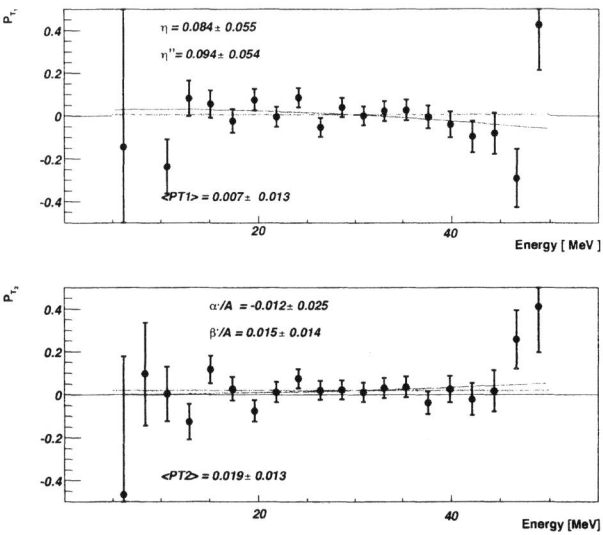

FIGURE 3. Energy dependence of the two transverse polarization components of the positrons from polarized muon decay at the moment of interaction (Preliminary result).

Here the question arises: Could we have detected a nonzero signal if there is a nonzero transverse polarization? The answer is yes:
By making use of the fact that positrons hitting the magnetized foil off the symmetry axis have a component of the longitudinal polarization in the direction of the electron polarization or opposite to it we can simultaneously measure the longitudinal polarization. The preliminary result is $P_\mathrm{L} = 1.06 \pm 0.15$.

REFERENCES

1. W. Fetscher, H.-J. Gerber and K.F. Johnson, Phys. Lett. **173B** 102 (1986).
2. H. Burkard *et al.*, Phys. Lett. **160 B** (1985) 343.
3. W. Fetscher and H.-J. Gerber, in *Precision Tests of the Standard Electroweak Model*, ed. P. Langacker, World Scientific, Singapore, 1995
4. I. Barnett *et al.*, to be published in NIM A (2000).

Measuring the Michel Parameter ξ'' in Polarized Muon Decay

X. Morelle[†,1], N. Danneberg[‡], J. Deutsch[*], J. Egger[†], W. Fetscher[‡],
F. Foroughi[†], J. Govaerts[*], C. Hilbes[‡], K. Kirch[‡], P. Knowles[¶],
K. Köhler[‡], A. Kozela[‡], J. Lang[‡], Y.W. Liu[†], R. Medve[*],
O. Naviliat[§], A. Ninane, R. Prieels, P. Van Hove[*]

[†] *Paul Scherrer Institut, Switzerland*
[‡] *Institut für Teilchenphysik, ETHZ, Switzerland*
[*] *Université catholique de Louvain, Belgium*
[¶] *Université de Fribourg, Switzerland*
[§] *Université de Caen, France*

Abstract. The study of pure leptonic electroweak processes, such as muon decay, is appropriate to test the Standard Model of the electroweak interactions as well as some of its extensions since the radiative corrections are negligible (and understood) at this level of precision. Unlike most of the Michel parameters used to describe muon decay, the uncertainty on the experimental value of ξ'' that fits the Standard Model V-A interaction, is large ($\xi'' = 0.65 \pm 0.36$) [1]. In order to improve the accuracy of this value, we have measured the longitudinal polarisation, P_L, of positrons emitted in the decay of polarized muons as a function of the positron energy. In fact, the value of P_L, at the positron endpoint energy equal to unity in the Standard Model, decreases for high energy positrons emitted antiparallel to the muon spin if the combination of Michel parameters $\xi''/\xi\xi' - 1$ deviates from the Standard Model value, i.e. from zero. We plan to improve the precision on ξ'' to 0.5%. Preliminary results are described.

INTRODUCTION

The observables of polarized muon decay (see fig. 1) ($\vec{k}_e, \vec{P}_e(P_L, P_{T_1}, P_{T_2})$) can be expressed as different combinations of the so-called "Michel parameters". The measured values of these parameters are all well in agreement with the predictions of the V-A theory with errors smaller than a few percent [2]. One exception is the ξ'' parameter ($\xi'' = 0.65 \pm 0.36$) [1] which governs the angular and energy dependency of the positron longitudinal polarization P_L in polarized muon decay:

[1)] Contact person: Xavier.Morelle@psi.ch

$$P_L(x, \cos\theta) = \xi' \left[1 + \underbrace{\frac{P_\mu \xi \cos\theta (2x-1)}{(3-2x) + P_\mu \xi \cos\theta (2x-1)}}_{enhancement\,factor} \left(\frac{\xi''}{\xi\xi'} - 1\right) \right], \quad (1)$$

where P_μ is the muon polarization, θ is the angle between the positron momentum and the muon spin, x is the relative energy of the positron, and ξ, ξ', and ξ'' are Michel parameters.

If the combination of Michel parameters $\left(\frac{\xi''}{\xi\xi'} - 1\right)$ differs from zero, the value of P_L will have a maximal deviation from unity when the Michel positron is emited at $\theta = 180°$ and $x \simeq 1$ ($E_{max} = 52.8$ MeV). This reflects the energy and angular dependency of the enhancement factor multiplying the combination of the Michel parameters shown in Eq. 1 (see fig. 2).

Therefore, in order to display such a possible effect, one has to measure the dependency of the positrons' longitudinal polarization on the energy. Our goal is to reach a precision on $\left(\frac{\xi''}{\xi\xi'} - 1\right)$ of 0.005 (whose value presently syands at -0.35±0.33 [1]).

By performing a relative measurement of the longitudinal polarization rather than an absolute one, we have only to determine the enhancement factor from the positron emission asymmetry (see the dots in fig. 2) for polarized and unpolarized muons. Even though the emission angle is not measured, the enhancement factor, which is an integral over a nonmeasured angular range which sets the sensitivity level of our experiment, is determined experimentally.

The relative positron longitudinal polarization $R(P_\mu)$ reads,

$$R(P_\mu) = \frac{P_L(\mu\ polarized)}{P_L(\mu\ unpolarized)} = 1 + \frac{A_{exp}}{1 + A_{exp}} \left(\frac{\xi''}{\xi\xi'} - 1\right), \quad (2)$$

where A_{exp} is the energy dependent experimental value of the asymmetry which enters the enhancement factor, which is obtained from:

$$A_{exp} = \frac{N(P_{\mu\ (polarized)})}{N(P_{\mu\ (unpolarized)})} - 1, \quad (3)$$

where $N(P_\mu)$ is the number of positrons entering the polarimeter for the same number of muons having a polarization P_μ. Close to the endpoint, the value of A_{exp} is negative.

In our set-up, muons arrive polarized to about 95% on the target. They keep 99.8% of their initial polarization if they are stopped in aluminium. In order to obtain unpolarized muons in similar conditions, we stop them in sulfur, where they are depolarized up to the value of 10% of the initial value, even in a large magnetic field environment (1 kG).

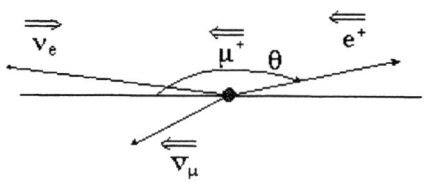

Figure 1. Muon decay in its rest frame. The positron has its maximal energy if both neutrinos are emitted at 180° with respect to the positron momentum.

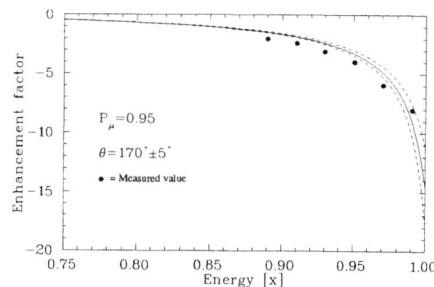

Figure 2. The enhancement factor as a function of the emitted positron energy, for a given emission angle (170°) and a muon polarization of 95%. Dots are measured values of the enhancement factor for energy bins of 1 MeV.

DISCUSSION

The relevant combination of Michel parameters may also be expressed as a function of the muon decay coupling constants [2]. The observable then takes the form

$$\frac{\xi''}{\xi\xi'} - 1 = 4\left[2\frac{|g_{RL}^V|^2}{|g_{LL}^V|^2} + \frac{|g_{RR}^V|^2}{|g_{LL}^V|^2} + \frac{1}{4}\frac{|g_{RR}^S|^2}{|g_{LL}^V|^2} + 4\frac{|g_{RL}^T|^2}{|g_{LL}^V|^2} + 2\mathcal{R}\left(\frac{g_{RL}^S}{g_{LL}^V}\frac{g_{RL}^{T*}}{g_{LL}^{V*}}\right)\right] \quad (4)$$

The upper indices of these coupling constants refer to the three possible Lorentz covariant structures (vector, scalar, tensor), while their lower indices correspond to the chirality of the e^+ (resp. μ^+) taking part in the corresponding interaction. In the standard model (V-A electroweak interactions), the only nonvanishing effective coupling constant is $g_{LL}^V = 1$. The present limits on all coupling constants coming from measurements of other observables do not provide any useful constraints on our Michel parameter combination (see fig. 3).
In the next sections, the discussion is organized into two parts. In the first part, we retain only the coupling constants $g_{RL}^{S,T}$ and assume that the others vanish. In the second part, we consider the four vector parameters $g_{\alpha,\beta}^V(\alpha, \beta = R,L)$ assuming that no scalar and no tensor effective interactions contribute (left-right symmetric model). The uncertainty on $(\xi''/\xi\xi'-1)$ is dominated by the last two terms of Eq. 4 which could contribute as much as 0.17, compared to 0.029 for the sum of the first three terms!

Scalar and tensor couplings

In Fig. 4, the constraints to be obtained in the g_{RL}^S vs. g_{RL}^T parameter plane are displayed, assuming a result of $(\xi''/\xi\xi' - 1) = 0 \pm 0.005$, and that only g_{RL}^S and g_{RL}^T are real and assumed to be non vanishing.

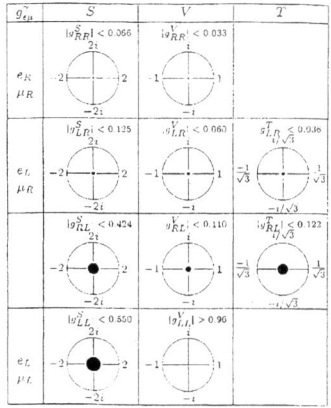

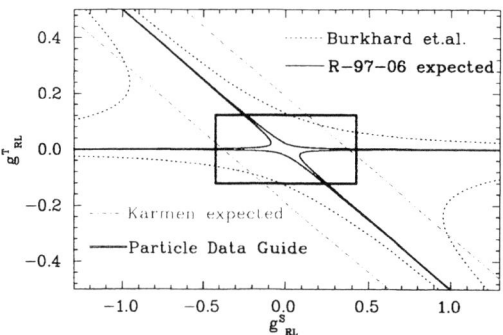

Figure 3. The 90% confidence limits for the various coupling constants, where the allowed values are inside the shaded regions. The g parameters have maximal possible values of 2,1,and $1/\sqrt{3}$ for S,V,T respectively.

Figure 4. Constraints on $g_{RL}^{S,T}$ stemming from the existing measurement of Burkhard et al. [1], the Review of Particle Properties and those expected from the KARMEN experiment at 1% precision and this experiement (R-97-06). The limits are shown at 1.64σ (90%).

General left-right symmetric models

The second scenario, in which all scalar and tensor effective coupling constants are taken to vanish $g_{\alpha\beta}^{S,T} = 0$ (α, β = R,L) but not the vector ones $g_{\alpha\beta}^V$ (α, β = R,L), corresponds to general left-right symmetric models. The parameters of such models are [3–5]: the mass-squared ratio $\delta_M = (M_1/M_2)^2$, where M_1 and M_2 are the masses of the known "left-handed" and the hypothetical "right-handed" physical gauge bosons,respectively, $t = \tan\zeta$, ζ being the mixing angle between these two gauge boson mass eigenstates (the recent 95 % C.L. lower bound of 720 GeV/c^2 on the mass of such a right-handed gauge boson set by the D0 [6] and CDF [7] collaborations from $p\bar{p}$ collider experiments, albeit assuming a more restricted parameter-space then the one we assume here, would require from our experiment a level of precision we cannot reach), the ratio r of the two gauge coupling constants and ν_l ($l = e, \mu$), the ratio of specific combinations of the leptonic Cabibbo-Kobayashi-Maskawa matrix elements U_{li}^L and U_{li}^R ($i = 1, 2, 3$; $l = e, \mu, \tau$) associated to each chirality sector, and corresponding to massive neutrino mixing. More explicitly, one has:

$$\delta_M = \frac{M_1^2}{M_2^2} \qquad t = \tan\zeta \qquad r = \frac{g_R^W}{g_L^W} \qquad \nu_l = \frac{\sum_i^l |U_{li}^R|^2}{\sum_i^l |U_{li}^L|^2} \qquad l = e, \mu \qquad (5)$$

Using these notations, the combination of Michel parameters to be measured takes the form,

$$\frac{\xi''}{\xi\xi'} - 1 \cong 4\nu_e r^2 \left[2t^2 + \nu_\mu r^2 \delta_M^2\right] \qquad (6)$$

therefore, our experiment will provide new constraints on the two terms of Eq. 4. It is further illuminating to compare Eq. (6) to the constraints provided by the positron asymmetry measurement of Ref. [8] which combines the purely leptonic process of muon decay with semi-leptonic one of the polarized muon producing pion decay. Assuming the condition $V_{ud}^R = 0$ in that case, the positron asymmetry at its energy end point is related to the quantity,

$$|P_\mu| \xi \frac{\delta}{\rho} - 1 \simeq -2\nu_\mu r^2 \left[t^2 + \nu_e r^2 \delta_M^2\right]. \qquad (7)$$

If our experiment provides a nonvanishing value for $\frac{\xi''}{\xi\xi'} - 1$, a comparison to the positron asymmetry value (Eq. 7) will allow us to conclude that the null result of asymmetry measurements is due to $\nu_\mu \approx 0$, i.e., to the massiveness of the right-handed muon neutrino.

THE EXPERIMENT

We are performing a relative measurement of the longitudinal polarization of the positron as a function of the positron energy. The positron polarization is analyzed using rate asymmetries of annihilation ($e^+ + e^- \to \gamma + \gamma$) and Bhabha scattering ($e^+ + e^- \to e^+ + e^-$) events on spin-polarized electrons for particles having their spin parallel or anti-parallel, and this for polarized and unpolarized muons. To take advantage of the large value of the enhancement factor for high energy positron, we select positrons having at least an energy equal to 90% of their maximal energy [9].

The incoming polarized muons are stopped in a 1.5 mm thick target of pure aluminium or of sulfur if we want to depolarize it. During the experiment the muon polarization is monitored by measuring the rate asymmetries in 3 telescopes surrounding the target.

Following the decay of the muon, the positrons have to pass a first magnet, which accepts only particles emitted in the angular range of $\theta = 160°$ to $180°$ relative to the muon spin. Collimators placed inside this magnet in addition select positrons in an energy window between 45 and 52.8 MeV, the positron energy end-point.

The first magnet is followed by a second one which has a homogeneous magnetic field of 2 Teslas. Three planes of silicon strip detectors placed in this magnet measure the spiraling trajectory of each positron. The tracking information (x,y,z are known within 1mm for each plane) given by this detector allows us to reconstruct the positron momentum with a resolution of 1.12 MeV/c (resp 1.19 MeV/c) for unpolarized muon (resp. polarized muon). The energy reconstruction efficiency of the silicon tracker is about 80%.

The presence of a third magnet is required to focus the positron beam onto the polarimeter. The longitudinal polarization of the positrons is measured by use of annihilation in flight and Bhabha scattering events of positrons on polarized electrons. The spin dependent parts of the cross sections of these two channels has opposite signs in the region of interest. The simultaneous measurement of both types of events will indicate and allow to control possible systematic effects. The polarimeter (see fig. 5) comprises 2 scintillators (used as trigger counters for the acquisition); 5 multi-wire proportional chambers (used to track the particle); 2 magnetized Vacoflux foils (49% iron, 49% cobalt, 2% nickel) magnetized in opposite directions and placed alternatively at +45° and -45° relative to the positron beam axis so that positrons can interact with polarized electrons (the foils are sandwiched between three of the five wire chambers); 2 planes of hodoscopes (used to discriminate an annihilation from a Bhabha or Michel event); and a wall of 127 BGO crystal scintillators (collecting in each cell the deposited energy of the incoming particles (one electron and one positron, 2 γ's, or the Michel positron)). The tracks which are reconstructed with 91.5% efficiency from the wire chambers allow to locate the interaction point on the magnetized foil. Additional geometrical cuts greatly suppress background from bremsstrahlung events.

Using the measured event rates we can infer the asymmetry in the two foils for both event types as a function of the positron energy and obtain from this information the longitudinal polarization of the incoming positron:

$$P_L = \frac{a}{AP_{e^-}} \qquad \text{with } a = \frac{n_{\uparrow\uparrow} - n_{\uparrow\downarrow}}{n_{\uparrow\uparrow} + n_{\uparrow\downarrow}} \qquad \text{and } A = \frac{\sigma_{\uparrow\uparrow} - \sigma_{\uparrow\downarrow}}{\sigma_{\uparrow\uparrow} + \sigma_{\uparrow\downarrow}} \quad (8)$$

where a is the asymmetry of the measured rate (see Fig. 6), A the theoretical analysing power and P_{e^-} the electron polarization in the magnetized foils. We should point out that we don't have to know A and P_{e^-} with high precision since we are doing a relative measurement of P_L.

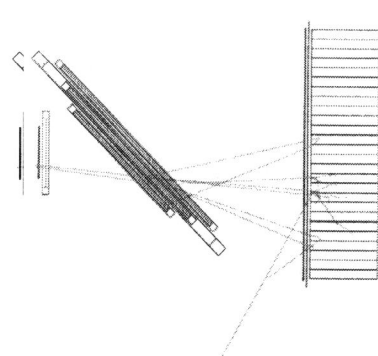
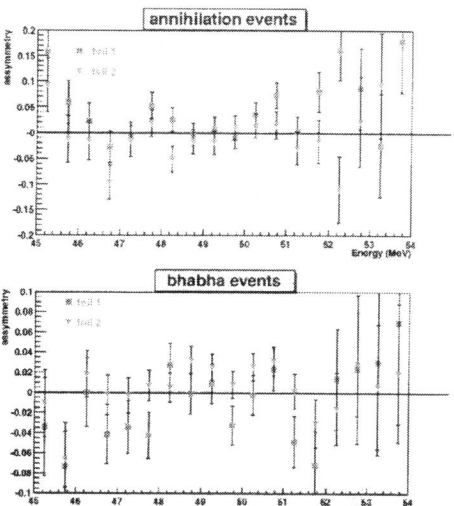

Figure 5. The polarimeter with the 5 multi-wire proportional chambers, the 2 vacoflux foils, the hodoscope and the BGO wall. One annihilation and one Bhabha event from the Monte Carlo simulation are shown.

Figure 6. The asymmetry for annihilation and Bhabha events is shown as a function of the energy of the positrons for statistics of only seven hours running time. The asymmetry is of the opposite sign for each foil due to the opposite magnetization, and for annihilation and Bhabha events because of the opposite analysing power of the two processes. The asymmetry is moreover twice as large for annihilation than for Bhabha processes, as is also predicted by the theory.

CONCLUSION

During our last production run, we took enough statistics to reach a good precision on the ξ'' parameter. We had the expected energy selection and resolution given by the silicon strip detector so that we are able to calculate the dependency of the enhancement factor versus the energy. We took data with all possible different geometries (foils at $\pm$ 45°, with both aluminum and sulfur targets, and with both electron polarization directions in the foils). Investigation of the backgrounds and the systematic effects is in progress. We expect to publish a new value for ξ'' in the near future.

REFERENCES

1. H. Burkhard et al., *Phys. Lett.* **160B**, 343 (1985).
2. W. Fetscher and H.-J. Gerber, *in Precision Tests of the Standard Electroweak Model*, ed. P. Langacker (World Scientific, Singapore,1995),pp. 657-705.
3. P. Langacker and S. Uma Sankar, *Phys. Rev.* **D40**, (1989) 1569.
4. P. Herczeg, *Phys. Rev.* **D34**, (1986) 3449.
5. J. Govaerts, Assymetry-Longitudinal Polarization Correlation in Muon Decay, a Reappraisal, *UCL-IPN-96-R02 Internal Report* (July 1995).
6. S. Abachi et al., *Phys Rev. Lett.* **76**, (1996) 3271; *Phys. Lett.* **B358**, (1995) 405.
7. F. Abe et al., *Phys. Rev. Lett.* **74**, (1995) 2900.
8. A. Jodidio et al., *Phys. Rev.* **D34**, (1986) 1967; (E) *ibid* **D37**, (1988) 237.
9. P. Van Hove, *L'expérience MELPOMENE: une nouvelle approche à la polarimetrie des positrons dans la désintégration des muons.*, Ph.D. dissertation, UCL, December 2000, in print.

Physics Beyond SM at RHIC with Polarized Protons

A. Ogawa[†], V. L. Rykov[§1] and N. Saito[‡2]

[†]*Penn State University, University Park, PA 16802, USA*
[§]*Wayne State University, Detroit, MI 48202, USA*
[‡]*RIKEN–BNL Research Center, Upton, NY 11973, USA*

Abstract. The capabilities of RHIC with polarized protons to test the Lorentz structure of electroweak interactions and also the properties of MSSM Higgs, should it be discovered, are discussed.

INTRODUCTION

RHIC-Spin experiment is about being started. It has the solid program[3], mainly concentrated around the measurements of proton spin dependent structure functions. For the last years, few attempts have been undertaken to also evaluate RHIC-Spin capabilities for testing physics beyond the Standard Model (SM) [1–3]. Since the recent suggestion [4] on a considerable increase of RHIC luminosity in pp mode along with a sizeable energy increase, RHIC-Spin potential in this area became looking even more promising.

In this report, we update our earlier study [2] of RHIC-Spin capabilities to explore the Lorentz structure of electroweak interactions and also provide some estimates on whether Higgs sector could be reachable at RHIC. In our considerations, we assume $\sqrt{S} = 500$ GeV and the luminosity of 0.8 fb^{-1}/year for RHIC before an upgrade (RHIC-500), and $\sqrt{S} = 650$ GeV and the luminosity of $\sim$20 fb^{-1}/year for the after upgrade (RHIC-650).

LORENTZ STRUCTURE OF QUARK ELECTROWEAK CURRENT

Lorentz structure of electroweak interactions had always been and remains to be the focus of many precise nuclear and particle physics experiments. Over the

---

[1]) Supported in part by the U.S. Department of Energy Grant DE-FG0292ER40713.
[2]) This work has been done partly within the framework of RIKEN RHIC-Spin project.
[3]) See plenary talk by N. Saito in this proceedings.

last few years, the first measurements of the proton's weak magnetism have been accomplished [5]. Much attention had also been paid to test the universality of the electroweak interactions, in general, and to evaluating electroweak dipole moments of quarks and leptons, in particular. The current experimental constraints to magnetic and electric dipole moments of quarks and leptons, except electrons and muons, are still quite far from the values, predicted in the SM. If any of these moments were found to be nonzero and above SM's expectations, this would be a clear signal of a new physics beyond SM [6,7].

The most stringent experimental constraints, applicable to all components of quark and τ-lepton dipole moments, come from the recent analyses [6,7] of electroweak data from high energy colliders. In these analyses, it has been assumed that theories beyond the SM, emerging at some characteristic energy scale above W/Z mass, have effects at low energies $E \leq M_{W,Z}$, and these effects can be taken into account by considering a Lagrangian that extends the SM Lagrangian, L_{SM}: $L = L_{SM} + L_{eff}$. To preserve the consistency of the low energy theory, it had also been assumed that L_{eff} is SU(3)⊗SU(2)⊗U(1) gauge invariant. Phenomenologically, this extension is equivalent to an introduction of a tensor coupling of fermions to gauge bosons which had also been discussed in Refs. [8,2,3]:

$$L_{eff}^{charged} = \frac{g}{2\sqrt{2}\cdot\Lambda}\left\{\bar{q}_d\sigma^{\mu\nu}(f_T^+ + f_T^-\gamma_5)q_u\partial_\nu W_\mu^- + \bar{q}_u\sigma^{\mu\nu}(f_T^{*+} - f_T^{*-}\gamma_5)q_d\partial_\nu W_\mu^+\right\} \quad (1)$$

In Eq. (1), representing the charged-current part[4] of L_{eff}, g is the electroweak coupling constant, Λ is the energy scale of the "full strength" tensor interactions, and the asterisk denotes the complex conjugate. The notations q_u and q_d are for the "upper" (u and c)[5] and "lower" (d,s,b) quarks, respectively. The CP- and T-invariance of model (1) is broken if any or all "formfactors" $f_T^\pm$ were complex[6].

With the coupling (1), a number of prohibited in the SM spin asymmetries must show up in the annihilation of polarized $q\bar{q} \to W^\pm/Z^0/\gamma \to l\bar{l}$ [2,3]. For polarized proton collisions, particularly interesting would be single-spin asymmetries with transverse polarization, arising from the interference of SM's and tensor couplings, because: a) these asymmetries are strongly suppressed in SM and b) they are expected to have a good sensitivity to anomalous interactions due to quite strong correlations between the proton spin and polarizations of high-x valence quarks, that participated in gauge boson production [9]. The triple-vector correlations, $\propto(\boldsymbol{k}\cdot[\boldsymbol{\zeta}_q^\perp \times \boldsymbol{p}_q])$, give rise to the "left-right" asymmetry A_N, while nonzero "up-down" asymmetry A_T comes from P- and CP-violating[7] two-vector correlations, $\propto(\boldsymbol{\zeta}_q^\perp \cdot \boldsymbol{k})$. In the formulae above, $\boldsymbol{p}_q$ and $\boldsymbol{\zeta}_q^\perp$ are for the momentum and transverse polarization of an incident quark, and $\boldsymbol{k}$ is the momentum of a final lepton. The

---
[4] With omitted quadratic terms, proportional to $g^2W_\mu W_\nu$; the neutral-current L_{eff} looks similarly.
[5] t-quark is virtually not reachable at RHIC.
[6] In this paper: $\gamma_5 = -i\gamma^0\gamma^1\gamma^2\gamma^3$ and $\sigma^{\mu\nu} = \frac{1}{2}(\gamma^\mu\gamma^\nu - \gamma^\nu\gamma^\mu)$.
[7] In this particular model for $L_{eff}^{charged}$.

respective asymmetries $\hat{a}_{N,T}$ in $W^{\pm}$ production at the quark interaction level, integrated over the phase space of final leptons, would be as follows:

$$\hat{a}_N \simeq \frac{3\pi}{16} \cdot \frac{M_W}{\Lambda} \cdot \text{Re}\{f_T^+ \mp f_T^-\} / \left\{1 + \frac{M_W^2}{4\Lambda^2}(|f_T^+|^2 + |f_T^-|^2)\right\} \qquad (2)$$

$$\hat{a}_T \simeq \frac{3\pi}{16} \cdot \frac{M_W}{\Lambda} \cdot \text{Im}\{f_T^- \mp f_T^+\} / \left\{1 + \frac{M_W^2}{4\Lambda^2}(|f_T^+|^2 + |f_T^-|^2)\right\} \qquad (3)$$

Formulae (2,3) have been obtained, using the solutions of Ref. [2] and with the assumptions that, the SM part of $q\bar{q}W$-interactions as well as the lepton coupling to $W^{\pm}$ were purely V-A. The upper and lower signs correspond to W^+ and W^- productions, respectively.

TABLE 1. RHIC-Spin sensitivity to $q\bar{q}W$ tensor coupling. Constraints (2σ) in columns 2-3 are for the asymmetries at quark ($\hat{a}$) and proton ($A \simeq \hat{a}/2$ [9,3]) interaction levels. RHIC sensitivities (2σ) to $A_{N,T}$ are shown in columns 4-5, assuming a 70% proton beam polarization.

Constraints from Ref. [6]			RHIC sensitivity in (W^+/W^-) modes	
$(\|f_T^+\|^2 + \|f_T^-\|^2) \cdot M_W^2/\Lambda^2$	$\hat{a}_{N,T}$	$A_{N,T}$	RHIC-500	RHIC-650
<0.15	$\lesssim(20\text{-}30)\%$	$\lesssim(10\text{-}15)\%$	$\sim(1.5/3.0)\%$	$\sim(0.2/0.4)\%$

The experimental constraint on "formfactors" $f_T^{\pm}$ in charged current $u \leftrightarrow d$ transitions, converted to the convenient for our goals representation, is shown in Table 1, along with the respective limits on spin asymmetries in $pp^{\uparrow} \to W^{\pm} + X \to l\bar{l} + X$ processes. These limits then are compared with the RHIC sensitivities to $A_{N,T}$. One can observe that the current limits on f_T in quark sector could be lowered by a factor of 5-10 at RHIC-500, and by almost an order of magnitude more after the upgrade to RHIC-650. And what is probably even more important, the real and imaginary parts of $f_T^{\pm}$ could be constrained individually from four independent measurements of A_N and A_T in W^+ and W^- productions.

WILL HIGGS PHYSICS BE REACHABLE AT RHIC?

In the recent years, extensive efforts for cornering Higgs bosons resulted in the significantly shrunken area of the still allowed Higgs sector parameters. The actual discovery of Higgs particle(s) is expected to occur in not so remote future. Then, the focus will be shifted to studying their properties. Does RHIC with *polarized* protons have capabilities to contribute in this study?

The production cross section of Higgs strongly depends on the helicities of initial gluons and quarks. As a result, measurements of spin-correlations in polarized pp collisions may allow us, for example, to determine CP-parities of Higgs states. Potentially, an interference between Higgs production and other SM processes in collisions of polarized gluons and quarks can generate a number of interesting single- and double-spin asymmetries[8], including those sensitive to CP-violation in Higgs

---

[8] Similar to the ones described, for example, in Refs. [10].

sector. In some cases, the polarization may help to improve signal-to-background ratio compared to unpolarized particle collisions.

Currently, the data favor low mass Higgs, probably just around $\sim$100 GeV, and for the Minimal Supersymmetric extension of the Standard Model (MSSM), the versions with large $\tan\beta \gtrsim 10$ seems as taking preferences[9]. Let's imagine for a moment that, just before RHIC-Spin upgrade, a neutral scalar[10] has been discovered with the mass $M_h \sim$115-120 GeV, and its characteristics were consistent with the MSSM Higgs for $\tan\beta \simeq 30$. The estimated production cross section, σ_h, for such boson(s) at RHIC-650 would be $\sim$0.5 pb with approximately equal contributions of $gg \to h$ and $q\bar{q} \to h$ subprocesses[11]. For an integrated luminosity of 20 fb/year, one may expect a yield of $\sim 10^4$ h/year.

The main decay modes of low mass Higgs scalars are: $h \to b\bar{b}$ with the branching of $\sim$90%, and $h \to \tau^+\tau^-$ with the branching of $\sim$8-9%. At the early stage, both these modes have been determined as hardly be suitable for the Higgs searches. However, as it was pointed out in Ref. [13], $h \to \tau^+\tau^-$ decay might become useful "to provide confirming evidence for a signal found in other modes"[12]. In this report, as a first step toward understanding the Higgs-at-RHIC problem, we will start with the evaluation of only these two modes with the highest branchings, keeping in mind of course, that future studies could reveal better ways for the Higgs sector exploration at RHIC[13].

The main obstacle for using $b\bar{b}$ channel at hadron colliders is the large QCD background from $gg \to b\bar{b}$. At RHIC-650, the estimated with PYTHIA production cross section $\sigma^{b\bar{b}}_{QCD}$ for $b\bar{b}$-pairs in the mass interval of 115$\times$(1$\pm$10%) GeV is $\gtrsim 10^3$ pb. Resulting from cross section estimates Higgs signal-to-background ratio, σ_h/σ_{QCD}, at $\lesssim 5\cdot 10^{-4}$ does not seem too encouraging, particularly for studying Higgs properties. Potentially, an interference between $gg \stackrel{QCD}{\to} b\bar{b}$ and $gg \stackrel{h}{\to} b\bar{b}$ could improve this ratio up to $\sim \sqrt{\sigma_h/\sigma_{QCD}} \sim 10^{-2}$. Unfortunately, in the ultra-relativistic limit of massless b-quark, these two channels do not interfere. As a result, the actual interference term is additionally suppressed by a factor $m_b/M_h \sim 0.03\text{-}0.04$, which brings it back to the same low level of $\lesssim 5\cdot 10^{-4}$.

The background to $\tau^+\tau^-$ pairs from Higgs decays predominantly comes from the Drell-Yan process: $q\bar{q} \to \gamma/Z^0 \to \tau^+\tau^-$. For RHIC-650, PYTHIA estimates the Drell-Yan cross section for $\tau^+\tau^-$ at $\sim$0.25 pb in the τ-pair mass interval of 115$\times$(1$\pm$10%) GeV. Taking into account $\sim$8% branching of Higgs to $\tau^+\tau^-$, the signal-to-background ratio in this mode would be $\sim$15%. With the detection efficiency for high-mass τ-pairs at about 20%, the event rate in the mass interval above should be expected at $\sim 10^3$ τ-pairs/year. For this rate, $\sim$15% excess of

---

[9]) See, for example, [11,12] and references therein.
[10]) Or scalars; for large $\tan\beta$, the theory allows for all three MSSM neutral bosons, h^0, H^0 and A^0, to be sitting simultaneously in the mass region from $\sim$100 to 130 GeV [12].
[11]) Notation h is used here as a generic name for either h^0, H^0 or A^0.
[12]) See also Ref. [12], page 29 for the similar qualification.
[13]) See, for example, J. F. Gunion and T. C. Yan, *Phys. Rev. Lett.*, **71** (1993) 488.

events due to Higgs decays should be detectable well above the statistical fluctuations. Then, for example the *CP*-parity of a detected Higgs boson could be determined by measuring the sign of the contributing to the cross section double-spin correlation ($\zeta_q^\perp \cdot \zeta_{\bar{q}}^\perp$) in collisions of transversely polarized protons. However, the statistics of just $\sim 10^2$ of $h \to \tau^+\tau^-$ decays/year may not be sufficient for measuring double-spin asymmetries. Unfortunately, we cannot count on potentially more sensitive single-spin ones which could arise from an interference of two competing $q\bar{q} \to \tau^+\tau^-$ channels. This interference will be vanishingly small because Drell-Yan pairs are mainly produced by light quarks, while the respective part of Higgs cross section will predominantly be due to $b\bar{b}$-annihilation.

CONCLUSION

It has been shown that RHIC with polarized protons will have highly competitive capabilities for hunting anomalies in the Lorentz structure of electroweak interaction due to physics beyond SM.

The low-mass Higgs might be reachable at high luminosity RHIC-650, although finding appropriate modes to study polarization phenomena in Higgs sector will be a quite challenging task.

Authors appreciate the valuable discussions with D. Boer, D. A. Cinabro, S. Dawson, R. L. Jaffe, J. S. Lange, T. Maruyama and J. Soffer. One of us (VLR) is thankful to T. M. Cormier for the support of his participation in SPIN2000 Symposium.

REFERENCES

1. P. Taxil and J.-M. Virey, *Phys. Lett.* **B404** (1997) 302, **B441** (1998) 376, *Phys. Rev.* **D55** (1997) 4480; J. Murata, Report to the RHIC Spin Collaboration, Oct. 2000 (unpublished).
2. V. L. Rykov, *Proc. of SPIN-98 Symposium (September 1998)*, p. 450–452; *hep-ex/9908050*.
3. S. Kovalenko, I. Schmidt, J. Soffer, *hep-ph/9912529*.
4. T. Roser, *Private communications*, and *www.rhichome.bnl.gov/RHIC/luminosity/upgrade/*.
5. B. Mueller et al. (SAMPLE Collaboration), *Phys. Rev. Lett.*, **78** (1997) 3824.
6. R. Escribano, E. Masso, *Nucl. Phys.*, **B249** (1994) 19.
7. R. Escribano, E. Masso, *Phys. Lett.*, **B301** (1993) 419; *Phys. Lett.*, **B395** (1997) 369; T. G. Rizzo, *Phys. Rev.*, **D56** (1997) 3074; K. Ackerstaff et al. (OPAL Collaboration), *Z. Phys.*, **C74** (1997) 403; M. Acciarri et al. (L3 Collaboration), *Phys. Lett.*, **B426** (1998) 207; G. A. Gonzalez-Sprinberg, A. Santamaria, J. Vidal, *hep-ph/0002203*.
8. G. L. Kane, G. A. Ladinsky and C.-P. Yuan, *Phys. Rev.*, **D45** (1992) 45.
9. J. Soffer, *Nucl. Phys. (Proc. Suppl.)*, **64** (1998) 143.
10. E. Akawa et al., *hep-ph/9912373*; J. I. Illana, *hep-ph/9912467*, and references therein.
11. P. Igo-Kemenes in Rev. of Part. Phys., *Eur. Phys. J.*, **C15** (2000) 274; J. Erler, *hep-ph/0010153*; T. Affolder et al. (CDF Collaboration), *hep-ex/0010052*.
12. M. Carena, J. S. Conway, H. E. Harber, J. D. Hobbs, *hep-ph/0010338*.
13. J. F. Gunion, H. E. Haber, G. Kane and S. Dawson, *The Higgs Hunter's Guide*, Addison-Wesley Publishing, 1990, p. 168.

3. SPIN STRUCTURE OF NUCLEONS

Measurements of the Spin Structure Function g_1 of the Proton and the Deuteron

Uta Stösslein* on behalf of the HERMES Collaboration

Nuclear Physics Laboratory, University of Colorado, Boulder, Colorado 80309-0446, USA
E-mail: uta.stoesslein@ifh.de

Abstract. New HERMES data are presented on the spin structure function g_1 of the proton in an extended kinematic range, $0.0021 < x < 0.021$ and $0.1\,\text{GeV}^2 < Q^2 < 1.2\,\text{GeV}^2$, and of the deuteron in the standard kinematic range, $0.021 < x < 0.85$ and $Q^2 > 0.8\,\text{GeV}^2$. Combined with CERN and SLAC data, for both nucleons it is found that the structure function ratio g_1/F_1 remains independent of Q^2.

Measurements of the structure function g_1 in inclusive deep-inelastic scattering (DIS) of longitudinally polarized lepton beams from longitudinally polarized targets have been used successfully for more than a decade to reveal the quark contribution to the nucleon spin, $\Delta\Sigma$. Although a wealth of data on polarized nucleon structure functions already exists [1], the range of small Bjorken-x has yet to be explored and the experimental precision is still unsatisfactory for the deuteron (neutron). Measurements at small x will reduce the extrapolation uncertainty for the integral $\int g_1 dx$ thus improving the accuracy in the determination of $\Delta\Sigma$. The combination of proton and deuteron (neutron) data is important for an extraction of flavor separated quark contributions, and for tests of the Bjorken sum rule.

The centre-of-mass energies presently available to spin experiments kinematically correlate the data at small Bjorken-$x < 0.01$ with low photon virtualities of $Q^2 < 3\,\text{GeV}^2$. However, exploring this region for spin dependent quantities has an interest in its own, allowing one to make tests of the validity range of perturbative QCD (pQCD), possibly by studying scaling violations.

The results presented here are based on the 1997 (proton target) and 1999 (deuteron target) data taking periods of the HERMES experiment at DESY. HERMES uses the 27.6 GeV positron (electron) beam of the HERA storage ring. The beam is transversely self-polarized. Longitudinal polarization in the interaction region is obtained by means of spin rotators. For the data reported the average beam polarization is 0.55 ± 0.02 (syst.). The polarized longitudinal gas target is confined in a tubular open-ended, cooled storage cell which is mounted inside the beam pipe and fed by an atomic beam source of nuclear-polarized Hydrogen (p) or Deuterium (d). The average target polarization is 0.88 ± 0.04 (syst.) for p and 0.81 ± 0.06 (preliminary syst.) for d.

The HERMES detector [2] is a forward spectrometer with a dipole magnet providing an integrated field of 1.3 Tm. A horizontal iron plate shields the beam line against the field thus dividing the spectrometer into two identical halves with a minimum vertical acceptance of ± 40 mrad. The acceptance extends to ± 140 mrad vertically and ± 170 mrad horizontally. Tracking is accomplished by 42 drift chamber planes in each detector half, resulting in an angular resolution of $\delta\Theta \leq 0.6$ mrad and a momentum resolution of $\delta p/p \leq 1.5\,\%$. The trigger for DIS events is formed by a coincidence of signals from three hodoscope planes with those from a lead-glass calorimeter with a required energy deposit of at least 1.4 GeV. Particle identification is achieved using a transition-radiation detector, a preshower hodoscope, the calorimeter, and a gas threshold Čerenkov counter (1997 data only). For the kinematic range $x > 0.021$ (< 0.021) the positron identification efficiency exceeds 99\% (97\%) with an estimated hadron contamination of less than 0.5\% (1.0\%).

The spin structure function $g_1(x, Q^2)$ is determined from the structure function ratio g_1/F_1 using assumptions about $F_1(x, Q^2)$. The ratio g_1/F_1 is approximately equal to the virtual photon asymmetry A_1 and is measured via the longitudinal cross section asymmetry $A_{\|}$. The function $F_1(x, Q^2) \simeq F_2(x, Q^2)/[2x(1 + R(x, Q^2))]$ is evaluated by parameterizations for F_2 [3], the unpolarized structure function, and for $R = \sigma_L/\sigma_T$ [4], the ratio of longitudinal to transverse virtual-photon absorption cross sections. For more analysis-related details see Ref. [5].

In a new analysis of the 1997 proton data the kinematic range covered is extended to $x < 0.021$ and $Q^2 < 0.8\,\text{GeV}^2$ complementing the previous HERMES publication [5] on the spin proton structure function $g_1^p(x, Q^2)$. To access the small x and low Q^2 data, the upper limit in y, the relative energy transfer in the laboratory frame, is increased from 0.85 to 0.91. This requires good understanding of the positron detection at low momenta, i.e. to control the trigger efficiency and the steeply rising contributions of background and of radiative events. Most relevant contributions to experimental systematic uncertainties originate from normalization, background corrections, acceptance cut variations, and scattering angle systematics. The systematic uncertainties arising from radiative and smearing corrections have been minimized using an iterative procedure for the extraction of the asymmetry at the Born cross section level. The total systematic uncertainty is about 9\%, and 14\% at the lowest x value.

The behavior of the proton structure function ratio g_1/F_1, see Fig. 1a), as a function of x shows a smooth transition to values close to zero at small values of $x < 0.021$. Comparing HERMES and SMC data no significant Q^2 dependence within the experimental uncertainties can be seen; the SMC small x ($Q^2 > 1\,\text{GeV}^2$) data are confirmed for the first time and with improved precision by HERMES. In Fig. 1b) the Q^2 dependence of the spin structure function $g_1^p(x, Q^2)$, shifted to common x values, is shown for world data [1,6,7] with $Q^2 > 1\,\text{GeV}^2$, the HERMES data [5] for $x > 0.021$, and the new small x and low Q^2 data points. The large beam

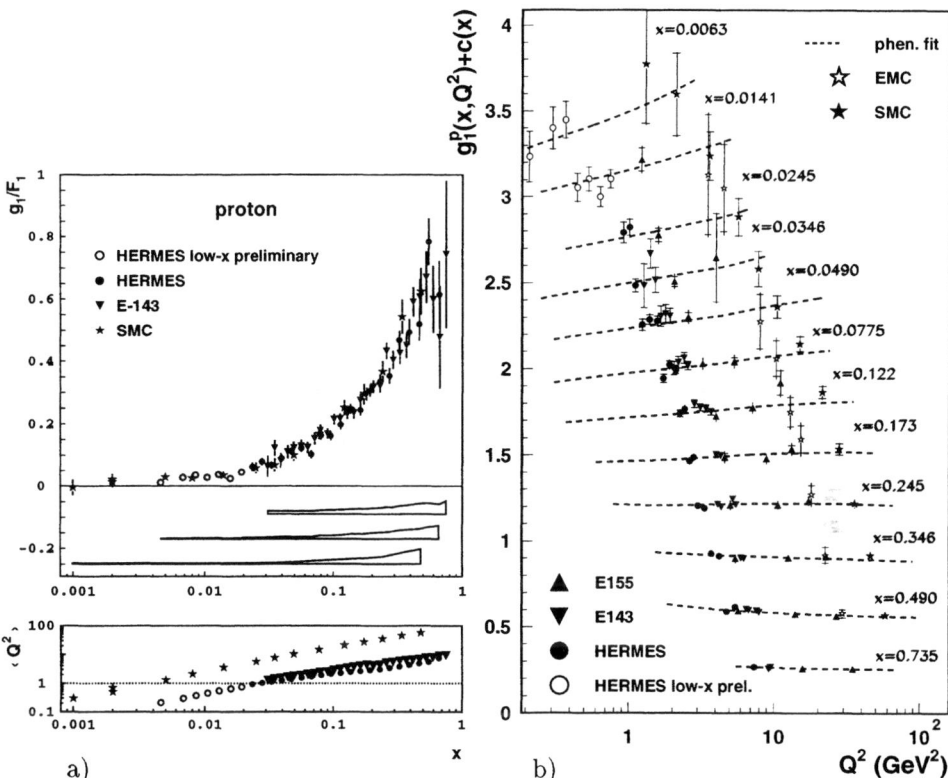

FIGURE 1. a) The structure function ratio g_1^p/F_1^p of the proton as a function of Bjorken-x measured at $\langle Q^2 \rangle$ as indicated in the lower panel. The bands show the systematic uncertainties. b) The spin structure function g_1^p of the proton [7] as a function of Q^2 but shifted to common x values as labeled. The phenomenological fit of g_1^p is based on parameterizations, see text.

energy of SMC (EMC) leads, in every x bin, to a lever arm in Q^2 of about one order of magnitude, whereas the recently published E155 data [8] lies in between. A fair description of the Q^2 dependence of g_1 is achieved based on phenomenological fits of g_1^p/F_1^p [8] and F_1^p [3,4]. It allows conclusions to be drawn on the Q^2 dependence of g_1^p: almost no or weak negative scaling violations at $x > 0.15$, significant positive scaling violations for $0.07 < x < 0.15$, and a first evidence for positive scaling violations at lower x values. Despite the availability of low Q^2 data for g_1^p/F_1^p and g_1^p, conclusions about possible higher twist contributions can not be drawn so far, since there is no clear concept yet for QCD analyses at these low Q^2 values. From the observed Q^2 independence of the ratio g_1^p/F_1^p it can be concluded that the scaling violations of g_1^p are very similar to those of F_1^p.

Since 1998 HERMES has taken data with a longitudinally polarized deuteron target. A first inclusive analysis of about 10^6 DIS events in the HERMES standard kinematic range, $x > 0.021$ and $Q^2 > 0.8\,\mathrm{GeV}^2$, is presented here. The measured x dependence of the structure function ratio g_1^d/F_1^d for the deuteron is shown in Fig. 2a) for the HERMES data only and in Fig. 2b) in comparison with recent $Q^2 > 1\,\mathrm{GeV}^2$ data [1] of CERN and SLAC indicating a Q^2 independence of the deuteron g_1/F_1 measurements presently available. Although only about 1/6 of the full data set collected is analyzed so far, the statistical precision is already comparable. The systematic uncertainty of this measurement is about 0.01 (0.03 at the edges). It is rising towards small x values mainly due to the lack of knowledge of the zero-crossings of g_1^d/F_1^d at $0.02 < x < 0.06$. Additionally, a possible contribution may occur from the still unknown tensor structure function b_1^d. In this analysis $b_1^d = 0$ is assumed. A final HERMES analysis of g_1^d will rely on a high statistics measurement including a first measurement of b_1^d using a separate data set with large tensor polarization taken in the year 2000.

A compilation of preliminary and published measurements of g_1 and xg_1 for protons, deuterons and neutrons is given in Figs. 3 a) and b), respectively. So far the proton data are of highest precision. More precise data at small x will open the field of small x physics of spin structure functions and presumably shed light on their partonic interpretation.

I would like to thank all HERMES colleagues. Special thanks go to my co-workers Yvo Gärber, Lara De Nardo and Christoph Weiskopf for their contributions and the good team work in obtaining and understanding the data.

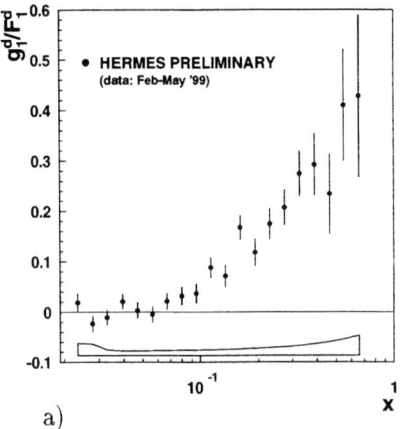
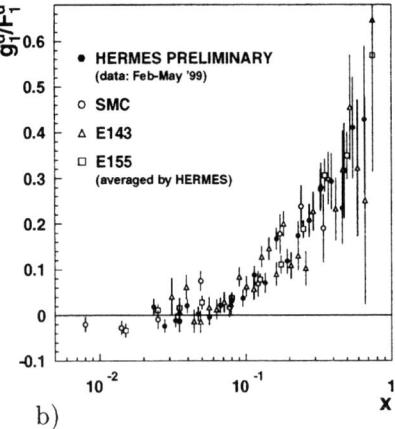

FIGURE 2. a) The structure function ratio g_1^d/F_1^d of the deuteron as a function of Bjorken-x measured at average $\langle Q^2 \rangle$ values. The band shows the systematic uncertainty. b) Same data but compared with results from SLAC and CERN, shown with the statistical errors only.

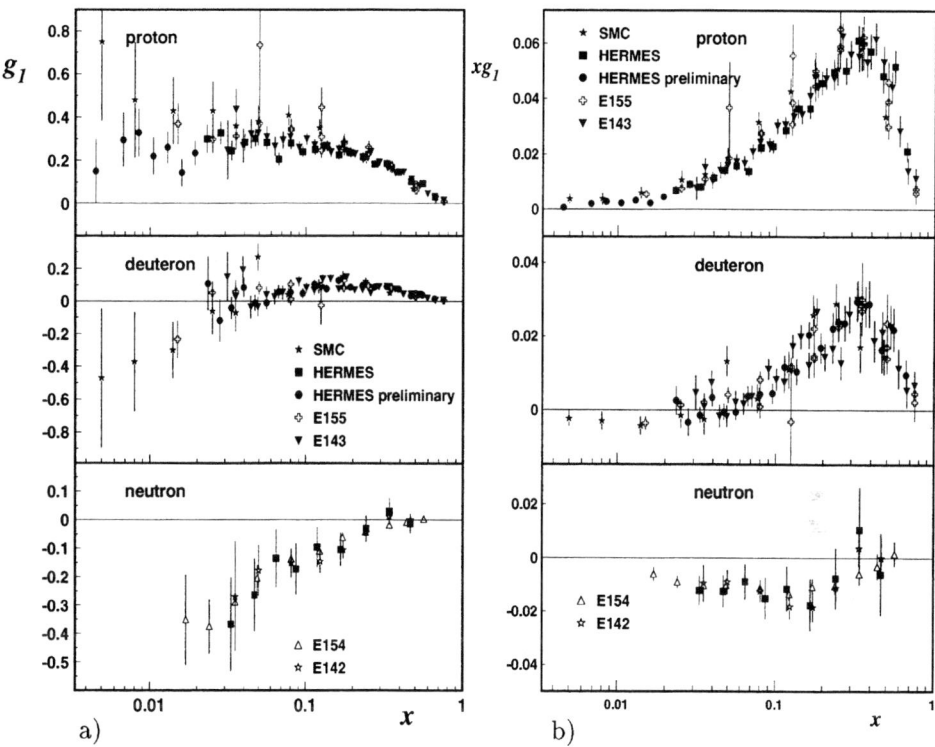

FIGURE 3. Compilation of world data on the nucleon spin structure function a) g_1 and b) xg_1 in dependence of Bjorken-x for measured average Q^2 values (not shown). The error bars are combined from statistical and systematic uncertainties.

REFERENCES

1. For a compilation of polarized data see: http://durpdg.dur.ac.uk/HEPDATA.
2. HERMES Coll., K. Ackerstaff et al., Nucl. Instr. and Meth. **A 417**, 230 (1998).
3. F_2^p : H. Abramowicz and A. Levy, hep-ph/9712415 (1997). For $W < 2.2$ GeV^2 or $Q^2 < 0.5$ GeV^2: A. Bodek et al., Phys. Rev. **D20** 1427, (1979).
   F_2^d : NMC (P. Amaudruz et al.), Nucl. Phys. **B371** (1992) 3.
4. L. W. Withlow et al., Phys. Lett. **B 250**, 193 (1990).
5. HERMES Coll., A. Airapetian et al., Phys. Lett. **B 442**, 484 (1998).
6. Recently published 4 additional data points at very small x and $Q^2 > 0.01$ GeV^2 values are not shown here: SMC, B. Adeva et al., Phys. Rev. **D 60**, 072004 (1999).
7. All g_1^p values were calculated using g_1^p/F_1^p [1] and F_1^p evaluated by F_2^p [3] and R [4].
8. E155 Coll., P.L. Anthony et al., SLAC-PUB-7994, hep-ph/0007248 (2000); Fit using $Q^2 > 1$ GeV^2 data: $g_1^p/F_1^p = x^{0.7}(0.817 + 1.014x - 1.489x^2)(1 - 0.04/Q^2)$.

Measurement of the Polarised Quark Distributions in the Nucleon at HERMES

Thore Lindemann[a] on behalf of the HERMES Collaboration

[a] *DESY, Deutsches Elektronen Synchrotron, 22603 Hamburg, Germany*

Abstract. The inclusive and semi-inclusive double spin asymmetries for positively and negatively charged hadrons were measured by HERMES using a polarised positron beam and polarised ^{3}He (1995), ^{1}H (1996/97) and ^{2}D (1998-2000) targets. Preliminary results in the kinematical range $0.023 < x < 0.6$ and 1 GeV$^2 < Q^2 < 10$ GeV^2 for the asymmetries taken with the polarised deuterium target in the spring of 1999 are presented. Using the whole set of data from 1995 to spring 1999, the polarised quark distributions were extracted as a function of x for up $(u+\bar{u})$ and down $(d+\bar{d})$ flavours, and for valence and sea quarks. The up quark polarisation is positive and the down quark polarisation is negative in the measured range. The polarisation of the sea is compatible with zero.

The HERMES experiment was mainly designed for inclusive and semi-inclusive polarised DIS measurements. The results reported here are based on data taken using the 27.5 GeV beam of longitudinally polarised positrons in the HERA storage ring at DESY, incident on longitudinally polarised ^{3}He(1995) [1], ^{1}H(1996/97), and ^{2}D(1998-2000) [2] internal gas targets. Particle identification is accomplished with the HERMES spectrometer [3] using a lead-glass calorimeter, a preshower counter, a transition radiation detector, and a threshold Čerenkov detector which was replaced by a dual radiator Ring Imaging Čerenkov (RICH) detector [4] in 1998.

Semi-inclusive polarised deep inelastic scattering allows the determination of the polarised quark distributions $\Delta q_f(x, Q^2) = q_f^\uparrow(x, Q^2) - q_f^\downarrow(x, Q^2)$ for each quark flavour $f = u, \bar{u}, d, \bar{d}, s, \bar{s}$. The function $q_f^{\uparrow(\downarrow)}$ represents the distribution of quarks of flavour f with spin parallel (antiparallel) to the nucleon spin. Here, x is the Bjorken scaling variable $x = Q^2/2M_p\nu$, Q^2 and ν are the negative squared four-momentum and energy of the exchanged virtual photon in the nucleon rest frame. Provided factorisation of the scattering and the fragmentation process holds, the cross section for the production of a particular hadron h can be written as

$$\sigma^h(x, Q^2, z) \propto \sum_f e_f^2\, q_f(x, Q^2)\, D_f^h(z, Q^2). \tag{1}$$

In Eq. (1) e_f denotes the quark charge in units of the elementary charge e, $q_f(x, Q^2)$ are the unpolarised quark distributions of flavour f, and $D_f^h(z, Q^2)$ are the fragmentation functions which are defined as the probability that a struck quark q_f fragments into a hadron h with fractional energy $z = E_h/\nu$, where E_h is the energy of the hadron. The sum is over the up, down, and strange quark and antiquark flavours. The asymmetry of the spin-dependent virtual photon nucleon absorption cross sections is defined as $A_1 = (\sigma_{1/2} - \sigma_{3/2})/(\sigma_{1/2} + \sigma_{3/2})$, with the projection $\frac{1}{2}$ ($\frac{3}{2}$) of the total photon and nucleon spin along the direction of the photon momentum.

According to this the semi-inclusive asymmetry A_1^h is related to the polarised and unpolarised quark distributions and fragmentation functions by

$$A_1^h(x, Q^2, z) = \mathcal{C_R} \cdot \frac{\sum_f e_f^2 \Delta q_f(x, Q^2) D_f^h(z, Q^2)}{\sum_f e_f^2 q_f(x, Q^2) D_f^h(z, Q^2)}, \qquad (2)$$

where $\mathcal{C_R} = (1 + R(x, Q^2))/(1 + \gamma^2)$ with $\gamma^2 = Q^2/\nu^2$ and $R = \sigma_L/\sigma_T$ the ratio of the photo absorption cross sections for longitudinally and transversely polarised virtual photons on a nucleon. The inclusive asymmetry can be similarly expressed as $A_1(x, Q^2) = \mathcal{C_R} \cdot (\sum_f e_f^2 \Delta q_f(x, Q^2))/(\sum_f e_f^2 q_f(x, Q^2))$. In Eq. (2) it has been assumed that the contribution of the spin structure function g_2 is negligible. The measured spin asymmetries $A_1^h(x, Q^2, z)$ were integrated in each x bin over the corresponding Q^2-range and in the z-range from 0.2 to 1 to yield $A_1^h(x)$. The results for the inclusive and semi-inclusive asymmetries A_1^h at HERMES on polarised ^{3}He and hydrogen can be found in Ref. [5]. The preliminary results for the inclusive and semi-inclusive asymmetries on deuterium are shown in Fig. 1.

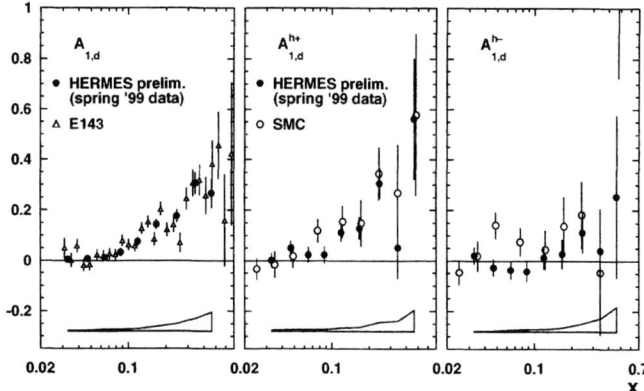

FIGURE 1. Preliminary results for asymmetries on deuterium data taken in spring 1999. The plots are, from left to right, the inclusive, semi-inclusive h^+, and semi-inclusive h^- asymmetries. Inclusive asymmetries are compared with SLAC E143 [6] and semi-inclusive asymmetries with SMC [7] measurements. All points are presented at the measured x and Q^2. The error bars represent the statistical errors. The error bands show the HERMES systematic uncertainties.

The expression (2) may be rearranged by defining a quark purity for each event as

$$P_f^h(x) = \frac{e_f^2 \, q_f(x) \int_{0.2}^1 D_f^h(z)\,dz}{\sum_{f'} e_{f'}^2 \, q_{f'}(x) \int_{0.2}^1 D_{f'}^h(z')\,dz'}. \quad (3)$$

P_f^h represents the probability that a detected event originated from a struck quark of flavour f in the nucleon. Using this purity formalism, Eq. (2) can be written in a matrix form as a system of equations

$$\vec{A}(x) = \mathcal{P} \cdot \vec{Q}(x). \quad (4)$$

In Eq. (4) the components of the vector $\vec{A}(x)$ are the measured inclusive and semi-inclusive charged hadron asymmetries on the ^{3}He, hydrogen, and deuterium targets. The vector $\vec{Q}(x)$ contains the quark and antiquark polarisations for each quark flavour and the matrix $\mathcal{P}$ depends on the unpolarised quark distributions q_f, the fragmentation functions D_f^h, and the correction term $\mathcal{C_R}$. The quark purities relevant for the HERMES experiment have been estimated with a Monte Carlo simulation of polarised DIS. The simulation uses the LUND string fragmentation model [8], CTEQ Low-Q^2 unpolarised parton distributions [9], and a model of the HERMES detector. The free parameters of the LUND fragmentation model were tuned to fit the measured hadron multiplicities.

In solving Eq. (4) constraints were imposed on the sea polarisation to improve the statistical significance. As was done in Ref. [5] it is assumed that the polarisation $\Delta q_s(x)/q_s(x)$ of the sea is independent of flavour, i.e.

$$\frac{\Delta q_s(x)}{q_s(x)} \equiv \frac{\Delta u_s(x)}{u_s(x)} = \frac{\Delta d_s(x)}{d_s(x)} = \frac{\Delta s(x)}{s(x)} = \frac{\Delta \bar{u}(x)}{\bar{u}(x)} = \frac{\Delta \bar{d}(x)}{\bar{d}(x)} = \frac{\Delta \bar{s}(x)}{\bar{s}(x)}. \quad (5)$$

The flavour decomposition was obtained by solving Eq. (4) for the vector

$$\vec{Q} = \left(\frac{\Delta u(x) + \Delta \bar{u}(x)}{u(x) + \bar{u}(x)}, \frac{\Delta d(x) + \Delta \bar{d}(x)}{d(x) + \bar{d}(x)}, \frac{\Delta q_s(x)}{q_s(x)} \right). \quad (6)$$

Fig. 2(a) shows the HERMES results for the quark polarisations obtained from data taken from 1995 to spring 1999. As it is seen, the up quark polarisation is positive and the values increase with increasing x. The down quark polarisation is negative with no evident x dependence. The sea polarisation is compatible with zero over the measured range of x.

For $x > 0.3$ the statistical error of the extracted sea polarisation $\Delta q_s(x)/q_s(x)$ is larger than the positivity limit $|\Delta q_s(x)/q_s(x)| \leq \mathcal{C}_R^{-1}$. Therefore, in the region $x > 0.3$, the sea polarisation was set to zero and the corresponding effect on the results for the non-sea polarisation was included in their systematic uncertainties. The flavour decomposition has been repeated with different unpolarised parton distributions [10], with the independent fragmentation model [11], and with different parametrisations for the fit of the LUND string model to HERMES multiplicities. A systematic uncertainty is assigned for the variation in the resulting quark polarisations.

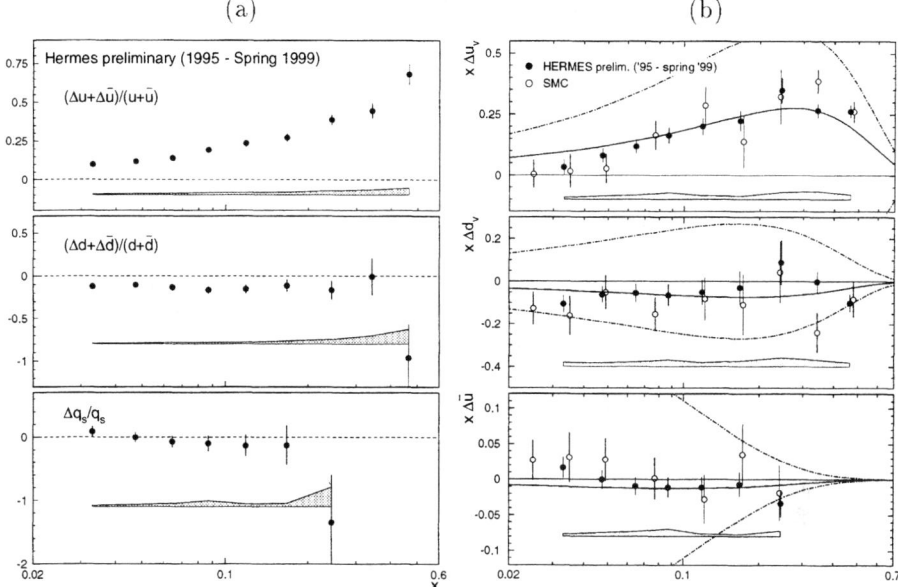

FIGURE 2. (a) The flavour decomposition $(\Delta u(x) + \Delta \bar{u}(x))/(u(x) + \bar{u}(x))$, $(\Delta d(x) + \Delta \bar{d}(x))/(d(x) + \bar{d}(x))$ and $\Delta q_s(x)/q_s(x)$ of the quark polarisation as a function of x. The sea polarisation is assumed to be flavour independent in this analysis. (b) The polarised quark distributions from HERMES at $Q^2 = 2.5$ GeV^2 separately for the valence quarks $x\Delta u_v(x)$, $x\Delta d_v(x)$, and the sea quarks $x\Delta \bar{u}(x)$. The distributions are compared to results from SMC, extrapolated to $Q^2 = 2.5$ GeV^2. The solid lines indicate the positivity limit and the dashed lines are the parametrisation from Ref. [12] (Gluon A, LO). All error bars shown are statistical errors, and the bands indicate the systematic uncertainties of HERMES.

For the up and down quarks, the systematic uncertainties are dominated by those of the beam and target polarisation measurements. For the sea quarks the systematic uncertainty is dominated by the uncertainty in the fragmentation functions.

Polarised quark distributions $\Delta q_f(x)$ were determined by multiplying the quark polarisations $\Delta q_f(x)/q_f(x)$ with the unpolarised quark distributions from Ref. [9] determined at $Q^2 = 2.5$ GeV^2. It was assumed that the polarisation is independent of Q^2 within the Q^2-range of measurement. Fig. 2(b) compares the polarised quark distributions $x\Delta u_v(x)$, $x\Delta d_v(x)$ and $x\Delta \bar{u}(x)$ as measured by HERMES, obtained by solving Eq. (4) for the vector $\vec{Q} = \left(\frac{\Delta u(x) - \Delta \bar{u}(x)}{u(x) - \bar{u}(x)}, \frac{\Delta d(x) - \Delta \bar{d}(x)}{d(x) - \bar{d}(x)}, \frac{\Delta q_s(x)}{q_s(x)} \right)$, with results from SMC [7]. Note that the HERMES definition of the flavour independence of the sea polarisation differs slightly from the sea model chosen by SMC. The SMC results were derived under the assumption that $\Delta u_s = \Delta d_s = \Delta s = \Delta \bar{u} = \Delta \bar{d} = \Delta \bar{s}$. It has been verified that the HERMES results are insensitive to the differences between this two assumptions. The positivity limit and a parametrisation [12] derived from inclusive spin asymmetries are included in Fig. 2(b). Within statistical and systematical uncertainties the results of both experiments are consistent, though the high precision HERMES ^{3}He, hydrogen, and deuterium data has yielded more sensitivity to Δu_v, Δd_v, and $\Delta \bar{u}$.

Conclusion and Outlook

The last five years of HERMES data taking have provided a large amount of polarised deep-inelastic scattering data on ^{3}He, hydrogen, and deuterium. The polarised measurements presented here yield inclusive and semi-inclusive charged hadron asymmetries using data taken until spring 1999. For deuterium, a factor of 6 more polarised DIS data was taken from summer 1999 until summer 2000.

The polarisation for up and down flavours, and for valence and sea quarks have been extracted. Using the preliminary HERMES results the precision of the polarised valence d-quark distribution could significantly be improved compared to the previous publication [5].

With the complete set of high statistics data on deuterium, the precision of the polarised quark distributions will be significantly improved compared to this analysis.

Before the deuterium run began in 1998, the threshold Čerenkov detector was replaced by a Ring Imaging Čerenkov (RICH) detector. This new detector identifies π, K, and p particles over nearly all momenta accepted by the spectrometer. As kaon asymmetries provide direct sensitivity to strange quark polarisation in the nucleon, future analysis including the RICH detector should provide a first direct measurement of the light and strange sea polarisations in the nucleon.

REFERENCES

1. De Schepper D. et al., *Nucl. Instr. Meth.* **A419**, 16 (1998).
2. Lenisa P., these proceedings.
3. Ackerstaff K. et al., *Nucl. Instr. Meth.* **A417**, 230 (1998).
4. Aschenauer E.C. et al., *Nucl. Instr. Meth.*, to be published.
5. Ackerstaff K. et al., *Phys. Lett.* **B464**, 123 (1999).
6. Abe K. et al., *Phys. Rev.* **D58**, 112003 (1998).
7. Adeva B. et al., *Phys. Lett.* **B420**, 180 (1998).
8. Ingelmann G., Edin A., Rathsman J., *DESY Report* **96-057**, (1996).
9. Lai H.L. et al., *Phys. Rev.* **D55**, 1280 (1997).
10. Glück M. et al., *Z. Phys.* **C67**, 433 (1995).
11. T. Sjöstrand, Comp. Phys. Comm. **82**, 74 (1994).
12. Gehrmann T. and Stirling W.J., *Phys. Rev.* **D53**, 6100 (1996).

Polarized structure functions of the deuteron

S. Kumano

Department of Physics, Saga University, Saga 840-8502, Japan

Abstract. Physics of spin-1 hadron is an unexplored topic in the high-energy region although spin-1/2 physics has been well investigated in the last decade. It is important to test our knowledge of hadron spin structure in a quite different field of spin physics. We discuss tensor structure functions, which do not exist for the spin-1/2 nucleon, in lepton scattering and in hadron reactions such as the polarized proton-deuteron Drell-Yan process.

INTRODUCTION

High-energy spin physics is entering into a new era in the sense that the difference between a naive quark-model prediction and experimental data is now roughly understood and that much detailed studies are in progress with the completion of the RHIC facility. So far, spin-structure studies have been focused on the spin-1/2 nucleon. For higher-spin hadrons, we know that different spin physics exists such as the tensor structure in the deuteron. In order to test our understanding of the hadron structure, it is important to investigate other spin quantities. In this sense, spin-1 hadrons, for example the deuteron, should be suitable for future investigations.

In this paper, the present status is explained for the deuteron spin structure in the high-energy region. First, polarized electron-deuteron scattering is discussed briefly in connection with the deuteron spin physics. In particular, new polarized structure functions are introduced. Then, as an alternative method, the polarized proton-deuteron Drell-Yan process is discussed for finding the new distributions.

SPIN-1 STRUCTURE IN ELECTRON SCATTERING

A general formalism of the deep inelastic electron-deuteron scattering is discussed in Ref. [1]. It suggests that there exist additional structure functions in comparison with the electron-proton scattering due to the spin-1 nature of the deuteron. We find that there are eight independent amplitudes for $\gamma(h_1) + d(H_1) \rightarrow \gamma(h_2) + d(H_2)$, where $h_{1,\,2}$ and $H_{1,\,2}$ are helicities, by using momentum conservation, parity invariance, and time-reversal invariance. Four of them are usual structure functions: F_1, F_2, g_1, and g_2, which exist in the spin-1/2 proton. In addition, there are four new structure functions: b_1, b_2, b_3, and b_4. The b_3 and b_4 are higher-twist functions and b_2 is related to b_1 by the "Callan-Gross type" relation $b_2 = 2xb_1$ in the Bjorken scaling limit, so that the most essential part of new physics is contained in b_1. A phenomenological sum rule $\int dx\, b_1(x) = 0$ was proposed [2] if there is no tensor polarization for antiquark distributions; however, it does not mean that $b_1(x)$ itself vanishes. This sum rule is similar to the Gottfried sum rule in a way [3]. Because there is no solid reason why the antiquark tensor polarization vanishes, the violation of the above sum rule could suggest a finite tensor polarization as the Gottfried-sum-rule violation indicated a finite $\bar{u}/\bar{d}$ asymmetry.

There are three twist-2 structure functions: F_1, g_1, and b_1, and they are related to electron-deuteron scattering cross sections as

$$F_1 \propto \langle d\sigma \rangle,$$
$$g_1 \propto d\sigma(\uparrow, +1) - d\sigma(\uparrow, -1),$$
$$b_1 \propto d\sigma(0) - [d\sigma(+1) + d\sigma(-1)]/2. \tag{1}$$

Here, $d\sigma(\uparrow, +1)$ and $d\sigma(\uparrow, -1)$ indicate spin parallel and anti-parallel cross sections, respectively. On the other hand, $\sigma(0)$ and $d\sigma(\pm 1)$ indicate cross sections with the deuteron spin state 0 and ± 1. This fact means that the electron does not have to be polarized for measuring b_1. There are a number of predictions on the x dependence of b_1, but we cannot discuss the model results because of the space limitation. The interested reader may find a list of papers in the reference section of Ref. [4]. At this stage, b_1 is expected to be quite small, typically $b_1/F_1 \sim$ a few percent or much less. However, our experience of g_1 implies that the predictions could be wrong. In particular, the tensor structure in the high-energy region is still an unexamined topic experimentally. The HERMES experiment for measuring b_1 is in progress and its results will be reported in the near future. Therefore, the b_1 physics could become one of popular topics in hadron physics.

PROTON-DEUTERON DRELL-YAN PROCESS

In addition to the lepton scattering studies, it is, in principle, possible to investigate the same spin physics in hadron facilities. However, there are only a few theoretical studies on the spin-1 physics in hadron collisions. So far, the polarized proton-deuteron (pd) Drell-Yan reaction has been investigated in connection with the tensor structure of the deuteron [4,5].

Imposing the conditions of Hermiticity, parity conservation, and time-reversal invariance, we found many structure functions in the pd Drell-Yan. Even after integrating the cross section over the transverse momentum of the virtual photon $\vec{Q}_T$ or after taking the limit $Q_T \to 0$, there are 22 functions [4]:

unpolarized : $W_{0,0}$, $W_{2,0}$

L/T polarized : $V_{0,0}^{LL}$, $V_{0,0}^{TT}$, $V_{2,0}^{LL}$, $V_{2,0}^{TT}$, $U_{2,1}^{TU}$, $U_{2,1}^{UT}$, $U_{2,1}^{TL}$, $U_{2,1}^{LT}$, $U_{2,2}^{TT}$

tensor polarized : $V_{0,0}^{UQ_0}$, $V_{0,0}^{TQ_1}$, $V_{2,0}^{UQ_0}$, $V_{2,0}^{TQ_1}$, $U_{2,1}^{TQ_0}$, $U_{2,1}^{LQ_1}$, $U_{2,1}^{TQ_2}$, $U_{2,1}^{UQ_1}$,

$U_{2,2}^{UQ_2}$, $U_{2,2}^{TQ_1}$, $U_{2,2}^{LQ_2}$. (2)

The subscripts of $F_{L,M}^{pol_p\, pol_d}$ ($F=W$, V, or U) indicate that it is obtained by the integration $\int d\Omega\, Y_{LM}\, d\sigma/(d^4Q\, d\Omega)$. The superscripts indicate the polarization states of the proton and deuteron. The notations U, L, and T indicate unpolarized, longitudinally polarized, and transversely polarized states. The quadrupole polarizations Q_0, Q_1, and Q_2 are specific in the reactions with a spin-1 hadron, and they are associated with the quadrupole terms: Q_0 for the term $3\cos^2\beta_B - 1 \sim Y_{20}$, Q_1 for $\sin\beta_B \cos\beta_B \sim Y_{21}$, and Q_2 for $\sin^2\beta_B \sim Y_{22}$ with the polarization angle of the deuteron β_B. The structure functions in the categories "unpolarized" and "L/T polarized" exist in the proton-proton (pp) Drell-Yan. The new functions are listed as "tensor polarized" and they are associated with the spin-1 deuteron.

In defining the spin asymmetry, we should be careful about the tensor contributions. If it is defined in the usual way, for example $A_{LL}^{usual} = [\sigma(\uparrow,-1) - \sigma(\uparrow,+1)]/[\sigma(\uparrow,-1) + \sigma(\uparrow,+1)]$, the denominator is not equal to the unpolarized cross section because of

$$2 <\sigma> = \sigma(\uparrow,+1) + \sigma(\uparrow,-1) + \frac{1}{3}\left[2\sigma(\uparrow,0) - \sigma(\uparrow,+1) - \sigma(\uparrow,-1)\right]. \quad (3)$$

In order to avoid the above tensor contribution, the unpolarized cross section is used for the denominator in Ref. [4] as a "theoretical" definition of the spin asymmetry. Using the obtained pd cross section with the structure functions in Eq.(2), we find the following spin asymmetries:

$$<\sigma>,\ A_{LL},\ A_{TT},\ A_{LT},\ A_{TL},\ A_{UT},\ A_{TU},$$
$$A_{UQ_0},\ A_{TQ_0},\ A_{UQ_1},\ A_{LQ_1},\ A_{TQ_1},\ A_{UQ_2},\ A_{LQ_2},\ A_{TQ_2}. \quad (4)$$

Those in the first line exist in the pp reaction. However, the ones in the second line are specific for the pd Drell-Yan, and they are related to the tensor polarizations Q_0, Q_1, and Q_2. For example, A_{UQ_0}, A_{UQ_1}, and A_{UQ_2} are given as

$$A_{UQ_0} = \frac{1}{2<\sigma>}\left[\sigma(\bullet,0_L) - \frac{\sigma(\bullet,+1_L) + \sigma(\bullet,-1_L)}{2}\right] = \frac{2V_{0,0}^{UQ_0} + (\frac{1}{3} - \cos^2\theta) V_{2,0}^{UQ_0}}{2W_{0,0} + (\frac{1}{3} - \cos^2\theta) W_{2,0}},$$

$$A_{UQ_1} = \frac{\sigma(\bullet,I_1) - \sigma(\bullet,I_3)}{2<\sigma>} = \frac{\sin\theta\cos\theta\sin\phi\, U_{2,1}^{UQ_1}}{2W_{0,0} + (\frac{1}{3} - \cos^2\theta) W_{2,0}},$$

$$A_{UQ_2} = \frac{1}{2<\sigma>}\left[\frac{\sigma(\bullet,\phi_B=0) + \sigma(\bullet,\phi_B=\pi)}{2} - \frac{\sigma(\bullet,\phi_B=\pi/2) + \sigma(\bullet,\phi_B=3\pi/2)}{2}\right]$$

$$= \frac{\sin^2\theta\cos 2\phi\, U_{2,2}^{UQ_2}}{2W_{0,0} + (\frac{1}{3} - \cos^2\theta) W_{2,0}}. \quad (5)$$

Here, θ and ϕ are polar and azimuthal angles of the ℓ^+ momentum. The solid circle indicates that the proton is unpolarized. The three tensor polarizations are illustrated in Fig.1.

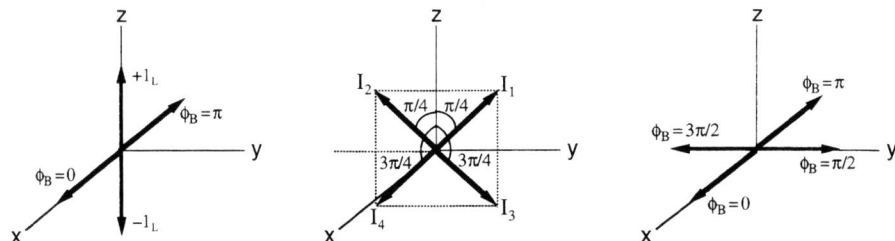

FIGURE 1. Q_0, Q_1, and Q_2 polarizations [4].

The Q_0 polarization asymmetry is in principle the same as the one in defining the tensor structure function b_1. On the other hand, Q_2 is associated with the tensor asymmetry in the transverse plane. The Q_1 polarization is very peculiar. The deuteron is polarized with the angle 45° or 135° with respect to the longitudinal axis, and the cross section difference is taken. This kind of peculiar asymmetry does not exist, of course, for the proton. This "intermediate" (according to Ref. [4]) polarization asymmetry could lead to a new unique field of spin physics. A parton-model analysis suggests that the asymmetry Q_0 should be related to b_1 type tensor distributions as [4]

$$A_{UQ_0} = \frac{\sum_a e_a^2 \left[f_1^a(x_1)\, \bar{b}_1^a(x_2) + \bar{f}_1^a(x_1)\, b_1^a(x_2) \right]}{\sum_a e_a^2 \left[f_1^a(x_1)\, \bar{f}_1^a(x_2) + \bar{f}_1^a(x_1)\, f_1^a(x_2) \right]}, \qquad (6)$$

where f_1^a and b_1^a are unpolarized and tensor-polarized distributions for the quark a, and $\bar{f}_1^a$ and $\bar{b}_1^a$ are those for the antiquark. The momentum fractions are given by x_1 and x_2 in the proton and the deuteron. For example, b_1^a is given by $b_1^a = [q_a(0) - (q_a(+1) + q_a(-1))/2]/2$. We mentioned the sum rule $\int dx b_1(x) = 0$. However, this relation crucially depends on the antiquark tensor polarization $\bar{b}_1^a$, which cannot be well probed in the electron scattering. One of the major advantages of using the pd reaction is to find $\bar{b}_1^a$ rather easily, at least according to the formalism. The situation is similar to the case that the detailed x dependence of $\bar{u}/\bar{d}$ was clarified by the Fermilab-E866 Drell-Yan experiment although the difference was originally indicated by the NMC data [3]. For example, in the large-x_F region, Eq.(6) becomes

$$A_{UQ_0}(\text{large } x_F) \approx \frac{\sum_a e_a^2\, f_1^a(x_1)\, \bar{b}_1^a(x_2)}{\sum_a e_a^2\, f_1^a(x_1)\, \bar{f}_1^a(x_2)}. \qquad (7)$$

Therefore, the antiquark tensor distributions $\bar{b}_1^a$ can be determined if the unpolarized distributions are well known in the proton and deuteron.

Another advantage of the pd Drell-Yan is that the $\bar{u}/\bar{d}$ asymmetry can be investigated for the transversity distributions [5]. Because they have chiral-odd property, they cannot be investigated in the inclusive electron scattering. The pd/pp Drell-Yan ratio is given by [5]

$$R_{pd} \equiv \frac{\Delta_{(T)}\sigma_{pd}}{2\Delta_{(T)}\sigma_{pp}} = \frac{\sum_a e_a^2 \left[\Delta_{(T)}q_a(x_1)\Delta_{(T)}\bar{q}_a^d(x_2) + \Delta_{(T)}\bar{q}_a(x_1)\Delta_{(T)}q_a^d(x_2)\right]}{2\sum_a e_a^2 \left[\Delta_{(T)}q_a(x_1)\Delta_{(T)}\bar{q}_a(x_2) + \Delta_{(T)}\bar{q}_a(x_1)\Delta_{(T)}q_a(x_2)\right]}$$

$$\rightarrow \frac{1}{2}\left[1 + \frac{\Delta_{(T)}\bar{d}(x_2)}{\Delta_{(T)}\bar{u}(x_2)}\right]_{x_2 \rightarrow 0} \quad \text{for } x_F \rightarrow 1, \quad (8)$$

where $\Delta_{(T)} = \Delta$ or Δ_T depending on the longitudinal or transverse polarization, and $\Delta_{(T)}\sigma$ indicates the polarized cross-section difference. In the second line, $x_F \rightarrow 1$ limit is taken together with the assumption $\Delta_{(T)}u_v(x \rightarrow 1) \gg \Delta_{(T)}d_v(x \rightarrow 1)$. Leading-order numerical results are shown in Fig.2, where $r_{\bar{q}} \equiv \Delta_{(T)}\bar{u}/\Delta_{(T)}\bar{d}$. It obviously indicates that the ratio is suitable for finding the polarized light antiquark distributions, particularly in the large-x_F region.

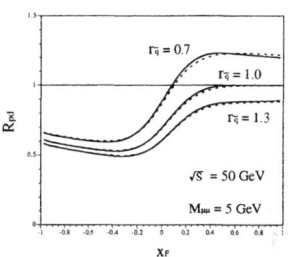

FIGURE 2. pd/pp Drell-Yan ratio [5].

In this way, we found that the pd Drell-Yan is an alternative way of studying the deuteron spin structure to the electron scattering. Because other studies of spin-1 physics are not discussed in this paper, the author hopes that the interested reader will look at the papers in the extensive lists of Ref. [4,5].

SUMMARY

We explained the possibilities of investigating the polarized deuteron. Exploring a new field of hadron spin, namely the tensor structure, we should be able to test our knowledge of high-energy spin physics. Because it has not been well investigated yet, we may encounter an unexpected result. Depending on the HERMES measurement in the near future, this kind of topic could become an interesting area of hadron physics. We also expressed the importance of polarized proton-deuteron reactions, which are possible not only in collider experiments but also in fixed-target ones.

REFERENCES

1. P. Hoodbhoy, R. L. Jaffe, and A. Manohar, *Nucl. Phys.* **B312**, 571 (1989).
2. F. E. Close and S. Kumano, *Phys. Rev.* **D42**, 2377 (1990).
3. S. Kumano, *Phys. Rep.* **303**, 183 (1998).
4. S. Hino and S. Kumano, *Phys. Rev.* **D59**, 094026 (1999); **D60**, 054018 (1999).
5. S. Kumano and M. Miyama, *Phys. Lett.* **B497**, 149 (2000).

Extraction of g1 in the Resonance Region

Renee Fatemi

*University of Virginia
on behalf of the CLAS Collaboration*

Abstract. Spin asymmetries are extracted from polarized electron - nucleon scattering data taken at The Thomas Jefferson Accelerator Facility in 1998. The integral of g_1 is shown for Q^2 of 0.3-1.3 GeV^2 for the proton and 0.15-1.1 GeV^2 for the deuteron.

INTRODUCTION

Since the development of the "spin crisis" over twenty years ago there has been considerable effort to measure precisely the spin structure functions, g_1 and g_2, in the deep inelastic region. The results are in close agreement with the Bjorken Sum Rule, which predicts the difference in the integrals of g_1 for the proton and the neutron. It is very difficult to predict the low Q^2 behavior of these structure functions however because of the strong interaction between the valence quarks. At low Q^2 the resonance region is expected to make an important contribution to the integrals. Moreover, polarized electron - polarized nucleon scattering at low Q^2 will yield important information on the structure of the nucleon resonances. For this reason a series of polarized electron - polarized nucleon experiments has been proposed and run at Jefferson Lab in recent years. EG1, the first polarized target experiment to run in the CEBAF Large Acceptance Spectrometer (CLAS), was designed with exactly these purposes in mind.

EXPERIMENTAL SETUP

Over 3 billion events were accumulated during the first run of EG1 by scattering the Jefferson Lab polarized electron beam off polarized proton and deuteron targets. The scattered electron and hadron tracks were reconstructed using the three sets of drift

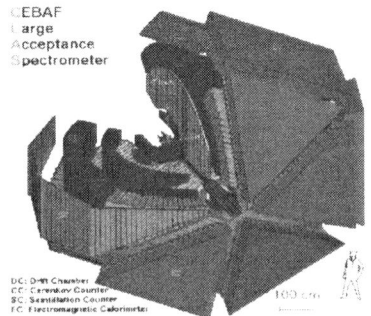

FIGURE 1. CLAS detector

chambers that make up the inner layers of the CLAS detector. Located behind the drift chambers are the Cerenkov detector and Electron Calorimeter, which are used for electron identification. The large acceptance of the CLAS detector package, nearly 2π in phi and $8° - 240°$ in theta, allowed us to simultaneously take inclusive and exclusive data for large ranges of Q^2 and W and several final state channels. The solid state dynamically polarized target design was optimized to match the CLAS acceptance. The 5 Tesla superconducting magnet, target chamber and exit window were designed to allow detection of particles scattered up to $50°$ and nearly $360°$ in phi. Radial windows also allowed detection of large angle scattering. The polarization was an average of 55-65% for the $N_{15}H_3$ target and 20-30% for the $N_{15}D_3$ target.

FIGURE 2. Polarized target with exit window removed and radial windows shielded.

THEORY

In the deep inelastic region many features of polarized electron-nucleon scattering are well understood. In the scaling limit the Bjorken Sum Rule[1] equates the difference in the integral of g_1 for the proton and neutron to the weak axial vector coupling constant.

$$\int_0^1 g_1^p(x)dx - \int_0^1 g_1^n(x)dx = \frac{g_A}{6}$$

This well tested sum rule can then be related to the spin distributions of the quarks inside the nucleon. At low Q^2 this relation isn't valid due to the increasingly large contribution of gluon spin and quark angular momentum to the integral. This is a region where α_s is changing rapidly and hence it is very difficult to formulate a strong prediction for the behavior of the spin structure functions in the resonance region. At low Q^2 the Gerasimov-Drell-Hearn Sum Rule[2,3] for absorption of real photons may provide a guide.

$$\int_{\pi thres}^{\infty} (\sigma^{1/2} - \sigma^{3/2}) \frac{d\nu}{\nu} = \frac{2\alpha\pi^2 k_n^2}{m_n^2}$$

The GDH sum rule relates the difference between spin aligned and anti-aligned polarized photon-nucleon cross-sections to k_n, the magnetic moment of the nucleon. If it is correct to assume that the virtual photon-nucleon scattering cross-sections connect smoothly to the real ones then it is possible to say that g_1 is limited by the Bjorken sum rule at high Q^2 and by the GDH sum rule at $Q^2=0$.

RESULTS

The calculation of g_1 first requires the extraction of the experimental asymmetry $A_{\|}$.

$$A_{\|} = \frac{N^{\uparrow\uparrow} - N^{\uparrow\downarrow}}{N^{\uparrow\uparrow} + N^{\uparrow\downarrow} - BG} * \frac{RC}{P_B P_T}$$

$A_{\|}$ is simply the raw asymmetry with several corrections applied. The three most important corrections are the background subtraction (BG), radiative corrections(RC), and the polarization factor($P_B P_T$). The background correction removes the nitrogen and helium contribution to the asymmetry leaving only those events that scattered from nucleons. The radiative corrections account for energy lost before, after and during the electron-nucleon interaction. The polarization correction simply divides the asymmetry by the product of the actual beam and target polarizations. This product is extracted from the elastic peak asymmetry in the data after the background subtraction has been completed.

$A_{\|}$ is related to the two physics asymmetries A_1 and A_2 by the equation $A_{\|} = D(A_1 + \eta A_2)$. D and n are kinematic coefficients related to R, Q^2 and ν. In fig.3 $A_{\|}/D$ is shown for the proton and deuteron data in order to demonstrate the resolution of the W spectrum in the resonance region. The blue shaded region represents the systematic error with only statistical errors shown by the error bars. The light red dotted lines show the theoretical expectation for only the resonance contribution,. The blue and green curves are parametric fits to the existing world data for A_1 and A_2.

A_2, F_1 and R are needed in order to calculate g_1 and therefore it is important to solve for A_1 and A_2 independently. There are two possible ways to obtain A_1 and A_2 individually. The first is to polarize the nucleons transversely to the beam and thus obtain $A_\perp$ which also can be related to A_1 and A_2. Unfortunately it is not possible to run a transversely polarized target in CLAS at this time. The second method is to use different kinematic ranges, and thus different coefficients D and η, in order to separate A_1 and A_2. Again there was not sufficient data in this first run to make a statistically significant separation, although this will be possible once the second EG1 run is completed in April of 2001. It is necessary then to use a model to simulate A_2 and then extract A_1 from the data and use these respective physics asymmetries to calculate g_1. The model used to simulate A_2 is a parameterization to the current g_1 and g_2 world data with the restrictions given by g_2^{ww} used to determine the resonance behavior of A_2. This model also provides fits to world data for F_1 and R, which are used in calculating D and g_1.

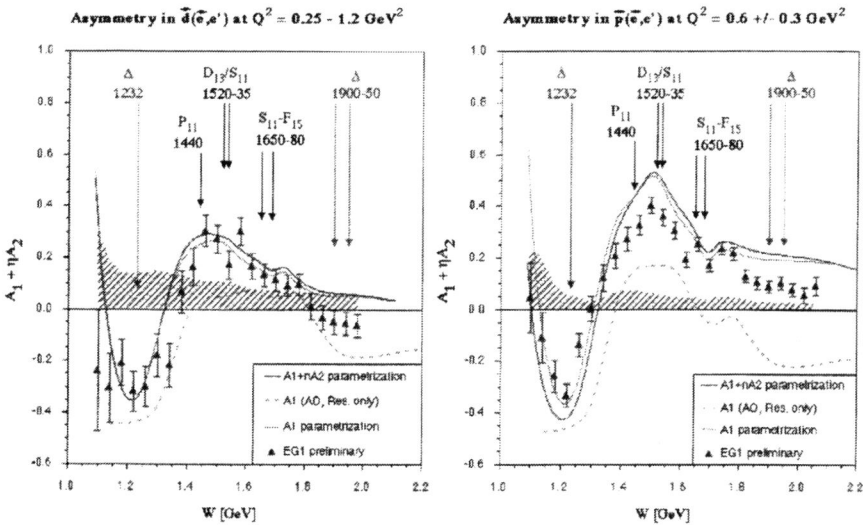

FIGURE 3. A1 + η A2 of the proton and deuteron show resolution of data in resonance region.

After extracting A_1 and modeling A_2 and F_1 it is possible to calculate and integrate g_1 over x = 0 to 1.

$$g_1 = \frac{v^2}{v^2+Q^2}\left(A_1 + \frac{Q}{v}A_2\right)F_1$$

In fig.4 the first moment of g_1 is shown for the deuteron and proton. The blue triangles are the actual data points with errors that reflect the statistical and systematic errors added in quadrature. For these data points g_1 is integrated up to W=2 GeV for the lower Q^2 and up to 2.2 for the higher Q^2 points. The pink triangles represent the data with the deep inelastic contribution added in order to complete the integral. This deep inelastic contribution is calculated using the same parametrization to the world data used in calculating A_2, F_1, and R. At low Q^2 the error is dominated by errors due to radiative corrections and at higher Q^2 the error is dominated by the background subtraction. The dark black line at the top right corner of fig.4 shows the pQCD extrapolation of DIS data. The light blue and green dotted lines are predictions of g1 made by Soffer[4] and Burkert and Ioffe[5] respectively. The dotted red line below the data is a calculation of the resonance contribution to g_1 using the AO model[6]. AO contains a parameterization for the helicity amplitudes of all measured resonance transitions. It is based on the analysis of pion and eta unpolarized electroproduction data, and parameterizes the energy dependence and Q2 dependence of resonance amplitudes. The most important observation is that the resonance contribution is very large at low Q^2 and levels off very quickly as Q^2 increases, at around 0.75 GeV^2 for the proton.

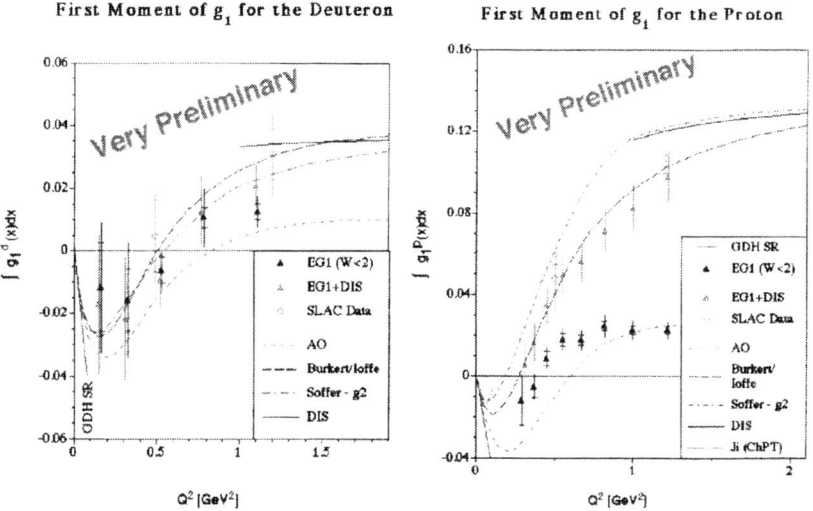

FIGURE 4. Integral of g1 for the proton and deuteron compared to theoretical expectation.

CONCLUSION

The first run of EG1 has contributed significantly to the spin structure function world data in the resonance region. Preliminary results show a rich structure of resonances in the spin asymmetries at low Q^2. Data also show that the behavior of the first moment of g_1 is dominated by these resonances at low Q^2. Data from the second run of EG1, which is currently running in Hall B at TJNAF, will extend the results to higher and lower Q^2 and reduce the statistical error bars by a factor of four.

ACKNOWLEDGMENTS

This work was supported in part by a grant from the U.S. Department of Energy and by an AEP grant from the University of Virginia.

REFERENCES

1. Bjorken, J.D., *Phys. Rev.* **148**, 1467 (1966) and *Phys. Rev. D* **1**, 1376 (1970)
2. Drell, S.D., and Hearn, A.C., *Phys.Rev.Lett.* **16**, 908 (1966).
3. Gerasimov, S.B., *Sov. J. Nucl. Phys.* **2**, 430 (1966)
4. Soffer, J., and Teryaev, O.V., *Phys. Rev. D* **51**, 25 (1995)
5. Burkert, V.D., and Ioffe, B.L., *Phys. Lett. B* **296**, 223 (1992)
6. Burkert, V., and Li, Zh., *Phys. Rev. D* **47**, 46 (1993)
7. E155, K.Abe et al., *Phys. Rev. D* **58**, 112003 (1998)

Preliminary Results for the Spin Structure Function g_2 from SLAC E155X

R. Prepost, for the E155X Collaboration

*Department of Physics, University of Wisconsin
Madison, WI 53706, USA*

Abstract. We have measured the spin structure functions g_2^p and g_2^d over the kinematic range $0.02 \leq x \leq 0.8$ and $0.6 \leq Q^2 \leq 20$ (GeV/c)^2 by scattering 29.1 and 32.3 GeV longitudinally polarized electrons from transversely polarized NH$_3$ and ^{6}LiD targets. The statistical errors are approximately three times smaller than previous measurements. Preliminary results for xg_2 are compared with the twist-2 Wandzura-Wilczek calculation. The twist-3 matrix element d_2 is compared with various theoretical models.

INTRODUCTION

The deep inelastic spin structure functions of the nucleons, g_1 and g_2, depend on the spin distribution of the partons and their correlations. The transverse structure function $g_T = g_1 + g_2$ has a simple parton interpretation in terms of the transverse polarization of the quark spins, which is proportional to quark masses. However, g_2 is also sensitive to higher twist effects such as quark-gluon correlations [1] and is not easily interpreted in pQCD where such effects are not included. By interpreting g_2 using the operator product expansion (OPE) [1,2], it is possible to study contributions to the nucleon spin structure beyond the simple quark parton model.

The structure function g_2 can be written:

$$g_2(x, Q^2) = g_2^{WW}(x, Q^2) + \overline{g_2}(x, Q^2) \tag{1}$$

where

$$g_2^{WW}(x, Q^2) = -g_1(x, Q^2) + \int_x^1 \frac{g_1(y, Q^2)}{y} dy.$$

$$\overline{g_2}(x, Q^2) = -\int_x^1 \frac{\partial}{\partial y}\left(\frac{m}{M}h_T(y, Q^2) + \xi(y, Q^2)\right)\frac{dy}{y}$$

where x is the Bjorken scaling variable and Q^2 is the absolute value of the virtual photon four-momentum squared. The twist-2 term g_2^{WW} was derived by Wandzura and Wilczek [3] and depends only on the well-measured g_1 structure function. The function $h_T(x,Q^2)$ is an additional twist-2 contribution [4] that depends on the transverse polarization density in the nucleon, but is suppressed by the quark-to-nucleon mass ratio. The twist-3 part, ξ, comes from quark-gluon correlations and is the main focus of our study.

EXPERIMENT

The recent SLAC experiment E155x made new measurements of g_2 for the proton and deuteron in February to May of 1999. The statistical errors are approximately three times smaller than previous measurements [5–7]. We used the 120 Hz SLAC electron beam with a longitudinal polarization of $(83 \pm 3)\%$ at energies of 29.1 and 32.3 GeV and a typical current of 25 nA. The beam helicity direction was randomly chosen pulse-by-pulse. We used transversely polarized NH_3 and ^6LiD targets as sources of polarized protons (average polarization 75%) and deuterons (average polarization 20%). Scattered electrons were detected in three independent spectrometers centered at 2.75°, 5.5°, and 10.5°.

The asymmetry $A_\perp$ for each kinematic bin was formed using

$$A_\perp = \frac{C_1}{f P_t f_{RC} \cos(\phi)} \left(\left(\frac{N_L - N_R}{N_L + N_R} \right) \frac{1}{P_b} + A_{EW} \right) + A_{RC}$$

where N_L and N_R are the measured counting rates from the two beam helicities, including small corrections for pion and charge symmetric backgrounds, P_b is the beam polarization, P_t is the target polarization, f is the target dilution factor, C_1 includes a small correction for polarized ^{15}N, ϕ is the angle between ep spin and scattering planes, and A_{EW} corrects for the electroweak asymmetry ($\approx 8 \times 10^{-5} Q^2$) The factors f_{RC} and A_{RC} correct for radiative tails, which can be large at low x. Preliminary radiative correction calculations were made using the g_2^{WW} model and the E155 fit to g_1. The calculations will be improved by iterating on a model fit to g_2.

RESULTS FOR g_2

We used our measured $A_\perp$ values together with a fit to g_1 to obtain values for g_2. Fig. 1 shows preliminary results for xg_2 as a function of Q^2 for the proton and deuterium in several x bins. There are up to 6 points in each plot, corresponding to the two energies and 3 spectrometers. The Q^2 dependence is small and consistent with the $g2_{WW}$ calculation shown as the solid black line.

Preliminary results for the three spectrometers were averaged together using the Q^2 dependence of $g2_W W$. The results at $Q^2 = 5(\text{GeV}/c)^2$ are shown in Fig. 2. The proton results are clearly different than zero, and exhibit an x-dependence similar to that of the g_2^{WW} model, although there appear to be statistically significant

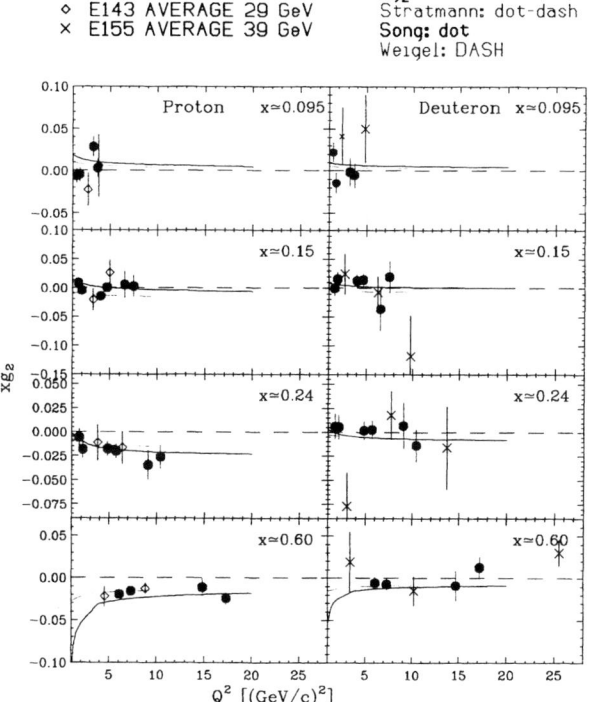

FIGURE 1. Preliminary results for xg_2 as a function of Q^2 for the proton and deuterium in several x bins.

differences from this model, possibly indicating non-zero twist-3 contributions are important. The data are in qualitative agreement with the bag model calculation of Stratmann [8] and the chiral calculation of Weigel [9], but are considerably more negative than the model of Song [4]. The deuteron data have larger errors than the proton data, but also indicate significantly negative values at negative x, and are in qualitative agreements with g_2^{WW}, Stratmann [8], and Weigel [9].

The twist-3 matrix element d_2 is given by

$$d_2 = \int_0^1 x^2 \Big[2g_1(x,Q^2) + 3g_2(x,Q^2)\Big] dx.$$

Preliminary values averaged over all world data are 0.0023 ± 0.0014 (proton) and 0.0077 ± 0.0043 (neutron) which are less than two standard deviations from zero. These results are a factor of 3-4 improvement over previous data.

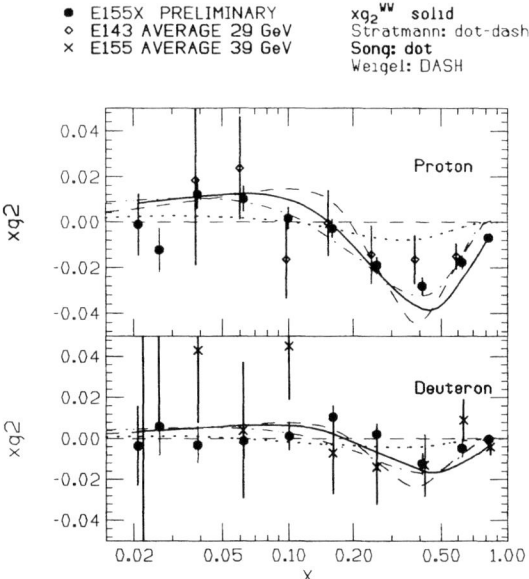

FIGURE 2. Preliminary values of the structure function xg_2 averaged over all three spectrometers (solid circles), and data from E143 [5] (diamonds) and E155 [7] (stars). The errors are statistical; the systematic errors are negligible. Also shown is our twist-2 g_2^{WW} at the average Q^2 of this experiment at each value of x (solid curves) and the calculations of Stratmann [8] (dot-dash curves), Weigel [9] (dashed curves), and Song [4] (dotted curves).

These data can also be used to test the Burkhardt-Cottingham sum rule:

$$\int_0^1 g_2(x)dx = 0.$$

We obtain for this integral: -0.005 ± 0.012 (proton) and 0.001 ± 0.017 (deuteron). Agreement with the prediction of the sum rule is excellent.

SUMMARY

The present measurements of $g_2(x, Q^2)$ are more precise by a factor 3-4 compared to previous data. We find that g_2^p is positive for $x < 0.1$ and negative for $x > 0.2$. There is reasonable agreement with the twist 2 g_2^{WW} model and the twist 3 value d_2 is zero to within 2 SD although consistent with the bag model of ref. [8] et al. and the chiral model of ref. [9] et al.

ACKNOWLEDGMENTS

Work supported by the National Science Foundation and the Department of Energy contract DE–AC03–76SF00515.

REFERENCES

1. E. Shuryak and A. Vainshtein, *Nucl. Phys.* B **201**, 141 (1982).
2. R. Jaffe and X. Ji, *Phys. Rev.* D **43**, 724 (1991).
3. S. Wandzura and F. Wilczek, *Phys. Lett.* B **72**, 195 (1977).
4. X. Song, *Phys. Rev.* D **54**, 1955 (1996).
5. E143 collaboration: K. Abe *et al.*, *Phys. Rev. Lett.* **76**, 587 (1996); *Phys. Rev.* D **58**, 112003 (1998).
6. SMC collaboration: D. Adams *et al.*, PLB **336**, 125 (1994); *Phys. Lett.* B **396**, 338 (1997).
7. E155 collaboration: P. Anthony *et al. Phys. Lett.* B **458**, 529 (1999).
8. M. Stratmann, *Z. Phys.* C **60**, 763 (1993).
9. H. Weigel, L. Gamberg, and H. Reinhart, *Phys. Rev.* D **55**, 6910 (1997).

Neutron Spin Structure and the Extended GDH Sum Rule at Low Q^2

Todd Averett
for the Jefferson Lab E94-010 collaboration

College of William and Mary
Williamsburg, VA 23187, USA

I INTRODUCTION

Over the past several decades, the spin structure of the nucleon has been a topic of central interest in nuclear physics. The primary goal is to understand the origin of the nucleon spin in terms of its fundamental constituents, quarks and gluons. It is also reasonable to assume that at small Q^2, the quark-gluon picture will begin to fail and a hadron-like description will be needed. Understanding the transition from quark-like to hadron-like behavior is an important step in obtaining a complete description of the spin structure of the nucleon. In the recently completed Jefferson Lab experiment E94-010, spin-dependent cross sections for the reaction $^3\vec{H}e(\vec{e},e')$ were measured in the quasi-elastic, resonance, and into the deep-inelastic regions. Using these cross-sections, the extended Gerasimov-Drell-Hearn (GDH) sum was obtained as a function of Q^2 in the range 0.1-1.0 GeV^2. The Q^2 evolution of this sum rule allows us to study the spin structure of the neutron at low Q^2 in this transition region.

The Gerasimov-Drell-Hearn (GDH) sum rule [1] is derived at $Q^2 = 0$ using the low-energy theorem and dispersion relations for the forward Compton scattering amplitudes. It relates the integral of the spin dependent cross sections to the anomalous magnetic moment of the nucleon and is considered to be a theoretically solid prediction. The sum rule is expressed as follows,

$$\int_{\nu_{thr}}^{\infty} \left[\sigma_{1/2}(\nu) - \sigma_{3/2}(\nu)\right] \frac{d\nu}{\nu} = -\frac{2\pi^2 \alpha}{M^2}\kappa^2 \qquad (1)$$

where $\sigma_{1/2}$ and $\sigma_{3/2}$ are the spin-dependent cross sections for real photon absorption, with the total helicity of the photon-nucleon system indicated by the subscripts. The quantity κ is the anomalous magnetic moment of the nucleon and ν is the photon energy.

It is possible to extend the left-hand side of the GDH sum rule to the $Q^2 > 0$ region by replacing the real-photon cross sections with the corresponding transverse virtual photon cross sections, $\sigma_{1/2}^T$ and $\sigma_{3/2}^T$,

$$\int_{\nu_{thr}}^{\infty} \left[\sigma_{1/2}^{T}(\nu) - \sigma_{3/2}^{T}(\nu)\right] \frac{d\nu}{\nu} \equiv 2 \int_{\nu_{thr}}^{\infty} \sigma^{TT}(\nu) \frac{d\nu}{\nu} \tag{2}$$

However, it is not possible to rigorously extend the right hand side of equation 1 in this case. At best, phenomenological models for the virtual photon cross sections can be used to predict the behavior of the integral, which must match the GDH sum rule prediction at $Q^2 = 0$.

Recently, Ji and Osborne [2] were able to obtain a sum rule that is valid at all Q^2 and which reduces to the GDH sum rule at the real photon point,

$$\int_{\nu_{el}}^{\infty} \frac{G_1(\nu, Q^2)}{\nu} d\nu = \frac{S_1(Q^2)}{4} \tag{3}$$

In the above expression, $G_1(\nu, Q^2)$ is one of the nucleon spin structure functions, and is measurable experimentally. The term $S_1(Q^2)$ is the forward virtual Compton scattering amplitude. At large and intermediate Q^2, twist expansions and lattice QCD can be used to calculate $S_1(Q^2)$, and at low Q^2, where hadronic degrees of freedom dominate, Chiral Perturbation Theory can be applied.

One remarkable feature of this extended sum rule is that as $Q^2 \to \infty$, equation 3 reduces to the fundamental Bjorken sum rule [3] (for the case of proton minus neutron):

$$\int_0^1 \left[g_1^p(x) - g_1^n(x)\right] dx = \frac{1}{6} g_A \tag{4}$$

where $g_1^{p,n}(x)$ are the proton and neutron deep-inelastic spin structure functions, and g_A is the axial coupling constant. It is important to note that Ji and Osbourne's extended GDH sum rule provides a theoretically rigorous tool for studying the spin structure of the nucleon that is valid from $Q^2 = 0$ to $Q^2 = \infty$. It is particularly useful for studying the dynamics of the nucleon spin at low Q^2 where a clear theoretical picture does not exist.

II EXPERIMENT E94-010

In experiment E94-010, electrons with longitudinal polarization of 70% were scattered from ^{3}He which was either longitudinally or transversely polarized with respect to the incoming electron momentum. Data was collected at six different beam energies, $E_{beam} = 0.86, 1.7, 2.6, 3.4, 4.3$, and 5.1 GeV, and the scattered electrons were detected (inclusively) in one of two magnetic spectrometers. The spectrometers were located on either side of the beam line at a scattering angle of $\theta = 15.5°$. The experiment took place in Hall A at Jefferson Lab and the complete kinematic coverage is shown in Figure 1.

The ^{3}He nuclei were polarized by spin exchange collisions with atomically polarized Rubidium atoms in a two-chambered glass target cell. Rubidium was contained in the upper chamber and polarized by the technique of optical pumping. Laser

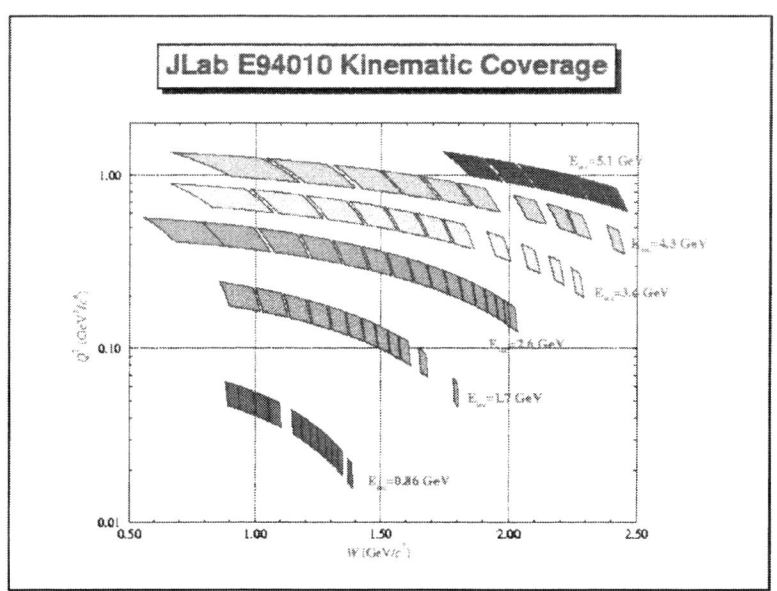

FIGURE 1. Kinematic coverage of experiment E94-010, where W is the invariant mass of the photon-nucleon system. Each box represents one kinematic point where data was collected.

light at a wavelength of $\lambda = 795$ nm was produced using 3 laser diode arrays with a total power output of roughly 100 Watts. The electron beam passed through the lower chamber which was 40 cm in length with thin end-windows ($\sim 120\mu$m) to minimize interactions between the beam and the glass. The target contained roughly 10 standard atmospheres of gaseous ^{3}He with an in-beam polarization of $30-40\%$. Target polarization was measured using two independent methods: NMR with adiabatic fast passage, and EPR (Electron Paramagnetic Resonance).

III PRELIMINARY RESULTS

The data analysis for this experiment is underway and the results presented here are preliminary. Differential cross sections were measured at each kinematic point for all possible combinations of beam and target polarization directions. The consistency of the data was checked by measuring the ^{3}He elastic cross section and asymmetry and comparing these to world data. Using the measured inelastic cross sections, the spin structure functions g_1 and g_2 were obtained for ^{3}He, as well as the virtual photon cross section difference σ^{TT} as defined in equation 2.

Data for $\sigma^{TT}(\nu)$ are shown in Figure 2. Notice the large negative value in the region of the delta resonance which will give a large negative contribution to the GDH integral.

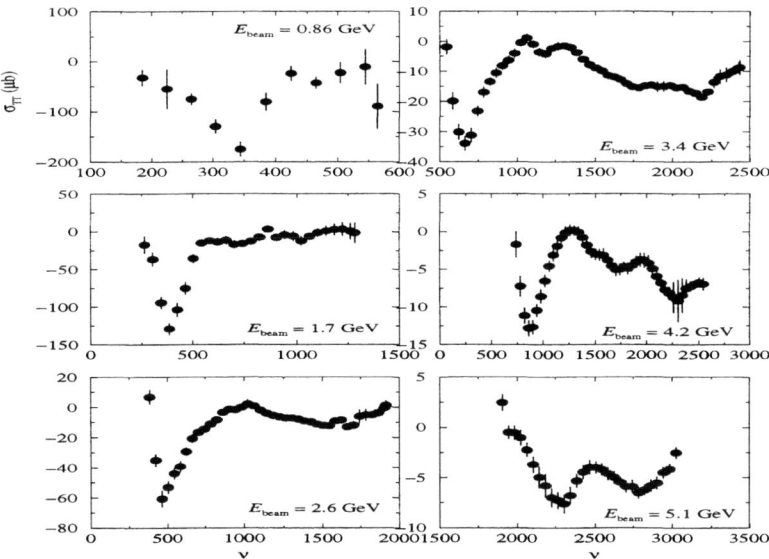

FIGURE 2. The cross section difference $\sigma^{TT} = [\sigma^T_{1/2} - \sigma^T_{3/2}]/2$ for ^{3}He is shown here versus the photon energy ν in MeV. The errors are statistical only. At all but the highest energy setting, the left-most dip is the delta resonance which is large and negative.

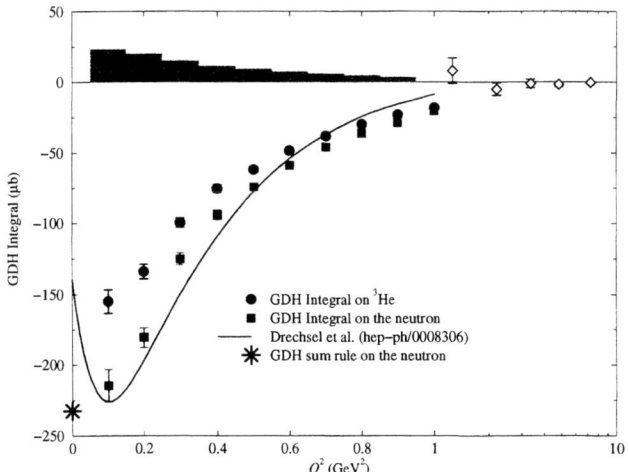

FIGURE 3. The GDH integral versus Q^2 for the neutron and ^{3}He as measured by E94-010. Error bars are statistical and the systematic error is indicated by the band. The high Q^2 HERMES data are shown by the diamonds and the Q^2 scale is logarithmic for that data only. The $Q^2 = 0$ value predicted by the GDH sum rule is shown by the asterisk, and the solid curve is the model of Drechsel et al.

A   Extraction of Neutron Results from ^{3}He

In its ground state, the ^{3}He nucleus is primarily S-wave, with relatively small S' and D-wave components. J. Friar et al. [4] calculated that the effective neutron polarization in ^{3}He is 87% and the proton is about -3%, which means that ^{3}He behaves as an effective polarized neutron target. Recently, C. degli Atti and S. Scopetta [5] studied the extraction of the neutron asymmetries and the generalized GDH sum from a polarized ^{3}He target. Although in the resonance region at low Q^2, the extracted neutron spin structure function g_1 could differ significantly from g_1 for the free neutron, the extracted neutron GDH sum does not differ significantly from the free neutron GDH sum. This is largely because the dominant nuclear effect is Fermi motion, which is effectively averaged out in the integration.

To calculate the GDH sum at fixed Q^2, interpolations were made between the measured kinematic points. Figure 3 shows the GDH sum in equation 2 as a function of Q^2 for the neutron and ^{3}He in the measured region only. No extrapolation to the unmeasured high energy region was made. Also shown is a phenomenological model of Drechsel et al. [6] and the high-Q^2 Hermes data [7].

IV   CONCLUSION AND FUTURE OUTLOOK

A first measurement was made of the spin dependent cross sections for the process $^3\vec{H}e(\vec{e}, e')$ in the region $0.1 < Q^2 < 1.0$ GeV^2. This data cover the quasi-elastic, resonance and into the deep-inelastic regions. The preliminary results indicate a large negative contribution from the delta region and, for the first time, we have observed the negative trend of the GDH sum towards the expected value at $Q^2 = 0$. Final results from this experiment are expected soon. A future Jefferson Lab experiment, E97-110 (Cates, Chen, Garibaldi et al.) will soon measure the cross sections for this process at scattering angles as low as 6° which will allow extraction of the GDH sum down to $Q^2 = 0.01$ GeV^2 with photon energies up to $\nu = 4.5$ GeV.

REFERENCES

1. S.B. Gerasimov, Sov. J. Nucl. Phys. **2** (1966) 430; S.D. Drell and A.C. Hearn, Phys. Rev. Lett. **16** (1966) 908.
2. X. Ji and J. Osbourne, hep-ph/9905410.
3. J.D. Bjorken, Phys. Rev. **148** (1966) 1467.
4. J. Friar et al., Phys. Rev. **C42** (1990) 2310.
5. C. degli Atti and S. Scopetta, Phys. Lett. **B44** (1997) 223.
6. D. Drechsel et al., hep-ph/0008306.
7. K. Ackerstaff et al., hep-ex/9809015.

Single-Spin Azimuthal Asymmetries in Semi-Inclusive Electro-Production of Pions at HERMES

Pasquale Di Nezza, for the HERMES Collaboration

INFN - Laboratori Nazionali di Frascati
via Enrico Fermi 40, I-00044 Frascati, Italy

Abstract. Single-spin azimuthal asymmetries in semi-inclusive pion production in longitudinally polarized deep-inelastic scattering have been measured for the first time in the HERMES experiment. A significant target-spin asymmetry amplitude of the order of 2% was observed both for π^+ and π^0. The π^- asymmetry is consistent with zero. The corresponding analyzing power in the $sin\phi$ moment has been determined as a function of x and $P_\perp$. These results can be interpreted as an effect of a chiral-odd spin distribution function in combination with the T-odd fragmentation function that is sensitive to the transverse polarization of the fragmenting quark.

INTRODUCTION

At leading order QCD the description of the nucleon involves three sets of Distribution Functions (DF). Two of them, the unpolarized DFs, f_1^q, related to the parton momentum distributions, and the polarized DFs, g_1^q, which represent the distribution of quark helicities in a longitudinally polarized nucleons, are well known. The third set of DFs, h_1^q, called transversity, remains hiterto unmeasured; it is the distribution of transverse quark spin in a transversely polarized nucleon with respect to its (infinite) momentum [1,2]. The function h_1^q is *chiral-odd* and, since the strong and the electromagnetic interactions conserve chirality, is not directly observable in inclusive lepton-nucleon scattering experiments. It can be measured through spin asymmetries containing two chiral-odd functions: in Drell-Yan with two DFs and in semi-inclusive DIS with a DF and a fragmentation function (FF). The latter, called Collins fragmentation function ($H_1^\perp$), describes how the probability of producing a hadron depends on its direction with respect to the direction of the transverse polarization of the struck quark. It has a time-reversal-odd structure (T-odd) resulting from the final-state interactions in the fragmentation process, rather than from any fundamental violation of time-reversal invariance [3]. An observable single-spin dependence is predicted to appear in the dependence of the cross section on the azimuthal angle ϕ_c between the spin axis of a *transversely* polarized target

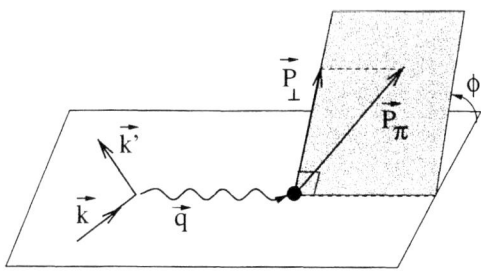

FIGURE 1. Kinematic planes and definition of the Collins angle for hadron production in semi-inclusive DIS of leptons off a nucleon.

and the plane spanned by the momenta of the virtual photon and the pion. This angle is known as the Collins angle [4].

Recently HERMES has measured semi-inclusive pion production from a *longitudinally* polarized proton target [5]; in this case the Collins angle becomes the azimuthal angle ϕ of the pion around the virtual photon direction, with respect to the lepton scattering plane, see Fig.1.

KINEMATICS AND DATA SELECTION

The kinematics of the process is illustrated in Fig.1. Here k and k' are the 4-momentum and E and E' the energies of the incoming and outgoing leptons, P_π is the 4-momenta of the produced pion, $\vec{q}$ is the 4-momentum transfer with $q^2 = -Q^2$, $W^2 = 2M\nu + M^2 - Q^2$ is the invariant mass of the photon-proton system and $x = Q^2/2M\nu$ is the Bjorken variable. E_π is the pion energy, M is the proton mass and $z = E_\pi/\nu$ is the pion fractional energy referred to the photon.

The data were taken during the 1996 and 1997 running periods of the HERMES experiment using a longitudinally polarized ^{1}H internal gas target in the 27.5 GeV longitudinally polarized positron beam of the HERA ring at DESY. A detailed description of the HERMES spectrometer can be found in [6].

The kinematical requirements on the scattered positron used in this analysis are $1 < Q^2 < 15$ GeV^2, $W > 2$ GeV, $0.023 < x < 0.4$ and $y = \frac{\nu}{E} < 0.85$. Charged pions were identified in the energy range $4.5 < E_{\pi^\pm} < 13.5$ GeV. The neutral pions were reconstructed by the two decay photons collected by the electromagnetic calorimeter after imposing the cuts on the single cluster energy $E_\gamma > 1.0$ GeV and on the invariant mass range $0.100 < M_{\gamma\gamma} < 0.170$ GeV, respectively.

Acceptance effects were minimized and exclusive pion production was suppressed by imposing the requirement $0.2 < z < 0.7$. The limit $P_\perp > 50$ MeV, where $P_\perp$ is defined with respect to the virtual-photon direction, was applied to a pion to allow an accurate measurement of the angle ϕ.

Measurements were performed with all combinations of beam and target helicities, opening the possibility of evaluating single- and double-spin terms in the cross

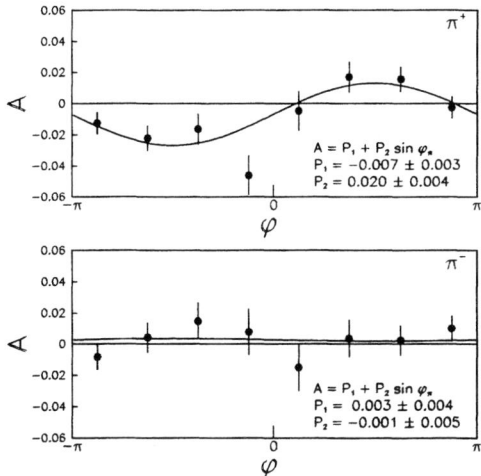

FIGURE 2. Spin-azimuthal asymmetry in the cross section for π^+ (top) and π^- (bottom). Error bars represent the statistical uncertainties. The curves are sinusoidal fits to the data.

section. The average hydrogen target polarization in the 1996 and 1997 HERMES running periods was 0.86 with a fractional uncertainty of 5%.

RESULTS

The ϕ dependence of the single target-spin asymmetry in the cross section is defined as:

$$A(\phi) = \frac{N^{\rightarrow}L^{\leftarrow} - N^{\leftarrow}L^{\rightarrow}}{N^{\rightarrow}L^{\leftarrow}P^{\leftarrow} + N^{\leftarrow}L^{\rightarrow}P^{\rightarrow}}. \tag{1}$$

Here N is the number of DIS events where the positron is accompanied by a pion, corresponding to the polarization values P and the dead-time corrected luminosities L, all averaged over the two beam helicities. The $\rightarrow$ and $\leftarrow$ symbols refer to the two target spin states.

The asymmetry (1) is shown as a function of the variable ϕ in Fig.2 for both π^+ and π^-. The π^+ asymmetry clearly shows a sinusoidal behavior while the π^- asymmetry is consistent with zero within the experimental errors. These findings are in agreement with the expectation based on the u-quark dominance in the valence region for the proton [4,7,8]. The average values of these asymmetries are $0.022 \pm 0.005(stat) \pm 0.003(syst)$ for the π^+ and $-0.002 \pm 0.006(stat) \pm 0.004(syst)$ for the π^-.

The various contributions to the ϕ dependent spin asymmetry are isolated by extracting moments of the cross section weighted by corresponding ϕ dependent functions W. This can be done by calculating the analyzing powers for the case of unpolarized beam and longitudinally polarized target. It can be written as:

$$A_{UL}^{W} = \frac{\frac{L^{\rightarrow}}{L_P^{\rightarrow}} \sum_{i=1}^{N^{\rightarrow}} W(\phi_i) - \frac{L^{\leftarrow}}{L_P^{\leftarrow}} \sum_{i=1}^{N^{\leftarrow}} W(\phi_i)}{\frac{1}{2}[N^{\rightarrow} + N^{\leftarrow}]}.$$

Hera L_P corresponds to the dead-time corrected luminosity, averaged with the magnitude of the target polarization. The weighting functions $W(\phi) = sin\phi$ and $W(\phi) = sin2\phi$ are expected to provide sensitivity to the Collins fragmentation function discussed above. Analyzing powers were extracted by integrating over the spectrometer acceptance in the kinematic variables y and z. Corrections were applied for the effects of the spectrometer acceptance, based on a Monte Carlo simulation.

The single target-spin related term $A_{UL}^{sin2\phi}$ was found to be consistent with zero within errors both for π^+ and π^-. The average values are $-0.002 \pm 0.005(stat) \pm 0.010(syst)$ for the π^+ and $-0.005 \pm 0.006(stat) \pm 0.005(syst)$ for the π^- respectively.

The term $A_{UL}^{sin\phi}$, integrated over the transverse pion momentum $P_\perp$ is presented as a function of Bjorken-x in Fig.3 (left) for both neutral and charged pions. The π^- behavior is compatible with zero and not shown in this figure. The fact that the dependence for π^0 and π^+ are very similar, is explained by u-quark dominance and, at the same time, indicates strong suppression of the Collins fragmentation functions associated to the other quarks (unfavored). $A_{UL}^{sin\phi}$ for π^+ and π^0 rises almost linearly from $x \simeq 0.05$ to $x \simeq 0.3$. This suggests that the sea contribution does not dominate the effect, in agreement with an existing interpretation of the single-spin asymmetry as being associated with valence quark contributions [9].

In Fig.3 (right) the analyzing power $A_{UL}^{sin\phi}$ is plotted as a function of $P_\perp$ of the detected pion, after integration over x. For π^+ and π^0 it shows an increasing behavior with increasing $P_\perp$ as long as $P_\perp$ remains below the typical hadronic mass of 1 GeV. This behavior can be understood by considering the dominant role of the intrinsic quark transverse momentum $k_\perp$ [10].

In all situations discussed above, the main contributions to the systematic uncertainties are coming from the uncertainty in the target polarization (5%) and from false asymmetries resulting from the spectrometer acceptance.

CONCLUSIONS

Single-spin azimuthal asymmetries have been measured for the first time in the production of pions in deep-inelastic scattering using a longitudinally polarized hydrogen target. The analyzing power involving the $sin\phi$ moment shows a sizeable asymmetry for π^+ and is consistent with zero for π^- within errors. The preliminary π^0 data are close to the π^+ ones. These effects measured as a function of x and $P_\perp$ are in qualitative agreement with the theoretical predictions based on the Collins model of fragmentation. The appearance of these asymmetries can be interpreted

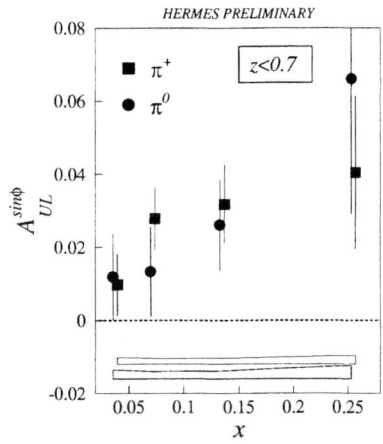 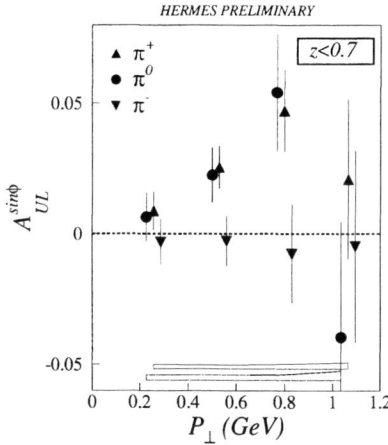

FIGURE 3. Target spin analyzing power in the $sin\phi$ moment as a function of Bjorken-x (left) and as a function of $P_\perp$ of the detected pion (right). Error bars show the statistical uncertainties and the bands represent the systematic uncertainties. The π^0 data are preliminary, the $\pi^\pm$ data have been published in Ref. [5].

as an effect of chiral-odd distribution functions coupled with a T-odd fragmentation function [11]. This offers a mean to measure the transversity distribution using a transversely polarized target in a simple way, e.g. as suggested in Ref. [12].

REFERENCES

1. J.P.Ralston and P.E.Soper, Nucl. Phys. **B152**, 109 (1979).
2. R.Jaffe and X.Ji, Nucl. Phys. **B375**, 527 (1992).
3. G.Gasiorowicz, *Elementary Particle Physics*, John Wiley & Sons, New York, 1966.
4. J.Collins, Nucl. Phys. **B396** 161 (1993).
5. HERMES Coll., A.Airapetian et al. Phys.Rev.Lett. **84** (2000) 4047.
6. HERMES Coll., K.Ackerstaff et al., Nucl Instr. & Meth. **A417** 230 (1998).
7. M.Anselmino and F.Murgia, Phys. Lett. **B483** (2000) 74.
8. A.Schäfer and O.Teryaev, Phys. Rev. **D61** 7903 (2000).
9. M.Anselmino, M.Boglione and F.Murgia, Phys. Lett. **B362** 164 (1995).
10. K.A.Oganessyan et al., hep-ph/0010063
11. P.J.Mulders and R.D.Tangerman, Nucl.Phys. **B461**, 197 (1996).
12. V.A.Korotkov, W.-D.Nowak and K.A.Oganessyan, hep-ph/0002268

Estimation of the proton transversity from azimuthal asymmetries in DIS

A.V. Efremov[1]

Joint Institute for Nuclear Research, Dubna, 141980 Russia

Using the preliminary experimental data from DELPHI [1] on the left right asymmetry in fragmentation of transversely polarized quarks ("Collins effect") and the theoretical calculation of the proton transversity distribution in the effective chiral quark soliton model [2] we explain the recently observed spin azimuthal asymmetries in semi-inclusive hadron production on longitudinally (HERMES) and transversely (SMC) polarized targets *with no free parameters*. (See Figure 1.) For more details see [3].

On this basis we state that the proton transversity distribution could be successfully measured in future DIS experiments with *longitudinally* polarized target simultaneously with measurement of the spin gluon distribution $\Delta g(x)$.

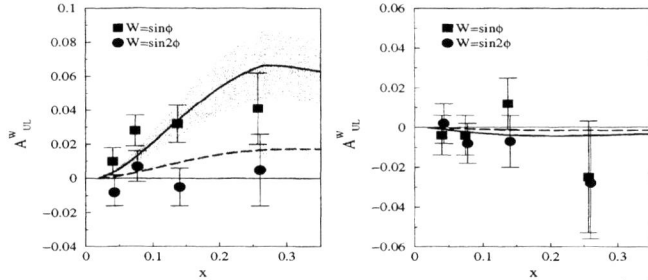

Figure 1. Single spin azimuthal asymmetry for π^+ (**left**) and π^- (**right**): $A_{UL}^{\sin\phi}$ (squares) and $A_{UL}^{\sin 2\phi}$ (circles) as a functions of x (HERMES data). The solid ($A_{UL}^{\sin\phi}$) and the dashed lines ($A_{UL}^{\sin 2\phi}$) correspond to the chiral quark-soliton model calculation at $Q^2 = 4\,GeV^2$. The shaded areas on the left figure represent the experimental uncertainty in the value of analyzing power of the Collins effect. For π^- (right) both asymmetries are compatible with zero.

REFERENCES

1. A.V.Efremov, O.G.Smirnova and L.G.Tkatchev, *Nucl. Phys. (Proc. Suppl.)* **74** (1999) 49 and **79** (1999) 554. hep-ph/9812522.
2. P.V.Pobylitsa and M.V.Polyakov, *Phys. Lett.* **B 389** (1996) 350.
3. A.V.Efremov, K.Goeke, M.V.Polyakov and D.Urbano, *Phys. Lett.* **B478** (2000) 94; hep-ph/0001119.

---

[1] Supported by Russian Foundation for Basic Research under the Grant 00-02-16696.

Spin physics with the HERMES experiment using hadron identification by the Ring Imaging Cherenkov Counter

Yasuhiro Sakemi
on behalf of the HERMES collaboration

*Department of Physics, Tokyo Institute of Technology,
2-12-1 Oookayama, Meguro-ku, Tokyo 152-8551, Japan*

Abstract. The HERMES experiment uses the 27.5 GeV beam of longitudinally polarized electrons of the HERA accelerator together with an internal gas target to study the spin structure of the nucleon. The ability to identify hadrons in coincidence with the scattered electron is crucial for the determination of the contributions of the different quark flavors to the nucleon spin. For this purpose a Ring Imaging CHerenkov (RICH) detector was developed. It is able to identify pions, kaons and protons over the relevant momentum range of 2-15 GeV/c by using two radiators, silica aerogel and C_4F_{10} gas. This detector was installed in May 1998 in the HERMES experiment. It is the first RICH detector in an actual experiment that uses aerogel as a radiator.

I INTRODUCTION

The HERMES experiment uses the 27.5 GeV beam of longitudinally polarized electrons of the HERA accelerator together with an internal gas target to study the spin structure of the nucleon [1]. Hadrons that are detected in coincidence with the scattered electron contain information on the nature of the parton that was hit in the scattering process. This method, called 'flavor tagging', in combination with the use of different target gases, allows the determination of the contributions of the different quark flavors to the spin of the nucleon [2]. However, the momentum of most hadrons produced in HERMES lies between 2 and 10 GeV, a region in which hadron identification with standard particle identification techniques was not feasible. As a result of the development of new clear silica aerogels with low indices of refraction this difficult momentum region can now be covered [3]. The combination of aerogel with a heavy gas (C_4F_{10}) in a dual radiator RICH makes it possible to cover the entire kinematic range of HERMES for hadron momenta from 2 to 15 GeV/c.

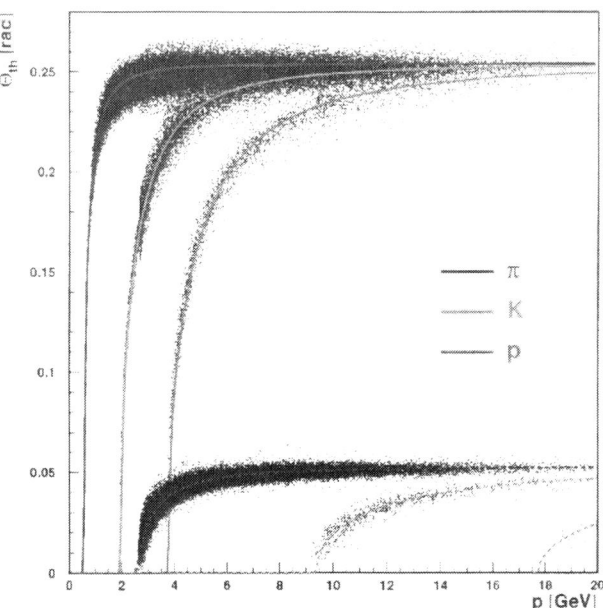

FIGURE 1. Momentum dependence of the Cherenkov angles θ_{th} for electrons, pions, kaons and protons for aerogel and C_4F_{10} radiators. The top curves show θ_{th} for the aerogel radiator and the bottom curves for the C_4F_{10} radiator. The points represent the experimental data.

II THE HERMES RICH DETECTOR

Due to its location at the HERA storage ring the HERMES spectrometer (including the RICH) consists of two symmetric halves, above and below the beam pipes [1]. To obey the spatial constraints within the HERMES spectrometer the existing boxes of the threshold Cherenkov counters were re-used as radiator boxes for the RICH. The index of refraction of the aerogel was chosen to be $n_{aero} = 1.03$, which in combination with the heavy gas C_4F_{10} (n = 1.0014) allows the coverage of the kinematic range from 2 to 15 GeV/c. The momentum dependence of the Cherenkov angles θ_{th} for pions, kaons and protons expected for this radiator combination are shown in Fig.1 with top curves for the aerogel radiator and bottom curves for the C_4F_{10} radiator.

A Aerogel Radiator

The aerogel was produced by Matsushita Electric Works, Ltd.(Osaka, Japan) in form of quadratic tiles ($114 \times 114 \times 11.3$ mm^3) [4]. The measured average index of refraction for the aerogel is $n_{aero} = 1.0304 \pm 0.0004$. The aerogel tiles are stacked

5 tiles deep, 5 tiles vertically and 17 tiles horizontally, in total 850 tiles for both radiators (corresponding to a total volume of 125 liter aerogel) [5]. The aerogel tiles are held in an aluminum frame with a 1 mm aluminum entrance window and a 3.2 mm lucite exit window. This volume is continuously flushed with dry nitrogen. To reduce background from stray photons, the individual tile stacks are separated from one another by black, opaque tedlar sheets that absorb photons crossing the stack boundaries. Rayleigh-scattering of the Cherenkov photon inside the aerogel itself is another major background source. As the Rayleigh-scattered light is centered at shorter wavelength, lucite with its transmission cutoff at 290 nm was chosen for the exit window. The basic geometry of the aerogel radiator with the entrance and exit windows, mirror array, and photomultiplier matrix is shown in Fig. 2.

B Photon Detector

Each photon detector consists of 1934 Philips XP1911/UV green enhanced 0.75 inch photomultiplier tubes (PMTs) arranged in a hexagonal grid that covers an area of 60 × 120 cm^2 centered at the focal point of the mirror. The PMT array is located close to the HERMES spectrometer magnet, where the fringe field reaches 100 Gauss. To achieve sufficient magnetic shielding, the PMTs were individually wrapped with thin μ-metal tubes and the matrix plates holding the PMTs were constructed from high permeability (C-1008) steel. Only 38 % of the area of the photon detector plane is covered directly by the PMT photo-cathodes. Cones of aluminized mylar foil, inserted for every PMT, increased the overall aerial coverage of the detector to 91 %. The photon detectors are read out using a LeCroy PCOS4 acquisition system. This system is characterized by high input sensitivity (3000 e$^-$) and high amplification (4.3 μV/e$^-$). The PCOS4 system does not provide any analog information; the response is digital - the PMT was 'hit' or not. As a result, the RICH data for a given event consist of a list of numbers of fired PMT in the top and bottom halves.

C Mirror Array

Each mirror array consists of eight individual mirror segments. Due to the position of the RICH in the spectrometer, it was important to keep the mirrors as thin as possible in terms of radiation length. Hence, the mirror segments were fabricated from a graphite fiber composite, coated with an epoxy film yielding an optically smooth surface, and finally aluminized. The surface was specified to have a roughness of less than 5 nm and a reflectivity above 0.85 (300-600 nm). A carbon fiber frame with adjustable three point mounts holds the mirror segments to form a mirror array with the total dimensions of 252.4 × 79.4 cm^2.

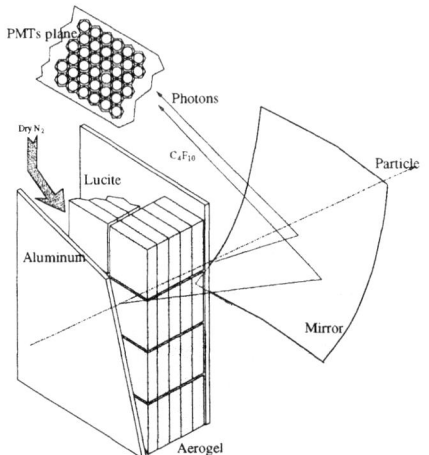
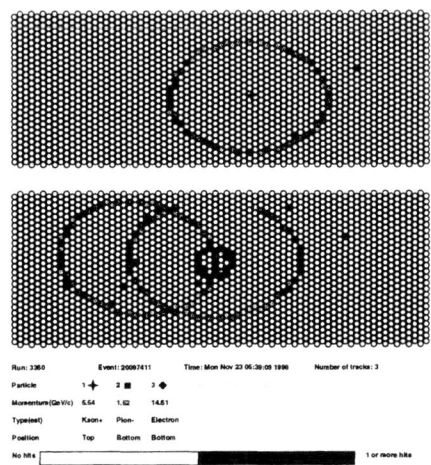

FIGURE 2. Basic geometry of aerogel radiator, mirror array and photomultiplier matrix. The mirror array consists of 8 individual mirrors.

FIGURE 3. HERMES RICH event display for an event with a 14.6 GeV electron and a 1.5 GeV pion in the lower half and a 5.5 GeV kaon in the upper half of the detector. See text for detailed description.

III DETECTOR PERFORMANCE

Since the detected 'rings' are not circles but in general asymmetrically deformed ellipses due to the imaging system, a ring fitting procedure is not feasible for an accurate reconstruction of the Cherenkov angles. The Cherenkov angle corresponding to each PMT hit is instead reconstructed from the tracking information and the position of the PMT [6]. For each hit, the Cherenkov angle θ_{th} is calculated under which the photon would have been emitted. This results in spectra of reconstructed Cherenkov angles for each track and radiator. An average angle $<\theta>$ is calculated from the angles inside a window around the theoretical angle θ_{th} for the given particle type. These average angles are converted into Gaussian likelihoods $L(<\theta>)$ for each particle hypothesis: $L(<\theta>) = exp(-\frac{(\theta_{th} - <\theta>)^2}{2\sigma^2_{<\theta>}})$.

For particles with $\beta \approx 1$ that are not affected by acceptance effects, on average 12 gas hits and 10 aerogel hits are observed. However, due to their sizes most aerogel rings are affected by the finite sizes of mirror, photon detector and aerogel tiles.

Due to the small size of the gas rings compared to the granularity of the photon detector one gas hit corresponds to about two photoelectrons, while for the larger aerogel rings one PMT hit represents one photoelectron. Fig.3 shows the HERMES RICH event display for an event with an electron and a pion in the lower half and a kaon in the upper half of the detector. The solid black dots indicate which PMTs have been fired. The markers in the ring centers indicate the point where the reflected track itself would hit the photon detector. The ellipsoids are the result of direct ray tracing indicating where hits are expected for this event.

The mirror alignment was made with the data itself. The position of a spherical mirror with known radius is unambiguously determined by the position of the mirror center. The alignment is done with a scan of the possible center positions for the mirror arrays leading to a mirror alignment that optimizes the detector resolutions. At each aerogel/gas/aerogel boundary the effect of the two refractions on the photon angle cancels. But as the photon is emitted inside an aerogel tile, the effect of the last aerogel/gas boundary does not cancel, and it was corrected in this analysis. The inclusion of the aerogel boundary refraction shifts the reconstructed angle towards smaller values. After the mirror alignment and the aerogel boundary correction, the single photon resolution is within 10 % of the value of about 7 mrad expected from Monte Carlo simulations.

The particle identification capabilities of the HERMES RICH are shown in Fig.1. The points represent the experimental data and the top/bottom curves show the calculated Cherenkov angle for each particle type. It can be seen that the pions, kaons, and protons are separated clearly over a wide momentum range from 2 to 15 GeV/c. The experimental data agree with the expected values showing that the reconstruction algorithm works well in the analysis.

IV CONCLUSION AND OUTLOOK

The HERMES RICH detector has been operated routinely as a part of the HERMES experiment since its installation in May 1998. After the mirror alignment, the single photon resolution of the detector agrees with Monte Carlo predictions within 10 %. It is the first RICH detector in an actual experiment that uses clear silica aerogel with a low index of refraction as a radiator. The combination of aerogal with C_4F_{10} gas in this dual radiator RICH makes it possible to cover the entire kinematic range of HERMES to separately identify pions, kaons, and protons. The identification of kaons in particular will permit a direct measurement of the strange sea polarization.

REFERENCES

1. K. Ackerstaff et al., *Nucl. Inst. Meth.* **A417** (1998) 230.
2. K. Ackerstaff et al., *Phys. Lett.* **B464** (1999) 123-134.
3. I. Adachi et al., *Nucl. Inst. Meth.* **A355** (1995) 390.
4. H.Yokogawa and M.Yokoyama, *J.Non-Cryst. Solids* **186** (1995) 23.
5. R.De Leo et al., *Nucl. Inst. Meth.* **A440** (2000) 338.
6. E. Cisbani et al., *HERMES internal note* **97-003**

Deeply Virtual Compton Scattering at HERMES

Moskov Amarian[1]
(for the HERMES Collaboration)

INFN - Sezione di Roma, Gruppo Sanita',
viale Regina Elena, 299, 00161 - Rome, Italy

Abstract. Single-spin asymmetries in the hard exclusive electroproduction of real photons have been measured for the first time. The data have been accumulated by the HERMES experiment at DESY using polarized and unpolarized hydrogen gas target internal to the 27.5 GeV polarized positron beam.

Inclusive deep-inelastic lepton-nucleon scattering experiments played a significant role in our understanding of the internal structure of the nucleon. Recent progress in this field is related to exclusive reactions and the *skewed parton distributions (SPD's)* which take into account the dynamical correlations between partons of different momenta. The SPD-framework can be used to evaluate a wide range of observables, such as electromagnetic form factors, inclusive parton distributions and exclusive meson-production cross sections.

One of the cleanest channels that can be used to measure SPD's is deeply virtual Compton scattering (DVCS), i.e. the observation of a multi-GeV photon radiated from a single quark in deep-inelastic lepton-nucleon scattering, when highly virtual photon (with large Q^2) is absorbed and the squared four-momentum transfer between initial- and final state nucleon, t, is small. This reaction is unique, as the produced real photon carries direct information about partonic structure of the nucleon without being distorted by hadronization processes. Another particular interest to DVCS measurements was triggered by its connection to the spin structure of the nucleon. In [1] it has been argued that DVCS provides information on the total angular momentum ($J = 1/2(\Delta\Sigma) + L$) of the partons in the nucleon. In view of the non-trivial spin structure of the nucleon, independent information on the angular momentum decomposition of the nucleon spin is highly desirable. It should be noted, though, that the exact relation between DVCS measurements and spin decomposition of the nucleon is still a subject of continued theoretical debate. Significant progress has been made in the last few years in our understanding of not

---
[1] On leave from Yerevan Physics Institute, 375036 - Yerevan, Armenia

only DVCS, but also other exclusive processes in the deep-inelastic domain [2–5]. However, a full overview of theoretical papers related to this subject is out of scope of our presentation.

On the other hand, experimental information on DVCS is almost absent. Recently, both collider experiments at HERA, ZEUS and H1 [8,7], have presented an excess of the real γ-production rate over background, which was interpreted as evidence of DVCS in the range of very low x_{Bj} (the Bjorken scaling parameter). In this paper we present the first measurement of a single-spin asymmetry in DVCS, which was obtained by the HERMES collaboration using the HERA polarized positron beam. Both, unpolarized and polarized (spin averaged) hydrogen target data have been included in this analysis.

In order to access the DVCS amplitudes either one has to select a kinematical domain where the Bethe-Heitler (BH) amplitudes are suppressed, or one has to use the interference between the BH and DVCS processes. In the present experiment we exploit the latter option. Following [6] the cross section for leptoproduction of real photons can be written as

$$\frac{d\sigma}{d\phi dt dQ^2 dx_{Bj}} = \frac{x_{Bj} y^2}{32(2\pi^4)Q^4} \frac{|\tau_{BH} + \tau_{DVCS}|^2}{\left(1 + 4x_{Bj}^2 m^2/Q^2\right)^{1/2}}, \qquad (1)$$

where τ_{BH}, τ_{DVCS} are the BH and DVCS amplitudes, $y = \nu/E_e$ is the fraction of the incoming lepton energy carried by the virtual photon, Q^2 its negative four-momentum squared, and m the proton mass. The full expression with all terms included is very lenghty, therefore it is omitted here. However interference term can be expressed as follows

$$(\tau_{BH}^* \tau_{DVCS} + \tau_{DVCS}^* \tau_{BH})_{pol} \sim e_l P_l \left[-\sin\phi \cdot \sqrt{\frac{1+\epsilon}{\epsilon}} \mathrm{Im}\tilde{M}^{1,1} + \sin 2\phi \cdot \mathrm{Im}\tilde{M}^{0,1} \right] \qquad (2)$$

In this equation the quantities $\tilde{M}^{\lambda,\lambda'}$ are linear combinations of the DVCS helicity amplitudes $M_{h,h'}^{\lambda,\lambda'}$ with λ, λ' (h, h') representing the helicities of the initial and final-state photon (target nucleon). $e_l = \pm 1$ (lepton charge) and $P_l = \pm 1$ (lepton helicity) of the incident lepton show the sensitivity of the interference term to the difference between electron and positron scattering and spin direction of incoming beam respectively. ϕ is defined as the angle between the lepton scattering plane and the plane defined by the virtual and real photon. The dependence of the cross section on ϕ gives rise to azimuthal asymmetries. The kinematical quantity ϵ represents the longitudinal-transverse polarization of the virtual photon.

Since $|\tau_{DVCS}|^2$ is very small and $|\tau_{BH}|^2$ does not depend on the lepton beam helicity, the interference term given above will dominate measurements of the cross-section asymmetry with respect to the lepton beam helicity P_l. From the equation above it can be seen that such measurements provide information on the imaginary part of the DVCS amplitudes $\tilde{M}^{1,1}$ and $\tilde{M}^{0,1}$.

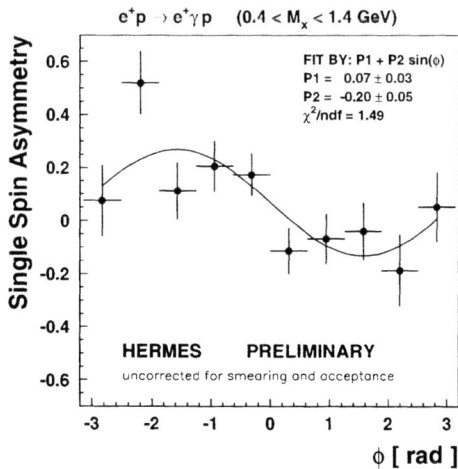

FIGURE 1. Single-spin asymmetry as a function of azimuthal angle ϕ.

The HERMES detector [9] is a forward spectrometer that identifies charged particles in the scattering angle range of $0.04 < \theta < 0.22$ rad. The electromagnetic lead-glass calorimeter and preceding scintillator counters with preshower were used to measure the energy deposition of a cluster without a track in the wire chamber detectors, that was identified as a photon.

The exclusive photons observed in the present analysis can be contaminated by decay photons from π^0 mesons. This contamination has been estimated by comparing exclusive π^+ and π^0 mesons observed within the HERMES acceptance in the range $0 < M_x < 2$ GeV, where $M_x^2 = (q + P_p - P_\gamma)^2$ with q, P_p and P_γ being four-momenta of virtual gamma, target nucleon and produced real photon respectively. It was found that the π^+ production rate is 5 times higher than that of the π^0's. Then in the same M_x-domain the real photon rate is 10 times higher than π^+ rate. Hence, the contamination of real photons due to π^0 decay is at most 2%.

A better measure of the sensitivity of the data to DVCS can be obtained from the ϕ-dependence of the cross-section asymmetry with respect to the beam helicity. In order to have a full coverage in the azimuthal angle ϕ and in order to ensure near collinearity of the virtual photon and the produced photon, the following requirement was introduced. Only data were selected with $0.015 < \theta_{\gamma\gamma^*} < 0.07$, where $\theta_{\gamma\gamma^*}$ represents the angle between the direction of the virtual photon and the produced real photon. The minimum value of $\theta_{\gamma\gamma^*}$ is dictated by the granularity of the calorimeter, and has been derived from a Monte-Carlo study. The obtained single-spin asymmetry is displayed in Fig.1. In this figure data have been selected with a missing mass between 0.4 and 1.4 GeV, i.e. including both the proton and the $\Delta(1232)$-resonance. As the data show a clear azimuthal asymmetry, which can only be caused by the DVCS subprocess, this result provides strong evidence for the observation of photons due to deeply virtual Compton scattering. Average values of relevant kinematical variables in this measurement are: $<Q^2> = 2.5\ GeV^2$, $<-t> = 0.2\ GeV^2$ and $<x_{Bj}> = 0.1$. The contribution of the BH-subprocess to the single-spin asymmetry shown in Fig.1 has been estimated using the aforementioned Monte-Carlo calculation. The false asymmetry due the BH subprocess was thus estimated to be at the level of only 2-3% of the measured asymmetry. Further checks of this result were carried out by dividing the acceptance of the HERMES spectrometer into four quadrants and extracting the single-spin asymmetry for each piece separately. In a similar fashion the single-spin asymmetry has also been determined for 3 different (with roughly the same number of events) time periods. Taken together all of these checks imply that the results are stable to within 5% of the measured asymmetry. In order to be able to compare the ϕ-dependence of the single-spin asymmetries for different mass bins, we introduce the $sin\phi$-weighted single-spin azimuthal asymmetry:

$$A_l = \frac{2 <sin\phi>_{LU}}{<|P_B|>} = \frac{2 \int_0^{2\pi} d\phi\ (d\sigma/d\phi)\ sin\phi}{<|P_b|> \int_0^{2\pi} d\phi\ (d\sigma/d\phi)}, \qquad (3)$$

where $|P_b|$ represents the average absolute value of the beam polarization as measured with lepton type l. The superscript LU means polarized beam and unpolarized target.

In Fig.2 (left panel) the measured value of the $sin\phi$-moment of the azimuthal asymmetry is plotted versus missing mass M_x for two helicity states of the positron beam. Also shown are the data for both helicity states together, weighted by the integrated luminosities for each helicity state. As can be seen the sign of the asymmetry is clearly opposite for the two beam helicities. Also, the beam-spin averaged data are consistent with zero, as it is expected. In fact, the beam-spin averaged data can be used as a measure of the false-asymmetry of the data, which – averaged between 0 and 2 GeV – yields a false asymmetry of 5%.

As the data in Fig.2 (left panel) for the two beam-helicity states represent the same physics information, it is natural to combine the two results. To this end, we define the difference of the sine-weighted single-spin azimuthal asymmetry for the two beam-helicity states by:

$$A_{LU}^{sin\phi} = \frac{2 \int_0^{2\pi} d\phi\ (d\sigma^+/d\phi - d\sigma^-/d\phi)\ sin\phi}{<|P_b|> \int_0^{2\pi} d\phi\ (d\sigma^+/d\phi + d\sigma^-/d\phi)}, \qquad (4)$$

where the superscripts +,- refer to the two beam-helicity states. The data for this sine-weighted asymmetry $A_{LU}^{sin\phi}$ are presented in Fig.2 (right panel) versus missing mass. A large asymmetry is observed in the missing mass region close to $M_x = m_p$. It is of interest to note that the asymmetry observed in the M_x-bins below m_p is

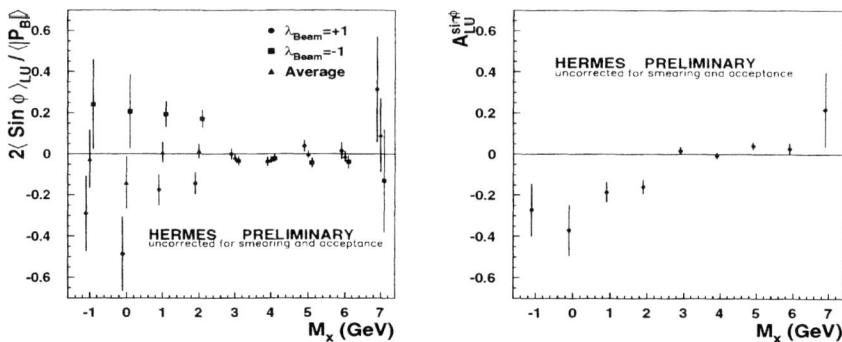

FIGURE 2. Sin-weighted azimuthal asymmetry for each beam-helicity state and averaged (left panel). The same for two beam-helicity states combined (right panel).

within errors the same as the asymmetry near m_p, which is to be expected as the observed strength below m_p is entirely due to the limited energy resolution of the experiment. If we combine the sine-weighted asymmetry $A_{LU}^{sin\phi}$ in the full M_x region that corresponds to exclusive photon production, a value of 0.20 ± 0.05 (stat) is found.

In summary, the single-spin asymmetry for the hard exclusive electroproduction of photons has been measured for the first time. A large value of asymmetry is observed in the exclusive limit, i.e. for $M_x \approx m_p$. The observed azimuthal asymmetry has the $sin\phi$ dependence and beam-helicity dependence expected for the interference amplitude between deeply virtual Compton scattering and the Bethe-Heitler process. Further information on DVCS in deep-inelastic lepton scattering can be obtained by also studying the lepton-charge asymmetry. Togeteher with the single-spin asymmetry measurement this will allow to access imaginary and real parts of all leading-, as well as non-leading amplitudes.

REFERENCES

1. X.Ji, Phys. Rev. D55 (1997) 7114 [hep-ph/9609381].
2. J.Collins, L.Frankfurt and M.Strikman, Phys. Rev. D56 (1997) 2982 [hep-ph/9611433].
3. A. Radyushkin, Phys. Rev. D56 (1997) 5524 [hep-ph/9704207].
4. X.Ji and J.Osborne, Phys. Rev. D58 (1998) 094018.
5. J.Collins and A. Freund, Phys. Rev. D59 (1999) 074009
6. M. Diehl et al.,Phys. Lett. B 411 (1997) 193.
7. P.R.Saull,ZEUS Collaboration, hep-ex/0003030.
8. E.Lobodzinska, Talk presented at the International Workshop on skewed Parton Distributions and Lepton-Nucleon Scattering, DESY, Hamburg, Sep. 11-12, 2000.
9. K.Ackerstaff et al., Nuc. Instr. Meth. A 417 (1998) 230.

Polarized gluon distributions from high-p_T pair hadron productions in polarized deep inelastic scattering

Teruya Yamanishi* [1], Dong Yu-Bing† [2], and Toshiyuki Morii‡ [3]

*Department of Management Science,
Fukui University of Technology, Fukui 910-8505, Japan
†Institute of High Energy Physics, Academia Sinica,
Beijing 100039, P. R. China
‡Faculty of Human Development and
Graduate School of Science and Technology,
Kobe University, Kobe 657-8501, Japan

Abstract. To study the polarized gluon density $\Delta g(x)$ in the nucleon, we propose the high–p_T pair charmed hadron production process in polarized ℓp scattering. The double spin asymmetry A_{LL} for this process is a good observable for testing the models of $\Delta g(x)$.

INTRODUCTION

Recently, new parametrization sets of the polarized parton distribution functions (pol–PDFs) were obtained from fitting to experimental data of polarized deep inelastic scattering (pol–DIS) with high precision for various targets [1]. The next–to–leading order QCD analyses on the x and Q^2 dependences of the structure function g_1 brought about information on the first moment of the polarized gluon Δg. However, there are large uncertainties in Δg extracted from g_1 alone. Knowledge of Δg is still limited because it is very difficult to directly extracting those information from existing data.

So far, a number of interesting proposals for studying longitudinally polarized distributions of gluons were presented: direct prompt photon production in polarized proton–polarized proton collisions [2], open charm–, J/ψ–, and pair–hadron production [3] in polarized lepton scattering off polarized nucleon targets. Recently, HERMES group at DESY reported the first measurement of the polarized gluon

---

[1] yamanisi@ccmails.fukui-ut.ac.jp
[2] dongyb@alpha02.ihep.ac.cn
[3] morii@kobe-u.ac.jp

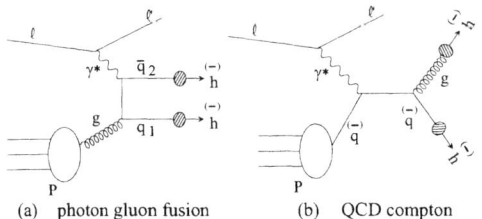

FIGURE 1. Lowest order Feynman diagrams for the high–p_T pair–hadron production.

distribution from di–jet analysis of semi–inclusive processes in pol–DIS, though only one data point is given as a function of Bjorken x [4].

Here, we propose a new process, the charmed hadron pair–production, in order to get an accurate information on the polarized gluon distribution in the nucleon. In general, the high–p_T pair–hadrons are produced via photon–gluon fusion (PGF) and QCD Compton at the lowest order of QCD (Figure 1). The PGF gives us a direct information on the polarized gluon distribution in the nucleon, while QCD Compton becomes background to the signal process for extracting the polarized gluon distribution. For ordinary hadron–pair productions, QCD Compton contribution is not necessary small and hence it is not so easy to extract the behavior of the polarized gluon unambiguously from those processes. However, for the charmed hadron pair–production, we can safely neglect the contribution of the QCD Compton process since the charm content in the proton is extremely small. Thus, we can expect the process proposed here is rather effective for extracting the polarized gluon.

DOUBLE SPIN ASYMMETRY FOR HIGH–P_T PAIR CHARMED HADRON PRODUCTIONS

Let us consider the process, $\gamma^* + N \to h_c + \bar{h}_{\bar{c}} + X$. An interesting observable is the double spin asymmetry defined by

$$A_{LL} = \frac{d\sigma_{-+} - d\sigma_{++}}{d\sigma_{-+} + d\sigma_{++}} = \frac{d\Delta\sigma}{d\sigma} , \quad (1)$$

where $d\sigma_{-+}$ means the cross section for the lepton and nucleon with the negative and positive helicity, respectively. The polarized differential cross section can be calculated by

$$\frac{d^3\Delta\sigma}{d\eta dz_1 dz_2} = \Delta g(\eta, Q^2) \int_{z_1}^{1} \frac{d\xi_1}{\xi_1} \int_{z_2}^{1} \frac{d\xi_2}{\xi_2} \frac{d^3\Delta\hat{\sigma}}{d\eta dz_c dz_{\bar{c}}} \\ \times D_c^{h_c}(\xi_1, Q^2) \, D_{\bar{c}}^{\bar{h}_{\bar{c}}}(\xi_2, Q^2) . \quad (2)$$

Here

$$\eta = \frac{\hat{s}}{s}, \quad z_1 = \frac{P \cdot p_1}{P \cdot q}, \quad z_2 = \frac{P \cdot p_2}{P \cdot q}, \quad (3)$$

and at the partonic level,

$$z_c = \frac{P \cdot p_q}{P \cdot q} = \frac{z_1}{\xi_1}, \quad z_{\bar{c}} = \frac{P \cdot p_{\bar{q}}}{P \cdot q} = \frac{z_2}{\xi_2}. \quad (4)$$

In eq.(2), Δg and $D_{c \, (\bar{c})}^{h_c \, (\bar{h}_c)}$ denote the longitudinally polarized gluon distribution and fragmenation function of a charm quark to a charmed hadron, respectively.

In this work, we take the D^{*+-} meson as the charmed hadron because its fragmentation function is well studied recently [5]. By using typical examples of the parametrization model [6] of polarized gluon distribution functions, we have calculated the double spin asymmetry A_{LL} which is shown in Figure 2.

SUMMARY

We have calculated the double spin asymmetry for the high-p_T pair charmed hadron production as another way for measuring the polarized gluon distribution. The process is dominated by the photon–gluon fusion and the contribution of virtual photon absorption and QCD Compton scattering to this process is very small.

As shown in Figure 2, the process is quite promising for extracting the polarized gluon distribution in the proton.

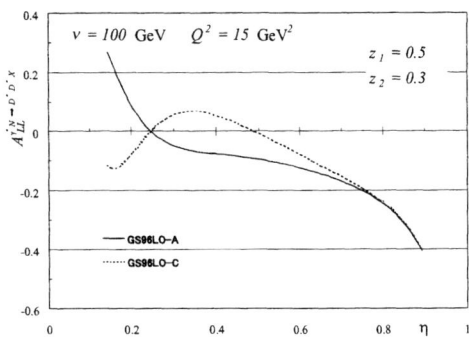

FIGURE 2. The double spin asymmetry A_{LL} for $\gamma^* + p \rightarrow D^{*+-} + D^{*+-} + X$. The solid and dashed lines are for GS96LO-A and GS96LO-C parametrizations of the polarized gluon distribution model, respectively.

REFERENCES

1. Gordon, L.E., Goshtasbpour, M., and Ramsey, G. P., *Phys. Rev.* **D58**, 094017 (1998); Ramsey, G. P., *Prog. Part. Nucl. Phys.* **39**, 599 (1997); Leader, E., Sidrov, A. V., and Stamenov, D. B., *Phys. Rev.* **D58**, 114028 (1998); *Phys. Lett.* **B445**, 232 (1998); **B462**, 189 (1999); Goto, Y., et al.: Asymmetry Analysis Collaboration, *Phys. Rev.* **D62** 34017 (2000).
2. Craigie, N. S., Hidaka, K., Jacob, M., and Renard, F. M., *Phys. Rep.* **99**, 69 (1983).
3. Bravar, A., von Harrach, D., and Kotzinian, A., *Phys. Lett.* **B421** 349 (1998).
4. Airapetian, A., et al.: HERMES Collaboration, *Phys. Rev. Lett.* **84** 2584 (2000).
5. Binnewies, J., Kniehl, B. A., and Kramer, G., *Phys. Rev.* **D58** 014014 (1998).
6. Gehrmann, T., and Stirling, W. J., *Phys. Rev.* **D53** 6100 (1996).

The Shape and Experimental Tests of the Q^2-Invariant Polarized Gluon Asymmetry

Gordon P. Ramsey[1,2]

*Physics Department, Loyola University, Chicago, IL 60626 and
High Energy Physics Division, Argonne National Lab, Argonne, IL 60439*

Abstract. The absence of "valence-gluon" degrees of freedom combined with an examination of radiative QCD diagrams leads to an implication that the gluon spin asymmetry in a proton, defined as $A_G(x,Q^2) = \frac{\Delta G(x,Q^2)}{G(x,Q^2)}$, should be approximately Q^2 invariant. The condition for scale invariance completely determines the x-dependence of this asymmetry, which satisfies constituent counting rules and reproduces the basic results of the Bremsstrahlung model originated by Close and Sivers. This asymmetry can be combined with the measured unpolarized gluon density, $G(x,Q^2)$ to provide a prediction for $\Delta G(x,Q^2)$. Existing and proposed experiments can test both the prediction of scale-invariance for $A_G(x,Q^2)$ and the nature of ΔG itself.

I INTRODUCTION

The spin-weighted gluon density, $\Delta G(x,Q^2)$, is of fundamental importance in understanding the dynamics of hadron structure. Numerous experiments have been proposed [1–3] to determine this distribution experimentally. Measurements of the deep-inelastic scattering asymmetry $A_1(x,Q^2)$ for protons, neutrons and deuterons yield data from which polarized quark distributions may be inferred, but the shape and size of the polarized gluon density has not been determined. However, the constituent quark model provides a framework for predicting an essential feature of $\Delta G(x,Q^2)$. To understand this, we assume that the spin structure of proton does not have a significant component representing a valence or "constituent" gluon polarization. Hadronic spin observables at small Q^2 conform to the non-relativistic quark model in which spin degrees of freedom are associated with constituent quarks. This assumption does <u>not</u> imply that $\Delta G \to 0$ at low Q^2. In fact, when the spin structure of the constituent quarks is resolved by inelastic scatter-

---
[1] Talk given at the Spin Physics Symposium (SPIN 2000), 16-21 October 2000, Osaka, Japan. Based upon work done with F. Close and D. Sivers.
[2] Work supported by the U.S. Department of Energy, Division of High Energy Physics, Contract W-31-109-ENG-38. E-mail: gpr@hep.anl.gov

ing measurements, this approach yields a variation of Close-Sivers Bremsstrahlung model [4] which displays a maximal gluon polarization at $x = 1$.

In a positive helicity proton, we define the gluon polarization asymmetry as

$$A_G(x,t) \equiv \Delta G(x,t)/G(x,t), \tag{1}$$

where the evolution variable $t \equiv \ln[\alpha_s(Q_0^2)/\alpha_s(Q^2)]$. It is assumed that the same factorization prescription is used to define all of the densities in equation (1). Since there are no overwhelming theoretical arguments favoring any single model for ΔG, we consider a more direct argument for its shape in terms of this asymmetry.

Our results follow from the observation that, in the absence of a "constituent" gluon, both $G(x,t)$ and $\Delta G(x,t)$ exhibit scaling violations which can be associated with measurements "resolving" radiative diagrams. The diagrams leading to positive and negative helicity gluons are the same. This implies that the relative probability of measuring a gluon of either helicity does not depend upon t. Thus, the gluon polarization asymmetry is predicted to be scale invariant: $\partial A_G(x,t)/\partial t = 0$. This would not be true if there were a valence gluon, since the shape of A_G would then depend upon the relative amount of valence and radiated gluons. It is reasonable to choose $t = 0$ to coincide with a typical hadronic scale, $Q^2 = m_H^2$. The scale-invariance assumption provides the x dependence of $A_G(x)$, which satisfies several important physical constraints:

- it obeys the constituent-counting rules,

- for large x, where quark distributions dominate the gluon distribution, the predicted asymmetry coincides with the original QCD-Bremsstrahlung model of Close and Sivers [4]. At other values of x, it corresponds to a natural extension of the QCD-Bremsstrahlung approach by allowing for radiation from both quarks and gluons, and

- for small x, where the gluon distribution is expected to dominate the quark distributions, the scale-invariant asymmetry arises as a natural asymptotic limit, independent of the starting point.

Thus, the arguments originally presented in ref. [4] can be combined with existing parametrizations of polarized and unpolarized quark distributions to provide a quantitative estimate for $\Delta G(x, Q_0^2)$ at any convenient reference scale.

II THE SHAPE OF $A_G(X)$

In 1977, Close and Sivers [4] proposed that the quark *sea* should be polarized and that *gluons* should exhibit a polarization of in the same direction as the proton. This was based on perturbative QCD and the theoretical understanding that valence quarks are polarized in the same sense as the proton. This happens since the γ_μ coupling of the quark-gluon vertex conserves quark helicity when quark masses are neglected. Thus, when a gluon is radiated by a quark, its helicity has the same sign as the creating quark.

A phenomenological picture of the proton assumes that at low $Q^2 \leq m_P^2$, a proton consists of three "valence" quarks, surrounded by radiated gluons and $q\bar{q}$ pairs. The Bremsstrahlung mechanism used in ref. [4] supplies a significant fraction of gluons in a proton found at low to medium values of Q^2. From a reference scale where the constituent quark picture is applicable, the QCD evolution equations can be used to generate a prediction for the quark and gluon distributions at higher Q^2.

The requirement that $A_G(x,t)$ has no t-dependence implies that

$$\frac{\partial A_G}{\partial t} = \frac{1}{G}\left[\frac{\partial \Delta G}{\partial t} - A_G(x,t)\frac{\partial G}{\partial t}\right] = 0. \qquad (2)$$

The t-dependence of the gluon distributions is given by the corresponding DGLAP evolution equations. [5] Combining the DGLAP equations with equation (2) gives

$$A_G = \frac{\frac{\partial \Delta G}{\partial t}}{\frac{\partial G}{\partial t}} = \left[\frac{\Delta P_{Gq} \otimes \Delta q + \Delta P_{GG} \otimes \Delta G}{P_{Gq} \otimes q + P_{GG} \otimes G}\right]. \qquad (3)$$

This follows since the diagrams which determine the t evolution of the distributions are the same as those which distribute the spin information to the gluons.

Since ΔG has not been measured, equation (3) can be converted into a non-linear equation for $A_G(x)$ by inserting $\Delta G(x,t) = A_G(x) \cdot G(x,t)$ into the convolution,

$$A_G = \left[\frac{\Delta P_{Gq} \otimes \Delta q + \Delta P_{GG} \otimes (A_G \cdot G)}{P_{Gq} \otimes q + P_{GG} \otimes G}\right]. \qquad (4)$$

An equation in this form can be solved iteratively. We first observe that for a given value of x, the distributions in the DGLAP equations enter only in the range $[x,1]$. Then, for a large enough x ($x \geq 0.6$), the gluon distributions on the right side of (4) can be neglected. Now, the polarized DIS data are consistent with the constituent counting rule result that $\lim_{x\to 1} A_1(x,Q^2) \approx \lim_{x\to 1} \Delta u_v(x,Q^2)/u_v(x,Q^2) = 1$. There exist parametrizations of the helicity-weighted quark distributions [6] which incorporate this result to reproduce all of the existing data. Thus, we make an initial approximation

$$\lim_{x\to 1} A_G^0 = \left[\frac{\Delta P_{Gq} \otimes \Delta u_v}{P_{Gq} \otimes u_v}\right]. \qquad (5)$$

in terms of the flavor non-singlet quark distributions, valid for large x. We can then define the iterative approximation:

$$A_G^{n+1} = \left[\frac{\Delta P_{Gq} \otimes \Delta q + \Delta P_{GG} \otimes (A_G^n \cdot G)}{P_{Gq} \otimes q + P_{GG} \otimes G}\right], \qquad (6)$$

which should converge for large enough n. It is important to note that (6) determines the form of $A_G(x)$ from the three distributions, $\Delta q(x,t)$, $q(x,t)$ and $G(x,t)$, extracted from data. The spin-weighted gluon asymmetry is then determined explicitly by $\Delta G(x,t) = A_G(x) \cdot G(x,t)$.

At small-x, we can also argue that $\frac{\partial A_G}{\partial t} = 0$. We can parametrize the asymmetry in the form $A_G(x,t) = A_G(x) + \epsilon(x,t)$, where $\epsilon(x,t)$ is a correction term which necessarily vanishes at large t. Then,

$$\frac{\partial A_G(x,t)}{\partial t} = \frac{\partial \epsilon(x,t)}{\partial t} = \frac{1}{G}\left[\frac{\partial \Delta G}{\partial t} - A_G(x,t)\frac{\partial G}{\partial t}\right]. \tag{7}$$

Now, insert the expression for $A_G(x,t)$ in terms of $A_G(x)$ and ϵ into (7) to get

$$\frac{\partial \epsilon(x,t)}{\partial t} = -\frac{\epsilon(x,t)}{G(x,t)} \cdot \frac{\partial G(x,t)}{\partial t}. \tag{8}$$

This implies that $\epsilon(x,t) \cdot G(x,t)$ is scale invariant. Thus, at small-x, the growth in G predicted by the evolution equations ensures that $\epsilon \to 0$ at large t and that A_G maintains its x-dependent shape asymptotically in t.

For the starting distributions in eq. (5) and the iterations of eq. (6), we use the polarized quark distributions outlined by GGR [6] and the CTEQ4M unpolarized distributions. [7] The evolution was performed in LO, since the NLO contributions to the splitting kernels, calculated in ref. [8], are most dominant at small-x, where the asymmetry is the smallest. Work is in progress to ensure that the effects of NLO are not significant for the ratio $\frac{\Delta G}{G}$. The iteration is relatively stable and converges within a couple of cycles. Some models of $G(x)$ converge more uniformly than others. The resulting shape of $A_G(x)$ is shown in Figure 1. This shape implies a larger polarized gluon distribution than the xG model of GGRA [6], so spin asymmetries which depend upon ΔG are enhanced.

III   EXPERIMENTAL TESTS OF $A_G(X)$ AND ΔG

There are a number of existing and planned experiments are suitable for measuring either $A_G(x)$ or a combination of $\Delta G(x,Q^2)$ and $G(x,Q^2)$. The HERMES experimental group at DESY has measured the longitudinal cross section asymmetry $A_\parallel$ in high-p_T hadronic photoproduction. [1] From this and known values of $\frac{\Delta q}{q}$ from DIS, a value for $A_G(x_G)$ can be extracted. Here, $x_G = \hat{s}/2M\nu$ is the nucleon momentum fraction carried by the gluon. Our corresponding value at $x_G = 0.17$ is within one σ of the quoted value of $A_G = 0.41 \pm 0.18$ (stat.) ± 0.03 (syst.).

Both direct-γ production and jet production at RHIC provide the best means of extracting information about $\Delta G(x,Q^2)$ and $G(x,Q^2)$ separately. [2,9] The kinematic regions of STAR and PHENIX can determine A_G over a suitable range of x_{Bj} to test this model of the gluon asymmetry. Coupled with additional direct measurements of $A_G(x_G)$ from HERMES, an appropriate cross check of $\Delta G(x)$ and $G(x)$ can be made. Since this model of $\frac{\Delta G}{G}$ implies a larger polarized glue than the GGRA model used in ref. [9], all of the asymmetries for direct-γ and jet

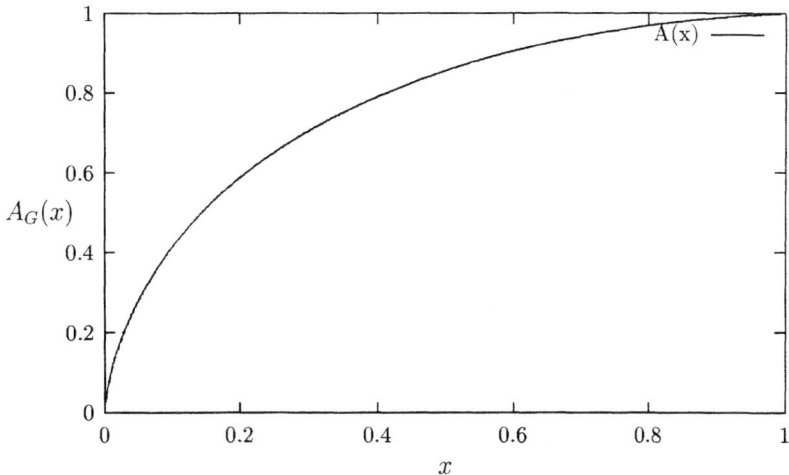

FIGURE 1. The gluon asymmetry $\frac{\Delta G}{G}$ plotted as a function of x

production should be enhanced, making them easier to distinguish from the other parmetrizations of ΔG.

The COMPASS group at CERN. [3] plans to extract A_G from the photon nucleon asymmetry, $A_{\gamma N}^{c\bar{c}}(x_G)$ in open charm muo-production, which is dominated by the photon-gluon fusion process. This experiment should be able to cover a wide kinematic range of x_G as a further check of this model. The combination of these experiments will be a good test of the assumptions of our gluon asymmetry model and a consistency check on our knowledge of the gluon distribution in the nucleon and its polarization.

REFERENCES

1. A. Airapetian, *et.al.*, **84**, 2584 (2000). Also see U. Stösslein, these proceedings.
2. G. Bunce, N. Saito, J. Soffer and W. Vogelsang, hep-ph/0007218.
3. F. Bradamante, Prog. Part. Nucl. Phys. **44**, 339 (2000).
4. F. E. Close and D. Sivers, Phys. Rev. Lett. **39**, 1116 (1977).
5. G. Altarelli and G. Parisi, Nucl. Phys. **B126**, 298 (1977); Dokshitzer; V.N. Gribov and L.N. Lipatov, Yad. Fiz.,**15**, 781 (1972) and Sov. J. Nucl. Phys.,**15**, 438 (1972).
6. L. E. Gordon, M. Goshtasbpour and G. P. Ramsey Phys. Rev. **D58**, 094017 (1998), (hep-ph/9803351).
7. CTEQ Collaboration, H. L. Lai *et al.*, Phys. Rev. **D51**, 4763 (1995).
8. W. Vogelsang, Acta. Phys. Polon. **B29**, 1189 (1998), (hep-ph/9805295).
9. L. E. Gordon and G. P. Ramsey, Phys. Rev. **D59**, 074018 (1999).

Measurement of the Gluon Polarization in the Proton at PHENIX

Yuji Goto, for the PHENIX Collaboration

RIKEN BNL Research Center, Upton, NY 11973, USA

Abstract. We plan to measure the gluon polarization in the proton at PHENIX using RHIC polarized proton collisions. Many channels of physic signals will be detected for this purpose in both the central arms and the muon arms of the PHENIX detector. Sensitivities to the gluon polarization in prompt photon, pion, and heavy-flavor production measurements are discussed.

INTRODUCTION

In polarized lepton–nucleon deep inelastic scattering (DIS) experiments, the quark spin have been found to contribute only about 10–30% of the nucleon spin. The remaining fraction needs to be explained by other carriers. We need to know the contribution of the gluon to the nucleon spin. In the polarized DIS experiments, the integrated gluon polarization has been obtained, $\Delta g = 1.0^{+1.0}_{-0.3}(\text{stat})^{+0.4}_{-0.2}(\text{sys})^{+1.4}_{-0.5}(\text{th})$, by the global QCD analysis by the SMC group [1]. The sensitivity to the gluon polarization is not good enough, because the measurements are sensitive only in the next-to-leading order (NLO). The polarized DIS experiments require Q^2 evolution, or semi-inclusive measurements to extract the gluon polarization.

Polarized proton collisions, which we will perform at RHIC/PHENIX [2], are sensitive to the gluon polarization in the leading order (LO). By measuring double longitudinal asymmetry, $A_{LL} = (d\sigma_{++} - d\sigma_{+-})/(d\sigma_{++} + d\sigma_{+-})$, where $d\sigma_{++}$ ($d\sigma_{+-}$) represents the cross section with parallel (antiparallel) beam helicity, the gluon polarization is extracted.

In the PHENIX experiment, we plan to measure the gluon polarization with many channels. The PHENIX detector system consists of two central arms and two muon arms. In the central arms, each of which covers $|\eta|<0.35$ and $\pi/2$ azimuthal angle, we will measure prompt photon and neutral/charged pion production using a fine-segment electro-magnetic calorimeter (EMCal), tracking chambers and particle-ID detectors. The segmentation of the EMCal is 0.01 radian in both the azimuthal and polar directions. It shows very good π^0 identification and very good π^0/γ discrimination up to 40GeV/c for orthogonally incident photons, and up to

25GeV/c even for photons with incident angle of 20 degree [3]. The muon arms which cover $1.2<|\eta|<2.4$ and 2π azimuthal angle consist of muon-tracking chambers and muon-ID detectors. By using both the central arms and muon arms, we will measure open and bound-state heavy-flavor production by detecting single leptons, di-lepton pairs and electron–muon ($e\mu$) pairs.

PROMPT PHOTON MEASUREMENT

The dominant process for prompt photon production is the gluon Compton process, $gq \to \gamma q$. We can expect a clear theoretical interpretation of the gluon distribution and the gluon polarization in the polarized proton collisions. The asymmetry, A_{LL}, can be factorized by the relation, $A_{LL}(p_T) = \Delta g(x_g, Q^2)/g(x_g, Q^2) \cdot A_1^p(x_q, Q^2) \cdot a_{LL}^{gq \to \gamma q}(\cos\theta^*)$, to the gluon polarization $\Delta g(x_g, Q^2)/g(x_g, Q^2)$, the quark polarization $A_1^p(x_q, Q^2)$ obtained by the polarized DIS experiments, and the calculable parton-level asymmetry of the gluon Compton process $a_{LL}^{gq \to \gamma q}(\cos\theta^*)$.

The measurement is experimentally challenging because there are many backgrounds, mainly from the two-photon decays of π^0 and η. The PHENIX detector has good background reduction capability with the fine segmentation of the EMCal. To estimate sensitivities of the measurements, we have performed PYTHIA simulations [4]. Results using GRV94 LO parton distribution functions (PDF) [5] and GS95 NLO polarized PDF [6] are shown.

In one-year run with full luminosity, 320pb^{-1}, at $\sqrt{s}$=200GeV, we expect 120K prompt photons in p_T>10GeV/c. We will have much enough statistics up to p_T=30GeV/c to distinguish between three models of GS95 NLO polarized PDF with different integrated gluon polarization values. The p_T region corresponds to the x_g region from 0.1 to 0.3. The raw ratio of experimental background to the prompt photon yield is 50% to 200% in the p_T region from 10 to 30GeV/c at $\sqrt{s}$=200GeV. Due to the fine-segment EMCal, this ratio can be reduced by invariant mass reconstruction of two photons and an isolation cut down to less than 20% [3].

PION MEASUREMENT

Asymmetry measurement of the jet production will provide us the gluon polarization information through quark–gluon and gluon–gluon reactions. At PHENIX, the asymmetry of pion production will be measured in the central arms as an alternative to the jet measurement in the limited acceptance. We will have high statistics of this measurement in the first year of the RHIC polarized proton run in 2001. With the fine-segment EMCal, π^0s are clearly identified by invariant mass reconstruction. If we assume 10% luminosity in one year at $\sqrt{s}$=200GeV, we will obtain 400M π^0s in p_T>2GeV/c. We will have sensitivity up to p_T=10GeV/c to distinguish between the GS95 NLO models.

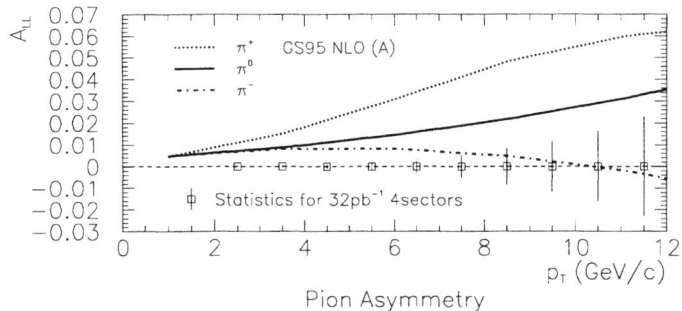

FIGURE 1. Asymmetries of π^0, π^+ and π^- with estimated errors of the A_{LL} measurement of π^0 in one year of 10% luminosity, 32pb^{-1} at $\sqrt{s}$=200GeV.

Because of the different fragmentation functions from specific partons to π^0, π^+ and π^-, the asymmetries of each pion are different. Figure 1 shows asymmetries of π^0, π^+ and π^- with estimated statistical errors of the A_{LL} measurement of π^0 in one year of 10% luminosity, 32pb^{-1}, at $\sqrt{s}$=200GeV. This measurement is similar to the semi-inclusive plus-charged and minus-charged hadron measurement in the polarized DIS experiment performed in the HERMES and SMC experiments [7], where they measured flavor decomposition of the quark polarization. We will add more information of the flavor decomposition of the quark polarization, in addition to the gluon polarization in this measurement.

How can we analyze these data ? We want to obtain decomposition of the quark polarization for each flavor and the gluon polarization. Asymmetries of the pion measurements can be written by

$$A_{LL}^h(p_T) = \frac{\sum_{f_1 f_2 f_3} \int dx_1 \int dx_2 \Delta f_1(x_1) \Delta f_2(x_2) d\Delta\hat{\sigma}(f_1 f_2 \to f_3 X) D_{f_3}^h(z)}{\sum_{f_1' f_2' f_3'} \int dx_1 \int dx_2 f_1'(x_1) f_2'(x_2) d\hat{\sigma}(f_1' f_2' \to f_3' X) D_{f_3'}^h(z)} \quad (1)$$

for $f = u, \bar{u}, d, \bar{d}, s, \bar{s}, g$, by omitting correspondence between pion's p_T and energy fraction z in the fragmentation functions $D_f^h(z)$, and Q^2 scales. If we define purity functions by

$$P_{f_1 f_2}^h(x_1, x_2, p_T) = \frac{\sum_{f_3} f_1(x_1) f_2(x_2) d\Delta\hat{\sigma}(f_1 f_2 \to f_3 X) D_{f_3}^h(z)}{\sum_{f_1' f_2' f_3'} \int dx_1 \int dx_2 f_1'(x_1) f_2'(x_2) d\hat{\sigma}(f_1' f_2' \to f_3' X) D_{f_3'}^h(z)} \quad (2)$$

like semi-inclusive analysis of the polarized DIS experiment, we can rewrite the asymmetries in much simpler form

$$A_{LL}^h(p_T) = \sum_{f_1 f_2} \int dx_1 \int dx_2 P_{f_1 f_2}^h(x_1, x_2, p_T) \cdot \frac{\Delta f_1(x_1)}{f_1(x_1)} \cdot \frac{\Delta f_2(x_2)}{f_2(x_2)} \quad (3)$$

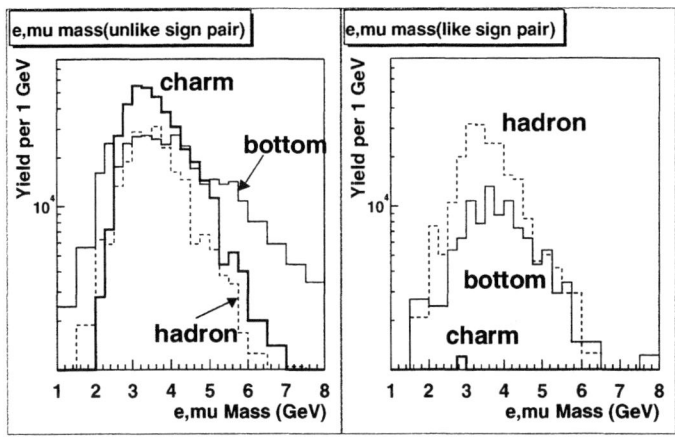

FIGURE 2. Simulated invariant mass spectra of unlike-sign and like-sign $e\mu$ pairs.

as a convolution of polarizations and purity functions. The purity functions (2) are more complicated than those of the semi-inclusive analysis of the polarized DIS experiment, and have many uncertainties in the fragmentation functions, PDFs, p_T-z relations, and scales. Our other data, for instance the prompt photon asymmetry, will also be written in the same form, and all data should be combined for the decomposition.

HEAVY-FLAVOR MEASUREMENT

By using both the central arms and muon arms, we will measure open and bound-state heavy-flavor production. Since the dominant process for the heavy-flavor production is the gluon fusion process, $gg \to c\bar{c}$ and $gg \to b\bar{b}$, the measurement of A_{LL} is sensitive to the gluon polarization.

In our heavy-flavor measurements, $e\mu$-pair coincidence measurement of the open heavy-flavor production is the most sensitive to the gluon polarization, because of low background rate. Figure 2 shows simulated invariant mass spectra of unlike-sign and like-sign $e\mu$ pairs. The unlike-sign pair spectrum shows $c\bar{c}$ production is dominant at around 3–4GeV/c^2, and $b\bar{b}$ production is dominant at more than 5GeV/c^2. Background from hadron (π and K) decay can be evaluated by the like-sign pair spectrum. As a result, we can evaluate the A_{LL} of the $e\mu$ pair by combining the $c\bar{c}$ production asymmetry and the $b\bar{b}$ production asymmetry.

We also have a large yield of single electrons from the open heavy-flavor production in the central arms. We will be able to begin the measurement in the first year of the RHIC polarized proton run. Many backgrounds come from the photon conversion and the Dalitz decay of π^0. Most of them can be rejected with a multiplicity vertex detector (MVD) by detecting an accompanying electron/positron.

TABLE 1. Summary of the estimated statistical errors and predicted values of the A_{LL} and the x_g range to be covered by each channel at $\sqrt{s}$=200GeV. The statistical errors show the absolute values for 10% luminosity 32pb^{-1} for channels indicated by *), and full luminosity 320pb^{-1} for other channels.

Probes	A_{LL} Errors	A_{LL} Predictions			x_g Range
		GS(A)	GS(B)	GS(C)	
prompt γ	0.006	0.05	0.03	0.001	0.1 − 0.3
π^0 / $\pi^\pm$	0.001*)	0.02	0.01	0.0001	0.05 − 0.2
$J/\psi \to \mu\mu$	0.006	0.01	0.01	0.002	0.005 − 0.01
$c\bar{c} \to eX$	0.001*)	−0.02	−0.01	−0.0003	0.005 − 0.2
$c\bar{c} \to e\mu X$	0.006	−0.05	−0.02	−0.0001	0.005 − 0.2
$b\bar{b} \to e\mu X$	0.006	0.01	0.01	−0.0004	0.01 − 0.3

On the other hand, if we select such background events with the MVD, we will observe the asymmetry of the jet production. This data is also useful for the jet study and the background study.

SUMMARY

We will perform the measurement of the gluon polarization in the proton at PHENIX using many channels. As a summary, table 1 shows the sensitivity of each channel to the gluon polarization measurement and the x_g range at $\sqrt{s}$=200GeV. These channels cover complementary kinematic ranges of the gluon which in total spread over $0.005 < x_g < 0.3$.

REFERENCES

1. B. Adeva et al., Phys. Rev. D **58**, 112002 (1998).
2. D.P. Morrison, Nucl. Phys. A **638**, 565c (1998); N. Saito, Nucl. Phys. A **638**, 575c (1998).
3. A. Bazilevsky, RIKEN Rev. **28**, 15 (2000).
4. T. Sjöstrand, Comp. Phys. Commun. **82**, 74 (1994).
5. M. Glück, E. Reya, and A. Vogt, Z. Phys. C **67**, 433 (1995).
6. T. Gehrmann and W.J. Stirling, Phys. Rev. D **53**, 6100 (1996).
7. B. Adeva et al., Phys. Lett. B **420**, 180 (1998); K. Ackerstaff et al., Phys. Lett. B **464**, 123 (1999).

Extracting $\Delta G(x)$ from the $\vec{p} + \vec{p} \to \gamma + \text{jet} + X$ Reaction with the STAR detector at RHIC

S. W. Wissink, for the STAR Collaboration

Department of Physics, Indiana University, and
Indiana University Cyclotron Facility, Bloomington, Indiana 47408 USA

Abstract. Two decades of polarized deep inelastic scattering experiments have revealed that the quarks within a proton carry only a small fraction of the proton's spin, but provide few clues as to the source of the missing components. In particular, the role played by gluons remains largely unexplored. In this talk, I discuss an experimental program that will allow a *direct* extraction of the gluon helicity distribution function $\Delta G(x)$, via a measurement of the spin correlation parameter A_{LL} in the reaction $\vec{p} + \vec{p} \to \gamma + \text{jet} + X$. Simulations indicate that a precise determination of the gluonic contribution to the proton's spin, represented by the integral ΔG, will require access to values of x_{gluon} as low as ~ 0.01. I will show that this can be achieved using the STAR detector at RHIC, when equipped with suitable electromagnetic calorimetry extending to a pseudorapidity of $\eta = 2$.

INTRODUCTION

Understanding the origin of the nucleon's spin in terms of its partonic contributions remains an important, though so far elusive, goal of hadronic physics. Polarized structure functions for the nucleon, obtained from studies of deep inelastic scattering (DIS) of polarized leptons on hydrogen, deuterium, and ^{3}He [1-3], have yielded quark helicity distributions that are in good agreement among the various experiments, but fall well below those needed to account fully for the proton spin. It is therefore essential that the other expected contributors to the nucleon's spin receive similar attention. In particular, a direct measurement of the gluon helicity distribution $\Delta G(x, Q^2)$ for the proton is a critical component in this effort; evaluation of its integral $\Delta G(Q^2)$ to an accuracy of $\sim \pm 0.5$ would clarify both the role of the gluons and that of the quarks in forming the proton's spin, and thereby place important constraints on the unknown orbital contributions.

To date, much of what is known about $\Delta G(x)$ has been gleaned from observed scaling violations in the polarized DIS work. As an example of this approach, consider the NLO perturbative QCD analysis of the SMC group [1], in which the parton helicity distributions are extracted in the Adler-Bardeen factorization scheme. In this work, the integral of $\Delta G(x)$ when evolved to $Q^2 = 1$ (GeV/c)^2 is given by

$$\Delta G \equiv \int_0^1 \Delta G(x) dx = 1^{+1.9}_{-0.6} \qquad (1)$$

where the quoted errors include both experimental and theoretical uncertainties summed in quadrature.

One can do much better with a more direct probe of the gluonic spin. Specifically, one would like to: (i) use a reaction dominated by a leading-order coupling to a gluon that has a clean experimental signature; (ii) use a probe that is itself highly polarized so that spin *correlations* (two-spin asymmetries) can be studied; and (iii) work at sufficiently high energy and momentum transfer ($p_T > 10$ GeV/c) that higher-twist contributions become negligible and pQCD can be reliably applied to the hard partonic scattering. One such process is photon-gluon fusion, in which virtual photons combine with gluons to yield $q\bar{q}$ pairs, resulting in di-jets, open charm, or (at lower energies) high p_T charged hadron pairs detected in the final state. This technique is currently being exploited by COMPASS [4], HERMES [5], and other fixed-target groups; but a full exploration of the critical low-x regime (and freedom from several ambiguities in interpretation) would require use of a high-energy polarized ep collider, such as polarized HERA would offer.

ADVANTAGES OF USING $\vec{p} + \vec{p} \to \gamma + $ jet

Another means of determining $\Delta G(Q^2)$, and the one we have chosen to pursue, is direct photon production via gluonic Compton scattering, *i.e.*, $q + g \to q + \gamma$, studied in high-energy collisions of longitudinally polarized protons. In this case, we benefit from the beautiful polarized DIS studies, and use the (calibrated) polarized quarks within one proton as the probe of polarized gluons within another. The correlation between these two spins—the dependence of the *partonic* cross sections on whether the $\vec{q}$ and $\vec{\gamma}$ are parallel or anti-parallel—can be calculated in pQCD, and is large. Finally, the outgoing high-p_T photon and its coincident away-side jet provide the needed clean signature of the underlying process, and also allow for reconstruction of the initial state parton kinematics. Thus, the primary advantages we see in using this reaction to extract ΔG can be summarized as:

- **The $qg \to q\gamma$ subprocess can be well isolated.** In LO pQCD, the main background process $q\bar{q} \to \gamma g$ only contributes at about the 10% level [6], with NLO calculations suggesting little change in sensitivity.

- **The partonic spin correlation $\hat{a}_{LL}$ approaches unity at back angles**, *i.e.*, when the γ is emitted along the direction of the incident quark. This is also the region where the scattering cross section reaches its maximum value.

- **Quark polarizations are large (>30%)**, especially for $x_q > 0.2$ [1], and also quite well measured over a broad range of x_q.

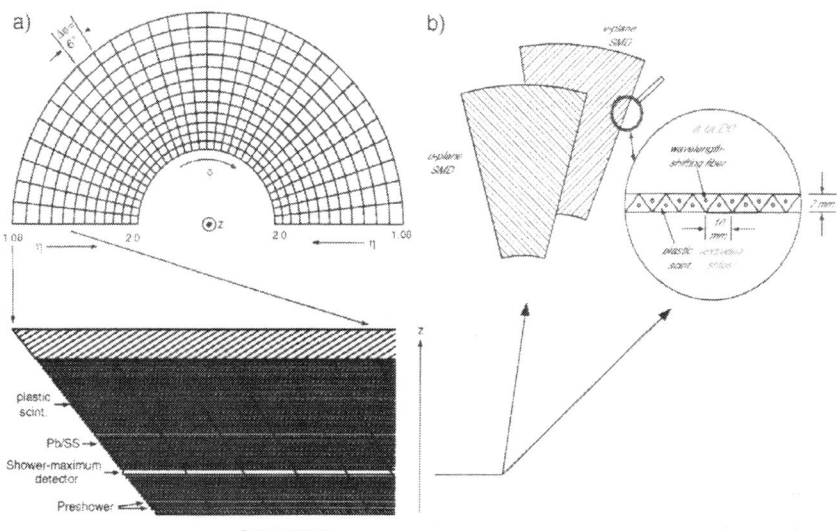

FIGURE 1. Features of the STAR Endcap EMC. *a)* View along the beam direction of the upper half, showing the 360 projective towers and (below) a slice along the depth indicating the location of the preshower and SMD layers. *b)* Schematic of a 30° sector of the SMD, with a close-up of the triangular scintillating strip structure.

- **The momentum fractions carried by the colliding partons can be determined**, on an event-by-event basis, via detection of γ–jet coincidences. While the photon energy and direction need to be measured, only the jet *direction* must be known [7].

To realize these advantages, of course, requires that a suitable detector be located at an appropriate facility. With the addition of an endcap electromagnetic calorimeter (EEMC [8], see Fig. 1), we believe the STAR detector at RHIC is such a device. Colliding beams of polarized protons will be available at RHIC over the energy range $50 \leq \sqrt{s} \leq 500$ GeV, with luminosities near 10^{32} cm^{-2}s^{-1} and polarizations near 70% anticipated by 2002. The large acceptance of STAR allows one to determine accurately the parameters of the away-side jet, spawned by the final-state quark, over a broad kinetic range. To obtain information on the coincident photon, the EEMC must be added to the previously envisioned barrel EMC for STAR, providing photon detection out to $\eta = 2$, and extending the kinematic reach of STAR into the critical region $0.01 < x_{gluon} < 0.1$ which is expected to dominate ΔG. The importance of the EEMC can be understood intuitively by realizing that the most useful collisions will be those that are *highly asymmetric* at the partonic level, in which high-x (and highly polarized) quarks in one proton collide with the low-x (but very abundant) gluons in the other, thereby boosting all outgoing particles along the direction of the incident quark. Because this is also (in the c.m. frame) the direction in which the cross section and spin asymmetry peak, detection of photons at forward angles (large pseudorapidity) is crucial.

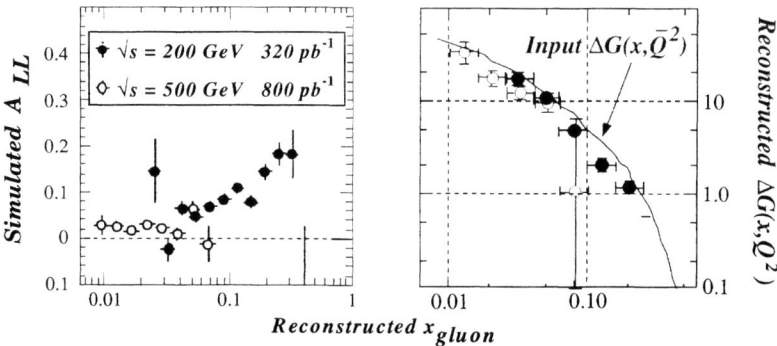

FIGURE 2. Simulated values of A_{LL} for all $\vec{p}+\vec{p} \to \gamma + \text{jet} + X$ events that satisfy anticipated software conditions, and a comparison of the reconstructed and input values for $\Delta G(x)$. The integrated luminosities shown correspond to those expected in two 10-week runs at STAR.

SIMULATION RESULTS

To make these statements more quantitative, detailed simulations have been carried out [7,8]. The event generator PYTHIA [9] was used to describe features of the "hard" pQCD processes, while the "soft" parton helicity distributions (including $\Delta G(x)$) were taken from the NLO analysis of Gehrmann and Stirling [10], after being evolved to $Q^2 = p_T^2/2$. The spin dependences predicted in LO pQCD for the hard scattering were put in by hand for each event. These events were converted to 'data' by accounting for the acceptance and resolution limitations of the EMC (barrel plus endcap) and other STAR detectors. The simulated data were then subjected to the same analysis (acceptance criteria) envisioned for use in analysis of real data, including algorithms for photon isolation cuts and/or jet reconstruction in some cases. Events passing these conditions were used to form spin asymmetries. Finally, these asymmetries were used to reconstruct the physics quantities of interest, assuming that all 'observed' $\gamma + \text{jet}$ events were due to $qg \to q\gamma$ processes in which $x_q > x_g$. By comparing the reconstructed to the input physics values, one can gauge the importance of various effects [7] that have been neglected in the current analysis scheme.

One such comparison, for $\Delta G(x)$, is illustrated in Fig. 2, where the error bars indicate the statistical precision that would be achieved in two 10-week runs at STAR under expected running conditions. Values for both the spin asymmetry (left-hand figure) and for x_g are reconstructed purely from the 'data.' We see that with this model input (set A of ref. [10], evolved to $Q^2 \sim 50$ GeV^2), the spin correlation A_{LL} is predicted to be quite small for $x_g < 0.1$; yet precise data in this region is crucial towards constraining the integral ΔG, simply because the gluon number is so large here. We also note a roughly 20% systematic underestimate of the input values for $\Delta G(x)$, due primarily to approximations made in the reconstruction algorithm. We believe most of this discrepancy will be removed in later analyses which incorporate

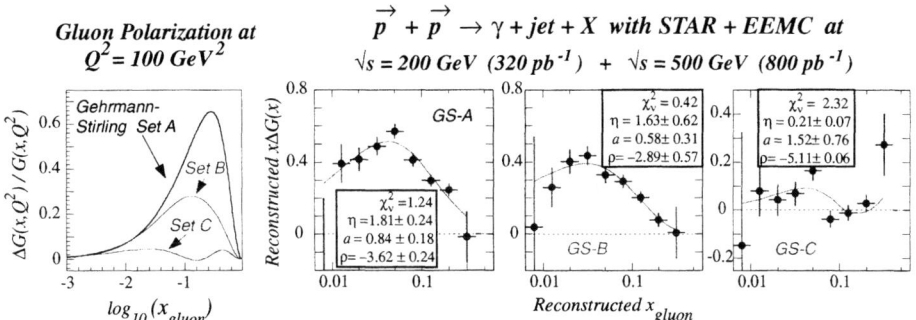

FIGURE 3. Reconstructed values for the gluon helicity distribution $\Delta G(x)$ for the three model predictions [10] shown at left. The curves on the right-hand plots are parameterized fits to the reconstructed values of $\Delta G(x)$; the parameter η corresponds to the first moment ΔG.

higher-order QCD effects and some iterative corrections.

Our main goal in these simulation studies, though, is to estimate how well the integral ΔG can be determined using STAR and the EEMC at RHIC. As indicated above, this will depend critically on the treatment of the low-x_g region; thus, rather than simply extrapolating the reconstructed values of $\Delta G(x)$ numerically to $x = 0$, we have fit them using a standard parameterization [10,7]. In Fig. 3 we show the reconstructed data and their fits for three model inputs [10], and the magnitude and error of the integral ΔG, represented by the variable η. While the error is seen to depend strongly on the model assumed for $\Delta G(x)$, an uncertainty of order 0.3–0.4 seems achievable. After taking into account additional uncertainties due to background subtraction and less-than-ideal γ/π discrimination [7], these simulations suggest that data of this sort taken with STAR will determine the integral ΔG to a statistical plus extrapolation uncertainty $< \pm 0.5$, for a wide range of possible x-dependences.

REFERENCES

1. B. Adeva *et al.*, Phys. Rev. D **58**, 112001 and 112002 (1998).
2. K. Abe *et al.*, Phys. Lett. **B405**, 180 (1997).
3. A. Airapetian *et al.*, Phys. Lett. **B442**, 484 (1998).
4. A. Bravar, D. von Harrach, and A. Kotzinian, Phys. Lett. **B421**, 349 (1998).
5. A. Airapetian *et al.*, Phys. Rev. Lett. **84**, 2584 (2000).
6. C. Bourrely, J. Soffer, F. M. Renard, and P. Taxil, Phys. Rep. **177**, 319 (1989).
7. L. C. Bland, *hep-ex*/9907058 (1999).
8. L. C. Bland, W. W. Jacobs, J. Sowinski, E. J. Stephenson, S. E. Vigdor, and S. W. Wissink, *An Endcap EMC for STAR: Conceptual Design*, STAR Note 401 (1999).
9. T. Sjöstrand, Comp. Phys. Commun. **82**, 74 (1994).
10. T. Gehrmann and W. J. Stirling, Phys. Rev. D **53**, 6100 (1996).

$\vec{p}\vec{p} \to W^{\pm}X$ Asymmetries with STAR at RHIC

S.E. Vigdor, for the STAR Collaboration

Dept. of Physics and Indiana University Cyclotron Facility, Bloomington, Indiana, USA

Abstract. W boson production in polarized pp collisions provides one of the cleanest ways to probe the flavor-dependence of sea antiquark polarizations in the nucleon. The relevance of such measurements, and their advantages over semi-inclusive deep inelastic scattering, for distinguishing among competing models for the origin of the sea are pointed out. Simulations of the expected performance of the STAR detector for W production are then described.

1. MODELS FOR THE SEA QUARK FLAVOR DEPENDENCE

A fundamental issue in understanding the structure of the nucleon is an assessment of the roles of different mechanisms in producing the $q\bar{q}$ sea. Gluon splitting should lead to roughly equal $u\bar{u}$ and $d\bar{d}$ contributions (with a possible small imbalance from Pauli blocking) and to quite similar contributions from all flavors of sea quarks and antiquarks to the proton spin. In contrast, if the emission and reabsorption of Goldstone bosons, by either nucleons or quarks, is a dominant mechanism, one would expect a preponderance of $\bar{d}$ over $\bar{u}$ in the proton sea (since there is greater opportunity for $u \to d\pi^+$ than for $d \to u\pi^-$ transitions), and one would also expect the *anti*quarks to have little spin preference, since they arise inside pseudoscalar mesons.

A large $\bar{d}$-$\bar{u}$ imbalance in the proton sea has indeed been clearly observed in FNAL experiment E866 [1], in at least qualitative agreement with pion cloud models [1,2]. However, if one takes seriously the $1/N_c$ expansion approach to QCD, then this unpolarized flavor imbalance arises at one higher order in the expansion than the *polarized* flavor imbalance, i.e., $(\Delta \bar{u} - \Delta \bar{d}) \sim N_c(\bar{d} - \bar{u})$, implying a quite large flavor-dependence of sea antiquark polarizations in a polarized proton [3]. This expectation is borne out by predictions within the chiral soliton model [3-5], wherein quarks are bound in a collective pion field, in a manner consistent with the large-N_c limit. This non-perturbative QCD model gives rise naturally to dynamical chiral symmetry breaking, and also accounts qualitatively for the E866 results [1]. Its predictions for both $\bar{d}(x)$ - $\bar{u}(x)$ and $\Delta\bar{u}(x)$ - $\Delta\bar{d}(x)$ are shown in Fig. 1, both at a low mass scale appropriate to the model [4] and evolved [5] to $\mu^2 = 25$ GeV^2.

These model considerations highlight the interest in finding clear experimental signatures for the flavor-dependence of $\bar{q}$ polarizations. As shown by the comparison of calculations [4] and measurements in Fig. 1(c,d), semi-inclusive deep inelastic scattering (SIDIS) asymmetries, as studied by SMC [6] and HERMES [7], provide only limited sensitivity to this flavor dependence, diluted by both the fragmentation functions for the outgoing hadrons and the charge-squared weighting preference for u

over d sensitivity. The current SIDIS uncertainties do not permit distinguishing between small and large values of $(\Delta\bar{u} - \Delta\bar{d})$. Even with improved data, the flavor dependence may well be masked by fragmentation function uncertainties. In contrast, W production at RHIC provides strong sensitivity to $(\Delta\bar{u} - \Delta\bar{d})$, far outweighing other relevant uncertainties in polarized parton distribution functions [see Fig. 1(e,f)]. This is because the $W^{\pm}$ production is dominated by flavor-dependent quark-antiquark collisions: $u + \bar{d} \rightarrow W^+$ vs. $d + \bar{u} \rightarrow W^-$. These weak production processes are parity-violating, permitting measurement of large *single*-spin longitudinal asymmetries, with strong sensitivity to the relevant antiquark helicity in the polarized proton [9].

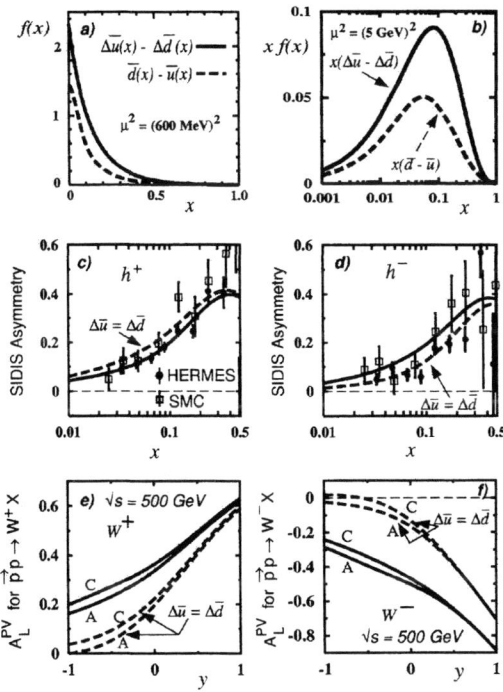

FIGURE 1. The flavor-dependence of sea antiquark polarizations in a polarized proton, as predicted by the chiral soliton model [4,5] (frames *a* and *b*), and the effects of this flavor-dependence in semi-inclusive deep inelastic scattering (frames *c*, *d*) and in $W^{\pm}$ production in polarized *pp* collisions (frames *e,f*). The figures are taken from Refs. [4,5]. The semi-inclusive predictions, with (solid) and without (dashed) flavor-dependence, are compared to data from HERMES [6] and SMC [7]. The W production calculations are shown with (solid) and without (dashed) flavor-dependence superimposed on two different sets of Gehrmann-Stirling [8] parton helicity distributions.

2. SIMULATIONS OF $W^{\pm}$ PRODUCTION WITH STAR

The produced W boson will be detected by its decay into a charged lepton plus a neutrino, with only the former detectable in STAR ($e^{\pm}$) or PHENIX ($e^{\pm}$ or $\mu^{\pm}$). In STAR, the electron detection will exploit electromagnetic calorimetry (EMC) over the pseudorapidity range $-1 \leq \eta \leq 2$ (when the barrel and endcap EMC's are completed in 2003) and tracking in the time projection chamber (TPC). The expected η distributions for the daughter electrons in STAR are shown in Fig. 2 for a nominal 10-week run at the expected luminosity ($L = 2 \times 10^{32}$ cm^{-2}s^{-1}) for $\sqrt{s}$ = 500 GeV collisions of polarized proton beams. The difference seen between e^+ and e^- distributions arises from simple features of the $W^\pm$ production and decay kinematics. When the colliding quark and antiquark have different Bjorken x-values, the W momentum in the collider frame is in the direction of the higher-x parton (usually the quark). Both $W^\pm$ are produced left-handed. The parity-violating decay is characterized by preferential emission of the daughter lepton in the W rest frame: *along* the direction of the W^+ spin and (by *CP*) *opposite* the direction of the W^- spin. The net result is a directional correlation between daughter and parent W (hence, quark) that is strong for W^- and weak for W^+. For W^- production, in particular, an e^- detected in STAR's endcap EMC thus preferentially probes asymmetric collisions of a d quark contributed by the proton headed toward the endcap with a $\bar{u}$ from the other proton.

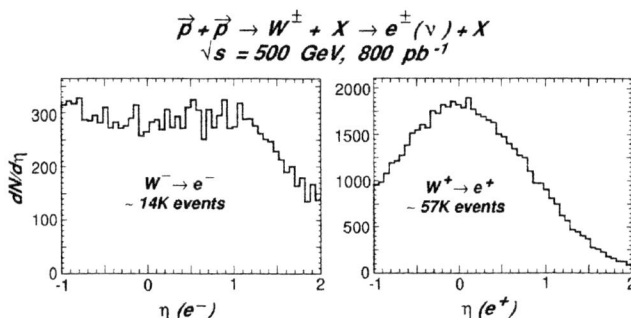

FIGURE 2. Simulated pseudorapidity distributions expected in STAR for $e^\pm$ daughters from W decay.

These kinematic conditions are also reflected in the simulated single-spin parity-violating asymmetries shown *vs.* η of the electron in Fig. 3. Here the simulations, carried out by L. Bland, have used flavor-*independent* antiquark helicity distributions from Ref. [8]. Four such asymmetries will be measured: for W^+ and W^- production, and with respect to spin flip of either beam in each case. While the two beams provide equivalent sensitivity at mid-rapidity, one preferentially provides the quark and the other the antiquark as one moves away from η=0. The extent of this preference, shown in the bottom frames of Fig. 3, is nearly complete for W^- leading to electrons in the STAR endcap. Hence, in this region one of the two W^- asymmetries provides a direct measure of d quark polarization, and the other of $\bar{u}$ polarization in a longitudinally polarized proton. In the W^+ case, the preference is weaker, but still the two asymmetries provide largely complementary sensitivities. The simulations in Fig. 3 indicate the statistical precisions achievable in a 10-week run with STAR.

The quark and antiquark x-ranges sampled are illustrated in Fig. 4. Electrons detected in the endcap sample $\Delta d/d$ over the range $0.1 \leq x_d \leq 0.6$ and $\bar{u}$ polarization for $0.05 \leq x \leq 0.2$. Positrons reach the endcap from either comparably asymmetric

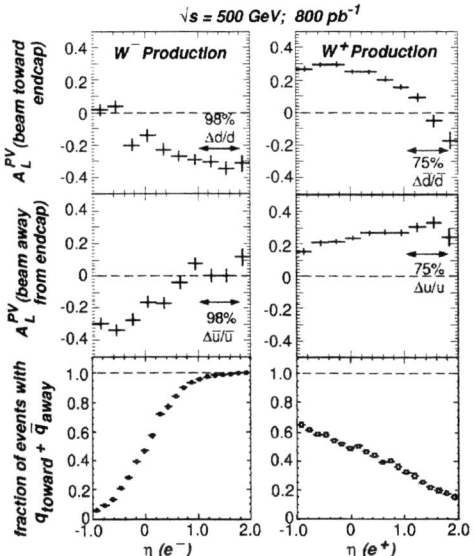

FIGURE 3. Simulated single-spin parity-violating analyzing powers for W^- (left frames) and W^+ (right) production at $\sqrt{s} = 500$ GeV, plotted vs. pseudorapidity of the daughter $e^\pm$ detected in STAR. The upper (middle) frames show the sensitivity to helicity flip of the proton beam headed toward (away from) the endcap, while the lower frames show the fraction of events where the colliding quark comes from the former beam. The averages of these fractions over the endcap region are summarized by the percentages specified in the upper four frames. The error bars reflect counting statistics only.

$u_{toward} \bullet \bar{d}_{away}$ or reasonably symmetric $\bar{d}_{toward} \bullet u_{away}$ collisions. For either sign, the asymmetric parton collisions probe the region $x < 0.1$ where the chiral soliton model predicts a large flavor-dependence of the antiquark polarizations. The energy and pseudorapidity of the electron or positron provide *event-by-event* information by which we can constrain the colliding parton x-values and their assignment to quark *vs.* antiquark [10]. This is illustrated in Fig. 5, based on a reconstruction neglecting the *transverse* momentum (q_T) with which the W is produced, as well as its intrinsic width. Treating the process as two-body fusion followed by two-body decay permits direct extraction of initial parton $x_{1,2}$ values with good resolution ($\sigma_{x\text{-}min} \approx 0.01$, $\sigma_{x\text{-}max} \approx 0.02$) from p_T and η of the electron. The approximation works well except for electrons at mid-rapidity or near the highest accessible p_T, where the W's small longitudinal momentum invalidates neglect of q_T.

The most significant background for W detection is from the far more abundant charged hadron yield. This will be suppressed by more than 3 orders of magnitude, yielding a W signal/hadron background ratio in excess of unity for $e^\pm p_T > 20$ GeV/c, via a combination of cuts on: (*i*) isolation of the detected particle from jet fragments; (*ii*) absence of an accompanying jet at opposite azimuth (replacing the missing energy cut usually applied with hermetic detectors); and (*iii*) EMC response. The latter cut is usually made by comparing transverse energy measured in the EMC to p_T deduced from the corresponding TPC track. But in the endcap region, where the p_T resolution from the TPC deteriorates, and at very high p_T, where jets often give a charged and a neutral hadron in the same calorimeter tower, the e/h discrimination relies increasingly on the EMC response alone. In particular, the comparison of energy depositions in the preshower, shower-maximum detector (SMD) and post-shower layers with that in the full EMC tower, together with SMD shower profile analysis, should still suppress charged hadrons by an order of magnitude. The TPC resolution remains adequate to distinguish e^- from e^+ cleanly over the entire relevant momentum range, up to $\eta \approx 1.7$.

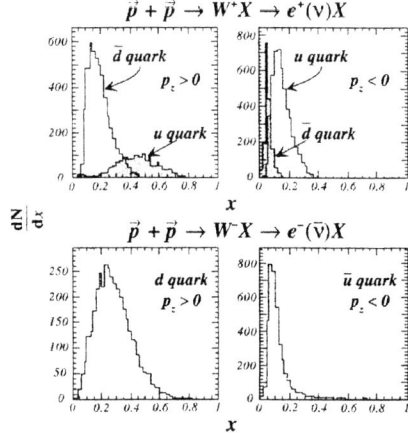

FIGURE 4. The distribution of x-values for quarks and antiquarks probed by $W^\pm$ production yielding $e^\pm$ in the STAR endcap region.

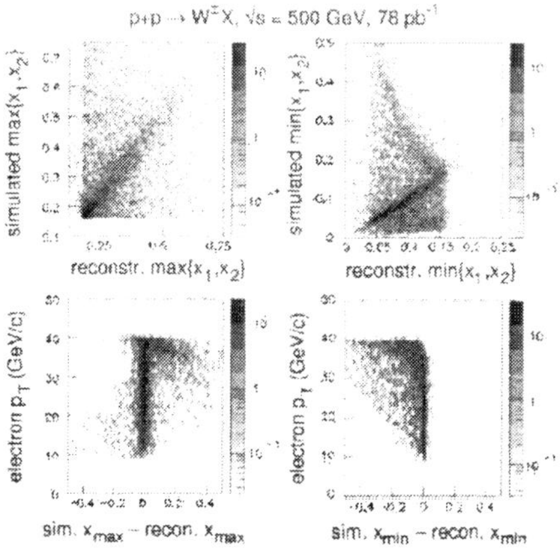

FIGURE 5. Comparisons of simulated with reconstructed x-values for the colliding quark and antiquark in $W^\pm$ production, taken from Ref. [10]. The reconstruction neglects transverse momentum of the W. The bottom frames include only $e^\pm$ with $\eta > 0.2$.

STAR is thus well equipped for $W^\pm$ production via detection of $e^\pm$ daughters over the range $-1 < \eta_e \leq 1.7$. These events are easily triggered by the high energy deposition in a single EMC tower. Measurement of four single-spin parity-violating asymmetries for $pp \to W^\pm X$ in a 10-week run at $\sqrt{s} = 500$ GeV will allow extraction of $\Delta\bar{u}/\bar{u}$ vs. $\Delta\bar{d}/\bar{d}$ with typical uncertainties $\sim \pm 0.05$ in bins of width $\delta x_q \approx 0.02$, for x between 0.05–0.15. This is precisely the range where the chiral soliton model suggests large and flavor-dependent antiquark polarizations. *Valence* quark polarizations $\Delta u/u$ and $\Delta d/d$ can be determined simultaneously, permitting an important crosscheck of the results against polarized deep inelastic scattering analyses. We still need a *complete* simulation, incorporating the W production and all relevant background processes, the full TPC response (including deteriorating resolution at $\eta > 1$ and pileup at high luminosity), and the full EMC response (including all layers), in order to tune our identification, reconstruction and analysis algorithms, and plan future extensions to the program.

REFERENCES

1. E.A. Hawker *et al.*, *Phys. Rev. Lett.* **80**, 3715(1998); J.C. Peng *et al.*, *Phys. Rev.* **D58**, 092004 (1998).
2. S. Kumano and J.T. Londergan, *Phys. Rev.* **D44**, 717(1991); S. Kumano, *Phys. Rep.* **303,** 183 (1998).
3. D.I. Diakonov *et al.*, *Phys. Rev.* **D56**, 4069 (1997); P.V. Pobylitsa *et al.*, *Phys. Rev.* **D59**, 034024 (1999); M. Wakamatsu and T. Watabe, *Phys. Rev.* **D62**, 017506 (2000).
4. B. Dressler *et al.*, *Eur. Phys. J.* **C14**, 147 (2000).
5. B. Dressler *et al.*, *hep-ph*/9910464 (1999).
6. B. Adeva *et al.*, *Phys. Lett.* **B420**, 180 (1998).
7. K. Ackerstaff *et al.*, *Phys. Lett.* **B464**, 123 (1999).
8. T. Gehrmann and W.J. Stirling, *Phys. Rev.* **D53**, 6100 (1996).
9. C. Bourrely and J. Soffer, *Phys. Lett.* **B314**, 132 (1993).
10. L.C. Bland, *hep-ex/0002061*.

Future Transversity Measurements at RHIC

Matthias Grosse Perdekamp*, Akio Ogawa[†]

*RIKEN BNL Research Center Upton, New York, USA
[†]Pennsylvania State University, University Park, Pennsylvania, USA

Abstract. The PHENIX and STAR collaborations at Brookhaven National Laboratory will probe the spin structure of the proton in polarized proton-proton collisions at the Relativistic Heavy Ion Collider (RHIC). Initial data taking is planned for 2001 with polarized proton beams at $\sqrt{s} = 200$ GeV. Double spin asymmetries with longitudinal polarization will give access to the proton gluon polarization distributions and parity violating single spin asymmetries will provide information on quark and anti-quark polarizations. Measurements of transverse spin asymmetries will permit the extraction of the currently unknown transversity distributions δq. We present first studies of experimental sensitivities for single transverse spin asymmetries at RHIC.

INTRODUCTION

High energy, deeply inelastic lepton-nucleon and hadron-hadron scattering cross sections can be described with the help of three independent nucleon helicity amplitudes. Measurements of the nucleon structure functions $F_1(x, Q^2)$ -the helicity average- and $g_1(x, Q^2)$ -the helicity difference-, have explored the helicity conserving part of the cross sections with great experimental accuracy. In contrast, no information is presently available on the helicity flip amplitude. The absence of experimental measurements is a consequence of the chiral-odd nature of the helicity flip amplitude and the related "transversity quark distributions", $\delta q(x, Q^2)$, which prevents the appearance of helicity flip contributions at leading twist in inclusive DIS experiments. Transversity distributions were first discussed by Ralston and Soper [1] in Drell-Yan scattering of two transversely polarized hadrons. In Drell-Yan processes the transverse double spin asymmetry, A_{TT}, is proportional to $\delta q \delta \bar{q}$ with even chirality.

Transverse single spin asymmetries $A^{\perp}$ (e.g. unpolarized leptons on transversely polarized nucleon targets) in *semi-inclusive* DIS and pp scattering may offer an alternative way to observe helicity flip contributions at leading twist. This possibility relies on the presence of quark fragmentation functions, $H_1^{\perp}$, which are sensitive to the quark polarization in the final state and possess the necessary negative chirality. The asymmetries $A^\perp$ are proportional to $\sum_q \delta q \times a_i^f \times H_1^\perp$, where a_i^f are the transversity dependent partonic initial-final-state asymmetries which can be calculated from pQCD.

For example, Collins suggested that in semi-inclusive single pion production the quark spin direction might be reflected in the azimuthal distribution of a final state pion [2]. Collins further demonstrated that the symmetry properties of the process do not require the proposed fragmentation function $H_1^\perp$ to be identical to zero. The current interest in transversity distributions results from a recent HERMES result on azimuthal single spin asymmetries [3] and a preliminary SMC result [4], which suggest that Collins's function $H_1^\perp$ and the transversity distribution function δq in fact are different from 0; see reference [5] for a detailed discussion.

In this paper we discuss a proposal by Collins, Heppelmann and Ladinsky [6] and more recently, Jaffe, Jin and Tang [7] to utilize two meson interference fragmentation in order to access the transversity distributions.

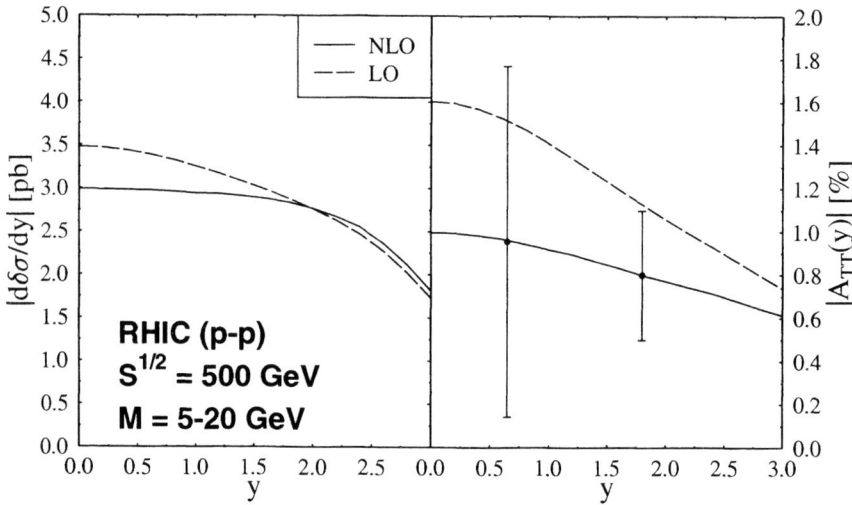

FIGURE 1. Projected sensitivities for transverse double spin asymmetries in Drell-Yan. The plots are taken from reference [8] and show for leading and next to leading order calculations the maximum possible polarized cross section versus dimuon rapidity y and the (right plot) maximum double spin asymmetries A_{TT} compared to statistical errors. The statistical errors correspond to an integrated luminosity of $\int L dt = 800\,\mathrm{pb}^{-1}$ at $\sqrt{s} = 500\,\mathrm{GeV}$.

I TRANSVERSITY AT RHIC

Originally the transverse double spin asymmetry, A_{TT}, in the Drell-Yan process was viewed as a good candidate for a measurement of the transversity distribution functions at RHIC, $A_{TT} \sim \delta q \delta \bar{q}$. Unfortunately, a recent analysis [8] estimates A_{TT} to about 1% with statistical errors comparable to the asymmetry itself for a projected measurement at RHIC. The sensitivities for A_{TT} in Drell-Yan are shown in Fig. (1), from reference [8], for the acceptance of the PHENIX experiment.

At present the proposal of Collins et.al. [6] and Jaffe et.al. [7] to utilize chiral odd two pion interference fragmentation processes appears to be the most promising approach to measure transversity at RHIC. In order to access the transversity distribution functions through this channel, it will be necessary to know the associated fragmentation functions. While currently unmeasured it should be possible in principle to extract this functions from existing e^+e^- data at LEP.

The relevant process at RHIC is pion pair production in pp scattering with one proton transversely polarized. For example, in the ρ/σ invariant mass region interference occurs between two pions in a superposition of s-wave and p-wave states. The spin analyzing power of this process is different from 0 in intervals of only a few 100 MeV above and below the ρ-mass and changes sign at the ρ-mass [9]. Therefore, it will be important for RHIC experiments to have sufficiently high invariant mass resolution to observe the invariant mass dependence of the analyzing power. The invariant mass resolution for pion pairs in the ρ-mass region is shown in the left plot of Fig. (2) for the example of the PHENIX experiment. The RMS of the distribution is 12 MeV – 15 MeV for STAR – which easily meets the demands for an experimental determination of transversity distributions at RHIC.

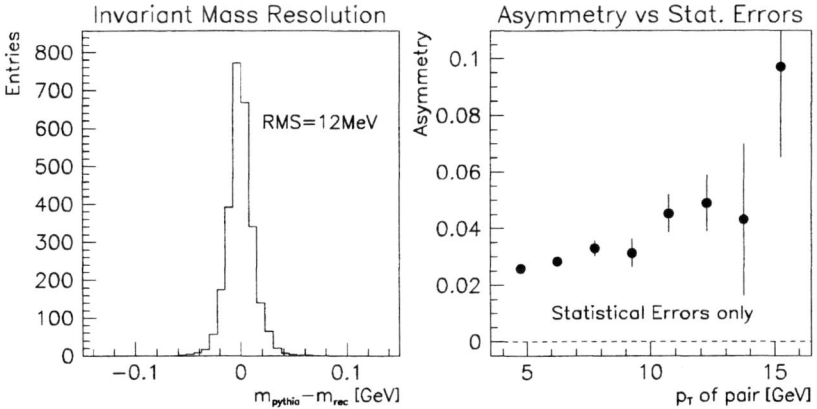

FIGURE 2. Simulation studies of two pion production in the ρ-mass region for the PHENIX detector. Left plot: The invariant mass resolution for pion pairs in the ρ-mass region. Right plot: Projected transverse single spin asymmetries versus the transverse momentum of the pion pair. The statistical errors correspond to an integrated luminosity of $\int Ldt = 32 \, \text{pb}^{-1}$ at $\sqrt{s} = 200 \, \text{GeV}$.

In STAR transversity measurements rely on the large acceptance, $|\eta| < 2.0$, time projection chamber for tracking and the electromagnetic calorimeter for triggering purposes. The PHENIX measurement will use the central detector arms which cover the pseudo rapidity interval $|\eta| < 0.35$. A combination of tracking chambers will give good momentum resolution: $\Delta p/p \approx 2\%$ at $p = 10\,\text{GeV}$. The electromagnetic calorimeter in combination with the ring imaging Cherenkov Counter will provide pion identification over are large momentum range, $4 < p_\pi < 12\,\text{GeV}$.

In order to estimate experimental sensitivities a first study was carried out at the event generator level including PHENIX detector acceptances and a parametrizations of the PHENIX central arm momentum resolution. The results using an integrated luminosity of $32\,\text{pb}^{-1}$ (corresponding to one week of polarized proton running at RHIC) are compared to asymmetry projections from Tang [9] in Fig. (2). The error bars shown in the plot represent statistical errors only. The asymmetries in Fig. (2) were obtained using upper bounds for the relevant distribution and fragmentation functions and represent an optimistic upper limit [9]. More detailed studies with different model assumptions and a full simulation of the detector response are underway. Nevertheless, the high rates at RHIC and the excellent momentum resolution in STAR and PHENIX are well suited for a transversity measurement in two pion interference fragmentation and a run with $320\,\text{pb}^{-1}$ integrated luminosity has been added to of the spin physics run plan [10].

REFERENCES

1. Ralston J., Soper D.E., *Nucl. Phys.* **B152**, 109 (1979).
2. Collins J.C., Nucl. Phys. **B396**, 161 (1993).
3. Airapetian A. et al., *Phys. Rev. Lett.* **84**, 4047 (2000).
4. Bravar A., *Nucl. Phys. (Proc. Suppl.)* **B79**, 520 (1999).
5. K.A. Oganessyan, N. Bianchi, E. De Sanctis, W.D. Nowak, hep-ph/0010261.
6. Collins J.C., Heppelmann S.F. and Ladinsky G.A., *Nucl. Phys.* **B420**, 565 (1994); Collins J.C. and Ladinsky G.A., Preprint PSU-TH-114, hep-ph/9411444; Collins J.C., Proceedings of the RHIC Spin Workshop, October 6 - 8, 1999, p. 158.
7. Jaffe R.L., Jin X., Tang J. et al., *Phys. Rev. Lett.* **80**, 1166 (1998);
8. Martin O. et al., Phys. Rev. **D60**, 117502 (1999). *Phys. Rev.* **D57**, 5920 (1998).
9. Tang J., Preprint hep-ph/9807560 and Tang J., Thesis, MIT (1999).
10. D. Boer et.al., RHIC Spin Physics Program, White Paper submitted to the DNP Town Meeting at Jefferson Laboratory, December 1-4, 2000.

Positivity Constraints in Spin Physics

Jacques Soffer

Centre de Physique Théorique
CNRS Luminy Case 907
13288 Marseille Cedex 09 France

Abstract. We will emphasize the relevance of *positivity* in spin physics, which puts non-trivial model independent constraints on spin observables. These positivity conditions are based on the positivity properties of density matrix or Schwarz inequalities for transition matrix elements in processes involving several particles carrying a non-zero spin. We will illustrate this important point by means of several examples chosen in different areas of particle physics.

POLARIZED TOTAL CROSS SECTIONS

Let us consider the spin-dependent total cross sections for the scattering of two spin-1/2 particles

$$a_{s=1/2} + b_{s=1/2} \to anything, \tag{1}$$

which are, for example, $pp, \bar{p}p, p\Lambda, pn, etc...$. The reaction (1) can be described in terms of three independent observables which are the unpolarized cross section

$$\sigma_{tot} = 1/2[\sigma_{tot}(+,+) + \sigma_{tot}(+,-)] = 1/2[\sigma_{tot}(\uparrow,\uparrow) + \sigma_{tot}(\uparrow,\downarrow)], \tag{2}$$

and the two asymmetries

$$\Delta\sigma_L = \sigma_{tot}(+,-) - \sigma_{tot}(+,+) \quad and \quad \Delta\sigma_T = \sigma_{tot}(\uparrow,\downarrow) - \sigma_{tot}(\uparrow,\uparrow). \tag{3}$$

Here +(-) denote the longitudinally polarized (or helicity) states of a and b and $\uparrow (\downarrow)$ their transversely polarized states.

If $\vec{P}_a$ and $\vec{P}_b$ are the polarization unit vectors of a and b, the polarized total cross sections corresponding to Eq.(1) are [1]

$$\sigma_{tot}(\vec{P}_a, \vec{P}_b) = Tr(\mathcal{M}\rho). \tag{4}$$

Here $\mathcal{M}$ is the forward scattering amplitude for the elastic reaction

$$a + b \to a + b, \tag{5}$$

and ρ is the 4x4 density matrix $\rho = 1/4(1 + \vec{P}_a\vec{\sigma}) \otimes (1 + \vec{P}_b\vec{\sigma})$, where $\vec{\sigma} = (\sigma_1, \sigma_2, \sigma_3)$ stands for the three 2x2 Pauli matrices. $\mathcal{M}$ is also a 4x4 matrix, Hermitian and *positive*, which implies that all its principal minors (*i.e.* subdeterminants of $\mathcal{M}$ with diagonal elements) must be *positive*. In terms of the three observables defined above, we get two trivial conditions, *i.e.* $|\Delta\sigma_i| \leq 2\sigma_{tot}, (i = L, T)$ and one non-trivial positivity bound [2], namely

$$|\Delta\sigma_T| \leq \sigma_{tot} + \Delta\sigma_L/2 . \tag{6}$$

In the case of proton-proton scattering we have checked that this rigorous bound is fulfill for $p_{lab} \geq 1 GeV/c$, where $\Delta\sigma_L$ and $\Delta\sigma_T$ have been measured [3]. For $0.5 \leq p_{lab} \leq 1 GeV/c$, σ_{tot} is of the order of $20mb$ or so, $\Delta\sigma_L$ is around $-30mb$, so by using Eq.(6) one gets $|\Delta\sigma_T| \leq 5mb$, with some errors, an interesting limit which must be satisfied by the data.

1/2 + 1/2 → 1/2 + 1/2 SCATTERING

A scattering process involving spinning particles and described by n *complex* amplitudes is completely determined in terms of $(2n - 1)$ *real* functions, up to an over-all phase. Since there are n^2 possible measurements for this reaction, we must have $(n-1)^2$ independent quadratic relations between the n^2 observables. For illustration, let us first consider the simplest case of an exclusive two-body reaction with a spin-0 and a spin-1/2 particle, namely $0 + 1/2 \rightarrow 0 + 1/2$. As an example $\pi N \rightarrow \pi N$ is described in terms of two amplitudes, the non-flip f_+ and the flip f_-, so we have four observables [1], the differential cross section $d\sigma/dt$, the polarization P and two rotation parameters R and A. There is one well known quadratic relation, that is $P^2 + A^2 + R^2 = 1$. For the reaction $1/2 + 1/2 \rightarrow 1/2 + 1/2$ and more specifically for nucleon-nucleon scattering, we have five amplitudes, therefore twenty five observables [1] and sixteen quadratic relations between them. For the derivation one considers a 5x5 matrix of the observables which is positive Hermitian and the final results can be found in Ref. [4]. These relations are very useful to check the data and when one observable is not measured, it is set to zero and the equality becomes an inequality. In particular one finds the following very simple condition

$$A_{LL}^2 + D_{NN}^2 \leq 1 , \tag{7}$$

between the two-spin correlation parameters [1] A_{LL} and D_{NN}. This condition turns out to be very usefull because if one of the parameters is close to one, the other one is bounded to be near zero. In the reaction $\bar{p}p \rightarrow \bar{\Lambda}\Lambda$, one has found in a certain kinematic region $A_{LL} = -1$, so before making any measurement one can conclude that $D_{NN} \sim 0$, which is also a no-go theorem for some theoretical considerations [5].

DEEP INELASTIC SCATTERING

As we have seen above positivity is playing an important role in constraining spin-dependent observables, in particular by providing a bound for the transverse asymmetry in polarized deep inelastic scattering (DIS). It is a well-known condition established long time ago [6] and based on an extensive study by Doncel and de Rafael [7], written in the form

$$|A_2| \leq \sqrt{R} , \tag{8}$$

where A_2 is the usual transverse asymmetry and $R = \sigma_L/\sigma_T$ is the standard ratio in DIS of the cross section of longitudinally to transversely polarized off-shell photons.

It reflects a non-trivial positivity condition one has on the photon-nucleon forward helicity amplitudes. However a stronger bound than Eq. (8) was established in Ref. [8],

$$|A_2| \leq \sqrt{R(1 + A_1)/2} , \tag{9}$$

where A_1 denotes the asymmetry with longitudinally polarized nucleon. It is worth noting that if A_1 and A_2 are measured with a good precision, Eq. (9) can be used as a lower bound for R, a rather non-trivial constraint. The above result can be generalized to each quark flavor separately. As we will see it leads to sensitive tests for parton distributions and sometimes gives a hint about higher twist terms. This may be achieved by considering a fictitious "photon" coupled to only one flavor and we get analogously to Eq. (9)

$$|A_2^f| \leq \sqrt{R^f(1 + A_1^f)/2} . \tag{10}$$

Let us denote by $q_f(x)$ the unpolarized quark distribution of flavor f and by $\Delta q_f(x)$ the corresponding polarized quark distribution. It was shown in Ref. [8], to the leading twist approximation, that Eq. (10) can be turned into a bound for $\Delta q_f(x)$, which reads

$$|\int_x^1 \frac{dz \Delta q_f(z)}{z}| \leq \sqrt{\frac{q_f(x) + \Delta q_f(x)}{2} \cdot \int_x^1 \frac{dz q_f(z)}{z}}. \tag{11}$$

We stress that unlike Eq. (9), which is a new and rigorous positivity bound, Eq. (11) is valid only to leading twist and provided that the target mass approximation for R, dominates over the contribution of perturbative QCD.

This bound was tested and we found that it is strongly dependent on the parametrization one uses for the quark distributions. Especially sensitive is the case of d-quark, because of its negative polarization. While the inequality is satisfied by the GRSV distributions [9], the GS distributions [10] exhibit a tiny violation of the inequality in the case of d-quark for $0.4 \leq x \leq 0.8$ (see Fig.1), but not for the u-quark as reported in Ref. [8]. We also anticipate a violation in the case of the strange quark, whose polarization, although losely known, is expected to be negative.

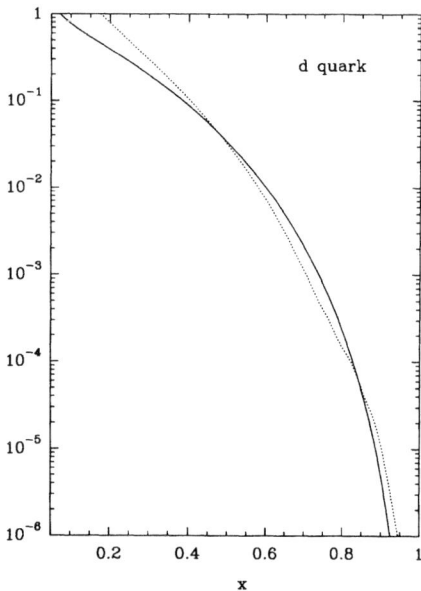

FIGURE 1. Test of the positivity bound using the GS distribution for d-quark at $Q^2 = 1 GeV^2$. The solid curve corresponds to the *l.h.s.* of Eq. (11) and the dotted curve to the *r.h.s.*

THE QUARK TRANSVERSITY DISTRIBUTIONS

Let us consider quark-nucleon elastic scattering

$$q_{h'} + N_H \to q_h + N_{H'} , \qquad (12)$$

where we have indicated the helicity of each particle. In the forward direction, only three helicity amplitudes survive $\mathcal{A}(++;++)$, $\mathcal{A}(+-;+-)$ and $\mathcal{A}(++;--)$. The quark distributions $q(x)$ and $\Delta q(x)$ are related to the first two and the third one is related to the transversity distribution $h_1^q(x)$, which measures the difference of probabilities for finding transverse spin aligned and anti-aligned with the transverse nucleon spin. This distribution is as fundamental as the helicity distribution $\Delta q(x)$, but so far it has not been measured yet. Of course in the absence of relativistic effects one has $h_1^q(x) = \Delta q(x)$ and as for $\Delta q(x)$, there is the obvious constraint

$|h_1^q(x)| \leq q(x)$, which must be always valid. However a straightforward use of positivity leads to the following non-trivial bound [11]

$$2|h_1^q(x)| \leq q(x) + \Delta q(x) . \qquad (13)$$

We show on Fig.2 the region allowed by Eq.(13), which is indeed smaller than the one resulting from the trivial constraint mentioned above. The inequality holds for all quark flavors and separately for their corresponding antiquarks. It was proved in several recent works [12] that the inequality is preserved under QCD evolution, that is, if it is assumed to be satisfied at a given energy scale Q_0^2, it will hold at any scale $Q^2 \geq Q_0^2$.

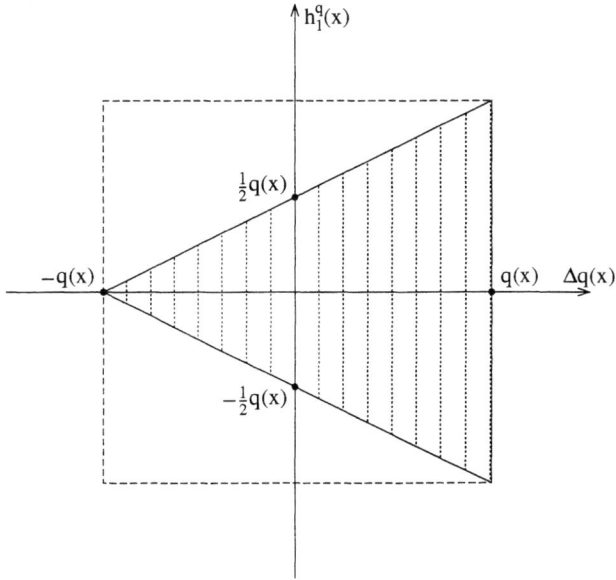

FIGURE 2. The hatched area represents the domain allowed by positivity (see Eq.(13)).

POLARIZED FRAGMENTATION FUNCTIONS

The structure of a hadron H is probed through the usual parton ($f = q, \bar{q}, g$) distributions $f_H(x, Q^2)$, which are extracted from DIS for a space-like energy scale ($Q^2 \leq 0$), but it can also be studied by means of the hadron fragmentation functions $D_f^H(z, Q^2)$. They represent the probability to find the hadron H with a fraction z of the momentum of the parent parton f, at a given value of Q^2, in the time-like region ($Q^2 \geq 0$). When H is a baryon and more specifically a Λ, it is interesting to study the fragmentation of a longitudinally polarized parton into a longitudinally polarized Λ which is described in terms of

$$\Delta D_f^\Lambda(z, Q^2) = D_{f(+)}^{\Lambda(+)}(z, Q^2) - D_{f(+)}^{\Lambda(-)}(z, Q^2) , \qquad (14)$$

where $D_{f(+)}^{\Lambda(+)}(z, Q^2)$ $[D_{f(+)}^{\Lambda(-)}(z, Q^2)]$ is the probability to find a Λ with positive [negative] helicity in a parton f with positive helicity. Clearly, one obtains the unpolarized fragmentation function D_f^Λ by taking in Eq.(14), the sum instead of the difference. For the fragmentation of a transversely polarized parton into a transversely polarized Λ, one uses Eq.(14) to define $\Delta_T D_f^\Lambda(z, Q^2)$ in complete analogy with $\Delta D_f^\Lambda(z, Q^2)$.

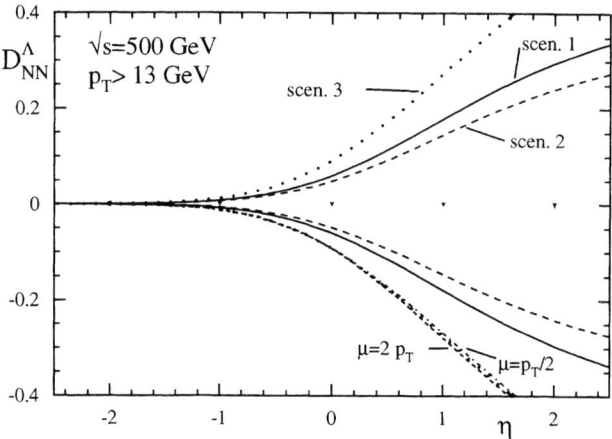

FIGURE 3. Various domains of D_{NN}^Λ allowed by positivity (see Eqs.(13,16), taken from Ref.[13]).

These polarized fragmentation functions can be tested by studying spin transfers in pp collisions. The forthcoming polarized pp collider at BNL RHIC will allow to undertake a vast spin physics programme at center of mass energies up to $\sqrt{s} = 500 GeV$. We are interested in the reaction $\vec{p}p \rightarrow \vec{\Lambda} X$, where both the initial p and the produced Λ are transversely polarized and we consider the spin transfer parameter D_{NN}^Λ defined as follows

$$D_{NN}^\Lambda(\eta, p_T) = (d\sigma_{\uparrow\uparrow} - d\sigma_{\uparrow\downarrow})/(d\sigma_{\uparrow\uparrow} + d\sigma_{\uparrow\downarrow}) \;, \quad (15)$$

where η and p_T are the rapidity and the transverse momentum of the outgoing Λ. $d\sigma_{\uparrow\uparrow}$ $[d\sigma_{\uparrow\downarrow}]$ denotes the cross section where the p and Λ spin have the same [opposite] sign. η is positive for a Λ in the direction of $\vec{p}$ and p_T is large. D_{NN}^Λ is directly related to the $h_1^q(x, Q^2)$'s and the $\Delta_T D_f^\Lambda$'s. Needless to say that we have no experimental information on these quantities, however we can derive some bounds on D_{NN}^Λ, using positivity arguments. A similar constraint to Eq.(13) holds for hadron fragmentation functions [13], namely

$$2|\Delta_T D_q^H(z, Q^2)| \leq D_q^H(z, Q^2) + \Delta D_q^H(z, Q^2) \;. \quad (16)$$

By saturating Eq.(13) and the above inequality, we get an estimate for an upper and lower bounds for D_{NN}^Λ which are displayed in Fig.3, where small error bars are due to the expected statistical accuracy.

We presented here only a few cases to exemplify the usefulness of positivity in spin physics, for lack of time, but the list is not exhaustive. For completeness let us just mention, a positivity bound for the longitudinal gluon distribution in a nucleon [14], some positivity constraints for off-forward parton distributions [15], bounds on transverse momentum dependent distribution and fragmentation functions [16], etc....

REFERENCES

1. Bourrely C., Leader E., and Soffer J. *Phys. Rep.* **59**, 95 (1980).
2. Soffer J., and Wray D., *Phys. Lett.* **B43**, 514 (1973).
3. Yokosawa A., *Phys. Rep.* **64**, 47 (1980).
4. Bourrely C., and Soffer J., *Phys. Rev.* **D12**, 2932 (1975).
5. Richard J.M., *Phys. Lett.* **B369**, 358 (1996).
6. Christ N., and Lee T.D., *Phys. Rev.* **143**, 1310 (1966).
7. Doncel M.G., and de Rafael E., *Nuovo Cimento* **4A**, 363 (1971).
8. Soffer J., and Teryaev O.V., *Phys. Lett.* **B490**, 106 (2000).
9. Gluck M., Reya E., Stratmann M., and Vogelsang W., *Phys. Rev.* **D53**, 4775 (1996).
10. Gehrmann T., and Stirling W.J., *Phys. Rev.* **D53**, 6100 (1996).
11. Soffer J., *Phys. Rev. Lett.* **74**, 1292 (1995).
12. Vogelsang W., *Phys. Rev.* **D57**, 1886 (1998); Bourrely C., Soffer J., and Teryaev O.V., *Phys. Lett.* **B420**, 375 (1998); Martin O., Schaefer A., Stratmann M., and Vogelsang W., *Phys. Rev.* **D57**, 3084 (1998); *Phys. Rev.* **D60**, 117502 (1999) and references therein.
13. de Florian D., Soffer J., Stratmann M., and Vogelsang W., *Phys. Lett.* **B439**, 176 (1998).
14. Soffer J., and Teryaev O.V., *Phys. Lett.* **B419**, 400 (1998).
15. Pire B., Soffer J., and Teryaev O.V., *Euro. Phys. J.* **C8**, 103 (1999).
16. Bacchetta A., Boglione M., Henneman A., and Mulders P.J., *Phys. Rev. Lett.* **85**, 712 (2000).

Twist-2 Polarized Fragmentation Function in the Open Charm Production in DIS

Yuri I. Arestov

Institute for High Energy Physics
142284 Protvino, Moscow Region, Russia

Abstract. To extract the polarized fragmentation function G_1, he transmitted polarization parameter D_{LL} has been considered in the semi-inclusive leptoproduction process $e_\uparrow^- + p \to e^- + \Lambda_{c\uparrow}^+ + X$ with both longitudinally polarized electron and charmed lambda. The polarization transfer $\hat{d}_{LL}$ for the lepton-gluon subprocess $e_\uparrow^- + g \to e^- + Q_\uparrow \bar{Q}$ has been carefully studied, and it appears to be sizeable. The quantitative estimates of $\hat{d}_{LL}$ has been made at the gluon momentum fraction $x_g = 0.2$.

The polarized charmed lambda production is considered in SIDIS reaction

$$e_\uparrow^- + p \to e^- + \Lambda_{c\uparrow}^+ + X \tag{1}$$

with the longitudinally polarized lepton beam shown in fig. 1 in the LO approximation. The polarization transmission parameter D_{LL} is defined as

$$D_{LL} = \frac{\sigma_{++} + \sigma_{--} - \sigma_{+-} - \sigma_{-+}}{\sigma_{++} + \sigma_{--} + \sigma_{+-} + \sigma_{-+}}, \tag{2}$$

where the subscripts $\{++\}$ etc. relate to the helicity states of the lepton and Λ_c^+.

In the absense of the initial polarization, Λ_c^+ may be polarized only transversely to the production plane. A longitudinal component of the Λ_c^+ polarization vector may arise due to the longitudinal lepton polarization in the initial state.

The polarization transmission coefficient which can be measured experimentally relates to the fragmentation function $G_1(z, \mu^2)$ (FF) through the following expression:

$$D_{LL} \sim G(x_g) \cdot \hat{d}_{LL} \cdot G_1, \tag{3}$$

where $G(x_g)$ is the gluon distribution in the (unpolarized) proton and $\hat{d}_{LL}$ is the polarization transmission coefficient in the heavy quark pair production in the lepton-gluon scattering

$$e^-_\uparrow + g \to e^- + Q_\uparrow \bar{Q}. \tag{4}$$

In order to plan measuring the unknown FF G_1, it would be instructive to know the range of the values of $\hat{d}_{LL}$ in the subprocess (4).

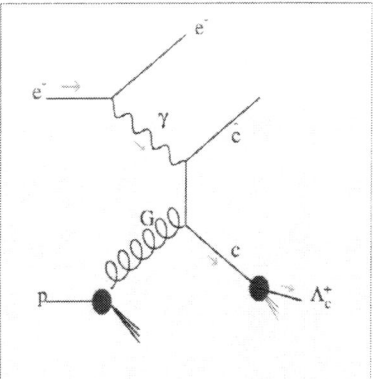

FIGURE 1. The SIDIS reaction (1).

The matrix element of the reaction (4) incorporates the contributions of u-channel ($Q\bar{Q}$ configuration in fig. 1), t-channel with the permuted Q's and the interference term. Below these contributions are referred as uu, tt and tu terms.

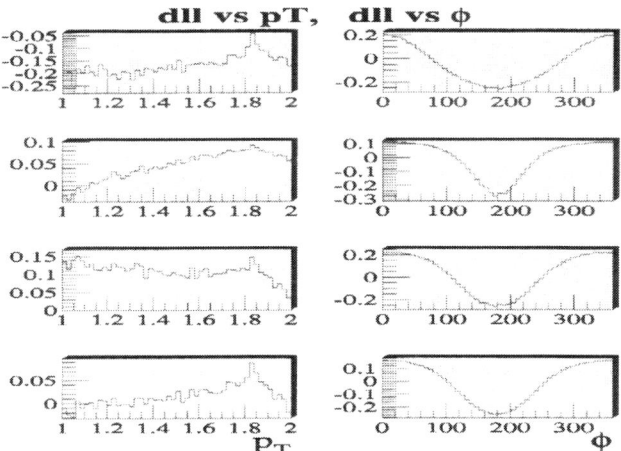

FIGURE 2. $\hat{d}_{LL}$ versus p_T and ϕ for (from up to down) tt, uu, tu and total terms.

As is seen from fig. 2, the d_{LL} properties may be quite different depending on t or u-channel that is on the $Q\bar{Q}$ configuration in the diagram. The azimuthal angle ϕ

counts from the lepton scattering plane. The total ϕ-dependence integrated over p_T exhibits a remarkable behaviour with a deep minimum at $\phi = 180^0$ and positive maximum at $\phi = 0^0$ (360^0). The maximum of the absolute value of d_{LL} may reach 0.3.

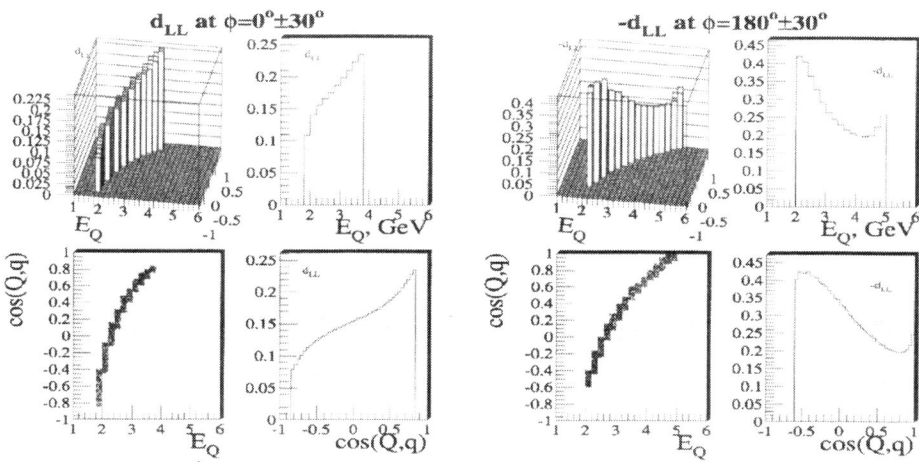

FIGURE 3. $\hat{d}_{LL}$ dependence on the quark energy E_Q and (Q,q) angle (see text).

Apart from p_T dependence it can be interested to track down the $\hat{d}_{LL}$ dependence on E_Q, the quark energy, and on $\cos(Q,q)$, the cosine of the scattering angle in respect to the virtual photon direction. In fig. 3 the corresponding two- and one-dimensional plots are shown in two regions: $\phi = 0 \pm 30^0$ and $\phi = 180 \pm 30^0$.

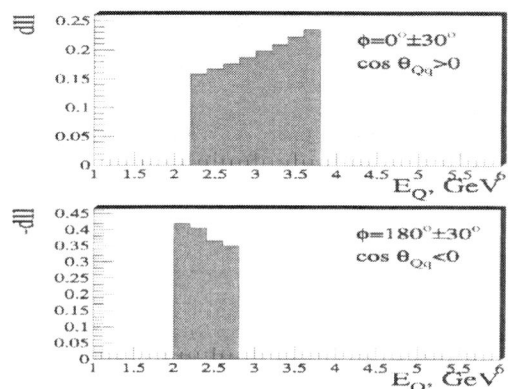

FIGURE 4. The quark energy dependence of $\hat{d}_{LL}$ in the forward and backward hemispheres.

Finally, fig. 4 presents the quark energy E_Q dependence of the polarization coefficient $\hat{d}_{LL}$ integrated in separate hemispheres in respect to the virtual photon momentum. It is seen that at the reasonable values of the quark energy, the coefficient $\hat{d}_{LL}$ appears to be sizeable.

From the above consideration, it follows that the model expectations for the fragmentation function G_1 (see (3)) may be quite reasonable because the underlying subprocess exhibits the large values of $\hat{d}_{LL}$ in some regions of the phase space.

Polarized gluon distribution function of nucleon in diffractive lepto-production of charmonium

Arata Hayashigaki[1] and Katsuhiko Suzuki[2]

Department of Physics, Faculty of Science, University of Tokyo, Tokyo 113-0033, Japan

Abstract. We investigate the longitudinal double spin asymmetry in the diffractive lepto-production of charmonium off the proton, and clarify its relation to the polarized gluon distribution ΔG of the proton. The asymmetry is found to be finite even in the forward limit, if we properly deal with the Fermi motion of charm quarks and transverse momenta of the exchanged gluons. Our calculation for the photo-production of J/ψ indicates the asymmetry is of $O(10^{-3}) \sim O(10^{-2})$ at HERA collider energy and of $O(10^{-2})$ at fixed-target energy (HERMES). The asymmetry for the ψ' is even large, of $O(10^{-1})$, due to the large Fermi motion of $c\bar{c}$ in ψ'.

INTRODUCTION

Diffractive lepto-production of vector-mesons has been investigated extensively to study the quark and gluon dynamics of the proton. In particular, the diffractive process of heavy vector-meson productions such as J/ψ and ψ' can supply us direct information on the gluon distribution, because the process predominantly couples to the gluons and not to light quarks inside the proton. On the basis of perturbative QCD (PQCD) model, we have described well the lepto-production of the J/ψ by assuming dominance of colorless two-gluon exchange in the Regge limit of the process, as shown in Fig.1. Indeed in the Regge limit, the amplitude $\mathcal{M}$ can be decomposed into the following three parts: $\mathcal{M} = \phi_\gamma^* \otimes \sigma_{pc\bar{c}} \otimes \phi_\psi$, where (i)$\phi_\gamma$; dissociation $\gamma^{(*)} \to c\bar{c}$, (ii)$\sigma_{pc\bar{c}}$; elastic scattering $c\bar{c}p \to c\bar{c}p'$ and (iii)ϕ_ψ; hadronization $c\bar{c}g \to J/\psi$ (or ψ')g. We can evaluate the part (i) within PQCD. The part (ii) contains the upper hard part calculable within PQCD and the lower blob, which includes both non-perturbative structure of the proton and complicated gluon dynamics between the exchanged two-gluons. For the last part (iii), we use the wave function of J/ψ evaluated with realistic non-relativistic quark model. Here we address the diffractive processes might be good probes to extract the gluon distribution in small x region accessible at HERA energy by three following reasons:

---

[1] E-mail: arata@nt.phys.s.u-tokyo.ac.jp. This research was supported by the Japan Society for Promotion of Science (JSPS).
[2] E-mail: ksuzuki@nt.phys.s.u-tokyo.ac.jp. This research was supported by the JSPS.

First, the cross section in the forward limit, the momentum transfer $\Delta = 0$, is sensitive to the small x distribution of the gluon, because it is proportional to squared gluon density. Here one can naively expect that the diffractive heavy quarkonium production at $\Delta = 0$ is related to the gluon distribution function. Second, we have less theoretical ambiguities in this process, because we need not consider the color-octet contribution in hadronization to the charmonium. Third, in the photoproduction the end point contribution in the longitudinal motion of real photon fluctuation could be strongly suppressed by convolution with the wave function of the charmonium in the final state. This property might guarantee the factorization even for the transversely polarized photon.

In this work, we study double spin asymmetry of the diffractive process, in which both lepton beam and target are longitudinally polarized, in order to determine the gluon spin-dependent distribution function inside the proton. The gluon contribution to the proton spin is a important subject related to proton spin problem. In particular, direct and detailed determination of the x-dependence of the polarized gluon distribution is necessary to extract the gluon contribution to the proton spin. The polarized diffractive process is one of the unique methods to do it. In fact, there are experimental efforts to measure this process at HERMES, although the data are still preliminary, and it could be possible at polarized HERA and eRHIC in future. There have been also several theoretical works to study the gluon polarization by the polarized diffractive $J/\psi(1S)$ production [1-3]. The authors of ref. [2] calculated the asymmetry at $\Delta = 0$ with perturbative two gluon exchange in the light-cone gauge, assuming the collinear scattering between $c\bar{c}$ and the target proton. Here the collinear approximation means evaluation of diagrams with gluon transverse momentum neglected. Unfortunately, the resulting asymmetry vanishes if one assumes the static approximation for the charmonium wave function, in which the Fermi motion of quarks are entirely neglected. The asymmetry is negligible even if a part of the Fermi motion correction is incorporated. Goloskokov [3] also found that the asymmetry vanishes at $\Delta = 0$ with the phenomenological Pomeron-proton vertex and the static approximation.

However, when we deal with the small x region of the asymmetry, the longitudinal and transverse components of the gluon momentum may be of the same order and thus the transverse degrees of freedom of the gluon must be taken into account. Here the amplitude should be calculated based on k_T-factorization scheme. Furthermore, several studies on the diffractive J/ψ production with the realistic wave function have shown that there are substantial Fermi motion corrections to the production cross section [4]. We have recently shown that the standard color-dipole approximation is not adequate for the description of the $\psi'(2S)$ production due to its large transverse size [5]. Here the color-dipole approximation means truncating a expansion due to the separation length of the charm quark pair up to the leading order by assuming the separation is very small. All these results suggest that the static approximation and the color-dipole approximation are no longer reliable for charmonium productions, in particular for ψ'. Motivated by these considerations, we evaluate the spin asymmetry without the collinear approximation. We explicitly incorporate the polarization tensor of the gluons with transverse momenta. We also deal with the Fermi motion of the charm quarks in the wave function properly. Such effects are especially crucial for ψ' case, and in fact enhance the resulting asymmetry. In next section, we will present the formulation to calculate the spin asymmetry.

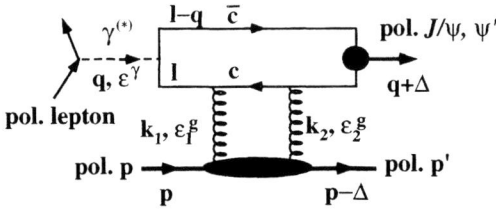

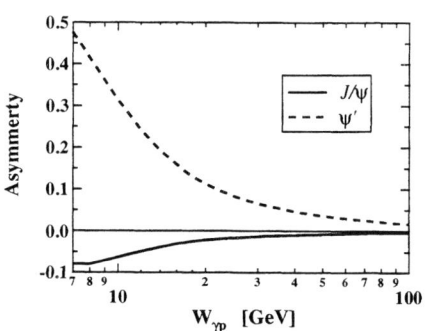

FIGURE 1. The Feynman diagrams contributing to polarized diffractive lepto-production of charmonium. To leading order, there are four diagrams independent in the way of attaching two gluon-lines to a quark line and an anti-quark line respectively. We show above diagram as one of them.

FIGURE 2. The W-dependence of the spin asymmetry in photo-production ($Q^2 = 0$).

I SPIN ASYMMETRY OF POLARIZED DIFFRACTIVE PROCESS

In proceeding to calculate the amplitude, we adopt light-cone perturbation theory [6] useful for studying processes with large transverse momenta. The labeling of respective particle's momenta, helicity and spin is shown in Fig.1. Here we consider zero momentum transfer ($\Delta = 0$) in t-channel (the forward limit of the process), and assume dominance of the target-spin non-flip process (S-channel helicity conservation (SCHC)). Then the helicity amplitude for $\gamma(q, \lambda) + p(p, S(= \pm 1/2)) \to \psi(q, \lambda) + p(p, S)$ reads

$$\mathcal{M}_{\lambda, S(\pm)} \propto \sum_{\lambda_1, \lambda_2} \delta_{\lambda_1, \lambda_2} \int dk_T^2 M_{\lambda, \lambda_1, \lambda_2}(k_T^2) \left[\frac{\partial(xG(x, k_T^2))}{\partial k_T^2} \pm \lambda_1 \frac{\partial(x\Delta G(x, k_T^2))}{\partial k_T^2} \right], \quad (1)$$

where λ denotes helicities of the photon and the charmonium. $\lambda_{1,2}$ are those of the exchange gluons. k_T is transverse momentum of the exchanged gluons ($k_1 = k_2$). $G(x, k_T^2)$ and $\Delta G(x, k_T^2)$ are usual unpolarized and polarized gluon parton distributions, respectively. Here x means longitudinal momentum fraction carried by the gluon. The helicity amplitude $M_{\lambda, \lambda_1, \lambda_2}$ of the upper part in Fig.1 for hard subprocess $\gamma(q, \lambda) + g(k_1, \lambda_1) \to \psi(q, \lambda) + g(k_1, \lambda_1)$ is given by the convolution of ϕ_γ, a perturbative $c\bar{c}$-dipole interaction and ψ wave function as mentioned before. The forward differential cross section for the polarized lepto-production of charmonium is described as $(d\sigma_\lambda/dt)_{t=0} \sim |\mathcal{M}_{\lambda,\pm}|^2$.

In the γ-p system, we can express the spin asymmetry A as $A(\gamma = L \text{ or } T) = [d\sigma(\gamma, S(+1/2)) - d\sigma(\gamma, S(-1/2))]/[d\sigma(\gamma, S(+1/2)) + d\sigma(\gamma, S(-1/2))]$, where "$L$" and "$T$" denote the "longitudinal" and "transverse" polarizations of the photon, respectively. We obtain $A(\gamma = L) = 0$ and $A(\gamma = T(+1)) = -A(\gamma = T(-1))$, as expected by the kinematical symmetry. So we discuss only the asymmetry induced by $(+1)$ component of the transversely polarized photon in the later calculation. Naively the asymmetry is rearranged with the gluon helicities (± 1) like

$$A(\gamma = T) = \frac{[M_{g(+1)} \otimes G][M_{g(+1)} \otimes \Delta G] - [+1 \to -1]}{[M_{g(+1)} \otimes G][M_{g(+1)} \otimes G] + [+1 \to -1]}, \qquad (2)$$

where $\otimes$ indicates the convolution of various integral variables. Here the amplitude $\mathcal{M}_{g(\pm 1)} = M_{g(\pm 1)} \otimes (\Delta)G$ is equivalent to

$$\mathcal{M}_{T(+1),g(\pm 1)} \propto \int dz \int db^2 \int_0^{Q^2_{eff}} dk_T^2 \alpha_s(k_T^2) \frac{(\Delta)f(x,k_T^2)}{k_T^4}$$
$$\times K_0(b\sqrt{z(1-z)Q^2 + m_c^2}) H_{g(\pm 1)}(z,b,k_T^2) \frac{\phi_\psi(z,b)}{z(1-z)}, \qquad (3)$$

where we define $Q^2_{eff} = z(1-z)Q^2 + m_c^2$ with $q^2 = -Q^2$, the strong coupling constant α_s and the Sudakov variable z, which parameterizes the momentum of the charm quark, $l = (zq^+, q^-, \vec{l}_T)$. The variable b represents the transverse size of the color-dipole. K_0 is the modified Bessel function, which corresponds to a part of the photon wave function calculated within PQCD [4]. On the other hand, we use the non-relativistic potential model to describe two-body $(c\bar{c})$ bound state of charmonium (S-state), which is a non-perturbative object. Indeed, after evaluating the wave function with the Cornell potential [7], we can deduce a light-cone wave function $\phi_\psi(z,b)$ following the prescription of [4,6]. $H_{g(\pm 1)}(z,b,k_T^2)$ are functions related to the $c\bar{c}g$-interaction amplitude. For the gluon exchange with $(+1)$ helicity

$$H_{g(+1)}(z,b,k_T^2) = \frac{W^2}{k_T^2}(1-2z)[1 - J_0(k_T b)]. \qquad (4)$$

For the gluon exchange with (-1) helicity

$$H_{g(-1)}(z,b,k_T^2) = \frac{W^2}{k_T^2}(1-2z)[1 - J_0(k_T b)] + \frac{(1-2z)}{z(1-z)} J_0(k_T b), \qquad (5)$$

where J_0 indicates Bessel function of the first kind. In Eq.(3), $f(x,k_T^2)$ can be approximately related to the unintegrated gluon density [5], $(\Delta)f(x,k_T^2) = k_T^2 \partial[x(\Delta)G(x,k_T^2)]/\partial k_T^2$, where the momentum fraction of the gluon x is identified with $x \sim Q^2_{eff}/W^2$ at large W. Furthermore, in evaluation of the k_T^2-integral in Eq.(3) we introduce the infrared separation scale Q^2 and perform the integral in the same manner as [5]. After performing all integrals in Eq.(3), we can see that Eq.(2) is proportional to x, because the term proportional to W^2/k_T^2 in the numerator of the asymmetry disappears by the cancellation between $(+)$ and $(-)$ components of the gluon helicity.

II NUMERICAL RESULTS AND SUMMARY

In Fig.2, we show the numerical results of the asymmetry in the photo-production ($Q^2 = 0$) using the GRV [8] and GRSV [9] parameterizations as inputs of the unpolarized and polarized gluon distributions, respectively. The W-dependence of the asymmetries show opposite signs between J/ψ(solid line) and ψ'(dashed line). At the W range relevant for HERA collider energy, these asymmetries are of $O(10^{-2}) \sim O(10^{-3})$, which are very small. At a fixed-target energy like HERMES, the asymmetries are of $O(10^{-2})$ for J/ψ and of $O(10^{-1})$ for ψ'. We note if we take $\vec{k}_T = \vec{0}$ in Eq.(2), the l_T-dependence appears from next-to-next leading order of W^2/k_T^2 in Eq.(4) and Eq.(5).

To summarize, we have studied the longitudinal double-spin asymmetry to extract the polarized gluon distribution ΔG in the polarized diffractive leptoproduction of J/ψ and the first excited state ψ'. ΔG could be extracted from the asymmetry, which has a finite value dependent on the transverse momentum of the exchanged gluons in the diffractive process of transversely polarized photon. The asymmetry, however, disappears for the longitudinally polarized photon. Furthermore, we have calculated the asymmetry with realistic ψ wave functions which incorporate longitudinal and transverse Fermi motion effects. The asymmetry at $W = 90$ GeV is a very small value, $O(10^{-3}) \sim O(10^{-2})$, in J/ψ production, and similar for ψ' production. But ψ' might have rather larger value $O(10^{-1})$ than J/ψ near $W \sim 10$ GeV (HERMES), due to the large Fermi motion of $c\bar{c}$.

REFERENCES

1. M.G. Ryskin, *Phys. Lett.* **B403**(1997) 335.
2. M. Vänttinen and L. Mankiewicz, *Phys. Lett.* **B440**(1998) 157; *Phys. Lett.* **B434**(1998) 141.
3. S.V. Goloskokov, *Eur. Phys. J.* **C11**(1999) 309.
4. L. Frankfurt, W. Koepf and M. Strikman, *Phys. Rev.* **D54**(1996) 3194; *Phys. Rev.* **D57**(1998) 512.
5. K. Suzuki, A. Hayashigaki, K. Itakura, J. Alam and T. Hatsuda, *Phys. Rev.* **D62**(2000) 031501(R).
6. G.P. Lepage and S.J. Brodsky, *Phys. Rev.* **D22**(1980) 2157.
7. E.J. Eichten and C. Quigg, *Phys. Rev.* **D52**(1995) 1726; E. Eichten *et al.*, *Phys. Rev.* **D21**(1980) 203.
8. M. Glück, E. Reya and A. Vogt, *Eur. Phys. J.* **C5**(1998) 461.
9. M. Glück, E. Reya, M. Strikmann and W. Vogelsang, *Phys. Rev.* **D53**(1996) 4775.

Determination of polarized parton distribution functions

M. Hirai *[1], H. Kobayashi †[2], and M. Miyama ‡[3]
(Asymmetry Analysis Collaboration)

*Department of Physics, Saga University, Saga 840-8502, Japan
†RIKEN BNL Research Center, Upton, NY 11973-5000, U.S.A.
‡Department of Physics, Tokyo Metropolitan University, Tokyo 192-0397, Japan

Abstract. We study parametrization of polarized parton distribution functions in the α_s leading order (LO) and in the next-to-leading order (NLO). From χ^2 fitting to the experimental data on A_1, optimum polarized distribution functions are determined. The quark spin content $\Delta\Sigma$ is very sensitive to the small-x behavior of antiquark distributions which suggests that small-x data are needed for precise determination of $\Delta\Sigma$. We propose three sets of distributions and also provide FORTRAN library for our distributions.

INTRODUCTION

Experimental data on polarized structure functions g_1 have been accumulated for the last several years. The data with the proton, deuteron, and ^{3}He targets are now available and new data are expected to be given by RHIC at BNL, COMPASS at CERN, and other facilities in the near future. In the light of such progresses and future projects, we should summarize present knowledge of polarized parton distribution functions by using all available data at this stage. For this purpose, we formed the group called Asymmetry Analysis Collaboration (AAC) and tried to determine the polarized parton distribution functions.

In this paper, we explain our analysis to determine optimum polarized distributions by χ^2 fitting to the experimental data of spin asymmetry A_1. In addition to the discussions based on the work in Ref. [1], we introduce a FORTRAN library for our distributions as a recent progress.

---

[1] 98td25@edu.cc.saga-u.ac.jp
[2] hyuki@bnl.gov
[3] miyama@comp.metro-u.ac.jp

PARAMETRIZATION

We determine initial polarized parton distributions at $Q^2 = 1.0$ GeV^2 ($\equiv Q_0^2$). Considering the counting rule, we adopt the following functional form:

$$\Delta f_i(x, Q_0^2) = A_i \, x^{\alpha_i} \left(1 + \gamma_i \, x^{\lambda_i}\right) f_i(x, Q_0^2). \tag{1}$$

Here, Δf_i and f_i represent the polarized and unpolarized parton distributions, respectively and A_i, α_i, γ_i, and λ_i are free parameters. These parameters are constrained by the positivity condition, $|\Delta f_i(x)| \leq f_i(x)$. We simply apply it not only for the leading-order (LO) case but also for the next-to-leading-order (NLO) case although it is valid only in LO. Furthermore, we assume SU(3) flavor-symmetric sea because we don't have enough data to extract flavor dependence of polarized sea. Under this assumption, the first moments of Δu_v and Δd_v can be fixed by the axial charges for octet baryon, F and D. Therefore, we should determine Δu_v, Δd_v, $\Delta \bar{q}$, and Δg distributions and total number of the parameters becomes 14.

As the experimental data to which the parameters are fitted, we chose spin asymmetry A_1 instead of g_1 since A_1 is closer to direct observable in experiment rather than g_1. A_1 is given by

$$A_1(x, Q^2) \simeq \frac{g_1(x, Q^2)}{F_1(x, Q^2)} = g_1(x, Q^2) \frac{2x[1 + R(x, Q^2)]}{F_2(x, Q^2)}, \tag{2}$$

where, R is the ratio of longitudinal to transverse cross-sections. In our analysis, A_1 is calculated by using F_2 obtained by GRV98 distributions, g_1 obtained by our parametrized distributions, and R taken from the SLAC-1990 analysis. Then, $\chi^2 = \sum [A_1^{\text{data}}(x, Q^2) - A_1^{\text{calc}}(x, Q^2)]^2 / [\sigma^{\text{data}}(x, Q^2)]^2$ is calculated and is minimized by the subroutine MINUIT. Here, A_1^{data}, A_1^{calc}, and σ^{data} indicate the experimental A_1, calculated A_1, and experimental error, respectively.

RESULTS AND DISCUSSIONS

From the χ^2 analysis, we obtain the results with $\chi^2 = 322.6$ in the LO case, and $\chi^2 = 300.4$ in the NLO case for 375 data points. The NLO χ^2 is significantly smaller than the LO one. It suggests that the NLO analysis is necessary for precise analysis. The obtained A_1 for the proton and for the neutron at $Q^2 = 5.0$ GeV^2 are shown in Fig.1 with the experimental data. Although the comparison is not straightforward because the data are taken at various Q^2, the obtained parameters reproduce the data well both in the LO and the NLO case. Figure 2 show the obtained polarized parton distribution functions at $Q^2 = 1.0$ GeV^2. In our analysis, the first moment of Δu_v (Δd_v) is fixed by positive (negative) value and the obtained distribution becomes positive (negative). Furthermore, antiquark and gluon distributions become negative and positive, respectivly.

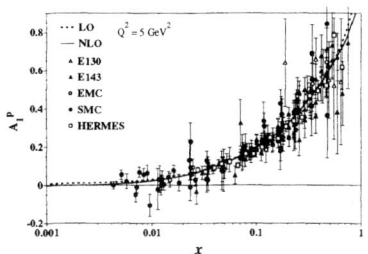 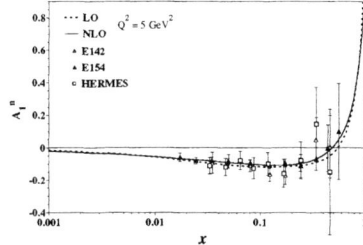

FIGURE 1. Spin asymmetries A_1 for the proton (left figure) and neutron (right figure).

For these distributions, quark spin content becomes $\Delta\Sigma = 0.201$ for LO and $\Delta\Sigma = 0.051$ for NLO at $Q^2 = 1.0$ GeV^2. The NLO $\Delta\Sigma$ is significantly smaller than the LO one. It is also smaller than the generally quoted values in other analyses that range in $0.1 \sim 0.3$. The difference of $\Delta\Sigma$ comes mainly from the difference of $\Delta\bar{q}$ in the small-x region. Because we don't have a small-x data at this stage, we should fix the small-x behavior of $\Delta\bar{q}$ from theoretical suggestions. We perform χ^2 fitting by fixing the parameter $\alpha_{\bar{q}}$, which controls the small-x behavior of $\Delta\bar{q}$, to $\alpha_{\bar{q}} = 1.0$ and 1.6 according to the predictions of the Regge theory and the perturbative QCD. These values are rather larger than the value $\alpha_{\bar{q}} = 0.32 \pm 0.22$ which is obtained by the previous NLO analysis. As a result, we get $\Delta\Sigma = 0.241$ ($\chi^2 = 305.8$) for $\alpha_{\bar{q}} = 1.0$ and $\Delta\Sigma = 0.276$ ($\chi^2 = 323.5$) for $\alpha_{\bar{q}} = 1.6$. The obtained $\Delta\Sigma$ are larger than the previous NLO result and are within the usually quoted range. From these results, we find that $\Delta\Sigma$ is very sensitive to the small-x behavior of the antiquark distributions. In order to determine the quark spin content precisely, the data in the small-x region are needed. Because χ^2 for the $\alpha_{\bar{q}} = 1.0$ case is close to the one for the previous NLO, this set can be also considered as a good parametrization. Therefore, we propose three sets of distributions, LO, NLO with free $\alpha_{\bar{q}}$ (NLO-1), and NLO with $\alpha_{\bar{q}} = 1.0$ (NLO-2), as AAC polarized distribution functions.

Finally, we briefly comment on the error estimate of the obtained distributions [2]. The uncertainties of our distributions can be estimated by using the error

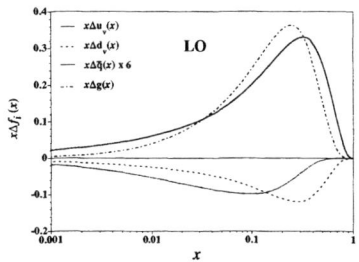 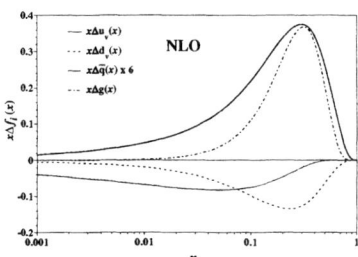

FIGURE 2. Obtained polarized distributions at $Q^2 = 1.0$ GeV^2.

matrix which is obtained by MINUIT. The 1-σ boundary of the distribution F is given by

$$(\delta F)^2 = \sum_i \sum_j \frac{\partial F}{\partial a_i} V_{ij} \frac{\partial F}{\partial a_j}, \qquad (3)$$

where, V_{ij} is the error matrix element for parameters a_i and a_j. This analysis is in progress and will be reported elsewhere.

LIBRARY

For practical application of our distributions, we provide a library program which is given as FORTRAN subroutine AACPDF(ISET, Q2, X, POLPDF, STRUCT). The users should call this subroutine in their programs with the input parameters ISET, Q2, and X. This library contains three sets of distributions, LO, NLO-1, and NLO-2, and the parameter ISET designates which set is used. When ISET=1, 2, or 3, the returned values are those in the LO, NLO-1, or NLO-2, respectively. The parameters Q2 and X specify Q^2 and Bjorken-x at which the distributions are calculated. The allowed ranges are 1.0 GeV$^2 \leq Q^2 \leq 10^6$ GeV^2 and $10^{-9} \leq x \leq 1.0$. The obtained values are returned by the arrays POLPDF(-3:3) and STRUCT(3). The available polarized parton distribution functions and the structure functions are listed in Table 1. It should be noted that the returned values are the distributions and the structure functions multiplied by x.

The distribution values are obtained by interpolating the grid data which are provided as DATA statements in the subroutine. The number of the grid points is 23 for the variable Q^2 and 68 for the variable x. Because the Q^2 dependence of the distributions is almost linear function of $t \equiv \ln Q^2$, we simply use the linear interpolation of t for the variable Q^2. If t' is in the range $t_i \leq t' < t_{i+1}$, the distributions at $t = t'$ are approximated by

$$f(t') \approx \frac{t_{i+1} - t'}{t_{i+1} - t_i} f(t_i) + \frac{t' - t_i}{t_{i+1} - t_i} f(t_{i+1}). \qquad (4)$$

TABLE 1. Available distributions and structure functions.

I	POLPDF(I)	STRUCT(I)
-3	$\bar{s}$ quark [$x\Delta\bar{s}(x,Q^2) = x\Delta s(x,Q^2)$)]	—
-2	$\bar{d}$ quark [$x\Delta\bar{d}(x,Q^2) = x\Delta d_{sea}(x,Q^2)$]	—
-1	$\bar{u}$ quark [$x\Delta\bar{u}(x,Q^2) = x\Delta u_{sea}(x,Q^2)$]	—
0	gluon [$x\Delta g(x,Q^2)$]	—
1	u-valence quark [$x\Delta u_v(x,Q^2)$]	$xg_1(x,Q^2)$ for proton
2	d-valence quark [$x\Delta d_v(x,Q^2)$]	$xg_1(x,Q^2)$ for neutron
3	s quark [$x\Delta s(x,Q^2) = x\Delta\bar{s}(x,Q^2)$]	$xg_1(x,Q^2)$ for deuteron

On the other hand, the x dependence of the distributions is rather complicated. Therefore, we use the cubic spline interpolation for the variable x. If x' is in the range $x_i \leq x' < x_{i+1}$, the distributions at $x = x'$ are approximated by

$$f(x') \approx f(x_i) + B_i(x' - x_i) + C_i(x' - x_i)^2 + D_i(x' - x_i)^3. \tag{5}$$

Here, B_i, C_i, D_i are the spline coefficients which can be calculated by the grid data. By using these interpolation methods and the grid data, the AACPDF returns the values of the polarized parton distribution functions and the structure functions g_1 at the specified Q^2 and x point. We verified that the distributions obtained by the library reproduce well the results in Ref. [1].

AAC library can be downloaded from our web page [3]. The subroutine AACPDF is in the file " aac.f ". This file also includes the subroutine SPLINE and the function ISERCH which are used in the AACPDF.

SUMMARY

The polarized parton distribution functions have been determined by χ^2 fitting to A_1 experimental data. As a result, we found that the NLO χ^2 is significantly smaller than the LO one. It implies that the NLO analysis is important. In the NLO analysis, the obtained $\Delta\Sigma$ is rather smaller than the usually quoted values, and the differences come from the small-x behavior of $\Delta\bar{q}$. In order to fix $\Delta\Sigma$ precisely, the small-x measurements are required. We propose three sets of AAC distributions, LO, NLO-1 ($\alpha_{\bar{q}}$: free), and NLO-2 ($\alpha_{\bar{q}} = 1.0$ fixed). The FORTRAN library is provided for calculating the AAC distributions numerically. This library is useful for practical application of the AAC distributions and available at our web site [3].

ACKNOWLEDGEMENTS

M.H. and M.M. were supported by the JSPS Research Fellowships for Young Scientists and by the Grant-in-Aid from the Japanese Ministry of Education, Science, and Culture. This talk is based on the work with Y. Goto, N. Hayashi, H. Horikawa, S. Kumano, T. Morii, N. Saito, T.-A. Shibata, E. Taniguchi, and T. Yamanishi.

REFERENCES

1. Asymmetry Analysis Collaboration, Y. Goto, et al., *Phys. Rev.* **D 62**, 034017 (2000).
2. Asymmetry Analysis Collaboration, research in progress.
3. http://spin.riken.bnl.gov/aac/

Solving the nucleon spin puzzle based on the chiral quark soliton model

Masashi Wakamatsu

Department of Physics, Faculty of Science
Osaka University, Toyonaka, Osaka 560, Japan

Abstract. An incomparable feature of the chiral quark soliton model as compared with many other effective models like the MIT bag model is that it can give reasonable predictions not only for quark distributions but also for antiquark distributions. This will be exemplified by the argument on the positivity constraint for $\bar{u}(x) + \bar{d}(x)$ as well as the Soffer inequality for quark and antiquark distributions. We also explain how the model can resolve the so-called "nucleon spin puzzle" without assuming large gluon polarization at the low energy scale.

INTRODUCTION

Undoubtedly, the EMC measurement in 1988 and the NMC measurement in 1991 are two of the most striking findings in the recent experimental studies of nucleon structure functions [1,2]. A prominent feature of the chiral quark soliton model (CQSM) is that it can simultaneously explain the above two big discoveries in no need of artificial fine-tuning [3,4]. What is the chiral quark soliton model, then? First of all, it is a relativistic field theoretical model effectively incorporating the idea of large N_c QCD [5]. For large enough N_c, a nucleon is thought to be a composite of N_c valence quarks and infinitely many Dirac sea quarks bound by the self-consistent pion field of hedgehog shape. After canonically quantizing the spontaneous rotational motion of the symmetry breaking mean field configuration, we can perform nonperturbative evaluation of any nucleon observables with full inclusion of valence and Dirac sea quarks [3]. It is this incomparable feature of the model that enables us to make a reasonable estimation not only of quark distributions but also of *antiquark* ones, as we shall show later. Finally, but most importantly, only 1 parameter of the model was already fixed by low energy phenomenology, which means that we can give *parameter-free predictions* for parton distributions function at the low renormalization scale.

CQSM AND TWIST-2 PDF

For obtaining quark distribution functions, we need to evaluate nucleon matrix elements of quark bilinear operators with light-cone separation. By using the path integral formulation of the CQSM, such nonlocality effects in time as well as spatial coordinates can be treated in a consistent manner [6,7].

The following novel N_c dependencies follow from the theoretical structure of the model, i.e. the mean-field approximation and the subsequent perturbative treatment of collective rotational motion [6,8] :

$$u(x) + d(x) \sim N_c \left[O(\Omega^0) + 0 \right] \sim O(N_c^1), \qquad (1)$$
$$u(x) - d(x) \sim N_c \left[0 + O(\Omega^1) \right] \sim O(N_c^0), \qquad (2)$$
$$\Delta u(x) + \Delta d(x) \sim N_c \left[0 + O(\Omega^1) \right] \sim O(N_c^0), \qquad (3)$$
$$\Delta u(x) - \Delta d(x) \sim N_c \left[O(\Omega^0) + O(\Omega^1) \right] \sim O(N_c^1) + O(N_c^0). \qquad (4)$$

Because of the peculiar spin-isospin correlation embedded in the hedgehog mean field, there is no leading-order N_c contribution to the isovector unpolarized distribution as well as to the isoscalar longitudinally polarized one, in contrast to the other combinations. This especially means that the isoscalar or flavor-singlet axial charge is parametrically smaller than the isovector one, in conformity with the EMC observation.

NUMERICAL RESULTS

In Fig.1, we summarize our parameter-free predictions for the twist-2 PDF at the model energy scale. We emphasize that *seeds* of all the success of the model are already contained in these four figures. Here, the functions in the negative x region should be interpreted as antiquark distributions according to the rule [6] :

$$u(-x) \pm d(-x) = -[\bar{u}(x) \pm \bar{d}(x)] \qquad (0 < x < 1), \qquad (5)$$
$$\Delta u(-x) \pm \Delta d(-x) = \Delta\bar{u}(x) \pm \Delta\bar{d}(x) \qquad (0 < x < 1). \qquad (6)$$

The long-dashed curves peaked around $x \simeq 1/3$ are the contributions of N_c valence quarks, while the dash-dotted curves represent those of Dirac-sea quarks. The sum of these two contributions are denoted by solid curves.

The crucial importance of the Dirac-sea contribution is most clearly seen in the isoscalar unpolarized distribution. Here, the "valence-quark-only" approximation leads to positive $u(x) + d(x)$ in the negative x region, thereby violating the positivity of the antiquark distribution [6]. On the other hand, if we include the vacuum polarization of Dirac-sea quarks, the positivity constraint for the antiquark distributions holds properly. The effect of Dirac-sea quarks is very important also for the isovector unpolarized distribution function [9,10]. Especially interesting here is the fact that $u(x) - d(x) > 0$ in the negative x region, which means that $\bar{u}(x) - \bar{d}(x) < 0$ for the physical value of x, just as required by the NMC measurement. In fact, after taking account of the scale-dependence by means of the DGLAP equation, the theory turns out to successfully explain the NMC data for the unpolarized nucleon structure functions $F_2^p(x)$ and $F_2^n(x)$ [11].

Turning to the longitudinally polarized distributions, one observes very different x dependencies between the isoscalar and isovector ones. One interesting feature of the isoscalar distribution is its sign change in the small x region. It has been shown that this sign change is just what is required by the recent experimental data for the longitudinally polarized structure functions of the deuteron [11]. Turning to the isovector distribution, we notice that the effect of Dirac-sea quarks has a peak of positive sign around $x \simeq 0$. What is remarkable here is the positivity in the negative x region. It means that anti-quark distributions are isospin asymmetric also for the longitudinally polarized distributions [11]. It is interesting to point out that some support is already given to this unique prediction of the CQSM by several semi-phenomenological and/or semi-theoretical analyses [12].

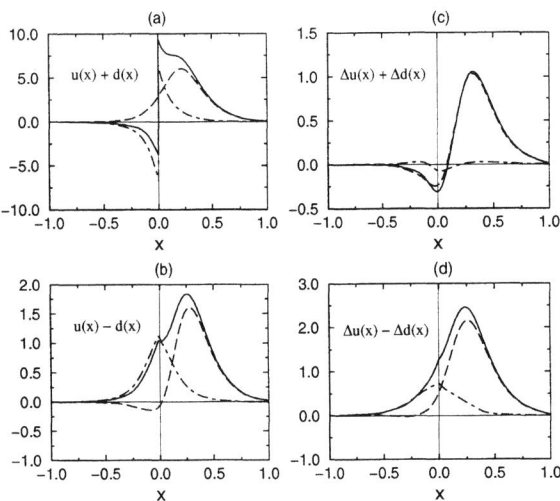

FIGURE 1. The theoretical predictions of the CQSM for the unpolarized distributions $u(x) + d(x)$ and $u(x) - d(x)$ as well as for the longitudinally polarized distributions $\Delta u(x) + \Delta d(x)$ and $\Delta u(x) - \Delta d(x)$.

To complete the list of twist-2 PDF, we need another distribution function $\delta q(x)$, usually called the transversity distribution. It is known that this distribution function must satisfy the so-called Soffer inequality [13] :

$$| \pm \delta q(x)| \leq \frac{1}{2} \left(\pm q(x) + \Delta q(x) \right) \qquad (x > 0, \ x < 0). \tag{7}$$

Now the question is whether the predictions of the CQSM fulfill this inequality or not. Fig.2 show that, if one includes the vacuum polarization effects properly, the Soffer inequality is well satisfied for both of u-quark and d-quark. On the other hand, if one ignores the Dirac-sea contributions, the Soffer inequality is badly broken for the antiquark distributions. An important lesson learned from this observation is that the field theoretical nature of the model, that is, the proper inclusion of the vacuum polarization effects, plays essential roles in giving reasonable predictions for antiquark distributions. Another lesson is that the frequently-used saturation Ansatz of the Soffer inequality for estimating $\delta q(x)$ is not justified.

Also noteworthy is another consequence of the soliton picture of the nucleon. Shown in Fig.3 are the spin and the orbital angular momentum distribution functions at the model energy scale [14]. One notices that the Dirac-sea contribution to the orbital angular momentum distribution function is sizably large and peaked around $x \simeq 0$. Among others, large support in the negative x region suggests that

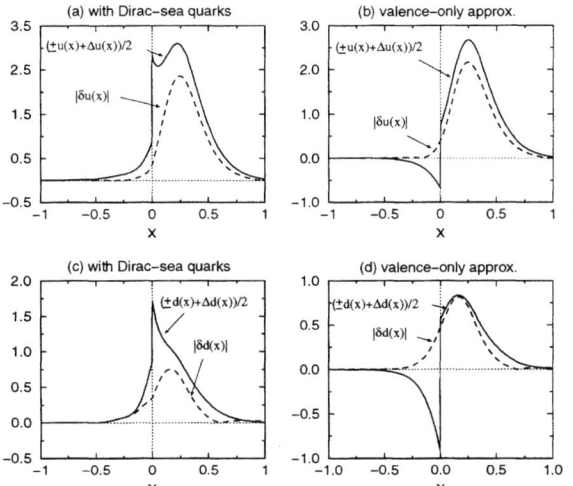

FIGURE 2. The theoretical check of Soffer inequality. The distributions in the negative x region denote the antiquark distributions.

seizable amount of orbital angular momentum is carried by antiquarks. After integration over x, one also finds that only about 35% of the total nucleon spin comes from the quark spin, while the remaining 65% is due to the orbital angular momentum of quark and antiquarks [14,3]. It is interesting to see that the dominance of the orbital angular momentum part over the intrinsic spin one is also indicated by the recent lattice QCD simulation [15].

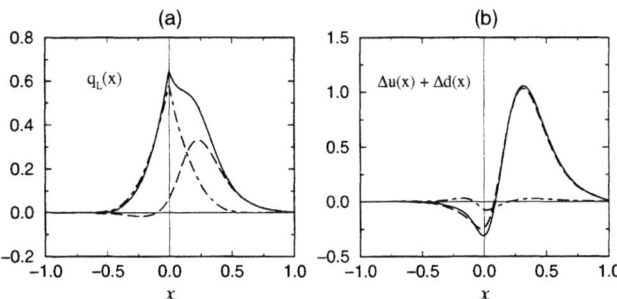

FIGURE 3. (a) The theoretical predictions of the CQSM for the quark and antiquark orbital angular momentum distribution functions $q_L(x)$ and (b) the isosinglet quark polarization $\Delta u(x) + \Delta d(x)$. The curves have the same meaning as in Fig.1.

The spin and orbital angular momentum contents of the nucleon are of course scale-dependent quantities. We recall that, at the NLO with the gauge-invariant factorization scheme, $\Delta \Sigma$ has a weak scale dependence mainly at low Q^2. The theoretical value $\Delta \Sigma = 0.31$ obtained at $Q^2 = 10\,\text{GeV}^2$ is qualitatively consistent with the recent SMC result, $\Delta\Sigma_{SMC}^{exp} = 0.22 \pm 0.17$ [16].

CONCLUSION

In summary, an incomparable feature of the CQSM as compared with many other effective models like the MIT bag model is that it can give reasonable predictions also for the *antiquark distribution functions* as exemplified by the argument on the positivity constraint for $\bar{u}(x)+\bar{d}(x)$ and also on the Soffer inequality for antiquarks. It has been emphasized that parton distribution functions evaluated at the model energy scale contain all the *seeds* of the success of the model in explaining existing experimental data given at the high energy scale. It naturally explains the excess of $\bar{d}$ sea over the $\bar{u}$ sea in the proton. The most puzzling observation, i.e. unexpectedly small quark spin fraction of the nucleon can also be explained in no need of large gluon polarization at the low energy scale.

As a further unique prediction of the model, we pointed out the possibility of large *isospin asymmetry* of the *spin-dependent sea-quark distributions*, which seems to be a natural consequence of the large N_c-counting rule, but appears inconsistent with the naive "meson cloud convolution model". Then, if this large asymmetry of the longitudinally polarized sea is experimentally established, it would offer a strong evidence in favor of nontrivial spin-isospin correlation imbedded in the "large N_c chiral soliton picture" of the nucleon.

The talk is based on the collaborations with T. Watabe and T. Kubota.

REFERENCES

1. EMC Collaboration, J. Ashman et al., Phys. Lett. **B206** (1988) 364-370 ; Nucl. Phys. **B328** (1989) 1-35.
2. NMC Collaboration, P. Amaudruz et al., Phys. Rev. Lett. **66**, 2712 (1991).
3. M. Wakamatsu and H. Yoshiki, Nucl. Phys. **A524** (1991) 561.
4. M. Wakamatsu, Phys. Rev. **D46** (1992) 3762.
5. D.I. Diakonov, V.Yu. Petrov, and P.V. Pobylitsa, Nucl. Phys. **B306** (1988) 809.
6. D.I. Diakonov, V.Yu. Petrov, P.V. Pobylitsa, M.V. Polyakov, and C. Weiss, Nucl. Phys. **B480** (1996) 341 ; *ibid.*, Phys. Rev. **D56** (1997) 4069.
7. M. Wakamatsu and T. Kubota, Phys. Rev. **D60** (1999) 034020.
8. M. Wakamatsu and T. Watabe, Phys. Lett. **B312** (1993) 184.
9. M. Wakamatsu and T. Kubota, Phys. Rev. **D57** (1998) 5755.
10. P.V. Pobylitsa, M.V. Polyakov, K. Goeke, T. Watabe and C. Weiss, Phys. Rev. **D59** (1999) 034024.
11. M. Wakamatsu and T. Watabe, Phys. Rev. **D62** (2000) 017506.
12. T. Morii and T. Yamanishi, Phys. Rev. **D61** (2000) 057501 ; M. Glück and E. Reya, Mod. Phys. Lett. **A15** (2000) 883 ; D.de Florian and R. Sassot, Phys. Rev. **D62** (2000) 094025 ; R.S. Bhalerao, hep-ph/0003075.
13. J. Soffer, Phys. Rev. Lett. **74** (1995) 1292.
14. M. Wakamatsu and T. Watabe, Phys. Rev. **D62** (2000) 054009.
15. N. Mathur, S.J. Dong, K.F. Liu, L. Mankiewics, and N.C. Mukhopadhyay, hep-ph/9912289.
16. SMC Collaboration, B. Adeva et al., Phys. Rev. **D58** (1998) 112001.

4. SPIN PHYSICS AND HADRONS

Measurement of Λ^0 polarization in ν_μ CC interactions in NOMAD

Dmitry V. Naumov
for the NOMAD Collaboration

Laboratory of Nuclear Problems, Joint Institute for Nuclear Research, 141980, Dubna, Russia
e-mail: naumov@thsun1.jinr.ru

Abstract. The Λ^0 polarization in ν_μ charged current interactions has been measured in the NOMAD experiment. The event sample (8087 reconstructed Λ^0's) is more than an order of magnitude larger than that of previous bubble chamber experiments, while the quality of event reconstruction is comparable. We observe negative polarization along the W-boson direction which is enhanced in the target fragmentation region: $P_x(x_F < 0) = -0.21 \pm 0.04(\text{stat}) \pm 0.02(\text{sys})$. In the current fragmentation region we find $P_x(x_F > 0) = -0.09 \pm 0.06(\text{stat}) \pm 0.03(\text{sys})$. These results provide a test of different models describing the nucleon spin composition and the spin transfer mechanisms. A significant transverse polarization (in the direction orthogonal to the Λ^0 production plane) has been observed for the first time in a neutrino experiment: $P_y = -0.22 \pm 0.03(\text{stat}) \pm 0.01(\text{sys})$. The dependence of the absolute value of P_y on the Λ^0 transverse momentum with respect to the hadronic jet direction is in qualitative agreement with the results from unpolarized hadron-hadron experiments.

I INTRODUCTION

A study of the Λ^0 polarization in (anti)neutrino nucleon DIS is motivated by several reasons. First, a possible longitudinal polarization in the *target fragmentation region* can be related to the polarized nucleon strangeness which is a conclusion (in the framework of quark parton model assuming $SU(3)_F$ symmetry and vanishing gluon spin contribution to the nucleon spin) of DIS experiments with both polarized beam and target [1]. The authors [2] attempt to explain the negative sign of the nucleon strangeness and predict that it could manifest itself as a *negative* (aligned in the opposite direction to the W exchange boson) longitudinal polarization of Λ^0 hyperons produced in (anti)neutrino nucleon DIS. Second, a measurement of the longitudinal polarization in the *current fragmentation region* provides a test of different models for the Λ^0 spin structure [3–5] with a *clean flavour separation* provided by the nature of neutrino interactions. Last but not least, transverse polarization of Λ^0 hyperons has been observed for a long time in unpolarized hadron-hadron experiments [6], and was never observed in (anti)neutrino nucleon DIS experiments [7]. This surprising feature challenges experimental and theoretical efforts in this field. The neutrino nucleon DIS is again exceptional due to the different Λ^0 production mechanisms in the target and current fragmentation regions in contrast to pp scattering. The quark and di-quark fragmentations are believed to be the subject of the current and target fragmentation regions respectively. Therefore, it is possible to study the transverse polarization in connection with the different mechanisms of Λ^0 production.

II EXPERIMENTAL PROCEDURE AND ANALYSIS

The active part of the NOMAD detector consists of 44 drift chambers located in a 0.4 Tesla magnetic field. The drift chambers serve as a nearly isoscalar target for neutrino interactions and as a tracking medium. These drift chambers provide an overall efficiency for charged track reconstruction of better than 95% and a momentum resolution of approximately 3.5% in the momentum range of interest (less than 10 GeV/c). Reconstructed tracks are used to determine the event topology (the assignment of tracks to vertices), to reconstruct the vertex position and the track parameters at each vertex and, finally, to identify the vertex type (primary, secondary, V^0, etc.).

Λ^0 hyperons appear in the detector as two charged tracks with opposite charges emerging from a common vertex separated from the primary interaction vertex (V^0-like signature). These events correspond to $\Lambda^0 \to p\pi^-$ decay. The background to Λ^0 decays consists of $K_S^0 \to \pi^+\pi^-$, $\bar\Lambda^0 \to \bar p\pi^+$ decays, $\gamma \to e^+e^-$ conversions, and random combinations of tracks wrongly labeled as V^0s. To identify Λ^0 hyperons we first apply some quality cuts to reject as much as possible of the combinatorial background, γ-conversions, and secondary interactions, we then perform a kinematic fit with energy and momentum constraints for each V^0 for the final resolution of V^0-like particles. As a result we obtained 8087 reconstructed and identified Λ^0 hyperons with about 4% background contamination in our data sample [8]. This sample is used for the polarization analysis reported below.

The Λ^0 polarization is measured through the *asymmetry* in the angular distribution of the protons in the parity violating decay process $\Lambda^0 \to p\pi^-$. In the Λ^0 rest frame the decay protons are distributed as: $\frac{1}{N}\frac{dN}{d\Omega} = \frac{1}{4\pi}(1 + \alpha_\Lambda \mathbf{P} \cdot \mathbf{k})$, where $\mathbf{P}$ is the Λ^0 polarization vector, $\alpha_\Lambda = 0.642 \pm 0.013$ [9] is the decay asymmetry parameter and $\mathbf{k}$ is the unit vector along the decay proton direction. A fit of the raw angular distributions of the decay protons in the data can only be performed after correction for the detector acceptance. To take into account the detector acceptance, and smearing of the angular distributions we developed a new 3-dimensional method for the polarization analysis [8]. We used its 1-dimensional option for the results reported below because of its better applicability to the samples with low statistics as is the case in the study of the polarization dependence on different kinematic variables.

The axes are defined as follows (in the Λ^0 rest frame):

- $\mathbf{n_x} = \mathbf{e_W}$, where $\mathbf{e_W}$ is the reconstructed W-boson direction;

- $n_y = e_W \times e_T / |e_W \times e_T|$ axis is orthogonal to the Λ^0 production plane
- $n_z = n_x \times n_y$.

III RESULTS AND DISCUSSION

Table 1 displays the results for the polarization of Λ^0 hyperons in our sample as a function of x_F. We observe negative longitudinal ("P_x") and transverse ("P_y") polarizations of Λ^0's which are enhanced in the target fragmentation region. Note that transverse polarization has never been observed before in (anti)neutrino nucleon DIS experiments. It is believed that the origin of Λ^0 polarization is different in the target and in the current fragmentation regions. Therefore it can be useful to study Λ^0 polarization at $x_F < 0$ and $x_F > 0$ separately.

TABLE 1. Dependence of the Λ^0 polarization on x_F in ν_μ CC events (statistical errors only).

Selection	Entries	$<x_F>$	Λ^0 Polarization		
			P_x	P_y	P_z
full sample	8087	−0.18	−0.15 ± 0.03	−0.22 ± 0.03	−0.04 ± 0.03
$x_F < 0$	5608	−0.36	−0.21 ± 0.04	−0.26 ± 0.04	−0.08 ± 0.04
$x_F > 0$	2479	0.21	−0.09 ± 0.06	−0.10 ± 0.06	0.02 ± 0.06

A Target fragmentation region

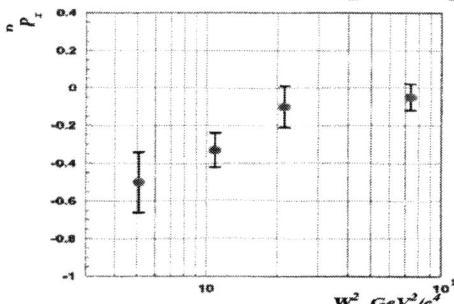
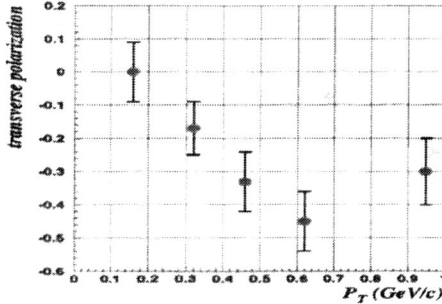

FIGURE 1. Longitudinal polarization of Λ^0 as a function of W^2 for $x_F < 0$

FIGURE 2. Transverse polarization of Λ^0 as a function of P_T for $x_F < 0$

1 Longitudinal polarization

The dependence of the longitudinal polarization of Λ^0 on W^2 at $x_F < 0$ is shown in Fig. 1. Large negative P_x is observed at small W^2, while at larger W^2 the longitudinal polarization vanishes. Such an effect can be interpreted as a manifestation of the polarized nucleon strangeness due to the larger probability for Λ^0 at small W^2 to include an s−quark originally present in the nucleon, while at larger W^2 the s−quarks (presumably unpolarized) are also created in the fragmentation process. The same dependence of P_x on Q^2 is observed.

2 Transverse polarization

We have performed a study of the dependence of the transverse polarization on the Λ^0 transverse momentum with respect to the jet direction (p_T) in the target fragmentation region and found it to be in qualitative agreement (both sign and shape) with that found in unpolarized hadron-hadron collisions [6]. Also, we observed no dependence of P_y on W^2. These features make possible to conclude that the origin of the transverse polarization is in the fragmentation process.

B Current fragmentation region

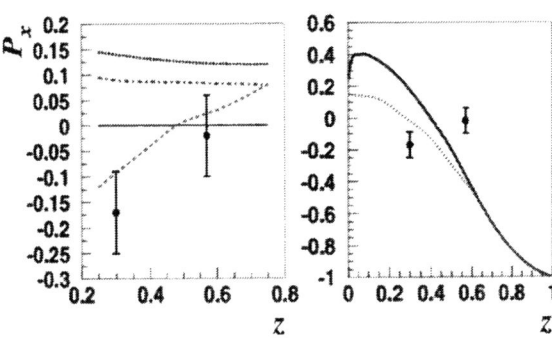
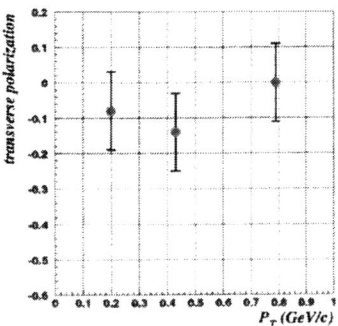

FIGURE 3. Longitudinal polarization of Λ^0 as a function of z in comparison with predictions from [4] (left) and [5] (right)

FIGURE 4. Transverse polarization of Λ^0 for $x_F > 0$ as a function of P_T

1 Longitudinal polarization

Measurement of the longitudinal Λ^0 polarization in the current fragmentation region provides a test of different models for the Λ^0 spin structure. A comparison of our data to theoretical calculations [4] and [5] performed for different models of the Λ^0 spin content (details can be found in [4,5]) is presented in Fig. 3. One can draw the conclusion that naive quark parton model is favoured by our measurement while still some progress in theoretical calculations is expected.

2 Transverse polarization

Transverse polarization of Λ^0 in the current fragmentation region in neutrino-nucleon DIS is related to the quark fragmentation processes, therefore it is crucial to look for its p_T dependence. Fig. 4 displays such a dependence.

C Target nucleon effects

Imposing a cut on the total charge (Q_{tot}) of the event we can study Λ^0 polarization on different target nucleons. We select $\nu_\mu p$ ($\nu_\mu n$)-like events requiring $Q_{tot} \geq 1$

($Q_{tot} \leq 0$) with purity of the selection 76% (85%). The results are summarized in Table. 2. There is a strong dependence of the polarization vector on the target nucleon. We attribute it to the different contribution of the polarization transfer from Σ^*, Ξ, Σ^0 to Λ^0 during their decays into Λ^0 in the final state.

TABLE 2. *The dependence of the Λ^0 polarization on the type of target nucleon.*

Target	Entries	Λ^0 Polarization		
		P_x	P_y	P_z
"proton"	3472	-0.26 ± 0.05	-0.09 ± 0.05	-0.07 ± 0.05
$x_F < 0$	2407	-0.29 ± 0.06	-0.10 ± 0.06	-0.09 ± 0.06
$x_F > 0$	1065	-0.23 ± 0.09	-0.06 ± 0.09	-0.02 ± 0.10
"neutron"	4615	-0.09 ± 0.04	-0.30 ± 0.04	-0.03 ± 0.05
$x_F < 0$	3201	-0.16 ± 0.05	-0.37 ± 0.05	-0.07 ± 0.05
$x_F > 0$	1414	0.01 ± 0.08	-0.11 ± 0.08	0.04 ± 0.09

IV CONCLUSION

Our measurements indicate for many interesting phenomena are hidden in the nucleon and in the fragmentation process.

ACKNOWLEDGEMENTS

Many thanks to the SPIN2000 organizers for making possible me to participate in the symposium. Special thanks to the NOMAD collaborators making possible this work to appear.

REFERENCES

1. J.Ashman et al., [EMC Collaboration], *Phys. Lett.* **B206** (1988) 364; *Nucl. Phys.* **B328** (1989) 1; D.Adams et al. [SMC Collaboration], *Phys. Rev.* **D56** (1997) 5330; B.Adeva et al., [SMC Collaboration], *Phys. Lett.* **B420** (1998) 180; K.Abe et al., [E143 Collaboration], *Phys. Rev.* **D58** (1998) 112003
2. J.Ellis, D.Kharzeev, A.Kotzinian, *Z. Phys.* **C69** (1996) 467; J.Ellis, M.Karliner, D.E.Kharzeev and M.G.Sapozhnikov, hep-ph/9909235
3. M.Burkardt and R.L.Jaffe, *Phys. Rev. Lett.* **70** (1993) 2537; R.L.Jaffe, *Phys. Rev.* **D 54** (1996) R6581 C.Boros and Z-t. Liang, *Phys. Rev.* **D57** (1998) 4491; D.de Florian, M.Stratmann and W.Vogelsang,*Phys. Rev. Lett.* **81** (1998) 530; B.Ma and J.Soffer, *Phys. Rev. Lett.* **82** (1999) 2250
4. A.Kotzinian, A.Bravar, D.von Harach, *Eur. Phys. J.* **C2** (1998) 329
5. B.Ma, I.Schmidt, J.Soffer and J.Yang, hep-ph/0001259
6. see review J.Félix, *Mod. Phys. Lett.* **A14** (1999) 827
7. G.T.Jones et al., *Z. Phys.* **C28** (1985) 23; S.Willocq et al., *Z. Phys.* **C53** (1992) 207; D.DeProspo et al., *Phys. Rev.* **D50** (1994) 6691; V.Ammosov et al., *Nucl. Phys.* **B162** (1980) 205; D.Allasia et al., *Nucl. Phys.* **B224** (1983) 1
8. P.Astier et al., [NOMAD Collaboration], *Nucl. Phys.* **B588** (2000) 3
9. Review of Particle Properties, *Eur. Phys. J.* **C15** (2000)

Hyperon Polarization in Inclusive Hadronic Production

Y. Kanazawa and Yuji Koike

Department of Physics, Niigata University, Ikarashi, Niigata 950-2181, Japan

Abstract. A QCD formula for the polarization in the large-p_T Λ hyperon production in the unpolarized nucleon-nucleon collision at large x_F is derived. We focus on the mechanism in which the chiral-odd spin-independent twist-3 quark distribution $E_F(x,x)$ becomes the source of the transversely polarized quarks fragmenting into the polarized Λ. A simple model estimate for that contribution shows the possibility that it gives rise to a sizable Λ polarization.

It is a well known experimental fact that the hyperons produced in the unpolarized nucleon-nucleon collisions are polarized transversely to the production plane [1,2]. In this letter we focus on the polarization of the Λ hyperon production with large transverse momentum in pp collision

$$N(P) + N'(P') \to \Lambda(l, \vec{S}_\perp) + X. \tag{1}$$

Ongoing experiment at RHIC is expected to provide more data on the polarization. The nonzero Λ polarization in this process requires a presence of particular quark-gluon correlation (higher twist effect) and/or the effect of transverse momentum either in the unpolarized nucleon or the fragmentation function for Λ. According to the generalized QCD factorization theorem, the polarized cross section for this process consists of two kinds of twist-3 contributions:

$$(A) \quad E_a(x_1, x_2) \otimes q_b(x') \otimes \delta D_{c \to \Lambda}(z) \otimes \hat{\sigma}_{ab \to c}, \tag{2}$$

$$(B) \quad q_a(x) \otimes q_b(x') \otimes D^{(3)}_{c \to \Lambda}(z_1, z_2) \otimes \hat{\sigma}'_{ab \to c}. \tag{3}$$

Here the functions $E_a(x_1, x_2)$ and $D^{(3)}_{c \to \Lambda}(z_1, z_2)$ are the twist-3 quantities representing, respectively, the unpolarized distribution in the nucleon and the fragmentation function for the transversely polarized Λ hyperon, and a, b and c stand for the parton's species. Other functions are twist-2; $q_b(x)$ the unpolarized distribution (quark or gluon) and $\delta D_{c \to \Lambda}(z)$ the transversity fragmentation function for Λ. The symbol $\otimes$ denotes convolution. $\hat{\sigma}_{ab \to c}$ *etc* represents the partonic cross section for the process $a + b \to c + anything$ which yields large transverse momentum of the parton c.

Note that (A) contains two chiral-odd functions E_a and $\delta D_{c\to\Lambda}$, while (B) contains only chiral-even functions.

In this report, we derive a QCD formula for the polarized cross section (1) from the (A) term in the kinematic region $|x_F| \to 1$, using the valence quark-soft gluon approximation proposed by Qiu and Stermann [3]. Employing this approximation, they reproduced the E704 data for the single-transverse spin asymmetries in the pion production at $x_F \to 1$ reasonably well. The fact that the perturbative QCD description for the pion production is valid as low as $l_T \sim 1$ GeV encouraged us to apply the method to the polarized Λ hyperon production (1) for which the data exist only in the relatively small l_T region. At large $x_F > 0$, which mainly probes large x and small x' region, the cross section is dominated by the particular terms in (A) which contain the derivatives of the *valence* twist-3 distribution $E_{Fa}(x,x)$. The reason for this observation is the relation $|\frac{\partial}{\partial x} E_{Fa}(x,x)| \gg E_{Fa}(x,x)$ owing to the behavior of $E_{Fa}(x,x) \sim (1-x)^\beta$ ($\beta > 0$) at $x \to 1$. We thus keep only the terms with the derivative of E_{Fa} for the valence quark (*valence quark-soft gluon approximation*).

The polarized cross section for (1) is a function of three independent variables, $S = (P+P')^2 \simeq 2P\cdot P'$, $x_F = 2l_\parallel/\sqrt{S}$ ($= (T-U)/S$), and $x_T = 2l_T/\sqrt{S}$. $T = (P-l)^2 \simeq -2P\cdot l$ and $U = (P'-l)^2 \simeq -2P'\cdot l$ are given in terms of these three variables by $T = -S\left[\sqrt{x_F^2 + x_T^2} - x_F\right]/2$ and $U = -S\left[\sqrt{x_F^2 + x_T^2} + x_F\right]/2$. In this convention, production of Λ in the forward hemisphere in the direction of the incident nucleon ($N(P)$) corresponds to $x_F > 0$. Since $-1 < x_F < 1$, $0 < x_T < 1$ and $\sqrt{x_F^2 + x_T^2} < 1$, $x_F \to 1$ corresponds to the region with $-U \sim S$ and $T \sim 0$.

In the valence quark-soft gluon approximation, the cross section for the (A) term reads,

$$E_l \frac{d^3 \Delta\sigma^A(S_\perp)}{dl^3} = \frac{\pi M \alpha_s^2}{S} \sum_{a,c} \int_{z_{min}}^1 \frac{dz}{z^3} \delta D_{c\to\Lambda}(z) \int_{x_{min}}^1 \frac{dx}{x} \frac{1}{xS + U/z}$$
$$\times \int_0^1 \frac{dx'}{x'} \delta\left(x' + \frac{xT/z}{xS + U/z}\right) \varepsilon_{lS_\perp pn}\left(\frac{1}{-\hat{u}}\right)\left[-x\frac{\partial}{\partial x} E_{Fa}(x,x)\right]$$
$$\times \left[G(x')\delta\hat{\sigma}_{ag\to c} + \sum_b q_b(x')\delta\hat{\sigma}_{ab\to c}\right], \qquad (4)$$

where p and n are the two light-like vectors defined from the momentum of the unpolarized nucleon as $P = p + M^2 n/2$, $p\cdot n = 1$ and $\varepsilon_{lS_\perp pn} = \varepsilon_{\mu\nu\lambda\sigma} l^\mu S_\perp^\nu p^\lambda n^\sigma \sim \sin\phi$ with ϕ the azimuthal angle between the spin vector of the Λ hyperon and the production plane. The invariants in the parton level are defined as $\hat{s} = (p_a + p_b)^2 \simeq (xP + x'P')^2 \simeq xx'S$, $\hat{t} = (p_a - p_c)^2 \simeq (xP - l/z)^2 \simeq xT/z$, $\hat{u} = (p_b - p_c)^2 \simeq (x'P' - l/z)^2 \simeq x'U/z$. The lower limits for the integration variables are $z_{min} = \frac{-(T+U)}{S} = \sqrt{x_F^2 + x_T^2}$ and $x_{min} = \frac{-U/z}{S+T/z}$. $q_b(x')$ is the unpolarized quark distribution, and $G(x')$ is the unpolarized gluon distribution. $\delta\hat{\sigma}_{ag\to c}$ and $\delta\hat{\sigma}_{ab\to c}$ are partonic cross sections for the quark-gluon and quark-quark processes, respectively. $E_F(x,x)$ is the soft gluon component of the unpolarized twist-3 distribution defined as

$$E_{Fa}(x,x) = \frac{-i}{2M} \int \frac{d\lambda}{2\pi} e^{i\lambda x} \langle P|\bar{\psi}^a(0) \not{n} \gamma_{\perp\sigma} \left\{ \int \frac{d\mu}{2\pi} gF^{\sigma\beta}(\mu n)n_\beta \right\} \psi^a(\lambda n)|P\rangle. \tag{5}$$

The summation for the flavor indices of $E_{Fa}(x,x)$ is to be over u- and d- valence quarks, while that for the twist-2 distributions is over u, d, $\bar{u}$, $\bar{d}$, s, $\bar{s}$. $\delta\hat{\sigma}_{ab\to c}$ and $\delta\hat{\sigma}_{ag\to c}$ can be obtained from the $2\to 2$ cut diagrams. The result reads

$$\delta\hat{\sigma}_{qq'\to q} = \left(\frac{\hat{s}\hat{u}}{\hat{t}^2}\right)\left[\frac{2}{9}+\frac{1}{9}\left(1+\frac{\hat{u}}{\hat{t}}\right)\right], \quad \delta\hat{\sigma}_{q\bar{q}'\to q} = \left(\frac{\hat{s}\hat{u}}{\hat{t}^2}\right)\left[\frac{7}{9}+\frac{1}{9}\left(1+\frac{\hat{u}}{\hat{t}}\right)\right],$$

$$\delta\hat{\sigma}_{qq\to q} = -\left(\frac{\hat{s}}{\hat{t}}\right)\left[\frac{10}{27}+\frac{1}{27}\left(1+\frac{\hat{u}}{\hat{t}}\right)\right], \tag{6}$$

for $\delta\hat{\sigma}_{ab\to c}$, and

$$\delta\hat{\sigma}_{ag\to c} = \frac{9}{8}\left(\frac{\hat{s}\hat{u}}{\hat{t}^2}\right) + \frac{9}{8}\left(\frac{\hat{u}}{\hat{t}}\right) + \frac{1}{8} + \left[\frac{1}{4}\left(\frac{\hat{s}\hat{u}}{\hat{t}^2}\right) + \frac{1}{72}\right]\left(1+\frac{\hat{u}}{\hat{t}}\right). \tag{7}$$

We now present a simple estimate of the Λ polarization. To this end we use a model for $E_F(x,x)$ introduced in Ref. [7]. It is based on the comparison of the explicit form (5) with the transversity distribution

$$\delta q_a(x) = \frac{i}{2}\varepsilon_{S_\perp \sigma pn} \int \frac{d\lambda}{2\pi} e^{i\lambda x} \langle PS|\bar{\psi}^a(0) \not{n}\gamma_\perp^\sigma \psi^a(\lambda n)|PS\rangle, \tag{8}$$

where $\varepsilon_{S_\perp \sigma pn} \equiv \varepsilon_{\mu\sigma\nu\lambda}S_\perp^\mu p^\nu n^\lambda$. We make an ansatz

$$E_{Fa}(x,x) = K_a \delta q_a(x), \tag{9}$$

with a flavor-dependent parameter K_a which simulates the effect of the gluon field with zero momentum in $E_F(x,x)$. We note that even though $E_F(x,x)$ is an unpolarized distribution, the quarks in $E_F(x,x)$ is "transversely polarized" which eventually fragments into the transversely polarized Λ. The relation (9) is in parallel with the ansatz originally introduced in [3]

$$G_{Fa}(x,x) = K_a q_a(x). \tag{10}$$

Here we assume the same parameter for K_a in (9) and (10) [7]. Since we are only interested in the estimate of order of magnitude, we do not pay much attention to the scale dependence of each distribution and fragmentation function and use those functions at the scale $1 \sim 2$ GeV which is a typical size of the transverse momentum of the produced Λ and pion. For the unpolarized distribution $q_a(x)$ and $G(x)$, we use the GRV LO distribution at the scale $\mu = 1.1$ GeV $(= l_T$ of Λ in R608 data

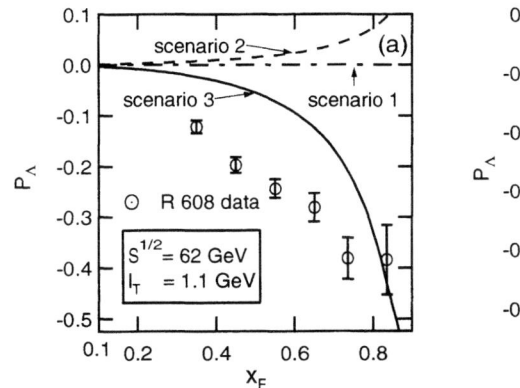 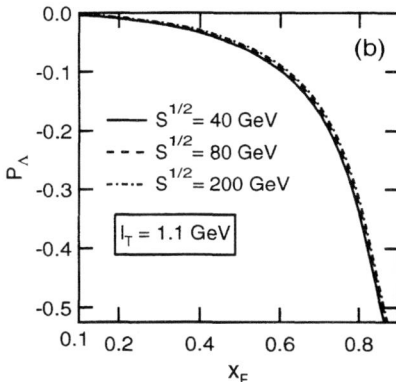

FIGURE 1. (a): The R608 data for the polarization P_Λ of the produced Λ hyperon in the unpolarized proton-proton collision at $\sqrt{S} = 62$ GeV [2]. The transverse momentum of the Λ is $l_T = 1.1$ GeV. The curves show the calculated polarization with three scenarios for $\delta D_{c\to\Lambda}$. (b): The Λ polarization in the unpolarized proton-proton collision at $\sqrt{S} = 40, 80, 200$ GeV with $l_T = 1.1$ GeV. Scenario 3 is used for $\delta D_{c\to\Lambda}$.

below) [4]. For the transversity distribution $\delta q_a(x)$, we use the GRSV helicity distribution $\Delta q_a(x)$ (LO, standard scenario) [5] assuming $\delta q_a(x) = \Delta q_a(x)$ at the scale $\mu = 1.1$ GeV. For the fragmentation function of the unpolarized and transversely polarized Λ, we use, respectively, the fragmentation function of the unpolarized and longitudinally polarized Λ, $D_{c\to\Lambda}(z)$ and $\Delta D_{c\to\Lambda}(z)$ (three scenarios for ΔD), given by de Florian et al. [6] with the assumption that $\delta D_{c\to\Lambda}(z) = \Delta D_{c\to\Lambda}(z)$ at the scale $\mu = 1.1$ GeV. Following our recent paper [7], we determine $K_{u,d}$ to fit the FNAL E704 data of the single-transverse spin asymmetry in the pion production [8] using $G_{Fa}(x,x)$ with (10) at the scale $\mu = 1.5$ GeV and the fragmentation function of the pion given at $\mu = 2$ GeV in [9]. The result is $K_u = -K_d = 0.06$. The obtained Λ polarization is shown in Fig. 1(a) for the three scenarios of $\delta D_{c\to\Lambda}(z)$ in [6] with the CERN R608 data [2]. The scenario 1 corresponds to the expectation from the naive non-relativistic quark model, where only strange quarks can fragment into a polarized Λ. In our approximation, $E_F(x,x) = 0$ for the s-quark and thus the polarization is zero in this scenario. The scenario 2 is based on the assumption that the flavor-dependence of $\Delta D_{c\to\Lambda}(z)$ is the same as that of the polarized structure function g_1^Λ as proposed in Ref. [10]. In the scenario 3, three flavors of quarks equally fragment into the polarized Λ; $\Delta D_{u\to\Lambda} = \Delta D_{d\to\Lambda} = \Delta D_{s\to\Lambda}$. As is clear from Fig. 1(a), the scenario 3 gives rise to increasing polarization at large x_F comparable to the data, while the scenario 2 gives small polarization with opposite sign. For a complete understanding on the hyperon polarization, combined analysis of both (A) and (B) contributions is necessary.

In Fig. 1(b), we plotted the polarization from the term (A) for various values of

$\sqrt{S}$ at $l_T = 1.1$ GeV with scenario 3 for $\delta D_{c\to\Lambda}$. One sees that the result is almost independent of the value of $\sqrt{S}$ in this kinematic region. This tendency is the same as the experimental data.

A different approach to the Λ polarization introduces the so-called T-odd distribution or fragmentation functions with the intrinsic transverse momentum instead of twist-3 distributions introduced here. Similarly to (A) and (B), this approach starts from the factorization assumption for the two types of contributions to the polarization; (i) $h_1^\perp(x, \mathbf{p}_\perp) \otimes q(x') \otimes \delta D(z) \otimes \hat{\sigma}$, (ii) $q(x) \otimes q(x') \otimes D_{1T}^\perp(z, \mathbf{k}_\perp) \otimes \hat{\sigma}'$, where $h_1^\perp$ represents distribution of a transversely polarized quark with nonzero tranverse mometum inside the unpolarized nucleon, and $D_{1T}^\perp$ represents a fragmentation function for an unpolarized quark fragmenting into a transversely polarized Λ with the transverse momentum ("polarizing function"). Anselmino *et al.* fitted the experimental data for the Λ polarization assuming the above (ii) is the sole origin of the polarization [11]. We expect from the present study, however, that the large portion of the Λ polarization should be ascribed to the twist-3 distribution in the unpolarized nucleon and $\delta D(z)$ which should be related to the above contribution (i). It is interesting to explore the connection between the present approach and that in [11].

To summarize, we have derived a cross section formula for the polarized Λ production in the unpolarized nucleon-nucleon collision at large x_F. A simple model estimate for this contribution suggests a possibility that the contribution from the soft gluon pole gives sizable Λ polarization.

REFERENCES

1. G. Bunce et al., Phys. Rev. Lett. **36** (1976) 1113; K. Heller et al., Phys. Rev. Lett. **41** (1978) 607; E. J. Ramberg et al., Phys. Lett. **B338** (1994) 403; P. M. Ho et al., Phys. Rev. Lett. **65** (1990) 1713; A. Morelos et al., Phys. Rev. Lett. **71** (1993) 2172.
2. A. M. Smith et al., Phys. Lett. **B185** (1987) 209.
3. J. Qiu and G. Sterman, Phys. Rev. **D59** (1999) 014004.
4. M. Glück, E. Reya and A. Vogt, Z. Phys. **C67** (1995) 433.
5. M. Glück, E. Reya, M. Stratmann and W. Vogelsang, Phys. Rev. **D53** (1996) 4775.
6. D. de Florian, M. Stratmann and W. Vogelsang, Phys. Rev. **D57** (1998) 5811.
7. Y. Kanazawa and Y. Koike, Phys. Lett. **B478** (2000) 121; **B490** (2000) 99.
8. D.L. Adams et al., Phys. Lett. **B261** (1991) 201; **B264** (1991) 462; A. Bravar et al., Phys. Rev. Lett. **77** (1996) 2626.
9. J. Binewise, B.A. Kniehl and G. Kramer, Z. Phys. **C65** (1995) 471.
10. M. Burkardt and R. L. Jaffe, Phys. Rev. Lett.**70** (1993) 2537.
11. M. Anselmino, D. Boer, U. D'Alesio and F. Murgia, hep-ph/0008186.

Polarization of hyperons in photon induced reaction at high energy

K. Suzuki[a1], N. Nakajima[b], H. Toki[b] and K.-I. Kubo[c]

[a] *Department of Physics, University of Tokyo, Tokyo 113-0033, Japan*
[b] *RCNP, Osaka University, Osaka 567-0047, Japan*
[c] *Tokyo Metropolitan College of Aeronautical Engineering, Tokyo 116-0003, Japan*

Abstract. We study the origin of the spin polarizations of hyperons observed in unpolarized hadron collision at high energy, and its consequence for the photon induced reaction. We assume the polarization is generated by the non-perturbative hadronization process with the scalar interaction. Non-trivial spin dependence of the production cross section arises from the interference term of hadronization amplitudes. Results are in good agreement with experiments. We then apply it to the photoproduction of hyperons and find large polarizations in unpolarized γN collision.

INTRODUCTION

Against a naive expectation that spin effects become less important at high energies [1], significant transverse polarizations in inclusive hyperon productions [2] and large analyzing powers in π, K, η productions from a transversely polarized nucleon [3] have been observed at low transverse momentum p_T and high Feynman x_F ($= 2p_L/\sqrt{s}$ in CM, p_L is the longitudinal momentum of the observed hadron). Such unexpected spin phenomena have attracted considerable experimental and theoretical interests. Since transverse spin asymmetry calculated by perturbative QCD is very small, we may naively expect that the polarization originates from some non-perturbative mechanism.

Let us consider the high energy photon scattering off a nucleon. If the polarizations were due to the spin structure of the beam nucleon [4], they would vanish in the photon induced reaction. On the other hand, we expect the non-vanishing polarization if the mechanism relies on the non-perturbative effects in the hadron production process, because the hadronization is independent of the beam. Therefore, the photon induced reaction will provide a valuable constraint on the generation of the polarization. Here, we discuss how the polarization emerges in the photoproductions which are accessible at future facilities. We focus on the production of Λ and Σ hyperons in this work.

---

[1] e-mail address: ksuzuki@nt.phys.s.u-tokyo.ac.jp, work supported by JSPS

ORIGIN OF POLARIZATIONS

Before going to discuss the hyperon photoproduction we briefly introduce our model which is designed to describe the inclusive particle production for low p_T and high x_F, in which the observed polarizations are significant. In ref. [5] Yamamoto et al. constructed a simple relativistic model for recombinations of quarks and/or diquarks to produce a final state hadron in terms of the quark distribution function of the incident hadron and wave function of the final state hadron. In this model, polarizations are brought by the scalar confining color force through the hadronization process in purely non-perturbative way. It was shown that this model reproduces the phenomenological rule of DeGrand and Miettinen (DM) [6], and provides polarizations in good agreement with experiments. In ref. [7], it is also demonstrated that the single spin asymmetry of the pions produced by transversely polarized proton-proton collision is well reproduced by taking into account the large Bjorken-x behavior of the nucleon spin structure function.

In this model, we assume that a fast *valence* quark (or diquark=quark-pair) from the beam proton picks up slow quarks created by the string breaking in the hadronization process in order to form a final state hadron, e.g. Λ. Such a direct recombination process which involves maximum numbers of valence quarks from the beam hadron is dominant for the particle production with high x_F over the standard fragmentation processes. In this sense, our approach is viewed as a model of the T-odd fracture function [8]. Apparently, gluons of the incoming nucleon gives negligible contributions to the high x_F hadron production, because the gluons have relatively small momentum fractions in the nucleon.

Based on the above picture, we further assume, (*a*) Each parton which participates in this reaction has the intrinsic transverse momentum distribution. (*b*) Quark and diquark are combined by the scalar interaction in the hadronization process. (*c*) SU(6) spin-flavor symmetry for the initial and final state hadron structure. Details are found in ref. [5,9].

Under these assumptions, dominant contribution to Λ production in the proton-proton collision comes from the recombination process of a spin-0 ud diquark $(ud)^0$ from the beam proton and the strange quark s generated by the pair creation. On the other hand, the spin-1 pair of ud quarks and the strange quark s will recombine to produce Σ^0.

By choosing the x-axis as the beam direction and z as the orientation of the Λ spin in the final state, the Λ production probability in the infinite momentum frame is given by [5]

$$S_{p \to \Lambda} = \int [dx_i \, dy_i \, dz_i/x_i] G_\Lambda |M(x_i, y_i, z_i)|^2 f_s(x_2, y_2, z_2) \, G_{(ud)/p}(x_1, y_1, z_1) \, \Delta^4 \, \Delta^3 \quad (1)$$

where M represents the elementary hadronization amplitude of $(ud)^0$ and s. $G_{(ud)/p}(x_1, y_2, z_3)$ is the $(ud)^0$ diquark wave functions of the incoming proton with x_i being the longitudinal momentum fraction and y_i, z_i the transverse momentum fractions. We assume the factorized form; $G_{(ud)/p}(x_1, y_2, z_3) = q_{(ud)/p}(x_1) \, e^{-y_1^2} \, e^{-z_1^2}$, where $q_{(ud)/p}(x)$ is the $(ud)^0$ diquark distribution function of the proton taken from ref. [10], while we use the Gaussian distribution for the transverse components with their average values chosen to be 400MeV [5,9]. $f(x_2, y_2, z_2)$ denotes the momentum distribution of the picked-up quark. We take the Gaussian form as $f(x_2, y_2, z_2) = e^{-y_2^2} e^{-z_2^2} \theta(x_1 - x_2)$ as suggested by the Schwinger mechanism. G_Λ is the light-cone Λ wave function.

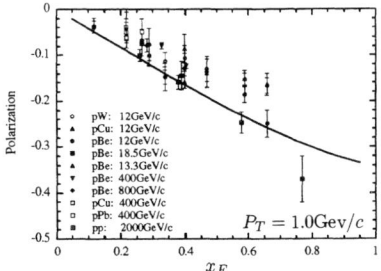

FIGURE 1. Λ polarization in pp collision

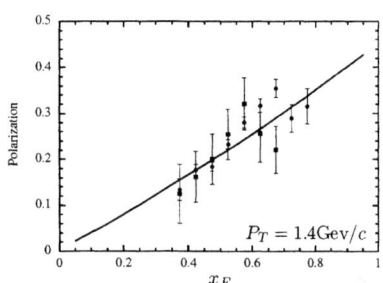

FIGURE 2. Σ polarizations in pp collision

To calculate the elementary amplitude M, we use the scalar interaction between partons, which may be responsible for the quark confinement. Since the hadronization is purely non-perturbative process, we must add all possible diagrams to calculate the hadronization probability. We find that the interference between the leading and higher order terms does generate the non-trivial spin dependence of the hadron production probability, although each diagram square shows no spin-dependence at all. The spin dependence of the cross section provided by the interference term is, in fact, what we need to account for the experimental data.

After the simple algebra, we obtain an expression for the polarizations;

$$P_N(\Lambda) = \frac{\int [dx_i \, dy_i \, dz_i/x_i] G_\Lambda \, \sigma_{dep} \, f(y_2, z_2) \, G_{(ud)/p} \, \Delta^4 \Delta^3}{\int [dx_i \, dy_i \, dz_i/x_i] G_\Lambda \, \sigma_{ind} \, f(y_2, z_2) \, G_{(ud)/p} \, \Delta^4 \Delta^3} , \qquad (2)$$

where the spin-dependent cross section is

$$\sigma_{dep} = -R_0 \, (x_F x_4 x_2) \left(\frac{x_F x_4 y_2 - x_2 y_4}{x_F x_3 x_1} \bar{p}_t \right) .$$

Coefficient R_0 in eq. (2) is determined by 'unknown' underlying dynamics of the confinement force, and here simply assumed to be a constant parameter. The spin dependent part of the cross section is defined as $\sigma_{dep} = \sigma(\Lambda \uparrow) - \sigma(\Lambda \downarrow)$, by assuming the Λ spin orientation is determined by the spin of the strange quark. This term originates from the interference of the hadronization amplitudes. However, this spin-dependence purely comes from the dynamical effect in the hadron formation process, and does not contribute to the static properties of hadrons. Similar expressions can be written for the Σ^0 [9], which consists of the spin-1 ud diquark and a strange quark s due to the SU(6) symmetry.

Since the model parameters in the wave functions are taken from other phenomenologies, we have two parameters to be fixed by experiments, that is, R_0 and R_1, for spin-0 and spin-1 diquarks, respectively. We shall fix R_0 by the Λ polarization data and R_1 by Σ, respectively. We also calculate the polarizations of other hyperons to check the consistency of our approach. All these calculations are consistent with experiments [5].

PHOTON INDUCED PROCESSES

Now we come back to the photon induced reaction. It is well known that the real photon can be approximated by the vector mesons, ρ, ω and ϕ, since the real photon ($Q^2 = 0$) gets enough time to turn into a hadronic $q\bar{q}$ system. It is easy to apply our model to the photoproduction $\gamma + N \to \Lambda + X$. The parton distribution of the incoming nucleon is now replaced with the quark distribution of the photon $G_{q/\gamma}(x_1, y_1, z_1)$. The u (or d or s) quark from the photon picks up the diquark through the hadronization in this case. Note that both spin-0 and spin-1 diquark recombination processes contribute to the $\gamma + N \to \Lambda + X$ reaction, according to the SU(6) wave function of hyperons.

Let us consider sub-processes in detail. The process $\gamma \to u, d \to \Lambda$ looks like the Λ production by the pion beam, $\pi + p \to \Lambda + X$. On the other hand, $\gamma \to s \to \Lambda$ process is similar to $K^- + p \to \Lambda + X$. In experiments, the former process gives somewhat negative polarization, and the latter is positive. Our calculations reproduce both behavior within errors with the parameters fixed previously. However, relative magnitudes of spin-0 and spin-1 diquark recombination processes are not well determined within our model and needs another parameter. In principle, relative magnitude could be fixed by the experiment, e.g. Ξ production cross section at large x_F. However, the data are poor at present, and we still have ambiguities for their relative magnitudes. Nevertheless, our model definitely predicts $P_\Sigma - P_\Lambda > 0$, where P_Λ, P_Σ are polarizations of Λ and Σ in the unpolarized photon-nucleon collision.

We show P_Λ and P_Σ in Fig.3 and 4. We use GRV parametrization of the quark distribution of the real photon at the scale $\mu^2 = 1 \text{GeV}^2$ [11]. Data are taken from [12]. Considerable polarization is obtained for the Λ production, while Σ polarization is small due to the large cancellation between sub-processes. As mentioned above, we have certain ambiguity to fix the relative contributions of the spin-0 and spin-1 diquark processes. If we change it slightly, Λ polarization becomes consistent with zero(similar to Fig.4), while Σ^0 shows a large positive polarization.

Recently interesting data also come out from NOMAD on the Λ polarizations in the neutrino induced reaction $\nu + p \to \mu + \Lambda + X$ [13]. It is found that only the polarizations in the target fragmentation region is significant. Our model can be applied to such a kinematical domain, and further study is in progress.

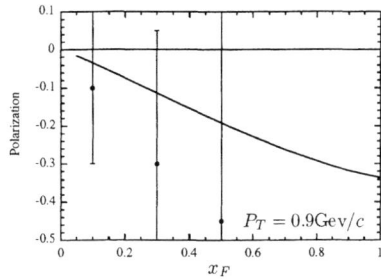
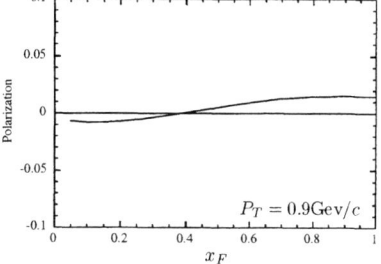

FIGURE 3. Λ polarization in γN reaction.   **FIGURE 4.** Σ polarization in γN reaction

CONCLUSION

We have studied the polarization phenomena in the hadron-hadron collision as well as the photon induced reaction. We have constructed the model which accounts for the polarization of the hyperons produced by the hadron-hadron collision assuming the polarization is brought by the scalar interaction in the hadronization process. We have found sizeable polarizations in the hyperon photoproduction which can be accessible in future experiments.

REFERENCES

1. Kave G.L., Pumplin J., Repko W., *Phys. Rev. Lett.* **41** (1978) 1689
2. Bunec G. et al., *Phys. Rev. Lett.* **36** (1976) 1113
   Recent experimental references are found in ref. [5].
3. Adams D.L.,et al., *Phys. Lett* **B264** (1991) 463
4. Kanazawa Y. and Koike Y., in these proceedings
5. Yamamoto Y., Kubo K.-I., and Toki H., *Prog. Theor. Phys.* **98** (1997) 95
   Kubo K.-I., in these proceedings
6. DeGrand T.A., Miettinen H.I., *Phys. Rev.* **D23** (1981) 1227
7. Suzuki K., Nakajima N., Toki H., Kubo K.-I., *Mod. Phys. Lett.* **A14** (1999) 1403
   Nakajima N. et al., *Nucl. Phys.* **A663** (2000) 573
8. K.S. would like to thank D. Boer for pointing out this.
9. Nakajima N., Suzuki K., Toki H., Kubo K.-I., to be published (2000); proceedings of APCTP Workshop on Strangeness in Nuclear Physics, Seoul, Korea, February 1999
10. Ekelin S. and Fredriksson S., *Phys. Lett.* **B95** (1985) 373
11. Glück M., Reya E.and Vogt A., *Phys. Rev.* **D46** (1992) 3986
12. Abe K. et al., *Phys. Rev.* **D29** (1984) 1877
13. Naumov D., in these proceedings; Astier P., et al., *Nucl. Phys.* **B588** (2000) 3

Longitudinal and Transverse Lambda Polarization at Hermes

Stefan Bernreuther[*][1]
on behalf of the HERMES -Collaboration

*Department of Physics, Tokyo Institute of Technology
2-12-1 Oookayama, Meguro-ku, Tokyo 152-8551, Japan*

Abstract. Spin transfer to Λ^0 Hyperons in semi-inclusive deep inelastic scattering of longitudinally polarized positrons from unpolarized nucleons has been investigated by HERMES. The result for the longitudinal spin transfer, which is dominated by scattering from u quarks in deep inelastic scattering, is compatible with zero and therefore in agreement with the Naive Quark-Parton Model expectation. Furthermore, Λ^0 polarization transverse to the production plane has been studied in quasi-real photoproduction, where the scattered leptons are not detected due to their small Q^2 value. The observed transverse polarization is positive and increases with the transverse momentum of the Λ^0. The value for $\bar{\Lambda}^0$ polarization is slightly negative, but consistent with zero due to limited statistical precision.

INTRODUCTION

The HERMES experiment at DESY in Hamburg [1], which started data taking in 1995, measures the spin structure of nucleons by deep inelastic scattering (DIS) of longitudinally polarized leptons from polarized internal gas targets, and using both inclusive and semi-inclusive measurements. However, the experiment also performs other semi-inclusive measurements, using both polarized and unpolarized targets, where various hadrons resulting from fragmentation of the struck quark are identified using the detector's excellent particle identification capabilities.

Among these hadrons the Λ^0 hyperon is unique: its decay products (proton and pion) can be identified cleanly, and parity violation in the weak decay couples the direction of the decay products with the spin orientation of the Λ^0. This offers the possibility to obtain information about spin-dependent effects in the fragmentation process. Moreover, such measurements can also provide information about the spin structure of the Λ^0 itself.

---

[1] Supported by the Japan Society for the Promotion of Science (JSPS).

LONGITUDINAL SPIN TRANSFER IN ELECTROPRODUCTION OF Λ^0

For a longitudinally polarized lepton beam and an unpolarized target, the Λ^0 polarization along its momentum axis is given in the quark parton model by [2,3]:

$$P_\Lambda = P_{\text{Beam}} D(y) D_{LL'}^\Lambda = P_{\text{Beam}} D(y) \frac{\sum_q e_q^2 q(x) \Delta D_q^\Lambda(z)}{\sum_q e_q^2 q(x) D_q^\Lambda(z)}. \qquad (1)$$

Here x is the Bjorken scaling variable, y is the fraction of the lepton beam energy transferred to the virtual photon, and z is the fraction of the photon's energy carried by the outgoing hadron. P_{Beam} is the polarization of the incoming lepton beam and $D(y) \approx y(2-y)/(1+(1-y)^2)$ is the virtual photon depolarization factor. Finally, $q(x)$ is the quark density in the nucleon, e_q^2 is the square of the quark's charge, and $D_q^\Lambda(z)$ and $\Delta D_q^\Lambda(z)$ denote respectively the unpolarized and polarized fragmentation functions for Λ's produced from a quark of flavor q.

Following ref. [2], the component of the longitudinal spin transfer to the Λ^0 along a spin quantization axis L' is defined as

$$D_{LL'}^\Lambda \equiv \frac{\vec{P}_\Lambda \cdot \hat{L}}{P_{\text{Beam}} D(y)} = \frac{\sum_q e_q^2 q(x) \Delta D_q^\Lambda(z)}{\sum_q e_q^2 q(x) D_q^\Lambda(z)}. \qquad (2)$$

The subscripts L and L' denote the fact that the spin is transferred from a polarized photon to a polarized Λ and that the two longitudinal spin quantization axes may be different. In DIS, the charge factor for the up quark leads to scattering from the nucleon's u quark content as the dominant process; the spin transfer can thus be approximated by $D_{LL'}^\Lambda(z) \approx \Delta D_u^\Lambda(z)/D_u^\Lambda(z)$. Furthermore, if one assumes that the helicity of the struck quark is conserved in the fragmentation process, one may then relate the spin transfer coefficient to the fractional polarization of the u quarks in the Λ: $D_{LL'}^\Lambda(z) \approx \Delta u^\Lambda/u^\Lambda$. In the naive quark parton model, the spin of the Λ is entirely due to the strange quark, and the up and down quark spin contributions are zero. On the other hand, assuming SU(3) flavor symmetry, the quark distributions and fragmentation functions for the Λ can be related to those in the proton. If data on hyperon decays and polarized structure functions of the nucleon are interpreted in the framework of SU(3) symmetry, one can estimate $\Delta u^\Lambda \approx \Delta d^\Lambda \approx -0.2$ [2,4].

The data used in this analysis were obtained in two three-week running periods, one in each of 1996 and 1997, which were dedicated to measurements with unpolarized targets of hydrogen, deuterium, ^{3}He and nitrogen. As no nuclear effects were observed within the limited statistics of this measurement, the data collected on the various targets have been added together. The sample of semi-inclusive Λ^0 events covers a kinematic range of $Q^2 > 0.8$ GeV, $0.02 < x < 0.5$ and $z \geq 0.2$. Using a moment method [5–8], and two data sets with opposite beam polarization (to avoid asymmetries caused by acceptance effects), the following expression for the spin transfer coeffcient may be obtained:

$$D_{LL'}^{\Lambda} = \frac{1}{\alpha \langle P_{Beam}^2 \rangle} \cdot \frac{\sum_{i=1}^{N_\Lambda} P_{Beam,i} \cos\theta_{P,i}}{\sum_{i=1}^{N_\Lambda} D(y_i) \cos^2\theta_{P,i}}, \qquad (3)$$

This expression depends only on measureable quantities: $\langle P_{Beam}^2 \rangle$ is the luminosity weighted average of the beam polarization squared, N_Λ is the total number of Λ^0's in the sample and $\cos\theta_{P,i}$ is the angle in the Λ^0 center of mass frame between the momentum of the decay proton and the Λ^0 spin quantization axis. The parameter $\alpha = 0.642 \pm 0.013$ describes the asymmetry of the parity violating Λ^0 decay.

Figure 1 shows the result for the spin transfer coefficient $D_{LL'}^\Lambda$. The measured value is compared to three scenarios following [9], each of which is motivated by the assumption of quark helicity conservation in the fragmentation process together with some model of the Λ spin structure. Scenario 1 corresponds to the prediction of the naive Quark Parton Model (QPM), with $\Delta s^\Lambda = 1$, $\Delta u^\Lambda = \Delta d^\Lambda = 0$; scenario 2 corresponds to the SU(3) model proposed in [2,4] with $\Delta u^\Lambda \approx \Delta d^\Lambda \approx -0.2$; and scenario 3 represents an 'extreme' picture where all three light quarks contribute equally to the Λ^0 spin.

Assuming helicity conservation in the fragmentation process, the HERMES measurement favors the naive QPM of the Λ spin structure. However, the present statistical precision is not yet sufficient to exclude the other scenarios.

TRANSVERSE POLARIZATION OF Λ^0 AND $\bar{\Lambda}^0$ FROM QUASI-REAL PHOTOPRODUCTION

It is a well known, but poorly understood, fact that hyperons produced in unpolarized proton-proton collisions are significantly polarized, in the direction perpendicular to the production plane. The HERMES experiment has recently explored this phenomenon, in the quasi-real photoproduction of Λ^0 and $\bar{\Lambda}^0$. The transverse polarization can be calculated as follows [5]:

$$\alpha P_{\Lambda,n} = \frac{\frac{1}{N_\Lambda^T} \sum_{i=1}^{N_\Lambda^T} \cos\theta_{P,i}^T + \frac{1}{N_\Lambda^B} \sum_{i=1}^{N_\Lambda^B} \cos\theta_{P,i}^B}{\frac{2}{N_\Lambda} \sum_{i=1}^{N_\Lambda} \cos^2\theta_{P,i}} \qquad (4)$$

The direction of the extracted polarization $P_{\Lambda,n}$ is the normal direction $\hat{n} = \hat{p}_{Beam} \times \hat{p}_\Lambda$, where $\hat{p}_{Beam}$ and $\hat{p}_\Lambda$ are unit vectors pointing along the momentum directions of the lepton beam and the Λ^0 respectively. The symbols N_Λ^T and N_Λ^B represent the numbers of Λ^0's in the top and bottom half of the detector, and N_Λ

represents the sum $(N_\Lambda^T + N_\Lambda^B)$. Finally, $\cos\theta_{P,i}$ denotes the corresponding angles between the proton and the Λ^0 polarization axis.

As in the case of the longitudinal polarization transfer analysis described above, data collected in 1996 and 1997 from several unpolarized targets were used. In this case, however, detection of the scattered beam positron was not required, to increase statistics. Consequently, the data on transverse Λ^0 polarization are concentrated in the quasi-real photoproduction regime ($Q^2 \approx 0$) where the cross-section is largest. As can be seen from Fig.2, the data show a significant positive polarization, increasing with the Λ^0's transverse momentum. For $\bar{\Lambda}^0$ the polarization tends to be slightly negative, however the limited statistics does not allow for a binning in momentum. The error bars in the figure are statistical only; the systematic uncertainty on the measurement is $\delta P_n = \pm 0.025$.

FIGURES

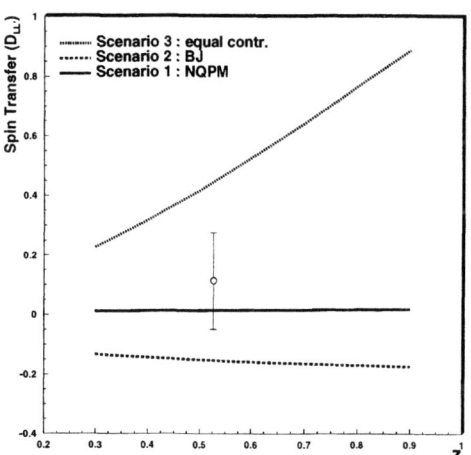

FIGURE 1. Longitudinal spin transfer versus the fractional energy z of the Λ^0, compared to the three scenarios described in the text.

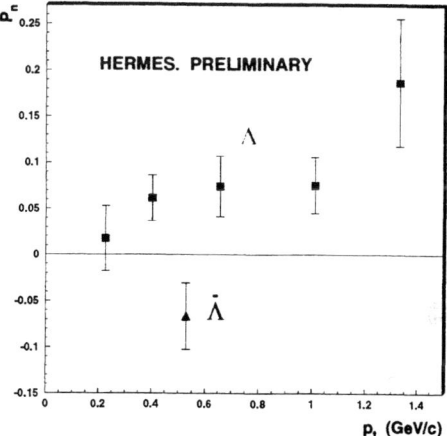

FIGURE 2. Transverse Λ^0 and $\bar{\Lambda}^0$ polarization. The error bars are statistical only; the systematic uncertainty is $\delta P_n = \pm 0.025$.

OUTLOOK

The results presented here are based on a data set containing 10×10^6 inclusive DIS events, collected in 2 dedicated periods of about 3 weeks each in 1996 and 1997. However HERMES has already collected another 9×10^6 DIS events with polarized targets and about 17×10^6 with several unpolarized targets from 1998 till 2000. Additionally the particle identification for this data has been significantly improved by replacing the spectrometer's threshold Cherenkov Counter by a Ring-Imaging

Cherenkov Counter (RICH) before starting data taking in 1998. Furthermore for the period beyond 2000, HERMES is preparing a detector upgrade with two round silicon counters (Lambda Wheels) 50 cm downstream of the internal gas target [10]. This will increase the acceptance for low momentum pions from the Λ^0 decay, and will also enable future exploration of Λ^0 production in the target fragmentation region.

REFERENCES

1. K.Ackerstaff et al.(HERMES Coll), NIM **A417**, 230 (1998).
2. R.L.Jaffe, Phys. Rev. **D54**, R6581 (1996).
3. P.J.Mulders and R.D.Tangermann, Nucl. Phys. **461**, 197 (1996)
4. M.Burkardt and R.L.Jaffe, Phys. Rev. Lett. **70**, 2537 (1993).
5. S. Belostotski and O. Grebenyuk, Hermes Internal Communication.
6. G.Schnell, Ph.D. Thesis, New Mexico State University (1999).
7. S.Bernreuther, Ph.D. Thesis (in German), Friedrich-Alexander Universität Erlangen-Nürnberg (1999), DESY-Thesis-2000-003.
8. K.Airapetian et al. (HERMES Coll.), hep-ex/9911017, DESY-99-151 (1999).
9. D.de Florian, M. Stratmann and W. Vogelsang, Phys. Rev. **D57**, 5811 (1998).
10. M.Amarian et al., HERMES Internal Report 97-032, (1997).

Spin transfer in high energy fragmentation processes

Liang Zuo-tang, and Liu Chun-xiu

Department of Physics, Shandong University, Jinan, Shandong 250100, China

Abstract. We point out that measuring longitudinal polarizations of different hyperons produced in lepton induced reactions are ideal to study the spin transfer of the fragtmenting quark to produced hadron in high energy hadronization processes. We briefly summarize the method used in calculating the hyperon polarizations in these processes, then present some of the results for e^+e^- and e^-p or νp reactions obtained using two different pictures for the spin structure of hyperon: that drawn from polarized deep inelastic lepton-nucleon scattering data or that using SU(6) symmetric wave functions. The results show in particular that measurements of such polarizations should provide useful information to the question of which picture is more suitable in describing the spin effects in the fragmentation processes.

The talk was a summary of a series of papers [1-3] by C. Boros and ourselves. Spin transfer in high energy fragmentation process is defined as the probability for the polarization of the fragmenting quark to be transferred to the produced hadron. It is one of the important issue in connection with the spin effects in high energy fragmentation processes which have attracted much attention recently [4]. The problem contains the following two questions: (1) Will the polarization of the fragmenting quark be retained in the fragmentation process? (2) What is the relationship between the spin of the quark and that of the hadron which contains this quark? Clearly, the answers to these questions depend not only on the hadronization mechanism but also on the spin structure of hadrons. Study of such effects provide useful information for the spin structure of hadron and spin dependence of high energy reactions. There exist now two distinctively different pictures for the spin contents of the baryons: the static quark model picture using SU(6) symmetric wave function [hereafter referred as SU(6) picture], and the picture drawn from the data for polarized deep inelastic lepton-nucleon scattering (DIS) and SU(3) flavor symmetry in hyperon decay [hereafter referred as DIS picture]. It is particular interesting to ask which picture is suitable to describe the relationship between the polarization of the fragmenting quark and that of the produced hadron which contains this quark. Obviously, the answer to this question is essential in the description of the puzzling hyperon transverse polarization observed already in the

1970s in unpolarized hadron-hadron reactions [5].

It has been pointed out that [1,2] measurements of the longitudinal Λ polarization in e^+e^- annihilations at the Z^0 pole provide a very special check to the validity of SU(6) picture in connecting the spin of the constituent to the polarization of the hadron produced in the fragmentation processes. This is because the Λ polarization in this case obtained from the SU(6) picture should be the maximum among different models. There are now data with reasonably high statistics available from both ALEPH [6] and OPAL [7] Collaborations. Their results show that the SU(6) picture seems to agree better with the data [6,7] compared with the DIS picture. This is rather surprising since the energy is very high at LEP thus the initial quarks and anti-quarks produced at the annihilation vertices of the initial e^+e^- are certainly current quarks and current anti-quarks rather than the constituent quarks used in describing the static properties of hadrons using SU(6) symmetric wave functions. It is thus interesting and instructive to make further checks in experiments by making complementary measurements. For this purpose, we have made a systematic study of hyperon polarizations in different lepton-induced reactions using the SU(6) or the DIS picture. The results we obtained can be used as further check of the pictures and now we give a brief summary of the calculation method and the obtained results.

We first summarize the calculation method by taking $e^+e^- \to H_i + X$ as an example.

Since the longitudinal polarization P_{H_i} of the hyperon H_i in the inclusive process $e^+e^- \to H_i + X$ originates from the longitudinal polarization P_f of the initial quark q_f^0 (where the subscript f denotes its flavor) produced at the annihilation vertex of the initial state e^+e^-, we should consider the H_i's which have the following different origins separately.

(a) Hyperons which are directly produced and contain the initial quarks q_f^0's originated from the annihilations of the initial e^+ and e^-;

(b) Hyperons which are decay products of other heavier hyperons which were polarized before their decay;

(c) Hyperons which are directly produced but do not contain any initial quark q_f^0 from e^+e^- annihilation;

(d) Hyperons which are decay products of other heavier hyperons which were unpolarized before their decay.

It is clear that hyperons from (a) and (b) can be polarized while those from (c) and (d) are not. We obtain therefore,

$$P_{H_i} = \frac{\sum_f t^F_{H_i,f} P_f \langle n^a_{H_i,f} \rangle + \sum_j t^D_{H_i,H_j} P_{H_j} \langle n^b_{H_i,H_j} \rangle}{\langle n^a_{H_i} \rangle + \langle n^b_{H_i} \rangle + \langle n^c_{H_i} \rangle + \langle n^d_{H_i} \rangle}. \tag{1}$$

Here P_f is the polarization of the initial quark q_f^0, and is determined by the electroweak vertex; $\langle n^a_{H_i,f} \rangle$ is the average number of the hyperons which are directly produced and contain the initial quark of flavor f; $\langle n^b_{H_i,H_j} \rangle$ is the average number of H_i hyperons coming from the decay of H_j hyperons which are polarized;

P_{H_j} is the polarization of the hyperon H_j before its decay; $\langle n_{H_i}^a \rangle (\equiv \sum_f \langle n_{H_i,f}^a \rangle)$, $\langle n_{H_i}^b \rangle (\equiv \sum_j \langle n_{H_i,H_j}^b \rangle)$, $\langle n_{H_i}^c \rangle$ and $\langle n_{H_i}^d \rangle$ are average numbers of hyperons in group (a), (b), (c) and (d) respectively; $t_{H_i,f}^F$ is the probability for the polarization of q_f^0 to be transferred to H_i in the fragmentation process and is called the polarization transfer factor, where the superscript F stands for fragmentation; t_{H_i,H_j}^D is the probability for the polarization of H_j to be transferred to H_i in the decay process $H_j \to H_i + X$ and is called decay polarization transfer factor, where the superscript D stands for decay. $t_{H_i,f}^F$ is equal to the fraction of spin carried by the f-flavor-quark divided by the average number of quark of flavor f in the hyperon H_i. This fractional contribution to the hyperon spin from f-flavor-quark is different in the above-mentioned SU(6) or the DIS picture. The results in the SU(6) picture can easily be obtained from the wave functions. In the DIS picture, the fractional contribution of quarks of different flavors to the spin of a baryon in the $J^P = \frac{1}{2}^+$ octet is extracted from $\Gamma_1^p \equiv \int_0^1 g_1^p(x) dx$ obtained in deep-inelastic lepton-proton scattering experiments and the constants F and D obtained from hyperon decay experiments. The way of doing this extraction is now in fact quite standard. A brief summary can, e.g., be found in the Appendix of [1]. The results can be found e.g. in [1,2]. The decay polarization transfer factor t_{H_i,H_j}^D is determined by the decay process and is independent of the process in which H_j is produced. They can be extracted from the materials in Review of Particle Properties (see e.g. [8,1,2]).

The average numbers of the hyperons of different origins mentioned above are determined by the hadronization mechanism and should be independent of the polarization of the initial quarks. Hence, we can calculate them using a hadronization model which give a good description of the unpolarized data for multiparticle production in high energy reactions. Presently, such calculations can only be carried out using a Monte-Carlo event generator. We used the Lund string fragmentation model [9] implemented by JETSET in our calculations.

The method described above was first applied to $e^+e^- \to \Lambda + X$ at the Z^0 pole. We recall that among all the $J^P = \frac{1}{2}^+$ hyperons, Λ is most copiously produced. Furthermore, the spin structure of Λ in the $SU(6)$ picture is very special, which makes it play a very special role in distinguishing the SU(6) and the DIS pictures. In the $SU(6)$ picture, spin of Λ is completely carried by the s valence quark, while the u and d quarks have no contribution. Since the initial s quark produced in the annihilation of the initial e^+e^- takes the maximum negative polarization, $|P_\Lambda|$ obtained using the SU(6) picture is the maximum among all the different models. In contrast, in the DIS picture, the s quark carries only about 60% of the Λ spin, while the u or d quark each carries about -20%. The resulting $|P_\Lambda|$ should be substantially smaller than that obtained in the $SU(6)$ picture. Comparing the maximum with experimental results provide us a good test of the validity of the picture.

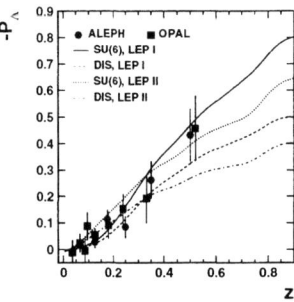 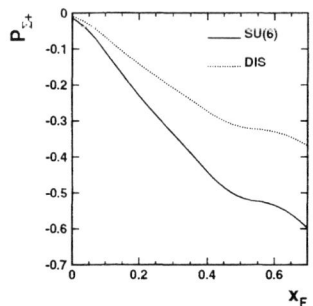

Fig.1: Longitudinal Λ polarization, P_Λ, in $e^+e^- \to \Lambda + X$ at LEP I and LEP II energies as a function of z. The data of ALEPH and those of OPAL are taken from [6] and [7] respectively.

Fig.2: Longitudinal Σ^+ polarization, P_{Σ^+}, in $\nu_\mu + p \to \mu^- + \Sigma^+ + X$ at $p_{inc} = 500 \text{GeV}/c$ as a function of x_F.

Using the method described above, we obtained the longitudinal polarization of Λ as shown in Fig.1. A comparison with the ALEPH data [6] and the OPAL data [7] shows that the data [6,7] of both groups agree better with the calculated results based on the $SU(6)$ picture. But, these available data [6,7] are still far from accurate and enormous enough to make a decisive conclusion. Further complementary measurements are needed. We therefore made a systematic study of hyperon produced in different lepton-induced reactions and obtain in particular the following results which can be used as further checks of the pictures.

1. Λ polarization in different subsamples of events in $e^+e^- \to \Lambda + X$.

We think it would be interesting to measure Λ polarization in events where the following criteria are satisfied: (i) Λ is the leading in one direction; (ii) the leading particle in the opposite direction is K^+. We expect that such Λ's should mainly have the origin (a) mentioned above.

Using the event generator JETSET, we showed that the Λ's from (a) contribute indeed substantially higher in these events than they do in the average events and the obtained $|P_\Lambda|$ is also much higher [2]. This should be easily be check in experiments.

2. Energy dependence of P_Λ in $e^+e^- \to \Lambda X$.

To see the energy dependence, we calculated P_Λ in $e^+e^- \to \Lambda X$ at LEP II energy. The results are also shown in Fig.1. We see a significant energy dependence due to that of P_f of the initial quark q_f^0. This can also be checked.

3. Longitudinal polarization of other $J^P = \frac{1}{2}^+$ hyperons.

The production rates for other octet hyperons are smaller than that for Λ so the statistic errors should be larger for the polarizations of these hyperons. On the other hand, decay contributions from heavier hyperons to these hyperons are also much less significant than that in case of Λ. Hence, the contaminations from the decay processes are much smaller. These conclusions can easily be checked using a Monte-Carlo event generator for e^+e^- annihilation into hadrons. On the other hand, we see also that the contribution from heavier hyperon decays is also much smaller. For example, for Σ^+'s, the decay contribution takes only about 7% of the total rate. The situations for Σ^-, Ξ^0, and Ξ^- are similar to that for Σ^+. Most of them are directly produced. Hence, the theoretical uncertainties in the calculations for these hyperons are much smaller. The study of polarizations of these hyperons should provide us with good complementary tests of different pictures. We thus calculated the longitudinal polarizations of all these $J^P = (1/2)^+$ hyperons[2], i.e., Σ^+, Σ^-, Ξ^0 and Ξ^-, in e^+e^- annihilation at LEP I and LEP II energies. We found that they are all polarized and the polarizations are different for different hyperons. They are also different in SU(6) picture and the DIS picture which can indeed be used as complementary check to the picture.

4. Hyperon polarization in the current fragmentation region in lepton-nucleon deeply inelastic scatterings.

The advantages of using hyperons in lepton-nucleon deeply inelastic scatterings are the following: First, we can study here not only longitudinal polarization transfer but also to check whether it is the same for transverse polarization case. Second, flavor separation is, in some cases, automatically. But it will be more difficult to reach the same statistics as that in e^+e^- annihilation at the Z^0 pole.

We calculated the hyperon polarization in different reactions and different kinematic regions [3]. We found it is partcularly interesting that Λ polarization in these reactions are usually small and also the differences from different pictures. On the other hand, polarizations of Σ are larger and the differences between different pictures are also larger. Hence, it should be more sensitive to use Σ as a check to different pictures than to use Λ in these reactions. As an example, we show P_Σ in $\nu_\mu + p \to \mu^- + \Sigma^+ + X$ in Fig.2.

This work was supported in part by the National Science Foundation of China and the Chinese Education Ministry.

REFERENCES

1. C. Boros, and Liang Zuo-tang, Phys. Rev. **D57**, 4491 (1998).
2. Liu Chun-xiu and Liang Zuo-tang, Phys. Rev **D 62**,094001 (2000).
3. Liu Chun-xiu, and Liang Zuo-tang, in preparation (2000).
4. See, e.g., the references we cited in our publications [1-3].
5. A summary of data can be found in e.g., K. Heller, in proceedings of the 12th International Symposium on High Energy Physics, Amsterdam, 1996.
6. ALEPH-Collaboration; D. Buskulic et al., Phys. Lett. **B 374** (1996) 319.
7. OPAL-Collaboration; Euro. Phys. J. **C2**, 49-59 (1998).
8. G.Gustafson and J.Häkkinen, Phys. Lett. **B303**, 350 (1993).
9. B. Anderson, G. Gustafson, G. Ingelman, and T. Sjöstrand, Phys. Rep. **97**, 31 (1983); T. Sjöstrand, Comp. Phys. Comm. **39**, 347 (1986).

Top Quark Physics at the LHC

Lars Sonnenschein

RWTH Aachen IIIB, D-52056 Aachen, Germany

Abstract. The proton proton collider LHC will operate at $\sqrt{s} = 14\,\text{TeV}$ center of mass energy which gives rise to a cross section of about 800 pb for the production of $t\bar{t}$ events. This leads, already for a start-up luminosity of $\mathcal{L} = 10^{33}\,\text{cm}^{-2}\text{s}^{-1}$, to the huge production rate of $8 \cdot 10^6$ $t\bar{t}$ events per year, corresponding to an integrated luminosity of $10\,\text{fb}^{-1}$. Simulations including the CMS detector response show the expected precision in the determination of the top quark mass. Moreover, the measurement of the $t\bar{t}$ spin correlation and the W boson helicity will allow to test the standard model interactions. Higgs scenarios and rare top decays are also considered.

DETERMINATION OF THE TOP QUARK MASS

The top quark mass in combination with the W boson mass allows to test the standard model and to constrain the Higgs boson mass. Thus it is very important to measure the mass of the top as precise as possible. Present combined results from the Tevatron lead $m_t = 174.3 \pm 3.3(\text{stat}) \pm 3.9(\text{syst})\,\text{GeV}$ [1]. At the LHC about $3.5 \cdot 10^6$ semileptonic decaying $t\bar{t}$ events, corresponding to $10\,\text{fb}^{-1}$ allow the precise measurement of the top quark mass based on CMS detector simulations.

To select the events, one isolated charged lepton and missing transverse energy, each with $E_\perp > 20\,\text{GeV}$ is required. At least four jets above $E_\perp > 40\,\text{GeV}$ have to be reconstructed within a pseudo rapidity of $|\eta| < 2.4$. Two of them have to be tagged as b jets with $E_\perp > 50\,\text{GeV}$. The p_z of the neutrino, coming from the leptonically decaying W boson which is imposed to have a transverse mass below 100 GeV, can be obtained exploiting the W mass constraint. To reduce the dependence on the calorimeter energy scale the two four vectors of the non b jets assumed to come from one W boson are rescaled to match its mass. The b jets are assigned to the reconstructed W bosons in minimizing the longitudinal momenta of the resulting tops. To reduce combinatorical background the two reconstructed tops have to fulfil $\cos\theta_{t\bar{t}} \leq -0.8$ in the plane transverse to the beam axis and the two top masses should not differ more than 25 GeV. About 5 000 events remain leading to 250 MeV statistical error of the top mass in the hadronic decay chain (see fig. 1, left). Between the generated and the reconstructed top mass appears a linear dependence such that masses and errors have to be scaled with a factor of 1.015. Background processes are negligible. Among the most important systematics (see

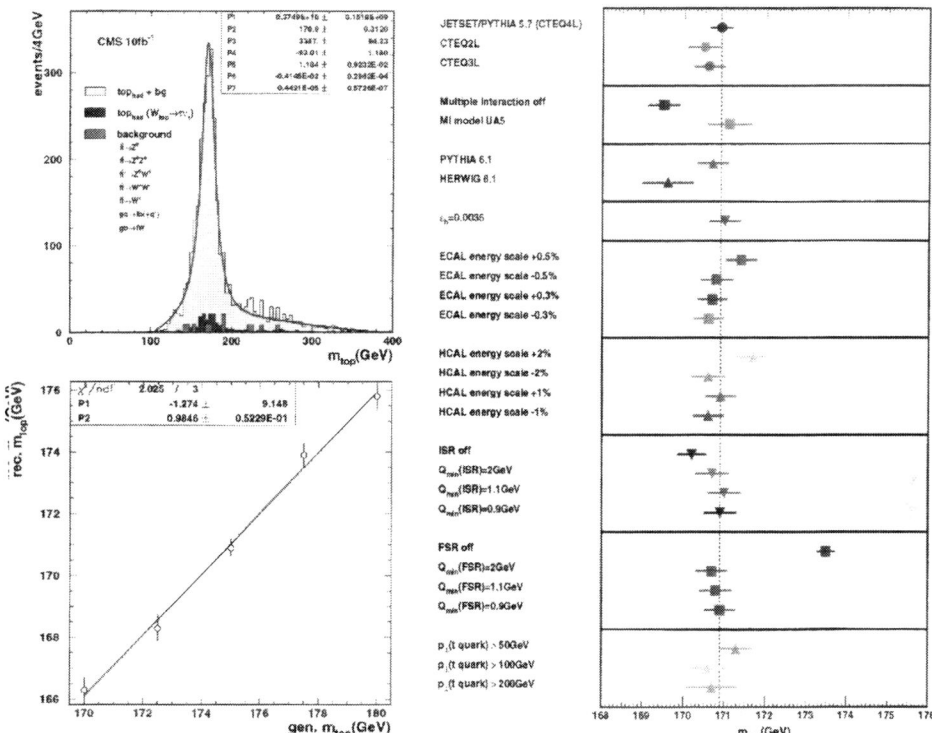

FIGURE 1. To the left the reconstructed top with the hadronically decaying W, including background processes and the linear dependence between the reconstructed and the generated top. To the right the most important systematic uncertainties. HERWIG has to be compared to PYTHIA without multiple interactions, since this feature is not implemented in HERWIG yet.

fig. 1, right) the uncertainty in the prediction of the top $p_\perp$ spectrum may dominate [2] (p. 429). With the variation of parameter values in a reasonable range a total error in the top mass of about 900 MeV can be achieved. Another possibility to determine the top mass via $t\bar{t} \to \ell + J/\Psi + X$ is described in [2] (p. 451).

HELICITY OF THE W BOSON

A precise measurement of the W boson helicity provides a unique test of the standard model and decreases systematic uncertainties in other measurements. The distribution of the helicity angle θ^* of W decay leptons is given by the differential cross section

$$\frac{1}{N_{tot}}\frac{dN}{d\cos\theta^*} = \frac{3}{8}\frac{1}{1+f}(1-\cos\theta^*)^2 + \frac{3}{4}\frac{f}{1+f}\sin^2\theta^*, \qquad f = \frac{m_t^2}{2m_W^2} \qquad (1)$$

in the standard model (see fig. 2).

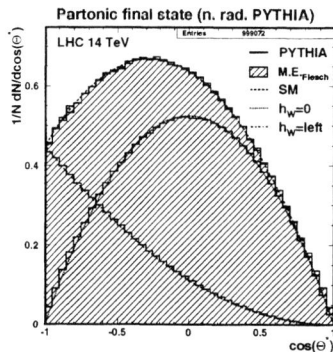

FIGURE 2. The cosine of the helicity angle θ^* of the lepton in the rest frame of the parent W boson. The left circular ($h_W = -1$) and the longitudinal ($h_W = 0$) polarized contributions to the standard model helicity of the W boson are drawn separately and as superposition. With $m_t = 175\,\text{GeV}$ and $m_W = 80.41\,\text{GeV}$ about 70 % of the W bosons are expected to be longitudinal polarized and about 30 % are expected to be left circular polarized. A right circular polarized contribution would reveal in a term proportional to $(1 + \cos\theta^*)^2$ in formula 1.

The observable is obtained from the leptonically decaying W boson in the semileptonic $t\bar{t}$ decay channel. The events are reconstructed as before but with 20 GeV jet cuts. The influence of different parton density functions, different event generators, i.e. PYTHIA and HERWIG, the variation of the top and W masses and of the Peterson fragmentation parameter ϵ_b in a wide range yield systematic errors which are certainly overestimated. The reconstructed decomposition of helicity states and the corresponding errors are

$$\left.\begin{array}{l} h(\text{left}) = 0.304 \pm 0.017(\text{stat}) \pm 0.046(\text{syst}) \\ h(\text{long}) = 0.672 \pm 0.027(\text{stat}) \pm 0.058(\text{syst}) \\ h(\text{right}) = 0.024 \pm 0.017(\text{stat}) \pm 0.060(\text{syst}) \end{array}\right\} \quad (\mathcal{L} = 10\,\text{fb}^{-1})\,.$$

$T\bar{T}$ SPIN CORRELATION

The $t\bar{t}$ spin correlation allows the direct measurement of the top quark spin. In contrast to the event generator PYTHIA which does not include the spin correlation (see fig. 3, left), the matrixelement [3], including the correlation was implemented in PYTHIA 5.7 (see fig. 3, right). In the dileptonic $t\bar{t}$ decay channel the helicity angles θ^* of the leptons are given by the double differential cross section

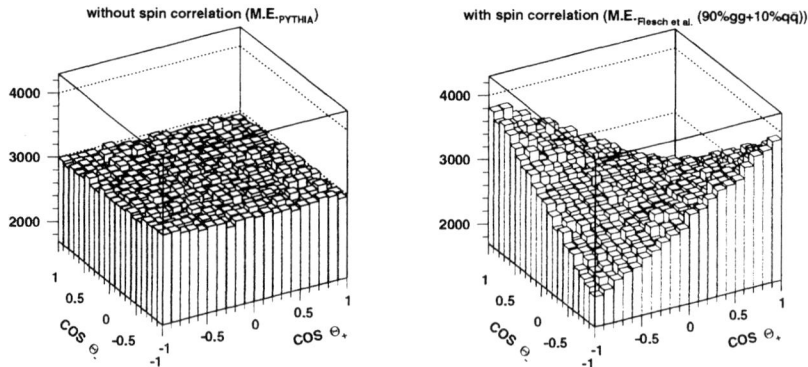

FIGURE 3. The lepton angles θ^* in the helicity basis without and with spin correlation.

$$\frac{1}{N_{\text{tot}}}\frac{d^2 N}{d\cos\theta^*_{\ell+} d\cos\theta^*_{\ell-}} = \frac{1}{4}(1 - \mathcal{A}\cos\theta^*_{\ell+}\cos\theta^*_{\ell-}) \qquad (2)$$

with the asymmetry $\mathcal{A}$ in the helicity basis defined as the normalized difference between like-spin and unlike-spin top pairs. At the LHC where the gg fusion dominates with about 87 % the asymmetry is expected to be $\mathcal{A} = 0.31 \pm 0.002(\text{stat}) \pm 0.03(\text{syst})$.

For the event selection two b jets and two oppositely charged leptons and missing transverse energy above $E_\perp > 20\,\text{GeV}$ each are required [2] (p. 472). Up to 16 ambiguities arise in the reconstruction of the event topology [4]. After correction for the detector acceptance and dilution due to combinatorical background results

$$\mathcal{A} = 0.309^{+0.039}_{-0.040}(\text{stat}) \pm 0.031(\text{syst}), \qquad (\mathcal{L} = 30\,\text{fb}^{-1}). \qquad (3)$$

The dominating contribution to the systematics comes from the different parton density functions due to different slopes and fractions of the gluon densities.

HIGGS SECTOR

If a neutral Higgs boson in the mass range above the $t\bar{t}$ production threshold, e.g. $m_H = 400\,\text{GeV}$ would be found, its CP properties have to be investigated [5]. Here a CP violating Higgs without coupling to vector bosons was used [3]. It can not be distinguished in the invariant $t\bar{t}$ mass spectrum from a pseudoscalar Higgs of the minimal supersymmetric model since the couplings to the top quark are of the same order. Both yield a Higgs width of $\Gamma_H \simeq 14.5\,\text{GeV}$ in contrast to the standard model case, where the width is $\Gamma_H \simeq 27\,\text{GeV}$ and the Higgs signal disappears in the $t\bar{t}$ background. The events were selected in the semileptonic decay channel. In half a year running at high luminosity ($\mathcal{L} = 34\,\text{cm}^{-2}\text{s}^{-1}$) a Higgs boson without vector boson coupling would be observable with a significance $\sigma := N_S/\sqrt{N_B} \simeq 14$ (see fig. 4). Therefore the precise knowledge of the $t\bar{t}$ background shape is crucial.

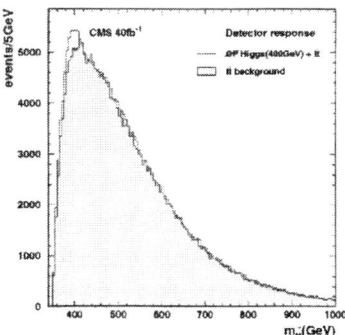
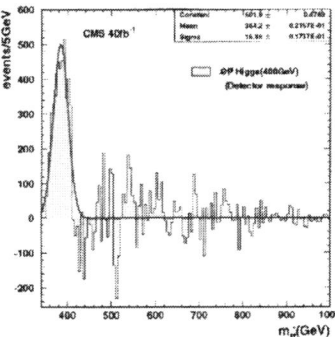

FIGURE 4. To the left the reconstructed invariant $t\bar{t}$ mass of signal plus background in comparison to the background only. To the right the signal after subtraction of the background.

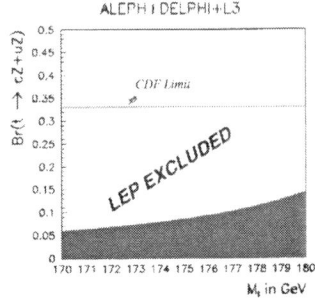 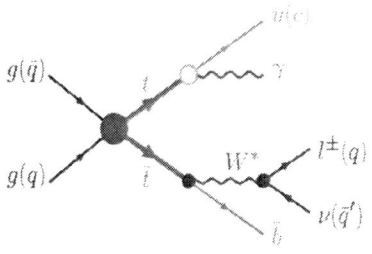

FIGURE 5. To the left the limits on the branching ratio $Br(t \to cZ + uZ)$. To the right the Feynman graph of the $t\bar{t}$ production with a FCNC transition.

RARE TOP DECAYS

Flavour changing neutral current (FCNC) transitions are highly suppressed in the top quark decay of the standard model. Beyond, a two Higgs doublet model would yield $Br(t \to \gamma(Z)q) \sim 10^{-8}$ and a R parity violating supersymmetric model $Br(t \to \gamma(Z)q) \sim 10^{-4}$. The current CDF results restrict the branching ratios to $Br(t \to \gamma c) + Br(t \to \gamma u) < 3.2\%$ and $Br(t \to Zc) + Br(t \to Zu) < 33\%$ (at 95% CL). The excluded branching ratio $Br(t \to cZ + uZ)$ from LEP is shown in fig. 5 on the left.

The matrix element for the $2 \to 5$ process (see fig. 5, right) was implemented in PYTHIA 5.7 for the investigation [2] (p. 479). One isolated γ with $E_\perp > 75\,\mathrm{GeV}$ and one isolated lepton with $E_\perp > 15\,\mathrm{GeV}$ within $|\eta| < 2.5$ are required. In addition at least two jets with $E_\perp > 20\,\mathrm{GeV}$ have to be found within $|\eta| < 4.5$. The highest $E_\perp$ jet has to be tagged as b jet and no additional jets with $E_\perp > 50\,\mathrm{GeV}$ should appear. The invariant mass of the reconstructed b jet and the W boson should differ less than 25 GeV from the top mass.

The considered background processes $t\bar{t}, W+\mathrm{jets}, WW, W\gamma$ and single top production are not exhaustive and point of further investigations. Assuming an integrated luminosity of $100\,\mathrm{fb}^{-1}$ and a top mass of 173.8 GeV the expected sensitivity of CMS to FCNC top decays is $Br(t \to c\gamma) + Br(t \to u\gamma) < 4 \cdot 10^{-5}$ and $Br(t \to cZ) < 1.8 \cdot 10^{-4}$ with 99% confidence level.

REFERENCES

1. D. E. Groom et al., Eur. Phys. J. **C15** (2000) 1.
2. G. Altarelli, M. L. Mangano et al., Geneva, Switzerland: CERN 2000-004 (2000).
3. W. Bernreuther, M. Flesch, P. Haberl, Phys. Rev. **D58**, (1998), 114031.
4. V. Šimàk, P. Homola, J. Valenta, R. Leitner, ATL-COM-PHYS-99-073, (1999).
5. W. Bernreuther, A. Brandenburg, M. Flesch, hep-ph/9812387, (1998).

Total and Differential Cross Sections and Polarization Effects in pp Elastic Scattering at RHIC

Wlodek Guryn[1]

for PP2PP (R7) collaboration[9]
Brookhaven National Laboratory, Upton, NY 11973, USA

Abstract. We shall describe an approved experiment [1] at the Relativistic Heavy Ion Collider (RHIC) Total and Differential Cross Sections and Polarization Effects in pp Elastic Scattering at RHIC, PP2PP experiment. By measuring single spin asymmetries such as A_N, double spin correlation parameter A_{NN}, and cross section difference $\Delta\sigma_T = \sigma_{tot}(\uparrow,\downarrow) - \sigma_{tot}(\uparrow,\uparrow)$, we will be able to determine the helicity amplitudes ϕ_i. Those amplitudes are not well known now, so their systematic study at RHIC will help to understand the spin structure of the nucleon and of the exchanged mediators of the force.

I INTRODUCTION

The Relativistic Heavy Ion Collider RHIC at Brookhaven National Laboratory, operating with polarized protons, with beam polarization 70% and luminosity $L \approx 10^{32}$ cm^{-2}s^{-1}, opens up an entirely new energy range for the study of spin dependence in proton-proton (pp) elastic scattering. Thus enabling us to measure the spin-dependent parameters of elastic pp scattering at much higher $\sqrt{s}$ than the highest energy data to date, which is $\sqrt{s} = 24$ GeV.

In discussing the polarization data, the s-channel helicity amplitudes [2] for NN elastic scattering ϕ_i ($i = 1-5$) are used. It is somewhat more convenient to express these in combinations that explicitly exhibit the t-channel exchange characteristics at high energy: $N_0 = 1/2(\phi_1 + \phi_3)$, $N_1 = \phi_5$, $N_2 = 1/2(\phi_4 - \phi_2)$, $U_0 = 1/2(\phi_1 - \phi_3)$, $U_2 = 1/2(\phi_4 + \phi_2)$.

The N and U amplitudes correspond respectively to natural and unnatural-parity exchanges; the subscripts 0, 1 and 2 correspond to the total s-channel helicity flip involved. The analyzing power A_N can be expressed in terms of these amplitudes as:

---

[1] Supported under Prime Contract between Brookhaven Science Associates and the Department of Energy No.DE-AC02-98CH10886

$$A_N \frac{d\sigma}{dt} = -2\frac{(\hbar c)^2}{16\pi K^2}\text{Im}[(N_0 - N_2)N_1^*], \quad (1)$$

where the spin-averaged differential cross section is

$$\frac{d\sigma}{dt} = \frac{(\hbar c)^2}{16\pi K^2}[|N_0|^2 + 2|N_1|^2 + |N_2|^2 + |U_0|^2 + |U_2|^2], \quad (2)$$

where $K = \sqrt{s(s-4m^2)}$. The double-spin asymmetry parameter is expressed as:

$$A_{NN}\frac{d\sigma}{dt} = 2\frac{(\hbar c)^2}{16\pi K}\text{Re}[U_0 U_2^* - N_0 N_2^* + |N_1|^2]. \quad (3)$$

The difference in the total cross-sections as a function of the initial transverse polarization states is:

$$\Delta\sigma_T = \{\sigma_{tot}(\uparrow,\downarrow) - \sigma_{tot}(\uparrow,\uparrow)\} = \frac{(\hbar c)^2}{K}\text{Im}(N_2 - U_2). \quad (4)$$

Elastic pp scattering with polarized protons covers a large spectrum of interesting physics. In the Coulomb Nuclear Interference (CNI) region, the electromagnetic and hadronic amplitudes are of comparable magnitude in the very forward direction, resulting in a small but significant asymmetry A_N in pp scattering near the point of maximum interference [3]. The interference between the hadronic non-flip and the electromagnetic spin-flip amplitudes gives rise to this asymmetry, which is expected to be about 4-5%. An up to date status of calculation can be found in [4].

Since the spin-dependent asymmetries are very sensitive to the quantum numbers of the exchanged object in the process of elastic scattering [5], the PP2PP experiment has a unique opportunity to shed light and possibly provide sufficient detail to help the theoretical understanding of long-range QCD interactions. In view of the recent theoretical and experimental evolution in the picture of diffraction, the complex structure at the constituents' scale of the Pomeron should be probed also at the level of helicity amplitudes. For example an additional triple-gluon O-exchange might provide the necessary phase difference with P-exchange to obtain an essentially energy-independent spin asymmetry. Current data are too scarce and imprecise to allow any definite conclusion on the large-$|t|$ behavior of A_N. New A_N measurements from RHIC will help settle this question.

With both RHIC beams polarized, A_{NN} will also be measured up to rather large $|t|$ values. A new calculation of A_{NN}, which includes C-odd exchange in the CNI region has been done [6] and can be verified at RHIC. It will also be possible to investigate the puzzling observations of large differences between parallel and antiparallel spin cross sections observed at around 12 GeV/c [7]. From these measurements, it appears that two protons interact harder when their spins are parallel.

All these endorse an experimental program to answer open questions such as: The precision measurement of A_N will address the issue of non-vanishing spin-flip contributions in diffraction. The precision measurement of A_{NN} will address

RHIC Intersection Region with PP2PP Basic CB Setup

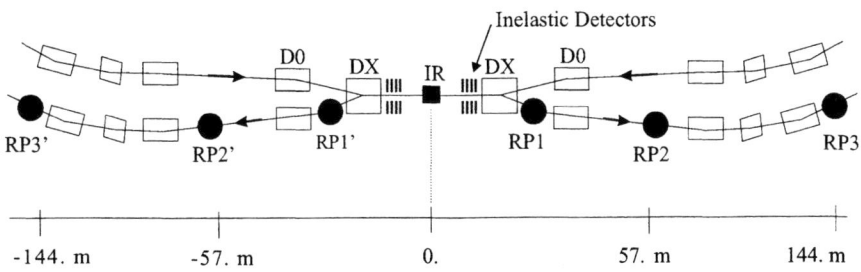

FIGURE 1. Layout of the PP2PP experiment.

the issue of existence of an C-odd exchange, the "Odderon". Does hard scattering produce large spin effects? What are the details of the long-range, non-perturbative QCD interaction between protons, and how does it approach the perturbative QCD regime at large $|t|$?

II THE EXPERIMENT

The two protons collide at the interaction point (IP), and since the scattering angles are small, protons follow trajectories determined by the lattice of the accelerator until they reach the detector, where the positions of the scattered particles with respect to the reference orbit are measured. That position y at the detector, is given by the transport equation:

$$y = a_{11}y^* + L_{eff}\theta_y^*, \qquad (5)$$

where y^* is the position at the IP and θ_y^* is the scattering angle at the IP. The optimum condition for the experiment is to have $a_{11} = 0$ and L_{eff} as large as possible, since the answer is then independent of the coordinate at the IR in the transverse plane of the accelerator. With the above condition satisfied, rays that are parallel to each other at the interaction point are focused onto a single point at the detector, commonly called "parallel to point focusing". The expression for the y coordinate at the detection point then simplifies to $y = L_{eff}\theta_y^*$ and the scattering angle is determined just from the measurement of the displacement alone. In our case, $L_{eff} = 198$ m results in the smallest measured four-momentum-transfer squared $|t|_{min} \approx 10^{-4}\,(\text{GeV}/c)^2$.

We intend to measure elastic and diffractive scattering in two experimental conditions:

1. In medium $|t|$ region, $0.006 < |t| < 1.3\,(\text{GeV}/c)^2$, no modifications to the accelerator tune and setup are required. In that setup, the RP1 and RP2 detectors give full $|t|$-coverage, and the DX dipole magnet of the RHIC lattice will be used for the momentum analysis of the scattered protons.

2. In small $|t|$ regime, for measurements reaching into the Coulomb region, $|t| \approx 10^{-4}(\text{GeV}/c)^2$, a special accelerator tune is required, which is also incompatible with high-luminosity running at the other IPs around the ring. For that reason, these special (week-long) runs for PP2PP are deferred until the second year of RHIC operation. Two weeks of special runs per year are required.

The basic detection elements of the experiment are Roman Pots (RP), located downstream from the IP. These devices allow the insertion (with UHV bellows) of Silicon strip detectors inside the RHIC beam pipe to measure scattered protons. The layout of the experiment with the three RP locations (with different $|t|$ coverages) is shown in Figure 1. The geometrical acceptance is determined by the beam pipe size for large $|t|$ and the t_{min} cutoff for small $|t|$ due to the beam emittance.

UHV bellows allow for transverse motion of the pot-detector system. In order to obtain the required t-resolution at the smallest $|t|$-values, $\delta t/t \approx 2\%$ the position of the detectors has to be known to better than 100 μm (0.004"), which imposes strict demands on the mechanics of the system in view of the pressure differential.

In order to keep the distance between beam, the bottom of the pot facing the beam is made as thin as possible, ≈ 0.5 mm, i.e. the point where the pot bottom starts intercepting the outer edge of the circulating beam, typically at 15-20 σ_{beam}. Also the closest sensitive part of the detector is ≈ 0.5 mm from its edge.

Single sided ac-coupled Si detectors will be mounted in Roman Pots. Their active area is approximately 7.5× 5.0 cm^2 with a characteristic strip pitch of 100 μm. Two versions exist, an "x-view" detector (vertical strips for measurement of horizontal coordinates), and a "y-view" detector (horizontal strips for vertical coordinates).

The SVX-IIe chips designed and tested by the $D\emptyset$ Collaboration for their microstrip vertex detector [8] will be used for readout. Each has 128 input channels, and may be daisy-chained together in the readout. Each channel contains a preamplifier stage, an analog pipeline, and an ADC.

A set of scintilaltors covering pseudorapidity range $2.5 < \eta < 5.5$ will be used to tag diffractive events and will allow to measure signgle and double diffractive cross sections.

Our simulations show [1]show that the errors of the elastic scattering parameters σ_{tot}, ρ, b due to uncertainties in the parameters describing the experimental, given 4×10^6 events collected, are expected to be $\Delta \sigma_{tot} = 200$ μbarn, $\Delta \rho = 0.005$, $\Delta b = 0.02$. The major sources of error are from the t_{min} cutoff. We have also simulated the A_N in the CNI region. The result is shown in Figure 2.

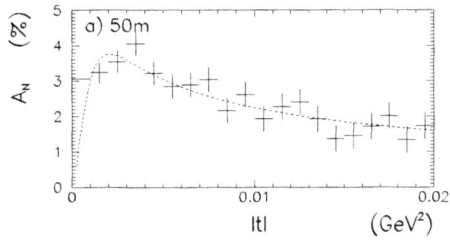

FIGURE 2. A_N as function of $|t|$. The + are simulation and the line is calculation [3].

III ACKNOWLEDGMENTS

The author would like to thank his collaborators [9], whose ongoing work contributed to this paper, and the organizers of the SPIN2000 Symposium for organizing and conducting a great conference.

REFERENCES

1. W. Guryn et al., PP2PP proposal "Total and Differential Cross Sections and Polarization Effects in pp Elastic Scattering at RHIC" (unpublished).
2. M. Jacob and G.C. Wick, Ann. Phys. 7 (1959) 404; M.L. Goldberger, M.T. Grisaru, S. MacDowell, D.Y. Wang, Phys. Rev. 120 (1960) 2250.
3. J. Schwinger, Phys. Rev. 69 (1946) 681; ibid. 73 (1948) 407; B.Z. Kopeliovich and L.I. Lapidus, Sov. J. Nucl. Phys. 19 (1974) 114; N.H. Buttimore, E. Gotsman, and E. Leader, Phys. Rev. D18 (1978) 694; C. Bourrely and J. Soffer, Nuovo Cim. Lett. 19 (1977) 569; R. Jakob and P. Kroll, Z. Phys. A344 (1992) 87.
4. N. Buttimore, B.Z. Kopeliovich, E. Leader, J. Soffer, T.L. Trueman, Phys. Rev. D59 (1999) 114010-1.
5. E.L. Berger, A.C. Irving and C. Sorensen, Phys. Rev. D17 (1978) 2971.
6. E. Leader and T.L. Trueman, Phys. Rev. D61 (2000) 077504.
7. D.G. Crabb et al., Phys. Rev. Lett. 41 (1978) 1257; E.A. Crosbie et al., Phys. Rev. D23 (1981) 600; I.P. Auer et al., Phys. Rev. D34 (1986) 1; D.G. Crabb et al., Phys. Rev. Lett. 65 (1990) 3241.
8. R. Lipton, Nucl. Instr. and Methods A 418 (98) 85-90.
9. B. Chrien, R. Gill, W. Guryn, D. Lynn, A. Rusek, M. Sakitt, S. Tepikian, Brookhaven National Laboratory, USA; J. Bourotte, M. Haguenauer, Ecole Polytechnique/IN2P3-CNRS, Palaiseau, France; N. Akchurin, C. Newsom, Y. Onel, University of Iowa, Iowa City, USA; A. A Bogdanov, V.A. Kaplin, A. Karakash, S. Nurushev, M.F. Runtzo, M.N. Strikhanov, Moscow Engineering Physics Institute (MEPHI), Moscow, Russia; M. Rijssenbeek, C. Tang, SUNY Stony Brook, USA; K. De, A. Vartapetian, University of Texas at Arlington, USA; G. N. Dudkin, I. V. Glavanakov, Yu. F. Krechetov, G. A. Naumienko, A. P. Potylitsyn, Tomsk Nuclear Physics Institute, Tomsk, Russia; R. Giacomich, A. Penzo, P. Schiavon, Universita di Trieste and Sezione INFN, Italy.

New Approaches to the *pp* Total Cross Section Measurements at Polarized Colliders

A.A. Bogdanov[1], S.B. Nurushev[1,2], A. Penzo[3], M.F. Runtzo[1], O.V. Selyugin[4], M.N. Strikhanov[1], A.N. Vasiliev[2]

1 Moscow Engineering Physics Institute, Kashirskoe Ave. 31, 115409 Moscow, Russia
2 Institute for High Energy Physics, 142284 Protvino, Moscow Region, Russia
3 Universite di Trieste and sezione INFN, Triest, Italy
4 BLTPh, JINR, Dubna, Russia

Abstract. There are suggestions to extract the *pp* total cross section, $\sigma_T(pp)$, at polarized colliders by measuring the analyzing power, $A_N(t)$, or so called the factor of merit, $M(t) = A_N^2(t) d\sigma/dt$. This talk is devoted to a detailed study of these new concepts.

1 THE SPECIFIC FEATURES OF THE ELASTIC DIFFERENTIAL *PP* CROSS SECTION.

Assume we can independently measure with high accuracy two differential cross sections for *pp* elastic scattering: the Coulomb one and the nuclear cross section. Then we extrapolate both cross section to the crossing point, $|t_0|$, which is related directly to the total cross section

$$\sigma_{tot} \cong \frac{8\pi\alpha}{|t_0|}. \qquad (1)$$

Therefore, by knowing $|t_0|$ one can determine σ_{tot}. We do not know any use of this relation in high energy colliders. The possible explanation of this situation might be the experimental difficulty in measuring the Coulomb cross section.

We look now for the additional specific features of the *pp* differential cross section, which can be used for extraction of total cross section. It is well known that elastic pp-differential cross section at the CNI region can be presented as a sum of three terms

$$\frac{d\sigma^{el}(t)}{dt} = \frac{d\sigma_c}{dt} + Int + \frac{d\sigma_n}{dt}. \qquad (2)$$

The first and third terms were discussed above. The second term, *Int*, means the interference between the Coulomb and nuclear interactions. It gives two additional points $-t_{ic}$ and $-t_{in}$ where *Int* is equal either to the Coulomb or to the nuclear cross section respectively. Together with the new value, -t2, coming from the extremum of the function $t^2 \dfrac{d\sigma^{el}(t)}{dt}$ they are presented in the Table1. A very interesting situation appears in the elastic *pp*-scattering. Since *Int* has a negative sign at $t=t_{in}$ this term completely compensates the nuclear scattering. Therefore, at this point only the Coulomb scattering survives which can used for normalization of the counting rates.

At $\sqrt{s}$=541 GeV $-t_{ic}$=4.2×10^{-3} (GeV/c)^2. This is an accessible point in the pp2pp experiment. Since at this point we have the pure nuclear scattering we can use the measurement at this point for checking some hypotheses such as the change of the slope parameter or oscillation of the differential cross section.

TABLE 1. TABLE 1. The specific points –t in the elastic pp differential cross section. The general formulae for –t as well as their magnitudes for the top RHIC energy are presented.

No	labels	Expression	value at √s=500 GeV	Comments
1	$-t_0=$	$\dfrac{8\pi\alpha(\eta c)^2}{\sigma_T}$	1.1·10^{-3} (GeV/c)^2	Coulomb=Nucl.
2	$-t_{ic}=$	$\dfrac{4\pi\alpha(\eta c)^2}{\sigma_T \rho}$	4.2·10^{-3} (GeV/c)^2	Coulomb=Int.
3	$-t_{in}=$	$\dfrac{16\pi\alpha\rho(\eta c)^2}{\sigma_T(1+\rho^2)}$	3·10^{-3} (GeV/c)^2	Int.=Nucl.
4	$-t2=$	$\dfrac{8\pi\alpha\rho(\eta c)^2}{\sigma_T(1+\rho^2)}$	1.5·10^{-3} (GeV/c)^2	Extremum in $t^2\cdot(d\sigma/dt)$

2 THE NEW APPROACHES TO σ_{tot} MEASUREMENTS AT POLARIZED COLLIDERS

In the following we discuss the possibility of extraction of $\sigma_T(pp)$ from the measurements of the analyzing power, $A_N(t)$, and also a factor of merit, $M(t)$. We borrow the corresponding formulae from paper, N. Buttimore et al., Phys. Rev. D59 (1999) 114010-1. But as it is usually accepted, omit in the fitting procedure all the spin dependent terms.

The expression for $A_N(t)$ in our approach looks like

$$A_N(t) = C_0 \frac{\sigma_T(1-\rho\varphi\alpha)(-t)^{3/2}}{1 + C_1(\rho+\alpha\varphi)\sigma_T + C_2(1+\rho^2)\sigma_T^2 |t|^2} \qquad (3)$$

Here $$C_0 = \frac{\mu_p - 1}{8\pi\alpha m_p(\eta c)^2} = 26.7735 GeV^{-3} mb^{-1},$$

$$C_1 = \frac{1}{4\pi\alpha(\eta c)^2} = -27.9972 GeV^{-2} mb^{-1}$$,and

$$C_2 = \frac{1}{[8\pi\alpha(\eta c)^2]^2} = 195.9609 GeV^{-4} mb^{-2}.$$ The μ_p and m_p are the proton magnetic moment and mass, respectively. Assuming that the ρ parameter is known from $\frac{d\sigma^{el}(t)}{dt}$, we van define the total cross section, σ_T, by one parameter fit to the experimental data.

From the same paper we extracted the formula for the factor of merit

$$M(t) = \overline{C}_0 \frac{\sigma^2(1-\rho\varphi)^2 e^{bt}|t|}{1 + C_1\sigma_T(\rho + \alpha\varphi)|t| + C_2(1+\rho^2)\sigma_T^2 t^2} \quad (4)$$

Here $\overline{C}_0 = \frac{(\mu_p - 1)^2}{16\pi m_p^2(\eta c)^2} = 0.1867 GeV^{-4} mb^{-1}$. Other parameters were defined earlier. For small t $bt<<1$, $\alpha\varphi \approx \alpha\varphi\rho \approx 0$. M has a maximum at the point t_M, which is

$$-t_M = \frac{8\pi\alpha(\eta c)^2}{\sigma_T\sqrt{1+\rho^2}} \quad (5)$$

Therefore the alternative to the fitting procedure is the extraction of σ_T from this relation. Ii is seen that t_M is not sensitive to the magnitude of ρ in the energy range of interest.

3 EXTRACTION OF THE σ_T (pp) FROM E704 DATA

The E704 collaboration at Fermilab measured the analyzing power, $A_N(t)$, in elastic pp-scattering at the initial momentum of the polarized beam, p_{in}=200 GeV/c in the region $2 \cdot 10^{-3} \leq |t|[(GeV/c)^2] \leq 4 \cdot 75 \cdot 10^{-2}$. We should fix the parameter ρ at $\sqrt{s}$=19.4 GeV. This parameter was measured at FNAL at p_{lab}=199 GeV/c and equaled ρ=-0.034±0.014. After inserting this number into formula (3) and fitting to the 6 experimental points on $A_N(t)$, we got

$$\sigma_T(pp) = (37.8 \pm 8.1) mb, \quad (6)$$

with χ^2=1.49 for ndf=5. This number is to be compared to the experimental value $\sigma_T(pp)$=38.9±0.7 measured at $\sqrt{s}$=23.0 GeV at ISR. The compatibility of these two values proves the correctness of the new method, though the statistics of E704 data do not allow to reach the precision of the standard technique. For that one needs to improve the E704 statistics by two orders of magnitude. Such drastic improvements of statistics can be made at RHIC by using a jet target. Therefore, $A_N(t)$ measurement in the CNI region at the fixed target mode (FTM) at RHIC is very desirable.

We turn now to the factor of merit, M(t). This is not the direct observable as $A_N(t)$ is, but it can be calculated using known data on $A_N(t)$ and $d\sigma^{el}(t)/dt$ at $\sqrt{s}$=19.4 GeV. Putting b=12 CeV^{-2} and ρ=-0.034 we made a fit to M(t) and got

$$\sigma_T = (22 \pm 40) mb, \qquad (7)$$

at χ^2=3.8 for ndf=5. We are not able to make any conclusion at such precision of σ_T extraction.

4 EXTRACTION OF THE $\sigma_T(pp)$ FROM "SIMULATED" pp2pp DATA AT RHIC

One of the important tasks of pp2pp experiment at polarized RHIC is to measure the analyzing power, $A_N(t)$, for elastic pp-scattering in CNI region. AN(t) was simulated at $\sqrt{s}$=500 GeV by using relation (3) with ρ=0. The beam polarization was set to be 70%, the running luminosity was assumed to be $2 \cdot 10^{29}$ cm$^{-2}\cdot$s^{-1}. In order to optimize the left-right difference in counting rates with vertically polarized beam, the events produced in the azimuthal region of $|\cos\phi|>1/\sqrt{2}$ were accepted. In such conditions the running time of about 3.7 hours will be required to collect $2.5 \cdot 10^6$ events. We took these simulated data as the "experimental measurements" and made a fit to the function given in equation (3) with ρ=0. At χ^2/ndf=26.64/18, we got

$$\sigma_T(pp) = (58.64 \pm 1.97) mb. \qquad (8)$$

This value has to be compared to the experimental data on $\sigma_T(\bar{p}p)$ obtained at $S\bar{p}pS$ at $\sqrt{s}$=541 GeV:63.0±1.5mb and at Tevatron at $\sqrt{s}$=546 GeV 61.26±0.93 (CDF). Consistency is very good providing that a new approach is workable and becoming competitive in precision with standard technique.

Now we are going to use the factor of merit for extraction of σ_T from the "simulated" pp data. For that we reconstructed the pp elastic differential cross section at $\sqrt{s}$=541 GeV from the UA4/UA2 results on $\bar{p}p$ assuming that the nuclear and Coulomb parts of the cross sections are equal, while the interfering term is different and has a different sign. The restored by such a way $d\sigma^{el}(t)/dt$ was multiplied by the A^2_N (T) and the experimental data on M(t) are restored. Applying formula (4) to these data, one gets σ_T=42.1±4.9 mb, with χ^2/ndf=26/18=1.42. Though this value is much smaller than the expected one and the error bar is twice bigger than in the $A_N(t)$ approach, nevertheless, M(t) also works in the right direction. One needs only more precise experimental data.

5 SINGLE SPIN FLIP CONTRIBUTION TO THE $\sigma_T(pp)$

Now we attempt to estimate the single spin flip contribution to the extracted value of $\sigma_T(pp)$. For that we use the E704 results on $A_N(t)$, the FNAL data for $d\sigma^{el}/dt$ and reconstruct the function

$$\psi(t) = \frac{m_p \cdot \sqrt{-t}}{\sigma_T} A_N(t) \frac{d\sigma^{el}}{dt}. \qquad (9)$$

The explicit from of function $\psi(t)$ is taken from paper N. Buttimore et al. (see above). Putting the numerical values σ_T=38mb, ρ=-0.034, μ_p=2.793, α=1/137, we got the final formula for the fit to the $\psi(t)$ at $\sqrt{s}$=19.4 GeV

$$\psi(t) = -0.00654 + 0.0073 I_s + 7.7(0.034 I_s + R_s)|t| \qquad (10)$$

Two-parametric fit to the $\psi(t)$ lead to the following results: $I_5 = 1.63 \pm 0.31$ and $R_5 = -0.05 \pm 0.02$. The position of the $A_N(t)$ maximum will be changed by the factor

$$\Delta t / t = \Delta_s = \frac{8(\rho I_s - R_s)}{(\mu_p - 1)(\sqrt{3} - (\rho + \alpha\phi))}, \qquad (11)$$

so numerically the magnitude of $\sigma_T(pp)$ will be changed by the same factor or ≈4%. Therefore, for the precise determination of the $\sigma_T(pp)$, let, say, of order 1%, one needs to make a better measurement of the parameters I_5 and R_5.

CONCLUSIONS

The suggestion to extract the $\sigma_T(pp)$ from the precise measurement of the analyzing power, $A_N(t)$, in the elastic *pp*-scattering at Coulomb-Nuclear Interference region becomes very attractive. At present there are several restrictions coming from the following factors: 1) precision of beam polarization measurement, 2) *t*-resolution of apparatus, 3) ambiguity in single spin flip term. In order for a new approach to $\sigma_T(pp)$ extraction to be competative to standard techniques at RHIC, we must find a way of improving precisions for items 1-3 listed above.

Measurement of Analyzing Powers and Spin Correlation Coefficients for Elastic pp Scattering

Frank Bauer for the EDDA[1] Collaboration

I. Institut für Experimentalphysik, Universität Hamburg,
Luruper Chaussee 149, 22761 Hamburg, Germany

Abstract. The EDDA experiment at the cooler synchrotron COSY measures four polarization observables of the elastic proton-proton scattering. Excitation functions of the analyzing power A_N and the three spin correlation parameters A_{SS}, A_{NN} and A_{SL} are acquired during beam acceleration for projectile momenta from 1.0 to 3.3 GeV/c, using a polarized atomic beam target. While the analyzing power measurements have been recently completed yielding about 850 data points, first results for the three spin correlation coefficients have been obtained at fixed energies with the polarized COSY beam. The results are compared to predictions of phase shift analyses.

INTRODUCTION

The proton-proton elastic scattering is fundamental to the understanding of the strong interaction. Excitation functions and angular distributions for cross sections and polarization observables of the pp interaction, in particular the elastic channel, provide the data base for phase shift analysis serving as input to calculations of effective NN interactions in nuclei, and to test meson exchange models. At present, the long and medium range part of the NN interaction seems to be well studied. The data are well represented by phase shift solutions and meson exchange models. However, at higher energies precise data, especially of polarization observables, can be useful to improve and extend phase shift analyses.

---

[1] M. Altmeier[1], F. Bauer[2], J. Bisplinghoff[1], M. Busch[1], K. Büßer[2], T. Colberg[2], L. Demirörs[2], O. Diehl[1], H. P. Engelhardt[1], P. D. Eversheim[1], O. Eyser[2], O. Felden[3], R. Gebel[3], M. Glende[1], J. Greiff[2], F. Hinterberger[1], E. Jonas[2], H. Krause[2], C. Lehmann[2], T. Lindemann[2], J. Lindlein[2], B. Lorentz[3], R. Maier[3], R. Maschuw[1], A. Meinerzhagen[1], C. Pauly[2], D. Prasuhn[3], H. Rohdjeß[1], D. Rosendaal[1], P. von Rossen[3], N. Schirm[2], V. Schwarz[1], W. Scobel[2], H. J. Trelle[1], K. Ulbrich[1], E. Weise[1], A. Wellinghausen[2], T. Wolf[2], R. Ziegler[1]
(1) Inst. f. Strahlen- und Kernphysik, University of Bonn
(2) I. Inst. f. Experimentalphysik, University of Hamburg
(3) Inst. f. Kernphysik, FZ Jülich

EXPERIMENTAL SETUP

The EDDA experiment at the cooler synchrotron COSY [1] of the Forschungszentrum Jülich is designed to measure excitation functions of the unpolarized differential cross section $d\sigma/d\Omega$ [2], of the analyzing power A_N [3] and the spin correlation parameters A_{SS}, A_{NN} and A_{SL} with a high relative accuracy over a wide momentum range from 1.0 up to 3.3 GeV/c.

EDDA was conceived as an internal target experiment using the recirculating COSY beam. Data collection proceeds during synchrotron acceleration in a multipass technique, so that a complete excitation function is measured in each acceleration cycle. Statistical accuracy is obtained by averaging over many cycles.

The EDDA detector (shown in Fig. 1) was designed to provide a fast trigger for coplanar two-prong events (i. e. $\Phi_2 - \Phi_1 = 180°$) that fulfill the kinematic correlation of elastic pp scattering. It consists of two cylindrical hodoscope layers with a large solid angle coverage ($35° \leq \Theta_{c.m.} \leq 90°$). The inner layer (H) is composed of scintillating fibers which are helically wound in opposing directions around the beam pipe in four layers. The outer layer consists of scintillator bars (B) which are running parallel to the beam axis; they are surrounded by scintillator semi-rings (R).

For the measurements of the analyzing power and the spin correlation coefficients a polarized atomic beam target was used. The polarization is prepared with two permanent sixpole magnets and an RF-transition unit. The direction of the polarization in the interaction region is aligned with a magnetic guide field of about 1 mT. The target thickness was $1.8 \cdot 10^{11}$ atoms/cm^2 resulting in a luminosity for the analyzing power measurements of $8 \cdot 10^{27}$ cm^{-2} s^{-1}. An effective polarization of 70-75% could be achieved.

The atomic beam has a diameter of 12 mm (FWHM). The resolution of the vertex reconstruction - obtained by a kinematic fit to the points of incidence of inner and outer layer - is better than 2 mm.

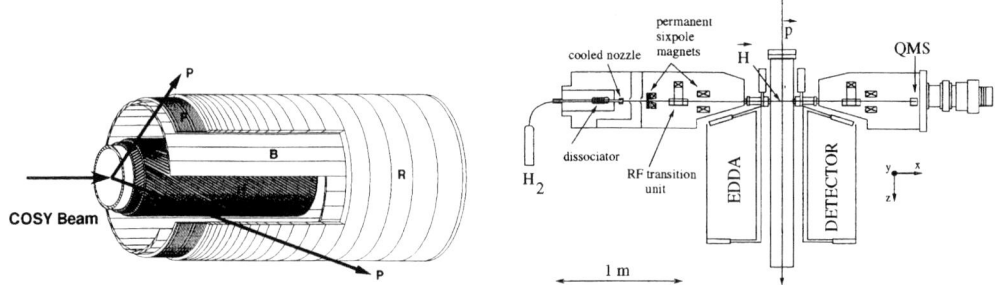

FIGURE 1. Scheme of the EDDA detector (left) and its combination with the polarized atomic beam target (right). The direction normal to the accelerator plane is denoted y.

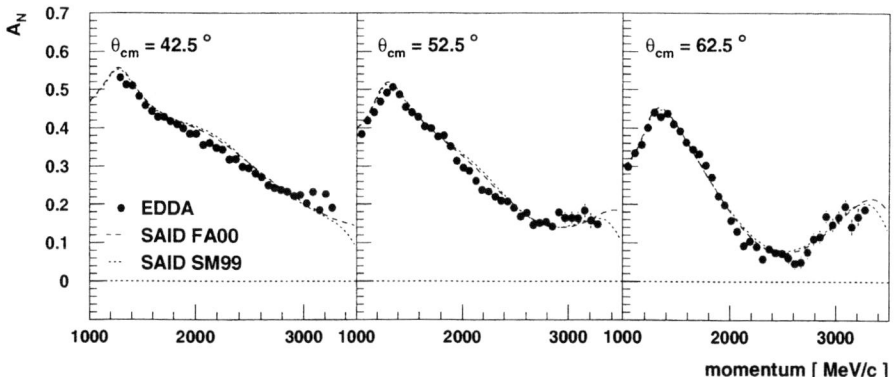

FIGURE 2. Three out of 12 excitation functions of the analyzing power A_N for $\Delta p = 60$ MeV/c and $\Delta\Theta_{c.m.} = 5°$ bins. Also shown are the most recent solutions from the phase shift analyses of the VPI group [4] SAID SP99 (dotted line) and FA00 (dashed line).

ANALYZING POWER

The analyzing power was measured using the polarized atomic beam target and the unpolarized COSY beam. The measurements were performed in cycles of about 13 s duration. The direction of the polarization was changed cyclewise from $+x$ to $-x$, $+y$ and $-y$. The analyzing power is determined from top-bottom (left-right) count rate asymmetries for a target polarization in x (y) direction. The absolute target polarization values are established by normalizing the observed asymmetry for one momentum bin $\Delta p = 60$ MeV/c around the energy of 730 MeV to a precise angular distribution of the analyzing power from McNaughton et al. [5].

Results with different polarization states were combined to apply a correction for false asymmetries [6]. The largest systematic error arises from inelastic reactions that accidentally fulfill the signature of elastic pp scattering. The remaining background was estimated, guided by Monte Carlo simulations, to be mostly $\leq 2\%$ and only at highest energies and most backward angles up to 4.5 %.

Excitation functions with about 850 data points have been deduced and are published [3]. The comparison to the recent phase shift solutions of the VPI group SAID SP99 [4](not including our data) shows (see Fig. 2) agreement in the general size and momentum dependence, but also systematic deviations in the momentum range 1800-2500 MeV/c, where other data are scarce. This discrepancy is smaller for the SAID solution FA00, which does include our data and results in lower values of χ^2 per datum.

SPIN CORRELATION COEFFICIENTS

The spin correlation coefficients A_{SS}, A_{NN} and A_{SL} are measured using the polarized atomic beam target and the polarized COSY beam [7]. The target polarization is changed with every cycle between $\pm x$, $\pm y$ and $\pm z$. The beam polarization is changed cyclewise from $+y$ to $-y$, thus 12 different combinations of polarizations result. Because of the Φ-dependence of the polarized differential cross section, one obtains a modulation of count rates depending on the polarization configuration. For analysis the detector is devided in four Φ-segments (centered at 45°, 135°, 225° and 315°). The Φ-dependence of the cross section leads then to count rate asymmetries in the four sectors that can be used to determine the spin correlation coeffcients as well as the target and beam polarization if the analyzing power is known. Here we used our results [3].

A first measurement was carried out at a momentum (energy) of 1430 MeV/c (772 MeV). This is the highest energy in the energy range of COSY where data for all three spin correlation coefficients from other experiments exist. Our data are in good agreement with the other data as well as the phase shift solution FA00, showing that the determination of the spin correlation coefficients works (see Fig. 3).

Subsequently, the measurements have been extended to higher energies. Data collection proceeds during beam acceleration and the following flat top. The result from a flat top measurement at 3100 MeV/c (2300 MeV) is shown in Fig. 4. At this energy the world data bases for A_{SS} and A_{SL} are rather sparse. Other data exist

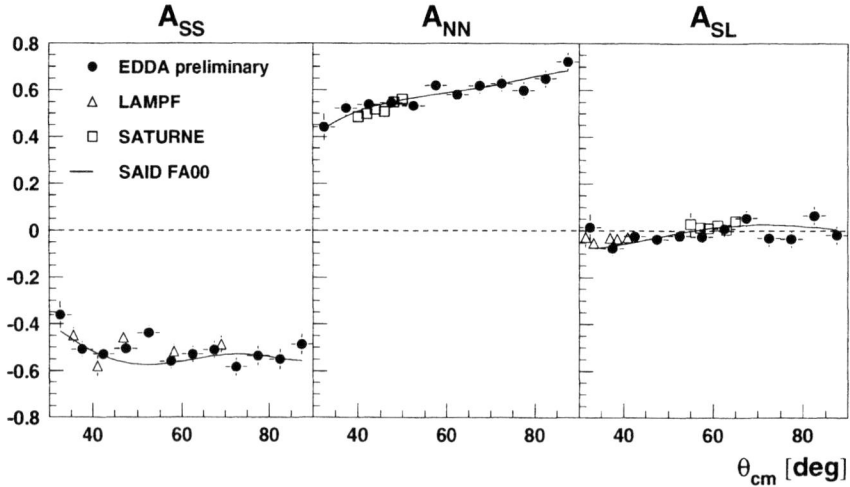

FIGURE 3. The spin correlation coefficients A_{SS}, A_{NN} and A_{SL} at 1430 MeV/c. Closed symbols, this experiment; open symbols, experiments at LAMPF and SATURNE ([8-11]). The SAID solution FA00 is given as a solid line.

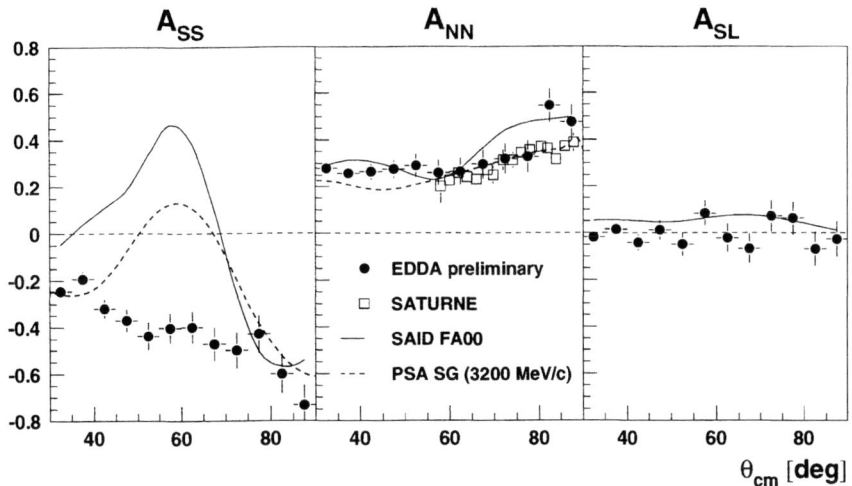

FIGURE 4. The spin correlation coefficients A_{SS}, A_{NN} and A_{SL} at 3100 MeV/c. The SAID solution FA00 is given as a solid line, a solution of the Saclay-Geneva group [12] as the dashed line

only for A_{NN}, which are in good agreement with our measurements. Also shown are predictions of phase shift analyses of the VPI [4] and the Saclay-Geneva [12] group. From the A_{SS} results, the limited predictive power of phase shift solutions not sufficiently supported by experimental data can be seen. These results are still preliminary.

The EDDA collaboration gratefully acknowledges the great support received from the COSY accelerator group. This work is supported by the BMBF and by the Forschungszentrum Jülich.

REFERENCES

1. R. Maier *Nucl. Instrum. Methods* **A390**, 1 (1997).
2. D. Albers et al. *Phys. Rev. Lett.* **78**, 1652 (1997).
3. M. Altmeier et al. *Phys. Rev. Lett.* **85**, 1819 (2000).
4. R. A. Arndt et al. *Phys. Rev. C* **62**, 034005 (2000), SAID solutions SP99 and FA00.
5. M. W. McNaughton, et al. *Phys. Rev. C* **41**, 2809 (1990).
6. G. G. Ohlsen and P. W. Keaton *Nucl. Instrum. Methods* **109**, 41 (1973).
7. F. Hinterberger *these proceedings*.
8. W. R. Ditzler et al. *Phys. Rev. D* **29**, 2137 (1984).
9. J. Bystricky et al. *Nucl. Phys.* **B262**, 727 (1985).
10. G. Glass et al. *Phys. Rev. C* **45**, 35 (1992)
11. A. de Lesquen et al. *Nucl. Phys.* **B304**, 673 (1988).
12. J. Bystricky, C. Lechanoine-Leluc, and F. Lehar *Eur. Phys. J.* **C4**, 607 (1998).

A_N for Inclusive $\pi^\pm$ Production at 21.6 GeV/c from C and LH$_2$

H. Spinka*, C. Allgower*, T. Kasprzyk*, K. Krueger*,
D. Underwood*, A. Yokosawa*, G. Bunce[†], H. Huang[†], Y. Makdisi[†],
T. Roser[†], M. Syphers[†], N.I. Belikov[‡], A.A. Derevschikov[‡],
Yu.A. Matulenko[‡], L.V. Nogach[‡], S.B. Nurushev[‡], A.I. Pavlinov[‡],
A.N. Vasiliev[‡], M. Bai[||], S.Y. Lee[||], Y. Goto[¶], N. Hayashi[¶],
T. Ichihara[¶], M. Okamura[¶], N. Saito[¶], H. En'yo[§], K. Imai[§],
Y. Kondo[§], Y. Nakada[§], M. Nakamura[§], H.D. Sato[§], H. Okamura**,
H. Sakai**, T. Wakasa**, V. Baturine[††], A. Ogawa[††],
V. Ghazikhanian[‡‡], G. Igo[‡‡], S. Trentalange[‡‡], and C. Whitten[‡‡]

Argonne National Laboratory, Argonne, IL 60439, USA
[†] *Brookhaven National Laboratory, Upton, NY 11973, USA*
[‡] *Institute for High Energy Physics, Protvino, Russia*
[||] *Indiana University, Bloomington, IN 47405, USA*
[¶] *RIKEN, The Institute of Physical and Chemical Research, Japan*
[§] *Department of Physics, Kyoto University, Japan*
** *Department of Physics, University of Tokyo, Japan*
[††] *Pennsylvania State University, University Park, PA 16802, USA*
[‡‡] *UCLA, Los Angeles, CA 90095, USA*

Abstract. Measurements of the spin asymmetry A_N for the inclusive reactions $\vec{p}C \to \pi^\pm X$, pX and $\vec{p}p \to \pi^\pm X$ are presented. Comparisons of hydrogen and carbon target results, and of data at 21.6 GeV/c and 200 GeV/c are given.

The inclusive production of charged pions and protons has been measured at the BNL AGS with an extracted polarized proton beam at 21.6 GeV/c. Targets of carbon, CH_2, and liquid hydrogen were used. One of the primary goals was to provide new data to assist in the understanding of single spin asymmetries at large x_F for hyperons and mesons. Another goal was to obtain data useful for a polarimeter at the RHIC injection energy.

Many experiments have measured the polarization of inclusively produced hyperons over a wide range in beam energy and for many target nuclei [1]. Large spin effects have been observed, increasing with x_F, but approximately independent of energy or target. These measurements are at relatively low P_T, where it is generally difficult to apply perturbative QCD.

Polarized proton and antiproton beams at 200 GeV/c have been used to measure asymmetries in charged pion production at large x_F (and low P_T). Large single spin asymmetries were observed, increasing with x_F [2]. There are also data with polarized proton beams at 11.75 [13] and at 13.3 and 18.5 GeV/c [14,15]. However, the systematics of inclusive pion production are less well established than for inclusive hyperon production.

Many models have been developed to try to understand the Fermilab E704 data in Ref. [2]. These include quark recombination models [3], orbital angular momentum of quarks or flavor currents [4,5], intrinsic P_T effects [6], twist-3 quark-gluon correlation functions and higher twist effects [7–9], and quark transversity effects [10–12]. These models have not made predictions at significantly lower beam momenta, such as in this experiment.

The large analyzing power observed at 200 GeV/c suggested that inclusive pion production might be useful for a polarimeter at RHIC or other high energy accelerators. One purpose of these measurements was to ascertain whether the large analyzing powers persisted to lower energy. Note that a competing technique, pp elastic scattering in the Coulomb-nuclear interference region, has an expected analyzing power roughly a factor 5 - 10 smaller.

The experimental layout [16], shown schematically in Fig. 1, consisted of two parts. One was a magnetic spectrometer to detect the charged pions and the associated production target. The spectrometer consisted of three trigger scintillation counters, four scintillation hodoscopes, and a gas threshold Cerenkov counter. The dipole magnet gave a P_T kick of 0.95 GeV/c for the carbon target data (Nov., 1997) and 0.86 GeV/c for the liquid hydrogen target results (Mar., 1999). The spectrometer was complemented with a set of beam veto counters to reject beam halo particles, and with an up-down pair of scintillation counter telescopes used to monitor the incoming beam intensity. The inclusive trigger consisted of a coincidence among the three trigger counters, and a requirement of a signal in three of the four hodoscope X-measuring planes, where X is in the direction of the magnet bend. For the π^+ data, a signal was also required from the Cerenkov counter. The trigger was vetoed by an OR of all the beam halo counter signals. The elements of each hodoscope plane were encoded and recorded event by event, along with ADC and TDC values for the trigger and Cerenkov counters, by a dedicated computer. Various scaled quantities were recorded for each spill.

The other part of the experiment was a beam polarimeter based on pp elastic scattering for $-t = 0.1$ to 0.2 GeV^2/c^2. Narrow CH_2 and C targets were used, in order to minimize energy loss for the recoil protons. The polarimeter included left-right symmetric recoil arms, each consisting of three scintillation counters, a wedge-shaped degrader, and a veto counter. The conjugate forward particles were

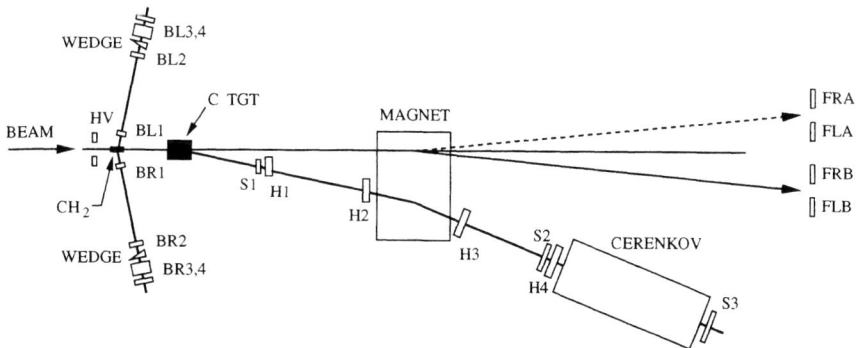

FIGURE 1. Diagram of the experimental layout, including the targets, magnetic spectrometer, elastic detectors, and beam veto counters (not to scale).

detected in a single scintillation counter. Two pairs of counters were required, since the forward scattered protons and the beam passed through the field of the spectrometer analyzing magnet (FRA, FLA during π^- and FRB, FLB during π^+ measurements). The trigger was a coincidence of the three counters from either the left or right recoil arm, and ADC and TDC values were read out separately for each counter to a dedicated computer. Again, the events were rejected by an OR of the beam veto counters. Various scaled quantities were also recorded spill by spill. Beam intensities used were about 10^7/spill.

The elastic data were analyzed to require both forward and recoil particles to be detected, with cuts on TDC and ADC values. Backgrounds for the CH_2 target data due to the presence of carbon nuclei were $\sim 2-4\%$, and due to empty target events were less than 0.3 %. The analyzing power for this polarimeter was approximately 0.040 ± 0.005 from Ref. [17] and data at other energies. The derived beam polarizations were 0.27 ± 0.06 and 0.40 ± 0.03 for the carbon and hydrogen target inclusive data, respectively. The quoted errors are statistical only. There is an additional uncertainty of $\pm 12\%$ due to knowledge of the pp elastic analyzing power.

Various cuts were applied to the inclusive events to select good $\pi^\pm$ candidates. A tracking requirement insured that at least one combination of hits from each hodoscope matched a possible particle trajectory. The resulting tracks were extrapolated to the target, and were removed if they were too far from the nominal beam center. The times of flight from counter S_1 to S_3 and to the Cerenkov counter were required to correspond to a high momentum pion or proton. A cut was also applied to the Cerenkov ADC value to reject K^+ events. Nearly all the remaining events exhibit a tight correlation of P_T versus x_F; those far from the majority or those that gave nonphysical kinematical quantities were rejected. Events passing all these cuts were binned in x_F, and were identified as either pions or protons via

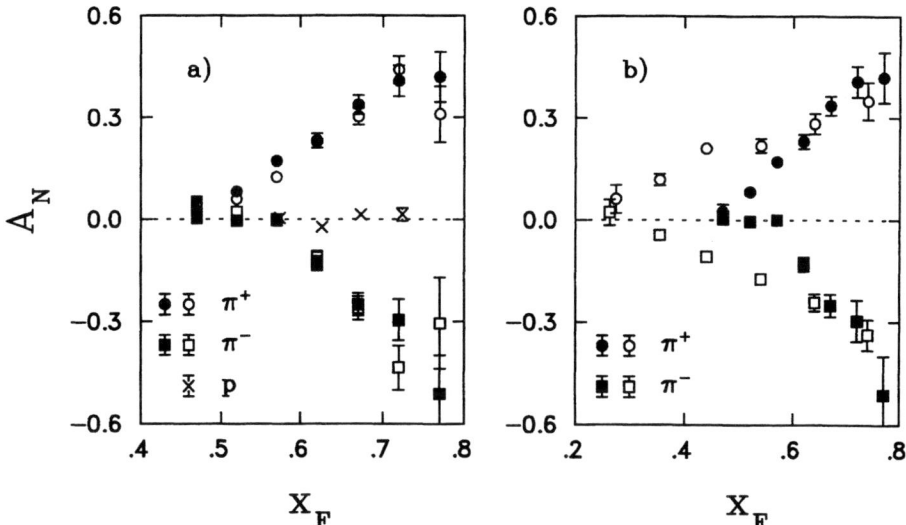

FIGURE 2. Preliminary analyzing powers for inclusive $\pi^\pm$ production as a function of x_F. a) A comparison of results for hydrogen (solid circles and squares) and for carbon (open circles and squares, and crosses) targets at 21.6 GeV/c. b) A comparison of hydrogen target results from this experiment at 21.6 GeV/c (solid circles and squares) and data from Ref. [2] at 200 GeV/c.

the Cerenkov counter ADC.

Backgrounds from target empty runs were found to be very small. Relative cross sections were derived from the good $\pi^\pm$ candidates, and compared to published values [18] after normalization to one x_F point. Good agreement was observed as a function of x_F, indicating little contamination of the data by kaons, accidental coincidences, etc.

Preliminary results from this experiment are shown in Fig. 2. The carbon target data have been published [16], and a final paper on the experiment is being written. A comparison with results from Fermilab E704 is provided in Fig. 2b, though the P_T range differs somewhat from the 21.6 GeV/c measurements. It can be seen that:

- Large, and approximately symmetric, A_N values are observed in inclusive $\pi^\pm$ production by polarized protons at 21.6 GeV/c.

- A_N seems to be the same for carbon and hydrogen targets at most values of x_F.

- The shapes of A_N as a function of x_F differ at 21.6 and 200 GeV/c, in contrast to the case of inclusive hyperon production.

- $A_N(\vec{p}p \to \pi^+ X)$ and $A_N(\vec{p}p \to \pi^- X)$ are not perfectly mirror symmetric, at either 21.6 or 200 GeV/c. For example, A_N for π^- seems consistent with zero to larger x_F than for π^+.

- $A_N(\vec{p}C \to pX)$ is consistent with zero.

As noted earlier, many model predictions exist for the 200 GeV/c results. It will be interesting to see whether such predictions can be extended down to 21.6 GeV/c, and to faithfully reproduce all the observations above. Finally, values of $A_N(\vec{p}p \to K^+X)$ at a few x_F will probably be extracted from the data collected. These may be able to further test predictions and, with the $\pi^\pm$ inclusive results presented here, provide a better understanding of the mechanism(s) producing the large spin effects seen.

REFERENCES

1. K. Heller in *Proceedings of the 12th International Symposium on High Energy Spin Physics, Amsterdam, 1996*, edited by C.W. de Jager et al., (World Scientific, 1996), p. 23.
2. D.L. Adams et al., Phys. Lett. **B264**, 462 (1991); A. Bravar et al., Phys. Rev. Lett., **77**, 2626 (1996).
3. Y. Yamamoto, K. Kubo, and H. Toki, Prog. Theor. Phys. **98**, 95 (1997).
4. Z. Liang and T. Meng, Z. Phys. **A344**, 171 (1992); C. Boros, Z. Liang, and T. Meng, Phys. Rev. Lett. **70**, 1751 (1993).
5. S.M. Troshin and N.E.Tyurin, Phys. Rev. **D52**, 3862 (1995); ibid. **D54**, 838 (1996).
6. D. Sivers, Phys. Rev. **D41**, 83 (1990); ibid. **D43**, 261 (1991).
7. A.V. Efremov and O.V. Teryaev, Phys. Lett. **B150**, 383 (1985).
8. M. Anselmino, M. Boglione, and F. Murgia, Phys. Lett. **B362**, 164 (1995); M. Anselmino and F. Murgia, Phys. Lett. **B442**, 470 (1998).
9. J. Qiu and G. Sterman, Phys. Rev. **D59**, 014004 (1998).
10. J. Collins, Nucl. Phys. **B396**, 161 (1993).
11. M. Anselmino, M. Boglione, and F. Murgia, Phys. Rev. **D60**, 054027 (1999).
12. M. Boglione and E. Leader, Phys. Rev. **D61**, 114001 (2000).
13. W.H. Dragoset et al., Phys. Rev. **D18**, 3939 (1978).
14. B.E. Bonner et al., Phys. Rev. **D41**, 13 (1990).
15. S. Saroff et al., Phys. Rev. Lett. **64**, 995 (1990).
16. K. Krueger et al., Phys. Lett. **B459**, 412 (1999).
17. D.G. Crabb et al., Nucl. Phys. **B121**, 231 (1977).
18. T. Eichten et al., Nucl. Phys. **B44**, 333 (1972).

5. SPIN PHYSICS WITH PHOTONS AND ELECTRONS

Spin effects in diffractive hadron photoproduction

S.V. Goloskokov

Bogoliubov Laboratory of Theoretical Physics, Joint Institute for Nuclear Research, Dubna 141980, Moscow region, Russia

Abstract. We study spin asymmetries in diffractive $Q\bar{Q}$ and vector meson production which are sensitive to the spin-dependent part of the two-gluon-nucleon coupling. It is found that the A_{ll} and A_{lT} asymmetry in diffractive reactions can be used to study polarized gluon distributions of the proton.

I INTRODUCTION

Investigation of the structure of hadrons, is a problem of considerable interest now. The inclusive reaction can be used to study ordinary parton distributions. However, it is difficult to distinguish events with a single outgoing proton or jet in a fixed target experiment like COMPASS [1]. In this case, at small x the diffractive events will contribute together with nondiffractive one. The measured asymmetry can be written in the form

$$A_{exp} = \frac{\Delta\sigma_{ND} + \Delta\sigma_D}{\sigma_{ND} + \sigma_D} = A(1-R) + A_D R, \quad R = \frac{\sigma_D}{\sigma_{ND} + \sigma_D}. \tag{1}$$

Here $A = \Delta\sigma_{ND}/\sigma_{ND}$ and $A_D = \Delta\sigma_D/\sigma_D$. It can be shown that the ratio R should increase with $x \to 0$. The integrated over x R ratio has been found at HERA to be about 20–30% [2]. This means that diffractive events might be important in extraction of asymmetry at small x from experiment. The diffractive hadron photoproduction can be expressed in terms of skewed parton distribution (SPD) in the nucleon $\mathcal{F}_\zeta(x)$ [3]. Investigation of such diffractive reactions should play a keystone role in future study $\mathcal{F}_x(x)$ at small x. In the diffractive charm quark production including J/Ψ reactions, the predominant contribution is determined by the two-gluon exchange (gluon SPD). Analysis of these reactions should throw light on the gluon structure of the proton at small x [4,5].

To study spin effects in the diffractive hadron production, one must know the structure of the two-gluon coupling with the proton at small x. The QCD-inspired diquark model generates the spin-dependent ggp coupling [6] of the following form:

$$V_{pgg}^{\alpha\alpha'}(p,t,x_P,l_\perp) = (\gamma^\alpha p^{\alpha'} + \gamma^{\alpha'} p^\alpha)B(t,x_P,l_\perp) + 4p^\alpha p^{\alpha'} A(t,x_P,l_\perp)$$
$$+ \epsilon^{\alpha\beta\delta\rho} p_\delta \gamma_\rho \gamma_5 D(t,x_P,l_\perp). \quad (2)$$

The first two terms of the vertex (2) are symmetric over α, α' indices. The structure $(\gamma^\alpha p^{\alpha'} + \gamma^{\alpha'} p^\alpha)B(t)$ in (2) determines the spin-non-flip contribution. The term $p_\alpha p_{\alpha'} A(r)$ leads to the transverse spin-flip in the vertex which does not vanish in the $s \to \infty$ limit. The single spin transverse asymmetry predicted in the models [6,7] is about 10% for $|t| \sim 3\text{GeV}^2$ which is of the same order of magnitude as has been observed experimentally [8]. These model approaches give for the ratio $\alpha = A/B \leq 0.1 \text{GeV}^{-1}$

The asymmetric structure in (2) is proportional to $D\gamma_\rho\gamma_5$ and can be associated with ΔG. It should give a visible contribution to the double spin longitudinal asymmetry A_{ll} [9]. The value of this structure is not well known now from our model estimations.

In this report, we shall analyze spin effects caused by the structures A and B. It will be shown here that such effects will be small in the A_{ll} asymmetry. The double spin asymmetry for a longitudinally polarized lepton and a transversely polarized proton is predicted to be not small and mainly determined by the A term in (2). Such asymmetry should be used to study this structure in the ggp coupling.

II DIFFRACTIVE HADRON PRODUCTION AND SPD

Let us study the diffractive J/Ψ production at high energies and fixed momentum transfer. The fractions of the momenta of proton carried by the Pomeron, $x_P \sim (m_J^2 + Q^2 + |t|)/W^2$ is small at high energies. The $\gamma^* \to J/\Psi$ transition amplitude is described by a nonrelativistic wave function [4,10]. Gluons from the Pomeron are coupled with the single and different quarks in the $c\bar{c}$ loop. The spin-average and spin dependent cross sections of the J/Ψ leptoproduction with parallel and antiparallel longitudinal polarization of a lepton and a proton are determined by the relation

$$\frac{d\sigma(\pm)}{dQ^2 dy dt} = \frac{1}{2}\left(d\sigma(\rightleftarrows) \pm d\sigma(\rightrightarrows)\right) = \frac{|T^\pm|^2}{32(2\pi)^3 Q^2 s^2 y}. \quad (3)$$

For the spin-average amplitude square we find [11]

$$|T^+|^2 = s^2 N\left((2-2y+y^2)m_J^2 + 2(1-y)Q^2\right)\left[|\tilde{B} + 2m\tilde{A}|^2 + |\tilde{A}|^2|t|\right]. \quad (4)$$

Here the term proportional to $(2-2y+y^2)m_J^2$ represents the contribution of the virtual photon with transverse polarization. The $2(1-y)Q^2$ term describes the effect of longitudinal photons. The N factor in (4) is normalization, and the $\tilde{A}$ and $\tilde{B}$ functions are expressed through the integral over transverse momentum of the gluon. The function $\tilde{B}$ is determined by

$$\tilde{B} = \frac{1}{4\bar{Q}^2} \int \frac{d^2 l_\perp (l_\perp^2 + \vec{l}_\perp \vec{\Delta}) B(l_\perp^2, x_P, ...)}{(l_\perp^2 + \lambda^2)((\vec{l}_\perp + \vec{\Delta})^2 + \lambda^2)[l_\perp^2 + \vec{l}_\perp \vec{\Delta} + \bar{Q}^2]}$$

$$\sim \frac{1}{4\bar{Q}^4} \int_0^{l_\perp^2 < \bar{Q}^2} \frac{d^2 l_\perp (l_\perp^2 + \vec{l}_\perp \vec{\Delta})}{(l_\perp^2 + \lambda^2)((\vec{l}_\perp + \vec{\Delta})^2 + \lambda^2)} B(l_\perp^2, x_P) = \frac{1}{4\bar{Q}^4} \mathcal{F}_{x_P}^g(x_P, t, \bar{Q}^2). \quad (5)$$

and connected with the gluon SPD [12]. Here $\bar{Q}^2 = (m_J^2 + Q^2 + |t|)/4$. The $\tilde{A}$ function is determined by the similar integral. The spin-dependent amplitude square looks like

$$|T^-|^2 = s|t|(2-y)N\left[|\tilde{B}|^2 + m(\tilde{A}^\star\tilde{B} + \tilde{A}\tilde{B}^\star)\right] m_J^2. \quad (6)$$

The asymmetry $A_{ll} = \sigma(-)/\sigma(+)$ depends on the ratio of the spin-flip to the non-flip parts of the coupling (2) $\alpha_{flip} = \tilde{A}(t)/\tilde{B}(t)$ which has been found to be about 0.1. The predicted asymmetry at HERMES energies is shown in Fig. 1. The contribution of the spin-dependent A term in (2) to the double spin A_{ll} asymmetry of the J/Ψ production does not exceed two per cent for the momentum transfer $|t| \leq 1\text{GeV}^2$. Sensitivity of the asymmetry to α is rather weak. At HERA energies, the asymmetry will be negligible.

For the diffractive $Q\bar{Q}$ leptoproduction the spin-average and spin-dependent cross section can be written in the form

$$\frac{d^5\sigma(\pm)}{dQ^2 dy dx_P dt dk_\perp^2} = \left(\frac{(2-2y+y^2)}{(2-y)}\right) \frac{C(x_P, Q^2)}{\sqrt{1 - 4k_\perp^2 \beta/Q^2}} N(\pm). \quad (7)$$

Here $C(x_P, Q^2)$ is a normalization function which is common for the spin average and spin dependent cross section. The $N(\pm)$ functions are expressed through the

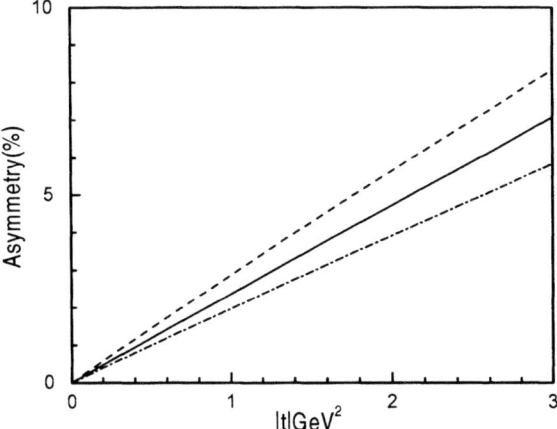

FIGURE 1. The A_{ll} asymmetry of the J/Ψ production at HERMES: solid line -for $\alpha_{flip} = 0$; dot-dashed line -for $\alpha_{flip} = -0.1$; dashed line -for $\alpha_{flip} = 0.1$

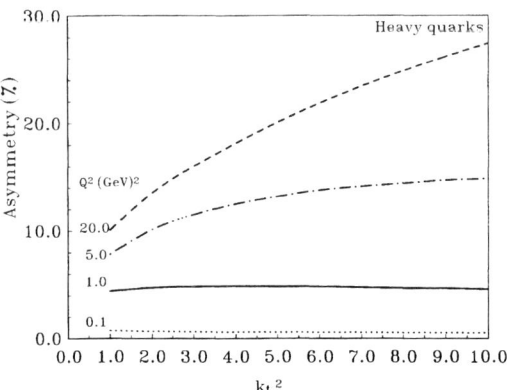

FIGURE 2. The predicted Q^2 dependence of the A_{lT} asymmetry for the $c\bar{c}$ production at COMPASS for $\alpha = 0.1 \text{GeV}^{-1}$, $x_P=0.1$, $y=0.5$

same skewed gluon distributions $\mathcal{F}^g_{x_P}(x_P, t, \bar{Q}_1^2)$ as for vector meson production but at a different scale $Q_1^2 = m_Q^2 + k_\perp^2$ (k is a quark momentum). Note that x_P is not fixed in this reaction and usually $x_P \leq 0.1$ The predicted asymmetry is quite small and does not exceed 1%. It has a week dependence on the $\alpha = \tilde{A}/\tilde{B}$ ratio. Moreover, the A_{ll} asymmetry is predicted to vanish for $Q^2 \to 0$ as $A_{ll} \propto Q^2/(Q^2 + Q_0^2)$ with $Q_0^2 \sim 1 \text{GeV}^2$.

The A structure in (2) should contribute to the A_{lT} asymmetry with longitudinal lepton and transverse proton polarization. The calculation of this asymmetry is similar to the analysis of A_{ll}, which has been carried out before. It has been found that the A_{lT} asymmetry is not small and proportional to the $\alpha = \tilde{A}/\tilde{B}$ ratio. The A_{lT} asymmetry is proportional to the scalar production of the proton spin vector, and the jet momentum $A_{lT} \propto (s_\perp \cdot k_\perp) \propto \cos(\phi_{Jet})$ and the asymmetry integrated over the azimuthal jet angle ϕ_{Jet} is zero. We have calculated the A_{lT} asymmetry for the case when the proton spin vector is perpendicular to the lepton scattering plane and the jet momentum is parallel to this spin vector. The predicted asymmetry is large and shown in Fig. 2. The reason for the large value of A_{lT} is that we do not find here a small coefficient x_P as for the A_{ll} asymmetry [13].

III CONCLUSION

In the present report, the polarized cross section of the diffractive hadron leptoproduction at high energies has been studied. The two-gluon exchange model with the spin-dependent gg-proton coupling (2) has been used. We consider all the graphs where the gluons from the Pomeron couple to a different quark in the loop and to the single one. This provides a gauge-invariant scattering amplitude.

Our calculations show that the contribution of the structure A in (2) to A_{ll} is smaller than 1-2%. Not small effects in the double spin A_{ll} asymmetry should be determined by the $\Delta G \propto D\gamma_\rho\gamma_5$ term of the vertex (2). The results obtained here show that diffractive asymmetry in the $Q\bar{Q}$ production vanishes as $Q^2 \to 0$. We can conclude that most likely such effects do not provide additional problems in extracting ΔG from the A_{ll} asymmetry because the COMPASS experiment plans to study the open charm production at small Q^2 [1].

It is shown that the gluon SPD $\mathcal{F}^g_{x_P}(x_P)$ and connected with ΔG distribution $\mathcal{G}^g_{x_P}(x_P)$ at the small $x_P \sim (m_V^2 + Q^2)/W^2$ can be studied from the double spin asymmetry in the vector meson photoproduction. The contributions of the quark SPDs are non-negligible for x of about 0.1 where the HERMES and COMPASS experiments will operate. Thus, in the case of the ϕ production the strange quark SPD might be studied in addition to the gluon one.

It is found here that the A_{lT} asymmetry of the diffractive heavy quark production is predicted to be not small, about 10-20%. It can give direct information about the spin-dependent structure A in the ggp coupling. A similar contribution to A_{lT} in the vector meson production vanishes because of the integration over $k_\perp$. The structure, which is proportional to x_p in the A_{lT} asymmetry of the vector meson production will be studied later.

We can conclude that important information on the spin–dependent SPD at small x can be obtained from double spin asymmetries in diffractive hadron photoproduction reactions.

The author is grateful to the Organizing Committee of SPIN2000 for financial support. This work was supported in part by the Russian Fond of Fundamental Research, Grant 00-02-16696.

REFERENCES

1. COMPASS Collaboration, Baum G., et al, *Proposal COMPASS*, CERN/SPSLC 96-16, (1996).
2. H1 Collaboration, Aid S. et al, *Z. Phys.* **C69**, 27 (1995).
3. Radyushkin A.V., *Phys. Rev*, **D56**, 5524 (1997);
   Ji X., *Phys. Rev* **D55**, 7114 (1997).
4. Ryskin M.G., *Z. Phys* **C57**, 89 (1993).
5. Brodsky S.J. at al, *Phys. Rev.* **D50**, 3134 (1994).
6. Goloskokov S.V., Kroll P., *Phys. Rev.* **D60**, 014019 (1999).
7. Goloskokov S.V., Kuleshov S.P., Selyugin O.V., *Z. Phys.* **C50**, 455 (1991).
8. Peaslee D.C., et al, *Phys. Rev. Lett.* **51**, 2358 (1983);
   Fidecaro G., at al, *Phys. Lett.* **B105**, 309 (1981).
9. Bartels J., Gehrmann T., Ryskin M.G., *Eur. Phys.J.* **C11**, 325 (1999).
10. Diehl M., *Eur. Phys. J.* **C4**, 497 (1998).
11. Goloskokov S.V., *Eur. Phys. J.* **C11**, 309 (1999).
12. Diehl M., Feldmann T., Jakob R., Kroll P., *Eur. Phys. J.* **C8**, 409 (1999).
13. Goloskokov S.V., *Mod. Phys. Lett.* **A12**, 173 (1997).

Nucleon resonances in polarized ω photoproduction

Yongseok Oh [a], Alexander I. Titov [b], and T.-S. H. Lee [c]

[a] *Institute of Physics and Applied Physics, Yonsei University, Seoul 120-749, Korea*
[b] *Bogolyubov Laboratory of Theoretical Physics, JINR, Dubna 141980, Russia*
[c] *Physics Division, Argonne National Laboratory, Argonne, Illinois 60439*

Abstract.
   The role of the nucleon resonances (N^*) in ω photoproduction is investigated by using the resonance parameters predicted by Capstick and Roberts. The contributions from the nucleon resonances are found to be significant in various spin asymmetries. In particular, we found that a crucial test of our predictions can be made by measuring the parity asymmetry and beam-target double asymmetry at forward scattering angles.

   The constituent quark models predict a much richer nucleon excitation spectrum than what has been observed in pion-nucleon scattering [1]. This has been attributed to the possibility that a lot of the predicted nucleon resonances (N^*) could couple weakly to the πN channel [2]. Therefore it is necessary to search for the nucleon excitations in other reactions to resolve the so-called "missing resonance problem." Electromagnetic production of vector mesons (ω, ρ, ϕ) is one of such reactions and is being investigated experimentally, e.g., at LEPS of SPring-8, TJNAF, ELSA-SAPHIR of Bonn and GRAAL of Grenoble.
   The role of the nucleon excitations in vector meson photoproduction was studied recently by Zhao et al. [3,4] within an $SU(6) \times O(3)$ constituent quark model. With the meson-quark coupling parameters adjusted to fit the existing data, they found that the single polarization observables are sensitive to the nucleon resonances.
   We are motivated by the predictions by Capstick and Roberts [5,6] based on the constituent quark model which accounts for the configuration mixing due to the residual quark-quark interactions [7] and the $^3P_0$ model [8] for the meson decay channels. Thus it would be interesting to see how these predictions differ from those of Refs. [3,4] and whether it can be tested experimentally.
   We focus on ω photoproduction in this work, simply because its non-resonant reaction mechanisms are fairly well established [9,10]. This reaction is dominated by diffractive process at high energies and by one-pion exchange at low energies which may be assumed as the dominant part of non-resonant background and may be used as a starting point for investigating the N^* effects.

We assume that the non-resonant (background) invariant amplitude has the form

$$I_{fi}^{bg} = I_{fi}^{P} + I_{fi}^{ps} + I_{fi}^{N}, \tag{1}$$

where I_{fi}^{P}, I_{fi}^{ps}, and I_{fi}^{N} denote the amplitudes due to the Pomeron, pseudoscalar-meson exchange, and direct and crossed nucleon terms, respectively. The four-momenta of the incoming photon, outgoing ω, initial nucleon, and final nucleon are denoted as k, q, p, and p' respectively, which defines $t = (p - p')^2 = (q - k)^2$, $s \equiv W^2 = (p + k)^2$, and the ω production angle θ by $\cos\theta \equiv \mathbf{k} \cdot \mathbf{q}/|\mathbf{k}||\mathbf{q}|$.

For the Pomeron exchange, which governs the total cross sections and differential cross sections at low $|t|$ in the high energy region, we follow the Donnachie-Landshoff model [11]. For the details of this model, we refer to, e.g., Refs. [12,13]. The pseudoscalar-meson exchange amplitude is calculated from the effective Lagrangian of Refs. [9,10] with slightly modified cut-off parameters $\Lambda_\pi = 0.6$ GeV and $\Lambda_{\omega\gamma\pi} = 0.7$ GeV.

We evaluate the direct and crossed nucleon terms from the Lagrangian,

$$\mathcal{L}_{VPP} = -g_V \bar{P} \left(\gamma_\mu V^\mu - \frac{\kappa_V}{2m_p} \sigma^{\mu\nu} \partial_\nu V_\mu \right) P, \tag{2}$$

where P stands for the proton Dirac spinor and V denotes γ or ω. When $V = \gamma$, we have $g_\gamma = e$ and $\kappa_\gamma = 1.79$. For the ωNN coupling, i.e., when $V = \omega$, we take $g_{\omega NN} = 10.35$ and $\kappa_\omega = 0$, which are determined in a study of πN scattering and pion photoproduction [14]. The ωNN vertices are dressed by the form factor, $\Lambda_N^4/[\Lambda_N^4 + (p^2 - m_p^2)^2]$, where p is the four momentum of the off-shell nucleon with $\Lambda_N = 0.5$ GeV.

In order to estimate the nucleon resonance contributions we make use of the quark model predictions on the resonance photo-excitation ($\gamma N \to N^*$) and the resonance decay ($N^* \to \omega N$) reported in Refs. [5,6] using a relativised quark model. Referring the detailed description of our resonant model to Ref. [15], here we discuss the main results of our investigation. The resonant amplitude is defined via N^* production amplitude $\mathcal{M}_{\gamma N \to N^*}$ and decay amplitude $\mathcal{M}_{N^* \to N'\omega}$:

$$I^{N^*} \propto \sum_{J,M_J} \mathcal{M}_{N^* \to N'\omega} \mathcal{M}_{\gamma N \to N^*} / \left(\sqrt{s} - M_R^J + \frac{i}{2} \Gamma^J(s) \right), \tag{3}$$

where M_R^J is the mass of an N^* with spin quantum numbers (J, M_J) and $\Gamma^J(s)$ is the energy dependent total decay width [16]. Since the most nucleon resonances we are dealing with are missing resonances, there is no information for their total decay widths. Therefore we rely on the averaged decay widths of N^* listed in Particle Data Group [17] and take $\Gamma^J(M_R^J) \simeq 300$ MeV. The amplitudes $\mathcal{M}_{\gamma N \to N^*}$ and $\mathcal{M}_{N^* \to N'\omega}$ are related to the corresponding transition amplitude as $\mathcal{M}_{\gamma N \to N^*} \propto A_{M_J}$ and $\mathcal{M}_{N^* \to N'\omega} \propto \sum G(J, L, S)$, where the resonance parameters are taken from Refs. [5,6]. In this study, we consider 12 positive parity and 10 negative parity nucleon

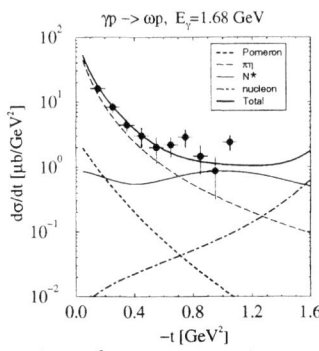

FIGURE 1. Differential cross sections of $\gamma p \to p\omega$ reaction as a function of t at $E_\gamma = 1.68$ GeV. Data are taken from Ref. [18].

resonances up to spin-9/2. Three of them were seen in the πN channel with four-star rating, five of them with two-star rating, and one of them with one-star rating. (See Ref. [15].) The majority of the predicted N^*'s are "missing" so far. Here we should also mention that we are not able to account for the resonances with the predicted masses less than the ωN threshold, since their decay vertex functions with an off-shell momentum are not available yet in the model of Refs. [5,6].

As an example for the role of nucleon resonances, we present our results for the differential cross section of ω photoproduction at $E_\gamma = 1.68$ GeV in Fig. 1. We also found that the data could be described to a very large extent for $E_\gamma \leq 5$ GeV. One can see that the contributions due to the N^* excitations (dotted line) and the direct and crossed nucleon terms (dot-dashed line) help bring the agreement with the data at large angles. Close inspection of the resonance part shows that the contributions from $N\frac{3}{2}^+(1910)$ and $N\frac{3}{2}^-(1960)$ are the largest at $W = 1.79 \sim 2.12$ GeV. In Ref. [5], the $N\frac{3}{2}^-(1960)$ is identified as a two star $D_{13}(2080)$ resonance of PDG, while the $N\frac{3}{2}^+(1910)$ is a missing resonance. In the study of Ref. [3], the authors found that $F_{15}(2000)$ dominates. This resonance is identified with $N\frac{5}{2}^+(1995)$ in Ref. [5] and is found to be not so strong in our calculation. The difference between the two calculations reflects the difference of the employed quark models.

Since it is difficult to test our predictions by considering only the angular distributions, we turn to the spin variables. We first examine the single spin observables [19,20]. Our predictions for photon asymmetry (Σ_x), target asymmetry (T_y), recoil nucleon asymmetry (P_y), and vector-meson tensor asymmetry ($V_{z'z'}$) are shown in Fig. 2. We find that the N^* excitations change the predictions from the dotted curves to the solid curves. The dashed curves are obtained when only the $N\frac{3}{2}^+(1910)$ and $N\frac{3}{2}^-(1960)$ are included in calculating the resonant part of the amplitude. Although our predictions are different from those of Ref. [3], we confirm their conclusion that the single polarization observables are sensitive to the N^* excitations but mostly at large scattering angles.

In order to probe the role of the nucleon resonances in ω photoproduction, we

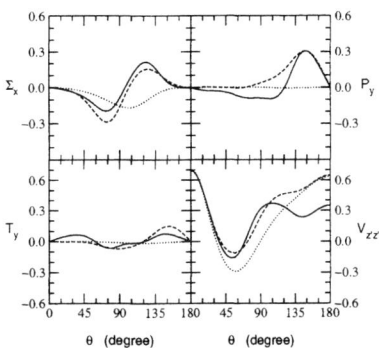

FIGURE 2. Single asymmetries at $E_\gamma = 1.7$ GeV. The dotted curves are calculated without including N^* effects, the dashed curves include contributions of $N\frac{3}{2}^+(1910)$ and $N\frac{3}{2}^-(1960)$ only, and the solid curves are calculated with all N^*.

address two polarization observables that are sensitive to the N^* contributions *at forward scattering angles*. The first one is the parity asymmetry P_σ [21]. At forward scattering region where the one-pion exchange is dominant, one expects $P_\sigma = -1$. Thus any deviation from this value will be only due to N^* excitation and Pomeron exchange, since the contribution from the direct and crossed nucleon terms is two or three orders in magnitude smaller at $\theta = 0$ (see Fig. 1). Our predictions for P_σ are shown in Fig. 3 (left panel). We show the results from calculations with (solid curve) and without (dotted curve) including the N^* contributions. The difference between them is striking and can be unambiguously tested experimentally. Here we also find that the $N\frac{3}{2}^+(1910)$ and $N\frac{3}{2}^-(1960)$ contributions are dominant. By keeping only these two resonances in calculating the resonant part of the amplitude, we obtain the dashed curve which is not too different from the full calculation (solid curve). Another asymmetry which is sensitive to the N^* excitations at forward scattering angles is the beam-target double asymmetry (C_{zz}^{BT}) [19]. Given in the right panel of Fig. 3 are our predictions on C_{zz}^{BT} at $\theta = 0$ as a function of invariant mass W. The striking difference between the solid curve and dotted curve is due to the N^* excitations. Again, the $N\frac{3}{2}^+(1910)$ and $N\frac{3}{2}^-(1960)$ give the dominant contributions (dashed curve).

In summary, we have investigated the role of nucleon resonances in ω photoproduction, especially in the resonance region. It was found that their role is important in the differential cross sections at large angles and some spin asymmetries can be used to identify the role of the nucleon resonances at forward scattering angles where precise measurements might be more favorable because the cross sections are peaked at $\theta = 0$. Experimental test of them will be a useful step toward resolving the so-called "missing resonance problem" or distinguishing different quark model predictions.

Acknowledgements. This work was supported in part by the Brain Korea 21 project of Korean Ministry of Education, Russian Foundation for Basic Research

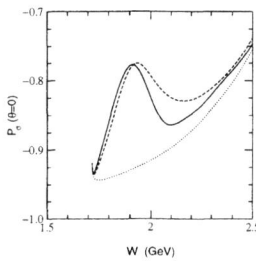 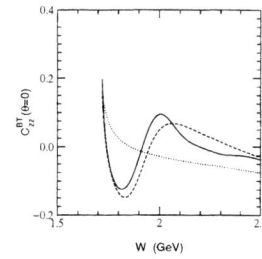

FIGURE 3. Parity asymmetry P_σ at $\theta = 0$ (left panel) and beam-target asymmetry C^{BT} (right panel) as a function of W. Notations are the same as in Fig. 2.

under Grant No. 96-15-96426, and U.S. DOE Nuclear Physics Division Contract No. W-31-109-ENG-38.

REFERENCES

1. N. Isgur and G. Karl, Phys. Lett. **72B**, 109 (1977); Phys. Rev. D **18**, 4187 (1978); **19**, 2653 (1979), **23**, 817(E) (1981); R. Koniuk and N. Isgur, *ibid.* **21**, 1868 (1980).
2. See, e.g., S. Capstick, Florida State Univ. Report, nucl-th/0011082.
3. Q. Zhao, Z. Li, and C. Bennhold, Phys. Lett. B **436**, 42 (1998); Phys. Rev. C **58**, 2393 (1998).
4. Q. Zhao, Nucl. Phys. **A675**, 217 (2000); nucl-th/0010038, Phys. Rev. C (in print).
5. S. Capstick, Phys. Rev. D **46**, 2864 (1992).
6. S. Capstick and W. Roberts, Phys. Rev. D **49**, 4570 (1994).
7. S. Godfrey and N. Isgur, Phys. Rev. D **32**, 189 (1985); S. Capstick and N. Isgur, *ibid.* **34**, 2809 (1986).
8. See, for example, A. Le Yaouanc, L. Oliver, O. Pene, and J. C. Raynal, *Hadron Transitions in the Quark Model* (Gordon and Breach, New York, 1988).
9. P. Joos *et al.*, Nucl. Phys. **B122**, 365 (1977).
10. B. Friman and M. Soyeur, Nucl. Phys. **A600**, 477 (1996).
11. A. Donnachie and P. V. Landshoff, Nucl. Phys. **B244**, 322 (1984); **B267**, 690 (1986); Phys. Lett. B **185**, 403 (1987); **296**, 227 (1992).
12. J.-M. Laget and R. Mendez-Galain, Nucl. Phys. **A581**, 397 (1995).
13. M. A. Pichowsky and T.-S. H. Lee, Phys. Rev. D **56**, 1644 (1997).
14. T. Sato and T.-S. H. Lee, Phys. Rev. C **54**, 2660 (1996).
15. Y. Oh, A. I. Titov, and T.-S. H. Lee, nucl-th/0006057, Phys. Rev. C (in print).
16. T. Yoshimoto, T. Sato, M. Arima, and T.-S. H. Lee, Phys. Rev. C **61**, 065203 (2000).
17. Particle Data Group, D. E. Groom *et al.*, Eur. Phys. J. C **15**, 1 (2000).
18. F. J. Klein, Ph.D. thesis, Bonn Univ. (1996); SAPHIR Collaboration, F. J. Klein *et al.*, πN Newslett. **14**, 141 (1998).
19. A. I. Titov, Y. Oh, S. N. Yang, and T. Morii, Phys. Rev. C **58**, 2429 (1998).
20. M. Pichowsky, Ç. Şavkli, and F. Tabakin, Phys. Rev. C **53**, 593 (1996).
21. K. Schilling, P. Seyboth, and G. Wolf, Nucl. Phys. **B15**, 397 (1970).

Polarization Phenomena in Vector Meson Photoproduction on Nucleons near Threshold

H. Babacan, T. Babacan, A. Gokalp, O. Yilmaz

Physics Department, Middle East Technical University, 06531 Ankara, Turkey

Abstract. We propose and develop a model for ρ^0 and ω photoproduction on nucleons near threshold based on t-, s-, and u-channel amplitudes. In t-channel we consider $\pi-$ and $\sigma-$ exchanges and we take the coupling constants (g_v, g_t) in (s+u) nucleon contributions as free parameters. We fit our model to the available experimental differential cross-section data of $\gamma + p \to p + V$ reactions. We find that different models which are equivalently good for the description of the differential cross-section on proton targets may be differentiated by their predictions of the beam asymmetry on proton and neutron targets and of the differential cross-section on neutron targets.

The photoproduction of ρ^0- and ω-mesons from proton targets near threshold region have been studied employing the mechanisms of pseudoscalar and scalar t-channel exchanges, (s+u) nucleon contributions, intermediate nucleon resonances, and Pomeron exchange [1,2]. However, the existing experimental data [3] about photoproduction of ρ^0- and ω-mesons from proton targets do not allow to establish definitely the relative role of these different possible mechanisms. Since in vector meson photoproduction reactions we may hope to find the missing nucleon resonances, it is crucial to understand the background production mechanism. For this purpose, we propose a model for ρ^0- and ω-photoproduction on nucleons near threshold involving t-, s-, and u-channel amplitudes, and we do not consider Pomeron exchange since it is expected to play a dominant role at higher energies. The mechanisms in our model are illustrated in Fig. 1.

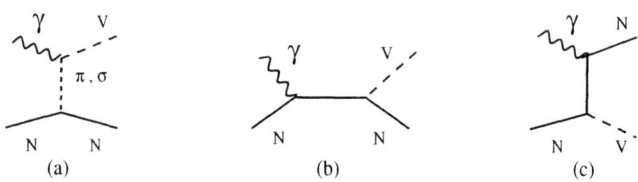

FIGURE 1. Mechanisms of the model for vector meson photoproduction: (a) t-channel exchanges, (b) and (c) s- and u-channel intermediate nucleon state contributions.

We describe the t-channel exchanges by the effective Lagrangians [4]

$$\mathcal{L}_{V\pi\gamma} = \frac{e}{M_V} g_{V\pi\gamma} \epsilon^{\mu\nu\alpha\beta} \partial_\mu V_\nu \partial_\alpha A_\beta \pi - ig_{\pi NN} \bar{N} \gamma_5 N \quad , \tag{1}$$

$$\mathcal{L}_{V\sigma\gamma} = \frac{e}{M_V} g_{V\sigma\gamma} [\partial^\alpha V^\beta \partial_\alpha A_\beta - \partial^\alpha V^\beta \partial_\beta A_\alpha] + g_{\sigma NN} \bar{N} N \sigma \quad , \tag{2}$$

and we regularize the resulting one-meson-exchange amplitudes by the factors

$$F_{\phi NN} = \frac{\Lambda_\phi^2 - M_\phi^2}{\Lambda_\phi^2 - t} \quad , F_{V\phi\gamma} = \frac{\Lambda_{V\phi\gamma}^2 - M_\phi^2}{\Lambda_{V\phi\gamma}^2 - t} \quad , \tag{3}$$

where $\phi = \pi^0, \sigma$ and we use $\Lambda_\pi = 0.7$ GeV, $\Lambda_\sigma = 1$ GeV $\Lambda_{V\pi\gamma} = 0.77$ GeV, and $\Lambda_{V\sigma\gamma} = 0.9$ GeV [2,4]. The coupling constants $g_{V\pi\gamma}$ are obtained from the experimental partial widths of vector meson radiative decays as $g_{\rho\pi\gamma} = 0.54$ and $g_{\omega\pi\gamma} = 1.82$. Moreover, we use $g_{\pi NN}^2/4\pi = 14.0$, $g_{\sigma NN}^2/4\pi = 8.0$, and $g_{\rho\sigma\gamma} = 2.71$, $g_{\omega\sigma\gamma} = 0.18$ [1,4]. The (s+u) intermediate nucleon-terms amplitude is obtained from the effective Lagrangian

$$\mathcal{L} = \bar{N}[g_v \gamma_\mu V^\mu - \frac{g_t}{2M_N} \sigma_{\mu\nu} \partial^\nu V^\mu] N + e\bar{N}[Q_N \gamma_\mu A^\mu - \frac{\kappa_N}{2M_N} \sigma_{\mu\nu} \partial^\nu A^\mu] N \tag{4}$$

where $Q_N = 1(0)$ for proton (neutron). We determine the coupling constants (g_v, g_t) by fitting the experimental differential cross-section data on proton targets. We, furthermore, also consider in our fits the coupling constants $g_{V\sigma\gamma}$ that are estimated recently on the basis of the experimental partial widths of the radiative decays $V \to \pi\pi\gamma$ [5,6]. In Fig. 2 and Fig. 3, we compare three different models for ω photoproduction with the coupling constants $(g_v = -2.54, g_t = 0.0, g_{\omega\sigma\gamma} = 0.182)$, $(g_v = 0.88, g_t = 0.52, g_{\omega\sigma\gamma} = 0.13)$, $(g_v = 1.01, g_t = 0.42, g_{\omega\sigma\gamma} = -0.27)$. In these models we use the coupling constant $g_{\omega\sigma\gamma} = 0.182$ estimated on the basis of vector meson dominance [1] and $g_{\omega\sigma\gamma} = 0.13$ and $g_{\omega\sigma\gamma} = -0.27$ estimated from the experimental branching ratio of the radiative decays $V \to \pi\pi\gamma$ [5,6]. The vector

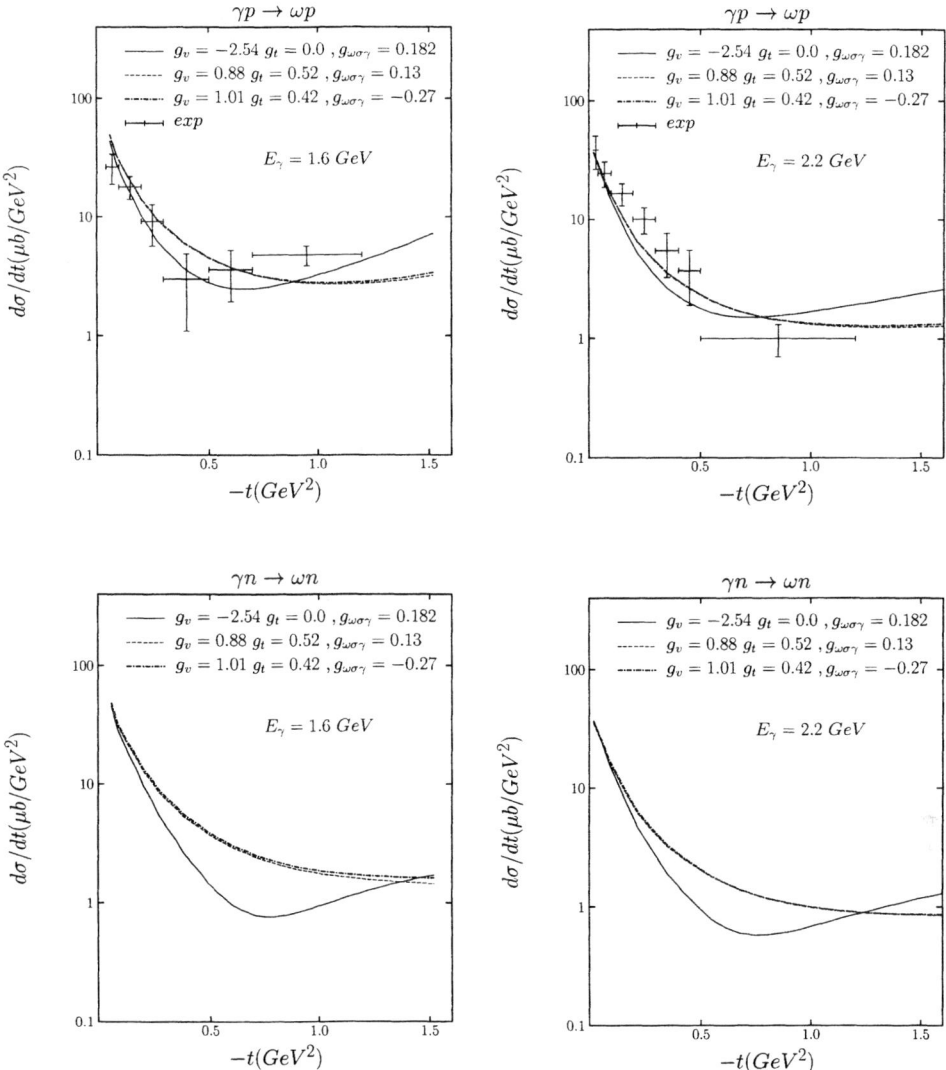

FIGURE 2. Fits of different models for differential cross-sections on proton targets and their predictions for differential cross-sections on neutron targets at $E_\gamma = 1.6$ GeV and $E_\gamma = 2.2$ GeV.

and tensor coupling constants (g_v, g_t) are obtained by performing a fit to the experimental differential cross-section data on proton targets. In our fit, we use $E_\gamma = 1.6$ GeV and $E_\gamma = 2.2$ GeV data together to determine the coupling constants (g_v, g_t), and then using these coupling constants we show the resulting differential cross-sections on proton targets and the predicted differential cross-sections on neutron targets in Fig. 2 for $E_\gamma = 1.6$ GeV and $E_\gamma = 2.2$ GeV. In Fig. 3 we plot

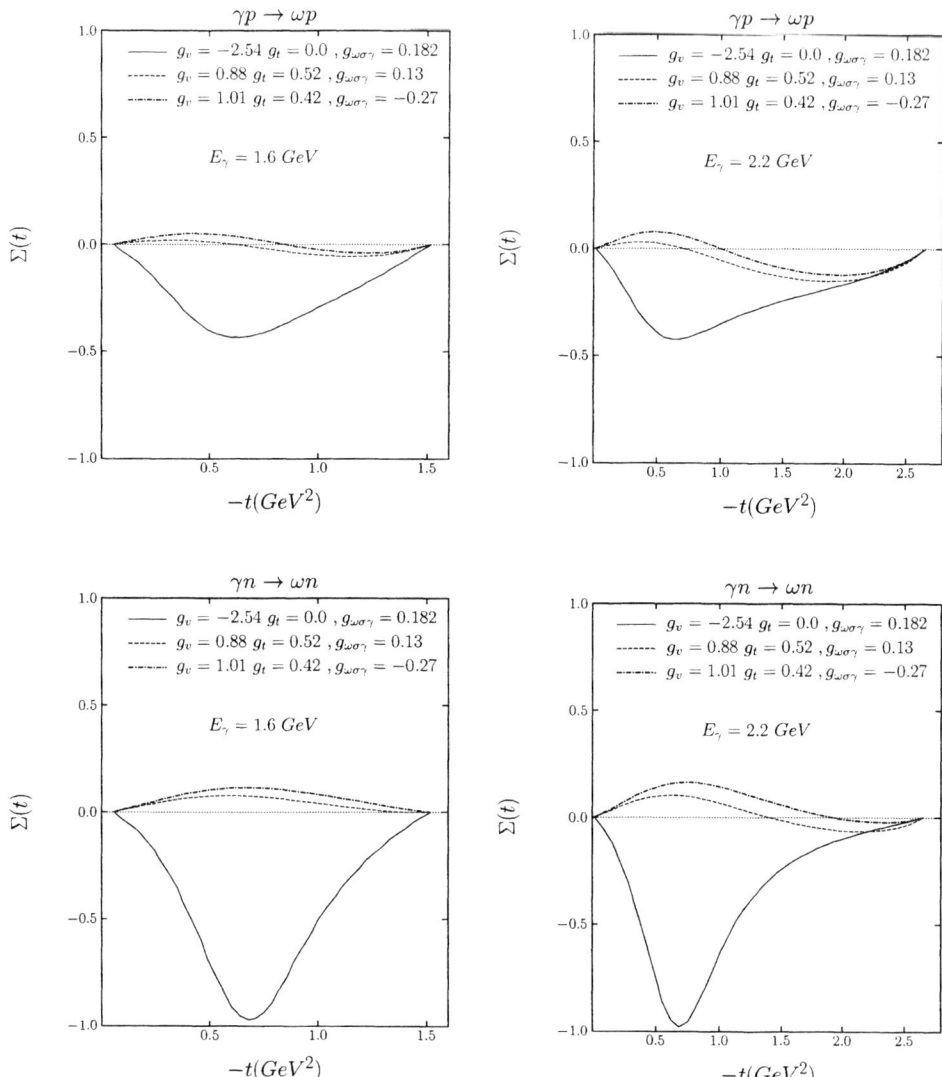

FIGURE 3. Predictions of different models for beam asymmetry on proton and neutron targets at $E_\gamma = 1.6$ GeV and $E_\gamma = 2.2$ GeV.

the predicted beam asymmetries by different models on proton and neutron targets as a function of t. We observe that these three models with different contributions of the above mentioned mechanisms are almost equivalently good for the description of the differential cross-sections of $\gamma p \to p\omega$ reactions near threshold. However, the predictions of these models for the beam asymmetry on proton and neutron targets as well as the differential cross-section for neutron target are quantitatively different, and they may lead to differentiation between different models. Work along these lines for ρ production is in progress and the preliminary results are in accordance with the above conclusion.

ACKNOWLEDGMENT

We thank M. P. Rekalo for suggesting this problem to us and for his guidance during the course of our work.

REFERENCES

1. B. Friman, M. Soyeur, Nucl. Phys. Rev. **A600**(1996)477.
2. Y. Oh, A. I. Titov, T. -S. H. Lee, nucl-th/0004055.
3. Aachen-Berlin-Bonn-Hamburg-Heidelberg-Munchen Collaboration, Phys. Rev. **175** (1968) 1669.
4. A. I. Titov, T. -S. H. Lee, H. Toki, O. Streltsova, Phys. Rev. **C60**(1999) 035205.
5. A. Gokalp, O. Yilmaz, Phys. Rev. **C 62**(2000) 093018.
6. A. Gokalp, O. Yilmaz, Phys. Lett. **B464**(2000) 69.

Hard Exclusive Meson Production at HERMES

D.Ryckbosch*, on behalf of the HERMES collaboration

*Department of Subatomic Physics, University of Gent, B-9000 Gent, Belgium

Abstract. An overview is given of recent results on hard exclusive production of vector and pseudoscalar mesons, obtained by the HERMES experiment. Cross sections for ρ^0-production by longitudinal photons are presented. For pion-production the azimuthal asymmetry for a longitudinally polarized proton target and $-t$-distribution are discussed.

INTRODUCTION

Of late the exclusive production of mesons (and photons) has received quite a lot of interest. This is mainly due to the recent advancement in the perturbative QCD (pQCD) description of such processes in terms of so-called Skewed Parton Distributions (SPD). These generalizations of the standard parton distributions provide a unified framework to discuss the hard exclusive production of various mesons and Deeply Virtual Compton Scattering (DVCS). While a reasonable amount of data exists for the electroproduction of vector mesons, there is much less information available for the pseudoscalar mesons. This is largely a consequence of the lower relevant cross sections. To distinguish the exclusive channel from the large non-exclusive background then requires an excellent energy resolution.

While the HERMES spectrometer was not designed and constructed specifically for a study of exclusive reactions (rather its forte is the study of semi-inclusive DIS reactions) its energy resolution is good enough to begin a systematic investigation of the hard exclusive electroproduction of the light mesons. First preliminary results are presented here.

SKEWED PARTON DISTRIBUTIONS

A major breakthrough in the description of exclusive reactions was the proof that the leading order amplitude for DVCS and exclusive meson production by longitudinal photons can be factorized in a hard scattering part and a soft nonperturbative nucleon structure part [1]. This latter can be parametrized in terms of 4

Skewed Parton Distributions which in the notation of Ji [2] are known as $H, \tilde{H}, E$ and $\tilde{E}$. Here H and E are averages over quark helicity, while $\tilde{H}$ and $\tilde{E}$ are helicity dependent. These distributions are functions of three variables: x, ξ and t. The momentum transfer to the target nucleon is given by $t = \Delta^2$, while ξ stands for its longitudinal fraction, and x is the average longitudinal momentum fraction of the parton. There exist important connections to the standard parton distributions:

$$H^q(x, 0, 0) = q(x)$$
$$\tilde{H}^q(x, 0, 0) = \Delta q(x)$$

The SPDs are furthermore directly related to the elastic formfactors of the nucleon, providing important boundary conditions.

The DVCS amplitude depends on all four SPDs (see contribution by M.Amarian to these proceedings). On the other hand one can use the production of either vector mesons or pseudoscalar mesons as a helicity-filter on the SPDs: the vector meson production depends only on the unpolarized distributions H and E, while the production of pseudoscalar mesons depends only on the polarized distributions $\tilde{H}$ and $\tilde{E}$. Of particular interest is the case of π^+-production. Here the presence of the pion pole at low $-t$ may make the contribution of $\tilde{E}$ dominate, whereas it is usually much smaller than the contribution from $\tilde{H}$.

EXCLUSIVE ρ^0-PRODUCTION

A large body of data exists for ρ^0-production at high W, coming from the HERA collider experiments. Fixed target experiments like E665 and NMC have provided data at somewhat lower W. All these experiments have mainly taken data at low values of x_{Bj}, meaning that they are sensitive to the skewed gluon distributions. In contrast to that, the average x_{Bj} probed in the HERMES experiment is of the order 0.3, increasing the sensitivity to the quark distributions.

The exclusive character of ρ^0-production in HERMES is shown in the left panel of fig.1. The small non-diffractive background can be accounted for by using standard DIS Monte Carlo simulations (properly normalized in the non-exclusive region) as shown by the histogram in the figure.

To deduce the longitudinal cross section from the total cross section is fairly easy for vector mesons. In this case it is possible to determine the polarization of the meson from the angular distribution of its decay particles. Assuming s-channel helicity conservation then leads to the required ratio of longitudinal to transverse photon cross sections (see contribution by E.Kinney to this proceedings.) The resulting longitudinal cross section for ρ^0-production is shown in fig.1, together with SPD calculations by Vanderhaeghen et al. It is clear that at HERMES kinematics the quark exchange mechanism dominates the cross section in this model.

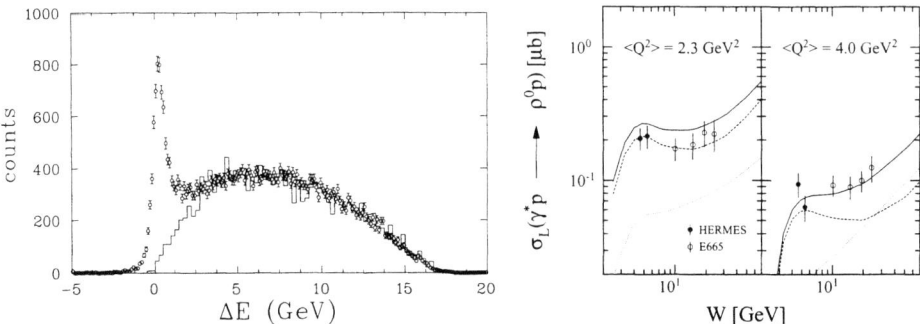

FIGURE 1. Left panel: ΔE-spectrum for ρ^0-production. The histogram shows an estimate of the non-diffractive background. Right panel: Cross section as compared to calculations of ref. [3] for quark exchange (dashed lines) and gluon exchange (dotted lines) mechanisms.

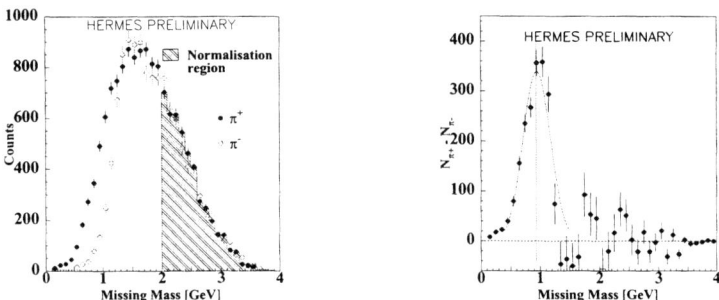

FIGURE 2. Left: Measured π^+ and π^- yield as function of missing mass. Right: Difference of π^+ and (normalised) π^- yield as function of missing mass.

EXCLUSIVE π^+-PRODUCTION

Exclusivity

While in the case of vector meson production the separation of the exclusive channel from the non-exclusive background was straightforward, this is much more difficult for pseudo-scalar mesons. In this case the non-exclusive background completely dominates the spectrum. This is shown in the left panel of figure 2 for charged pions. To determine the non-exclusive background in the π^+ spectrum it was assumed that it has the same shape as the π^- spectrum, where for a proton target there is no exclusive signal. The resulting net spectrum for π^+ production is then given in the right hand panel of figure 2. It shows a clear peak at the missing energy of the proton ground state with a width as expected from the results of simulations of the resolution of the spectrometer. Deducing cross sections for exclusive pion production by longitudinal photons is not possible in the same way as it was for vector mesons and requires a full Rosenbluth separation. Recently it

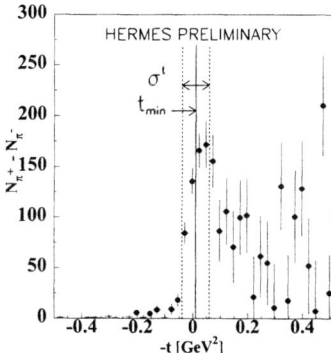

FIGURE 3. Exclusive π^+ yield as function of t.

was shown that it is possible to operate the HERMES detector with a lower beam energy, so a Rosenbluth separation may well be performed in the future.

The $-t$-distribution

It is expected [4] that the pion-pole term which appears in the $\tilde{E}$ SPD may dominate π^+ production, especially at low $\mid t \mid$. The $-t$-distribution as measured at HERMES is shown in fig.3. There are several theoretical calculations based on SPDs [3,5] for this $-t$-distribution. Once the data have been corrected for smearing and acceptance effects a comparison will be made.

Single spin azimuthal asymmetries

Apart from the cross sections and the $-t$-distribution there is yet another quantity that can be compared to calculations based on skewed parton distributions: azimuthal asymmetries. Calculations exist [6] for the case of a *transverse* target polarization and sizeable asymmetries are predicted. With the HERMES spectrometer it was possible to measure azimuthal asymmetries for a *longitudinal* target polarization (data taking with a transverse target is expected to start in 2001). In figure 4 the $\sin(\phi)$-moment of the azimuthal asymmetry is plotted as a function of the missing mass E_m, ϕ being the azimuthal angle of the pion around the virtual photon direction, with respect to the lepton scattering plane. At $E_m \gtrsim 2$ GeV this moment has been interpreted in terms of the transversity distribution (see contribution by P. Di Nezza to these proceedings). At the missing mass corresponding to exclusive pion production there is a clear change in the behaviour: for both π^+ and π^0 production the asymmetries become very much larger than in the semi-inclusive case. In the case of π^+-production the asymmetry even changes sign as

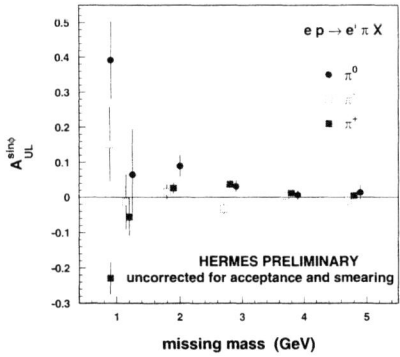

FIGURE 4. The $\sin(\phi)$-moment of the azimuthal asymmetry for longitudinally polarized hydrogen target as a function of missing mass. Only statistical errors are shown.

the exclusive region is reached. For π^- production, where an exclusive reaction is impossible, the value for the asymmetry is compatible with zero at low E_m.

SUMMARY

Clear evidence for exclusive meson production in the HERMES experiment is shown. As an example of vector meson production the longitudinal ρ^0 cross section is shown. For pseudoscalar mesons the first exclusive production peak for positive pions off hydrogen is also presented. Also the single spin azimuthal asymmetry displays a prominent exclusive signal. In these cases the HERMES data are compared to theoretical calculations using Skewed Parton Distributions as input.

REFERENCES

1. Collins, J.C., Frankfurt, L., and Strikman, M., *Phys. Rev. D* **56**, 2982 (1997)
2. Ji, X., *J. Phys. G* **24**, 1181 (1998)
3. Vanderhaeghen, M., Guichon, P.A.M., and Guidal, M., *Phys. Rev. D* **60**, 094017 (1999)
4. Frankfurt, L.L., Polyakov, M.A., and Strikman, M., hep-ph/9808449 (1998)
5. Piller, G. et al., *Eur. Phys. J. C* **10**, 307 (1999)
6. Frankfurt, L.L., Polyakov, M.V., Strikman, M., and Vanderhaeghen, M., *Phys. Rev. Lett.* **84**, 2589-2592 (2000)

Spin Structure Function of Virtual Photon and Polarized Parton Distributions[1]

Ken Sasaki* and Tsuneo Uematsu†

*Department of Physics, Faculty of Engineering, Yokohama National University
Yokohama 240-8501, Japan
†Department of Fundamental Sciences, FIHS, Kyoto University, Kyoto 606-8501, Japan

Abstract. The structure function $g_1^\gamma(x, Q^2, P^2)$ and parton distributions of polarized virtual photon target are investigated in pQCD up to the next-to-leading order (NLO). It is found that the NLO corrections significantly modify the leading log result of g_1^γ, in particular, both at small and large x. Parton distributions, predicted entirely up to NLO, are factorization-scheme-dependent. Analysis is made in four different factorization schemes. The scheme dependence is clearly seen in the $n = 1$ moments and the large-x behaviors of quark parton distributions. Through the analysis of the scheme dependence of parton distributions, the non-vanishing result of the first moment of g_1^γ is understood to be connected to QED and QCD axial anomaly.

1. Introduction

In polarized e^+e^- collision experiments expected to be performed in the future linear colliders, we can measure the spin-dependent structure function $g_1^\gamma(x, Q^2, P^2)$ of virtual photon, where $-Q^2$ ($-P^2$) is the mass squared of probe (target) photon (Fig.1). There are at least three significant features in the study of g_1^γ. Firstly, as in the case of the unpolarized photon structure function $F_2^\gamma(x, Q^2, P^2)$ [1], g_1^γ with the kinematical region $\Lambda^2 \ll P^2 \ll Q^2$, where Λ is the QCD scale parameter, can be calculated completely up to the next-to-leading order (NLO) by pQCD, in contrast to the case of real photon target ($P^2 = 0$) in which there exist in NLO non-perturbative pieces. Thus the study of the virtual photon structure functions such as F_2^γ and g_1^γ may provide a unique test of QCD.

Secondly, the first moment of g_1^γ has relevance to QCD axial anomaly, which has played an important role in the QCD analysis of nucleon spin structure functions. In the case of photon target, not only QCD but QED axial anomaly also appears

---
[1] Presented by K. Sasaki

on the stage. So it is interesting to see how QCD and QED axial anomalies have to do with g_1^γ. Thirdly, the behaviors of parton distributions in virtual photon can also be predicted entirely up to NLO, but they are factorization-scheme dependent. In particular, the prescriptions to treat the axial anomaly are different from scheme to scheme. The polarized virtual photon target, therefore, serves as a good testing ground for the study of factorization-scheme dependence of parton distributions.

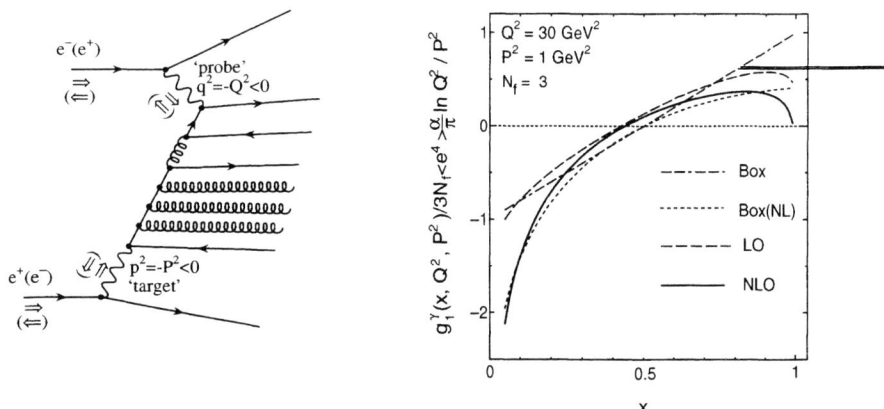

FIGURE 1. Two-photon process in polarized e^+e^- collision for $\Lambda^2 \ll P^2 \ll Q^2$ and the polarized photon structure function $g_1^\gamma(x, Q^2, P^2)$ for $Q^2 = 30$ GeV^2 and $P^2 = 1$ GeV^2 with $N_f = 3$. LO, NLO and Box (NL) denote QCD LO, NLO and Box-diagram (non-leading) results.

2. pQCD Calculation of $g_1^\gamma(x, Q^2, P^2)$

Applying the OPE supplemented by the renormalization group method or equivalently DGLAP type parton evolution equations, we find that the n-th moment of $g_1^\gamma(x, Q^2, P^2)$ for the kinematical region $\Lambda^2 \ll P^2 \ll Q^2$ is given by [2]

$$\int_0^1 dx x^{n-1} g_1^\gamma(x, Q^2, P^2) = \frac{\alpha}{4\pi} \frac{1}{2\beta_0} \left[\sum_{i=+,-,NS} L_i^n \frac{4\pi}{\alpha_s(Q^2)} \left\{ 1 - \left(\frac{\alpha_s(Q^2)}{\alpha_s(P^2)} \right)^{\lambda_i^n/2\beta_0+1} \right\} \right.$$
$$\left. + \sum_{i=+,-,NS} \mathcal{A}_i^n \left\{ 1 - \left(\frac{\alpha_s(Q^2)}{\alpha_s(P^2)} \right)^{\lambda_i^n/2\beta_0} \right\} + \sum_{i=+,-,NS} \mathcal{B}_i^n \left\{ 1 - \left(\frac{\alpha_s(Q^2)}{\alpha_s(P^2)} \right)^{\lambda_i^n/2\beta_0+1} \right\} \right.$$
$$\left. + \mathcal{C}^n + \mathcal{O}(\alpha_s) \right] \quad (1)$$

where $\alpha_s(Q^2)$ is the QCD running coupling constant, L_i^n, $\mathcal{A}_i^n$, $\mathcal{B}_i^n$ and $\mathcal{C}^n$ are computed from the 1- and 2-loop anomalous dimensions [3,4] as well as from 1-loop coefficient functions, and λ_i^n ($i = +, -, NS$) denote the eigenvalues of 1-loop anomalous dimension matrices. In Fig.1 we have shown $g_1^\gamma(x, Q^2, P^2)$ evaluated from Eq.(1) by inverse Mellin transform for $Q^2 = 30$ GeV^2 and $P^2 = 1$ GeV^2 with $N_f = 3$ [5].

We find that the NLO corrections significantly modify the leading log result of g_1^γ, in particular, both at small and large x.

3. Sum Rule for $g_1^\gamma(x, Q^2, P^2)$

The polarized structure function g_1^γ of real photon ($P^2 = 0$) has been shown to satisfy a remarkable sum rule, $\int_0^1 dx g_1^\gamma(x, Q^2) = 0$, which holds true in all orders of $\alpha_s(Q^2)$ in QCD [6–8]. Now we ask what happens to the case of virtual photon. Taking the $n \to 1$ limit, we find that the first three terms except the fourth, $C^{n=1}_\gamma$, in the square brackets of Eq.(1) vanish and that the sum rule remains finite but non-vanishing. In fact, we can go a step further to the $\mathcal{O}(\alpha_s)$ QCD corrections and we obtain [2]

$$\int_0^1 dx g_1^\gamma(x, Q^2, P^2) = -\frac{3\alpha}{\pi} \sum_{i=1}^{N_f} e_i^4 \left(1 - \frac{\alpha_s(Q^2)}{\pi}\right)$$
$$+ \frac{6\alpha}{\pi \beta_0} (\sum_{i=1}^{N_f} e_i^2)^2 \frac{\alpha_s(P^2) - \alpha_s(Q^2)}{\pi} + \mathcal{O}(\alpha_s^2). \tag{2}$$

which is in agreement with the one obtained by Narison, Shore and Veneziano [7], apart from the overall sign for the definition of g_1^γ. In fact, through the analysis of the factorization-scheme-dependence of polarized parton distributions, we see that the first term of the sum rule (2) is coming from QED axial anomaly and the second is from QCD axial anomaly[2].

4. Polarized Parton Distributions in Virtual Photon and Factorization Scheme Dependence

The moments of g_1^γ can be expressed, in terms of the moments of parton distributions and coefficient functions, as

$$g_1^\gamma(n, Q^2, P^2) = \Delta \vec{q}^\gamma(n, Q^2, P^2) \cdot \Delta \vec{C}^\gamma(n, Q^2) \tag{3}$$

with $\Delta \vec{q}^\gamma = (\Delta q_S^\gamma, \Delta G^\gamma, \Delta q_{NS}^\gamma, \Delta \Gamma^\gamma)$ and $\Delta \vec{C}^\gamma = (\Delta C_S^\gamma, \Delta C_G^\gamma, \Delta C_{NS}^\gamma, \Delta C_\gamma^\gamma)$, where $\Delta q_S^\gamma(\Delta q_{NS}^\gamma)$, ΔG^γ, $\Delta \Gamma^\gamma$ are the flavor-singlet (non-singlet)-quark, gluon, and photon distributions, respectively, in polarized virtual photon with mass $-P^2$, and $\Delta C_S^\gamma(\Delta C_{NS}^\gamma)$, ΔC_G^γ, and ΔC_γ^γ are the corresponding coefficient functions. Again, with the kinematical region $\Lambda^2 \ll P^2 \ll Q^2$, parton distributions $\Delta \vec{q}^\gamma$ can be predicted completely up to NLO. Although g_1^γ is a physical quantity and thus unique, there remains a freedom in factorization of g_1^γ into $\Delta \vec{q}^\gamma$ and $\Delta \vec{C}^\gamma$. Given the formula Eq.(3), we can always redefine $\Delta \vec{q}^\gamma$ and $\Delta \vec{C}^\gamma$ as follows:

---

[2] This notion was first pointed out by the authors of Ref. [7]

$$g_1^\gamma = \Delta\vec{q}^\gamma|_a \cdot \Delta\vec{C}^\gamma|_a , \qquad \text{with} \quad \Delta\vec{q}^\gamma|_a = \Delta\vec{q}^\gamma Z_a \text{ and } \Delta\vec{C}^\gamma|_a = Z_a^{-1}\Delta\vec{C}^\gamma . \qquad (4)$$

We have studied the factorization-scheme-dependence of parton distributions in polarized virtual photon. We report, in this talk, the results of our study in four different schemes, namely, 1) $\overline{\text{MS}}$, 2) chirally invariant (CI), 3) Adler-Bardeen (AB), 4) off-shell (OS) (For the detailed description of each factorization scheme, see [5].) The treatment of axial anomaly is different from scheme to scheme. In $\overline{\text{MS}}$ scheme, QCD and QED anomaly effects reside in the quark distributions, and we find

$$\Delta q_S^\gamma(n=1,Q^2,P^2)|_{\overline{\text{MS}}} = \left[-\frac{\alpha}{\pi} 3(\sum_{i=1}^{N_f} e_i^2)\right]\left\{1 - \frac{2}{\beta_0}\frac{\alpha_s(P^2) - \alpha_s(Q^2)}{\pi}n_f\right\} \qquad (5)$$

$$\Delta q_{NS}^\gamma(n=1,Q^2,P^2)|_{\overline{\text{MS}}} = \left[-\frac{\alpha}{\pi} 3(\sum_{i=1}^{N_f} e_i^4 - (\sum_{i=1}^{N_f} e_i^2)^2/N_f)\right]\left\{1 + \mathcal{O}(\alpha_s^2)\right\} \qquad (6)$$

$$\Delta C_{G,\,\overline{\text{MS}}}^{\gamma,\,n=1} = 0 , \quad \Delta C_{\gamma,\,\overline{\text{MS}}}^{\gamma,\,n=1} = 0 . \qquad (7)$$

On the other hand, in CI, AB, and OS schemes, QCD (QED) anomaly appears in the gluon (photon) coefficient function and not in the quark distributions. With $a = $CI, AB, and OS, we find

$$\Delta q_S^\gamma(n=1,Q^2,P^2)|_a = \Delta q_{NS}^\gamma(n=1,Q^2,P^2)|_a = 0 \qquad (8)$$

$$\Delta C_{G,\,a}^{\gamma,\,n=1} = -(\sum_{i=1}^{N_f} e_i^2)\frac{\alpha_s(Q^2)}{2\pi} , \quad \Delta C_{\gamma,\,a}^{\gamma,\,n=1} = -\frac{3\alpha}{\pi}(\sum_{i=1}^{N_f} e_i^4)\left(1 - \frac{\alpha_s(Q^2)}{\pi}\right) \qquad (9)$$

As for the gluon distribution, all four schemes give the same result up to NLO,

$$\Delta G^\gamma(n=1,Q^2,P^2) = \frac{12\alpha}{\pi\beta_0}(\sum_{i=1}^{N_f} e_i^2)\frac{\alpha_s(Q^2) - \alpha_s(P^2)}{\alpha_s(Q^2)} \qquad (10)$$

It is interesting to note that the contribution to the sum rule (2) is coming from quark distributions in $\overline{\text{MS}}$ scheme, while in other three schemes the sum rule is derived from the gluon and photon contributions. In fact, we see from Eqs.(9-10) that $[\Delta C_{G,\,a}^{\gamma,\,n=1}\Delta G^\gamma(n=1,Q^2,P^2) + \Delta C_{\gamma,\,a}^{\gamma,\,n=1}]$ gives the sum rule (2). Also it is clear now that the sum rule is the consequence of axial anomalies and that the first term $(\sum_{i=1}^{N_f} e_i^4)$ is coming from QED and the second $((\sum_{i=1}^{N_f} e_i^2)^2)$ from QCD axial anomaly.

By performing the inverse Mellin transform of the moments, parton distributions are reproduced numerically as functions of x. We present in Fig.2 our results for the singlet-quark and gluon distributions predicted by the above four factorization-schemes. It is noted that the NLO quark distribution in $\overline{\text{MS}}$, CI and AB schemes diverge as $[-\ln(1-x)]$ for $x \to 1$, while the one in OS remains finite and approaches a constant value. Gluon distribution vanishes as $[-\ln x]$ for $x \to 1$.

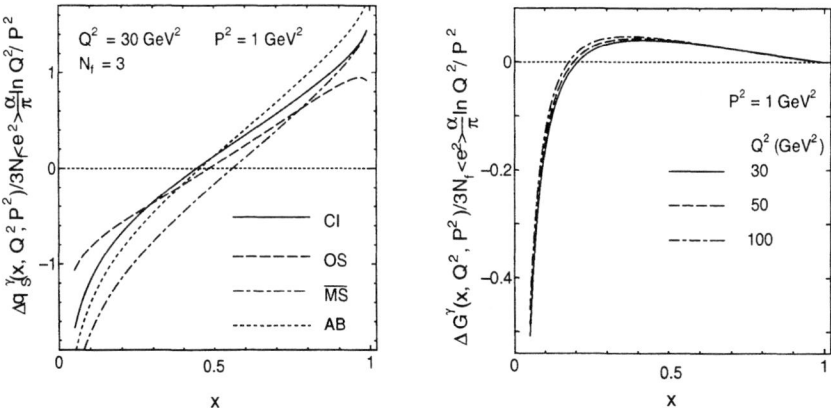

FIGURE 2. Polarized singlet-quark distribution $\Delta q_S^\gamma(x, Q^2, P^2)$ in four different factorization schemes and the polarized gluon distribution $\Delta G^\gamma(x, Q^2, P^2)$ in virtual photon for $Q^2 = 30$ GeV^2 and $P^2 = 1$ GeV^2 with $N_f = 3$

5. Concluding Remarks

We have studied the spin structure function $g_1^\gamma(x, Q^2, P^2)$ and the polarized parton distributions in virtual photon for the kinematical region $\Lambda^2 \ll P^2 \ll Q^2$, which are perturbatively calculable up to NLO in QCD. The first moment of g_1^γ is non-vanishing in contrast to the case of real photon target. This is the consequence of both QED and QCD axial anomalies. We also studied the factorization-scheme-dependence of polarized parton distributions. The scheme dependence is clearly seen in the $n = 1$ moments and the large x-behaviors of the quark distributions.

REFERENCES

1. Uematsu, T., and Walsh, T. F., *Nucl. Phys.* **B199**, 93 (1982).
2. Sasaki, K., and Uematsu, T., *Phys. Rev.* **D59**, 114011 (1999).
3. Mertig, R., and van Neerven, W. L., *Z. Phys.* **C70**, 637 (1996).
4. Vogelsang, W., *Phys. Rev.* **D54** 2023 (1996).
5. Sasaki, K., and Uematsu, T., *Phys. Lett.* **B473**, 309 (2000); hep-ph/0007055 (2000).
6. Efremov, A. V., and Teryaev, O. V., *Phys. Lett.* **B240**, 200 (1990).
7. Narison, S., Shore, G. M., and Veneziano, G., *Nucl. Phys.* **B391**, 69 (1993).
8. Bass, S. D., Brodsky, S. J., and Schmidt, I., *Phys. Lett.* **B437**, 417 (1998).

Polarization Observables in Wide Angle Compton Scattering

Bogdan B. Wojtsekhowski, for the RCS 99-114 Collaboration

Thomas Jefferson National Accelerator Facility,
12000 Jefferson Ave., Newport News, VA 23606, USA

INTRODUCTION

The physics of Wide Angle Compton Scattering (WACS) from the proton was addressed in a pioneering experiment [1] performed at Cornell in 1979. The process was analyzed in the pQCD framework [2]. The general features of pQCD predictions were found in agreement with the experiment. However, detailed calculations [3–5] got a cross section smaller by a factor of five-ten than from the experiment, for most models of nucleon distribution amplitudes. This raised a question of the origin of the scaling and the appropriate mechanism of WACS.

Recently WACS attracted considerable attention because of its connection to the novel concept of skewed parton distributions [6]. The progress of theory [4–8] and expected soon an experiment [9] will allow clarification of the mechanism of WACS process at $s, -t, -u$ in the range of 5 –10 GeV^2.

Experiment E99-114 [9] will be done at photon energy up to 6 GeV and large angles (from 60 up to 130 degrees in cms). In comparison with the only existing previous measurement [1] the counting rate in our experiment will be 300 times higher. Such gain in luminosity results from a combination of the 100% duty factor of CEBAF, high segmentation and resolution of the photon detector, and successful use of the mixed photon/electron beam.

The dominant mechanism of WACS will be investigated through s and t dependence of the cross section and by using polarization transfer from photon to proton. Presently, we are studying the possibility to measure the effect of photon linear polarization.

THE EXPERIMENT

The photon beam produced in a Cu radiator of 1 mm thickness and the primary electron beam will pass through a 15 cm long liquid hydrogen target. The typical photon intensity will be of $4 \cdot 10^{12}$ Hz. The scattered photons will be detected in a calorimeter assembled from 704 lead glass blocks. The acrylic Cherenkov veto counters upstream of the calorimeter and the deflection magnet will be used for discrimination of the electrons from elastic scattering. The recoil proton will be analyzed in the High Resolution Spectrometer (HRS). Kinematical correlation between the proton and scattered photon is a key signature for selecting events of Compton scattering.

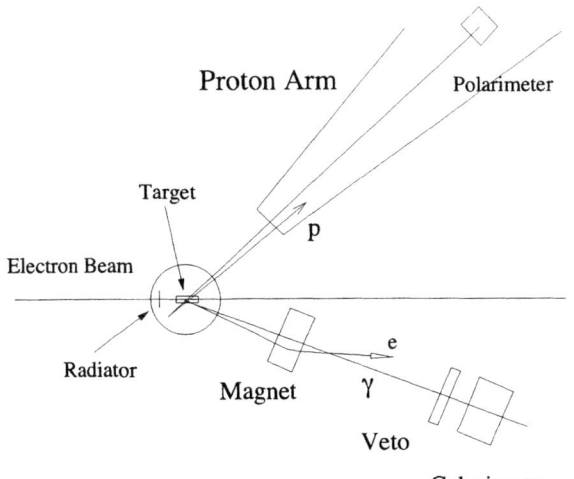

FIGURE 1. The overview of experimental setup.

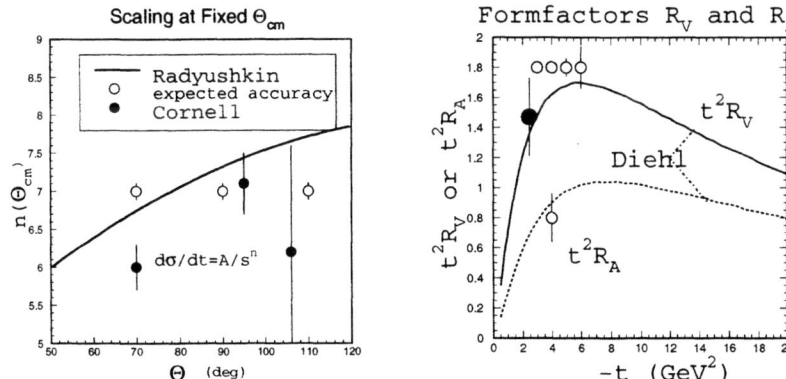

FIGURE 2. Expected accuracy for n and formfactors $R_{V,A}$.

The performance of the described technique was tested at the required luminosity in 1999. The construction of the full scale apparatus should be completed by fall 2001. The polarized photons will have 70% circular polarization and the same intensity as unpolarized ones. The recoil proton polarization at proton momentum

3 GeV/c will be measured by polarimeter in the HRS focal plane for which the Figure of merit is of 1%.

In the pQCD limit the scaling parameter for WACS can be found from a general rule [10]. For fixed scattering angle in cms the cross section $d\sigma/dt \sim s^{-n}$, with $n = 6$. In the soft overlap mechanism the predictions for n range from 6 to 7.5. Figure 2 shows the expected accuracy of the data and predictions for the scaling parameter and WACS formfactors [6,7].

THE POLARIZATION TRANSFER ASYMMETRY

The polarization observables are well known for their sensitivity to the reaction mechanism. In case of WACS the principal difference between polarization observables in the pQCD and the soft overlap mechanisms related with gluon exchange in the hard kernel.

The reaction amplitude is pure real in the soft overlap mechanism, so the A_{LL}, which is defined as: $(\sigma(+-) - \sigma(--))/(\sigma(+-) + \sigma(--))$, has the same positive sign and large value as the one in the Klein-Nishina $e\gamma \to e\gamma'$ process. The predicted A_{LL} in the pQCD approach becomes negative at back angles for most parton distributions [4,5].

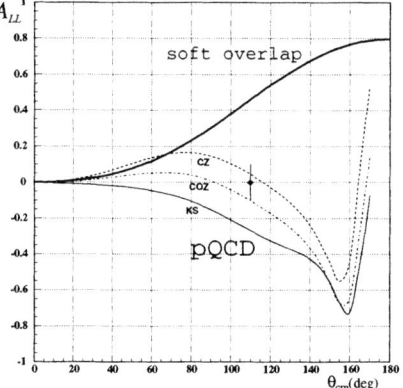

FIGURE 3. A_{LL} for different models and expected accuracy of experiment [9].

THE SIGMA ASYMMETRY

When linear polarized photons are used, one can analyze the difference in cross sections for reaction planes parallel and perpendicular to the photon polarization plane: $\Sigma = (\sigma(\|) - \sigma(\perp))/(\sigma(\|) + \sigma(\perp))$. The Σ asymmetry is defined by both photon spin-flip and non-spin-flip amplitudes of reaction. The calculation of Σ was done using the pQCD approach [5] where for large scattering angles it was found large and positive. The soft overlap calculation leads to zero value of Σ [11].

The analysis of experimental possibilities, which is outlined below, concentrates on $E_\gamma = 4$ GeV and $\theta_{cms} = 90°$.

Measurement of the Σ asymmetry for WACS with apparatus like [9] takes about 200 hours per kinematical point and requires a photon intensity of a few 10^{10} Hz (for 100% photon polarization). Let us discuss how to achieve such photon intensity. There are two ways to study the Σ asymmetry.

The first requires production of a high–intensity, high–energy photon beam. This is possible by using coherent bremsstrahlung of 10–11 GeV electrons from a crystal radiator. Recently this technique was reviewed [12], where a collimated photon beam was designed with intensity 10^8 Hz. The use of larger photon flux will require often replacement of the radiator. Because the radiator has thickness less than 10^{-3} radiation length, which is much less than the Hydrogen target itself, the electron beam should be removed from a target by a chicane to the main beam dump or to a local one.

The second way is based on the technique of small angle electron scattering. The differential cross section of reaction $H(e,e'\gamma)p$ at small electron scattering angle has only two components: σ_T and σ_{TT}. Its ratio has $cos(2\phi)$ dependence on the angle between the electron scattering plane and the momentum of the final photon.

The Σ asymmetry will be found as: $(\sigma_{\phi=90°} - \sigma_{\phi=0°})/(\sigma_{\phi=90°} + \sigma_{\phi=0°})$.

The electron spectrometer must to be organized on the beam line. It can be based on a quadrupole magnet or a combination of four dipoles, which forms the toroidal type system. An electron scattered at a few degrees with an energy of 20–30% of the initial beam energy should be detected. In the case under consideration, the initial electron energy is 6 GeV, the final energy 2 GeV, and the scattering angle 6°. In any proposal of the tagger (our electron spectrometer is essentially a tagger, which records both the energy of the final electron and its scattering plane) the key question is the counting rate of the detector. The effective flux of photons at a given detector rate characterizes the perspective of the technique. The large contribution to the rate at small (but nonzero) angles comes from inelastic electron scattering. In our case 80% of the rate is due to inelastic and 20% is from the radiative tail of elastic scattering. Because the four momentum transfer is very low, the rate of electrons defined by an effective flux of the photons the J_γ and the total cross section of photon absorption in GeV region, which is $\sim 10^{-28}\,cm^2$. The same effective flux defines the rate of the Compton Scattering process. With an electron spectrometer rate of $1 \cdot 10^6$ Hz and a 15 cm liquid hydrogen target we come to an effective flux of photons $J_\gamma \sim 2 \cdot 10^{10}$ Hz. Assuming 10 msr solid angle and 20% momentum acceptance of the tagger, the experiment will require 50 μA of beam intensity.

It is interesting to note that when this technique is applied for measurement of Σ for π° photoproduction, the detection of the scattered electron can be omitted and the electron momentum reconstructed from the proton and pion parameters.

I am grateful to my RCS collaborators and specially to A. Nathan for important discussions.

REFERENCES

1. M. A. Shupe et al., *Phys. Rev.* **D19** 1929 (1979).
2. G. R. Farrar and H. Zhang, *Phys. Rev. Lett.* **65** 1721 (1990).
3. A. S. Kronfeld and B. Nizic, *Phys. Rev.* **D44** 3445 (1991).

4. M. Vanderhaeghen, et al., *Nucl. Phys.* **A622** 144c (1997).
5. T. Brooks and L. Dixon, *Phys. Rev.* **D62** (2000)114021.
6. A. Radyushkin, *Phys. Rev.* **D58** (1998) 114008.
7. M. Diehl, Th. Feldman, R. Jakob, and P. Kroll, *Eur. Phys. J.* **C8** 409 (1999).
8. P. Kroll, hep-ph/9908242.
9. C. Hyde-Wright, A. M. Nathan, and B. Wojtsekhowski (co-spokespersons), JLab experiment E99-114.
10. S. J. Brodsky and G. Farrar, *Phys. Rev. Lett.* 31 1953 (1973).
11. M. Diehl, private communication.
12. R. Jones, Intense Beams of Polarized and Nearly Monochromatic Photons from Coherent Bremsstrahlung, http://zeus.phys.uconn.edu/halld/.

Spin effects in the fragmentation of transversely polarized and unpolarized quarks

M. Anselmino[a], D. Boer[b], U. D'Alesio[c] and F. Murgia[c]

[a] Dipartimento di Fisica Teorica, Università di Torino and
INFN, Sezione di Torino, Via P. Giuria 1, I-10125 Torino, Italy

[b] RIKEN-BNL Research Center
Brookhaven National Laboratory, Upton, NY 11973, USA

[c] Istituto Nazionale di Fisica Nucleare, Sezione di Cagliari
and Dipartimento di Fisica, Università di Cagliari
C.P. 170, I-09042 Monserrato (CA), Italy

Abstract. We study the fragmentation of a transversely polarized quark into a non collinear ($k_\perp \neq 0$) spinless hadron and the fragmentation of an unpolarized quark into a non collinear transversely polarized spin 1/2 baryon. These nonperturbative properties are described by spin and $k_\perp$ dependent fragmentation functions and are revealed in the observation of single spin asymmetries. Recent data on the production of pions in polarized semi-inclusive DIS and long known data on Λ polarization in unpolarized p–N processes are considered: these new fragmentation functions can describe the experimental results and the single spin effects in the quark fragmentation turn out to be surprisingly large.

INTRODUCTION

Several large and puzzling single spin asymmetries in high energy inclusive processes are experimentally well known since a long time and new ones have just been or are being measured. These effects are absent in massless perturbative QCD dynamics and they depend on new and interesting aspects of nonperturbative QCD; as such, they deserve a careful study, both theoretically and experimentally.

We consider here spin properties of quark fragmentation processes, which have been suggested in the literature [1–3], and address the question of whether or not they might explain some single spin asymmetries observed in inclusive processes: in particular we look at the recently observed azimuthal dependence of the number of pions produced in polarized semi-inclusive DIS, $\ell p^\uparrow \to \ell \pi X$ [4,5], and at the longstanding problem of the polarization of Λ's produced in unpolarized p–p and p–n interactions, $pN \to \Lambda^\uparrow X$ [6]. Both these unexpected spin dependences should originate in the fragmentation process of a quark, polarized in the first case ($q^\uparrow \to \pi X$) and unpolarized in the second one ($q \to \Lambda^\uparrow X$).

QUARK ANALYSING POWER

The inclusive production of hadrons in DIS with transversely polarized nucleons, $\ell N^\uparrow \to \ell h X$, is the ideal process to study the so-called Collins effect, *i.e.* the spin and $\boldsymbol{k}_\perp$ dependence of the fragmentation process of a transversely polarized quark, $q^\uparrow \to hX$. In such a case, in fact, possible effects in quark distribution functions [7], which require initial state interactions [8,9], are expected to be negligible.

If one looks at the $\gamma^* N^\uparrow \to hX$ process in the γ^*–N c.m. frame, the elementary interaction is simply a γ^* hitting head on a transversely polarized quark, which bounces back and fragments into a jet containing the detected hadron. The hadron $\boldsymbol{p}_T$ in this case coincides with its $\boldsymbol{k}_\perp$ inside the jet; the fragmenting quark polarization can be computed from the initial quark one.

The spin and $\boldsymbol{k}_\perp$ dependent fragmentation function for a quark with momentum $\boldsymbol{p}_q$ and a *transverse* polarization vector $\boldsymbol{P}_q$ ($\boldsymbol{p}_q \cdot \boldsymbol{P}_q = 0$) which fragments into a hadron with momentum $\boldsymbol{p}_h = z\boldsymbol{p}_q + \boldsymbol{p}_T$ ($\boldsymbol{p}_q \cdot \boldsymbol{p}_T = 0$) can be written as:

$$D_{h/q}(\boldsymbol{p}_q, \boldsymbol{P}_q; z, \boldsymbol{p}_T) = \hat{D}_{h/q}(z, p_T) + \frac{1}{2} \Delta^N D_{h/q^\uparrow}(z, p_T) \frac{\boldsymbol{P}_q \cdot (\boldsymbol{p}_q \times \boldsymbol{p}_T)}{|\boldsymbol{p}_q \times \boldsymbol{p}_T|} \quad (1)$$

where $\hat{D}_{h/q}(z, p_T)$ is the unpolarized fragmentation function. Notice that – as required by parity invariance – the only component of the polarization vector which contributes to the spin dependent part of D is that perpendicular to the q–h plane; in general one has:

$$\boldsymbol{P}_q \cdot \frac{\boldsymbol{p}_q \times \boldsymbol{p}_T}{|\boldsymbol{p}_q \times \boldsymbol{p}_T|} = P_q \sin \Phi_C, \quad (2)$$

where $P_q = |\boldsymbol{P}_q|$ and we have defined the *Collins angle* Φ_C.

When studying single spin asymmetries one considers differences of cross-sections with opposite transverse spins; by reversing the nucleon spin all polarization vectors, including those of quarks, change sign and the quantity which eventually contributes to single spin asymmetries is:

$$D_{h/q}(\boldsymbol{p}_q, \boldsymbol{P}_q; z, \boldsymbol{p}_T) - D_{h/q}(\boldsymbol{p}_q, -\boldsymbol{P}_q; z, \boldsymbol{p}_T) = \Delta^N D_{h/q^\uparrow}(z, p_T) \frac{\boldsymbol{P}_q \cdot (\boldsymbol{p}_q \times \boldsymbol{p}_T)}{|\boldsymbol{p}_q \times \boldsymbol{p}_T|} \quad (3)$$

which implies the existence of a *quark analysing power* for the fragmentation process $q \to h + X$:

$$\begin{aligned} A^h_q(\boldsymbol{p}_q, \boldsymbol{P}_q; z, \boldsymbol{p}_T) &= \frac{D_{h/q}(\boldsymbol{p}_q, \boldsymbol{P}_q; z, \boldsymbol{p}_T) - D_{h/q}(\boldsymbol{p}_q, -\boldsymbol{P}_q; z, \boldsymbol{p}_T)}{D_{h/q}(\boldsymbol{p}_q, \boldsymbol{P}_q; z, \boldsymbol{p}_T) + D_{h/q}(\boldsymbol{p}_q, -\boldsymbol{P}_q; z, \boldsymbol{p}_T)} \\ &= \frac{\Delta^N D_{h/q^\uparrow}(z, p_T)}{2 \hat{D}_{h/q}(z, p_T)} \frac{\boldsymbol{P}_q \cdot (\boldsymbol{p}_q \times \boldsymbol{p}_T)}{|\boldsymbol{p}_q \times \boldsymbol{p}_T|} \equiv A^h_q(z, p_T) \frac{\boldsymbol{P}_q \cdot (\boldsymbol{p}_q \times \boldsymbol{p}_T)}{|\boldsymbol{p}_q \times \boldsymbol{p}_T|}. \end{aligned} \quad (4)$$

This results in a single spin asymmetry [10]:

$$A_N^h(x,y,z,\Phi_C,p_T) = \frac{d\sigma^{\ell+p,\boldsymbol{P}\to\ell+h+X} - d\sigma^{\ell+p,-\boldsymbol{P}\to\ell+h+X}}{d\sigma^{\ell+p,\boldsymbol{P}\to\ell+h+X} + d\sigma^{\ell+p,-\boldsymbol{P}\to\ell+h+X}}$$

$$= \frac{\sum_q e_q^2 \, h_{1q}(x) \, \Delta^N D_{h/q}(z,p_T)}{2\sum_q e_q^2 \, f_{q/p}(x) \, \hat{D}_{h/q}(z,p_T)} \, \frac{2(1-y)}{1+(1-y)^2} \, P \sin\Phi_C \,, \quad (5)$$

where P is the transverse (with respect to the γ^* direction) proton polarization.

We wonder how large the quark analysing power can be. Such a question has been addressed in Ref. [10], where recent data on A_N^π [4,5] were considered. We refer to that paper for all the details and only outline the main procedure here. Under some realistic assumptions and using isospin and charge conjugation invariance Eq. (5) gives ($i = +, -, 0$):

$$A_N^{\pi^i}(x,y,z,\Phi_C,p_T) = \frac{h_i(x)}{f_i(x)} \, A_q^\pi(z,p_T) \, \frac{2(1-y)}{1+(1-y)^2} \, P \sin\Phi_C \quad (6)$$

where:

$$i = + : \quad h_+ = 4h_{1u} \qquad f_+ = 4f_{u/p} + f_{\bar{d}/p} \quad (7)$$
$$i = - : \quad h_- = h_{1d} \qquad f_- = f_{d/p} + 4f_{\bar{u}/p} \quad (8)$$
$$i = 0 : \quad h_0 = 4h_{1u} + h_{1d} \qquad f_0 = 4f_{u/p} + f_{d/p} + 4f_{\bar{u}/p} + f_{\bar{d}/p} \,. \quad (9)$$

The f's are the unpolarized distribution functions and the h_1's are the transversity distributions. Notice that the above equations imply – at large x values – $A_N^{\pi^+} \simeq A_N^{\pi^0}$ as observed by HERMES [11].

We bound the unknown transversity distributions saturating Soffer's inequality [12]

$$|h_{1q}| \leq \frac{1}{2}(f_{q/p} + \Delta q) \,, \quad (10)$$

and, by comparing with SMC data [5]

$$A_N^{\pi^+} \simeq -(0.10 \pm 0.06) \sin\Phi_C \,, \quad (11)$$

we obtain the significant lower bound for pion valence quarks:

$$|A_q^\pi(\langle z\rangle,\langle p_T\rangle)| \gtrsim (0.24 \pm 0.15) \qquad \langle z\rangle \simeq 0.45 \,, \quad \langle p_T\rangle \simeq 0.65 \,\text{GeV}/c \,. \quad (12)$$

A similar result is obtained by using HERMES data [4], although their transverse polarization is much smaller. These lower bounds of the quark analysing power are remarkably large and indeed the Collins mechanism might be (at least partly) responsible for several other observed single transverse spin asymmetries [8].

QUARK POLARIZING FRAGMENTATION FUNCTIONS

We consider now the possibility that an unpolarized quark fragments into a transversely polarized hadron [2,3]; in analogy to Eq. (1) we write

$$\hat{D}_{h\uparrow/q}(z, \boldsymbol{k}_\perp) = \frac{1}{2} \hat{D}_{h/q}(z, k_\perp) + \frac{1}{2} \Delta^N D_{h\uparrow/q}(z, k_\perp) \frac{\hat{\boldsymbol{P}}_h \cdot (\boldsymbol{p}_q \times \boldsymbol{k}_\perp)}{|\boldsymbol{p}_q \times \boldsymbol{k}_\perp|} \tag{13}$$

for an unpolarized quark with momentum $\boldsymbol{p}_q$ which fragments into a spin 1/2 hadron h with momentum $\boldsymbol{p}_h = z\boldsymbol{p}_q + \boldsymbol{k}_\perp$ and polarization vector along the $\uparrow = \hat{\boldsymbol{P}}_h$ direction. $\Delta^N D_{h\uparrow/q}(z, k_\perp)$ (denoted by $D_{1T}^\perp$ in Ref. [2]) is a new *polarizing fragmentation function*.

This reflects into a possible polarization of hadrons inclusively produced in the high energy interaction of unpolarized nucleons. Indeed, it is well known since a long time that Λ hyperons produced with $x_F \gtrsim 0.2$ and $p_T \gtrsim 1$ GeV/c in the collision of two unpolarized nucleons are polarized perpendicularly to the production plane, as allowed by parity invariance; a huge amount of experimental information, for a wide energy range of the unpolarized beams, is available on such single spin asymmetries [6]:

$$P_\Lambda = \frac{d\sigma^{pN \to \Lambda^\uparrow X} - d\sigma^{pN \to \Lambda^\downarrow X}}{d\sigma^{pN \to \Lambda^\uparrow X} + d\sigma^{pN \to \Lambda^\downarrow X}}. \tag{14}$$

By taking into account intrinsic $\boldsymbol{k}_\perp$ in the hadronization process, and assuming that the factorization theorem holds also when $\boldsymbol{k}_\perp$'s are included [1], one obtains

$$\frac{E_\Lambda d^3\sigma^{pN \to \Lambda X}}{d^3\boldsymbol{p}_\Lambda} P_\Lambda = \sum_{a,b,c,d} \int \frac{dx_a\, dx_b\, dz}{\pi z^2} d^2\boldsymbol{k}_\perp\, f_{a/p}(x_a)\, f_{b/N}(x_b)$$
$$\times \hat{s}\, \delta(\hat{s} + \hat{t} + \hat{u}) \frac{d\hat{\sigma}^{ab \to cd}}{d\hat{t}}(x_a, x_b, \boldsymbol{k}_\perp)\, \Delta^N D_{\Lambda\uparrow/c}(z, \boldsymbol{k}_\perp) \tag{15}$$

We use the above equation, together with a simple parameterization of the polarizing fragmentation functions, to see whether or not one can fit the experimental data on Λ and $\bar{\Lambda}$ polarization. All details can be found in Ref. [3] and some results are shown here in Fig. 1.

The data can be described with remarkable accuracy in all their features: the large negative values of the Λ polarization, the increase of its magnitude with x_F, the puzzling flat $p_T \gtrsim 1$ GeV/c dependence and the $\sqrt{s}$ apparent independence; data from p–p processes are in agreement with data from p–Be interactions and also the tiny or zero values of $\bar{\Lambda}$ polarization are well reproduced. The resulting functions $\Delta^N D_{\Lambda\uparrow/q}$ are very reasonable and realistic.

We conclude by stressing that a systematic phenomenological approach towards the description and prediction of single transverse spin asymmetries, based on perturbative QCD dynamics and nonperturbative quark properties, is now possible and worth being developed.

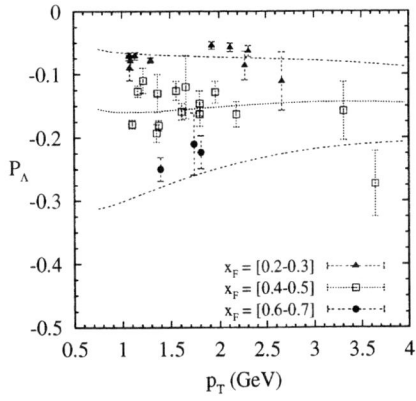 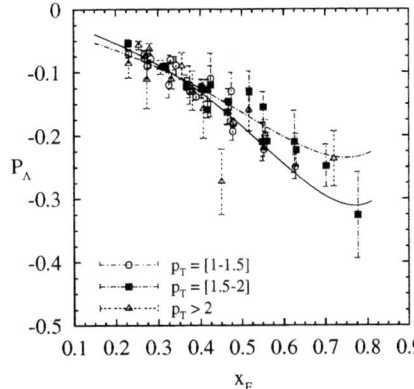

FIGURE 1. Our best fit to P_Λ data from p–Be reactions (a partial collection from Ref. [6]) as a function of p_T (on the left) and of x_F (on the right). For each x_F-bin, the corresponding theoretical curve is evaluated at the mean x_F value in the bin. The two theoretical curves, on the right, correspond to $p_T = 1.5$ GeV/c (solid) and $p_T = 3$ GeV/c (dot-dashed).

REFERENCES

1. J.C. Collins, *Nucl. Phys.* **B396**, 161 (1993)
2. P.J. Mulders and R.D. Tangerman, *Nucl. Phys.* **B461**, 197 (1996); *Nucl. Phys.* **B484**, 538 (1997) (E)
3. M. Anselmino, D. Boer, U. D'Alesio and F. Murgia, e-Print Archive: hep-ph/0008186, submitted for publication in *Phys. Rev.* **D**
4. H. Avakian (on behalf of the HERMES collaboration), *Nucl. Phys.* **B79** (Proc. Suppl.), 523 (1999); HERMES Collaboration, A. Airapetian *et al.*, *Phys. Rev. Lett.* **84**, 4047 (2000)
5. A. Bravar (on behalf of the SMC collaboration), *Nucl. Phys.* **B79** (Proc. Suppl.), 520 (1999)
6. For a review of data see, *e.g.*, K. Heller, in Proceedings of Spin 96, C.W. de Jager, T.J. Ketel and P. Mulders, Eds., World Scientific (1997); or A.D. Panagiotou, *Int. J. Mod. Phys.* **A5**, 1197 (1990)
7. D. Sivers, *Phys. Rev.* **D41**, 83 (1990); **D43**, 261 (1991)
8. M. Anselmino, M. Boglione and F. Murgia, *Phys. Rev.* **D60**, 054027 (1999)
9. M. Anselmino, M. Boglione, J. Hansson and F. Murgia, *Eur. Phys. J.* **C13**, 519 (2000)
10. M. Anselmino and F. Murgia, *Phys. Lett.* **B483**, 74 (2000)
11. Talk by P. Di Nezza in these Proceedings (SPIN2000, Osaka, Oct. 16-21, 2000)
12. J. Soffer, *Phys. Rev. Lett.* **74** (1995) 1292

Precise Measurement of the Spin-Dependent Transverse Asymmetry in Quasielastic $^3\vec{\text{He}}(\vec{e}, e')$ and the Neutron Magnetic Form Factor

J.-O. Hansen
for the Jefferson Lab E95-001 Collaboration

Jefferson Lab, 12000 Jefferson Avenue, Newport News, Virginia 23606

Abstract. We report a measurement of the transverse asymmetry $A_{T'}$ in $^3\vec{\text{He}}(\vec{e}, e')$ quasielastic scattering with high statistical and systematic precision at Q^2-values from 0.1 to 0.6 $(\text{GeV}/c)^2$. Using a state-of-the-art Faddeev calculation, the neutron magnetic form factor G_M^n is extracted for $Q^2 = 0.1$ and 0.2 $(\text{GeV}/c)^2$ with an experimental uncertainty of less than 2%. The results are in excellent agreement with those recently obtained at NIKHEF and Mainz using a deuterium target.

INTRODUCTION

The electromagnetic form factors of the nucleon have been a long-standing subject of interest in nuclear and particle physics. They describe the distribution of charge and magnetization within nucleons and thus allow sensitive tests of nucleon models and Quantum Chromodynamics at low energies.

The neutron form factors are known with much less precision than the proton form factors because of the lack of free neutron targets. As experimental capabilities have improved over the past decade, much effort has been directed towards increasing the precision of the neutron data. Considerable attention has been devoted to the precise measurement of the magnetic form factor, G_M^n [1].

Until recently, most data on G_M^n had been deduced from elastic and quasielastic (QE) electron-deuteron scattering [2–5]. By measuring the ratio of the $d(e, e'n)$ to $d(e, e'p)$ cross sections at QE kinematics, uncertainties in G_M^n of 1-3% have been achieved [3–5]. Unfortunately, there is a significant disagreement between these results (cf. Fig. 2), and further data are desirable to help clarify the situation.

Precision data on G_M^n can also be obtained from the inclusive reaction $^3\vec{\text{He}}(\vec{e}, e')$ at QE kinematics. In comparison to deuterium experiments, this technique employs a different target and relies on polarization degrees of freedom. It is thus subject

TABLE 1. Kinematics of the quasielastic measurements.

Q^2	$(GeV/c)^2$	0.1	0.193	0.3	0.4	0.5	0.6
E	(GeV)	0.778	0.778	1.727	1.727	1.727	1.727
E'	(GeV)	0.717	0.667	1.559	1.506	1.453	1.399
θ	(deg)	24.44	35.50	19.21	22.62	25.80	28.85

to completely different systematics. Pilot experiments using this technique were carried out in 1990-93, and a result for G_M^n was extracted [6]. In this talk, we report the first precision measurement of G_M^n using a polarized ^{3}He target [7].

Polarized ^{3}He is useful for studying the neutron electromagnetic form factors because of its unique spin structure: The ^{3}He ground state is dominated by a spatially symmetric S wave in which the proton spins cancel and the spin of the ^{3}He nucleus is carried by the unpaired neutron [8]. In electron scattering, the spin-dependent properties of ^{3}He can be studied by measuring the spin-dependent asymmetry, defined as $A \equiv (\sigma_+ - \sigma_-)/(\sigma_+ + \sigma_-)$, where the subscript $+(-)$ refers to the helicity of the incident electrons, and σ is the differential cross section. In terms of nuclear response functions R_k, the asymmetry can be written [9]

$$A = -\frac{\cos\theta^* \nu_{T'} R_{T'} + 2\sin\theta^* \cos\phi^* \nu_{TL'} R_{TL'}}{\nu_L R_L + \nu_T R_T}, \tag{1}$$

where the ν_k are kinematic factors, and θ^* and ϕ^* are the polar and azimuthal angles of target spin with respect to the 3-momentum transfer vector $\mathbf{q}$. By orienting the target spin at $\theta^* = 0$, i.e. parallel to $\mathbf{q}$, one selects the transverse asymmetry $A_{T'}$ (proportional to $R_{T'}$).

Because the ^{3}He nuclear spin is carried mainly by the neutron, $R_{T'}$ contains a dominant neutron contribution at the QE peak and is essentially proportional to $(G_M^n)^2$, similar to elastic scattering from a free neutron. This picture has been confirmed in several theoretical studies [10–12]. Thus, the inclusive asymmetry $A_{T'}$ in the vicinity of the ^{3}He QE peak is highly sensitive to $(G_M^n)^2$.

JEFFERSON LAB EXPERIMENT E95-001

The experiment was carried out in Hall A at the Thomas Jefferson National Accelerator Facility (JLab), using a longitudinally polarized continuous wave electron beam of 10 μA current incident on a high-pressure polarized ^{3}He gas target [14]. The target was polarized by spin-exchange optical pumping at a density of 2.5×10^{20} nuclei/cm^3 using rubidium as the spin-exchange medium. The beam and target polarizations were approximately 70% and 30%, respectively. The kinematics are detailed in Table 1.

Scattered electrons were observed in the two Hall A High Resolution Spectrometers, HRSe and HRSh. Both spectrometers operated in single-arm mode. The HRSe was set for QE kinematics while the HRSh detected elastically scattered

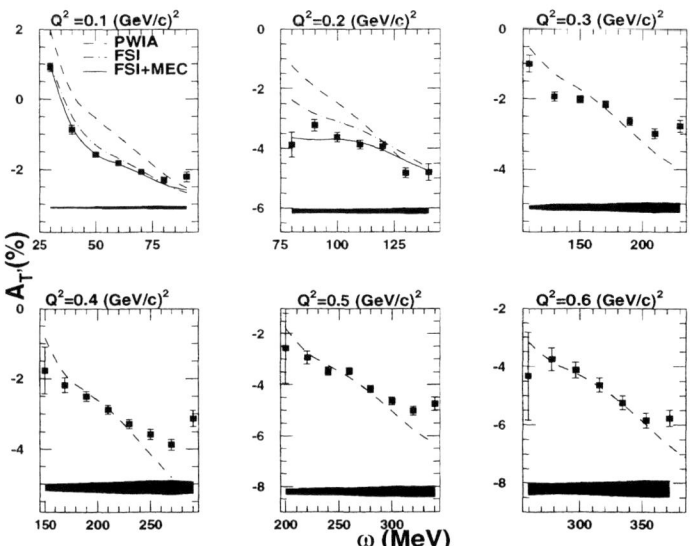

FIGURE 1. Results for the asymmetry $A_{T'}$ as a function of the electron energy transfer ω and the squared four-momentum transfer Q^2. The curves are explained in the text.

electrons. The elastic measurement allowed monitoring of the product of the beam and target polarizations, $P_t P_b$, with better than 3% precision. A more detailed description of the experiment and analysis has been published elsewhere [7].

The measured asymmetries $A_{T'}(\omega)$ are shown in Fig. 1. The error bars indicate the statistical uncertainty. The total experimental systematic uncertainty is depicted as an error band in each panel and amounts to 2% for $Q^2 = 0.1$ and 0.2 (GeV/c)^2, dominated by the error in $P_t P_b$, and 5% for the remaining Q^2 values, dominated by the uncertainty in the radiative corrections.

Also shown in Fig. 1 are the results of several calculations: PWIA [11] (dashed lines), Faddeev with final-state interaction (FSI) corrections [12] (dash-dotted lines), and Faddeev with both FSI and meson-exchange current (MEC) corrections [13] (solid lines). The latter two calculations are non-relativistic. All theory results are based on the Höhler nucleon form factor parametrization [15] and were averaged over the spectrometer acceptances using a Monte Carlo simulation. The advanced calculations [12,13] are not available for $Q^2 > 0.2$ (GeV/c)^2 because relativistic corrections become too large in that regime for the results to be reliable.

EXTRACTION OF THE FORM FACTOR

To extract G_M^n for the two lowest Q^2 kinematics, predictions for $A_{T'}(G_M^n)$ were generated using the full Faddeev calculation [13] and averaged over a 30 MeV bin

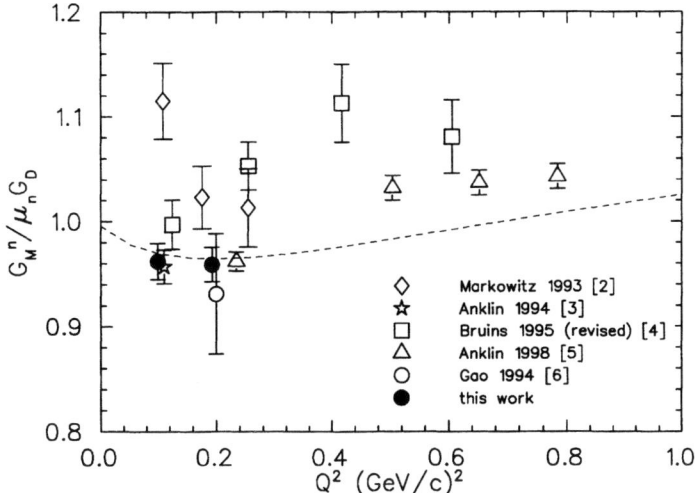

FIGURE 2. Recent results for the neutron magnetic form factor G_M^n in dipole units, $G_D = (1 + Q^2/0.71)^{-2}$. The dashed curve represents the Höhler parametrization [15].

around the QE peak. G_M^n was varied by multiplying the Höhler functional form with a constant factor. The extracted G_M^n corresponds to the value at the central Q^2 of each kinematics for which the predicted $A_{T'}$ agrees with the experimental number in the center 30 MeV bin. Our results are given in Fig. 2 and Tab. 2. The error bars in the figure are the quadrature sum of the statistical and experimental systematic uncertainties. Also shown in Fig. 2 are the results of the recent deuterium experiments [2-5]. As can be seen, our data are in agreement with those recently measured by Anklin et al. [3,5] at NIKHEF and Mainz.

The theoretical uncertainty in extracting G_M^n was estimated [7] to be 1.9% and 2.6% at $Q^2 = 0.1$ and 0.2 $(\text{GeV}/c)^2$, respectively, dominated by MEC and relativistic corrections. These uncertainties can be reduced once fully relativistic calculations become available and MEC corrections are further improved.

To extract G_M^n from our asymmetry data at $Q^2 = 0.3$ to 0.6 $(\text{GeV}/c)^2$, a fully relativistic 3-body calculation that includes FSI and MEC corrections is required, which is presently not at hand. Efforts are underway to extend the full calculation [13] to higher Q^2 [16].

TABLE 2. Extracted values for G_M^n. The uncertainties are statistical and experimental systematic, respectively.

Q^2 $(\text{GeV}/c)^2$	$G_M^n/G_M^n(Dipole)$	Uncertainties
0.1	0.962	±0.014 ±0.010
0.2	0.959	±0.013 ±0.010

CONCLUSIONS

In conclusion, we have measured the asymmetry $A_{T'}$ in inclusive QE $^3\vec{\text{He}}(\vec{e},e')$ scattering at Q^2-values from 0.1 to 0.6 (GeV/c)^2 with very high precision. The neutron magnetic form factor G_M^n has been extracted at $Q^2 = 0.1$ and 0.2 (GeV/c)^2 with $\lesssim 2\%$ experimental accuracy. Our G_M^n data agree with the recent measurements of Anklin et al. [3,5] on deuterium. The present experiment provides the first precision data on G_M^n using a fundamentally different experimental approach than previous experiments. Thus it is a significant step towards understanding the discrepancy among the existing data sets in the low-Q^2 region.

ACKNOWLEDGEMENTS

We thank the Hall A technical staff and the JLab Accelerator Division for their outstanding support during this experiment. This work was supported in part by the U.S. Department of Energy under contract no. DE-AC05-84ER40150 (JLab) and other grants, DOE/EPSCoR, the U.S. National Science Foundation, the Science and Technology Cooperation Germany-Poland and the Polish Committee for Scientific Research, the Ministero dell'Università e della Ricerca Scientifica e Tecnologica, the French Commissariat à l'Énergie Atomique, Centre National de la Recherche Scientifique, and the Italian Istituto Nazionale di Fisica Nucleare. Numerical calculations were performed at the U.S. National Energy Research Scientific Computer Center (NERSC) and at the NIC in Jülich.

REFERENCES

1. H. Gao, plenary talk, SPIN2000, these proceedings.
2. P. Markowitz et al., *Phys. Rev. C* **48**, R5 (1993).
3. H. Anklin et al., *Phys. Lett.* **B336**, 313 (1994).
4. E.E.W. Bruins et al., *Phys. Rev. Lett.* **75**, 21 (1995); B. Schoch, priv. comm.
5. H. Anklin et al., *Phys. Lett.* **B428**, 248 (1998).
6. H. Gao et al., *Phys. Rev. C* **50**, R546 (1994); *Nucl. Phys.* **A631**, 170c (1998).
7. W. Xu et al., *Phys. Rev. Lett.* **85**, 2900 (2000).
8. B. Blankleider and R.M. Woloshyn, *Phys. Rev. C* **29**, 538 (1984).
9. T.W. Donnelly and A.S. Raskin, *Ann. Phys. (N.Y.)* **169**, 247 (1986).
10. R.-W. Schulze and P. U. Sauer, *Phys. Rev. C* **48**, 38 (1993).
11. A. Kievsky, E. Pace, G. Salmè, M. Viviani, *Phys. Rev. C* **56**, 64 (1997).
12. S. Ishikawa et al., *Phys. Rev. C* **57**, 39 (1998).
13. V.V. Kotlyer, H. Kamada, W. Glöckle, J. Golak, *Few-Body Syst.* **28**, 35 (2000).
14. J.S. Jensen, Ph.D. Thesis, California Institute of Technology, 2000 (unpublished).
15. G. Höhler et al., *Nucl. Phys.* **B114**, 505 (1976).
16. W. Glöckle, private communication.

Single π^0 Electroproduction from CLAS Data at Jefferson Lab

K. Joo* for the CLAS collaboration

Thomas Jefferson National Accelerator Facility, Newport News, Virginia 23606, U.S.A.

Abstract. New measurements of the electroproduction of the $\Delta(1232)$ resonance through the $p(e,e'p)\pi^o$ and $p(\vec{e},e'p)\pi^o$ reactions have been performed. The data were taken with the CEBAF Large Acceptance Spectrometer (CLAS) at Jefferson Lab using polarized and unpolarized incident electron beam. Cross sections were measured simultaneously with continuous coverage over a large range of four-momentum transfer Q^2=(0.3-1.2 GeV^2). Decay angular distributions in the $p\pi^o$ center-of-mass were obtained over the full range of $\cos\theta_{c.m.}$ and $\phi_{c.m.}$. The high statistical accuracy of this data set is expected to provide strong constraints on dynamic models of the $N \to \Delta$ transition form factors.

Introduction

The determination of $R_{EM} = E_{1+}/M_{1+}$ and $R_{SM} = S_{1+}/M_{1+}$ in the region of the $\Delta(1232)$ resonance has been the aim of a considerable number of experiments and theoretical activities in the past. Even though theoretical models have become more refined, most previous measurements have large systematic, statistical errors and significant kinematic limitations. A new program using CLAS at Jefferson Lab/Hall B has been inaugurated to vastly improve the systematic and statistical precision and cover a wide kinematic range in four momentum transfer Q^2 and invariant mass W, as well as the full angular range of the resonance decay into the $p\pi^o$ final state. In this report, some of the first preliminary results including the electron beam asymmetry will be presented.

Experimental Setup and Data Analysis

The data were taken using an polarized electron beam at 100 % duty factor incident on a liquid hydrogen target. Scattered electrons and protons were detected by CLAS. A schematic view of CLAS is shown in Figure 1. CLAS is a toroidal magnetic spectrometer divided into six identical sectors. Each sector has three drift chambers to determine the momentum and charge of the charged particles, 48 time-of-flight scintillator paddles to determine the masses of the hadrons, a threshold Čerenkov detector to distinguish between electrons and negatively charged pions, and a calorimeter to detect electromagnetic showers from electrons and photons.

To obtain differential cross sections, CLAS geometrical acceptance, tracking efficiency and resolution were simulated using a GEANT model of the detector geometry which incorporated the magnetic field map and the surveyed positions of detector elements (including target position relative to coils). Software fiducial cuts were used to define the solid angle for electrons and hadrons, excluding regions of low Čerenkov efficiency or multiple scattering from magnetic coils. Radiative corrections were also calculated using a Monte-Carlo method based on the Mo and Tsai formula [1], without using the peaking approximation. To obtain beam asymmetries, two electron beam helicities were utilized. The average electron beam polarization was 70 %.

Once differential cross sections and beam asymmetries were obtained, an energy independent multipole analysis were performed. In the region of the $\Delta(1232)$ resonance, one may expect that only $s-$ and $p-$ waves with $J \leq 3/2$ contribute. Therefore, the partial wave expansion in the differential cross section contains seven unknown multipoles, E_{0+}, M_{1-}, E_{1+}, M_{1+}, S_{0+}, S_{1+} and S_{1-}. Furthermore, one can retain only terms containing M_{1+} assuming that the $\gamma N \Delta$ transition is dominated by a M_{1+} magnetic dipole transition. By utilizing these two approximations, R_{EM} and R_{SM} can be determined unambiguously by mapping the angular dependence of the differential cross sections.

Preliminary Results

Preliminary results for R_{EM} and R_{SM} are shown in Figure 2 for Q^2=0.40,0.52, 0.65, 0.75 and 0.91 GeV^2 compared to recent model calculations [2–4] and previous measurements [6–8]. The errors shown throughout are statistical only. R_{EM} is small and negative with a weak Q^2 dependence, while R_{SM} exhibits a strong Q^2 dependence with a trend towards increasingly negative values. The trend of the data is qualitatively described by quark models that include pion degrees of freedom. These results are in contrast to previous data which give ambiguous results for the sign of R_{EM} and show no clear Q^2 dependence of R_{SM}. Preliminary results for $Re(S_{0+}M_{1+})$, $Re(S_{1+}M_{1+})$, $Im(S_{0+}M_{1+})$ and $Im(S_{1+}M_{1+})$ are also shown in Figure 3. These multipole measurements are compared with theoretical calculations [5,4]. $Im(S_{0+}M_{1+})$ and $Im(S_{1+}M_{1+})$ are obviously very sensitive to the phase relation between the different multipoles since they would vanish identically if the multipoles were strictly in phase. The $Im(S_{1+}M_{1+})$ term would vanish if the multipoles contain contributions from the $\Delta(1232)$ resonance only. Any deviation from zero would indicate the existence of additional non-resonant contribution to the S_{1+} multipole.

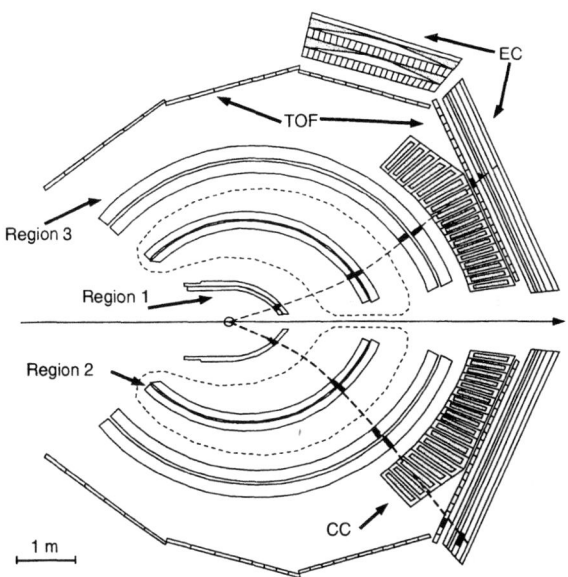

FIGURE 1. Horizontal midplane cut through the CLAS detector at beam line elevation showing two charged particles traversing the drift chambers (Region 1,2,3) in opposite sectors. Outside of Region 3 are time-of-flight (TOF) counters, calorimeters (EC), and Čerenkov counters (CC). The 2 T toroidal magnetic field is contained within the boundary surrounding Region 2.

ACKNOWLEDGMENTS

This work was supported in part by DOE contract DEFG02-97ER41018. Thanks go to Lee Cole Smith and Ralph Minehart of the University of Virginia for their help in the data analysis, and to the staffs at Jefferson Lab/Hall B and the CLAS collaboration for their hard work in supporting the experiment.

REFERENCES

1. L. W. Mo and Y. Tsai, *Rev. Mod. Phys.* **41**, 205 (1969)
2. A. Silva, D. Urbano, T. Watabe, M. Fiolhais and K. Goeke, hep-ph/9904326
3. S.S. Kamalov and Shin Nan Yang, *Phys. Rev. Lett.* **83**, 4494 (1999)
4. T. Sato and T.-S.H. Lee, *Phys. Rev.* C **54**, 2660 (1996) and T.-S.H. Lee, private communication, 2000.
5. D. Drechsel, O. Hanstein, S.S. Kamalov, L. Tiator, private communication, 2000.
6. R. Beck *et al*, *Phys. Rev. Lett.* **78**, 606 (1997)
7. G. Blanpied *et al*, *Phys. Rev. Lett.* **79**, 4337 (1997)
8. C. Mertz *et al*, nucl-ex/9902012, H. Schmieden, nucl-ex/9909006

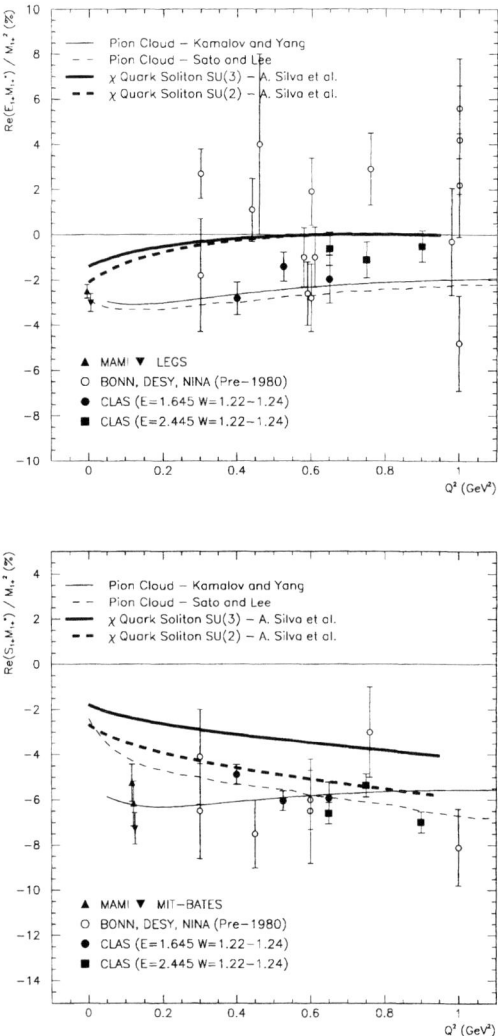

FIGURE 2. Preliminary multipole ratios vs. Q^2 at electron beam energies E=1.645, 2.445 GeV. Curves are from recent model calculations[2,3,4]. Errors shown are statistical only. Top: Electric quadrupole $\mathrm{Re}(E_{1+}^* M_{1+})/|M_{1+}|^2$ compared to previous measurements[6,7]. Bottom: Coulomb quadrupole $\mathrm{Re}(S_{1+}^* M_{1+})/|M_{1+}|^2$ compared to previous measurements[8].

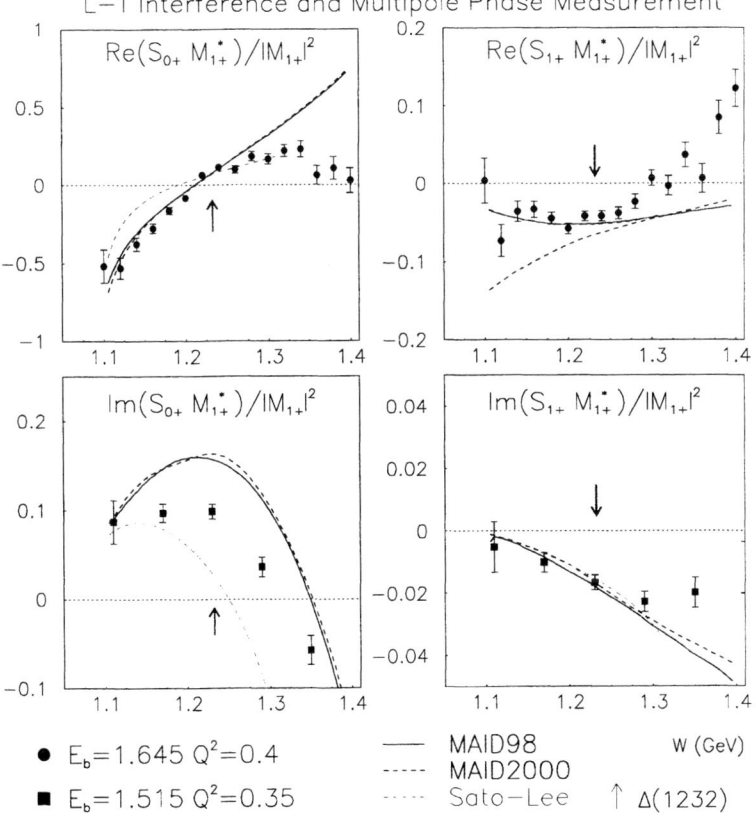

FIGURE 3. Preliminary results from an energy independent multipole analysis. Curves are from recent model calculations[4,5]. Errors shown are statistical only. Top left: $\mathrm{Re}(S_{0+}^* M_{1+})/|M_{1+}|^2$ vs. W (GeV) at $Q^2=0.40$ GeV^2. Top right: $\mathrm{Re}(S_{1+}^* M_{1+})/|M_{1+}|^2$ vs. W (GeV) at $Q^2=0.40$ GeV^2. Bottom left: $\mathrm{Im}(S_{0+}^* M_{1+})/|M_{1+}|^2$ vs. W (GeV) at $Q^2=0.35$ GeV^2. Bottom right: $\mathrm{Im}(S_{1+}^* M_{1+})/|M_{1+}|^2$ vs. W (GeV) at $Q^2=0.35$ GeV^2.

Results and Status of Inelastic ed-scattering Experiments at the Internal Polarized Deuterium Targets of VEPP-3

V.N. Stibunov*, M.V. Dyug†, B.A. Lazarenko†, A.Yu. Loginov*,
S.I. Mishnev†, D.M. Nikolenko†, A.V. Osipov*, I.A. Rachek†,
R.Sh.Sadykov†, Yu.V. Shestakov†, A.A.Sidorov*, D.K. Toporkov†,
and S.A. Zevakov†

† *Budker Institute of Nuclear Physics, Novosibirsk, 630090, Russia*
* *Nuclear Physics Institute at Tomsk Polytechnical University, Tomsk, 634050, Russia*

Abstract. A brief description of the investigation on the polarization phenomena in the reactions $\vec{d}(e,pn)e'$, $\vec{d}(e,pp)e'\pi^-$ and $\vec{d}(e,e'p)n$ at the Novosibirsk 2 GeV electron storage ring VEPP-3 is given. Some experimental technique improvements are described. The plan of the experiments on inelastic $e\vec{d}$-scattering are presented.

INTRODUCTION

The polarization observables measured in the inelastic exclusive ed-scattering provide the information on the spin-dependent properties of the deuteron. Such measurements gives the additional information on the nucleon-nucleon interaction, the non-nucleon degrees of freedom, the excitation of $\Delta(1232)$-resonance.

The measurements of the polarization observables in the reactions $\vec{d}(e,pn)e'$, $\vec{d}(e,pp)e'\pi^-$ and $\vec{d}(e,e'p)n$ have been performed in Novosibirsk at 2 GeV VEPP-3 electron storage ring using an internal tensor-polarized deuterium target [1].

Here we will focus on the experimental results obtained for the first two reactions. The study of these processes is performed at small electron scattering angles, that almost corresponds to the reactions with real photons.

DESCRIPTION OF EXPERIMENTS

The measurements have been performed by means of the two identical two-arms detectors, detecting two protons or a proton and a neutron in coincidence [2]. The

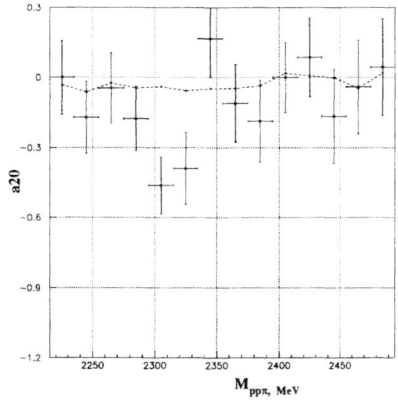
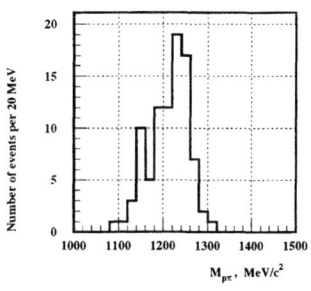

FIGURE 1. a_{20} - component of the tensor target asymmetry as a function of the invariant mass $pp\pi^-$-system. The curve is calculated in the model, described in [3].

FIGURE 2. The fast proton-pion effective mass distribution for the events with the cuts 2280 MeV $\leq M_{pp\pi^-} \leq$ 2320 MeV.

detector arms were placed symmetrically around the electron-beam axis at the polar angle of 75° with respect to the beam line. The degree of tensor polarization for deuterium nuclei in storage cell and integral luminosity were determined from the elastic ed-scattering measurements [4]. During the data acquisition a direction and sign of the target polarization were changed at regular intervals. The tensor target asymmetry components were calculated combining the differences in counting rates [1,5].

The reaction $\vec{d}(e, pp)e'\pi^-$ was studied for the case when both final protons had the momentum more then 300 MeV/c. Their transverse components were large also. The cross section and the tensor target asymmetry components were measured as the functions of the proton-proton invariant mass and the slow and the fast proton momenta [5]. These observables were also presented as the functions of the photon energy, the proton-proton-pion invariant mass, and the fast or slow proton-pion masses [5–7]. The calculation of the cross section and target asymmetry were made within spectator model using the relativistic elementary pion photoproduction amplitude and the realistic deuteron wave function models [5–7]. These calculations had not described the obtained experimental results.

Figure 1 shows the measured and calculated tensor target asymmetry component a_{20} as a function of the invariant mass $M_{pp\pi^-}$ of the $pp\pi^-$-system. The target asymmetry a_{20} has the sharp dip near the mass $M_{pp\pi}$=2310 MeV. From a Monte-Carlo simulation of the reaction we have concluded that this dip is associated with a $\Delta(1232)$-isobar production [6]. The fast proton-pion invariant mass distribution of the events from the dip region is presented in Figure 2. Recently the effects of the pion-nucleon and nucleon-nucleon re-scattering on the observables of the $d(\gamma, ppi)\pi^-$ -reaction were calculated in [3] within diagrammatical approach. A relativistic invariant form of the pion photo-production and pion scattering operators were used in these calculation. The experimental a_{20} -target asymmetry and theoretical one obtained taking into account the influence of the re-scattering effects are noticeably different for this kinematical region (Figure 1), as for the experimental and theoretical reaction yields have an nearly equal values. These experimental results obtained with insufficient statistical accuracy had stimulated the new measurements at VEPP-3.

A number of the major technical developments have been achieved for these new experiments. In order to decrease an electron beam cross section and to be able to use a small-aperture cell the electron beam optics in the experimental section of VEPP-3 ring was modified. This storage cell was fabricated from the 30 μm aluminum foil and cooled by the circulating liquid nitrogen. The new intensity Atomic Beam Source (ABS) has been constructed to fill storage cell target by polarized deuterium atoms [8]. The new polarized target based on the new ABS was thicker then that used in the previous experiment in the hundred times. The detector has the two identical two-arms systems. Each of the systems consists of an electron calorimeter on one side of the electron beam and a hadron scintillator hodoscope on the other side. For the registration of the particle tracks the different

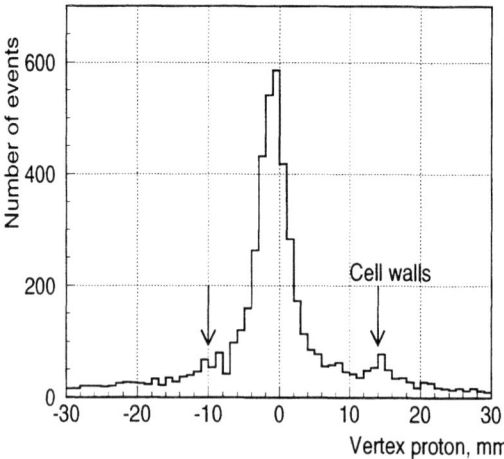

FIGURE 3. The distribution of the vertex coordinate transverse to the electron beam for a fragment of the Run2000 *pp*-events.

sets of drift chambers are employed. From April, 1999 to March, 2000 the target and the detector were used in the measurement of the polarization observables in the elastic ed-scattering and inelastic process $\vec{d}(e,pp)e'\pi^-$ [9]. The serious problem caused by the background that produced by stray particles the cell walls has been successfully solved for this reaction channel also (Figure 3). The analysis of the accumulated data is in progress.

The obtained in [1] results of the measurement of the tensor analyzing power T_{20}- and T_{22}-components for the deuteron disintegration and the theoretical predictions within the simple models don't agree. In general, these experimental results are consistent with those theoretical predictions which take into account final-state interaction, meson-exchange currents and other corrections which complicate the analysis and are far from being well understood yet. Therefore the found discrepancies between theory and experiment for the components T_{20} and T_{22} remain unexpained so far. The new more precise measurements are need in order to check the models of few-body systems, while the existing data are of insufficient quality.

PERSPECTIVES

We started the preparation of the detector system for the new experiments at VEPP-3 using the last development of the super-thin internal target technique. The scheme of the experiments is presented in Figure 4. There will be two pairs

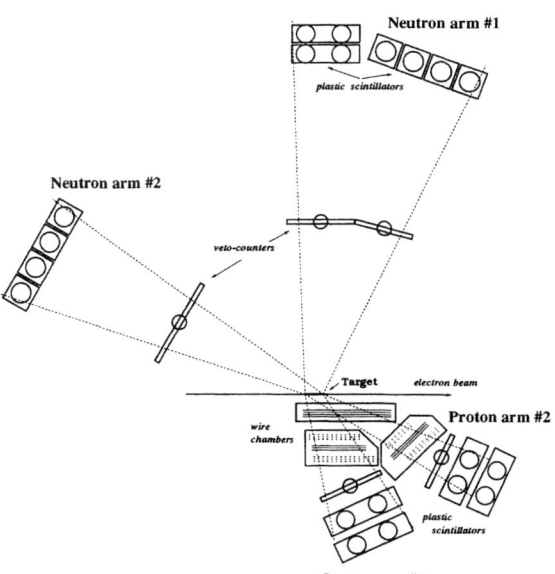

FIGURE 4. Schematical side view at the new experimental set-up for a new asymmetry measurement in the disintegration channels of the polarized deuteron.

of the detector arms. Two proton detecting arms will be placed below the electron beam line and two neutron detecting arms are above beam line. Degree of tensor polarization of the target atoms will be determined by the same method as in the previous ed-scattering experiment at VEPP-3, using the new "low-Q" polarimeter. The aim of the proposed investigations is an experimental determination of the tensor analyzing power components T_{20}, T_{21} and T_{22} of the pion production and the deuteron photo-disintegration at much higher statistical accuracy and with smaller systematical uncertainties than in the previous experiments. We have estimated the statistical accuracy for the measurement of T_{20} of the photo-disintegration by one of the detection arm pair corresponding to the center of mass angle $\Theta_{cm} = 80^0$. The estimation was performed for the total electron beam integral 50 kC, assuming the target thickness $3 \times 10^{13} at/cm^2$ and target polarization $P_{zz}=+0.4/-0.8$. We have obtained that a drastical improvement in statistical accuracy is expected, which should result in rigorous test of the theoretical models.

ACKNOWLEDGMENT

We indebted to all members of the Novosibirsk "DEUTERON" collaboration who participated in these experiments. The work was supported in part by the RFBR under Grants 98-02-17993, 98-02-17949 and INTAS under Grant 96-0424.

REFERENCES

1. Mishnev, S.I., Nikolenko, D.M., Popov, S.G., et al., *Phys. Lett.* **B302**, 23-28 (1993).
2. Gilman, et al., *NIM* **A327**, 277 (1993). Theunissen, J.A.P. et al, *NIM* **A348**, 61-72 (1994).
3. Loginov, A.Yu., Sidorov, A.A., and Stibunov, V.N., *Physics of Atomic Nuclei* **63**, 391-398 (2000).
4. Gilman, R., Holt, R.J., Kinney, E.R.,et al., *Phys. Rev. Letters* **65**, 1773 - 1776 (1990).
5. Loginov, A.Yu., Osipov, A.V., Sidorov, A.A., et al., *JETF Letters* **67**, 770 - 777 (1998).
6. Stibunov, V.N., Loginov, A.Yu., Osipov, A.V., et al., *Few-Body System Suppl.* **10**, 507 -510 (1999).
7. Osipov, A.V., Loginov, A.Yu., Sidorov, A.A., et al.,"Target Asymmetry in the $d(e,pp)e'\pi^-$-Reaction at High Proton Momenta", in*Barions'98*, edited by D.W.Menze et al., World Scientific Publ. Co. Pte. Ltd., Singapore, 1999, pp. 567-570.
8. Isaeva L.G., Lazarenko, B.A., Mishnev, S.I., et al., "Status of the Novosibirsk Cryogenic Atomic Beam Source", in *SPIN 98 Proceedings*, edited by N.E.Tyurin, et al., World Scientific Publ. Co. Pte. Ltd., Singapore, 1999, pp. 631-633.
9. Rachek, I.A., Arenhoevel, H., Barkov, L.M., et al., "Recent results from the internal polarized deuterium target experiment at the electron storage ring VEPP-3", in *Nuclear Physics at Storage Rings*, edited by H.- O.Meyer and P.Schwandt, AIP Conference Proceedings 512, New York,2000, pp. 362-364.

Measurement of spin correlation parameters in the Δ region for the $^1\vec{H}(\vec{e},e')$ reaction

L. D. van Buuren[a,b], D. Szczerba[c], R. Alarcon[d],
Th. S. Bauer[b,e], D. Boersma[b,e], J. F. J. van den Brand[a,b],
H. J. Bulten[a,b], M. Ferro-Luzzi[a,b], D. W. Higinbotham[b,f],
S. Klous[a,b], H. Kolster[a,b], J. Lang[c], F.A. Mul[a], D. Nikolenko[h],
B. E. Norum[f], I. Passchier[b], H. R. Poolman[a,b], I. Rachek[h],
M. C. Simani[a,b], E. Six[d], H. de Vries[b], K. Wang[f], Z.-L. Zhou[i]

[a] Faculty of Sciences, Vrije Universiteit, NL-1081 HV, Amsterdam, The Netherlands
[b] NIKHEF, NL-1009 DB Amsterdam, The Netherlands
[c] Institut für Teilchenphysik, ETH, CH-8093 Zürich, Switzerland
[d] Department of Physics and Astronomy, ASU, Tempe, AZ 85287, USA
[e] Physics Dpt., UU, NL-3508 TA Utrecht, The Netherlands
[f] Dpt. of Physics, UVa, Charlottesville, VA 22901, USA
[g] TJNAF, Newport News, VA 23606, USA
[h] BINP, Novosibirsk, 630090, Russian Federation
[i] Laboratory for Nuclear Science, MIT, Cambridge, MA 02139, USA

Abstract. Sensitivity to the quadrupole transition form factors ($C2$ and $E2$) of the N-Δ excitation can be obtained by measuring double-polarization electromagnetic observables in the Δ-region. Such measurements have been carried out with the internal target facility at NIKHEF. Radiative corrections are an important ingredient of the data interpretation. They need to be evaluated carefully, taking into account their spin-dependence, in order to compare the data to theoretical models of pion electroproduction. From these comparisons the $E2/M1$ and $C2/M1$ ratios can be extracted.

I INTRODUCTION

A topic of interest in particle physics concerns the spin content of the nucleon, in particular the contribution of orbital angular momentum, of which little is known. It has been argued that the failure of the quark model to properly describe fundamental observables (such as G_A/G_V, the SU(3) decay ratio (D+F)/(D-F) and the $\pi N \Delta$ coupling constant), might be removed if one allows for a substantial D-state admixture in the wave functions of the baryons [1]. This issue can be experimentally addressed by measuring double-polarization electromagnetic observables in the Δ region, which are sensitive to the quadrupole transition form factors ($C2$ and $E2$) of the N-Δ excitation, hence to a possible D-state admixture in the nucleon and Δ wave function. Considerable work has been done at a four-momentum transfer $Q^2 = 0$, using polarized real photons [2,3], constraining the $E2/M1$ ratio. Here, we report on a measurement of the longitudinal spin correlation parameter $A_\parallel$ (with momentum transfer parallel to the nuclear polarization) and the sideways spin correlation parameter $A_\perp$ (with momentum transfer perpendicular to the nuclear polarization in the scattering plane) in the Δ region for the $^1\vec{\mathrm{H}}(\vec{e},e')$ reaction, at an average Q^2 of 0.11 GeV/c^2. The N-Δ quadrupole form factor $E2$ ($C2$) enters $A_\parallel$ ($A_\perp$) via an interference with the dominant magnetic ("spin-flip") dipole form factor M1.

II EXPERIMENTAL SETUP

The experiments were carried out at the internal target facility at NIKHEF, Amsterdam. Longitudinally polarized electrons [4] were accelerated to 720 MeV in a medium energy accelerator and in-

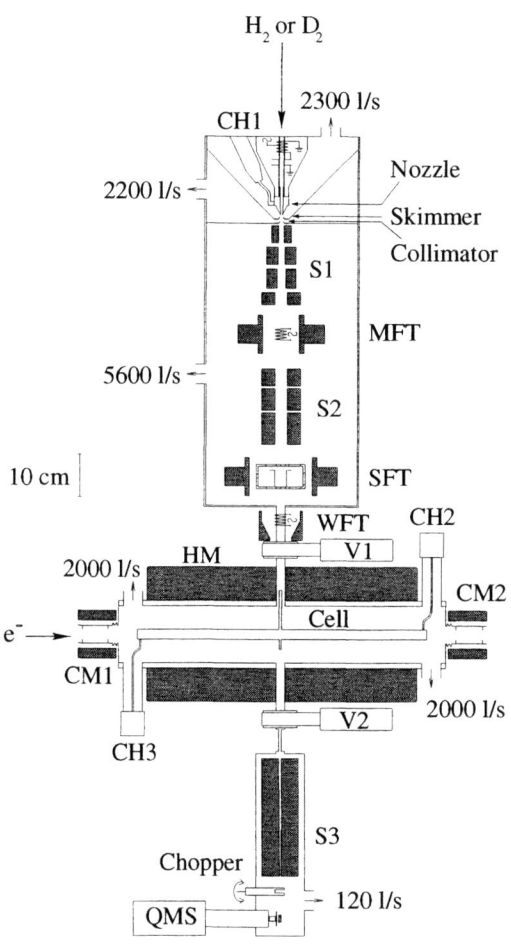

Figure 1 Internal target. Pumping speeds for H$_2$. CH1, CH2, CH3: coldheads; S1, S2, S3: sextupole magnets; MFT, SFT, WFT: medium, strong and weak field transition units; V1,V2: valves; HM: target holding field magnet; CM1, CM2: compensation magnets; QMS: quadrupole mass spectrometer.

jected into the Amsterdam pulse stretcher storage ring. The electron polarization was maintained by using a Siberian snake and measured regularly with a Compton backscattering polarimeter [5]. The polarized electron beam was guided through a storage cell, in which polarized gas ($^{1,2}\vec{H}$ or $^{2}\overleftrightarrow{H}$) was injected by an atomic beam source (ABS, see section 3). The scattered electrons were detected in a large acceptance magnetic spectrometer (96 msr) and the ejected hadrons in a TOF detector (250 msr). Exclusive and inclusive data were collected with both polarized hydrogen and deuterium targets. Here, we only discuss data obtained for the $^{1}\vec{H}(\vec{e}, e')$ reaction.

III INTERNAL TARGET

In the ABS, molecular gas (H_2 or D_2) is dissociated into atoms by a 27.1 MHz RF discharge. A beam is formed by the cooled nozzle, skimmer and collimator (fig. 1). Two sets of sextupole magnets (de) focus atoms with electron spin (anti) parallel to the magnetic field into the target feed tube. Nuclear polarization ($^{1,2}\vec{H}$ or $^{2}\overleftrightarrow{H}$) is obtained by inducing RF transitions between selected hyperfine states. In our configuration, vector polarized hydrogen or deuterium as well as tensor polarized deuterium can be produced. Systematic errors in the measured asymmetries are reduced by flipping the polarization every 8 seconds. The polarized atomic beam is injected into a 60 cm long, 15 mm diameter storage cell, fabricated from 30 μm thin aluminium foil and cooled to $\sim$ 70 K. A magnetic field is produced in the target region to provide a target polarization axis. This axis can be oriented in any direction in the scattering plane. A small fraction of the direct atomic beam flows into a Breit-Rabi-type Polarimeter (BRP) via a sample tube. The BRP, which consists of a sextupole magnet followed by a quadrupole mass spectrometer, measures the electron polarization of the atomic beam and is used to tune the RF transition units.

The main improvements, with respect to the previous internal target experiments at NIKHEF with tensor-polarized deuterium [6], were the implementation of permanent sextupole magnets, a more powerful pumping system in the ABS beam-formation chamber and in the target region, the use of a longer and colder storage cell to increase the target thickness up to 1.2×10^{14} nuclei/cm^2 and the ability to run with $^{1,2}\vec{H}$ (polarization up to 0.71) as well as $^{2}\overleftrightarrow{H}$ [7-9].

IV ANALYSIS

To be able to compare our data to theoretical models ([10,11]), background and radiative contributions have to be taken into account. The background contributions were determined from frequent measurements without polarized gas in the storage cell and subtracted from the raw data (fig. 2).

To take into account the contributions of events which have radiated one or more photons before or after scattering, the structure functions predicted from the theoretical models were implemented in the code POLRAD [12]. This code calculates the spin-dependent radiative contributions of both elastic and inelastic processes. The corrected cross sections are folded with our detector acceptance using a Monte-Carlo technique. The calculations for the longitudinal and sideways asymmetries of the elastic radiative tail (which is the main radiative contribution to the asymmetry in the Δ-region), folded with our detector acceptance are shown in fig. 3. By comparing the asymmetries obtained from the data to the ones obtained in the Monte-Carlo, (model dependent) values for the multipoles $E2$ and $C2$ can be extracted.

V RESULTS AND OUTLOOK

By monitoring the experimental asymmetry at the elastic peak continuously and comparing it to the known elastic asymmetry, the product of the electron beam and target polarizations is known precisely. Using this product as a normalization and by ap-

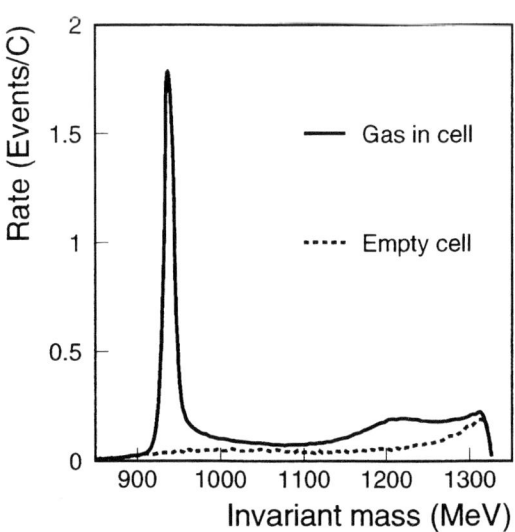

Figure 2 Rate plotted versus invariant mass for measurements with and without gas in the storage cell.

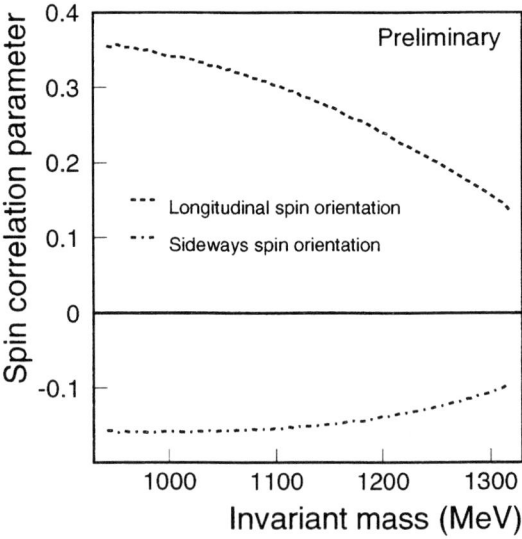

Figure 3 Spin correlation parameters for the elastic radiative tail for longitudinal and sideways spin orientations.

plying a linear interpolation between the predictions of the models for different $E2$ and $C2$ strengths in the Δ-region (1170 − 1270 MeV) best fits can be obtained for the ratios $E2/M1$ and $C2/M1$ at $Q^2 = 0.11(\text{GeV/c})^2$ using the least squares method. Systematic errors arising from the normalization, the spin dependent radiative corrections and the Monte-Carlo method have to be taken into account. These results together with the previous data and the upcoming results from Bates, MAMI, ELSA and JLAB will constrain the $E2$ and $C2$ form factors.

This work was supported in part by the Stichting voor Fundamenteel Onderzoek der Materie (FOM), which is financially supported by the Nederlandse Organisatie voor Wetenschappelijk Onderzoek (NWO) and the Swiss National Foundation.

REFERENCES

1. S.L. Glashow, Physica **96 A** (1979) 27.
2. R. Beck et al., Phys. Rev. Lett **78** (1997) 606.
3. G. Blanpied et al., Phys. Rev. Lett. **79** (1997) 4337.
4. B.L. Militsyn et al., Nucl. Instr. and Meth. **A 427** (1999) 46.
5. I. Passchier et al., Nucl. Instr. and Meth. **A 414** (1998) 444.
6. Z.-L. Zhou et al., Nucl. Instr. and Meth. **A378** (1996) 40.
7. D. Szczerba et al., to be published in Nucl. Instr. and Meth. **A**.
8. L.D. van Buuren et al., Nucl. Phys. **A663-664** (2000) 1049c.
9. L.D. van Buuren for the 97-01 collaboration, in *Nuclear Physics at Storage Rings*, pages 356–358, 2000, AIP Conference Proceedings.
10. D. Drechsel et al., Nucl. Phys. **A645** (1999) 145.
11. T. Sato and T-.S.H. Lee, to be submitted to Phys. Rev. C (2000).
12. I. V. Akushevich and N. M. Shumeiko, J. Phys. **G 20** (1994) 513.

6. SPIN PHYSICS IN NUCLEI

Precise determination of the spin-transfer coefficient $K_{NN'}$ for $\vec{n}p$ elastic scattering at 187 MeV

S. W. Wissink, S. Choi, W. A. Franklin, W. W. Jacobs,
T. Peterson, T. Rinckel, J. Sowinski, E. J. Stephenson,
M. Wolanski, and H. Yang

Indiana University Cyclotron Facility, Bloomington, Indiana 47408 USA

Abstract. We have measured the normal-component spin-transfer observable $K_{NN'}$ for $\vec{n}p$ elastic scattering to a precision of ±0.02 at 15 angles in the far-backward region ($105° \leq \theta_{cm} \leq 172°$) at a neutron bombarding energy of 187 MeV. This kinematic regime is of particular interest in that the scattering is mediated predominantly through exchange of a single charged pion, and hence is expected to depend sensitively on the strength of the $\pi^{\pm}NN$ coupling constant g_c^2, a quantity whose precise value remains highly controversial. This controversy has been heightened by new np absolute cross section data, also taken at back angles, which appear to favor the higher values of g_c^2 obtained in earlier partial-wave analyses, but disagree with most recent predictions. After reviewing the current situation, I will describe the experimental apparatus used in our measurement and our data acquisition and analysis procedures, paying particular attention to potential sources of systematic error and their effect on the data. Our primary result is that these new values for $K_{NN'}$, in contrast with the recent cross section data, agree most closely with the predictions of modern NN potential models in which a relatively *weak* πNN coupling is used.

BACKGROUND AND MOTIVATION

A quantitative understanding of the free nucleon–nucleon (NN) force remains an important goal of nuclear physics. While the strong interaction contributions should, in principle, be calculable within the framework of QCD, our best descriptions of most low- and intermediate-energy nuclear phenomena are presently obtained from *effective* theories of the interaction, in which baryons and mesons serve as efficient, collective degrees of freedom; hence, there is great interest in determining the values of the constants that necessarily appear in these theories. In meson-based models, the one-pion exchange process plays a fundamental role, as it dominates the long-range hadronic force, especially the tensor terms, with a strength determined primarily by the size of g_π^2, the πNN coupling constant.

This coupling therefore plays a crucial role in these models for predictions of NN and πN scattering behavior, and for reproducing known ground state properties of few-nucleon systems. More generally, g_π^2 is a key parameter in NN interaction models used as input to microscopic nuclear reaction or structure calculations; at a more basic level, accurate knowledge of this coupling is needed in low-energy theorems of charged pion photo- or leptoproduction, and to gauge the extent, via the Goldberger-Treiman discrepancy [1], to which chiral-symmetry breaking is present in the πN interaction.

Yet despite its obvious importance, new determinations of both the neutral [2] and charged [3] pion coupling constants, based largely on analysis of NN and πN scattering data, respectively, have yielded values for g_π^2 that are 5–7% *smaller* than values determined prior to 1987. Perhaps more disturbing is that this weaker coupling appears to be incompatible with the strengths required in meson-exchange models to produce the correct deuteron quadrupole moment and asymptotic D/S-state ratio [4], and with values extracted from new, precise np cross section data [5]. To help resolve this issue, we decided to look in detail at how the values extracted for g_π^2 depend on the spin observable data set. We found that the spin-transfer coefficients for NN elastic scattering (the diagonal terms in particular) are very sensitive to the magnitude of this coupling, especially at low momentum transfer, $q ¡ 0.8$ fm^{-1}, where contributions from multi-pion or heavy meson exchange are expected to be small. This enhanced sensitivity stems from the similarities in spin structure between the one-pion-exchange potential and particular amplitudes in the NN interaction which these observables can isolate [6]. The strong influence of these specific observables on the determination of g_π^2 has also been established empirically through an extensive study of local (single-energy) phase shift analyses by Bugg and Machleidt [7] from which subsets of the database were excluded.

An examination of the current database revealed that there was very little data between 150 and 400 MeV for any spin transfer observable at low momentum transfer. We therefore launched a series of precise measurements of pp spin transfer observables at IUCF [6], motivated by the above arguments. These pp data have been shown to impose severe constraints on the size of the *neutral* pion coupling, and have provided strong support for models of the NN interaction in which a weaker coupling is employed [6]. Because these same arguments should also apply to *charged* pion exchange in back-angle $\vec{n}p$ spin transfer, we decided to investigate this process as a possible means of constraining the value of g_c^2.

EXPERIMENTAL TECHNIQUES AND ANALYSIS

The two previous measurements of $K_{NN'}$ in a comparable energy range [8,9] suffered from poor statistics ($\delta K_{NN'} \sim 0.1$) and large systematic error. This is understandable, given that studying spin transfer in the np system requires a true triple scattering experiment initiated with a polarized proton beam, in order to (1) produce the polarized neutron beam, (2) scatter it from a proton target, and then

FIGURE 1. Top view of the detector arrangement used during 'zero-degree' mode running, showing the neutron hodoscopes and the forward proton tracking and polarimetry.

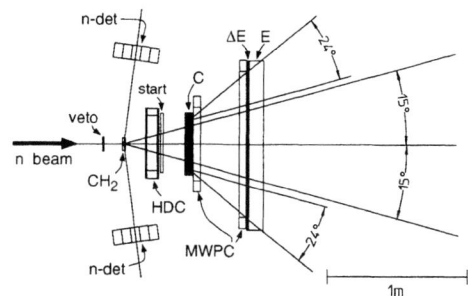

(3) determine the polarization of the outgoing proton. For our measurement, step 1 was carried out using the IUCF Polarized Neutron Facility, in which an intense, highly polarized proton beam ($I = 150$–500 nA, $P_p \approx 0.75$) incident upon a 10-cm long liquid deuterium cell was used to produce the secondary neutron beam. After collimation through 2 m of steel and concrete, neutron fluxes of up to 6×10^6 n/s, with polarizations $P_n \approx 0.6$, were typically available. The mean neutron kinetic energy was determined to be 187 MeV, with a FWHM spread of about 8 MeV.

For this study, the neutrons impinged on a 1.4 cm thick solid CH_2 target. All valid np scattering events required detection of both outgoing nucleons. Two different detector geometries were used: a 'zero-degree' mode, in which recoil protons were detected at angles $|\theta_p^{lab}| \leq 15°$, and a larger-angle mode for $10° \leq \theta_p^{lab} \leq 40°$. The former configuration is shown schematically in Fig. 1. In both cases, the trajectory of the forward-going proton was tracked using 8 planes of alternating x and y horizontal drift chambers ('HDC' in Fig. 1). Thin scintillator planes provided fast timing (to start the HDC's) and charged particle identification, as well as crude time-of-flight information. The timing of the start scintillator relative to the phase-compensated cyclotron rf signal was used to select out only the highest energy neutrons from the secondary beam.

The normal (vertical) polarization of the scattered proton flux was determined via p-C elastic scattering. Downstream of the HDC stack, a graphite target (either 5.0 or 2.5 cm thick, depending on the mode of operation) was followed by two pairs of x-y multiwire proportional chambers (MWPC) and two large planes (ΔE,E) of plastic scintillator. The MWPC's were used to determine the proton trajectory *after* scattering in the C target, while the ΔE–E scintillator combination was sufficiently thick to stop all protons of interest. By combining this information offline to determine the recoil proton's initial energy (after correcting for energy loss), we could select only those events in which the p-C scattering was close to elastic, and thus would exhibit a substantial (and well calibrated) analyzing power. The entire proton polarimeter was modeled very closely after the one used successfully on the IUCF K600 spectrometer [10], so extensive calibration of the effective analyzing power of the apparatus was not required.

The low-energy recoil neutrons ($T_n \sim 0.8$–60 MeV) were detected in a highly segmented ($\Delta\theta = \Delta\phi = \pm 2°$) scintillator hodoscope array, comprised of 60 plastic blocks with individual phototube readout. In zero-degree mode, a set of 30 blocks was positioned on each side of the beam, while a single set of 60 was used for the larger angle running. The depth of the individual hodoscope blocks was varied with the (lab) neutron angle, to maximize efficiency while minimizing the active volume of scintillator; at the most forward angles, a 10-cm depth provided detection efficiencies of 10–20%, while as θ_n^{lab} approached 90° (and $T_n \to 0$), a 2-cm depth maintained very high efficiency yet allowed us to set very low thresholds.

Offline, track reconstruction through the HDC's determined the scattering angle of the detected charged particle, and ensured that it had originated from the known beam-target interaction region. Precise neutron timing relative to the proton start signal provided crisp n–γ discrimination, with clear peaks from 'prompt' photons, as well as a flat background, due primarily to thermal neutron capture, lying under the neutron events of interest. Requirements on these and other free-scattering correlations, imposed in software between the angles and energies of the two detected nucleons, reduced quasi-free contributions from $^{12}C(n,pn)$ processes to less than 1% of the measured yields. These contributions were reduced further by subtracting data taken on a carbon secondary target. Our final step in isolating the free np scattering events was to use the relative n-p timing spectra, corrected for kinematic effects, to account for remaining accidental coincidences in our scattering yields.

Analysis of the proton polarimeter information paralleled very closely that used on the K600 spectrometer. MWPC hits were used to establish the proton direction after passing through the graphite analyzer; combined with the HDC track, we could extract the p+C scattering angle, and also verify that the two tracks intersected within the analyzer. Knowing the complete proton trajectory, as well as its initial energy, energy losses throughout the apparatus could be calculated, but only by *assuming* that the proton had scattered elastically from a ^{12}C nucleus in the graphite target. To check this assumption, measured and calculated energy depositions in the ΔE and E scintillators were compared, and events showing large differences in either scintillator were rejected. All accepted events with p+C scattering angles between 6° and 22° were then sorted by neutron spin state (up or down) and scattering direction (left or right). $K_{NN'}$ and other observables were deduced from these four spin-sorted yields.

RESULTS AND CONCLUSIONS

Preliminary values extracted for $K_{NN'}$ in this work are presented in Fig. 2. The desired statistical precision has been achieved, resulting in total errors for $K_{NN'}$ (statistical plus systematic) of about ±0.02 for 15 angle bins extending from $\theta_{c.m.} = $ 105° to 172°. (Some of the more forward-angle points will be rebinned in our final analysis.) Normalization uncertainties are very small in this case, as both P_n and A_{pol} are well calibrated, and $K_{NN'}$ itself is (fortuitously) quite close to zero. Note that over the angle range from about 135° to 160° the data shown represent the average of three largely independent measurements of $K_{NN'}$: from the left-

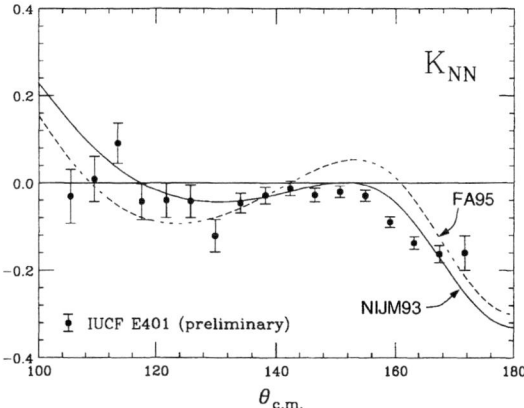

FIGURE 2. Preliminary data for the spin transfer observable $K_{NN'}$ for np elastic scattering at 187 MeV. The two curves are predictions from partial wave analyses of the NN database by different groups, and are described in more detail in the text.

and right-scattered protons in 'zero-degree' mode, and the large-angle data. The excellent agreement found among these three data sets demonstrates that most potential sources of systematic error are well under control.

Our data are seen to be in very good agreement with the predictions of the Nijmegen partial wave analysis [2] (solid line in Fig. 2). This is especially true in the angle range where our results are most robust, and where there are clear differences among the PWA predictions, such as that by the VPI-GW group [11] (dashed line). It is significant that this is the same region where the Uppsala differential cross section data [5] differ substantially in shape from the Nijmegen PWA predictions, a discrepancy which translates fairly directly into the differing values found for the $\pi^{\pm}NN$ coupling constant ($g_\pi^2 \approx 13.6$ in NIJM93 vs. 14.5 in [5]). In contrast to the Uppsala data, our study of spin transfer in the np system, much like our pp results, provides strong support for partial-wave analyses and NN potential models which incorporate a relatively *weak* pion-nucleon coupling.

REFERENCES

1. M. L. Goldberger and S. B. Treiman, Phys. Rev. **110**, 1178 (1958).
2. V. G. J. Stoks, R. A. M. Klomp, M. C. M. Rentmeester, and J. J. de Swart, Phys. Rev. C **48**, 792 (1993).
3. R. A. Arndt, R. L. Workman, and M. M. Pavan, Phys. Rev. C **49**, 2729 (1994).
4. R. Machleidt and F. Sammarruca, Phys. Rev. Lett. **66**, 564 (1991).
5. J. Rahm et al., Phys. Rev. C **57**, 1077 (1998).
6. S. W. Wissink et al., Phys. Rev. Lett. **83**, 4498 (1999).
7. D. V. Bugg and R. Machleidt, Phys. Rev. C **52**, 1203 (1995).
8. D. Spalding et al., Phys. Rev. **58**, 1338 (1967).
9. S. Clough et al., Phys. Rev. C **21**, 988 (1980).
10. W. W. Wissink, in *Spin and Isospin in Nuclear Interactions*, edited by S. W. Wissink, C. D. Goodman, and G. E. Walker (Plenum, New York, 1991), p. 253.
11. R. A. Arndt, I. I. Strakovsky, and R. L. Workman, Phys. Rev. C **50**, 2731 (1994).

Neutron Densities in ^{120}Sn Observed by Polarized Proton Scattering

H.Sakaguchi*, H.Takeda*, T.Taki*, M.Yosoi*, M.Itoh*, T.Kawabata*,
T.Ishikawa*, M.Uchida*, N.Tsukahara*, T.Noro[†], M.Yoshimura[†],
H.Fujimura[†], H.Yoshida[†], E.Obayashi[†], A.Tamii**, and H.Akimune [††]

*Grad. School of Phys., Kyoto Univ., Kyoto 606-8501, Japan
[†] Research Center for Nuclear Phys., Osaka Univ., Osaka 567-0047, Japan
** Grad. School of Phys., Univ. of Tokyo, Tokyo 113-0033, Japan
[††] Dep. of Phys., Konan Univ., Kobe 658-8501, Japan

Abstract. Cross sections, analyzing powers and spin rotation parameters of proton elastic scattering from ^{58}Ni and ^{120}Sn have been measured at intermediate energies. By elastic scattering off $N \simeq Z$ nulcei like ^{58}Ni at intermediate energies we can study medium modification of the nucleon-nucleon (NN) interaction inside the nucleus, because proton distributions in target nuclei are constrained by charge distributions measured by electron scattering and neutron distributions can be assumed to be the same as proton's. In order to explain our experimental data of ^{58}Ni at large scattering angles, it was found to be necessary to use experimental densities deduced from charge densities measured by electron scattering and to modify the coupling constants and the masses of exchanged σ and ω mesons in the RIA, assuming linear dependencies of meson propeties to nuclear densities. Parameters of the medium effect have been searched to reproduce the data.

For $N \neq Z$ nuclei, neutron density distribution can be extracted from the elastic scattering, assuming the same medium modifications fixed by the ^{58}Ni data and using proton distributions obtained from charge distributions. We have searched neutron density distribution so as to reproduce ^{120}Sn data at the proton incident energy of 300 MeV. Deduced neutron distribution has an increase at the nuclear center, which is consistent with the $3s_{1/2}$ orbit wave function as expected in ^{120}Sn. At energies other than 300 MeV, experimental data of ^{120}Sn have been also well reproduced by the neutron distribution obtained at 300 MeV.

I INTRODUCTION

Protons at intermediate energies are considered to be a suitable probe to extract internal information of nuclei because of the large mean free path in nuclear medium. For $N \simeq Z$ nuclei ambiguities due to the target nuclear structure are relatively small in the elastic scattering, because proton distributions in target nuclei are constrained by charge distributions measured by the electron scattering and

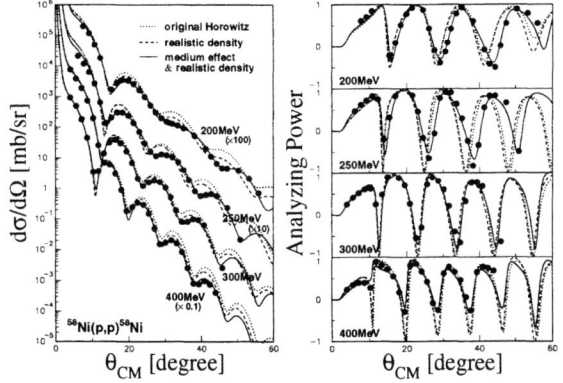
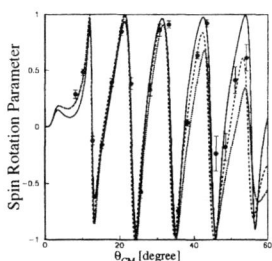

FIGURE 1. Elastic scattering of polarized protons off ^{58}Ni. Dotted curves show the original RIA calculation of Horowitz et al.. Broken curve are the similar calculation with realistic densities deduced from the electron scattering. The solid curves indicates the calculation with medium modified NN interaction and the realistic densities.

FIGURE 2. Spin rotation parameters for the elastic scattering of protons off ^{58}Ni at 300 MeV. Curves in the figure are the same as in Fig. 1.

neutron distributions can be assumed as same as proton distributions for $N \simeq Z$ nuclei. Thus the proton elastic scattering at intermediate energies has been used to discern various microscopic approaches for nuclear interactions. Among various models relativistic impulse approximations (RIA) have successfully explain elastic scattering, especially polarization observables. Since these observables are directly related to the coupling constants and the masses of exchanged mesons in the RIA formalism, we can study the medium modification of nucleon–nucleon (NN) interactions related to the modification of hadron properties in nuclear medium.

For $N \neq Z$ nuclei we can not expect that neutron distribution has the same shape as protons. However, the elastic scattering is sensitive to both NN interactions in nuclear medium and density distributions of target nucleus. Neutron density distribution can be extracted from the elastic scattering, assuming the same medium modifications obtained by the data for $N \simeq Z$ nuclei and using proton distributions obtained from charge distributions.

We first fix medium modification parameters by using the elastic scattering off ^{58}Ni, which is the heaviest nucleus with almost equal neutron and proton numbers. Then, we can extract neutron density distributions for ^{120}Sn from the proton elastic scattering.

II MEDIUM MODIFICATION OF NN INTERACTION

Cross sections and polarization observables of elastic scattering of polarized protons off ^{58}Ni are shown in Fig. 1,2. Dotted curves in the figure show the original RIA calculations of Horowitz et al. [3]. They can explain the analyzing powers but fail to explain cross section at backwards. This tendency remains, even if we use density distributon deduced from the charge distribution. As reported in detail in our previous report [2], in order to explain the elastic scattering data, it was necessary to modify NN interactions in nuclear medium. We modified the coupling constants and the masses of exchanged σ and ω mesons in RIA calculations, assuming linear dependencies to nuclear densities as follows;

$$g_j^2, \bar{g}_j^2 \longrightarrow \frac{g_j^2}{1+a_j\rho(r)/\rho_0}, \frac{\bar{g}_j^2}{1+\bar{a}_j\rho(r)/\rho_0}, \qquad (1)$$

$$m_j, \bar{m}_j \longrightarrow m_j\left[1+b_j\rho(r)/\rho_0\right], \bar{m}_j\left[1+\bar{b}_j\rho(r)/\rho_0\right], \qquad (2)$$

where index j indicates σ or ω mesons and $\rho_0 = 0.1934\,\text{fm}^{-3}$ is a normal density. Best fit parameters are searched at each energy. With this modification of NN interaction we have succeeded to explain differential cross sections (espetially at backward angles), analyzing powers and spin rotation parameters [2].

III NEUTRON DENSITY DISTRIBUTIONS

Experimental results of cross sections, analyzing powers and spin rotation parameters of ^{120}Sn at 300 MeV are plotted in Fig. 3, 4 as solid circles. For $N \neq Z$ nuclei it can not be expected that neutron distributions have the same shape as proton. However, assuming the same density dependence parameters as ^{58}Ni for NN interactions and using proton distributions deduced from the electron scattering data, neutron distributions can be extracted from our elastic scattering data.

In order to search the neutron distribution we have used a sum of Gaussians type distribution;

$$\rho_n(r) = \frac{N}{2\pi^{3/2}\gamma^3}\sum_i \frac{Q_i}{1+2R_i^2/\gamma^2}\left(e^{-(r-R_i)^2/\gamma^2}+e^{-(r+R_i)^2/\gamma^2}\right). \qquad (3)$$

Normalization condition $\int \rho_n(r)\,d^3\vec{r} = N$ results in a constraint for Q_i; $\sum Q_i = 1$. Q_i are searched so as to reproduce ^{120}Sn data at 300 MeV, while width γ and position R_i of each Gaussian are fixed with the values listed in a reference [4].

Deduced neutron distributions are displayed in Fig. 5, 6. Hatched area is a superposition of possible densities with good reduced chi-squares $\chi_\nu^2 \, (= \chi^2/\nu)$;

$$\chi_\nu^2 - \chi_{\nu\min}^2 \leq 1, \qquad (4)$$

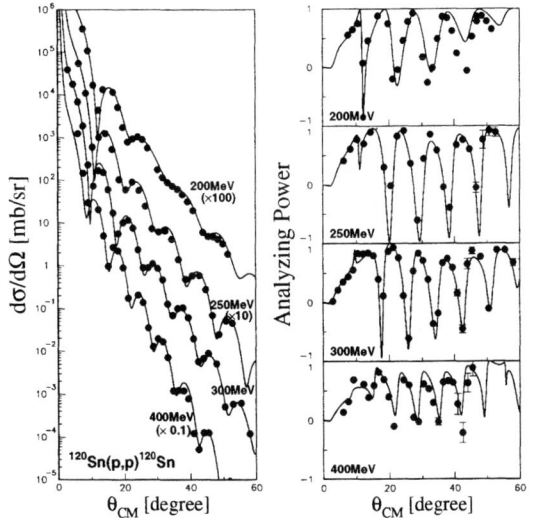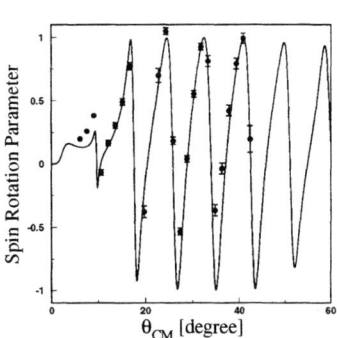

FIGURE 3. Cross sections and analyzing powers of proton elastic scattering off ^{120}Sn at 200, 250, 300 and 400 MeV. Neutron densities are searched to explain the scattering of 300 MeV data and the same neutron density are used for all the 4 energies.

FIGURE 4. Spin rotation parameters of proton elastic scattering off ^{120}Sn at 300 MeV.

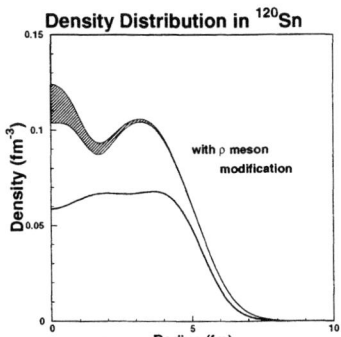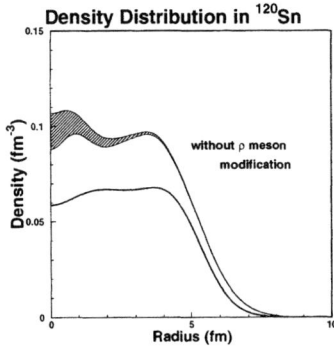

FIGURE 5. Deduced neutron distribution in ^{120}Sn with medium effects for ρ-mesons in NN interaction

FIGURE 6. Deduced neutron distribution in ^{120}Sn without medium effects for ρ-mesons in NN interaction

where ν is the number of degrees of freedom, $\chi^2_{\nu\min}$ is the minimum value of χ^2_ν. Deduced neutron distribution has an increase at nuclear center. This result may be a reflection of the wave function of neutrons in $3s_{1/2}$ orbit as expected to be occupied in ^{120}Sn nuclei. Solid lines in Fig. 3, 4 are the calculations using the best fit neutron density.

^{120}Sn data at energies other than 300 MeV are also explained by the same neutron distribution deduced at 300 MeV. Figure 3 shows cross sections and analyzing powers at all energies. The obtained density distribution of neutrons depends on the medium modification of the ρ-meson part in the NN interaction. If we introduce medium effects of ρ-mesons by scaling law, rms radii of protons and neutrons are almost equal and the density distribution of neutrons is shown in Fig. 5. But, if we switch off the medium effect, the rms radius of neutron distribution is larger than that of protons by about 0.1 fm and the density distribution becomes as shown in Fig. 6 .

IV SUMMARY

We have measured cross sections, analyzing powers and spin rotation parameters of proton elastic scattering from ^{58}Ni and ^{120}Sn at intermediate energies at RCNP. Original RIA calculations based on free *NN* interactions can not explain experimental data of ^{58}Ni, especially cross sections at large scattering angles. We have introduced medium effects into the RIA calculation and successfully explained the data.

Using obtained *NN* interactions in nuclear medium, we have searched neutron density distribution so as to reproduce ^{120}Sn data at 300 MeV. Deduced neutron distribution has an increase at nuclear center, which is consistent with $3s_{1/2}$ orbit wave function as expected in ^{120}Sn. At energies other than 300 MeV experimental data of ^{120}Sn are also well reproduced with the obtained neutron distribution.

REFERENCES

1. M. Yosoi *et al.*, *11th Int. Symp. on High Energy Spin Physics*, AIP Conference Proceedings No. 343, p. 157.; M. Yosoi *et al.*, RCNP Annual Report 1994, p. 147.
2. H. Sakaguchi *et al.*, Phys. Rev. C57 (1998) 1749; T. Taki *et al.*, RCNP Annual Report 1997, p. 56.
3. D. P. Murdock and C. J. Horowitz, Phys. Rev. C35(1442)1987; C. J. Horowitz *et al.*, *Computational Nuclear Physics 1* (Springer-Verlag, Berlin, 1991), Chap. 7.
4. H. de Vries *et al.*, Atomic Data and Nuclear Data Tables 36(1987)495.

Spin-flip probability for the ^{26}Mg(^{3}He,t)^{26}Al*(1$^+$; 1.058 MeV) reaction at 177 MeV

S. Sakoda[a], H. Sakai[a], A. Tamii[a], T. Ohnishi[a], K. Yako[a],
M. Hatano[a], Y. Maeda[a], T. Uesaka[b], M.B. Greenfield[c],
M.N. Harakeh[d], A.M. van den Berg[d], V.M. Hannen[d],
B. Krüsemann[d], R.G.T. Zegers[d], M.A. de Huu[d],
D. Frekers[e], S. Rakers[e], H.J. Wörtche[e], F. Ellingh[e],
M. Hagemann[f], J. Heyse[f], N. Blasi[g], F. Camera[g]

[a] Department of Physics, University of Tokyo, Bunkyo, Tokyo 113-0033, Japan
[b] Department of Physics, Saitama University, Saitama 338-8570, Japan
[c] Division of Natural Sciences, International Christian University, Mitaka, Tokyo 181-8585, Japan
[d] Kernfysisch Versneller Instituut, Zernikelaan 25, 9747 AA Groningen, The Netherlands
[e] Institut für Kernphysik, Universität Münster, D-48149 Münster, Germany
[f] Vakgroep Subatomaire en Stralingsfysica, Rijksuniversiteit Gent, B-9000 Gent, Belgium
[g] Dipartimento di Fisica, Universita di Milano and INFN sez. Milano, I-20133 Milano, Italy

Abstract. Spin-flip probabilities and the differential cross sections for the ^{26}Mg(^{3}He,t)^{26}Al*(1$^+$;1.058 MeV) reaction have been measured at $E_{^3{\rm He}}$=177 MeV. This is the first spin-flip probability data for the (^{3}He,t) reaction in the energy region above 20 MeV/A. The data are compared with distorted wave Born approximation calculations by using the ^{3}He-n effective interaction extracted from the differential cross sections.

INTRODUCTION

For the study of the (^{3}He,t) reaction, it is very important to determine the effective interaction between the ^{3}He nucleus and the neutron in the target nucleus. However, the ^{3}He-n effective interaction has never been extracted from the spin-flip probability which is sensitive to the spin dependent part but only from the differential cross sections because there was only one measurement of the spin-flip probability for the (^{3}He,t) reaction at the low energy region [1]. There is no data in the energy region above 20 MeV/A where the reaction mechanism is simple.

To extract the information about the spin dependent part of the ^{3}He-n effective interaction, we have measured the spin-flip probability for the ^{26}Mg(^{3}He,t)^{26}Al*(1^+;1.058 MeV) reaction at $E_{^3He}$=177 MeV by applying the Bohr theorem (Schmidt method).

Schmidt method

By applying the Bohr theorem [2], the spin-flip probability can be measured without polarized beams nor a polarimeter.

According to the Bohr theorem, the change of the magnetic substate between a projectile and an ejectile is related to the magnetic substate of the residual nucleus by the parity conservation law. Table 1 shows the case of the ^{26}Mg(0^+)(^{3}He,t)^{26}Al*(1^+) reaction. The magnetic substate m_y of ^{26}Al*(1^+) is restricted to $\pm 1(0)$ in the case of spin-flip (non-spin-flip) reactions. This means that the spin-flip probability of the (^{3}He,t) reaction can be obtained from the population in the magnetic substate of ^{26}Al*.

The magnetic substate population can be measured by detecting the γ-rays emitted from the ^{26}Al* nuclei. In the case of $1^+ \to 0^+$ γ-decay, γ-rays are emitted in the direction of the quantum axis only when the magnetic substate of ^{26}Al* is ± 1. Therefore, we can obtained the spin-flip probability by measuring the γ-rays emitted perpendicularly to the scattering plane.

This method was first applied by F.H. Schmidt et al. [3] to the measurement of the spin-flip probability in inelastic proton scattering, and is called Schmidt method. We apply this method to the reaction

$$^{26}Mg(^3He, t\gamma)^{26}Al^*(1^+; 1.058 MeV \to 0^+; 0.228 MeV)$$

at $E_{^3He}$=177 MeV and at $\theta_{cm} = 0.7 - 4°$. The branching ratio of the γ-decay is very close to 100 %.

TABLE 1. Allowed spin substates by the Bohr theorem.

J^π	$\frac{1}{2}^+$	0^+		$\frac{1}{2}^+$	1^+		
particle	^{3}He	^{26}Mg	$\to$	t	^{26}Al*	$\|\Delta m_y\|$	
m_y	+1/2	0		+1/2	0	0	
	+1/2	0		−1/2	± 1	0,2	spin-flip
	−1/2	0		+1/2	± 1	2,0	spin-flip
	−1/2	0		−1/2	0	0	

EXPERIMENT

The experiment was performed at the KVI in the Netherlands. A ^{3}He^{++} beam accelerated up to $E_{^3\text{He}}$=177 MeV by the AGOR super-conducting cyclotron was bombarded an enriched ^{26}Mg foil with a thickness of 4.66 mg/cm^2. Scattered tritons were momentum analyzed by the Big-Bite Spectrometer and were detected at the focal plane. Emitted γ-rays were detected by a GSO(Gd$_2$SiO^5:Ce) scintillation counter with a dimension of 35 × 35 × 10mm^3 in coincidence with the tritons. The GSO counter was placed perpendicular to the reaction plane at a distance of 60 mm from the target. In order to obtain the angular distribution of the emitted γ-rays, four BaF$_2$ scintillation counters were installed around the target.

The spin-flip probability is proportional to the ratio of the number of the detected γ-rays to the number of the tritons populating the 1^+ state (1.058 MeV). The spin-flip probability S is written by

$$S = \frac{8\pi}{3} \frac{Y_\gamma}{Y_t} \frac{1}{\varepsilon_\gamma \Delta \Omega_\gamma}, \quad (1)$$

where Y_t is the number of singles events of tritons, Y_γ is the number of coincidence γ-rays (1^+; 1.058MeV $\to$ 0^+; 0.228MeV) with an energy of 830 keV, and $\varepsilon_\gamma \Delta \Omega_\gamma$ is the detection efficiency and the solid angle of the γ-ray detector.

We also measure the differential cross section for the ^{26}Mg(^{3}He,t)^{26}Al*(1^+;1.058 MeV) at θ_{cm}=0-34°.

DWBA CALCULATION

Distorted wave Born approximation (DWBA) calculations were performed.

The ^{3}He-n effective interaction determined by van der Werf et al. [4] to fit the differential cross sections in the energy region from 66 to 90 MeV. The effective interaction is expressed in terms of single Yukawa potentials ($Y(r)$) as

$$V_{eff}(r) = \left\{ V_{\sigma\tau} (\boldsymbol{\sigma}_1 \cdot \boldsymbol{\sigma}_2) Y\left(\frac{r}{R_{\sigma\tau}}\right) + V_{T\tau} S_{12} r^2 Y\left(\frac{r}{R_{T\tau}}\right) \right\} (\boldsymbol{\tau}_1 \cdot \boldsymbol{\tau}_2), \quad (2)$$

where S_{12} is the usual tensor operator. The parameters, R and V, are $V_{\sigma\tau}$=-3.0 MeV, $V_{T\tau}$=-6.5 MeV·fm^{-2}, $R_{\sigma\tau}$=1.415 fm, and $R_{T\tau}$= 0.878 fm.

The optical-model parameters are obtained by extrapolating the ^{3}He global optical potential for $E_{^3\text{He}}$=90-120 MeV [5] to $E_{^3\text{He}}$=177 MeV. The DWBA calculation with the optical-model parameters reproduces the elastic scattering cross section very well.

Transition density for the 1^+ state is Brown and Wildenthal [6] in a full $0d_{5/2}, 0d_{3/2}, 1s_{1/2}$ basis.

RESULTS AND DISCUSSION

The experimental data and the calculation result are shown in Figure 1. The Left panel shows spin-flip probability, the right panel shows the differential cross section, the solid lines show the result of the DWBA calculation. The differential cross sections are not reproduced by the calculation at $\theta_{cm}=5-20°$. The spin-flip probabilities are not reproduced at all the measured angles.

We discuss the reason of the disagreement, especially for the spin-flip probability.

Firstly the spin-flip probability is insensitive to the optical potential. Actually the agreement between the data and the calculation is completely destroyed for the differential cross sections by slightly changing the volume terms of the optical-model parameters, while the spin-flip probability is not so much affected.

Next we consider the effective interaction. The effective interaction is for the energy range of 66 – 90 MeV and may not be appropriate at 177 MeV. The spin-flip probability is sensitive to the ratio $V_{T\tau}/V_{\sigma\tau}$. We change the strength of $V_{T\tau}$ keeping the same $V_{\sigma\tau}$ strength. The results are shown in Figure 1. The calculation with $V_{T\tau}/V_{\sigma\tau}=2.0$ fm^2 well reproduces the spin-flip probability but does not reproduce the differential cross section, especially at the second peak. Both the spin-flip probability and the differential cross section can not be simultaneously reproduced by changing the effective interaction.

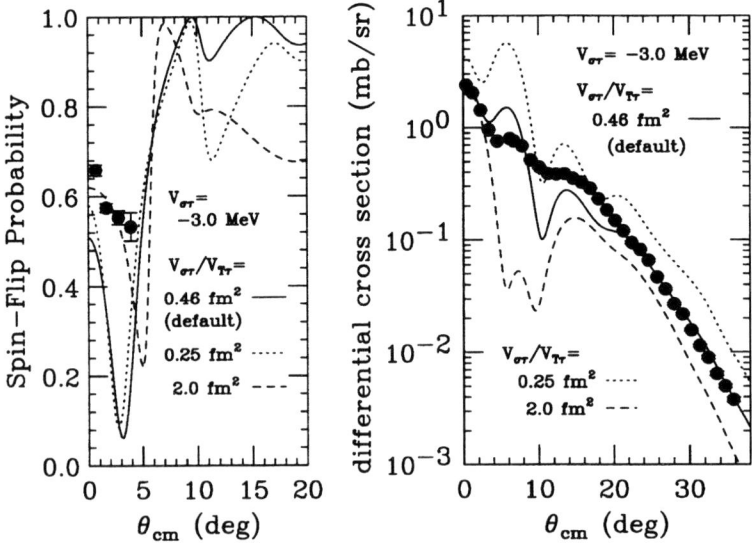

FIGURE 1. DWBA calculations with changing the value of $V_{T\tau}$ (the ratio $V_{T\tau}/V_{\sigma\tau}$). Solid lines show the calculation with given values from Ref. [4] ($V_{T\tau}/V_{\sigma\tau}=0.46$). Dotted and dashed lines show the calculation with $V_{T\tau}/V_{\sigma\tau}=0.25$ and 2.0, respectively.

SUMMARY

The spin-flip probabilities and the differential cross sections for the ^{26}Mg(^{3}He,t)^{26}Al (1^+; 1.058 MeV) reaction have been measured at 177 MeV. This is the first measurement of the spin-flip probability for the (^{3}He,t) reaction in the energy region above 20 MeV/A. DWBA calculations were performed by using the effective interactions parameterized by van der Werf *et al.* from the differential cross section. The calculations do not reproduce both the spin-flip probabilities and differential cross sections simultaneously.

ACKNOWLEDGMENTS

We would like to express our appreciation to the continuous help of the staff of the KVI facility. This project is supported by the Ministry of Education, Science, Sports and Culture of Japan with the Grant-in-Aid for Science Research No. 10044058.

REFERENCES

1. H. Sakai, T. Aoyama, S. Fujitaka, M.N. Harakeh, K. Kubota, H. Okamura, H. Otsu, Y. Satou, M. Tanaka, T. Uesaka, and T. Wakasa, Nucl. Phys. **A 588** (1995) 479
2. A. Bohr, Nucl. Phys. **10** (1959) 486.
3. F. H. Schmidt R.E. Brown, J.B. Gerhart and W.A. Lolasinski, Nucl. Phys. **52** (1964) 353.
4. S.Y. van der Verf, S. Brandenburg, P. Grasdijk, W.A. Sterrenburg, M.N. Harakeh, M.B. Greenfield, B.A. Brown and M. Fujiwara, Nucl. Phys. **A496** (1989) 305.
5. M. Hyakutake, *et al.*, Nucl. Phys. **A333** (1980) 1.
6. B.A. Brown and B.H. Wildenthal, At. Data Nucl. Data Tables 33 (1985) 347.

Study of Isospin Structure of 1^+ Spin States in ^{58}Ni and ^{58}Cu by the Comparison of ^{58}Ni(p,p') and ^{58}Ni$(^3$He,$t)^{58}$Cu Reactions

H. Fujita[*,1], Y. Fujita[*], G.P.A. Berg[‡,2], Y. Shimbara[*],
A.D. Bacher[†], C.C. Foster[†], K. Hara[‡], K. Harada[*], K. Hatanaka[‡],
J. Jänecke[**], K. Katori[*,3], T. Kawabata[§], T. Noro[‡],
D.A. Roberts[**], H. Sakaguchi[§], T. Shinada[*],
E.J. Stephenson[†], T. Taki[§], H. Ueno[*,4], and M. Yosoi[‡]

[*]*Department of Physics, Osaka University, Toyonaka, Osaka 560-0043, Japan*
[†]*IUCF, Indiana University, Bloomington, IN 47408, USA*
[‡]*RCNP, Osaka University, Ibaraki, Osaka 567-0047, Japan*
[**]*Department of Physics, University of Michigan, Ann Arbor, MI 48109, USA*
[§]*Department of Physics, Kyoto University, Sakyo, Kyoto 606-8224, Japan*

Abstract. High-resolution measurements of proton inelastic and charge-exchange (^{3}He,t) reactions at $0°$ on ^{58}Ni were performed. These reactions mainly cause isovector spin flip transitions to M1 states in ^{58}Ni and GT states in ^{58}Cu. The "level-by-level" comparison of analogous M1 and GT transition strengths allows the study of the isospin symmetry structure of highly-excited analog states in ^{58}Ni and ^{58}Cu and identification of isospin T ($T = 1$ and 2) based on the different sensitivities of inelastic and charge exchange reactions.

I ANALOGOUS GT AND M1 TRANSITIONS

A Gamow-Teller (GT) transition is characterized by the quantum numbers $\Delta L = 0$, $\Delta S = 1$ ($J^\pi = 1^+$) and $\Delta T = 1$. GT states are excited favorably in a (^{3}He,t) charge exchange (CE) reaction. The GT final states excited by the (^{3}He,t) reactions on $T_0 = 1$ nucleus ^{58}Ni can have isospin values T_f are 0, 1 and 2.

---

[1] email address: hfujita@lns.sci.osaka-u.ac.jp
[2] On leave from IUCF, Indiana
[3] RIKEN, Wako, Saitama 351-0198, Japan
[4] RIKEN, Wako, Saitama 351-0198, Japan

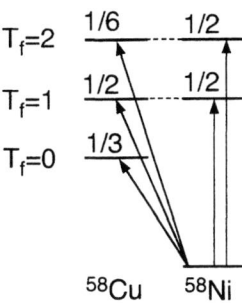

FIGURE 1. The squared isospin CG coefficients of GT and M1 states in ^{58}Cu and ^{58}Ni.

In an inelastic (IE) proton scattering at $0°$, 1^+ states can also be excited. These are called M1 states. Only $T_f = 1$ and 2 are allowed for M1 states excited by the ^{58}Ni(p,p') reaction. These M1 states are analogous states of $T_f = 1$ and 2 GT states in ^{58}Cu. Therefore, under the assumption of charge symmetry of nuclear interactions, the excitation energies of 1^+ states with $T_f = 1$ and 2 excited by ^{58}Ni$(^3$He,$t)$ and ^{58}Ni(p,p') reactions should show good correspondence if Q-values and the energy shift from Coulomb interactions are taken into account.

In hadron reactions like $(^3$He,$t)$ and (p,p') at intermediate incident energies and at $0°$, there is a good proportionality between the strengths of GT and M1 transitions and cross sections [1,2]. Because of the similarity of hadron CE and IE reactions, the main difference of GT and M1 transition strengths, and thus the difference of cross sections come from different isospin Clebsch-Gordan (CG) coefficients assuming pure $\sigma\tau$ interaction [3].

In Fig 1, the squared CG coefficients of GT and M1 transitions for allowed T_f are shown for the target ^{58}Ni. For the GT and M1 transitions to $T_f = 1$ states, the squared CG coefficients are both 1/2. However, for the transitions to $T_f = 2$ states, they are 1/6 and 1/2 for GT and M1 transitions, respectively. This suggests that in the IE reaction $T_f = 2$ states are three times more favorably excited than in the CE reaction compared with $T_f = 1$ states. Excited states with isospin $T_f = 0$ are observed only in CE reaction. In (n,p)-type CE reaction like $(t,^3$He$)$ and $(d,^2$He$)$ on ^{58}Ni, only 2 is allowed as a T_f value of GT final states in ^{58}Co.

II EXPERIMENTS

Since the isospin symmetry structure of 1^+ states should be studied on a level-by-level base, good energy resolution is needed. This can be realized by the application of *focus* and *lateral dispersion matching* techniques using high resolution magnetic spectrometers. For a large dispersion spectrometer like Grand Raiden, dispersion of the beam line for *lateral dispersion matching* may prevent precise horizontal scattering angle measurement. By applying the *angular dispersion matching* technique,

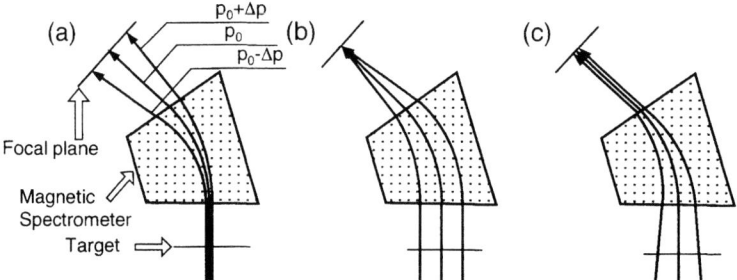

FIGURE 2. Ion trajectories under different matching conditions of a beam line and a magnetic spectrometer. Each of them show the situation (a) when achromatic beam transportation is used; (b) when *lateral dispersion matching* is realized in transporting the beam; (c) when both *lateral dispersion matching* and *angular dispersion matching* are realized. No angle spread of the beam is assumed in the figures so as to show the behaviors of beams with slightly different momentum $\pm \Delta p$ from p_0.

good horizontal angle resolution can be realized. These techniques are illustrated in Fig. 2. Trajectories of the beam with slightly different momenta, p_0 and $p_0 \pm \Delta p$, and assuming no horizontal angle spread in a spectrometer at 0° are shown under the condition that (a) the beam is achromatically transported, (b) *lateral dispersion matching* is realized and (c) both *lateral* and *angular dispersion matchings* are realized.

The 0° ^{58}Ni(p,p') experiment was performed at IUCF by using the K600 spectrometer [4]. The incident proton energy was 160 MeV. In the 0° (p,p') reaction, the incident beam passes through the spectrometer with inelastically scattered particles and is stopped in a special Faraday cup behind the detector system. Therefore, the low excitation energy region cannot be observed and beam halo can cause a large background. The cyclotron and the beam line was tuned to minimize the background from the beam during the experiment. The *lateral dispersion matching* technique was applied [5] and an energy resolution of 35 keV was achieved. As shown in Fig. 3 (a), a low background spectrum was obtained. The excited states at 8.46, 8.60, 8.67, 9.07 and 9.15 MeV are also reported in ^{58}Ni(γ,γ') experiments of F. Bauwens et al. [6] and identified as M1 states.

The ^{58}Ni$(^3\mathrm{He},t)$ experiment was performed at RCNP using the Grand Raiden spectrometer and the new beam line WS course at 150 MeV/nucleon. This beam line is designed to realize both *lateral* and *angular dispersion matchings* efficiently [7]. By applying matching techniques, high energy resolution of 50 keV was achieved. In this experiment, the energy spread of the incident beam was estimated to be 250 keV. An energy spectrum of corresponding to the ^{58}Ni(p,p') spectrum is shown in Fig. 3 (b). The fine structure of the GT resonance in the energy region $E_x = 8.5 - 9.5$ MeV was clearly observed.

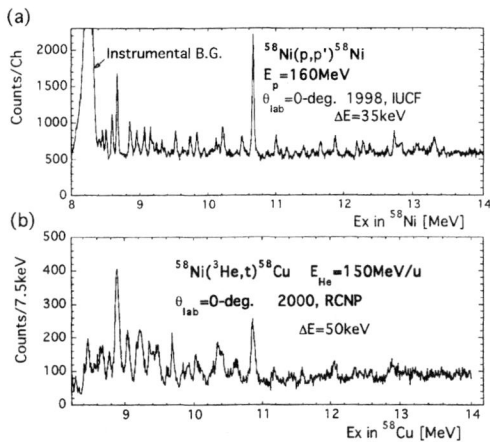

FIGURE 3. Energy spectra from $E_x = 8 \sim 14$ MeV are shown for the ^{58}Ni(p,p') (a) and the ^{58}Ni$(^3$He,$t)$ (b) reaction.

III COMPARISON OF EXPERIMENTAL RESULTS

In an earlier work, isospin distributions of GT states in ^{58}Cu were studied by comparing ^{58}Ni$(^3$He,$t)$ and various other reactions [8]. With the high energy resolution ^{58}Ni(p,p') and ^{58}Ni$(^3$He,$t)$ spectra taken at 0°, more direct and detailed level-by-level comparison of analogous M1 and GT transitions can be performed in order to study the isospin symmetry structure.

From the comparison of ^{58}Ni(p,p') and ^{58}Ni$(^3$He,$t)$ spectra, which are shown in Figs. 3 (a) and (b), correspondence of excitation energies of states is found. For example, the 10.67 MeV state in Fig. 3 (a) corresponds to the 10.85 MeV state in Fig. 3 (b) and the 8.67 MeV state to the 8.88 MeV state. Correspondences of excitation energies suggest that isospin symmetry structure exists even at these high excitation energies and each pair of excited states are analogous GT and M1 states. These analogous GT and M1 states should have isospin values either $T_f = 1$ or 2.

The ratios of events for each pair of analogous states are shown in Fig. 4. For the 10.67 MeV state in ^{58}Ni(p,p') spectrum, which was assigned to be $T_f = 2$ in Ref. [9], the ratio of 3 is given. As discussed above, $T_f = 2$ states are three times more strongly excited in IE reaction than in the CE reaction compared to $T_f = 1$ states. Therefore, the ratios of other $T_f = 2$ states should also be approximately 3, while the ratios for $T_f = 1$ states should nearly be 1.

The calculated ratios are clearly divided into two groups. The group of states in the 8.6 - 10.5 MeV region in (p,p') spectrum show the ratios of about 1. On the other hand, the states above 10.67 MeV show ratios larger than 1, namely about 3. These ratios show good agreement with the predicted values from isospin CG

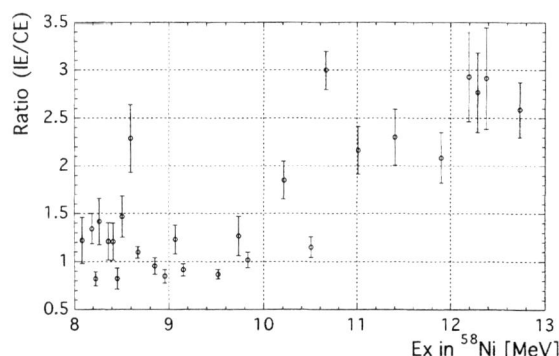

FIGURE 4. Empirical ratios between cross sections of identified analogous states. From the isospin CG coefficients, expected ratios are 1 and 3 for $T_f = 1$ and 2 states, respectively. The horizontal axis shows the excitation energy in (p,p'). Peak fitting was done by software to deduce the number of events of the peaks for each spectrum.

coefficients, which suggests that the states in the first group is of $T_f = 1$ nature and those in the second group is of $T_f = 2$ nature. As expected from isospin symmetry energy, $T_f = 2$ states are found at higher excitation energies compared to $T_f = 1$ states.

It was reported that three states with the excitation energies of 9.83, 10.20 and 10.49 MeV in the (p,p') spectrum were observed in the ^{58}Ni(e,e') experiment [9] and that corresponding analogous states were observed in the ^{58}Ni$(t,^3$He$)^{58}$Co experiment [10], which suggests these states have $T_f = 2$. In our present analysis these states show the ratios of about 1, suggesting that they are of $T_f = 1$ nature. A possible explanation of these reduced ratios may be the existence of unresolved $T_f = 0$ states near $T_f = 2$ states in the ^{58}Ni$(^3$He,$t)$ spectrum.

REFERENCES

1. T.N. Taddeucci et al., Nucl. Phys. A **469**, (1987) 125.
2. Y. Fujita et al., Phys. Rev. C **59**, (1999) 90.
3. Y. Fujita et al., Proc. Int. Conf. GR2000, Osaka, June 2000, to be published in Nucl. Phys. A, and references therein.
4. Y. Shimbara et al., OULNS Annual Report 1998, p.86.
5. G.P.A. Berg et al., IUCF Annual Report 1993, p.106.
6. F. Bauwens et al., Phys. Rev. C. **62**, (2000) 024302.
7. Y. Fujita et al., Nucl. Instrm. and Meth. B **126**, (1997) 274.
8. Y. Fujita et al., Phys. Lett. B **365**, (1996) 29.
9. W. Mettner et al., Nucl. Phys. A **473**, (1987) 160.
10. F. Ajzenberg-Selove et al., Phys. Rev. C **31** (1985) 777.

Momentum Transfer Dependence of Spin Isospin Modes in Quasielastic $(\vec{p}, \vec{n})$ Reactions

T. Wakasa,[a] H. Sakai,[b] M. Ichimura,[c] K. Hatanaka,[a] H. Okamura,[d]
K. Kawahigashi,[e] A. Tamii,[b] H. Otsu,[f] Y. Nakaoka,[b] T. Ohnishi,[g]
K. Yako,[b] K. Sekiguchi,[b] T. Yagita,[h] J. Kamiya,[a] S. Sakoda,[b]
K. Suda,[d] H. Kato,[b] M. Hatano,[b] and Y. Maeda[b]

[a] *Research Center for Nuclear Physics (RCNP), Ibaraki, Osaka 567-0047, Japan*
[b] *Department of Physics, University of Tokyo, Tokyo 113-0033, Japan*
[c] *Faculty of Computer and Information Sciences, Hosei University, Tokyo 184-8584, Japan*
[d] *Department of Physics, Saitama University, Saitama 338-8570, Japan*
[e] *Department of Information Science, Kanagawa University, Kanagawa 259-1293, Japan*
[f] *Department of Physics, Tohoku University, Miyagi 980-8578, Japan*
[g] *Institute of Chemical and Physical Research (RIKEN), Saitama 351-0198, Japan*
[h] *Department of Physics, Kyushu University, Fukuoka 812-8581, Japan*

Abstract. A complete set of polarization transfer coefficients has been measured for quasielastic $(\vec{p}, \vec{n})$ reactions on ^{2}H and ^{12}C at a bombarding energy of 346 MeV and laboratory scattering angles of 16°, 22°, and 27°. The spin-longitudinal ID_q and spin-transverse ID_p polarized cross sections are deduced. The theoretically expected enhancement in the spin-longitudinal mode is observed. The observed ID_q is consistent with the pionic enhanced ID_q evaluated in a distorted wave impulse approximation (DWIA) calculation employing a random phase approximation (RPA) response function. On the contrary, the theoretically predicted quenching in the spin-transverse mode is not observed. The observed ID_p is not quenched, but rather enhanced in comparison with the DWIA+RPA calculation.

INTRODUCTION

The role of the pion (π) and rho-meson (ρ) in the nuclear spin-isospin response functions is one of the most interesting subjects of nuclear physics. The spin-isospin dependent residual interaction is often given by the $\pi+\rho+g'$ model. In this model [1] with a standard value of g', the spin-longitudinal interaction becomes moderately attractive for $q > 0.8$ fm^{-1}, while the spin-transverse interaction remains repulsive for the wide range of q. Thus the relevant isovector spin response functions in the quasielastic region are expected to show an enhanced ratio of the spin-longitudinal response function R_L relative to the spin-transverse response function R_T [1].

In this article, we present the measurements of a complete set of polarization transfer coefficients for quasielastic $(\vec{p}, \vec{n})$ reactions on ^{2}H and ^{12}C at $T_p = 346$ MeV and laboratory scattering angles of $\theta_{\text{lab}} = 16°$, $22°$, and $27°$ which correspond to $q_{\text{lab}} \simeq 1.2$, 1.7 and 2.0 fm^{-1} at the quasielastic peak [2]. The measured polarization transfer coefficients and cross sections are used to separate the cross sections into non-spin, spin-longitudinal, and spin-transverse polarized cross sections. The experimental polarized cross sections will be compared with theoretical calculations in frameworks of a distorted wave impulse approximation (DWIA) and a random phase approximation (RPA).

EXPERIMENT AND DATA REDUCTION

The data presented here were obtained with the Neutron Time-Of-Flight (NTOF) facility [3] at the Research Center for Nuclear Physics (RCNP), Osaka University. Detailed descriptions concerning the NTOF facility and the neutron detection system can be found in elsewhere [3–5].

The (p, n) unpolarized double differential cross section I in the NA laboratory frame can be separated into four polarized cross sections ID_i as

$$I = ID_0 + ID_q + ID_n + ID_p , \tag{1}$$

where D_i are the polarization observables introduced by Bleszynski et al. [6]. In a plane wave impulse approximation (PWIA) with eikonal and optimal factorization approximations, ID_i are expressed as

$$\begin{aligned}
ID_0 &= 8CK(2J_A+1)N_{\text{eff}}(|A^\eta|^2 R_0 + |C_2^\eta|^2 R_n) , \\
ID_n &= 8CK(2J_A+1)N_{\text{eff}}(|B^\eta|^2 R_n + |C_1^\eta|^2 R_0) , \\
ID_q &= 8CK(2J_A+1)N_{\text{eff}}(|E^\eta|^2 R_q + |D_1^\eta|^2 R_p) , \\
ID_p &= 8CK(2J_A+1)N_{\text{eff}}(|F^\eta|^2 R_p + |D_2^\eta|^2 R_q) ,
\end{aligned} \tag{2}$$

where C is the transformation factor required to obtain R_i defined by the intrinsic (internal) states of the target and the residual A-body system, K is the kinematical factor, J_A is the target spin, N_{eff} is the effective neutron number, A^η–F^η are the components of the optimal-frame t matrix, and R_i are the nuclear spin response functions. The formalism to derive Eq. (2) is given in Ref. [7]. The isospin degree of freedom neglected in Ref. [7] is properly accounted for in Eq. (2).

The (p, n) reaction on ^{2}H can be treated in PWIA, and we deduce its response function experimentally by using Eq. (2) with $N_{\text{eff}} = 1$. The distortion effects for nuclear targets are treated in a eikonal approximation, and they are expressed as an effective neutron number N_{eff} in Eq. (2). In this approximation, the spin-direction dependence of the distortion effects is neglected. As a result, the N_{eff} values are independent of the spin direction. However, Kawahigashi et al. [8] pointed out that the distortion effects significantly depend on the spin direction and that the use of N_{eff} in an eikonal approximation is a quantitatively poor approximation. Therefore we compare directly the experimental ID_i with DWIA+RPA calculations.

RESULTS AND DISCUSSIONS

$^2\text{H}(\vec{p},\vec{n})$

Figure 1 shows the spin-longitudinal R_L and spin-transverse R_T response functions for ^{2}H. Recently, Itabashi, Aizawa, and Ichimura [9] calculated the response functions for the $^2\text{H}(\vec{p},\vec{n})$ reaction including the final state interaction between two protons. The results are shown in Fig. 1 with the solid curves. The theoretical calculation can reproduce the shape of both spin-longitudinal and spin-transverse response functions very well, while the magnitude is somewhat underestimated.

The spin-transverse response function R_T^0 of the quasielastic electron scattering on ^{2}H has been reported by Dytman et al. [10]. The definition of R_T^0 is described in detail in Ref. [11], and the result at $q = 2.0$ fm^{-1} is compared with the present $(\vec{p},\vec{n})$ spin-transverse response R_T in Fig. 1. The (e, e') spin-transverse response function agree fairly well with the corresponding $(\vec{p},\vec{n})$ one, while it is slightly smaller than the theoretical calculation. Note that there is the contribution from the meson exchange current [12,13] to R_T^0 of the (e, e') scattering, which is neglected in the theoretical calculation. Furthermore, there are uncertainties for the reaction mechanisms such as the shadowing effect since we have deduced R_i on the assumption of $N_{\text{eff}} = 1$. These uncertainties might be responsible for the discrepancy between the theoretical and experimental spin response functions.

$^{12}\text{C}(\vec{p},\vec{n})$

Figure 2 compares the experimental polarized cross sections ID_q and ID_p for ^{12}C with DWIA+RPA calculations. The RPA calculations are performed without the commonly used universality ansatz ($g'_{NN} = g'_{N\Delta} = g'_{\Delta\Delta}$), namely all of the g's are treated independently. The nonlocality of the mean field is treated by an effective mass m^* with radial dependence of

$$m^*(r) = m_N - \frac{f_{\text{WS}}(r)}{f_{\text{WS}}(0)}[m_N - m^*(0)], \tag{3}$$

where f_{WS} is the Woods-Saxon radial form factor. The formalism of DWIA calculations is described in Ref. [8].

The solid curves in Fig. 2 are the results of DWIA calculations with the RPA response functions employing $(g'_{NN}, g'_{N\Delta}, g'_{\Delta\Delta}) = (0.6, 0.3, 0.5)$ and $m^*(0) = 0.7 m_N$.

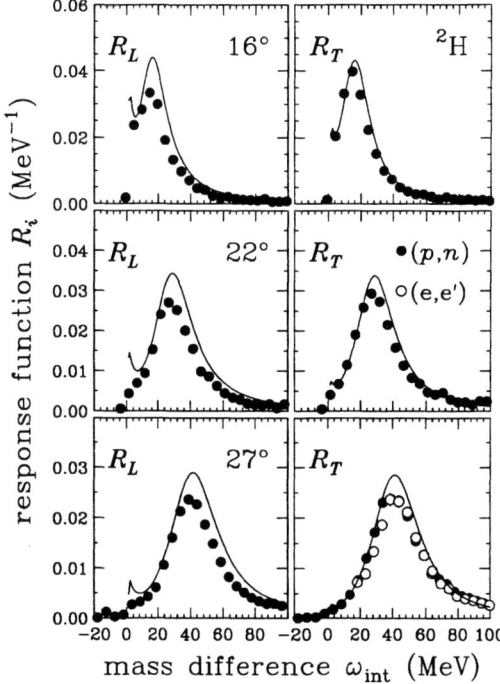

FIG. 1. The experimental spin-longitudinal R_L (left panels) and spin-transverse R_T (right panels) response functions for the $^2\mathrm{H}(\vec{p},\vec{n})$ reaction at $T_p = 346$ MeV and $\theta_{\mathrm{lab}} = 16°$, $22°$, and $27°$. The solid curves are the theoretical predictions from Itabashi, Aizawa, and Ichimura [9]. The open circles are R_T^0 of the quasielastic electron scattering on ^{2}H [10].

The dashed curves are the DWIA results with the free response functions employing $m^*(0) = m_N$.

In the spin-longitudinal mode, the calculations reasonably reproduce the observed ID_q at large angles of $22°$ and $27°$, which is consistent with the predicted enhancement of R_L in this momentum-transfer region. The present calculations indicate that the smaller $g'_{N\Delta}$ ($\simeq 0.3$) compared with g'_{NN} ($\simeq 0.6$) and the smaller effective mass at the center ($m^* \simeq 0.7 m_N$) are preferable. The calculation at $16°$ is slightly larger than the observed one. This might mean that the effective interaction in this momentum-transfer region is not so attractive as is expected in the $\pi + \rho + g'$ model.

In the spin-transverse mode, at all three angles, the experimental ID_p is significantly larger than the calculated one. The RPA correlation quenches ID_p as is predicted, while the experimental results are significantly enhanced. Recent theoretical investigations [14] indicate that the two-step contribution is more effective for ID_p than ID_q.

ACKNOWLEDGMENTS

We are grateful to K. Nishida and A. Itabashi for their helpful correspondence. This work is supported in part by the Grants-in-Aid for Scientific Research

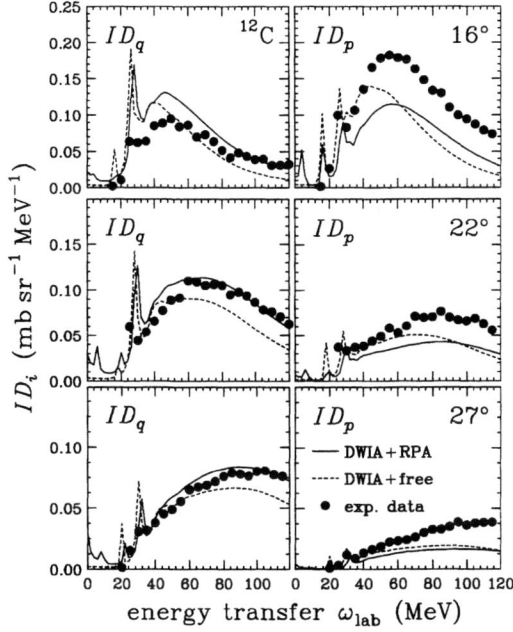

FIG. 2. The spin-longitudinal ID_q (left panels) and spin-transverse ID_p (right panels) polarized cross sections for the $^{12}\text{C}(\vec{p},\vec{n})$ reaction at T_p = 346 MeV and θ_lab = 16°, 22°, and 27°. The solid and dashed curves represent the results of DWIA calculations with RPA and free response functions, respectively.

Nos. 6342007, 02640215, 04402004, 05640328, 10304018, 12640294, and 12740151 of the Ministry of Education, Science, Sports and Culture of Japan.

REFERENCES

1. W. M. Alberico, M. Ericson, and A. Molinari, Nucl. Phys. **A379**, 429 (1982).
2. T. Wakasa et al., Phys. Rev. C **59**, 3177 (1999).
3. H. Sakai et al., Nucl. Instrum. Methods Phys. Res. A **369**, 120 (1996).
4. H. Sakai et al., Nucl. Instrum. Methods Phys. Res. A **320**, 479 (1992).
5. T. Wakasa et al., Nucl. Instrum. Methods Phys. Res. A **404**, 355 (1998).
6. E. Bleszynski, M. Bleszynski, and C. A. Whitten, Jr., Phys. Rev. C **26**, 2063 (1982).
7. M. Ichimura and K. Kawahigashi, Phys. Rev. C **45**, 1822 (1992).
8. K. Kawahigashi et al., Phys. Rev. C, submitted.
9. A. Itabashi, K. Aizawa, and M. Ichimura, Prog. Theor. Phys. **91**, 69 (1994).
10. S. A. Dytman et al., Phys. Rev. C **38**, 800 (1988).
11. K. Nishida and M. Ichimura, Phys. Rev. C **51**, 269 (1995).
12. T. Suzuki, Nucl. Phys. **A495**, 581 (1989).
13. J. Carlson and B. Schiavilla, Phys. Rev. C **49**, R2880 (1994).
14. Y. Nakaoka and M. Ichimura, Prog. Theor. Phys. **102**, 599 (1999).

Relativistic calculations of quasielastic proton-nucleus spin observables using a complete Lorentz invariant description of the NN scattering matrix

B.I.S. van der Ventel, G.C. Hillhouse, and P.R. De Kock

Physics Department, University of Stellenbosch, Stellenbosch 7600, South Africa

Abstract. Effective-mass-type medium effects are investigated by calculating complete sets of spin observables for quasifree proton-nucleus scattering. Results are presented for a ^{40}Ca target between 500 and 200 MeV. The principle conclusion is that the use of the incomplete five-term parameterization of the NN scattering matrix (the SPVAT form or the IA1 representation) is to be avoided since it can severely overestimate the importance of effective-mass-type medium effects on quasifree spin observables.

The modification of nucleon properties inside the nuclear medium remains a fundamental open question in nuclear physics. The quasielastic scattering of polarized protons from a nucleus is an attractive reaction since it simulates the scattering of the incoming nucleon from *one nucleon* from the nuclear surface [1].

The first relativistic analysis of quasielastic scattering was done in Ref. [2]. The relativistic plane wave impulse approximation (RPWIA) of Ref. [2] reduces quasielastic scattering to a two-body process where the interacting nucleons are described by Dirac spinors. The nuclear medium is incorporated via the use of effective projectile and nucleons masses in the spinors. Values for the effective masses were calculated microscopically in Ref. [3]. The calculations of Refs. [2,3] all employed a five-term parameterization of the NN scattering matrix ($\hat{F}$), referred to as the IA1 representation or the SPVAT form. Despite obtaining encouraging results these calculations were flawed from the outset since a five-term parameterization is necessarily ambiguous. This motivated the use of a general Lorentz invariant representation (called the IA2 representation) of $\hat{F}$ in the calculation of quasifree spin observables [4]. The IA2 representation of $\hat{F}$ shows that when parity and time-reversal invariance, together with charge symmetry are enforced, $\hat{F}$ contains, in fact, 44 independent invariant amplitudes. Five of these amplitudes are solely determined by fitting to free NN scattering data, while the remaining 39 amplitudes may be obtained by solving the Bethe-Salpeter equation using a meson-exchange model for the NN force with pure pseudovector pion-nucleon coupling. The five amplitudes mentioned above are identical to the standard SPVAT amplitudes and therefore *IA2 contains IA1 as a special case*. In the special case of using free nucleon masses IA2 and IA1 give identical results and this fact was used as a non-trivial check on our formalism. It follows therefore that *IA1 neglects 39 additional amplitudes which must appear on the grounds of very general symmetry principles*. The question immediately arises if neglecting these amplitudes will have a dramatic effect on the spin observables and therefore if their inclusion in IA2 will then necessarily lead to a better description of the data?

We first compare IA2 to IA1 calculations for the reaction $^{40}\text{Ca}(\vec{p},\vec{p}')$ at $T_{lab} = 500$ MeV and $\theta_{lab} = 19°$ (see Fig. 1). This was the first reaction were the so-called 'quenching effect' in A_y was observed which was hailed as a 'relativistic signature' [2]. For this reaction we chose the effective mass combination $M_1/M = M_2/M = 0.8$ [1] for both representations. It is clear that the IA1 representation describes the data very well and that it is quenched relative to the free mass calculation. Using *the same* effective masses in the IA2 calculation does not lead to such a large quenching effect, with the result that IA2 does not give a good description of the data. There is also no clear relativistic signature as for IA1. However, the question immediately arises if there does not exist another effective mass combination which will give a good description of the A_y data? Furthermore, will this combination still give a good description of the other observables? To address these questions we introduce effective mass bands which show the variation of the spin observables when the effective masses are varied over the range $(0.5;0.5) \leq (M_1/M, M_2/M) \leq (1.0;1.0)$. The quasielastic data are fairly constant with energy transfer, hence results will be presented *at the quasielastic peak*, i.e. as a function of laboratory scattering angle. We first consider Fig. 2. The data point lies just on the boundary of the IA2 band, hence we *can* find a particular effective mass combination which will accurately describe A_y. We also see that the effective mass combination which reproduces A_y is different from the free mass and it is *quenched* relative to the free mass. In this respect IA1 and IA2 are in complete agreement. Similarly, since the data lie inside the IA1 band, there is also an effective mass combination which can accurately describe A_y using IA1 (as was seen in Fig. 1). In addition, the IA1 band is wider than the IA2 band which illustrates that IA1 exaggerates medium effects on A_y. We also see that the optimal effective mass combinations [2] are close to the free mass calculation for both representations. The consequence is that, even though it gives a very good description of the other observables, it cannot describe A_y. All the reactions we have studied (see Table I in Ref. [4]) indicate that IA2 does not unanimously do better than IA1. However, a straightforward comparison of IA2 to IA1 in this sense is too simplistic. Instead we should rather see whether IA2 changes our conclusions as to what we will interpret as a medium effect. This fact was already alluded to in the analyzing power for the reaction $^{40}\text{Ca}(\vec{p},\vec{p}')$ at $T_{lab} = 500$ MeV and $\theta_{lab} = 19°$. In Fig. 3 we show $D_{\ell's}$ and $D_{s'\ell}$ for the reaction $^{40}\text{Ca}(\vec{p},\vec{p}')$ at $T_{lab} = 200$ MeV and $\theta_{lab} = 24°, 37°$ and $48°$. We notice that none

---

[1] M_1 and M_2 are projectile and target effective masses respectively
[2] that combination of effective masses which best describe *a complete set of data*

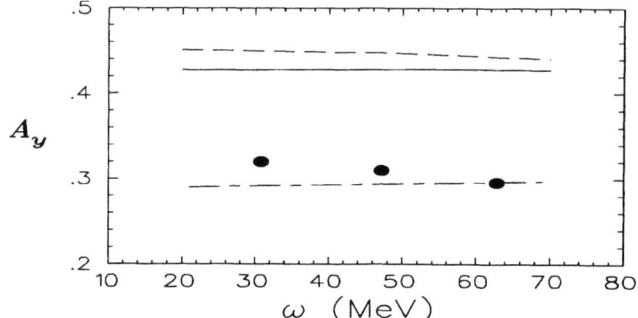

FIGURE 1. Analyzing power (A_y) for the reaction ^{40}Ca$(\vec{p},\vec{p}')$ at T_{lab} = 500 MeV and θ_{lab} = 19°. The solid line represents the IA2 calculation, the dashed line the IA2 free mass calculation and the long-dashed–short-dashed line the IA1 calculation. The data are from Ref. [5].

of the data points for $D_{\ell's}$ lie within the IA2 effective mass band, but that all three are contained in the IA1 band. This does not mean that IA1 is superior to IA2, but illustrates how IA1 may exaggerate the medium effect on the spin observables. For $D_{s's}$ we clearly see how small the medium effect on this observable really is (the IA2 band is very narrow) and yet IA1 predicts a strong medium effect on this observable.

In conclusion we have seen that the IA1 representation may overestimate the effective-mass-type medium effects on quasiefree observables. In order to give a true interpretation of medium effects one has to use a general Lorentz invariant representation of the NN scattering matrix. Even though the RPWIA based on the IA2 representation of the NN scattering matrix is the most advanced treatment of effective-mass-type medium effects to date, it still fails to describe a complete set of spin observable; the glaring examples being the prediction of A_y and D_{nn} for the $(\vec{p},\vec{p}')$ reaction as the energy is lowered from 500 MeV to 200 MeV. We have developed a relativistic distorted wave impulse approximation for quasielastic proton-nucleus scattering and are currently implementing the numerical code associated with this formalism [7].

REFERENCES

1. O. Häusser et al., Phys. Rev. C **43**, 230 (1991).
2. C.J. Horowitz and D.P Murdock, Phys. Rev. C **37**, 2032 (1988).
3. G.C. Hillhouse and P.R. De Kock, Phys. Rev. C **49**, 391 (1994).
4. B.I.S. van der Ventel, G.C. Hillhouse, and P.R. De Kock, Phys. Rev. C **62**, 024609 (2000).
5. T.A. Carey et al, Phys. Rev. Lett. **53**, 144 (1984).
6. C.L. Hautala, Ph.D. thesis, Ohio University, 1998 (unpublished).
7. G.C. Hillhouse, Ph.D. thesis, University of Stellenbosch, 1999 (unpublished).

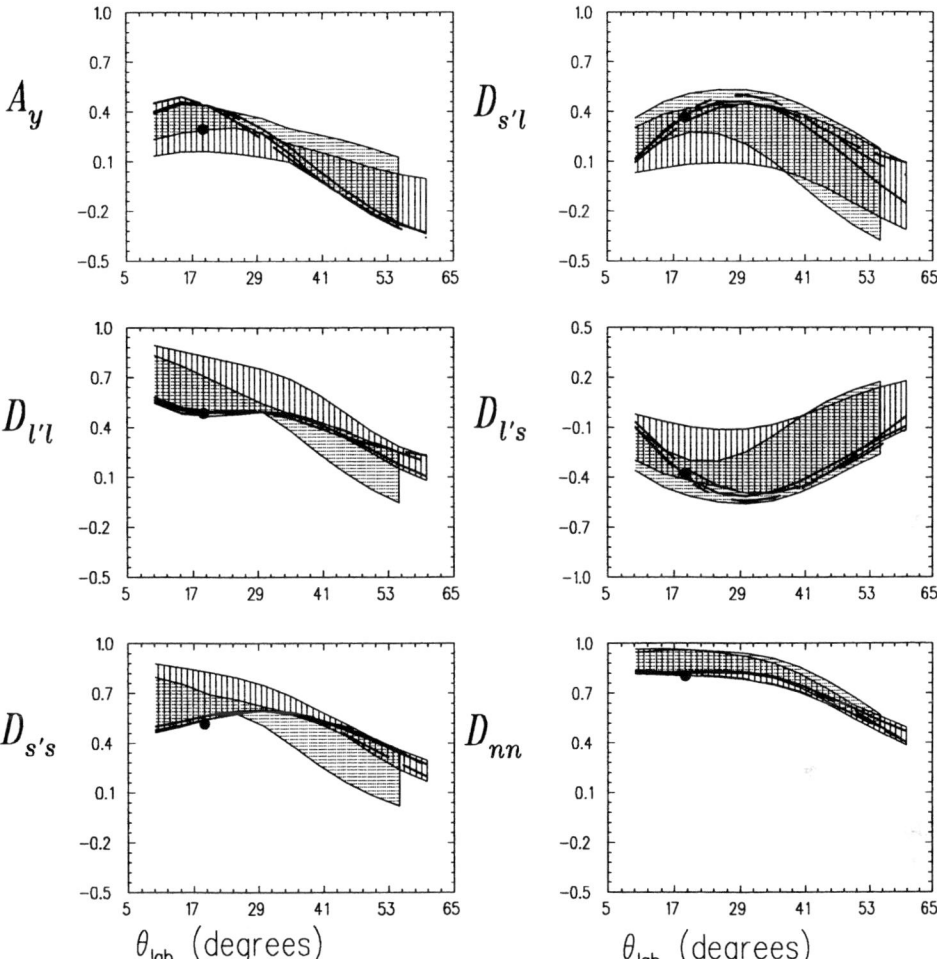

FIGURE 2. Values of A_y and polarization transfer observables $D_{i'j}$ for ^{40}Ca$(\vec{p},\vec{p}\,')$ at $T_{lab} = 500$ MeV. Solid and dashed lines represent the calculations with optimal effective masses in, respectively the IA2 and IA1 representations. The hatched bands denote the range of values which results from varying M_1/M and M_2/M over the full range (see text): The straight line hatch pattern denotes the IA1 model, the dotted hatch pattern the IA2 model. The long-dashed–short-dashed lines represent the free mass values. Data (at $\theta_{lab} = 19°$) are from Ref. [5].

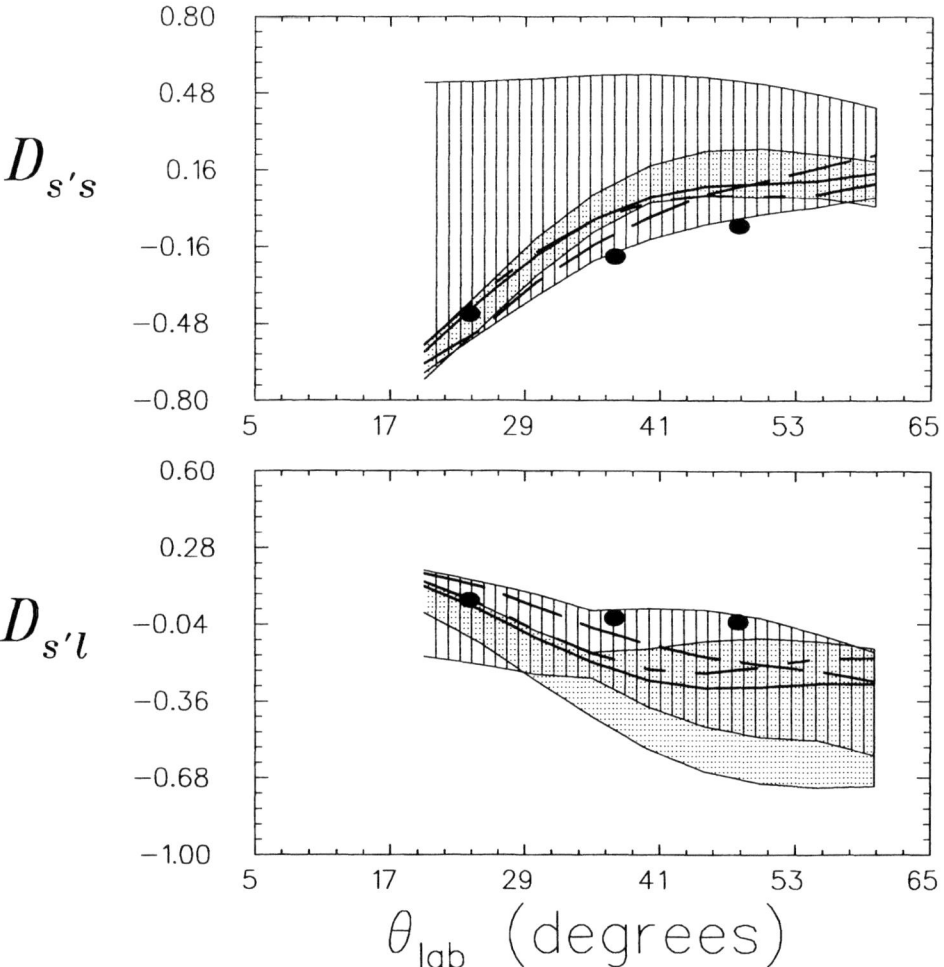

FIGURE 3. Same as Fig. 1 but for the reaction $^{40}\text{Ca}(\vec{p},\vec{p}\,')$ at $T_{lab} = 200$ MeV and only the observables $D_{s's}$ and $D_{s'l}$. The data at $24°, 37°$ and $48°$ are from Ref. [6].

Calculation of the complete set of spin transfer coefficients including one- and two-step processes in (p, nx) reaction at 346 MeV

Kazuyuki Ogata*, Yukinobu Watanabe[†], Sun Weili[‡],
Michio Kohno[||], and Mitsuji Kawai*

*Department of Physics, Kyushu University, Fukuoka 812-8581, Japan
[†]Department of Advanced Energy Engineering Science, Kyushu University, Kasuga, Fukuoka 816-8580, Japan
[‡]Department of Applied Quantum Physics and Nuclear Engineering, Kyushu University, Fukuoka 812-8581, Japan
[||]Physics Division, Kyushu Dental College, Kitakyushu 803-8580, Japan

Abstract. The semiclassical distorted wave (SCDW) model is extended to calculate the complete set of spin transfer coefficients for (p, nx) reaction process at 346 MeV including one- and two-step processes. The results are compared with the experimental data and analyzed in terms of the contribution of each step process, the bare NN forces on which the effective interactions are based, and the in-medium modification on them.

INTRODUCTION

Pre-equilibrium processes, which consist of multistep compound (MSC) and multistep direct (MSD) ones, play an important role in nuclear reactions. The MSD process is dominant at intermediate energies and has been studied by many quantum mechanical models [1-3] and methods of simulation [4-6]. We have studied the pre-equilibrium MSD process by a semiclassical distorted wave model, hereafter referred to as the SCDW model, and applied it to the calculation of the double differential inclusive cross sections, DDXs, in $(p, p'x)$ and (p, nx) processes at intermediate energies, say, 60-200 MeV, taking account of up to three-step process [7]. The results have been in overall good agreement with experimental data except at very forward angles and at large momentum transfers.

Recently we extended the model to calculate the depolarization D_{NN} for $(p, p'x)$ and (p, nx) reactions including one- and two-step processes [8]. And we found the following results: (a) The calculated D_{NN} in $^{58}\text{Ni}(p, p'x)$ at 80 MeV agrees with the experimental data semi-quantitatively, (b) the two-step contribution to D_{NN}

is appreciable where the two-step DDX is comparable to the one-step one, and (c) the effect of the in-medium modification on the NN effective interaction is clearly seen in the calculated D_{NN} for ^{90}Zr(p,nx) process. These are very interesting and important results since they mean that the detailed and systematic study about the spin observables can provide an knowledge about the mechanism of the nuclear reaction and the effective interaction in the nuclear medium.

Nowadays, active measurements [9] for the complete set of spin transfer coefficients are made at Research Center for Nuclear Physics (RCNP) in Osaka university for quasi-free (p,nx) reaction processes at about 350 MeV incident energy. Since the spin transfer coefficient D_{ij} contains important information on various transition densities and effective interactions, it is very interesting to analyze the experimental data systematically and extract such information. The quantitative estimation of the the multistep contributions will be particularly important.

Under these circumstances, we extended the SCDW model to calculate D_{ij} at about 346 MeV including one- and two-step processes, and compared the results with the experimental data. The dependence of the calculated DDX and D_{ij} on the effective interactions and the in-medium modification on them is discussed. In this paper, we show only the outline of the formulation and a typical result. Complete ones and detailed discussions will be shown in a forthcoming paper.

FORMALISM

The starting point of the SCDW model is the DWBA series expansion of the reaction T-matrix elements as in the previous quantum mechanical models. The spin transfer coefficient D_{ij} is defined by

$$D_{ij} = \frac{\text{Tr}\text{Tr}'[\mathcal{T}\sigma_i \mathcal{T}^\dagger \sigma_j]}{\text{Tr}\text{Tr}'[\mathcal{T}\mathcal{T}^\dagger]}, \tag{1}$$

where $\sigma_i(\sigma_j)$ is the Pauli spin matrix for the leading particle, LP, in the initial (final) channel and $\mathcal{T}$ is the two-by-two matrix of which each component is the reaction T-matrix element with the specified spin directions of the incoming and outgoing LPs. Tr stands for the trace of the spin substates of LP and Tr' that for the initial and the final nuclear states. First we make the "never-come-back" assumption in each step of the MSD process. Then each of the numerator and the denominator of Eq. (1) is given by an incoherent sum of the contributions of individual steps.

To calculate the one-step contribution to the numerator of D_{ij}, we make use of the local density Fermi gas (LFG) model for the nucleus and the local semiclassical approximation (LSCA) [7] to the distorted waves and also we neglect the spin orbit coupling in the distorting potential. The last approximation is justified by the fact that the difference in the numerical results for D_{ij} by the PWIA and DWIA calculations are quite small. With these approximations, one can obtain

$$\text{Tr}\text{Tr}'[\mathcal{T}\sigma_i\mathcal{T}^\dagger\sigma_j]^{1\text{step}} = \frac{1}{(2\pi)^3}\int d\boldsymbol{R}\,|\chi_f(\boldsymbol{R})|^2\,|\chi_i(\boldsymbol{R})|^2 \iint_{k_\alpha<k_F(\boldsymbol{R}),k_\beta>k_F(\boldsymbol{R})} d\boldsymbol{k}_\alpha d\boldsymbol{k}_\beta$$
$$\times \sum_{\nu_2}\sum_{m_2,m_2'} \text{Tr}\big[\mathcal{M}\sigma_i\mathcal{M}^\dagger\sigma_j\big]$$
$$\times \delta(\boldsymbol{k}_\beta - \boldsymbol{k}_\alpha + \boldsymbol{k}_f(\boldsymbol{R}) - \boldsymbol{k}_i(\boldsymbol{R}))\,\delta(\varepsilon_\beta - \varepsilon_\alpha + Q_{\alpha\beta} - \omega), \quad (2)$$

where the χ, ε, k, and $k(\boldsymbol{R})$ are the distorted waves, the kinetic energies of the struck target nucleon, the asymptotic and the local wave numbers of LP, respectively, the subscripts i and α (f and β) standing for the initial (final) state. ω is the energy transfer and $Q_{\alpha\beta}$ is the reaction Q value. m_2 (m_2') is the z-component of the spin of the struck nucleon in the initial (final) channel and ν_2 is that of the isospin. $\mathcal{M}$ is the two-by-two matrix of which each component is the half-off-shell matrix element of the in-medium NN effective interaction in coordinate representation, with the specified spin directions of LP. The resultant formula for the denominator of D_{ij}, which is proportional to DDX, is just the same as Eq. (2), with $\mathcal{M}\sigma_i\mathcal{M}^\dagger\sigma_j$ replaced by $\mathcal{M}\mathcal{M}^\dagger$. In fact, the derivation of Eq. (2) is quite similar to that of DDX formula described in Ref. [8]. It should be noted that this is possible since we neglect the spin orbit coupling in the distorting potential as mentioned above.

The extension of Eq. (2) to two- and three-step processes can be made with the assumptions of the LFG model and the LSCA just the same as in one-step process, and the additional assumption of the eikonal approximation [10] to the intermediate Green functions. This is possible since we neglect the spin flip of LP during the propagation in the intermediate channel. We have numerically verified the validity of this approximation. The explicit formula for two-step process will be shown in a forthcoming paper. We emphasize that the axis of the spin quantization is fixed throughout multistep processes since the SCDW model is described in the coordinate representation, which greatly simplifies the calculation.

In this study, the effect of relativity is included minimally by the modification on the kinematics, and the multiplication of the Möller factor to transform the two-body transition matrix in the colliding two-nucleon (NN) c.m. frame to that in the nucleon-nucleus (NA) one.

RESULTS AND DISCUSSIONS

We adopted the global optical potentials based on Dirac phenomenology of Hama et al. [11,12] for proton and Ishibashi et al. [13] for neutron. We used for the effective two-nucleon interaction in the nuclear medium the G matrix parameterized in coordinate representation by Melbourne group [14,15] based on Bonn-B NN potential [16] with the optical model modification [17], which we refer to as the Melbourne G matrix. In the calculation of D_{ij}, we followed the Madison convention. Other input data are the same as in Ref. [8].

The calculated D_{ij} in ^{40}Ca(p,nx) at 346 MeV and 22° is compared with the experimental data [9] in Fig. 1 as a function of the energy transfer ω. The corre-

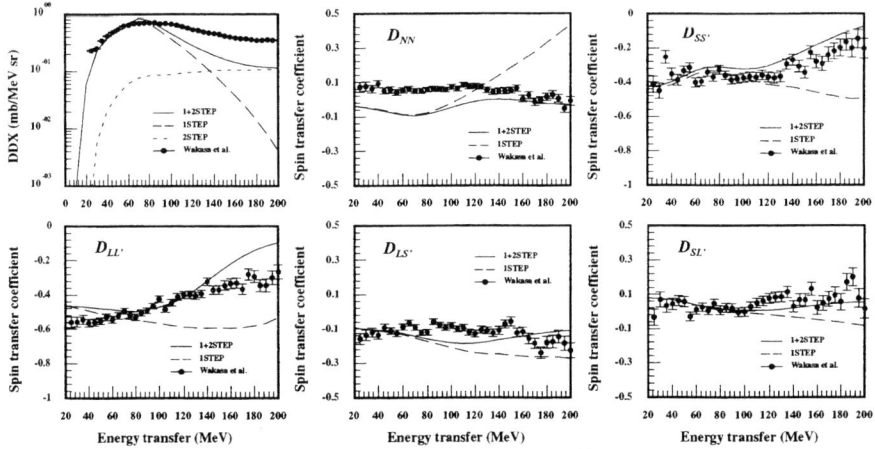

FIGURE 1. Calculated and measured [9] DDX and D_{ij} in ^{40}Ca(p,nx) at 346 MeV and 22°.

sponding result for DDX is also plotted. The dashed and solid lines represent the results including one- and one- plus two-step processes, respectively. The dotted line in the DDX panel is the two-step cross section. It should be noted that D_{ij} including only two-step process lacks physical meaning and is not plotted. The calculated DDX agrees with the data at small ω including the quasi-free peak position but underestimates them at large ω. On the other hand, the theoretical and experimental results for D_{ij} are in overall good agreement in all energy transfer regions, except a considerable disagreement in D_{NN} at small ω. One sees from Fig. 1 that the two lines in each panel for D_{ij} are quite different from each other at large ω, which shows the importance of two-step process for D_{ij} in this region.

The disagreement in DDX at large ω makes the agreement in D_{ij} look fortuitous. However, since D_{ij} is a ratio of the expectation values of two quadratic forms of the T-matrix elements as shown in Eq. (1), it is possible that the calculated D_{ij} comes out right even if the individual expectation values, including DDX, do not. The fact that all components of the spin transfer coefficients agree with data at large ω strongly suggests that this is indeed the case in this angular region. In addition, it was found that the results of DDX and D_{NN} to 40° emission process both agree with the preliminary experimental data [18] quite well at all ω. Although this is not a direct evidence that the results of the SCDW model at 22° are correct, it enforces the validity of the model. In any case, it is interesting to clarify the reason for the disagreement in DDX at 22° and large ω.

The discrepancy in D_{NN} at small ω might be due to the neglect of the nuclear correlation in this study. From the results of the analyses, however, D_{ij} is found to be quite sensitive to the choice of the bare NN forces on which the effective interactions are based, the way of calculating the G matrix, and also the in-medium modification on the effective interaction. In fact, by using the Melbourne G matrix without the optical model modification [17] on the Bonn-B potential [16], we can obtain a rather good agreement in D_{NN}, although not sufficient to reproduce the experimental data. Thus it seems to be too early yet to conclude that the disagreement in D_{NN} is due to the effects of the correlation. Not only to clarify the validity and limit of the SCDW model but also to extract important information on the effective interaction, more extensive and systematic studies, especially on the angular distribution of DDX and D_{ij}, will be necessary. At present, however, the experimental data are not enough for that purpose. The measurements for the complete set of spin transfer coefficients at about 40° must be very important.

SUMMARY

The semiclassical distorted wave (SCDW) model is extended to calculate the spin transfer coefficient D_{ij} at about 350 MeV incident energy. The calculated and measured D_{ij} in ^{40}Ca(p,nx) at 346 MeV and 22° are in overall good agreement except a considerable discrepancy in D_{NN} at small energy transfer ω. The two-step contribution was found to be fairly large at large ω. The disagreement in D_{NN} above might be due to the neglect of the nuclear correlation. However, it will be too early to conclude it since D_{ij} is quite sensitive to the effective interaction and the in-medium modification on it. More extensive and systematic studies on D_{ij}, especially in middle angular region, will be necessary.

REFERENCES

1. Feshbach, H. et al., Ann. of Phys. **125**, 429-476 (1980).
2. Nishioka, N. et al., Ann. of Phys. **183**, 166-187 (1988).
3. Tamura, T. et al., Phys. Rev. C **26**, 379-404 (1982).
4. Chiba, S. et al., Phys. Rev. C **53**, 1824 (1996).
5. Ono, A. et al., Prog. Theor. Phys. **87**, 1185 (1992).
6. Tanaka, E.I. et al., Phys. Rev. C **52**, 316 (1995).
7. Watanabe, Y. et al., Phys. Rev. C **59**, 2136-2151 (1999) and the references therein.
8. Ogata, K. et al., Phys. Rev. C **60**, 054605-1-054605-11 (1999).
9. Wakasa, T. et al., Phys. Rev. C **59**, 3177-3195 (1999).
10. Kawai, M., and Weidenmüller, H.A., Phys. Rev. C **45**, 1856-1862 (1992).
11. Hama, S. et al., Phys. Rev. C **41**, 2737-2755 (1990).
12. Cooper, E.D. et al., Phys. Rev. C **47**, 297-311 (1993).
13. Ishibashi, K. et al., "Neutron-Incident Phenomenological Dirac Optical Model Potential," in *Proceedings of a Specialists Meeting on Nucleon-Nucleus Optical Model up to 200 MeV, 1996*, Bruyeres-le-Chatel, France, 1997, pp. 91-99.
14. Dortmans, P.J., and Amos, K., Phys. Rev. C **49**, 1309-1314 (1994).
15. von Geramb, H.V. et al., Phys. Rev. C **44**, 73-80 (1991).
16. Machleidt, R. et al., Phys. Reports **149**, 1-89 (1987).
17. von Geramb, H.V. et al., Phys. Rev. C **58**, 1948-1965 (1998).
18. Wakasa, T. et al., RCNP E131 Collaboration.

Relativistic plane wave model for complete sets of spin transfer observables for exclusive proton-induced knockout reactions

[a]S. M. Wyngaardt, [a]A. A. Cowley, [a]G. C. Hillhouse, [a]B. I. S. van der Ventel and [b]J. Mano

[a]*Department of Physics, University of Stellenbosch, Stellenbosch, 7600*

[b]*Department of Electrical Engineering, Osaka Prefectural College of Technology, Osaka, 572-0017, Japan*

Abstract. A relativistic plane wave model is presented for the calculation of complete sets of $(p, 2p)$ spin transfer observables. We examine the sensitivity of these observables to different kinematic prescriptions, nuclear medium modifications of the nucleon mass, as well as different forms of the πNN vertex.

INTRODUCTION

In recent years considerable effort has been devoted to the measurement and interpretation of complete sets of $(p, 2p)$ spin transfer observables (**STO**). At present only non-relativistic Schrödinger based models [1,2] exist for the prediction of complete sets of STO's. The recent success of Dirac based descriptions of elastic and inelastic $(\vec{p}, \vec{p}')$ spin transfer observables in describing data has lead to the development of relativistic models for $(p, 2p)$ reactions. To date relativistic models have only been applied to calculations of unpolarized triple differential cross sections and analyzing powers [3–5]. Presently both relativistic and non-relativistic models fail to reproduce the analyzing power for the knockout of s-state protons. A proper treatment of medium modifications of the free NN interactions could possibly remedy the situation. Further insight can be gained by comparing model predictions of complete sets of spin transfer observables to data. An alternative relativistic model is presented which is based on a technique used in electron scattering work [6]. Nuclear distortion effects on the scattered protons are currently ignored in order to first understand the role of various model parameters. Our model should be seen as a first step to fully understand the effects of relativity and medium modifications on the scattering observables. The work presented in this paper links up with the experimental $(p, 2p)$ programs at the Research Center for Nuclear Physics (Japan) and the National Accelerator Center (South Africa).

RELATIVISTIC PLANE WAVE IMPULSE APPROXIMATION MODEL

The $(p, 2p)$ spin transfer observables $D_{i'j}$ are calculated from polarized transition amplitudes $T_{LJ}(\mu_{a'}\mu_{b'}, \mu_a, M)$ which in a relativistic plane wave model are defined by the expression

$$T_{LJ}(\mu_{a'}, \mu_{b'}, \mu_a, M) = \bar{u}(k_{a'}, \mu_{a'})\bar{u}(k_{b'}, \mu_{b'}) t_{NN}(|\vec{x} - \vec{x}'|) u(\vec{k}_a, \mu_a) \Phi^B_{LJM}(-\vec{k}_C) , \quad (1)$$

where the symbols a, a', b' and C represent the projectile, the two ejectiles scattered to the left and right of the scattering beam and the residual nucleus respectively. We use μ_α and $\vec{k}_\alpha$ to represent the spin projection and momentum of the particle represented by the label α. The Dirac-spinor for a particle with momentum $\vec{k}$ and spin projection μ is represented by $u(\vec{k}, \mu)$. The orbital angular momentum (L) and total angular momentum quantum number (J) specify the orbital in which the bound proton is located. The symbol M is used to represent the total angular momentum projection quantum number of the bound proton. We assume the IA1 representation of the NN interaction [7]. Using trace algebra one can show that

$$\sum_{\mu_{b'} M} |T_{LJ}(\mu_{a'}, \mu_{b'}, \mu_a, M)|^2 \sim \sum_{i,j=S}^{T} (F^i)^*(\vec{q}) F^j(\vec{q}) Tr \left[\left(\frac{\slashed{P}_{a'} + m}{2m} \right) \left(\frac{1 + \gamma^5 \slashed{S}_{a'}}{2} \right) \lambda_1^j \right.$$
$$\left. \left(\frac{\slashed{P}_a + m}{2m} \right) \left(\frac{1 + \gamma^5 \slashed{S}_{a'}}{2} \right) \bar{\lambda}_1^i \right]$$
$$Tr \left[\left(\frac{\slashed{P}_{b'} + m}{2m} \right) \lambda_{2j} \Phi^B_{LJM}(-\vec{k}_C) \Phi^B_{LJM}(-\vec{k}_C) \bar{\lambda}_{2i} \right]$$
(2)

where P_α represents the four momentum of particle α and S_α is the corresponding four component spin vector. We use F^i to represent the NN scattering amplitude which is obtained from the relativistic Love-Franey model [8]. The symbol λ^i represents the 4×4 Lorentz covariant matrices $\{1, \gamma^\mu, \gamma^5, \gamma^5\gamma^\mu, \sigma^{\mu\nu}\}$ [9]. Upper and lower component radial bound state wave functions are generated via the self consistent Dirac-Hartree approximation [10]. Due to uncertainty in the calculation of the NN center-of-mass scattering angle and effective laboratory kinetic energy, both the initial and final energy prescriptions are included into our model. In the case of the initial energy prescription we use the four momentum of the projectile and the left scattered proton in the laboratory system to calculate these quantities. While for the final energy prescription we use the four momenta of the two outgoing protons. The Brown-Rho scaling relationships are used to include nuclear medium effects on the scattered protons as well as the NN interaction [11]. With medium effects included into the calculation the free proton mass m_p of the projectile and scattered protons as well as the proton mass and σ, ρ and ω-meson masses used in the calculation of the NN amplitudes which appear in Eq. (2) are replaced by effective masses which are given by the Brown-Rho scaling relations. With medium-modifications on the NN interaction we also allow for a choice between a pseudoscalar and pseudovector πNN coupling by replacing the free πNN pseudoscalar coupling constant with the medium modified pseudovector coupling constant

$$g_{\pi NN}^{PV} = g_{\pi NN}^{free} \chi \tag{3}$$

where $\chi = 0.7$ represents the scaling factor for the nucleon mass in the nuclear medium.

RESULTS

Calculations were performed for an incident energy of 400 MeV, scattering angle pair $(32.5°, 50°)$ and knockout from a $2s_{\frac{1}{2}}$ state. In fig.(1) comparisons are made between calculations of complete sets of spin transfer observables with medium effects on the scattered protons and the NN interaction to calculations done with free parameters. These plots also reflect the differences between medium-modified predictions with a pseudoscalar to that with a pseudovector coupling. In fig. (2). we show comparisons of predicted spin transfer observables calculated with the initial energy prescription to calculations done with the final energy prescription.

CONCLUSIONS AND SUMMARY

According to our results it seems that the analyzing powers are not very sensitive to medium effects. Therefore other observables such as D_{nn} would be a better choice to search for signatures of nuclear medium effects. We also see that the spin transfer observables are insensitive to the different kinematic prescriptions used. Some refinement to the present model are clearly needed. These include an extension of the model to include distorted wave functions. We also need to replace the IA1 representation of the NN amplitudes with a more general IA2 representation: see the contribution of BIS van der Ventel.

REFERENCES

1. N. S. Chant and P. G. Roos, Phys. Rev. C 15, 57, (1977).
2. Y. Kudo and K. Miyazaki, Phys. Rev. C 34, 1192, (1986).

3. E. D. Cooper and O. V. Maxwell, Nuclear Phys. A 493 (1988), 468.
4. Y. Ikebata, Phys. Rev. C 52, 890 (1995).
5. J. Mano, M. Arima, Y Kudo and H. Tsunoda, Prog. Theo. Phys., Vol 96, no. 5 (1996).
6. W. Greiner, Quantum Electrodynamics (Springer-Verlag, Berlin,1992).
7. J. A. McNeil, L. Ray and S. J. Wallace, Phys. Rev. C 27,2123 (1983).
8. C. J. Horowitz, Phys. Rev. C 31, 488 (1985).
9. W. Greiner, Relativistic Quantum Mechanics (Springer-Verlag, Berlin, 1992).
10. K. Langanke, J. A. Maruhn and S. E. Koonin, Computational Nuclear Physics 1, (Springer-Verlag, Berlin Heidelberg, 1991).
11. G. E. Brown and M. Rho, Phys. Rev. Lett. 66,2720 (1991).

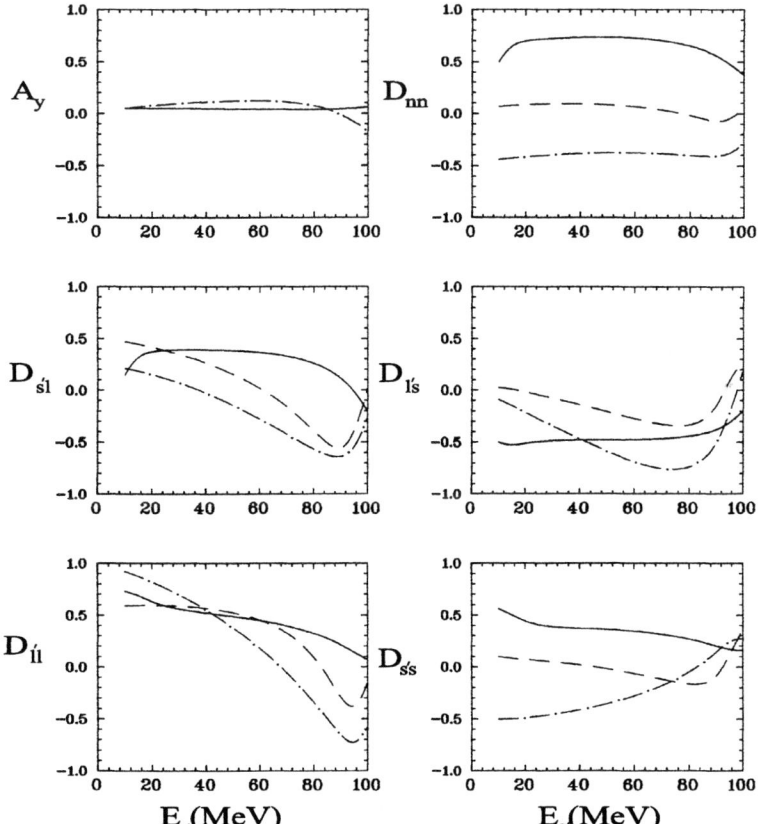

FIGURE 1. Plots comparing complete sets of $(p, 2p)$ spin transfer observables for the $2s_{\frac{1}{2}}$ state with medium effects and a pseudoscalar πNN coupling (dashed-doted line), medium effects and a pseudovector coupling (dashed line) to free calculations with free parameters (solid black line).

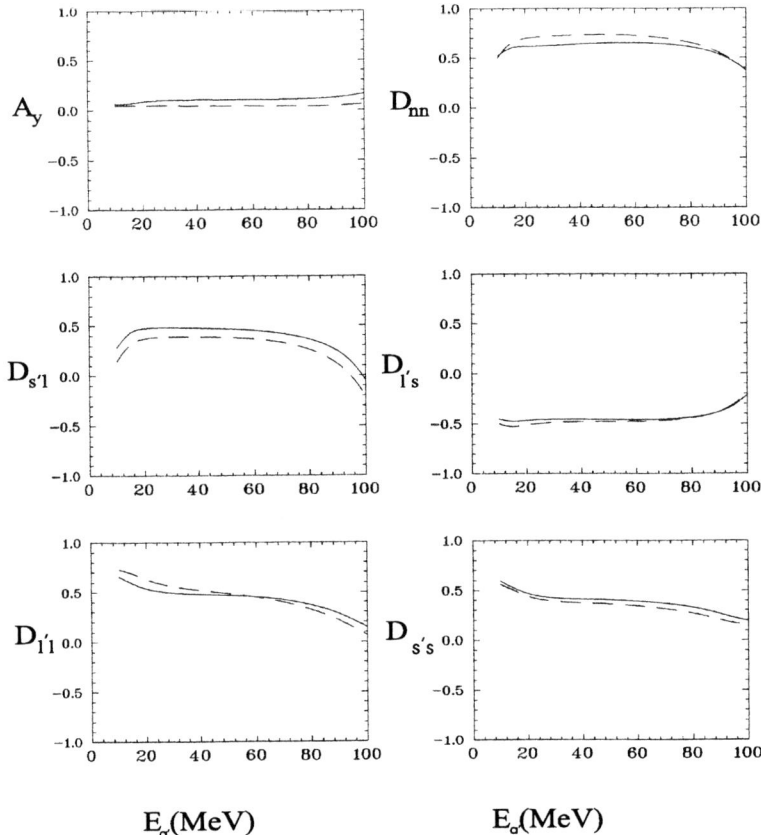

FIGURE 2. Plots comparing complete sets of $(p,2p)$ spin transfer observables for the $2s_{\frac{1}{2}}$ state calculated with the initial energy prescription (solid line) to predictions obtained with the final energy prescription (dashed line).

Spin-dependent Effective Interaction Studied by the ^{12}C, ^{28}Si$(\vec{p}, \vec{p}')$ Reactions at Zero Degrees

A. Tamii, T. Kawabata,[a] H. Akimune,[b] I. Daito,[c] Y. Fujita,[d]
M. Fujiwara,[e] K. Hatanaka,[e] K. Hosono,[f] F. Ihara,[e] T. Inomata,[e]
T. Ishikawa,[a] M. Itoh,[a] M. Kawabata,[e] M. Nakamura,[a] T. Noro,[e]
E. Obayashi,[e] H. Sakaguchi,[a] H. Takeda,[a] T. Taki,[a] H. Toyokawa,[g]
H.P. Yoshida,[e] M. Yoshimura,[e] and M. Yosoi[e]

Department of Physics, University of Tokyo, Tokyo 113-0033, Japan
[a] *Department of Physics, Kyoto University, Kyoto 606-8502, Japan*
[b] *Department of Physics, Konan University, Kobe 658-8501, Japan*
[c] *Center for Integrated Research in Science and Engineering*
Nagoya University, Nagoya 464-8601, Japan
[d] *Department of Physics, Osaka University, Toyonaka, Osaka 560-0043, Japan*
[e] *Research Center for Nuclear Physics, Osaka University, Ibaraki, Osaka 567-0047, Japan*
[f] *Himeji Institute of Technology, Himeji, Hyogo 678-1297, Japan*
[g] *Japan Synchrotron Radiation Research Institute, Kamigori, Hyogo 678-1298, Japan*

Abstract. Polarization transfer (PT) coefficients in the ^{12}C, ^{28}Si(p,p') reactions at a scattering angle of 0° are measured at $E_p = 392$ MeV. The strength of the isoscalar central spin-dependent part (V_σ) of the effective nucleon-nucleon interaction is studied by using the sensitivity of the PT coefficients for the excitation of the 1^+, $T=0$ state in ^{12}C to the V_σ strength. It is found that distorted wave impulse approximation (DWIA) calculations employing the Franey and Love effective interaction well reproduce the measured PT coefficients.

I INTRODUCTION

The effective nucleon-nucleon (NN) interaction in nuclear medium has been a subject of much interest over the past decades. Nucleon-nucleus scattering in an intermediate energy region has been extensively studied for extracting the strength of each effective interaction term taking advantage that distorted wave impulse approximation (DWIA) provides a good starting point to describe the reaction mechanism in that energy region. For example, isovector central interactions have been studied by measuring the (p, n) cross sections [1], and isovector tensor interactions are studied by the (p,n) polarization transfer coefficients at $0°$ [2,3] and the (p,p') cross sections at $0°$ [4]. However, the strength of the isoscalar spin-dependent interaction (V_σ) is considered to be very small and has not been well determined before. Love and Franey extracted the parameters of the effective interactions to reproduce the NN scattering data. But they noted that V_σ was poorly determined [5].

In this paper, we report an attempt to study the V_σ strength by measuring the PT coefficients for isoscalar $M1$ transitions (1^+, $T=0$) in inelastic proton scattering at $0°$. The study is based on the property that the PT coefficients for 1^+ excitation at $0°$ are sensitive to the strength ratio between central spin-dependent interactions and knock-on exchange tensor interactions as described in the next section. In addition, we discuss a model independent signature of spin transfers, *total spin transfer*.

II PT COEFFICIENTS AT $0°$ AND THE EFFECTIVE INTERACTION

In plane wave impulse approximation (PWIA), the NN scattering amplitude can be written in the form

$$M(q) = A + \frac{1}{3}(B + E + F)\boldsymbol{\sigma}_1 \cdot \boldsymbol{\sigma}_2 + C(\boldsymbol{\sigma}_1 + \boldsymbol{\sigma}_2) \cdot \hat{\mathbf{n}}$$
$$+ \frac{1}{3}(E - B)S_{12}(\hat{\mathbf{q}}) + \frac{1}{3}(F - B)S_{12}(\hat{\mathbf{Q}}) , \quad (1)$$

where S_{12} is the usual tensor operator. The unit vectors are defined as

$$\hat{\mathbf{q}} = \frac{\mathbf{k}_f - \mathbf{k}_i}{|\mathbf{k}_f - \mathbf{k}_i|} , \quad \hat{\mathbf{Q}} = \frac{\mathbf{k}_i + \mathbf{k}_f}{|\mathbf{k}_i + \mathbf{k}_f|} , \quad \text{and } \hat{\mathbf{n}} = \hat{\mathbf{Q}} \times \hat{\mathbf{q}} . \quad (2)$$

In this paper, we concentrate on $M1$ transitions from a 0^+ ground state of an $N=Z$ ($T_0=0$) nucleus. In this case, non-vanishing PT coefficients at $0°$ are written by [6]

$$D_{SS} = D_{NN} = \frac{-|F|^2}{|F|^2 + 2|B|^2} , \quad (3)$$

$$D_{LL} = \frac{|F|^2 - 2|B|^2}{|F|^2 + 2|B|^2} , \quad (4)$$

neglecting the $\Delta L=2$ contributions. From Eq. 3, one can see that PT coefficients are determined by the ratio $|F|/|B|$. From Eq. 1, the central spin-dependent interaction (the second term) is proportional to $B+E+F$ ($=2B+F$ at $0°$), and the knock-on exchange tensor interaction (the final term) is proportional to F-B. By a simple calculation, it can be shown that both D_{SS} and D_{LL} become $-1/3$ if the central spin-dependent interaction is dominant ($|2B+F| \gg |F$-$B|$), while D_{SS} and D_{LL} become $-2/3$ and $+1/3$, respectively, if the knock-on exchange tensor interaction is dominant ($|2B+F| \ll |F-B|$). In the case of 1^+, $T=0$ transitions at $0°$, the PT coefficients are closely related to the relative strength of the two interactions since these transitions are considered to be mediated by V_σ (central spin-dependent interaction) and V_τ^T (tensor interaction through knock-on exchange). By using this property, the strength of V_σ relative to that of V_τ^T can be studied by measuring the PT coefficients at $0°$ for 1^+, $T=0$ states.

Next, a model-independent signature of spin transfers, called *total spin transfer* (Σ), can be formed from the PT coefficients as

$$\Sigma \equiv \frac{3-(D_{SS}+D_{NN}+D_{LL})}{4} = \frac{3-(2D_{SS}+D_{LL})}{4} \quad \text{(at } 0°\text{)}. \tag{5}$$

Under the conservation of parity invariance, Σ becomes zero for $\Delta S=0$ transitions and unity for $\Delta S=1$ transitions [7] provided that the spin-orbit (LS) interaction is negligible (this is the case at the scattering angle of $0°$). Spin-flip properties of excited states can be studied by using this model-independent signature.

III EXPERIMENT

The experiment was performed at the Research Center for Nuclear Physics using the high resolution spectrometer *Grand Raiden* [8] and a focal plane polarimeter *FPP*. [9]. Horizontally polarized proton beams with an energy of 392 MeV were tuned to have no sizable halo. The primary beam was transported inside the spectrometer and was extracted from the focal plane. Details of the experiment are described in Refs. 10 and 11.

IV RESULTS AND DISCUSSION

Excitation spectra for ^{12}C and ^{28}Si targets are shown in Fig. 1. Low lying discrete states are clearly observed without background events.

The measured PT coefficients and statistical errors for the discrete peaks are summarized in Tab. 1. Distorted wave impulse approximation (DWIA) calculations are performed using the program code DWBA91 [12]. The 425 MeV parameter set of the Franey and Love (FL) effective interaction [13] is employed as an elementary NN interaction. The target wave functions by Cohen and Kurath [14] are employed for ^{12}C, and those for ^{28}Si are calculated with the program code OXBASH [15] in the sd-shell configuration space using the USD interaction by Wildenthal [16]. The results are listed in the Tab. 1.

In Fig. 2, the measured PT coefficients (hatched region) for excitation of the 1^+, $T=0$ state are compared with DWIA calculations. The horizontal axis is the artificially changed V_σ strength normalized to the original V_σ strength in the FL interaction. It is found that the PT coefficients are surprisingly well reproduced by the original V_σ strength (see 1.0 in the horizontal axis).

TABLE 1. Measured PT coefficients with statistical errors and the results of DWIA calculations for ^{12}C and ^{28}Si targets.

nucleus	E_X	$J^\pi;T$	D_{SS}^{exp}	D_{LL}^{exp}	Σ^{exp}	D_{SS}^{th}	D_{LL}^{th}	Σ^{th}
^{12}C	7.65	$0^+;0$	1.00±0.06	1.07±0.07	-0.02±0.04	1.00	1.00	0.00
^{12}C	12.7	$1^+;0$	-0.63±0.07	0.74±0.08	0.88±0.04	-0.64	0.62	0.92
^{12}C	15.1	$1^+;1$	-0.18±0.02	-0.58±0.02	0.99±0.01	-0.21	-0.50	0.98
^{28}Si	9.5	$1^+;0$	-0.04±0.13	0.98±0.15	0.52±0.08	-0.59	0.63	0.89
^{28}Si	10.6	$1^+;1$	-0.07±0.13	-0.85±0.14	1.00±0.08	-0.14	-0.60	0.97
^{28}Si	11.5	$1^+;1$	-0.17±0.06	-0.50±0.06	0.96±0.04	-0.20	-0.55	0.99
^{28}Si	12.3	$1^+;1$	-0.05±0.12	-0.37±0.13	0.87±0.08	-0.15	-0.60	0.97
^{28}Si	13.3	$1^+;1$	-0.22±0.13	-0.41±0.15	0.96±0.09	-0.08	-0.66	0.95
^{28}Si	14.0	$1^+;1$	-0.31±0.08	-0.63±0.09	1.06±0.05	-0.06	-0.69	0.95

As has been explained in Sec. II, Σ become zero for non-spin-flip ($\Delta S=0$) transitions and unity for spin-flip ($\Delta S=0$) transitions. By taking average in the solid angle, Σ deviates from unity for $T=0$ spin-flip excitation due to relatively large contribution of the LS interaction in contrast to the case of $T=1$ spin-flip excitation. The prediction of Σ is satisfactory for all the states except the 1^+, $T=0$ state on ^{28}Si, for which the measured Σ largely deviates from the DWIA prediction.

ACKNOWLEDGEMENTS

The authors gratefully acknowledge H. Sakai and T. Suzuki for valuable discussions. This research was supported in part by the Ministry of Education, Science, Sports and Culture of Japan with the Grant-in-Aid for Scientific Research No.

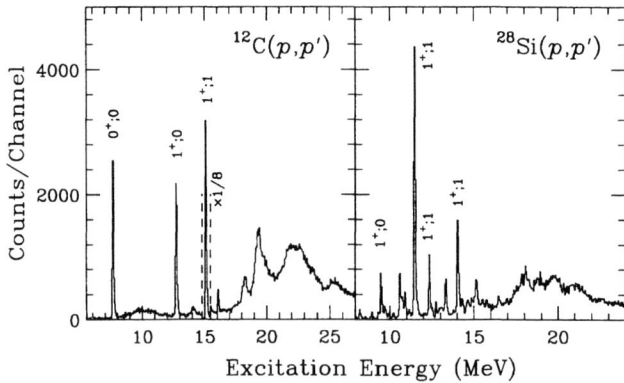

FIGURE 1. Excitation energy spectra of proton inelastic scattering at $0°$ from ^{12}C (left panel) and from ^{28}Si (right panel).

07454051 and by the Japan Society for the Promotion of Science.

REFERENCES

1. For example, see C.D. Goodman et al., Phys. Rev. Lett. **44**, 1755 (1980); T.N. Taddeucci et al., Phys. Rev. C **25**, 1094 (1982).
2. T.N. Taddeucci et al., Phys. Rev. Lett. **52**, 1960 (1984).
3. T. Wakasa et al., Phys. Rev. C **51**, R2871 (1995).
4. Y. Sakemi et al., Phys. Rev. C **51**, 3162 (1995).
5. W.G. Love and M.A. Franey, Phys. Rev. C **24**, 1073 (1981).
6. J.M. Moss, Phys. Rev. C **26**, 727 (1982).
7. T. Suzuki, Prog. Theor. Phys. **103**, 859 (2000).
8. M. Fujiwara et al., Nucl. Instrum. Methods A **422**, 1999 (484).
9. M. Yosoi et al., *AIP Conf. Proc. 11th Int. Symp. on High Energy Spin Physics*, ed. K. J. Heller and S. L. Smith (AIP, New York, 1995) pp. 157.
10. A. Tamii et al., Phys. Lett. B **459**, 61 (1999).
11. M. Yosoi et al., in these proceedings.
12. J. Raynal, computer code DWBA91, 1981 (NEA 1209/02).
13. M. A. Franey and W. G. Love, Phys. Rev. C **31**, 488 (1985).
14. S. Cohen and D. Kurath, Nucl. Phys. **73**, 1 (1965).
15. B.A. Brown et al., MSU cyclotron laboratory report No. 524 (1986).
16. B.H. Wildenthal, Prog. Part. Nucl. Phys. **11**, 5 (1984).

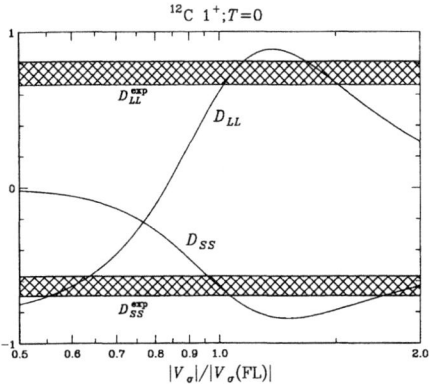

FIGURE 2. PT coefficients for excitation of the 1^+, $T=0$ state on ^{12}C. The shaded areas show the experimental results with 1σ statistical error. The solid curves are the result of DWIA calculations using Franey and Love effective interaction with artificially changed V_σ strength.

Conventional and Non-conventional Medium Effects in (p,p′) Reactions

E.J. Stephenson[a] and F. Sammarruca,[b]

[a] *Indiana University Cyclotron Facility, Bloomington, IN 47408 USA*
[b] *University of Idaho, Moscow, ID 88434 USA*

Abstract. We report the results of an investigation of the medium effects in (p,p′) reactions on nuclear targets at 200 MeV. A model using a high-quality one-boson-exchange potential for the free interaction, and including Pauli blocking, nuclear binding, and the effects of strong relativistic mean fields, provides overall a good description of natural and unnatural parity excitations. Isolated discrepancies are observed for the unnatural-parity polarization transfer coefficients. These are not relieved by changes to the meson properties of the in-medium one-boson-exchange potential.

SUMMARY OF THE MODEL

The properties of the nuclear medium are often modelled through the use of an effective nucleon-nucleon (NN) interaction that varies with the local nuclear density. Here we summarize one such model that has been constructed for the purpose of exploring non-conventional medium effects through a comparison to sets of polarization observables for (p,p′) reactions at 200 MeV. A more detailed discussion and list of references may be found in Refs. 1 and 2.

The free NN interaction is calculated from a one-boson-exchange (OBE) potential of a type similar to the Bonn interaction, with pseudovector coupling at the πNN vertex. The parameters of this potential have been adjusted, in particular in the scalar meson sector [1], to produce a high quality fit ($\chi^2/\nu \sim 1$) to NN data below 350 MeV.

Conventional medium effects have been included using effective masses within an infinite nuclear matter framework. For a Brueckner-Hartree-Fock (BHF) calculation, these effects include Pauli blocking and nuclear binding. A Dirac-Brueckner-Hartree-Fock (DBHF) calculation adds the effects of strong relativistic mean fields. The effective masses are calculated self-consistently as a function of the local nuclear density. The DBHF calculation reproduces the binding energy and density of nuclear matter.

A mathematical transformation is made to convert the effective density-dependent interaction into a Yukawa expansion suitable for use in Distorted-Wave Impulse Approximation calculations of (p,p') reactions. These calculations are made with the LEA program for natural parity transitions and DWBA86 for unnatural parity transitions. In both cases, the transition density is adjusted to match (e,e') data for each excited state. LEA offers a better way to incorporate the electric and magnetic formfactors, and DWBA86 contains the finite-range exchange needed to handle the tensor exchange amplitudes. The distortions are computed from the folding model using the same interaction that is used for the DWIA transition potential. This choice may overestimate the effects of the repulsion present in the nuclear interior for the DBHF case. Nevertheless, the DBHF treatment represents the most realistic treatment of conventional medium effects for (p,p') reactions.

RESULTS FOR NATURAL PARITY STATES

Isoscalar, natural-parity transitions are the most sensitive to conventional medium effects since these effects make large changes to the isoscalar central and spin-orbit parts of the effective interaction. For such transitions, an examination of the cross section and analyzing power is sufficient for an evaluation. Typical examples of calculations are shown in Fig. 1. The dotted curves are based on the free interaction at all densities, the dashed include BHF medium effects, and the solid include DBHF medium effects.

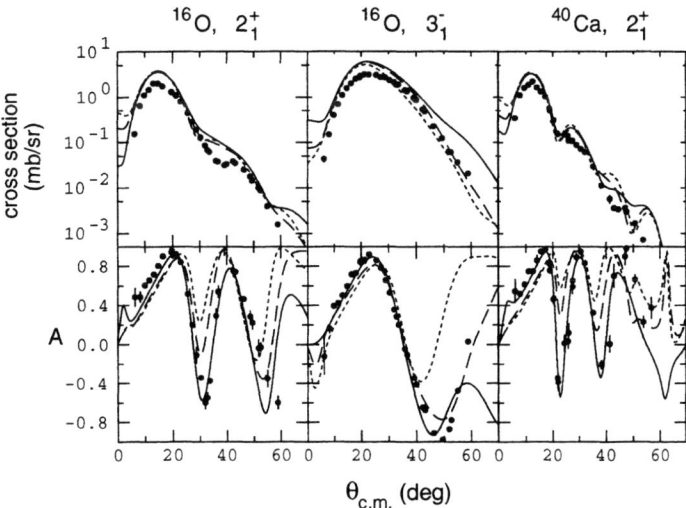

FIGURE 1. Measurements of the cross section and analyzing power for the (p,p') reaction leading to the 2^+ and 3^- states in ^{16}O at 6.917 and 6.130 MeV, and the 2^+ state in ^{40}Ca at 3.904 MeV. The curves are described in the text.

The 2^+ transitions shown in Fig. 1 are representative of low-spin states with a significant diffraction pattern, while the 3^- transition with a single analyzing power oscillation is similar to higher spin states. The calculated cross section is generally too large by 30-50%, but otherwise tends to track the shape of the measured angular distributions. This problem, present at 200 MeV, goes away with increasing bombarding energy. The DBHF cross section is also too large at large momentum transfer, particularly for states of high spin; this is a consequence of the spin-orbit amplitude being overestimated by the relativistic calculation. The DBHF calculation reproduces the analyzing power well, particularly near the maximum of the cross section. The free and BHF calculations, which contain too little repulsion in the nuclear interior, produce analyzing powers that are too positive. The 3^- transition shows little difference among the calculations because its formfactor is peaked at lower nuclear densities. Thus, comparing the 2^+ and 3^- calculations shows that this treatment of medium effects appears to have the correct variation with changing nuclear density. Despite the issues noted above, the DBHF model gives a good overall description of these transitions.

RESULTS FOR UNNATURAL PARITY STATES

States with unnatural parity and a "stretched" spin configuration offer more dynamically independent polarization observables and an opportunity to compare combinations of them with the size of particular spin-dependent pieces of the effective interaction in both the isoscalar and isovector channels. The isovector channel has been of considerable interest because these terms in the effective interaction arise in simple ways from the exchange of π and ρ mesons. A set of calculations similar to those in Fig. 1 is shown in Fig. 2 for the $\Delta T=1$, 4^- transition in ^{16}O.

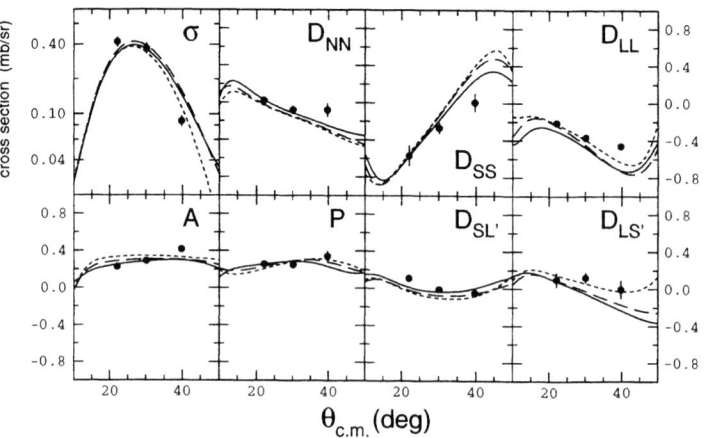

FIGURE 2. Measurements of the cross section and polarization observables for the 4^+, T=1 state at 18.98 MeV in ^{16}O. The curves are described in the text.

In this case, there is little difference among the free, BHF, and DBHF calculations. This reflects the fact that the density dependence is much smaller for the isovector parts of the effective interaction and the higher spin of this particle-hole transition $(p_{3/2}^{-1} d_{5/2})$ pushes it into regions of smaller nuclear density. In general, the agreement with the calculation here is excellent, with differences appearing only at the largest angle. When these differences are referred to the individual terms in the effective interaction, they indicate the need for a smaller spin-longitudinal ($\sigma_{1q}\sigma_{2q}$) and a larger transverse ($\sigma_{1n}\sigma_{2n}$) amplitude.

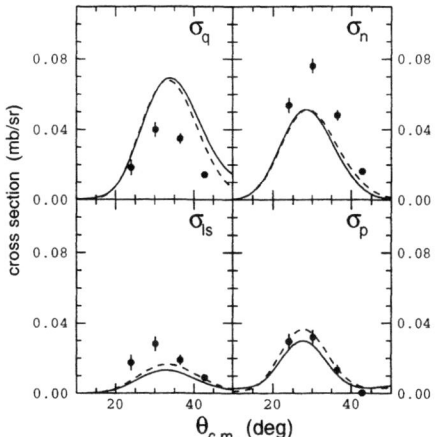

FIGURE 3. Measurements of the polarized cross sections for the 6^-, T=1 state at 14.35 MeV in ^{28}Si. In the plane wave impulse approximation, σ_{ls} is associated with the spin-orbit term in the effective interaction. The remaining cross sections are associated with the tensor terms for each axis as σ_q with the momentum transfer $\mathbf{q} = \mathbf{k}_{in} - \mathbf{k}_{out}$, σ_n with the normal $\hat{n}$ to the reaction plane, and σ_p with $\hat{p} = \hat{n} \times \hat{q}$. The calculations were generated using the Rapp (solid) and DBHF (dashed) effective interactions.

The differences noted for the ^{16}O case are much more pronounced for the 6^- transition in ^{28}Si, as shown in Fig. 3. Since all three calculations (free, BHF, and DBHF) are similar, only the DBHF is shown as the dashed curves. Instead of the polarization transfer coefficients, Fig. 3 shows the combinations which, as cross sections, are associated in plane wave with the squares of products of particular amplitudes in the NN effective interaction with the transition formfactor. Since the formfactor is constrained by electron scattering data, such differences are usually interpreted as deficiencies in the medium-dependent effective interaction. The overestimate of σ_q suggests too much pion-like attraction and the underestimate of σ_n too little repulsion in particular parts of the tensor interaction. Even the spin-orbit term σ_{ls} seems too small. While similarities in the differences (Fig. 2 and Fig. 3) suggest a common explanation in new medium-effect physics, the differences between nuclei would also point to problems with either structure or reaction mechanism.

Problems also appear for the isoscalar parts of the effective interaction. There the tensor parts of the interaction, which are particularly small, would need to be larger in order to explain measurements for the diagonal polarization transfer coefficients that are not as positive as predicted.

NON-CONVENTIONAL MEDIUM EFFECTS

New medium effects have been proposed based on the notion that a partial chiral restoration of the meson masses takes place at full nuclear density. This idea is also supported by calculations based on a consideration of quark condensates. In the isovector channel, a lowering of the ρ-meson mass would reduce the tensor attraction apparent in the $\sigma_{1q}\sigma_{2q}$ amplitude. While some preliminary analyses indicated that such a change would bring the calculations closer to the data, when such effects are included in a density-dependent manner, the changes in the predictions are much smaller than what is needed to even approach the ^{28}Si results. A model which combines the Brown-Rho scaling concept for meson masses with a realistic nuclear matter framework was suggested by Rapp et al. [3]. The predictions of this interaction are shown by the solid curves in Fig. 3. They are only slightly different from the predictions of the DBHF model.

PROSPECTS

The DBHF model described here is the first occasion on which a high-quality description of NN scattering data has been combined with a description of the nuclear medium that matches the properties of nuclear matter in order to explain a wide spectrum of (p,p') observables using the best available descriptions of the nuclear transitions. The quality of the results exceeds previous efforts across a broad range of nuclear states and demonstrates the necessity of using the best available ingredients in the model.

The search for new physics in the nuclear medium remains inconclusive because of a lack of sensitivity in the (p,p') data to changes suggested in the properties of the OBE mesons in the medium. Large changes involving both vector and scalar mesons do result in noticeable sensitivity, but any improvement in the spin-flip states typically deteriorates the quality of the predictions for the natural parity transitions. At the same time, there remain a number of systematic discrepancies which suggest the need to redistribute the strength among the various spin-dependent pieces of the interaction.

On the other hand, it is possible that the origin of these differences should be sought in more conventional ingredients of the model. Further investigations are underway [4] to examine the changes in spin dependence that occur when the Pauli exclusion operator is not treated within the usual angle-average approximation. In addition, extensions of the model to include coupling of nucleons to the Δ may change the spin characteristics of the in-medium repulsion.

The authors acknowledge financial support from the U.S. National Science Foundation under grant NSF-PHY-9602872 (E.S.) and from the Department of Energy under contract No. DE-FG03-00ER41148 (F.S.).

REFERENCES

1. F. Sammarruca, E.J. Stephenson, and K. Jiang, Phys. Rev. C **60**, 064610 (1999).
2. F. Sammarruca et al. Phys. Rev. C **61**, 014309 (2000).
3. R. Rapp, R. Machleidt, J.W. Durso, and G.E. Brown, Phys. Rev. Lett. **82**, 1827 (1999).
4. F. Sammarruca X. Meng, and E.J. Stephenson, Phys. Rev. C **62**, 014614 (2000).

Spin-dipole resonances in $^{16}O(\vec{p}, \vec{p}')$ reaction

T. Kawabata, H. Akimune[a], H. Fujimura[b], H. Fujita[c], Y. Fujita[c],
M. Fujiwara[b,d], K. Hara[b], K. Hatanaka[b], D. Hirooka[b], K. Hosono[e],
T. Ishikawa, M. Itoh, J. Kamiya[b], M. Nakamura, T. Noro[b],
E. Obayashi[b], H. Sakaguchi, Y. Shimbara[c], H. Takeda, T. Taki,
A. Tamii[f], H. Toyokawa[g], N. Tsukahara, H. Ueno[c], M. Uchida,
T. Wakasa[b], K. Yamasaki[a], Y. Yasuda, H.P. Yoshida[b] and M. Yosoi[b]

Department of Physics, Kyoto University, Kyoto 606-8502, Japan
[a]*Department of Physics, Konan University, Kobe, Hyogo 658-8501, Japan*
[b]*Research Center for Nuclear Physics (RCNP), Ibaraki, Osaka 567-0047, Japan*
[c]*Department of Physics, Osaka University, Toyonaka, Osaka 560-0043, Japan*
[d]*Advanced Science Research Center, Japan Atomic Energy Research Institute, Tokai-mura, Ibaraki 319-1195, Japan*
[e]*Department of Engineering, Himeji Institute of Technology, Hyogo 678-1297, Japan*
[f]*Department of Physics, University of Tokyo, Tokyo 113-0033, Japan*
[g]*Japan Synchrotron Radiation Research Institute, Hyogo 679-5198, Japan*

Abstract. Complete sets of observables in proton inelastic scattering from ^{16}O have been measured at forward angles including $0°$. Obtained spectra were decomposed into spin-flip and non spin-flip spectra by polarization transfer coefficients and spin-dipole strength splitting into several peaks was identified in the giant resonance region of $E_x = 19 \sim 27$ MeV in ^{16}O.

INTRODUCTION

Study of spin-isospin excitation modes is an important subject in nuclear physics. Gamow-Teller resonances (GTR; $\Delta T = 1$, $\Delta S = 1$, $\Delta L = 0$) were systematically studied by charge exchange reactions like (p,n) and (^{3}He, t). The quenching problem of GT strength is extensively discussed from the view point of mixing of 2p-2h configurations or ΔN^{-1} excitations into 1p-1h configurations. On the other hand, spin-dipole resonances (SDR; $\Delta T = 1$, $\Delta S = 1$, $\Delta L = 1$) have not been studied fully up to present. Strength distributions of three different spin states of SDR ($J^\pi = 2^-, 1^-$ and 0^-) can give a strong constraint on nuclear models, but only several experimental studies of SDR states for light nuclei were performed by measuring polarization observables or decaying particles from highly excited states in (d,^{2}He) and (p,n) reactions [1,2] and these results are not still satisfactory.

Excitations of SDR states in ^{16}O are interesting from the view point of not only nuclear physics but also astrophysics. Recently Langanke et al. [3] proposed a new detection scheme of supernova neutrinos via neutral current excitation. Supernova neutrinos with $E_\nu = 10 \sim 25$ MeV predominantly excite 1^- and 2^- giant resonance states, and high energy gamma-rays emitted in the de-excitation processes can be detected as signals of neutrino. It is important to study SDR distribution on ^{16}O in order to establish this new detection scheme.

Measurements of polarization transfer observables at forward angles enable to decompose spin modes in excitation spectra unambiguously [4]. Thus we have measured complete sets of observables in proton inelastic scattering from ^{16}O aiming at studies of SDR states. In this paper, our experimental procedures and results will be reported.

EXPERIMENTAL PROCEDURE

The experiment has been performed at RCNP, Osaka University by using a 392 MeV polarized proton beam accelerated by $K = 400$ MeV ring cyclotron. The extracted beam was achromatically transported from ring cyclotron to the target. Scattering particles were momentum analyzed by the high-resolution spectrometer Grand Raiden [5], and the polarization was measured by a focal plane polarimeter. The measurements were performed for three different beam polarizations, whose directions make about 90° of angles each other, in order to measure complete sets of spin observables. One of the beam polarizing direction was selected as perpendicular to the scattering plane and other two directions were in the scattering plane to avoid mixing between horizontal and vertical polarization transfer observables. Additionally, in measurements of in-plane beam polarization, a special dipole magnet for spin rotation (DSR) was employed to measure all components of polarization of scattering particles. The DSR magnets is a bending magnet of $+18°$ or $-17°$, and proton spin precesses about $\pm 45°$ against the momentum direction in the horizontal plane depending on the magnet polarity and proton energy.

Generally, it is difficult to make an oxygen target due to its gaseous property. Metal oxides are commonly used as oxygen targets since their preparation in a chemical process is relatively easy. The solid compositions of BeO and Li_2O are conveniently used in many experiments. The events from metal bases, however, should be measured and subtracted to extract the oxygen events in case of metal oxide targets. This procedure often causes problem concerning statics and quality of spectra. Although oxygen gas kept in a cell is also used as a target, it is difficult to obtain a thickness enough to measure polarization transfer observables with high statistics. Moreover, background subtraction procedures are needed again at forward angles. In order to overcome these difficulties, we have developed a self supporting windowless ice target system for this experiments [6]. An ice sheet with

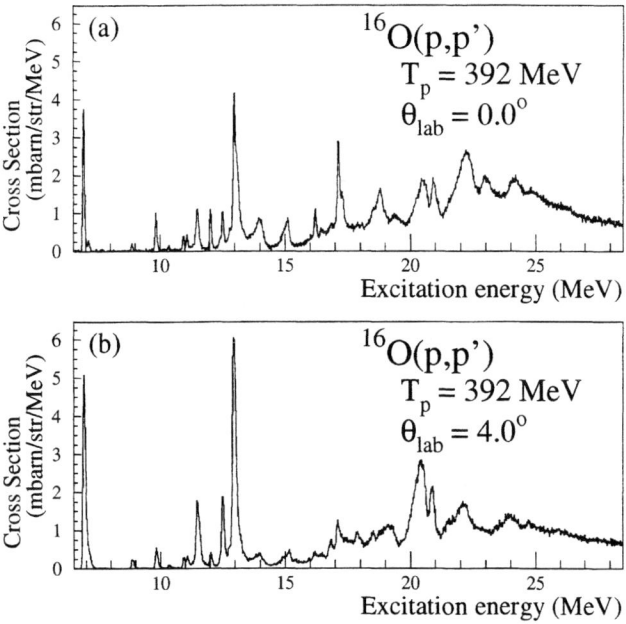

FIGURE 1. Spectra for ^{16}O(p,p') at (a) 0° and (b) 4°, respectively.

a thickness of 10 ~ 30 mg/cm² made of pure water is cooled under 140 K by liquid nitrogen where sublimation loss of ice can be neglected in vacuum chamber. Then we can assume H$_2$O as a pure oxygen target since hydrogen contaminations in H$_2$O can be removed due to a large difference of kinematical effects between oxygen and hydrogen. We have succeeded in obtaining background free oxygen spectra by using this new ice target.

RESULT AND DISCUSSION

Figure 1 shows spectra of ^{16}O(p,p') at (a) 0.0° and (b) 4.0° respectively. The energy resolution was 80 ~ 150 keV (FWHM). The spectra consist of discrete levels at lower excitation energies and broad resonance bumps at higher energies. Here, we introduce new quantity *total spin transfer* [7] as follow;

$$\Sigma = \frac{3 - (D_{SS} + D_{NN} + D_{LL})}{4}. \qquad (1)$$

This quantity takes a value of unity for spin-flip and zero for non spin-flip transitions at forward angles where the spin-orbit term in the NN effective interaction is negligible. By using this quantity, the spin-flip cross section of $\Sigma \cdot d\sigma/d\Omega$ and the non spin-flip cross section of $(1 - \Sigma) \cdot d\sigma/d\Omega$ can be obtained. The spectra of $\Sigma \cdot d\sigma/d\Omega$ and $(1-\Sigma) \cdot d\sigma/d\Omega$ are shown in figure 2. Spin-flip and non spin-flip transitions are shown as hatched and white spectra in the figures, respectively. Spin-flip transitions occupy major parts near the unnatural parity states at 12.53 and 12.97 MeV. Large non spin-flip component near the 12.97 MeV states in zero-degree spectrum is due to the neighboring 1^- state at 13.09 MeV which peaks at 0°. Narrow spin-flip resonances at 16.21 MeV, 17.14 MeV and 18.79 MeV are identified as 1^+ since they are forward peaking. Several bumps exist in the region of $E_x = 20 \sim 24$ MeV at 0° while resonant structures are not evident in the non spin-flip spectrum at 4°. The non spin-flip spectrum has the same shape as a gamma absorption spectrum [8]. Dipole resonances are expected to be enhanced at 0° via Coulomb excitations in (p,p') reaction at 392 MeV. Thus, our observation is consistent with the expectation. Spin-flip resonance states are observed at 20.3 MeV, 20.9 MeV,

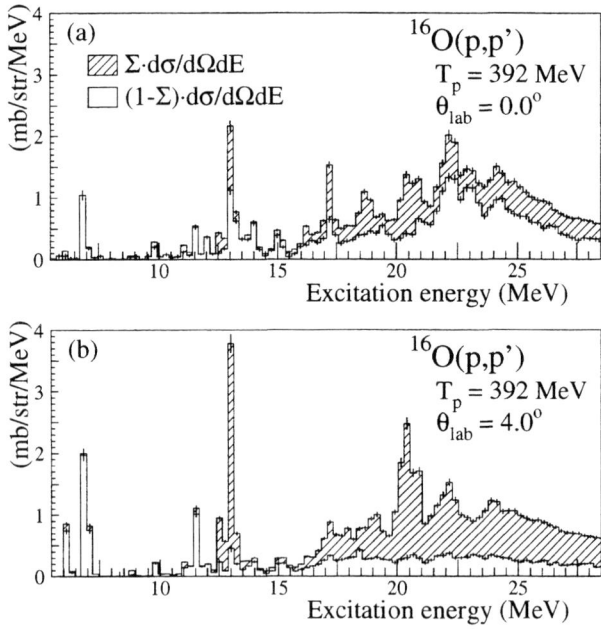

FIGURE 2. Spin-flip (hatched) and non spin-flip (white) spectra for ^{16}O(p,p') at (a) 0° and (b) 4°, respectively.

22.2 MeV, and 23.9 MeV. Since all the states decrease at 8° again, these states are identified to be excited with spin-dipole transitions. Further studies are in progress in order to determine J^π modes from polarization transfer observables.

REFERENCES

1. H. Okamura *et al.*, Phys. Lett. B **345**, 1 (1995).
2. T. Inomata *et al.*, Phys. Rev. C **57**, 3153 (1998).
3. K. Langanke, P. Vogel and E. Kolbe, Phys Rev. Lett. **76**, 2628 (1996).
4. A. Tamii *et al.*, Phys. Lett. B **459**, 61 (1999).
5. M. Fujiwara *et al.*, Nucl. Instr. Meth. A **422**, 484 (1999).
6. T. Kawabata *et al.*, Nucl. Instr. Meth. A (in press).
7. H. Sakai *et al*, in proceedings of *Internatinal Symposium on New Facet of Spin Giant Resonaces in Nuclei*, (World Scientific, Singapore 1998) pp.29.
8. J. Ahrens *et al*, Nucl. Phys. A **251**, 479 (1975).

Measurement of Single and Double Spin-Flip Probabilities in Inelastic Deuteron Scattering on ^{12}C and ^{28}Si at 270 MeV

Y. Satou[a], S. Ishida[a], H. Sakai[b], H. Okamura[b], H. Otsu[a],
N. Sakamoto[a], T. Uesaka[b], T. Wakasa[a], T. Ohnishi[b], T. Nonaka[b],
G. Yokoyama[b], K. Sekiguchi[b], K. Yako[b], S. Fukusaka[b],
T. Ichihara[a], T. Niizeki[c], K. S. Itoh[a] and N. Nishimori[d]

[a] *The Institute of Physical and Chemical Research (RIKEN), Wako, Saitama 351-0198, Japan*
[b] *Department of Physics, University of Tokyo, Bunkyo, Tokyo 113-0033, Japan*
[c] *Department of Physics, Tokyo Institute of Technology, Oh-okayama, Tokyo 152-0033, Japan*
[d] *Japan Atomic Energy Research Institute (JAERI), Ibaraki, 319-11, Japan*

Abstract. We report measurements of the deuteron single and double spin-flip probabilities (SFPs), S_1 and S_2, for the $(\vec{d},\vec{d'})$ reaction on ^{12}C and ^{28}Si at 270 MeV over an excitation energy range of 4–24 MeV and a scattering angular range of 2.5–7.5°. The SFP S_1 exhibits large positive values for known spin-flip states, while it has values close to zero for non-spin-flip states. The SFP S_2 is close to zero for all the measured excitation energy region. The results for the 1^+ (12.71 MeV) and 2^+ (4.44 MeV) states in ^{12}C are well described by the microscopic DWIA calculations. The deuteron SFPs provide signatures of isoscalar single and double spin-flip nuclear excitations.

INTRODUCTION

The study of spin-flip processes in deuteron inelastic scattering at intermediate energies (E/A > 100 MeV) is an important extension to the corresponding study using the nucleon projectile such as the $(\vec{p},\vec{p'})$ and $(\vec{p},\vec{n})$ reactions. Stringent selectivity to the isoscalar transitions of the (d,d') reaction provides a good opportunity to investigate the isoscalar spin-flip mode of nuclear excitation [1]. Much less is known about this mode due to the lack of efficient probes as well as to the weakness of the effective nucleon-nucleon (NN) interaction in the relevant channel. Furthermore, spin-1 nature of the deuteron may offer a unique capability to probe double spin-flip transitions, such as the proposed double GT state [2,3].

In (d,d'), one can define single and double SFPs, S_1 and S_2, as fractions of deuterons undergoing spin-flip by 1 and 2 units along an axis normal to the reaction plane. S_1 will be a signature of single spin-flip excitations, as S_{nn} is in (p,p'), and S_2 a signature of double spin-flip excitations. In terms of polarization observables they are given by the relations:

$$S_1 = \frac{1}{9}(4 - P^{y'y'} - A_{yy} - 2K^{y'y'}_{yy}), \qquad (1)$$

$$S_2 = \frac{1}{18}(4 + 2P^{y'y'} + 2A_{yy} - 9K^{y'}_{y} + K^{y'y'}_{yy}), \qquad (2)$$

where one (two) indices stand for the vector (tensor) polarization, and lower (upper) indices stand for the incident (outgoing) beam. The determination of S_1 and S_2 requires vector and tensor polarized beams and vector and tensor polarimeters.

The primary aim of this work is to explore the possibility of using the deuteron single and double SFPs as probes of isoscalar single and double spin-flip excitations. We have measured the $(\vec{d},\vec{d'})$ reaction on ^{12}C and ^{28}Si at $E_d = 270$ MeV. A focal plane polarimeter, which was specifically designed to measure both vector and tensor components of the deuteron polarization, was utilized. The ^{12}C and ^{28}Si targets were chosen as they provide typical isoscalar spin-flip 1^+ states.

EXPERIMENT

The experiment was performed at RIKEN accelerator research facility (RARF). The polarized deuteron beam from the Ring Cyclotron with an energy of 270 MeV and an average current of 10 nA was led through a beam twister and a beam swinger onto ^{12}C (87.2 mg/cm^2) and ^{28}Si (58.1 mg/cm^2) targets. The vector and tensor polarizations with maximum values of (p_y, p_{yy})=(0,0), (0,-2), (2/3,0) and (-1/3,1) were used, the magnitudes were measured using the $\vec{d}+p$ elastic scattering at 270 MeV [4] to be 60–70 % of the ideal values. Scattered deuterons were detected using the SMART spectrometer [5], which was instrumented with a multiwire drift chamber (MWDC) for track reconstruction and two plastic scintillation detectors (SC1 and SC2) for triggering. It had a solid angle of 5.0 msr and a momentum acceptance of 4%. Data were acquired at a swinger angle of 5°.

The polarizations of outgoing deuterons were measured in a deuteron polarimeter DPOL installed in the focal plane of SMART. Both vector and tensor components of outgoing deuterons were determined by utilizing, as the analyzing reactions, the $\vec{d}$+C elastic scattering [6] and the charge exchange ^{1}H$(\vec{d},2p)$ reaction [7], respectively. The deuterons from the primary scattering were incident on the analyzer target, behind MWDC, consisting of a 2.5-cm thick CH$_2$ block bracketed by SC1 and SC2 (1.0 and 0.5 cm thick). A counter hodoscope consisting of two layers of plastic scintillation counter array, HOD and CM, ($\sim 2\times$ 2 m^2) located 4 m behind the analyzer target was used to detect proton pairs produced by the ^{1}H$(\vec{d},2p)$ reaction, and deuterons from the $\vec{d}$+C elastic scattering. An iron absorber between HOD and CM was used to discriminate between deuterons and protons.

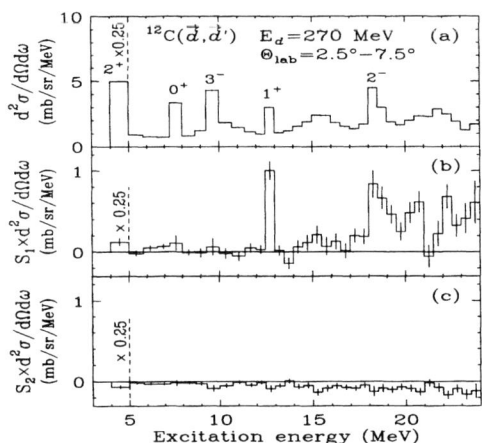

FIGURE 1. Excitation energy spectra for the $^{12}C(\vec{d},\vec{d'})$ reaction at $E_d = 270$ MeV integrated over $\Theta_L = 2.5°-7.5°$. (a) The double differential cross section. (b) The cross section multiplied by S_1. (c) The cross section multiplied by S_2.

RESULTS AND DISCUSSIONS

Results for the ^{12}C target are shown in Fig. 1. The error bars are statistical ones. Figure 1 (a) shows the double differential cross sections as a function of the ^{12}C excitation energy over $\Theta_L = 2.5°-7.5°$. The well known isoscalar 1^+ (12.71 MeV) state is clearly excited among natural parity 2^+ (4.44 MeV), 0^+ (7.65 MeV) and 3^- (9.64 MeV) states. Isovector states such as the 1^+ state at 15.11 MeV are entirely absent due to the isoscalar selectivity of the (d,d') reaction. Figure 1 (b) shows the cross section multiplied by S_1. In this spectrum, the 1^+ spin-flip state stands out, while the other natural parity states, which are predominantly non-spin-flip transitions, are suppressed. This shows that S_1 is indeed a sensitive signature of isoscalar spin-flip transitions. The state at 18.3 MeV has been assigned as $(J^\pi, T) = (2^-, 0)$ from the S_{nn} measurement in $(\vec{p},\vec{p'})$ [8]. Present result shows consistency with its identification as an isoscalar spin-flip state. Figure 1 (c) plots the cross section multiplied by S_2. The spectrum is consistent with zero and almost no structure can be seen in it.

Figure 2 shows results for the ^{28}Si target, (a) the differential cross section, (b) single and (c) double spin-flip cross sections. The single spin-flip cross section exhibits the largest strength at the energy bin where an isoscalar 1^+ (9.50 MeV) state is known. Above 10 MeV in excitation energy, the spin-flip cross section fluctuates at around 0.4 mb/sr/MeV, suggesting that the isoscalar spin-flip strengths are fragmentary existing in this energy region. The double spin-flip cross section is found to be almost consistent with zero.

To understand the results on S_1 and S_2, microscopic DWIA calculations [9] were performed for the 2^+ (4.44 MeV) and 1^+ (12.71 MeV) states in ^{12}C. The transition

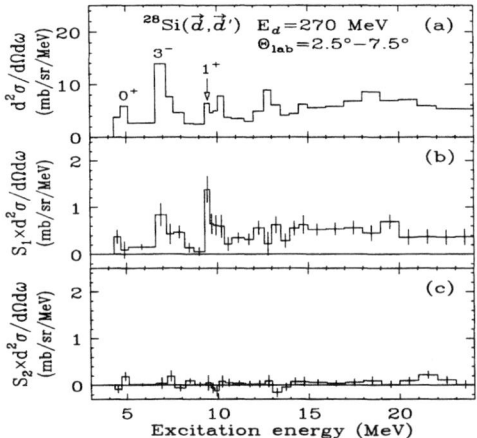

FIGURE 2. The same as Fig.1 but for ^{28}Si.

density was calculated using the Cohen and Kurath wave functions [10]. The optical potential was obtained by fitting the elastic scattering data. The projectile-nucleon effective interaction was the free deuteron-nucleon (dN) t-matrix obtained from the three nucleon Faddeev calculations [11] using the CD-Bonn potential for the nucleon-nucleon interaction [12].

In Fig. 3 (a), experimental angular distributions of S_1 and S_2 for the 2^+ state are compared with calculated results. The solid (dashed) lines represent DWIA (PWIA) results. The experimental results are well reproduced by the DWIA calculations. Finite S_1 values predicted in DWIA are mainly due to the spin-orbit distortion in the optical potential. The experimental S_1 values for the 1^+ state (see Fig. 3 (b)) are large at small scattering angles and decrease with increasing the angle. The S_2 data are close to zero. The data are qualitatively well described by the DWIA calculations. The dot-dashed line for S_1 for the 1^+ state represents a theoretical curve corresponding to the transferred orbital, spin and total angular momenta of $LSJ = (011)$, calculated as given in Ref. [13] using the dN impulse approximation amplitudes. The agreement between the DWIA and this simple IA estimate at forward angles ($\Theta \leq 10°$) indicates that in this angular region S_1 is relatively insensitive to distortion effects and also to details of the transition density, instead it is mainly determined by the impulse approximation amplitudes. Together with small values predicted for non-spin-flip transitions, this feature of S_1, which is qualitatively consistent with results obtained for S_{nn} in $(\vec{p}, \vec{p}')$ [14], makes measurement of S_1 at forward angles a unique tool identifying the presence of isoscalar spin strengths in the spectrum. With regard to S_2, present calculations show that the optical distortion has little influence on this observable at forward angles. This may be useful in exploiting S_2 at forward angles as a signature of double spin-flip excitations.

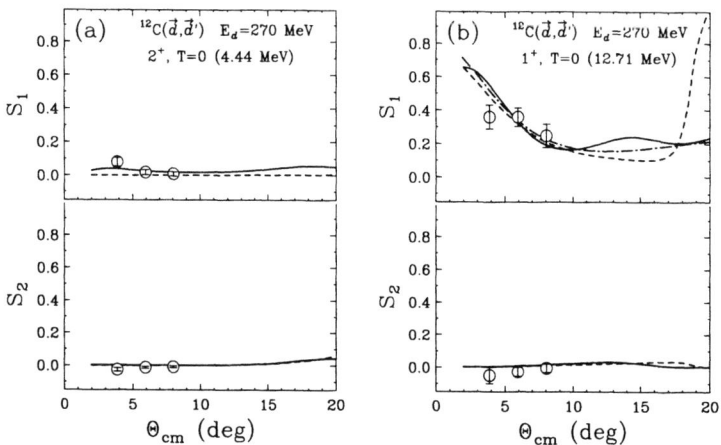

FIGURE 3. Spin-flip probability S_1 and S_2 vs center-of-mass scattering angle for the (a) 2^+ (4.44 MeV) and (b) 1^+ (12.71 MeV) states in ^{12}C. The solid (dashed) lines are DWIA (PWIA) calculations. The dot-dashed line for the 1^+ state (shown only for S_1) is an IA estimate using the dN amplitudes for the $LSJ=(011)$ transition.

ACKNOWLEDGMENTS

We thank Dr. H. Kamada for providing us with the deuteron-nucleon amplitudes. This work was supported financially in part by the Grant-in-Aid for Scientific Research No. 04402004 of Ministry of Education Science and Culture of Japan.

REFERENCES

1. M. Morlet et al., Phys. Lett. B 247 (1990) 228.
2. P. Vogel et al., Phys. Lett. B 212 (1988) 259.
3. N. Auerbach, L. Zamick and D. C. Zheng, Ann. Phys. 192 (1989) 77.
4. N. Sakamoto et al., Phys. Lett. B 367 (1996) 60.
5. T. Ichihara et al., Nucl. Phys. A 569 (1994) 287c.
6. B. Bonin et al., Nucl. Instr. and Meth. A 288 (1990) 389.
7. S. Kox et al., Nucl. Instr. and Meth. 346 (1994) 527.
8. K. W. Jones et al., Phys. Lett. B 128 (1983) 281.
9. J. Van de Wiele, A. Willis, M. Morlet, Nucl. Phys. A 588 (1995) 829.
10. S. Cohen and D. Kurath, Nucl. Phys. 73 (1965) 1.
11. W. Glöckle, H. Witala, D. Hüber, H. Kamada, J. Golak, Phys. Rep. 274 (1996) 107.
12. R. Machleidt, F. Sammarruca and Y. Song, Phys. Rev. C 53 (1996) 1483.
13. T. Suzuki, Nucl. Phys. A 577 (1994) 167c.
14. S. J. Seestrom-Morris et al., Phys. Rev. C 26 (1982) 2131.

DWIA Calculations for Inelastic Scattering of Deuterons at $E_d = 400$ MeV

Y.Hirabayashi*, T. Suzuki[†] and M.Tanifuji[‡]

*Center fo Information and Multimedia Studies, Hokkaido Univ.,
[†]Department of Physics, Tokyo Metropolitan Univ.
[‡]Department of Physics, Hosei Univ.

Abstract. A new type of DWBA calculation for inelastic scattering of deuterons is formulated based on the sudden approximation. Cross section and polarization observables are expressed in terms of the scattering matrices for the constituent nucleons with the same target transitions. The formalism is applied for an excitation of isoscalar states in ^{12}C using DWBA amplitudes for the nucleon-nucleus scattering. Excellent agreement with existing data has been obtained for natural parity transitions, while some discrepancies remain for unnatural parity ones. The latter could be related to the choice of the effective nucleon-nucleon interaction. A relationship between observables for deuteron-nucleus and proton-nucleus scattering was found, which are well satisfied in the existing data. Details will be given elsewhere [1].

Sudden Approximation

Recently a new series of polarization data has been obtained for inelastic scattering of deuterons from nuclei, which is expected to shed a light on the reaction mechanism and on the structure of isoscalar excitations of the target [2,3]. For a scattering of weakly bound composite particles such as deuterons, an adiabatic picture should be valid, suggesting an impulse type approximation. Here the internal motion of the projectile is frozen while the latter interacts with the target. This is known as the sudden approximation and has been applied to transfer reactions and elastic scattering [4,5].

In the sudden approximation, each nucleon in the insident deuteron is distorted under the nuclear optical potential, giving rise to the deuteron-nucleus wave function in the incident channel

$$\Psi_{dA}^{(+)} \sim \int d\vec{k}\, a(\vec{k}) [\phi_{k_n}^{(+)} \phi_{k_p}^{(+)}] \cdot \Psi_A, \qquad (1)$$

where $a(\vec{k}) = a_S(\vec{k}) + a_D(\vec{k}) S_{12}(\hat{k})$ denotes a momentum space wave function of the deuteron with S- and D-waves, and neutron/proton distorted wave functions $\phi^{(+)}$ are characterized by the momenta $\vec{k}_n = \frac{1}{2}\vec{k}_d + \vec{k}$ and $\vec{k}_p = \frac{1}{2}\vec{k}_p - \vec{k}$. We simplify the calculation of the transition matrix by 1) calculating only the single collision term, and 2) by evaluating nucleon distorted waves $\phi^{(+)}$ at $\vec{k}=0$ where $a(\vec{k})$ is strongly peaked. The resulting deuteron-nucleus scattering matrix is given as the matrix element in the deuteron spin space as:

$$T_{dA} = \langle \chi_{d'}|\hat{M}_d|\chi_d\rangle, \qquad \hat{M}_d = F(q)(\hat{M}_n + \hat{M}_p) + \cdots, \qquad (2)$$

where $F(q)$ is the deuteron monopole form factor, the dots denote terms proportional to the quadrupole form factor, and $\hat{M}_n$ and $\hat{M}_p$ are the nucleon-nucleus scattering matrices for the same transition of the target. The latter are evaluated at half incident energy of the deuteron.

The deuteron-nucleus scattering observables are defined by:

$$I_0 K_\alpha^\beta = \text{Tr}_d[\hat{M}_d \mathcal{P}_\alpha \hat{M}_d^\dagger \mathcal{P}_\beta], \quad I_0 \equiv \text{Tr}_d[\hat{M}_d \hat{M}_d^\dagger] \qquad (3)$$

where I_0 is proportional to the cross section, and $\mathcal{P}_\alpha$ are the spin weight factors. As the matrix $\hat{M}_d$ is expressed in terms of $\hat{M}_n$ and $\hat{M}_p$, the trace Tr_d in the deuteron spin space is transformed to that in the neutron and proton spin spaces. The resultant matrix consists of the *direct* terms which involve $\hat{M}_n \hat{M}_n^\dagger$ or $\hat{M}_p \hat{M}_p^\dagger$, and the *crossed* terms involving $\hat{M}_n \hat{M}_p^\dagger$ and $\hat{M}_p \hat{M}_n^\dagger$. The direct terms can be expressed in terms of the nucleon-nucleus scattering observables defined in a similar manner as eq.(3).

Calculation

We consider inelastic scattering of polarized deuterons at $E_d = 400$MeV on ^{12}C nucleus leading to the isoscalar 2^+(4.4MeV), 3^-(9.6MeV) and 1^+(12.7MeV) states. We calculate cross sections, A_y, and other polarization observables. The y-direction is taken in the normal to the scattering plane, while z is taken in the direction of the momentum transfer $\vec{q}$. We first calculate nucleon-nucleus amplitude at 200MeV in DWBA and then use sudden approximation formula to obtain deuteron-nucleus observables. The ^{12}C wave functions are those of ref. [8], and the proton-nucleus optical potential was taken from ref. [7]. The neutron optical potential is the same as those for the proton without Coulomb interaction. We use the effective interaction of ref. [9] which reasonably reproduces the proton-nucleus data except for some observables. The deuteron form factors have been obtained from ref. [10]. We compare our calculation with the DWIA calculation [6]. Note that in DWIA the distortion effect is applied to the deuteron center-of-mass motion, while in the sudden approximation each nucleon independently feels the nulcear potential in conformity with the impulse approximation picture. We also effectively include the double collision correction by the momentum transfer shift of 7%. [5]

Figures 1 and 2 show the results of our calculation(solid lines) together with those of DWIA(dashed lines) for 3^- and 1^+ excitations. For the 3^- state all the observables are nicely reproduced in the calculation. Agreement of a similar quality has been obtained also for the observables of the 2^+ state. For the 1^+ state, however, our calculation fails to reproduce A_{yy}, while DWIA is succesful. This difference is present already in the plane wave limit. For P^y in 1^+, on the other hand, our calculation shows a much better agreement than DWIA.

In these calculations the contribution of the deuteron D-state is small at angles $\theta \leq 20°$ where experimental data exist. We checked the nucleon DWBA input by using another optical potential parameters. The results are similar in quality as those obtained above. Thus the failure in the calculation of A_{yy} for the 1^+ state should originate from other sources, one of which may be the choice of the interaction.

Relations between d-A and p-A Observables

It has been noted(ref. [2]) that the analysing powers in deuteron-nucleus scattering are very similar to those of the corresponding nucleon-nucleus scattering when plotted against momentum transfer(Fig.3, left). Our formulation is especially suited for the study of this kind of relationship.

For the isoscalar excitations in ^{12}C, we may assume that the proton- and neutron-nucleus scattering matrices $\hat{M}_{p,n}$ are similar aside from that they operate in different spin spaces. This is indeed well satisfied except at very forward angles where the effect of the Coulomb interaction is important. Under this assumption the crossed terms in the sudden approximation formula give an equal contribution as the direct terms. As noted before, therefore, the deuteron-nucleus scattering observables are entirely expressed in terms the nucleon-nucleus ones. If we further omit the deuteron D-state contribution we obtain

$$I_0\Big|_{dA} = \frac{1}{2}I_0(9 + K_x^x + K_y^y + K_z^z)\Big|_{NA} \cdot |F(q)|^2 \qquad (4)$$

$$I_0 A_y\Big|_{dA} = I_0(3A_y + P^y)\Big|_{NA} \cdot |F(q)|^2 \qquad (5)$$

$$I_0 P^y\Big|_{dA} = I_0(A_y + 3P^y)\Big|_{NA} \cdot |F(q)|^2 \qquad (6)$$

$$I_0 A_{yy}\Big|_{dA} = I_0(-K_x^x + 2K_y^y - K_z^z)\Big|_{NA} \cdot |F(q)|^2 \qquad (7)$$

$$I_0 K_y^y\Big|_{dA} = 2I_0(1 + K_y^y)\Big|_{NA} \cdot |F(q)|^2, \qquad (8)$$

etc. As the quantity I_0 is related to the cross section, the above formula give direct relations between $d - A$ and $N - A$ observables in terms of the single factor

$$f(q) \equiv 4\frac{I_0(NA)}{I_0(dA)} \cdot |F(q)|^2 = \frac{8}{3}\left(\frac{k_f}{k_i}\right)_d \left(\frac{k_i}{k_f}\right)_N \left(\frac{\mu_d}{\mu_N}\right)^2 \frac{\sigma(NA)}{\sigma(dA)} \cdot |F(q)|^2 \qquad (9)$$

For instance, for the analysing power we obtain

$$A_y|_{dA} = f(q)\frac{1}{4}(3A_y + P^y)|_{NA} \simeq f \cdot A_y|_{NA}, \qquad (10)$$

where the last relation is satisfied if $P^y \simeq A_y$ for nucleon-nucleus scattering.

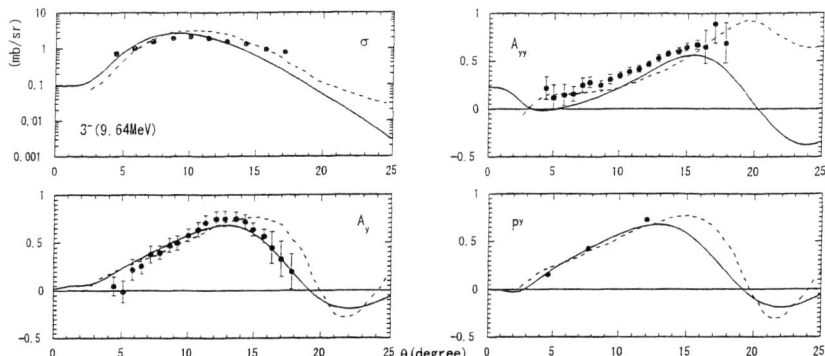

FIGURE 1. Cross section and polarization observables for $(d,d')^{12}\mathrm{C}(3^-)$ at 400MeV in the sudden approximation. Dashed lines show the DWIA results.

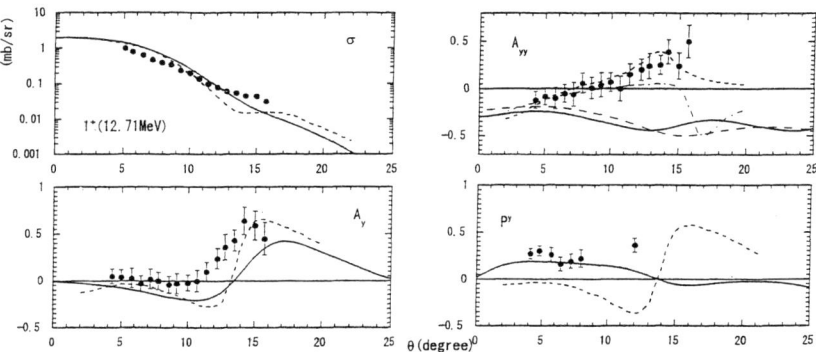

FIGURE 2. Similar to Fig.1 but for $(d,d')^{12}\mathrm{C}(1^+)$. Long-dashed (dash-dotted) lines for A_{yy} show the plane wave results in the present (ref. [6]) calculation.

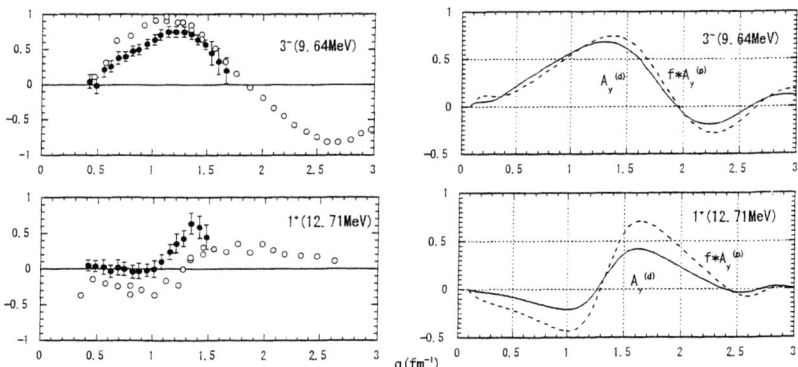

FIGURE 3. Left: Comparison of the experimental data of A_y for (d,d')(filled circles) and (p,p')(open circles) at the same momentum transfer for the 3^- and 1^+ states. Right: Comparison of A_y for (d,d') in the sudden approximation (solid lines) and $f * A_y$ for (p,p')(dashed lines).

Figure 3(right) shows a comparison of the calculated $A_y(d,d')$ and $f \cdot A_y(p,p')$. The equality is well satisfied for natural parity (non-spin-flip) transitions where the relation $P^y \simeq A_y$ holds for (p,p'). The latter relation is violated for the 1^+ state, however, and the last relation in eq.(10) does not hold as seen in the figure.

We obtain another relation from eqs.(4)-(8):

$$R \equiv (\frac{4}{3} + \frac{2}{3} A_{yy} - K_y^y)/f = 1. \qquad (11)$$

This may be used as a signature of the validity of the sudden approximation. (If one eliminates factor f using the expression for K_{yy}^{yy}, one obtains a relation $S_2 = 0$ for the double-spin-flip transition of ref. [2], which is a natural consequence of the single-collision calculation.) Figure 4 plots the calculated ratio R for the 1^+ state showing that the relation is well satisfied. Similar results are obtaind for 2^+ and 3^-. We plot also the experimental data for 1^+ and find that it satisfies the relation rather well, although each of the observables involved has not been reproduced in the present calculation. This suggests that the basic picture of the present calculation is not in contradiction with data, but the origin of the failure for A_{yy} for 1^+ should be found in the input parameters, e.g., the choice of the effective interaction.

REFERENCES

1. Hirabayashi, Y., Suzuki, T., and Tanifuji, M., to be published.
2. Morlet, M. et al., *Phys. Lett.* **247B**, 228(1990); *Phys. Rev.* **C46**, 1008(1992).
3. Satou, Y., contribution to the present volume and references therein.
4. Tanifuji, M., *Nucl. Phys.* **58**, 81(1964); Butler, S.T., *Nature* **207**, 1349(1965).
5. Sakuragi, Y., and Tanifuji, M., *Nucl. Phys.* **A560**, 945(1993).
6. Van de Wiele, J., Willis, A., and Morlet, M., *Nucl. Phys.* **A588**(1995).
7. Jones, K.W. et al., *Phys. Rev.* **C33**, 17(1986).
8. Cohen, S. and Kurath, D., *Nucl. Phys.* **73**, 1(1965); Millener, D.J. and Kurath, D., *Nucl. Phys.* **A255**, 315(1978).
9. Love, W.G. and Franey, M.A., *Phys. Rev.* **C24**, 1073(1981); *ibid.* **C31**, 488(1985).
10. Tamura, K., *Nucl. Phys.* **A536**, 597(1992) and private c

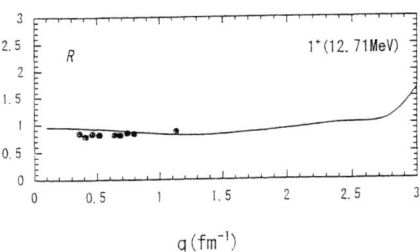

FIGURE 4. The ratio R for $(d,d')^{12}\text{C}(1^+)$ as a function of momentum transfer. The experimental data (using calculated factor f is also plotted.

Nuclear Spectroscopy By Means Of $(\vec{p},\alpha)$ Reactions On Magic And Near Magic Nuclei: ^{122}Sn$(\vec{p},\alpha)^{119}$In

P.Guazzoni[*], M.Jaskola[ø&], V.Yu.Ponomarev[+&], L.Zetta[*], Y.Eisermann[¶], G.Graw[¶], R.Hertenberger[¶], A.Vitturi[†], J.N.Gu[‡], G.Staudt[♪]

[*]*Dipartimento di Fisica dell'Università and Istituto Nazionale di Fisica Nucleare - I20133 Milano, Italy*
[ø]*Soltan Institute for Nuclear Studies - Warsaw, Poland*
[+]*Bogoliubov Laboratory of Theoretical Physics, JINR - Dubna, Russia*
[¶]*Ludwig-Maximillians Universität München - D85748 Garching, Germany*
[†]*Dipartimento di Fisica dell'Università and Istituto Nazionale di Fisica Nucleare - I35131 Padova, Italy*
[‡]*Institute of Modern Physics, Academia Sinica - Lanzhou, People Republic of China*
[♪]*Physikalisches Institut der Universität - D72076 Tübingen, Germany*

Abstract. $(\vec{p},\alpha)$ reactions on near magic nuclei was used to assign unambiguously spin and parity to higher excited states using the homology concept, in the regions A~208 and A~90. Here some preliminary results, both experimental and theoretical for the ^{122}Sn$(\vec{p},\alpha)^{119}$In reaction are presented.

INTRODUCTION

The weak coupling model has been used in spectroscopic studies of $(\vec{p},\alpha)$ reactions on near magic target nuclei (odd-A nuclei) having one proton (or one neutron) outside a complete filled magic shell (core) to investigate the spectator role of the unpaired nucleon. The simplest form of this model foresees that a class of states in odd-A nuclei will arise from the coupling of the odd particle with a basically undisturbed state of the (A-1) even-even nucleus. The coupling of the odd spectator particle with an excited state of the core generates a multiplet of homologous states with spin J ranging from $|J_P - J_C|$ to $(J_P + J_C)$ where J_P (J_C) is the spin of the particle (core). According to this approach the excitation probability for the members of the multiplet is proportional to $(2J + 1)$.

We have experimentally investigated via $(\vec{p},\alpha)$ reactions the concept of homologous states in the regions with A ~ 208 [1] and A ~ 90 [2,3] and detailed theoretical studies on the properties of homologous states in these A regions [2-4] were also done in the framework of the shell model. In order to investigate the spectator role of the unpaired nucleon in a different mass region we have studied the ^{122}Sn$(\vec{p},\alpha)^{119}$In reaction and we plan to also study the ^{123}Sb$(\vec{p},\alpha)^{120}$Sn reaction. The

---

[&] guest researcher of the Istituto Nazionale di Fisica Nucleare, Sezione di Milano

measured ^{122}Sn$(\vec{p},\alpha)^{119}$In energy spectrum is compared with the theoretical predictions of nuclear structure calculations in the framework of the Quasiparticle-Phonon Model (QPM), which accounts for the interaction between simple and complex configurations of nuclear excitations.

EXPERIMENTAL PROCEDURE AND RESULTS

The angular distributions of cross sections $\sigma(\theta)$ and analyzing powers $A_y(\theta)$ of the triton pick-up reaction ^{122}Sn$(\vec{p},\alpha)$ ^{119}In have been measured using the 26 MeV proton beam of the Garching HVEC MP Tandem, the polarized source, the Q3D magnetic spectrograph and the light ion focal plane detector with cathode periodic readout [5]. Transitions to 25 levels of ^{119}In were observed up to an excitation energy of ~ 2.6 MeV.

A DWBA analysis of $\sigma(\theta)$ and $A_y(\theta)$ was carried out assuming a semimicroscopic triton cluster pickup mechanism. The calculations have been done in finite range approximation with the code TWOFNR [6], using a proton-triton interaction potential of Gaussian form. Table a) summarizes the optical model parameters for the proton incident and α exit channels, and the geometric parameters used for evaluating the bound state wave function of the transferred triton cluster. Fig.1 shows the comparison between experimental and calculated $\sigma(\theta)$ and $A_y(\theta)$ for some ^{119}In levels.

TABLE a). Optical model parameters

	V_r MeV	r_r fm	a_r fm	W_v MeV	r_v fm	a_v fm	W_d MeV	r_d fm	a_d fm	V_{so} MeV	r_{so} fm	a_{so} fm	r_c fm
p	52.0	1.25	0.65				8.0	1.3	0.6	6.0	1.25	0.70	1.25
α	179.8	1.42	0.57	39.9	1.42	0.57							1.30
BS		1.45	0.4										

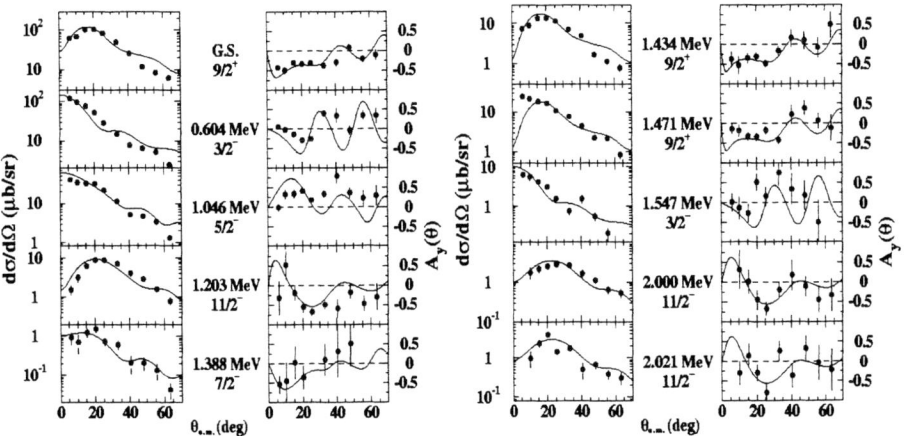

FIGURE 1. The angular distributions of $\sigma(\theta)$ and $A_y(\theta)$ for some of the observed levels.

Exploiting the J^π dependence of $\sigma(0)$ and $A_y(\theta)$ an unambiguous attribution of spin and parity was made for 20 of the 25 observed levels, and comparing the experimental results to Nuclear Data Sheet (NDS)[7] we have done 7 new J^π attributions, as shown in table b).

TABLE b). ^{119}In levels

Adopted Levels [7]		Present Experiment		
E_x (keV)	J^π	E_x (MeV)	J^π	$\sigma_{int}(\mu b)$
0.0	9/2$^+$	0.000	9/2+	85.341
311.37	1/2$^-$	0.310	1/2$^-$	19.268
604.18	3/2$^-$	0.604	3/2$^-$	49.134
654.27	1/2$^+$,3/2$^+$			
720.60	(7/2,9/2,11/2$^-$)			
788.26	1/2$^+$,3/2$^+$			
941.43	5/2$^+$			
1025.02	(7/2,9/2,11/2$^-$)			
1044.44	5/2$^-$	1.046	5/2$^-$	29.021
1050.21	5/2$^+$			
1143.00	(9/2,11/2,13/2$^+$)			
1203.71	7/2$^-$			
		1.203	11/2$^-$	9.461
1353.10	(9/2$^+$)			
1388.39	7/2$^-$	1.388	7/2$^-$	1.186
1436.44	7/2$^+$,9/2$^+$	1.434	9/2$^+$	12.764
1474	7/2$^+$,9/2$^+$	1.471	9/2$^+$	15.998
1553	1/2$^-$,3/2$^-$	1.547	3/2$^-$	4.230
1624.94	1/2,3/2,5/2			
1649	3/2$^+$,5/2$^+$			
1729	1/2$^-$,3/2$^-$			
1769.97	1/2$^+$,3/2,5/2			
1806.59	1/2$^+$,3/2,5/2			
1837	1/2$^-$,3/2$^-$			
1920.99	1/2$^+$,3/2$^+$			
		1.929		3.595
		1.960	5/2$^+$	1.860
1979	7/2$^+$,9/2$^+$			
		2.000	11/2$^-$	1.936
2021.35	(9/2$^-$,11/2$^-$)	2.021	11/2$^-$	1.530
2050	1/2$^-$,3/2$^-$			
2064.15	1/2$^+$,3/2,5/2$^-$			
2104.30	(9/2,11/2,13/2$^+$)	2.104	13/2$^+$	8.367
2126.93	(9/2$^-$,11/2$^-$)			
		2.193		3.333
		2.247	11/2$^-$	8.738
2272				
		2.308	11/2$^-$	1.267
2338.04	1/2$^+$,3/2$^+$			
2343	1/2$^-$,3/2$^-$			
2359.54	(9/2,11/2,13/2)			
2367.95	1/2$^+$,3/2$^+$			

Adopted Levels [7]		Present Experiment		
E_x (keV)	J^π	E_x (MeV)	J^π	$\sigma_{int}(\mu b)$
		2.372	$5/2^+$	4.426
2389.18	$(9/2^-,11/2^-)$			
2410	$5/2^-,7/2,9/2^+$	2.413	$(5/2^-,7/2^+)$	5.118
2422.54	$(9/2^-,11/2^-)$			
		2.430	$11/2^-$	2.590
2460	$5/2^-,7/2,9/2^+$	2.469	$(5/2^-,9/2^+)$	4.954
2487.21	$(9/2^-,11/2^-,13/2^-)$	2.488	$11/2^-$	4.399
2502	$1/2^-,3/2^-$	2.509	$3/2^-$	1.083
2520.3	$(9/2,11/2,13/2^+)$			
2524.57	$1/2,3/2,5/2$			
2526.6	$(9/2,11/2,13/2)$			
2554.40	$1/2,3/2,5/2$	2.554	$5/2^-$	3.818
2564.64	$(11/2^-,13/2^-)$			
2564.8	$1/2^+,3/2,5/2$			
		2.598	$5/2^+$	3.884
2618	$7/2^+,9/2^+$			

THEORETICAL CALCULATIONS

The spectrum of excited states in ^{119}In has been calculated within the Quasiparticle-Phonon Model. The ground and excited states of ^{119}In have been described by wave functions of the form:

$$\Psi^\nu(JM) = \left\{ C^\nu(J)\alpha^+_{JM} + \sum_{j\lambda i} S^\nu_{j\lambda i}(J)\left[\alpha^+_j Q^+_{\lambda i}\right]_{JM} + \sum_{j\beta_1\beta_2 I} \frac{D^\nu_{j\beta_1\beta_2}(J)\left[\alpha^+_j \left[Q^+_{\beta_1} Q^+_{\beta_2}\right]_I\right]_{JM}}{\sqrt{1+\delta_{\beta_1\beta_2}}} \right\} \left| ^{118}Cd \right\rangle_{g.s.} \quad (1)$$

where: α^+_{jm} is a quasiparticle (qp) creation operator, Q^+_β is a phonon (ph) creation operator, $\left| ^{118}Cd \right\rangle_{g.s.}$ is the ground state wave function of the neighboring even-even core.

Square brackets in the equation (1) denote angular momentum coupling, i.e. $\left[\alpha^+_j Q^+_\beta\right]_{JM} = \sum_{m\mu} C^{JM}_{jm\lambda\mu} \alpha^+_{jm} Q^+_{\lambda\mu}$, where $C^{JM}_{jm\lambda\mu}$ are the Clebsch-Gordan coefficients. Quasiparticles with shell quantum numbers $jm \equiv (n,l,j,m)$ have half-odd-integral angular momenta. The index β of the phonon operator means a combination of (λ,μ,i) where i is used to distinguish between one-phonon states of the same multipolarity λ, but with different excitation energies.

In the actual calculations we have used several of the lowest QRPA states for each multipolarity $\lambda = 2 \div 7$ corresponding to natural parity excitations of the ^{118}Cd core nucleus. Calculations in ^{119}In up to 3 MeV excitation energy have been performed in a practically complete basis which includes [qp × 1ph] configurations up to 4 MeV, and [qp × 2ph] configurations up to 6 MeV, not violating the Pauli principle. Coefficients C, S, and D in the equation (1), and energy eigenvalues are obtained by diagonalizing the model Hamiltonian on the set of wave functions (1). The index ν is used to distinguish between different excited states with the same J^π. For more details of the QPM application to the excited states description in odd-mass nuclei, refer to [8-10].

An advantage of the QPM calculations in comparison with shell-model ones is that within the QPM it is possible to carry out the calculations employing rather complete single-particle bases. In the present calculations single-particle levels of the Wood-Saxon potential (for neutrons and protons) from $1s_{1/2}$ to narrow quasi-bound levels in the continuum are used. The summations in Eq. (1) are taken over all these levels j. Thus no additional effective charges are needed to reproduce the collectivity of the lowest vibrational states. The shell model calculations employ only several-single particle levels near the Fermi surface and realistic calculations for medium-heavy nuclei with open shell, like ^{119}In, become very complex. In fig.2 the spectrum from the QPM calculations is shown, compared with experimental and NDS [7] data.

Calculations of the cross sections for the ^{122}Sn$(\bar{p},\alpha)^{119}$In reaction using the internal microscopic structure of any excited state of ^{119}In from QPM are in progress.

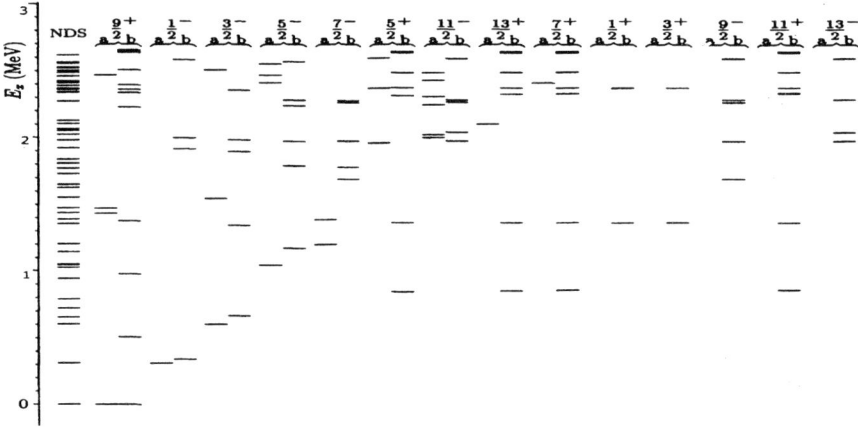

FIGURE 2. Energy spectrum for ^{119}In: the first column shows NDS [7] levels; in columns **a** are shown the levels observed in the $(\bar{p},\alpha)$ reaction, and in columns **b** the states from QPM calculations.

REFERENCES

1. Guazzoni, P., et al., *Phys. Rev.* **C49**, 2784-2787 (1994).
2. Guazzoni, P., et al., *Zeit. Phys.* **A356**, 381-391 (1997).
3. Guazzoni, P., et al., *Eur. Phys. J.* **A1**, 365-378 (1998).
4. Gu, J.N., et al., *Phys. Rev.* **C55**, 2395-2406 (1997).
5. Zanotti, E., et al., *Nucl. Instr. and Meth.* **A310**, 706-707 (1991).
6. Igarashi, M., *Computer Code TWOFNR* (1977), unpublished.
7. Ohya, S., and Kitao, K., *Nuclear Data Sheets* **89**, 345-480 (2000)
8. Vdovin, A.I., et al., *Sov. J. Part. Nucl.* **16**, 105-120 (1985)
9. Gales, S., et al. *Phys. Rep.* **166**, 125-193 (1988)
10. Bryssinck, J., et al., *Phys. Rev.* **C62**, 014309_1- 12 (2000)

Measurement of analyzing power for $pp \to pp\pi^0$ reaction at 392 MeV

Y. Maeda, M. Segawa, H.P. Yoshida, M. Nomachi[a], Y. Shimbara[a],
Y. Sugaya[a], K. Yasuda[b], K. Tamura[c], T. Ishida[d] and T. Yagita[d]

Research Center for Nuclear Physics, Osaka University, Ibaraki, Osaka 567-0047, Japan
[a] *Department of Physics, Osaka University, Toyonaka, Osaka 560-0043, Japan*
[b] *The Wakasa Wan Energy Research Center, Fukui 914-0192, Japan*
[c] *Physics Division, Fukui Medical University, Fukui 910-1193, Japan*
[d] *Department of Physics, Kyushu University, Fukuoka 812-8581, Japan*

Abstract. The angular dependence of analyzing power for $pp \to pp\pi^0$ reaction was measured at 392 MeV using a polarized proton beam and liquid hydrogen target. It is demonstrated that meson exchange model fails to reproduce observed behavior of the analyzing power.

Pion-production near the threshold gives us important informations about the low-energy strong-interaction physics.

Recently several experimental studies of $NN \to NN\pi$ reaction near the pion-production threshold have been performed. The total cross section of $pp \to pp\pi^0$ reaction was measured at IUCF [1] very precisely. It has been pointed out that large contribution of the short-range s-wave pion-production amplitude Ss, in which final protons are in the S state, is necessary to reproduce this data. The importances of heavy-meson exchange and off-shell property of πN interaction were indicated [2–4]. The theoretical investigations based on the chiral perturbation theory were also performed [5] and give that the sign of s-wave pion-production amplitude is opposite to the sign predicted in Ref. [2–4]. Therefore the systematic comprehension for s-wave pion-production mechanisms is not established yet.

All of mentioned theoretical models considered lowest partial wave amplitude Ss and discussed only the total cross section of π^0 production. Perspective theoretical approches to pion-production mechanisms were proposed by RCNP group [6] and Jülich group [7] in the meson exchange picture. These groups study relevant pion production channels ($pp \to pp\pi^0$, $pp \to pn\pi^+$, $pp \to d\pi^+$ and $pn \to pp\pi^-$) in consistent way. The higher partial waves were taken into account as well as s-wave pion-production mechanisms. These models enable one to predict the differential cross section and polarization observables of pion production above the threshold and discuss about the relative sign and strength of partial wave amplitude. The RCNP model works well from near-threshold to the Δ energy region for all pion production reaction.

Recently the integrated polarization observables of $\vec{p}\vec{p} \to pp\pi^0$ reaction were also measured at IUCF using the polarized beam and polarized target between 325 and 400 MeV [8](Fig.1). These observables were integrated by the emitted angle and the pion energy. The integrated analyzing power Ay is expressed by the product of p-wave amplitude Pp and s-wave amplitude Ps in which final protons are in the P-state. The calculation of RCNP model predicts significantly smaller value of Ay than the experimental one, typically by a factor of 4 [6]. This result suggested that large Ps amplitude is necessary to explain these data. One may conclude that origin of strong s-wave pion-production amplitude corresponding to Ss and Ps is not yet clear.

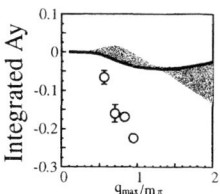

FIGURE 1. Integrated analyzing power. q_{max} is maximum momentum of pion in C.M. system. The data is taken from Ref. [8]. The line shows model calculation [6].

The global structure of angular dependence of Ay is given by

$$Ay \frac{d\sigma}{d\Omega_\pi} = \sqrt{2} Ps \times Pp\, P_{11}(\cos\theta_\pi)\sin(\delta) - \frac{\sqrt{5}}{10} Ss \times Sd\, P_{21}(\cos\theta_\pi)\sin(\delta'), \qquad (1)$$

where the $P_{ij}(x)$ is the associated Legendre function, $d\sigma/d\Omega_\pi$ is the differential cross section of pion, δ is the phase shift of initial nucleons state and θ_π is the scattered angle of pion in center of mass system. From this expression one sees that conribution of the Ss amplitude is vanished in integrated Ay, but in nonitegrated Ay (1), the term of P_{21} gives the structure which changes sign at $\theta_\pi = 90°$ and it's magnitude is determined by the Ss and Sd amplitude. On the other hand, P_{11} has symmetric structure and it's magnitude is proportion to the Ps and Pp amplitudes. Therefore we can deduced the relative strength of these partial wave amplitudes from experimental data on angular dependence of Ay.

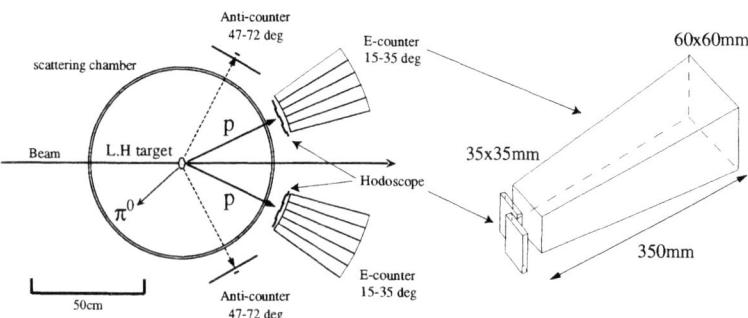

FIGURE 2. Experimental setup. The shadow region shows Liquid Hydrogen target. Solid arrows show detected protons, whereas dotted arrows show recoil protons from pp-elastic scattering. The schematic view of hodoscope and E-counter are also shown on the right hand side.

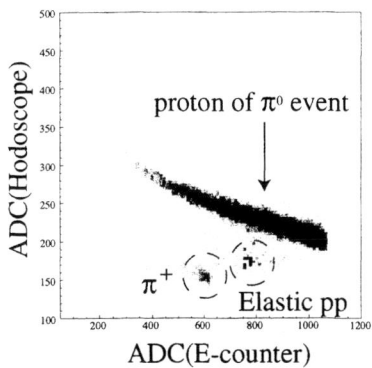

FIGURE 3. Particle identification.

For above purpose, we measured the $pp \to pp\pi^0$ reaction using the 392 MeV polarized proton beam from Ring Cyclotron at RCNP, Osaka, University. The view of the experimental setup are shown in Fig.2. Detecting two outgoing particles, four kinematical variables (θ_1, θ_2, E_1, E_2) are measured; $\phi_1, \phi_2 = 0$ are assumed on the basis of coplanarity. The number of measured variables is enough to determine the kinematics in final state. The energy and momentum of the pion can be calculated from four-momentum of the two protons.

The Liquid hydrogen target was used. The target system was developed by the Kyushu University group [9]. The target thickness was 8.55mm and the container windows were made of 12.5μm thick aramid foils. The energy of the proton was measured by plastic scintillators (E-counter) which can stop protons up to 250 MeV kinetic energy. The direction of out-going particle are determined by the plastic scintillator Hodoscopes which were putted in front of the E-counter. The energy resolution of E-counter is better than 3%(FWHM). The angular resolution of the hodoscope is 2°. Due to a nuclear reaction between an incident proton and a nucleus in the scintillator, there are events which do not deposite the full energy at the E-counter. For the mesurement of the proton with 100-200MeV kinetic energy, the efficiency of full-energy deposit is 85-60% obtained from the measurement of the proton-proton elastic scattering. Anti-coincidence counters were set up to reduce accidental coincidence events from the pp elastic scattering by detecting a recoil proton. The beam polarization was monitored by detecting the elastic proton-proton scattering from the liquid hydrogen target. The value of -0.3571(SAID) is used for analyzing power of the pp scattering. During the experiment, the beam polarization was 65~75%.

The typical particle identification spectrum obtained from the ADC of the E-counter and the hodoscope is shown in Fig.3. The proton of π^0 event is identified and the background events, which are the elastic pp and π^+ events, are also shown.

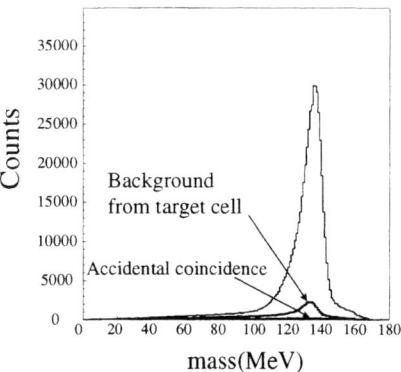

FIGURE 4. The missing-mass spectrum of $pp \to ppX^0$

These background events are cutted in this spectrum.

The missing-mass spectrum is shown in Fig.4. A clear peak at around π^0 mass is observed. The background event due to the accidental coincidence and the background from the target cell are also shown. The number of events from a target cell is estimated from the measurement of hydrogen gas target. Finally the missing mass spectra with 135.6MeV (MEAN) and 9.4MeV (FWHM) are obtained after subtracting background event. The width of the missing-mass spectrum is determined basically by the energy resolution. The tail distribution comes from the event which dose not deposite full energy in E-counter. The event with the missing-mass 100-150 MeV are selected. The amount of the events which do not deposite full energy is estimated 5% at the selected region. The uncertainty of Ay which comes from this amount is less than 2%.

In the center of mass system, three body final state kinematics of $pp \to pp\pi^0$ reaction is characterized by three variables: the pion angle(θ_π), relative momentum of final two protons(k) and it's direction(θ_k). In this measurement, the range for these variables are $\theta_\pi=0°$-$180°$, $\theta_k=40°$-$140°$ and $k=152$-230MeV/c, which cover 0.4% of phase space for π^0 production.

In the data analysis, the kinematical variables are reconstructed for event by event. The yields are obtained at θ_π for each polarization direction of the incident beam. The bin-size for the pion angle is $20°$. The analyzing power is obtained by

$$Ay = \frac{(Y_\pi)\uparrow - (Y_\pi)\downarrow}{P\downarrow(Y_\pi)\uparrow + P\uparrow(Y_\pi)\downarrow},$$

where the arrows indicate the polarization of proton beam, Y_π and P show the normalized pion yields and beam polarization, respectively.

The experimental results of angular dependence of Ay are shown in Fig.5. These data are averaged by the $\theta_{pp}=40°$-$140°$ and integrated by the k with the following three regions: (a)152-178 MeV/c, (b)178-204 MeV/c and (c)204-230 MeV/c. The error bars show the statistical errors.

In Fig.5, the calculated results of RCNP model [6] are also shown. The angular dependence of the experimental Ay has more symmetrical behavior than the calculated result and the absolute value of experimental Ay is larger than the calculated one. The experimental data suggests that the term of the P_{11} function dominates over the term of the P_{21} function and the strength of the $SsSd$ is much smaller than the one of the $PpPs$. Therefore this model has problems with the description of the relative strength of partial wave. Our data of Ay indicates that the large strength of Ps amplitude is necessary and this indication is consistent with IUCF data.

The partial wave amplitude is calculated by the overlap integral of the wave function of pion and nucleons, and Yukawa function which is characterized by the mass of the exchanged meson. The wave function of S-state, in which final nucleons are in the $^1S_0$ state, has a large strength at the short-range region($\sim$0.5fm). Therefore the heavy-meson exchange plays an important rule in the Ss amplitude. On the other hand, the wave function of P-state is pushed out from the short-range region due to the centrifugal barrier and does not have a strength at the short-range region. Therefore the long-range interaction such as the pion-exchange mechanisms is required to increase the Ps amplitude in the model.

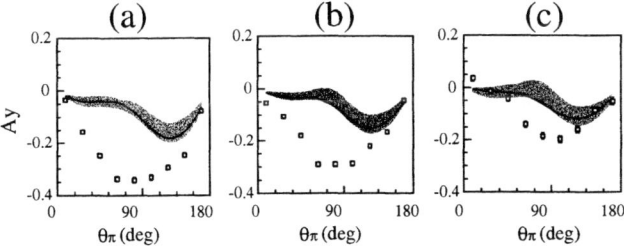

FIGURE 5. The angular dependence of Ay of the $pp \to pp\pi^0$ reaction. Solid line shows calculated result with Paris potential. The shadow region shows uncertainty coming from the potential models which are Reid, Argonne and Bonn potential.

REFERENCES

1. H.O. Meyer et al., Phys. Rev. Lett. **65**, 2846 (1990); Nucl. Phys. **A539**, 633 (1992).
2. T.-S.H. Lee and D. Riska, Phys. Rev. Lett. **70**, 2237 (1993).
3. C.J. Horowitz, H.O. Meyer and D.K. Griegel, Phys. Rev. **C49**, 1337 (1994).
4. E. Hernández and E. Oset, Phys. Lett. **B350**, 158 (1995).
5. T. Sato et al., Phys. Rev. **C56**, 1246 (1997) and references therein.
6. K. Tamura, Y. Maeda and N. Matsuoka, Nucl. Phys. **A663-664** 457c-460c(2000).
7. C. Hanhart, J. Haidenbauer, O. Krehl and J. Speth Phys. Lett. **B444** 25-31(1998).
8. H.O. Meyer et al., Phys. Rev. Lett. **81**, 3096 (1998).
9. K. Sagara et al., in RCNP Annual Report 1995,p.158.

Energy Dependence of ^{12,13}C$(p,\pi^-)^{13,14}$O$_{g.s.}$ Reactions in the Δ_{1232} Resonance Region

J. Kamiya[a], K. Hatanaka[a], T. Noro[a], T. Wakasa[a], Y. Maeda[a],
H.P. Yoshida[a], E. Obayashi[a], D. Hirooka[a], K. Tamura[b], H. Sakai[c],
A. Tamii[c], K. Yako[c], Y. Maeda[c] and H. Okamura[d]

[a] *Research Center for Nuclear Physics, Osaka University, Ibaraki, Osaka 567-0047, Japan*
[b] *Department of Physics, Fukui Medical University, Fukui 910-1193, Japan*
[c] *Department of Physics, University of Tokyo, Bunkyo, Tokyo 113-0033, Japan*
[d] *Department of Physics, Saitama University, Saitama 338-8570, Japan*

Abstract. Theoretical investigations and measurements have been performed to reproduce the experimental data for the ^{12,13}C(p, π^-) reactions near threshold energies and to investigate the role of the delta resonance in the (p, π^-) reactions to ground states. In order to clarify the reaction mechanism, it is inevitable to study the energy dependence of the cross section and the asymmetry. In the present paper, angular distributions of cross sections and analyzing powers were measured for ^{12,13}C(p, $\pi^\pm$) reactions leading to ground states of the residual nuclei at incident energies of 250, 300 and 350 MeV. Experimental results are compared with the full-range DWBA calculations.

INTRODUCTION

Proton induced π production on nuclei, for instance (p,π) reactions, involves a large momentum transfer, because a large fraction of the incident energy is used to create the pion mass. Thus, (p,π) reactions have been studied as a potentially valuable probe for high momentum components of the nuclear wave functions. The mechanism of the (p,π) reactions is considered to be a two-nucleon process(NN $\rightarrow$ NNπ), which means the incident proton interacts only with one of the nucleons in the target nucleus [1]. In contrast to the situation for (p, π^+) reactions, only a single two-nucleon process, involving interaction with a target neutron, can contribute to (p, π^-) reactions. When the configurations of the initial and final [2p (protons) - 1h (neutron)] states connected by (p, π^-) transitions are known, the shell model orbit of the struck neutron is uniquely determined. Because of the simplicity and the striking selectivity for high spin states, (p, π^-) reactions have been used as a spectroscopic tool mainly in the threshold region in the 1980's [2–5]. However, experimental results were explained only qualitatively. Recently, DWBA calculations within the two-nucleon model succeeded in explaining the overall features of the angular distributions of differential cross sections and analyzing powers for stretched state transitions [6–8].

Transitions to ground states also show the dominance of the two nucleon process in (p, π^-) reactions in the threshold region, although experimental data are rather limited. The cross sections on ^{14}C targets are twice as large as those on ^{13}C, which corresponds to the neutron occupancy number in the valence $p_{\frac{1}{2}}$ shell [9]. Furthermore, striking sign change was observed for the angular distributions of analyzing powers. For the case of ^{13,14}C, the analyzing powers are positive and exhibit similar angular distributions, while it has large negative values for ^{12}C. Although these experimental results strongly suggest the dominance of the two-nucleon process, Vigdor et al. pointed out that semi-classical considerations lead to opposite signs of the analyzing powers [10]. Calculations were performed by Kume et al. [11] on the two-nucleon pion-production model which was successfully applied to stretched state transitions near threshold energies. The analyzing powers were found to be greatly influenced by pion distortion effects and to exhibit a clear isotope dependence. They could not reproduce the experimental data by their calculations, although they obtained stable numerical results rather independent of the choice of the pion-nucleus optical potentials and the other parameters. Unlike for transitions to high-spin states, the large momentum and angular momentum transfers are not well matched in the ground state transitions in carbon isotopes. Core polarization effects have been studied but could not reduce the discrepancies between theory and experiments [12]. In order to clarify the reaction mechanism, it was pointed that the energy dependence of the cross section and the asymmetry for these ground state transitions should be studied. In the near-threshold region, the two-nucleon process is complicated because of the interference between the s- and p-wave rescattering contribution. At higher energies, the s-wave rescattering contribution is expected to become less important. They compared the calculated energy dependence of the differential cross sections with experimental data for ^{13}C$(p, \pi^-)^{14}$O$_{g.s.}$ reaction. The DWBA results predicted a resonance structure around 300 MeV [13]. The available experimental data points are too limited to clearify discrepancies between experimental and theoretical results. There have been neither experimental data around 300 MeV nor complete sets of angular distributions of cross sections and analyzing powers which are very important to study pion distortions.

The present experiment was motivated by these theoretical backgrounds [13] and the lack of data. We measured a wide range of angular distributions of cross sections and analyzing powers for the (p, π^-) reactions on ^{12,13}C leading to oxygen ground states at 250, 300 and 350 MeV. For comparison, we also measured those of the (p, π^+) reactions under the same conditions. Experimental results are compared with full-range DWBA calculations.

RESULTS AND DISCUSSIONS

Measurements were performed at the RCNP cyclotron facility. Polarized protons from the atomic beam type polarized ion source [14] were accelerated by the AVF and the Ring cyclotrons. Pions were momentum analyzed by the Large Acceptance Spectrometer and detected by the focal plane counter system. Because cross sections of (p, π^-) reactions are very small(several to several tens of nb/sr), a second level trigger technique was employed to improve the signal/noise ratio. Signals from the first x-plane of VDC were utilized as second level triggers. Pions were identified by the time of flight method. The energy resolution was bout 400 keV FWHM.

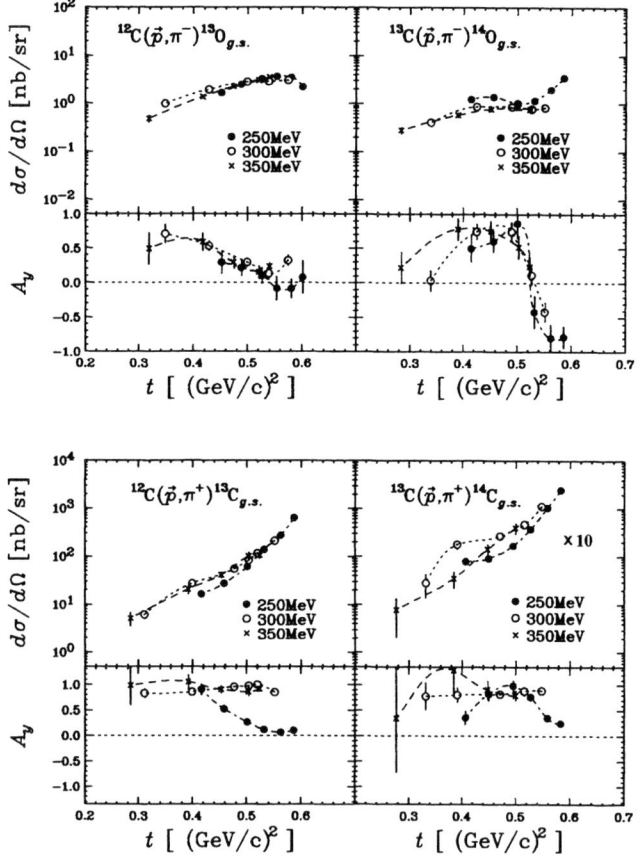

FIGURE 1. t-distributions of differential cross sections and analyzing powers for ^{12,13}C(p, $\pi^{\pm}$) reactions leading to ground states of the residual nuclei at incident energies of 250, 300 and 350 MeV. The upper panel shows results for (p, π^-) reactions, the lower panel those for (p, π^+) reactions. Cross sections for the ^{13}C(p, π^+) reaction are multiplied by a factor 10.

Distributions of measured differential cross sections and analyzing powers are shown in Fig. 1. They are plotted as function of the relativistically invariant Mandelstam variable t, which is defined as the square of the four-momentum transfer. The advantage of these t-plots is that the nuclear structure effects are fixed in first order and distributions reflect the reaction mechanism only. Curves in the figure are guides to eyes. From the figure, it can be seen that the overall features for the (p, π^-) reactions do not change in the present energy region. On the other hand, analyzing powers for the (p, π^+) reactions are very different at 250 MeV from those at higher energies. It is very interesting that analyzing powers of the (p, π^+) reactions have the maximum values of unity at all angles at 300 and 350 MeV.

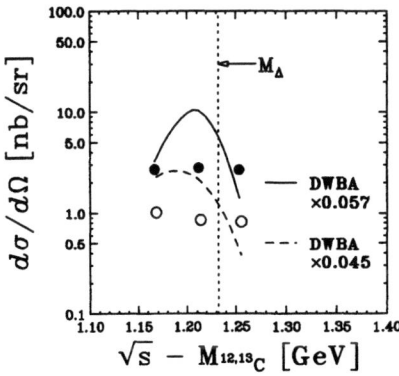

FIGURE 2. The energy dependence of the cross sections at a constant value of $t=0.5$ (GeV/c)^2 plotted vs the center of mass energies ($\sqrt{s}$-m$_{12,13C}$). Filled and open circles are results on ^{12}C and ^{13}C, respectively. The solid curve shows calculations for the ^{12}C(p, π^-) reaction which is multiplied by 0.057. The dashed curve shows calculations for the ^{13}C(p, π^-) reaction which is multiplied by 0.045.

Figure 2 shows the energy dependence of the cross sections at a constant value of $t=0.5$ (GeV/c)^2 plotted vs the center of mass energies ($\sqrt{s}$-m$_{12,13C}$). Curves in the figure show DWBA calculations by Nose-Togawa [15]. Distorted waves for incident protons were generated with optical potentials which describe elastic scattering. The MSU optical potential was adopted for pions. For nuclear wave functions of ^{12,13}C, 1p-shell wave functions of Cohen-Kurath were employed. In the calculations, the reaction was assumed to proceed dominantly through the delta resonance. The s-wave rescattering process was neglected and only p-wave rescattering processes were included in the calculations, since the present incident proton energies are well above the thresholds. Contrary to DWBA results, the experimental cross sections seem to have no peaks around the Δ_{1232} resonance. For positive pion production process p+p $\to$ p+n+π^+, the dominant channel is pp($^1D_2$) $\to$ NΔ($^5S_2$) $\to$ NN($^3S_1$) + π^+(p-wave), where s-wave intermediate NΔ and final NN states are involved. For negative pion production, the short-range nature of the (p,π^-) process also favors the relative S-state for final two protons ($^1S_0$). If we assume the dominant final ($^1S_0$) channel and also the pion orbital angular momentum of 0 or 1, only $N\Delta(^3P_0)$ intermediate state is allowed. The phase shift analysis of the $\pi^- pp(^1S_0)$ $\rightarrow$ pn angular distribution extracted from the ^{3}He(π^-,pn)n data suggests that the reaction proceeds via $\pi^- pp(^1S_0) \rightarrow$ pn ($^3D_1$, T=0), where the intermediate $N\Delta$ state is forbidden [16]. The present results confirm that the non-resonant process dominates the (p, π^-) reaction leading to ground states.

In summary, we have measured angular distributions of cross sections and analyzing powers for ^{12,13}C(p, $\pi^\pm$) reactions leading to ground states in the residual nuclei at incident energies of 250, 300 and 350 MeV. This is the first measurement covering a large angular range of cross sections and analyzing powers in the Δ_{1232} region. Experimental results were compared with the full-range DWBA calculations based on the two-nucleon model. The energy dependence of the cross sections, it was confirmed that the non-resonant process dominates the ^{12,13}C(p, π^-) reactions leading to ground states.

ACKNOWLEDGMENTS

We thank Professors H. Toki and K. Kume for helpful dicussions and Dr. N. Nose-Togawa for the results of her calculations which she made available to us along with many helpful comments. This experiment was performed under the Program No. E125 at RCNP.

REFERENCES

1. Vigdor, S.E. et al., *Nucl. Phys.* **A396**, 61c (1983).
2. Vigdor, S.E. et al., *Phys. Rev. Lett.* **49**, 1314 (1982).
3. Green, M.C. et al., *Phys. Rev. Lett.* **53**, 1893 (1984).
4. Cao, Z-J. et al., *Phys. Rev.* **C35**, 625 (1987).
5. Throwe, T.G. et al., *Phys. Rev.* **C35**, 1083 (1987).
6. Kume, K., *Nucl. Phys.* **A504**, 712 (1989).
7. Kume, K., *Nucl. Phys.* **A511**, 701 (1990).
8. Nose, N. et al., *Phys. Rev.* **C53**, 2324 (1996).
9. Jacobs, W.W. et al., *Phys. Rev. Lett.* **49**, 855 (1982).
10. Toki, H. and Kubo, K.-I., *Phys. Rev. Lett.* **54**, 1203 (1985).
    Vigdor, S.E. et al., *Phys. Rev. Lett.* **54**, 1204 (1985).
11. Kume, K. and Nose, N., *Nucl. Phys.* **A528**, 723 (1991).
12. Nose, N. and Kume, K., *Phys. Rev.* **C45**, 2879 (1992).
13. Nose-Togawa, N. et al. *Phys. Rev.* **C57**, 2502 (1998).
14. Hatanaka, K. et al. *Nucl. Instr. Meth.* **A348**, 575 (1997).
15. Nose-Togawa, N., *private communication*.
16. Hahn, H. et al. *Phys. Rev.* **C53**, 1074 (1996).

Effective charge anomaly in neutron-rich nuclei revealed from spin-polarized RI beam experiments

H. Ogawa[1], K. Sakai[1], H. Ueno[2], T. Suzuki[1], K. Asahi[1,2],
H. Miyoshi[1], M. Nagakura[1], K. Yogo[1], A. Goto[1], T. Suga[1],
T. Honda[1], N. Imai[3], Y.X.Watanabe[2], K. Yoneda[2], A. Yoshimi[2],
N. Fukuda[3], N. Aoi[3], Y. Kobayashi[2], W.-D. Schmidt-Ott[4],
G. Neyens[5], S. Teughels[5], A. Yoshida[2], T. Kubo[2], and M. Ishihara[6]

[1] *Department of Physics, Tokyo Institute of Technology, Oh-okayama 2-12-1, Meguro-ku, Tokyo 152-8551, Japan*
[2] *RIKEN, Hirosawa 2-1, Wako-shi, Saitama 351-0198, Japan*
[3] *Department of Physics, University of Tokyo, Hongo 7-3-1, Bunkyo-ku, Tokyo 113-0033, Japan*
[4] *Zweites Physikalisches Institut, Der Universität Göttingen, Bunsenstrasse 7-9, D-37073 Göttingen, Germany*
[5] *Instituut voor Kern- en Stralingsfysica, University of Leuven, Celestijnenlaan 200 D, B-3001 Leuven, Belgium*
[6] *RIKEN-BNL Research Center, Brookhaven National Laboratory, Upton, NY 11973, USA*

Abstract. The electric quadrupole moment of ^{17}B was measured by means of spin polarized fragment beams produced by the projectile fragmentation reaction. The experimental quadrupole moment of ^{17}B was very close to that of neutron p-shell closed nucleus ^{13}B. This result indicates that the neutron effective charge for ^{17}B is considerably smaller than the value commonly used in the sd-shell region.

INTRODUCTION

The nuclear spin polarization of fragments has been observed in the projectile fragmentation reaction at intermediate energy $E/A = 40 - 110$ MeV/u [1,2]. By using the projectile fragmentation-induced nuclear polarization technique, nuclear moments of light-mass unstable nuclei have been measured at the RIKEN Accelerator Research Facility during the last decade, incorporating with the β-ray detected nuclear magnetic resonance(β-NMR) method. The nuclear moments of B isotopes in the neutron-rich region have been measured systematically, and at present the magnetic moments μ of ^{14}B, ^{15}B [3] and ^{17}B [4] and the electric quadrupole moments Q of ^{14}B and ^{15}B [5] were determined.

In this paper, we report on measurements of the electric quadrupole moments of B isotopes. It is plausible that the quadrupole moment becomes larger, as the number of extra neutrons added to neutron p-shell closed nucleus ^{13}B increases, because valence neutrons attract a ^{13}B core and induce deformation of a core. However the experimental data of quadrupole moment for ^{15}B, $|Q(^{15}B)| = 38.01$ mb, was found very close to that of $|Q(^{13}B)| = 36.93$ mb [6]. This suggests that the influence of two extra neutrons on the core is very small, or in other words, the neutron effective charge e_n is quenched in this nucleus. In order to clarify the behavior of the neutron effective charge also appears in the more neutron-rich nucleus ^{17}B($I^\pi = 3/2^-$, $T_{1/2} = $ 5.08 ms) which has four valence neutrons in the sd shell, the measurement of the quadrupole moment for ^{17}B was performed.

EXPERIMENT

A beam of spin-polarized ^{17}B was produced using the fragment separator RIPS [7] through the fragmentation of ^{22}Ne at an energy of 110 MeV/nucleon in a 1.07 g/cm^2 thick ^{93}Nb target by selecting a finite emission angle and a momentum of fragments. The polarized fragments were transported to the β-NMR apparatus, as shown in Fig.1, located at the final focus of RIPS, and were implanted in a Mg single-crystal stopper which has an electric field gradient q. After the implantation, an oscillating radio frequency field B_1 was applied perpendicular to the static magnetic field B_0. Then the β rays emitted from ^{17}B fragments in the stopper were detected with

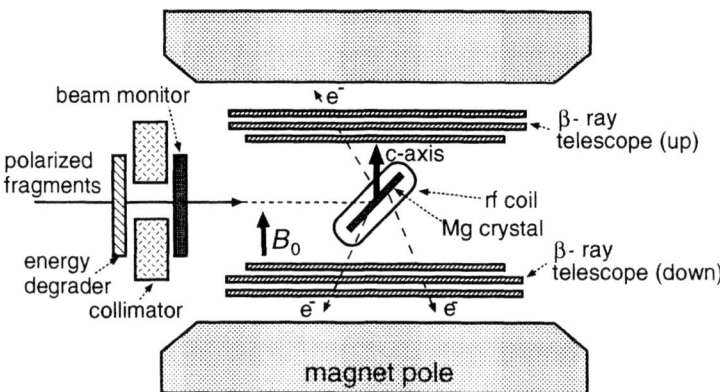

FIGURE 1. Schematic drawing of the β-NMR apparatus. Polarized fragments are implanted in a Mg single-crystal stopper to which a static magnetic field B_0 is applied. The direction of the crystal c-axis is chosen parallel or perpendicular to the B_0 field. An oscillating magnetic field is applied with an rf coil around the stopper. The β rays are counted with β-ray telescopes in order to detect the spin-flip transition through change in the β-ray up/down asymmetry.

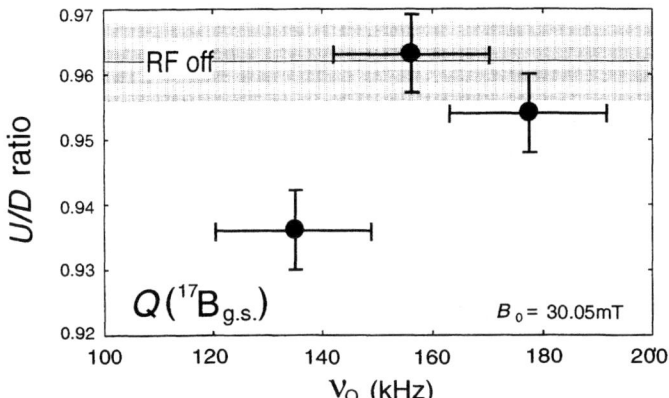

FIGURE 2. β-NMR spectrum obtained in the measurement of quadrupole moment for ^{17}B implanted in a Mg single crystal with the c-axis perpendicular to the magnetic holding field B_0. The up/down ratio of β-ray intensities is plotted as a function of $\nu_Q (\equiv eqQ/h)$ of the B_1 field.

two sets of plastic scintillator telescope which were located above and below the stopper. When the frequency of the oscillating field is swept to meet the transition frequency described below, the nuclear magnetic resonance is observed as the change in the up/down ratio of β-ray counting rates. The transition frequency between magnetic substates m and $m+1$ of the nuclear spin I, with the magnetic moment $g\mu_N I$, is given by the first-order perturbation theory as

$$\nu_{m,m+1} = \frac{g\mu_N B_0}{h} - \frac{3eqQ(3\cos^2\beta - 1)(2m+1)}{8I(2I-1)h}, \qquad (1)$$

where β denotes the angle of the principal axis (the c-axis in the case of Mg) of the electric field gradient with respect to the magnetic holding field B_0.

In the early stage of this experiment, the measurement was performed with the Mg c-axis aligned at an angle of $\beta = 90°$ so that the region of ν_Q twice as wide as that with $\beta = 0°$ was covered. In Fig. 2 we show the NMR spectrum obtained with a broad-bin frequency sweep at an angle of $\beta = 90°$. The NMR dip was observed in the region of $\nu_Q = 121 - 149$ kHz. We have also performed measurements with finer ν_Q bin widths and the angle of the c-axis at $\beta = 0°$ for more precise measurement. The ν_Q value obtained in the present analysis is $|\frac{eqQ}{h}(^{17}\text{B in Mg})| = 138.1 \pm 4.7$ kHz. From this the quadrupole moment for ^{17}B is tentatively determined as $|Q(^{17}\text{B})| = 38.8 \pm 1.5$ mb by taking an electric field gradient q from Ref. [8]. The details of the experimental procedure and analysis will be presented in the forthcoming paper.

DISCUSSION

As was observed for the quadrupole moment of ^{15}B, the obtained quadrupole moment for ^{17}B is found to be again close to that for $N = 8$ shell-closed isotope ^{13}B. Based on the weak-coupling model, the ground state of ^{17}B consists of the ^{13}B core and four extra neutrons in the sd shell. In shell model, the quadrupole moment is expressed as $Q = e_p Q_p + e_n Q_n$, where e_p and Q_p (e_n and Q_n) are the effective charge and the matter quadrupole moment for proton(neutron), respectively. Taking the experimental quadrupole moment of ^{13}B, $|Q(^{13}B)| = 36.93$ mb [6], as proton part $e_p Q_p$ and the matter quadrupole moment of the neutron for ^{17}B deduced from shell model calculation [9,10], $Q_n = 60.5$ mb, the neutron effective charge e_n of ^{17}B is obtained as $e_n(^{17}B) = [Q(^{17}B) - e_p Q_p]/Q_n = 0.03$. This e_n value is substantially smaller than $e_n \approx 0.5$ used to describe the sd-shell nuclei near the β-decay stability region [11].

For further discussions, shell model calculations in the $0\hbar\omega$ model space with the effective interactions PSDMK and PSDWBT [9,10,12] based on the OXBASH code [10] were performed. In Fig.3, the experimental quadrupole moments for the odd-mass B isotopes are compared with the shell model predictions. The calculated quadrupole moments with effective charges $e_p = 1.3$ and $e_n = 0.5$ [11] increase rapidly with mass number A for $A > 13$, reflecting the increasing contribution of neutrons in the sd orbit. On the contrary, the experimental quadrupole moments stay almost constant against the mass number, and as a result the shell model calculations overestimate $Q(^{17}B)$ by 60%. The small experimental $Q(^{17}B)$ suggests

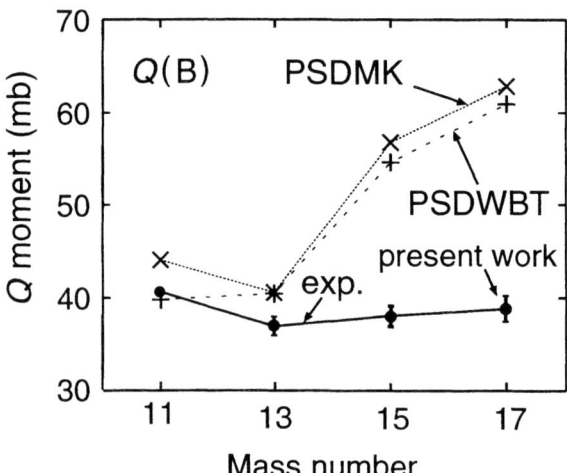

FIGURE 3. Comparison between the experimental(exp.) and theoretical quadrupole moments for odd-mass B isotopes. The experimental $Q(^{17}B)$ is considerably smaller than the shell model calculations using the effective interaction PSDMK and PSDWBT with $e_p = 1.3$ and $e_n = 0.5$.

that the core-polarization effect is hindered in this nucleus. This quenching feature of the E2 core-polarization effect has been also indicated in the previous results for $Q(^{14}B)$, $Q(^{15}B)$ [5], and $Q(^{18}N)$ [13]. It may be suggested that the core polarization induced by excess neutrons tends to diminish as the degree of neutron richness increases.

SUMMARY

The electric quadrupole moment of ^{17}B was measured by using a spin polarized radioactive isotope beam. From the measured quadrupole coupling constant for ^{17}B in Mg, $|Q(^{17}B)| = 38.8 \pm 1.5$mb is preliminary deduced. The obtained $Q(^{17}B)$ is substantially smaller than the shell model predictions assuming a standard value $e_n = 0.5$ for neutron effective charge normally employed in the sd-shell region. It is interesting that the experimental value of $Q(^{17}B)$ is found very close to that of $Q(^{13}B)$ and, together with the previous results for $Q(^{14}B)$, $Q(^{15}B)$, and $Q(^{18}N)$, reveals a tendency that the neutron effective charge becomes gradually quenched as the degree of neutron excess develops.

One of the authors, (H.O.), acknowledges support from the Japan Society for the Promotion of Science for Young Scientists. This work was partly supported by a Grant-in-Aid in Scientific Research of the Japanese Ministry of Education, Science and Culture.

REFERENCES

1. Asahi K. et al., *Phys. Lett.* **B 251**, 488 (1990).
2. Okuno H. et al., *Phys. Lett.* **B 335**, 29 (1994).
3. Okuno H. et al., *Phys. Lett.* **B 354**, 41 (1995).
4. Ueno H. et al., *Phys. Rev.* **C 53**, 2142 (1996).
5. Izumi H. et al., *Phys. Lett.* **B 366**, 51 (1996).
6. $Q(^{13}B)$ was deduced from the moment ratio $Q(^{13}B)/Q(^{12}B)$ [14] and the most recent value $Q(^{12}B)$ [8].
7. Kubo T. et al., *Nucl. Instrum. Methods Phys. Res.* **B 70**, 309 (1992).
8. Minamisono T. et al., *Phys. Rev. Lett.* **69**, 2058 (1992).
9. Warburton E.K. and Brown B.A., *Phys. Rev.* **C46**, 923 (1992).
10. Brown B.A., Etchegoyen A., and Rae W.D.M., *Computer code OXBASH*, MSU Cyclotron Laboratory Report No. 524 (1986).
11. Brown B.A. and Wildenthal B.H., *Ann. Rev. Nucl. Part. Sci.* **38**, 29 (1988).
12. Millener D.J. and Kurath D., *Nucl. Phys.* **A255**, 315 (1975).
13. Ogawa H. et al., *Phys. Lett.* **B451**, 11 (1999).
14. Haskell R.C. and Madansky L., *J. Phys. Soc. Japan. Suppl.* **34**, 167 (1973).

Role of Deuteron Internal Variables in the ^{3}He$(d,p)^4$He Reaction

T. Uesaka,[a] H. Sakai,[b] H. Okamura,[a] A. Tamii,[b] Y. Satou,[c]
N. Sakamoto,[c] T. Ohnishi,[c] T. Wakasa,[d] K. Itoh,[e] K. Sekiguchi,[b]
K. Yako,[b] K. Suda,[a] S. Sakoda,[b] J. Nishikawa,[a] M. Hatano,[b]
H. Kato,[b] Y. Maeda,[b] T. Saito,[b] N. Uchigashima,[b] T. Wakui,[c] and
S. Yamamoto[f]

[a] *Department of Physics, Saitama University, Saitama 338-8570, Japan*
[b] *Department of Physics, University of Tokyo, Hongo, Bunkyo, Tokyo 113-0033, Japan*
[c] *RIKEN (The Institute of Physical and Chemical Research), Saitama 351-0198, Japan*
[d] *Research Center for Nuclear Physics, Osaka University, Osaka 567-0047, Japan*
[e] *Tandem Center, Tsukuba University, Ibaraki 305-8577, Japan*
[f] *Department of Physics, Toho University, Funabashi, Chiba 274-8510, Japan*

Abstract. Polarization observables for the ^{3}He$(\vec{d},p)^4$He reaction have been measured at E_d =140, 200, and 270 MeV. This work aims to investigate the high momentum component of deuteron D-state wave function and the mechanism for the one-nucleon transfer reaction at intermediate energies. Data are compared with impulse approximation calculation based on the three-nucleon scattering amplitudes.

I INTRODUCTION

The high momentum component of deuteron wave function provides a unique opportunity to reveal the short-range behavior of nucleon-nucleon (NN) interaction [1,2]. Especially, the D-state wave function, which arises from the existence of the tensor component in NN interaction, manifests its importance in the short-range ($r < 1$ fm) or high-momentum ($k > 1$ fm^{-1}) region. The $^3\vec{\mathrm{He}}(\vec{d},p)^4$He reaction is expected to be an effective probe to the deuteron D-state wave function because of its strong spin-selectivity [3].

We have measured the vector (A_y) and tensor (A_{yy} and A_{xx}) analyzing powers at E_d= 140, 200, and 270 MeV. In this paper, we discuss how the deuteron structure affects the polarization observables, by comparing the experimental data with impulse approximation calculations.

II EXPERIMENTAL PROCEDURE

The experiment was carried out at RIKEN Accelerator Research Facility (RARF). A polarized deuteron beam was supplied by the high-intensity polarized ion source [4] and accelerated up to 140–270 MeV in the accelerator-complex consisting of AVF and Ring cyclotrons. The polarization of deuteron beam was measured with the beam-line polarimeter which is based on the $d+p$ elastic scattering [5,6].

A cryogenic ^{3}He gas target with a density of 6.6×10^{20} cm^{-3} was bombarded by the polarized deuteron beam. Entrance and exit windows of the target were 6 μm-thick Harvar foils. The scattered protons were momentum-analyzed in the magnetic spectrometer SMART [7] and detected at the focal plane. The detector system consisted of an eight-plane multi-wire drift chamber and three plastic scintillators.

III RESULTS AND DISCUSSIONS

Figure 1 shows data of vector and tensor analyzing powers at three energies as a function of a scattering angle in the center of mass system. The statistical uncertainties are less than 0.02.

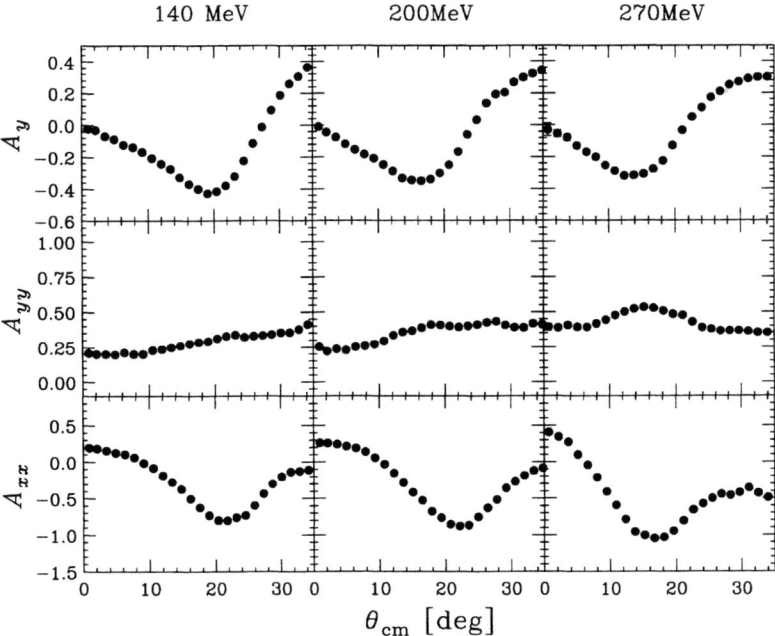

FIGURE 1. Vector (A_y) and tensor analyzing powers (A_{yy} and A_{xx}) for the ^{3}He$(d,p)^4$He reaction at E_d=140, 200, and 270 MeV.

Recent theoretical work by Bochum group revealed that impulse picture based on the three-nucleon amplitude is valid in describing the polarization observables for the reaction in the energy region considered [8]. Namely, the incident deuteron scatters off a nucleon in ^{3}He and the other two nucleons (2N) are left as spectators (see Fig. 2). As is usual in impulse approximation for nucleon scattering at forward angles, the scattering amplitude is calculated for a specific value of momentum of the spectator. However, unlike the usual cases, the value of k_{2N} affects the scattering amplitude considerably when a large momentum transfer is involved. In Ref. [8], it is assumed that $k_{2N} = k_{\tilde{d}}$ giving a large probability to form ^{4}He nucleus in the final state (referred as Model I).

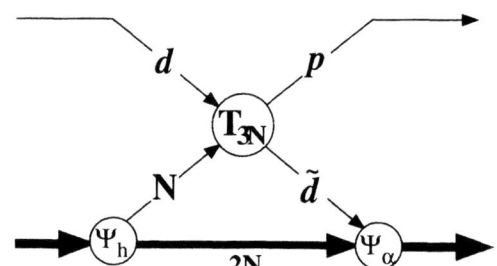

FIGURE 2. Graph of the 3N impulse approximation.

For further development of this model, it is necessary to investigate the relevance of the kinematics. We employ a different model, referred as Model II in the rest of this paper. Since a ^{4}He nucleus has a large binding energy of 28.3 MeV compared with that of ^{3}He (7.7 MeV), one can expect larger probability in finding high momentum component in ^{4}He than in ^{3}He. To put it more precisely, we calculate a typical internal momentum Q which is defined as $Q = \sqrt{2\mu B}$. Here μ is a reduced mass of the system and B is its binding energy. For ^{4}He and ^{3}He, they are obtained as $Q_\alpha = 230$ MeV/c and $Q_h = 98$ MeV/c, respectively. Here we neglect small binding energies of two-nucleon subsystems in both cases. To take into account this difference of Q properly, we choose k_{2N} to fulfill the condition $K_\alpha/Q_\alpha = K_h/Q_h$, where K_α and K_h are the internal momenta in ^{4}He and ^{3}He, respectively. In Table 1, internal momenta in ^{4}He (K_α) and ^{3}He (K_h) at $\theta = 0°$ calculated with Model I and II are shown.

TABLE 1. Internal momenta in ^{4}He (K_α) and ^{3}He (K_h) at $\theta = 0°$ calculated with Model I and II.

	Model I		Model II	
E_d	K_α	K_h	K_α	K_h
140 MeV	92 MeV/c	0 MeV/c	27 MeV/c	64 MeV/c
200 MeV	115 MeV/c	0 MeV/c	34 MeV/c	80 MeV/c
270 MeV	137 MeV/c	0 MeV/c	41 MeV/c	96 MeV/c

We employ one-nucleon exchange (ONE) model to calculate the three-nucleon amplitude in stead of the Faddeev amplitude in Ref. [8]. With the help of ONE model, deuteron internal variables, i.e., internal momentum k_{pn} and internal angle θ_{pn} are defined as

$$k_{pn} = k'_p - \frac{1}{2}k'_d, \tag{1}$$

$$\theta_{pn} = \cos^{-1}\left(\frac{k_{pn} \cdot k'_d}{k_{pn} k'_d}\right). \tag{2}$$

Here, k'_d and k'_p are momenta of the incident deuteron and the scattered proton in the three-nucleon subsystem, respectively. It should be noted that both k'_d and k'_p depend on the momentum of 2N cluster. ONE can also provide direct relation between tensor analyzing powers for the ^{3}He$(d,p)^4$He reaction and the deuteron wave function as

$$A_{yy} = \frac{-\sqrt{2}u(k_{pn})w(k_{pn}) + \frac{1}{2}w^2(k_{pn})}{u^2(k_{pn}) + w^2(k_{pn})}, \tag{3}$$

$$A_{xx} = \frac{-\sqrt{2}u(k_{pn})w(k_{pn}) + \frac{1}{2}w^2(k_{pn})}{u^2(k_{pn}) + w^2(k_{pn})} \left(1 - 3\sin^2\theta_{pn}\right), \tag{4}$$

where $u(k_{pn})$ and $w(k_{pn})$ are the deuteron radial wave functions for S- and D-states, respectively. Note that A_{yy} depends only on the internal momentum through $u(k_{pn})$ and $w(k_{pn})$. To examine the relevance of the internal angle clearly, a ratio of A_{xx} to A_{yy} is introduced because it is independent from the internal momentum and a function of the internal angle as $A_{xx}/A_{yy} = 1 - 3\sin^2\theta_{pn}$. Accordingly, we can discuss the k_{pn}- and θ_{pn} dependences separately, by using A_{yy} and A_{xx}/A_{yy}.

In Fig. 3, calculations made with Model I (broken lines) and II (solid lines) are compared with the experimental data. It is clearly seen that magnitudes of tensor analyzing powers A_{yy} are sensitive to the value of 2N momentum. Model II gives better description than Model I. On the other hand, A_{xx}/A_{yy} is almost dependent the value of 2N momentum. Both models (I and II) fail to reproduce the ratio in the whole angular range except for $\theta = 0°$ where A_{xx} is identical to A_{yy}. Dot-dash lines in the figure represent the results with Faddeev amplitude from Ref. [8]. The Faddeev calculations give better description than Model I and II, especially at lower energies.

IV CONCLUSION

We have measured the vector (A_y) and tensor (A_{yy} and A_{xx}) analyzing powers at E_d= 140, 200, and 270 MeV. The data are analyzed by using impulse approximation calculation based on three-nucleon amplitudes. We discuss the k_{pn}- and θ_{pn} dependences separately by using A_{yy} and A_{xx}/A_{yy}. It has been shown that the value of the momentum of "spectator" 2N cluster affects tensor analyzing power A_{yy} substantially, but not the ratio A_{xx}/A_{yy}.

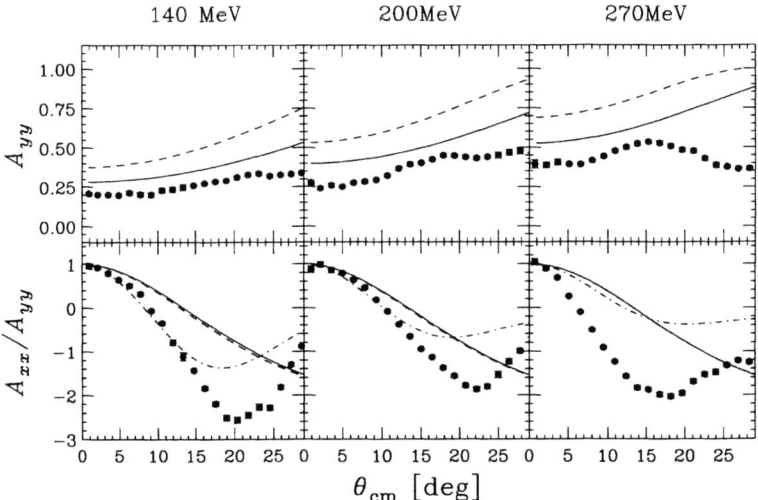

FIGURE 3. Vector (A_{yy}) and tensor analyzing powers (A_{yy} and A_{xx}) for the ^{3}He$(d,p)^4$He reaction at E_d=140, 200, and 270 MeV. Solid and broken lines represent one-nucleon calculation of Model I and II, respectively. Dot-dash lines in the lower panel are calculations with Faddeev amplitudes. Details of theoretical lines are explained in the text.

ACKNOWLEDGEMENTS

The authors thank Dr. H. Kamada and Bochum group for fruitful and valuable discussions. They also would like to express their gratitude to all the staff of RIKEN Accelerator Research Facility. One of the authors (T.U.) would like to acknowledge the financial support of the Special Researcher's Basic Science Program of RIKEN.

REFERENCES

1. A.P. Kobushkin, Phys. Lett. B **421** (1998) 53 and references therein.
2. L.S. Azhgirey et. al., Phys. Lett. B **391** (1997) 22 and references therein.
3. T. Uesaka et al., Phys. Lett. B **467** (1999) 199.
4. H. Okamura *et al.*, AIP Conf. Proc. **293** (1994) 84.
5. N. Sakamoto *et al.*, Phys. Lett. B **367** (1996) 60.
6. T. Uesaka *et al.*, in *14th International Spin Physics Symposium*, Osaka, Japan, October 2000.
7. T. Ichihara et al., Nucl. Phys. **569** (1994) 287c.
8. H. Kamada et al., Prog. Theor. Phys. **104** (2000) 703.

Tensor Analyzing Power A_{yy} in Deuteron Breakup on Hydrogen and Nuclei at Large Transverse Momenta of Proton

V.P.Ladygin*[1], L.S.Azhgirey*, S.V.Afanasiev*, V.V.Arkhipov*,
V.K.Bondarev*,†, Yu.T.Borzounov*, G.Filipov*,$, L.B.Golovanov*,
A.Yu.Isupov*, V.I.Ivanov*, A.A.Kartamyshev#, V.A.Kashirin*,
A.N.Khrenov*, V.I.Kolesnikov*, V.A.Kuznezov*, N.B.Ladygina*,
A.G.Litvinenko*, S.G.Reznikov*, P.A.Rukoyatkin*,
A.Yu.Semenov*, I.A.Semenova*, G.D.Stoletov*, A.P.Tzvinev*,
N.P.Yudin&, V.N.Zhmyrov* and L.S.Zolin*

*Joint Institute for Nuclear Researches, 141980 Dubna, Russia
†St.-Petersburg State University, 198350 St.-Petersburg, Russia
$Institute of Nuclear Research and Nuclear Eenergy, 1784 Sofia, Bulgaria
#Russian Scientific Center "Kurchatov Institute", 123182 Moscow, Russia
&Moscow State University, 117234 Moscow, Russia

Abstract. The data on the tensor analyzing power A_{yy} in the deuteron breakup reaction on hydrogen and nuclei (^9Be and ^{12}C) obtained at 9 and 4.5 GeV/c up to proton transverse momenta of ~ 900 MeV/c are presented. The data demonstrate an approximate independence on the A-value of the target, as well as on the deuteron initial momentum, when they are plotted versus variables α and p_T, that describe the internal motion of nucleons in the deuteron. The results suggest that a deuteron structure function at short distances may depend on more than one independent variable unlike an ordinary wave function.

INTRODUCTION

Recent measurements of the polarization observables in deuteron inclusive breakup on hydrogen and nuclei [1–7] have shown that their behaviour at large internal momenta can not be reproduced by the models using conventional deuteron wave functions (DWFs). The account of the mechanisms [8]

---

[1] E-mail address: ladygin@sunhe.jinr.ru

additional to Impulse Approximation (IA) or non-nucleonic degrees of freedom [9] improves the agreement with the data, however, some problems in the description still persist.

In this respect, the measurements of the polarization observables at an non zero emission angle is of greate interest [10]. On the one hand, the mechanism of the deuteron breakup reaction under such kinematical conditions is significantly simplified, because the diagrams of the rescattering and virtual pion production are suppressed. Hence, the breakup process can be treated as a interfering sum of diagrams describing direct fragmentation and hard scattering of the deuteron nucleon on the target nucleon. Such a simple approach called relativistic hard scattering (RHS) model [11] allows to describe the data on the differential cross section using the standard DWFs [12]. On the other hand, the observables obtained at non zero angle can be sensitive to the relativistic effects in the deuteron, namely, to the dependence of its internal structure at large internal momenta on more than one variable [13].

The results of our recent measurements of the tensor analyzing power A_{yy} in the breakup of 9 GeV/c deuterons with the emission of protons at angle of 85 mr [14] have shown deviation on the calculations performed within RHS using conventional DWFs [10]. In this report we present data on A_{yy} in breakup of 9 GeV/c and 4.5 GeV/c deuterons on hydrogen and nuclei with the transverse momenta of secondary protons up to ~ 900 MeV/c.

EXPERIMENT

The experiment has been performed using a tensorially polarized deuteron beam of the Synchrophasotron of JINR. The tensor polarization of the beam has been measured from the asymmetry of protons from the deuteron breakup on nuclear targets,
$d + A \to p + X$ at a zero angle and the momenta $p_p \sim 2/3 p_d$ [15]. The vector polarization of the beam has been measured from the asymmetry of quasi-elastic pp scattering on CH_2 target [16]. The tensor and vector polarizations, p_{zz} and p_z, were $p_{zz}^+ = 0.798 \pm 0.002(stat) \pm 0.040(sys)$, $p_{zz}^- = -0.803 \pm 0.002(stat) \pm 0.040(sys)$, $p_z^+ = 0.231 \pm 0.014(stat) \pm 0.012(sys)$ and $p_z^- = 0.242 \pm 0.014(stat) \pm 0.012(sys)$, respectively.

A slowly extracted deuteron beam with a typical intensity of $\sim 5 \cdot 10^8 \div 10^9$ $\vec{d}$/spill was directed onto a liquid hydrogen target of 30 cm in length or onto nuclear targets with varied length. The data at 9 GeV/c of the deuteron initial momentum were obtained at secondary proton emission angles of 85, 130 and 160 mr and proton momenta between 4.5 and 7.0 GeV/c on hydrogen and carbon, while the data at 4.5 GeV/c were obtained only at $\sim$80 mr on beryllium target. The separation of the protons and inelastically scattered deuterons was done by the measurements of their time-of-flight (TOF) over a base line of ~ 34 m. The residual background was completely eliminated by the requirement that particles are detected at least in two prompt TOF windows.

RESULTS

The data on A_{yy} obtained at 9 GeV/c of the deuteron initial momentum in this experiment are shown in Fig.1 by the triangles (160mr), squares (130mr) and circles (85mr) along with the data obtained at a zero emission angle (diamonds) [2] versus proton momentum in the laboratory. The empty and solid symbols correspond to carbon and hydrogen target, respectively.

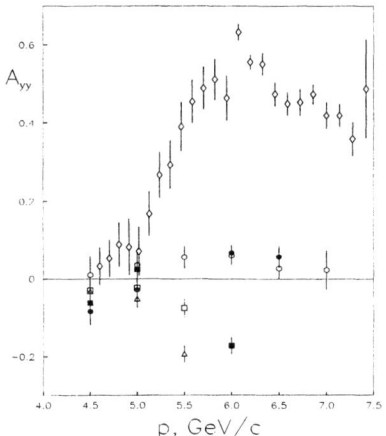

FIGURE 1. A_{yy} data obtained at 9 GeV/c and 85(circles), 130(squares) and 160 mr(triangles) of the proton emission angle. The solid and empty symbols correspond to the hydrogen and carbon targets, respectively. Diamonds are the data obtained at a zero angle [2].

One can see that the A_{yy} data from our experiment do not depend on the atomic number of the target, and hence multiple scattering processes play a minor role and the obtained information reflects the internal deuteron structure. A_{yy} has a positive sign at small emission angles of the proton, but at large angles it changes the sign. Even at relatively small momenta of the proton the significant dependence of A_{yy} on transverse momentum p_T is observed.

The observed features of the A_{yy} data suggest that an adequate description of the data may be achieved by using a deuteron structure function that depends at short distances on more than one variable. This can be, for example, longitudinal momentum fraction $\alpha = (E_p + P_p)/(E_d + P_d)$, where E_d and E_p are the energies, and p_d and p_p the momenta of the incoming deuteron and detected proton, respectively, and transverse momentum p_T.

The measurement of the tensor analyzing power A_{yy} at different initial energies, but at the same α and p_T could clarify the question of deuteron structure dependence on two internal variables. For that reason the measurement of A_{yy} at 4.5 GeV/c and ~ 80 mr on beryllium target have been performed.

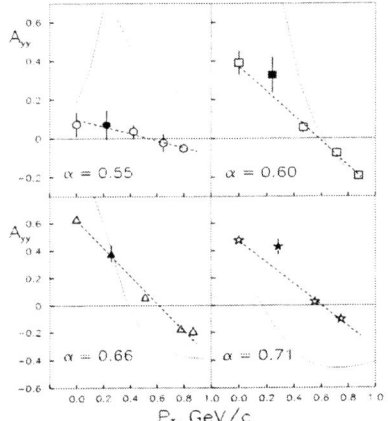

FIGURE 2. A_{yy} data obtained at 9 GeV/c and 4.5 GeV/c on carbon and beryllium are shown by the open and solid symbols, respectively. The solid curves are the calculations in the framework of RHS model using Paris DWF [17].

The results of measurements at the fixed longitudinal momenta fractions α are shown in Fig.2 as a function of the transverse momentum p_T. The open and solid symbols are the data obtained at 9 GeV/c and 4.5 GeV/c on carbon and beryllium targets, respectively. One can see, that both sets of the data do not contradict each other and in the firs approximation can be fitted as linear functions of p_T (dashed lines). A_{yy} has a positive value at small p_T and changes the sign at p_T higher than ~ 600 MeV/c independently on the longitudinal momentum fraction α. Such a behaviour contradicts the predictions of the RHS model [10] with Paris DWF [17]; those are shown in Fig.2 by the solid curves.

To summarize:
-New data on A_{yy} in deuteron breakup at 9.0 GeV/c and 4.5 GeV/c are obtained up to transverse momenta 900 MeV/c.
-The data demonstrate an approximate independence on the A-value of the target, as well as on the initial energy of the deuteron, when they are plotted at the fixed α as a function of transverse momentum p_T.
-The A_{yy} data demonstrate an approximate linear dependence upon the transverse momentum of the protons, p_T. Such a behaviour definitely contradicts the predictions of the RHS model with conventional DWFs.
-The observed features of the A_{yy} data suggest that a deuteron structure function at short distances, where relativistic effects are significant, may depend

on more than one variable [13].

Acknowledgement

Authors thank the Director of LHE A.I.Malakhov and vice-Director of LHE V.N.Penev for their permanent help. Authors would like to express their thanks to the accelerator crew and POLARIS staff. They are grateful to L.V.Budkin, V.P.Ershov, V.V.Fimushkin, A.S.Nikiforov, Yu.K.Pilipenko, V.G.Perevozchikov, E.V.Ryzhov, A.I.Shirokov and O.A.Titov for their assistance during the experiment. Authors thank A.N.Prokofiev and A.A.Zhdanov for the lending of the scintillation counters for the polarimeter. They thank V.A.Karmanov for helpful comments. One of authors (V.P.L.) is grateful to the Organizing Committee of SPIN2000 Conference for the financial support and kind hospitality, and to Russian Academy of Science for the travel grant.

REFERENCES

1. Perdrisat C.F. et al., *Phys. Rev.Lett.* **59**, 2840 (1987); Punjabi V. et al., *Phys. Rev.* **C39**, 608 (1989).
2. Ableev V.G. et al., *Pis'ma Zh.Eksp.Teor.Fiz.* **47**, 558 (1988) ; *JINR Rapid Comm.* **4[43]-90**, 5 (1990).
3. Aono T. et al., *Phys.Rev.Lett.* **74**, 4997 (1995).
4. Azhgirey L.S. et al., *Phys.Lett.* **B387**, 37 (1996).
5. Cheung E. et al., *Phys.Lett.* **B284**, 210 (1992).
6. Nomofilov A.A. et al., *Phys.Lett.* **B325**, 327 (1994).
7. Kuehn B. et al., *Phys.Lett.* **B334**, 298 (1994); Azhgirey L.S. et al., *JINR Rapid Comm.* **3[77]-96**, 23 (1996).
8. Lykasov G.I., *Part. and Nucl.* **24** , 140 (1993).
9. Kobushkin A.P., *J.Phys. G: Nucl.Part.Phys.* **19**, 1993 (1993); *Phys.Lett.* **B421**, 53 (1998).
10. Azhgirey L.S., and Yudin N.P., *Yad.Fiz.* **57**, 160 (1994).
11. Blankenbecler R. et al.,*Phys.Rev.* **D15**, 3321 (1977); *Phys.Rev.***C22**, 2433 (1980); Azhgirey L.S. et al., *Yad.Fiz.* **48**, 87 (1988).
12. Azhgirey L.S. et al.,*Nucl.Phys.* **A528**, 621 (1991).
13. Karmanov V.A. and Smirnov A.V.,*Nucl.Phys.* **A575**, 520 (1994); Carbonell J. and Karmanov V.A.,*Nucl.Phys.* **A581**, 625 (1994)
14. Afanasiev S.V. et al.,*Phys.Lett.* **B434**, 21 (1998); Ladygin V.P. et al., *Few Body Systems Suppl.***10**, 451 (1999); Azhgirey L.S. et al., *Phys.Atom.Nucl.* **62**, 1673 (1999).
15. Zolin L.S. et al.,*JINR Rapid Comm.* **2[88]-98**, 27 (1998).
16. Azhgirey L.S. et al.,*PTE* **1**, 51 (1997).
17. Lacombe M. et al.,*Phys.Lett.* **B101**, 139 (1981).

Polarization transfer for the ^{2}H(d,p)^{3}H reaction at $\theta = 0°$ at a very low energy

Tatsuya Katabuchi, Kohei Kudo, Kazuyuki Masuno,
Tomoyuki Iizuka, Yasuo Aoki, and Yoshihiro Tagishi

*Institute of Physics and Tandem Accelerator Center,
University of Tsukuba, Ibaraki 305-8577, Japan*

Abstract. The polarization transfer coefficient $K^{y'}_y$ has been first measured for the ^{2}H(d,p)^{3}H reaction at a scattering angle of 0° for incident deuteron energies below 90 keV. The polarization of the emitted protons from the reaction was measured with a proton polarimeter using the p - ^{28}Si elastic scattering. As a result, the value of $K^{y'}_y = 0.09 \pm 0.13$ was obtained.

INTRODUCTION

The fusion reactions ^{2}H(d,p)^{3}H and ^{2}H(d,n)^{3}He at the sub-Coulomb energy region have been studied for both the nuclear reaction mechanism of the four-nucleon system and application in fusion energy research. Some researchers have pointed out that the reactions initiated by two polarized deuterons in parallel to each other would be suppressed [1]. The suppression would reduce undesired neutrons created from the simultaneous reaction ^{2}H(d,n)^{3}He in D-^{3}He plasma, thus leading to "neutron-lean" fusion reactors [1], [2]. However, this suggestion has aroused controversy because, for the ratio of the polarized to the unpolarized cross sections, values derived from several models conflict [3]- [10]. Hence, for the controversy, spin-dependent experiments for these reactions are important. Researchers have measured the cross section, analyzing powers and ejectile polarization induced by unpolarized deuterons for the reactions [11]- [15]. However, any double-scattering experiments using polarized deuteron beam for these reactions at very low energies, which have been desired for more understanding, have not been previously performed. In the present work, we have first measured the polarization transfer coefficient $K^{y'}_y$ for the ^{2}H(d,p)^{3}H reaction at a scattering angle of 0° at $E_d \leq 90$ keV.

In any reaction of spin-1 projectile and spin-$\frac{1}{2}$ ejectile, the polarization of ejectile at a scattering angle of $\theta = 0°$ is expressed in simple form using the polarization transfer coefficients under the following particular condition [16]. If the spin quantization axis is oriented at an angle of 54.7° with respect to the incident beam direction, the polarization component of ejectile along the y' axis, $p_{y'}$, where the y' axis is perpendicular to the incident beam direction and in a plane given by the spin quantization axis and the incident beam direction, is written using the polarization transfer coefficient $K_y^{y'}$ as follows [16]:

$$p_{y'}(0°) = \sqrt{\frac{2}{3}} p_3 K_y^{y'}(0°), \qquad (1)$$

where p_3 is the vector polarization of the incident beam at the polarized ion source. According to Eq.1, we obtained $K_y^{y'}$ for the ^{2}H(d,p)^{3}H reaction at 0° angle from measurements of the polarization $p_{y'}$ of the emitted protons and the polarization p_3 of the incident deuterons.

EXPERIMENT

The experiments were performed with a 90-keV polarized deuteron beam from a Lamb-shift polarized ion source at the Tandem Accelerator Center of the University of Tsukuba (UTTAC) [17]. In order to measure the polarization of protons from the ^{2}H(d,p)^{3}H reaction, a proton polarimeter using the p - ^{28}Si elastic scattering was used. We should stress two characteristics in the present work that allow polarization measurements at the very low energy. First, the polarized ion source at UTTAC utilizes a spin filter for nuclear polarization, so we measured the incident beam polarization by a quench-ratio method [18] without using any nuclear reactions. We have tested the accuracy of this method with nuclear reactions and estimate it to be correct within 2 % at UTTAC. Second, a silicon solid-state detector (SSD) was used as an analyzing ^{28}Si-target of the proton polarimeter, so the background events were reduced by requiring coincidence between the target and detector SSD's of the polarimeter. As a result, the double-scattering events were separated from the background well and counted with a good signal-to-background ratio.

In the present work, deuterated-polyethylene $(CD_2)_n$ was used as a deuterium target. The $(CD_2)_n$ target was backed with foil of aluminum to keep the target from a rise of temperature during the bombardments, thus reducing rapid depletion of deuterium caused by ion irradiation. The targets were prepared by pouring a solution of $(CD_2)_n$ in 100°C-xylene onto horizontal aluminum foil and then evaporating the xylene from the foil at a temperature of 75°C. In the present work, the targets with a deuterium thickness of approximately 0.3 mg/cm^2 corresponding to a $(CD_2)_n$ thickness of 10 μm were prepared on 15-μm foil of aluminum. The mean range of 90-keV deuterons in $(CD_2)_n$ is 1.34 μm [19]. Hence, the incident deuteron beam was stopped in the $(CD_2)_n$ target. On the other hand, protons emitted from the ^{2}H(d,p)^{3}H reaction at 0° have a large energy of 3.44MeV, so they penetrated the $(CD)_2$ target and the aluminum backing toward the proton polarimeter. The mean reaction energy in the $(CD_2)_n$ target at an incident deuteron energy of 90keV was estimated to be 68 keV from the reaction cross sections [11] and the stopping powers of $(CD_2)_n$ for deuterons [19].

The polarization of the ^{2}H(d,p)^{3}H protons emitted at 0° through the $(CD)_2$ target and the aluminum backing was measured by the polarimeter using the p - ^{28}Si elastic scattering. The polarimeter consisted of two silicon solid-state detectors. The first detector with an active area of 28×28 mm^2 and a thickness of 450 μm (HAMAMATSU S5377-03) was placed at a scattering angle of 0° as a silicon target to analyze the polarization of the protons. The accepted angle of the detector was ±6.5° around 0° angle. The p - ^{28}Si elastic scattering has a large analyzing power near 115° lab. angle below $E_p = 3$ MeV [20]. For the detection of protons scattered at the first detector, the second detector with an active area of 48×48 mm^2 and a thickness of 300 μm (HAMAMATSU S4276) was placed at 115° lab. angle with respect to the first detector. The analyzing powers for scattering energies ranging from 2 MeV to 3 MeV between 95° and 135° lab. angles were used in this polarimeter. The analyzing power A_y of the polarimeter was measured with 3-MeV polarized protons from a tandem accelerator (Pelletron 12UD) at UTTAC. The measured analyzing power was -0.44 ± 0.03. The detection efficiency of the polarimeter was 9.7×10^{-6}.

The polarization transfer coefficient $K_y^{y'}$ was obtained by measuring each cross section of the incident deuteron beams with the $m_I = +1$ and $m_I = -1$ deuteron magnetic substates from the polarized ion source for the spin-quantization axis oriented at an angle of 54.7° to the beam direction. The spin-quantization axis was controlled by a Wien-filter system of the polarized ion source. The magnetic substate sequence of the deuteron beam was changed every 5000×10^{-10} C beam charges. The beam current was approximately 200nA and the counting rate of the true events was approximately 6 counts for an hour.

RESULT AND DISCUSSION

As a result, the polarization transfer coefficient $K_y^{y'}$ was determined to be 0.09 ± 0.13. The error includes statistics for proton polarization and uncertainties associated with the ion source and the proton polarimeter. We compare the present result with values calculated using 16 transition amplitudes for the ^{2}H(d,p)^{3}H reaction determined by Lemaître and Schieck in Ref. [9]. In the analysis, they assumed the energy dependence of the matrix elements to result entirely from the Coulomb penetrability in the entrance channel. And the energy-independent parts of the matrix elements were determined from a fit to Legendre expansion coefficients of the experimental data for the cross section, analyzing powers, and proton polarization induced by unpolarized deuterons for $E_d < 500$keV. According to their analysis, we calculated the polarization transfer coefficient $K_y^{y'}$ using the transition amplitudes. The calculated results for incident energies of 90 and 10 keV and the present experimental datum are shown in Fig.1 as solid and dashed lines and a solid circle. As can be seen in Fig.1, in the calculation, $K_y^{y'}$ is not largely dependent on incident deuteron energy. The polarization transfer coefficient $K_y^{y'}$ calculated from the transition amplitudes determined by Lemaître and Schieck is found to be close to the present experimental datum within the error.

SUMMARY

In summary, we have measured the polarization transfer coefficient $K_y^{y'}$ for the ^{2}H(d,p)^{3}H reaction at a scattering angle of $\theta = 0°$ at an incident energy of $E_d = 90$ keV. In the experiments, it was measured with the Lamb-shift type polarized ion source and the proton polarimeter using the p - ^{28}Si elastic scattering. As a result, the value of $K_y^{y'} = 0.09 \pm 0.13$ was obtained and found to be close to the calculated values from the reaction amplitudes which have been previously determined by Lemaître and Schieck.

ACKNOWLEDGEMENTS

We would like to thank Professor K. Sagara of Kyushu University, Japan, for providing useful information in target fabrication. We are indebted to the staff of the Tandem Accelerator Center, University of Tsukuba for their support.

REFERENCES

1. R. M. Kulsrud, H. P. Furth, and E. J. Valeo, Phys. Rev. Lett. **49**, 1248 (1982)
2. G. W. Shuy, A. E. Dabiri, and H. Gurol, Fusion Technol. **9**, 459 (1986)
3. H. M. Hofmann and D. Fick, Phys. Rev. Lett. **52**, 2308 (1984)
4. K. F. Liu and J. S. Zhang, Phys. Rev. Lett. **55**, 1649 (1985)
5. D. Fick and H. M. Hofmann, Phys. Rev. Lett. **55**, 1650(1985)
6. J. S. Zhang and K. F. Liu, Phys. Rev. Lett. **57**, 1410 (1986)
7. S. Abu-Kamar, M. Igarashi, R. C. Johnson and J. A. Tostevin, J. Phys. **G14**, L1 (1988)
8. H. Paetz. gen. Schieck, B. Becker, R. Randermann, S. Lemaître, P. Niessen, R. Reckenfelderbaumer and L. Sydow, Phys. Lett. **B276**, 290 (1992)
9. S. Lemaître and H. Paetz. gen. Schieck, Ann. Phys. **2**, 503 (1993)
10. J. S. Zhang, K. F. Liu, and G. W. Shuy, Phys. Rev. **C60**, 054614 (1999)
11. R.E.Brown and N.Jarmie, Phys. Rev. **C41**, 1391 (1990)
12. Y.Tagishi, N. Nakamoto, K. Katoh, J. Togawa, T. Hisamune, T. Yoshida, and Y. Aoki Phys. Rev. **C46**, R1155 (1992)
13. K. A. Fletcher, Z. Ayer, T. C. Black, R. K. Das, H. J. Karwowski, E. J. Ludwig, and G. M. Hale, Phys. Rev. **C49**, 2305 (1994)
14. Y. Tagishi, T. Katabuchi, K. Mizukoshi, N. Yamada, M. Yamaguchi, N. Kawachi, and Y. Aoki, Nucl. Instr. Meth. in Phys. Res. **A202**, 436 (1998)
15. P. Kozma and P. Bém, Nucl. Phys. **A442**, 17 (1985)

16. T. B. Clegg, D. D. Armstrong, R. A. Hardekopf, and P. W. Keaton, Jr., Phys. Rev. C8, 922 (1973)
17. Y. Tagishi and J. Sanada, Nucl. Instrum. Methods 164, 411 (1979)
18. G. G. Ohlsen, J. L. McKibben, G. P. Lawrence, P. W. Keaton, Jr., and D. D. Armstrong, Phys. Rev. Lett. 27, 599 (1971)
19. J. Biersack and J. F. Ziegler, TRIM89, version-5.1 (1989)
20. G. Hempel, A. Hofmann and K. Kilian, Nucl. Instr. & Meth. 105, 91 (1972)

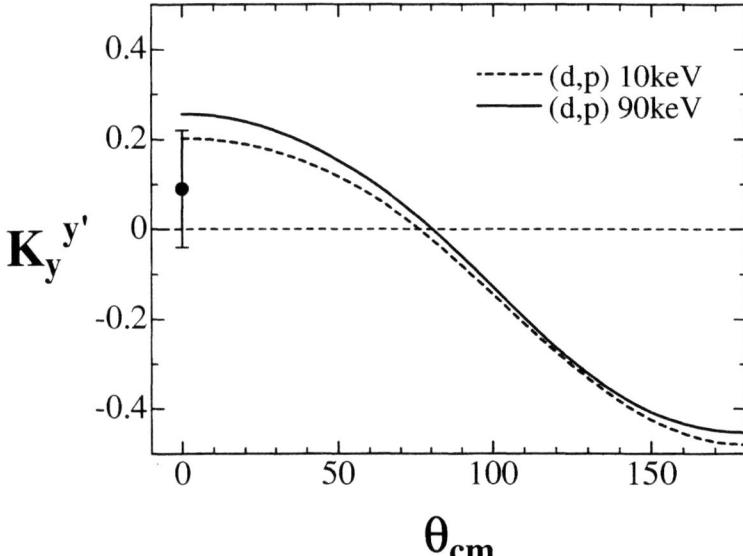

FIGURE 1. The present datum of the polarization transfer coefficient $K_y^{y'}$ and calculated value from the transition amplitudes determined by Lemaître and Schieck in Ref.9.

Calculation of Low Energy ^{2}H$(d,p)^3$H Reaction by the Four-Body Faddeev-Yakubovsky Equation

Eizo Uzu*, Shinsho Oryu*, and Makoto Tanifuji†

*Department of Physics, Faculty of Science and Technology, Science University of Tokyo,
Yamazaki 2641, Noda, Chiba 278-8510, Japan
†Department of Physics, Hosei University,
Fujimi 2-17-1, Chiyoda-ku, Tokyo 102-8160, Japan

Abstract. Cross sections and analyzing powers for ^{2}H$(d,p)^3$H reactions are calculated at $E_d = 30$ keV by the four-body Faddeev-Yakubovsky equation. The PEST potential is adopted for inter-nucleon forces. New results for the vector and tensor analyzing powers are similar to our previous ones which employ the Yamaguchi potentials. However, the new calculation gives larger differential cross section, improving the agreement with the experimental data. By the use of the solution, the polarization transfer coefficients $K_y^{y'}$, $K_{xz}^{y'}$, $K_{xx}^{y'}$, $K_{yy}^{y'}$ and $K_{zz}^{y'}$, and the suppression ratio σ_{pol}/σ_0 are calculated.

INTRODUCTION

In this decade we have investigated low-energy ^{2}H$(d,p)^3$H and ^{2}H$(d,n)^3$He reactions by the Faddeev-Yakubovsky (FY) equations. In the ^{2}H$(d,p)^3$H reaction at $E_d = 30$ keV, the measured vector and tensor analyzing powers are well reproduced [1,2] by calculations with the Yamaguchi-type N-N potential, although the calculation underestimates the differential cross section.

Here we will investigate if the discrepancy in the differential cross section is improved by adopting realistic potentials. For the first attempt, we choose the PEST [3] potentials for the N-N interactions. In the following, numerical results for this potential are presented and are discussed in comparison with our previous results [2].

FADDEEV-YAKUBOVSKY EQUATIONS

The FY equations for the four-identical particle system are described by coupled-channel equations, where two channel, say α channel and β one, are considered.

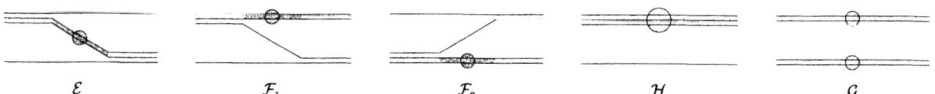

FIGURE 1. Typical diagrams which appear in the Faddeev-Yakubovsky equations.

The former describes the [3+1] channel and the latter the [2+2] one. The explicit form of the equation is given as

$$\begin{pmatrix} \mathcal{M}_{\alpha\alpha} & \mathcal{M}_{\alpha\beta} \\ \mathcal{M}_{\beta\alpha} & \mathcal{M}_{\beta\beta} \end{pmatrix} = \begin{pmatrix} \pm\mathcal{E} & \mathcal{F}_1 \\ 2\mathcal{F}_1^t + 2\mathcal{F}_2^t & 0 \end{pmatrix} + \begin{pmatrix} \pm\mathcal{E} & \mathcal{F}_1 \\ 2\mathcal{F}_1^t + 2\mathcal{F}_2^t & 0 \end{pmatrix} \begin{pmatrix} \mathcal{H} & 0 \\ 0 & \mathcal{G} \end{pmatrix} \begin{pmatrix} \mathcal{M}_{\alpha\alpha} & \mathcal{M}_{\alpha\beta} \\ \mathcal{M}_{\beta\alpha} & \mathcal{M}_{\beta\beta} \end{pmatrix}, \quad (1)$$

where the $\mathcal{M}$'s are the scattering amplitudes, $\mathcal{E}$ the four-body potential for the $\alpha \to \alpha$ transition, and the plus sign corresponds to the boson system, the minus sign to fermions, $\mathcal{F}_1$ signifies the $\beta \to \alpha$ transition, $\mathcal{F}_1^t$ and $\mathcal{F}_2^t$ the $\alpha \to \beta$ one, and subscripts 1 and 2 distinguish the two channels in the [2+2] subsystem. $\mathcal{H}$ and $\mathcal{G}$ are the propagators for the [3+1] and [2+2] subsystems, respectively (see Fig. 1).

INPUT

For the two-nucleon interactions we employ the PEST potential [3] in the $^1S_0$, $^3S_1$-$^3D_1$, $^3P_0$, $^1P_1$, $^3P_1$ and $^3P_2$ states. We take rank 1 for the $^1S_0$ and $^3S_1$-$^3D_1$ states, rank 2 for the $^3P_0$, $^1P_1$, and $^3P_1$ states, and rank 3 for the $^3P_2$ states. The $^3P_2$ state is coupled to the $^3F_2$ one, and after solving the coupled channel Lippmann-Schwinger equation, we neglect $^3F_2$ elements of the amplitude.

The amplitudes of the [3+1] and [2+2] subsystems are treated using a modified Hilbert-Schmidt expansion method. We take 24 terms in the [3+1] subsystem and 12 terms in the [2+2] subsystem, respectively.

The convergence of the four-body amplitudes with respect to the spin and parity of the [3+1] subsystem is confirmed by taking up to $7/2^\pm$, which is consistent to our previous calculation employing the Yamaguchi potentials. In the [2+2] subsystem, all of the allowed states are included. The convergence of the total spin and parity of the four-body system is confirmed by taking up to $4^\pm$.

The FY equations are solved by the method of continued fraction [4] without taking into account the Coulomb force. However, Coulomb corrections are considered by including a Coulomb penetration factor in the cross section and by introducing Coulomb phase shifts in the scattering amplitude.

RESULTS AND DISCUSSIONS

The calculated differential cross section and analyzing powers are presented in Fig. 2, where the solid lines show our new results by the PEST potentials, and

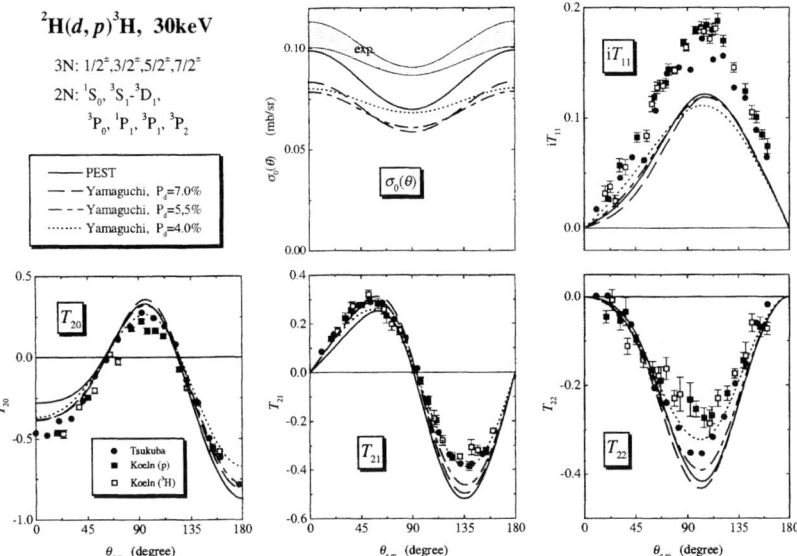

FIGURE 2. Comparison of unpolarized cross section σ_0, vector analyzing power iT_{11}, and tensor analyzing powers T_{20}, T_{21}, and T_{22} between the calculation and the measurement for $^2\text{H}(d,p)^3\text{H}$ reactions at $E_d = 30\text{keV}$. The cross section data [5] are shown by the shaded area between the two thin solid lines. In the vector and tensor analyzing powers, the solid circles are for the 30keV Tsukuba data [6] and the open and solid squares for the 28keV Köln data [7]. The solid lines are calculated by the PEST potentials. Others are our previous results by the Yamaguchi potentials, the dashed lines by $P_d = 7.0\%$, the dash-dotted lines by $P_d = 5.5\%$, and the dotted lines by $P_d = 4.0\%$.

other lines show our previous ones [2] by the Yamaguchi potentials. The new results of the vector and tensor analyzing powers are similar to our previous ones by $P_d = 7.0\%$. However in the differential cross section, the new calculation is larger than the previous ones, and the discrepancy between the measured and the calculated is much improved.

The predictions of polarization transfer coefficients $K_y^{y'}$, $K_{xz}^{y'}$, $K_{xx}^{y'}$, $K_{yy}^{y'}$ and $K_{zz}^{y'}$ are shown in Fig. 3, where the definitions of the lines are the same as those in Fig. 2. The well-known relationship, $K_{xx}^{y'}+K_{yy}^{y'}+K_{zz}^{y'} = 0$, is satisfied by the calculation. The measured $K_y^{y'}$, which is reported by T. Katabuchi, is 0.09 ± 0.08 at $\theta \approx 0°$, and the sign is opposite to that of the new calculation.

The calculated suppression ratio for the PEST potential is shown in Fig. 4 as a function of Θ, where Θ is the angle between the incident momentum and the polarization axis. The ratio we obtain is 0.86 at $\Theta = 0°$ and 1.06 at $\Theta = 90°$, that is the cross section is not much suppressed by the polarization. It is interesting that

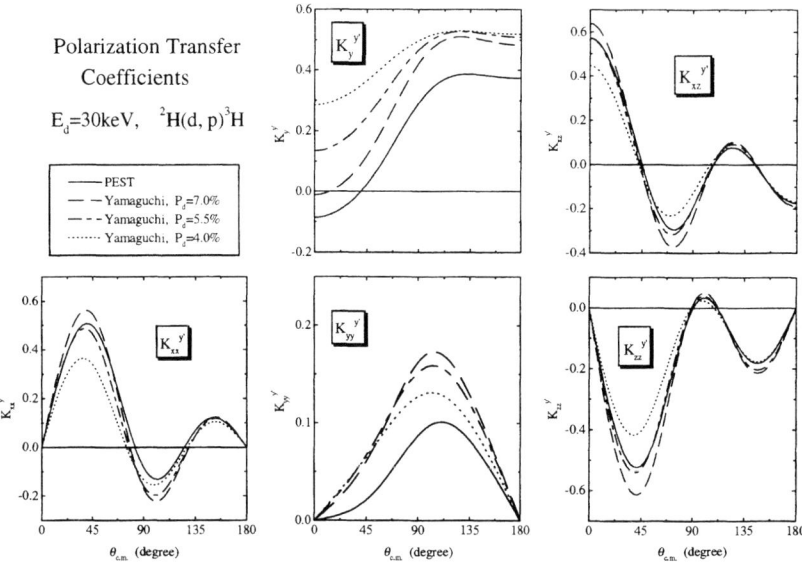

FIGURE 3. Predictions of the polarization transfer coefficients. The definitions of the lines are the same as those in Fig. 2.

the calculated suppression ratio strongly depends on the nuclear force employed, as seen in Fig. 4.

As seen above, there are still some discrepancies between the calculated quantities, say cross section and vector analyzing power, and the measured ones. They will be due to the next causes: 1. insufficient ranks in the two-body subsystem, 2. the lack of the higher partial waves in the two-body subsystem, 3. the absence of the three-body force, and 4. insufficient modification for the Coulomb force. Improved calculations for the points 1. and 2. are now in progress.

Acknowledgement. The calculations in this work are carried out by computer systems of RCNP Computer Center in Osaka University, Computer Center of National Institute for Fusion Science, Computer Center of RIKEN, and Frontier Research Center for Computational Science in Science University of Tokyo.

REFERENCES

1. E. Uzu, H. Kameyama, S. Oryu, and M. Tanifuji, Few-Body Sys. **22**, 65 (1997); E. Uzu, S. Oryu, and M. Tanifuji, Prog. Theor. Phys. **90**, 937 (1993); E. Uzu, S. Oryu, and M. Tanifuji, AIP Conference Proceedings 334 on Few-Body Problems in Physics, p447-p450, (Williamsburg, Virginia, 1994); E. Uzu, S. Oryu, and M. Tanifuji, Few-

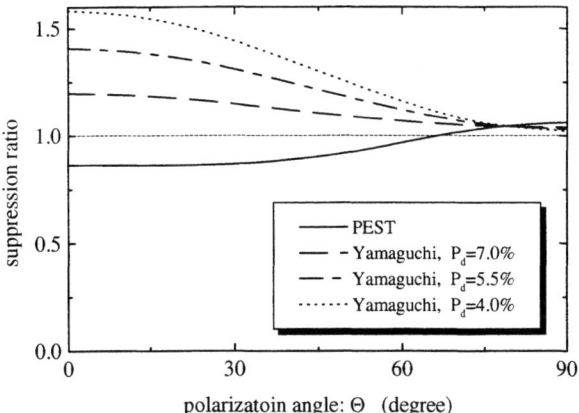

FIGURE 4. Prediction of the suppression ratio $\sigma_{p}ol/\sigma_0$. Θ indicates the angle between the incident momentum and the polarization axis. The definitions of the lines are the same as those in Fig. 2.

Body Systems Suppl. **99**, 97 (1995); S. Oryu, E. Uzu, T. Hino, and S. A. Sofianos, Proceedings of the European Conference on Advances in Nuclear Physics and Related Areas, p534-p539, (Thessaloniki, Greece, 1999).
2. E. Uzu, S. Oryu, and M. Tanifuji, Few-Body Sys. Suppl. **12**, 398 (2000).
3. J. Haidenbauer and W. Plessas, Phys. Rev. **C30**, 1822 (1984); ibid. **C32**, 1424 (1985).
4. J. Horáček and T. Sasakawa, Phys. Rev. **A28**, 2151 (1983); ibid., **A30** (1984) 2274; ibid., **C32**, 70 (1985).
5. R.E. Brown and N. Jarmie, Phys. Rev. C41, 1391 (1990).
6. Y. Tagishi, N. Nakamoto, K. Katoh, J. Togawa, T. Hisamune, T. Yoshida, and Y. Aoki, Phys. Rev. C46, R1155 (1992).
7. B. Becker, R. Randermann, B. Polke, R. Lemaître, R. Reckenfelderbäumer, P. Niessen, G. Rauprich, L. Sydow, and H. Paetz gen. Schieck, Few-Body Sys. **13**, 19 (1992).

Measurement of Analyzing Power T_{20} in Elastic Electron-Deuteron Scattering in the Momentum Transfer Range of 0.3 - 0.8 $(GeV/c)^2$

H. Arenhövel[a], L.M. Barkov[b], S.L. Belostotsky[c], V.F. Dmitriev[b], M.V. Dyug[b], R.J. Holt[d], C.W.de Jager[e], E.R. Kinney[f], B.A. Lazarenko[b], S.I. Mishnev[b], D.M.Nikolenko[b], V.V. Nelyubin[c], A.V. Osipov[g], V.G. Popov[b], D.H. Potterveld[h], I.A. Rachek[b], R.Sh. Sadykov[b], Yu.V. Shestakov[b], V.N. Stibunov[g], D.K. Toporkov[a], V.V. Vikhrov[c], H.de Vries[e], and S.A. Zevakov[b].

[a] *IKPJGU, Mainz, Germany,* [b] *BINP, Novosibirsk, Russia,* [c] *PINP, St.-Petersburg, Russia,* [d] *UI, Urbana-Champaign, IL, USA,* [e] *NIKHEF, Amsterdam, The Netherlands,* [f] *CU, Boulder, CO, USA,* [g] *INP, Tomsk, Russia,* [h] *ANL, Argonne, IL, USA*

Abstract. The results on the measurement of the analyzing power component T_{20} in elastic ed scattering are presented. An internal storage cell target technique was used in the VEPP-3 electron ring in Novosibirsk for this measurement. The measurement has been performed in the range of momentum transfer poorly investigated but important from theoretical point of view because of the node of the monopole charge form factor of the deuteron. The comparison of the other experimental data and some theories are discussed.

The measurement of the polarization observables in the elastic electron-deuteron scattering together with the differential cross section measurement gives a possibility for complete experimental determination of the electromagnetic form factors of the deuteron (charge monopole, quadrupole and magnetic). Starting from 80th several laboratories (BATES, BINP, Bonn University, NIKHEF, TJNAF) have performed such experiments in a different modifications.

Here we present recent results on the T_{20} analyzing power measurement which were obtained in the experiment with the use of internal polarized deuterium target in the storage ring VEPP-3, Novosibirsk, at 2 GeV electron energy.

A new Atomic Beam Source has been designed and constructed to fill the storage cell target by polarized atoms [1]. It has a conventional design of the dissociator and high-frequency transitions units. The main difference of this ABS from the other modern sources is the usage of the superconductive sextupole magnets instead of permanent ones which allow to provide a magnetic pole-tip field three times larger than that for permanent magnets and consequently a higher flux of polarized atoms is available from the source. The electron beam optics in the experimental section

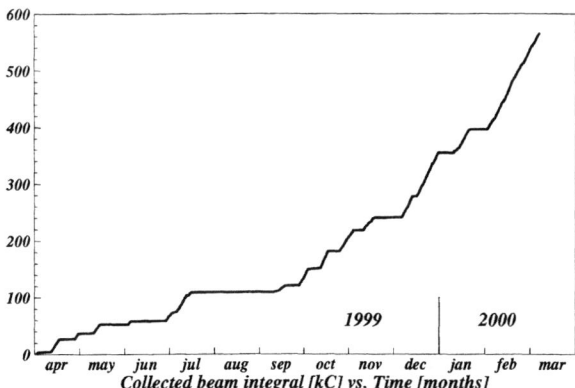

FIGURE 1. Collected beam integral.

of VEPP-3 ring was modified to get the possibility to use a small-aperture storage cell. An aluminum (0.03mm thick) storage cell has an elliptic cross section 13×24 mm, the length of 400mm and coated by 'drifilm'. The cell is cooled by liquid nitrogen to get more dense target.

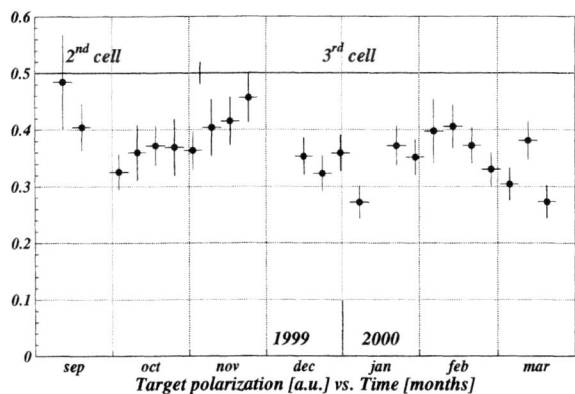

FIGURE 2. Polarization of the target vs time measured with Low-Q polarimeter.

The particle detector (partly described in [2]) has two identical systems. For the registration of the particle tracks different sets of drift chambers are employed. Each of the systems has of a segmented CsI + NaI electron calorimeter (total thickness is 16 radiation lengths) on one side of the electron beam and a hadron scintillation hodoscope (3 layers of plastic scintillators with a total thickness of 26cm) on the other side. Angular acceptance of each electron arm is $\theta = 20° \ldots 30°$ and $\Delta\phi = 60°$ and that for each hadron arm is $\theta = 60° \ldots 70°$ $\Delta\phi = 60°$. The detection system permits to identify the $D(e,e'd)$, $D(e,e'p)n$ and $D(e,pp)e'\pi^-$ events.

In March 1999 all the equipment was assembled at the VEPP-3. The electron beam current integral of 566 kC was collected during almost one year run, as shown in Fig.1 A large amount data in both elastic and inelastic scattering channels was collected.

The tensor polarization of the atomic beam injected into the storage cell is monitored by the Breit-Rabi polarimeter and polarization is very high, about 98%.

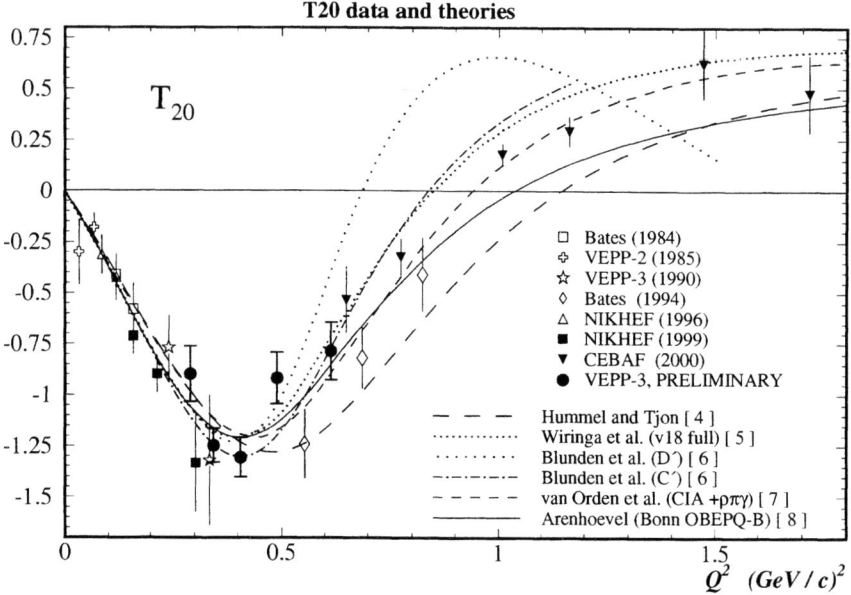

FIGURE 3. World data on T_{20} measurement in elastic ed scattering versus the square of momentum transfer and some theoretical predictions.

However, the degree of the polarization in the storage cell lower due to the depolarization effects, which couldn't be numerically accounted for. Moreover, the polarization is getting worse during the run because of degradation of the cell surface. To control the average polarization of the target during the run so called Low-Q polarimeter based on the measurements of the asymmetry in elastic ed scattering was installed at a small angle (average $\theta_e = 9°$) where the absolute measurement of the analyzing power T_{20} is available [3]. The polarization of the target versus the time extracted with the use of Low-Q polarimeter is shown in Fig.2.

The cross section of elastic ed scattering falls dramatically as compared with the

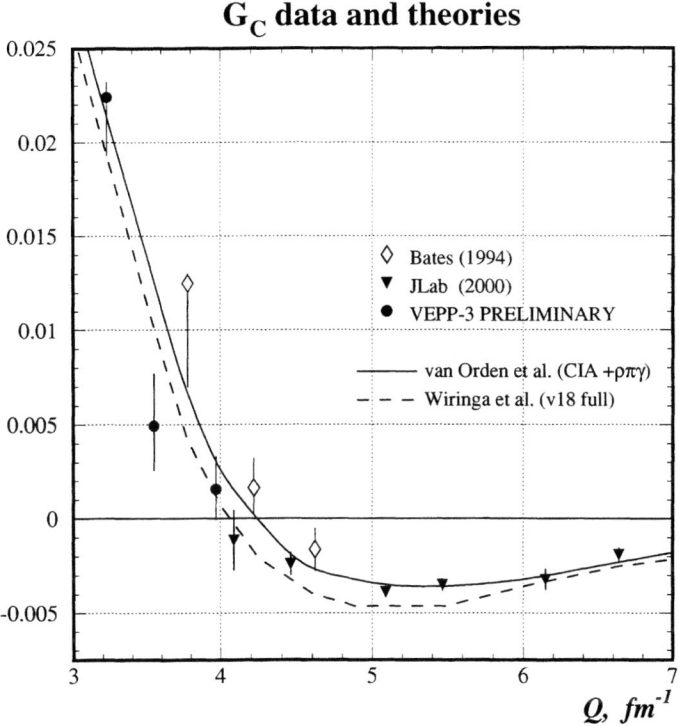

FIGURE 4. The data on G_c measurement in elastic ed scattering versus the momentum transfer and some theoretical predictions.

quasi-elastic one when the momentum transfer is increased. Therefore to extract the events of the elastic scattering among the lot of background events three kinematical correlations and two particle identification methods were applied:
* e-d polar angles correlation , * e-d azimuthal angles correlation, * scattering angle – deuteron energy correlation, * E-ΔE method, * time of flight method.

The preliminary results obtained for the tensor analyzing power component T_{20} are shown in Fig.3 together with the existing world data and with predictions of several theoretical models. These data are provided by one of the detection system that is sensitive mostly to T_{20} (the other one is sensitive mostly to T_{21} component). Small corrections were applied to calculate T_{20} at the conventionally accepted angle $\Theta_e = 70°$. The shown error bars include statistical errors only. The measurements of T_{20} were performed in the range of momentum transfer where it reaches the minimum and where the experimental data are poor. Analysis of the available theories has shown that the theoretical description of T_{20} given by Wiringa *et al.* (v18full) and Blunden *et al.* (C') ensure almost the same good fit to our experimental data. It seems that prediction for T_{20} given by van Orden *et al.* (

CIA + $\rho\pi\gamma$) provides the best description of the experimental data in the whole range of momentum transfer available so far.

The data on the tensor analyzing power component T_{20} in elastic ed scattering together with the data on the unpolarized elastic differentional cross section provide a possibility to extract each form factor separately. The results for the monopole G_c charge form factor shown in Fig.4. The results show a node located at a lower value than inferred from the Bates measurements and agree well with the recent measurements at JLab [9].

This work was supported by Russian Foundation for Basic Researches, grants 98-02-17949 and 98-02-17993, INTAS, grant 96-0424 and State Scientific and Technical Programm (Nuclear Physics).

REFERENCES

1. L.G. Isaeva *et al.*, Proc. of 13th Int. Symp. on High Energy Spin Phys., Protvino, Russia, Word Scientific, p.631 (1998).
2. Theunissen J.A.P. *et al.*, Nucl. Instrum. and Methods, **A 348**, 61 (1994).
3. M. Ferro-Luzzi *et al.*, Phys. Rev. Lett. **77**, 2630 (1996).
4. E. Hummel and J. Tjon, Phys. Rev. **C42**, 423 (1990).
5. R.B. Wiringa, V.G.J. Stoks and R. Schiavilla, Phys. Rev. **C51**, 38 (1995).
6. P.G. Blunden, W.R. Greenberg and E.L. Lomon, Phys. Rev. **C40**, 1541 (1989).
7. J.W. Van Orden, N. Devine and F. Gross, Phys. Rev. Lett. **75**, 4369 (1995).
8. H. Arenhövel, F. Ritz and T. Wilbois, arXiv:nucl-th/9910009 v2 1 Dec 1999.
9. D. Abbot *et al.* arXiv:nucl-ex/0001006 17 Jan 2000.

Difference between $nd\ A_y$ and $pd\ A_y$ below 16 MeV

T. Fujita[1], K. Sagara[1], K. Shigenaga[1], K. Tsuruta[1], T. Yagita[1], T. Nakashima[1], N. Nishimori[2], H. Nakamura[3] and H. Akiyoshi[4]

[1]*Department of Physics, Kyushu University, Fukuoka, Japan*
[2]*FEL lab., Japan Atomic Energy Research Institute, Tokai, Japan*
[3]*Kitakyushu National College of Technology, Kokura, Japan*
[4]*RIKEN, Wakou, Saitama, Japan*

Abstract. Analyzing power $A_y(\theta)$ of nd scattering was accurately measured at E_n =12 and 16 MeV. Polarization of the n-beam produced by spin transfer $^2\text{H}(d,n)^3\text{He}$ reaction at 0° was determined in a separate experiment using $^4\text{He}(n,n)$ scattering. The present $nd\ A_y(\theta)$ data are consistent with the existing $nd\ A_y(\theta)$ data below E_n = 14.1 MeV. It was confirmed that the difference between $nd\ A_y$ and $pd\ A_y$ becomes large above 14 MeV and the assumption on slow-down effects by Coulomb force breaks down. The observed large difference in $A_y(\theta)$ requires systematic calculations with exact treatment of Coulomb force to study charge symmetry in nuclear force.

INTRODUCTION

Large discrepancy between experiment and calculation for $A_y(\theta)$ of Nd (nd and pd) scattering below 30 MeV has been well known as A_y puzzle for about 15 years. Witała and Glöckle [1] showed that both $Nd\ A_y(\theta)$ and $NN\ A_y(\theta)$ can be excellently reproduced by slightly modifying ^3P_j-state NN interactions. Since Coulomb force is not included in their calculation, their modification includes too strong charge independence breaking (CIB) in ^3P_j-state NN interactions. Their attempt, nevertheless, suggests a possibility to find effects of charge symmetry breaking (CSB) in ^3P_j-state NN interactions from the difference between $nd\ A_y$ and $pd\ A_y$.

About 15 % of the difference between nn and pp scattering lengths arises from CSB and about 10 % of the difference between ^3H and ^3He binding energies comes from CSB. The difference between $nd\ A_y$ and $pd\ A_y$ around their maxima is about 10-20 %, and 10-15 % of the difference may be expected to come from CSB.

To study CSB in $Nd\ A_y$, therefore, experimental accuracy of about 1% is necessary. We have already systematic $pd\ A_y(\theta)$ data of about 0.7 % accuracy in the energy range of 2-18 MeV [2,3] . Accurate $nd\ A_y(\theta)$ data are necessary, together with exact pd calculations.

EXPERIMENT

Three kinds of experiments were made; Calibration of a d-beam polarimeter, measurement of the polarization of a n-beam produced by $^2\mathrm{H}(d,n)^3\mathrm{He}$ reaction at $0°$, and measurement of nd A_y.

Preparation of a *d*-beam polarimeter

Polarization of the n-beam, P_y^n, produced by $^2\mathrm{H}(d,n)^3\mathrm{He}$ reaction at $0°$ is expressed as

$$P_y^n = \frac{3}{2} P_y^d K_y^y(0°)/[1 - P_{yy}^d A_{zz}(0°)/4] \tag{1}$$

where P_y^d and P_{yy}^d are vector and tensor polarizations of the d-beam, respectively, and $K_y^y(0°)$ and $A_{zz}(0°)$ are the spin transfer coefficient and the tensor analyzing power, respectively, of $^2\mathrm{H}(d,n)^3\mathrm{He}$ reaction at $0°$. P_y^n is proportional to P_y^d, whereas P_y^n is not so sensitive to P_{yy}^d since $P_{yy}^d A_{zz}(0°)/4$ is about -0.07.

We adopted $^3\mathrm{He}(d,p)^4\mathrm{He}$ reaction to measure the d-beam polarizations. In a separate experiment, we measured the analyzing powers $A_y(\theta)$ and $A_{yy}(\theta)$ of this reaction at $E_d = 6.6$-16.0 MeV around the angle where A_y takes the maximum value. The tensor polarization of the d-beam was determined by a $^3\mathrm{He}(d,p)^4\mathrm{He}$ polarimeter whose $A_{zz}(0°)$ had been already calibrated using $^{16}\mathrm{O}(d,\alpha)$ reaction. The vector polarization of the d-beam, produced in a Lamb-shift ion source, was evaluated from the tensor polarization.

In the n-beam production, P_y^d was precisely measured using the maximum of $A_y(\theta)$ of $^3\mathrm{He}(d,p)^4\mathrm{He}$ reaction, and P_{yy}^d was coursely measured using $A_{yy}(\theta)$ at the same angle.

Determination of *n*-beam polarization

The polarization of the n-beam was determined using $^4\mathrm{He}(n,n)^4\mathrm{He}$ scattering whose $A_y(\theta)$ becomes the theoretically maximum ($=1$) near $\theta = 110°$ and around $E_n = 12$ MeV, and slowly varies with the energy and angle.

A polarized d-beam from the Kyushu university tandem accelerator was injected into a D_2 gas target to produce a polarized n-beam, as shown in Figure 1. The d-beam was then lead into a $^3\mathrm{He}$ gas target to measure the d-beam polarizations using $^3\mathrm{He}(d,p)^4\mathrm{He}$ reaction. The d-beam was stopped on the bottom of the $^3\mathrm{He}$ gas target cell. Thin Ta foils were used for the windows of the gas target cells and also Ta plate was used for the beam stopper in order to reduce production of unfavorable neutrons and gamma rays by high Coulomb barrier.

A liquid helium target of 25 mm in diameter and 50 mm in height was placed about 50 cm downstream from the n-beam production unit. Walls of the helium cryostat were made thin to reduce multiple scattering of neutrons. The target cell had a stainless steel wall of 0.3 mm in thickness, and the cell was surrounded by a copper wall of 1 mm in thickness at the liquid nitrogen temperature and kept in vacuum by a stemless steel wall of 1 mm in effective thickness.

The bottom of the target cell was sealed by a quartz glass through which scintillation light produced in the target by n-He scattering and other reactions came out and was guided to a photo-multiplier tube placed in the atmosphere.

Neutrons scattered by the helium target were detected by eight liquid scintillation detectors (NE213) on the left and right sides of the n-beam at four angles, as shown in Figure 2. The scintillators had the size of 3 inches in diameter and 6 inches in height and were placed about 1 m apart from the target. Lead bricks and Paraffin blocks were placed between the scintillators and the n-beam production unit to reduce neutron and gamma fluxes. The helium target was in normal liquid state since no bubbles were seen in the target through the glass.

The light output from a recoil He and the time of flight of a neutron from the target to a scintillator were measured to specify the events of $^4\text{He}(n,n)^4\text{He}$ scattering. Measurements were made at $E_n = 12$ MeV and 16 MeV by flipping the sign of the n-beam polarization, and angular distribution of left-right asymmetry was derived.

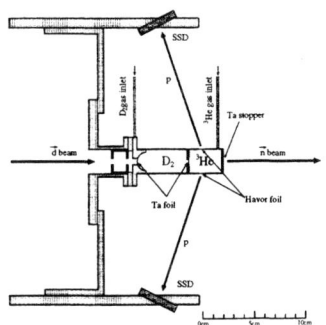

FIGURE 1. polarized n-beam production unit.

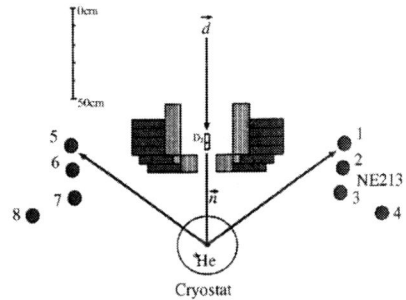

FIGURE 2. Setup to determine the n-beam polarization.

At 12 MeV, the maximum of $A_y(\theta)$ of $^4\text{He}(n,n)^4\text{He}$ was assumed to be 1 because the maximum is estimated to be above 0.9995 from the existing n-^{4}He phase shifts. The n-beam polarization was determined from the assumption and the maximum of the measured left-right asymmetry.

At 16 MeV, we searched new phase shifts which reproduced the measured asymmetry and calculated cross section from the existing phase shifts. The maximum value of $A_y(\theta)$ was treated as an additional parameter to be searched. Several sets of existing phase shifts were chosen as the initial values and we obtained nearly the same results, from which maximum value of $A_y(\theta)$, and as a result the n-beam polarization, were determined.

Finally we obtained $K_y^y(0°)$ and $A_{zz}(0°)$ of the ^{2}H$(d,n)^3$He reaction. The obtained $A_{zz}(0°)$ agrees with existing data. Present $K_y^y(0°)$ at 16 MeV agrees with new data by Cologne group [4], however, disagrees with previous TUNL data [5].

In the following experiments of nd $A_y(\theta)$, we used the same n-production unit and the same d-beam polarimeter as in the determination of the n-beam polarization. Therefore the n-beam polarization was correct even if $K_y^y(0°)$ and $A_{zz}(0°)$ of the ^{2}H$(d,n)^3$He reaction were incorrect.

Measurement of n-d A_y

Measurements of nd $A_y(\theta)$ at E_n =12 and 16 MeV were made using the same setup except for the target which was a deuterated scintillator NE232 ($(CD_2)_n$). An intense and pulsed polarized d-beam from RIKEN injector cyclotron was used. Additional experiments were made using Kyushu university tandem accelerator.

Very accurate data for nd $A_y(\theta)$ at E_n =12 and 16 MeV were obtained (see Figure 3). The data at 12 MeV agree well with TUNL data [6].

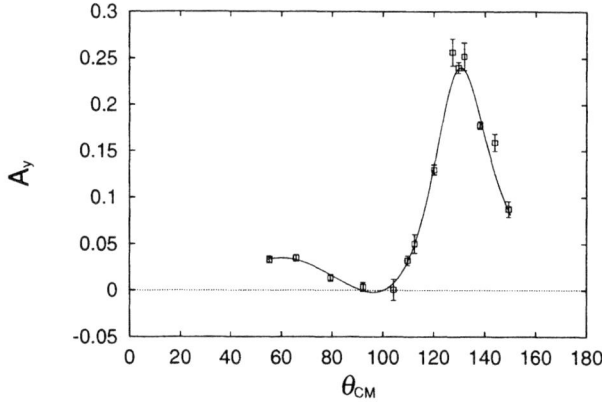

FIGURE 3. Measured nd $A_y(\theta)$ at 16 MeV and Legengre fit to data.

RESULTS AND CONCLUSION

The present nd $A_y(\theta)$ data and previous ones [6–10] and pd $A_y(\theta)$ [2,3] were fitted by Legendre polynomials to derive the maximum values, $A_y^{max}(nd)$ and $A_y^{max}(pd)$, respectively.

Figure 4 shows that $A_y^{max}(nd)$ and $A_y^{max}(pd)$ form nearly parallel lines below 12 MeV. The phenomena could be explained if effective $E_{cm}(pd)$ was assumed to be smaller than $E_{cm}(nd)$ by a constant (energy shift) due to slow-down effect by Coulomb force.

The present data at 16 MeV revealed that $A_y^{max}(nd)$ continuously increases with energy at least up to 16 MeV, whereas $A_y^{max}(pd)$ reduces increment above 14 MeV. Hence the difference between $A_y^{max}(nd)$ and $A_y^{max}(pd)$ becomes substantially large above 14 MeV, and apparently disagrees with the slow-down hypothesis.

It is very interesting to see whether energy dependence of the difference between $A_y^{max}(nd)$ and $A_y^{max}(pd)$ is fully explained by the effect of Coulomb force alone. When systematic pd calculations with exact treatment of Coulomb force are made, some conclusion on CSB will be obtained from the present data.

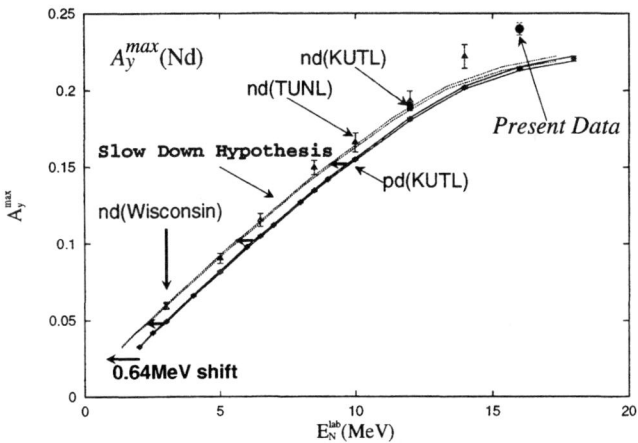

FIGURE 4. $A_y^{max}(nd)$ and $A_y^{max}(pd)$. The dashed line is obtained by shifting $A_y^{max}(pd)$ by 0.64 MeV of E_p.

REFERENCES

1. H. Witała and W. Glöckle, *Nucl. Phys.* **A528**, 48 (1991).
2. K. Sagara et al., *Phys. Rev.* **C50**, 576 (1994).
3. S. Shimizu et al., *Phys. Rev.* **C52**, 1193 (1995).
4. W. von Witsch et al., *Phys. Rev.* **C57**, 2104 (1998).
5. P. W. Lisowski et al., *Nucl. Phys.* **A242**, 298 (1975).
6. C. R. Howell et al., *Few Body Systems* **2**, 19 (1987).
7. J. E. McAninch et al., *Phys. Lett.* **B307**, 13 (1993)
8. W. Tornow et al., *Phys. Lett.* **B257**, 273 (1991).
9. W. Tornow et al., *Phys. Rev. Lett.* **49**, 312 (1982)
10. W. Tornow et al., *Phys. Rev.* **C27**, 2439 (1983)

Measurement of Cross Section and Analyzing Powers for dp Scattering at Intermediate Energies and Three – Nucleon Force Effects

K. Sekiguchi*[1], H. Sakai*[†], H. Okamura[††], N. Sakamoto[†],
A. Tamii*, T. Uesaka[††], T. Wakasa[‡] Y. Satou[†], T. Ohnishi[†],
K. Yako*, S. Sakoda*, K. Suda[††], H. Kato*, Y. Maeda*,
M. Hatano* and J. Nishikawa[†]

* *Department of Physics, University of Tokyo, Tokyo 113-0033, Japan*
[†] *The Institute of Physical and Chemical Research, Saitama 351-0198, Japan*
[††] *Department of Physics, Saitama University, Saitama 338-8570, Japan*
[‡] *Research Center for Nuclear Physics, Osaka University, Osaka 567-0047, Japan*

Abstract. Precise measurements of the cross sections and deuteron analyzing powers (A_y, A_{yy}, A_{xx} and A_{xz}) for the d–p elastic scattering has been made at $E_d = 140, 200$ and 270 MeV at RIKEN Accelerator Research Facility. The obtained results are compared with the Faddeev calculations based on modern nucleon–nucleon forces together with Tucson-Melbourne type and Urbana–Argonne type of three nucleon forces.

INTRODUCTION

A current interest has been focused on the three–nucleon force effects in the nucleon–deuteron (Nd) elastic scattering at intermediate energies ($E/A \geq 60$ MeV). This is partly because a solution of the Faddeev equation in which three–nucleon force (3NF) is taken into account has become available based on realistic modern nucleon–nucleon (NN) interactions up to $E/A \leq 250$ MeV. Thus it is now possible to compare the calculated results with the precise experimental data. One of the old version 3NF models is the Fujita–Miyazawa type [1]. This is a two–pion exchange 3NF model. A delta (Δ) particle is generated in the intermediate state. Lately more refined 3NF models such as Tucson–Melbourne (TM) type [2] and Urbana–Argonne (UR) type [3] have been available for the calculations. Both types of 3NFs have explained well the binding energies of three– and four–nucleon systems.

---

[1] Electronic Address: kimiko@nucl.phys.s.u-tokyo.ac.jp

We have reported in Ref. [4,5] that the significant discrepancy exists where the cross section takes minimum between the data and the theoretical predictions considering NN forces only for the d–p elastic scattering at $E_d = 270$ MeV. Recently it is shown that this discrepancy can be removed if 3NF is included in the calculations [6,7]. However the results of the tensor analyzing powers have suggested some deficiencies of the 3NF model in its spin part [5].

In order to study the dynamical aspects of the 3NF effects, we have made precise measurements of the cross sections and all deuteron analyzing powers (A_y, A_{yy}, A_{xx} and A_{xz}) for the d–p elastic scattering not only at 270 MeV but also at 140 and 200 MeV, covering the wide angular range $\theta_{c.m.} = 10° - 180°$.

EXPERIMENT

The experiment was performed at the RIKEN Accelerator Research Facility using vector and tensor polarized deuteron beams at 140, 200 and 270 MeV. The direction of the polarization axis was controlled by a Wien Filter system prior to acceleration [8]. The beam polarization was monitored by using d–p elastic scattering [4] and it was 60 – 80% of the theoretical maximum values during the experiments. A CH_2 target was bombarded and either scattered deuteron or recoil proton was momentum analyzed by the magnetic spectrometer SMART [9] depending on the scattering angle and detected at the focal plane.

RESULTS

Fig.1 shows the experimental results of the cross sections and the vector analyzing power A_y and the tensor analyzing power A_{yy} with open circles. The errors are statistical ones only. The statistical errors are within 1.3% for the cross sections and ±0.02 or less for A_y, ±0.01 or less for A_{yy}, ±0.02 or less for A_{xx} and ±0.03 or less for A_{xz}, respectively. The uncertainties of the target thickness and the charge collection of the beam were estimated by comparing the measured cross sections for the $^1H(p,p)$ scattering with the calculated values by the program code SAID [10]. The systematic error due to these uncertainties is estimated to be 2% at most.

DISCUSSION

Three theoretical predictions of the cross section and analyzing powers A_y and A_{yy} in terms of Faddeev theory are shown in Fig.1 together with the observed values. The solid curves (ND-TM) are the calculations with including TM 3NF and the dashed curves (ND-UR) are those with UR 3NF [11]. The dotted curves (ND-NN) are the theoretical predictions which consider NN force only. The Argonne v_{18} (AV18) potential [12] is used as NN interaction in these calculations. Note that all these calculations do not include Coulomb force. The theoretical predictions with and without TM 3NF based on CD BONN [13], Nijimegen I and II [14] potentials, which are not shown here, have also been reported by Glöckle et al. [5,15]. They have obtained almost the same results for all these observables so that it could be considered that the calculated results are independent of NN interactions.

Compared with the observed cross section, the ND-NN calculation underestimates the data at backward angles. In particular the discrepancy is clearly seen where the cross section takes minimum. The agreement of the ND-TM and ND-UR calculations with the observed values is excellent except for forward angles. Thus the discrepancy in the cross section minimum can be removed by adding 3NF.

For the vector analyzing power A_y, the ND-TM and ND-UR calculations well reproduce the data at 200 and 270 MeV. At 140 MeV, however, both of the 3NF models overestimate the data slightly.

For the tensor analyzing powers A_{yy}, the difference between the data and the calculations without 3NF becomes clearer as an energy increases and this difference is described by neither the ND-TM calculations nor the ND-UR ones. The agreement is even deteriorated when 3NF is included and it is interesting that the effects by these two 3NFs are in the opposite direction.

The successful agreement of angular distribution of the cross section shows that the strengths of TM and UR type of 3NFs are right size, although spin dependent part of them may have some deficiencies.

SUMMARY

Cross sections and deuteron analyzing powers A_y, A_{yy}, A_{xx} and A_{xz} for the d-p elastic scattering have been measured at $E_d = 140$, 200 and 270 MeV which cover the wide angular range of $\theta_{c.m.} = 10° - 180°$. Highly accurate data have been obtained. These results are compared with the Faddeev calculations with and without the Tucson-Melbourne 3NF or the Urbana–Argonne 3NF. The large difference where the cross section takes minimum can be removed by including 3NF, although the analyzing power data are not always explained by adding 3NF.

To conclude, the discrepancy in the cross section minimum is considered to be one of the clearest signatures of 3NF effects. However, the insufficient reproduction of analyzing powers might indicate some ambiguities in the spin-dependent part of these 3NF models.

ACKNOWLEDGMENT

We would like to thank H. Witała, W. Glöckle and H. Kamada for their strong theoretical support. We would also like to thank S. Nemoto and P. U. Sauer for their useful comments on theoretical issues. We would also like to express our appreciation to the continuous help of the staff of RIKEN Accelerator Research Facility.

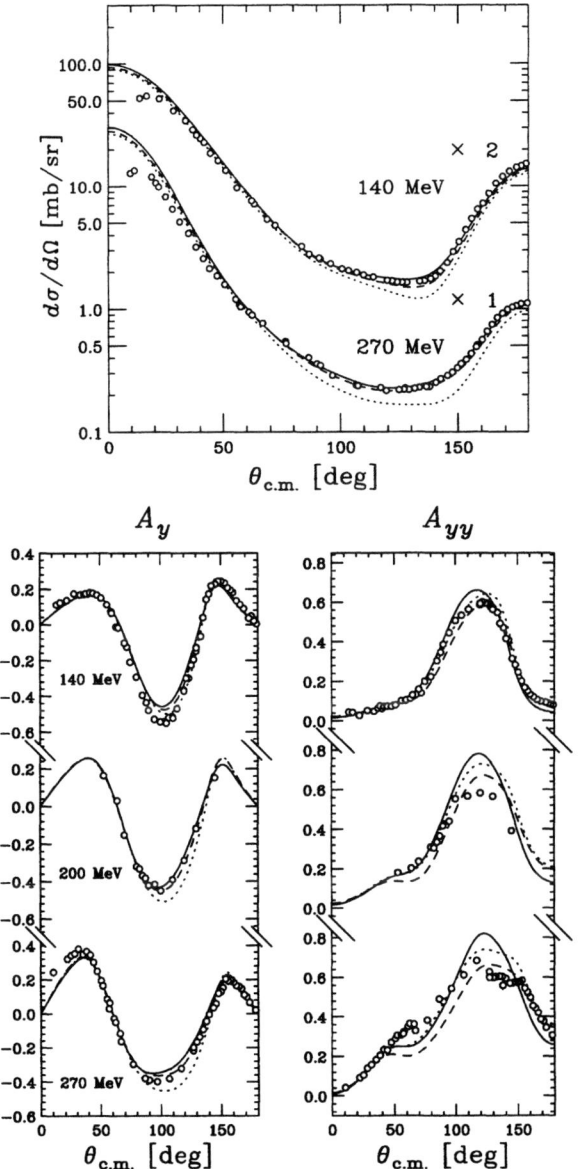

FIGURE 1. Cross sections and vector and tensor analyzing powers A_y, A_{yy} for the d–p elastic scattering at $E_d = 140$, 200 and 270 MeV. The open circles are the experimental results. The dotted curves are the Faddeev calculations in which the Argonne v_{18} potential is used as NN interactions. The solid and dashed curves are the Faddeev calculations including the Tucson-Melbourne and Urbana–Argonne 3NF, respectively.

REFERENCES

1. J. Fujita and H. Miyazawa, *Prog. Theo. Phys.* **17** 360 (1957).
2. S.A. Coon and W. Glöckle, *Phys. Rev.* **C23** 1790 (1981).
   S.A. Coon et al., *Nucl. Phys.* **A317**, 242 (1979).
3. B. S. Pundliner, R. Pandharipande, J. Carlson, Steven C. Pieper and R. B. Wiringa, *Phys. Rev.* **C56**, 1720 (1997).
4. N. Sakamoto, H. Okamura, T. Uesaka et al., *Phys. Lett. B* **367**, 60 (1996).
5. H. Sakai, K. Sekiguchi, H. Witała et al., *Phys. Rev. Lett.* **84**, 5288 (2000).
6. H. Witała et al., *Phys. Rev. Lett.* **81**, 1183 (1998).
7. S. Nemoto et al., *Phys. Rev.* **C58**, 2599 (1998).
8. H. Okamura et al., AIP Conf. Proc. **293**, 84 (1994),
   H. Okamura et al., *ibid.* **343**, 123 (1995).
9. T. Ichihara et al., *Nucl. Phys.* **A569**, 287c (1994).
10. R. A. Arndt and L. D. Roper, *Scattering Analysis Interactive Dial-In program* (SAID), Virginia Polytechnic Institute and State University (unpublished).
11. H. Witała et al. (to be published).
12. R. B. Wiringa, V. G. J. Stoks, and R. Schiavilla, *Phys. Rev.* **C51** 38 (1995).
13. R. Machleidt, F. Sammarruca, R. Schiavilla, *Phys. Rev.* **C53**, R1483 (1996).
14. V. J. G. Stoks, R. A. M. Klomp, C. P. F. Terheggen, J. J. de Swart, *Phys. Rev.* **C49** 2950 (1994).
15. W. Glöckle, H. Witala, D. Hüber, H. Kamada, and J. Golak, *Phys. Rep.* **274** 107 (1996).

Measurement of differential cross sections and vector analyzing powers for the $\vec{n}d$ scattering at 250MeV

Y. Maeda[a], H. Sakai[a], K. Hatanaka[c], H. Okamura[b], A. Tamii[a],
T. Wakasa[c], K. Yako[a], K. Sekiguchi[a], S. Sakoda[a], J. Kamiya[c],
K. Suda[b], H. Kato[a], M. Hatano[a], D. Hirooka[c], T. Saito[a],
N. Uchigashima[a], M.B. Greenfield[d], J. Rapaport[e]

[a] Department of Physics, University of Tokyo, Bunkyo, Tokyo 113-0033, Japan
[b] Department of Physics, Saitama University, Urawa, Saitama, 338-9570, Japan
[c] Research Center for Nuclear Physics (RCNP), Ibaraki, Osaka 567-0047, Japan
[d] Division of natural Science, International Christian University, Mitaka, Tokyo 181-8585, Japan
[e] Department of Physics, Ohio University, Athens, Ohio, 45701-2979, USA

Abstract. The differential cross sections and vector analyzing powers for the $\vec{n}d$ elastic scattering at $E_n = 250$ MeV have been measured at $\theta_{cm} = 85° - 180°$. Our aims are to study the three-nucleon force (3NF) without ambiguities from the Coulomb effect. The data are compared to the results of Faddeev calculations using modern nucleon-nucleon (NN) forces with and without 3NF. It is found that the calculations better reproduce the cross sections by including the 3NF but still underestimate the data.

INTRODUCTION

Recently the three-nucleon force (3NF) has been actively studied from both theoretical and experimental sides. 3NFs arise naturally in the standard meson exchange picture. One of the most well-known 3NF models is the Fujita-Miyazawa type [1] in which a nucleon is excited to a delta particle as an intermediate state in a 2π-exchange diagram among three nucleons.

We have measured angular distributions of the following observables for the $\vec{d}p$ elastic scattering at $E_d = 270$ MeV at RIKEN : differential cross sections, deuteron vector and tensor analyzing powers and the deuteron tensor to proton vector polarization transfer coefficients [2]. Faddeev calculations which include two-body NN forces alone significantly underestimate the cross section in the backward angular region. Adding the Tucson-Melbourne (TM) 3NF leads to an excellent agreement with the data. The deuteron vector analyzing power is also well reproduced. These results are considered as clear signatures of 3NF effects. However the inclusion of the TM-3NF rather deteriorates the agreement of tensor analyzing powers, which implies deficiencies in spin dependence of the TM-3NF.

In addition to that, since the inclusion of the Coulomb interaction into the calculation is difficult, 3NF effects have been discussed by the comparison between the $\vec{d}p$ data and the $\vec{d}n$ Faddeev calculations. The Coulomb effect is certainly dominant in the very forward angular region and causes large discrepancies between the $\vec{d}p$ and the $\vec{d}n$ differential cross sections. However it was pointed out that the Coulomb effect may remain even in the backward scattering region [3] where 3NF effects are expected to be large. To study 3NF in a Coulomb-free system and to reach a decisive conclusion about 3NF, we have measured the differential cross sections and vector analyzing powers for the $\vec{n}d$ elastic scattering at 250MeV.

EXPERIMENT

We have carried out the measurement at the (n,p) facility [4] at the Research Center for Nuclear Physics (RCNP). Figure 1 shows a schematic layout of the facility.

The polarized neutron beam was produced by the ^{7}Li$(\vec{p},\vec{n})$ reaction at $E_p = 250$ MeV. The differential cross section for the ^{7}Li$(p,n)^7$Be(g.s. and 0.4 MeV) reaction at $\theta_{cm}=0°$ is 27.0±0.8 mb/sr [5]. The thickness of the ^{7}Li target was 580 mg/cm^2.

We put the clearing magnet just downstream of the ^{7}Li target with inclining the magnet to the floor by 24°. The primary proton beam was bent by 23.2° by this magnet after passing through the ^{7}Li target and then transported to a beam dump in the floor.

We obtained an almost mono-energetic neutron beam with a flux of 2.7×10^5 neutrons/cm^2·s at the $\vec{n}d$ target position, i.e. 98 cm downstream of the ^{7}Li target. Because the polarization of the proton beam was 0.6 and D_{NN} of this production reaction is −0.28, the polarization of neutron beam was about 0.16.

We used deuterized polyethylene (CD$_2$) [6] as deuteron targets. Since the neutron beam was spread, a large target size was required (about 37 mm × 47 mm) to obtain a reasonable counting rate. The targets were contained in the MWDC box, called the target chamber. The target chamber has 6 planes of MWDC and the reaction points are determined by their tracking information. We used four CD$_2$ targets. Each target has a thickness of 220 mg/cm^2. These CD$_2$ targets were separated from one another by wire planes to distinguish in which target the reaction occurred. With this method, we could attain better energy resolution by correcting for the deuteron energy loss in the CD$_2$ target ($\Delta E = 1.4$ MeV) in part, than the case with using one thick target. Recoil deuterons were momentum analyzed by the Large Acceptance Spectrometer (LAS) and were detected at the focal plane. At the entrance of the LAS, we installed another MWDC (Front End Chamber, FEC) and we determined reaction angles by using information from the target chamber and FEC. The resolution of scattering angle was 0.1°. The measurement covered $\theta_{cm} = 85°$ – 180° angular region where the effects of 3NF are expected to be large.

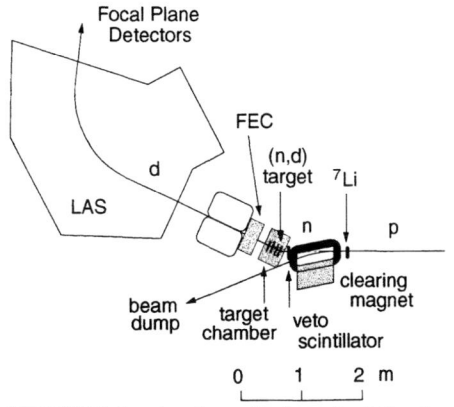

FIGURE 1. A schematic layout of the (n,p) facility at RCNP. The accelerated proton beam is injected to the ^{7}Li target from the right side of this figure. This proton beam is bent by the clearing magnet to the beam dump (not shown). The produced neutrons pass through a scintillator, which is used as a charged particle veto, and bombard the $\vec{n}d$ targets in the target chamber. Recoil deuterons are tracked by two MWD-Cs (the target chamber and FEC) and momentum analyzed by LAS.

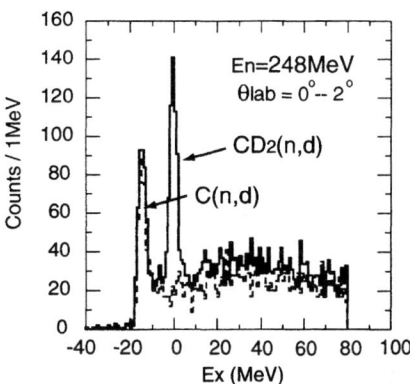

FIGURE 2. Excitation energy spectrum for the $CD_2(n,d)$ (solid lines) and $C(n,d)$ (dashed lines) reactions at $E_n = 248$ MeV.

We also used C targets (180 mg/cm^2 × 4) and measured the $C(n,d)$ reaction to subtract the events of the C target from those of the CD_2 target. Figure 2 shows the spectra of (n,d) reactions at $\theta_{LAB} = 0° - 2°$. Two sharp peaks are due to the ^{2}H(n,d)n and ^{12}C$(n,d)^{11}$B$(g.s.)$ reactions, respectively. In order to obtain the value of the neutron flux and the acceptance of the LAS, we measured the $\vec{n}p$ elastic scattering with CH$_2$ (190 mg/cm^2 × 4) targets.

RESULTS AND DISCUSSIONS

The preliminary data of the differential cross sections and the vector analyzing powers are shown in Figure 3 by solid circles. We determined the amount of neutrons at the target position by comparing the results of $\vec{n}p$ measurements and the calculations with the program code SAID [8], then we extracted the absolute values of the differential cross sections.

The dashed and dotted curves are the Faddeev calculations [7] of $\vec{n}d$ elastic scattering with using the CD-BONN potential (the total angular momenta j≤6 are included) [9] and Argonne v_{18} potential (AV18) [10] as NN interactions, respectively. From these two calculations, we can consider that the effect due to the ambiguity of

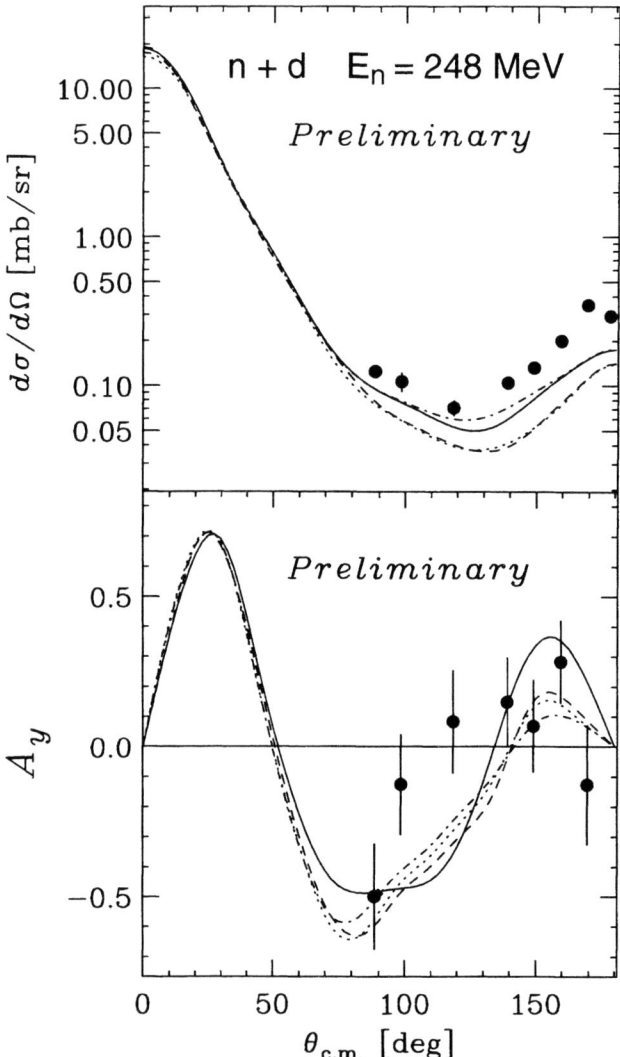

FIGURE 3. Differential cross sections and vector analyzing powers for the $\vec{n}d$ elastic scattering at $E_n = 248$ MeV. The solid circles are the results of this experiment. The statistical errors are shown in the figures. The calculations with CD-BONN potential (dashed curves), AV18 potentials (dotted curves), CD-BONN potential including TM-3NF (solid curves) and AV18 potential including Urbana 3NF (dot-dashed curves) are also shown.

NN interaction is small. The solid curves are the calculations with the CD-BONN potential (j≤5) including Tucson-Melbourne 3NF [11] and the dot-dashed curves are with the AV18 potential including Urbana 3NF [12]. We can recognize that the calculations including 3NF slightly better reproduce the data of cross sections. But the calculations still underestimate the data by 30%. These discrepancies may be considered as an indication of relativistic effects [13,14] which is not accounted in the present calculations.

The statistical errors of the vector analyzing powers are still large to be compared with the calculations because only one third of the total data are analyzed.

We have also measured $\vec{p}d$ elastic scattering at $E_p = 250$ MeV [15]. The data are now under analysis. We can extract the Coulomb effect by directly comparing the $\vec{n}d$ and the $\vec{p}d$ results.

Acknowledgement The authors would like to thank Dr. H. Kamada for his valuable calculations. This project is supported by the Ministry of Education, Science, Sports and Culture of Japan with the Grant-in-Aid for Science Research No. 10304018.

REFERENCES

1. J.I. Fujita and H. Miyazawa, *Prog. Theor. Phys.* **17**, 360 (1957).
2. H. Sakai et al., *Phys. Rev. Lett.* **84**, 5288 (2000).
3. Y. Koike, private communication.,
   N. Sakamoto et al., *Phys. Lett.* B **367**, 60 (1996).
4. K. Yako et al., RCNP Annual Report 1999.
5. T.N. Taddeucci et al., *Phys. Rev.* C **41**, 2548 (1990).
6. Y. Maeda et al., poster session at this conference.
7. H. Kamada, private communication.
8. CNS DAC online Services, http://gwdac.phys.gwu.edu/
9. R. Machleidt et al., *Phys. Rev.* C **53**, R1483 (1996).
10. R.B. Wiringa et al., *Phys. Rev.* C **51**, 38 (1995).
11. S.A. Coon et al., *Nucl. Phys.* A **317**, 242 (1979).
12. J.L. Forest et al., *Phys. Rev.* C **60**, 014002 (1999).
13. H. Rohdjeß et al., *Phys. Rev.* C **57**, 2111 (1998).
14. H. Witała et al., *Phys. Rev.* C **57**, 2111 (1998).
15. D. Hirooka et al., poster session at this conference.

Forward Scattering Amplitudes and Contributions of Three-Nucleon Forces in nd Elastic Scattering

S. Ishikawa*, M. Tanifuji*, and Y. Iseri[†]

*Department of Physics, Hosei University, Fujimi 2-17-1, Chiyoda, Tokyo 102-8160, Japan
[†]Department of Physics, Chiba-Keizai College, Todoroki 4-3-30, Inage, Chiba 263-0021, Japan

Abstract. In neutron-deuteron scattering, four total cross sections are shown to form a complete set for the determination of the imaginary parts of the forward scattering amplitudes by means of the optical theorem. Contributions of three-nucleon forces to the forward amplitudes, which are decomposed into scalar and tensor components in spin space, are numerically studied.

I INTRODUCTION

Spin observables of the three-nucleon scatterings are expected to provide significant information of three-nucleon forces (3NF). The most frequently used models of 3NF in the recent theoretical investigations are based on the exchange process of two pions among three nucleons, which is related to the spin-tensor components of nucleon-nucleon (NN) interaction as well as the spin-scalar ones. This characteristic should be reflected on spin structure of nucleon-deuteron scattering amplitudes. However, in general, expressions of the scattering observables are given in quadratic form of the scattering amplitudes, which means that the observables are results of interference between the amplitudes. So, it is not so easy to get clear-cut information on the spin structure of the amplitudes from the scattering observables. Here, we notice the use of the optical theorem, in which total cross section corresponding to an initial spin state described by spin density matrix ρ_i, is related with imaginary part of the forward scattering amplitude as,

$$Im\left[Tr\left(\rho_i M(0°)\right)\right] = \frac{k}{4\pi}\sigma^{tot}_{\rho_i}. \qquad (1)$$

An adequate choice of the spin density matrices will lead to the determination of matrix elements of the forward scattering amplitude by measuring the corresponding cross sections. In this paper, firstly, we will present a complete set of total cross sections to determine the imaginary part of the neutron-deuteron (nd)

forward scattering amplitudes. Next, we decompose the amplitudes by spin space operators to see how these amplitudes are related with the particular components of the nuclear force. Finally, we will show some numerical results to show how these amplitudes are affected by 3NF.

II ND FORWARD ELASTIC SCATTERING AMPLITUDE

The elastic scattering of spin-1/2 and spin-1 particles is described by $6 \times 6 = 36$ scattering matrix elements, among which 12 components are independent due to the parity conservation and time reversal invariance. At the forward angle, we have four independent amplitudes to be non-zero:

$$M_1 = \langle \tfrac{1}{2}, 1 | \boldsymbol{M} | \tfrac{1}{2}, 1 \rangle_{\theta=0},$$
$$M_2 = \langle -\tfrac{1}{2}, 1 | \boldsymbol{M} | -\tfrac{1}{2}, 1 \rangle_{\theta=0},$$
$$M_3 = \langle -\tfrac{1}{2}, 1 | \boldsymbol{M} | \tfrac{1}{2}, 0 \rangle_{\theta=0} = \langle \tfrac{1}{2}, 0 | \boldsymbol{M} | -\tfrac{1}{2}, 1 \rangle_{\theta=0},$$
$$M_4 = \langle \tfrac{1}{2}, 0 | \boldsymbol{M} | \tfrac{1}{2}, 0 \rangle_{\theta=0}. \qquad (2)$$

Therefore, the measurements of four independent cross sections will determine the imaginary parts of these amplitudes. We will consider the following four cross sections: the one for the scattering of unpolarized neutrons by unpolarized deuterons (σ_0^{tot}), that of unpolarized neutrons by tensor (t_{20}) polarized deuterons (σ_{20}^{tot}), the cross section asymmetries for the scattering of vector polarized neutrons by vector polarized deuterons, $\Delta\sigma_L$ and $\Delta\sigma_T$, where the spin quantization axis is parallel to the beam direction in the former while is perpendicular in the latter [1]. With these cross sections, we obtain the following:

$$\alpha \mathrm{Im}(M_1) = \sigma_0^{tot} + \frac{1}{\sqrt{2}}\sigma_{20}^{tot} - \frac{1}{2}\Delta\sigma_L,$$
$$\alpha \mathrm{Im}(M_2) = \sigma_0^{tot} + \frac{1}{\sqrt{2}}\sigma_{20}^{tot} + \frac{1}{2}\Delta\sigma_L,$$
$$\alpha \mathrm{Im}(M_3) = -\frac{1}{\sqrt{2}}\Delta\sigma_T,$$
$$\alpha \mathrm{Im}(M_4) = \sigma_0^{tot} - \sqrt{2}\sigma_{20}^{tot}, \qquad (3)$$

where $\alpha = \frac{4\pi}{k}$ with k being the magnitude of the nd relative momentum. Therefore, the measurements of these cross sections shall determine the imaginary part of the nd elastic amplitudes completely.

Next we will consider a representation of the nd forward amplitude by spin operators. Analyses of the spin structure of the amplitude by the invariant amplitude

[2,3] give that the forward amplitude consists of two scalar and two tensor amplitudes. From this consideration, we write the scattering amplitude at the forward angle as the following equation.

$$\boldsymbol{M}(\theta=0) = S_0 + S_\sigma\,(\boldsymbol{s}_n \cdot \boldsymbol{s}_d) + W_D[\boldsymbol{s}_d \otimes \boldsymbol{s}_d]_0^2 + W_T[\boldsymbol{s}_n \otimes \boldsymbol{s}_d]_0^2. \quad (4)$$

Using Eqs. (2) and (3), the imaginary parts of the above amplitudes are give by the cross sections, σ_0^{tot}, $\Delta\sigma_L$, $\Delta\sigma_T$, and σ_{20}^{tot}, as follows:

$$\alpha\mathrm{Im}(S_0) = \sigma_0^{tot},$$
$$\alpha\mathrm{Im}(S_\sigma) = -\frac{1}{3}\left(\Delta\sigma_L + 2\Delta\sigma_T\right),$$
$$\alpha\mathrm{Im}(W_D) = \frac{3}{\sqrt{2}}\sigma_{20}^{tot},$$
$$\alpha\mathrm{Im}(W_T) = -\left(\Delta\sigma_L - \Delta\sigma_T\right). \quad (5)$$

There is an advantage for the representation of the scattering amplitude in the form of Eq. (4). If we consider a simple folding potential between the neutron and the deuteron neglecting antisymmetrizations and other reaction mechanisms, and assuming that the NN interaction consists of spin-independent and spin-spin central components, and tensor one, the relation between $S_0 \sim W_T$ and the nuclear force components turns to be rather straightforward. In the first order approximation, it is easily shown that S_0 and S_σ are provided by the spin-independent and spin-spin NN interactions, respectively, with the S-state component of the deuteron internal wave function, W_D is derived from the NN central interactions with the D-state component, and W_T from the NN tensor interaction with the deuteron S-state component. Therefore, the measurements of σ_0^{tot}, $\Delta\sigma_L$, $\Delta\sigma_T$, and σ_{20}^{tot} would provide rather pure information on the respective interactions.

It is well known that there is a correlation between calculated values of triton binding energy (B_3) and those of the nd s-wave doublet scattering length, which has been known as "Phillips relation". This means that if we use a nuclear force model to reproduce the binding energy, the same interaction reproduces the correlated observable. Therefore these observables do not give any further information on 3NF than B_3. In this context, we introduce another representation for the spin-scalar amplitudes due to the channel-spin, which, in the present case, takes the values of 1/2 (the doublet scattering) or 3/2 (the quartet scattering):

$$S(1/2) = \sqrt{2}\,(S_0 - S_\sigma),$$
$$S(3/2) = 2S_0 + S_\sigma. \quad (6)$$

III NUMERICAL RESULTS

The total cross sections for the complete set at low incident energies are obtained by the Faddeev calculation, in which the NN force (2NF) is fixed to the Argonne V_{18}

potential (AV18) [4] while the 3NF is the 2π exchange Brazil model (BR-3NF) [5] with the cut-off parameter adjusted so as to reproduce the empirical triton binding energy. Due to the 2π exchange mechanism, the BR-3NF is expected to contribute to not only scalar nuclear forces but also tensor ones in the spin space. In order to demonstrate the role of the tensor forces, a fictitious Gaussian 3NF which is the spin independent force (GS-3NF),

$$V_G = V_0^G \exp\{-(\frac{r_{21}}{r_G})^2 - (\frac{r_{31}}{r_G})^2\} + (c.p.), \qquad (7)$$

is examined with $r_G = 1.0$ fm and $V_0^G = -45$ MeV, which are fixed so as to reproduce the empirical triton binding energy.

Numerical calculations of the low energy nd scattering are performed by solving the 3N Faddeev equation in the coordinate space [6–8]. In the present calculation, 3N partial wave states for which 2NF and 3NF act, are restricted to those with total NN angular momenta $j \leq 2$. The total 3N angular momentum (J) is truncated at $J = 19/2$.

In Fig. 1, $\mathrm{Im}(S_{1/2})$, $\mathrm{Im}(S_{3/2})$, $\mathrm{Im}(W_D)$, and $\mathrm{Im}(W_T)$ are shown by cross sections, $\sigma_A = \alpha \mathrm{Im}(A)$, where A is $S_{1/2}$ etc. as a function of the neutron incident energy up to 15 MeV. In the figure, one can see the following characteristics of the calculated amplitudes. $\mathrm{Im}(W_D)$, which is expected to be sensitive to the deuteron D-state, is

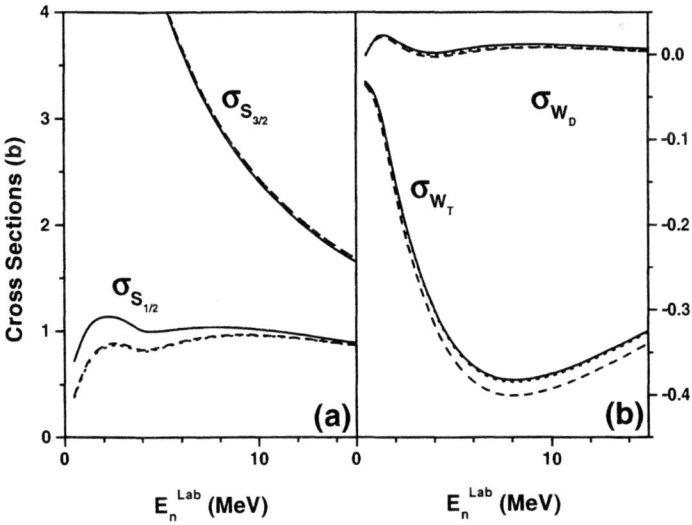

FIGURE 1. The total cross sections of the nd scattering as a function of incident neutron energy in laboratory system for AV18 (solid lines), AV18+BR-3NF (dashed lines), and AV18+GS-3NF (dotted lines). The dashed line and the dotted one are overlapped with each other for $\sigma_{S_{1/2}}$.

small compared to other three cross sections. The effect of the 3NF on the scalar amplitude for the quartet state is very small, while that for the doublet state is remarkable, particularly at low incident energies. In the latter, however, we cannot distinguish the effect of the BR-3NF from that of the GS-3NF as is expected from the Phillips relation. On the other hand, the effect of 3NF on the tensor amplitude W_T has an interesting feature. One can see that the effect of the BR-3NF on $\mathrm{Im}(W_T)$ is visible at large incident energies. However that of the GS-3NF is almost negligible. This is what to be expected from the consideration that W_T is directly related with the NN tensor interaction.

IV SUMMARY AND CONCLUSION

We have shown that the four non-vanishing independent forward amplitudes in the nd elastic scattering consist of two scalar amplitudes and two tensor amplitudes, which are respectively related to the two central interactions and the two tensor ones, and that one can determine experimentally the imaginary parts of these scattering amplitudes by measuring four cross sections, σ_0^{tot}, $\Delta\sigma_L$, $\Delta\sigma_T$, and σ_{20}^{tot}. By the Faddeev calculation, the spin dependence of the 3NF contribution is investigated and it is found that one of the tensor amplitude, W_T, is most sensitive to the 3NF except for very low energies and it provides the information of the tensor effect of the 3NF. These predictions will be encouraging the measurements of the total cross sections to obtain a significant information of the interaction between three nucleons.

REFERENCES

1. Witała, H., Glöckle, W., Golak, J., Hüber, D., Kamada, H., and Nogga, A., *Phys. Lett. B* **447**, 216 (1999).
2. Tanifuji, M., and Yazaki, K., *Prog. Theor. Phys.* **40**, 1023 (1968).
3. Tanifuji, M., Ishikawa, S., and Iseri, Y., *Phys. Rev. C* **57**, 2493 (1998).
4. Wiringa, R.B., J. Stokes, V.G., and Schiavilla, R., *Phys. Rev. C* **51**, 38 (1995).
5. Robilotta, M.R., and Coelho, H.T., *Nucl. Phys.* **A460**, 645 (1986).
6. Sasakawa, T., and Ishikawa, S., *Few-Body Syst.* **1**, 3 (1986); Ishikawa, S., and Sasakawa, T., *ibid.* **1**, 143 (1986).
7. Ishikawa, S., *Nucl. Phys.* **A463**, 145c (1987).
8. Ishikawa, S., Wu, Y., and Sasakawa, T., *Proceedings of the Few-Body Problems in Physics*, edited by F. Gross, AIP Conference Proceedings **334**, New York, 1995, p. 840.

7. POLARIZED BEAMS AND POLARIMETERS

Polarized Proton Beam Development at COSY with EDDA as a Fast Internal Polarimeter

F. Hinterberger for the COSY Team[1] and the EDDA Collaboration[2]

ISKP, University of Bonn, Nussallee 14-16, D-53115 Bonn, E-mail: fh@iskp.uni-bonn.de

Abstract. Polarized protons in the Cooler Synchrotron COSY encounter five imperfection and nine intrinsic depolarizing resonances during the acceleration from 300 to 3300 MeV/c. When crossing imperfection resonances vertical correction dipoles are excited in order to enhance the average vertical displacement and thereby the resonance strength to result in a complete spin flip without loss of polarization. When crossing intrinsic resonances a rapid vertical tune jump is applied to minimize polarization losses. In order to find the optimum machine parameters a novel and fast method was developed to measure the internal beam polarization as a function of the beam momentum in the vicinity of a depolarizing resonance as well as in the full acceleration ramp. Using very thin internal CH_2- and/or C-fiber targets the polarization is deduced from the left-right asymmetry of fast scaler rates. To this end the EDDA detector is used. This detector consists of two cylindrical scintillation hodoscope layers covering about 87 % of 4π for pp elastic scattering. The effective analyzing power of the fast method is obtained by a special calibration procedure using a "slow but proper" EDDA-style measurement of the elastic pp scattering asymmetries. For this calibration precise analyzing power excitation functions measured by EDDA became available in time.

---

[1] COSY Team: U.Bechstedt[a], J. Dietrich[a], R. Gebel[a], K. Henn[a], A. Lehrach[b], B. Lorentz[a], R. Maier[a], D. Prasuhn[a], P. von Rossen[a], A. Schnase[a], H. Schneider[a], R. Stassen[a], H. Stockhorst[a], R. Tölle[a]
[a] Forschungszentrum Jülich, Germany, [b] Brookhaven National Laboratory, USA
[2] EDDA Collaboration: M. Altmeier[a], F. Bauer[b], J. Bisplinghoff[a], K. Büßer[b], M. Busch[a], T. Colberg[b], L. Demirörs[b], O. Diehl[a], F. Dohrmann[b], H.P. Engelhardt[a], P.D. Eversheim[a], O. Eyser[b], O. Felden[a], R. Gebel[c], M. Glende[c], J. Greiff[b], R. Groß-Hardt[a], F. Hinterberger[a], R. Jahn[a], E. Jonas[b], H. Krause[b], C. Lehmann[c], T. Lindemann[b], J. Lindlein[b], B. Lorentz[c], R. Maier[c], R. Maschuw[a], A. Meinerzhagen[a], O. Nähle[a], D. Prasuhn[c], C. Pauly[b], H. Rohdjeß[a], D. Rosendaal[a], P. von Rossen[c], N. Schirm[b], V. Schwarz[a], W. Scobel[b], H.J. Trelle[a], E. Weise[a], A. Wellinghausen[b], T. Wolf[b], R. Ziegler[a]
[a]University of Bonn, [b]University of Hamburg, [c]Forschungszentrum Jülich
Supported by BMBF and Forschungzentrum Jülich

POLARIZED PROTON BEAM IN COSY

The paper presents the current status of the polarized proton beam development at the cooler synchrotron COSY [1] using the internal target and the detector of the EDDA experiment [2] as a fast polarimeter. It refers to previous work on polarized proton beam development [3] and fast polarimetry [4].

The cooler synchrotron COSY [1] accelerates protons from 300 MeV/c (45 MeV) up to 3300 MeV/c (2500 MeV). The layout is shown in Fig. 1. COSY is a race-track synchrotron with two long straight sections for internal target experiments. Electron cooling as well as stochastic cooling [5] can be used to improve the beam quality, i.e. the phase space density of the beam. At the moment four internal and three external target stations are installed at COSY.

The starting point of the polarized proton beam is a colliding beams source [6]. The polarized H^- beam is preaccelerated in the Jülich cyclotron up to 45 MeV and injected into COSY via stripping injection. The internal target experiment EDDA [2] serves as a fast polarimeter for the polarization development. In addition, there are a low-energy polarimeter in the injection beam line, an external high-energy polarimeter and another internal polarimeter [7].

The lattice is characterized by two arcs with six unit cells and two long straight sections. By adjusting telescopic optics the straight sections can be matched horizontally as well as vertically to a phase advance of 2π. Then supersymmetries 2 and 6 can be achieved and have been used previously [3]. However, in the normal

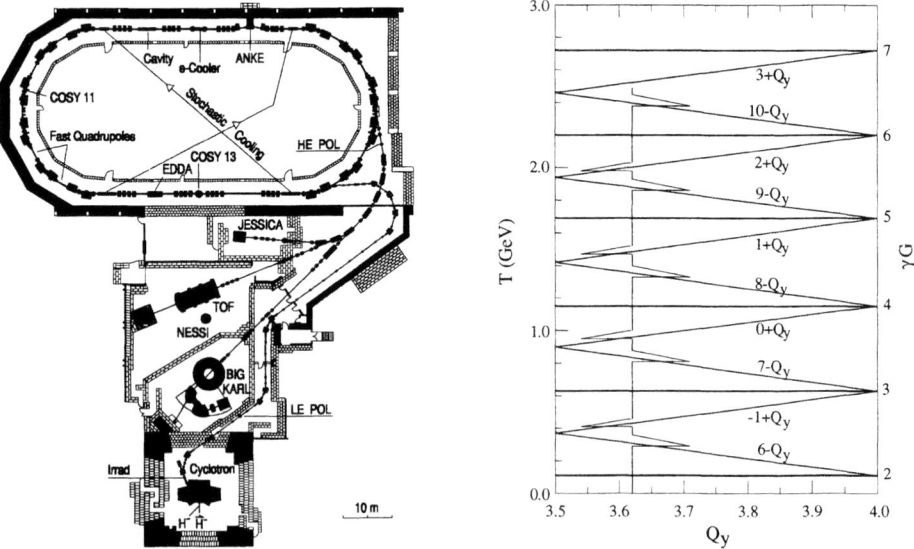

FIGURE 1. Layout of the cooler synchrotron COSY (left). Depolarization resonances (right).

mode of operation supersymmetry is not reached due to installations breaking this symmetry.

In Fig. 1 the scheme of depolarization resonances in COSY is plotted as a function of the kinetic energy T and the vertical betatron tune Q_y. On the right axis the corresponding spin tune γG is indicated. There are five imperfection resonances at γG equal to 2,3,4,5, and 6 and nine intrinsic resonances in between. The imperfection resonances are crossed by exciting one of the vertical steerer magnets during resonance crossing. By this method the resonance strength is enhanced and the polarization is flipped adiabatically without polarization loss. The intrinsic resonances are crossed by fast tune jumps using two pulsed quadrupoles. The sequence of alternating tune jumps is indicated schematically in the plot. By a careful adjustment of those tune jumps the intrinsic resonances can be passed with negligible polarization losses. Between the intrinsic resonances the vertical betatron tune is about 3.62.

EDDA AS A FAST INTERNAL POLARIMETER

The EDDA experiment is designed for a high precision measurement of proton–proton elastic scattering excitation functions ranging from 0.5 to 2.5 GeV laboratory kinetic energy [2]. To this end an internal target and the internal proton beam of COSY are used. A continuous energy variation is achieved by measuring during the beam acceleration. Each machine cycle lasting about 10 s yields a complete excitation function. In order to get statistics many thousand cycles are accumulated in a multipass technique. Unpolarized differential cross sections have been measured as a function of energy using very thin (4 μm × 5 μm) CH_2 fiber targets and Carbon fiber targets of 5 or 7 μm diameter for background subtraction. For the measurement of the analyzing power A_N [8] and the polarization correlation parameters A_{NN} A_{SS}, and A_{SL} [9] an internal polarized atomic hydrogen beam is used as target [10].

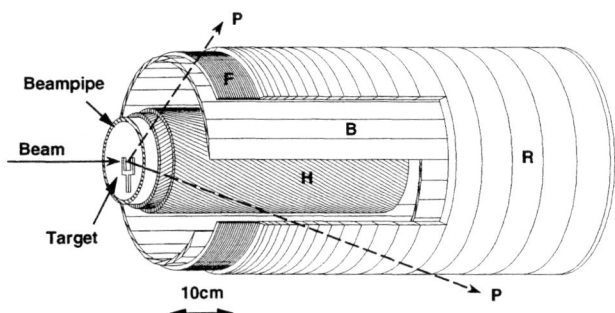

FIGURE 2. Scheme of the EDDA detector. Targets: CH_2 fiber, C fiber and polarized atomic hydrogen beam; H: inner hodoscope of helically wound scintillating fibers; B: scintillator bars; R: scintillator semi-rings; F: semi-rings from scintillating fibers.

The EDDA detector (see Fig. 2) consists of two cylindrical hodoscopes. The inner hodoscope is made up of scintillating fibers which are helically wound in opposing directions around the beam pipe. The outer hodoscope is comprised of scintillator semi-rings and bars. The semi-rings close to the target are made from scintillating fibers. The trigger is based on the coplanarity and the kinematic correlation of the two outgoing protons. EDDA can be used as a very fast polarimeter by counting only the signals from the outer left and right semi-rings. The CH_2 and C fiber targets are moved at a preselected time of the acceleration ramp into the beam using a fast linear motor. Background is suppressed to some degree by requiring kinematic coincidence between the signals from the left and right semi-rings. Since the method of counting two-prong events with fast scalers does not require any digitizing of data and event reconstruction it is dead-time free and very fast.

The fast polarimetry was calibrated in a dedicated calibration run by measuring the effective analyzing power $A(p, \Theta)$ of CH_2 and C fiber targets. To this end, exclusive elastic proton-proton scattering data were taken simultaneously but due to data acquisition dead-time with much reduced beam currents and countrates. The calibration data were taken as a function of beam momentum p during the acceleration of a polarized COSY beam. For the calibration precise elastic pp analyzing power data $A_{pp}(p, \Theta)$ measured by EDDA with an unpolarized COSY beam and a polarized atomic beam target became available in time [8]. The effective analyzing power $A(p, \Theta)$ was deduced from

$$A(p, \Theta) = A_{pp}(p, \Theta) \frac{\epsilon(p, \Theta)}{\epsilon_{pp}(p, \Theta)}. \tag{1}$$

Here, ϵ is the "fast but inclusive" two-prong ring scaler asymmetry and ϵ_{pp} the "slow but exclusive" elastic pp scattering asymmetry. The effective analyzing power of the CH_2 targets is roughly a factor of two larger than that of the C targets. But the CH_2 targets have the disadvantage that the hydrogen content and therewith the effective analyzing power slowly decreases due to radiaton damage. Therefore, we decided to use only C fiber targets of 5 or 7 μm diameter for the polarized beam development.

PROTON BEAM POLARIZATION

In Fig. 3 the beam polarization is plotted as a function of the beam momentum between 1.3 and 3.3 GeV/c. Beam intensities of 2 to $4 \cdot 10^9$ polarized protons in the ring have been achieved. The depolarization resonances are indicated by dashed lines. At low momenta the beam polarization is about 80 %, at high momenta, that means above the 8- intrinsic resonance, it is about 75 %. Such a measurement over a wide momentum range can be performed within half an hour. For a small momentum range in the vicinity of a depolarization resonance it can be even faster. There, the polarization can be measured with sufficient accuracy within

5-10 minutes. As a consequence fast turn-around times are possible in fine tuning the machine.

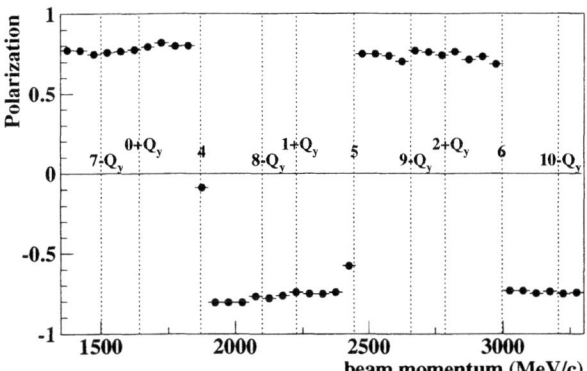

FIGURE 3. Proton beam polarization as a function of beam momentum. The dashed lines indicate the position of imperfection and intrinsic resonances.

A nice feature of the method to move the horizontal fiber target from below into the beam is the possibility to study the beam polarization as a function of the vertical betatron amplitude. This can be done by scanning the beam polarization as a function of the vertical position of the fiber target. Such measurements were performed several times during the beam development. Fortunately, for the final machine setup no vertical position dependence was found within errors.

Concluding, the EDDA detector has been calibrated and used as fast polarimeter for beam development at COSY in the momentum range 1.3 − 3.3 GeV/c. The imperfection resonances are crossed by exciting one of the vertical steerer magnets, intrinsic resonances by exciting fast tune jumps. The proton polarization achieved is about 80 % at low momenta and 75 % at high momenta.

REFERENCES

1. R. Maier, Nucl. Instr. and Meth. A 390, 1 (1997)
2. D. Albers et al. Phys. Rev. Lett 78, 1652 (1997)
3. A. Lehrach et al., Nucl. Instr. and Meth. A 439, 26 (2000)
4. V. Schwarz et al., Proc. 13th Int. Symp. on High Energy Spin Physics, Protvino (1998), eds. N.E. Tyurin et al., World Scientific Singapure (1999) 560
5. D. Prasuhn et al., Nucl. Instr. and Meth. A 439, 26 (2000)
6. P.D. Eversheim et al., Proc. 12th Int. Symp. on High Energy Spin Physics, Amsterdam (1996), eds. C.W. de Jager et al., World Scientific Singapure (1997) 306
7. F. Bauer et al., Nucl. Instr. and Meth. A 431, 385 (1999)
8. M. Altmeier et al., Phys. Rev. Lett. 85, 1819 (2000)
9. F. Bauer, these Proceedings
10. P.D. Eversheim et al., Nucl. Phys. A626, 117c (1997)

Spin-flipping with an rf-dipole and a full Siberian snake*

A.M.T. Lin[1]**, B.B. Blinov[1], Ya.S. Derbenev[1], T. Kageya[1],
D.Yu. Kantsyrev[1(a)], A.D. Krisch[1], V.S. Morozov[1(a)], J.R. Murray[1],
D.W. Sivers[1(b)], V.K. Wong[1], K. Yonehara[1],
V.A. Anferov[2], C.M. Chu[2], P. Schwandt[2], B. von Przewoski[2],
V.N. Grishin[3], V.L. Solovianov[3],
K. Jacobs[4], and G.T. Zwart[4]

[1] *Spin Physics Center, University of Michigan, Ann Arbor, MI 48109-1120*
[2] *Indiana University Cyclotron Facility, Bloomington, IN 47408-0768*
[3] *IHEP, Protvino, Russia 142 284*
[4] *MIT-Bates Linear Accelerator Center, 21 Manning Ave., Middleton, MA 01949*

Abstract.
We recently used a vertical-field rf-dipole magnet to study the spin-flipping of a 120 MeV horizontally polarized proton beam stored in the presence of a nearly-full Siberian snake in the IUCF Cooler Ring. The spin was flipped by ramping the rf-dipole's frequency through an rf-induced depolarizing resonance. After optimizing the frequency ramp parameters, we used multiple spin-flips to measure a maximum spin-flip efficiency of 86.5±0.5% in April 2000, and 92.5±0.5% in June 2000. The spin-flip efficiency was apparently limited by the maximum achievable current in the rf-dipole. This result indicates that spin-flipping a stored polarized proton beam should be possible in high energy rings such as RHIC (and perhaps HERA in the future), where Siberian snakes are utilized and the dipole rf-flipper-magnets should be quite practical. During the June 2000 run, a new *faster* technique of locating the rf depolarizing resonance frequency was developed.

---

* Supported by research grants from the U.S. Department of Energy and the U.S. National Science Foundation.
** E-mail: alilin@umich.edu
(a) Also at: Moscow State University, Moscow, Russia.
(b) Also at: Portland Physics Institute, Portland, OR 97201, USA.

INTRODUCTION

Siberian snake experiments in the IUCF Cooler Ring have been instrumental in understanding the spin dynamics of a polarized proton beam in a storage ring. In this past year, we studied spin-flipping of a polarized proton beam at the Cooler Ring. Frequent reversals of the beam polarization direction should help to significantly reduce systematic errors in a scattering experiment's spin asymmetry measurements. We previously used an rf solenoid to spin-flip, with 97 ± 1% efficiency, a horizontally polarized proton beam stored in the Cooler Ring containing a Siberian snake. [1] However, a solenoid's spin rotation decreases linearly with energy because of the Lorentz contraction of its $\int B \cdot dl$; thus, a solenoid would be impractical for spin-flipping in high energy rings. On the other hand, a transverse rf dipole's spin rotation remains constant, independent of energy and thus should be more practical at high energies. As a first step, we earlier succeeded in spin-flipping efficiently with an rf-dipole in the absence of a Siberian snake. [2]

In a storage ring, each proton's spin precesses around the Stable Spin Direction (**SSD**) with a frequency of:

$$f_s = f_c \, \nu_s ; \tag{1}$$

where f_c is the circulation frequency, and ν_s is the **spin tune**, which is the number of spin precessions during one turn around the ring. With a Siberian snake in the ring, the spin tune is:

$$\nu_s = \frac{1}{\pi} cos^{-1} \left(cos(\pi G \gamma) cos(\pi s) \right) ; \tag{2}$$

where $G = (g-2)/2 = 1.792847$ is the proton's anomalous magnetic moment, γ is its Lorentz energy factor, and s is the **snake strength**. For a full (100%) snake, $s = 1$, and then $\nu_s = \frac{1}{2}$, independent of energy.

Spin-flipping with a full Siberian snake

One can spin-flip a horizontally polarized proton beam in a ring with a Siberian snake using a depolarizing resonance induced by a vertical-field rf-dipole; similarly, a horizontal-field rf-dipole can spin-flip a vertically polarized beam. A resonance occurs when the rf magnetic field's frequency is synchronized with the spin tune and the circulation frequency according to:

$$f_r = f_c(k \pm \nu_s); \tag{3}$$

where k is an integer. This resonance condition allows *coherent* kicks to build up and rotate the spin about the vertical axis by 180° causing a **spin-flip**.

Spin-flipping parameters

The rf resonance strength ϵ is given by:

$$\epsilon = \frac{\theta_s}{\pi} = \frac{Ge \int B \cdot dl}{2\pi m_p v} \quad \text{for an rf dipole}, \qquad (4)$$

where θ_s is the rf-dipole's spin rotation angle, m_p is the proton's mass and v is the proton's velocity. By varying the rf dipole frequency from below to above the resonance frequency f_r, one can *cross* the depolarizing resonance. After this crossing, the final beam polarization, P_f, is related to the initial beam polarization, P_i, by the **Froissart-Stora** form:

$$P_f = P_i \left\{ 2 \, exp\left[\frac{-(\pi \epsilon f_c)^2}{\Delta f / \Delta t}\right] - 1 \right\}, \qquad (5)$$

where ϵ is the resonance strength, and $\Delta f/\Delta t$ is the resonance crossing rate for the frequency range Δf during a ramp time Δt.

Three distinct conditions for the frequency variation rate $\Delta f/\Delta t$ are:

Rate	Polarizations	Effect
Fast crossing:	$P_f \approx P_i$	Little or no depolarization
Medium crossing:	$P_i > P_f > -P_i$	Depolarization
Slow crossing (adiabatic):	$P_f \approx -P_i$	Spin-flip

The **spin-flip efficiency** is defined as:

$$\eta = \frac{-P_f}{P_i}. \qquad (6)$$

To maximize the spin-flip efficiency, one must carefully choose $\epsilon, \Delta f,$ and Δt.

For a description of the apparatus used in this experiment, including the rf-dipole, the IUCF Cooler Ring, the Siberian snake, and the polarimeter, the reader is referred to Ref. [3]. The 120 MeV horizontally polarized proton beam in the Cooler Ring was obtained using the new Cooler Injector Polarized Ion Source (CIPIOS) and the new Cooler Injection Synchrotron. The beam polarization was about $(50-60)\pm 2\%$ after the 7 MeV Linac. At 120 MeV, the circulation frequency in the Cooler Ring was $f_c = 1.59784$ MHz.

With a nearly-full Siberian snake in the Ring, from Eq. (2), the spin tune ν_s is very near but not exactly equal to $\frac{1}{2}$. Therefore, Eq. (3) implies that two closely spaced rf depolarizing resonances, f_r^- and f_r^+, exist:

	April 2000 run	June 2000 run
Center of 2 resonances	$1.5 f_c = 2.39676$ MHz	$0.5 f_c = 0.79892$ MHz
f_r^-	$f_c(2 - \nu_s)$	$f_c(1 - \nu_s)$
f_r^+	$f_c(1 + \nu_s)$	$f_c(0 + \nu_s)$

With a snake strength s of 1.01, ν_s was about 0.505; thus, the f_r^- resonance was:
slightly below $1.5 f_c$ for k = 2 for April 2000 run; and
slightly below $0.5 f_c$ for k = 1 for June 2000 run.

APRIL 2000 RESULTS

In the April 2000 run, the f_r^- resonance was first located using the much stronger rf solenoid. Once found, this resonance was mapped out by the radial polarization response to varying the rf dipole's frequency, as shown in Fig 1.

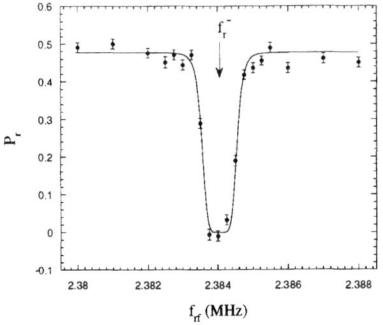

Fig. 1. The measured radial proton polarization plotted against the rf-dipole's frequency; arrow shows f_r^-. [3]

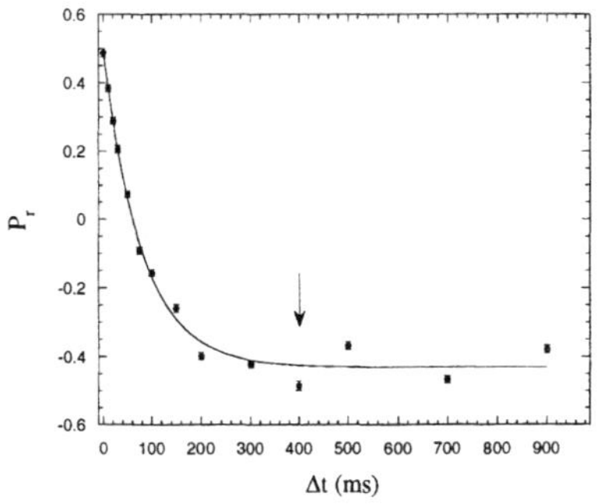

Fig. 2 Radial polarization plotted against the ramp time for a fixed frequency range of $\Delta f = 10$ kHz. [3]

The parameter Δt was then optimized as shown in Fig. 2; it indicated that *spin flips with* **slow** *rf ramp times*.

In addition, remnant polarization of the beam was measured after various number of flips. From these graphs, a maximum spin-flip efficiency of 86.5 ±0.5 % was obtained.

JUNE 2000 RESULTS

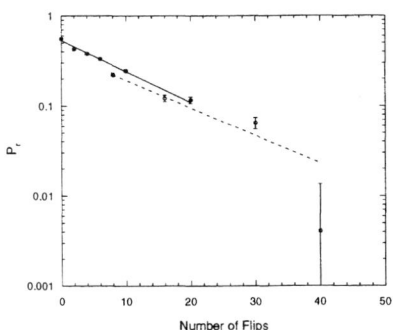
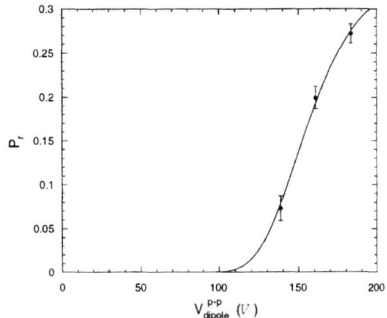

Fig. 3 Radial proton polarization plotted against number of spin-flips. The two fitted lines gave a spin-flip efficiency η of about $92.5 \pm 0.5\%$.

Fig. 4 Maximum radial polarization after ten spin-flips plotted against the peak-to-peak dipole voltage.

In June 2000, CE-69 enhanced the rf dipole strength by sweeping through a lower resonant frequency of $f_r^- = f_c(1 - \nu_s)$. The data for multiple spin-flips is shown in Fig. 3; the two curves were fitted to $P_n = P_i \cdot \eta^n$, where η is the efficiency and n denotes the number of flips. The maximum spin-flip efficiency was found to be increased to $92.5 \pm 0.5\%$.

Fig. 4 shows how the maximum spin-flip efficiency grows with the rf dipole's strength; a higher spin-flip efficiency should be possible with a higher strength.

During the June run, a new method was developed to locate the rf depolarizing resonance as outlined below:

- first make a large frequency sweep range, containing the spin-flip;
- then divide the frequency sweep range into two equal sweeps;
- continue subdividing the range, always keeping the ones containing the spin-flip.

This method was much more efficient and has the advantage of observing the actual spin-flips, instead of a depolarizing dip.

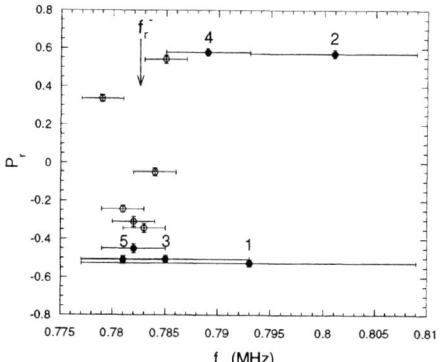

Fig. 5 Radial proton polarization plotted against the rf-dipole's central frequency, with widths indicating the frequency range; arrow shows f_r^-.

Such a process is shown graphically in Fig. 5 by points 1 to 5, which were done sequentially. Notice also that, once roughly located, the resonance was mapped out by (small) equal frequency ranges of ± 2 kHz around the resonance. The arrow indicates the resonance frequency f_r^- thus obtained; the same value was later obtained by mapping the polarization at many fixed frequency points around the resonance.

We thank the staff of IUCF for the successful operation of the Cooler Ring.

REFERENCES

1. B.B. Blinov et al., Phys. Rev. Lett. **81**, 2906 (1998); V.A. Anferov et al., Proc. 13th Intl. Symposium on High Energy Spin Physics, eds. N.E. Tyurin et al., (World Scientific, Singapore, 1999) p. 503.
2. *Spin flipping a stored polarized proton beam with an rf dipole*, V.A. Anferov et al., Phys. Rev. ST–AB **3**, 041001 (2000).
3. *Spin-flipping with an rf-dipole and a full Siberian snake*, B.B. Blinov et al., Phys. Rev. ST–AB **3**, 104001 (2000).

Crossing a Coupling Spin Resonance With an RF Dipole[1]

M. Bai, T. Roser

Brookhaven National Laboratory, Upton, NY 11973, U.S.A

Abstract. In accelerators, due to quadrupole roll errors and solenoid fields, the polarized proton acceleration often encounters coupling spin resonances. In the Brookhaven AGS, the coupling effect comes from the solenoid partial snake which is used to overcome imperfection resonances. The coupling spin resonance strength is proportional to the amount of coupling as well as the strength of the corresponding intrinsic spin resonance. The coupling resonance can cause substantial beam polarization loss if its corresponding intrinsic spin resonance is very strong. A new method of using an horizontal RF dipole to induce a full spin flip crossing both the intrinsic and its coupling spin resonances is studied in the Brookhaven's AGS. Numerical simulations show that a full spin flip can be induced after crossing the two resonances by using a horizontal RF dipole to induce a large vertical coherent oscillation.

I INTRODUCTION

In an accelerator, particles undergo betatron oscillations in both horizontal and vertical planes while they circulate around the machine. In a perfect machine, both oscillations are independent of each other. However, this independence can be broken if there is any quadrupole roll errors or solenoid fields. In this case, the horizontal motion is coupled to the vertical oscillation. Unlike the uncoupled case, the frequency spectrum of the betatron oscillation in either of the two transverse plane then consists of two components ν_1 and ν_2 given by [1,2]

$$\nu_1 = \frac{1}{2}(\nu_x + \nu_z) + \frac{1}{2}\sqrt{(\nu_x - \nu_z)^2 + \Delta Q^2_{min}} \rightarrow \nu_x; \text{ without coupling} \quad (1)$$

$$\nu_2 = \frac{1}{2}(\nu_x + \nu_z) - \frac{1}{2}\sqrt{(\nu_x - \nu_z)^2 + \Delta Q^2_{min}} \rightarrow \nu_z; \text{ with couple} \quad (2)$$

where ν_x and ν_z are the unperturbed horizontal and vertical tunes. ΔQ_{min} is the minimum tune split between the two eigen tunes when $\nu_x = \nu_z$ and is proportional to the coupling strength [3]. With weak coupling,

---

[1] The work was performed under the auspices of the US Department of Energy

$$\nu_1 \simeq \nu_x \qquad (3)$$
$$\nu_2 \simeq \nu_z. \qquad (4)$$

In the Brookhaven AGS, the main coupling source comes from the solenoid partial snake which is used to overcome the imperfection spin resonances in the AGS [4]. The minimum tune split ΔQ_{min} from the 5% partial snake is about 0.015.

In a coupled machine, in addition to the intrinsic spin resonance at $G\gamma = kP \pm \nu_2$ ($G\gamma = kP \pm \nu_z$ without coupling) [5], the vertical betatron oscillation also drives a coupling spin resonances at $G\gamma = kP \pm \nu_1$ [3]. The strength of the coupling resonance ϵ_{ν_x} is proportional to the amount of the coupling and it is given by

$$\epsilon_{\nu_x} \propto C_x \sqrt{\varepsilon_u} \epsilon_{\nu_z} \qquad (5)$$

where ϵ_{ν_z} is the strength of the adjacent intrinsic spin resonance and C_x is the coupling coefficient. For a fully coupled machine, $\nu_x = \nu_z$ and $C_x = 1$. For a decoupled machine, $C_x = 0$. ε_u is the beam emittance in the eigen direction [6] and equals the horizontal beam emittance if $C_x = 0$.

In the AGS, there are four strong intrinsic spin resonances at $0+\nu_z$, $12+\nu_z$ and $36\pm\nu_z$ [5]. Traditionally in the AGS, the beam polarization loss at the coupling resonances is minimized by separating the horizontal and vertical tunes. The coupling resonances around these four strong intrinsic resonances can produce about 35% polarization losses with the normal AGS polarized proton setting [7,8]. In order to achieve 70% polarization in the AGS, one needs to minimize the polarization loss at the coupling resonances. Since they are adjacent to the intrinsic resonances, it is very difficult to use the vertical RF dipole [9] to obtain full spin flips at both the intrinsic and the coupling resonances.

Analogous to the method of using a vertical RF dipole at the intrinsic spin resonance, one should also expect to obtain a full spin flip by inducing a strong artificial resonance if the intrinsic and its coupling spin resonances are fully overlapped. Because of the coupling effect, the two spin resonances can never be brought closer than the minimum tune split ΔQ_{min}. However, ΔQ_{min} in general is small and a full spin flip still should be achievable if the induced resonance is strong enough. In a fully coupled machine, the unperturbed tunes are equal and the intrinsic and the coupling resonances are equally strong and located on either side of the unperturbed betatron tune at a distance of half of ΔQ_{min}.

Unlike using a vertical RF dipole to obtain a vertical coherence in an uncoupled machine [10], the vertical coherence is excited by a horizontal RF dipole instead in a fully coupled machine. This can be understood by solving the differential equation of a coupled driven oscillator

$$x'' + (2\pi f\nu)^2 x + qz = A\cos(2\pi f\nu_m\theta)$$
$$z'' + (2\pi f\nu)^2 z + qx = 0. \qquad (6)$$

Here f is the revolution frequency, ν_m is the modulation tun, q is the coupling strength and A is the amplitude of the driving term. When the modulation tune

ν_m equals the unperturbed tune ν, the solution of Eq. 6 gives $x = 0$ and a pure vertical oscillation $z = \frac{A}{q}cos2\pi f\nu_m\theta$.

Fig. 1 shows numerical spin tracking results at $G\gamma = 36 + \nu_z$. The dotted line shows the result with the nominal AGS tune setting ($\nu_x = 8.8, \nu_z = 8.7$) and no correction scheme for the intrinsic spin resonance. In this case, the depolarization at the coupling resonance is obvious. The solid line is the result of using a horizontal RF dipole with the horizontal and vertical betatron tunes set at 8.7. Due to the coupling from the solenoid partial snake, the two betatron tunes are split by 0.0144. The horizontal RF dipole tune was set to 0.3. With a horizontal RF dipole amplitude 28.0 G-m, a full spin flip was achieved.

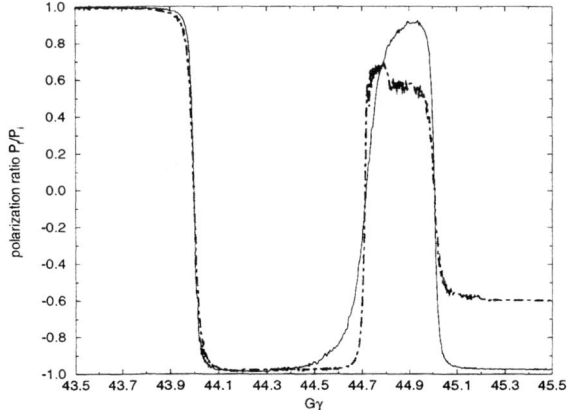

FIGURE 1. The two curves are the calculated polarization ratio P_f/P_i as a function of energy. The solid line is for the case of a fully coupled machine and a horizontal RF dipole was used to obtain an adiabatic vertical coherence. The dotted line is the result of a weakly coupled machine with the two betatron tunes set 0.1 apart. No correction scheme was used at the $G\gamma = 36 + \nu_z$ intrinsic spin resonance. For both cases, the horizontal and vertical emittance are 20π mm-mrad and 10π mm-mrad respectively.

II EXPERIMENTAL RESULTS

The method of using a horizontal RF dipole to excite a vertical coherence to cross the coupling spin resonance was tested in the AGS during the 2000 RHIC polarized proton commissioning run. The polarized H^- beam was pre-accelerated in the 200 MeV LINAC and then stripped and injected into the Booster. It was then injected into the AGS at $G\gamma = 4.7$ and then accelerated up to $G\gamma = 46.5$. In the AGS, the nominal tune setting is $\nu_x = 8.8$ and $\nu_z = 8.7$.

During the experiment, the AGS skew quadrupoles were all set to 17 A. Due to a hardware limit, the partial snake strength at $G\gamma = 36 + \nu_z$ is actually only about 3.5% instead of 5%. The combined effect of the skew quadrupoles and the weaker snake gave a smaller minimum tune spilt ΔQ_{min} of 0.007. The horizontal RF dipole was set in the middle of the two betatron tunes ν_1 and ν_2. The turn by turn beam position monitor data confirmed that a vertical coherence was excited without horizontal response as shown in the two left plots of Fig. 2. The horizontal response was not zero once the RF dipole tune deviated from the average of the two eigen tunes as shown on the right of Fig. 2.

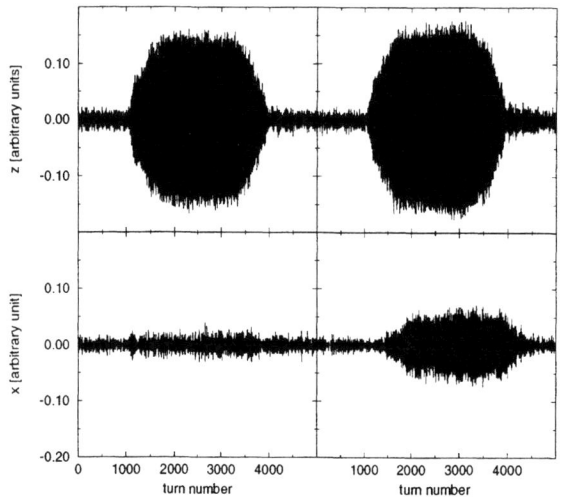

FIGURE 2. The top and bottom plots on the left are vertical and horizontal turn-by-turn beam position data when the horizontal RF dipole modulation tune $\nu_m = \frac{1}{2}(\nu_1 + \nu_2)$. As shown, no horizontal coherence was excited. The two plots on the right correspond to the case where the horizontal modulation tune $\nu_m \neq \frac{1}{2}(\nu_1 + \nu_2)$ and the horizontal coherence was no longer zero.

Table 1 shows the comparison of the measured beam asymmetries of using vertical RF dipole, no correction and using horizontal RF dipole at $G\gamma = 36 + \nu_z$. Comparing the measured asymmetry when using the horizontal RF dipole with the case of no correction, it is clear that the horizontal RF dipole did help to recover the beam polarization. However, the excited coherence was not optimized and about 70% beam emittance growth was observed. Because of limitations of the AGS sextupole power supplies, we could not achieve small chromaticities in both planes and obtain a fully adiabatic excitation. This is the most likely reason that the horizontal RF dipole did not recover 100% beam polarization as expected.

TABLE 1. measured asymmetry

	measured asymmetry (x10^{-3})	condition
1	1.50± 0.04	with vertical RF dipole
2	1.25± 0.1	with horizontal RF dipole
3	0.067± 0.063	no correction

III CONCLUSION

It has been demonstrated in the AGS that in a fully coupled machine, a vertical coherence can be excited by an horizontal RF dipole. Although beam polarization was improved, we think the residual polarization loss was due to the not fully adiabatic beam motion.

IV ACKNOWLEDGEMENT

We would like to thank Dr. L. Ahrens, Dr. E. D. Courant, W. J. Glenn, Dr. H. Huang, Dr. A. Lehrach, Dr. A. Luccio, Dr. W. Mackay, V. Ranjbar, Dr. N. Tsoupas, Dr. W. van Asselt for the fruitful discussions. We also would like to thank K. Zeno and D. Warburton for their great help. This work is performed under the auspices of Department of Energy of U.S.A.

REFERENCES

1. S. Y. Lee, *Accelerator Physics*, World Scientific Pub. Singapore, 1999.
2. D. A. Edwards, M. J. Syphers, *An Introduction To The Physics of High Every Accelerators*, Wiley-Interscience Pub. 1993.
3. S. Y. Lee, *Spin Dynamics and Snakes in Synchrotrons*, World Scientific Pub. Singapore, 1997.
4. T. Roser, *Partial Siberian Snake Test at the Brookhaven AGS*, in High Energy Spin Physics: 10th International Symposium, ed. T.Hasegawa, et al., Nagoya, Japan, 1992, (Univesal Academic Press, Inc.,1992), p.429.
5. E. D. Courant, R. D. Ruth, *The Acceleration of Polarized Protons in Circular Accelerators*, BNL report, BNL 51270, 1980.
6. D. A. Edwards, L. C. Teng, *Parametrization of LINEAR Coupled Motion in Periodic Systems*, IEEE Trans. on Nucl. Sc. 20, 885 (1973).
7. H. Huang et al., *Preservation of Proton Polarization by a Partial Siberian Snake*, Phy. Rev. Letters. 73, 2982 (1994).
8. H. Huang et al., *Polarized Proton Beam in the AGS*, Proceedings of 13th international symposium in High Energy Spin Physics, P.492 (1998).
9. M. Bai et al., *Overcoming Intrinsic Spin Resonances with an rf Dipole*, Physical Review Letters 80, 4673(1998).
10. M. Bai, et al., *Experimental Test of Coherent Betatron Resonance Excitations*, Physical Review E, 5(1997).

Beam Polarization Distributions for the Relativistic Heavy Ion Collider *

A. Lehrach, A.U. Luccio, W.W. MacKay, T. Roser

Brookhaven National Laboratory, Upton, NY 11973, USA

Abstract. In a high energy accelerator the polarization distribution of the beam can vary substantially across the beam. This decreases the amount of polarization provided to experiments. A method has been developed to calculate the beam polarization distribution based on adiabatic particle and spin excitation by inducing coherent betatron oscillations using an ac dipole. In earlier studies we calculated the beam polarization distribution in Brookhaven's Relativistic Heavy Ion Collider (RHIC) at fixed energy to compare spin motion on and away from strong spin resonances [1]. In this paper we present calculations of the beam polarization distribution during spin resonances crossing in RHIC.

INTRODUCTION

In order to investigate the polarization distribution of the beam in a circular accelerator, the invariant spin field has to be calculated. The invariant spin field $\vec{n}(\vec{z})$ [2] depends on the position $\vec{z}$ of the particle in the six dimensional phase space. A particle with the initial spin $\vec{s}_i$ at the phase space position $\vec{z}_i$ has the final spin $\vec{s}_f$ after it has been transported to the phase space point $\vec{z}_f$ during one turn in a storage ring. If $T_{t=1}$ is the one turn spin transfer matrix, then for every phase space point $\vec{z}_i$ a spin field vector $\vec{n}(\vec{z}_i)$ exists such that

$$\vec{n}(\vec{z}_f) = T_{t=1}\,\vec{n}(\vec{z}_i). \tag{1}$$

The spin follows the invariant spin field if the motion of the spin is adiabatic. A particle spin vector with an initial angle ϕ with respect to the corresponding invariant spin field vector $\vec{n}(\vec{z}_i)$ will be rotated around $\vec{n}(\vec{z}_i)$ every time it comes close to $\vec{z}_i$. If the spin motion is adiabatic, the angle ϕ is constant. The time averaged polarization at $\vec{z}_i$ will be parallel to $\vec{n}(\vec{z}_i)$ with a magnitude of one if the spin vectors are parallel to the invariant spin field. The maximum time averaged polarization P_{lim} of the beam is given by the average of the invariant spin field vectors $\langle \vec{n} \rangle$ over the beam, and therefore depends on the phase space distribution

---

*) Work performed under the auspices of the U.S. Department of Energy.

of the beam [3]. A large angle spread of $\vec{n}(\vec{z})$ leads to low maximum time averaged polarization. Thus the invariant spin field has to be calculated to determine the amount of maximum beam polarization delivered to experiments. A method to calculate the invariant spin field is stroboscopic averaging [4], which is based on multi-turn tracking and averaging of the spin viewed stroboscopically from turn to turn at one position in the ring. The invariant spin field has also been studied using a method called adiabatic anti-damping [3], which is very similar to the method presented here. In our study, the motion of the particle and spin is adiabatically excited with coherent betatron oscillations using an ac dipole [1]. Spin and particle motion were calculated using the spin tracking program SPINK [5].

CALCULATION OF THE INVARIANT SPIN FIELD

A controlled betatron oscillation is introduced by an ac dipole to calculate the invariant spin field. The ac dipole is slowly energized to the desired field and back to zero field in a way that the resulting orbit and spin motion are adiabatic. Adiabaticity is maintained during this process if the spin and particle return to their initial positions after de-energizing this device. In order to get a large coherent amplitude Z_{coh}, the modulation tune of the ac dipole, defined as the oscillation frequency of the ac dipole field divided by the revolution frequency of the particles in the accelerator, has to be close to the fractional betatron tune [6]:

$$Z_{coh} = \frac{1}{2} \beta_z \frac{B_{ac}\, l}{B\rho} \frac{1}{2\pi\delta}. \tag{2}$$

In Eq. 2, β_z is the betatron amplitude at the location of the ac dipole, δ the difference between the modulation tune of the ac dipole field and the fractional betatron tune, l the length of the ac dipole, B_{ac} the amplitude of the ac dipole field, and $(B\rho)$ the magnetic rigidity of the beam. If the spin of the particle on the closed orbit is started parallel to the corresponding spin field vector[†], and the ac dipole excites the motion of the spin adiabatically, then the spin of the particle remains parallel to the invariant spin field during the excitation. In this case one particle spin tracking is sufficient to calculate the invariant spin field for the entire phase space, as shown in the next section.

INVARIANT SPIN FIELD DURING ACCELERATION

In the presented simulation, the design RHIC lattice with a vertical tune of $\nu_y = 29.18$ was chosen without field errors and misalignments of the magnets [7]. The motion of the invariant spin field has been investigated during crossing the strongest intrinsic resonance in the energy range of RHIC. The amplitude of the ac dipole field was increased in 5000 turns starting at $\gamma G = 416$, kept constant at the

---

[†] The $\vec{n}$ axis on the closed orbit is calculated using stroboscopic averaging [4]

desired field strength during spin resonance crossing for 13600 turns, and decreased back to zero amplitude in 5000 turns just before reaching $\gamma G = 430$. The particle and spin return adiabatically to their initial position. To determine the maximum beam polarization that can be provided to experiments, the invariant spin field is investigated at the interaction point of the PHENIX experiment.

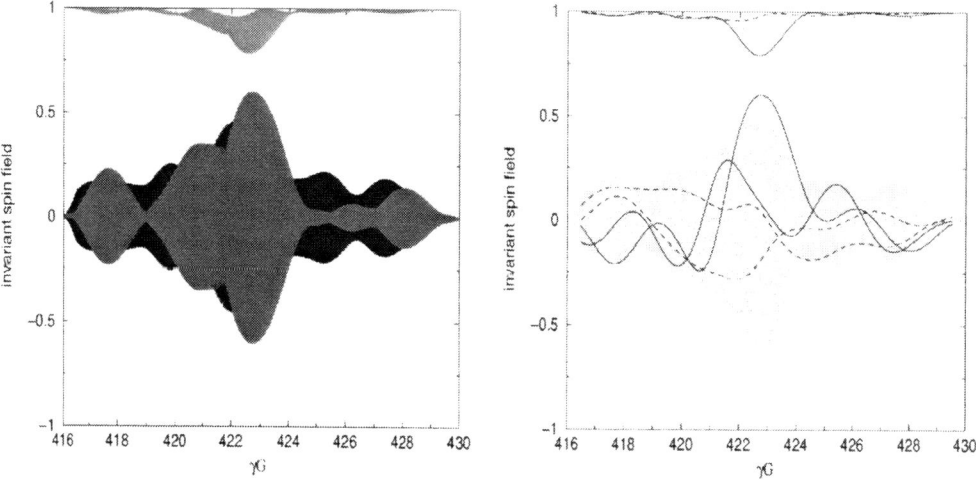

FIGURE 1. Components of the invariant spin field versus energy in units of γG. The left plot shows the invariant spin field for a particle on a phase space ellipse with a normalized vertical emittance of 5π mm mrad. The upper curve is the vertical component of the invariant spin field, the two lower curves are the horizontal and longitudinal components. The lines in the plot on the right side show the invariant spin field for two selected vertical phase space points with maximum orbit (solid line) and angular displacement (dashed line).

Fig. 1 shows, that spin resonances in this energy range are overlapping. The invariant spin field is strongly depending on the position in vertical phase space. For different phase space points the motion of the invariant spin field is quite different. In this energy range the maximum time averaged polarization of the beam with a normalized vertical emittance of 5π mm mrad will be higher than 0.75 for any vertical phase space distribution. Fig. 2 shows the dependence of the invariant spin field on the vertical emittance of the beam. As expected, the excitation of the invariant spin field is increasing with vertical emittance; whereas the width (full width half maximum) of the invariant spin field distribution at spin resonances is constant. Up to a vertical excitation of 20π mm mrad the spin motion stays adiabatic, even during crossing the strongest intrinsic resonance when the oscillation of the invariant spin field is very large with a vertical component continuously flipping sign. The beam energy to deliver a polarized beam to experiments has to be chosen carefully in this energy range. Even for a beam with a normalized vertical emittance of 20π mm mrad it is possible to get a maximum time averaged polariza-

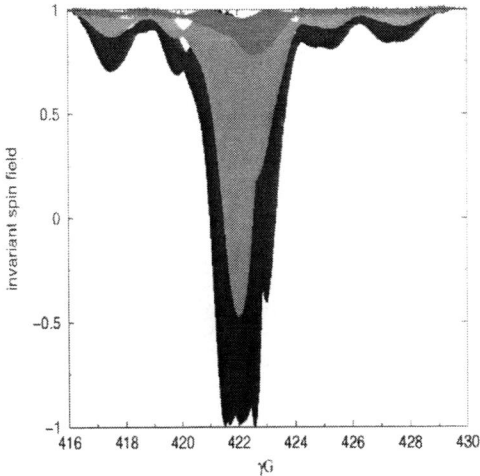

FIGURE 2. Vertical component of the invariant spin field versus energy in units of γG for particles on different phase space ellipse corresponding to a normalized vertical emittance of 5π mm mrad (upper curve), 10π mm mrad (curve in the middle), and 20π mm mrad (lower curve).

tion higher than 0.9, if polarized beam is provided to experiments away from spin resonances, at energies around $\gamma G = 418.5$ or 426.5.

POLARIZATION LOSSES DURING SPIN RESONANCE CROSSING

Polarization losses occur if the spin vector is not able to follow the motion of the invariant spin field adiabatically during spin resonance crossing. Thus the angle ϕ between the spin vector and its corresponding invariant spin field vector is increasing. In this calculation the coherent amplitude of the betatron oscillation was increased to observe non-adiabatic spin motion. Since the invariant spin field is vertical after turning off the ac dipole, the maximum time averaged polarization equals one. Therefore polarization losses for different cases can be investigated by comparing the vertical component of the spin vector at $\gamma G = 430$. Fig. 3 shows the non-adiabatic spin motion for different vertical particle excitations. The spin vector starts to oscillate around the corresponding invariant spin field vector during crossing of the strongest intrinsic resonance, and remains non-parallel to the invariant spin field after turning off the ac dipole. As can be seen, the polarization losses are increasing with beam excitation and even spin flips are possible.

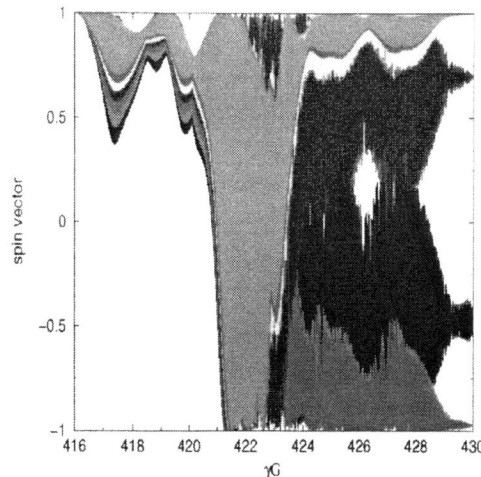

FIGURE 3. Vertical component of the spin vector versus energy in units of γG for particles on equally spaced phase space ellipse of normalized vertical emittances between 20 and 40π mm mrad.

CONCLUSION AND OUTLOOK

Polarization distributions for RHIC have been calculated by exciting the motion of particle and spin with an ac dipole field. It has been shown that the maximum time averaged polarization is depending on the vertical phase space distribution of the beam and strongly decreasing with the vertical beam size. The simulation indicates that some energy ranges in the vicinity of spin resonances should be excluded from being used for experiments at RHIC. Furthermore polarization losses during spin resonance crossing have been investigated.

This studies will be continued to learn more about beam polarization distributions and spin resonance crossing in RHIC. In addition spin tunes will be calculated by analyzing spin oscillation of particles with non-parallel initial spin direction to the corresponding invariant spin field vector.

REFERENCES

1. A. Lehrach et al., DESY-Proc-1999-03, 210 (1999).
2. Ya. Derbenev, A.M. Kondratenko, Sov. Phys. JETP 37(6), 968 (1973).
3. M. Vogt et al., Proc. of EPAC 98, Stockholm, 1362 (1998).
4. G.H. Hoffstätter et al., Phys. Rev. E 54, 4240 (1996).
5. A.U. Luccio et al., Proc. of PAC 99, New York, 1578 (1999).
6. M. Bai et al., Phys. Rev. E 56, 6002 (1997).
7. Design Manual - Polarized Proton Collider at RHIC, BNL (1998); http://www.agsrhichome.bnl.gov/RHIC/Spin/.

Using the amplitude dependent spin tune to study high order spin–orbit resonances in storage rings

D.P. Barber*, G.H. Hoffstätter* and M. Vogt[†]

Deutsches Elektronen–Synchrotron (DESY), Hamburg, Germany
[†] *University of New Mexico, Albuquerque, NM 87131, USA*

Abstract. We define the amplitude dependent spin tune and illustrate its use for identifying spin–orbit resonances and evaluating their strength.

INTRODUCTION

A proper understanding of the spin–orbit resonance structure at high energy in storage rings can only be obtained with a correct definition of the "spin tune". This requires establishing a proper coordinate system for "measuring" spin precession and that, in turn, requires the notion of the "invariant spin field". This paper illustrates that programme. More details can be found in [1–6]. Our calculations were made with the spin–orbit tracking code SPRINT [5,6]. The algorithms in SPRINT are *non-perturbative*. SPRINT has extensive facilities for the *dynamical* variation of the reference energy and tunes so that it has been possible to carry out realistic tracking simulations of acceleration.

THE INVARIANT SPIN FIELD

The transverse and longitudinal motion of particles in storage rings is described in terms of three pairs of canonical coordinates $\vec{u} = (q_1, p_1, q_2, p_2, q_3, p_3)$. The independent variable is the distance along the ring l. There is a corresponding classical Hamiltonian $h_{\text{orb}}(\vec{u}; l)$. In distorted rings $\vec{u}$ describes motion with respect to the resulting closed orbit. In the absence of spin flip, spin motion for electrons and protons moving in electric and magnetic fields is described by the T-BMT equation [1] $d\vec{S}/dl = \vec{\Omega} \times \vec{S}$ where $\vec{S}$ is the rest frame spin expectation value of the particle ("the spin") and $\vec{\Omega}$ depends on the electric and magnetic fields, the velocity and the energy so that it depends on $\vec{u}$ and l.

As a first step in setting up a coordinate system for spin we attach a laboratory space 3–vector $\hat{f}(\vec{u};l)$ of fixed unit length, to every point $(\vec{u};l)$. At this stage $\hat{f}$ is a definite but freely chosen smooth vector function of $\vec{u}$ and l. The rate of change of $\hat{f}$ along some path in $(\vec{u}, l)$ space is $\frac{d\hat{f}}{dl} = \frac{\partial \hat{f}}{\partial l} + \sum_{k=1}^{3} \frac{dq_k}{dl} \frac{\partial \hat{f}}{\partial q_k} + \frac{dp_k}{dl} \frac{\partial \hat{f}}{\partial p_k}$. Then along a particle trajectory, and in terms of a Poisson bracket, the equation of motion takes the form $\partial \hat{f}/\partial l + \{\hat{f}, h_{\text{orb}}\} = \vec{F}_{\hat{f}}(\vec{u};l)$. Since, by choice, $||\hat{f}||$ is invariant, the motion of $\hat{f}$ must be a rotation so that $\vec{F}_{\hat{f}}$ must have the form $\vec{G}(\vec{u};l) \times \hat{f}$. We now choose $\hat{f}$ so that it obeys the T–BMT equation: $d\hat{f}/dl = \vec{\Omega} \times \hat{f}$ along particle orbits. Moreover we require that it reflects the periodicity of the magnet structure by being 1–turn periodic in l, i.e. $\hat{f}(\vec{u}; l+C) = \hat{f}(\vec{u};l)$ where C is the ring circumference. We denote this special choice by $\hat{n}(\vec{u};l)$. Except at the spin–orbit resonances to be discussed later, $\hat{n}(\vec{u};l)$ is unique.

Thus $\hat{n}(\vec{u};l)$ is a pre–established 1–turn periodic *vector field* on $(\vec{u},l)$ obeying the T–BMT equation. For one turn $\hat{n}(\vec{M}(\vec{u};l); l+C) = \hat{n}(\vec{M}(\vec{u};l); l) = R_{3\times 3}(\vec{u};l)\hat{n}(\vec{u};l)$ where $\vec{M}(\vec{u};l)$ is the new phase space vector after one turn starting at $\vec{u}$ and l and $R_{3\times 3}(\vec{u};l)$ is the corresponding spin transfer matrix. If a spin $\vec{S}$ is followed along an orbit, the scalar product $\vec{S}\cdot\hat{n}$ of $\vec{S}$ and the local $\hat{n}$ is invariant since both vectors obey the T–BMT precession equation. Thus with respect to the local pre–established $\hat{n}$ the motion of $\vec{S}$ is very simple, namely a precession around $\hat{n}$. On the closed orbit $\hat{n}(\vec{u};l)$ becomes $\hat{n}(\vec{0};l)$ which we denote by $\hat{n}_0(l)$. Obviously $\hat{n}_0(l+C) = \hat{n}_0(l)$. It is given by the real unit eigenvector of the 1–turn 3×3 spin transport matrix on the closed orbit.

Examples of the invariant spin field at 800 GeV for a HERA proton optic with a suitable arrangement of Siberian Snakes are shown in figure 1. In these particular simulations the protons only execute stable linear vertical betatron motion of fixed amplitude. Each picture shows the locus, on the surface of a sphere, of the tip of the $\hat{n}$ vector as the betatron phase varies at a point on the ring where $\hat{n}_0$ is vertical. The parameters are shown in the captions. For each picture in figure 1, the phase space coordinates of a particle are not 1–turn periodic but at a fixed position on the ring ("azimuth") they lie on a closed elliptical curve at positions depending on its vertical betatron phase. Likewise a spin at some $\vec{u}$ set parallel to $\hat{n}$ and tracked, is not 1–turn periodic but on tracking it turn to turn, it lies on the closed curve, parametrised by the orbital phase, of the field $\hat{n}$. Thus, like the invariant orbital ellipses, the curves of figure 1 are invariant when tracked from turn to turn and we therefore call $\hat{n}(\vec{u};l)$ the *invariant spin field*. As the amplitude is increased, the invariant spin field becomes convoluted, especially near the spin–orbit resonances to be discussed below. For motion with one degree of freedom, the loci on the sphere are closed as, for example, in figure 1. For more than one degree of freedom, the phase space coordinates lie on invariant tori and the loci of $\hat{n}$ do not close in general although the field $\hat{n}$ is still an invariant of the 1–turn spin–orbit map. If the spins for an ensemble of particles distributed uniformly around the phase space ellipses for figure 1, are all set initially parallel to $\hat{n}_0$ and then tracked, the beam polarisation at that azimuth oscillates. If they are set parallel to $\hat{n}$, the beam polarisation is stationary. The maximum *stationary* beam polarisation that can be reached is $P_{\lim}(l) = ||\int d^6u\, w_{\rm st}(\vec{u};l)\hat{n}(\vec{u};l)||$ where $w_{\rm st}(\vec{u};l)$ is the normalised stationary phase space density. For motion on a vertical betatron ellipse $P_{\lim}$ is just given by the average of $\hat{n}$ over the betatron phase [3]. On the 64π mm mrad ellipse $P_{\lim}$ is much smaller than for the 4π mm mrad ellipse — it pays to devise ways to keep the spread of $\hat{n}$ small. $P_{\lim}$ should be calculated before carrying out simulations of acceleration. If $P_{\lim}$ is small such a simulation is not worthwhile. Note that for $\vec{u} \neq \vec{0}$, the constraint $\hat{n}(\vec{u};l+C) = \hat{n}(\vec{u};l)$ for the invariant spin field

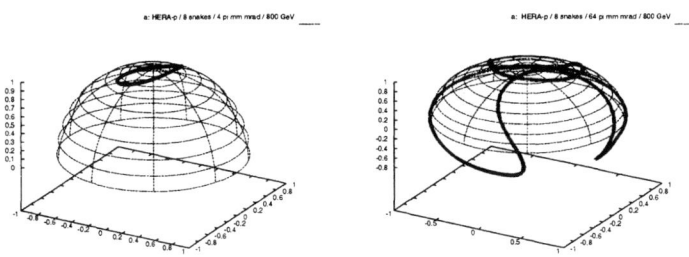

FIGURE 1. The field $\hat{n}$ in HERA–p calculated with SPRINT *on* the 4π mm mrad (left) and the 64π mm mrad (right) ellipses at $800\ GeV$. A normalised emittance of 4π mm mrad $\equiv$ "$1 - \sigma$".

is obviously *not* equivalent to the closure condition $\vec{N}(\vec{u};l) = R_{3\times 3}(\vec{u};l)\vec{N}(\vec{u};l)$. In fact, the calculation of the real $\hat{n}(\vec{u};l)$ is computationally nontrivial and requires either "stroboscopic averaging" [2], Fourier analysis as in SODOM-II [7] or "adiabatic anti–damping". All three algorithms are non–perturbative and are implemented in the code SPRINT [5,6].

Although we have concentrated on protons and have introduced the invariant spin field as an essential geometrical object it was first motivated by Derbenev and Kondratenko [8,9], for providing semiclassical spin quantisation axes when calculating radiative spin flip for electrons.

THE AMPLITUDE DEPENDENT SPIN TUNE

To complete the construction of our coordinates for describing spin motion, two other unit vectors $\hat{n}_1(\vec{u};l)$ and $\hat{n}_2(\vec{u};l)$ are attached to all $(\vec{u},l)$ such that the sets $(\hat{n}_1, \hat{n}_2, \hat{n})$ form local orthonormal coordinate systems at all points in phase space at each l. Like $\hat{n}$, $\hat{n}_1$ and $\hat{n}_2$ are 1–turn periodic in l: $\hat{n}_i(\vec{u};l+C) = \hat{n}_i(\vec{u};l)$ for $i \in \{1,2\}$. But unlike $\hat{n}$ they do not obey the T-BMT equation. As pointed out above the motion of $\vec{S}$ is a precession around $\hat{n}$. Now, with the basis vectors $\hat{n}_1$ and $\hat{n}_2$ we have a way to quantify the rate of spin precession around $\hat{n}$: it is the rate of rotation of the projection of $\vec{S}$ onto the $\hat{n}_1, \hat{n}_2$ plane. Except for the uninteresting case of running on orbital resonance, the fields $\hat{n}_1(\vec{u}; l)$ and $\hat{n}_2(\vec{u}; l)$ can be chosen so that the rate of precession is constant and independent of the starting orbital phases [1-6]. The number of precessions per turn "measured" in this way is called the spin tune ν. The spin tune depends only on the orbital amplitudes — a tune depending in some way on phases would hardly be a useful quantity since it would have to change as the phases advance.

Spins are particularly strongly perturbed, and the locus of $\hat{n}$ is then expected to be very convoluted, when the spin tune is near resonance with the orbital tunes: $\nu(J_1, J_2, J_3) = k_0 + k_1 Q_1 + k_2 Q_2 + k_3 Q_3$ where the Q's are the amplitude dependent tunes of the orbital modes, the k's are integers and the J's are orbital amplitudes. The spin tune on the closed orbit, $\nu(0,0,0)$, is the number of precessions per turn of an arbitrary spin around $\hat{n}_0(l)$. We denote it by ν_0. Note that contrary to common practice our expression for resonance does not contain ν_0. Indeed, that is the whole point of having a clean definition of spin tune as we now illustrate. Figure 2 (left) shows the dependence of the spin tune on orbital amplitude (=

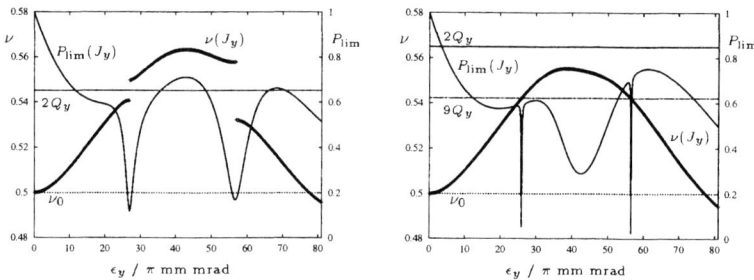

FIGURE 2. The amplitude dependent spin tune ν and P_{lim} on phase space ellipses with normalised vertical emittance ϵ_y as calculated with SPRINT for HERA–p at 805 GeV. Left: vertical tune $Q_y = 32.2725$, right: $Q_y = 32.2825$.

enclosed normalised emittance) for purely vertical betatron motion in HERA–p at 805 GeV with a suitable arrangement of snakes [3]. On the design orbit, i.e. at zero amplitude, ν is $1/2$ as expected. But it deviates from $1/2$ as the amplitude increases and at 27π mm mrad it jumps symmetrically across the resonant value $2Q_y$. After increasing further, ν then decreases and at a normalised emittance of 56π mm mrad it jumps back across the resonant value $2Q_y$. So ν never actually hits the resonant value but as one can see P_{lim} becomes small around the resonant amplitudes as the locus of $\hat{n}$ becomes convoluted and extends over the whole unit sphere. Thus the behaviours of ν and P_{lim} are mutually consistent. Figure 2 (right) shows the behaviour of ν when Q_y is increased. The second order resonance can no longer be crossed but 9th order resonant behaviour occurs instead. These curves illustrate just how complicated spin motion can be at very high energy. Such phenomena could obviously not be seen without a properly defined spin tune. For example, a "fake spin tune" erroneously extracted from the complex eigenvalues of $R_{3\times 3}$ shows no correlation with dips in P_{lim}. That is no surprise since that "fake tune" depends on the orbital phase and is therefore unsuitable for describing long term spin–orbit coherence. With the properly defined ν, the proximity to spin–orbit resonances can be properly judged and the changes in orbital tunes needed to avoid resonances can be properly estimated.

An especially satisfying aspect of these concepts is that they have provided a way to generalise the application of the Froissart–Stora formula [11] for the polarisation loss when passing through resonances. In particular, the size of a resonant jump in ν, $\Delta\nu$, for a high order resonance, is a measure of the strength of the resonance and using the facilities in SPRINT it has been possible to parametrise polarisation loss *with respect to* $\hat{n}$, when varying various machine parameters *dynamically* through such high order resonances, in terms of a generalised Froissart–Stora formula [5,6], containing $\Delta\nu$. More details on these calculations will be published elsewhere.

SUMMARY AND CONCLUSION

The use of a coordinate system based on the invariant spin field and of the properly defined spin tune are indispensable for a clear understanding of spin–orbit resonant behaviour in storage rings. Their use allows high order resonances to be cleanly identified, their strengths to be determined and misconceptions based on false definitions of spin tune to be avoided.

REFERENCES

1. D.P. Barber et al., five articles in proc. ICFA workshop "Quantum Aspects of Beam Physics", Monterey, U.S.A., 1998, World Scientific (1999). DESY report 98-96, Los Alamos archive: physics/9901038 – physics/9901044.
2. K. Heinemann and G.H. Hoffstätter, *Phys.Rev.* E **54**(4) 4240 (1996).
3. D.P. Barber, G.H. Hoffstätter and M. Vogt, proc. 13th International Symposium on High Energy Spin Physics (SPIN98), Protvino, Russia, 1998, World Scientific (1999).
4. G.H. Hoffstätter, M. Vogt and D.P. Barber, Phys. Rev. ST Accel. Beams **11**(2) 114001 (1999).
5. G.H. Hoffstätter, Habilitation Thesis, Technical University of Darmstadt (2000). Accepted for publication by Springer.
6. M. Vogt, Ph.D. Thesis, University of Hamburg (2000). To be published.
7. K. Yokoya, DESY report 99-006 (1999), Los Alamos archive: physics/9902068.
8. Ya. S. Derbenev and A. M. Kondratenko, *Sov.Phys. JETP.* **37** 968 (1973).
9. D.P. Barber and G. Ripken, *Handbook of Accelerator Physics and Engineering*, Eds. A.W. Chao and M. Tigner, World Scientific (1999). DESY report 99-095 (1999), Los Alamos archive: physics/9907034.
10. K. Heinemann and D.P. Barber, *Nucl. Inst. Meth.* accepted for publication. DESY report 98-145 (1998), Los Alamos archive: physics/9901045.
11. M. Froissart and R. Stora, *Nucl. Inst. Meth.* **7** 297 (1960).

The polarized electron beam at ELSA

M. Hoffmann[a], W. v. Drachenfels[a], F. Frommberger[a], M. Gowin[a],
K. Helbing[b], W. Hillert[a], D. Husmann[a], J. Keil[a], T. Michel[b],
J. Naumann[b], T. Speckner[b], G. Zeitler[b]

[a] *Physikalisches Institut der Universität Bonn, Nussallee 12, D-53115 Bonn, Germany;*
[b] *Physikalisches Institut der Universität Erlangen-Nürnberg, Erwin-Rommel-Str. 1, D-91058 Erlangen, Germany;*

Abstract.
The future medium energy physics program at the electron stretcher accelerator ELSA of Bonn University mainly relies on experiments using polarized electrons in the energy range from 1 to 3.2 GeV.

To provide a polarized beam with high polarization and sufficient intensity a dedicated source has been developed and set into operation. To prevent depolarization during acceleration in the circular accelerators several depolarizing resonances have to be corrected for. Intrinsic resonances are compensated using two pulsed betatron tune jump quadrupoles. The influence of imperfection resonances is successfully reduced applying a dynamic closed orbit correction in combination with an empirical harmonic correction on the energy ramp.

In order to minimize beam depolarization, both types of resonances and the correction techniques have been studied in detail. It turned out that the polarization in ELSA can be conserved up to 2.5 GeV and partially up to 3.2 GeV which is demonstrated by measurements using a Møller polarimeter installed in the external GDH[1]-beamline.

I INTRODUCTION

At ELSA [1] external fixed target experiments with longitudinally polarized electrons or circularly polarized photons (produced by Bremsstrahlung) are carried out. The first one is the GDH experiment, which just has started with data acquisition. Because self polarization of the beam by Sokolov-Ternov effect can not be used due to the operation mode of ELSA, a polarized electron source is used. The polarized electron beam is preaccelerated in a linac and a fast cycling booster (50 Hz). After injection into the main ring further acceleration up to 3.5 GeV is possible (Fig. 1).

---

[1] The GDH collaboration is named after the authors of the so-called Drell-Hearn-Gerasimov sum rule.

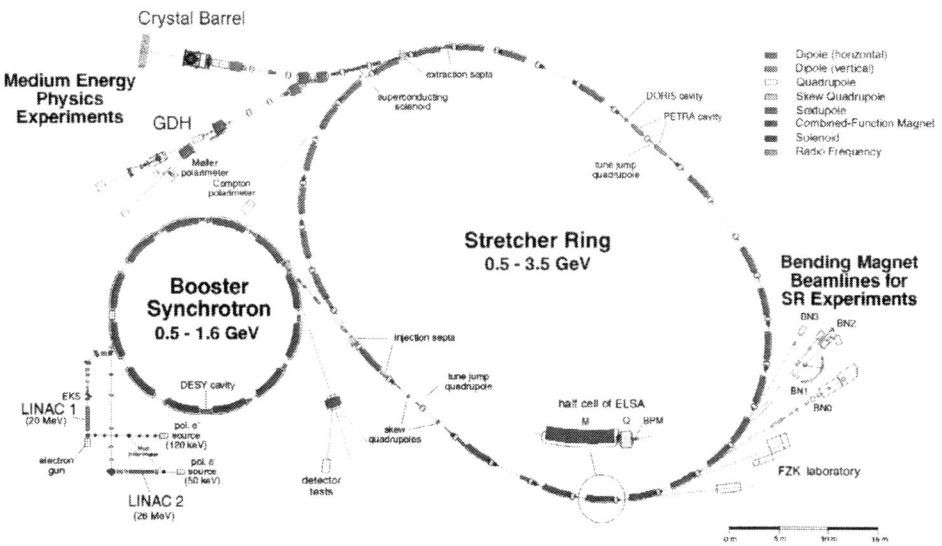

FIGURE 1. The ELSA facility at Bonn University.

II THE POLARIZED ELECTRON SOURCE

A pulsed low energy beam of polarized electrons is produced in a newly developed polarized electron gun [2]. In order to enhance the overall efficiency it operates with a new pulsed injector linac which requires an injection energy of 50 keV, a pulse length of 1 μs and a repetition rate of 50 Hz [3]. The inverted high voltage structure of the gun permits to vary the distance between cathode and anode and allows to adjust the perveance of the gun. For medium energy experiments, the gun is operated in space charge limitation, emitting a peak current of 100 mA in rectangular 1 μs long electron pulses. Using a Be-InGaAs/Be-AlGaAs superlattice photocathode [4] a polarization of 80 % and a corresponding quantum efficiency of 0.4 % was obtained. The photocathode lifetime during operation is higher than 3000 hours [5].

III DEPOLARIZING RESONANCES

In ELSA depolarization is caused mainly by intrinsic and imperfection resonances [6]. The spin originally oriented nearly perpendicular to the accelerator plane precesses around the direction of the magnetic field in the bending magnets. During circulation of the particles in the synchrotron only the polarization component parallel to the guiding field is conserved while the other components are lost. Horizontal magnetic fields cause polarization loss in the case of a resonance with the spin precession frequency. Imperfection resonances are caused by closed orbit displacements in the focusing quadrupoles. Intrinsic resonances are driven by the vertical betatron motion of the electrons, characterized by the vertical betatron tune Q_z.

The depolarization caused by linear crossing of an isolated resonance is quantified by the Froissart-Stora-Formula [7]

$$\frac{P_f}{P_i} = 2e^{-\frac{\pi|\epsilon|^2}{2\alpha}} - 1 \quad , \tag{1}$$

where ϵ is the resonance strength and α the crossing speed. Small polarization losses can be obtained for a small resonance strength ϵ or a high crossing speed ($\alpha = \frac{\dot{\gamma}a \mp \dot{Q}_z}{\omega_0}$ for intrinsic resonances). In this formula the influence of synchrotron oscillations and radiation is neglected. To take these effects into account we use a modified Froissart-Stora-Formula which essentially describes the depolarization after independently crossing of a depolarizing resonance and the two first order synchrotron satellites:

$$\frac{P_f}{P_i} = \left(2e^{-\frac{\pi|\epsilon|^2}{2\alpha}} - 1\right)\left(2e^{-\frac{\pi|\epsilon_s|^2}{2\alpha}} - 1\right)^2 \quad , \tag{2}$$

where ϵ_s is the resonance strength of the first order synchrotron satellites.

Spin tracking studies show that this description is sufficient to explain the behavior of depolarization at resonances at higher energies in ELSA. Especially a total spin flip ($\frac{P_f}{P_i} = -1$) can not be observed in ELSA at energies higher than 1.6 GeV.

Three techniques are used to avoid depolarization: Intrinsic resonances are crossed fast with help of two **pulsed betatron tune jump quadrupoles** which shift the vertical betatron tune. The strengths of imperfection resonances must be reduced to avoid depolarization. This is done with a **dynamic correction of the closed orbit** during the energy ramp. For further reduction of the remaining resonance strengths **harmonic corrections** are applied.

A Tune jumps

For intrinsic resonances a high resonance crossing speed can be achieved by fast shifting the vertical betatron tune Q_z using pulsed quadrupoles. Before crossing the resonance, the vertical betatron tune is shifted fast by the tune jump quadrupoles. The tune jump system at ELSA consists of two quadrupoles which can be pulsed up to 500 A in 4–14 μs [11–13]. This corresponds to a tune shift of $\Delta Q_z = 0.1$. The vertical betatron tune is shifted back to its original value within 4 to 20 ms before the next resonance is crossed.

B Closed orbit correction

The closed orbit is measured with a BPM system consisting of 28 monitor stations. After correction with 19 vertical and 21 horizontal corrector magnets the

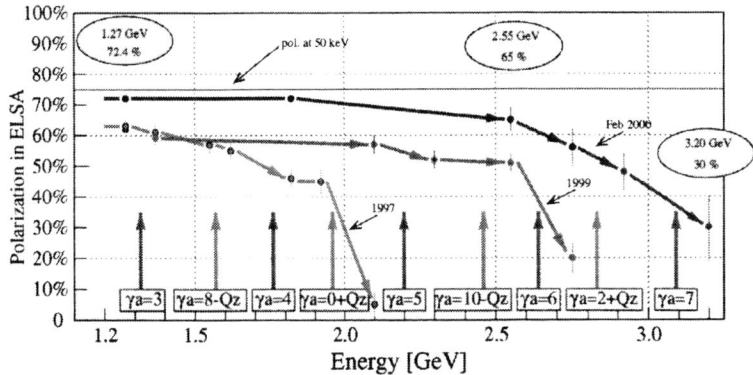

FIGURE 2. Achieved polarization in ELSA. The black arrows indicate the energies of intrinsic resonances ($Q_z = 4.431$), the gray ones the position of the imperfection resonances.

remaining horizontal and vertical distortions (determining the depolarization) are smaller than 0.2 mm (rms) [8–10].

The closed orbit is measured and corrected at the energies of the imperfection resonances. Between these energies a linear interpolation of all corrector kick angles is done. With this interpolation a dynamic closed orbit correction during the energy ramp is achieved. After this correction most of the imperfection resonances become so weak that no significant depolarization is observed.

C Harmonic correction

A further reduction of the strength of each imperfection resonance can be achieved by harmonic correction. The closed orbit harmonics relevant for a single resonance are corrected by modifying the vertical closed orbit. For each resonance two modulation parameters have to be found empirically by measurements of the polarization of the extracted beam.

IV THE GDH-MØLLER POLARIMETER

All polarization measurements were performed with the GDH-Møller polarimeter in the external beamline [14]. This two-arm polarimeter consists of a target system, a spectrometer dipole magnet and a detector system. The target system is composed of several changeable polarized foils with different orientations of the magnetization relative to the electron beam which enable the measurement of all three vector components of the electron beam polarization. Both Møller scattered electrons are energy separated in the dipole magnet and are identified in coincidence in the detector system.

V POLARIZATION OF THE EXTRACTED BEAM

Polarized electrons could be successfully accelerated to higher energies by means of the corrections for depolarizing resonances. We observe a polarization of $P \approx 72\%$ up to energies of 2 GeV, which decreases to $P \approx 65\%$ at 2.55 GeV and drops to $P \approx 30\%$ at 3.2 GeV (see Fig.2). An external current of max. 3 nA could be delivered to the GDH-Tagger target. Optimization of the polarization levels at higher energies is subject to future studies at ELSA.

VI SUMMARY

At ELSA a new polarized electron source has been developed and set into operation. A polarization of $P = 80\%$, $QE = 0.4\%$ and a current of 100 mA were obtained. Polarized electrons have been successfully accelerated to higher energies. Several techniques for correction of depolarizing resonances have been implemented. The longitudinal and transversal polarization components were measured with a Møller polarimeter installed in the extraction beamline. We now can provide a polarized electron beam over the full energy range of ELSA.

REFERENCES

1. K.H. Althoff et al., Part. Acc. 27 (1990) 101
2. W. Hillert et al., Proc. Low Energy Polarized Electron Workshop, St. Petersburg, (1998) p. 115
3. W. Hillert et al., Proc. GDH2000, World Scientific, Singapure (2000)
4. T. Nakanishi et al., Proc. Low Energy Polarized Electron Workshop, St. Petersburg, (1998) p. 118
5. W. Hillert et al., Proc. Low Energy Polarized Electron Workshop, Nagoya, this issue
6. S. Nakamura et al., Nucl. Instr. & Meth. A 441 (1998) 93
7. M. Froissart and R. Stora, Nucl. Instr. & Meth. 7 (1960) 297
8. J. Keil, PhD thesis, BONN-IR-00-09 (2000)
9. J. Dietrich, J. Keil, I. Mohos, Proceedings of the 4th European Workshop on Beam Diagnostics and Instrumentation for Particle Accelerators, Daresbury 1999
10. J. Dietrich, J. Keil, I. Mohos, Proceedings of the 18th Particle Accelerator Conference, New York 1999
11. C. Steier et al., Proceedings of the 18th Particle Accelerator Conference, New York 1999
12. C. Steier, PhD thesis, BONN-IR-99-07 (1999)
13. M. Hoffmann, diploma thesis, BONN-IB-98-10 (1998)
14. B. Kiel, PhD thesis, Universität Erlangen (1999)

Electron beam polarization with the Compton Polarimeter at JLab

T. Pussieux* for the Compton Polarimeter collaboration

*CEA Saclay, DAPNIA/SPhN, 91191 Gif sur Yvette Cedex, France.
http://www.jlab.org/compton

Abstract. We built and operated a new kind of Compton polarimeter to measure the electron beam polarization of the Thomas Jefferson National Accelerator Facility (Virginia, USA) to 3% total error within an hour. The heart of this polarimeter is the coupling of a High Finesse monolithic Fabry-Pérot cavity to the particle accelerator. Its purpose is to amplify a primary 300 mW laser beam to increase the luminosity at the Compton interaction point. The measured Finesse and amplification gain of the cavity are $F = 26000$ and $G = 7300$. We have used this facility during the HAPPEX (April-July 1999) experiment.

INTRODUCTION

Among all the possibilities to measure the polarization of an electron beam, (Mott and Moller polarimeters), Compton polarimetry presents the advantage of providing a measurement while the main experiment is running. Our group has completed the installation of the Jlab Hall A Compton polarimeter in feb 1999. It has given its first results for the HAPPEX experiment [1] with a 4 GeV electron beam of 50 μA. Whereas Compton polarimeters operating at higher energies (SLAC, HERA) or higher currents (NIKHEF) use a high power laser photon source, the challenge of the Compton polarimeter at Jlab was to operate a Fabry-Perot cavity to amplify the photon beam in order to achieve a statistical precision of 1% within 1 hour.

I THE COMPTON POLARIMETER

The hearth of this polarimeter is a high power ($P_L \simeq 1.4 kW$) Fabry-Perot cavity injected by a 300 mW NdYAG infra-red laser [2]. The light inside the Cavity is circuraly polarized, and its helicity can be reversed thanks to a rotatable $\lambda/4$ plate. The polarization for both helicity is $P_\gamma^{R,L} = \pm 99.3 \pm 1.1\%$ [3].

A magnetic chicane steers the electron beam to the center of the optical Cavity. Backscattered photon are dectected by a $PbWO_4$ calorimeter [4]. Counting rates for events with photon energy k' above $k_s' \simeq 30 MeV$ are recorded by a scaler. For a small fraction of events (1 %), the energy deposited in the calorimeter k'_r is also measured thanks to a charge ADC.

The beam polarization P_e is extracted from the Compton experimental asymmetry, A_{exp}, via $P_e = \dfrac{1}{P_\gamma <A_c>} A_{exp}$, where $<A_c>$ is the analyzing power (A.P.). In practise we measure a raw conting rate asymmetry $A_{raw} = \dfrac{r^+ - r^-}{r^+ + r^-}$ given by

$$A_{exp} = \left[\left(1 + \frac{B}{S}\right) A_{raw} + \frac{B}{S} A_B + A_F^p\right]. \tag{1}$$

In order to measure the background over signal ratio $\frac{B}{S} \simeq 0.3$ and the background asymmetry A_B, we take data alternatively with the cavity on ($\simeq$ 5 minutes) and off ($\simeq$ 2 minutes). To reduce systematic errors associated with the luminosity asymmetries A_F^p, the photon polarization is set alternativly right or left.

In addition to the determination of the Light polarisation, the two steps to extract the polarization are the determinaitions of the experimental asymmetry A_{exp} from the measured raw asymmetry A_{raw} and of the analyzing power $<A_c>$.

II MEASUREMENT OF THE EXPERIMENTAL ASYMMETRY A_{EXP}

For each photon helicity state (R, L), a raw asymmetry $A_{raw}^{R,L}$ is determined using the measured rates (normalized to the beam intensity) $(r_1^{R,L})$ at 30 Hz when the cavity is On. If all the parameters were stable between two consecutive photon helicity, then averaging the two raw asymmetries $A_{raw}^{R,L}$ cancels out the false asymmetries (background and luminosity)

$$A_{exp} \simeq \left(1 + \frac{B}{S}\right) \frac{1}{2} \left[A_{raw}^R - A_{raw}^L\right]. \tag{2}$$

The background dilution was estimated by using the rate r_0 measured when the cavity is off $1 + \dfrac{B}{S} = \dfrac{r_1}{r_1 - r_0}$. With this method, we have measured experimental asymetries close to $A_{exp} \simeq$ 1.2 % with a relative statistical error of 1.4% for a typical one hour run. Variations with time of the rates and the background asymmetry result in a contribution to the relative systematic error of 0.5 % for the dilution and 0.5 % for the background asymmetry.

There is also a contribution to the systematic error from the non complete cancellation of the luminosity asymmetry between the two photon helcity states. False luminosity asymmetry can be traced back to the fact that we are crossing two beams with small transverse sizes ($\simeq 100\mu m$) in the vertical direction (y). We thus expect our luminosity to be very sensitive to the vertical electron beam position. For a non optimal crossing (e.g. off-centering by 50 μm), the luminosity exhibits a sensitivy to the electron beam vertical position $\frac{1}{r}\frac{\partial r}{\partial Y}$ that can reach the 0.1%/μ

level. This effect, combined with the fact that there might be systematic different beam positions for the two electron helcity $\Delta y = y^+ - y^-$ of the order of 100 nm, could results in a relative false asymmetry close to 1%. To keep this systematics at a small level, roughly 3% of the data for which the beams were off-centered by more than 50μ were rejected. Again, in practise, a false asymmetry is computed for each photon polarization state

$$A_p^{R,L} = \sum_{p=x,y,\theta_a,\theta_y} \left(\frac{1}{r}\frac{\partial r}{\partial p}\right)^{R,L} \Delta p^{R,L} \quad (3)$$

and the residual systematic error to the experimental asymmetry using Eq. 2 is given by $A_p^F \simeq 1/2\left(A_p^R - A_p^L\right)$. This is the largest contribution ($\simeq 1.2\%$) to the relative systematics error on the experimental asymmetry.

III DETERMINATION OF THE ANALYZING POWER

In terms of backscattered photon energy k', the A.P. is defined by

$$<A_C> = \frac{\int_0^{k'_{max}} \epsilon(k')\frac{d\sigma_c^+}{dk'} - \int_0^{k'_{max}} \epsilon(k')\frac{d\sigma_c^-}{dk'}}{\int_0^{k'_{max}} \epsilon(k')\frac{d\sigma_c^+}{dk'} + \int_0^{k'_{max}} \epsilon(k')\frac{d\sigma_c^-}{dk'}} \quad (4)$$

where $k'_{max} = 190 MeV$ is the Compton edge, $\epsilon(k')$ is the calorimeter efficiency and $\frac{d\sigma_c^{\pm}}{dk'}$ the known Compton helcity dependent cross-sections [2]. In practise, we only measure the energy deposited in the calorimeter k'_r, which differs from the backscattered photon energy k' for various reasons (statistical fluctuations, non-linearity, ...). The A.P. has thus to be computed using this measured energy

$$<A_C> = \frac{\int_0^\infty \epsilon(k'_r)\frac{d\sigma_c^+}{dk'_r} - \int_0^\infty \epsilon(k'_r)\frac{d\sigma_c^-}{dk'_r}}{\int_0^\infty \epsilon(k'_r)\frac{d\sigma_c^+}{dk'_r} + \int_0^\infty \epsilon(k'_r)\frac{d\sigma_c^-}{dk'_r}} \quad (5)$$

In order to determine the dependance of the cross-section $\frac{d\sigma_c^{\pm}}{dk'_r}$ and the efficiency ϵ with the reconstructed energy k'_r we use the background substracted ADC spectrum. This spectrum is first corrected for the non-linearity of the electronics, and then calibrated in energy using the compton edge to get the energy reconstructed spectrum. We then fit this energy reconstructed spectrum by $p_{(k'_r,\sigma_r^s)}(k'_r)\left[\frac{d\sigma_c}{dk'}\otimes g_{(a,b,c)}(k')\right]$ where the Compton cross-section $\frac{d\sigma_c}{dk'}$ is convoluted by a gaussian resolution $g_{(a,b,c)}$ with $\sigma = a \oplus b/\sqrt{k'} \oplus c/k'$ and the calorimeter efficiency is parametrized by an erf function centered around the threshold k'^s_r with a slope proportionnal to σ_r^s. With this procedure we extract for each run the A.P., $<A_C> \simeq 1.7\%$. To estimate the systematic error associated with this procedure, we allow the parameters of the fit (e.g the three parameters for the resolution

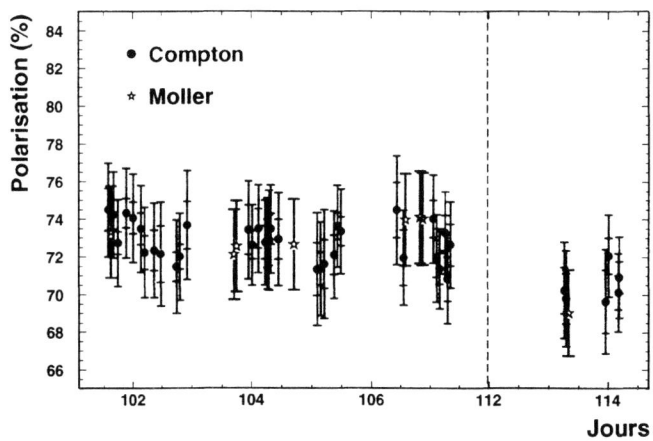

FIGURE 1. Compton and Møller measurement as a function of time.

(a, b, c) and the two parameters of the response function $k_r'^s, \sigma_r^s)$ to vary within a large range and assign as a systematic error on the A.P. the maximum absolute difference between the A.P. determined using the nominal parameters of the fit and the ones determined with each combinaition of the parameters in the above range. This contributes to a relative systematics errors of 1.9 %, and is largely dominated by the sensitivity of the A.P. to the threshold value. Eventually, we have added a 1% contribution to the relative systematic error associated with the non-linearity of the electronics, along with a 1% contribution to take into account the calibration procedure. This ends up with a total relative systematics errors of 2.4 % associated to the A.P.

IV RESULTS

The 40 polarization measurement are given on figure 1. There are in good agreement with the Møller measurements. The typical total error is around 3.3%. In addition to these absolute measurements, the Compton was able to monitor the polarization at the 1.5 % and ruled out possible large variation of the polarization between two Moller measurements.

REFERENCES

1. HAPPEX Collab (K.A. Aniol et al.) Phys.Rev.Lett. **82** :1096 -1100 (1999)
2. J. P. Jorda et al., Nucl. Instrum. Meth. **A412** (1998) 1.
3. N. Falletto, PhD Thesis, Grenoble University, (1999), unpublished.
4. D. Neyret et al., ton polarimeter," Nucl. Instrum. Meth. **A443**, 231 (2000)

Instrumentation for the Polarization Transfer Experiment in Proton Inelastic Scattering at 0°

M. Yosoi[a], H. Akimune[b], I. Daito[c], H. Fujimura[a], Y. Fujita[d],
M. Fujiwara[a], K. Hatanaka[a], K. Hosono[e], T. Inomata[f],
T. Ishikawa[g], M. Itoh[g], M. Kawabata[d], T. Kawabata[g],
M. Nakamura[g], T. Noro[a], E. Obayashi[a], H. Sakaguchi[g],
H. Takeda[g], T. Taki[g], A. Tamii[h], H. Toyokawa[i], M. Uchida[g]
H. P. Yoshida[a] and M. Yoshimura[a]

[a] *Research Center for Nuclear Physics, Osaka University, Ibaraki, Osaka 567-0047, Japan*
[b] *Department of Physics, Konan University, Kobe 658-8501, Japan*
[c] *Center for Integrated Research in Science and Engineering, Nagoya University, Nagoya, 464-8601, Japan*
[d] *Department of Physics, Osaka University, Toyonaka, Osaka 560-0043, Japan*
[e] *Department of Engineering, Himeji Institute of Technology, Hyogo, 678-1297, Japan*
[f] *Wakasa Wan Energy Research Center, Tsuruga, Fukui 914-0192, Japan*
[g] *Department of Physics, Kyoto University, Kyoto 606-8502, Japan*
[h] *Department of Physics, University of Tokyo, Hongo, Tokyo 113-0033, Japan*
[i] *Japan Synchrotron Radiation Research Institute, Hyogo 679-5198, Japan*

Abstract. We have constructed an experimental ensemble to measure polarization transfer observables in proton inelastic scattering at zero degrees. Excited states can simultaneously be measured for the excitation energy from 5 MeV to 35 MeV at E_p=400 MeV. A fast data acquisition system has been developed to endure the large background event rate originated by the primary beam itself. We have found the effective analyzing power of the focal plane polarimeter slightly depends on the treatment of two-ray events.

INTRODUCTION

Inelastic scattering of energetic hadrons at extremely forward angles including 0° is one of the suitable tools to investigate the effective interaction and the small ΔL collective modes in nuclear excitations. Until recently, however, such studies using proton inelastic scattering were hindered by difficulties in determining the spin-transfer (ΔS) decomposition of the measured cross sections. The advent of focal-plane polarimeters has changed this situation. Now all polarization transfer

(PT) obsevables D_{ij} can be measured along with the cross section $d\sigma/d\Omega$, and not only the spin-dependent interactions can be precisely determined but also the spin transfer into the excited states become able to be decomposed. At 0°, especially, the quantity of total spin transfer Σ, defined as $\Sigma = (3 - (D_{NN} + D_{SS} + D_{LL}))/4$, is unity for spin-flip ($\Delta S=1$) transitions and zero for non-spin-flip ($\Delta S=0$) transitions [1].

We have constructed a system to measure PT observables of zero-degree inelastic scattering at RCNP, consisting of the *Grand Raiden* spectrometer [2], the focal-plane polarimeter (FPP) [3] and the external beam dump. We report here on the detection system and some results obtained from the experiment of the FPP calibration.

BEAM POLARIMETRY AND THE FPP

In the polarization transfer measurements, it is important to control the spin direction of both the beam and the scattered particles. Two super-conducting solenoids installed in the transport line between the injection AVF cyclotron and the ring cyclotron are used to give the two independent in-plane polarized beam. The beam polarimetry system consists of a low energy polarimeter (LBLP) placed in the injection line and two high energy polarimeters (HBLP) in the beam line after the ring cyclotron. The p-C elastic and p-p scattering from polyethylene sheets are employed for the LBLP and for the HBLP, respectively.

Momenta of scattered particles from the primary target are analyzed by the spectrometer *Grand Raiden*. This spectrometer has a very high momentum resolution ($p/\Delta p = 37,000$) and we have achieved the 13 keV (FWHM) resolution at 300 MeV with the dispersion matching technique. The *Grand Raiden* has a special dipole magnet for spin rotation called DSR downstream from the main QSQDMD magnet system, which can make other two focal planes with +18° deflection and -17° one. Although, usually, we can not measure the longitudinal component of the polarization at the focal plane, this makes it possible to measure all in-plane polarization observables with the same accuracy.

The FPP is the most important equipment for the spin transfer measurement. As shown schematically in Fig. 1, the FPP consists of a thick carbon analyzer slabs, two pairs of multi-wire proportional chambers (MWPC's) and two plastic-scintillator hodoscopes (X and Y). It is placed just downstream from the standard focal plane counters (two multiwire vertical drift chambers (MWDC's) and two trigger scintillators), which determine the positions and angles of incident particles. The large MWPC3 and MWPC4 with 1.4m width of effective areas measure re-scattered particles from thick carbon slabs. The MWPC1 and MWPC2 are mainly used to make the second-level trigger more effective.

In order to measure the D_{ij} at 0° for inelastic scattering as low excitation as possible, all large wire chambers have holes in their frames near the effective areas (see Fig.1), through which we can let a vacuum pipe for the primary beam pass.

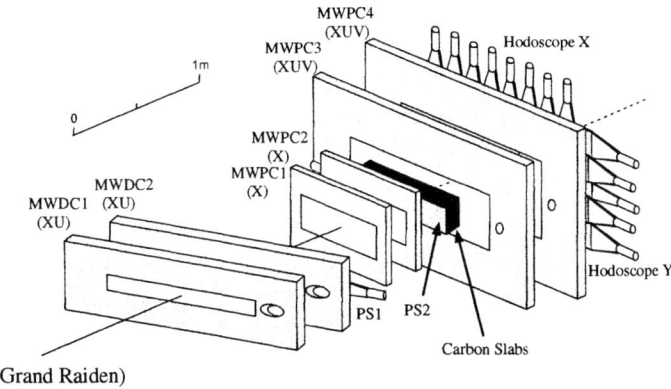

FIGURE 1. Schematic layout of the focal plane detectors.

CALIBRATION OF THE EFFECTIVE ANALYZING POWER OF THE FPP

Accurate values of the analyzing power (A_y) in inclusive proton-carbon scattering are required in determining the polarization of protons incident upon the FPP. There exist a lot of data at SIN, TRIUMF, LAMPF, etc. [4–6] for the intermediate energies. Those data were parametrized as a function of the laboratory scattering angle and the mean energy at the carbon. There remains, however, slight discrepancies among different global fittings.

We have used single scattering of a faint polarized beam itself to calibrate the effective analyzing power (A_y^{eff}) at several energies and for various carbon thicknesses. The faint beam has been obtained by decreasing the beam intensity by about 1/100,000 using the attenuators downstream from the polarized ion source. The beam polarization has been measured periodically by the LBLP when the attenuator was moved aside. Since the systematic uncertainty of the beam polarization was rather large in the above method, we have normalized A_y^{eff} value by another run using the usual intensity beam and employing the elastic scattering at the angle with very large A_y at some energy points. In this case, we can get the incident protons with nearly maximum polarization (~ 1).

It should be noted that the most part of inclusive scattering from carbon is (p,2p) reaction. In the case using large detectors, probabilities that two protons hit in the first wire chamber are increased, while in the small acceptance case, mostly the only forward proton hits the detector. We have first rejected all multi-cluster events in wire chambers due to the simplicity of the analysis. Next, some corrections for the two-ray events have been made using the hodoscope information. The both results of the angle-averaged effective analyzing power A^c over 5°-20° are shown in Fig.2, together with calculations using parameters of the LAMPF(solid) [4], the TRIUMF(dashed) [5] and the SIN(dotted and dot-dashed) [6]. The present data

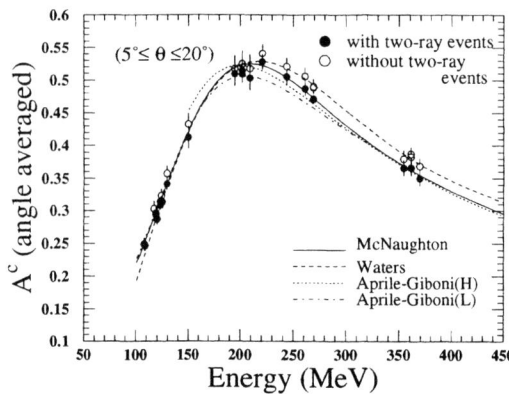

FIGURE 2. Angle-averaged effective analyzing power A_c. Curves are calculated using several global parametrizations (see text). Data of the present measurements are plotted by closed (open) circles with (without) multi-cluster corrections.

rejecting multi-cluster events are plotted as open circles and almost agree with the fitting of Waters *et al.* [5]. The data with corrections for two-ray events are presented as closed circles. Apparently, effective analyzing powers decrease by 2-5 % and are reproduced well by the McNaughton's fitting [4]. In fact, the chamber size was about 1 m in the TRIUNF case, on the other hand it was about 60 cm in the SIN and LAMPF case. It should be mentioned that the vertical acceptance of our system is not so large that the difference between two analysis methods is small in the measured values of up-down asymmetry.

POLARIZATION TRANSFER EXPERIMENT AT 0°

The detector setup at the focal plane for the measurement of proton inelastic scattering at 0° is shown in Fig. 3. The primary beam is transported inside the spectrometer at the highest momentum side and stopped at the external beam dump after passing through the focusing magnet. In the measurement of zero-degree inelastic scattering, large background events originated by the primary beam itself restrict the rate of the data acquisition (DAQ). Fortunately, vertical focusing property of the focal plane let us successfully subtract the background events in offline analysis. We have developed a fast DAQ system [7]. The most important point is to exclude the CAMAC actions in the data flow and we have succeeded it by the sophisticated DAQ programming and the development of some extra modules. Finally, the speed of the present DAQ system is achieved to be about 1 MByte/s. A fast second-level trigger using the LeCroy PCOSIII read out systems, the Memory

Lookup Unit (LRS2372) and the Universal Logic Module (LRS2366) has also been developed to reduce the trigger rate, which rejects the small angle scattering events in the second scattering from the carbon analyzer within a few micro seconds.

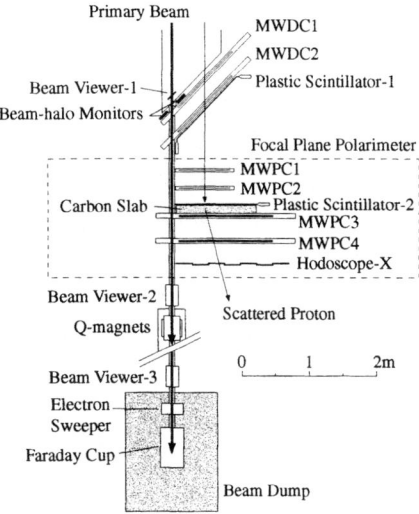

FIGURE 3. Experimental setup of the detectors at the focal plane of the *Grand Raiden* for the measurement of polarization transfer observables in proton inelastic scattering at 0°.

We have first succeeded such PT measurements in ^{12}C(p,p') reaction at 0° [8]. It has been found that the excitation strengths are well decomposed into the spin-flip part and the non-spin-flip one as well as the small isoscalar spin-dependent interaction is quantitatively determined. Some experimental results and analyses are presented in other part of this proceedings [9].

REFERENCES

1. Sakai, H., *Proc. Int. Symp. on New Facet of Spin Giant Resonances in Nuclei*, eds. H. Sakai, H. Okamura and T. Wakasa (World Scientific, Singapore, 1998) p29: T. Suzuki, *Prog. Theor. Phys.* **103**, 859 (2000).
2. Fujiwara, M., et al., *Nucl. Instr. Meth.* **A422**, 484 (1999).
3. Yosoi, M., et al., *AIP Conf. Proc.* **343**, *11th Int. Symp. on High Energy Spin Physics*, eds. K.J. Heller and S. Smith (AIP, New York, 1995) p157.
4. McNaughton, M.W., et al., *Nucl. Instr. Meth.* **A241**, 435 (1985).
5. Waters, G., et al., *Nucl. Instr. Meth.* **153**, 401 (1978).
6. April-Giboni, E., et al., *Nucl. Instr. Meth.* **A215**, 435 (1983).
7. Tamii, A., et al., *IEEE Tr. Nucl. Sci.* **43**, 2488 (1996).
8. Tamii, A., et al., *Phys. Lett.* **B459**, 61 (1999).
9. Kawabata, T., et al., in this proceedings : Tamii, A., et al., in this proceedings.

Deuteron Polarimeter DPOL and Calibration of the System

H. Kato[a], Y. Satou[b], H. Sakai[a], A. Tamii[a], K. Sekiguchi[a], K. Yako[a],
S. Sakoda[a], M. Hatano[a], Y. Maeda[a], N. Sakamoto[b], K. S. Itoh[e],
T. Ohnishi[b], H. Okamura[c], T. Uesaka[c], K. Suda[c], J. Nishikawa[c]
and T. Wakasa[d]

[a] *Department of Physics, University of Tokyo, Bunkyo, Tokyo, 113-0033, Japan*
[b] *RIKEN Accelerator Research Facility, Wako, Saitama, 351-0198, Japan*
[c] *Department of Physics, Saitama University, Urawa, Saitama, 338-9570, Japan*
[d] *Research Center for Nuclear Physics (RCNP), Ibaraki, Osaka, 567-0047, Japan*
[e] *Department of Physics, Tsukuba University, Tsukuba, Ibaraki, 305-8571, Japan*

Abstract. We have calibrated the effective analyzing powers of the deuteron polarimeter, DPOL, in the deuteron energy region of $210 < E_d < 270$ MeV. The calibration was needed to analyze the polarization transfer measurement for the ^{12}C$(\vec{d},\vec{d'})$ reaction at 270 MeV whose aim is to search for Double Gamow-Teller resonances in ^{12}C in an excitation energy region of $E_x < 60$ MeV.

INTRODUCTION

The measurement of polarization transfer coefficients in the deuteron inelastic scattering is an excellent probe to explore isoscaler spin excitations in nuclei. In the $(\vec{d},\vec{d'})$ scattering, single spin-flip probability (S_1) and double spin-flip probability (S_2) in the y-direction can be defined as

$$S_1 = \frac{1}{9}(4 - P^{y'y'} - A_{yy} - 2K^{y'y'}_{yy}) , \qquad (1)$$

$$S_2 = \frac{1}{18}(4 + 2P^{y'y'} + 2A_{yy} - 9K^{y'}_{y} + K^{y'y'}_{yy}) , \qquad (2)$$

where $P^{y'y'}$ is the tensor polarizing power, A_{yy} is the tensor analyzing power, $K^{y'y'}_{yy}$ is the tensor to tensor polarization transfer coefficient and $K^{y'}_{y}$ is the vector to vector polarization transfer coefficient. We have mainly two purposes. One is to study the strength of the V_σ term of the effective nucleon-nucleon interaction. The effective interactions other than V_σ are well studied by using the (p,p') or (p,n') scattering. However, we have little information about the V_σ term because it is weak. Since a deuteron has a spin 1 and an isospin 0, only isoscaler transitions can take place in the (d,d') scattering. S_1 is a good signature of single spin-flip reaction ($\Delta S = 1$). By measuring the S_1 value, we can derive the information about the V_σ term. The other purpose is to search for the Double Gamow-Teller (DGT) resonances [1,2]. Lately, double giant resonances have been of considerable interest. Up to now double isobaric analogue states, double isovector dipole resonances and a dipole resonance built on an analogue state have been identified [3]. However, there is no report on the DGT resonance. The DGT resonance is a 2 phonon state in which GT resonances are doubly excited. $\Delta S = 2$ is a good signature for the DGT resonances. By measuring the S_2 value, we can search for the DGT resonances.

In order to obtain these spin-flip probabilities, polarization of ejectile deuterons must be measured. For this purpose we have constructed a focal plane Deuteron POLarimeter (DPOL) [4] at RIKEN.

Recently we have measured S_1 and S_2 for the highly excited continuum region of $E_x < 60$ MeV of the $^{12}\text{C}(\vec{d},\vec{d'})$ scattering at 270 MeV where the DGT resonances are predicted to exist. To derive S_1 and S_2 we needed calibrate the effective analyzing powers of the DPOL in the deuteron energy range of $210 < E_d < 270$ MeV.

DEUTERON POLARIMETER DPOL

The DPOL is located at the second focal plane of the spectrometer, SMART [5]. A full view of the DPOL is illustrated in Fig.1. The DPOL consists of two MWDCs, scatterers (SC), a hodoscope (HOD) and a calorimeter (CM). Incoming deuterons are scattered in the SCs and ejectile particles are detected by the second MWDC and the HOD. The SCs are composed of 4 plastic scintillators and 3 polyethylene plates. The thicknesses of each plastic scintillator and polyethylene plate are 0.5 cm and 1 cm, respectively. The HOD consists of 28 bars of plastic scintillators whose length is 220 cm. The distance between the SCs and the HOD is about 4 m. We can calculate the momentum of the scattered particles from their time of flight.

The DPOL utilizes the $^{12}\text{C}(\vec{d},d)$ reaction for the vector polarization measurement and the $^1\text{H}(\vec{d},2p)$ reaction for the tensor polarization measurement. The two reactions can be clearly distinguished by light outputs of scintillators in SCs. The vector and tensor polarizations of incoming deuterons can be measured simultaneously. This is why the DPOL is expected to have less systematic errors in the measurement. However, because the SCs are mixtures of hydrogen and carbon, the two reactions must be well separated from background events such as the $^1\text{H}(\vec{d},d)$ reaction or the $^{12}\text{C}(\vec{d},2p)$ reaction.

CALIBRATION OF THE DPOL

In the spherical coordinate system, the cross section σ for a polarized beam of spin 1 particle is written as

$$\sigma(\theta,\phi) = \sigma_0(\theta)[1 + Re(it_{11}e^{-i\phi})iT_{11}(\theta) + t_{20}T_{20}(\theta)$$
$$+ 2Re(t_{21}e^{-i\phi})T_{21}(\theta) + 2Re(t_{22}e^{-2i\phi})T_{22}(\theta)] , \quad (3)$$

where σ_0 is the cross section with an unpolarized beam, θ and ϕ are the scattering angle and the azimuthal angle in the laboratory system, respectively, t_{ij} is the polarization of the beam and T_{ij} is the analyzing power. In order to calibrate the analyzing powers, polarizations of deuteron should be known as well as the analyzing powers should be known when measuring the polarizations. In this calibration of the DPOL, a faint polarized deuteron beam ($\sim 10^5$ per second) was directly injected to the DPOL. We measured the beam polarization at 1 nA by using the beam-line polarimeter assuming that the polarization of the beam does not depend on the beam intensity. The beam energy was lowered to 250, 230 and 210 MeV by using the aluminum degraders.

At RIKEN, we can freely control the polarization axis of the beam. For example, when the polarization axis is set to the y-direction, we can simplify the Eq. (3) as

$$\sigma(\theta,\phi) = \sigma_0(\theta)[1 + \sqrt{3}p_y iT_{11}(\theta)cos\phi + t_{20}T_{20}(\theta)$$
$$- \frac{1}{2\sqrt{2}}p_{yy}T_{20}(\theta) - \frac{\sqrt{3}}{2}p_{yy}T_{22}(\theta)cos2\phi] , \quad (4)$$

where p_y and p_{yy} are polarizations in the cartesian description. By combining the data of 4 polarization modes produced at the ion source, the analyzing powers of the DPOL can been extracted from the ϕ-dependence of σ.

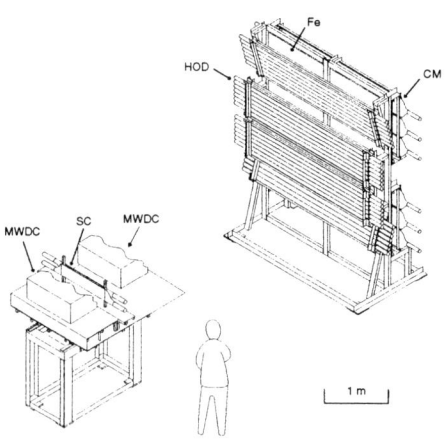

FIGURE 1. Full view of the Deuteron POLarimeter DPOL.

RESULT

The obtained angular distribution of the effective analyzing powers at a deuteron energy of $E_d = 250$ MeV is shown in Fig. 2. iT_{11} is the vector analyzing power of the $^{12}C(\vec{d},d)$ reaction. T_{20} and T_{22} are the tensor analyzing powers of the $^1H(\vec{d},2p)$ reaction. The relative energies of two protons are selected as $E_{rel} < 3$ and < 9 MeV in the analysis of T_{20} and T_{22}, respectively. The relative energy range is determined so as to get the highest figure of merit (FOM) of each analyzing power. The FOM(F_{ij}) is defined as,

$$F_{ij} = \sqrt{\int T_{ij}(\theta)^2 * \varepsilon(\theta) d\theta} ,$$

where ε, called the efficiency, is the ratio of the number of the detected reactions to that of the injected deuterons. The FOMs and thicknesses of scatterers are compared with those of the polarimeters at Saturne [6,7] in Table 1. The FOMs of the DPOL are smaller than those at Saturne due mainly to the thickness of scatterers. However, the DPOL has an advantage that the vector and tensor polarizations can be measured simultaneously, while at Saturne different polarimeter systems are used for the vector and tensor polarization measurements, respectively.

The energy dependence of the effective analyzing powers is shown in Fig. 3. The scattering angular ranges in the analysis of iT_{11}, T_{20} and T_{22} are selected as $5° < \theta < 14°$, $0° < \theta < 10°$ and $3° < \theta < 12°$, respectively.

SUMMARY

We have constructed the DPOL which utilizes the $^{12}C(\vec{d},d)$ reaction for the vector polarization measurement and the $^1H(\vec{d},2p)$ reaction for the tensor polarization measurement. We have calibrated the effective analyzing powers of the DPOL for

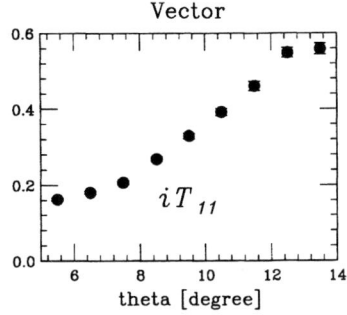
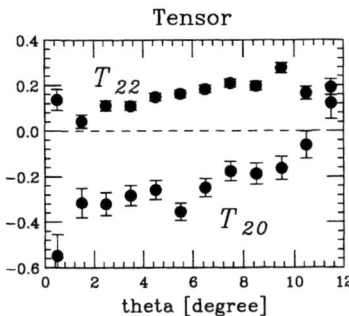

FIGURE 2. Angular distributions of vector and tensor analyzing powers at $E_d = 250$ MeV. The relative energy of two protons are selected as $E_{rel} < 3$ and < 9 MeV in the analysis of T_{20} and T_{22}, respectively.

TABLE 1. FOMs at $E_d = 250$ MeV and thicknesses of the scatterers.

	F_{11} (10^{-2})	F_{20} (10^{-2})	F_{22} (10^{-2})	thickness (g/cm^2)
DPOL	1.91	0.60	0.59	C:4.2 H:0.47
Saturne[a]	4.2	1.2	1.0	C:8.2 H:1.1

[a] F_{11} (F_{20}, F_{22}) is a value of the vector (tensor) polarimeter POMME (POLDER).

the deuteron energy range of $210 < E_d < 270$ MeV. The FOMs of the DPOL are smaller than those of Saturne. However, the DPOL has a merit that the vector and tensor polarizations can be measured simultaneously. Therefore we can reduce systematic errors in deducing the S_1 and S_2 values.

REFERENCES

1. P. Vogel et al., *Phys. Lett.* **B212**, 259 (1988).
2. N. Auerbach et al., *Ann. Phys.* **192**, 77 (1989).
3. S. Mordechai and C.F. Moore, *Nature* **352**, 393 (1991).
4. S. Ishida et al., *AIP Conf. Proc.* **343**, 182 (1995).
5. T. Ichihara et al., *Nucl. Phys.* **A569**, 287c (1988).
6. B. Bonin et al., *Nucl. Instr. and Meth.* **A288**, 389 (1990).
7. S. Kox et al., *Nucl. Instr. and Meth.* **A346**, 527 (1994).

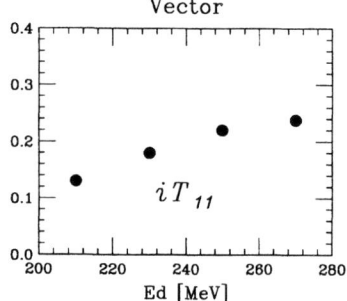
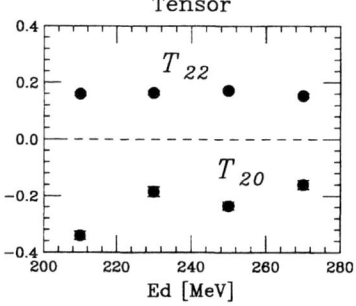

FIGURE 3. Energy dependence of the effective analyzing powers. The scattering angular range in the analysis of iT_{11}, T_{20} and T_{22} are selected as $5° < \theta < 14°$, $0° < \theta < 10°$ and $3° < \theta < 12°$, respectively.

Radiative Polarization in the BATES South Hall Ring

M.Korostelev and Yu.M.Shatunov

Budker Institute of Nuclear Physics, Novosibirsk, 630090, Russia

Abstract. This report describes some aspects of spin matching conditions at storage ring with Siberian snake to reach larger equilibrium polarization degree. An enhancement of the so-called kinetic polarizing mechanism by special polarizing wigglers is considered.

INTRODUCTION

It is good known that the spin dependence of the synchrotron radiation leads to polarization of an electron beam with a characteristic time τ to the equilibrium value P_{eq} [1]:

$$P_{eq} = -\frac{8}{5\sqrt{3}}\frac{\alpha_-}{\alpha_+}; \quad \tau^{-1} = \frac{5\sqrt{3}}{8}\frac{e^2\hbar\gamma^5}{m^2c^3}\alpha_+ \tag{1}$$

where

$$\alpha_- = \left\langle \frac{\hat{\mathbf{b}}}{|\rho|^3}(\hat{\mathbf{n}} - \mathbf{d}) \right\rangle; \quad \alpha_+ = \left\langle \frac{1}{|\rho|^3}[1 - \frac{2}{9}(\hat{\mathbf{n}}\hat{\mathbf{v}})^2 + \frac{11}{18}|\mathbf{d}|^2] \right\rangle \tag{2}$$

Here $\hat{\mathbf{n}}$ is a periodical spin solution, $\mathbf{d} = \gamma\frac{\partial \hat{\mathbf{n}}}{\partial \gamma}$ is a spin-orbit coupling vector, ρ is bending magnet curvature, $\hat{\mathbf{v}}$ and $\hat{\mathbf{b}}$ are unit vectors in the direction of particle velocity and magnetic field respectively. The average is taken over the ring azimuth θ and, generally, over beam distribution.

Experimentally this effect have been studied at many machines, when the $\hat{\mathbf{n}}$ vector coincides practically with the guiding field direction. Using one Siberian snake creates an unusual situation, when the direction of the beam polarization is in horizontal plane excepting the snake, and the scalar product of $\hat{\mathbf{b}}$ and $\hat{\mathbf{n}}$ vectors in α_- expression (2) is zero. However, the equilibrium polarization should differ from zero level due to so-called kinetic polarizing mechanism. Maximum polarization degree about $\sim 80\%$ may be achieved when $|\mathbf{d}| \approx 1.2$.

SPIN MATCHING CONDITIONS

In the energy range of few Gev the longitudinally polarized beams can be obtained with the use of Siberian Snake, a special kind of spin rotator, which rotates particle spin by 180° angle around a direction lying in the horizontal plane. This direction is called the snake axis. An insertion of Siberian snake with longitudinal snake axis provides automatically the longitudinal beam polarization on a ring azimuth opposite to the snake insertion. Such solenoidal snake was developed for BATES SHR [2].

At first-order approximation the **d** vector is orthogonal to $\hat{n}$: $\mathbf{d} = \mathrm{Re}(iD\hat{\eta}^*)$ where $\hat{\eta}$ is an eigen solution of spin motion equation orthogonal to $\hat{n}$. As it was shown in the paper [3], in general case the D function can be represented as a sum of two contributions: $D = D_\gamma + D_\beta$.

$$D_\gamma = -\frac{\pi}{2}\sin(\pi\nu_0) + i\nu_0(\pi - \int_0^\theta K_z d\theta') \qquad (3)$$

$$D_\beta = -\frac{\nu_0 \pi}{4\cos(\pi\nu_x)}\left[\cos(\pi\nu_0)\mathrm{Im}(e^{i\pi\nu_x}J(\theta)G_{Ix}^*) + i\mathrm{Im}(e^{i\pi\nu_x}G_{Iz}^*J(\theta))\right] \qquad (4)$$

where $G_{Ix,z} = f'_{Ix,z(out)} - f'_{Ix,z(in)}$ is the difference of the first mode Floquet function derivatives at the entrance of the first solenoid and at the exit from the second one and $J(\theta) = f_{Ix}\psi'_x - f'_{Ix}\psi_x$.

Here D_γ comes from the direct dependence of the vector $\hat{n}$ on the particle energy while D_β results from a jump of betatron amplitudes during an emission of quanta.

A spin matching condition in the case of one Siberian snake is $|D_\beta| = 0$, that can be satisfied if $f'_{Ix(in)} = f'_{Ix(out)}$ and $f'_{Iz(in)} = f'_{Iz(out)}$.

SOUTH HALL RING WITH SIBERIAN SNAKE

The Fig. 2 shows the location of the Siberian snake in the SHR. For $E = 1$ GeV the magnetic field of solenoids is 6.586 T. The snake scheme was chosen for the orbital matching. Unfortunaly the proper spin matching with existing snake is not possible, but the contribution of D_β can be considerably suppressed due to an optics tuning by nearest to Siberian snake pairs of regular machine quadrupole magnets.

After such corrections, the $|\mathbf{d}|$ behavior (solid line) along azimuth of SHR is shown in the Fig. 2b ($E = 1$ GeV), together with D_γ (dashed line) derived from (3). A behavior of eigen modes within Siberian snake is shown in Fig. 1.

All calculations here and later were performed with ASPIRRIN code [3]. The Fig. 3 present the dependence of τ and P_{eq} on the beam energy for the SHR. One can see that the polarization time is too long and the equilibrium polarization degree is small.

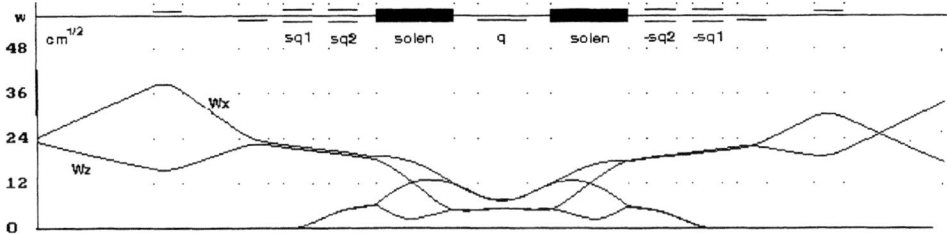

FIGURE 1. The Floquet functions nearby the Siberian snake.

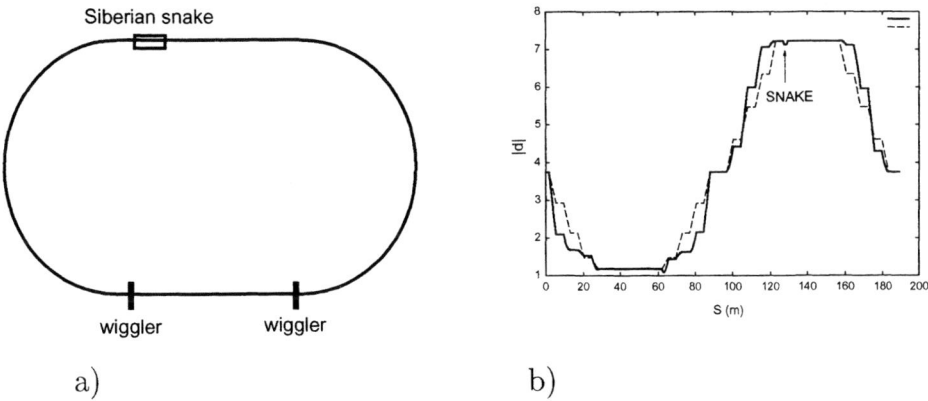

FIGURE 2. a) Position of Siberian snake in the SHR, b) $|d|$ along the azimuth of SHR

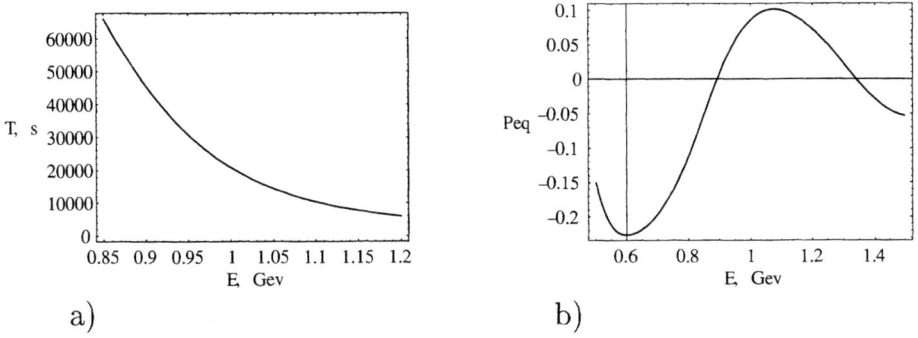

FIGURE 3. The dependence of polarization time (a) and equilibrium polarization degree (b) on the beam energy.

THE METHOD OF POLARIZATION ENHANCEMENT

It's good known, that asymmetric wigglers with high magnetic field poles accelerate the radiative polarization. But in the case of machine with Siberian snake an application of the asymmetric wiggler looks much more attractive. If we install such wiggler in the straight section opposite to the snake insertion (see Fig. 2) where $|d| \approx 1.2$, we get not only faster polarization but also an enhancement of P_{eq}.

In this paper we have considered two wigglers with three pairs of magnetic poles. The length of central pole is 10 cm and its field is 10 T. Both compensated poles are identical. The length and field each of them is 45 cm and -1.1111 T respectively. To compensate the distortion of Floquet functions induced by wiggler, due to strong enough radial focusing, thin quadrupole magnets on both sides of the center pole are needed, namely with -1.95 KGs/cm and 10 cm in length.

The disposition of both wigglers have mirror symmetry with respect to center of straight section. As it is seen from the Fig. 4 the wiggler is located after three quadrupole magnets Q6, Q7, Q8 following last arc magnet. Computer simulation shows that we can not restore the original machine optics symmetry by adjustment gradient field of only these (Q6, Q7, Q8) quadrupole magnets. Possible optical solution is in a shift of Q8 towards to center of straight section and one additional quadrupole magnet Q10 between Q7 and Q8.

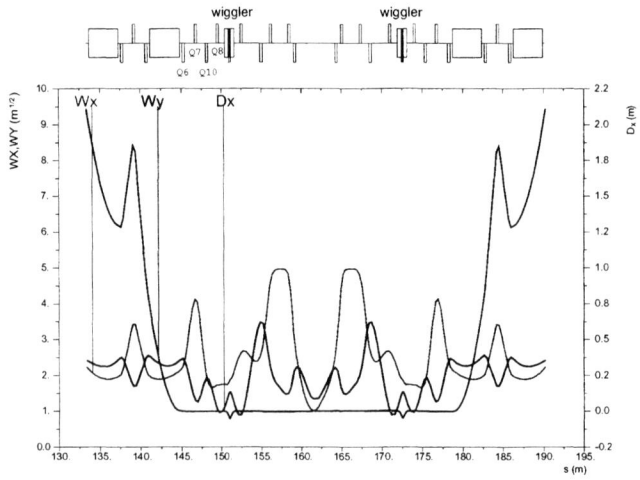

FIGURE 4. The Floquet and dispersion functions in the straight section with wigglers.

The calculation of P_{eq} and τ for the new lattice with two wigglers are shown in the Fig. 5. The data for the energy $E = 1$ GeV are presented in the Table 1, from where one can see that we can achieve 57% polarization degree in spite of discussed above snake imperfections.

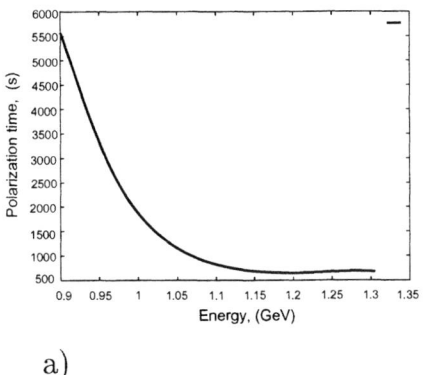
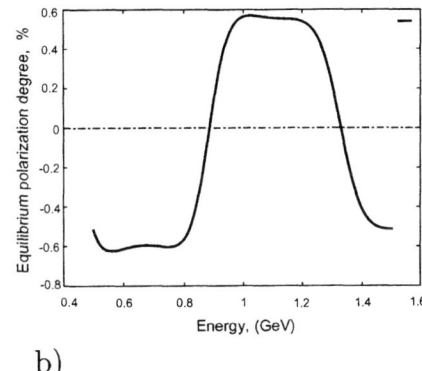

a)   b)

FIGURE 5. The dependence of polarization time (a) and equilibrium polarization degree (b) on the beam energy.

TABLE 1. The results of the computer simulation with ASPIRIN code for SHR

	result
Equilibrium polarization degree	57%
Relaxation time polarization	1871s
Betatron tunes ν_x	7.202
Betatron tunes ν_z	8.921
Spin tune ν_s	0.5

Hence the application of the polarizing wigglers can make the kinetic radiative polarization a real instrument for spin experiments at SHR.

REFERENCES

1. Ya. S. Derbenev and A. M. Kondratenko, Sov. Phys. JETP, **37**, (1973) 968.
2. S. Kovalsky, T. Zwart, P. Ivanov and Yu. M. Shatunov, *Spin Control System for the SHR at Bates linear accelerator center*, High Energy Spin Physics eleventh international symposium, Bloomington in 1994.
3. E. A. Perevedentsev, V. I. Ptitsin and Yu. M. Shatunov, *Calculation of spin resonances in storage rings*, V Workshop on High Energy Spin Physics, Protvino, September 1993.

Polarised $e^{\pm}$ at Hera: experience and expectations after the Luminosity Upgrade

D.P. Barber, M. Berglund, E. Gianfelice

Deutsches Elektronen-Synchrotron DESY, Notkestrasse 85, D-22603 Hamburg, Germany

Abstract. After a short summary of experience with $e^{\pm}$ polarisation at HERA, the impact of the collider Luminosity Upgrade on polarisation will be reviewed.

INTRODUCTION

HERA is a 6 km long $p/e^{\pm}$ double ring collider located at DESY in Hamburg. The proton and $e^{\pm}$ beams are accelerated up to 920 GeV and 27.5 GeV respectively and collide head–on at the Interaction Points (IP's) North and South, where the experiments H1 and ZEUS are located. These experiments started data taking in 1992. HERMES, which uses the longitudinally polarised $e^{\pm}$ beam on an internal polarised gas target, joined the collider experiments in 1994.

The HERA performance has greatly improved over the years. In particular, although the beam currents did not reach the design values, it was possible to attain the design luminosity through a reduction of the beam sizes at the IP's.

At the request of the physics community the feasibility of higher luminosity has been studied in the last few years. The resulting Luminosity Upgrade project was approved in December 1997 and is currently being realised.

EXPERIENCE WITH POLARISATION AT HERA

An integral part of the original HERA design was the provision of longitudinally spin polarised $e^{\pm}$ beams for the collider experiments. In a storage ring, $e^{\pm}$ beams can become spin polarised through the Sokolov–Ternov effect [1]. The polarisation direction is given by the periodic solution to the Thomas–BMT equation for the spin on the closed orbit, $\hat{n}_0(s)$, which is vertical in a perfectly planar ring. To provide the experiments with longitudinal polarisation, $\hat{n}_0(s)$ must be rotated into the longitudinal direction at the experiments by special magnet insertions called "spin rotators". At high energy, rotators must involve radial fields; this means that the ring is, by design, no longer planar everywhere. In a non–planar ring, where $\hat{n}_0(s)$ is not everywhere vertical and/or the beam has a finite vertical dimension, the stochastic photon emission causes the single particle spins to diffuse away from $\hat{n}_0(s)$ with a consequent decrease of polarisation. Spin rotators are therefore a source of spin diffusion; this can be partially neutralised by designing a "spin matched" optics [2,1].

Similarly, the unavoidable magnet misalignment and field errors lead to spin diffusion [1]. Simulations show that for a high energy storage ring, like HERA, even after that the closed orbit distortion has been minimised, the polarisation is very low and of no practical interest for the experiments. A dedicated minimisation of the $\hat{n}_0(s)$ distortion, $\delta\hat{n}_0(s)$, is then needed and was for the first time successfully applied at PETRA [3]. The method has been improved for HERA [4,5]; 8 closed vertical orbit bumps ("harmonic bumps") allow the 8 most important Fourier components of $\delta\hat{n}_0(s)$ to be minimised.

There were some doubts in the scientific community about whether large beam polarisation could be observed in high energy storage rings and whether it could be maintained in the presence of spin rotators. Transverse beam polarisation was observed for the first time at HERA in November 1991 and after June 1992 [5], with dedicated machine tuning, high transverse polarisation became a routine aspect of HERA operation. This success played a decisive role in the approval of the HERMES experiment and of the installation of a first pair of spin rotators of the Buon–Steffen type [6] during the 1993–1994 shut down to provide longitudinal polarisation to HERMES. The vertical bending magnets of the rotators were turned on for the first time on May 4, 1994 [7]. High longitudinal beam polarisation was then routinely delivered to HERMES and the doubts were seen to be unfounded.

In a collider such as HERA the interaction with the counter–rotating beam was also expected to be a source of trouble for polarisation. Indeed, since 1996 while the proton current and the specific luminosity have steadily increased, a clear correlation between $e^\pm$ polarisation and luminosity has been observed. By careful machine tuning it has nevertheless been possible to cope with the beam–beam interaction.

HERA–e is the only high energy $e^\pm$ ring delivering longitudinal spin polarisation.

THE HERA LUMINOSITY UPGRADE

Impact of the new IR design on polarisation

The aim of the Luminosity Upgrade is to increase the luminosity without spoiling the polarisation for HERMES and for H1 and ZEUS, for which two more rotator pairs have been manufactured in the meantime. Since the solution adopted has been described elsewhere (see for example [8]), we here only recall the aspects which have a direct impact on polarisation. In particular:
- The new interaction regions will no longer be mirror symmetric w.r.t. the IP's.
- Due to lack of space the anti–solenoids will be removed.

- The IR quadrupoles will be stronger. So too, will those in the arcs where the FODO phase advance will be increased in both planes from 60 to 72 degrees in order to reduce the $e^\pm$ horizontal emittance, ϵ_x.
- It is planned to operate the machine with a RF frequency offset of about 250 Hz to get a further ϵ_x reduction.

In the new design the betatron coupling resulting from the lack of anti–solenoids will be corrected by 4 independently powered skew quadrupoles per IP. Moreover, although the H1 solenoid, for example, would rotate a vertical spin by about 86 mrad at 27.5 GeV, at first sight there should be no distortion of $\hat{n}_0(s)$ since the nominal $\hat{n}_0(s)$ is longitudinal with the H1 and ZEUS spin rotators running. But we must nevertheless expect some reduction of the achievable polarisation due to the presence of two additional spin rotators and of the solenoids themselves [9].

The experiment solenoids have been treated as a perturbation and ignored when designing the new optics. However, subsequent local spin matching of the optics in the presence of the solenoids does not seem to be possible. A mutual compensation of the two solenoids is not feasible either, mainly because it would at least require the phase advance between the two IP's to be a multiple of π.

While the ZEUS solenoid ($B_{sol}L$=4.4 Tm) fits physically into the 3.9 m free space between the machine magnets, the H1 solenoid ($B_{sol}L$=7.6–8.3 Tm, L=7.3 m) overlaps with the machine magnets, namely with the long combined function superconducting magnet GO. Moreover it is longitudinally shifted by 1.1 m. Therefore the nominal particle velocity and $\hat{n}_0$ are not perfectly parallel to the solenoid field when entering it. The overlap produces a (mainly vertical) orbit distortion (Δz_{rms}=1.2 mm) and the longitudinal offset produces a residual $\hat{n}_0$ distortion ($\delta\hat{n}_{0,rms}$=8.8 mrad). The distortions due to the H1 solenoid were computed initially by solving the equations of particle motion for the transverse coordinates and the Thomas–BMT equation for the spin in a general 3–dimensional field. Later on symplectic/orthogonal maps produced by numerical integration of the equations of particle/spin motion in the measured fields were introduced into the existing codes SLIM/SLICK [10] and SITF/SITROS [11]. In parallel we also used interleaved slices of combined function magnets, solenoids and vertical correctors — the "sandwich" model. This approach does not need code changes, but of course leads to unphysical non–cancelling solenoid end fields. Nevertheless, the results of these different approaches are in reasonable agreement.

The orbit will be locally corrected by using the vertical dipole windings of the two closest machine magnets. The residual $\hat{n}_0$ distortion will be compensated by slightly asymmetric settings of the vertical bending magnets of the rotators on the left and on the right side of the IP North.

Fig. 1 shows polarisation vs. energy for the optics with 3 rotator pairs (linear calculations with SLIM): (a) ideal optics; (b) optics with H1 solenoid turned on ("sandwich" model); (c) with H1 solenoid turned on, after correcting the orbit, the coupling and the $\hat{n}_0$ distortion ("sandwich" model). The three dashed lines correspond to the polarisation related to each of the three degrees of freedom of the motion.

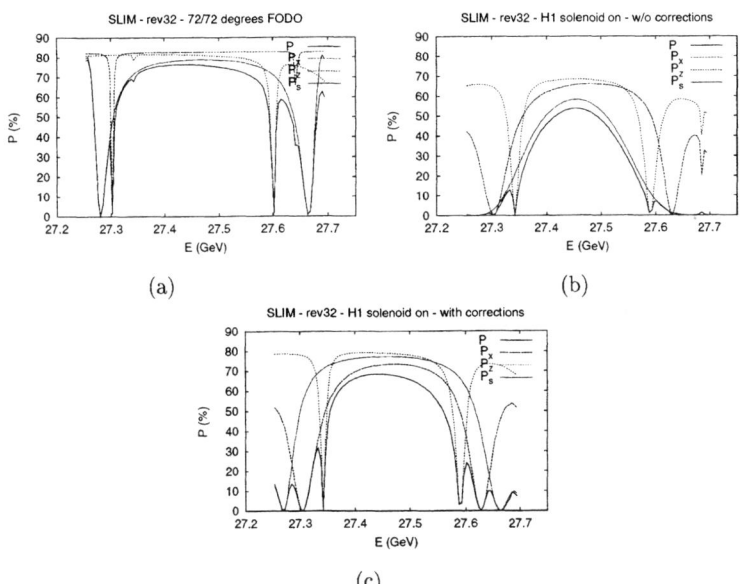

FIGURE 1. Polarisation vs. energy (linear spin motion calculations).

TABLE 1. Expected $\delta\hat{n}_0$ and polarisation in presence of random errors

after usual orbit correction			with harmonic bumps in addition		
$\delta\hat{n}_0$(mrad)	$P(\%)$	$P_x(\%)$ $P_z(\%)$ $P_s(\%)$	$\delta\hat{n}_0$(mrad)	$P(\%)$	$P_x(\%)$ $P_z(\%)$ $P_s(\%)$
35.6±11.1	12.3±11.8	73.5±4.1 72.5±4.4 11.3±11.9	16.9±3.7	67.5±4.8	71.5±4.1 70.8±5.3 67.6±5.8

Effects of random distortions

The impact of the unavoidable random alignment errors has been studied for the optics with 3 rotator pairs and, initially, without experiment solenoids. The assumed rms value of the horizontal and vertical quadrupole displacement is 0.3 mm with a 3 σ cut. In some cases a roll–angle error (0.35 mrad rms value) has also been introduced.

In Table 1 the expected rms value of $\delta\hat{n}_0$ and the polarisation, after usual orbit correction and after additional optimisation of the harmonic bumps, are quoted; the values are averaged over 6 seeds. The orbit has been corrected down to x_{rms}=0.76±0.06 mm and z_{rms}=0.81±0.14 mm.

In comparison with the old optics there is a larger $\hat{n}_0$ perturbation and the closed orbit must be better corrected (down to $\simeq 0.7$ mm) to ensure $P \geq 60\%$, after harmonic bump optimisation. We have also noticed that in some cases the polarisation is limited by the large horizontal dispersion around the IP's. Therefore it is sensible for the future to have a dispersion correction algorithm.

Recent polarisation studies

After the Luminosity Upgrade, routine running with longitudinal polarisation is expected to be more difficult than in the past. Therefore studies aiming to explore some of the issues were carried out before shutting down the machine last September. A spin matched 72°/72° optics was established and it was possible to get about 60% polarisation within 24 hours. Afterwards, the dependence of the polarisation on the RF frequency shift, at constant particle energy (by tuning the main dipole current), was measured. A slight drop of polarisation was observed but the effect of the RF shift was not as large as predicted by the corresponding preliminary calculations with SITROS to include nonlinear spin motion.

A dispersion correction algorithm based on singular value decomposition was also tested and the results look encouraging.

SUMMARY AND OUTLOOK

HERA demonstrated the possibility of getting longitudinal polarisation in a high energy electron ring. Machine tuning for polarisation required care, but was feasible and high longitudinal beam polarisation was routinely delivered to HERMES together with luminosity for H1 and ZEUS. The Luminosity Upgrade will have a large impact on HERA-e and on polarisation. The orbit will have to be very well corrected. The increase of the vertical incoherent beam–beam tune shift by about 24% will be quite challenging. Machine commissioning for polarisation will in practice be more difficult than in the past.

REFERENCES

1. D.P. Barber and G. Ripken, in *Handbook of Accelerator Physics and Engineering*, Eds. A.W. Chao and M. Tigner, World Scientific (1999), 153-165.
2. A.W. Chao and K. Yokoya, KEK TRISTAN Report, 81-7, (1981).
3. R. Rossmanith and R. Schmidt, *Nucl. Instrum. Methods* A **236**, 231 (1985).
4. D.P. Barber, H. Mais, G. Ripken and R. Rossmanith, DESY Report 85-044 (1985).
5. D.P. Barber et al., *Nucl. Instrum. Methods* A **338**, 166-184 (1994).
6. J. Buon and K. Steffen, DESY HERA Report 85-09 (1985)
7. D. P. Barber et al., *Phys. Lett.* B **343**, 436-443 (1995).
8. M. Seidel and F. Willeke, in *Proceedings of EPAC 2000*, 2000, 379-381.
9. D. P. Barber, et al., DESY Report 82-76 (1982).
10. A. W. Chao, *Nucl. Instrum. Methods* A **180**, 29-36 (1981)
    (SLICK is the thick lens version of SLIM from D.P. Barber).
11. J. Kewisch, DESY 85-109 (1985).

Stern-Gerlach Interaction in Fermion Beams

P. Cameron[a], M. Conte[b1], M. Ferro[b], G. Gemme[b], A.U. Luccio[a],
W.W. MacKay[a], M. Palazzi[b], R. Parodi[b], M. Pusterla[c]

(a) RHIC Project, Brookhaven National Laboratory, Upton, NY 11973, USA.
(b) Dipartimento di Fisica dell'Università di Genova, INFN Sezione di Genova,
Via Dodecaneso 33, 16146 Genova, Italy.
(c) Dipartimento di Fisica dell'Università di Padova, INFN Sezione di Padova,
Via Marzolo 8, 35131 Padova, Italy

Abstract. The Stern-Gerlach interaction, between a moving charged particle endowed with a magnetic moment and a radio-frequency e.m. field, is studied by means of a semi-classical approach. Theoretical results are presented, and a possible experimental check of this theory is discussed.

In the example of a charged particle with magnetic moment which travels inside a time varying electromagnetic field, the expression of the Stern-Gerlach force in the laboratory frame has been deduced [1,2] by means of some quite complicated calculations. In fact we had to start from the particle rest frame (x', y', z', t') where such a force assumes the usual form

$$\vec{f}'_{SG} = \nabla'(\vec{\mu}^* \cdot \vec{B}') = \tfrac{\partial}{\partial x'}(\vec{\mu}^* \cdot \vec{B}')\hat{x} + \tfrac{\partial}{\partial y'}(\vec{\mu}^* \cdot \vec{B}')\hat{y} + \tfrac{\partial}{\partial z'}(\vec{\mu}^* \cdot \vec{B}')\hat{z}, \qquad (1)$$

where $\vec{\mu}^* = g\frac{e}{2m}\vec{S}$ is the magnetic moment. The partial derivatives and the fields $\vec{E}, \vec{B}, \vec{E}', \vec{B}'$ are Lorentz boosted along the z-axis via the following transformations:

$$\tfrac{\partial}{\partial x'} = \tfrac{\partial}{\partial x}, \qquad \tfrac{\partial}{\partial y'} = \tfrac{\partial}{\partial y}, \qquad \tfrac{\partial}{\partial z'} = \gamma\left(\tfrac{\partial}{\partial z} + \tfrac{\beta}{c}\tfrac{\partial}{\partial t}\right), \qquad (2)$$

$$\vec{E}' = \gamma(\vec{E} + c\vec{\beta} \times \vec{B}) - \tfrac{\gamma^2}{\gamma+1}\vec{\beta}(\vec{\beta} \cdot \vec{E}), \qquad (3)$$

and

$$\vec{B}' = \gamma\left(\vec{B} - \tfrac{\vec{\beta}}{c} \times \vec{E}\right) - \tfrac{\gamma^2}{\gamma+1}\vec{\beta}(\vec{\beta} \cdot \vec{B}). \qquad (4)$$

Moreover, bearing in mind that the force transforms as

$$\vec{f}_\perp = \tfrac{1}{\gamma}\vec{f}'_\perp, \qquad \vec{f}_\parallel = \vec{f}'_\parallel, \quad \text{and} \quad f_z = f'_z, \qquad (5)$$

---

[1] Corresponding author: Mario.Conte@ge.infn.it

we obtain

$$\vec{f}_{SG} = \frac{1}{\gamma}\frac{\partial}{\partial x}(\vec{\mu}^* \cdot \vec{B}')\hat{x} + \frac{1}{\gamma}\frac{\partial}{\partial y}(\vec{\mu}^* \cdot \vec{B}')\hat{y} + \frac{\partial}{\partial z'}(\vec{\mu}^* \cdot \vec{B}')\hat{z} = f_x\hat{x} + f_y\hat{y} + f_z\hat{z} \qquad (6)$$

with

$$f_x = \mu_x^*\left(\frac{\partial B_x}{\partial x} + \frac{\beta}{c}\frac{\partial E_y}{\partial x}\right) + \mu_y^*\left(\frac{\partial B_y}{\partial x} - \frac{\beta}{c}\frac{\partial E_x}{\partial x}\right) + \frac{1}{\gamma}\mu_z^*\frac{\partial B_z}{\partial x}, \qquad (7)$$

$$f_y = \mu_x^*\left(\frac{\partial B_x}{\partial y} + \frac{\beta}{c}\frac{\partial E_y}{\partial y}\right) + \mu_y^*\left(\frac{\partial B_y}{\partial y} - \frac{\beta}{c}\frac{\partial E_x}{\partial y}\right) + \frac{1}{\gamma}\mu_z^*\frac{\partial B_z}{\partial y}, \qquad (8)$$

$$f_z = \mu_x^* C_{zx} + \mu_y^* C_{zy} + \mu_z^* C_{zz} \qquad (9)$$

where

$$C_{zx} = \gamma^2\left[\left(\frac{\partial B_x}{\partial z} + \frac{\beta}{c}\frac{\partial B_x}{\partial t}\right) + \frac{\beta}{c}\left(\frac{\partial E_y}{\partial z} + \frac{\beta}{c}\frac{\partial E_y}{\partial t}\right)\right], \qquad (10)$$

$$C_{zy} = \gamma^2\left[\left(\frac{\partial B_y}{\partial z} + \frac{\beta}{c}\frac{\partial B_y}{\partial t}\right) - \frac{\beta}{c}\left(\frac{\partial E_x}{\partial z} + \frac{\beta}{c}\frac{\partial E_x}{\partial t}\right)\right], \qquad (11)$$

$$C_{zz} = \gamma\left(\frac{\partial B_z}{\partial z} + \frac{\beta}{c}\frac{\partial B_z}{\partial t}\right). \qquad (12)$$

Let us consider a rectangular resonator, whose sides a, b, and c are respectively parallel to the x, y, and z-axes, and which is excited in the TE$_{011}$ mode. If the spin orientation is 50% parallel and 50% antiparallel to $\vec{B}_{\text{ring}} \parallel \hat{y}$, as commonly assumed for describing an unpolarized fermion beam, and if we choose $x = \frac{a}{2}$ and $y = \frac{b}{2}$ as beam coordinates, such a cavity is characterized [3] by the following parameters

$$K_c = \frac{\pi}{b}, \quad \omega = c\sqrt{\left(\frac{\pi}{b}\right)^2 + \left(\frac{\pi}{d}\right)^2}, \quad \beta_{\text{ph}} = \frac{d}{\pi}\sqrt{\left(\frac{\pi}{b}\right)^2 + \left(\frac{\pi}{d}\right)^2} \qquad (13)$$

with $v_{\text{ph}} = \beta_{\text{ph}} c =$ wave's phase velocity, and field components

$$\vec{B} = \begin{cases} 0 \\ -B_0\frac{b}{d}\cos\left(\frac{\pi z}{d}\right)\cos\omega t \\ 0 \end{cases}, \quad \vec{E} = \begin{cases} -\omega B_0\frac{b}{\pi}\sin\left(\frac{\pi z}{d}\right)\sin\omega t \\ 0 \\ 0 \end{cases}. \qquad (14)$$

Therefore the most important force-component is

$$f_z = \mu^*\gamma^2 B_0 b\left\{\frac{1}{\pi}\left[\left(\frac{\pi}{d}\right)^2 + \left(\frac{\beta\omega}{c}\right)^2\right]\sin\left(\frac{\pi z}{d}\right)\cos\omega t + \frac{2}{d}\left(\frac{\beta\omega}{c}\right)\cos\left(\frac{\pi z}{d}\right)\sin\omega t\right\}, \qquad (15)$$

whose integration over the cavity length d gives the following expression of the energy gained, or lost, by a fermion which crosses the cavity:

$$\Delta U = \int_0^d f_z dz = \mu^*\gamma^2 B_0 \frac{b\left(\frac{\pi}{d}\right)^2 + \left(\frac{\beta\omega}{c}\right)^2 - \left(\frac{\omega}{c}\right)^2}{\left(\frac{\pi}{d}\right)^2 - \left(\frac{\omega}{\beta c}\right)^2}\left[1 + \cos\left(\frac{\omega d}{\beta c}\right)\right], \qquad (16)$$

having carried out the substitution $\omega t = \frac{\omega z}{\beta c}$. The stationary wave conditions, pertaining to the TE$_{011}$ mode, imply that $d = \frac{1}{2}\beta_{\text{ph}}\lambda$; hence Eq. (16) becomes

$$\Delta U = \mu^* B_0 \frac{b}{d} \left[\beta^2 \gamma^2 \frac{\beta_{ph}^2 - 1}{\beta_{ph}^2 - \beta^2} + \frac{\beta_{ph}^2 \beta^2}{\beta_{ph}^2 - \beta^2} \right] \left(1 + \cos \frac{\beta_{ph}}{\beta} \pi \right), \qquad (17)$$

or in the ultrarelativistic limit ($\gamma \gg 1$ and $\beta \simeq 1$)

$$\Delta U \simeq \mu^* B_0 \frac{b}{d} \gamma^2 (1 + \cos \beta_{ph} \pi) = 2 \mu^* B_0 \frac{b}{d} \gamma^2 \quad (\beta_{ph} = \text{even integer}). \qquad (18)$$

This energy exchange has to be compared to the one caused by the electric field, whose evaluation [1,2] yields

$$\Delta U_E = \int_0^d eE_x \, dx = \left[ew B_0 \frac{bd}{\pi^2} \frac{\beta^2}{\beta_{ph}^2 - \beta^2} \sin \frac{\beta_{ph}}{\beta} \pi \right] x' \simeq \left[\frac{bd}{2\pi} \frac{\beta_{ph}}{\beta_{ph}^2 - 1} \frac{ew B_0}{\gamma^2} \right] x', \qquad (19)$$

where x' is the trajectory slope, for β_{ph} equal to an **integer** and for ultrarelativistic particles.

As far as a spin-splitter [1,2] is concerned, we recall that spin up particles receive (or loose) that amount of energy given by Eq. (18) at each rf cavity crossing. Simultaneously, spin down particles behave exactly in the opposite way, i.e. they loose (or gain) the same amount of energy turn after turn. The most important issue is that the transferred energies add up coherently, i.e. the final energy separation after N_{SS} revolutions is

$$\Delta U_{\uparrow\downarrow} = \sum \{\Delta U_\uparrow - (-\Delta U_\downarrow)\} = 4 \frac{b}{d} N_{SS} \mu^* B_0 \gamma^2 \simeq 4 N_{SS} \mu^* B_0 \gamma^2. \qquad (20)$$

Summing up the energy contributions (19) from the electric field gives

$$(\Delta U_E)_{tot} = \sum \Delta U_E = \kappa \sum x' = 0, \qquad (21)$$

since the sign of x' changes continuously due to the incoherence of betatron oscillations.

Recalling that the spin-splitter principle requires a repetitive crossing of N_{cav} cavities, and that after each revolution the particle experiences a deviation of its momentum spread

$$\zeta = \frac{\delta p}{p} = \frac{1}{\beta^2} \frac{\delta E}{E} \simeq \frac{N_{cav} \Delta U}{E} \simeq \frac{2\sqrt{3}}{3} N_{cav} \frac{B_0}{B_\infty} \gamma, \qquad (22)$$

having made use of Eq. (18) with $\beta_{ph} = 2$ and $B_\infty = \frac{mc^2}{\mu^*} \simeq 10^{16}$ T for (anti)protons. From Eq. (22) we find that the number of turns and the time Δt needed for attaining a momentum separation equal to $2(\Delta p/p)$ are respectively

$$N_{SS} = \frac{(\Delta p/p)}{\zeta} = \frac{1}{2 N_{cav} \gamma} \frac{B_\infty}{B_0} \left(\frac{\Delta p}{p} \right), \quad \text{and} \quad \Delta t = N_{SS} T_{rev}. \qquad (23)$$

TABLE 1. RHIC and HERA parameters

	RHIC	HERA
E(GeV)	250	820
γ	266.5	874.2
$\tau_{rev}(\mu s)$	12.8	21.1
$\frac{\Delta p}{p}$	4.1×10^{-3}	5×10^{-5}
N_{SS}	6.67×10^9	2.48×10^7
Δt	8.52×10^4 s $\simeq$ 23.7 h	523 s
μ^*	1.41×10^{-26} JT^{-1}	

TABLE 2. MIT-Bates parameters

τ_{rev}	634 nsec	b/d	$\simeq 1$
$\omega_{rev}/2\pi$	1.576 MHz	B_0	$\simeq 0.1$ T
$N_{electrons}$	$3.6 \times 10^8 \cdot 225 = 8.1 \times 10^{10}$	ω_{rf}/ω_{rev}	1820
γ	978.5	μ^*	9.27×10^{-24} JT^{-1}

Table 1 shows estimates for RHIC [4] and HERA [5] with

$$B_0 \simeq 0.1 \; T, \quad \text{and} \quad N_{cav} = 200.$$

Let us now evaluate a possible experimental test [6,7] of a polarimeter in the MIT-Bates [8] ring (see Table 2), loaded with 500 MeV polarized electrons. For a single rectangular cavity, with peak magnetic field B_0, and for a bunch train made up of N particles with polarization P, the average power transferred is the ratio between the total energy transfer of Eq. (18) and the revolution period, namely

$$W \simeq 2NP \frac{\mu^* B_0}{\tau_{rev}} \frac{b}{d} \gamma^2. \tag{24}$$

If we adopt a two coupled cavity scheme as a parametric amplifier, similar to the one proposed for a different [9] application, we may obtain a huge amplification of the small signal generated by the Stern-Gerlach interaction. Two cavities, tuned at the same frequency and coupled either in a symmetric or antisymmetric mode, can act as a parametric converter [10,11] provided that the frequency separation between the two modes is equal to the revolution frequency of the beam. With an initially empty level, the power transferred to this level is

$$W_2 = \frac{\omega_{rf}}{\omega_{rev}} W \simeq 2\,N\,P \frac{\mu^* B_0}{\tau_{rf}} \frac{b}{d} \gamma^2 \simeq 431\,P \text{ watt}, \tag{25}$$

where the cavity's period is $\tau_{rf} = 1/f_{rf}$ and $f_{rf} \sim 3$ GHz.

Comparing the energy exchanges (18) and (19) for $x' \simeq 1$ mrad, $\beta_{ph} = 2$ and $\lambda = 10$ cm in the MIT-Bates ring, we obtain:

$$r = \frac{\Delta U_E}{\Delta U} = \frac{x'}{8} \frac{\beta_{ph}^3}{\beta_{ph}^2 - 1} \frac{\lambda ec}{\mu^*} \frac{1}{\gamma^4} = 1.72 \times 10^{-4}, \tag{26}$$

i.e. the spurious signal, depending upon the electric interaction is negligible with respect to the measurable signal generated by the magnetic interaction. However, numerical simulations with spin-tracking [12] and cavity-designing [13] codes should be made by considering a real machine with a system of either rectangular or cylindrical cavities.

REFERENCES

1. M. Conte et al. "The Stern-Gerlach Interaction Between a Traveling Particle and a Time Varying Magnetic Field", INFN/TC-00/03. Available as e-Print: physics/0003069.
2. M. Conte, M. Ferro, G. Gemme, W.W. MacKay, R. Parodi and M. Pusterla, "The Time Varying Stern-Gerlach Interaction", submitted for publication on European Physical Journal A.
3. S. Ramo, J.R. Whinnery and T. Van Duzer, *Fields and Waves in Communication Electronics*, John Wiley and & Sons, New York, 1965.
4. M.A. Harrison, Proc. of EPAC96, p. 13, Sitges (Barcelona), 1996.
5. E. Gianfelice-Wendt, Proc. of EPAC98, p. 118, Stockholm, 1998.
6. P. Cameron et al., "An RF Resonant Polarimeter Phase 1 Proof-of-Principle Experiment", RHIC/AP/126, January 6 1998.
7. M. Ferro, Thesis of Degree, Genoa University, April 22 1999.
8. K.D. Jacobs et al., Proceedings PAC95, p. 327, Dallas, 1995.
9. Ph. Bernard, G. Gemme, R. Parodi and E. Picasso, Part. Acc., **61** (1998) [343]/79.
10. J.M. Manley, H.E. Rowe, Proc. of the IRE, 44 (1956) 904.
11. W.H. Louisell, *Coupled Modes and Parametric Electronics*, John Wiley & Sons, New York, 1965.
12. A. Luccio, "Numerical Spin Tracking in a Synchrotron. Computer Code SPINK - Examples (RHIC)", BNL-52481, September 18 1995.
13. P. Fernandes and R. Parodi, "Oscar2D User's Guide", INFN/TC-90/04.

Measurement of the Analyzing Power for Proton-Carbon Elastic Scattering in the CNI Region with a 22 GeV/c Polarized Proton Beam

The BNL-AGS E950 Collaboration

J. Tojo[1], I. Alekseev[2], M. Bai[3], B. Bassalleck[4], G. Bunce[3,5],
A. Deshpande[5], J. Doskow[6], S. Eilerts[4], D.E. Fields[4], Y. Goto[5],
H. Huang[3], V. Hughes[7], K. Imai[1], M. Ishihara[5], V. Kanavets[2],
K. Kurita[5], K. Kwiatkowski[6], B. Lewis[4], B. Lozowski[6], Y. Makdisi[3],
H.O. Meyer[6], B.V. Morozov[2], M. Nakamura[1], B. Przewoski[6],
T. Rinkel[6], T. Roser[3], A. Rusek[3], N. Saito[5,8], B. Smith[4],
D. Svirida[2], M. Syphers[3], A. Taketani[8], T.L. Thomas[4],
D. Underwood[9], D. Wolfe[4], K. Yamamoto[1], L. Zhu[1]

[1] *Dept. of Physics, Kyoto University, Kyoto 606-8502, Japan*
[2] *Institute of Theoretical and Experimental Physics, 117259 Moscow, Russia*
[3] *Brookhaven National Laboratory, Upton, NY 11973, USA*
[4] *Dept. of Physics and Astronomy, University of New Mexico, Albuquerque, NM 87131, USA*
[5] *RIKEN BNL Research Center, Upton, NY 11973, USA*
[6] *Indiana University Cyclotron Facility, Bloomington, IN 47405, USA*
[7] *Department of Physics, Yale University, New Haven, CT 06511, USA*
[8] *RIKEN(The Institute of Physical and Chemical Research), Wako, Saitama 351-0198, Japan*
[9] *Argonne National Laboratory, Argonne, IL 60439, USA*

Abstract. We have carried out the experiment BNL-AGS E950 to measure the analyzing power for proton-carbon elastic scattering in the Coulomb-Nuclear Interference (CNI) region with a 22 GeV/c polarized proton beam. Recoil carbons from 300 keV to a few MeV in the CNI region, were detected inside the AGS ring to identify proton-carbon elastic scattering. The preliminary results of the analyzing power measurement are presented.

Measurement of the analyzing power A_N for proton-carbon (pC) elastic scattering in the Coulomb-Nuclear interference (CNI) region provides an unique opportunity to study the spin dependence of hadronic interaction at small momentum transfer

t. The A_N in the range of $10^{-3} < -t < 10^{-2}$ (GeV/c)^2, originating from the interference between electro-magnetic (EM) spin-flip amplitude and hadronic spin-nonflip amplitude, is calculable [1,2],

$$A_N(t) = \frac{\sqrt{-t}\,(\mu_p - 1)\frac{t_c}{t}}{m_p\,\left(\frac{t_c}{t}\right)^2 + 1}, \qquad t_c = -\frac{8\pi Z\alpha}{\sigma_{tot}^{pC}}, \qquad (1)$$

where m_p is the proton mass, μ_p is the proton magnetic moment, and σ_{tot}^{pC} is the pC total cross section. In Eq.(1), the EM and hadronic form factors of proton and carbon, the Coulomb phase and the real-to-imaginary ratio of the hadronic spin-nonflip amplitude are neglected. Hadronic spin-flip amplitude has not been well understood and has been examined in several theoretical frameworks [3–5]. The amplitude contributes to the A_N through the interference with EM and hadronic spin-nonflip amplitudes and can be constrained by the precision measurement of the A_N.

The importance of the analyzing power measurement also has been arisen from the requirement for high energy proton polarimetry at RHIC, where acceleration of polarized proton beam from 24 GeV to 250 GeV has been developed [6]. The pC CNI analyzing power is expected to be nearly independent of beam energy. The figure of merit is large owing to the large cross section. The usage of "micro-ribbon carbon target" developed at IUCF [7] realizes pC scattering events at a high rate inside the RHIC ring for the fast polarization measurement. Those advantages are ideal for a high energy proton polarimeter for RHIC [8].

The experiment BNL-AGS E950 set the goals to measure the pC CNI analyzing power with a 22 GeV/c polarized proton beam circulating in the AGS ring and to establish the experimental base for the process to be used as RHIC polarimetry. One experimental difficulty is in identifying pC elastic scattering inside the AGS ring with an unknown background and noise environment. Scattered protons are difficult to detect inside the AGS ring due to very small scattering angles. Recoil carbons have a very low energy from 100 keV to a few MeV and recoil angles are concentrated around 89° with respect to the beam direction. Therefore, the pC elastic scattering has to be identified by detecting only low energy recoil carbons with kinematic constraints.

The experimental setup, consisting of a carbon target and two symmetric detector arms for recoil carbons, are schematically drawn in Fig. 1. Two types of ultra-thin carbon targets were prepared : (1) a "thin target", which was 6 μm wide, 3.7 μg/cm^2 thick and 3 cm long and (2) a "thick target", which was a bundle of micro-ribbon targets used to increase event rate. Those extremely thin targets make it possible to extend its life-time longer by reduction of the heating, to suppress the loss and the emittance growth of the beam and to decrease energy loss and multiple scattering of recoil carbons in the targets. The silicon strip detector (SSD) had a thickness of 400 μm and was segmented into 6 strips. Each strip was 4 mm wide and 10 mm high. The surface of the SSD had a special feature of quite thin Boron implantation with small aluminum wire electrodes so that low energy recoil carbons

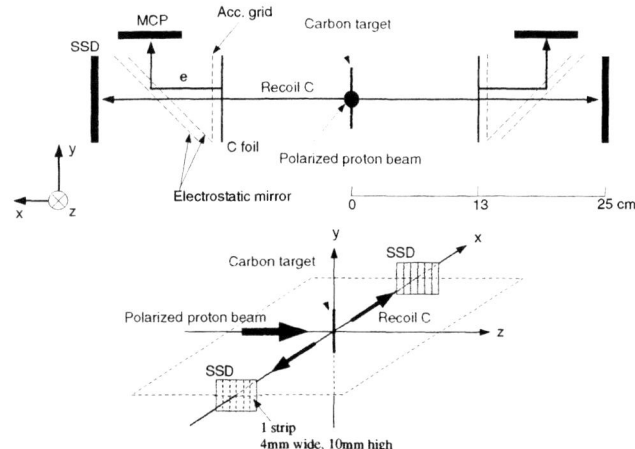

FIGURE 1. Schematic layout of the experimental setup (not scaled). Top : Beam view upstream of the target. Bottom : Definition of xyz axes (only SSDs are drawn).

could penetrate the surface without being hindered by the electrodes. The micro channel plate (MCP) assembly, consisting of two layers of MCPs and a single anode, had a sensitive area of 27 mm in diameter. It detected secondary electrons emitted from a 5 $\mu g/cm^2$ thin carbon foil primarily by the passage of recoil carbons. The secondary electrons were accelerated and reflected toward the MCP by both an accelerating grid and an electrostatic mirror. The system in each arm provides the measurements of (1) energy, timing and particle emission angle with respect to the beam direction from the SSD strips and (2) timing from the MCP and (3) time-of-flight (TOF) between the MCP and the SSD.

The experiment was carried out in March 1999. The transversely polarized proton beam was accelerated to 22 GeV/c. The polarization sign was flipped each spill in order to reduce systematic errors in the A_N measurement. The beam polarization P_B was independently measured by the AGS internal polarimeter to be about 40%. The beam was one bunch of 4-6×10^9 protons with a width of 6 ns and a revolution frequency of 370 kHz. The trigger was a coincidence between signals from any of the SSD strips and a beam bunch crossing (RF) in the AGS. The trigger rate was about 2×10^3/spill for the thick target and was about 10 times lower for the thin target. ADCs and TDCs of all detectors were taken with the ADC gate and the TDC start which were generated by the RF. The majority of data were taken with the thick target.

Signals from recoil carbons were clearly seen as a strong correlation between the energy E and the timing T obtained from each SSD strip shown in the left plot of Fig. 2. The center of the carbon locus was fitted as a function of energy $F(E)$ and the time deviation defined by the ratio $T/F(E)$ was used to select recoil carbon

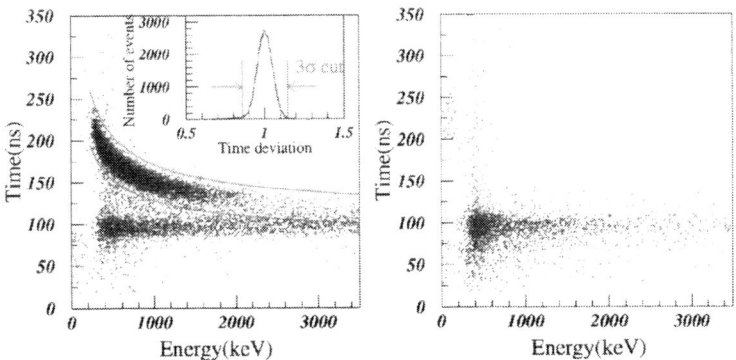

FIGURE 2. Correlation between energy and timing obtained from one of the SSD strips. Left : Correlation obtained from data taken with the thick target. Left-top : Definition of 3σ cut in time deviation. Right : Correlation obtained from data taken without the target.

events. The 3σ cut was defined in the time deviation. Physics backgrounds from $C^*(4.4\ \text{MeV})$ excitation and Δ production were not kinematically allowed to enter the detector acceptance. Other backgrounds were estimated to be less than 3 % using data taken without the target shown in the right plot of Fig. 2. Momentum transfer $-t$ was kinematically determined from both the energy E and the timing T. Energy loss of recoil carbons in materials were estimated from (1) the thickness of the "thick target" determined by comparing the multiple scattering width with that from the known thickness of the "thin target", (2) the known thickness of the electron emitter carbon foil and (3) the thickness of the SSD dead layer calibrated by carbons of known energies from 200 keV to 800 keV at the Tandem Van de Graaff in Kyoto University. The effects of those energy losses were corrected to determine the correct $-t$ value. The A_N was determined using a formula [9]:

$$A_N(t) = \frac{1}{P_B} \frac{\sqrt{N_L^\uparrow(t) \cdot N_R^\downarrow(t)} - \sqrt{N_L^\downarrow(t) \cdot N_R^\uparrow(t)}}{\sqrt{N_L^\uparrow(t) \cdot N_R^\downarrow(t)} + \sqrt{N_L^\downarrow(t) \cdot N_R^\uparrow(t)}}, \quad (2)$$

where $N_L^\uparrow(t)$ ($N_R^\downarrow(t)$) represents the number of recoil carbons at $-t$, detected in the right (left) detector arm with the beam polarization up (down). The preliminary result of the A_N as a function of $-t$ is shown in Fig. 3 with a theoretical curve from Eq.(1). Error bars are statistical only. Systematic errors are still under study. The deviation from the theory implies the possible contribution of hadronic spin-flip amplitude to the analyzing power, which will be studied in the further analysis.

Based on the successful result and the established technique, the pC CNI polarimeter for RHIC consisting of 4 SSDs without MCPs, was constructed and was installed in one of RHIC rings. During the RHIC polarized proton beam commissioning this year, the polarimeter was commissioned successfully [10].

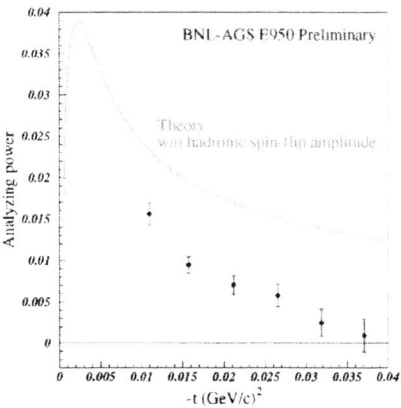

FIGURE 3. Preliminary result of the analyzing power for pC elastic scattering in the CNI region. Error bars are statistical only. The theoretical curve in Eq.(1) are drawn for comparison.

In summary, we have developed the experimental technique for the measurement of the pC CNI analyzing power, which is also applicable to the pC CNI polarimeter for RHIC. The pC elastic scattering was identified inside the AGS ring by detecting only low energy recoil carbons. The analyzing power was successfully measured for the first time with high statistical accuracy. Our data will be compared with a theory including the contribution of hadronic spin-flip amplitude. Hadronic spin-flip amplitude will be constrained from the further analysis.

REFERENCES

1. B.Z. Kopeliovich and L.I. Lapidus, Yad. Fiz. 19 (1974) 218.
2. N.H. Buttimore, E. Gotsman and E. Leader, Phys. Rev. D 18 (1978) 694.
3. "Hadron Spin-Flip at RHIC Energies", Proceedings of RIKEN BNL Research Center Workshop, Organized by E. Leader and T.L. Trueman, 1997, Vol.3, BNL-64724.
4. B.Z. Kopeliovich, "High-Energy Polarimetry at RHIC", hep-ph/98006055.
5. N.H. Buttimore, B.Z. Kopeliovich, E. Leader, J. Soffer and T.L. Trueman, Phys. Rev. D 59 (1999) 114010.
6. T. Roser et al., "Design Manual - Polarized Proton Collider at RHIC", 1998, BNL, http://www.agsrhichome.bnl.gov/RHIC/Spin/ .
7. W.R. Lozowski and J.D Hudson, Nucl. Instr. and Meth. A 303 (1991) 34.
8. "Physics of Polarimetry at RHIC", Proceedings of RIKEN BNL Research Center Workshop, Organized by D.E. Fields and K. Imai, 1998, Vol.10, BNL-65926.
9. H. Spinka, "Note on RHIC Polarimetry", Argonne National Laboratory Report ANL-HEP-TR-99-113 (1999).
10. H. Huang et al., "Commissioning of RHIC p-Carbon CNI Polarimeter", these proceedings.

Commissioning of RHIC p-Carbon CNI Polarimeter [1]

H. Huang[1], I. Alekseev[2], G. Bunce[3], A. Deshpande[3], D. Fields[4], K. Imai[5], V. Kanavets[2], K. Kurita[3], B. Lozowski[6], W. MacKay[1], Y. Makdisi[1], T. Roser[1], N. Saito[3], H. Spinka[7], D. Svirida[2], J. Tojo[5], D. Underwood[7]

(1) Brookhaven National Laboratory, Upton, NY 11973, USA
(2) ITEP, Moscow, 117259, Russia
(3) RIKEN BNL Research Center, Upton, NY 11973, USA
(4) University of New Mexico, NM , USA
(5) Kyoto University, Japan
(6) Indiana University, Bloomington, IN 47405, USA
(7) Argonne National Laboratory, 9700 Cass Ave., Argonne, IL 60493, USA

Abstract. An innovative polarimeter based on proton carbon elastic scattering in the Coulomb Nuclear Interference (CNI) region has been installed and commissioned in the Blue ring of RHIC during the first RHIC polarized proton commissioning. The polarimeter consists of ultra-thin carbon targets and four silicon detectors. All elements are in a 1.6 meter target chamber. It worked very well as a pivotal tool during the commissioning. It demonstrated that the RHIC snake magnets rotate spin to the expected direction. The results also show that polarized proton beam has been accelerated in RHIC and the polarization is maintained up to 29 GeV.

For polarization monitoring during RHIC polarized proton operation, a fast and reliable polarimeter is required that produces a polarization measurement with a 10% relative error within a few minutes. The idea of using $\vec{p} + C$ elastic scattering in the Coulomb-Nuclear Interference region (CNI) was suggested for the fast polarimeter several years ago. It is attractive because a significant (4%) analyzing power is predicted over the entire RHIC range. The figure of merit, at the maximum analyzing power at $-t = 0.003 GeV^2/c^2$, is excellent compared to other candidate processes. The sizable analyzing power, the large cross section and the advantages of a solid ribbon target make this process ideal for a fast primary polarimeter for RHIC.

An AGS Experiment (E950) has been done to measure the asymmetry in the p-C CNI process at 21.7 GeV, which is in the RHIC injection energy range. The exper-

---
[1] Work performed under the auspices of the U.S. Department of Energy.

imental results show that the recoil carbons from CNI scattering can be obtained with the Si detectors, and the asymmetry behavior follows the theory qualitatively [1]. A p-C CNI polarimeter was then installed in the Blue ring of RHIC and commissioned during the first RHIC polarized proton commissioning.

The RHIC p-C CNI polarimeter uses a carbon target, which is much simpler and cheaper than a hydrogen jet, and also easier to handle in the vacuum. Using a carbon target also results in the high luminosity required for fast polarization measurements. The very thin carbon target developed at IUCF [2] is crucial to the p-C CNI polarimeter: both for survival in the RHIC beam and to get the carbon nuclei out of the target in the CNI region where the recoil carbon carries only hundreds of keV kinetic energy. The target is only 100 atoms thick, which allows the slow carbon nuclei (150 keV) to escape and reduces the scattering rate to a level tolerable by detector electronics. It survives the beam due to its large surface to volume ratio. A potential disadvantage is that the number of carbon atoms can actually become depleted, so it is important to use the scattering time efficiently. However, this also works out well, with an expected lifetime of hours in the beam and measurements taking tens of seconds. The detail of the design of the RHIC p-C CNI polarimeter is given in Ref. [3].

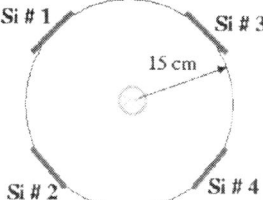

FIGURE 1. Cross section of the target chamber. Beam is going into the paper and hits the carbon target in the center of the beam pipe. The carbon target is $5.5\mu g/cm^2$ thick and $11.6\mu m$ wide.

In manipulation of polarization in RHIC, information on both vertical and horizontal components of beam polarization are needed. In addition, only one helical Siberian snake [4] has been installed this year. The stable spin direction is in the horizontal plane with snake on and in the vertical direction with snake off. The injected beam from AGS is in the vertical direction. It is necessary to have the snake off during injection and then have it adiabatically turned on so that beam polarization follows the stable spin direction and turns from vertical to horizontal. Four silicon detectors were installed obliquely, 15 cm from the target, as shown in Fig. 1, in order to measure both transverse components. When using these two pairs for either vertical or radial components, the analyzing power will drop by a factor of $\cos 45° = \sqrt{2}/2$. These detectors provide measurements of both vertical polarization (left-right)=((1+2)-(3+4)), and radial polarization (up-down)=((1+3)-(2+4)), with the detectors numbered as in Fig. 1.

The commissioning was done by injecting 6 proton bunches with alternating polarization, for example, ↑↓↑↓↑↓. There were about 3×10^{10} protons per bunch with a separation of about $2\mu s$. Injection energy was 24.3 GeV ($G\gamma = 46.5$). The AGS E950 readout, FERA ADCs and TDCs, were used for this run. The trigger was a coincidence between the beam bunch crossing timing and any delayed Si strip hits. Triggers were rejected if no silicon detector strip was above threshold.

The identification of a carbon band is shown in Fig. 2 as energy vs. time-of-flight plots, for each of 12 strips on one silicon detector. All 48 silicon strips worked beautifully. In fact, there is very little background as shown for the reconstructed mass in Fig. 3. An alpha peak can be seen clearly (presumably quasi-elastic p-alpha scattering). The energy was calibrated using observed carbon and alpha mass peaks. The commissioning showed that p-C CNI scattering can be identified with little background.

When the data are combined as discussed above to measure vertical and radial polarization, the vertical polarization of the injected beam was observed (see plot A of Fig. 4). Reversing the injected beam polarization resulted the beam polarization reversed. As a cross-check, unpolarized beam was injected and zero polarization was observed in RHIC. Based on the preliminary analyzing power for p-C CNI polarimeter and AGS internal polarimeter, the polarization at RHIC injection was about 19 ± 1 % (statistical error only), which was less than the polarization measured at the AGS extraction, 33 ± 1% (statistical error only). The polarization at injection is sensitive to the beam energy and betatron tunes. Due to the limited running time, no effort was devoted to explore the tune space and injection beam energy during the commissioning.

With one snake in the ring, it is necessary to turn it on after beam injection, so that the stable spin direction is then turned into the horizontal plane from vertical

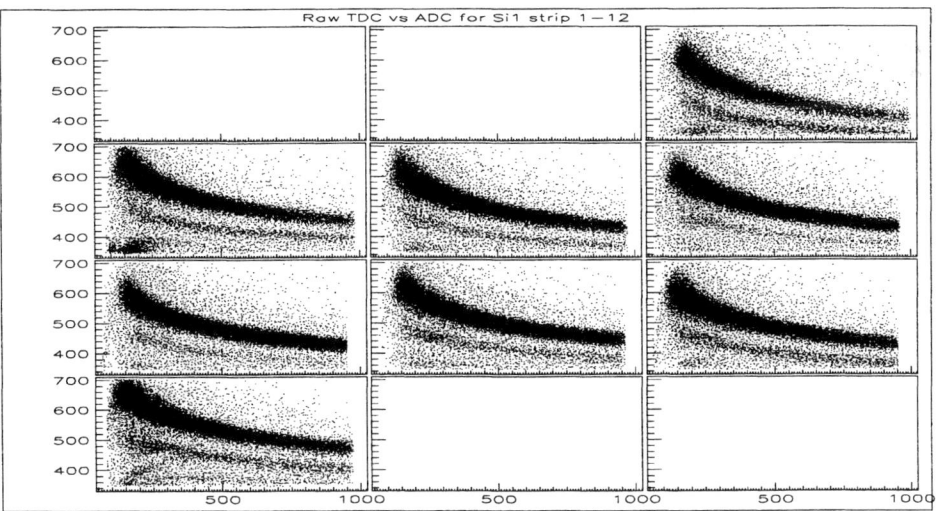

FIGURE 2. Scatter plot between the ADC values on the horizontal axis and the time of flight on the vertical axis for Si detector 1. Only center eight strips were used in the analysis.

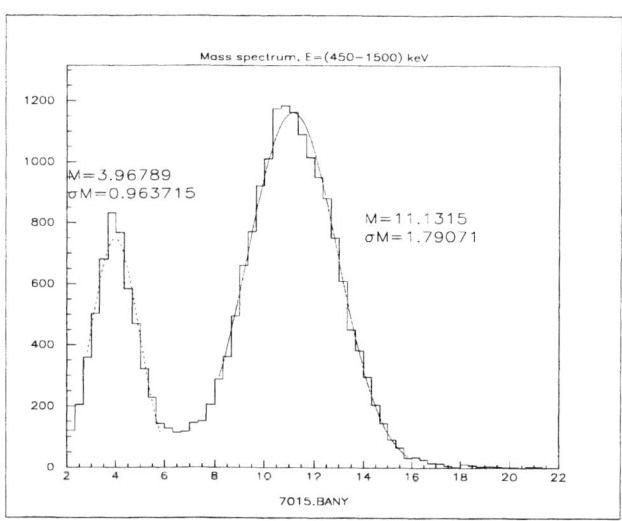

FIGURE 3. Mass spectrum. The horizontal axis is in the unit of GeV and vertical axis gives the event counts.

direction. The angle in the horizontal plane is determined by the beam energy, snake rotation axis and location of observation point relative to the snake. The polarimeter can only measure either vertical or radial components, so the beam energy was properly chosen to get radial component at the polarimeter. The expected radial polarization was observed after the snake was adiabatically turned on (see plot B of Fig. 4). This proved that the snake rotated the spin as expected. Polarized beam was accelerated to 25.1GeV ($G\gamma = 48$), past several resonances including a strong resonance with a strength of 0.02. Measured polarization reversed direction radially as expected and showed similar magnitude (see plot C of Fig. 4). Polarized beam was then accelerated to 29.2 GeV ($G\gamma = 55.7$), past many more resonances including two more strong resonances, and again was measured with the expected direction and no noticeable polarization loss (see plot D of Fig. 4). When polarized beam was accelerated to 31.6 GeV ($G\gamma = 60.3$), just above a much stronger resonance with strength of 0.07, no significant polarization was measured (see plot E of Fig. 4). Most likely the depolarization is due to the strong snake resonances because the vertical closed orbit was too large and the betatron tune was moving around during acceleration. No time was spent in the commissioning to correct them. Next year, our goal is to control the vertical betatron tune to better than .005 and the vertical closed orbit to better than 0.5 mm rms. Both of these requirements will be a major focus for the preparations for next run but are quite feasible based on the operational experience with gold beams.

With the E950 electronics, there was considerable dead time: readout of a full buffer took 1.5s. Next year with 60 bunches and higher bunch intensity, the DAQ system will be upgraded to wave form digitizer (WFD) readout, so there will be

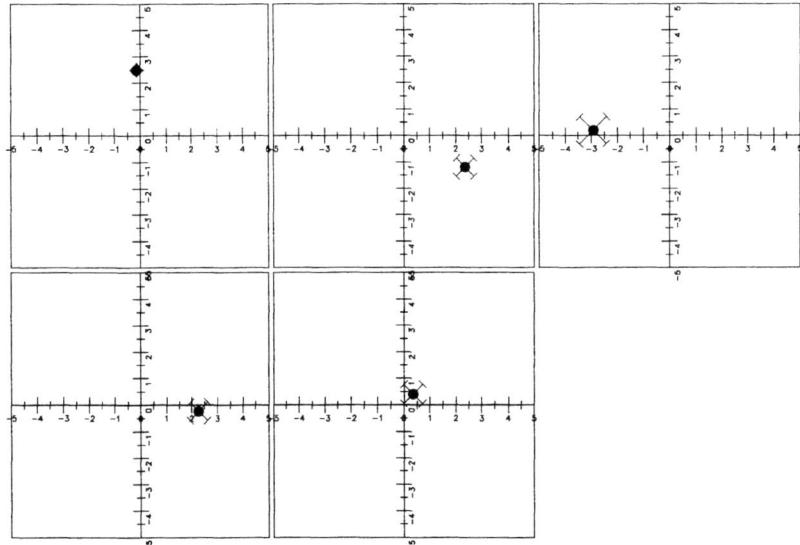

FIGURE 4. Plot of polarization vectors at various energies. Both horizontal and vertical axes are in units of asymmetry $\times 10^{-3}$. Top row from left to right: A) $G\gamma = 46.5$ and snake off; B) $G\gamma = 46.5$ and snake on; C) $G\gamma = 48$ and snake on; bottom row from left to right: D) $G\gamma = 55.7$ and snake on; E) $G\gamma = 60.3$ and snake on. The error bars reflect the fact that the asymmetries were derived from the 45° detectors.

no dead time. A prototype WFD module was tested during the commissioning run and worked as expected.

We have completed a successful test of a new relative polarimeter, which measures proton-Carbon elastic scattering at small angles(CNI region). The first successful use of a full helical Siberian snake was demonstrated. The p-C CNI polarimeter is ideal for high energy proton polarimetry: fast measurement, low cost and compact size. Another p-C CNI polarimeter will be installed in the Yellow ring before next run. To accommodate the higher beam intensity next year, the DAQ system will be upgraded by utilizing the WFD readout.

REFERENCES

1. J. Tojo, et al., *Measurement of the Analyzing Power for Proton-Carbon Elastic Scattering in the CNI Region with a 22 GeV/c Polarized Proton Beam*, these proceedings.
2. W.R. Lozowski and J.D. Hudson, NIM in Physics Research A334, 173(1993).
3. H. Huang, *RHIC p-C CNI polarimeter*, AIP Conf. Proc. 546, (AIP, New York, 2000), p. 440-447.
4. W.W. Mackay, et al., *Super-conducting Helical Snake Magnets: Design and Construction*, p.163, DESY-PROC-1999-03.

Deuteron Beam Polarimetry at JINR Accelerator Facility

Yu.K. Pilipenko, V.M. Slepnev, L.S. Zolin

Joint Institute for Nuclear Research, 141980 Dubna, Russia

Abstract. The spin physics program of the Laboratory of High Energies, JINR (Dubna), is based on use of the slowly extracted deuteron beam with momentum up toTwo-arm scintillation counter polarimeter with thin polyethylene target 9 GeV/c. To control operation of the polarized ion source POLARIS [1] as well as vector and tensor polarization of extracted deuteron beam, the following four types of polarimeters are used.

1. LOW ENERGY POLARIMETERS.

These are used to measure vector and tensor polarization of deuterons downstream the 10 MeV LINAC with $^4He(\vec{d},d)^4He$ and $^3He(\vec{d},p)^4He$ reactions correspondingly. The design of detector boxes of vector and tensor polarimeters (LEPV and LEPT) is shown in Fig.1. In both cases the gaseous target and detectors are combined in the hermetic box which can be brought in 10 MeV deuteron beam by means of a mechanical supporting device.

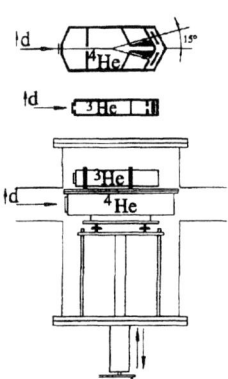

FIGURE 1. Detector boxes of low energy vector and tensor polarimeters, LEPV and LEPT. Depending on vector or tensor mode of beam polarization, the LEPV-box (^4He-filling) or the LEPT-box(^3He-filling) is inserted correspondingly in 10 MeV deuteron beam downstream LINAC.

The vector polarization of deuterons is defined by means of measuring left-right asymmetry of yields of alpha-particles at d^4He elastic scattering at the angle of 150° in c.m. frame. The collimated deuteron beam is scattered on the helium nuclei in

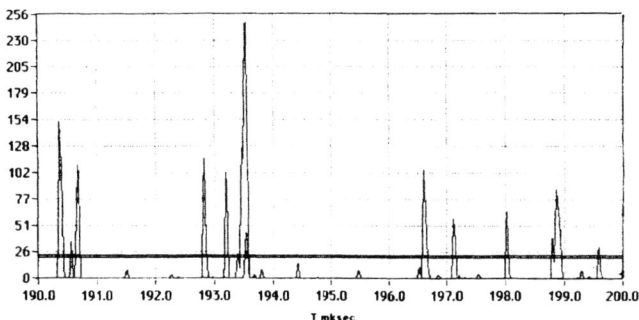

FIGURE 2. Si-detector signals of LEPT-polarimeter digitized by flash-ADC with $20ns$ step.

the target part of the detector box and recoiled alpha-particles are registered at the lab. angles of 15° with two thin ($55\mu m$) silicon detectors. The 4.5 MeV alpha particles are easily separated from deuterons, $\Delta E-$ losses of which are about 1.2 MeV in silicon. The beam polarization is related to left-right asymmetric $A_{LR} = \frac{N_L - N_R}{N_L + N_R}$ with the equation $p_z = \frac{2A_{LR}}{3A_y}$, where analyzing power of reaction $^4He(\vec{d},d)^4He$ is known with accuracy as low as 5% [2]: $A_y = 0.328 \pm 0.015$ at $T_d = 6.7$ MeV – the deuteron energy after the losses in the mylar window and in helium.

The detector box of LEPT–polarimeter is cylinder-shaped (ϕ 37 mm), the forward part of 80 mm is filled with ^3He at 1 atm. The fast protons (20 MeV) emitted at 0° in the reaction $^3He(\vec{d},p)^4He$ are detected with $500\mu m$ silicon detector after their crossing the filter of $2.3\frac{g}{cm^2}$, rejecting other lower energy particles. The tensor polarization is defined by difference of yields of the protons at polarized (N_p) or unpolarized (N_0) deuteron beam:

$$p_{zz} = 2\sqrt{2}\frac{I_0 N_p - I_p N_0}{T_{20} I_p N_0},$$

where I_p, I_0 – beam intensities are monitored by means of the dAu–scattering at thin gold–target. The tensor analyzing power in $^3He(\vec{d},p)^4He$ is known with uncertainty less than 2% [3]: $T_{20} = 1.254 \pm 0.014$ at $T_d=6$ MeV – deuteron energy in the center of the ^3He–target. To register the amplitude of signals of the silicon detectors a standard electronic line is used: *detector – cable line(30 m) – amplifier – ADC convertor*. First, during POLARIS testing and tuning, ADCs of conventional type with converting time of about $30\mu s$, were used. At LINAC burst of $300\mu s$, it limited an available count rate at 3-4 counts/burst. As result, it required about 3-hour data taking to measure a beam's polarization to within 5%. To overcome this disadvantage a new data acquisition system was developed [4]. The latter is based on using a fast ADC of flash type with the 32K memory. The system permits to increase count rate to 100 counts/burst. The ADC codes an input potential with

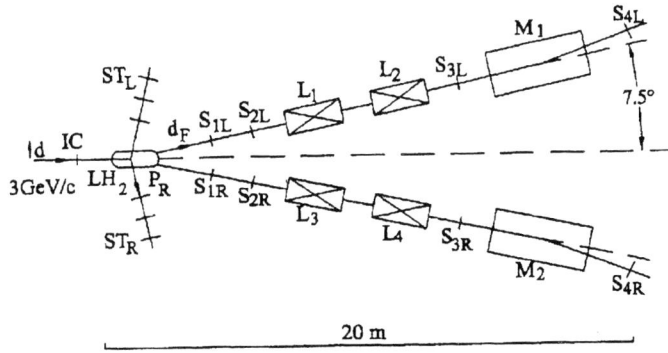

FIGURE 3. The two-arm ALPHA-polarimeter with momentum selection of dp-elastic scattering events. LH_2 is a liquid hydrogen target; IC is a ionization chamber beam monitor; $ST_{L,R}$ are recoil proton telescopes; S_{iL}, S_{iR} are scintillation counters of forward left-right spectrometers; M_i, L_i are magnetic dipoles and quadrupoles correspondingly.

the step of 20 ns thus covering $640\mu s (0.02\mu s \cdot 32K)$ time interval. Between two bursts all information is processed by the crate controller with SVGA monitor.

2. TWO ARM HIGH ENERGY POLARIMETER ALPHA

Two arm high energy polarimeter ALPHA with momentum selection of dp-elastic scattering events [5] measures vector p_z and tensor p_{zz} polarizations of slowly (400ms) extracted deuteron beam. The polarimeter operates with 10 cm liquid hydrogen target at the fixed deuteron momentum of 3 GeV/c.

The polarimeter setup is shown in Fig.3. Elastically scattered forward-deuterons are selected at coincidence with the recoil-protons according to kinematical constraint $d + p \rightarrow d(7.5°) + p_{recoil}(72°)$. Polarimeter azimuth angular acceptance corresponds to 4-momentum of $t = -0.150 \pm 0.017 (GeV/c)^2$. According to the precise Saclay data [6] for dp-elastic at P_d=3.0 GeV/c vector and tensor analyzing powers at selected t are equal to $A_y = 0.338 \pm 0.004$ and $A_{yy} = 0.598 \pm 0.016$.

The ALPHA-polarimeter is the most precise tool to control the p_z and p_{zz} of extracted deuteron beam. Its disadvantages are as follows: a) an extracted deuteron beam has to be adjust at fixed momentum, b) magnetic arms of polarimeter have to recollected at run with experimental setups located downstream the polarimeter, c) it takes an arbitrary long time to measure p_z, p_{zz} with a small uncertainty (about 3h to have $\Delta p_z(stat.) = 2\%$ at 10s acceleration cycle). The polarimeter is usually used for p_z, p_{zz}-measurements at the beginning and at the end of data taking run.

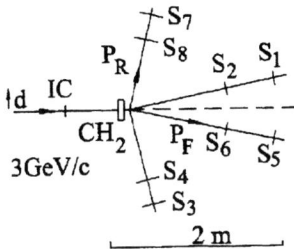

FIGURE 4. Two-arm scintillation counter polarimeter with thin polyethylene target (CH_2, $0.5 g/cm^2$) based on measurement of the left-right asymmetry in the elastic pp-scattering on hydrogen in polyethylene and on quasi-free protons in carbon nuclei.

3. TWO-ARM SCINTILLATION COUNTER POLARIMETER

Two-arm scintillation counter polarimeter with thin polyethylene target is used to continuously monitor vector polarization of extracted deuteron beam [7]. Polarimeters of this type are widely used at intermediate energies.

The polarimeter's geometry is shown in Fig.4. The vector analyzing power of CH_2–polarimeter was studied at the fixed arm angles of 14° at $P_d = 3 - 6$ GeV/c and of 8° at $P_d = 6 - 9$ GeV/c. The $A_y(CH_2)$ ranges from 0.466 ± 0.008 at 3 GeV/c to 0.138 ± 0.010 at 9 GeV/c [7]. At tensor polarization of deuteron beam the CH_2–polarimeter might be used only to watch stability of polarization by measuring the vector component in the beam.

4. TENSOR POLARIMETER

Tensor polarimeter based on using deuteron break-up reaction $dA \to p(0°)X$ measures tensor polarization of the deuteron beam at the fixed ratio of the proton to the deuteron momenta [8].

The studying of tensor analyzing power T_{20} in the deuteron break-up at Saclay and Dubna showed the next features of T_{20} in $dA \to p(0°)X$ [9], [10]:

- $|T_{20}|$ reaches its maximum at ratio of proton to deuteron momenta $P_p/P_d \simeq 2/3$ corresponding to internal momentum about $q=0.25$ GeV/c;

- A-dependence of T_{20} for light nuclei is not detected at q up to 0.24 within the said error limit and does not exceed 10% at $k \leq 0.27$;

- T_{20} is independent of deuteron momentum $2.5 \leq P_d \leq 9.0$ GeV/c and there are no theoretical reasons for revealing P_d-dependence at higher energies;

- $T_{20}(q)$ is constant at q range from 0.21 to 0.27 in $dC \to p(0°)X$ reaction: $T_{20}^{Saclay} = -0.825 \pm 0.025 \pm 0.025$, $T_{20}^{Dubna} = -0.820 \pm 0.040 \pm 0.033$

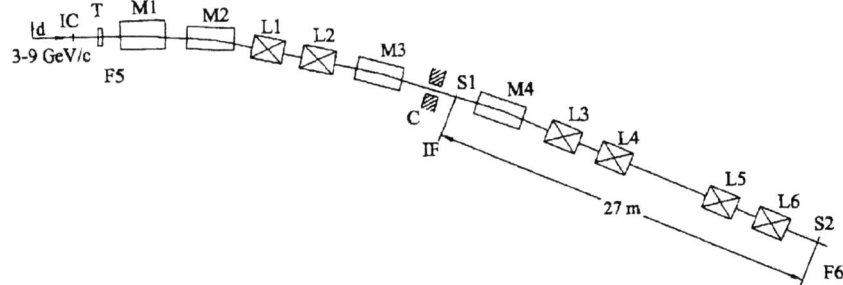

FIGURE 5. Setup of the polarimeter at 4V beam-line to measure tensor polarization of high energy deuteron beam by means of counting a yield of stripping protons at 0° in the deuteron breakup [8]. T is C(Be)-target; IC is an ionization chamber beam monitor; C is a collimator in intermediate focus IF; S1,S2 are scintillation counters for TOF-selection of protons, M_i, L_i are magnetic dipoles(M) and quadrupoles(L).

Tensor polarization is defined by difference of proton 0°-yields (n^+, n^-, n^0) at different beam polarizations:

$$p_{zz}^{\pm} = \frac{2\sqrt{2}}{T_{20}}(1 - \frac{n^{\pm}}{n^0}),$$

where $T_{20} = -0.82 \pm .04$ according to Saclay-Dubna data.

The measurement of tensor component in deuteron beam can be realized in the simplest way as measurement of count rate in a beam-line focus downstream the carbon target at beam-line momentum adjusted to $P_p = 0.65 P_d$. A typical arrangement is shown in Fig.5. At the beam-line momentum resolution of 2-3% an admixture of deuterons is about 6% that can be accounted with a corresponding correction of few % to p_{zz}. At more accurate measurements one can suppress deuterons using TOF-technique. The yield of stripping protons grows as P_d^2 at deuteron energy increase. It is rather high: $\frac{d\sigma}{dpd\Omega}(dC \to pX) \simeq 300$mb/(GeV/c sr) at $P_d = 9$, $P_p = 6$ GeV/c (so it allows us to reach $\Delta p_{zz}(stat.) \simeq 3\%$ at 10-min run using 4V beam line arrangement [8]).

In conclusion: considering an arbitrary simplicity of realization, a high count rate, and analyzing power energy independence, one can recommend using the deuteron breakup reaction as an effective express method to monitor the tensor polarization of deuteron beams at several GeV and higher energies.

REFERENCES

1. V.P.Ershov et al., *Intern. Workshop on Polarized Beams and Polarized Gas Targets*, Cologne, Germany June 6-9, 1995, p.193.

2. R.R.Cadmus and W.Haeberly, Nucl.Inst.Meth. **129**, 403 (1975).
3. W.Gruebler et al., Phys.Rev.,**C22** 2243 (1980).
4. S.N.Basylev et al., In: sl Proc. of the Intern. Workshop "Relativistic Nuclear Physics...", Slovak Republic, Stara Lesna, June 14-18, 1999, pp.163-167.
5. V.G.Ableev *et al.*, Nucl. Instrum. Methods Phys. Res., **A306**, 73 (1991).
6. V.Chazikhanian et.al., Phys. Rev.,**C43** 1532 (1991).
7. L.S.Azhgirey et al., Prib. Techn. Exper., No.**1**, 51 (1998).
8. L.S.Zolin et al., JINR Rapid Communications, N2[88]-98, pp.27-36.
9. R.Perdrisat et al., Phys. Rev.,**C39** 608 (1989).
10. T.Aono et al., Phys. Rev. Lett.,**74** 4997 (1995).

Absolute Calibration of the Deuteron Beam Polarization at Intermediate Energies via the $^{12}\text{C}(\vec{d}, \alpha)^{10}\text{B}^*(2^+)$ Reaction

K. Suda[*], H. Okamura[*], N. Sakamoto[†], A. Tamii[‡], T. Uesaka[*],
Y. Satou[†], T. Ohnishi[†], K. Sekiguchi[‡], K. Yako[‡], S. Sakoda[‡],
J. Nishikawa[*], H. Kato[‡], M. Hatano[‡], Y. Maeda[‡], and H. Sakai[‡]

[*]*Department of Physics, Saitama University, Saitama 338-8570, Japan*
[†]*The Institute of Physical and Chemical Research (RIKEN), Saitama 351-0198, Japan*
[‡]*Department of Physics, University of Tokyo, Tokyo 113-0033, Japan*

Abstract. The $^{12}\text{C}(\vec{d}, \alpha)^{10}\text{B}^*(2^+)$ reaction at $0°$ is proposed as a new standard for the deuteron beam polarization at an intermediate energy region. The absolute value of the polarization is measured using this reaction at 270 and 140 MeV deuteron energies for the first time. The obtained beam polarization is used for the calibration of the analyzing powers for the $\vec{d}$–p elastic scattering which we use as a polarimetry. The result is almost consistent with the data of the previous calibrations.

INTRODUCTION

In recent years, experiments have been performed using polarized deuteron beam at intermediate energies ($E_d \gtrsim 100$ MeV). Among those experiments, the study of the three–nucleon force (3NF) via the $\vec{d}$–p elastic scattering draws the interest [1]. Since the 3NF effects are small but persistent, especially for the spin-dependent part of the interaction, highly precise data are required to be compared with rigorous calculations. Accordingly, the accuracy of the beam polarization also becomes crucial. At the RIKEN Accelerator Research Facility (RARF), the $\vec{d}$–p elastic scattering is used to measure the deuteron beam polarization. The analyzing powers for the reaction were calibrated against the polarization measured by the $^{12}\text{C}(d,p)^{13}\text{C}$ or $^3\text{He}(d,p)^4\text{He}$ reaction at the exit of the injector AVF cyclotron [2,3]. At RARF, however, systematic uncertainty arises from the fact that the beam polarization and the analyzing powers for the d–p elastic scattering can not be measured simultaneously. Moreover, systematic uncertainties of the analyzing powers for the two reference reactions mentioned above are already significant in amount. Thus, it is desired to calibrate the analyzing powers using the reaction, the absolute value of which is unambiguously known. One of such reactions is the ^{16}O$(d,\alpha)^{14}$N$(0^-;$ 4.915 MeV) reaction and it was employed at low energies [4]. At intermediate energies, however, a measurement of this reaction is difficult because the energy resolution better than 200 keV is required and the 0^- state is populated only weakly. As an alternative, we propose to use the ^{12}C$(d,\alpha)^{10}$B reaction. In this reaction at $0°$, if the residual nucleus ^{10}B is in a natural parity state except for 0^+, it can be shown that tensor analyzing power T_{20} is identical to $1/\sqrt{2}$ as a consequence of the parity conservation. Among the levels of ^{10}B, the 2^+ state is advantageous because energy differences between the adjacent levels are larger than 1 MeV. Since the vector analyzing power vanishes at $0°$, it has no influence on the measurement of the tensor polarization, even if the beam has a component of vector polarization.

CALIBRATION OF ANALYZING POWERS FOR $\vec{d}$–p ELASTIC SCATTERING USING $(\vec{d},\alpha)$ REACTION

In order to confirm a usefulness of the (d,α) reaction at an intermediate energy region, we carried out the first measurement of the (d,α) reaction applying to the calibration of the analyzing powers for the d-p elastic scattering at $E_d = 270$ and 140 MeV at RARF. The deuteron beam was provided by the atomic beam type polarized ion source [5]. The sign of the tensor polarization t_{20} was changed every five seconds. We employed the spectrograph SMART [6] as a deuteron polarimeter to measure the ^{12}C$(d,\alpha)^{10}$B reaction at $0°$. The experimental conditions are summarized in Table. 1. The natural carbon foil was used as a target. The scattered α-particles are momentum analyzed by the SMART and detected at the focal plane. A typical excitation energy spectrum is shown in Fig. 1. The 2^+ state of interest is well separated from adjacent levels. The cross sections for the 2^+ state are 6 and 31 μb/sr at $0°$ for 270 and 140 MeV, respectively. The target contamination from ^{13}C and/or ^{16}O was negligibly small. The energy resolution $\Delta E \sim$ 600 keV is mainly due to the energy spread of the beam and the energy struggling of the particle in the target. The absolute value of the tensor polarization t_{20} was deduced from the asymmetry of the scattering. Since the measurement at absolute $0°$ is essentially impossible, the polynomial fit of the second order of scattering angle is used for four points taken at $\theta_{c.m.}=0°$–$4°$. It should be noted that the asymmetry changes considerably around $0°$. Thus, the higher angular resolution of

TABLE 1. Experimental condition for the (d,α) measurement

	270	140
beam energy (MeV)	270	140
target thickness ^{nat}C (mg/cm^2)	9.76	5.17
average beam intensity (nA)	70.5	112.5
data accumulation period (hour)	27.8	9.0

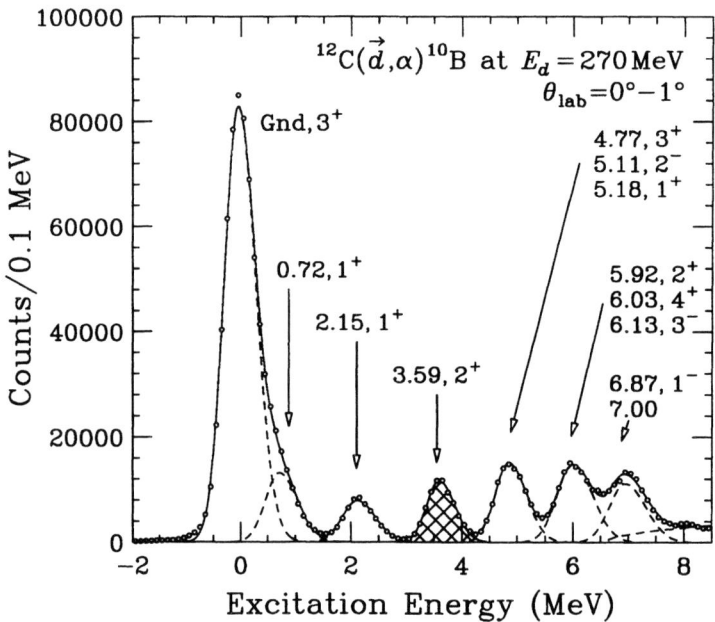

FIGURE 1. Excitation energy spectrum for the (d,α) reaction at $E_d=270$ MeV. The hatched peak indicates the 2^+ state of interest. The lines are the result of the Gaussian peak fit.

the spectrograph is required for an accurate determination of the polarization. For the spectrograph SMART, the vertical resolution of 3.8 mrad is sufficiently good as well as the horizontal one of 3.3 mrad.

Simultaneously with the measurement of the polarization, the asymmetry of the d–p elastic scattering was measured by the beam-line polarimeter using the same beam in order to reduce systematic uncertainty. A thin polyethylene (CH_2) film with a thickness of about 60 μm was used as a hydrogen target and it did not lose the quality of the beam. Both scattered deuterons and recoiled protons are detected in kinematical coincidence by four pairs of plastic scintillation counters placed in the direction of Left, Right, Up, and Down. Thus, together with the beam polarization, we obtained the analyzing power T_{20} and T_{22} for the d–p elastic scattering at $\theta_{c.m.} = 86.5°$ and $110.0°$ for 270 MeV and 140 MeV, respectively. Each angle is used for the polarization measurement at each energy [2,3].

RESULTS

The present results of the analyzing powers T_{20} and T_{22} for the d–p elastic scattering are plotted in Fig. 2 together with the data taken from Refs. [1–3,7]. The statistical errors are only shown. In comparison, the present data is almost con-

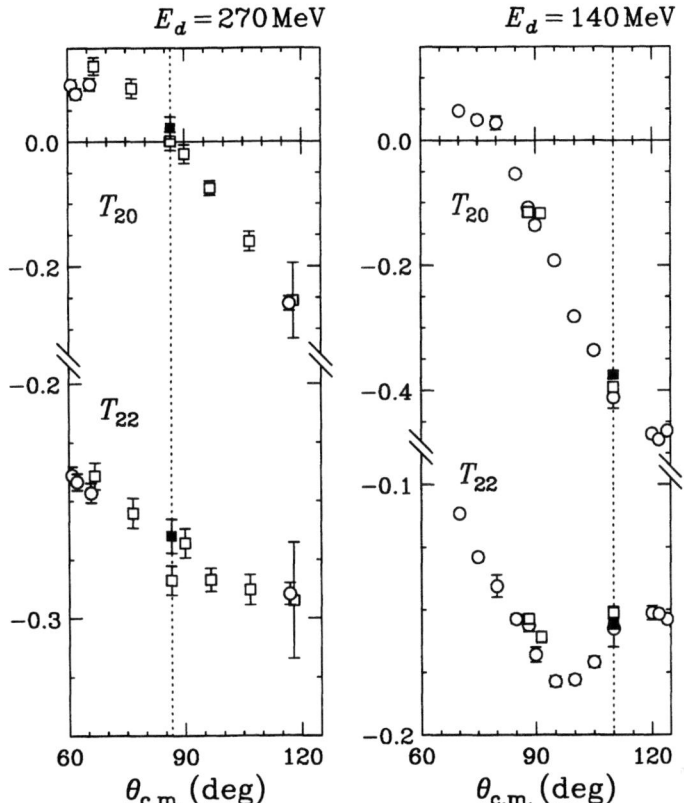

FIGURE 2. The analyzing powers T_{20} and T_{22} for the d–p elastic scattering at E_d=270 and 140 MeV. The solid squares represents the present results on both panels. The vertical dots shows the angle used to measure the beam polarization. In the left panel, open squares are taken from Ref. [2] and open circles from Ref. [1]. In the right panel, open squares are taken from Ref. [3] and open circles from Ref. [7].

sistent with the result of the previous calibrations [2,3]. Hence, it supports the usefulness of the (d,α) reaction for the standard of the deuteron beam polarization.

CONCLUSION

We propose to use the ^{12}C$(\vec{d},\alpha)^{10}$B reaction as a new standard for the deuteron beam polarization at an intermediate energy region. This reaction is used for the first time to measure the absolute value of the deuteron beam polarization. The calibration of the analyzing powers for the d–p elastic scattering is performed using the same beam at the same time with the polarization measurement in order to reduce systematic uncertainty. The result of the calibration is almost consistent with the previous calibrations. This suggests that $^{12}\text{C}(\vec{d},\alpha)^{10}\text{B}^*(2^+)$ reaction is useful at intermediate energies for the calibration of a deuteron polarimetry.

REFERENCES

1. H. Sakai et al., *Phys. Rev. Lett.* **84**, 5288 (2000).
2. N. Sakamoto et al., *Phys. Lett. B* **367**, 60 (1996).
3. T. Uesaka et al., *RIKEN Accel. Prog. Rep.* **33**, 153 (2000); contribution to this conference.
4. S. Kato et al., *Nucl. Instrum. Methods Phys. Res. A* **238**, 453 (1985).
5. H. Okamura, et al., *AIP Conf. Proc.* **293**, 84 (1994)
6. T. Ichihara et al., *Nucl. Phys.* **A569**, 287c (1994)
7. K. Sekiguchi, contribution to this conference.

A Spin Polarizer for Low Energy Radioactive Nuclear Beams

T. Shimoda*, S. Shimizu, E. Doumoto, M. Yagi, M. Asai,
M. Nakamura, Y. Hirayama, K. Horie, T. Shigematsu,
H. Izumi, M. Kawabata and N. Takahashi

Department of Physics, Graduate School of Science, Osaka University, Osaka 560-0043, Japan

Abstract. A versatile spin polarizer for radioactive nuclear beams is proposed for spectroscopic studies of exotic nuclei far from stability. The polarizer takes advantage of the polarized electron transfer process, as in the optically pumped polarized proton ion sources. This method is applicable for low energy beams with a wide variety of nuclear species. The feasibility of the polarizer has been experimentally investigated by using stable nuclear beams at the test stand of RCNP. The beams with 2+ charge were injected into the polarizer and the nuclear polarization of the 1+ charge ions was selectively measured. The polarization measurement was based on the method of beam-foil spectroscopy. The nuclear polarization was observed for 10 keV/amu ^{3}He, 3.4 keV/amu ^{14}N and ^{15}N beams to be $(3.89\pm0.76)\%$, $(3.04\pm0.11)\%$ and $(1.32\pm0.40)\%$, respectively with $P(Rb) = 70 - 80\ \%$. The transmission efficiency of the ^{14}N beam, (2+ ion → 1+ ion) was found to be $\sim 60\%$. The performance is very promising for practical applications.

INTRODUCTION

The nuclear spin orientation allows the measurements of the magnetic and quadrupole moments of the ground state of the β-decaying nuclei. The magnetic moment is sensitive to the configuration of the valence nucleons and the quadrupole moment is primarily sensitive to the spatial distribution of protons. They therefore provide insight into the nuclear structure of the loosely bound nuclei. Aiming at such spectroscopic studies of exotic nuclei far from stability, we proposed a versatile method to produce spin-polarized radioactive nuclear beams (RNB) of a wide variety of nuclear species [1]. The principle of the polarizer is the same as that in the optically pumped polarized proton ion-sources [2-4]. The procedure is as follows:

(i) The spin polarization of the outermost electron of Rb atom is produced by the laser optical pumping.

(ii) The radioactive nuclear beam passes through the Rb vapor and the nucleus picks up the polarized electron into its atomic state.

---

*) e-mail address: shimoda@phys.sci.osaka-u.ac.jp

(iii) The hyperfine interaction between the nucleus and the electron induces the nuclear polarization.

It should be noted that this method works, in principle, for any non-zero spin nuclear species. The cross section of the electron transfer process is expected to be highest at the beam energy of a few keV/amu. This method is, therefore, best suited to the ISOL based RNB's. The feasibility of the polarizer however is subject to the unknown problems inherent in the nature of RNB's; (a) the depolarization effect after the electron transfer process, and (b) the transmission efficiency of RNB. The latter problem depends on the electron pickup and stripping cross sections. As for the former problem, the electron is transferred mostly to the atomic excited states because of the lower ionization potential of Rb than that of the beam ion and the depolarization may occur during the transition to the lower atomic states. In the proton ion-sources this problem is overcome by applying high magnetic field (~ 2 T) so as to decouple the spin-orbit interaction in the hydrogen atom. This method is, however, not applicable for RNB, because such decoupling field is not always feasible, furthermore, the emittance blowup in high field must be serious [5] for RNB whose intensity is limited. However, it is expected that without high field some part of polarization remains in the atomic states [6]. It is worth noting that the polarization even on the order a few % enables the nuclear spectroscopic studies. The polarizer of this type is possibly a versatile tool. In order to investigate the feasibility of the polarizer, we have constructed a test stand of the polarizer at RCNP, Osaka University, and some successful results have been obtained with stable nuclear beams of ^{14}N. In the following the experimental method and the results are described.

EXPERIMENTAL SETUP

The test stand essentially consists of four parts; an ion-source to supply the beam, a Rb cell which contains high density Rb vapor, a polarimeter to measure the nuclear polarization, and a laser system for the optical pumping. The layout of the test stand is shown in Fig. 1. The beam of stable nucleus at a few keV/amu is supplied from a 2.45 GHz ECR ion-source. After the mass/charge state analysis, the beam goes through the Rb cell where Rb vapor is contained. The outermost electron in the Rb atom is spin polarized by the laser optical pumping. Thus the beam ion picks up the polarized electron in the collisions with the Rb atoms. In order to select ions that have picked up the electron, the beam goes to the electrostatic deflector and the beam direction is bent by 90°. The beam ions finally reach the polarimeter where the polarization of stable nucleus is measured through the atomic transition in the beam-foil interaction. The detail of the polarimeter is described in Ref. [7]. A low magnetic field (~ 10 Gauss) to preserve the nuclear polarization is applied in the whole region downstream of the Rb cell.

The Rb cell consists of two coaxial cylinders and the liquidized Rb is contained in the outer cylinder which serves as the reservoir for the Rb vapor at a saturated

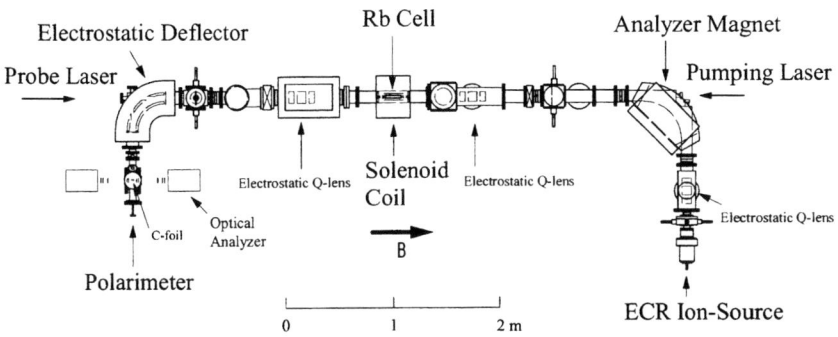

FIGURE 1. Layout of the test stand.

vapor pressure and the Rb vapor diffuses into the inner cylinder through thin slits. The Rb density is controlled by the Rb temperature. The Rb vapor is evacuated through the apertures at both ends of the inner cylinder. The beam goes through the inner cylinder. In the whole region of the cell a magnetic field is applied in the axial direction by a solenoid coil in order to cause Zeeman splitting for the optical pumping. The maximum field strength is 5 kG at the center of the cell. A pumping laser beam (tuned for the D_1 transition in Rb; 795 nm, 2.4 W) to polarize the Rb atom is supplied from the right hand side in Fig. 1 through the vacuum window on the analyzer magnet. The polarization axis is parallel to the laser beam direction. Another laser beam (780 nm, 18 mW) to monitor both the Rb polarization and the Rb thickness is shot from the left hand side in Fig. 1. These measurements are based on the Faraday rotation effect.

RESULTS

Figure 2. shows the Rb polarization as a function of the Rb thickness. The magnetic field strength at the Rb cell was 2 kG. It should be noted that very high polarization has been achieved at the density around 10^{13} - 10^{14} atoms/cm^2, which is related to the roughly estimated cross section on the order of 10^{-14} cm^2 for the electron pickup reaction. The decrease of polarization at higher Rb density is due to the relaxation at the cell wall and due to the "radiation trapping". If the electron pickup cross section is small, we need higher density of Rb and consequently higher laser power. The cross section was measured in the transmission experiment: Figure 3. shows the intensity of the ^{14}N ions after the Rb cell, resulting from the 4 keV/amu ^{14}N^{++} beam, as a function of the Rb thickness. The intensity was measured as an electric current on a Faraday cup placed downstream of the electrostatic deflector. The intensities are normalized to the incident beam intensity. The intensity of the ^{14}N$^+$ ion (N+) increases with increasing Rb thickness and that of the ^{14}N^{++}

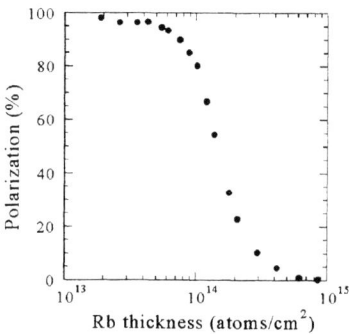

FIGURE 2. Polarization of Rb as a function of Rb thickness.

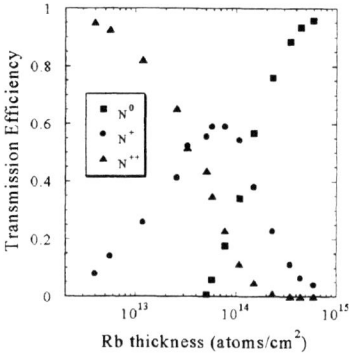

FIGURE 3. Transmission efficiency of N ions as a function of Rb thickness.

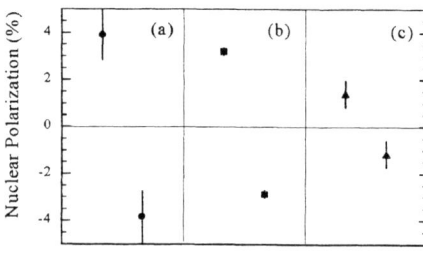

FIGURE 4. Nuclear polarization of (a) ^{3}He, (b) ^{14}N and (c) ^{15}N. The upper (lower) part shows the nuclear polarization with the pumping laser helicity +1 (−1).

(N++) decreases. This observation is due to the single electron pickup process. The intensity of the neutralized beam (N0), which is due to the subsequent electron pickup processes or double electron pickup process, was deduced by subtracting the N+ and N++ from the incident beam intensity. This process is significant at the Rb density higher than 10^{14} atoms/cm^2. Because of this process the ^{14}N$^+$ intensity is limited: The highest transmission efficiency for the ^{14}N$^+$ ion is approximately 60 % at the Rb density of 7×10^{13} atoms/cm^2. This efficiency is reasonably high for the practical application of the polarizer. It should be noted that the Rb polarization is sufficiently high at this density region: The laser we used is intense enough. From the data in Fig. 3. the cross section of the single electron pickup process was estimated to be 1.5×10^{-14} cm^2. The nuclear polarization of 10 keV/amu ^{3}He$^+$, 3.4 keV/amu ^{14}N$^+$ and 3.4 keV/amu ^{15}N$^+$ ions was measured by the method of beam-foil spectroscopy: The circular polarization of the fluorescence P_{atom} was observed for the atomic transitions 2s ^{3}S–3p ^{3}P (388 nm in He$^+$) and $2s^22p3s$ ^{3}P $- 2s^22p3p$ ^{3}D (568 nm in N$^+$), and the nuclear polarization $P_{nucleus}$ was evaluated as $P_{nucleus} = P_{atom}/A$, where A is the analyzing power which depends on the nuclear spin and atomic spins of the initial and final states. The Rb thickness was set at 7×10^{13} atoms/cm^2 and the Rb polarization was 70 - 80 %, which is somewhat smaller than that shown in Fig. 2, due to bad laser condition. The results are shown in Fig. 4. The upper (lower) part shows the results with the pumping laser helicity +1 (−1). The sign of the nuclear polarization is opposite for the opposite laser helicity, ensuring that we are not observing spurious polarization. Combining these results the nuclear polarization was determined to be $(3.89 \pm 0.76)\%$, $(3.04\pm0.11)\%$ and $(1.32\pm0.40)\%$, for ^{3}He, ^{14}N and ^{15}N, respectively. The smaller polarization for ^{15}N ($I = 1/2$) than for ^{14}N ($I = 1$) suggests that the atomic polarization we observed does not properly reflect the nuclear polarization. This may be because of the cascade atomic transitions leading to the initial state of the fluorescence. The nuclear polarization must be larger than that inferred from the beam-foil spectroscopy results.

The performance of the polarizer is very promising for practical applications with radioactive nuclear beams. The conditions to increase the polarization are under investigation, and the direct measurement of the nuclear polarization based on the β-NMR method is planned.

REFERENCES

1. Shimoda, T. and Shimizu, S., in *JHF Science*, KEK Proceedings 98-5, Tsukuba, 1998, p.337.
2. Mori, M. *et al.*, AIP Conf. Proc. **117**, 123 (1983).
3. Zelenski, A.N. *et al.*, *Nucl. Instru. and Meth.* **A334**, 285 (1993).
4. Tanaka, M. *et al.*, in *SPIN2000*, Osaka, 2000.
5. Ohlsen, G.G. *et al.*, *Nucl. Instru. and Meth.* **73**, 45 (1969).
6. Liu, C.-J. *et al.*, *Phys. Rev. Lett.* **64**, 1354 (1990).
7. Shimizu, S. *et al.*, *Rev. Sci. Instr.* **71**, 2045 (2000).

8. POLARIZED ION SOURCES AND POLARIZED TARGETS

Improved NMR System with Non-Resonant Cable Arrangement for Target Polarization Measurements

D. G. Crabb[a], S. Bültmann[a], G. Court[b], D. B. Day[a], M. Houlden[b], C. Keith[c], S. Penttilä[d] and Y. Prok[a]

[a] *University of Virginia, Charlottesville, VA 22901, USA*
[b] *University of Liverpool, Liverpool L69 7ZE, UK*
[c] *Jefferson Laboratory, Newport News, VA 23606, USA*
[d] *Los Alamos National Laboratory, Los Alamos, NM 87545, USA*

Abstract. The implementation of a NMR circuit designed to improve the detection of small signals and avoid the pitfalls of using a resonant cable is discussed. This new circuit combines a modification of the standard Liverpool Q- meter, together with a circuit of passive components, maintained at liquid helium temperatures. Results from the first tests are presented.

I INTRODUCTION

Over the past twenty years, the Liverpool Q-meter [1] has become a standard for measuring the polarization in solid targets. Originally designed and optimised for measuring the proton polarization at a field of 2.5T, where the proton resonance frequency is 106.5MHz, its use has been extended to 213 MHz operation as well as to the processing and measurement of the very small signals obtained with deuteron targets at frequencies of 16 or 32 MHz. In a number of recent experiments (for example reference [2]) using solid state polarized targets the systematic uncertainty in the target polarization measurement has been a major or even the dominant error on the final experimental result. In general the most important factor determining the size of the systematic uncertainty for proton targets is the system non-linearities. While for deuteron targets it is the signal- to-noise ratio of the thermal equilibrium (TE) signal used to provide the absolute polarization calibration factor. This talk describes two techniques for improving system performance when used with deuteron targets.

FIGURE 1. Insert target cells with NMR board.

II LIVERPOOL NMR

The Liverpool system is a constant current Q-meter which uses a resonant length $(n\lambda/2)$ cable to connect the target sampling coil to the RF signal processing module. This arrangement has the practical advantage that it allows the tuning capacitor, which is used to tune the coil and cable to resonance, to be mounted outside the target cryostat and adjusted at room temperature. However it has the serious disadvantage of generating a background signal superimposed on the nucleon signal which, in the case of deuteron targets, can be up to a factor of 100 larger than the signal itself. This background signal limits the RF gain which can be used in the signal processing module and so limits the signal-noise-ratio. In addition, the background signal size and shape are critically dependent on the cable parameters which are in turn temperature dependent. Thus background signal instability is another, and often very important, source of systematic uncertainty in the measurement [3]. This group therefore decided to try a different mode of operation which avoids the use of a resonant cable: by mounting the fixed-value capacitor and its associated circuit input components close to the sampling coil in the refrigerator, requiring that that the components operate at liquid helium temperatures. This is discussed in the next section.

III NEW NMR

In the new design, the circuit input components are mounted on a small (2cm x 1.2cm) circuit board and attached to the sampling coil as is shown in Figure 1. Two cables are then required inside the cryostat for each coil, one to supply a constant voltage to the input circuit and the second to feed the resultant resonance signal to the RF processing module. This mode of operation has the secondary advantage that it allows simple adjustment of the coil current over a wide range of values. The coil current can then be optimised to give the best possible signal-to-noise ratio under the operational conditions.

In the standard circuit arrangement the coil current can only be changed by changing the value of the constant current resistor, which is both inconvenient and lacking in a broad adjustment range.

A simple modification to the Liverpool Q-meter is also required and involves removing the input circuit board and connecting the sampling coil output directly to the first stage RF amplifier input. This disconnects the reference signal feed to the phase sensitive detector inside the RF module so that this connection must be then made externally.

The tuning of the circuit to resonance was, for these initial tests, carried out by changing the value of the capacitor on the circuit board at room temperature and by testing under liquid nitrogen. Fine tuning was achieved by changing the length of the sampling coil, before the insert was mounted in the cryostat. The components were specifically selected to have very low temperature coefficients so that there was very little change in tune going from room temperature to liquid helium temperatures. The phase tuning is very similar to that of the conventional system in that the length of the rf cable to the reference side of the phase detector is adjusted. But to maintain system linearity the RF voltage at the input to the phase detector must be maintained at its correct working value of 270 mV$_{rms}$.

The potential advantages and disadvantages of this new approach are as follows:

Advantages	**Disadvantages**
Eliminates $\lambda/2$ cable	No tuning *in-situ*
Shallower background curve	Tune outside cryostat
	-temperature effects
Use higher input rf level	
and more gain	Two cables per coil
Larger signal	Circuit sensitivity to:
Better signal/noise ratio	a) microwaves
Better stability	b) radiation damage

A Test Results

Initial tests were carried out to investigate how the resonance signal area varied with coil current using our standard mode resonant cable system. Results for proton and deuteron TE signals are shown in Figs. 2 & 3. These plots show the expected linear relationship between coil current (= RF level) and signal area and that there are no significant saturation effects at the highest currents used. Therefore in our particular circumstances, the signal-to-noise ratio for deuteron signals can be improved by at least factor of three, without any increase in systematic uncertainty, by increasing the coil current.

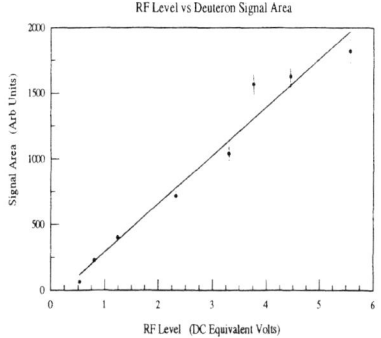
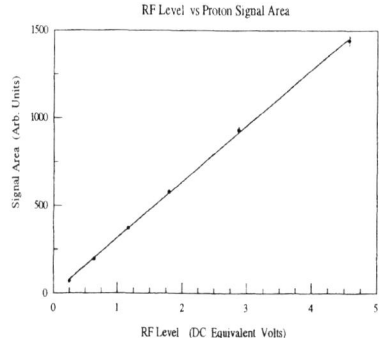

FIGURE 2. Deuteron TE area vs. RF input level.

FIGURE 3. Proton TE area vs. RF input level.

Tests on the new system were made in April and August 2000, using deuterated butanol and deuterated ammonia as target material. Our computer simulations of the circuit showed that a raw deuteron TE signal (ie one with no background subtraction and no digital noise reduction) should be observed. Though not shown here, it was indeed observed in the data.

A comparison between the standard system and the new system under the same operating conditions is shown in Fig 4. Both signal-noise ratio (S/N) and stability have been substantially improved. The S/N went from ~4:1 to 15:1 while the stability improvement is demonstrated by the fact that the measurement with the conventional system was taken twenty minutes after the baseline, while with the new system it was about 20 hours.

IV CONCLUSIONS

Tests of the new NMR arrangement have shown significant improvement in performance over the conventional system and which we believe can be further improved. Modern miniature components with superb temperature characteristics have contributed to this excellent performance and we are awaiting delivery of even better components. So far the only minor drawback is the neccesity of tuning the resonant circuit outside the crystat if one wants to avoid the chore of pulling the target insert in and out of the refrigerator to change components. For the future we will study small proton signals as well as investigate very small deuteron signals. We plan to incorporate such a system into a scattering experiment in 2001.

Our simulations of the new system shows that the background signal size is very much smaller than with the conventional system but is still finite. This arises

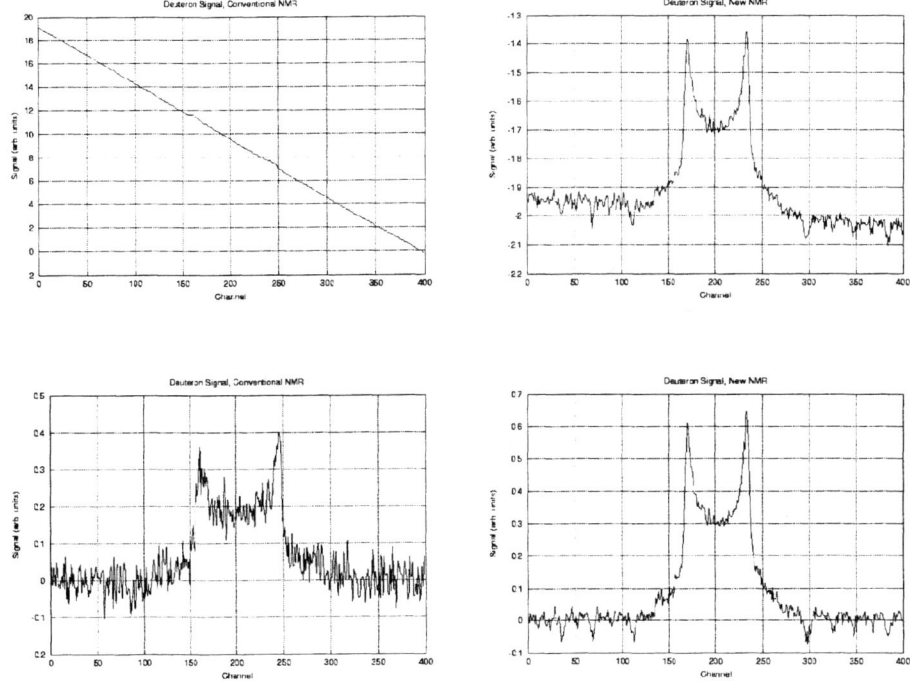

FIGURE 4. Comparison between the conventional and new NMR systems. The two left hand graphs for the conventional system show (top) the raw signal with only background subtracted, and (bottom) the resultant signal after fitting to the tails of the top curve and subtracting. Similarly for the new NMR in the right hand graphs.

because of the loading of the resonant LC circuit by the relatively low amplifier input impedance, which must, of necessity, be 50 ohms to ensure that the cryostat signal output cable is correctly matched. The background signal is theoretically eliminated if a very high impedance amplifier is used. It is therefore probable that with very small volume deuteron targets the best signal-to-noise ratio will be obtained by using the new system in conjunction with a high input impedance amplifier mounted in close proximity to the NMR coil. In most practical situations this would necessitate the use of an amplifier to be operated in the 1K to 4K temperature regime [4] which would also give the advantages of high gain stability and greatly reduced shot noise. However a hostile radiation environment would likely preclude such an arrangement.

V ACKNOWLEDGEMENTS

This work was supported in part by the US. Department of Energy under Grant number DEFGO2 - 96ER40950 and by an AEP grant at the University of Virginia.

REFERENCES

1. G. R. Court et al., Nucl. Instr. & Meth., **A324**, 433 (1993)
2. P. L. Anthony et al., Phys. Lett. **B493**, 19 (2000)
3. S. K. Dhawan et al., IEEE Trans. Nucl. Sci., **43**, 2128 (1996
4. P. Hautle, Proc. Workshop on NMR in Polarized Targets, Charlottesville, 1998, eds. S. Bueltmann and D. G. Crabb, p. 54

A High Intensity Stern-Gerlach Polarized Hydrogen Source for the Munich MP-Tandem Laboratory Using ECR Ionization and Charge Exchange in Cesium Vapour

R. Hertenberger, Y. Eisermann, A. Metz, P. Schiemenz and G. Graw

Ludwig-Maximilians Universität, Am Coulombwall 1, D-85748 Garching, Germany [1]

Abstract. The 14 year old Lamb-Shift hydrogen source of the Munich Tandem laboratory is presently replaced by a newly developed Stern-Gerlach type atomic beam source (ABS) with electron-cyclotron-resonance (ECR) ionization and subsequent double charge exchange in a supersonic cesium vapour jet target. The atomic beam source provides an intensity of $6.4*10^{16} \frac{atoms}{sec}$ of polarized hydrogen and of about $5*10^{16} \frac{atoms}{sec}$ of polarized deuterium. Beam intensities larger than 100 μA were observed for positive $\vec{H}^+$ and $\vec{D}^+$ ion beams after ECR ionization and intensities larger than 10 μA for negative $\vec{D}^-$ ion beams in three magnetic substates.

INTRODUCTION

The Munich 14 MV tandem accelerator in combination with the Q3D magnetic spectrograph enables nuclear structure studies with highest energy resolution. Nuclear spin polarized hydrogen projectiles eliminate inambiguities in the assignment of quantum numbers and enable in many cases complete spectroscopic information as in our recent study on ^{195}Pt and ^{196}Au indicating for the first time evidence for the existence of supersymmetry in a nuclear system. [1].

To optimize the experimental conditions our 14 year old Lamb-Shift hydrogen source is presently replaced by a newly developed Stern-Gerlach type atomic beam source (ABS) with electron-cyclotron-resonance (ECR) ionization and subsequent double charge exchange in a supersonic cesium vapour jet target. The goal is to produce negative ion beams with intensities around 10 μA and polarisations of about 80 %. Transfer experiments with an energy resolution of 3 keV FWHM at beam energies near 28 MeV will become feasible due to the improved brilliance

---

[1] Work is supported by the DFG under IIC4-Gr 894/2 and by the Beschleunigerlaboratorium der Universität und der Technischen Universität München.

of the ion beams. Multi-particle transfer reactions will profit decisively from the increase in intensity.

STERN-GERLACH ATOMIC BEAM SOURCE

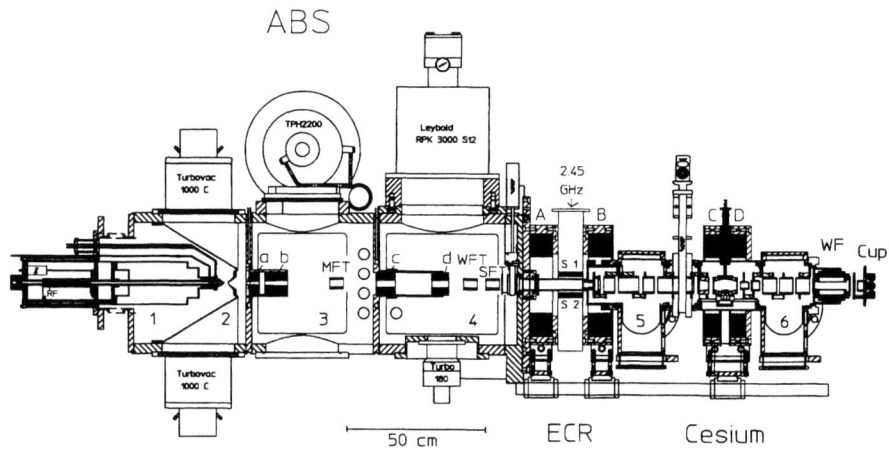

FIGURE 1. Vertical section of the Stern-Gerlach atomic beam source (ABS), the ECR ionizer (ECR), the electrostatic lens system and the cesium target (CS). For diagnostics of the positive ion beam a moveable wire scanner for beam profile measurements or a Faradaycup can be installed at the position of the cesium target. A compact Wien-filter (WF) allowed to investigate the composition of positive and negative ion beams on the test bench and to estimate the degree of depolarization.

The Stern-Gerlach ABS is designed similar to the HERMES and to the Madison/IUCF sources [2]. A vertical section is given in figure 1. The first of the four differentially pumped vacuum chambers houses the dissociator [3], the third and fourth the Stern-Gerlach sixtupoles of FeNdB permanent magnets [4]. The additional pumping in chamber 2 reduces scattering of beam atoms with molecules of the residual gas by about 30 %. The hydrogen flux through the dissociator is 2.2 $\frac{mbl}{s}$ at 320 Watts of 13.5 MHz rf power and a degree of dissoziation of 77 ± 5 %. Through the entrance of the ECR ionizer, an aperture of 10 mm in diameter, the ABS provides an intensity of 6.4*10^{16} $\frac{atoms}{sec}$ of polarized hydrogen and of about 5*10^{16} $\frac{atoms}{sec}$ of polarized deuterium [5].

POSITIVE ION BEAMS

For diagnostics of the positive ion beam a moveable wire scanner for beam profile measurements or a Faradaycup were installed at the position of the cesium vapour target. A compact Wien-filter allowed to determine the composition of the ion beam. Out of the ECR ionizer more than 100 μA of positively charged polarized H^+ and D^+ ions were observed in the acceptance of the cesium target at plasma currents of about 400 μA. To stabilize the ECR plasma 40 % of nitrogen were introduced into the ECR volume by a separate gas inlet. Ion beam mass spectra and profiles are presented in figure 2 in four subsections, in sections a-c for positive D ion beams, in d for negative $\vec{D}$ beams. Pictures a and b show two different optimizations for the ion optical setting with reference to the cesium cell. The polarized positive $\vec{D}^+$ beam in b has a maximum intensity of 45 μA in an emittance of $\epsilon < 1.6$ cm·rad·$\sqrt{eV}$. In a, an intensity of 35 μA corresponds to $\epsilon < 0.7$ cm·rad·$\sqrt{eV}$.

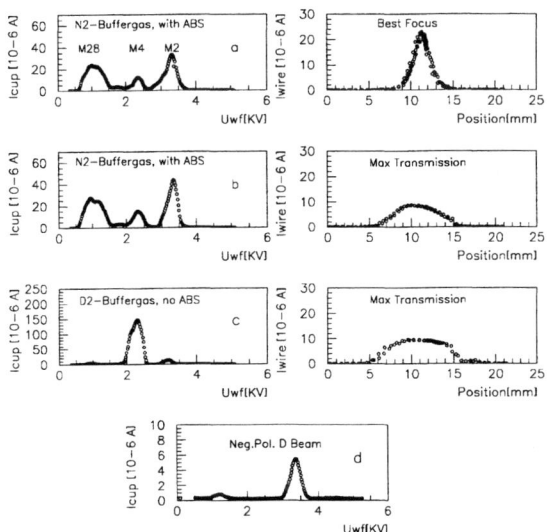

FIGURE 2. Wien-filter mass spectra (left) and beam profile scans (right) for positive polarized $\vec{D}^+$ ion beams (sections a and b) and unpolarized D_2 gas inlet (c). Best focus and max. transmission refer to different ion optical settings with respect to the cesium target. Part d shows the first observation of negative $\vec{D}^-$ ion beam. Its intensity of 5 μA is not yet optimized, intensities above 10 μA are meanwhile standard in 3 magnetic substates. The peaks labeled M2, M4 and M28 in the WF-scans refer to masses 2, 4 and 28, respectively.

The comparison of polarized and unpolarized beam composition in b and c allows to estimate the depolarization of the ion beam due to recombined residual D_2 gas in the ECR volume. In the polarized case b, the positive $\vec{D}^+$ ion beam (M2) dominates and the fraction of recombined D_2^+ ions (M4) is comparatively weak. The unpolarized case c was realized by ionization of D_2 gas, which was injected by a side inlet into the ECR volume. Here the D_2^+ peak dominates, the atomic fraction of the D^+ beam contributes with 10 % only. The amount of the injected D_2 gas was adjusted to the standard plasma current of 0.6 mA, as in a,b and d. From the comparison of the ratio of mass 4 to mass 2 peaks in b and c we conclude, that the contribution from recombination to beam depolarization is only 3 %. The depolarization due to hyperfine coupling in the magnetic holding fields of 1 kGauss during charge exchange is negligible for deuterium. For hydrogen it causes a reduction in polarization of about 15 %.

Bottom part d of figure 2 shows our first observation of a negatively charged polarized $\vec{D}^-$ ion beam with an intensity of 5 μA. With slight modifications of the cesium target intensities above 10 μA are meanwhile standard.

NEGATIVE $\vec{D}^-$ ION BEAM

The subsequent double charge exchange in the cesium jet target saturates at a cesium area density of $10^{15}\ \frac{atoms}{cm^2}$ [6]. This corresponds for our jet target to a temperature of 310 °C for the evaporation of liquid cesium at 120 Watts of heater power. At a pressure of 2.6 mbar the cesium vapour expands adiabatically in a hot nozzle. 30 % of the evaporated gas pass an 8 mm wide aperture and enter the ion charge exchange canal. A mass analysed negative deuterium ion beam of 13 μA (in 3 hyperfine substates) has been observed in stable operation over a period of 16

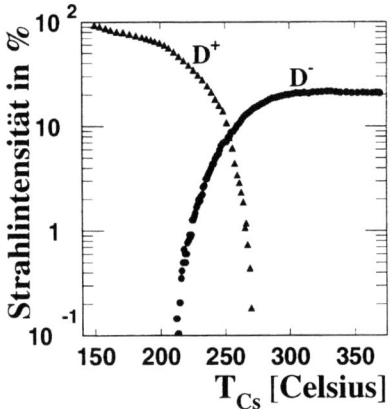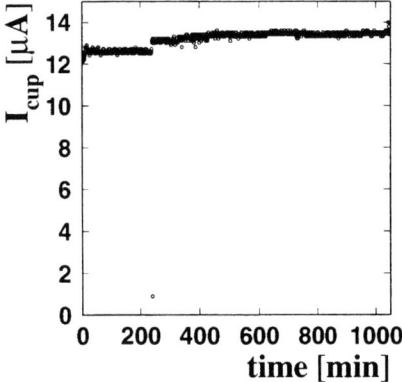

FIGURE 3. Left: D_1^+ and D_1^- beam intensities against cesium temperature. The intensities are given in units relative to the maximum positive beam intensity of 59 μA through the Wien-filter. Right: Test run with 13 μA of negative $\vec{D}^-$ beam intensity for about 16 hours at T_{cs}=310 °C.

hours using a compact Wien-filter with an acceptance similar to that of the tandem accelerator, see figure 3.

The left part of figure 3 shows the performance of the cesium vapour target. The decrease of the incoming positive deuterium ion beam with increasing vapour density (= increasing cesium temperature) starts at 150 °C, at temperatures of about 270 °C the positive beam vanishes completely. Negative ions start to built up around 220 °C. The negative ion beam increases up to about 310 °C where saturation occurs at a cesium area density of $10^{15} \frac{at}{cm^2}$. The efficiency for double charge transfer $D^+ \to D^-$ lies above 20 %. For a theoretical description of the cesium vapour target see [7].

SUMMARY

The new polarized hydrogen source is meanwhile installed at the Tandem accelerator. It runs reliably with intensities of negatively charged polarized $\vec{D}^-$ ion beams above 10 µA. The tuning of the polarization is presently ongoing. Beam time at the Tandem accelerator is herefore necessary. We use the elastic scattering of 22 MeV $\vec{D}$ on a ^{96}Zr target, ^{96}Zr($\vec{d}$,d$_0$). It has an analyzing power of about A_y=-0.4 with vanishing tensor polarization at a scattering angle of 56 degrees. Reliable values of the polarization will be available soon.

REFERENCES

1. A. Metz, et al., *Phys. Rev. Lett.* **83**, 1542 (1999)
   A. Metz, et al., *Phys. Rev.* **C 61**, 064313-1 (2000)
   Frankfurter Allgemeine Zeitung, 15. Sep. 1999
   Neue Züricher Zeitung, 2. Sep. 1999.
2. F. Stock, et al., *Proc. Int. Workshop on Polarized Beams and Polarized Gas Targets*, Cologne 1995, p. 260
   T. Wise, A.D. Roberts, W. Haeberli, *Nucl. Inst. Meth* **A336**, 410 (1993).
3. K.el. Abiary, *Diploma thesis*, LMU München, 1996, p.66.
4. P. Schiemenz, A. Ross, G. Graw, *Nucl. Inst. Meth* **A305**, 15 (1991).
5. R. Hertenberger et.al., *Rev. Sci. Inst.* **69,2**, 750 (1998).
6. A. S. Schlachter et al., *Phys. Rev.* **A22,6**, 2494 (1980).
7. C. Pertl, *Thesis*, LMU München, 2000.

The Polarized Internal Gas Target for ANKE at COSY-Jülich[1]

R. Brüggemann[1,†], R. Emmerich[1], R. Engels[1], H. Kleines[2], V. Koptev[3], P. Kravtsov[3], S. Lemaître[1,‡] J. Ley[1], B. Lorentz[4], S. Lorenz[5], M. Mikirtytchiants[4,#], M. Nekipelov[4,#], V. Nelyubin[3], H. Paetz gen. Schieck[1], F. Rathmann[5,§], J. Sarkadi[2], H. Seyfarth[4], E. Steffens[5], H. Ströher[4], A. Vassiliev[3], K. Zwoll[2]

[1] *Institut für Kernphysik, Universität zu Köln, 50937 Köln, Germany*
[2] *Zentrallabor für Elektronik, Forschungszentrum Jülich, 52425 Jülich, Germany*
[3] *High Energy Physics Dpt., St. Petersburg Nuclear Physics Institute, 188350 Gatchina, Russia*
[4] *Institut für Kernphysik, Forschungszentrum Jülich, 52425 Jülich, Germany*
[5] *Physikalisches Institut II, Friedrich-Alexander Universität, 91058 Erlangen, Germany*
[†] *Present address: Promatis AG, Stolberger Straße 200, 50933 Köln, Germany*
[‡] *Present address: Department of Physics and Astronomy, University of North Carolina-Chapel Hill, Chapel Hill, NC, 27599, USA*
[#] *PhD-Student from High Energy Physics Department, St. Petersburg Nuclear Physics Institute, 188350 Gatchina, Russia*
[§] *Now: Institut für Kernphysik II, Forschungszentrum Jülich, 52425 Jülich, Germany*

Abstract. A polarized internal hydrogen and deuterium gas target for the magnetic spectrometer ANKE in the COSY-Jülich storage ring is being developed. The polarized atomic beam will feed a storage cell. At present, the beam intensity measured with use of a compression tube is $6.4 \cdot 10^{16}$ H atoms per second. The nuclear polarization of the beam and the target gas will be investigated with a Lamb-shift polarimeter. First studies on the COSY beam properties have been performed with an aperture at the ANKE-target position. They will be extended with prototypes of cell tubes as soon as a new target chamber is available.

INTRODUCTION

Up to now studies of meson production at the magnetic spectrometer ANKE [1] in the COoler SYnchrotron COSY-Jülich [2] are carried out with the unpolarized proton beam incident on the unpolarized hydrogen or deuterium cluster-beam target [3] and on thin solid strip targets of higher atomic mass like carbon [4]. The first measurements of the non-mesonic proton-induced deuteron break-up will be performed with the unpolarized beam and the deuterium-cluster target [5]. A future field of research is accessible by extending these measurements to polarization

---

[1] work supported by BMBF (contracts RUS 99/686 and 06 ER 831), by DFG (contract 436 RUS 113/430), by FZ Jülich (FFE, contract 41149451), and by the Russian Ministry of Sciences.

observables, i. e. analyzing powers and spin correlations. To accomplish such studies, a polarized storage-cell gas target for the ANKE set-up is being developed. The polarized atomic beam source (ABS) will be utilized to feed a gas-storage cell. Thereby the luminosity will be increased by about two orders of magnitude compared to an atomic jet crossing the stored COSY beam. Storage cells are being routinely employed for nuclear physics experiments at storage-rings such as PINTEX at IUCF (Bloomington, IN, USA) [6] and HERMES at DESY (Hamburg, Germany) [7]. With $5 \cdot 10^9$ protons circulating in the COSY ring at a frequency of $1.6 \cdot 10^6$ s^{-1}, and a gas target thickness of $5 \cdot 10^{13}$ atoms per cm^2, a luminosity of $4 \cdot 10^{29}$ cm^{-2}s^{-1} will be obtained. The first experiments with the polarized gas target will be studies of the proton induced deuteron break-up reaction [8].

THE ATOMIC BEAM SOURCE

The ABS has been assembled and tested. Its layout is shown in Fig. 1. Due to space restrictions at the ANKE target area, only a vertical installation is possible. The slight tilting by about 6° may be required in order to avoid sputtering material from the discharge in the dissociator (1^2) to drizzle into the storage cell (7). Laboratory test runs will show, whether the inclination is needed.

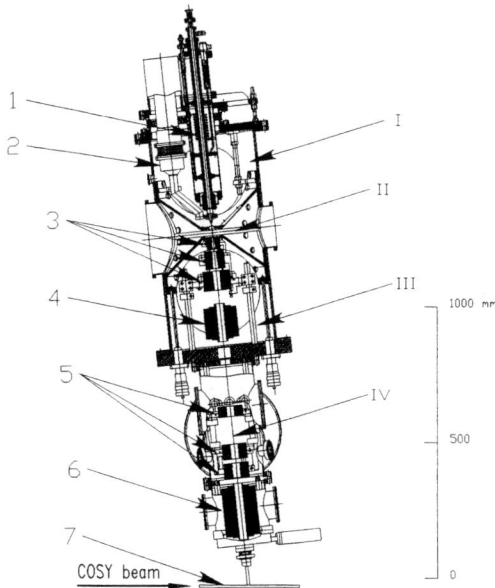

FIGURE 1. The atomic beam source for ANKE. (For explanations see text.)

The vacuum system: a powerful pumping system provides the necessary vacua in the chambers **I** to **IV**. Two Pfeiffer[3] TPH2200 turbopumps (combined H$_2$ pumping speed 5600 ℓ/s) are mounted on chamber **I**, each followed by a Pfeiffer TMH260 turbopump. The two TMH260 then are backed by a common third-stage TMH260 and finally by a Pfeiffer MD8 diaphragm pump. The smaller gas load in chamber **II** is pumped by a single set of TPH2200-TMH260-MD8. Two Leybold[4] Coolvac 3000 cryopumps (each of 3000 ℓ/s H$_2$ pumping speed and of 28 barℓ H$_2$ capacity at 10^{-6} mbar) are installed at the chambers **III** and **IV**. This pumping system (14400 ℓ/s in total) at a primary H$_2$ inlet flow of 1.5

---

[2] throughout this paper, label numbers in bold face refer to Fig. 1.
[3] Pfeiffer Vacuum GmbH, D-35614 Asslar, Germany.
[4] Leybold Vakuum GmbH, D-50968 Köln, Germany.

mbarℓ/s yields pressures of about 10^{-4}, 10^{-6}, 10^{-7}, and $5 \cdot 10^{-8}$ mbar in chamber **I**, **II**, **III**, and **IV**, respectively.

The dissociator (1): the plasma discharge is maintained by inductive coupling of 13.56 MHz from a Hüttinger[5] rf-power supply and an adapter network. The inner Duran 8330[6] discharge tube (14 mm outer diameter and 1.5 mm wall thickness) is cooled by a coaxial water flow. The nozzle at the lower end of the dissociator can be cooled to about 40 K by a Leybold RGS120 1-stage cryocooler (**2**) via a flexible connection and either a Cu heat bridge or a Ne heat pipe [9]. The details at the lower end of the dissociator are shown in Fig. 2. The transition from the cold nozzle to the outer dissociator tube via the shown components avoids any force on the discharge glass tube. Thermal contact to the lower part of the discharge tube is provided by a modified sliding rf connector. This special construction allows to extract and reinstall the glass tubes and the rf-generating elements without disassembling the heat bridge and nozzle.

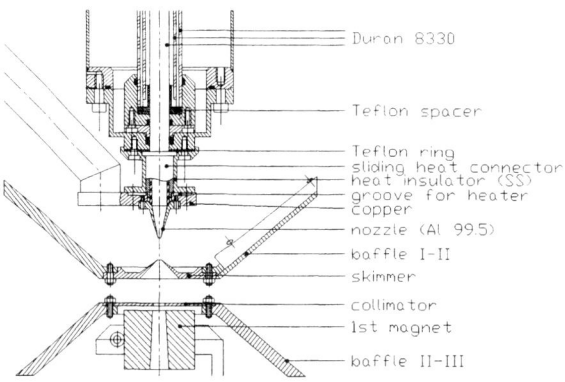

FIGURE 2. The region between the lower end of the dissociator and the first sextupole magnet (diameter of nozzle, skimmer, and collimator: 2, 3, and 8 mm, respectively; distance from skimmer entrance to first magnet 35 mm). Components not labeled are made from stainless steel (SS).

The permanent sextupole magnets (two groups, **3** and **5**): their dimensions and positions as shown in Table 1 have been obtaineded by Monte-Carlo trajectory calculations [10] to maximize the polarized beam intensity fed into the storage cell. Magnetic field and flux-density calculations have been performed with the MAFIA code in order to select appropriate NdFeB compounds for magnet production. The calculated pole tip fields B_o^{calc} agree very well with the B_o^{meas} measured after delivery[7] as is shown in the last columns of Table 1. The design considerations as well as the results of the field distribution measurements and their interpretation are found in a recent paper [11]. To protect the magnets from hydrogen, they were enclosed[8] by stainless steel cans with inner tubes of 0.2 mm thickness only in order to maximize the free magnet apertures.

---

[5]) Hüttinger Elektronik GmbH, D-79110 Freiburg, Germany.
[6]) glass brand name by Schott Glas (81% SiO_2, 13% B_2O_3, 4% Na_2O+K_2O, 2% Al_2O_3), identical to Pyrex, brand name by Corning Inc.
[7]) Vakuumschmelze GmbH, D-63412 Hanau, Germany.
[8]) welding with a low power pulsed Nd-YAG laser by Fraunhofer-Institut für Lasertechnik, D-52074 Aachen, Germany.

TABLE 1. The sextupole-magnet geometry and the calculated and measured values B_o.

Magnet no.	length [mm]	inner radius [mm]	outer radius [mm]	distance to next magnet [mm]	B_o^{calc} [T]	B_o^{meas} [T]
1[a]	40	$5/7^b$	20	10	1.633	1.630(4)[c]
2	65	$8/11^b$	32	10	1.641	1.689(2)
3	70	14	47	430^d	1.642	1.628(4)
4	38	15	47	102	1.564	1.583(3)
5	55	15	47	15	1.605	1.607(2)
6	55	15	47	300^e	1.605	1.611(3)

[a] distance to the nozzle 50 mm.
[b] $r_{entrance}/r_{exit}$ are given for the conical magnet.
[c] the values of this column result from the measured azimuthal field distributions [11].
[d] distance necessary to install the first rf-transition unit.
[e] distance to the storage-cell feeding tube, necessary to install the second rf-transition unit.

The rf-transition units: both units to provide purely polarized H atoms with $m_I = \pm 1/2$ have been designed and built at Universität Erlangen [12]. The medium-field unit (**4**) has been installed, tested, and is ready for operation. The weak- and strong-field unit (**6**) will be installed as soon as the Lamb-shift polarimeter (see below) is available for beam-polarization studies.

The slow control and interlock system: it is based on the consistent use of industrial technologies like SCADA (Supervisory Control and Data Acquisition), field bus (PROFIBUS DP), and programmable logic controllers (Siemens Simatic S7-300) [13]. The system will be extended to control the Lamb-shift polarimeter and the target-chamber components.

RESULTS OF ABS-BEAM STUDIES

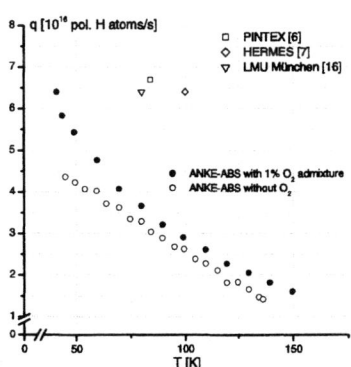

FIGURE 3. $\vec{H}$-beam intensity into the compression tube as function of the nozzle temperature (see text) compared with the values given for other sources in similar geometry.

With a crossed-beam quadrupole mass spectrometer [14] degrees of dissociation of the beam have been measured to be about 70% for a pure H_2 inlet flow of 1.0 mbarℓ/s. The ABS operation parameters have been determined by maximizing the atomic beam intensity into a compression-tube [15] with dimensions of the future storage-cell feeding tube (10 mm diameter and 100 mm length) and at the position given in Table 1. The resulting values of 1.3 mbarℓ/s for the H_2 inlet flow and 300 W for the rf power are close to those of other sources [6,7,16]. With a fixed distance of 35 mm between skimmer entrance and first magnet, the optimum nozzle-to-skimmer distance was found to be 14.8 mm.

With these values the H beam intensity into the compression tube has been measured as function of the temperature at the base of the nozzle. The results are shown in Fig. 3. The highest intensity of $(6.4 \pm 0.3) \cdot 10^{16}$ is close to the best values published for other sources. However, it is obtained with a lower input flux. Measurements of the temperature distribution along the nozzle are underway.

FUTURE DEVELOPMENTS

ABS beam: as suggested by Fig. 3, the beam intensity will be measured at lower nozzle temperatures possibly after installation of a 2-stage coldhead.

The Lamb-shift polarimeter: it has been designed, built, and tested at the Universität zu Köln [17] and will be transferred to Jülich around the coming turn of the year 2000/2001 and then first will be used to study the nuclear polarization of the ABS, thus enabling a fine tuning of the rf transition units, and later to investigate the polarization of the gas in the storage cell.

The target chamber and the storage cell: a new large target chamber is under development which will allow to extend the beam studies, already started with simple diaphragms, with storage-cell prototypes. The chamber also will house tailored beam position monitors near the cell. The time schedule aims at having the complete polarized gas-target set-up ready for installation at ANKE in 2002.

REFERENCES

1. Barsov, S. et al., *Nucl. Instrum. Methods*, in publication (2000).
2. Maier, R., *Nucl. Instrum. Method* **A390**, 1 (1997).
3. Khoukaz, A. et al., *Eur. Phys. J.* **D5**, 275 (1999).
4. Barsov, S. et al., *Nucl. Phys.* **A675**, 230c (2000).
5. Abasov, V. et al., *COSY Exp.Proposal No.* **20.3** (1999), spokesperson V. Komarov.
6. Rathmann, F. et al., *AIP Conf. Proc.* **421** 89 (1998).
7. Stewart, J. et al., *AIP Conf. Proc.* **421** 69 (1998).
8. Dshemuchadze, S.V. et al., *COSY Exp. Proposal No.* **20.1** (1991).
9. Vassiliev, A. et al., *AIP Conf. Proc.* **421** 479 (1998).
10. Lemaitre, S. et al., IKP Annual Report 1996, report Jül-3365 (1997), p.59.
11. Vassiliev, A. et al., *Rev. Sci. Instr.* **71**, 3331 (2000).
12. Lorenz, S., diploma work, Universität Erlangen (1999).
13. Kleines, H. et al., Proc. Int. Conf. on Accelerator and Large Experimental Physics Control Systems (icalepcs 99), Trieste, Italy, 1999. Ed. Synchrotron Trieste, Italy (2000), section TA2O03.
14. Mikirtytchiants, M., diploma thesis, St.Petersburg Technical State University (1999).
15. Nekipelov, M., diploma thesis, St.Petersburg Technical State University (1999).
16. Hertenberger, R. et al., Proc. Int. Workshop on Polarized Sources and Targets, Erlangen. Ed. A. Gute, S. Lorenz, E. Steffens (Universität Erlangen, 1999), p.52.
17. Engels, R., PhD work, Universität zu Köln.

Development of Polarized Negative Hydrogen Ion Source With Resonant Charge-Exchange Plasma Ionizer

A.S. Belov, S.K. Esin, L.P. Netchaeva, A.V. Turbabin, and G.A. Vasil'ev

Institute for Nuclear Research of Russian Academy of Sciences, 117312, Moscow, Russia

Abstract. Polarized negative hydrogen ion beam with peak current of 2.5 mA has been obtained from an atomic beam-type polarized ion source of Institute for Nuclear Research, Moscow. The intensity improvement has been achieved due to increase of efficiency of conversion of polarized hydrogen atoms into polarized negative ions. New converter for production of deuterium plasma with high density of unpolarized negative ions is described. Limitations of the method and possible improvements are discussed.

INTRODUCTION

A source of polarized negative hydrogen ions of Institute for Nuclear Research of Russian Academy of Sciences is atomic beam-type source in which polarized negative hydrogen ions are produced by resonant charge exchange between polarized hydrogen atoms and unpolarized negative deuterium ions in deuterium plasma. Initially, the source has been developed for production of pulsed polarized proton beam [1]. In the same paper it was proposed to enrich the deuterium plasma by D^- ions and use this D^- plasma ionizer for production of polarized H^- ions. This was some modification of W. Haberli idea [2] in which it was proposed to produce polarized H^- ions in collisions between thermal polarized hydrogen atoms and several keV energy unpolarized D^- ions (colliding beam ionizer).

The D^- plasma ionizer has several advantages in comparison with the colliding beam ionizer such as very large cross-section $\sigma_{ce} \sim 10^{-14}$ cm^2 for the resonant charge transfer due to low energy ~ 10 eV of colliding particles in plasma; elimination of a space charge problem (existing for an intense D^- ion beam with energy of several keV) because of plasma quasi-neutrality; a convenient possibility to transport plasma in longitudinal magnetic field of the ionizer practically without reduce of its density.

However, a possible destruction of the D^- and polarized H^- ions by plasma electrons inside the plasma column 30 cm in length in the ionization region can be a problem in the plasma ionizer. A cross-section of this electron detachment process at energy of incident electrons ~ 10 eV is ~ 4 10^{-15} cm^2. A life-time of H^- or D^- ion in the plasma with electron density 10^{11} cm^{-3} is about 10 μs due to the detachment process in collisions with the plasma electrons. This time can be compared with time of flight of D⁻ ion with typical energy of 5 eV through the ionization region which is about 15 μs. In order to avoid strong destruction of negative ions in plasma due to their stripping by plasma electrons it is necessary to generate plasma with low relative electron density: $n_e/n_{D^+} \ll 1$ or $n_{D^+} \sim n_{D^-}$. This means that ideal plasma for this kind of ionizer would be plasma consisting from D⁻ and D⁺ ions with relatively low density of electrons.

This kind of plasma injector has been developed at INR, Moscow [3]. Initially, only 1.2 mA of D⁻ ion pulsed current has been extracted from the ionizer. Respective polarized H⁻ ion current was 0.15 mA [3]. Later the D⁻ ion pulsed current was increased to 10 mA and polarized H⁻ ion current was increased to 1 mA level [4,5].

In this paper we describe next improvement in the plasma ionizer which allows to increase further intensity of the polarized H⁻ ion current produced.

IONIZER

The resonant charge-exchange plasma ionizer with the new two-stage converter is shown schematically in fig.1.

The ionizer works as follows. Polarized atomic hydrogen beam is injected into an ionization region through a bending magnet and an extraction electrode system. Deuterium plasma jet is injected into the ionization region from opposite direction. Polarized H⁻ ions are formed in the ionization region due to the resonant charge-exchange reaction: $H^0 \uparrow + D^- \rightarrow H^- \uparrow + D^0$.

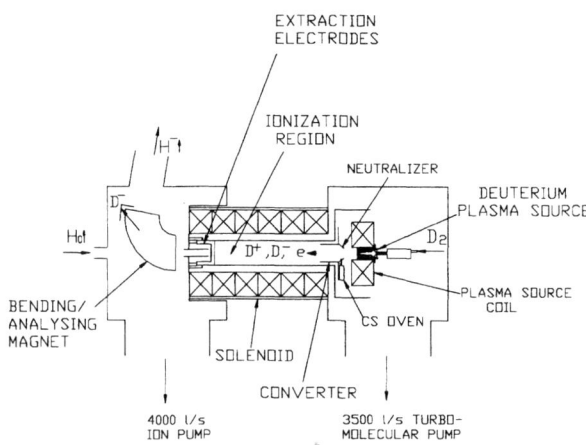

FIGURE 1. Schematic diagram of the resonant charge – exchange plasma ionizer with the two-stage converter.

The polarized H⁻ ions are confined in radial direction by longitudinal magnetic field of ≈1 kG created by the ionizer solenoid. Interaction with the plasma leads to increase of the polarized H⁻ ions energy from thermal to several eV value. During this process direction of the H⁻ ions movement is reversed, so they move then towards the acceleration electrode system.

The polarized H⁻ ions are extracted from plasma and accelerated to 20 keV energy together with unpolarized D⁻ ions and electrons in electric field produced by the three electrode ion-optical system. Then the different beams are separated spatially in the magnet mass-analyzer.

The polarized atomic hydrogen beam is generated by apparatus designed more then decade ago. It is described in detail in ref. [1].

The deuterium plasma is produced by an arc-discharge plasma source operating in longitudinal magnetic field about 800 G. The source produces plasma ion flux of several tenths of Amperes.

This intense plasma flux is enriched by negative ions in the two-stage converter. At first stage, positive ions of the plasma jet are converted into neutral atoms with eV energy in interaction with the neutralizer internal surface. The neutralizer has geometry of ringed blunt-nosed circular cone. Secondly, the hot atoms move from the neutralizer to the converter internal surface where the atoms are converted into negative ions. The converter tube is made from stainless steel; molybdenum sheet is installed inside the converter. Efficiency of the negative ions generation process is enhanced due to decrease of the converter surface work function by Cs atoms. The Cs atoms flux is only − 0.3 mg/hour, and it is produced by a Cs oven also shown in fig.1. The converter is heated to ~500°C. Transversal magnetic field in the converter area ~ 300 G (produced by electromagnet) decreases the plasma electron density and temperature.

RESULTS AND DISCUSSION

The new converter assembly produces deuterium plasma with relatively high D⁻ ion density in comparison with a previous converter design. The D⁻ ion flux in the present ionizer is proportional to the discharge current in the plasma source as shown in fig.2. For a given discharge current value parameters of feeding gas flux and extraction system voltages were changed to get maximum D⁻ ion current. The D⁻ ion beam was recorded by an ion collector installed downstream the bending magnet. The D⁻ ion beam with peak current of 42 mA has been obtained with discharge current of the plasma source of 460 A. The D⁻ ion density estimated from the extracted D⁻ ion current is $6 \cdot 10^{10}$ cm^{-3}.

Linear behavior of the dependence shows little influence of destruction of the D⁻ ions due to collisions with plasma electrons and positive ions in the ionization region in spite of relatively large plasma density. This is explained by low density of the plasma electrons. The electron current was measured to be typically ~ 400 mA. The corresponding electron density is -10^{10} cm^{-3}.

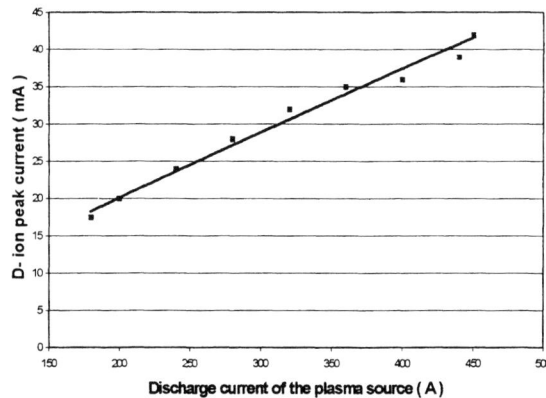

FIGURE 2. D⁻ ion peak current versus discharge current of the plasma source. Points are experimental data. Solid line is linear approximation of the data.

Respective characteristic time of the D⁻ ions destruction due to collisions with the plasma electrons ($\tau \sim 1/(n_e v_e \sigma_e)$) is estimated to be 50 μs which is significantly larger then the flight – time of the D⁻ ions through the ionization region (15 μs).

A mutual neutralization process between negative and positive ions becomes more important for plasma with low electron density. A cross-section for the mutual neutralization of the D⁻ ions with D⁺ ions for relative energy of 10 eV is very large: $\sigma_+ \sim 3 \cdot 10^{-14}$ cm⁻². However, characteristic time of the D⁻ ions destruction due to the mutual neutralization ($\tau \sim 1/(n_+ v_+ \sigma_+)$) for the given conditions is 200 μs. This means that destruction of negative ions by electrons remains the most important process even for the plasma with suppressed electrons density.

Polarized H⁻ ions are slower then unpolarized D⁻ ions because the H⁻ ions are formed from thermal polarized hydrogen atoms. However, we do not observe yet of saturation of polarized H⁻ ion current due to destruction of the H⁻ ions by plasma electrons even for maximum D⁻ ion densities achieved in the considered ionizer. Data for the polarized H⁻ ion current versus the D⁻ ion current are shown in fig.3.

Linear character of the dependence corresponds to simple formula for the polarized H⁻ ion current obtained with assumption that all H⁻ ions formed due to resonant charge – exchange reaction are transported without any losses to acceleration electrode system: $I_{H^-} = I_{D^-} \sigma_{ce} n_{Ho} L$, where I_{H^-} is the H⁻ ion current, I_{D^-} is the D⁻ ion current overlapping with the atomic hydrogen beam, σ_{ce} is the cross-section for the resonant charge transfer, n_{Ho} is averaged density of the polarized hydrogen atoms in the ionization volume, L is length of the ionization region.

Angle of the line shown in fig.3 is determined by averaged density of atomic hydrogen beam in the ionization volume. It can be changed in some degree depending of atomic beam apparatus operation.

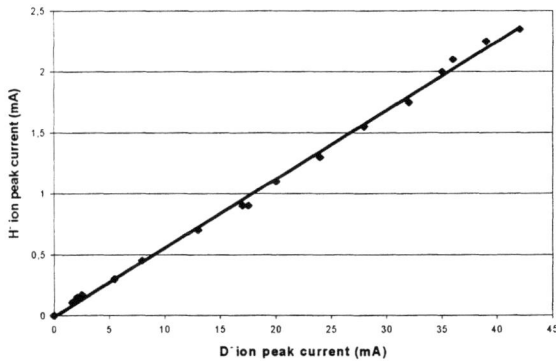

FIGURE 3. Polarized H⁻ ion peak current versus D⁻ ion peak current.

The best intensity of polarized H⁻ ion beam obtained so far from the ionizer described was 2.5 mA. Background H⁻ ion current from the ionizer (atomic hydrogen beam "off") was typically less than 5% from the total polarized H⁻ ion current. The H⁻ ion pulse duration was 150 µs and repetition rate can be changed up to 10 Hz.

It is worth note that obtained intensity is close to estimated in ref. [1] (2 mA) as an upper limit for this method (for given polarized atomic hydrogen beam intensity).

However, it is clear from the experimental results shown in fig.3 that it is possible to increase intensity of polarized H⁻ ion beam further by increase of the D⁻ ion density in the plasma ionizer.

Another possibility for the intensity increase is given by use of high-field sextupole magnets. In the present source conventional electromagnet sextupoles with magnetic field at the magnets pole tips up to 0.9T are used. As calculations show, sextupole magnets with magnetic field ~ 5 T could produce gain about 2.7 in atomic hydrogen beam density and respective increase in the polarized H⁻ ion beam current.

In conclusion, we summarize that polarized hydrogen ion source with resonant charge-exchange plasma ionizer is seemed to be very useful and promising technique for production of high-intensity polarized H⁻ ion beams for high-energy accelerators.

ACKNOWLEDGMENTS

This work was supported in part by Russian Foundation for Basic Research and by Indiana University Cyclotron Facility.

REFERENCES

1. Belov A.S., Esin S.K., Kubalov S.A., Kuzik V.E., Stepanov A.A., and Yakushev V.P., Nucl. Instr. and Meth. In Phys. Research **A255,** 442-459 (1987).
2. Haebeli W., Nucl. Instr. and Meth. **62,** 335 (1968).
3. Belov A.S., Dudnikov V.G., Kuzik V.E., Plokhinsky Yu.V., and Yakushev V.P., Nucl. Instr. and Meth. In Phys. Research **A333,** 256-259 (1993).
4. Belov A.S., Netchaeva L.P., Plokhinsky Yu.V., Vasil'ev G.A., Klenov V.S., Turbabin A.V., and Dudnikov V.G., "High-intensity source of polarized negative hydrogen ions with resonant charge-exchange plasma ionizer", *in Proc. of International Workshop on Polarized Beams and Polarized Gas Targets,* edited by H.P. gen. Schieck and L. Sydow, Cologne, Germaany, 1995, World Scientific, 1996, p.p. 218-223.
5. Belov A.S., Esin S.K., Netchaeva L.P., Plokhinsky Yu.V., Vasil'ev G.A., Klenov V.S., Turbabin A.V., Yakushev V.P., and Dudnikov V.G., Rev. of Sci. Instr. **67**, 1293-1295 (1996).

Development of polarized ³He ion source - From OPPIS to Spin-exchange

M. Tanaka[a*], T. Yamagata[b], K. Yonehara[c], Y. Arimoto[d], and N. Shimakura[e],

[a] Kobe Tokiwa College, Ohtani-cho 2-6-2, Nagata, Kobe 653-0838, Japan
[b] Department of Physics, Konan University, Okamoto 8-9-1, Higashinada, Kobe, 658-8501, Japan
[c] Department of Physic, University of Michigan, Ann Arbor, Michigan 48109, USA.
[d] Japan Synchrotron Radiation Research Institute (JASRI), Kouto 1-1-1, Mikazuki-cho, Sayo, Hyoto 679-5198, Japan
[e] Department of Chemistry, Niigata University, Ikarashi Nino-cho 8050, Niigata 950-2181, Japan

Abstract. A long history is presented on the polarized ³He ion source being developed at RCNP for nuclear physics research at an intermediate energy region. A particular emphasis is placed on how to solve serious problems encountered in each phase of the development, i.e., OPPIS (Optical Pumping Polarized Ion Source)[phase 1], EPPIS (Electron Pumping Polarized Ion Source) [phase 2], and SEPIS (Spin-exchange Polarized Ion Source)[phase 3].

INTRODUCTION

Polarization phenomena in nuclear physics have been an important tool to study both the nuclear structure and nuclear reaction mechanism from both the traditional and topical view points. Almost ten years ago, the RCNP ring cyclotron (K=400 MeV) started providing a variety of beams from proton to light heavy ions. With this cyclotron a polarized proton beam has also been dedicated to spin observable measurements. Meanwhile, over a decade ago a development of a polarized ³He ion source started at RCNP primarily to extend the territory of nuclear physics. This was a starting point of our long painful but most adventuresome journey.

---

*) e-mail address: tanaka@rcnp.osaka-u.ac.jp

OPPIS [PHASE 1]

We initiated to develop a polarized ^{3}He ion source based on a polarized electron capture by a fast ion [1] later named, "OPPIS" (Optical Pumping Polarized Ion Source). The reason why we chose this was because the OPPIS revealed great success in producing a polarized proton beam with high intensity and high polarization [2,3] relative to other methods such as an AB (Atomic Beam) or a direct ^{3}He pumping method [4]. The ^{3}He$^+$ OPPIS ion source applied to the ^{3}He polarization uses an electron capture between an incident fast ^{3}He^{2+} ion and a polarized alkali atom optically pumped [5–9]. After completion of a bench test device, we measured the ^{3}He$^+$ nuclear polarization by varying a hepp impact energy for optimization of the ^{3}He OPPIS. The result showed a gentle decrease with an incident ^{3}He^{2+} energy. This behavior was reasonably reproduced by the theory based on the semi-classical impact parameter method [10,11]. Observed absolute values of the ^{3}He nuclear polarization were, on the other hand, greatly reduced relative to the alkali atomic polarization. This was in striking contrast with the proton OPPIS, where almost no proton depolarization was observed with respect to the alkali polarization [3]. The origin of this reduction was understood in terms of an insufficient LS decoupling field for a polarized ^{3}He$^+$ ion; the LS decoupling field needed for hydrogen is only about 2 T, whereas that for ^{3}He$^+$ ion was estimated to be over 30 T [12].

In addition to the above discouraging result, we met another serious problem; an extracted beam intensity from the ^{3}He OPPIS [13] was greatly reduced. This was simply understood in terms of a transversal emittance growth due to fringing fields of the solenoidal coil at both ends [14].

EPPIS [PHASE 2]

This urged us to discover an alternating idea which uses multiple electron capture and stripping collisions between an incident ^{3}He$^+$ ion and a polarized Rb vapor [15]. Since this method uses multiple charge changing collisions between a he atom and ^{3}He$^+$ ion with almost no contribution from a ^{3}He^{2+} ion, the LS decoupling could be accomplished with a practically available magnetic field ($\sim$ a few T). This eventually results in the expectation that most of the alkali polarization would be converted to the ^{3}He nuclear polarization. We named this novel method an "electron pumping" from analogy of the optical pumping. After an enormous effort, we succeeded in experimentally proving validity of the electron pumping [16]. Later, we named this polarized source "EPPIS" (Electron Pumping Polarized Ion Source).

Another principal superiority of the EPPIS over the OPPIS is a suppression of emittance growth since a ^{3}He charge state is the same, i.e., +1 when the ion is incident on and emerging out of the solenoid coil. Nevertheless, there may be additional emittance growth induced by the multiple stripping and capture collisions under the strong magnetic field. This may further influence a spatial distribution of the ^{3}He nuclear polarization. To see these effects more closely, we carried out the Monte Carlo simulations. We could find that no sizable emittance growth occurs thanks to the fact that a penetrating ^{3}He spends a substantial time with a ^{3}He atomic state free from a magnetic field [17]. On the other hand, concerning the polarization distribution of the ^{3}He$^+$ beam, a hole having a less polarization the surrounding region was predicted, which was later named a "polarization hole" [18]. These findings would be beneficial not only to design a practical ion source but also to study plasma physics and probably astrophysics.

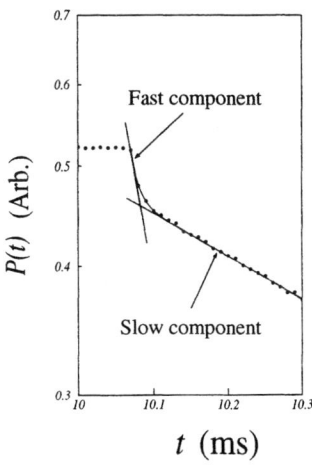

FIGURE 1. Rb polarization (arbitrary unit) is observed as a function of time. Pumping laser is switched off at t = 10.07 ms, where an external magnetic field is 4T, and Rb cell temperature is 125 °C.

In spite of great success in the EPPIS, we met a somewhat serious problem. The EPPIS requires a polarized Rb vapor with thickness higher than 10^{14} cm^{-2}, an order of magnitude higher than that required for the OPPIS ion source [15,16]. However, fabrication of highly polarized Rb vapor with such a high thickness is not easy. In addition, no basic study has been systematically done on this subject. In our work, we studied the relaxation mechanism by a time differential measurement with a chopped pumping laser [19]. One of the remarkable results was an obvious presence of a fast decaying component in addition to a well established slow component caused by the wall relaxation, diffusion, effusion and so on as shown in Fig. 1. This new component was understood in terms of formation of a sheath with higher polarization due to a radiation trapping predominated at high vapor thickness. We found that the radiation trapping and absorption of the pumping light make it difficult to attain a highly polarized Rb vapor homogeneously distributed, which is indispensable to the EPPIS.

SEPIS [PHASE 3]

This urged us to devise a new polarization principle which does not use a thick Rb vapor. A new idea which meets the above requirement was proposed after a detailed analysis of our measurement of the electron pumping [21]. Since this method uses an enhanced spin-exchange cross section at a low ^{3}He$^+$ incident energy, we named it SEPIS (Spin Exchange Polarized Ion Source). The atomic collision theory based on

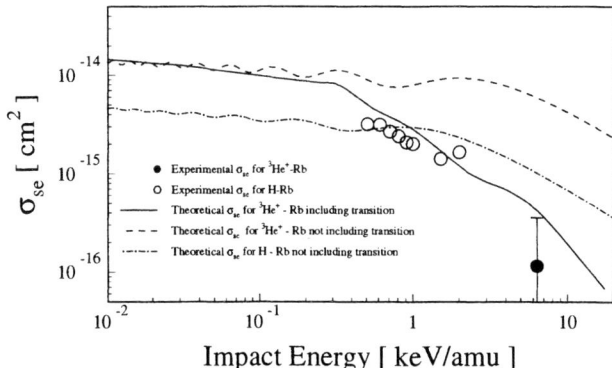

FIGURE 2. Observed and calculated spin-exchange cross sections for proton and hep incident on Rb atom

the semi-classical close-coupling method allowing inclusion of possible transitions unexpectedly predicted an anomalously large spin-exchange cross section between a ^{3}He$^+$ ion and a polarized Rb atom in particular at a ^{3}He$^+$ impact energy lower than a few 100 eV/amu (a solid curve in Fig. 2) [20]. This is in striking contrast with the behavior of the proton spin-exchange cross section. Since the SEPIS requires neither an un-practically large magnetic field nor an extremely thick Rb vapor, the ^{3}He ion SEPIS will hopefully be one of the most practical polarized ^{3}He ion sources in the next generation.

CONCLUSION AND FUTURE PROSPECT

After a long journey we have reached the SEPIS. We believe that the SEPIS is one of the most promising polarized ^{3}He ion source ever proposed. An application of the SEPIS should not be restricted in nuclear physics regime but must be extended to particle physics such as RHIC or DESY.

Everybody always asks us why we did not begin with the SEPIS ten years ago. Of course, this is true. But, it must be a universal truth that long redundancy is indispensable to a thorough change of concept.

Roma non uno die aediffcata est!

REFERENCES

1. L.W. Anderson, Nucle. Instr. Meth. **158** (199) 363.
2. Y. Mori, A. Takagi, K. Ikegami, S. Fukumoto,a and A. Ueno, J. Phys. Soc. Japan, Suppl. **55** (1986) 453.
3. A.N. Zelenski, C.D.P. Levy, P.W. Schmor, W.T.H. van Oers, and G. Dutto, *AIP conference Proceedings 293, 1994 ed. by L.W. Anderson, and W. Haeberli, p.173*.
4. M. Leduc, Coll. de Phys. **C6** (1990) 317.
5. M. Tanaka, T. Ohshima, K. Abe, K. Katori, M. Fujiwara, T. Itahashi, H. Ogata, and M. Kondo, Colloque de Physique **C6** (1990) 553.
6. M. Tanaka, T. Ohshima, K. Katori, M. Fujiwara. T. Itahashi, H. Ogata, and M. Kondo, Phys. Rev. **A41** (1990) 496.
7. M. Tanaka, T. Ohshima, K. Katori, M. Fujiwara, T. Itahashi, H. Ogata, and M. Kondo, Nucl. Instr. Meth. **A302** (1991) 460.
8. T. Ohshima, K. Abe, K. Katori, M. Fujiwara, T. Itahashi, H. Ogata, M. Kondo, and M. Tanaka, Phys. Lett. **B279** (1992) 163.
9. M. Tanaka, T. Ohshima, K. Katori, M. Fujiwara, T. Itahashi, H. Ogata, and M. Kondo, Hyperfine Interactions **74** (1992) 205.
10. M. Tanaka, T. Ohshima, K. Katori, M. Fujiwara, H. Ogata, M. Kondo, and N. Shimakura, Hyperfine Interactions **78** (1993) 251.
11. M. Tanaka, N. Shimakura, T. Ohshima, K. Katori, M. Fujiwara, H. Ogata, and M. Kondo, Phys. Rev. **A50** (1994) 1184.
12. B.H. Bransden, and C.J. Joachain, Physics of atoms and molecules, Longman Scientific and Technical, 1983, England.
13. T. Yamagata, M. Tanaka, K. Yonehara, Y. Arimoto, T. Takeuchi, M. Fujiwara, Yu.A. Plis, L.W. Anderson, and R. Morgenstern, Nucl. Instr. Meth. **A402** (1998) 199.
14. G.G. Ohlsen, J.L. McKibben, R.R. Stevenson Jr., and G.P. Lawrence, Nucl. Instr. Meth. **73** (1969) 45.
15. M. Tanaka, M. Fujiwara, S. Nakayama, L.W. Anderson, Phys. Rev. **A52** (1995) 392.
16. M. Tanaka, T. Yamagata, K. Yonehara, T. Takeuchi, Y. Arimoto, M. Fujiwara, Y. Plis, L.W. Anderson, and R. Morgenstern, Phys. Rev. **A60** (1999) R3354.
17. T. Takeuchi, T. Yamagata, K. Yonehara, Y. Arimoto, and M. Tanaka, Rev. Sci. Instr. **69** (1998)412.
18. Y. Arimoto, K. Yonehara, T. Yamagata, and M. Tanaka, Nucl. Instr. Meth. **A** (2000) in press.
19. K. Yonehara, T. Yamagata, Y. Arimoto, T. Takeuchi, and M. Tanaka, submitted to Nucl. Instr. Meth. **A** (2000) in reviewing.
20. Y. Arimoto, N. Shimakura, K. Yonehara, T. Yamagata, and M. Tanaka, to be submitted to Phys. Rev. **A** (2000)
21. Y. Arimoto, N. Shimakura, K. Yonehara, T. Yamagata, and M. Tanaka, Eur. Phys. J. **D8** (2000) 305.

The new LEGS Highly Polarized Frozen-Spin Solid HD Target Facility

X. Wei[1,¶], F. Lincoln[1], M. Lowry[1], T. Saitoh[2] and A. M. Sandorfi[1]

[1]*Brookhaven National Laboratory, Upton, NY 11973, USA*
[2]*Virginia Polytechnic Institute & State University, Blacksburg, VA 24061, USA*

Abstract. Three large highly H polarized frozen-spin HD targets (1.3 moles each in the solid phase) have been produced simultaneously at 20 mK and 15 Tesla with a new production facility located at Brookhaven National Laboratory. After *freezing* the hydrogen spins into alignment, the targets were cold transferred (at 2.5 K and 0.016 Tesla) to other cryostats with <2% loss in polarization. Highlights of the new target polarization facility are discussed.

INTRODUCTION

Development of highly polarized frozen-spin targets consisting of HD in the solid phase began at Syracuse University [1-4] as a joint effort with the Laser Electron Gamma Source (LEGS) collaboration. In the summer of 1999, a **S**trongly **P**olarized **H**ydrogen deuteride **ICE** (*SPHICE*) target was produced at Syracuse, cold-transferred into a storage cryostat, and successfully shipped 300 km to LEGS at Brookhaven National Laboratory (BNL) by truck. Following this, the *SPHICE* target production facility has been moved to BNL. Since then, two sets of three *SPHICE* targets have been made at BNL in preparation for a program of nucleon spin-physics experiments at LEGS.

PHYSICS MOTIVATION

The physics program at LEGS is centered on double-polarization experiments with both beam and target polarized. In particular, we are focusing on the spin structure sum rules for the nucleon, the forward spin-polarizability,

$$\gamma_o = \frac{1}{4\pi^2} \int \frac{\sigma^{1/2} - \sigma^{3/2}}{E_\gamma^3} dE_\gamma \qquad (1)$$

and the Gerasimov-Drell-Hearn,

$$-\frac{\alpha}{2m^2}\kappa^2 = \frac{1}{4\pi^2} \int \frac{\sigma^{1/2} - \sigma^{3/2}}{E_\gamma} dE_\gamma \quad . \qquad (2)$$

[¶] e-mail: xwei@bnl.gov

The right-hand side of both contain the difference of total reaction cross sections with the photon and nucleon spins anti-parallel and parallel, weighted by a power of the energy E. The integrals on the right-hand side of the sum rules will be measured at LEGS from π-threshold up to 470 MeV. The left-hand side of γ_o can be predicted from chiral perturbation theory, while the GDH integral is determined by the anomalous magnetic moment of the target, under the assumption of vanishing Compton spin-flip at infinite energy. The energy weighting of these integrals emphasizes low energies. At LEGS we will cover about 90% of the γ_o integral and 60% of the GDH integral.

SPHICE TARGETS

Most conventional polarized hydrogen targets are made from either Ammonia or Butanol. Although high polarizations have been achieved for the atomic hydrogen nuclei in these molecules, nonetheless atomic hydrogen represents only a small fraction of their molecular mass. The new LEGS target consists of molecular HD in the solid phase. In *SPHICE*, both the H and the D nuclei can be highly polarized and oriented at will, and over 85% of the target consists of the molecular species of interest, while effects from the remaining 15% can be removed with empty-target subtraction. (This is a marked advantage over experiments with more complex targets such as C_4H_9OH or NH_3 for which *empty-cell* measurements cannot sample the effects from the C, O or N nuclei.) A schematic drawing of the target is shown in figure 1. The target cell contains a copper ring with a left-handed thread inside and a right-handed thread outside for target transfer, an inner and an outer Mylar™ cup, and 3000 gold-coated aluminum $\varnothing = 38\mu m$ cooling wires.

SPHICE targets can be produced in several polarization combinations. The basic idea is to use a *brute force* method to polarize the proton and then transfer the polarization from the proton to the deuteron via RF forbidden adiabatic fast passage. The intrinsic relaxation times of HD are almost infinite at high B and low T. Although this is ideal once the target is polarized, a trace of ortho-H_2 has to be introduced into the HD system to reduce the initial relaxation time to allow polarization [5]. Ortho-H_2 decays to para-H_2 in 6.25 days, and provides an effective *relaxation switch*. After about

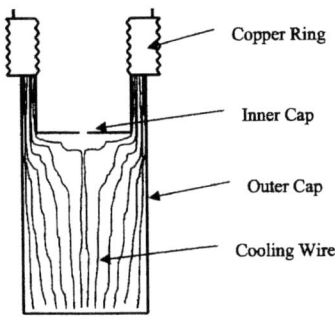

FIGURE 1. Schematic drawing of an HD target cell.

6 decay times, the o-H_2 concentration is only about 0.1% of its initial value and the *relaxation switch* is turned off. The aluminum wires in the target are needed to remove the heat from the ortho-H_2 to para-H_2 transition.

The stages involved in manufacturing SPHICE targets in various polarization combinations are shown in figure 2. When only H polarization is desired, one can just wait for about a month in the dilution refrigerator (at 15mK and 17T) to polarize H. When only D polarized targets are desired, the H polarization is transferred to D via RF Forbidden Adiabatic Fast Passage (FAFP). The cycle of polarizing H followed by FAFP polarization transfer from H to D is repeated twice. If polarization of both H and D is needed in the final products, repolarizing H after the second FAFP is then necessary. The final polarizations are different for each case because the relaxation times are getting longer and longer while H is being polarized and eventually H polarization growth will stop. When relaxation times are long enough, the targets can then be transferred to an in-beam cryostat for double polarization measurement or to a storage cryostat for long-term holding, with less than 2% relative loss in polarization. Long relaxation times in storage conditions (1.5K and 10T) provide the opportunity of making and using the SPHICE targets at different times and locations. In fact, in May 1999, one H-polarized HD target, produced at Syracuse University, was successfully cold transferred to a *Storage dewar* and transported 300 kilometers to BNL.

Two sets of 3 H-polarized SPHICE targets have been produced at BNL in 2000 and have been used for detector calibrations. At present reliability problems with two of the cryostats have delayed the start of the double-polarization physics program. These cryostats are currently being repaired, and replacements are also being designed.

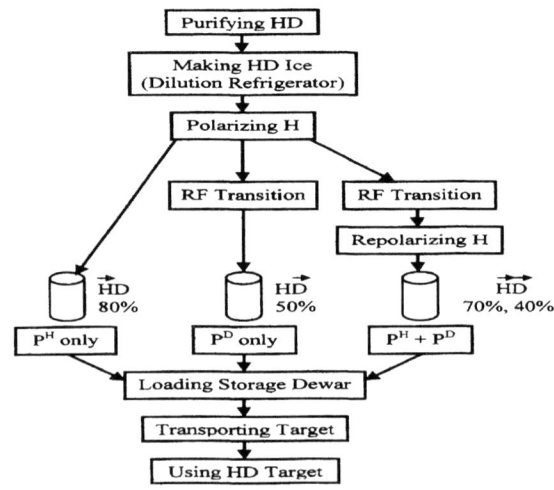

FIGURE 2. Alternate cycles in SPHICE target production.

$\pi^\pm$ PRODUCTION ON UNPOLARIZED HD TARGET

π^+-production from protons in HD is shown in figure 3 for 320 MeV beam energy, left panel. There are two components to this spectrum, the yield from the free proton, which produces the narrow peak, and the yield from the bound proton of deuterium, which shows up as a much broader structure. Empty target yields have been subtracted, but these are coincidence data in which we detect the recoil nucleon, so the contributions from the aluminum wires turns out to be very small. π^- production from the neutron is shown in the right panel, broadened by the Fermi motion in deuterium.

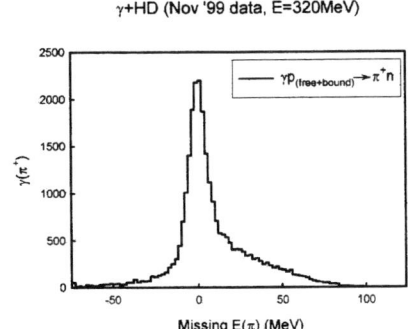
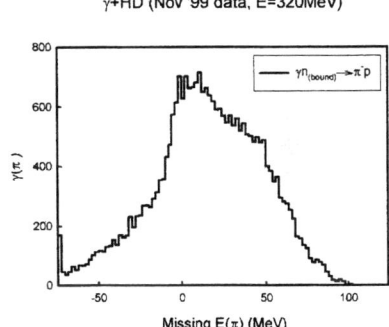

FIGURE 3. π^+ p missing-energy spectra from the proton in unpolarized HD (left panel), and π^- n spectra from the neutron (right panel).

The π^+ spectrum from the free proton can be approximated by assuming the yield from the bound neutron is the same as that from the bound proton and subtracting it, as shown by solid line in Figure 4. This assumption evidently works quite well and the tail on the *free proton* peak is essentially gone. The neutron spectrum from the right panel of figure 3 is reproduced in figure 4 as a dotted line, normalized so that it has the same area as the solid curve. This is why we are very enthusiastic about this target for neutron experiments. Although an over-determination of kinematics can be used with conventional complex targets to pull contributions from atomic hydrogen into a peak, this peak always sits on a large background from reactions in the accompanying heavier nuclei. Such backgrounds can be subtracted, but not without significantly affecting the propagated uncertainties. In HD the large backgrounds characteristic of complex targets are completely absent. Moreover, while Fermi motion of the single bound neutron in HD broadens the $\gamma n \to \pi^- p$ peak considerably, there are no unpolarizable neutrons in HD to obscure the signal. In contrast, the corresponding $\pi^- p$ spectrum from Butanol or Ammonia would completely blend into its background.

CONCLUSIONS AND IMPROVEMENTS

The main features of the new SPHICE target are high polarization (80% for hydrogen and 50% for deuterium), no dilution by unwanted extraneous nuclear species,

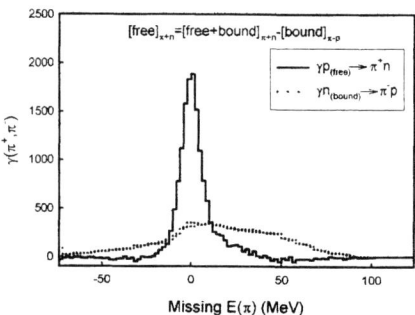

FIGURE 4. Missing-energy spectra from $p(\gamma,\pi^+ n)$ and $n(\gamma,\pi^- p)$.

a significantly higher accuracy for D polarization, and long relaxation times. The reliability problems with some of the cryogenic equipment that has delayed the start of experiments are being addressed, and measurements are expected to begin within the next year.

ACKNOWLEDGMENTS

This work is supported by the US Department of Energy under contract DE-AC02-98CH10886 and by the US National Science Foundation. The authors wish to thank Dr. S. Hoblit and Mr. H. Meyer for providing the missing energy spectrums and the entire LEGS Collaboration for participating the γ+HD experiment.

REFERENCES

1. A. Honig, Q. Fan, X. Wei, M. Breuer, J. P. Didelez, M. Rigney, M. Lowry, A. Sandorfi, A. Lewis and S. Whisnant, Proceeding of 12th Int'l Symp. On High Energy Spin Physics, Amsterdam, Sept. 1996. **World Scientific**, p. 365 (1997).
2. X. Wei, A. Honig, A. Lewis, M. Lowry, A. Sandorfi, S. Whisnant and J. P. Didelez, Physica B, 284-288, 2051 (2000).
3. A. Honig, X. Wei, F. Lincoln, M. Lowry, A. Sandorfi, A. Lewis, S. Whisnant, C. Commeaux, J. P. Didelez and C. Schaerf, Proceeding of Int'l Workshop on Polarized Sources and Targets, Erlangen, Sept. 29-Oct. 2, 1999, PST99, p.392 (2000).
4. A. Honig Q. Fan, X. Wei, A. M. Sandorfi and C. S. Whisnant, Nucl. Inst. Meth. A356, 39 (1995).
5. A. Honig, Phys. Rev. Lett. 19, 1009 (1967).

The HERMES Polarized Internal Deuterium Gas Target

P. Lenisa*

on behalf of the HERMES Collaboration

Università di Ferrara and INFN
44100 Ferrara - Italy

Abstract.
The HERMES target makes use of a storage cell internal to the HERA lepton ring in which polarized Deuterium atoms are injected from an Atomic Beam Source (ABS). A sample beam is extracted from the center of the storage cell to determine the atomic fraction and the atomic polarization using respectively a Target Gas Analyzer (TGA) and a Breit-Rabi Polarimeter (BRP). The target performance with Deuterium gas in 2000 will be presented.

INTRODUCTION

The HERMES experiment in the HERA electron storage ring at DESY is studying the spin structure of the proton and neutron via deep inelastic scattering of polarized electrons off the nucleons in a highly polarized internal gas target. The target was operated with Hydrogen gas in 1996 and 1997 and has been modified to use Deuterium gas since 1998. A schematic diagram of this target, which uses a storage cell to increase the target density, is shown in figure 1.

A beam of Deuterium atoms is generated in a surfatron based MW dissociator which forms part of the atomic beam source (ABS) [1]. These atoms are electron polarized by means of Stern-Gerlach separation in a sextupole magnet system. The electron polarization is transferred to the nucleons by means of adiabatic high frequency transitions that interchange the occupations of two different hyperfine states.

The beam of nuclear polarized atoms is injected into the centre of the thin-walled storage cell via a side tube and the atoms then diffuse to the open ends of the cell where they are removed by a high speed pumping system. The storage cell acts to confine the polarized atoms near the beam and therefore increases the target areal density 150 times the free beam. The cell (Fig. 2) is coated with Drifilm [2] in order to minimize wall interaction effects.

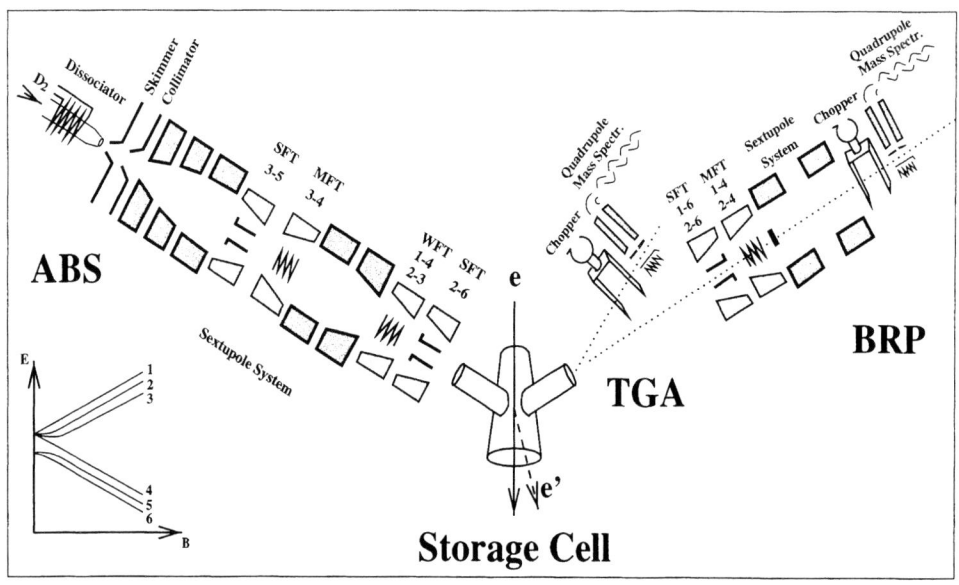

FIGURE 1. Schematic picture of the HERMES target.

A longitudinal magnetic field of approximately 330 mT is maintained over the cell volume by a superconducting magnet coil which results in a non uniformity of $\Delta B/B = \pm 1\%$. The field provides a quantisation axis for the spins and inhibits nuclear spin relaxation by decoupling nucleon and electron spins.

A second side tube is provided to sample the gas within the target cell. The beam emerging from this tube is analysed with a target gas analyser (TGA) to

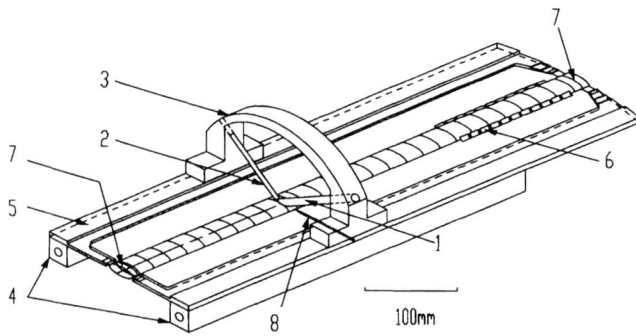

FIGURE 2. The HERMES storage cell: (1) injection tube; (2) sample tube; (3) bridge; (4) cooling rails; (5) holding frame; (6) cell extension; (7) connection to wake field suppressor; (8) unpolarized gas feed line.

determine its atomic fraction and a Breit-Rabi polarimeter (BRP) to measure its hyperfine occupation in order to calculate the atomic polarization [3]. During the atom diffusion process relaxation by wall and spin exchange collisions and wall recombination changes the polarization and the atomic fraction of the target gas. The atom polarization and atomic fraction values measured by the BRP and TGA must be corrected for these effects to obtain the absolute target polarization.

THE TARGET POLARIZATION

The target polarization P_T is calculated using the expression:

$$P_T = \alpha_0 \left(\alpha_r + (1 - \alpha_r)\beta\right) P_T^{atom}, \tag{1}$$

where α_0 is the initial atomic fraction, α_r is the fraction of atoms surviving recombination and β is the polarization of recombined molecules relative to the polarization of the atoms P_T^{atom}. In order to relate the above mentioned parameters, which are average values over the cell, with the corresponding measurement on the sample beam, so called sampling correction c_α and c_P have to be introduced. They are defined in the following way:

$$\alpha_r = c_\alpha \, \alpha_r^{TGA} \tag{2}$$

$$P_T^{atom} = c_P \, P_{BRP}. \tag{3}$$

The sampling corrections in equations (2) depend on the sensitivity of the BRP and the TGA to the different parts of the cell and are derived by means of Monte-Carlo simulations investigating the history of gas particles travelling through the storage cell [4].

In the special case of no recombination and no depolarization inside the storage cell, $c_\alpha = 1$ and $c_P = 1$ and:

$$\alpha_r = \alpha_r^{TGA} = 1 \tag{4}$$

$$P_T^{atom} = P^{BRP} \tag{5}$$

so that the formula for the target polarization becomes simply:

$$P_T = \alpha_0 P^{BRP} \tag{6}$$

TARGET PERFORMANCE IN 2000

Several hardware upgrades were implemented in 1999 to increase the target density. The cell cross section has been reduced from 29 × 9.8 mm to 21 × 8.9 mm and its temperature decreased from $90 K$ to $60 K$. The RF dissociator has been replaced by a surfatron based microwave dissociator. Thanks to improvements to the system, the HERMES Deuterium target in 2000 was running very stable with double atomic density compared to Hydrogen in 1996/1997. The magnetic holding field was operated at $B = 335 mT$ which resulted in a strong decoupling of electron end nucleon spin, as indicated by the decoupling factor $x = (B/B_c^D) = 28.6$ ($B_c^D = 11.7$ mT is the critical field for Deuterium), which is 4.33 times higher than for Hydrogen. The integrated areal target density was $D_T = 2.1 \times 10^{14}\ nucl.cm^{-2}$. The target polarization was, in standard operation, randomly flipped between P_z^+ and P_z^- (vector polarization). For certain studies or special measurements any combination of hyperfine states could be injected, e.g. the target has been run for two months with purely tensor polarized Deuterium atoms. All numbers quoted below refer to the 2000 data taking period with a polarized Deuterium target.

The measured average α_0^{TGA} was 0.95, where two sources for molecular contribution are taken into account: the ballistic flow of molecules coming from the ABS and the residual gas from the vacumm chamber around the storage cell.

Both the atomic fraction and the atomic polarization showed no temperature dependence trough the year 2000. Since both the recombination and depolarization mechanisms active in the cell depend exponentially on the temperature [5], this can be considered as an indication that the storage cell used in 2000 caused no measurable recombination and depolarization on the cell surface.

The hypothesis of no recombination has been confirmed with detailed calibration measurements which allow the separation of the detected molecular flow into its different components and the determination of the possible contribution coming from recombination. The result for the average atomic fraction surviving recombination, α_r^{TGA}, was $0.997 \pm 0.010\ (syst.)$.

The range for β is per definition between 0 (the nucleons loose their polarization completely during a recombination process) and 1 (all nucleons keep their polarization). The large uncertainty does not significantely contribute to the target polarization in equation (1) because α_r is very close to one.

A fit of the measured occupation numbers using a *rate equation model* which is based on a combination of well established spin relaxation models and a one-dimensional diffusion equation describing the particle transport through the storage cell [6], allows the determination of the injected polarization for $\vec{D}$ atoms, which was typically 90.5%, and the relaxation parameters with small uncertainties. The average polarization loss was determined to be less than 0.2% from spin exchange collisions and less than 0.5% from wall collisions, essentially confirming the hypothesis of negligible depolarization.

Studies on the surface coating performed at the end of the run evidenced the crucial rule played by a layer of heavy-water frozen on the cell surface. Some amount of water is in fact continously injected in the cell by the ABS where a small amount of Oxygen is added in order to improve the discharge performance. The cell surface lost its good quality after the first warming up. In the successive cooling down the cell exhibited strong recombination and depolarization which are indications of a severe damage of the Drifilm surface.

Although a final analysis is the moment not yet available, the projected error on the final target polarization is less than 4%.

PLANS FOR THE 2001 TRANSVERSE RUNNING

During the winter shutdown 2000/2001 the target will be modified and upgraded to start running transversally polarized hydrogen in 2001. Particular attention is being given to the design of the transverse magnet producing the holding field. Based on the 1996/1997 experience, an absolute value of the field higher than 300 mT should be desirable to inhibit depolarization due to spin-exchange collisions and wall depolarization [7]. Additional constraints concern field homogeneity, which has to be better than $1.4 \times 10^{-4}T$ in order to avoid the resonances induced by the RF field associated with the time structure of the lepton beam. The required homogeneity is higher than for the longitudinal field due to an additional class of hyperfine transitions not present for the longitudinal case [8]. At the same time a new sextupole magnet system is being implemented for the BRP which will increase the present statistical precision.

CONCLUSIONS

The HERMES target ran reliably with very stable operation during the 2000 running period. Due to improvements in the running conditions, the target density has been doubled with respect to the previous period. The knowledge gained during the running with Hydrogen has been successfully transferred to the Deuterium target allowing a fast and reliable analysis of the data. The collected data are consistent with having little or no recombination and depolarization inside the storage cell. During the winter shutdown 2000/2001 the target will be modified and upgraded to start running transversally polarized hydrogen in 2001 with improved statistical precision.

REFERENCES

1. N. Koch and E. Steffens, *Rev. Sci. Instrum.*, 70, 1999, p.1631.
2. G.E. Thomas et. al.; *Nucl. Instr. Meth. A* 257, 1987, pp. 32.
3. B. Braun; *AIP Conf. Proc. no.* 421 : *Pol. Beams and Pol. Gas Targets*, Ed. Roy J. Holt & Michael A. Miller, Urbana-Champaign 1997, pp. 156.
4. M. Henoch; *PST* 99 : *International Workshop on Polarized Sources and Targets*; A. Gute, S. Lorenz, E. Steffens; Univerität Erlangen-Nürnberg 1999, pp.472.
5. H. Kolster; *PhD. Thesis*, LMU München 1998.
6. C. Baumgarten; *PhD. Thesis*, Univerität München 2000.
7. J. Stewart; *HERMES Internal Note* 00 − 13
8. D. Reggiani, *Diploma Thesis*, Universitá di Ferrara, 1999.

Michigan Ultra-Cold Polarized Atomic Hydrogen Jet Target

B.B. Blinov[1], S.E. Gladycheva[1], T. Kageya[1], D.Yu. Kantsyrev[1],
A.D. Krisch[1], V.G. Luppov[1], V.S. Morozov[1], J.R. Murray[1],
R.S. Raymond[1], N.S. Borisov[2], V.V. Fimushkin[2], V.N. Grishin[3],
A.I. Mysnik[3], D. Kleppner[4]

[1]*Randall Lab. of Physics, University of Michigan, Ann Arbor, MI - 48109-1120, USA*
[2]*Joint Institute for Nuclear Research, RU-141980, Dubna, Russia*
[3]*Institute for High Energy Physics, RU-142284, Protvino, Russia*
[4]*Department of Physics, Massachusetts Inst. of Technology, Cambridge, Massachusetts 02139, USA*

Abstract. To study spin effects in high energy collisions, we are developing an ultra-cold high-density jet target of proton-spin-polarized hydrogen atoms. The target uses a 12 Tesla magnetic field and a 0.3 K separation cell coated with superfluid helium-4 to produce a slow monochromatic electron-spin-polarized atomic hydrogen beam, which is then focused by a superconducting sextupole into the interaction region. In recent tests, we studied a polarized beam of hydrogen atoms focused by the superconducting sextupole into a compression tube detector, which measured the polarized atoms' intensity. The Jet produced, at the detector, a spin-polarized atomic hydrogen beam with a measured intensity of about $2.8 \cdot 10^{15}$ H s^{-1} and a FWHM area of less than 0.13 cm^2. This intensity corresponds to a free jet density of about $1 \cdot 10^{12}$ H cm^{-3} with a proton polarization of about 50%. When the transition RF unit is installed, we expect a proton polarization higher than 90%.

To study spin effects in high energy collisions, we are developing an ultra-cold high-density jet target of proton-spin-polarized hydrogen atoms (Michigan Jet). The jet first uses a high magnetic field (12 T) and an ultra-cold (0.3 K) separation cell coated with a superfluid helium-4 film to produce a slow monochromatic electron-spin-polarized atomic hydrogen beam; an rf transition unit will then convert this beam into a proton-spin-polarized beam [1].

A layout of the Michigan Jet is shown in Fig. 1. Atomic hydrogen is produced in a room- temperature rf dissociator and guided to the ultra-cold separation cell coated with superfluid helium-4 to depress the surface recombination of atoms. The double walls of the cell form the mixing chamber of the dilution refrigerator. The cell's entrance and exit apertures are respectively located at about 95% and 50% of the 12 Tesla central magnetic field of the superconducting solenoid. After the hydrogen atoms are thermalized by

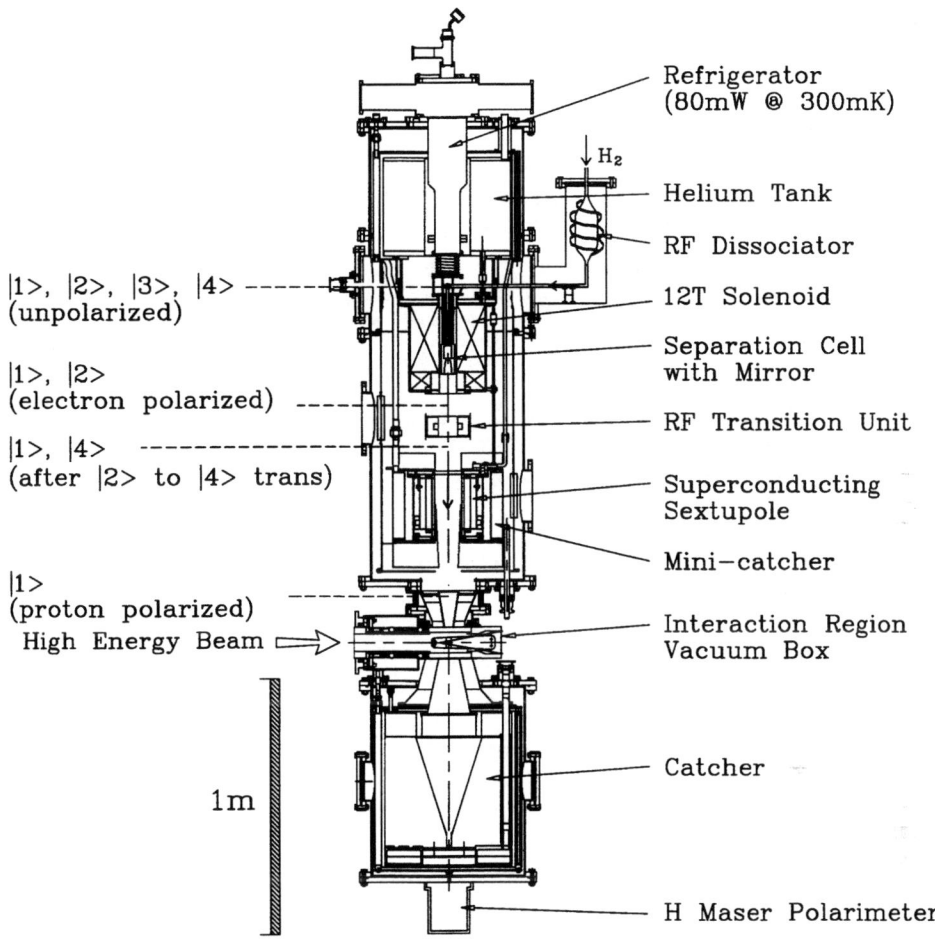

FIGURE 1. The layout of the Michigan ultra-cold jet.

collisions with the cell surface, the magnetic field gradient physically separates the atoms according to their electron-spin states. The atoms in the two lowest hyperfine states 3 and 4 are attracted toward the high field region and escape from the cell. They quickly recombine on bare surfaces and are cryopumped. The atoms in the two higher hyperfine states 1 and 2 are repelled toward the low field region and effuse from the exit aperture, forming an electron-spin polarized beam. To increase the jet density we use a polished gold-coated copper focusing mirror covered with a helium-4 superfluid film similar to the prototype mirror [2]. After an rf transition unit, which interchanges atoms in states 2 and 4, the beam passes through a superconducting sextupole.

The sextupole selects atoms in electron spin state +1/2 by focusing atoms in state 1 into the interaction region and defocusing atoms in state 4, which are then cryopumped. The proton-spin polarized beam then passes through the interaction region and is caught below by a cryopumping catcher. A maser polarimeter below the catcher monitors the beam proton polarization.

Most of the Michigan Jet parts have been fabricated and successfully tested. This hardware includes a 12 Tesla superconducting solenoid, a dilution refrigerator with a cooling power of almost 80 mW at 300 mK, a 20 cm long superconducting sextupole magnet with iron poles and a 10.5 cm diameter bore, a cryocondensation pump with a measured pumping speed of about $1.2 \cdot 10^7$ l/s ($4.2 \cdot 10^{26}$ atoms/(Torr s)), and a hydrogen maser polarimeter capable of monitoring the polarization to about ±2% in a few minutes.

In recent tests we studied a polarized beam of hydrogen atoms focused by the superconducting sextupole into a compression tube detector, which measured the polarized atoms' intensity. The measured compression tube signal versus focusing sextupole current is shown in Fig. 2. As we expected the sextupole curve shape corresponds to calculated beam velocity distribution [3].

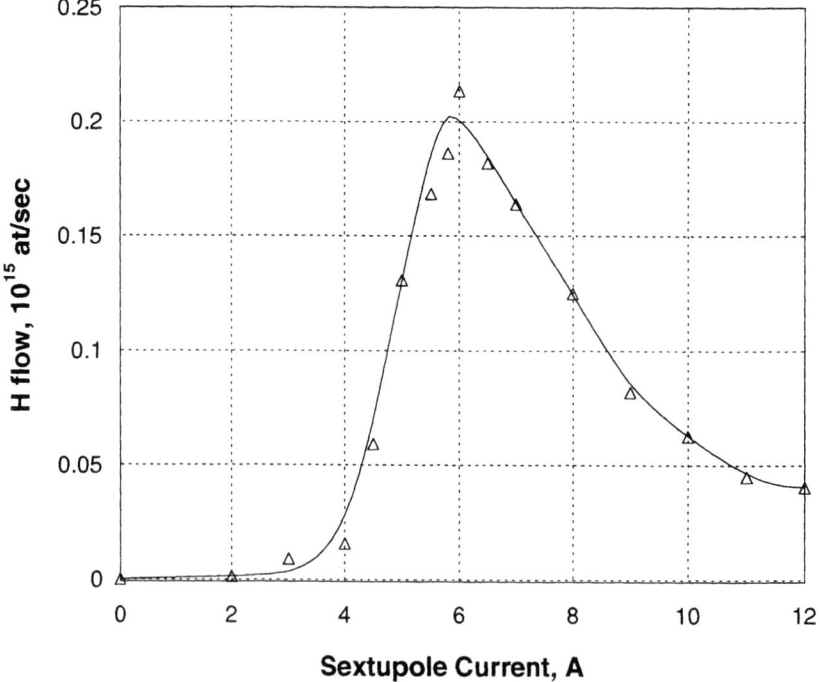

FIGURE 2. The measured compression tube signal versus the sextupole current. The line is a fit to the data.

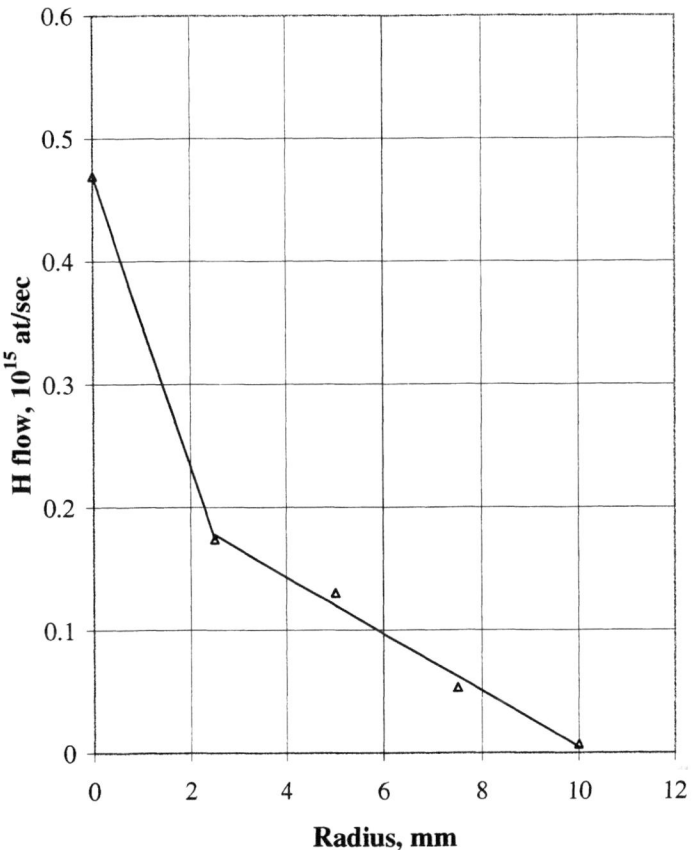

FIGURE 3. The measured radial beam distribution. The line is a fit to the data.

According to our Monte-Carlo simulations [3] the ultra-cold beam is very monochromatic due to the low temperature and the much larger magnetic field gradient acceleration along the solenoid axis. As a result we expected a small size of the focused beam. Indeed, the measured radial beam distribution shown in Fig. 3 gives a FWHM of only about 4 mm. A test of the new superfluid-^{4}He-coated parabolic mirror, attached to the separation cell, appeared to increase the beam intensity by a factor of about 3, as expected [4].

The highest measured spin-polarized atomic hydrogen jet intensity to the 12 mm by 2.5 mm compression tube slot is about $2.8 \cdot 10^{15}$ H s^{-1}. Within the 4 mm FWHM area of 0.13 cm^2, this intensity corresponds to a free jet density of $1 \cdot 10^{12}$ H cm^{-3}. The Jet's total thickness is about $6 \cdot 10^{11}$ H cm^{-2}. So far, the intensity is limited by the high insulation vacuum pressure due to the evaporation of the separation cell's helium film.

The electron-spin polarized beam has a proton polarization of about 50%. The electron polarization will be converted into proton polarization by adiabatic passage through an rf transition unit with a novel ring dielectric resonator that accepts the 6 cm beam diameter. A room temperature prototype rf unit was tested with a maximum measured transition efficiency of 97%. A preliminary design of the cryogenic rf unit has been fabricated and tested [5].

ACKNOWLEDGEMENTS

Supported by a research grant from the U.S. Department of Energy.

REFERENCES

1. V.G. Luppov et al., Status of the Mark-II Polarized Hydrogen Jet Target, in *Proceedings of the 13-th International Symposium on High-Energy Spin Physics,* Protvino, Russia, September 1998, edited by N.E. Tyurin et al, World Scientific, 1999, pp.409-411.
2. V.G. Luppov, W.A. Kaufman, K.M. Hill, R.S. Raymond, and A.D. Krisch, *Phys. Rev. Lett.* **71**, 2405 (1993).
3. V.G. Luppov et al., Status of the Mark-II Polarized Hydrogen Jet Target, in *Proceedings of the 12-th International Symposium on High-Energy Spin Physics,* Amsterdam, The Netherlands, September 1996, edited by C.W. de Jager et al, World Scientific, 1997, pp.434-437.
4. T. Kageya, Private communication.
5. R.S. Raymond, Development of a Large-bore Cryogenic 2-4 Transition Unit, in *Proceedings of the International Workshop on Polarized Sources and Targets,* edited by A. Gute et al., Erlangen, Germany, September 1999.

Development of a polarized proton target in a low magnetic field at high temperature

Takashi Wakui*, Michio Hatano†, Hideyuki Sakai†, Atsushi Tamii†, and Tomohiro Uesaka‡

*RIKEN, Wako, Saitama 351-0198, Japan
†University of Tokyo, Bunkyo-ku, Tokyo 113-0033, Japan
‡Saitama University, Urawa, Saitama 338-8570, Japan

Abstract. A proton polarizing system have been designed and constructed in order to develop a polarized proton solid target applicable to an experiment with a radioactive isotope beam. Protons are polarized by a method called microwave-induced optical nuclear polarization. Our goal is to attain a polarization of more than 30% in the target sample with a diameter of 20 mm and a length of 5 mm.

INTRODUCTION

The RI beam factory (RIBF) is currently being constructed at RIKEN. The RIBF will be capable of providing more than 3000 species of radioactive nuclei. At the facility, a polarized proton target will be used to investigate the structure of unstable nuclei through an experiment involving an inverse kinematics. In such an experiment, a proton target which can be polarized in low magnetic fields (100 ∼ 3000 Gauss) at high temperatures (77 ∼ 300 K) is required because it is necessary to detect low momentum recoiled protons at a large detection solid angle.

Protons in a single crystal of aromatic molecules can be polarized under such conditions by a method called microwave-induced optical nuclear polarization (MIONP) [1]. In this method, guest aromatic molecules doped in the single crystal are optically pumped to a higher singlet state by laser irradiation as shown in Fig. 1. The excited molecules have a probability of transition from the first excited singlet state to the lowest triplet state owing to the intersystem crossing [2]. Populations of sublevels in the triplet state are then different from each other due to the angular momentum selection rule [3,4]. Thus, the electron polarization due to the population difference between two sublevels can be obtained almost independent of the magnetic field strength and the temperature. Then, the population difference is transferred to proton polarization by means of the integrated solid effect (ISE) technique [5,6]. In this technique, the magnetic field is swept adiabatically through the electron spin resonance (ESR) line during the lifetime of the triplet state, and the pulsed microwave with a frequency corresponding to the energy between two sublevels is simultaneously applied. This technique enables efficient polarization transfer even in a low magnetic field where the efficiency of the conventional DNP method is greatly reduced.

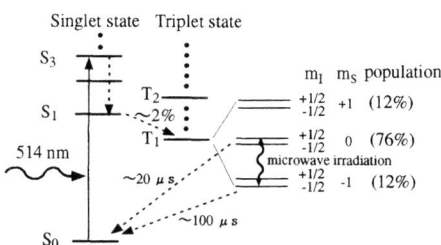

FIGURE 1. Energy level of pentacene. The lifetime of the first excited singlet state is 20 ns. The electron polarization due to the population difference between two sublevels is 73%.

Recently, Iinuma et al. reported a proton polarization of 32% in a single crystal of naphthalene doped with pentacene was achieved by using this method in a magnetic field of 3 kGauss at a temperature of 77 K [6]. We are planning to apply this method to a proton target for RI beam experiments. Figure 2 (a) shows a schematic of the polarizing system. The system consists of a C-type magnet, a laser system for the optical pumping, a microwave system for the ISE and an nuclear magnetic resonance (NMR) system for measuring the proton polarization. The current status of the system is presented here.

PREPARATION OF MATERIALS

As the guest aromatic molecule, we chose pentacene because it has a large population difference (73%) in the photo-excited triplet state [7]. Moreover, for crystallization, there are suitable molecules, naphthalene and p-terphenyl, as a host material of pentacene. Typical concentration of pentacene in naphthalene is 0.01 mol%, and in p-terphenyl 0.1 mol%. These values depend on whether a fine crystal can be made. This influence the relaxation time of protons.

In order to attain high proton polarization, it is necessary for the molecules to be of high purity. In this work, the molecules are purified by the zone-melting method. Molecules in a glass cell with a diameter of 20 mm are passed through 30 molten zones with the speed of about 20 mm/h. This is repeated for 2 or 3 times. After the purification, the target material is crystallized with the Bridgeman technique. The host molecule and small amount of pentacene contained in a glass tube are moved from the region of temperature above their melting point to that of below with a speed of 1 mm/h. After the crystallization, a section is cut from the crystal and placed in the polarizing system.

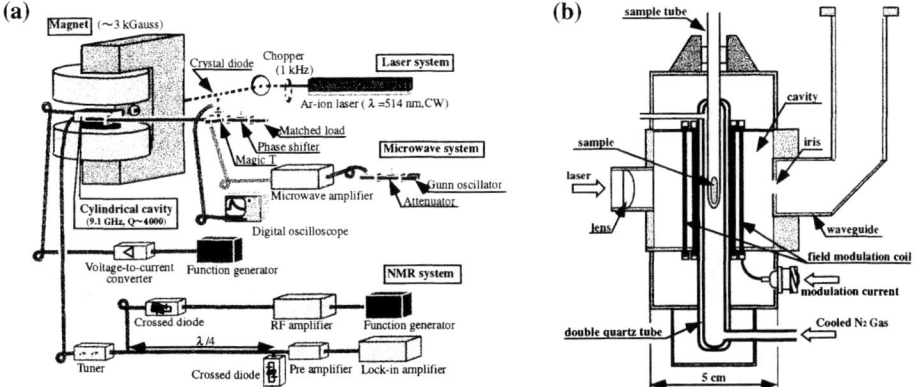

FIGURE 2. (a) Schematic of the proton polarizing system. The system consists of a magnet, a laser system, a microwave system and a NMR system. (b) Schematic view of the TE_{011} cylindrical cavity. The cavity contains set of field modulation coil. The quality factor of the cavity is about 4000 at room temperature.

MAGNET SYSTEM

The magnetic field produced by the C-type magnet is about 3 kGauss. The magnetic field inhomogeneity of 10^{-3} over the target size is required for obtaining a proton polarization of more than 30%. We designed a magnet which satisfies this condition. The diameter of the coil is 150 mm and the gap between the poles is 60 mm. The magnetic field inhomogeneity measured using a NMR teslameter at a field of about 3 kGauss is 3.5×10^{-4} over the target size. The field homogeneity is sufficient for polarizing protons.

LASER SYSTEM

We use an Ar-ion laser for the optical pumping. The wavelength of the main mode in the laser, 514 nm, corresponds to the energy of transition to the third excited singlet state in pentacene (Fig. 1). To obtain a high population difference in the lowest triplet state, a pulsed laser beam is required because the lifetime of the most populated level in the triplet state is shorter than that of other levels [8]. For pulsing the laser beam, we use an optical chopper. The pulse width and the repetition rate are 20 μs and 1 kHz, respectively. The pulsed laser beam is injected from behind the magnet to the target sample.

MICROWAVE SYSTEM

Since the frequency of the electron spin resonance (ESR) in pentacene is about 9.1 GHz under a magnetic field of 3 kGauss [8], the target sample is placed in a cylindrical microwave cavity. It is necessary that the cavity enables the laser irradiation for the optical pumping and the magnetic field sweep for the ISE, without impairing the quality factor. Moreover, the target sample has to be kept at a temperature ranging from 77 K to room temperature. We introduced a set of internal field modulation coil into TE$_{011}$ cylindrical cavity; it is placed parallel to the cylinder axis in the cavity, as shown in Fig. 2 (b). The coil is configured in a two-loop arrangement and is made from four Cu rods with a 2-mm diameter. The quality factor of the cavity is not impaired by the coil as long as the rods are kept parallel to the cylinder axis, and is about 4000 at room temperature. The coil configuration also permits the laser irradiation of the target sample. A quartz tube, on which the target sample is mounted, is placed into a double quartz tube placed among the coil rods along the cylinder axis. Cooled N$_2$ gas is introduced into the inner quartz tube when protons are polarized below room temperature.

The schematic diagram of the microwave system is shown in Fig. 2(a). A microwave from a Gunn oscillator is pulsed and amplified by a microwave amplifier. Its maximum output power is 20 W and the pulse duration is 10 μs. Pulsed microwaves are split with a magic tee, one arm of which is connected to the cylindrical cavity. The other arm (balance arm) is equipped with a phase shifter and a matched load. The impedance of the cavity is matched with that of waveguides by changing the diameter of the iris.

During the pulsed microwave irradiation, the magnetic field is swept through the entire width of an ESR spectrum, whose width is about 30 Gauss, in order to perform the ISE. The field sweep is done by applying a triangular wave of about 50 A$_{p-p}$ to the field modulation coil. The current is produced by a voltage-to-current conversion circuit using a high-power operational amplifier.

NMR SYSTEM

A pulsed NMR system is used for measuring the proton polarization. The schematic diagram of the NMR system is shown in Fig. 2(a). The RF signal produced by a function generator is pulsed and amplified by an RF amplifier, whose maximum output is 500 W. The frequency and the width of the RF pulse are 12.75 MHz and 2 μs, respectively. The power of the RF pulse is adjusted to a level corresponding to the 90° pulse. The pulse is passed through a tuner for impedance matching and applied to the RF coil. Then, the NMR signal due to the free induction decay (FID) of protons is picked up by the same RF coil. The picked-up signal is amplified by a pre-amplifier and its proton resonance frequency component is obtained using a lock-in amplifier. Figure 3 shows the FID signal of a single crystal of *p*-terphenyl doped with pentacene in the thermal equilibrium.

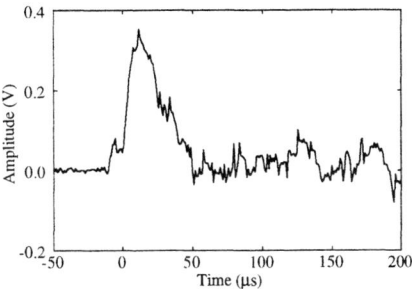

FIGURE 3. A FID signal of a single crystal of p-terphenyl doped with pentacene in the thermal equilibrium. We waited for 50 minutes before the measurement because the relaxation time of protons in p-terphenyl was about 10 minutes.

SUMMARY

We have designed and constructed a proton polarizing system in order to develop a polarized proton target for RI beam experiments. A single crystal of aromatic molecules are used as a target material. A combination of MIONP and ISE is used for polarizing protons. The matching and optimization of the proton polarizing system have been finished. We have just started to polarize protons.

ACKNOWLEDGEMENTS

The authors would like to thank Dr. I. Tanihata at RIKEN for his constant support of this study. We also wish to thank Dr. M. Iinuma and Dr. K. Takeda for valuable advice. One of us (T.W.) would like to acknowledge the Special Post-dctoral Reserchers Program.

REFERENCES

1. van Kesteren H. W., Wenckebach W. Th., and Schmidt J., *Phys. Rev. Lett.* **55**, 1642 (1985).
2. Hesselink W. H. and Wiersma D. A., *Phys. Rev. Lett.* **43**, 1991 (1979).
3. de Groot M. S., Hesselmann I. A. M., Schmidt J., and van der Waals J. H., *Mol. Phys.* **15**, 17 (1968).
4. Veeman W. S. and van der Waals J. H., *Mol. Phys.* **18**, 63 (1970).
5. Henstra A., Dirksen P., and Wenckebach W. Th., *Phys. Lett.* **A 134**, 134 (1988).
6. Iinuma M., Takahashi Y., Shake I., Oda M., Masaike A., and Yabuzaki T., *Phys. Rev. Lett.* **84**, 171 (2000).
7. Kim S. S. and Weissman S. I., *Rev. Chem. Intermed.* **3**, 107 (1979).
8. Van Strien A. J. and Schmidt J., *Chem. Phys. Lett.* **70**, 513 (1980).

Polarized nuclei in plastic scintillators: a new class of polarized targets

B. van den Brandt[‡,1], E.I. Bunyatova[*], P. Hautle[‡], J.A. Konter[‡]
S. Mango[‡] and I.B. Nemchonok[*]

[‡] *Paul Scherrer Institute, CH-5232 Villigen PSI, Switzerland*
[*] *Joint Institute for Nuclear Research, Dubna, Head P.O. Box 79, 101000 Moscow, Russia*

Abstract. Polarized scintillating targets are now routinely available: protons, deuterons or other nuclei in blocks of scintillating organic polymer, doped with the free radical TEMPO, are polarized dynamically in a field of 2.5 T in a vertical ^{3}He-^{4}He dilution refrigerator. A 19 mm diameter plastic lightguide transports the scintillation light from the sample in the mixing chamber to a photomultiplier outside the cryostat. Sizeable nuclear polarizations have been achieved newly in boron enriched polystyrene-based scintillating material. A scintillator target with high detection sensitivity for low energy neutrons has been so made available, in which both protons and boron nuclei are polarized.

INTRODUCTION

New possibilities for the measurement of spin-dependent observables in nuclear and particle physics are offered by the development of polarizable plastic scintillators. A polarized scintillating target is an instrument in which the hydrogen nuclei (or other nuclei of interest) in a piece of scintillator can be dynamically polarized at very low temperatures and the light produced in the scintillator by scattered particles can be forwarded to a photomultiplier at room temperature [1–5]. One such instrument has allowed e.g. to measure the neutron-proton spin correlation parameter at forward angles at 68 MeV [6] and the analyzing powers in $\pi\vec{p}$-scattering at 45-87 MeV [7], thanks to the coincident detection of the low energy recoil proton in the target itself. The background scattering could so be significantly suppressed, what is essential when measuring at forward angles.
The light output and the polarization achievable in a block of scintillator are determined by the doping and production process. The investigation of different

---

[1)] e-mail: Ben.vandenBrandt@psi.ch

preparation methods has led to a recipe for the production of solid, transparent and bubble-free samples of ca. 4 cm^3. A full account of the production process has been given elsewhere [3]. Satisfying degrees of proton polarization could be achieved in polyvinyltoluene (PVT) and polystyrene (PS) -based scintillators, but the scintillation properties were much better in the second ones.

POLARIZED PROTONS AND DEUTERONS IN PLASTIC SCINTILLATORS.

We achieve now routinely 80% proton polarization at 2.5 T and below 0.3 K in blocks of 18×18×5 mm of PS-based scintillator [4]. A nominal concentration of 2×10^{19} paramagnetic centres/gram was found to give high degrees of polarization in a reasonable time (s. table 1). For a sample in which 80 % and more polarization could be obtained, 85 minutes were enough to reach + 60% and 135 minutes to reach + 70%.

Table 1: Polystyrene-based scintillator with different concentrations of TEMPO

no.	spin conc. [p.c./g]	proton pol. [%]	$T_{1,p}^*$ [sec]
1	0.67×10^{19}	+67.5 / -71.6	334 (T = 1.2 K)
2	1.00×10^{19}	+74.7 / -72.4	255 (T = 1.2 K)
3	1.45×10^{19}	+76.0 / -67.5	273 (T = 1.1 K)
4	2.00×10^{19}	+83 / -84	235 (T = 1.1 K)

* in a magnetic field of 2.5 T

We measured the proton spin-lattice relaxation time $T_{1,p}$ of a series of samples with different concentrations of TEMPO at about 100 mK in magnetic fields of 0.4 and 0.8 T, to determine wether these samples can be used for a frozen spin target. Samples with a TEMPO concentration of 7×10^{18} p.c./g had at 90 mK a proton relaxation time of 70 hours in 0.4 T, and 230 hours in 0.8 T, but needed one day and more to reach full polarization, leading to the conclusion that operation in frozen spin mode with these samples should be chosen only if absolutely necessary.

Table 2: Deuterated PS-based scintillator with different TEMPO concentrations.

no.	spin conc. [p.c./g]	proton pol. [%]	deuteron pol. [%]	$T_{1,d}^*$ [sec]	$T_{1,p}^*$ [sec]	$T_{1,d}^{**}$ [hour]	$T_{1,p}^{**}$ [hour]
1	1.0×10^{19}	+79 / -77	+18 / -19	1448	-	50	53
2	1.5×10^{19}	+77 / -77	+22 / -25	845	-	32	27
3	2.0×10^{19}	+70 / -	+21 / -22	566	168	26	20

* in a magnetic field of 2.5 T at 1.15 K
** in a magnetic field of 0.8 T at 95 mK

In 98.7% deuterated polystyrene-based scintillator, prepared with the non deuterated additives p-terphenyl (1.5 wt.%) and 1,4-di-(2-(5-phenyloxazolil))-benzene

(0.15 wt.%), sizeable deuteron polarizations could be obtained at 2.5 T in blocks again of 18×18×5 mm [5] (s. table 2, which contains also a few measured relaxation times).

The scintillation characteristics of the blocks have been determined with a ^{90}Sr-source and with protons of energies ranging between 5 and 12 MeV at a tandem accelerator. The light output of the latest probes reaches up to 30% of the untreated reference material. Protons of energy down to 1.5 MeV can be detected, with an energy resolution degraded to about 25 % of the one of the untreated material and an almost unchanged timing resolution.

The light extracting system has been investigated with a raytracing program and the optimum shape and dimensions of the different components deduced. Accordingly, an improved target holder has been constructed, and the diameter of the 1150 mm long lightguide has been increased to 19 mm.

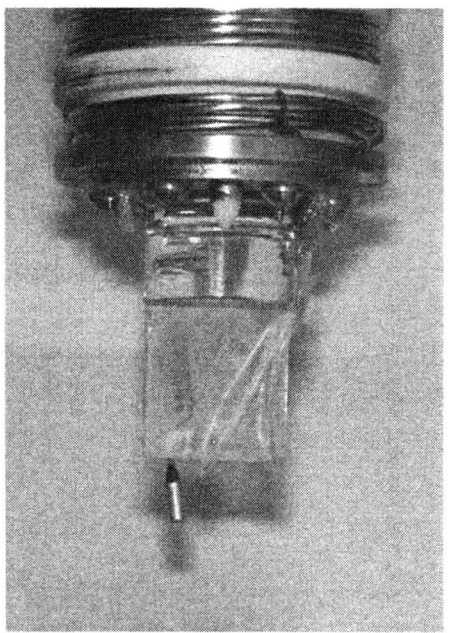

FIGURE 1. A polarizable scintillating block of 18×18×5 mm, mounted in the mixing chamber of the dilution refrigerator. A 1150 mm long, 19 mm diameter lightguide brings the scintillation light to a photomultiplier at room temperature at the top of the cryostat. Further details are given in the text.

The latest version of the instrument with a 18×18×5 mm scintillator mounted in the mixing chamber is shown in figure 1. The hydrogen-free light collecting quartz transition (diameter 19 mm) between the scintillator and the lightguide is visible above the scintillator around which the NMR coil (2 turns) is wound [8,3]. The microwave guide (with white PTFE cone) can be seen in the center of the figure and the end of the capillary for the ^{3}He rich mixture at the same height on the right. A calibrated rutheniumoxide thermometer is hanging from behind the scintillator. The microwave cavity is demounted; in place it enables the phase boundary in the liquid helium mixture to be situated at the level of the center of the scintillator. The lightguide that brings the scintillation light to a photomultiplier (Hamamatsu H6153-01) at the top of the cryostat is contained in a tight-fitting stainless steel tube inside the (white) ceramic bobbin, on which the continuous heatexchanger capillary is wound.

A reasonable polarization could be achieved (s. fig.2) and the base temperature attained was 65 mK. The system has been operated recently in a $\pi\vec{p}$- scattering [7] experiment over a period of two months, proving to be very robust and reliable. A $\gamma\vec{p}$ experiment requiring a larger volume target is planned for the near future [9].

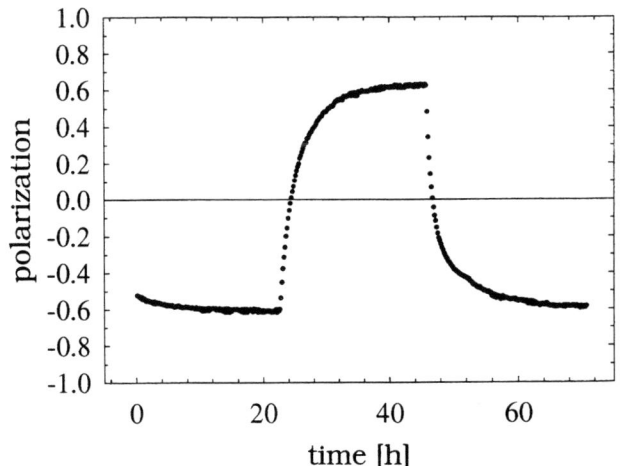

FIGURE 2. Time histogram of the proton polarization in two scintillating blocks of 18×18×5 mm, clamped together, mounted in the configuration shown in figure 1.

POLARIZED BORON IN PS-BASED SCINTILLATORS.

The addition of natural boron, containing 19.58% ^{10}B and 80.42% ^{11}B, to organic scintillators is known to improve the detection sensitivity for low energy neutrons by means of the reaction $^{10}_{5}$B + $^{1}_{0}$n → $^{7}_{3}$Li + $^{4}_{2}\alpha$ + 2.8 MeV, and to shorten the capture time. On the other side, appreciable proton polarizations can be achieved dissolving the free radical TEMPO in polystyrene-based scintillator. In order to obtain plastic scintillators with the above mentionned detection characteristics and in which the protons could be polarized, boron-loaded polystyrene-based scintillator has been used, containing the boron-organic compound o-carborane [10] (s. figure 3).

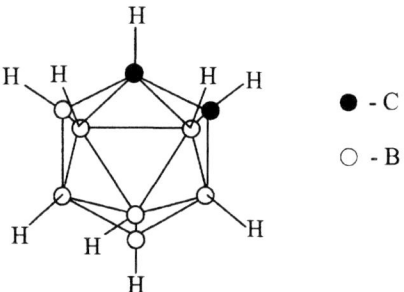

FIGURE 3. o-carborane.

The transmission and the photoluminescence in the visible range of a sample with 5% boron of natural isotopic composition are shown in figure 4 and 5; its light output has been measured to be 70 % of the unloaded scintillator.

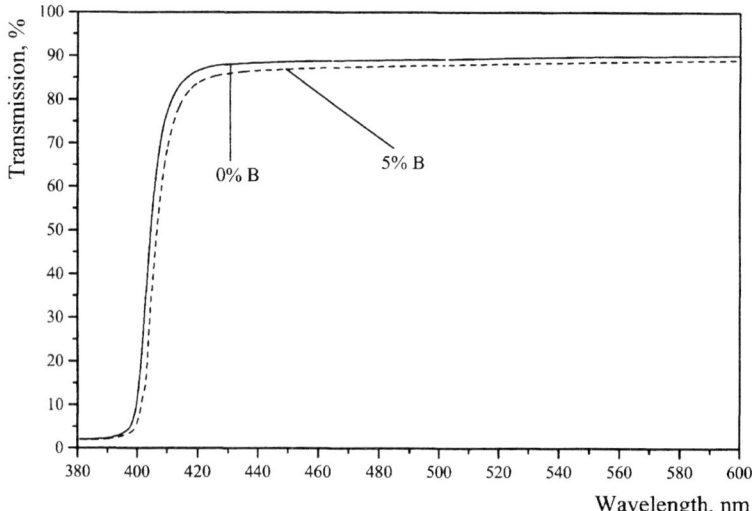

FIGURE 4. Transmission of a sample with 5% boron of natural isotopic composition.

The free radical TEMPO has been added and scintillating blocks have been produced respecting the procedure described earlier. These blocks have been polarized

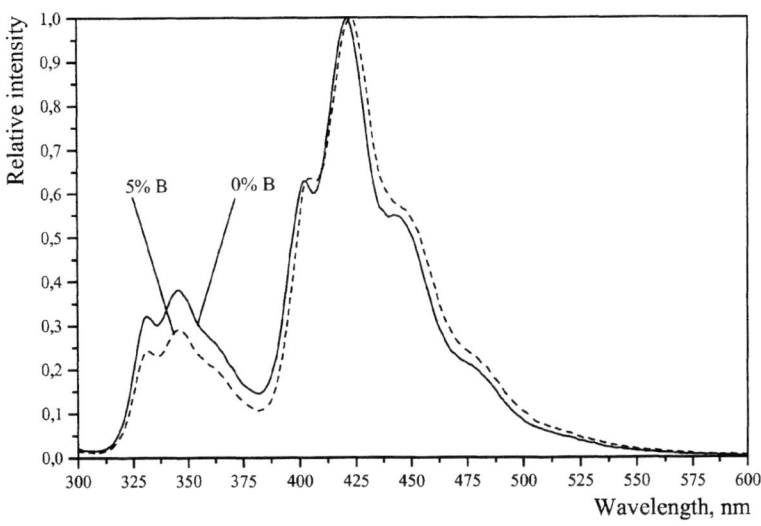

FIGURE 5. Photoluminescence of a sample with 5% boron of natural isotopic composition.

in 2.5 Tesla at dilution refrigerator temperatures. Proton polarizations above 80% could be achieved, while the ^{11}B polarization could be monitored but not measured. In fact the low nuclear concentration provided NMR T.E. signals too weak to be detected, while in the enhanced signals the satellite peaks were too weak to allow a reasonable evaluation of the polarization based on their intensity ratios. Assuming equal spin temperatures for the proton and the boron system, the ^{11}B polarization has to be close to 20%. However the polarized boron signals reveal some interesting features. We measure a smaller quadrupole splitting than Hill et al. in borhydrides [11], i.e. $\nu_Q^C = 560 kHz$ for the carbon and $\nu_Q^O = 400 kHz$ for the oxygen bond, compared to about 800 kHz. Furthermore, the ordering of the transitions is inverted, which means that the electric field gradient points antiparallel to the C-^{11}B and O-^{11}B bonds.

In any case it has been demonstrated that a high degree of proton polarization can be achieved in a boron enriched scintillator and that at least the ^{11}B nuclei, but almost certainly also the ^{10}B nuclei, can be polarized dynamically to an extent that we estimate appreciable. The way is open for the deployment of a polarized scintillating proton target with special detection features but a polarized boron background, or of a polarized boron scintillating target with a polarized proton background.

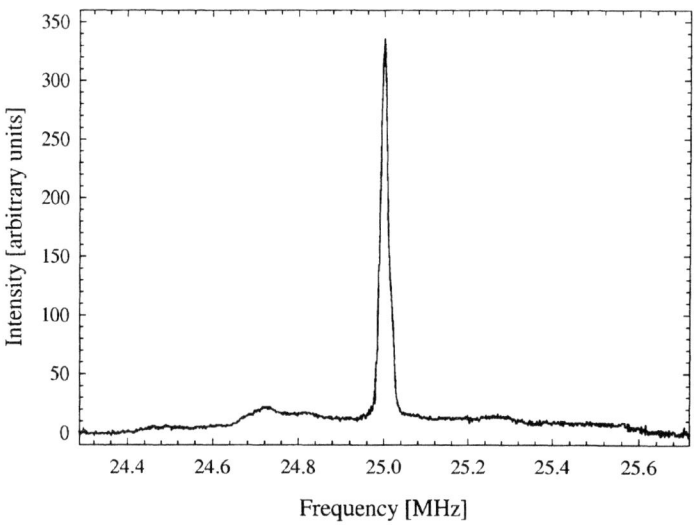

FIGURE 6. NMR signal of ^{11}B at 1.6 Tesla.

CONCLUSIONS

At the Paul Scherrer Institute scintillating polarized targets are now routinely available. Proton polarizations of more than 80 % and deuteron polarizations of 25% can be reached under optimum conditions in a ^{3}He-^{4}He dilution refrigerator with optical access, suited for nuclear and particle physics experiments. A high degree of proton polarization has been newly achieved also in boron enriched PS-based scintillator, together with an appreciable degree of boron polarization.

The financial support for one of us (E.I.B.) from the International Scientific Technical Center (contract N-608) is gratefully acknowledged.

REFERENCES

1. B. van den Brandt, E.I. Bunyatova, P. Hautle, J.A. Konter, S. Mango, Proc. of SPIN96, 12th Internat. Symp. on High-Energy Spin Physics, Sept. 10-14, 1996, Amsterdam, (World Scientific, Singapore, 1997), p. 238
2. B. van den Brandt, P. Hautle, J.A. Konter, S. Mango, E.I. Bunyatova, H. Denz, R. Meier, J. Jourdan, H. Wöhrle, Proc. of SPIN98, 13th Internat. Symp. on High-Energy Spin Physics, September 1998, Protvino, Russia, (World Scientific, Singapore, 1999), p. 442
3. B. van den Brandt, E.I. Bunyatova, P. Hautle, J.A. Konter, S. Mango, Nucl. Instr. and Meth. **A446** (2000) 592-599

4. B. van den Brandt, E.I. Bunyatova, P. Hautle, J.A. Konter and S. Mango, proc. PST'99, Sept. 29 - Oct. 2, 1999, Erlangen, p. 368-370
5. B. van den Brandt, E.I. Bunyatova, P. Hautle, J.A. Konter and S. Mango, proc. PST'99, Sept. 29 - Oct. 2, 1999, Erlangen, p.372-374
6. S. Buttazzoni *et al.*, PSI Annual Rep. **1** (1998) 23
7. R. Bilger *et al.*, PSI Annual Rep. **1** (1997) 22
8. B. van den Brandt, J.A. Konter, S. Mango, Nucl. Instr. and Meth. **A289** (1990) 526
9. J. Ahrens *et al.*, Polarized target asymmetries in threshold neutral pion photoproduction on the proton: Test of chiral dynamics, MAMI 1997
10. V.B. Brudanin, O.I. Kochetov, I.B. Nemchonok, A.A. Smolnikov, Preprint, JINR Dubna, No. R13-2000-195 (2000)
11. D. Hill, J. Hill, M. Krumpolc, 4th Internat. Workshop on Polarized Target Materials and Techniques, September 3-6, 1984, Bad Honnef, p.84

ABSTRACTS OF POSTERS

Helicity Dependence in Photodisintegration of the Deuteron

Bogdan Wojtsekhowski# and W.T.H. van Oers%,
for the DGNP collaboration

\# Thomas Jefferson National Accelerator Facility, Newport News, VA;
% TRIUMF/University of Manitoba, Winnipeg, Canada, R3T 2N2

Weak nucleon-nucleon forces were parameterized with weak meson-nucleon coupling constants about 20 years ago [1]. The interaction can be studied through parity violation in deuteron photodisintegration [2]. The threshold region allow to measure the combinations of weak couplings h_ρ and h_ω [3]. The progress in development high performance polarized electron source [4] allow to produce a record intensity of 70% circularly polarized beam of a few MeV photons ($\sim 10^{15}$ Hz). Almost 15 years after first attempt to measure PV in $D(\gamma,n)p$ [5] we had found that presently it is feasibile to study of the process in energy range from threshold up to 8 MeV by using CEBAF injector. Expected neutron yield is of $10^{11} - 10^{12}$ Hz. The systematics of measurement will be checked by using Compton scattering from atomic electrons.

1. B. Desplanques, J. Donoghue, and B. Holstein, Ann.Phys., **124** (1980) 449;
2. T. Oka, Phys. Rev. **D27** (1983) 523;
3. I.B. Khiplovich and R.V. Korkin, hep/nucl-th/0010032;
4. C. Sinclair, priv. communication; 5. E.D. Earle et al., Can.J.Phys., **66**(1988)534.

A new method of UCN production using a spatially alternating magnetic field with spin flips

K. Sakai, K. Asahi, H. Ogawa, A. Goto, K. Yogo, T. Suga, H. Miyoshi,
D. Kameda, M. Utsuro, K. Okumura, M. Hino, and A. Yoshimi

Neutrons with energies E below the Fermi effective potential $U_{eff} \sim 0.2$ μeV are called ultracold neutrons (UCN), which are totally reflected from a surface of material and thus can be stored in a bottle. The UCN in a bottle are of a great advantage for high-precision measurements in fundamental physics with neutrons, where the UCN density ρ_{UCN} takes the key role. We have proposed a new method for the production of UCN, which is based on the repeated application of a field-gradient force on a magnetic moment realized with a spatially alternating field and the correlated spin flips. To evaluate the feasibility and performance of the method, a computer simulation of neutron decelerating and transporting processes has been performed. The result indicates that the UCN density ρ_{UCN} in a storage bottle in the present method reaches quite large values as $\rho_{UCN} \approx 2.2 \times 10^3$ n/cm^3 in a realistic condition. An experiment to test the principle and to explore the technical ingredients of the present method is being conducted using the UCN beam from the supermirror turbine at KUR.

Experimental Sensitivity of Contact Interaction at RHIC

Jiro Murata

The Institute of Physical and Chemical Research, Wako, Saitama, Japan

One of the most vital tasks of an experiment using a polarized hadron collider in a new energy must be a study of physics beyond standard model(SM). In the past few years, several theoretical articles have been devoted showing that RHIC can reach a similar sensitivity of contact interaction for the TEVATRON, due to its polarized beam. The purpose of the present study is to explore the discovery potential for physics beyond SM, from the experimental perspective. A contact interaction(CI), which originates in interactions between quark- and lepton-sub constituents, is included in PYTHIA. The autor developed a plug-in program "POLBEYOND" for PYTHIA by which helicity-dependent matrix elements and spin asymmmetries for Drell-Yan and quark scattering process can be examined. The results on the expected single spin asymmetries at RHIC($\sqrt{s}$=500 GeV with 800 pb^-1) in $\mu^\pm$ pair production were obtained for various Λ, a model-independent scale parameter of the CI.

The Q^2–Dependence of the Generalised GDH Integral for the Proton

B. Seitz* for the HERMES Collaboration

* University of Alberta, Edmonton, AB, T6G 2J1, Canada

The Q^2–dependence of the generalised GDH integral for the proton has been measured in the range 1.2 GeV^2 < Q^2 < 12 GeV^2 by scattering longitudinally polarised positrons on a longitudinally polarised hydrogen gas target. HERMES has previously published a measurement of the deep inelastic portion of the generalized GDH integral. New results for the full integral were shown, including the resonance contribution for W < 2 GeV. The contributions of the nucleon resonance and deep inelastic regions have been evaluated separately. The latter has been found to dominate for Q^2 > 3 GeV^2. The total integral shows no significant deviation from a $1/Q^2$ behaviour in the measured Q^2 range indicating no effects due to either nucleon resonance excitations or non-leading twist. The combination of the data in the resonance and DIS regions provides the first measurement of the Q^2–dependence of the generalised GDH integral.

Measurement of the Gluon Polarization with the PHENIX Muon Arms

H. D. Sato* for the PHENIX Collaboration

*Department of Physics, Kyoto University, Japan

Polarized p-p collisions at RHIC with $50<\sqrt{s}<500$ GeV will provide us new information on the spin structure of the nucleon. Using the PHENIX detector, we plan to measure the double-longitudinal spin asymmetry (A_{LL}) for the heavy-flavor production which is sensitive to the gluon polarization ($\Delta G(x)$) in the proton.

We can measure A_{LL} for J/ψ's, identified with unlike-sign dimuon pairs, with a small experimental uncertainty ($\delta A_{LL} \sim 0.006$ with 320pb^{-1} luminosity and $\sqrt{s}$=200GeV). With the production mechanism of J/ψ understood, $\Delta G(x)$ can be extracted from A_{LL}.

Open heavy quarks can be identified with electron-muon pairs whose invariant mass is over 2 GeV/c^2 with a small background. Using both unlike-sign and like-sign pairs, we can obtain A_{LL} for both charm and bottom events separately, which constrain $\Delta G(x)$.

A new approach to parton-density evolution

Mehrdad Goshtasbpour[a,b] and Philip G. Ratcliffe[c,d]

[a] Department of Physics, Shahid Beheshti University, Tehran, Iran
[b] Center for Theoretical Physics and Mathematics, P.O.Box 11365-8486, Tehran, Iran
[c] Dip. di Scienze, Univ. degli Studi dell'Insubria, via Valleggio 11, 22100 Como, Italy
[d] Ist. Naz. di Fisica Nucleare—sezione di Milano, via Celoria 16, 20133 Milano, Italy

An approach to numerical evolution of the Altarelli-Parisi equations is outlined. The method consists of discretisation of the Bjorken x variable so that the resulting matrix equation may be solved exactly as a function of Q^2. The advantages are three-fold. The first is a major computational improvement, both increased speed and dramatically reduced storage space. The second gain is that data parametrisation may be achieved via data bins rather than a fixed functional form and thus (with the aid of the full correlation matrix), for example, appraisal of the low-x asymptotic behaviour is rendered more transparent and free of pollution from the higher x regions. Finally, the number of x bins used for parametrisation will certainly exceed the usual number of parameters and thus (in principle) allow a much improved fit.

Measurement of Transversal Handedness in 3π Diffractive Production

A.V.Efremov, Yu.I.Ivanshin, L.G.Tkatchev, R.Ya.Zulkarneev

Joint Institute for Nuclear Research, Dubna, Russia

I.Kachaev, A.Zaitsev

Institute for High Energy Physics, Serpukhov, Russia

Some years ago the concept of handedness (H) was introduced as a measur of polarization of parent partons or hadrons. It was defined as an asymmetry of a process probability W with respect to an axial vector projection and was shown to be proportional to polarization P. In this work the transversal handedness of diffractively produced pion triples $\pi^- + A \to (\pi^-\pi^+\pi^-) + A$ by π^- beam 40 GeV/c from a different nucleus A is studied. It is defined as

$$H_{T1} = \frac{W(\mathbf{Nn} > 0) - W(\mathbf{Nn} < 0)}{W(\mathbf{Nn} > 0) + W(\mathbf{Nn} < 0)} = \alpha P$$

where $\mathbf{N}$ is a normal to the production plane of the pion triple, $\mathbf{N} = (\mathbf{v}_{3\pi} \times \mathbf{v_b})$ ($\mathbf{v}_b$ and $\mathbf{v}_{3\pi}$ are velocities of the initial π^- beam and the 3 pion center of mass) and $\mathbf{n}$ is a normal to the "decay plane" of the triple in its center of mass. The value of the asymmetry, averaged over all nuclei is $H_{T1} = (5.96 \pm 0.21)\%$. The dependence of the handedness on atomic number, Feynmann variable, transversal momentum and on invariant mass of the neutral pion pair and of the pion triple were investigated. The former resembles the single spin asymmetry behaviour while the latter has rich resonance structure and show clear dynamic origin of the handedness. For more details see Phys. Atom. Nucl. **63** (2000) 445; nucl-th/9901005. (*Supported by RFBR Grant No. 98-02-16508.*)

Λ polarization in unpolarized hadron collisions

M. Anselmino[a], D. Boer[b], U. D'Alesio[c] and F. Murgia[c]

[a] *Dipartimento di Fisica Teorica, Università di Torino and INFN, Sezione di Torino, Via P. Giuria 1, I-10125 Torino, Italy*

[b] *RIKEN-BNL Research Center Brookhaven National Laboratory, Upton, NY 11973, USA*

[c] *Istituto Nazionale di Fisica Nucleare, Sezione di Cagliari and Dipartimento di Fisica, Università di Cagliari C.P. 170, I-09042 Monserrato (CA), Italy*

The fragmentation function of an unpolarized quark into a non collinear hadron may depend on the hadron spin and transverse momentum, $\boldsymbol{k}_\perp$. The data on the polarization of Λ, $\bar{\Lambda}$ baryons inclusively produced in unpolarized $p-N$ processes are used to determine a simple phenomenological expression for this new "polarizing fragmentation function", which describes remarkably well the experiments. More details can be found in the paper hep-ph//0008186 (Phys. Rev. **D**, in press), and in the talk by M. Anselmino in these Proceedings.

Nuclear Responses for Solar-ν, Supernova-ν and $\beta\beta - \nu$

H. Ejiri

RCNP, Osaka University, Ibaraki,Osaka, 567-0047, Japan
JASRI,SPring-8, Mikazuki-cho, Sayo-gun, Hyogo, 675-5918, Japan

Nuclear weak responses for neutrinos in nuclei, which are mainly axial vector spin isospin responses, are crucial for neutrino studies in nuclei. Nuclear spin isospin responses for solar-ν, supernova-ν, and $\beta\beta - \nu$ have been studied by using charge-exchange spin-flip reactions. ^{100}Mo is found to have large responses for these neutrinos. Then MOON(Mo Observatory Of Neutrinos)composed of supermodules of scintillators and ^{100}Mo isotopes is used for real-time studies of low energy solar-ν and supernova-ν and for spectroscopic studies of $\beta\beta - \nu$ with a ν-mass sensitivity of 0.03 eV.

1. H. Ejiri, Int. J. Mod. Phys. **E6** No1, 1 (1997).
2. H. Ejiri, Phys. Rep. 338, 265 (2000).
3. H. Ejiri et al, Phys. Rev. Lett. 85, 2917 (2000).

Search for T-violation in $K^+ \to \pi^0 \mu^+ \nu$ decay

Y. Asano representing KEK E246

Inst. of Applied Physics, Univ. of Tsukuba
Tsukuba, Ibaraki, 305-8573 Japan

Abstract. The experiment E246 looked for a T-violation by measuring the T-odd parameter in the decay of K^+, $\Im(\xi)$. A non-zero value of the parameter would indicate the T-violation. Its larger value, say 10^{-3}, would suggest some non-standard nature of the electro-weak interaction. The measurement is almost free from systematic errors, since a 12-fold symmetry of the spectrometer and a double ratio for the forward-going and backward-going π^0s cancel most of the systematic errors. By including the newly analyzed data taken in 1998, the central value of the parameter is almost zero within ± 0.01.

Study of Heavy Meson Production in NN Collisions with Polarized Beam and Target at COSY

F. Rathmann[1,2], M. Düren[1] P. Jansen[3], F. Klehr[3], S. Martin[2], H.O. Meyer[4], K. Rith[1] E. Steffens[1], H. Seyfarth[2], H. Ströher[2]

[1] *Physikalisches Institut II, Friedrich-Alexander Universität, 91058 Erlangen, Germany*
[2] *Institut für Kernphysik, Forschungszentrum Jülich, 52425 Jülich, Germany*
[3] *Zentralabteilung Technologie, Forschungszentrum Jülich, 52425 Jülich, Germany*
[4] *Department of Physics, Indiana University, Bloomington, Indiana, USA*

The study of near threshold meson production in pp and pd collisions involving polarized beams and polarized targets offers the rare opportunity to gain a better understanding of short range features of the nucleon-nucleon interaction. COSY presents itself as a unique environment to perform such studies above the pion threshold, once an appropriate detector system for the internal beam becomes available. The measurements of polarization observables of interest require a cylindrically symmetric large acceptance detector and because of the high momenta involved also magnetic separation. The magnetic spectrometer would consist of a superconducting toroid, providing fields around 3 T. A storage cell fed by a polarized atomic beam source will be used as a target.

The Observation of Double Strange Stable Dibaryons

Aslanyan P.Zh.[†*], Rikhvitskiy V.S.[*], Yemelyanenko V.N.[*]

[*] *Joint Institute for Nuclear Research, LHE Dubna, Moscow region, p.o. 141980*
[†] *Yerevan State University*

A few events, detected on the photographs of the JINR 2m propane bubble chamber exposed to a 10 GeV/c proton beam, were interpreted H dibaryons. A reliable identification of the above events needs a multivertex kinematic analysis which is in turn is feasible only using 4π-detectors and high measurement precisions of the sought objects. The 2 m propane bubble chamber is best suited for these purpose. These events are grouped into two following groups.

1. The first group is formed of three neutral, S=-2 stable dibaryons, the masses of which are below the $\Lambda\Lambda$ threshold.

2. The second group is formed of two neutral and three positively charged S=-2 heavy stable dibaryons. The masses of all the five dibaryons coincide within the limits of errors are over the $\Lambda\Lambda$, ΞN, $\Lambda\Sigma$ threshold.

TESLA-N: Polarized Electron-Nucleon Scattering at TESLA

Frank Ellinghaus[†] and Elke C. Aschenauer[†]
for the TESLA-N Study-Group

[†]*DESY Zeuthen, 15738 Zeuthen, Germany*

Measurements of polarized e-N scattering can be realized at the TESLA linear collider facility with projected luminosities that are about two orders of magnitude higher than those expected of other experiments at comparable energies. Longitudinally polarized electrons, accelerated as a small fraction of the total current in the e^+ arm of TESLA, can be directed onto a solid state target that may be longitudinally or transversely polarized. A large variety of polarized parton distribution and fragmentation functions can be determined with unprecedented accuracy, many of them for the first time. A main goal of the experiment is the precise measurement of the x- and Q^2-dependence of the unknown transversity distributions that will provide us with the full information on the nucleon's quark spin structure as relevant for high energy processes. The additional possibilities of using unpolarized targets and of experiments with a real photon beam turn TESLA-N into a versatile next-generation facility at the intersection of particle and nuclear physics.

A Search for Non-conventional Medium Effects in (p,p´) Reactions

E.J. Stephenson[a] and F. Sammarruca,[b]

[a]*Indiana University Cyclotron Facility, Bloomington, IN 47408 USA*
[b]*University of Idaho, Moscow, ID 88434 USA*

The inability of conventional many-body theories to fully explain (p,p′) polarization transfer measurements at 200 MeV, especially for the 6^-, T=1 state in ^{28}Si, has encouraged searching for new physics in the nuclear medium. Applications of the Brown-Rho scaling concept within a realistic nuclear matter framework did not yield any improvements. However, evidence has been reported [1] that vector meson spectral functions in the medium may be modified beyond a simple mass shift. We are in the process of confronting these ideas with (p,p′) data, while at the same time re-examining more standard approximations and the role of additional degrees of freedom in the baseline interaction.

1. B. Friman, M. Lutz, and G. Wolf, LANL preprint archive nucl-th/0003012.

Isoscalar, Isovector, Spin and Orbital Contributions in $M1$ Transitions

Y. Fujita[a], T. Adachi[a], G. P. A. Berg[b], B. A. Brown[c], H. Ejiri[b],
H. Fujita[a], H. Fujimura[b], K. Hara[b], K. Hatanaka[b], J. Kamiya[b],
K. Katori[a], T. Kawabata[d], S. Mizutori[a], P. von Neumann-Cosel[e], T.
Noro[b], A. Richter,[e] Y. Shimbara[a], T. Shinada[a], H. Ueno[a], A. Weiss[e],
M. Yoshifuku[a] and M. Yosoi[d]

[a] *Department of Physics, Osaka University, Toyonaka, Osaka 560-0043, Japan*
[b] *Research Center for Nuclear Physics, Osaka University, Ibaraki, Osaka 567-0047, Japan*
[c] *NSCL and Dep. of Physics and Astronomy, MSU, East Lansing, MI 48824, USA*
[d] *Department of Physics, Kyoto University, Sakyo, Kyoto 606-8224, Japan*
[e] *Institut für Kernphysik, TU Darmstadt, D-64289 Darmstadt, Germany*

Assuming isospin symmetry of structures in mirror nuclei, the four terms of the title in $M1$ transitions have been decomposed by comparing the strengths of analogous $M1$ transitions in $T_z = \pm 1/2$ nuclei and Gamow-Teller transitions, which are caused only by the isovector-spin term. For details, see Y. Fujita et al., Phys. Rev. C 59 (1999) 90 and Y. Fujita et al., Phys. Rev. C 62 (2000) 044314.

J^π Decomposition of a Bump at $E_x \simeq 7$ MeV in ^{12}N

S. Fukusaka[a], H. Sakai[a], K. Hatanaka[b], H. Okamura[c], T. Ohnishi[a],
S. Sakoda[a], K. Sekiguchi[a], A. Tamii[a], T. Wakasa[b] and K. Yako[a]

[a] *Department of Physics, University of Tokyo, Bunkyo, Tokyo 113-0033, Japan*
[b] *Research Center for Nuclear Physics (RCNP), Ibaraki, Osaka 567-0047, Japan*
[c] *Department of Physics, Saitama University, Urawa, Saitama 338-8570, Japan*

The ^{12}C$(\vec{d}, ^2\text{He})^{12}$B measurement by Okamura et al. has shown that the spin-dipole resonance bump at $E_x \simeq 7.5$ MeV in ^{12}B is most probably due mainly to the 2^- states. This fact disagrees with the widely accepted assignment as 1^-. Motivated by this work we studied the spin-parity of the mirror state located at $E_x \simeq 7$ MeV in ^{12}N by measuring a complete set of polarization transfer (PT) coefficients for the ^{12}C$(\vec{p}, \vec{n})^{12}$N reaction at $0°$. We employed a 300 MeV polarized proton beam and the NTOF facility at the Research Center for Nuclear Physics.

The PT coefficients enable us to disentangle the bump into spin-parity components as $1^-_{\Delta S=0} : 1^-_{\Delta S=1} : 2^-_{\Delta S=1} = 0.08\pm0.02 : 0.17\pm0.04 : 0.75\pm0.04$ in a plane-wave impulse approximation neglecting the contribution of the 0^- states. This result clearly indicates that the dominant component of the bump is due to the 2^- states and strongly supports the conclusion of Okamura et al.

Polarization Transfer Invariants at $0°$ in Nuclear Reactions

Toru Suzuki

*Department of Physics, Tokyo Metropolitan University,
Hachioji, Tokyo 192-0397, Japan*

Polarization transfer measurements at forward angles have been used to detect and extract strengths of spin-dependent excitations in nuclei. It was recently demonstrated experimentally[1] and proved theoretically[2] that a rotationally invariant quantity constructed from the polarization transfer coefficients for nucleon-nucleus scattering takes a unique value at $0°$ for unnatural parity transitions. For natural parity transitions the value of the invariant provides a measure of the spin-flip mixture in the transition.[2] The present contribution extends the above result to general inclusive nuclear reactions $A(a,b)B$ with arbitrary projectile/ejectile spins and target excitations. The result gives a relation between the set of invariant values and that of the target spin-flip transition strengths.

1) H.Sakai, Nucl.Phys.**A649**, 251c(1999); A.Tamii et al., Phys.Lett.**B459**, 61(1999).
2) T.Suzuki, Prog.Theor.Phys.**103**, 859(2000).

Systematic study of spin–isospin excitations in neutron rich light nuclei via the $(d, {}^2\text{He})$ reaction at 270 MeV

T. Ohnishi[1], H. Sakai[2], H. Okamura[3], T. Niizeki[4], K. Itoh[5],
T. Uesaka[3], Y. Satou[1], K. Sekiguchi[2], K. Yakou[2], S. Fukusaka[2],
N. Sakamoto[1], and H. Ohnuma[6]

[1] *The Institute of Physical and Chemical Research, Wako, Saitama, Japan*
[2] *Department of Physics, University of Tokyo, Tokyo, Japan*
[3] *Department of Physics, Saitama University, Urawa, Saitama, Japan*
[4] *Department of Home Economics, Tokyo Kasei University, Tokyo, Japan*
[5] *Tandem Accelerator Center, University of Tsukuba, Ibaraki, Japan*
[6] *Chiba Institute of Technology, Narashino, Chiba, Japan*

In order to study the nature of the neutron excess, the Gamow-Teller and spin-isospin dipole transitions have been measured by the $(d,{}^2\text{He})$ reaction for ${}^6\text{Li}$, ${}^9\text{Be}$, and ${}^{11}\text{B}$ targets at $E_d = 270$ MeV. The experimental $B(\text{GT})$ value is deduced from the measured cross section. In spite of $B(\text{SFD})$, the sum of the measured cross section at $\theta_{\text{c.m.}} = 4°$ is compared. In the ${}^{11}\text{Be}$ nucleus, the effect of the halo due to the large radial component of the halo wave function is observed.

Implications for the πNN coupling from spin transfer measurements in pp elastic scattering at 200 MeV

Scott W. Wissink, for the IUCF E367 Collaboration

Dept. of Physics and Indiana University Cyclotron Facility, Bloomington, Indiana, USA

Abstract. A detailed study of spin transfer in pp elastic scattering near 200 MeV has been carried out at the Indiana University Cyclotron Facility. The new data have much smaller uncertainties than all previous measurements, and span a kinematic range selected specifically to maximize sensitivity to the neutral πNN coupling constant g_0^2, a fundamental quantity in nuclear physics whose value remains highly controversial. Our results provide strong support for modern potential models of the NN interaction which use a relatively weak pion coupling ($g_0^2 \approx 13.6$), but disagree significantly with the predictions of models in which g_0^2 is ~ 14.4. Working in a one-boson-exchange framework, calculations suggest that most of these latter differences can be removed just by reducing the strength of g_0^2 from 14.4 to 13.6 for the long-range (higher partial wave) pion contributions in these models.

Isospin Identification for $A = 25$ Mirror Nuclei by High Resolution (p, p') and $(^3\text{He}, t)$ Experiments

Y. Shimbara, H. Fujita, Y. Fujita, T. Adachi, H. Fujimura*,
K. Harada, K. Katori, T. Shinada, H. Ueno, A. D. Bacher[†],
G. P. A. Berg*, C. C. Foster[†], K. Hara*, K. Hatanaka*, J. Jänecke[‡],
J. Kamiya*, T. Kawabata[§], S. Mizutori, T. Noro*, D. A. Roberts[‡],
E. J. Stephenson[†], M. Yoshifuku, and M. Yosoi[§]

Department of Physics, Osaka University, Toyonaka, Osaka 560-0043, Japan
[*] *Research Center for Nuclear Physics, Osaka University, Ibaraki, Osaka 567-0047, Japan*
[†] *Indiana University Cyclotron Facility, Bloomington, Indiana 47408, USA*
[‡] *Department of Physics, University of Michigan, Ann Arbor, Michigan 48109, USA*
[§] *Department of Physics, Kyoto University, Sakyo, Kyoto 606-8224, Japan*

Gamow-Teller and $M1$ states excited in $^{25}\text{Mg}(^3\text{He}, t)^{25}\text{Al}$ and $^{25}\text{Mg}(p, p')$ reactions at 0° and 450 MeV incident energy, respectively, have been measured and compared. Good symmetry structure in the mirror nuclei ^{25}Al and ^{25}Mg has been identified up to the highest measured excitation energy of $E_x \sim 16$ MeV.

Incident Energy Dependence of the Polarization Observables in Deuteron Elastic Scattering at $E_d = 50 \sim 700$ MeV

Yasunori Iseri* and Makoto Tanifuji[†]

* Department of Physics, Chiba-Keizai College, Inage, Chiba 263-0021, Japan
[†] Department of Physics, Hosei University, Chiyoda, Tokyo 102-8160, Japan

Abstract. In deuteron elastic scattering an energy dependence of the tensor analyzing powers is studied. The cross sections and some polarization oservables are calculated at several incident energies with the method of continuum-discretized coupled channels. A relationship between A_{yy} and A_{xx} has a form $A_{yy} \simeq -2A_{xx}$ at low incident energies ($E_d < 100$ MeV), but it seems to be $A_{yy} \simeq -A_{xx}$ at relatively high incident energies ($E_d \gtrsim 200$ MeV). The origin of the energy dependence has been discussed.

Tensor analyzing powers of ^{3}He(d,p)^{4}He reactions around 430keV resonance

M. Tanifuji* and H. Kameyama[†]

*Department of Physics, Hosei University, Fujimi, Chiyoda, Tokyo 102-8160, Japan
[†] Chiba-Keizai College, Todoroki, Inage, Chiba 263-0021, Japan

Abstract. In ^{3}He(d,p)^{4}He reactions, contributions of P waves and $1/2^+$ states to the 430keV resonance are investigated by the invariant amplitude method.

Measurement of H($\vec{d}$,³He)γ reaction using a large acceptance spectrograph

T. Yagita[1], K. Sagara[1], M. Kondo[1], S. Minami[1], T. Ishida[1],
K. Hatanaka[2], T. Wakasa[2], J. Kamiya[2], D. Hirooka[2],
T. Noro[2], H. P. Yoshida[2], E. Obayashi[2], K. Takahisa[2],
M. Yoshimura[2] and H. Akiyoshi[3]

[1] *Department of Physics, Kyushu University, Fukuoka, Japan*
[2] *Research Center for Nuclear Physics, Osaka, Japan*
[3] *RIKEN, Wakou, Saitama, Japan*

Abstract. A measurement of cross section and analyzing powers A_y, A_{yy} and A_{xx} of H($\vec{d}$,³He) reaction at $E_d = 200$ MeV is in progress at RCNP. The target is liquid hydrogen of about 1.5 mm in thickness. Since ³He particles from the reaction are concentrated in very forward angles in the laboratory frame, we use a large acceptance spectrograph (LAS) to measure the reaction simultaneously from $\theta_{cm} = 20°$ to $160°$ in the horizontal plane (A_y, A_{yy} and cross section) and in the vertical plane (A_{xx}). The preliminary data are compared with 3N Faddeev calculations.

Evidence for the Existence of Supersymmetry in Atomic Nuclei

A. Metz[1,2], J. Jolie[2], G. Graw[1], R. Hertenberger[1], J. Gröger[3],
C. Günther[3], N. Warr[2,4], Y. Eisermann[1]

[1] *Ludwig-Maximilians Universität, Am Coulombwall 1, D-85748 Garching, Germany*
[2] *Institut de Physique, University of Fribourg, Pérolles, CH-1700 Fribourg, Switzerland*
[3] *ISKP, University of Bonn, Nußallee 14-16, D-53115 Bonn, Germany*
[4] *Dept. of Physics and Astronomy, University of Kentucky, Lexington, KY 40506, USA*

Abstract. We found strong evidence for the existence of supersymmetry by studying the odd-odd nucleus ¹⁹⁶Au using the ¹⁹⁷Au($\vec{d}$,t), ¹⁹⁷Au(p,d) and ¹⁹⁸Hg($\vec{d}$,α) transfer reactions. High resolution ¹⁹⁶Pt(p,d)¹⁹⁵Pt and ¹⁹⁶Pt($\vec{d}$,t)¹⁹⁵Pt transfer experiments performed in parallel yielded at the same time an improved level scheme of ¹⁹⁵Pt. Using extended supersymmetry, a single fit of the six parameter eigenvalue expression with only 6 free parameters A, B, B', C, D and E yielded a complete description of all observed low-lying excited states in the four different nuclei forming the supermultiplet. The detailed comparison of the transfer amplitudes for the states up to 500 keV in the odd-odd member of the supermultiplet ¹⁹⁶Au using a semi-microscopic transfer operator, provides then evidence that this description is correct. For more detail see: A. Metz, et al., *Phys.Rev.Lett.* **83**, 1542 (1999) and *Phys.Rev.* **C 61**, 064313-1 (2000)

Spin Effects at Fragmentation of Polarized Deuterons into Pions

S. Afanasiev[a], V. Arkhipov[a], V. Bondarev[a], I. Daito[d], N. Doushita[c], S. Fukui[c], N. Horikawa[d], T. Iwata[c], A. Isupov[a], V. Kashirin[a], A. Khrenov[a], K. Kondo[c], V. Ladygin[a], A. Litvinenko[a], A. Malakhov[a], V. Penev[a,b], Yu .Pilipenko[a], S. Reznikov[a], P. Rukoyatkin[a], I. Rusanov[a,b], A. Wakai[d], L. Zolin[a]

[a] *Joint Institute for Nuclear Research, 141980 Dubna, Russia*
[b] *INPHE, 1784 Sofia, Bulgaria;* [c] *Department of Physics, and* [d] *Center for Integrated Research in Science and Engineering, Nagoya University, 464-8602, Japan*

Tensor analyzing power T_{20} in $\vec{d}A \to \pi(\theta = 0°)X$ has been measured at fragmentation of 9 GeV deuterons into pions with the momenta from 3.5 to 5.3 GeV/c at H-, Be-, and C-targets. This momentum range corresponds to region of cumulative pion production where the deuteron core structure up to $r \simeq 0.4$ fm is probed. The values of T_{20} are found to be small and positive in contradiction with the Impulse Approximation calculations assuming a direct mechanism of pion production on high momentum nucleon in the deuteron ($NN \to NN\pi$). In this report we present new data on tensor Ayy analyzing powers at non-zero angles ($\theta = 135, 180 mr$) with pion transverse momenta up to P_t=0.8 GeV/c. Ayy increases with rise of P_t in the cumulative region reaching the magnitude of -0.4 at P_t=0.7-0.8 GeV/c.

Quasi-periodicity of spin motion in storage rings — a new look at spin tune

D.P. Barber*, J. Ellison† and K. Heinemann*

Deutsches Elektronen–Synchrotron (DESY), 22603 Hamburg, Germany
†*University of New Mexico, Albuquerque, NM 87131, U.S.A.*

We show how spin motion on the periodic closed orbit of a storage ring can be analysed in terms of the Floquet theorem for equations of motion with periodic parameters. The spin tune on the closed orbit emerges as an extra frequency of the system which is contained in the Floquet exponent in analogy with the wave vector in Bloch wave functions for electrons in periodic atomic structures.

We then show how to analyse spin motion on quasi-periodic synchro-betatron orbits in terms of a generalisation of the Floquet theorem and find that if small devisors are controlled by applying a Diophantine condition, a spin tune can again be defined and that it again emerges as an extra frequency in a Floquet-like exponent.

We thereby obtain a deeper insight into the concept of "spin tune" and the conditions for its existence. The formalism suggests the use of Fourier analysis to "measure" spin tune during simulations of spin motion on synchro-betatron orbits.

A Vector and Tensor Polarimeter for Intermediate Energy Deuterons

Y. Satou[a], S. Ishida[a], H. Sakai[b], H. Okamura[b], H. Otsu[a],
N. Sakamoto[a], T. Uesaka[b], T. Wakasa[a], T. Ohnishi[b], T. Nonaka[b],
G. Yokoyama[b], K. Sekiguchi[b], K. Yako[b], S. Fukusaka[b],
T. Ichihara[a], T. Niizeki[c], K. S. Itoh[a] and N. Nishimori[d]

[a] *The Institute of Physical and Chemical Research (RIKEN), Wako, Saitama 351-0198, Japan*
[b] *Department of Physics, University of Tokyo, Bunkyo, Tokyo 113-0033, Japan*
[c] *Department of Physics, Tokyo Institute of Technology, Oh-okayama, Tokyo 152-0033, Japan*
[d] *Japan Atomic Energy Research Institute (JAERI), Ibaraki, 319-11, Japan*

Abstract. An intermediate energy Deuteron POLarimeter DPOL has been developed at RIKEN. It was designed to measure all components of the deuteron polarization by using the $\vec{d}+C$ elastic scattering and the charge exchange $^1H(\vec{d},2p)$ reaction in a CH_2 analyzer. The calibration in 1995 confirmed large vector analyzing powers for the former, and useful tensor analyzing powers for the latter reaction over 230 – 270 MeV.

RCNP (n,p) Facility

K. Yako[*], H. Sakai[*], A. Tamii[*], K. Sekiguchi[*], S. Sakoda[*],
Y. Maeda[*], M. Hatano[*], H. Kato[*], T. Saito[*], N. Uchigashima[*],
H. Okamura[†], K. Suda[†], M.B. Greenfield[‡],
T. Wakasa[∥], J. Kamiya[∥], D. Hirooka[∥], and K. Hatanaka[∥]

[*] *Department of Physics, University of Tokyo, Tokyo 113-0033*
[†] *Department of Physics, Saitama University, Saitama 338-8570*
[‡] *International Christian University, Tokyo 181-8585*
[∥] *Research Center for Nuclear Physics (RCNP), Osaka 567-0047*

An (n,p) facility has been constructed at the Research Center for Nuclear Physics. The polarized neutron beam, produced by the $^7Li(\vec{p},n)$ reaction, bombards the (n,p) targets placed in the multi-wire drift chamber (MWDC) box. The scattering position is detected by the MWDC and the momentum of the outgoing proton is analyzed by the Large Acceptance Spectrometer (see K. Yako et al. in RCNP annual report 1999). A commissioning run has been performed by using the np scattering. About 2×10^6/sec neutrons were produced in the target area of $20^H \times 30^W \text{(mm}^2\text{)}$. The typical n-beam polarization was 0.15. The energy resolution of the measurement was 1.4 MeV and the measured differential cross sections are consistent with the known values. By using this facility, the measurement for the $^{90}Zr(n,p)^{90}Y$ reaction at 300 MeV has been carried out as well as the nd elastic scattering measurement at 250 MeV.

ORIENTATION OF RADIOACTIVE NUCLEI

J. Dupák[1], M. Finger[2,3], M. Finger, Jr.[3], A. Janata[3], T.I. Kracíková[2,3],
N.A. Lebedev[3], M. Rotter[2], M. Slunečka[2,3], Yu.V. Yushkevich[3]

[1] *ISI AS Královopolská 147, CZ-612 64 Brno, Czech Republic*
[2] *Charles University, FMP, V Holešovičkách 2, CZ-180 00 Praha 8, Czech Republic*
[3] *Joint Institute for Nuclear Research, DLNP, RU-141980 Dubna, Russia*

Two complementary low temperature nuclear orientation facilities are being developed at JINR Dubna and at the Faculty of Mathematics and Physics of Charles University in Prague, respectively. These Facilities together with existing and planned JINR radioactive beam complexes YASNAPP and DRIBs will allow accomplish rich program in nuclear and solid state physics research. The method based on hyperfine interactions at low temperatures obtained with fast cooling top loading 3He-4He dilution refrigerators is used to orient radioactive nuclei. Using on-line coupling of the isotope production, mass separation and implantation of radioactive nuclei beams into a suitable host placed in the 3He-4He dilution refrigerator, studies of isotopes with life-times as low as 1 sec can be performed. With off-line coupling isotopes with life-times down to 1 hour can be oriented. The directional distribution and temperature dependence of the radiation from oriented nuclei as well as nuclear magnetic resonance on oriented radioactive nuclei reflected in the resonant destruction of observed nuclear radiation anisotropy in a radio-frequency field can be measured.

Suppressing Intrinsic Spin Harmonics in the AGS *

A. Lehrach[†], J.W. Glenn[†], T. Roser[†], V. Ranjbar[†,‡]

[†] *Brookhaven National Laboratory, Upton, NY 11973, USA*
[‡] *Indiana University, Bloomington, IN 47405, USA*

Over the last decade several improvements have been made to increase the polarization of the proton beam at the Brookhaven Alternating Gradient Synchrotron (AGS) [1]. A partial snake was installed to overcome all imperfection resonances in the energy range of the AGS [2]. The rf dipole concept to preserve polarization at strong intrinsic resonances has been demonstrated at the AGS for the first time [3]. Polarization losses have been observed at weak intrinsic resonances. One method to preserve polarization at weak intrinsic resonances is to change the betatron tune rapidly before crossing the resonance. Due to the non-adiabatic nature of this tune jump, the beam emittance will increase [4]. More efficient is a method called suppressing intrinsic spin harmonics, first successfully applied at the Cooler Synchrotron COSY [5]. In this presentation we describe how to use this method at the AGS [6].

Application of Internal Gas Target for Beam Polarization Measurement in the Electron Storage Ring

S.I. Mishnev[a], S.A. Nikitin[a], D.M. Nikolenko[a], I.Ya. Protopopov[a],
I.A. Rachek[a], Yu.M. Shatunov[a], A.N. Skrinsky[a], V.N. Stibunov[b],
G.M. Tumaikin[a], D.K. Toporkov[a], E.N. Zhilich[a]

[a]BINP, Novosibirsk, Russia; [b]INP, Tomsk, Russia

Abstract. First results on polarization measurement are presented. More sophisticated scheme of polarimeter is proposed. It is expected that the accuracy of the measurement of polarization of the circulated electron beam of about 20% will be achieved during 8 minuts of operation of this Möller polarimeter. The work was supported by Russian Foundation for Basic Researches, grants 98-02-17949 and 98-02-17993, INTAS, grant 96-0424 and State Scientific and Technical Programm (Nuclear Physics).

On feasibility of the experiments with a polarized deuteron beam and a polarized target at Charles University in relation with polarized fusion

Yu.A. Plis, on behalf of the collaboration:
Prague – Dubna – Kharkov – Moscow – Saclay

Joint Institute for Nuclear Research, RU-141980 Dubna, Russia

Abstract. There is an interest in the problem of polarized fusion with the neutron-free d^3He reaction. Up to now, the experimental data on the cross sections of two dd reactions, which produce neutrons at once or through secondary dt reaction, are absent for polarized deuterons. There is a relatively cheap way to carry out the experiments with polarized deuterons at the Charles University in Prague. A polarized deuteron beam with energy from 100 keV up to approximately 1 MeV may be produced on the Van de Graaff accelerator by the channeling of a deuteron beam through magnetized Ni single crystal foil, according M. Kaminsky [Phys. Rev. Lett. **23**, 819 (1969)]. This method permits to produce a polarized deuteron beam of an energy ≤ 1 MeV with a current of ~ 1 nA, vector polarization P_3 up to 2/3 and tensor polarization $P_{33} = 0$. It will be necessary to modify the existing polarized target at Charles University for work with a low energy deuteron beam [N.S. Borisov et al., Nucl. Instr. & Meth. **A 345**, 421 (1994)].

Synchrotron-sideband snake depolarizing resonances[e]

T. Kageya[a], V. Anferov[a], B. Blinov[a], C. Chu[b], Ya. Derbenev[a],
A. Krisch[a], S. Lee[b], W. Lorenzon[a], T. Rinckel[b], H. Sato[c],
P. Schwandt[b], D. Sivers[a], K. Sourkont[a], F. Sperisen[b],
B. von Przewoski[b], V. Wong[a], S. Youssof[a]

Abstract. We recently created a snake depolarizing resonance using an rf solenoid magnet in a ring containing a nearly 100 % Siberian snake. We found that the primary snake rf resonance also had two weaker synchrotron sidebands, which are second-order snake resonances; they were probably caused by the energy-dependent strength of the solenoid snake due to the Lorentz contraction of its longitudinal $\int B \cdot dl$. This was the first observation of an rf synchrotron-sideband depolarizing resonance in the presence of a nearly full Siberian snake.[d]

[a] Randall Lab of Physics, University of Michigan, Ann Arbor, MI 48109-1120, USA
[b] Indiana University Cyclotron Facility, Bloomington, IN 47408-0768, USA
[c] KEK, High Energy Accelerator Research Organization, Tsukuba, 305-0801, Japan
[d] B.B. Blinov et al., Phys. Rev. ST-AB **2**, 064001 (1999)
[e] Supported by research grants from the U.S. Department of Energy and U.S. NSF

Beam-line Polarimeter for Intermediate-Energy Deuteron

T. Uesaka,[a] H. Sakai,[b] H. Okamura,[a] A. Tamii,[b] Y. Satou,[c]
N. Sakamoto,[c] T. Ohnishi,[c] T. Wakasa,[d] K. Itoh,[e] K. Sekiguchi,[b]
K. Yako,[b] K. Suda,[a] S. Sakoda,[b]

[a] *Department of Physics, Saitama University, Saitama 338-8570, Japan*
[b] *Department of Physics, University of Tokyo, Hongo, Bunkyo, Tokyo 113-0033, Japan*
[c] *RIKEN (The Institute of Physical and Chemical Research), Saitama 351-0198, Japan*
[d] *Research Center for Nuclear Physics, Osaka University, Osaka 567-0047, Japan*
[e] *Tandem Center, Tsukuba University, Ibaraki 305-8577, Japan*

We have developed a beam-line polarimeter for intermediate energy deuterons at RIKEN Acclearator Research Facility. The $d + p$ elastic scattering is used as polarimetry. Recently, calibration measurement has been carried out at $E_d = 140$ and 200 MeV. The values of A_y (A_{yy}) are -0.519 ± 0.005 (0.541 ± 0.005) and -0.332 ± 0.005 (0.306 ± 0.006) at 140 MeV and 200 MeV, respectively.

Development of deuteron beam polarimeter at RCNP

T. Yagita[1], K. Sagara[1], M. Kondo[1], S. Minami[1], T. Ishida[1],
K. Hatanaka[2], T. Wakasa[2], J. Kamiya[2], D. Hirooka[2],
T. Noro[2], H. P. Yoshida[2], E. Obayashi[2], K. Takahisa[2],
M. Yoshimura[2] and H. Akiyoshi[3]

[1] *Department of Physics, Kyushu University, Fukuoka, Japan*
[2] *Research Center for Nuclear Physics, Osaka, Japan*
[3] *RIKEN, Wakou, Saitama, Japan*

Abstract. A polarimeter for a deuteron beam at 200 MeV has been developed at RCNP. The polarimeter uses the maximum values of A_{yy} and A_y of $d+p$ scattering. The polarimeter A_{yy} was calibrated using theoretically-known analyzing powers of $A_{yy} = -1/2(A_{zz} = 1)$ of $^{12}\text{C}(\vec{d},\alpha)^{10}\text{B}_{2+}$ reaction at 0°. The beam vector polarization p_y was estimated using the rf transition probabilities in the atomic ion source. The polarimeter A_y was evaluated from the measured asymmetry $p_y A_y$ and the estimated p_y. Uncertainties in the polarimeter A_y and A_{yy} are estimated to be about 2%.

Status of the HERMES Atomic Beam Source and Possible Improvements

A. Nass, N. Koch, M. Raithel and E. Steffens

Universität Erlangen - Nürnberg, E-Rommel Str. 1, 91052 Erlangen, Germany

To improve the flow rate of about $4.5 \cdot 10^{16} \vec{D}/\text{s}$ a new dissociator based on a microwave discharge at 2.45 GHz has been developed and installed into the HERMES Atomic Beam Source. The long term stability was found to be excellent. Compared with the Radio-Frequency dissociator (RFD) used so far the intensity was increased by 15% although the sextupole magnet system has been optimized for the RF-dissociator.

Furthermore the Carrier Jet method was tested, a new method of beam formation based on an expanded beam surrounded by an overexpanded carrier jet. Since there was a large deviation between the measurements and the results of the Navier-Stokes calculations Monte-Carlo simulations were performed. The results agree well with the measurements, showing that the method used in ref. does not provide a realistic description of beam formation in this flow range, in particular concerning the diffusive mixture of the two beams.

Polarization at the Nuclotron

V.Angelov, V.P.Ershov, V.V.Fimushkin, G.I.Gai, A.D.Kovalenko,
L.V.Kutuzova, A.I.Malakhov, V.A.Michailov, <u>Yu.K.Pilipenko</u>,
V.N.Penev, V.M.Slepnev, A.D.Stepanov, V.P.Vadeev, A.I.Valevich,
V.I.Volkov, L.S.Zolin, A.S.Belov[a]

Joint Institute for Nuclear Research, 141980 Dubna, Moscow reg. Russia
[a] *Institute for Nuclear Research RAN, Troitsk Moscow reg. Russia*

The cryogenic source of polarized deuterons ($\uparrow D^+$) POLARIS operates at the Dubna 10 GeV accelerator synhrophasotron. There is a project to make a polarized beam at the new superconducting accelerator nuclotron to continue a spin physics program. It is supposed to use the atomic beam stage of the source POLARIS, to modify an existing $\uparrow D^+$ charge exchange ionizer into $\uparrow D^-$ ionizer, to realize multiturn charge exchange injection in the nuclotron ring, to upgrade the polarimetry of low and high energy polarized beams. Description, parameters and some test results of the setups are presented.

Production of thick CD_2 targets for measurements of the $\vec{n}d$ scattering at 250MeV

Y. Maeda[a], H. Sakai[a], K. Hatanaka[c], H. Okamura[b], A. Tamii[a], T. Wakasa[c], K. Yako[a],
K. Sekiguchi[a], S. Sakoda[a], J. Kamiya[c], K. Suda[b], H. Kato[a], M. Hatano[a], D. Hirooka[c],
T. Saito[a], N. Uchigashima[a], M.B. Greenfield[d], J. Rapaport[e]

[a] *University of Tokyo,* [b] *Saitama University,* [c] *Research Center for Nuclear Physics (RCNP),*
[d] *International Christian University,* [e] *Ohio University*

In order to study the three-nucleon-force (3NF) effects in a Coulomb-free system, we have measured $\vec{n}d$ elastic scattering (see Y. Maeda *et al.*, in these proceedings). For the experiment, we produced self-supporting CD_2 sheets from the CD_2 powder obtained from C/D/N Isotopes Inc. We heated and pressed the CD_2 powder in a mold. The mold was evacuated and filled with nitrogen gas for several times during the process of heating in order to avoid the oxidation. And we kept heating at 145°C and pressing with 100 kgf/cm^2 for 60 hours. We succeeded in producing CD_2 targets with 0.3 - 4.0 mm thickness, 37×47 mm^2 area and ± 20 μm uniformity of the thickness.

The Bochum Polarized Target

G. Reicherz, S. Goertz, J. Harmsen, J. Heckmann, A. Meier, W. Meyer and E. Radtke

Experimentalphysik 1 AG, Ruhr-Universität Bochum, D-44780, Germany
This work is supported by the BMBF

The Bochum 'Polarized Target' group develops the target material ^6LiD for the COMPASS experiment at CERN. Several different materials like alcohols, alcanes and ammonia are under investigation. Solid State Targets are polarized in magnetic fields higher than $B = 2.5T$ and at temperatures below $T = 1K$. For the Dynamic Nuclear Polarization process, paramagnetic centers are induced chemically or by irradiation with ionizing beams. The radical density is a critical factor for optimization of polarization and relaxation times at adequate magnetic fields and temperatures. In a high sensitive EPR - apparatus, an evaporator and a dilution cryostat with a continuous wave NMR - system, the materials are investigated and optimized. To improve the polarization measurement, the Liverpool NMR - box is modified by exchanging the fixed capacitor for a varicap diode which not only makes the tuning very easy but also provides a continuously tuned circuit. The dependence of the signal area upon the circuit current is measured and it is shown that it follows a linear function.

Polarized Deuteron Target System for Low Energy $\vec{D}(\vec{d},p)T$ Measurement

I. Daito, H. Doushita[1], S. Hasegawa[1], N. Horikawa, S. Horikawa[1], T. Iwata[1], K. Kondo[1], Y. Miyachi, K. Mori[1], N. Takabayashi[1], T. Tojyo[1], A. Wakai

Center for Integrated Research in Science and Engineering, Nagoya University, Nagoya, Aichi, 464-8603, Japan
School of Science, Nagoya University, Nagoya, Aichi, 464-8603, Japan

Our group is constructing a polarized deuteron target system for a measurement of $\vec{D}(\vec{d},p)T$ reaction at low energies. If we assume that the incident energy of deuteron beam is 20 MeV, and its intensity is 1 nA, required specifications to the target system are the polarization is larger than 40 %, the target thickness is ~ 1 mg/cm^2, a cooling power of refrigerator is several mW at 200 mK. At present, the lowest temperature of 200 mK and the cooling power of 2.6 mW at 500 mK was achieved. On the other hand, the development of target material has been made. The polarization of 17 % was obtained to a electron irradiated - CD_2 foil with a doze of 1×10^{16} e$^-$/cm^2.

WORKSHOP ON POLARIZED ELECTRON SOURCE AND POLARIMETERS

Opening Address

Charles Y. Prescott

Stanford Linear Accelerator Center
Stanford CA 94309

Thank you for inviting me to open this Workshop on Polarized Electron Sources and Polarimeters, PES2000. I have been actively involved with polarized electrons for many years, and the experimental results using the polarized electrons at high energies have been important to the progress in High Energy Physics. I would like to give you a brief history of these historical developments.

Polarized electrons played a very important role in the development of the Standard Model. The story goes back to the 1960's when unification of the weak and electromagnetic forces was a theoretical issue. Steve Weinberg first discussed unification of these forces in 1966 (Model of Leptons). His discussion incorporated concepts of mixing (Salam) and origins of mass (Higgs). He incorporated the weak charged currents as purely V-A (left-handed couplings only), in agreement with the experiments. The coupling of the left-handed electron, e_L, to the Z, the carrier of the weak neutral force, is given by $g_L = T_{3L} - q \sin^2\Theta_w$, and for e_R is given by $g_R = T_{3R} - q \sin^2\Theta_w$, where $T_{3L,R}$ refers to the weak isospin assignment for e_L and e_R, q is the electron charge, and $\sin^2\Theta_w$ is the weak mixing parameter. Weinberg chose the simplest weak isospin assignment, where e_L and v_e were placed together in a doublet with $T_{3L} = \pm 1/2$ respectively, and e_R was in a singlet with $T_{3R} = 0$. This assignment naturally incorporated the purely left-handed weak charged current.

An alternative possibility, placing e_R in a doublet with a heavy neutrino (too heavy to be produced) was preferred by some. The first case, Weinberg's choice, predicted unequal couplings to the weak neutral currents, while the second predicted equality between g_L and g_R (parity conservation) in the couplings. Thus experimentally the search for parity non-conservation in the weak neutral current interaction became a test of Weinberg's ideas of weak-electromagnetic unification.

In the mid 1970's atomic parity violation experiments, specifically in the optical rotation of a laser beam passing through a bismuth vapor, came to an erroneous conclusion. Parity conservation was declared in the first measurements, but this was wrong. Only some years later was the atomic parity violation situation clarified experimentally. However, in 1978, an experiment at SLAC unequivocally demonstrated the unequal probability for inelastic scattering of polarized electrons, e_L versus e_R, from deuterium at high energies. Polarized electrons at high energies from an accelerator played a crucial role in establishing the validity of the Weinberg-Salam Model. The Weinberg-Salam Model later was incorporated in the Standard Model.

Today polarized electrons in $e^+ e^-$ annihilation at the Z pole have provided the most precise electroweak measurements (given the precise mass of the Z which establishes the electroweak parameters). Through virtual effects (loop and vertex corrections) the values of the Standard Model's heavy particles are indirectly determined. The SLD's A_{LR} measurement points to a light Higgs boson mass. Today, rumors of a light Higgs boson possibly seen at the LEP II detectors abound at CERN. This Higgs story has not yet reached its conclusion, but polarized electrons have played an important role in the lead up to the discovery of the Higgs boson.

Future linear colliders will use polarized electrons as an experimental tool. The physics goals incorporate polarized electrons as an important tool, and the results will be enhanced significantly by having the electrons polarized. Jefferson Lab has commissioned its polarized electron beams, and a rather impressive set of precision electronuclear measurements using the polarized electron beams will be presented next week in Osaka at the SPIN2000 Symposium. The impact of polarized electrons on nuclear physics and high energy physics has been important.

I have come to this Workshop to learn from your work. What you can teach us today will have a great influence on my field tomorrow. Keep up the good work. I am pleased to see you here in Nagoya. Welcome to PES2000.

Atomic and electronic engineering of p-GaAs-(Cs,O)-vacuum interface

V.E. Andreev[a], V.V. Bakin[b], A.N. Litvinov[b], A.A. Pakhnevich[b],
O.E. Tereshchenko[a], H.E. Scheibler[a], A.S. Jaroshevich[a], A.S. Terekhov[a,b]

[a]*Institute of Semiconductor Physics, Novosibirsk, 630090, Russia*
[b]*Novosibirsk State University, Novosibirsk, 630090, Russia*

Abstract. Preparation procedures generally used to prepare GaAs surface for subsequent activation to NEA state are analyzed. It is emphasizes that together with surface clearness other properties of surface are important, if extreme values of photocathode parameters are needed.

INTRODUCTION

Semiconductors with effective Negative Electron Affinity (NEA) play the key role in the development of laser-driven electron sources. The basic principles of NEA state formation are well understood, but the very important details of NEA technology and physics, which determine the ultimate values of parameters of NEA photocathodes, are still under the discussion. Among others, the influence of the initial atomic structure of the atomically clean semiconductor surface on the photocathode parameters is still unclear. For example, for silicon only atomically clean surface with (100) orientation and single-domain (2x1) reconstruction can be activated to NEA state [1]. Surfaces with other orientations can not be activated to NEA state. These data evidently demonstrate very strong influence of the atomic arrangements on the semiconductor surface, on the value of (Cs,O) dipole and, consequently, on the work function of the activated surface. In contrast to silicon, GaAs with any surface orientation and/or surface reconstruction can be activated to NEA state. Because of this finding little attention have been paid to the influence of the atomic structure of the GaAs surface on the resulting parameters of NEA emitters. There is a viewpoint that the clearness is the only parameter, which characterize the "quality" of the semiconductor surface before its activation to the NEA state. We consider this opinion is valid only in the case if photoemitters with extreme photoelectron escape probability, energy and angular spreads, polarization are not needed. If, on the other hand, extreme parameters of photoemitters are desirable, the momentum scattering and recombination of photoelectrons near the surface should be suppressed by proper preparation procedure. The goal of this paper is to formulate the list of requirements to the atomic structure of the NEA surface and to analyze the possibility to satisfy these requirements by means of the existing surface treatment procedures.

LIST OF REQUIREMENTS

To formulate the list of requirements to the surface treatment procedure one should start with the activation layer model. It has been shown [2] that for the GaAs(100) surface NEA state can be achieved for very broad range of thickness of the (Cs,O) activation layer. For thin (Cs,O) layers, when the total amount of Cs atoms lies within interval 1÷2 ML and the amount of O atoms is equal approximately 0.1÷0.3 ML, the dipole model of the activation layer seems to be adequate. For such activated surface the values of NEA are within 30÷200 meV energy interval. For thick activation layers when amounts of Cs and O atoms exceed 3 ML and 1 ML respectively, the activated GaAs(Cs,O) interface should be described as heterojunction [2]. The highest escape probability and the lowest photoelectron momentum scattering are achieved within the dipole layer region [2,3]

Therefore, requirement for the GaAs-surface, which is primary prepared for activation, should be formulated as follows: to provide the highest possible value of average surface dipole at the limited (≤1 ML) amount of Cs-atoms. This requirement means that for the minimization of the undesirable photoelectron scattering at the GaAs-(Cs,O) one should provide the highly ordered activation layer. It seems to be reasonable to assume, that the atomically clean and atomically flat semiconductor surface provides the best opportunity for the activation with the lowest probability of photoelectron scattering. Any kind of surface imperfections produces some disordering within the activation layer and increases the probability of photoelectron scattering during their escape. The primitive picture of the activated surface with different kinds of surface imperfections is shown in Fig.1. The surface impurities, which were not removed during the surface cleaning procedure, occupy some adsorption places and, therefore Cs-atoms can not be absorbed at these places. As a result, the highest possible value of average surface dipole decreases together with increase of photoelectron momentum scattering. It was experimentally shown [4] that for carbon contaminated GaAs surface the presence of 10^2 ML of foreign atoms leads to the measurable decrease of QE. Therefore, to eliminate the influence of the residual surface impurities on the photoelectron escape the amount of foreign atoms at the surface should not exceed one percent of monolayer.

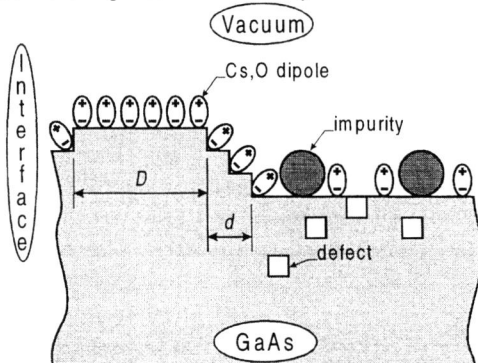

FIGURE 1. The model of (Cs,O)/GaAs interface.

Another important kind of surface imperfections is the surface roughness. There are several "elementary fragments", which are responsible for the surface morphology. Among them, the single- and multi-atomic steps seem to be very important. Near such steps (Cs,O) dipoles are directed not perpendicularly to the semiconductor surface. As a result the average dipole moment of the rough surface is lower than the dipole moment of the (Cs,O)-activated atomically flat surface. To eliminate this lowering of the dipole moment, the total area of atomically flat regions of the surface, which is proportional to $<D^2>$ should be much more than the total area of stepped regions, which is proportional to $<d^2>$. The meaning of D and d is shown in Fig.1. In addition to the misorientation of the surface dipoles, surface roughness leads to the elastic momentum scattering of emitted photoelectrons and subsequently broads their angular distribution. The photoelectron elastic scattering due to surface roughness can be eliminated if the value of $<D>$ is much more than the photoelectron momentum free path, which is limited by all others scattering processes. As far as we know the influence of surface roughness on the photoelectron escape from NEA-photocathodes is not experimentally proved up to now. There are only some speculations on this subject [5]. Nevertheless, some indirect arguments can be found in the literature. It is well known, that at early days of NEA-physics as-grown surfaces of GaAs epitaxial layers were activated to very high integral sensitivity, which often exceeded 2000 mkA/lm. The highest reported sensitivity [6] was equal to 3200 mkA/lm in reflection mode. Such values of sensitivity corresponds approximately to QE≈60% at λ=600 nm in reflection mode. If to take into account the reflection coefficient of GaAs, which exceeds 30 % within the actual spectral interval, one can conclude that every absorbed photon generates the emitted photoelectron! In this experiment [6] no chemical treatment was used prior to loading of GaAs layer grown by Liquid Phase Epitaxy (LPE) into the vacuum chamber. On the other hand, it is well known, that the as-grown surface of epitaxial GaAs layers, prepared by LPE, contains very large atomically flat areas. The fragment of such prepared surface, which was measured by atomic force microscope [7], is shown in Fig.2. One can see that single- and multi-atomic steps are separated by very broad atomically flat terraces with an average width of about 200÷300 nm. Therefore, one can conclude, that the very high QE, which was measured in [6], may be explained by the atomic flatness of the LPE-grown GaAs surfaces.

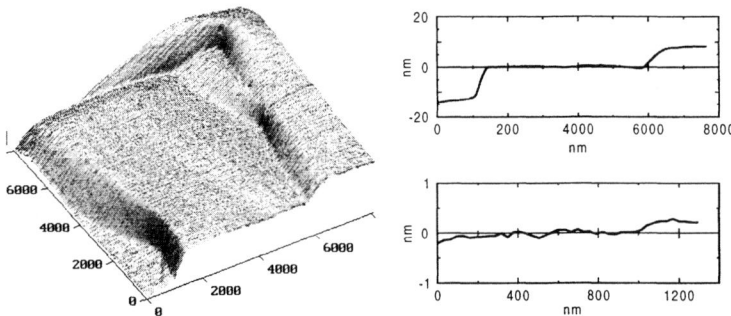

FIGURE 2. AFM image of the LPE-grown GaAs(100) surface and cross sectional view.

Another way of influence of surface disorder on the photoelectron escape from GaAs NEA-photocathode was found in our group. In our practice we use HCl-isopropanol surface chemical treatment to remove surface oxides [8,9]. This treatment enable ones to obtain atomically clean As-stabilized surface after low temperature (300÷400 °C) annealing of GaAs samples in UHV. Such prepared surfaces after (Cs,O) activation reproducibly demonstrate QE=10÷15% at λ=670 nm. If the same surface is annealed at high temperatures (510÷580 °C), the photocathode demonstrate usually QE=25÷28% at λ=670 nm in reflection mode. It is known, that high-temperature annealing converts As-stabilized GaAs(100) surface to Ga-stabilized one [8,9]. To understand the QE-difference between As-stabilized and Ga-stabilized surfaces, the evolution of atomic structures of both these surfaces along with Cs adsorption was investigated [10]. It was observed, that the adsorption of 0.1÷0.2 ML of Cs atoms on the As-stabilized GaAs(100) surface with (2×4)/c(2×8) reconstruction leads to complete disordering of the regular atomic structure of semiconductor surface [10]. On the other hand, after the adsorption of the same amount of Cs-atoms on the Ga-stabilized (4×2)/c(8×2) surface structure disordering was not detected by LEED. The disordering was observed after adsorption of 0.6÷0.7 ML of Cs atoms. It means, that atomic structure of Ga-stabilized GaAs(100) surface is more stable under the adsorption of cesium than the atomic structure of As-stabilized surface. This higher stability leads to the higher ordering of (Cs,O)-dipole layer. As a result, the activation of Ga-stabilized surface demonstrates highest QE values and lowest momentum scattering of emitted electrons.

In addition to the defects related to the surface roughness, the point defects of bulk atomic structure, such as As-vacancies (see Fig.1) play an important role in the photoelectron emission from NEA-photocathode. If the surface treatment procedure is not well designed, the concentration of such defects near the activated surface may be very high. These defects may influence the photoemission via several effects. (*i*) Such defects are effective centers of non-radiative electron recombination. As a result, some photoelectrons recombine via these defects. (*ii*) As-vacancies in GaAs are positively charged and compensate negatively charged acceptors [11]. This compensation shifts the potential profile within band bending region away from the optimum. This shift leads to the decrease of photoelectron escape probability. (*iii*) Randomly distributed charged vacancies near the surface increase the intensity of photoelectron elastic momentum scattering and provide additional broadening of photoelectron angular distribution. (*iiii*) If vacancies are located very near the activated surface, the (Cs,O) dipoles can be disturbed. Therefore, more detailed list of requirements for the GaAs(100) surface, prepared for (Cs,O) activation, which are needed to be satisfied if high QE and low momentum scattering of photoelectrons during their escape are desirable, is as follows. The GaAs(100) surface prepared for activation should be: (*i*) atomically clean; (*ii*) atomically flat; (*iii*) with low density of near-surface defects; (*iiii*) Ga-stabilized.

SURFACE TREATMENT PROCEDURES.

There are several surface treatment procedures, which were used to prepare GaAs surface for subsequent (Cs,O) activation.

Ion sputtering of surface contamination and subsequent thermal annealing may be used to prepare atomically clean, Ga-stabilized GaAs surface [12]. Nevertheless, this procedure is not used for preparation of NEA photocathodes, because of high density of near-surface defects, which are generated during ion bombardment. Unfortunately, thermal annealing can not decrease the density of these defects to the acceptable level.

Thermal desorption of surface oxides is very popular for the preparation of GaAs before (Cs,O) activation because of simplicity. Unfortunately, the simplicity is the only advantages of this method. Studies of GaAs surface oxide thermal desorption opened the following peculiarities of this process. (*i*) The temperature of GaAs oxide desorption depends on oxide composition. The arsenic oxides can be removed at 540°C while the gallium oxide (Ga_2O_3) is more stable and can be desorbed at the temperature above 580°C [13]. The surface decomposition and vacancy generation at so high temperatures is very probable. During the heat cleaning, some part of As_2O_3 is converted to Ga_2O_3 [13]. Therefore, high temperature heating can not be avoided. (*ii*) The thermal desorption of oxide does not occurred homogeneously: at the first moment the clean surface appears as a small holes in the oxide layer and the complete removal of oxide takes some time. During this time arsenic is desorbed from the clean regions of the surface while the oxidized regions are protected from desorption. Therefore, after the complete desorption of oxide layer the surface is very rough [14]. (*iii*) The thermal desorption of oxide layer does not lead to the removal of the carbon contaminations.

HCl-treatment and subsequent UHV annealing is very efficient, if the GaAs treatment and its subsequent transfer into UHV are preformed in oxygen-free atmosphere [8,9]. The only requirement, which is not fulfilled at the desirable quality, is the atomic flatness. After the HCl-based chemical treatment the surface contains large atomically flat terraces [15]. The only problem with this technique, which we have found up to now, is the following. To convert As-stabilized surface to Ga-stabilized the annealing at the temperatures above 520÷530 °C is needed. It was observed recently by means of ellipsometry [16], that at temperatures above 520 °C sharp decreasing of ellipsometric angle Δ takes place. The only way to explain this decreasing in our particular case is to assume that surface roughening or intense generation of As-vacancies starts at temperatures above 520 °C. Because both process are not desirable, the method to get Ga-stabilized GaAs(100) surface at temperature below 520 °C is needed.

Atomic hydrogen treatment of the GaAs surface was investigated in details. Several methods to produce atomic hydrogen were used: Electron Cyclotron Resonance (ECR) source [17] and decomposition of molecular hydrogen on the surface hot filament [18] or inside hot capillary [19]. Every kind of H^0 source provides the possibility to obtain atomically clean GaAs surface. Nevertheless, the atomic flatness of such prepared surfaces depends on the source used: while atomic hydrogen from hot capillary provides atomically flat surface [19], the usage of ECR source may lead to the surface roughening [17]. It has been shown, that the surface roughening during such treatment take place if the atomic hydrogen flow from ECR source is directed perpendicularly to the treated surface [20]. If the atomic hydrogen flow were directed at the glancing angle, the atomically flat surface could be obtained [19]. Because the GaAs surface treatment with atomic hydrogen is performed within 400 °– 500 °C temperature interval, the intensive thermal generation of As- or Ga- related point defects is not expected. The situation with surface stoichiometry of GaAs after the atomic hydrogen treatment is not absolutely clear. When ECR source was used, the Ga-stabilized (100) surface is obtained [17]. The treatment of GaAs by atomic hydrogen from hot capillary leaves the As-stabilized (100) surface. It is not clear at present can this surface be converted to Ga-stabilized by proper adjustment of hydrogen flow rate and GaAs temperature. Nevertheless atomic hydrogen treatment seems to be very promising tool for preparation of GaAs surface ideally suited for subsequent (Cs,O)-activation.

SUMMARY

The preparation of GaAs surface prior to (Cs,O)-activation can not be "reduced" to the surface cleaning The fit of several surface parameters to the desirable values is especially needed if photoemitter is prepared for the study of NEA-physics or if the photocathode with extreme set of parameters should be prepared for some particular application. The best instruments for the atomic and electronic engineering of the GaAs surface are the HCl-based treatment, H^0 treatment and UHV annealing at temperatures below ~500 °C.

ACKNOWLEDGMENTS

This work is partially supported by Ministry of Science of Russian Federation under the federal program "Surface Atomic Structures" (grant No. 107-25(00)-P), and by Ministry of Education of Russian Federation under the Federal program "Integration" through the Novosibirsk State University (project No. 274).

REFERENCES

1. P.M. Gundry, R. Holtom and V. Leverett, *Surf. Sci.* **43**, (1974) 647.
2. S. Pastuszka, D. Kratzmann, A. Wolf, D.A. Orlov, A.G. Paulish, H.E. Scheibler, A.S. Terekhov, O.E. Tereshchenko, "Elucidation of activation layer model by means of measurements of photoelectron energy distribution curves," in *Proceedings of the Seventh International Workshop on Polarized Gas Targets and Polarized Beams*, Ed. Roy J. Holt, M.A. Miller, American Institute of Physics, Urbana, (1997) p.493.
3. D.A. Orlov, M. Hoppe, D. Schwalm, A. S. Terekhov, U. Weigel and A. Wolf, "Investigation of transverse and longitudinal spreads on NEA GaAs-photocathodes with various GaAs(Cs,O)-interface structures," in *Proceedings of the International Workshop on Polarized Sources and Targets*, Ed. A. Gute, S. Lorenz, E. Steffens, Erlangen, (1999) p.280.
4. E.L. Nolle, S.N. Maximovskiy, S.A. Botnev, N.N. Loyko, A.E. Petrov, *Sol. St. Phys.* (in Russian) **23**, (1981) 2752.
5. B. Goldstein, *Surf. Sci.* **47**, (1975) 143.

6. L.I. Antonov, Yu.F. Biryulin, A.Ya. Vul, V.P. Denisov, L.G. Zabelina, P.P. Ichkitidze, A.I. Klimin, S.E. Kozlov, Yu.V. Shmartsev, *Pisma v ZhTF* (in Russian), **11**, (1985) 602.
7. A.V. Latyshev, V.L. Alperovich, N.S. Rudaya, private communication.
8. O.E. Tereshchenko, S.I. Chikichev, A.S. Terekhov, *J. Vac. Sci. Technol.* **A17**, (1999) 2655.
9. O.E. Tereshchenko, S.I. Chikichev, A.S. Terekhov, *Appl. Surf. Sci.* **142**, (1999) 75
10. O.E. Tereshchenko, H.E. Scheibler, S.N. Kosolobov, A.N. Litvinov, N.V. Kislyh, A.S. Terekhov, "Study of preparation procedure of GaAs(100) surfaces optimized for NEA-photocathodes," in *Proceedings of the International Workshop on Polarized Sources and Targets*, Ed. A. Gute, S. Lorenz, E. Steffens, Erlangen, 1999, p.284.
11. Ph. Ebert, *Surf. Sci. Rep.* **33**, (1999) 121.
12. M. Kamaratos, E. Bauer, *J. Appl. Phys.* **70**, (1991) 7564.
13. W. Mönch, Semiconductor surface and Interfaces, Springer, Berlin, 1995.
14. S. Goto, M. Yamada, Y. Nomura, Jpn. *J. Appl. Phys.* **34**, (1995) L1180.
15. Z. Song, S. Shogen, M. Kawasaki, and I. Suemune, *Appl. Surf. Sci.* **82/83**, (1994) 250.
16. S.I. Chikichev, private communication.
17. J.W. Elzey, P.F.A. Meharg and E.A. Ogryzlo, *J. Appl. Phys.* **77**, (1995) 2155.
18. Y. Ide and M. Yamada, *J. Vac. Sci. Technol.* **A12**, (1994) 1858.
19. H. Nagano, Zh. Qin, A. Jia, Y. Kato, M. Kobayashi, A. Yoshikawa, K. Takahashi, *J. of Cryst. Growth* **189/190** (1998) 265.
20. I. Suemune, A. Kishimoto, K Hamaoka, Y. Honda, Y. Kan, and M. Yamanishi, *Appl. Phys. Lett.* **56**, (1990) 2393.

Photoemission and STM, STS Study of Cs/*p*-GaAs(110)

T. Yamada, J. Fujii, and T. Mizoguchi

Faculty of Science, Gakushuin University, 1-5-1 Mejiro, Toshima-ku, Tokyo 171-8588 Japan

Abstract. Various stage of adsorbed Cs, i.e., one-dimensional (1D) lines, polygons and coherently c(4x4)-ordered polygons, on cleaved *p-GaAs*(110) are studied by scanning tunneling microscopy (STM) and scanning tunneling spectroscopies (STS) with an interest in relation to the photo-electron emission. It is understood that only the coherently c(4x4)-ordered Cs polygon surface (Cs coverage of 0.6ML and 0.7 ML) can emit photo-electrons due to a sufficient reduction of the local work function down to 1.3 eV to get a negative electron affinity state.

INTRODUCTION

p-GaAs is now practically used as a spin-polarized photoelectron source [1]. The crucial point of this source is surface treatments to get the negative electron affinity state in order to emit spin-polarized conduction electrons to the vacuum. Although the CsO_x/*p*-GaAs(001) has been used for a practical polarized photoelectron source, no atomic scale information about this surface has been reported. We tried to observe this surface by STM after various treatments according to the earlier described recipe for a practical photoelectron source, but we did not succeed in getting atomic scale images for the (001) surface [2].

On the other hand, it was found that the cleaved (110) surface of p-GaAs with only adsorbed Cs without oxygen is able to emit photoelectrons [3]. In this paper, experiments of Cs adsorption on *p*-GaAs(110) surface are reported. Although no study was done in relation to the photoemission, L.J.Whitman *et al.* [4] reported the STM study of Cs adsorption on GaAs(110). The purpose of this study is to find the relation between photo-electron emission and various surface configurations of Cs/*p*-GaAs(110). We studied to get information about not only topographic images with STM but also the band edges and especially the local work function with STS.

EXPERIMENTAL

A *p*-GaAs(001) wafer doped with Zn at a concentration of $\sim 10^{19}$ cm^{-3} was cleaved in an UHV chamber to get a clean *p*-GaAs (110) surface. Cs atoms were adsorbed on the (110) surface from a SAES Getters chromate dispenser after a sufficient outgas procedure. The deposition rate was about 0.08~0.2 ML/min in UHV (5×10^{-10} mbar during deposition). Here 1 ML (monolayer) is defined as one Cs atom per As atom on the (110) surface.

In order to observe a photoelectron emission, the 3.0x0.28 mm^2 area of the Cs/p-GaAs(110) surface which was biased at −9.2V was illuminated with linearly polarized GaAlAs laser light (λ =780 nm, 2 mW/mm^2). The photocurrent was measured with a pico-amperemeter.

All STM images were obtained in UHV (less than 6x10^{-11} mbar) at room temperature in constant current mode with a typical set point of 0.130 nA and −1.580 V. *I-V* spectra were taken by cutting off a feed back loop at a selected position of the STM image while the sample bias voltage is varied between −2 V to +2 V. The local work function was obtained from the tunneling current with varying separation, S, between a tip and a selected point of the sample surface (*I-S* spectra)[5],[6]~[8].

RESULTS AND DISCUSSION

STM topographic images of various stages of Cs/*p*-GaAs (110) are shown in Fig.1. The crystallographic direction [1-10] is oriented 45 degrees counterclockwise from the horizontal.
   (a) cleaved *p*-GaAs(110) surface (3 nm x 3 nm).
   (b) 0.1 ML Cs, 1D zigzag line (11 nm x 11 nm).
   (c) 0.23 ML Cs, 1D triple line (4 nm x 4 nm).
   (d) 0.35 ML Cs, non-coherently arranged polygons. (28 nm x 28 nm).
      It was proposed that each polygon consists of 5 Cs atoms [16]. No photocurrent was detected for (a) to (d).
   (e) 0.5 ML Cs, partially ordered Cs polygons (59 nm x 59 nm).
      The photocurrent of 0.18 nA/mm^2 was detected.
   (f) 0.6 ML Cs, c(4x4) coherently ordered Cs polygons (22 nm x 22 nm).
      The photocurrent of 240 nA/mm^2 was detected.
   (g) 0.7 ML Cs, layer by layer growth of c(4x4) ordered polygons (65 nm x 65 nm).
      The photocurrent of 360 nA/mm^2 was detected. If we assume the photocurrent emits uniformly from the illuminated surface, the quantum efficiency is calculated to be 3x 10^{-4} at this stage.
   (h) The thick rough surface of Cs grains of ~5 nm in size obtained with higher Cs deposition rate (40 nm x 40 nm). No photocurrent was detected from this stage.

In Fig.2 the results of *I-V* and *I-S* experiments are summarized in an energy band scheme. At the cleaved *p*-GaAs(110) ((a) in Fig1), the band gap is 1.4eV and there is a very slight band bending of 0.05 eV. Adsorbed Cs atoms on *p*-GaAs(110) form different configurations depending on the coverage as shown Figs.1 (b)~(g). All of them show reduced band gap of 0.9~1.2 eV, and the bottom of the conduction band are 0.45~0.6 eV above the Fermi level. The rough surface of Cs grains (Fig.1(h)) shows a metallic character without band gap.

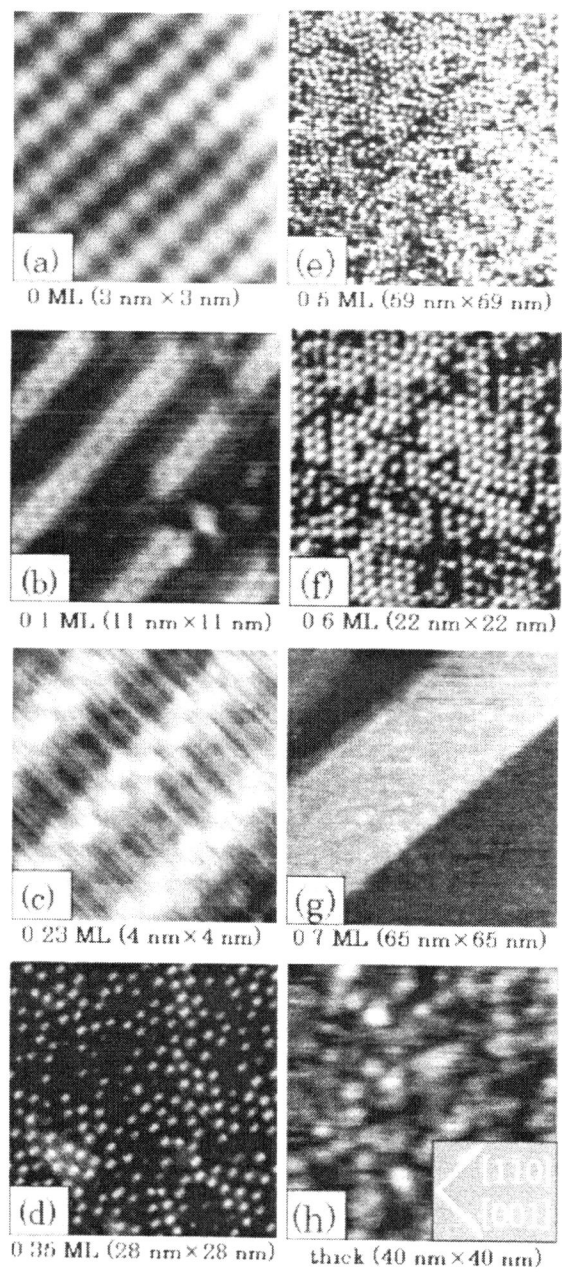

FIGURE 1. STM images from Cs/*p*-GaAs (110) surface.

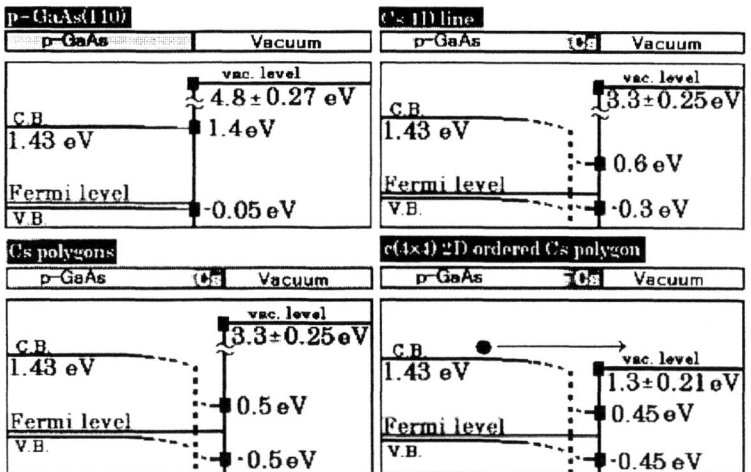

FIGURE 2. The energy band diagrams for the photoemission-process.

The local work function of the clean p-GaAs(110) surface is measured to be 4.8±0.27 eV, which well exceeds the energy of photo-excited conduction electrons (1.43 eV). The local work function at the position of Cs lines or Cs polygon in Figs.1 (b)~(e) become the same value of 3.3±0.25 eV. Only the local work function of coherently c(4x4) -ordered Cs-polygon surface (Figs.1(f) and (g)) is found to be remarkably reduced down to 1.3±0.21 eV which is less than the band gap of the bulk p-GaAs, i.e., 1.43 eV. It is understood that only the coherently c(4x4)-ordered Cs-polygon surface can emit photoelectrons due to the sufficient reduction of the local work function to get the negative electron affinity state.

There may be the possibility that a charge transfer and then the reduction of the local work function are enhanced by the c(4x4)-superlattice, which introduces a new gap at the zone boundary of the Brillouin zone so as to reduce the total energy of the electronic system.

REFERENCES

1. D. T. Pierce et al., *Appl. Phys. Lett.* **26**, 670 (1975).
2. D. E. Burgler, C. M. Schmidt, D. M. Schaller, F. Meisinger, R. Hofer, and H. J. Guntherodt, *Phys. Rev.* **B56**, 4149 (1997).
3. J. Fujii, K. Tamura, S. Ishizuka, T. Sato and T. Mizoguchi, *Surf. Rev. Lett.* **5**, 305 (1998).
4. L. J. Whitman, J. A. Stroscio, R. A. Dragoset, and R. J. Celotta, *Phys. Rev. Lett.* **66**, 1338 (1991).
5. R. Wiesendanger, *Scanning Probe Microscopy and Spectroscopy*, Cambridge Univ. press, 1994.
6. N. Garcia, C. Ocal and F. Flores, *Phys. Rev. Lett.* **50**, 2002 (1983).
7. J. F.J ia, K. Inoue, Y. Hasegawa, W. S. Yang, and T. Sakurai, *Phys. Rev.* **B58**, 1193 (1998).
8. G. Binnig, H. Rohrer, Ch. Gerber, and E. Weibel, *Appl. Phys. Lett.* **40**, 178 (1982).

Longitudinal and transverse energy distributions of electrons emitted from GaAs(Cs,O)

D.A. Orlov*, M. Hoppe*, U. Weigel*, D. Schwalm*,
A.S. Terekhov[†], A. Wolf*

*Max-Planck-Institut für Kernphysik,
69029 Heidelberg, Germany
[†]Institute of Semiconductor Physics,
630090 Novosibirsk, Russia

Abstract. Longitudinal and differential transverse energy distributions of electrons emitted from GaAs-photocathodes are studied at room (295 K) and low (90 K) temperatures. The obtained data prove that electron energy loss as well as elastic electron scattering are of crucial importance in the electron transfer through the GaAs(Cs,O)-vacuum interface.

An intense ultra cold electron source is being developed to study of electron ion interaction at the Heidelberg Test Storage Ring. The heart of the source is a GaAs(Cs,O) photocathode with negative electron affinity (NEA). The physics of the photoelectron escape process from GaAs(Cs,O), however, is still not well understood with the result that properties (longitudinal and transverse energy spreads of photoemitted electrons) of NEA-photocathodes can hardly be optimized for the application. In order to explore the conditions of the photoelectron escape process, detailed data on the longitudinal and transverse energy distributions of photoemitted electrons are required. On the basis of measurements of longitudinal energy distribution curves (EDCs) and mean transverse energies (MTEs) [1,2] of photoemitted electrons we developed a new method [3] to study differential transverse energy distributions $N(E_\perp, E_\parallel^f)$, i.e. the distribution of electrons as a function of their transverse energy $E_\perp$ with fixed longitudinal energy $E_\parallel^f$.

We used transmission mode GaAs photocathodes consisting of a double layer heterostructure GaAs/AlGaAs bonded to a transparent glass carrier. The emitting p^+-GaAs(100) layer was epitaxially grown to a thickness of $\approx 1.5\,\mu$m and doped with Zn to a concentration of holes of about 5×10^{18} cm^{-3}. A two-chamber photocathode preparation setup with a base pressure below 10^{-12} mbar allowed to activate photocathodes with (Cs,O) to quantum efficiencies of 20-25 % (in the reflection mode at 670 nm). The details of the photocathode preparation setup are described elsewhere [2]. After the activation the photocathode was transfered into the electron gun of the measurement setup, where it could be cooled to 90 K.

In the measurement setup the cathode was illuminated in the reflection mode with a 800 nm diode laser. The photoemitted electrons were accelerated to ≈ 20 eV, and guided by the longitudinal magnetic field to a retarding field analyzer, capable of measuring the longitudinal energy distribution of electrons. While the magnetic field B_0 at the cathode position was kept at a constant value of 0.05 T, the magnetic field $B_a = \alpha B_0$ at the analyzer was adjustable between $\alpha = 1.0$ and 2.5. A potential barrier in front of the photocathode was used to block electrons with longitudinal energies $E_\| < E_\|^f$ before the acceleration. Since $E_\perp/B$ is an adiabatic invariant [4] the effect of the adiabatically changing magnetic field on an electron is well known. The energy components of the electron at the cathode $(E_\perp, E_\|)$ and at the analyzer $(E_\perp', E_\|')$ are related by the following way: $E_\perp' = E_\perp + \Delta\alpha E_\perp$ and $E_\|' = E_\| - \Delta\alpha E_\perp$ with $\Delta\alpha = \alpha - 1$. The longitudinal EDCs of photoemitted electrons were measured at a uniform magnetic field ($\Delta\alpha = 0$) by differentiation of the retarding curve $I(U_{bias})$ measured by the analyzer. The mean transverse energy MTE of all photoelectrons with $(E_\| \geq E_\|^f)$ can be deduced from the average longitudinal energies $\overline{E_\|}$ of the distributions dI/dU_{bias} measured for different α, namely MTE=$\Delta\overline{E_\|}/\Delta\alpha$. The possibility of a fast control of the cutting energy $E_\|^f$ near the cathode was used for the measurements of differential transverse energy distributions $N(E_\perp, E_\|^f)$: Modulating $E_\|^f$ in time with a small amplitude $\Delta E_\|^f$ yields a modulation of the analyzer current which represents the signal only from electrons with initial longitudinal energies $E_\|$ near $E_\|^f$. The related ac - component of the analyzer current I^m was measured sensitively by synchronous lock-in detection. Recording $I^m(U_{bias})$ for fixed $E_\|^f$ and taking again the derivative dI^m/dU_{bias}, a signal proportional to the transverse energy distribution for electrons with $E_\| = E_\|^b$ could be obtained [3].

Figure 1 (left) shows the longitudinal EDCs measured from a GaAs(Cs,O) specimen at room (295 K) and low (90 K) temperatures when $E_\|^f < E_{vac}$, i.e. admitting all photoemitted electrons. The vacuum level E_{vac} was determined as the steepest point of the low-energy slope of the EDC. While the position of E_c at low temperature was found from an independent energy calibration measurement [5,6] on the same cathode, the position of E_c at 295 K (≈ 70 meV below E_c at 90 K) was determined by the known dependence of the band gap energy of GaAs on the temperature [7]. Similar to earlier research [2,5,6,8], it is seen that photoelectrons are distributed in a broad longitudinal energy range from E_c down to E_{vac}. The position of the low energy cut of the longitudinal EDCs is about the same at room and low temperatures (a small change of E_{vac} could be explained by a slight variation of the work function [9]). The position of the high energy cut is different due to changing the position of E_c. The high-energy tail (above E_c) originates from the Boltzmann distribution of electrons in the bulk of the semiconductor and falls off exponentially with a slope of 12 meV and 28 meV at low and room temperatures, respectively. A fine structure observed in the longitudinal EDC at 90 K (which consists of two peaks: one is 40 meV below E_c, another slightly above E_{vac}) could be explained by an efficient capture of electrons to the two-dimensional quantized subbands in the near-surface potential well [8]. Note that photoelectrons with the longitudinal energy below E_c in the EDCs can originate either from energy loss near the interface or from elastic electron scattering, which would lead to their emission over a broad range of angles, corresponding to reduction of the longitudinal energy. Thus the longitudinal EDC alone does not give sufficient information to clarify the relative importance of the elastic and inelastic scattering in the emission process.

The width of the longitudinal EDC, the value of $|\chi_{eff}| = E_c - E_{vac}$, and the quantum efficiency were found to be increased at low temperature in agreement with our previous studies [5,9]. We also found that on the one hand the MTEs of electrons emitted into vacuum with $E_\parallel \geq E_c$ *decrease* from 21 meV to 6 meV by cooling. Thus, by cooling the cathode and by blocking the longitudinally low energy part of the distribution it is possible to create an electron beam with very narrow transverse and longitudinal energy spreads. On the other hand it was shown that the MTE of all photoemitted electrons (for the same specimen) *increases* from 58 meV at 295 K to 72 meV at 90 K, that is opposite to that of thermionic cathodes. The raise of MTE could be explained by an increase of $|\chi_{eff}|$, which "opens" additional final states in the vacuum for electrons with higher transverse energies. Indeed, the maximum possible transverse energy of photoemitted electrons is about the entire

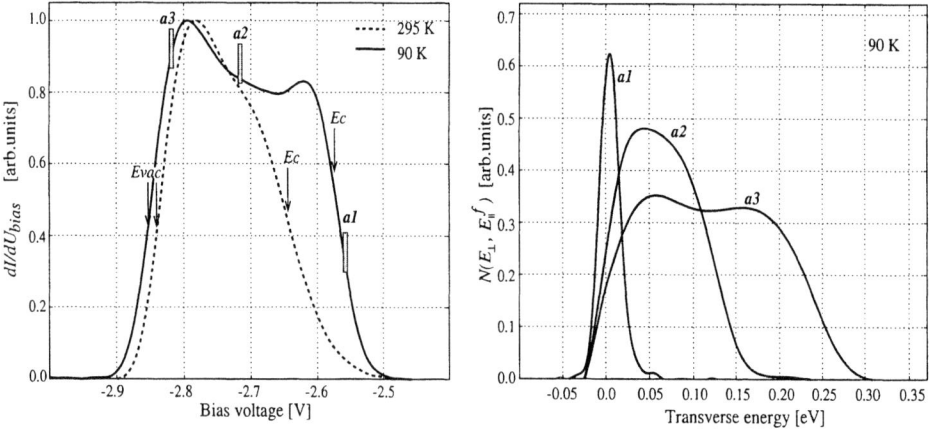

FIGURE 1. Left: Longitudinal EDCs at 295 K and 90 K. The vertical arrows indicate the positions of the vacuum level (E_{vac}) and the bottom of the conduction band in the bulk of the semiconductor (E_c). Right: Distributions of electrons as a function of their transverse energy for fixed longitudinal energy at 90 K. The positions of $E_\parallel^f$ for the curves a1, a2 and a3 are shown in the longitudinal EDC recorded at 90 K.

energy available from the surface potential step, that is the value of $|\chi_{eff}|$. It equals to 280 meV at 90 K for the used specimen, while at 295 K it is about 190 meV.

Distributions of electrons $N(E_\perp, E_\parallel^f)$ as a function of transverse energies for different $E_\parallel^f$ are shown in Fig.1 (right). It is seen that the Gaussian-like curve $a1$ ($E_\parallel^f$ of ≈ 20 meV above E_c) is narrow with a width (FWHM) of 25 meV limited by the resolution of the analyzer. The differential mean transverse energy dMTE (i.e, the MTE for electrons with fixed longitudinal energy) amounts to about 6 meV for this distribution. The line shapes of the $N(E_\perp, E_\parallel^f)$ and the values of the dMTE are about the same for all measured $E_\parallel^f$ above the E_c. Below E_c the transverse distribution broadens and the dMTE gradually increases. For curve $a3$ ($E_\parallel^f$ of 240 meV below E_c) the width of the distribution is already 230 meV and the dMTE is about 120 meV. The high energy cut of this distribution originated from electrons with total energy $E_\perp + E_\parallel \approx E_c$ is observed. That means these electrons pass through the band banding region without energy loss but are scattered elastically and emitted at a large angle to the surface normal. Electrons around $E_\perp \approx 0$ have suffered loss of total energy (of about 240 meV for $a3$-curve) but kept a low transverse energy. The main part of the transverse distribution corresponds to electrons which underwent elastic scattering and suffered energy loss as well.

In conclusion: a) The obtained data demonstrate that the cooling of a GaAs photocathode results in an increase of the negative electron affinity, of the quantum efficiency, and of the MTE for all photoemitted electrons. The MTE for longitudinally high-energy electrons ($E_\parallel^f \geq E_c$) falls off with decreasing temperature and equals about kT for all measured temperatures down to 90 K. b) The measurements of the differential transverse energy distributions prove that elastic as well as inelastic electron scattering are of crucial importance in the process of the electron transfer through the GaAs(Cs,O)-vacuum interface.

REFERENCES

1. Pastuszka S., Terekhov A.S, Kratzmann D., Schwalm D., and Wolf A., *Appl. Phys. Lett.* **71**, 2967-2969 (1997).
2. Pastuszka S., Hoppe M., Kratzmann D., Schwalm D., Wolf A., Jaroshevich A.S., Kosolobov S.N., Orlov D.A., Terekhov A.S., *J. Appl. Phys.* **88**, 6788-6800 (2000).
3. Orlov D.A., Hoppe M., Weigel U., Schwalm D., Terekhov A.S., Wolf A., to be published.
4. O'Neil T.M. and Hjorth P.G., *Phys. Fluids* **28**, 3241-3252 (1985).
5. Terekhov A.S. and Orlov D.A., *Proc. SPIE* **2550**, 157-164 (1995).
6. Drouhin H.-J., Hermann C., Lampel G., *Phys. Rev. B* **31**, 3859-3870 (1985).
7. Varshni Y.P., *Physica* **34**, 149-154 (1967).
8. Orlov D.A., Andreev V.E. and Terekhov A.S., *JETP Lett.* **71**, 151-154 (2000).
9. Terekhov A.S. and Orlov D.A., *JETP Lett.* **59**, 864-868 (1994).

Cesiumoxide-GaAs Interface and layer thickness in NEA surface formation

Sam D. Moré [1], Senku Tanaka [2], Shin-ichiro Tanaka [1*],
Tomohiro Nishitani [3], Tsutomu Nakanishi [3] and Masao Kamada [1,2]

[1] *UVSOR Facility, Institute for Molecular Science, Okazaki 444-8585, Japan*
[2] *Department of Structural Molecular Science, The Graduate University for Advanced Studies, Okazaki 444-8585, Japan*
[3] *Department of Physics, Nagoya University, Nagoya 464-8602, Japan*

Abstract. Difference in NEA activation processes on GaAs(100) has been investigated with high resolution photoemission spectroscopy. It was found that the NEA surface produced by different activation processes can be classified into three phases having characteristic chemical reactions of Cs and Oxygen with GaAs. A schematic phase diagram is presented to understand NEA activation processes.

INTRODUCTION

Negative electron affinity (NEA) surfaces have found applications as efficient photocathodes and the NEA surface of GaAs(100) and its superlattice is known to be a useful emitter of spin-polarized electrons with a high degree of polarization and efficiency. This can be achieved by the 'jo-jo'-technique [1], where Cs deposition and subsequent oxidation are repeated several times. The details of NEA surface formation, however, are not fully understood: previous reports [2-6] differ considerably both in describing the method of production as well as the underlying chemical and physical mechanism.

In this paper we present a systematic study of various NEA activation processes on GaAs(100) using photoemission spectroscopy. Analyzing the influence of the cesiation and oxidation on the bandbending and monitoring the electron yield from bulk GaAs(100), we have been able to distinguish three different phases of activation, which depend on the total thickness of the overlayer, the Cs:O ratio and the resulting chemical interaction with the substrate.

EXPERIMENTAL RESULTS AND DISCUSSION

The experimental details are similar to those described in previous publications [7,8]. The overall energy resolution was less than 0.17 eV (0.25 eV for phase III). All experiments were conducted at the BL 5A of the UVSOR facility in Okazaki, Japan.

Cs getters were obtained from SAES. A p-type Zn-doped GaAs wafer (5 x 10^{19} atoms/cm^3) was used as the substrate.

Fig. 1 shows the electron yield as a function of sample treatment. Curve (a) shows Cs deposition only. Curve (b) shows the so called Nagoya treatment; Cs is dosed until the electron yield decreases again, the yield raises under O expose. Curve (c) depicts the classical jo-jo treatment; the GaAs surface is alternately exposed to Cs and O, O lowers the electron yield, while Cs increases it. Curve (d) shows a mixed type of activation where during the initial Cs deposition small amounts of O have been added. This treatment involves larger Cs and O amounts. It is characteristic for saturation regimes of both O and Cs, resulting in the generation of overlayers with a larger thickness.

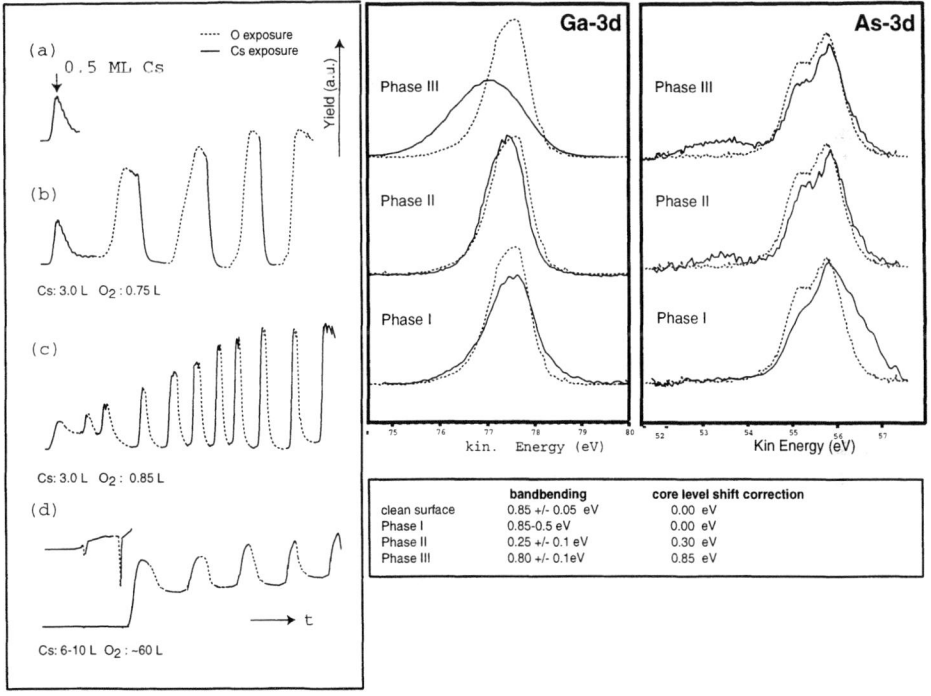

FIGURE 1 (LEFT). Electron yield as a function of sample treatment for different activation processes on GaAs(100): (a) Cs-only activation, (b) so-called Nagoya treatment, (c) classical jo-jo treatment, and (d) a mixed type of activation.
FIGURE 2 (RIGHT). Ga-3d and As-3d core-levels representative for different activation processes. Peak positions are corrected for bandbending. The dotted line shows the clean surface. Phase I corresponds to the activation (b), Phase II to the activation (c) and Phase III to the activation (d).

Fig. 2 shows the Ga-3d and As-3d photoemission spectra after the activation. All activations can achieve a high electron yield. The stability of the yield is however different with Phase III activations being frequently less stable than Phase I and Phase II activations.

The As-3d peak for Phase I is broadened towards higher kinetic energy, indicating donation of electrons by reaction with Cs. In Phase II the As-3d peak has split into a larger main peak and a smaller peak at 53-54 eV (47-46 eV binding energy), which marks the As=O double bond formation. The broadening to the higher kinetic energy side is less pronounced. This indicates that Phase II has a larger interface oxidation at the As than Phase I.

Phase III finally shows a strong interface oxidation. The direct Cs-interaction with the substrate seems to be negligible. The Ga and As core-level peaks for Phase III are much broader than the peaks of the previous phases. Also the halfwidth of the Cs-peak is about 25% broader. This indicates that the Cs in this phase consists of more different ionization stages than the one of Phase I. The chemical composition of the thicker overlayer is far more complex and heterogeneous than for the Phases I and II. The lower stability in terms of electron yield for the Phase III surface could be due to an ongoing chemical reaction inside such an overlayer.

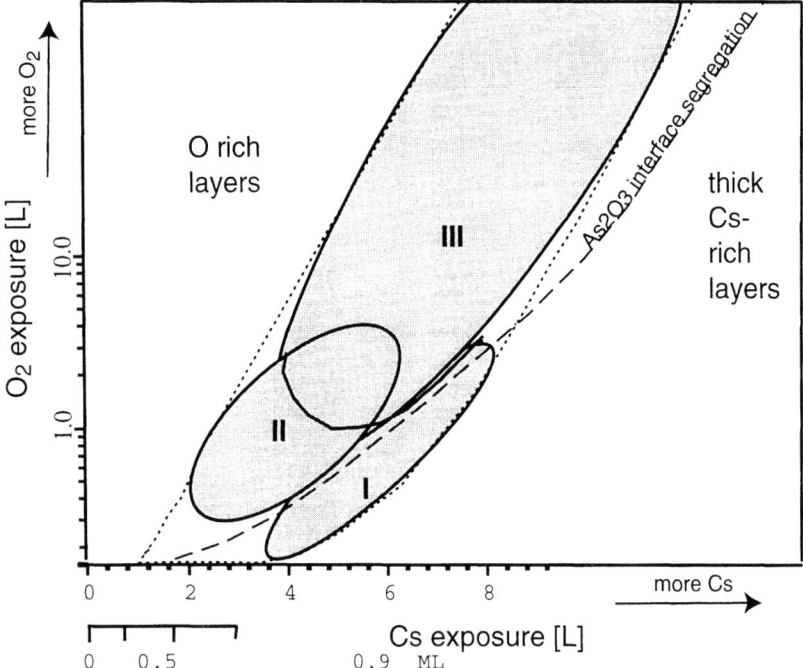

FIGURE 3. Semi-quantitative phase diagram over the different interface and overlayer conditions which have been found for the GaAs(100) system. Cs and O are both given as cumulative exposures. The different phases necessarily therefore overlap due to different sticking coefficients. Phases correspond to those described in the previous figures. Layers which have been exposed to the same amount of Cs but to lager amounts of O will therefore be thicker than layers which have been exposed to less oxygen. The resulting layer thickness increases therefore towards the upper right corner of the plot.

The Table below Fig. 2 gives the bandbending values obtained from the surface photovoltage experiment, which the photoemission data have been corrected for. A stronger oxidation of As is correlated with a decrease in bandbending in our experiment, which is in alignment with previous results from literature [2,3]. The Table shows a decrease in bandbending from Phase I to Phase III. None of the surfaces investigated in our study has shown therefore any increase in bandbending. This contrasts the dipole model as originally proposed [2]. No increase in bandbending but rather the decrease has been recently observed for activations on p-type GaN [9].

Fig. 3 shows a schematic phase diagram to understand different activation processes. The Nagoya treatment, classical jo-jo treatment and a mixed type of activation, which have been mentioned above, correspond to the Phases I, II and III in the diagram, respectively. These three phases have differences in the amounts of Cs and O, the Cs:O ratio and the overlayer thickness. The resulting chemical reactions of Cs and O with GaAs substrate are therefore very different among these three phases. During the technical use of NEA surfaces decayed surfaces are typically reactivated a few times before the substrate is re-cleaned and fresh overlayers are deposited. These surfaces may be also understood in this phase diagram. The detailed analysis of the chemical reactions in this phase diagram will be reported in near future [10].

CONCLUSIONS

Three different NEA activation processes on GaAs(100) have been characterized using high resolution photoemission spectroscopy. The photovoltage effect has been quantified and used to determine the amount of bandbending. Interface oxidation clearly depends on the activation scheme used and also on the absolute amounts of Cs and O the substrate is exposed to. A schematic phase diagram has been presented.

ACKNOWLEDGMENTS

This work was partially supported by a Grand-in-Aid for Scientific Research from the Ministry of Science, Education, Sports and Culture, Japan.

REFERENCES

* Present address: Department of Physics, Nagoya University, Nagoya 464-8602, Japan
1. C.Y. Su, W.E. Spiecer and I. Lindau, *J. Appl. Phys.* **54** (1983) 1413.
2. A. Alperovich, A.G. Paulish, A.S. Terekov, *Surf. Sci.* **331-333** (1995) 1250-1255.
3. M. Besancon, R. Landers and J. Juille, *J. Vac. Sci. Techn.* **A5** (1987) 2025.
4. N. Takahashi, S. Tanaka, M. Ichikawa, Y. Cai and M. Kamada, *J. Phys. Soc. Japan* **66** (1997) 2798.
5. T. Subbramanian, Jinming Cao and Y. Gao, *J. Vac Soc. Technol.* **A 10** (1992) 3158.
6. Q.-B. Lu, Y.-X. Pan and H. Gao, *J. Appl. Phys.* **68** (1990) 634.
7. S. Moré, S. Tanaka, S. Tanaka and M. Kamada, *Surf. Sci.* **454** (2000) 161.
8. M. Kamada, J. Murakami, S. Tanaka, S. D. Moré, M. Itoh, and Y. Fujii, *Surf. Sci.* (2000) 525.
9. C.I. Wu and A. Kahn, *J. Appl. Phys.* (1999).
10. S. More, S. Tanaka, Tomohiro Nishitani, Tsutomu Nakanishi, Kazutoshi Takahashi and Masao Kamada, in preparation.

Temperature Dependence of Electron Spin Dynamics

Yuri A. Mamaev[*], Yuri P. Yashin[*], Arsen V. Subashiev[**],
Anton N. Ambrajei[*] and Alexander V. Rochansky[*]

*Laboratory of Spin-Polarized Electron Spectroscopy, State Technical University,
195251, St.-Petersburg, Russia*
**State Technical University, 195251, St.-Petersburg, Russia*

Abstract. The results of systematic study of polarized electron emission from unstrained GaAs and strained GaAsP thin semiconductor layers are presented. The polarization spectra are shown to reflect the optical spin orientation of the electrons produced by the circularly polarized light excitation. The interpretation of the details of the polarization spectral and temperature dependencies shows the importance of valence band wrapping and correlation between the spin and the momentum of an electron for high-polarization photoemission. For thin overlayers the residual polarization losses occur in the band-bending region.

Unstrained GaAs

With the aim to clarify the photoelectron polarization loss mechanism, we have performed a set of experiments with unstrained GaAs overlayers. In general, polarization can be lost [1]: a) at the electrons creation under circular light excitation; b) during the thermalisation; c) in course of electrons transport to the band-bending region (BBR); d) in the BBR; e) during the escape to vacuum through the NEA surface barrier. The GaAs overlayers with the thickness 70, 100, 150, 500, 1000 and 1500 nm were MBE-grown at the top of n-doped GaAs wafer. An advantage of n-doped GaAs wafer is the elimination of the back scattering of photoexcited electrons from the overlayer/wafer interface, which leads to the lowering of an effective lifetime of electrons prior the emission. At the same time the study of electron emission from the overlayers with various thickness allowed us to investigate the emission of electrons under various lifetime conditions (the more thickness, the more lifetime). One should expect less depolarization at small lifetime. In the gradually doped GaAs overlayer the main part of a layer was low p-doped ($5 \cdot 10^{17}$ cm^{-3}), which should help to suppress Bir – Aronov – Pikus (BAP) spin relaxation mechanism, but the top 10 nm were heavily p-doped ($5 \cdot 10^{19}$ cm^{-3}) with the aim to achieve high quantum yield at NEA surface. The working GaAs overlayer was capped with As to prevent the layer pollution by the air. With the aim to study temperature dependence of electron spin dynamics, the electron polarization and yield spectral dependencies were measured at the temperatures 300 and 130 K for freshly activated samples and in the course of the photocathode surface degradation.

All measurements have been performed at the computer-controlled set-up [2] at the residual pressure $1\text{-}2\cdot10^{-10}$ Torr, circularly polarized light monochromaticity being equal $\Delta\lambda = 2$ nm. Thermal cleaning procedure (at the pressure not exceeding $3\text{-}5\cdot10^{-9}$ Torr) consisted of the sample heating with gradual increase of temperature up to 560 - 580 degrees centigrade, one hour exposition at 560 - 580 degrees centigrade and than cooling for one hour till room temperature. NEA has been achieved by cesium and oxygen deposition. During the course of experiments Mini - Mott detector was calibrated both with the energy – loss extrapolation procedure [3] and with GaAs/AlGaAs derivative standard for polarimeter calibration [4].

Fig. 1 shows the electron spin polarization $P(h\nu)$ and quantum yield $Y(h\nu)$ as a function of excitation energy for the GaAs unstrained sample with overlayer thickness 100 nm at room temperature. The maximum value of P is about 46% near the photothreshold. The typical features of $P(h\nu)$ and $Y(h\nu)$ curves are clearly seen, i.e. drastic enhancements of **P** near the photothreshold, then rather wide plateau, followed by decreasing of **P** when $h\nu$ is close to $E_g+\Delta$, where spin-orbit splitting $\Delta = 0.3$ eV. The wide plateau is a sequence of a suppression of the BAP spin relaxation mechanism due to low doping at the main part of the GaAs overlayer.

As it is evident from fig. 2, the maximum polarization value grows upon the reduction of GaAs overlayer thickness and for the most thin overlayer practically reaches 50%, which is the theoretical maximum value of P_{max} in equilibrium. At the thickness more than 400 nm the value of P_{max} for cool cathode exceeds one at room temperature. Such situation is typical as well for the bulk cathodes and is explained by fast averaging of the photoexcited electrons momenta and spin relaxation prior the emission into vacuum. One can see that at the thickness less than 400 nm $P_{max}(300K) > P_{max}(130K)$. This observation can be explained by more effective spin relaxation in BBR at 130K, since at low temperature the BBR quantum well is deeper, than at T=300K [5] and D'yakonov and Perel' (DP) mechanism is effective above mobility edge [6].

Data of Fig. 2 were obtained for freshly activated cathodes, i.e. $P(h\nu)$ and $Y(h\nu)$ curves were measured just after activation of NEA surface. The increase of the work function, which occurs upon the degradation of the sample surface, leads to the cut-off the photoelectrons, which are thermalised at the bottom of the conduction band. Hence, the only "hot" electrons, which have not yet undergo spin relaxation due to thermalisation, can be escaped into vacuum. Fig. 3 shows that in the case of photoexcitation at about 200 meV above the photo-threshold polarization is growing up upon the surface degradation, which is explained by high polarization of hot electrons. The GaAs valence band wrapping effect and the relationship between the spin and the momentum of an electron just after excitation can lead to the electron polarization of up to 67% [7]. It was first time illustrated by Mirlin [7], who measured the hot photoluminescence circular polarization of up to 35%.

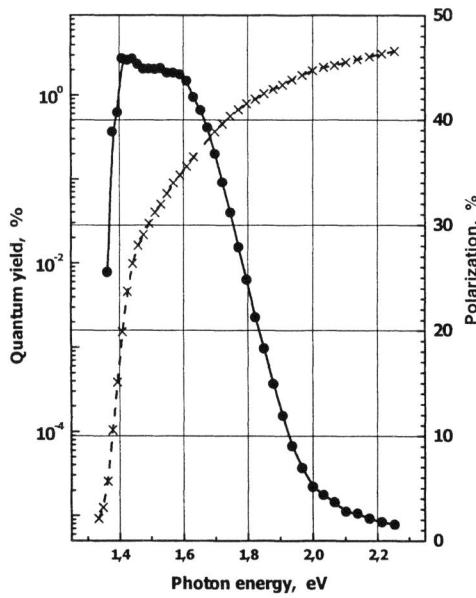

FIGURE 1. Electron spin polarization (circles) and quantum yield (crosses) as a function of excitation energy for the GaAs unstrained sample (overlayer thickness 100 nm) at room temperature.

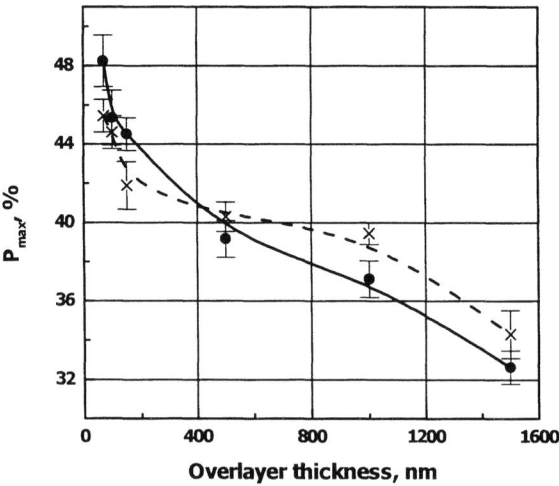

FIGURE 2. Polarization maximum values at various GaAs overlayer thickness. Freshly activated cathodes. Solid line and circles – room temperature; dash line and crosses – T = 130K.

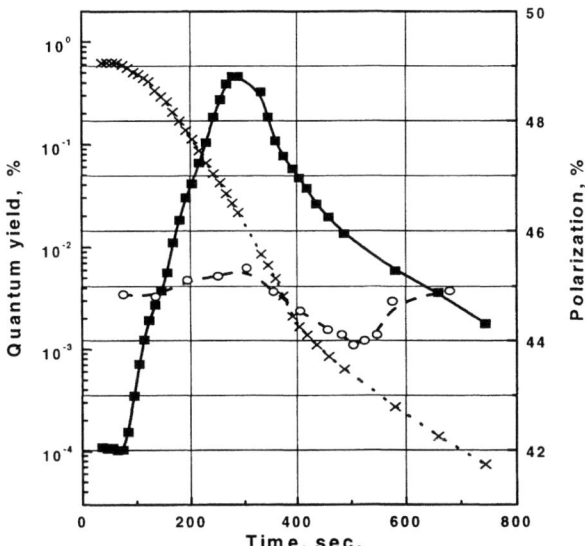

FIGURE 3. Polarization (circles and squares) and Quantum Yield (crosses) evolution upon the degradation of the GaAs sample surface at T = 130K for the near-threshold (open circles) excitation ($\Delta E = h\nu - E_g = 15$ meV) and excitation at 190 meV above the energy gap (solid squares).

Strained GaAsP

The same set of spectral and temperature experiments has been performed as well with the modified $GaAs_{0.95}P_{0.05}$ / $GaAs_{0.7}P_{0.3}$ strained MOCVD grown heterostructure [8]. The parameters of the sample have been improved on the base of X-ray, Raman and polarized photoluminescence studies of such structure [9]. The modification consisted of more sharp interfaces fabrication and gradual p-doping of 140 nm thick $GaAs_{0.95}P_{0.05}$ strained overlayer ($1 \cdot 10^{18}$ cm^{-3} at the main part of it, but $1 \cdot 10^{19}$ cm^{-3} at the top 20 nm).

Fig. 4 shows $P(h\nu)$ and $Y(h\nu)$ curves at room temperature. The $P(h\nu)$ dependence has a plateau of about 60 meV and drops drastically at $h\nu < 1.7$ eV. Both these features emphasise high quality of a cathode, namely high splitting of the heavy-hole and light-hole subbands due to high strain and, hence, weak influence of the band tails inter-mixture [9]. P_{max} for such freshly activated structure goes down upon the sample cooling, which is explained as well by more effective spin relaxation in a deeper BBR well. The comparison of the maximum polarization for the unstrained and strained thin layer cathodes showed close relaxation rate values in BBR while the surface recombination rate is reduced in the modulation doped structures.

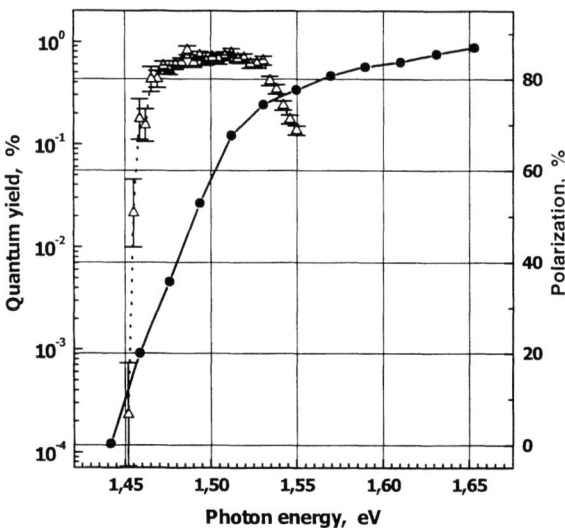

FIGURE 4. Electron spin polarization (triangles) and quantum yield (circles) as a function of excitation energy for the GaAsP strained sample at room temperature.

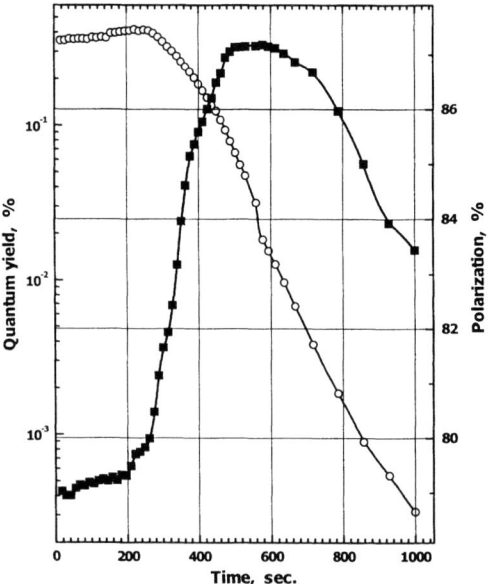

FIGURE 5. Polarization (squares) and Quantum Yield (open circles) evolution upon the degradation of the GaAsP sample at T = 130K. Excitation light energy 1,544 eV.

Fig 5 illustrates the polarization and quantum yield evolution upon the degradation of the GaAsP sample at T = 130K, excitation light energy being equal to 1,544 eV. A considerable increase of a polarization is explained, as in the case of GaAs, by high polarization of the "hot" electrons.

Conclusions

High polarization was achieved with specially designed and modified GaAs and GaAsP samples. Valence band wrapping and the relationship between the spin and the momentum of an electron are essential for high-polarization photoemission. For thin overlayers the most essential depolarization occurs due to spin relaxation in the band bending region.

ACKNOWLEDGEMENTS

This work was supported by the Russian State Program "Surface Atomic Structures" under grant 107-24(00), Russian Fond for Basic Research under grant 00-02-16775 and INTAS under grant 99-00125.

REFERENCES

1. Subashiev, A.V., Mamaev, Yu.A., Yashin, Yu.P. and Clendenin, J.E., *Phys. Low-Dim. Structures*, **1/2**, 1 (1999).
2. Yashin, Yu.P., Ambrajei, A.N. and Mamaev, Yu.A., *Instruments for Experimental Techniques*, **43**, #2, 245 (2000).
3. Gay, T.J., Khakoo, M.A., Brand, J.A., Furst, J.E., Meyer, W.V., Wijayaratna, W.M.K.P. and Dunning, F.B., *Rev. Sci. Instrum.*, **63 (1)**, 114 (1992).
4. Mulhollan, G. et al., " A Derivative Standard for Polarimeter Calibration" in Proceedings of the 16th IEEE Particle Accelerator Conference (PAC 95) and International Conference on High-energy Accelerators (IUPAP), Dallas, Texas, 1995, p. 1043.
5. Herman, C., Drouhin, H.-J., Lampel, G., et al., 'Photoelectronic processes in Semiconductors, Activated to Negative Electron Affinity", in *Spectroscopy of Nonequilibrium Electrons and Phonons*, edited by C.V. Shank and B.P. Zakharchenya, Elseiver Science, B.V.,1992, Ch. 9.
6. Subashiev, A.V., "Manifestation of the electron kinetics in BBR region in polarized electron emission", *in Proceedings. of the Low Energy Polarized Electron Workshop LE-98*, edited by Yu.A. Mamaev et al., SPES-Lab-Pub., St.Petersburg, 1998, p. 125.
7. Mirlin, D.N., "Electron momenta optical alignment in the GaAs-type semiconductors", in *Optical Orientation*, edited by F. Meier and B.P.Zakharchenya, North-Holland, Amsterdam, 1984, Ch. 4.
8. Mamaev, Yu.A, Yashin, Yu.P., Subashiev, A.V., Galaktionov, M.S., Yavich, B.S., Kovalenkov, O.V., Vinokurov, D.A. and Faleev, N.N., *Phys.Low-Dim.Structures*, **7**, 27 (1994).
9. Subashiev, A.V., Mamaev, Yu.A., Oskotskij, B.D., Yashin Yu.P. and Kalevich, V.K., *Semiconductors*, **33**, #11, 1182 (1999).

Latest Results from Time Resolved Intensity and Polarization Measurements at MAMI

J. Schuler[a], K. Aulenbacher[a], T. Baba[e], D. v. Harrach[a], H. Horinaka[d], T. Nakanishi[c], S. Okumi[c], E. Reichert[b], J. Roethgen[a] and K.Togawa[c].

[a] *Institut für Kernphysik, Johannes Gutenberg-Universität, D-55099 Mainz, Germany*
[b] *Institut für Physik, Johannes Gutenberg-Universität, D-55099 Mainz, Germany*
[c] *Department of Physics, Nagoya University, Nagoya 464-8602, Japan*
[d] *University of Osaka Prefecture, Osaka, Japan*
[e] *NEC Corporation*

Abstract. At the Institute of Nuclear Physics at Mainz we investigate time resolved intensity and polarization spectra in the photoelectron emission from different types of highly polarized p-type strained layer and superlattice photo cathodes used for polarized electron sources at accelerator facilities. After several improvements concerning time resolution and phase stability the observed response times decreased to 2-2.5ps which is the time resolution of our setup. Obtaining these short response times we decided to reinvestigate the emission process using simpler GaAs structures with different layer thickness. With the new data one may have to reconsider the escape process of the electrons for thin active layers. The measured response times of about 1ps or less for 100-200nm active layer thickness suggest some ballistic traveling of the electrons through the crystal without interaction. For these electrons energy loss and depolarization seem to take place only in the bent band region or on the crystal surface.

INTRODUCTION

At the Mainz Microtron MAMI we investigated time resolved intensity and polarization spectra in the photoemission from different types of superlattice photo cathodes to determine their response times, initial spin polarization and spin relaxation times. The shape of the electron bunch is obtained by a time resolved current measurement with a Faraday-Cup. To avoid bunch broadening by space-charge the bunch charge is reduced until the bunch duration is not effected by space-charge. Figure 1 shows a bunch duration measurement from a short period InGaAs-AlGaAs strained superlattice SLSA#4 with an active layer thickness of 90nm. This sample was developed by the group of T. Nakanishi, Nagoya University [5]. Due to various improvements concerning the time resolution of our setup the measured bunch duration of SLSA#4 decreased from 5ps to ~2.8ps. A bunch profile of SLSA#4 at ~0.01fC bunch charge is shown in figure 2. At this point we decided to reinvestigate our experiment with "simpler" GaAs structures. The results should help

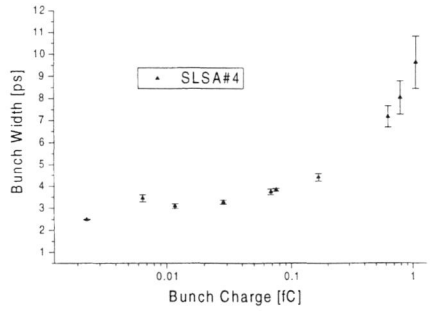
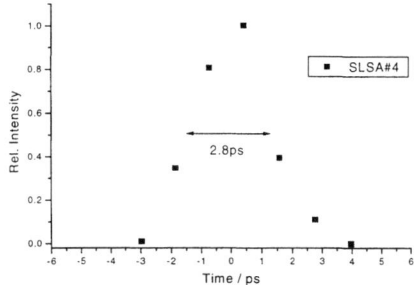

Figure 1. Bunch width vs. bunch charge to determine the minimum bunch duration.

Figure 2. Electron bunch shape from SLSA#4. Bunch width is 2.8ps (FWHM) at a bunch charge of ~0.01fC.

solving the question if the short response is a specific property of the superlattice or if it can be explained by a "conventional" model for the photoemission process.

EXPERIMENTAL SETUP AND RESULTS

Time resolved measurements on a picosecond time scale can be performed at our pulsed 100keV electron gun test facility [2][3]. The measurement principle is based on a radio-frequency streak method. The time structure of the emitted electron bunch is transformed into a transversal spatial profile by wobbling the beam over a narrow slit. For wobbling the beam we use a TM_{110} resonator cavity at the MAMI radio-frequency of 2.4493GHz. The photoemission is excited by a Kerr-lens mode locked Ti:Sapphire laser [4] with a repetition rate of 76.5406MHz which is the 32nd subharmonic of the MAMI-rf. The laser is phase-locked to the cavity-rf and time resolved measurements are obtained by shifting the phase between cavity and laser. The time resolution of the setup is limited by:

1. The geometrical resolution given by the slit width, the beam diameter on the slit and the deflection angle: ~0.8ps
2. Since the method is integrating over the measuring time the signal is broadened by a phase-jitter in the synchronization process: ~1ps.
3. The laser pulse duration after passing through an optical fiber: ~0.8ps
4. Some energy-spread of the emitted electrons causes a time of flight difference during the acceleration process: ~1-1.5ps

Point 4 is the most uncertain factor because it strongly depends on the surface conditions of the sample. Taking a value of 1.5ps for the energy-spread, the total time resolution results to ~2ps.

With this setup we performed time resolved intensity measurements with three different GaAs-samples. Table 1 shows a listing of the samples and their properties.

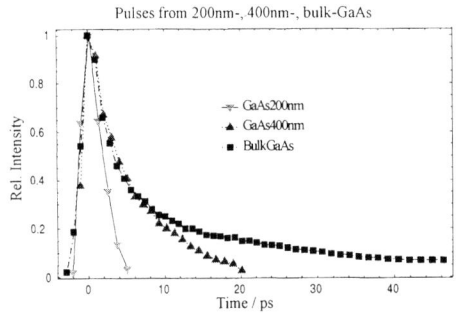

Figure 3. Intensity/current measurement normalized on maximum intensity.

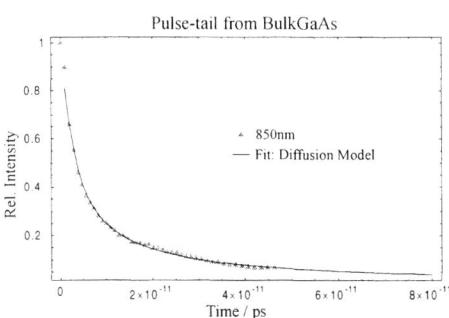

Figure 4. BulkGaAs. A diffusion model with $\alpha \cong 6000\,\text{cm}^{-1}$ and $D \cong 85\,\text{cm}^2/\text{s}$ is fitted to the tail.

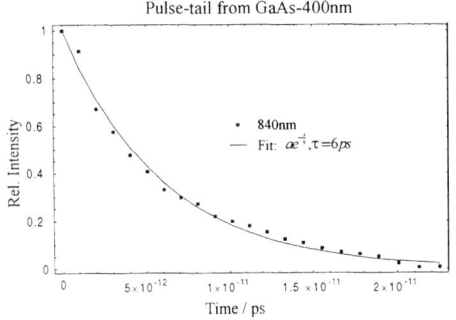

Figure 5. GaAs-400nm. An exponential function with a decay time of 6ps fitted to the tail.

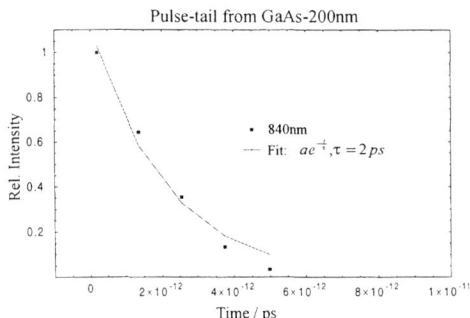

Figure 6. Bunch tail of GaAs-200nm. Exponential fit with 2ps decay time.

Table 1. Measured samples

Type	Doping	Doping concentration	Active layer thickness
BulkGaAs	p-doped	$2\text{-}3 \cdot 10^{19}\,\text{cm}^{-3}$	$\sim 3\,\mu\text{m}$
GaAs	p-doped	$3\text{-}5 \cdot 10^{18}\,\text{cm}^{-3}$	400nm
GaAs	p-doped	$3\text{-}5 \cdot 10^{18}\,\text{cm}^{-3}$	200nm

Figure 3 shows the bunch shapes of the samples. All values are normalized on maximum intensity. Figures 4-6 take a closer look on the bunch tails of each sample.

DISCUSSION

Comparison of the electron bunches in figure 3 suggests two kinds of electron transport processes contributing to the current signal: "Ballistic" electrons without or at the most a few number of interactions form the peak with escape times of 2-4.5ps even for the bulk structure. And diffusion electrons which may have undergone plural scattering are mainly located in the bunch tail. The fraction of diffusion electrons depends on layer thickness and doping concentration. Since the tail of the bulk structure should be dominated by diffusion electrons the shape of its tail is well described by a diffusion model [6] like shown in figure 4.

This diffusion model is not able to describe the tails of the 400nm and the 200nm structure. The fraction of diffusion electrons decreases with layer thickness why the response function of thin epilayers should be dominated by "ballistic" electrons. For those ballistic electrons the process can be described by an exponential decay function (figure 5+6).

CONCLUSION

After deconvoluting the time resolution we obtain response times (FWHM) for thin active layers (<200nm) of less than 1.5ps. Therefore depolarization processes with typical time constants of ~100ps [5][6] will not lead to a considerable depolarization of the emitted electrons. With the new data we suggest a ballistic travelling of the electrons through the epilayer to the bent band region almost without interaction. Considering the short response times spin relaxation for ballistic electrons may be caused by faster depolarization processes in the bent band region or on the crystal surface.

ACKNOWLEDGEMENT

This project is supported by the Deutsche Forschungsgemeinschaft (SFB443). The photocathode material was kindly put at our disposal by Prof. Y. Mamaev, St. Petersburg State Technical University, and Prof. T. Nakanishi, Nagoya University.

REFERENCES

[1] P. Hartmann et al, *NIM* **A379**, 15-20 (1996).
[2] P. Hartmann, *Aufbau einer gepulsten Quelle polarisierter Elektronen*, Dissertation, Shaker Verlag, Aachen, 1998.
[3] K. Aulenbacher, Ch. Nachtigall et al, *NIM* **A391** 498-506 (1997).
[4] COHERENT Mira900 D with Synchro-Lock unit.
[5] T.Nakanishi, "Highly polarized electrons from superlattice photocathodes", in Proceeding of 7th international workshop AIP, edited by R. Holt and H. Miller, AIP Conference Proceedings, New York, 1998, pp. 300-310.
[6] H. Horinaka et al, *Jpn. J. Appl. Phys.* **34** 6444-6447 (1995).
[7] P. Hartmann et al, *J. Appl. Phys.* **86** 2245 (1999).

Photo-luminescence Study of Superlattice Photocathode

Tetsuya MATSUYAMA, Masayasu MUKAI, Hiromichi HORINAKA, Kenji WADA, Tsutomu NAKANISHI [1], Shoji OKUMI [1], Kazuaki TOGAWA [1], Toshio BABA [2]

Department of Physics and Electronics, Osaka Prefecture University, 1-1 Gakuen-cho, Sakai, Osaka, 599-8531, Japan
[1] *Department of Physics, Nagoya University, Furo-cho, Chikusa-ku, Nagoya, Aichi, 464-8602, Japan*
[2] *Silicon SystemsResearch Labolatories, NEC Corporation, 1120, Simokuzawa, Sagamihara, Kanagawa, 220-1198, Japan*

Abstract. Luminescence with high circular polarization was observed in the InGaAs-AlGaAs strained layer superlattice which was designed as a spin polarized electron source. The maximum luminescence polarization of 65% was obtained around 1.60eV which corresponded to the lowest band gap. The dependence of luminescence polarization on the excitation photon energy reflected the valence band structure split by the internal strain and the quantum confinement effect. The spin polarization of conduction band electrons at the instant of excitation was found to be 94% from the lifetime and the spin relaxation time measured by the time-resolved luminescence measurement.

INTRODUCTION

Spin polarized electron sources have played very important roles in material sciences, nuclear and particle physics. We have developed semiconductor photocathodes emitting highly spin polarized electrons for use as electron gun of linear collider. In recent years, spin polarized electron sources have been also utilized to investigate spin devices such as spin filter using the semiconductor/ferromagnet junction.[1] Spin polarization of practical polarized electron sources is based on the selection rule for circularly polarized excitation.[2] To generate highly spin polarized electrons, it is necessary that the degeneracy of the valence band is removed and electrons are photoexcited only from the top valence band. There are three types of semiconductor photocathodes with high polarization; the strained thin layer[3], the superlattice[4] and the strained layer superlattice.[5] The strained layer superlattice is thought to be the photocathode with high polarization and high efficiency.

The method using Mott analyzer is standard for measurement of spin polarization of electrons. It can determine the spin polarization of extracted electrons directly. This method, however, has following disadvantages: (a) Measured spin polarization depends on the surface barrier because the spin polarization of electrons is relaxed during the stay in conduction band.[5] (b) This method is difficult to be applied to measurement of the initial spin polarization. The initial spin polarization denotes the spin polarization at the instant of excitation. (c) Surface treatment, high vacuum and high voltage are required to extract spin polarized electrons.

In this present work, we apply the spin dependent luminescence method to the InGaAs-AlGaAs strained layer superlattice designed as a highly spin polarized electron source. The spin dependent luminescence method is thought to be suitable for study of photocathode materials and structures because surface treatment and special equipments are not required and the initial spin polarization of conduction band electrons can be estimated.

STRUCTURE AND BAND STRUCTURE OF InGaAs-AlGaAs STRAINED LAYER SUPERLATTICE

Figure 1(a) shows the sample structure of InGaAs-AlGaAs strained layer superlattice. The superlattice composed of 18 periods of alternating InGaAs and AlGaAs layers was grown on the p-GaAs substrate ((100), $p=2 \times 10^{19}/cm^3$) after the deposition of a 50nm GaAs buffer layer and a 1μm AlGaAs barrier layer with aluminum fraction of 0.35. The well width and the indium fraction are 2nm and 0.15, respectively. The barrier width and the aluminum fraction are 3nm and 0.25, respectively. The total thickness of the superlattice layer is about 100nm. The lattice mismatch between alternating layers is about 1.1%. AlGaAs barrier layers are free of strain and the lattice mismatch between InGaAs well layer and AlGaAs barrier layer is completely accommodated by the elastic strain in the well layer.

Figure 1 (b) shows the energy band structure of InGaAs-AlGaAs strained layer superlattice at the Γ point calculated using a Kronig-Penny model which takes into account the effect of strain on the energy band structure of well layer. The lowest energy gap and the second lowest energy gap are estimated to be 1.60eV and 1.66eV at 10K, respectively. The valence band splitting of 60meV is large enough to excite the electrons with high polarization from the heavy hole band.

The spin polarized electrons in the conduction band recombine predominantly with holes in the top valence band because of the large splitting of valence band. The luminescence polarization coincides with the spin polarization of conduction band electrons. When the excitation photon energy is adjusted to the lowest energy gap, the only transition from the heavy hole band to the conduction band occurs. The spin polarization of electrons at the instant of excitation is ideally 100%. When the photon energy of excitation light exceeds the second lowest energy gap, the initial spin polarization is restricted to be 50% because the ratio of the transition intensities between the conduction band and the heavy hole band to that between the conduction band and the light hole band is 3:1.

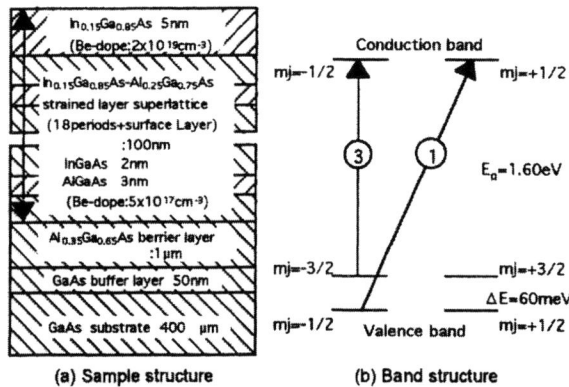

FIGURE 1. Sample structure and band structure of InGaAs-AlGaAs strained layer superlattice

EXPERIMENTAL AND DISCUSSION

As the first step, the luminescence polarization spectrum of the sample was measured. The light from a cw Ti:sapphire laser was converted into the circularly polarized light by a polarizer and a quarter wave plate, and focused on the surface of sample mounted in a cryostat. An optical chopper (chopping frequency 29Hz) was placed in front of the cryostat to make use of a lock-in amplifier for signal processing. The right and left circularly polarized light components of luminescence were selected by a quarter wave plate and a polarizer. The circularly polarized light components of luminescence were detected by a photomultiplier tube through a monochromator.

The photon energy of excitation laser was fixed at 1.64eV. Figure 2 shows the spectra of luminescence polarization calculated from the intensities of the right and left circularly polarized light components at 10K. The maximum luminescence polarization of 65% was obtained around 1.60eV which agreed well with the lowest energy gap of the sample at 10K.

As the next step, the luminescence polarization was measured as a function of the photon energy of excitation light. The measured photon energy of luminescence was fixed at the lowest band gap, 1.60eV, of the sample. The luminescence polarization strongly depends on the photon energy of excitation laser as shown in Fig. 3. When the photon energy of excitation laser exceeds than 1.66eV, the luminescence polarization is lower than 30%. The low polarization is thought to be due to the optical transition from the second valence band to the conduction band. As the photon energy of excitation light approaches the lowest band gap, the luminescence polarization increases to 65%.

The spin polarization of conduction band electrons is relaxed during stay in the conduction band. The measured luminescence polarization using the cw laser is smaller than the spin polarization of electrons at the instant of excitation due to the spin relaxation. It is important to measure the initial spin polarization for development of the highly spin polarized electron source. The luminescence polarization P is defined as following equation[2],

$$P = \frac{T_s}{\tau + T_s} P_i , \qquad (1)$$

where τ and T_s are the lifetime and the spin relaxation time of conduction band electrons, respectively. P_i is the initial spin polarization which defined as the spin polarization at the instant of excitation. If the spin relaxation time and the lifetime are determined experimentally, the initial spin polarization is able to be determined.

In order to determine the spin relaxation time and lifetime, the time-resolved luminescence was measured using a mode-locked Ti:sapphire laser and a streak scope with time resolution of 15ps. The photon energy of excitation laser was set at 1.63eV which agreed with the excitation photon energy of the highest luminescence polarization. The spin relaxation time T_s and the lifetime τ were determined by decay curves of luminescence polarization and the total luminescence intensity, respectively. T_s and τ were estimated to be 270ps and 120ps, respectively. Substituting P, T_s and τ into Eq.(1), the initial spin polarization P_i was determined to be 94%.

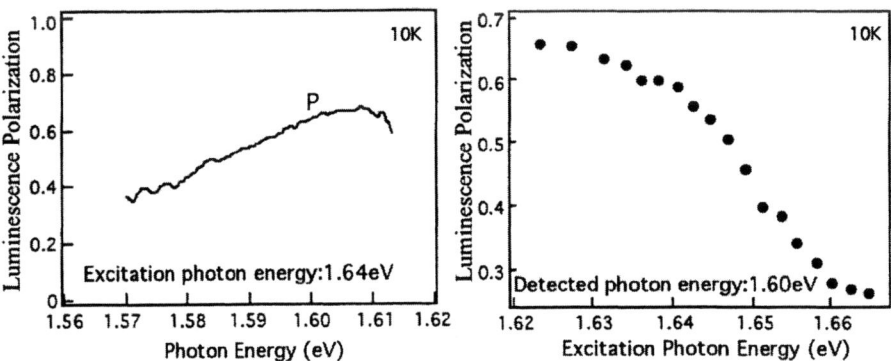

FIGURE 2. Spectrum of luminescence polarization at 10K

FIGURE 3. Excitation spectrum of luminescence polarization at 10K

CONCLUSION

Luminescence polarization of the InGaAs-AlGaAs strained layer superlattice reached 65% around 1.60eV which agreed with the estimated lowest energy gap of the superlattice. When the photon energy of excitation laser increased from 1.620eV to 1.665eV, the luminescence polarization decreased from 65% to 27%. The dependence of the luminescence polarization on the excitation photon energy reflected the valence band structure of the superlattice. From the result of time-resolved luminescence measurement, the spin polarization of conduction band electrons at the instant of excitation was estimated to be 94%.

REFERENCES

1. A. Filipe, H.J. Drouhin, G. Lample, Y. Lassailly, J. Nagle, J. Peretti, V.I. Safarov and A. Schuhl: *Phys. Rev. Lett.* **80** (1998) 2425.
2. D.T.Pierce and F.Meier: *Phys. Rev.* **B13** (1976) 5484.
3. T.Nakanishi, H.Aoyagi, H.Horinaka, Y.Kamiya, T.Kato, S.Nakamura, T.Saka and M.Tsubata: *Phys. Lett.* **A158** (1991) 345.
4. T.Omori, Y.Kurihara, T.Nakanishi, H.Aoyagi, T.Baba, T.Furuya, K.Itoga, M.Mizuta, S.Nakamura,Y.Takeuchi, M.Tsubata, and M.Yoshioka: *Phys. Rev. Lett.* **67**(1991) 3294.
5. T.Nakanishi, S.Okumi, K.Togawa, S.Nakamura, C.Suzuki, F.Furuta, T.Ida, K.Wada, T.Nishitani, M.Yamamoto, H.Kobayakawa, Y.Kurihara, H.Matsumoto, T.Omori, M.Yoshioka, K.Asano, H.Horinaka, K.Wada, T.Matsuyama, T.Baba, M.Mizuta, K.Kato and T.Saka: Proc. Low Energy Polarized Electron Workshop,St. Petersburg, 1998, p.118.

MeV Mott Polarimetry at Jefferson Lab

M. Steigerwald

Jefferson Lab, Newport News, Virginia 23606

Abstract. In the recent past, Mott polarimetry has been employed only at low electron beam energies ($\approx 100\,\text{keV}$). Shortly after J. Sromicki demonstrated the first Mott scattering experiment on lead foils at 14 MeV (MAMI, 1994), a high energy Mott scattering polarimeter was developed at Thomas Jefferson National Accelerator Facility (5 MeV, 1995). An instrumental precision of 0.5 % was achieved due to dramatic improvement in eliminating the background signal by means of collimation, shielding, time of flight and coincidence methods. Measurements for gold targets between $0.05\,\mu m$ and $5\,\mu m$ for electron energies between 2 and 8 MeV are presented. A model was developed to explain the depolarization effects in the target foils due to double scattering. The instrumental helicity correlated asymmetries were measured to smaller than 0.1 %.

INTRODUCTION

A Mott - polarimeter has been developed to measure the spin polarization of the electron beams produced by the sources of polarized electrons in the injector of Thomas Jefferson National Accelerator Facility. The polarimeter uses the counting rate asymmetry in the single elastic Mott scattering process which exists if the polarization vector is not parallel to the scattering plane. The Sherman - function determines the relation between measured asymmetry and the degree of polarization of the electron beam. Accurate polarimetry is ensured by addressing three concerns: First, the determination of the theoretical Sherman - function for the single elastic scattering process. Second, the correct measurement of the asymmetry for every target by the achievement of pure energy spectra and third, the understanding of the foil - thickness extrapolation to target thickness zero.

THE POLARIMETER SETUP

The present setup of the MeV polarimeter is shown in figure 1. Early versions of the polarimeter have been described in [1]. Since the kinetic energy of the electron beam is low between 2 and 8 MeV, the polarimeter uses a vacuum vessel to enclose the scattering target avoiding any spread of the electron beam in air

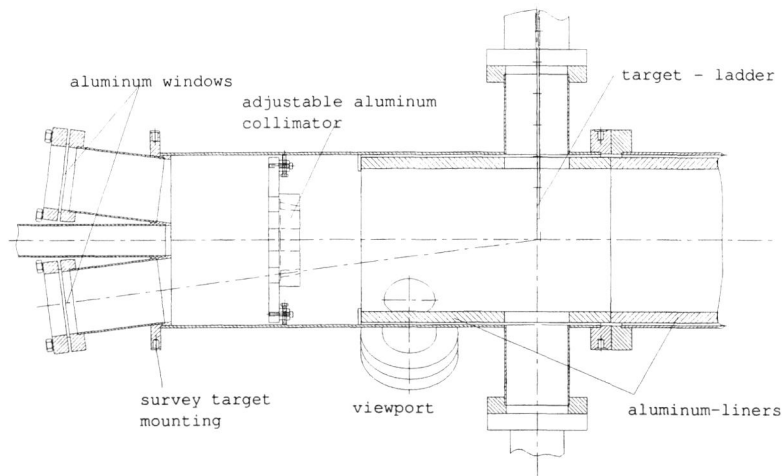

FIGURE 1. Cross section through the MeV Mott - polarimeter

(Fig. 1). The polarimeter's stainless steel chamber is directly connected to the vacuum beam line and can be isolated by means of a full metal valve. The electron beam is guided to the polarimeter target by a 12.5 degree dipole bend magnet, which is energized during a polarization measurement. The electrons enter the chamber from the left side of Figure 1 and hit the center of the target with a precision of 0.5 mm and an angle of less than 2 mrad. The beam spot may be observed by means of a CCD camera mounted to a side view port. It detects the transition radiation produced by the the electrons when they pass the target foil. The target ladder is mounted on a 600 mm linear drive that allows the selection of seventeen different targets. The target ladder holds 10 gold, 2 copper and 3 silver as well as a Cromox viewer and an empty target. The gold targets cover a thickness range of 500 Å to 5 µm. The polarimeter chamber has 4 ports that each support a 0.05 mm thin aluminum window, a detector assembly, and a lead collimator. Together with an additional adjustable aluminum collimator inside the vacuum chamber this arangement defines a scattering angle of 172.6 degrees and a solid angle of 0.18 msr. The collimators are chosen to accept only scattered electrons from a foil area that has a diameter of 3 mm. The scattered electrons penetrate the aluminum windows and reach the detectors. The main electron beam is dumped into a 21 mm thick aluminum plate which serves as the end flange of a 2500 mm long, 200 mm diameter aluminum tube connected directly to the polarimeter chamber. Aluminum liners are used to minimize photon production from electrons that hit the chamber walls after scattering from the target. In order to eliminate the Bremstrahlung photons which are produced by all types of electron

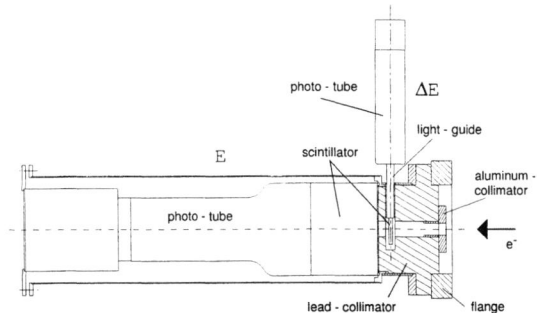

FIGURE 2. Cross - section of the detector arrangement

loss and field emission from nearby cryomodules a 50 to 100 mm lead shielding was constructed around the detector area.

Coincidence and time of flight methods

The analyzing power is only known for single elastic scattering, so it is necessary to separate the elastic scattered electrons from the inelastics. Hence, every detector assembly possesses two scintillator / photomultiplier assemblies (fig.2). The E detector serves as a stop detector and achieves an energy resolution of about 8 %. The phototube has an active cathode area of 75 mm. A cylindrical scintillator, 75 mm long with 75 mm diameter is glued directly to the photomuliplier window. Since the photon background crosses into the elastic electron peak of the detectors energy spectra and because it is impossible to shield against the photon background generated by the dump it was necessary to install a second detector assembly ΔE in front of the of the E detector assembly. The two detectors are run in coincidence. The ΔE detector assembly consists out of a 2 mm thin plastic plate and a 25 mm diameter phototube. A photon producing a signal is unlikely, so the ΔE detector serves only as a trigger. Finally, a time of flight methode was introduced in addition to the above described methods. Instead of illuminating the photemission gun with a laser pulse rate of 500 MHz which is the standard running mode the frequency is divided by four so that the electron bunchers possess a time gap of 8 ns. Hence, the electrons that are reflected at the aluminum dump that is 6 ns away from the target may be seperated from those that are scattered at the target by triggering on the laser frequency.

Figure 3 indicates the improvement made by using the different methods. By cleaning up the spectra so dramatically a background substraction is not necessary any more and therefore is the asymmetry measurement precision for one foil thickness limited to the statistical errorbar.

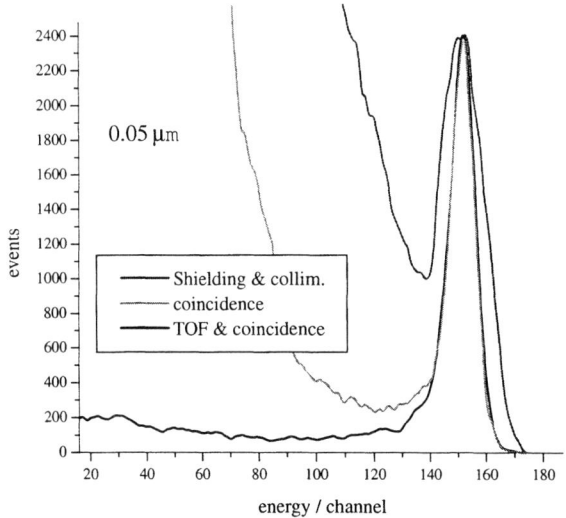

FIGURE 3. Energy spectra measured by means of the E detector

ANALAYZING POWER

The measured asymmetry is smaller for the thicker foil due to double scattering in the target foil. Therefore a foil thickness extrapolation has to be done to find the asymmetry for foil thickness zero that corresponds to the single elastic scattering process. In the past it was not clear which extrapolation function had to be used. This paragraph describes the calculation of the analyzing power for the single elastic scattering process as well as a new model that describes the dilution of the analyzing power with increasing foil thickness.

Single elastic scattering process

The analyzing power $S(\alpha)$ may be derived by solving the Dirac equation for a pure Coulomb potential [3] and by calculating the phase shifts of the scattering phases caused by the finite thickness of the nucleus [6]. Also important are the spin rotation functions $T(\alpha)$, $U(\alpha)$ as well as the cross section $\sigma(\alpha)$. These values are generously generated by Prof. Ch. Horowitz (Univ. of Indiana). They are defined by the scattering amplitudes f ang g:

$$\sigma(\alpha) = |f|^2 + |g|^2 \quad S(\alpha) = i\frac{fg^* - f^*g}{|f|^2 + |g|^2} \quad T(\alpha) = \frac{|f|^2 - |g|^2}{|f|^2 + |g|^2} \quad U(\alpha) = \frac{fg^* + f^*g}{|f|^2 + |g|^2}$$

The function f and g may be found in the literature [3] [5]. The screening of the Coulomb potential by the shell electrons is not taken into account in Horowitz's

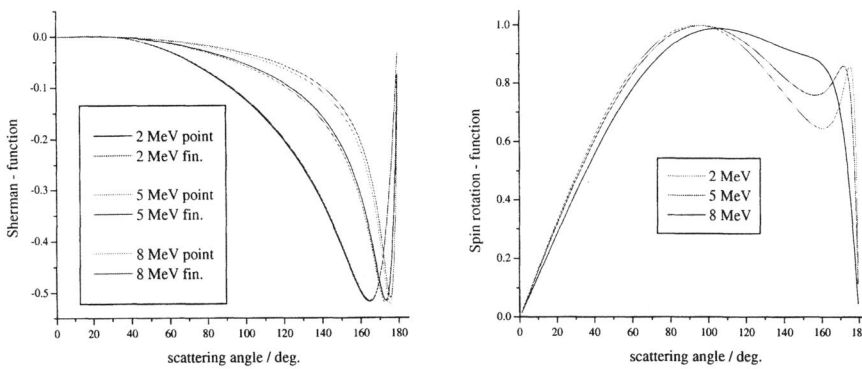

FIGURE 4. Sherman function and spin rotation function for 2, 5 and 8 MeV

claculation as well as the recoil of the center of mass and the charge distribution of the nucleus. The dominating uncertainty is caused by radiative corrections and is estimated to be lower than 1 %. Figures 4 show the results of the analyzing power and the spin rotation function for three different energies. Both solutions for point and finite size nucleus are indicated.

Double scattering

In order to explain the dilution of the asymmetry in target foils of finite thicknesses a model has been developed that follows Wegeners [4] exposition, but has been done numerically. In the following the number of big angle double and small angle multiple scattered electrons will be determined. The double scattering is calculated using cross - sections $\sigma(\alpha)$, Sherman functions $S(\alpha)$ and spin rotation functions $T(\alpha)$ and $U(\alpha)$ described in the upper section. The likelihood for triple scattering is neglected since it is very low at high electron energies in the MeV region. Hence, the following integral has to be determined

$$N = \int_{\theta=0}^{\pi} \int_{\varphi=0}^{2\pi} \int_{x_1=0}^{D} \int_{\vartheta=\theta_2}^{\theta_2+\Omega_\theta} 1(x_1\theta,\varphi) \cdot 2(x_2,\theta_2) \cdot E(x_1,x_2) \quad d\vartheta\, dl\, d\varphi\, d\theta$$

The dependencies on energy and number of protons in the nucleus is not indicated. The integration has to be carried out over the whole volume of the foil. Here θ and ϕ symbolize the polar - and azimuthal - angle of the first scattering and dl an integration step in foilthickness D. x_1 stands for the path between foil entrance and the first scattering, while x_2 symbolizes the path after the first scattering and exit of the foil. $1(x_1\theta,\varphi)$, $2(x_2,\theta_2)$ describe the first and second scattering processes, while $E(x_1, x_2)$ contains the impulse height spectra. $E(x_1, x_2)$ is achieved by Monte Carlo simulations and has different characteristics for foilthickness, the distances in the foil x_1 and x_2 and Energy and includes electron loss between the first and second scattering and electron absorption in the target. Therefore, it compensates the singularity that arises for scattering under 90 degrees when the electron travels parallel to the target surface and x_2 becomes infinit. The fourth integral is executed because of the change of $2(x_2, \theta_2)$ over the detector acceptance. The first scattering may be written as

$$1(x_1, \theta, \varphi) = I \cdot \frac{A_v \rho}{A} \cdot \sigma_1(\theta) \cdot x_1 \cdot \Omega_\varphi \cdot (1 + S(\theta_1) \cdot \vec{P_i} \hat{n}_1)$$

I is the incident electron current, A_v the Avogadro number, ρ the density of the target material, x_1 the path in the foil and Ω_φ the detector acceptance. The singularity of the modified Mott cross - section at a small angle is solved by means of GEANT simulations. Cross - sections for angles lower than the critical Møller angle are substituted by the results of the Monte Carlo simulation, which represents multiple scattering. The residual polarization after the first scattering is:

$$\vec{P}^\dagger = \frac{\left(\vec{P_i}\hat{n}_1 + S(\alpha)\right) \cdot \hat{n}_1 + T(\alpha) \cdot \hat{n}_1 \times \left(\vec{P_i} \times \hat{n}_1\right) + U(\alpha)\left(\hat{n}_1 \times \vec{P_i}\right)}{1 + \vec{P_i}\hat{n}_1 S(\alpha)}$$

The second scattering for the up (u) and down (d) detector results to

$$2(x_2, \theta_2)_{ud} = x_2 \cdot (1 + S(\theta_{2,ud}) \cdot \vec{P}^\dagger \hat{n}_{2,ud}) \cdot \sigma(\theta_{2,ud})$$

where θ_2 is the angle of the second scattering. The dependency of the cross section of the second scattering on the energy loss between first and second scattering is also taken into account.

Result

Figure 5 shows the number of electrons that hit the up and down detectors as a function of the first scattering angle θ_1. The numbers are calculated for an incident beam of $1\,\mu A$ and with an energy of 5 MeV having transverse polarization of 100 %. The target is a $5\,\mu m$ gold - foil. The 0 to 20 (160 to 180) degree peaks may be described as electrons that have a scattering with a first (second) small angle and a large second (first) scattering that possesses a high analyzing power. That produces a counting rate asymmetry between up and down detector. The peak at 90 degrees exists because the path x_2, and therefore the scattering probability $2(x_2, \theta_2)_{ud}$, becomes large. The integration over ϕ and the Sherman function of 4.3 % at 90 degrees lead to a small counting rate asymmetry between the up and down detectors. Writing the counting rate for up and down detector as the sum of polarization dependend portion A and independend portion U

$$N_{ud} = (1 \pm PS) \cdot A + U$$

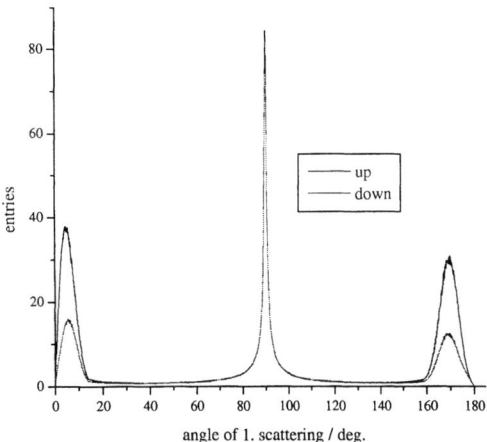

FIGURE 5. The polarization of the incident electrons is perpendicular to the scattering plane and its degree is 1. Number of electrons in the up and down detector in dependency of the first scattering angle

It follows the analyzing power in dependency of the foil thickness:

$$S(d) = S(0)\frac{1 + 0.00272 d^{0.866}}{1 + 0.23 d^{0.866} + 0.0729 d + 0.0146 d^2 + 0.00339 d^3}$$

The upper result is only correct if an energy cut is made at the center of the elastic peak. This is because the polarization dependent and independent portions have varying energy losses in the target material and have therefore different energy spectra.

CALIBRATION

Figure 6 compares the measurement (points with error bars) of foil thickness extrapolations for three different energies and the corresponding calculations (lines). The only free parameter in each case was the degree of spin polarization. Since the energy measurement results in an uncertainty of relative 0.3 % and each extrapolation to a statistical variance of 0.3 % all three extrapolations agree within the error bars. That leads to the conclusion that not only the form of the extrapolation function is understood but also that the calculation for the analyzing power agrees relatively well with each other. All investigations lead to the following result: The uncertainty of the extrapolated asymmetry at foil thickness zero is smaller than 0.5 %. The analyzing power for the single elastic scattering process is known to a

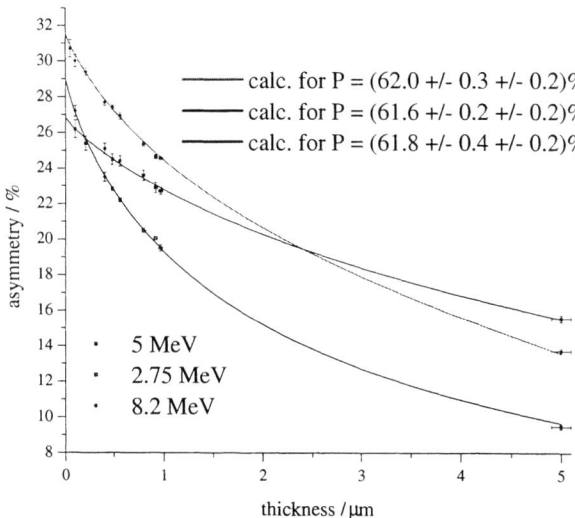

FIGURE 6. Foil - thickness extrapolation for 2.75, 5 and 8.2 MeV. Lines symbolize calculation and dots the measurement.

precision of 1 %. This results in a systematic uncertainty of the polarization measurement process of 1.1 %.

The instrumental asymmetry was determined to $(4 \pm 6) \cdot 10^{-4}$ by using an unpolarized electron beam. The asymmetries in both polarimeter arms agree within the statistical error bars. No significant dependency of the asymmetry was observed by varying the electron position on the target foil in an area of 10 mm diameter.

ACKNOWLEDGEMENT

This work was supported by the USDOE under contract DE-AC05-84ER40150. Special thanks to A. Day and B. Wojtsekhowski for their continous support.

REFERENCES

1. Price, J.S., *Proceedings of 13th International Symposium High Energy Spin Physics*, 554 (1998).
2. Sromicki, J., *Phys. Rev. Let.* **81(1)**, 57 (1999).
3. Sherman, N., *Phys. Rev.* **103(6)**, 1601 (1956).
4. Wegener, H., *Zeitschrift f. Physik* **151**, 252 (1958).
5. Mott, N.F., Massey, H.S.W., *Theory of Atomic Collision*, Oxford: Clarendon Press, 1965, ch. 6, pp. 23-26.
6. Horowitz, Ch. Private communication, (2000).

Polarized Source Performance and Developments at Jefferson Lab

M. Poelker, P. Adderley, J. Clark, A. Day, J. Grames, J. Hansknecht, P. Hartmann, R. Kazimi, P. Rutt, C. Sinclair, M. Steigerwald

Thomas Jefferson National Accelerator Facility
12000 Jefferson Ave., Newport News, VA 23606, USA

Abstract. The polarized photoinjector at Jefferson Lab continues to provide high average current, high polarization, high quality beam to nuclear physics Users in as many as three endstations simultaneously. Long lifetime operation has been obtained from two identical polarized guns. A new high power modelocked ti-sapphire laser has been constructed to enhance the effective operating lifetime of the photoinjector. Efforts to enhance beam polarization and reduce helicity correlated beam systematic effects are underway.

INTRODUCTION

Jefferson Lab is a cw electron accelerator with a recirculating linac design and beam energies to 6 GeV. There are three endstations and each can receive beam simultaneously. The beam current requirements in each hall vary significantly. Halls A and C can take up to 120 microA, Hall B typically receives only 1 to 4 nanoA. Multiple simultaneous Users and a large dynamic current range place strict demands on the photoinjector and laser systems.

The polarized beam program at Jefferson Lab began in December, 1997 with bulk GaAs and a peggy-style polarized electron gun. Performance was greatly enhanced when the gun was modified to accommodate improved pumping near the cathode/anode gap and other features described below [1]. Only high polarization photocathodes have been used since August, 1998. The most recent injector modification occurred during the summer of 1999 when the photoinjector was rebuilt to house two identical polarized guns, one in use for production beam delivery and one serves as a spare.

Gun performance improved significantly when non-evaporable getter (NEG) pumps were installed near the photocathode to provide more than 2000 L/s pump speed and a base pressure in the main gun chamber in the upper 10e-12 Torr range (Figure 1). But central to the success of the new gun design was the realization that during operation, a small portion of the total photoemitted gun current originates from the edge of the photocathode, beyond the region where the photocathode is directly illuminated with laser light. Photomission from large radius of the photocathode is probably a result of low level ambient light within the gun caused by stray laser light or spontaneous emission from the photocathode itself.

The Horizontal Gun at CEBAF

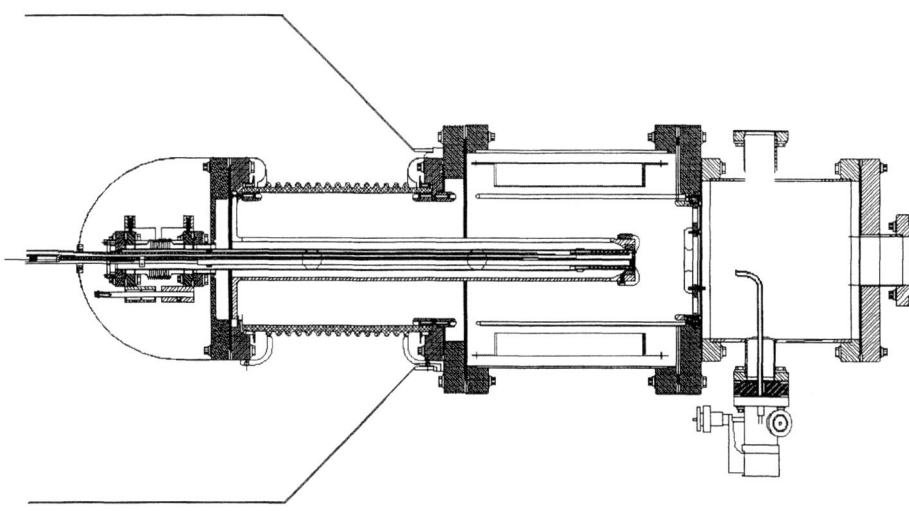

FIGURE 1. Schematic diagram of the Jefferson Lab polarized electron gun. There are two such guns installed in the photoinjector, one for production beam delivery and one spare available when necessary.

Electrons that originate from the edge of the photocathode follow extreme trajectories; some hit the anode plate, some travel further from the gun and hit beamline walls, situations that degrade vacuum pressure through stimulated desorption of gas and subsequent quantum efficiency degradation via ion backbombardment. The following precautions reduce the production of photoemission from large radius of the photocathode and help minimize deleterious effects should the circumstance persist; a) the edge of the photocathode wafer is anodized [2] to produce quantum efficiency of zero at large radius, b) large diameter vacuum beamline tubes (2.5 inch) are coated with NEG material to minimize electron stimulated desorption and provide improved pumping near the source of the gas source should electrons strike the beamline walls [3], and c) the short focal length electron optical elements of previous injector configurations (90 degree bend magnet and electrostatic bends of the z-style spin manipulator) were replaced with a 15 degree bend magnet and Wien-style spin manipulator.

The effectiveness of these features is illustrated by considering photoinjector performance during a recent 6 month period from January through June, 2000. During this period over 900 Coulombs were extracted, with approximately 2/3 delivered to the endstations. The average charge lifetime from both guns approached 200 Coulombs, where lifetime refers to the total charge extracted until the quantum efficiency falls to 1/e of the initial value. Ion backbombardment is the sole mechanism by which quantum efficiency degrades. The effective operating lifetime of the gun exceeds the charge lifetime because quantum efficiency degrades only at the location of laser spot and along a line radially inward to the electrostatic center of the cathode. The laser beam diameter (0.5 mm) is considerably smaller than the active area of the photocathode (6 mm) so it is possible to continue delivering beam to the endstations by moving the laser to fresh locations on the photocathode. The effective operating lifetime of the gun is also enhanced by occasional recesiations or heat treatment followed by full reactivations with cesium and nitrogen trifloride. Quantum efficiency degradation from ion backbombardment is completely restored following a heat treatment.

LASER SYSTEMS

Until very recently, only diode laser systems [4] were used to drive the Jefferson Lab photoinjector. The diode laser systems are composed of gain-switched diode seed lasers and diode optical amplifiers. Optical pulsewidths of ~ 50 ps are obtained with approximately 100mW average power at a pulse repetition rate of 499 MHz and 250mW at 1497 MHz. The diode laser systems have many appealing features including relatively low cost (~ 15K USD) and near maintenance-free, long lifetime operation. The laser output is "clean" with low amplitude noise and pulse-to-pulse timing jitter. The most appealing feature of the diode laser system is the ease by which the optical pulse train can be phase locked to the accelerator. This is because the pulse forming mechanism, gain switching, is independent of the diode laser cavity length. The pulse repetition rate and the phase of the optical output follows that of the sinusoidal rf signal applied to the diode seed laser.

Unfortunately, there are disadvantages of the diode laser systems. In particular, the relatively low output power of the systems (and the low quantum efficiency of strained layer photocathodes) makes it difficult to meet the high current beam requirements of the nuclear physics community. In addition, the manufacturer of the diode optical amplifier (Spectra Diode Labs) no longer sells this component and other vendors are unknown. Finally, the diode laser systems do not provide extensive wavelength tunability (only +/- 5 nm via temperature tuning) and as a result, new photocathodes may require completely new diode elements to reach the wavelength that provides peak polarization.

To overcome the limitations of the diode laser systems, a novel modelocked ti-sapphire laser has been constructed [5] and installed in the injector. The complete laser system composed of the new ti-sapphire laser and diode laser systems is shown in Figure 2. Diode laser systems continue to provide high-polarization beam to endstations that require < 30 microA. The ti-sapphire laser is used for endstations with higher beam current requirements

The ti-sapphire laser cavity has a double-fold geometry and is pumped with green light from a solid-state Verdi-10 laser (Coherent Inc.). The ti-sapphire laser can be configured with one or two output coupler mirrors; one output coupler mirror for high current beam delivery to one hall, and two output coupler mirrors when two halls require simultaneous high current beam. The cavity length is chosen to be 60.1 cm corresponding to a longitudinal mode spacing of 249.5 MHz, a subharmonic of the

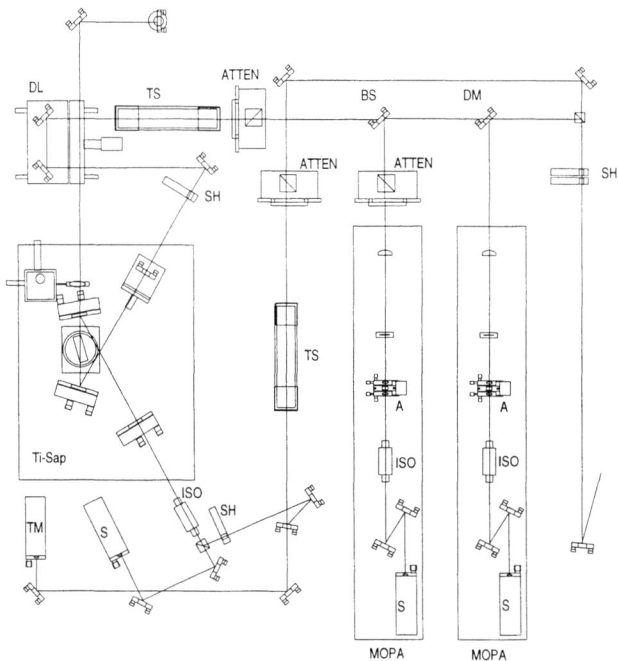

FIGURE 2. Schematic diagram of the Jefferson Lab drive laser system. MOPA, diode laser system; S, gain-switched diode laser; A, diode amplifier; ISO, optical isolator; Atten, optical attenuator; TS, telescope; SH, shutter; DM, dichroic mirror; BS, beamsplitter; DL, optical delay line; TM, tune mode laser diode.

accelerator cavity rf frequency. Modelocked pulsed operation of the ti-sapphire laser is obtained when light from a gain-switched diode laser is directed into the cavity and the repetition rate is set to 499 MHz. The best optical pulses (i.e., most stable) are obtained when the ti-sapphire and diode laser wavelengths are equal to within a few nanometers. The phase of the ti-sapphire laser optical pulse train is set by the phase of the rf signal used to drive the gain-switched diode laser, a situation that makes it a simple matter to lock the optical pulse train to the accelerator. Presently, the cumulative output power of the laser is ~ 800 mW with pump power of 5 W. The laser has been running nearly continuously at Jefferson Lab for approximately three months and although this laser requires more attention than diode laser systems, performance has been very good. Occasional mirror cleaning and adjustment of mirror alignment and cavity length are necessary to maintain optimum output power and pulse stability. In the future, feedback mechanisms will be implemented to reduce the necessity of making these occasional manual laser adjustments. Currently, experiments in two endstations receive 60 microA of high polarization beam using the ti-sapphire drive laser. The high power capability of this laser provides exceptionally long operating lifetimes, allowing weeks of uninterrupted high current beam delivery before actions must be taken to restore the photocathode quantum efficiency. Laboratory demonstration suggests output powers to 2 W are possible in the future.

ENHANCING BEAM POLARIZATION

To date, we have used only high polarization photocathodes from Spire Inc. (Bandwidth Semiconductor). These GaAs-on-GaAsP photocathodes are grown to the SLAC specification and the best samples have provided beam polarization 70 to 80%. Unfortunately, many samples provide lower beam polarization, and these wafers are rejected for use in the photoinjector. Polarization also varies from location to location across a single sample. Polarization within the 6 mm diameter active area can vary by +/-10%. There is speculation within the polarized source community that low polarization and polarization variation across a small sample may a be result of photocathode surface roughening caused by atomic hydrogen cleaning [6]. All photocathodes at JLAB are cleaned via atomic hydrogen [7] and tests are underway to investigate possible effects on beam polarization.

There is a considerable backlog of nuclear physics experiments awaiting beam time at Jefferson Lab. Higher beam polarization will reduce the amount of time required to achieve appropriate statistical accuracy and thereby hasten the process of conducting nuclear physics research. As such, it is very important to begin using photocathodes that provide polarization greater than 80%. To this end, we are testing photocathode material from the groups of Nakanishi, Terekhov and Mamaev.

REDUCING HELICITY CORRELATED BEAM SYSTEMATICS

During the spring of 1999, a parity violation experiment was conducted in Hall A (HAPPEX) using high polarization beam from a strained layer photocathode [8]. This experiment required control of helicity correlated charge asymmetry to less than one ppm and it was feared that strained layer photocathodes would be unacceptable because of the inherent quantum efficiency anisotropy associated with the presence of residual linearly polarized light in the two helicity states [9]. Fortunately helicity correlated charge asymmetry was kept acceptably small using a rotating halfwave plate downstream of the pockel cell. This device was used to orient the residual linear polarization in directions to produce equal beam current in the two helicity states.

Future nuclear physics experiments however will require more sophistication (i.e., tighter control of helicity correlated charge, position and beam energy asymmetries). Efforts are underway to minimize the problematic nature of strained-layer photocathodes by creating more highly circularly polarized laser light. An ellipsometer is being constructed to help properly orient the pockel cell with high accuracy and reproducibility with the hope that circular polarization of 99.9999% can be obtained. A test stand has been constructed to study the laser beam-steering properties of pockel cells and piezo-driven mirror mounts have been added to the laser table to counter the beam steering effects of the pockel cell. Finally, it may be possible to use Kerr cells rather than pockel cells as a means to eliminate helicity correlated position asymmetries.

ACKNOWLEDGMENTS

This work was supported by the USDOE under contract DE-AC05-84ER40150.

REFERENCES

1. Gun modifications have been described in other proceedings, for example; M. Poelker et.al., "Operation Experience with the Jefferson Lab Polarized Electron Source" in LE98 Proceedings, St. Petersburg, Russia, Sept. 1998, and C. Sinclair, "Performance of the Jefferson Lab Polarized Electron Source at High Average Current" in PST99 Proceedings, Erlangen, Germany, Sept., 1999.
2. B. Schwartz, F. Ermanis, and M.H. Brastad, "The Anodization of GaAs and GaP in Aqueous Solutions," J. Electrochem. Soc.: Solid State Science and Technology, Vol. 123, No 7, 1089-1097 (1976).
3. C. Benvenuti et al., J. Vac. Sci. Technol. A16, 148 (1998).
4. M. Poelker, "High Power Gain-Switched Diode Laser Master Oscillator and Amplifier," Appl. Phys. Lett. 67, 2762 (1995).
5. C. Hovater and M. Poelker, "Injection Modelocked Ti-sapphire Laser with Discreetly Variable Pulse Repetition Rates to 1.56 GHz," Nucl. Instrum and Meth A, 418, 280-184 (1998).
6. See Terekhov, these proceedings.
7. C. Sinclair, et al., "Atomic Hydrogen Cleaning of Semiconductor Photocathodes," Proc. 1997 Particle Accel. Conf., Vancouver, B.C. Canada, May 1997.
8. K.A. Aniol, HAPPEX Collaboration, "New Measurement of Parity Violation in Elastic Electron Proton Scattering and Implications for Strange Form Factors," submitted to Phys. Lett. B.
9. T. Maruyama, et al., "Recent Developments on Strained Photocathode Research," LE98 Proceedings, St. Petersburg, Russia, Sept. 1998.

New results from the Mainz polarized electron facilities

Kurt Aulenbacher* for the B2-collaboration:

K. Aulenbacher, H. Euteneuer, P. Jennewein, K.-H.Kaiser, H.J. Kreidel, D. v. Harrach, E. Reichert, J. Schuler, V. Tioukine, M. Wiessner, K. Winkler

Institut für Kernphysik der Universität Mainz, J.J. Becherweg 45, D-55099 Mainz

Abstract.
The polarized electron source at the Mainz microtron has improved it's performance in every parameter during the last year. Run times have been performed with simultaneously 20 Mikroamps of beam current and more than 80% of polarization. We demonstrate that the limited beam-current lifetime product is due to a transmission loss of about several 10^{-4}. The loss can be minimized by the usage of photocathodes with only partially photosensitive surface, thus resulting in a nearly one order of magnitude higher beam-current lifetime product.

I OPERATIONAL PARAMETERS IN 1999-2000

A Beam production

The source of polarized electrons is installed at the injection point of the MAMI accelerator (see fig:2). During the last year it has been used for about 2000 hours of runtime- about one third of the total accelerator time. Beam currents were up to $20\mu A$. Figure 1 shows the increase of the beam quality factor $P^2 * I$ in production runs at MAMI since the begin of polarized beam operation. The main improvements in the last year were:

- An increase of the beam-current lifetime product from 5 Coulomb to presently more than 10 Coulomb.

- By selecting the best available photocathodes from different wafers we have improved the polarization of the beam to more than 80%.

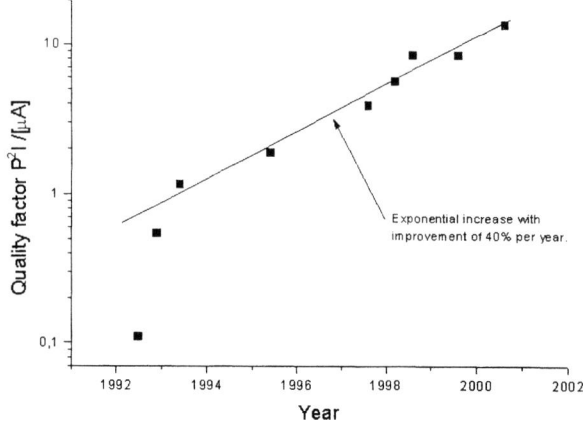

FIGURE 1. Improvement of beam quality factor (on target) in production runs at MAMI.

B Helicity correlated parameters

The helicity correlated beam intensity fluctuations have been minimized to a value below 10^{-5}. This has been done by an accurate angular orientation of the photocathode: The inhomogeneous strain relaxation [1] axis of the photocathode is oriented with respect to the remaining linear polarization components of the incoming light [3]. We have observed stability of the helicity correlated asymmetries over periods of days. Under this conditions we are able to determine the average value of the HCA much more precisely, so that we are optimistic to achieve the desired accuracy of correction $(2*10^{-7})$.

II PHOTOCATHODE LIFETIME INVESTIGATIONS

A Observations concerning life time limiting effects

The decay constant of quantum efficiency (q.e.) in our apparatus is roughly proportional to the beam current. Therefore from the observed 1/e decay time τ_{obs} one can form the beam-current lifetime product:

$$C = I * \tau_{obs}$$

In the recent production runs the value of C was about 10–15 Coulomb. The laser power is then sufficient to compensate for the decay for about 150 hours at a current of 20 Microampere. This is more than the typical production interval at MAMI (5-6 days). Therefore the availability of the source was virtually 100%.

For longer (permanent) run-time the availability would drop because of the time needed for photocathode reactivation. This time interval is about 2-4 hours, leading to a reduction of availability to 97%. However for the production of higher currents this reduction becomes larger, so there is interest in increasing the beam-current lifetime product C.

The present observations concerning the nature of C-limitation are the following:

Most decrease of q.e. is global, i.e. on the whole photocathode surface. The local effect which is caused by ion backbombardment [2] is very small, which is demonstrated in figure 3: Only $< 2\%$ of q.e. can be recovered by changing the laser spot location. The global effect is therefore of most concern in our apparatus at present time.

The figure 3 shows the q.e decay for an interval of one day after several days of operation at $23\mu A$. A remarkable feature of the curve is that q.e. decay continues while the beam production was stopped.

This effect could be explained by data from Gröbner et.al. [4]: Their experiment was done with soft X-rays from synchrotron-radiation impinging on a baked UHV stainless steel vacuum chamber. This conditions could be similar to the ones present in a polarized source when beam loss generates a spectrum of bremsstrahlung. Gröbner et. al. demonstrated that after the turn on of the X-rays several species of chemical active gasses (O_2, H_2O) are formed. The evaporation of water continues for hours even after the X-ray radiation is stopped, a fact that could explain the

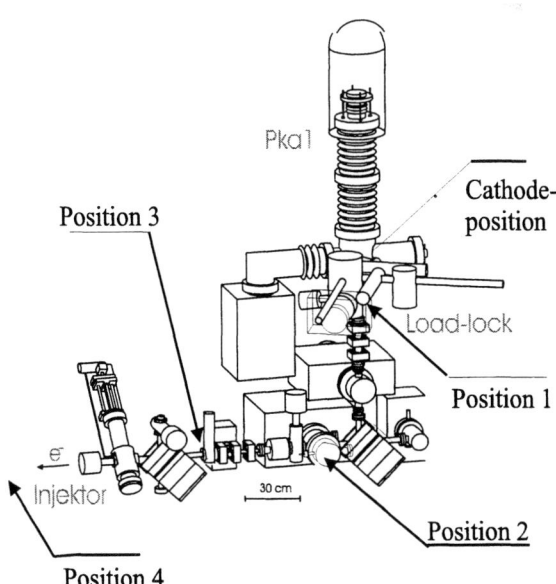

FIGURE 2. Compact set-up of polarized source in Hall-A of the MAMI accelerator

ongoing decay of the photocathode during a break in beam production.

This argument suggests that the observed global photocathode decay is still caused by beam loss, even though the losses are very small. Therefore a method to avoid beam losses could be more effective then for instance an increase in the effective pumping speed.

B Beam losses in low emittance electron sources

The electron beam in our source is started slightly (1.2 mm) out of the center of the source to minimize the effect of ion-backbombardment. The emitting area has 300 μm diameter. The normalized (rms-) emittance is $E_{norm}^{rms} = 0.05\,\pi\,mm\,mrad$.

The low emittance beam of the source guarantees that the beam has a smaller diameter than 3 mm FWHM everywhere in the beam line. Therefore beam losses on the vacuum pipe - which has a typical diameter of 30 mm - should be negligible.

On the other hand any amount of parasitic photoemission from the other areas of the 1 cm diameter photocathode will be treated differently by the electron optics. The focusing effect of the conical surrounding of the photocathode leads to a large loss for all electrons that are started on the edge of the photocathode[1].

C Comparison 'nude' and anodized cathode

We decided to test an 'anodized' photocathode (unstrained bulk GaAs) which was kindly produced for us by the CEBAF team. As reported in Erlangen [5] this process helped CEBAF -together with an optimization of the vacuum system - to achieve a much better beam-current lifetime product. (C= approx 100 C). In our experiment we wanted to separate the effects of vacuum improvement and transmission optimization by leaving the vacuum system unchanged.

For comparison we used 'nude' photocathodes (unstrained bulk GaAs and strained layer) were the full area (diameter 10mm) was photosensitive, and one where an outer ring was 'anodized' and therefore only had a 3mm diameter sensitive area in the center. The anodized photocathode allows to avoid the losses which are caused by parasitic photoemission from the outer ring. The anodized layer has no influence on the electron optics because it is very thin. All samples were tested with identical settings of the electron optics.

We compared the difference of transmission losses for both types of photocathode at a series of points (marked one to four in figure 2) in the beam line. Point one is an insulated piece of beam tube with 25 mm diameter at a distance d=0.1 m from the anode, where the loss current can be measured directly. At point two

---

[1] This is not the case for space charge limited operation with a pierce cone. However this mode of operation would require to have a very small area of the photocathode for a low emittance space charge limited operation. In this case the emission area -located at the center of the cathode- could be quickly destroyed by the ion backbombardment.

we measure with two small ionization chambers the current that is lost on a 8 mm diameter aperture in the differential pumping stage. Point 3 and 4 are emittance collimators each about 2.5 mm in diameter.

The results are summarized in table 1. It is clearly visible that the anodized photocathode offers superior conditions to avoid beam loss. Especially the loss close to the source is dramatically reduced. The observed increase in vacuum-level per emitted current in the source was at least one order of magnitude lower for the anodized cathode. The results with the 'nude' cathode do not depend on the cathode type (bulk or strained layer).

position	dia and distance	loss nude	loss anodized
1(tube)	$\phi = 25mm, d=0.1m$	$5*10^{-4}$ to 10^{-3}	10^{-6}
2(I-chamber)	$\phi = 8mm, d=2m$	$5*10^{-4}$	$< 5*10^{-6}$
3(Collimator)	$\phi = 2.5mm, d=2.5m$	$1.3*10^{-2}$	$1.1*10^{-3}$
4(Collimator)	$\phi = 2.5mm, d=3.8m$	$6*10^{-2}$	$1.8*10^{-2}$

TABLE 1. Transmission losses at various points of the beam line

Another experiment was done to observe the improvement in lifetime that can be achieved with the anodized photocathode. Because the limited amount of time available at MAMI for such an experiment the current was increased to initially 200 μA. Figure 4 shows the result:

After an initial sharp drop the quantum efficiency increases again and eventually reaches a plateau. This behavior is also observed in the operation with 'nude' photocathodes. After fifteen hours the q.e. starts to decrease. The decay constant at the end of the experiment is estimated to be 184 hours at a current of 175 μA.

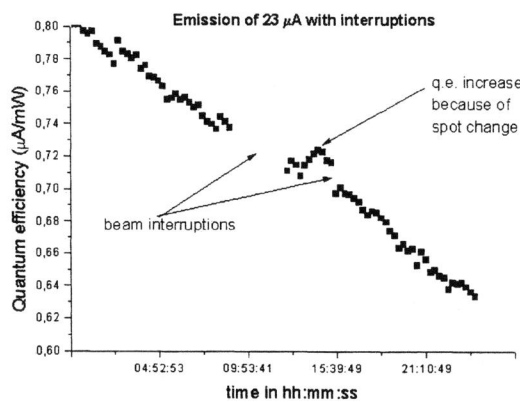

FIGURE 3. Lifetime in production with strained layer cathode: Effect of beam interruption

The total amount of charge that can be emitted from this photocathode can be extrapolated to be between 50 and 100 Coulomb in one 1/e decay from the start.

The uncertainty comes from the fact that we cannot completely exclude the possibility that the curve will show a stronger decay after longer time. However such an effect was not observed in the long term production runs with a 'nude' strained layer photocathode at 23 μA where a similar curve shape was observed.

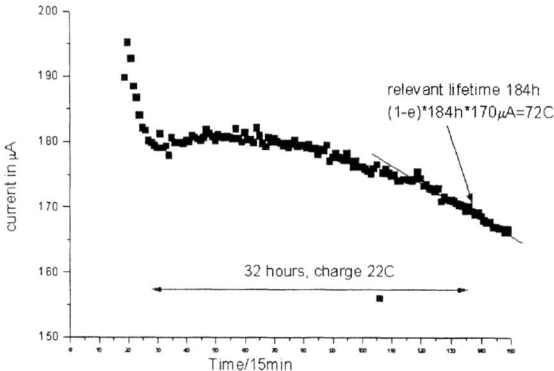

FIGURE 4. Decay of photocurrent of anodized cathode under constant laser illumination.

We therefore expect that the usage of anodized strained layer photocathodes will enable us to increase the beam-current lifetime product by a factor 3-10.

III ACKNOWLEDGMENTS

We want to thank Charlie Sinclair and his group at CEBAF for providing us with an anodized photocathode for the lifetime studies.

This work was supported by the Deutsche Forschungsgemeinschaft within the framework of the SFB 443.

REFERENCES

1. R. Mair et. al., Phys. Lett. **A 212** (1996) p. 231-236
2. K. Aulenbacher, C. Nachtigall et. al. NIM **A 391** (1997) p.498
3. K.Aulenbacher et. al. in: Y. Mamaev (ed.) *Proceedings of the LE98 Workshop*, St. Petersburg 1998. p.
4. O. Groebner et al. in: S. Turner (ed.) *CERN Accelerator school on Vacuum technology*, Geneva 1999, p.127-138
5. C. Sinclair et.al. in: A. Gute (ed.): *Proceedings PST 1999, Erlangen 1999*, p.222-p.229.

New Results from the MIT-Bates Polarized Source and the Test Beam Setup[†]

M. Farkhondeh, E. Tsentalovich, T. Zwart and E. Ihloff

*MIT-Bates Linear Accelerator Center
P. O. Box 846, Middleton, MA 0194, USA*

Abstract. During the summers of 1998 and 1999, the polarized source at MIT delivered over 300 Coulombs of high quality polarized pulsed beam to the SAMPLE parity violation experiment. This success in beam delivery signified achievements of long lifetime for the Bates polarized source in both high average and high peak current regimes. The injection requirements of high polarization beams for the South Hall Ring will be discussed. They are relatively modest for the storage mode but very demanding in peak current and repetition rate for the extraction mode. Because of these requirements, a 60 keV test beam setup with a Mott polarimeter was constructed at Bates and will be described here. This setup is independent of the main accelerator and has provided a platform for R&D in photocathodes and high power lasers. New results will be presented on high peak current photoemission from strained GaAsP samples using fiber coupled diode array laser systems on this test setup. A brief report will be presented on the photoemission requirements of a future e-p Collider that is currently under discussion in the nuclear and particle physics community in the US.

MIT-BATES POLARIZED SOURCE

The polarized source at Bates consists of a commercial CW laser system, a 60 kV diode electron gun, followed by a Wien filter for spin manipulation, a 300 kV acceleration column, a vertical transport line and an achromat with two 45° dipoles. The gun chamber and the associated instrumentation are located inside an elevated Faraday cage separated from the ground with corona rings. Two additional guns identical to the injector gun are prepared with certified photocathodes and are available as backups. A typical gun interchange requires about four days of downtime. The laser system consists of a 30W Ar laser pumping a CW Spectra-Physics Titanium-sapphire laser.

Polarized Beams for the SAMPLE Experiment

The SAMPLE experiment at Bates [1,2,3] measures a parity violating asymmetry between the yield of positive and negative helicity electrons scattered at backward angles from an unpolarized liquid hydrogen or deuterium target. Because the measured asymmetry is extremely small, $\sim 10^{-6}$, it is necessary to suppress any helicity-correlated differences in the properties of the electron beam that may cause false

---

[†] Work supported in part by the Department of Energy under a Cooperative agreement # BEFC294ER40818.

asymmetries in the measured yield. These include differences in the current, position, angle and the energy of the incident beam for the two helicities. For instance, the experiment required helicity-correlated beam position differences to be controlled at the level of better than 100 nm at the target for a beam that has a FWHM of about few millimeters. SAMPLE also required long lifetime from the source so that a minimum of two Coulombs a day on target is delivered for periods of several days between activations. Substantial improvements in the polarized source [4] for several years led to a successful data run of the SAMPLE experiment during the summers of 1998 and 1999. In those periods, high quality polarized beams were delivered to the experiment totaling over 300 Coulombs on the hydrogen and deuterium targets. The actual total charge from the source was over 900 Coulombs. Figure 1. shows the daily Coulomb delivery during the two periods of the experiment.

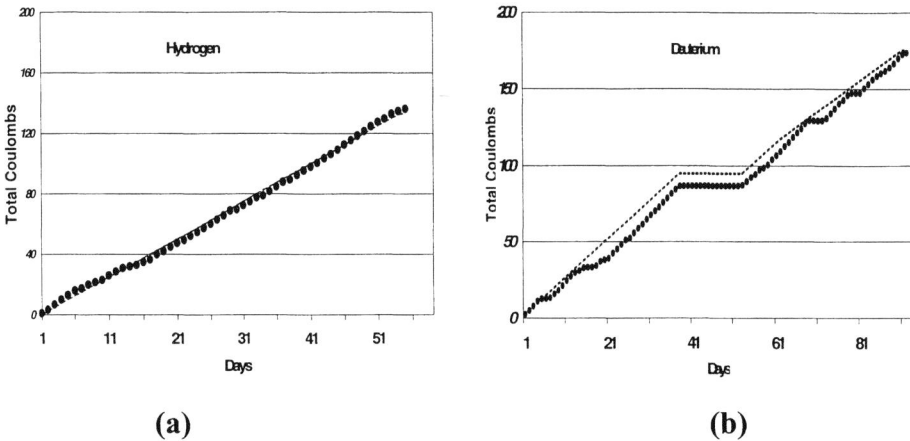

FIGURE 1. Total Coulombs shown vs. days during the SAMPLE a) hydrogen [1] and b) deuterium experiments [2].

Only a few photocathode heat cleanings and activations were needed during each summer; this signifies a very long photocathode lifetime of the order of ~150 Coulombs or about 25 days between activations. Peak currents of 8-10 mA and average currents of ~120 µA from the source were needed for SAMPLE. These average currents are comparable or higher than average currents delivered at Jefferson Lab or at Mainz which have CW machines operating with one to two orders of magnitude smaller peak currents. With a maximum of 3.5 W laser power available at the photocathode, quantum efficiencies (QE) of greater than 0.5% were needed. This requirement prevented the use of high polarization photocathodes, which have inherently low QE's. The new physics results from the deuterium experiment in 1999 [2,3] brought forth a new proposal for SAMPLE with an incident beam energy of 125 MeV. This newly approved experiment requires 150 Coulombs of high quality beam on a deuterium target. The purpose of this run is to measure the same quantity, but at different momentum transfer. The main polarized source injector will be left unchanged for this experiment. The data run will take place in 2001.

POLARIZED BEAMS FOR SOUTH HALL RING

The two modes of operation of the South Hall Ring (SHR) are the storage mode with stacking option and the extraction or pulse stretching mode. The parameters for high polarization beams ($P_e > 60\%$) required for each mode are listed in Table 1.

Table 1: SHR beam parameters

Beam Parameters	Storage Mode	Extraction Mode
Average current	25-200 mA* circulating	5-10 µA extracted
Injection Repetition rate	< 1 Hz	600 Hz (1800)**
Injection Peak current	1-5 mA with stacking	5-10 mA
Pulse duration	1.3 µs	1.3 µs

* Depending on the damping time of betatron oscillations during stacking.
** 1.8 kHz may be possible after an upgrade to the RF transmitters is completed.

The existing Ar-pumped Ti:sapphire laser with ~3.5 W of radiation is not adequate for high polarization requirements listed in Table 1. New lasers are therefore essential for SHR. As shown in the table, for the storage mode the technical challenge for the new laser system is greatly reduced by stacking option and by the low repetition rate of injection. For instance, the high power pulsed laser system used at the AMPS ring at NIKHEF would be adequate to provide high polarization beams for the internal physics program at Bates. With this laser, peak currents of 1-5 mA of high polarization beam would be injected to the linac for storage mode. With the linac capture efficiency of about 1/3, a peak current of 5-15 mA at low repetition rate would be needed from the polarized source. However, it is desirable to use a laser system that can meet the injection requirements for both the storage and extraction modes. In fact, a system that can meet the requirements of the extraction mode can easily satisfy the requirements of the storage mode.

The high peak current and high repetition rates of the extraction mode are very demanding for the polarized source. An attractive option for this mode is the use of commercial high power CW fiber-coupled diode lasers that can readily be operated in pulsed mode at high repetition rates. Specifically, we have tested the Spectra-Physics model OPC-BO60-mmm-FC that is rated for 60 Watts of unpolarized radiation at 848 nm. A reduction of a factor of two is expected when the light is subsequently polarized. This laser is potentially capable of meeting the requirements for both modes of operation for SHR. The drawback of fiber-based diode lasers is an inherently large emittance that is totally incompatible with our existing 15m-long laser transport system. With an emittance of ~100 mm-mr, the beam spot size after one meter grows to ~10mm, the diameter of the crystal. An alternative path to the photocathode is through an existing view port in the gun chamber that is at a 37° angle with respect to vertical and is intended for viewing the laser spot on the photocathode.

NEW INITIATIVES

During the past two years, motivated by the requirements of photoemission for SHR and guided by the performance of the source during the SAMPLE experiment, a series of new initiatives have been implemented as listed here.

- A clean room for UHV and photocathode work.
- An atomic hydrogen cleaning apparatus for surface cleaning of photocathodes.
- A 20-MeV transmission polarimeter on the linac for rapid polarization tests.
- A laser back scattering Compton polarimeter for the SHR.
- A high power flashlamp Ti:sapphire laser from NIKHEF.
- High power CW diode lasers at 850, 830 and 808 nm.
- A Test Beam Setup (TBSU) with a Mott polarimeter.

The Compton polarimeter was tested with unpolarized beam earlier this year and will be fully commissioned with polarized beam later this year. This polarimeter will be used for continuous monitoring of beam polarization for the storage mode. The TBSU facility provided for the first time the capability at Bates to characterize photocathode polarization and QE independent of the main accelerator.

Test Beam Set Up

The TBSU at Bates is a 60 keV beam line that consists of a Bates gun assembly, a beam transport line with a 90-degree magnetic bend, a Wien filter, a conventional Mott polarimeter and a beam dump. A schematic view of this setup is shown in Figure 2. The beam line controls and the Mott polarimeter data acquisition systems are based on the EPICS [5] software and VME hardware. The Mott polarimeter was calibrated by measuring the Mott asymmetry from a Bulk GaAs photocathode with a known polarization. This polarization was repeatedly measured with the 200 MeV S-line Møller polarimeter in 1999 during the SAMPLE experiment. Our gun assembly was designed to allow the interchange of guns while preserving the photocathodes.

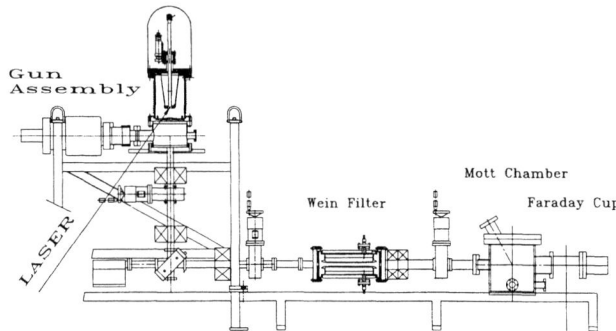

FIGURE 2. A view of the 60-keV test beam setup is shown. The direction of the laser striking the photocathode through a view port is also illustrated. The actual plane of the incident light is perpendicular to the view shown.

New Results

A strained GaAsP sample grown by SPIRE Corporation [6] was installed in one of the Bates guns and a photocathode was made. The polarization and QE were measured vs. wavelength using radiation from our exiting Ti:sapphire laser through a long fiber optic cable. The result is shown in Figure 3 (a). The 60-Watt diode laser [7] operating in pulse mode at a very low average power was also used on this photocathode. In pulsed mode, this laser provided peak power of as much as ~150W at 848 nm. A plot of peak current and QE vs. laser peak power is shown in Figure 3 (b). Peak currents as much as ~60 mA of unpolarized beams were achieved.

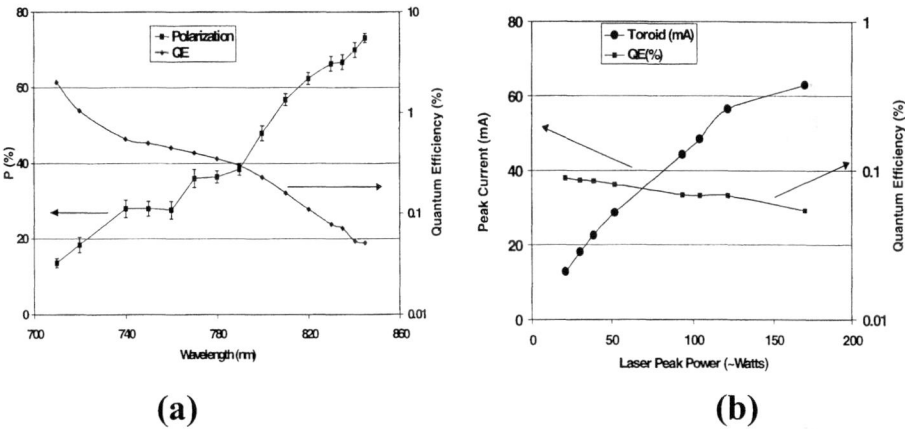

FIGURE 3. Photoemission results on TBSU from a strained GaAsP sample grown by SPIRE Co. [6]. Shown are a) Polarization and low power QE vs. wavelength and b) peak current and high power QE vs. laser peak power from the Spectra-Physics diode laser [7] at 848 nm.

This peak power corresponds to ~30 mA of high polarization beam in the injector and may be adequate for both modes of SHR. The radiation from the diode laser was introduced to the photocathode through a view port in the gun chamber. The axis of this port is at 37° with respect to the normal to the photocathode surface. Due to the large value of index of refraction of GaAs samples (n~ 4.5), the refracted light according to the Snell's law is within 8° to the normal. The polarization dilution of the incident light is therefore negligible in this setup. The diode assembly can therefore be used in the main injector with no modifications to the injector or the gun assembly.

At the present time, the production of the 850 nm diode laser by Spectra-Physics is discontinued and is not offered by any other vendors. However, Spectra-Physics and other vendors offer similar lasers at ~808 nm [8] where a significant market exists for pumping YAG lasers. With the new wavelength in mind, we have started testing strained GaAsP samples developed at St. Petersburg, Russia [9] which have high polarization near this wavelength.

POLARIZED SOURCE FOR AN E-P COLLIDER

There is a new initiative by the nuclear and particle physics community for a polarized electron- proton (ion) Collider with high luminosity and high center-of-mass energies. This topic was the subject of four workshops in the U.S. this year, and the last one was held at MIT in September [10]. An important component of such a collider is the ability to provide highly polarized electron beams at very high average currents. The community is planning a white paper for the next U.S. Long Range Plan in Nuclear Physics, due next year. Currently, two classes of colliders are considered.

- **Ring-Ring Option:** aided by the stacking option, this mode has modest requirements for the polarized source by stacking pulses of tens of mA to achieve circulating average currents of up to a few Amperes. However, it is still necessary to deal with spin manipulation in the storage ring.
- **Linac-Ring option**: This option has very demanding requirements for a polarized source. A CW source with 100-200 mA polarized beam is needed. This is about three orders of magnitude higher in average current over than is accomplished in the field today. A substantial level of R&D is needed to address these very high average currents in the polarized source and superconducting CW linac. This option may potentially allow achieving higher luminosities at higher center-of-mass energies. The 60 mA peak current from our test beam setup produced a general interest from the proponents of this option at the MIT workshop.

REFERENCES

1. B. Mueller, *et al.*, Phys. Lett. **78**, 3824 (1997).
2. D. T. Spayde, *et al.*, Phys. Lett. **84**, 1106 (2000).
3. R. Hasty, *et al.*, submitted to SCIENCE for publication.
4. M. Farkhondeh, *et al.*, Proceedings of 15th International Conf. on *Applications of Accelerators in Research and Industry* (AIP Conf. Proc. 475) AIP 1-56396-825-8/99, p.261.
5. Experimental Physics and Industrial Control System (EPICS), www.aps.anl.gov/epics/about.php.
6. Strained GaAsP sample with the "SLAC specification".
7. Spectra-Physics, Opto Power diode laser model OPC-DO60-mmm-FB with λ=848 nm.
8. Spectra-Physics, Opto Power diode laser model OPC-DO60-mmm-FC with λ=808 nm.
9. Laboratory of Spin-Polarized Electron Spectroscopy, State Technical University, St. Petersburg, Russia.
10. Second Workshop on Physics with a Polarized Electron Light-ion Collider (EPIC) September 14-16, 2000 MIT Cambridge, MA USA, http://mitbates.mit.edu/epic-workshop/.

The 50 kV inverted source of polarized electrons at ELSA

Wolfgang Hillert, Michael Gowin, and Bernhold Neff

University of Bonn, Physics Institute, Nussallee 12, 53115 Bonn, Germany

Abstract. The future medium energy physics program at the electron stretcher accelerator ELSA of Bonn University mainly relies on experiments requiring a beam of polarized electrons and a polarized target. To provide a polarized beam with high polarization and sufficient intensity a pulsed 50 kV inverted gun of polarized electrons has been set into operation. The gun is operated in space charge limitation, producing a peak current of 100 mA in rectangular $1\mu s$ long electron pulses. Photocathode lifetime during operation is higher than 3000 hours. Using a Be-InGaAs/Be-AlGaAs superlattice photocathode a polarization of 80 % and a corresponding quantum efficiency of 0.4 % could be obtained.

INTRODUCTION

Medium energy experiments requiring circularly polarized photons (produced by Bremsstrahlung of longitudinally polarized electrons) have started at the electron stretcher ELSA in Bonn [1]. To fullfill the demands of the experiments a new pulsed source of polarized electrons was developed [2] and set up. In order to enhance the overall efficiency it operates with a newly installed pulsed injector linac which requires an injection energy of 50 keV, a pulse length of $1\mu s$ and a repetition rate of 50 Hz [3]. In this paper, we will report on first measurements and our operation experience with the source.

GUN AND LOAD LOCK SYSTEM

A cross section of the gun and the load lock system is shown in Fig. 1. The cathode electrode is supported on three insulating macor rods, bolted to the upper flange of a bellow mounted on top of the gun chamber. High voltage is supplied to the cathode from the side via a pin-ball contact using a standard feedthrough which can be tilted without breaking the vacuum of the gun chamber. In this inverted configuration, the anode, the gun body and the load lock are on ground potential. The distance between cathode and anode can be varied from 45 to 70 mm. This permits to change the perveance of the gun and allows a space charge

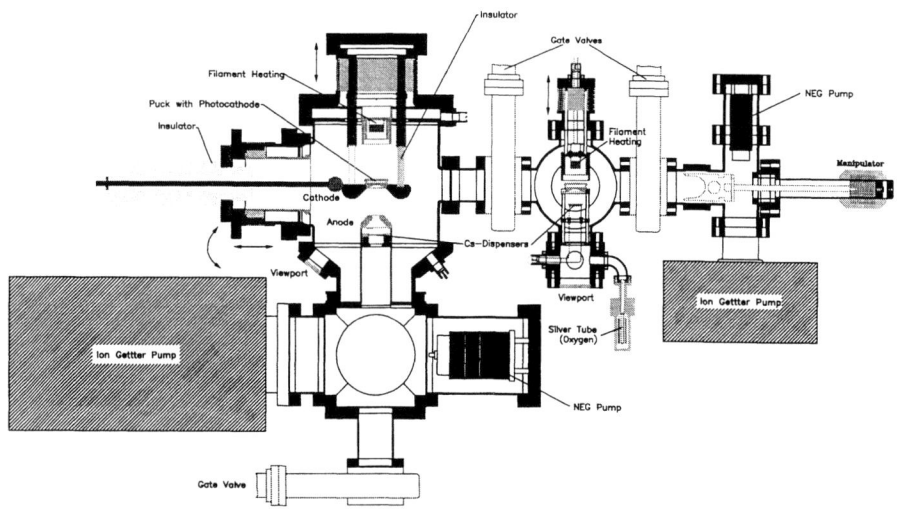

FIGURE 1. Setup of the 50 kV inverted gun and the load lock system.

limited operation over a wide range of currents. The range of operation can be enlarged using photocathodes with different photoemitting areas (see Fig. 2). To improve the gun vacuum and consequently the lifetime of the source heat cleaning and activation of the photocathode are carried out in the preparation chamber of the load lock system which is attached to the gun chamber. In addition this setup allows to exchange photocathodes without breaking the vacuum of the gun chamber.

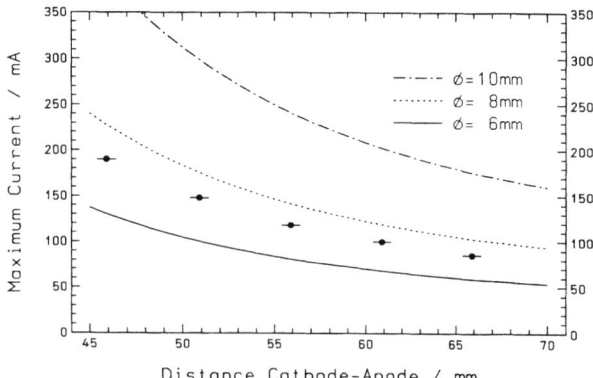

FIGURE 2. Simulated and observed space charge limitation. The simulation, represented by the lines, was carried out for three different photocathode diameters using the EGUN-code.

LASER SYSTEM

The light source is based on a tunable (700-900 nm) free running flashlamp-pumped 50 Hz Ti-Sapphire laser. The laser pulses (pulslength $10\,\mu s$) are chopped to $1\,\mu s$, fed via a 85 m long optical multimode fibre to the source and become circularly polarized after passing a linear polarizer and a pockels cell. A cw beam of polarized electrons of low intensity (typ. 100 pA), which is needed for the measurement of the beam polarization, is produced using a continous wave tunable (700-900nm) Ti-Sapphire laser. This laser is pumped by an argon vapor laser and can be fed into the fibre as well.

FIRST MEASUREMENTS

Maximum currents of up to 190 mA were obtained from a 8 mm diameter Be-InGaAs/Be-AlGaAs superlattice photocathode [5]. Electron emission could be varied from 85 mA to 190 mA (see. Fig. 2) by changing the cathode-anode-distance. A rectangular pulsstructure was obtained in all cases. The observed space charge limitation differ significantly from the calculated one which may be attributed to the different emission properties of a (cold) photocathode and a (hot) thermionic cathode, which is not implemented in the EGUN code [4] so far.

In Fig. 3 the wavelength dependance of polarization and quantum efficiency, obtained from Mott-scattering off thin gold foils, is presented. A maximum polariza-

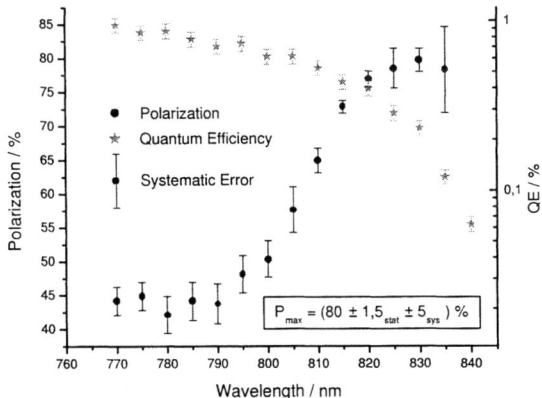

FIGURE 3. Polarization and quantum efficiency versus wavelength, obtained from Mott-scattering off thin gold foils. The error bars represent the statistical error only. A systematic error of 5 %, caused by an insufficient knowledge of the effective Sherman function, has to be taken into consideration for all data points.

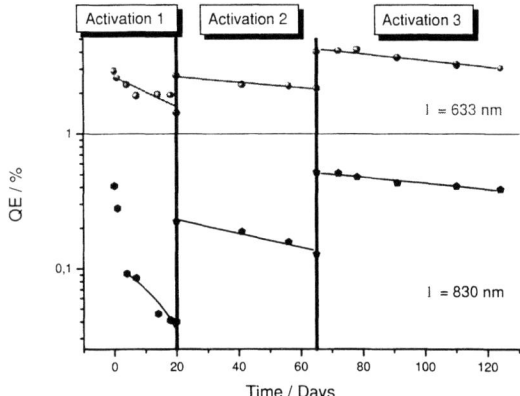

FIGURE 4. Observed decrease of quantum efficiency with time, measured at two different wavelengths.

tion of P = 80 % is achived at 830 nm, using a Be-InGaAs/Be-AlGaAs superlattice photocathode. The corresponding quantum efficiency was 0.2 %.

The degradation of the photocathode was determined by measuring the time dependent decrease of the quantum efficiency (QE) using a diode laser (wavelength 830 nm) and a He:Ne laser (wavelength 633 nm). We found lifetimes higher than 4500 hours for the last most successful activation of the photocathode (see Fig. 4). No siginificant decrease of the lifetime was observed during operation of the source.

CONCLUSIONS

A 50 kV source of polarized electrons has been set successfully into operation. A polarization of P = 80 %, QE = 0.4 % and a current of 100 mA were obtained. First experience showed a reliable operation and indicate a high source availability close to 100%.

REFERENCES

1. G. Anton et al., *Proposal*, Bonn, (1992).
2. W. Hillert et al., *Proc. Low Energy Polarized Electron Workshop*, St. Petersburg, 115, (1998).
3. W. Hillert et al., *Proc. GDH2000*, World Scientific, Singapore, in press.
4. W. B. Herrmannsfeldt, *SLAC PUB 331 UC 28*, (1988).
5. T. Nakanishi et al., *Proc. Low Energy Polarized Electron Workshop*, St. Petersburg, 118, (1998).

Polarized Electrons in Low Energy Electron Microscopy

Ernst Bauer

Department of Physics and Astronomy, Arizona State University, Tempe, AZ 85287-1504, USA

Abstract. The application of spin-polarized electron emission in low energy electron microscopy of ferromagnetic materials is reviewed and the possibilities of the spin-dependent reflectivity of slow electrons for efficient spin detection are described.

I. INTRODUCTION

The interaction of electrons with matter is spin-dependent due to the spin-orbit and spin-spin interaction. This fact has been used for some time in spin-polarized low energy electron diffraction (SPLEED) [1,2]. The spin information in SPLEED is contained in the asymmetry $A_{hk} = (I_{hk}^+ - I_{hk}^-)/(I_{hk}^+ + I_{hk}^-)$ where I_{hk}^+ and I_{hk}^- are the intensities of the diffracted beams (h,k) with opposite spin polarization direction **P**/P of the incident beam. In low energy electron microscopy (LEEM) only one diffracted beam, usually the specularly reflected beam (00) and its close environment can be used for imaging because of the electron-optical aberrations of the objective lens [3,4]. In the SPLEED energy range in which A_{hk} is large the intensities I_{hk}^+ and I_{hk}^- are usually small so that this energy range is not suitable for imaging. Large A values together with high intensities are in general only obtained at very low energies, typically below 10 eV. At these low energies all diffracted beams except the specular (00) beam are reflected into angles far off the optical axis, leaving only the (00) beam and its surrounding, that is approximate 180° scattering, for imaging. As a consequence, the reflected intensity contains no contribution from the spin-orbit interaction but only from the spin-spin interaction which makes spin-polarized LEEM (SPLEEM) [3,5,6] a powerful tool for the study of the spin distribution in matter.

If the spin distribution in the specimen produces a net magnetization **M** in regions larger than the resolution limit of the instrument, then the intensity of the reflected (00) beam contains a contribution I_M to the total intensity I_{00} which is proportional to **P**•**M**: $I_M^+ = I_{00}^+ - I_N$, where I_N is the nonmagnetic contribution. When the direction of **P** is reversed, the magnetic contribution is also reversed, $I_M^- = -I_M^+$, but not the nonmagnetic contribution I_N. Therefore, if images taken with opposite polarization are subtracted from each other, the resulting "asymmetry image" contains only magnetic information. Conversely, the sum image contains only microstructure information. The LEED pattern of the specimen that can be observed over a wide energy range gives information on the crystal structure. The fact that three complementary pieces of information, **M** distribution, microstructure and crystal structure, can be obtained from the same specimen region makes a SPLEEM instrument ideal for the study of the connection between magnetic structure and microstructure which is very pronounced in the thin films used in present and future magnetic sensor and storage media.

In the following the physical basis and the experimental realization of SPLEEM will be outlined first (Sect. II). Then the method is illustrated by some examples (Sect. III). A result of SPLEEM that may be important for efficient polarization determination is described next (Sect. IV), followed by a brief summary and outlook.

II. PHYSICAL BASIS AND EXPERIMENTAL REALIZATION

For the understanding of SPLEEM four aspects are important: the formation of the spin-polarized beam, the beam-specimen interaction, the image formation, that is the electron optics, and the image detection. The incident beam should have high polarization, high intensity, high stability and freely adjustable polarization direction. The last requirement has been fulfilled with a polarization manipulator consisting of a combined electrostatic-magnetic $90°$ beam deflector and a spin rotator lens [7]. The first three conditions are only poorly fulfilled in the two presently existing SPLEEM instruments. Both have no cathode airlock and preparation chamber nor the good vacuum which is now standard in high energy physics guns. They use standard unstrained GaAs sources which are activated in the gun chamber. Under these conditions polarization P and lifetime τ are low, typically $P \approx 0.2 - 0.25$ and $\tau \approx 1$ day – 1 week. The low polarization is presently the weakest point of SPLEEM. A large P value is important both for high magnetic signal, that is high asymmetry, and for good signal to noise ratio. If α is the spin-dependent fraction of the reflected intensity for $P = 1$, then $I_M = P\alpha I$ for $P < 1$ and $A = P\alpha/(1-\alpha)$. The noise is connected with the total signal, $N = I^{1/2}$, so that the magnetic signal to noise ratio is $I_M/N = P\alpha I/I^{1/2} = (P\alpha)^{1/2} I_M^{1/2}$ instead of $I_M/N = I_M^{1/2}$. α is primarily determined by the properties of the specimen. It can be optimized by proper choice of the beam energy at the specimen but is usually significantly less than 1. Therefore, high magnetic signal A and good signal/noise ratio I_M/N require large P. A good I_M/N ratio needs also a high intensity because of $I_M/N = P\alpha I^{1/2}$.

In addition to these conditions the incident beam must be focused into a fine spot in the back focal plane of the objective lens in order to obtain parallel illumination of the specimen. The angular aperture and the diameter of the beam in the back focal plane determine the diameter of the illuminated area and the divergence of the beam at the specimen, respectively. A fine spot in the back focal plane requires a small emitting area on the cathode because a large demagnification of the source is electron-optically impracticable.

The most important aspects of the beam-specimen interaction are the spin-dependent elastic reflectivity – which determines the image intensity and the magnetic contrast – and the spin-dependent inelastic mean free path which determines the spin-dependent sampling depth. The elastic reflectivity depends upon the band structure as illustrated in Fig.1 for the case of the Co(0001) surface. Due to the exchange splitting of the band structure the reflectivity of the surface for electrons with spin parallel and

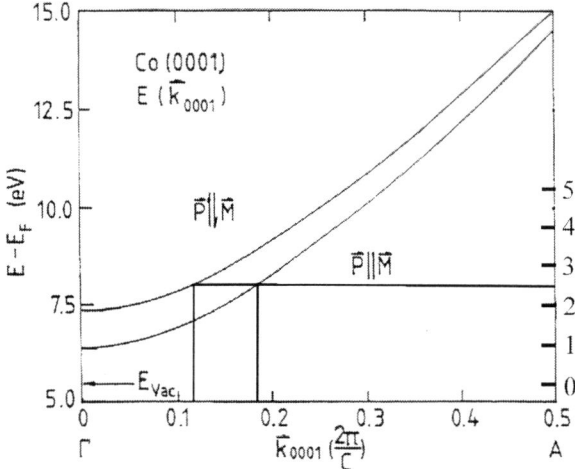

FIGURE 1. Band structure of Co along the [0001] direction. Left side energy from Fermi level, right side E_V from vacuum level. The two k values for $E_V = 2.5$ eV are indicated.

antiparallel to the magnetization **M** of the specimen is different. Co has in the [0001] direction a bandgap which extends above the vacuum level. In this bandgap Co is a reactive medium, that is the incident wave is strongly attenuated in the crystal and strongly reflected. When the energy is increased so that the majority spin states become accessible the electrons whose spin is parallel to the majority spin state can penetrate deeper into the crystal and loose energy by inelastic scattering. The electrons whose spin is parallel to the minority spin state are still strongly reflected until their energy is so high that they find allowed states in the crystal. Then they can also penetrate deeper into the crystal. As a consequence, between about 1 and 2 eV in the example of Fig.1, there is a strong contrast between regions with different spin orientation, that is magnetization **M**. At higher energies the contrast decreases significantly but still remains due to the different densities of states in the two bands. In thin films, however, strong contrast can be obtained even at higher energies due to the difference in the wave number $k = 2\pi/\lambda$ between the two states as indicated in Fig.1. At a given energy the interference condition $n\lambda/2 = d$ (n = 1,2,...) is fulfilled for different film thickness d. For a given thickness and energy the reflectivity is, therefore, different for the two spin orientations. This spin-dependent "quantum size contrast" will be discussed in more detail in Sect. IV.

Elastic scattering in the backward direction is weakened by phonon scattering and electron-hole pair creation. Plasmon excitation does not occur at the low energies used in SPLEEM. Although the energy transfer in phonon scattering is very small, the momentum transfer is large so that the electrons are scattered off the optical axis and are lost for imaging. Electron-hole pair creation is spin-dependent because the density of unoccupied minority spin states is significantly larger than that of the majority spin states. As a consequence, much more states are accessible for inelastic scattering of minority electrons than for majority electrons. Depending upon material and energy the mean free paths may differ as much as by a factor of 3 but both are small and decrease with the number of unoccupied d states [8]. Typical values are 0.3 – 1.5 nm. These short mean free paths make SPLEEM very surface-sensitive.

The image formation process is determined mainly by the objective lens. The electron-optical components further downstream – beam separator, transfer lens, diffraction lens and projective lens – which will be discussed below have little influence on the image quality. In the objective lens the incident electrons are decelerated from their full energy in the back focal plane – typically 5 – 20 keV – to the desired low energy at the specimen. The objective lens is thus a cathode lens or an immersion lens and the specimen is part of the lens. This type of lens has large chromatic and spherical aberrations which make it necessary to limit the angular aperture used for imaging to small angles. This is achieved by a "contrast" aperture in the back focal plane of the objective lens or in an image plane of this plane. The resolution is then determined by the aberrations and the diffraction at the aperture and depends upon energy and energy spread. For negligible energy spread the best lenses can achieve a resolution of several nm but in practice only 10 to several 10 nm are reached. For magnetic imaging electrostatic tetrode lenses are preferred because they produce no magnetic field at the specimen as uncompensated magnetic objectives do. The beam separator mentioned above is needed to separate the incident from the reflected (image) beam, the transfer lens to transfer the diffraction pattern in the back focal plane of the objective lens so that it can be imaged with the diffraction and projective lenses and these two lenses give also the possibility to change the magnification of the specimen image over a wide range.

The image is detected via multichannel plate image intensifier and a fluorescent screen by a CCD camera. Two images with opposite polarization are acquired immediately one after the other and subtracted on-line. In order to improve the signal/noise ratio up to 128 frames can be accumulated per image. Typical image acquisition times are presently 1 to several sec with an 8 bit camera. Significant improvements can be expected from a source with high polarization and from improvements of the image detection system that introduces significant additional noise and limits image recognition [9].

III. APPLICATION EXAMPLES

Except for the very first SPLEEM study in which the magnetic contrast was demonstrated with a Co(0001) single crystal surface [10] nearly all work done to-date is on *in situ* grown films during and after growth. Fig. 2 shows an example of the connection between magnetic domain structure and microstructure. The LEEM image (c) shows three thickness levels and gives some idea of the step distribution. The in-plane image (a) shows no influence of the local film thickness, the out-of-plane image (b) a strong influence of the step structure on the $\mathbf{M}_\perp$ distribution. The complete $\mathbf{M} = \mathbf{M}_\parallel + \mathbf{M}_\perp$ distribution is wrinkled [11].

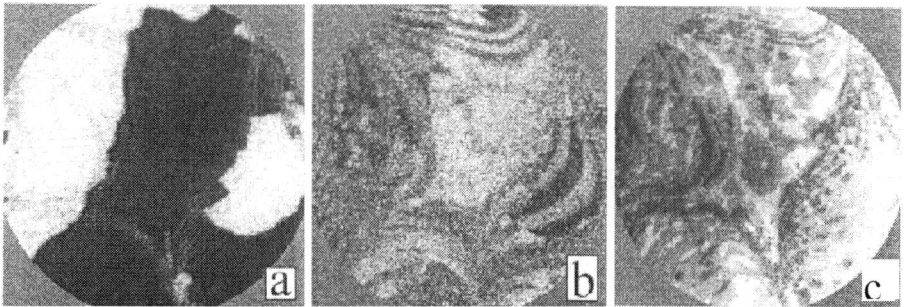

FIGURE 2. Images of a 5 monolayer thick Co film on W(110). a) **P** in-plane parallel to easy axis, b) **P** out-of-plane, c) LEEM image with quantum size contrast. Field of view 6μm diameter, electron energy 1.2 eV (a, b) and 3 eV (c).

A second example is the spin reorientation transition shown in Fig. 3. Below 4.2 monolayers **M** is out-of-plane, above 4.5 monolayers in-plane. A detailed SPLEEM study showed that the transition is initiated by domain wall broadening, not by rotation of the magnetization within the domains [12].

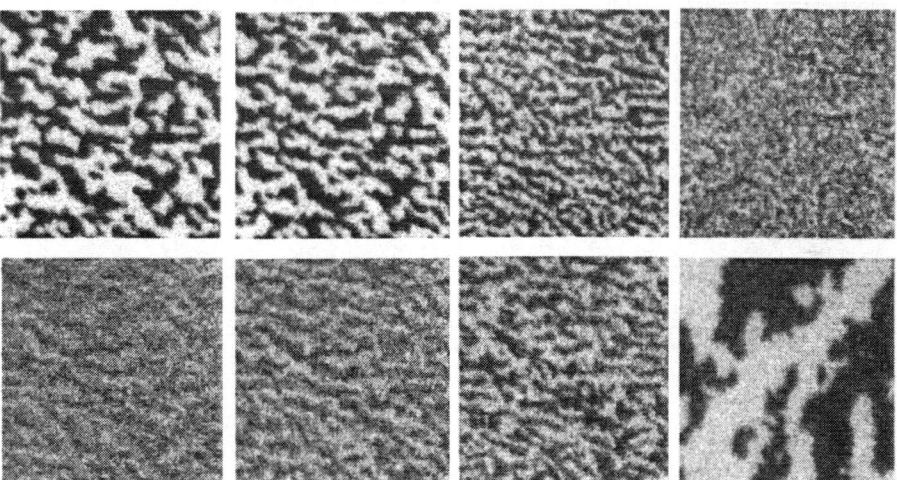

FIGURE 3. Spin reorientation transition in a Co film on Au(111). Top row: **P** out-of-plane, bottom row: **P** in-plane. Film thickness from left to right: 4.2, 4.3, 4.4 and 4.5 monolayers. Field of view 7×7 μm^2, electron energy 1.2 eV.

Another example is the study of the interlayer coupling between two magnetic films trough a nonmagnetic spacer layer which oscillates between ferromagnetic and antiferromagnetic coupling and is frequently accompanied by noncollinear coupling (90° coupling).Because of its low sampling depth SPLEEM allows to separate clearly the **M** distribution in the two layers and, thus, to study the coupling in detail as a function of microstructure [13].

IV. THE QUANTUM SIZE POLARIZATION DETECTOR

In thin planar films up to about 10 monolayers in thickness the wavelength of slow electrons is of the order of the film thickness. Accordingly standing waves can form in the film, provided that its surfaces are sufficiently reflecting and the electron wave is not attenuated too strongly in the layer. This is the case for Co films on W(110). Because of the exchange splitting in Co, electrons with spin up and spin down polarization have different wave lengths (see Fig. 1). Therefore, the standing wave condition is fulfilled for different energies in the two cases so that the asymmetry oscillates strongly with energy for a given film thickness as illustrated in Fig. 4 [14].

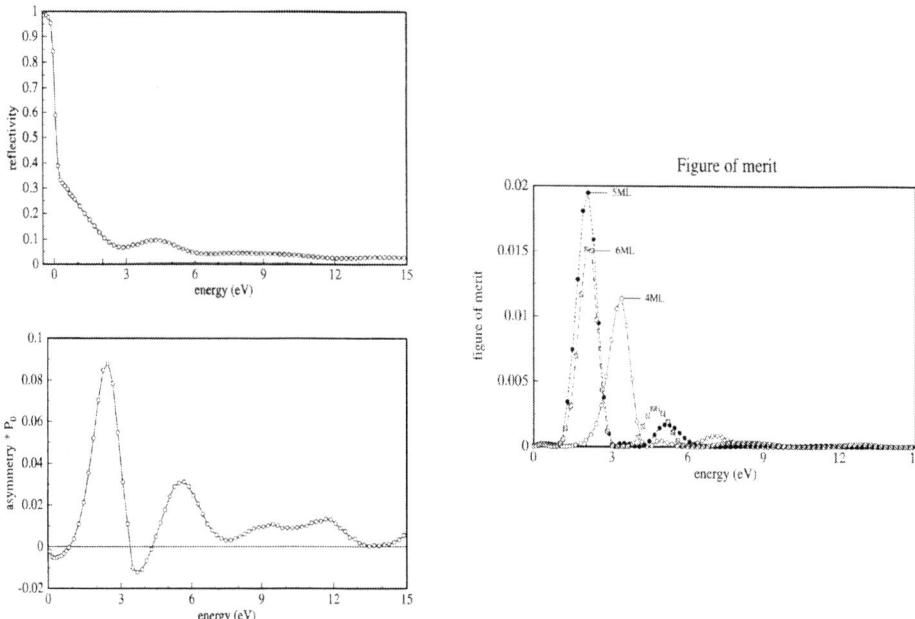

FIGURE 4. Reflectivity and asymmetry of a 5 monolayers thick Co film on W(110) (left side). Right side: figure of merit of a polarization detector consisting of 4, 5 or 6 Co monolayers, assuming a polarization of the incident beam of 0.2.

The effect is strongest at the lowest energies where W has a bandgap in the direction of the (110) surface normal and where the inelastic mean free path is largest. Near the first maximum of A the reflectivity is still relatively high so that the figure of merit of a spin detector IA^2 is high compared to all other spin polarimeters. However, calibration is necessary because the calculations of this effect [15] are not accurate enough. The Co layer may be coated with a thin metal layer which does not adsorb the typical UHV residual gases and thus be stabilized for long periods.

ACKNOWLEDGMENTS

The development of SPLEEM was stimulated by H. Poppa via an IBM cooperation agreement, subsequently supported by the DFG (Grant No. Ba 410/32-1) and finally by the NSF (Grant Nr. DMR-9818296). May people have contributed to it: M. Altman, J. Hurst, G. Marx, H. Pinkvos, H. Poppa, K. Wurm and foremost T. Duden.

REFERENCES

1. Kirschner, J., *Polarized Electrons at Surfaces,* Springer: Berlin, 1985
2. Feder, R., editor, *Polarized Electrons in Surface Physics,* World Scientific: Singapure, 1985
3. Bauer, E., Rep. Prog. Phys. **57**, 895 – 938 (1994)
4. Bauer, E. in: *Handbook of Microscopy, Methods II,* edited by S. Amelinckx et al, VCH: Weinheim, 1997, pp. 487 - 503
5. Bauer, E., in ref. 4, pp. 751 – 759
6. Duden, T. and Bauer, E., J. Electron Microscopy **47**, 379 –385 (1998)
7. Duden, T. and Bauer, E., Rev. Sci. Instrum. **66**, 2861 - 2864 (1995)
8. Siegmann, H.C., J. Phys.: Condens. Matter **4**, 8395 – 8434 (1992)
9. Duden, T. and Bauer, E., Surf. Rev. Lett. **5**, 1213 – 1220 (1998)
10. Altman, M.S. et al, MRS Symp. Proc. **232**, 125 – 132 (1991)
11. Duden, T. and Bauer, E., Phys. Rev. Lett. **77**, 2308 - 2311 (1996)
12. Duden, T. and Bauer, E., MRS Symp. Proc. **475**, 283 - 288 (1997)
13. Duden, T. and Bauer, E. Phys. Rev. B **59**, 474 –479 (1999)
14. Wurm, K. M.S. thesis, TU Clausthal, 1994
15. Scheunemann, T. et al, Solid State Comm. **104**, 787 – 792 (1997)

Electron Emission From Na/Fe(100) Surfaces By Deexcitation Of Spin-Polarized Helium Metastable Atoms

Yasushi Yamauchi, Mitsunori Kurahashi, and Taku Suzuki

National Research Institute for Metals, Sengen, Tsukuba 305-0047, Japan

Abstract. A pulsed helium metastable atom beam was generated by the pulsed nozzle-skimmer discharge and a high degree of spin polarization of the helium beam was obtained using the optical pumping method. Spin polarization of the outermost surface electron was detected for clean and sodium covered Fe(100) films deposited on MgO(100) by measuring the secondary electrons ejected by the irradiation of the spin-polarized helium metastable atom beam. Secondary electrons correspond to the Fermi level for a clean iron surface and those corresponding to the Na 3s level for the sodium-covered surface show a positive asymmetry, which indicates the negative polarization of electrons at these levels. The negative polarization of the Na 3s electrons rapidly decreases with the thickness increasing of the Na layer up to 0.5 ML but remains almost constant at a higher coverage (3ML).

INTRODUCTION

Slow helium metastable atoms impinging on a surface with a thermal energy do not penetrate into the subsurface layers. This characteristic produces an extreme surface sensitivity of metastable atom beams as a surface analysis probe. Metastable-atom deexcitation spectroscopy (MDS), which measures the secondary electron emission by the deexcitation of metastable atoms at surfaces, provides us information about electronic structures of the outermost surfaces. Metastable helium atoms have another feature, i.e., spin polarization, in which metastable atoms in the triplet state are unequally distributed into its three magnetic sublevels. This feature extends the capability of MDS as a spin-sensitive probe for the analysis of the spin-polarized electronic structures. Spin-polarized metastable-atom deexcitation spectroscopy (SPMDS) that was established at Rice University has been recognized as a unique method to investigate the magnetic properties of the outermost surfaces.[1-4]

When a He(2^3S) atom in one of the magnetic sublevels, $M_s = 1$ or -1, approaches a surface, the 1s hole of the He(2^3S) atom is filled by an electron which has an anti-parallel spin with respect to the remaining 1s electron irrespective of suffering from resonance ionization (high work function case) or electron attachment (very low work function case) and another electron is emitted via Auger-like transitions, i.e., Auger neutralization, Auger deexcitation or auto-detachment. The transition rates at the surface depend on the density of the surface electrons, which have anti-parallel spins with respect to the 1s electron of the He(2^3S) atom.[5] If surface electrons are spin-polarized along to the quantization axis of He(2^3S) defined by a certain magnetic field, the different populations and distributions of electrons for spin up or down result in an asymmetry of intensity for the emitted secondary electrons.

SPMDS has been applied to studies of clean surfaces of ferromagnetic samples and the effect of gaseous adsorption. Since work functions of surfaces studied in earlier research are high enough to cause resonance ionization followed by Auger neutralization, MDS spectra become a rather structureless self-convolution of the occupied density of states.[1-4] In the present study, Na is deposited on to Fe(100) to obtain a low work function surface and then the thickness of Na layer is increased to examine the thickness dependence of the magnetism of the paramagnetic overlayer on a ferromagnetic substrate.

EXPERIMENT

For the SPMDS we have developed a novel pulsed metastable atom beam source by modifying a conventional nozzle-skimmer DC discharge metastable atom source.[6] The total metastable flux in the pulse duration, which is equivalent to a continuous beam flux to be mechanically chopped, is estimated to be 1.7×10^{15} $s^{-1}sr^{-1}$. Spin polarization of the metastable atoms is performed by an optical pumping method.[7,8] The wavelength of the pumping laser diode is stabilized by a helium discharge cell. The degree of polarization is confirmed by a Stern-Gerlach measurement with a compact Rabi type magnet, which was also newly developed. A laser diode with a rating power of 25mW is enough to polarize most of the triplet atoms except 10 % of the singlet atoms when the laser frequency is stabilized. The polarity of the spin can be rapidly reversed by changing the helicity of the circularly polarized light.

The apparatus used in the present study is a scanning Auger electron microscope (cylindrical energy analyzer (CMA), PHI 590A) combined with the metastable helium atom beam source and the Stern-Gerlach analyzer.[9] Iron films of 99.99 % are deposited onto MgO(100) single crystal substrates by an electron bombardment evaporator in the measurement chamber. A radiative heater of 20 W made of a Ta filament is used to anneal the films up to 500 K. The mechanically measured thickness is greater than 30 nm. The pole figure of the x-ray diffraction of the iron film shows that the structure is Fe(bcc) and that Fe(100) is parallel to MgO(100) but that the [010] direction of Fe rotates towards the [011] direction of MgO. The sodium overlayer is deposited onto Fe(100) by a dispenser (SAES). Auger electron spectroscopy and its quantitative analysis are employed to estimate the Na thickness.

The residual magnetization of the iron film has been checked by a simple and qualitative Kerr effect experiment. The iron film showed a clear Kerr rotation and the direction residual magnetization was changed by a pulse magnetic field generated by a pulse current into a coreless coil. This coreless coil is introduced in the measurement chamber to assure the maximum residual magnetization and also to reverse the magnetization for checking the experiment validity. The iron films are magnetized both along to an easy magnetization axis in the sample plane and in the direction parallel or anti-parallel to the magnetic field of 100 mG, which secures the quantization axis of He(2^3S).

RESULTS AND DISCUSSION

SPMDS spectra for the clean and Na deposited Fe(100) surfaces, i.e., energy distributions of the secondary electrons obtained by CMA during the irradiation of spin-polarized He(2^3S) beam are shown in Fig. 1 (a). Two spectra, each taken at parallel and anti-parallel configurations between the He(2^3S) spin and the majority spin of the iron films, are overlaid at the same level. At a glance, differences are evident for the clean surface and for the Na deposited surfaces. The deviations in spectra between the opposite spins become clearer with the increasing thickness of the Na overlayer. These deviations are well reproduced even by changing the measurement order of the primary He(2^3S) spin polarity and/or the magnetization polarity of the iron films. Spin-dependent asymmetries, which are the intensity differences in the SPMDS spectra normalized by the total intensity, are plotted in Fig. 1 (b).

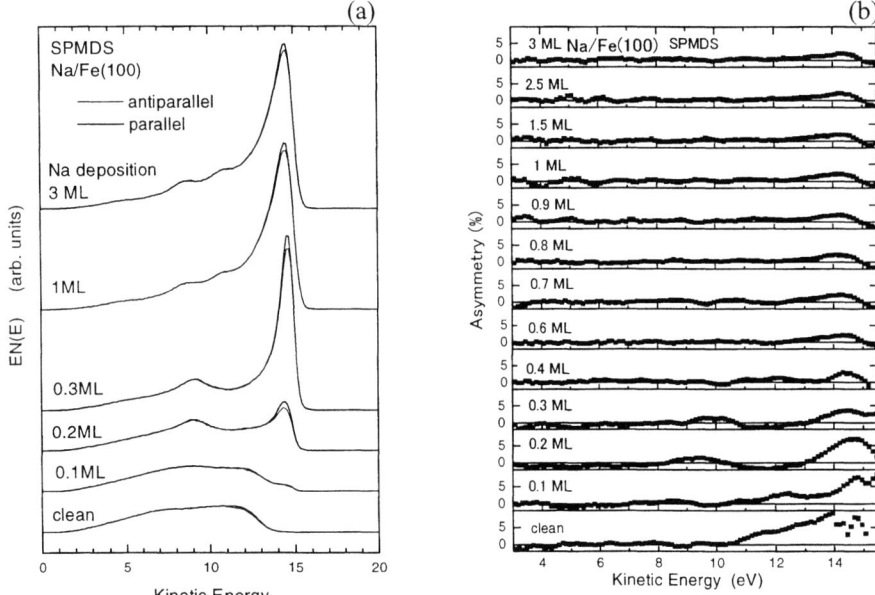

FIGURE 1. (a) SPMDS spectra for clean and Na covered Fe(100) surfaces, (b) Asymmetries of SPMDS spectra. Asymmetry $A=(I_p-I_a)/(I_p+I_a)$, where I_p and I_a denote intensities of SPMDS spectra obtained, respectively, at the parallel and anti-parallel spin configurations between He(2^3S) of the primary beam and the majority electron of the sample.

For the clean Fe(100) surface, a small but clear deviation between the two different spin configurations exists at the 13 eV shoulder. Figure 1 (b) clearly shows a positive asymmetry at 11 - 14 eV. The positive asymmetry indicates the negative polarization of electrons at the Fermi level. Theses results for the clean iron surface well reproduce our earlier study, which is consistent with other studies.[2-4]

While depositing Na onto the iron surface, the peak at 15 eV appears and the intensity drastically increases. (Fig. 1 (a)) The peak at 15 eV is due to the 3s-electron band, which Na atoms induce. The intensity of the peak has a maximum at the deposition of less than 1 ML then decreases. These behaviors have been observed for alkali-covered surfaces.[10,11] The dominant component (90%) of the primary metastable helium atoms is He(2^3S) in the present study and the remaining He(2^1S) atoms are converted to He(2^3S) on the low work function surfaces with a substantial probability before the Auger-like transitions take place.[11]

In Fig. 1 (a), the intensity of the peak at 15 eV for the parallel spin configuration (thick line) is obviously larger than for the anti-parallel spin configuration (thin line). The polarity of the asymmetry is the same as the clean surface. The absolute value of the intensity difference increases with the increasing amount of Na deposition and shows a maximum at depositions less than 1 ML (monolayer). In Fig. 1 (b), the asymmetry curves exhibit a sharp peak at 15 eV. The asymmetry peak is the largest at 0.1 ML then rapidly decreases until 0.4 ML while the SPMDS intensity peak at 15 eV increases. With an increasing thickness of the overlayer from 1 to 3 ML, both the SPMDS intensity peak and asymmetry peak at 15 eV retains an almost constant value of 2 %, which is fairly far from the stochastic error bars (not shown in the figure).

The magnetism of the paramagnetic overlayer on a ferromagnetic substrate was analyzed by applying the free-energy Ginzburg-Landau theory, and the correlation length equivalent to the lattice distance was reported for rare earth overlayers on transition metal substrates.[12] However, it is obvious that the thickness dependence of the asymmetries in Fig. 1 (b) can not be fitted by a single exponential profile.

At a submonolayer Na coverage, the adsorbed Na atom nearly releases its $3s$ electron to the substrate and creates a dipole moment, which lowers the local work function.[13] Therefore the $3s$ electron of the adsorbed Na strongly interacts with the Fe valence d and s electrons and may exhibit the nearly the same spin-polarization to the clean Fe(100) surface at the lower coverage. This interaction may rapidly decrease with the increasing thickness of the Na overlayer. On the contrary, the origin of the remaining constant asymmetry at a higher coverage is unclear. The thickness dependence of the asymmetry at much higher thickness up to tens of monolayers will be needed for further discussions on the constant asymmetry.

REFERENCES

1. Onellion, M., Hart, M. W., Dunning, F. B., and Walters, G. K., *Phys. Rev. Lett.* **52**, 380 (1984).
2. Hammond, M. S., Dunning, F. B., and Walters, G. K., *Phys. Rev.* **B45**, 3674 (1992).
3. Getslaff, M., Egert, D., Rappolt, P., Wilhelm, M., Steidl, H., Baum G., and Raith, W., *Surf. Sci.* **331-333**, 1404 (1995).
4. Ferro, P., Morioni, R., Salvietti, M., Canepa M., and Mattera, L., *Surf. Sci.* **409**, 212 (1998).
5. Penn, D. R., and Apell P., *Phys. Rev.* **B41**, 3303 (1990).
6. Yamauchi, Y., Kurahashi, M., and Kishimoto, N., *Meas. Sci. Technol.* **9**, 531 (1998).
7. Wallace, C. D., Bixter, D. L., Monroe, T. J., Dunning, F.B., and Walters, G. K., *Rev. Sci. Instrum.* **66**, 265 (1995).
8. Granitza, B., Salietti, M., Torello, E., and Mattera, L., *Rev. Sci. Instrum.* **66**, 4170 (1995).
9. Kurahashi, M., and Yamauchi, Y., *Surf. Sci.* **420**, 259 (1999).
10. Maus-Friedrichs, W., Wehrhahn, M., Dieckhoff, S., and Kempter, V., *Surf. Sci.* **237**, 257 (1990).
11. Hermmen, R., and Conrad, H., *Phys. Rev. Lett.* **67**, 1314 (1991).
12. Dowben, P. A., LaGraffe, D., Li, Dongqi, Miller, A., and Ling Zhang, *Phys. Rev.* **B43**, 3171 (1991).
13. Diehl, R. D., and McGrath, R. J., *Phys.: Condens. Matt.* **9**, 951 (1997).

Investigations of the Charge Limit Phenomenon in GaAs Photocathodes[1]

T. Maruyama*, J. E. Clendenin*, E. L. Garwin*, R. E. Kirby*,
G. A. Mulhollan*, R. Prepost[†], C. Y. Prescott*,
and A. V. Subashiev*

Stanford Linear Accelerator Center, Stanford, CA 94309, USA
[†]*University of Wisconsin, Madison, WI 53706, USA*

Abstract. The doping concentration dependence of the charge limit phenomenon is studied using a set of thin unstrained GaAs. The p-type doping concentration ranges from 5×10^{18} cm^{-3} to 5×10^{19} cm^{-3}. The surface photovoltage effect on photoemission is found to diminish rapidly with increasing doping concentration. The doping concentration effect on electron polarization is studied using high surface doped strained GaAs. The charge enhancement is explored using a direct, lateral surface charge sink in the form of a metallic grid overlaid on the surface.

INTRODUCTION

A fundamental concern in the operation of electron accelerators is the generation of intense beams with high polarization. The time structure of the electron beam determines whether the limiting factor in emission is laser power or surface charge limit [1]. The surface charge limit comes about due to the photon absorption that excites electrons from the valence to the conduction band. Some fraction of the electrons can be trapped near the surface which induces a rise in the electron affinity through the increased electrostatic potential. The high affinity causes a lower emission probability and thus less emitted charge at later times. The electron affinity can recover to the zero charge limit after electron-hole recombination. The charge limit phenomenon poses a serious problem for future linear colliders such as the JLC and the NLC. The NLC requires 95 micropulses having 2×10^{10} electrons per pulse in 0.7 ns with an interpulse spacing of 2.8 ns. Experience at SLAC has shown that it is possible to generate 16×10^{10} electrons in 2 ns pulse, but subsequent pulses may experience strong intensity damping. Several parameters can be varied to enhance the net charge output. Among these are increasing either the electric field at the photocathode surface (Shottky effect), or the surface charge dissipation

---

[1] This work was supported in part by Department of Energy Contract No. DE-AC03-76SF00515

rate. The latter is accomplished by raising the dopant concentration or by shunting the trapped charge.

DOPANT CONCENTRATION DEPENDENCIES

A series of measurements were performed using unstrained 100 nm thick GaAs to study the doping concentration dependence of the charge limit effect. The p-type doping concentrations were 5×10^{18} cm^{-3}, 1×10^{19} cm^{-3}, 2×10^{19} cm^{-3}, and 5×10^{19} cm^{-3} for a set of four samples. The experiments were performed using the 121 kV diode gun in the SLAC Gun Test Laboratory. Two excitation sources were used, two pulsed Ti:Sapphire lasers pumped by a frequency-doubled Nd:YAG laser, and a flash-lamp pumped Ti:Sapphire laser. The wavelength of the lasers was set at 850 nm. The two YAG-Ti lasers produced a 2 nsec pulse with energy up to 100 μJ. The time separation between the two lasers could be varied up to 60 nsec, providing a pump-probe technique. The Flash-Ti laser produced a longer pulse varying from 150 nsec to 350 nsec with energy up to 130 μJ.

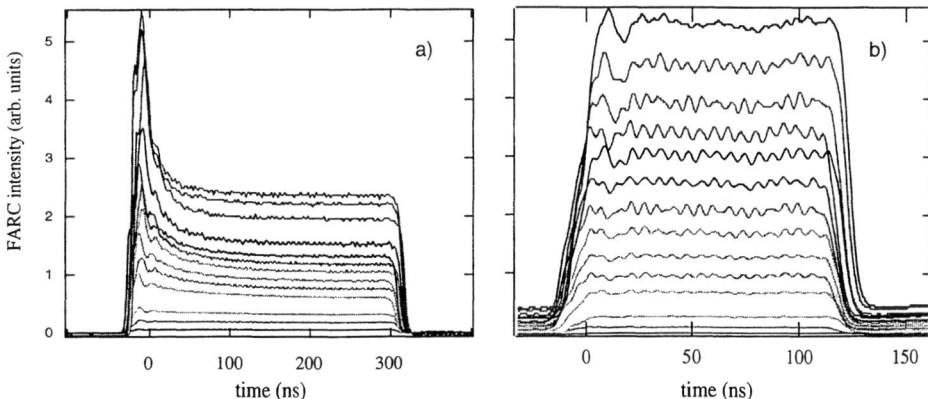

FIGURE 1. The temporal profiles of the emission pulses for a) the 5×10^{18} cm^{-3} sample and b) the 2×10^{19} cm^{-3} sample.

Figure 1(a) and 1(b) show the temporal profiles of the emission current pulses measured for a number of light pulse energies for the 5×10^{18} cm^{-3} sample and for the 2×10^{19} cm^{-3} sample, respectively. As seen in Figure 1(a), the emission current follows the rectangular laser shape when the laser energy is low. As the laser energy is increased, the emission peak at the start of the light pulse reflects retardation in the build up of the photovoltage due to the finite time of the charging of the surface states. As the photovoltage builds up, the emission current decreases exponentially. The suppression of the later portion of the emission pulse manifests the decrease of the surface escape probability with the growth of the photovoltage and its relaxation to the steady state. As the laser energy is increased, the photovoltage builds up more quickly and the emission current suppression is more pronounced. For the 1×10^{19} cm^{-3} sample, the charge limit behavior is much reduced. As seen in Figure 1(b), the 2×10^{19} cm^{-3} sample does not show the charge limit effect.

The laser pump-probe technique is used to explore the photovoltage effect produced by the electron pump pulse affecting the electron probe pulse. Figure 2 shows the probe electron pulse charge, Y_2, normalized by the charge without the pump pulse, Y_1, as a function of time separation between the two pulses. For the 5×10^{18} cm^{-3} sample (Figure 2(a)), the photovoltage produced by the pump pulse suppresses the probe pulse charge as much as 70% and the charge output recovers monotonically with increasing time separation ($\tau = 130$ ns). The recovery time constant for the 2×10^{19} cm^{-3} sample (Figure 2(b)) is $\tau = 8$ ns. No charge limit effect is observed for the 5×10^{19} cm^{-3} sample. This experiment suggests that high doping concentration of at least 2×10^{19} cm^{-3} should be sufficient to overcome the surface charge limit.

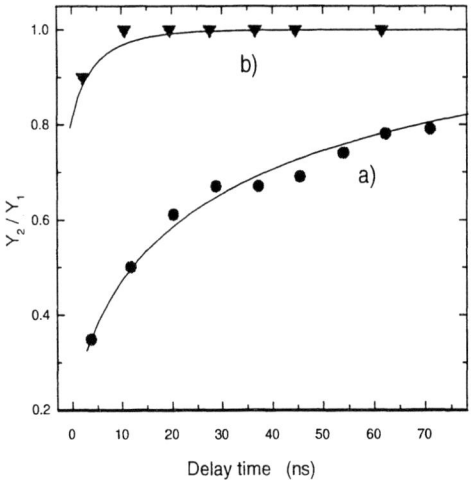

FIGURE 2. The normalized charge output as a function of the time separation for a) the 5×10^{18} cm^{-3} sample and b) the 2×10^{19} cm^{-3} sample.

HIGH SURFACE DOPING CONCENTRATION

While high doping concentration is necessary to overcome the charge limit, electron polarization will be significantly reduced as the doping level is increased. One way to achieve the high doping level without suffering from polarization loss is to dope only ~100 Å of the surface layer at $>2\times10^{19}$ cm^{-3}. To test this technique, polarization measurements were performed at the SLAC Cathode Test System using high surface doped strained GaAs. While the active layer is 1000 Å thick GaAs, the surface layer is doped at 4×10^{19} cm^{-3} and the rest at 3×10^{17} cm^{-3}. Two samples with the high doped layer thickness of 250 Å and 400 Å are used. The high doped layer thickness is reduced by an anodization technique [2]. Anodization is a reliable technique to remove thin layer from GaAs surface. Secondary-ion-mass-spectroscopy (SIMS) is used to calibrate the anodization rate (14.9 Å/V), to estimate the GaAs evaporation rate during heat-cleaning ($\sim$4 Å/hr), and to study the dopant (zinc) diffusion during heat-cleaning. Figure 3 shows the maximum polarization as a function of the high doped layer thickness. The polarization increases rapidly as the layer thickness is reduced. The 80% polarization seems achievable with a high doped layer thickness of about 50Å.

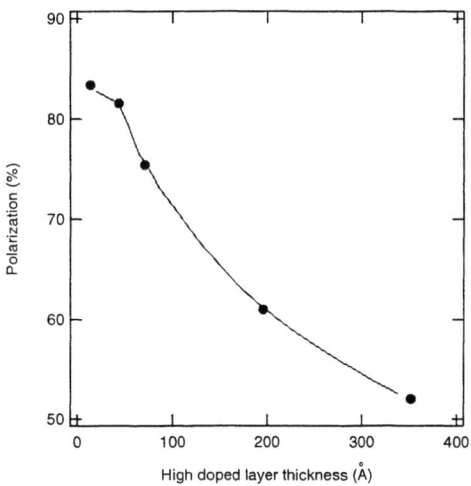

FIGURE 3. The maximum polarization as a function of the high doped layer thickness.

CHARGE BLEEDOFF GRID

The high surface doping technique seems promising, and is desirable if the charge limit problem can be solved by this technique alone. An alternative to the high surface doping technique is the addition of a direct, lateral surface charge sink in the form of a metallic grid overlaid on the surface. The creation of such a grid presents several challenges. First, the grid must be small enough that the spacing between the lines allows fast enough charge diffusion during the time of interest, a few nanoseconds. This requirement may be accomplished by using standard lithographic methods. The second requirement is that the grid material make good electrical contact, yet not react with the GaAs at the heat cleaning temperature,

600°C. Lastly, the material must be robust enough to withstand the heat cleaning process by having a very low vapor pressure at 600°C. The material that met all these requirements is tungsten. The grids were prepared on a 100 nm thick unstrained GaAs wafer with a doping concentration of 1×10^{18} cm^{-3} to deliberately exacerbate the surface charge buildup. The line width was 10μm with a 23 μm spacing which left 38% of the GaAs surface exposed.

Figure 3 shows the current density as a function of the laser intensity for the grid covered material (sold circles) and a no-grid sample (open triangles) prepared under the same conditions (same wafer and same lithography process). The peak charge measured at 862 nm was enhanced by about 30%, while at 790 nm a factor of two increase was achieved.

Although the enhancement in the current density is observed, the net current is much smaller for the grid covered material as 62% of the GaAs surface does not photoemit. Further improvements are expected from smaller line widths and closer spacing. The lithography technique has been developed at the Stanford Nanofabrication Facility, successfully fabricating a sample with a 2 μm line width and a 4 μm spacing. The charge limit measurements will be performed shortly.

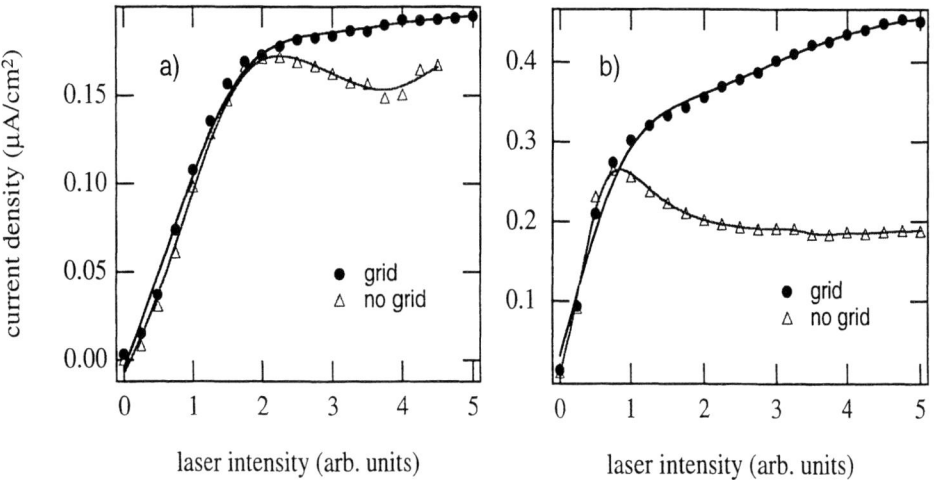

FIGURE 4. The charge density as a function of the laser intensity for a) 862 nm excitation and b) 790 nm excitation.

OUTLOOK

The charge limit effect is well understood in terms of the surface photovoltage effect. The high surface doping technique seems to achieve the beam intensity requirements of future linear colliders. The bleedoff grid is also promising for further increasing the charge output. However, a high intensity beam together with high polarization is yet to be demonstrated.

REFERENCES

1. M. Woods et. al, J. Appl. Phys. **9**, 2295 (1994),
   H. Tang in *Proceedings of the Workshop on Photocathodes for Polarized Electron Sourses for Accelerators*, SLAC-432 (January 1994),
   K. Togawa et. al, Nuclear Instr. Meth. **A 414**, 431 (1998),
   A. Herrera-Gómez, G Vergare and W. E. Spicer, J. Appl. Phys. **79**, 7318 (1996),
   M.H. Hecht, Phys. Rev. B **41**, 7918 (1991),
2. B. Schwartz, F. Ermanis and M. H. Brastad, J. Electrochem. Soc. **123**, 1089 (1976).

Polarized Electron Source for Japan Linear Collider

K. Togawa[1], T. Nakanishi[1], S. Okumi[1], C. Suzuki[1], F. Furuta[1], K. Wada[1], T. Nishitani[1], M. Yamamoto[1], H. Kobayakawa[2], Y. Takeda[2], Y. Takashima[2], H. Sugiyama[2], O. Watanabe[2], Y. Kurihara[3], H. Matsumoto[3], T. Omori[3], Y. Takeuchi[3], M. Yoshioka[3], H. Horinaka[4], K. Wada[4], T. Matsuyama[4], T. Saka[5], T. Baba[6] and T. Kato[7]

[1] Department of Physics, Nagoya University, Nagoya 464-8602, Japan
[2] Faculty of Engineering, Nagoya University, Nagoya 464-8603, Japan
[3] High Energy Accelerator Research Organization (KEK), Tsukuba 305-0801, Japan
[4] College of Engineering, University of Osaka Prefecture, Sakai 599-8531, Japan
[5] Daido Institute of Technology, Nagoya 457-8531, Japan
[6] Silicon System Research Lab., NEC Corporation, Sagamihara 220-1198, Japan
[7] New Materials Research Lab., Daido Steel Co. Ltd, Nagoya 457-8531, Japan

Abstract. Our collaboration group has been conducting the R&D works on the polarized electron source for Japan Linear Collider. The experimental results to produce the high-intensity multi-bunch beam from the superlattice photocathodes are summarized in this report.

INTRODUCTION

Japan Linear Collider (JLC) requires a highly spin-polarized ($\geq$80%) electron beam with high-intensity and multi-bunch structure. About 70 micro-bunches with a separation time of 2.8 ns must be produced at a 150 Hz repetition rate, and each bunch must include more than $\sim 2 \times 10^{10}$ electrons in ~700 ps bunch width at the gun [1]. The schematic diagram of JLC and its beam structure at the gun region is shown in Fig. 1. Such a beam is extracted from a GaAs-type semiconductor photocathode with a negative electron affinity (NEA) surface.

We have been conducting the developments of highly-polarized photocathodes, such as strained GaAs and superlattice. Nowadays, high polarization of $\geq$80% and high quantum efficiency (QE) of $\geq$0.3% are achieved simultaneously by using the strained-layer superlattice. Next subject is multi-bunch generation to satisfy the JLC requirements. In this report, the experimental results to produce the high-intensity multi-bunch beam from the superlattice photocathodes are summarized.

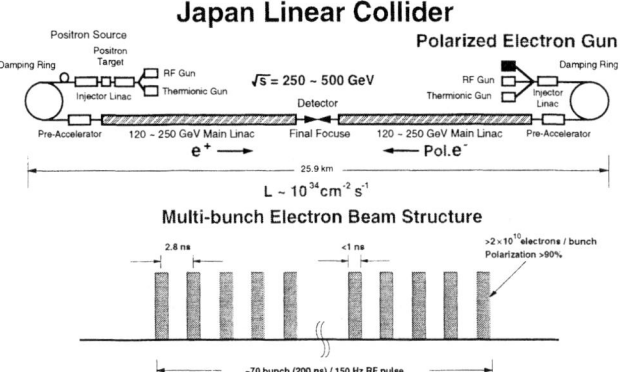

FIGURE 1. Schematic diagram of JLC and multi-bunch beam structure at the gun region.

SURFACE CHARGE LIMIT

In order to generate a polarized multi-bunch beam required by JLC, it is needed to overcome the "surface charge limit (SCL)" phenomenon in the NEA photocathode. The charge saturation level becomes lower than that of the space-charge-limit determined by the Child-Langmuir law [2]. The SCL mechanism is understood as follows [3]. Some of the excited electrons going toward the surface can escape to vacuum and become the extracted beam, however, the remainder of them are trapped at the band bending region (BBR) and become the surface charge. Due to the lack of a sufficient number of holes which can recombine with the trapped electrons in the BBR, the surface charge does not disappear so quickly. As a result, it prevents electrons in the later portion of excited electrons in the bunch from escaping to vacuum.

To overcome the SCL problem, the following two conditions must be satisfied simultaneously : (1) high escape probability of conduction-band electrons against the surface potential barrier and (2) high tunneling probability of valence-band holes against the surface band bending barrier.

SUPERLATTICE PHOTOCATHODE

We noticed that such conditions can be satisfied by superlattice structure [4]. The escape probability will become higher because the upward-shift of the conduction-band minimum (for example, ~150 meV for the GaAs-AlGaAs superlattice (SL#7 [5])) is as large as the amount of NEA. In the same way, the energy level of the valence-band hole is lower than that of the valence-band maximum of the well and the hole tunneling probability will be increased by this energy shift (~70 meV for SL#7). The schematic energy band diagram of the superlattice photocathode is shown in Fig. 2. Another effective method which can increase the hole tunneling probability is heavy p-doping, because the narrower the width of the BBR, the higher the tunneling probability becomes, and this induces a faster recombination of the trapped electrons with holes [6]. The modulation doping technique was employed for our superlattice photocathodes since high doping is necessary to achieve high QE and low doping is preferable to minimize depolarization in the electron emission process (4×10^{19} cm^{-3} at the 5 nm surface and 5×10^{17} cm^{-3} inside the ~100 nm superlattice for SL#7).

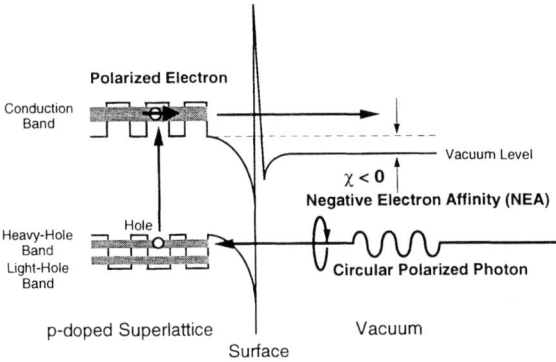

FIGURE 2. Enenrgy band diagram of the NEA superlattice photocathode

NANOSECOND MULTI-BUNCH EXPERIMENTS

The generation of a nanosecond multi-bunch beam has been achieved using the GaAs-AlGaAs superlattice (SL#7) [4]. This superlattice has been fabricated by NEC Corporation using the MBE method. The experiments have been performed using the 70 keV polarized electron source at Nagoya University. The quadruple-bunch laser (7ns bunch width, 25 ns bunch separation) has been used as a light source.

The charge saturation curves and the temporal beam profiles taken for different QE (monitored using a HeNe laser) are shown in Fig. 3, respectively. The pulse laser wavelength was tuned to 748 nm which gives the maximum polarization. For the high QE state, the bunches have the symmetrical shapes and the saturated peak current of ~1.6 A is limited by the space-charge-limit of the gun. All of the bunches keep the symmetrical shapes even for the degraded NEA state. The InGaAs-AlGaAs strained-layer superlattice (SLSA#2 [5]) has also shown similar behavior. These results indicate that the surface charge accumulated in those types of superlattice is negligible for the beam with peak current of ≥1 A and time scale of ~100 ns.

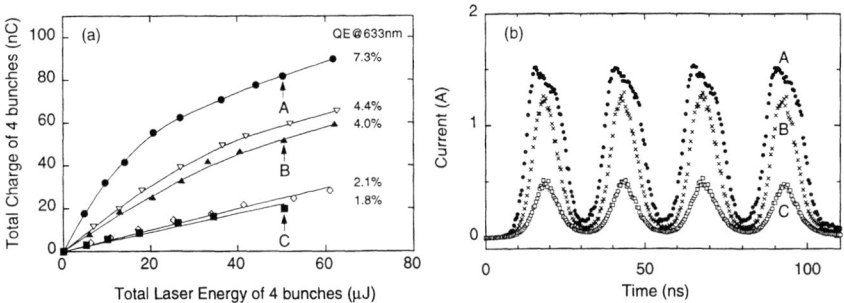

FIGURE 3. Charge saturation curves (a) and temporal beam profiles (b) measured for GaAs-AlGaAs superlattice using nanosecond quadruple-bunch laser.

SUB-NANOSECOND MULTI-BUNCH EXPERIMENTS

The experiment to produce a sub-nanosecond double-bunch beam has been performed using the GaAs-GaAsP strained-layer superlattice (SLSP#9). This superlattice has been fabricated by the MOCVD apparatus at Nagoya University [7]. The sample structure and the measured polarization and QE spectra are shown in Fig. 4.

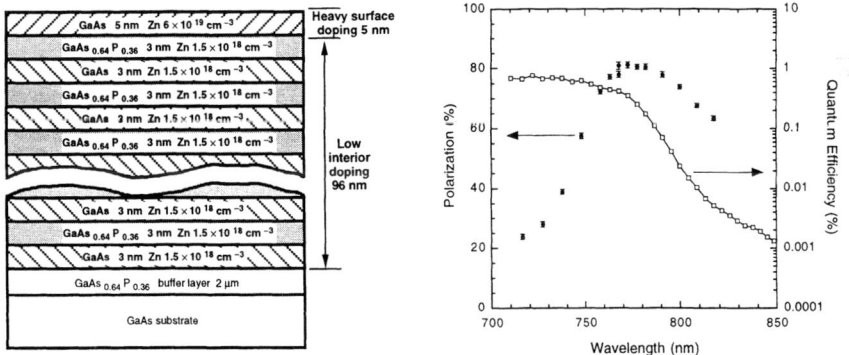

FIGURE 4. Structure of GaAs-GaAsP strained-layer superlattice (left) and measured polarization and QE spectra (right).

The sub-nanosecond double-bunch laser structure was made by using a fast Pockels cell [8]. The plane of linear polarization at the center part of the 7 ns original pulse was rotated by 90° by applying a 0.7 ns high voltage (~4 kV) pulse to the KD*P crystal of the Pockels cell. The modulated part was sliced off by a subsequent linear polarizer, and the sub-nanosecond bunch was remain. The delay path was set to give 2.8 ns bunch separation. The schematic view of this laser system and the time profile of the double-bunch laser light measured by a PIN photodiode are shown in Fig. 5. The pulse laser wavelength was tuned to 773 nm which gives the maximum polarization. The experiment has been performed at a gun bias voltage of 50 keV.

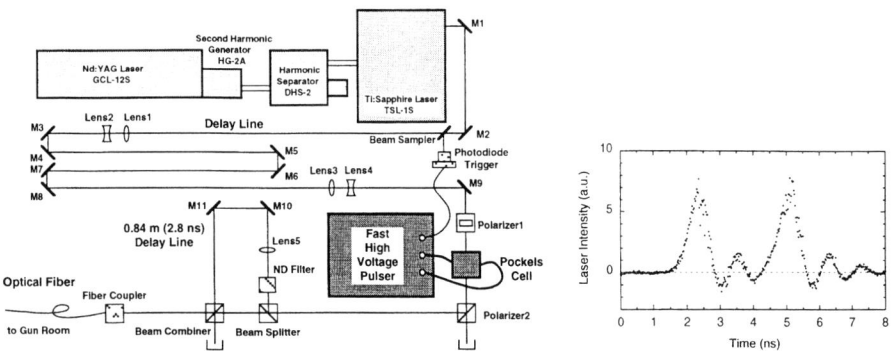

FIGURE 5. Sub-nanosecond double-bunch laser system (left) and time profile of the laser light (right).

The charge saturation curves taken for the first and the second bunches and the temporal beam profiles measured at several laser energies are shown in Fig. 6, respectively. The QE at 633 nm was 2.1%. Both data are in good agreement with the behavior governed only by the space-charge-limit phenomenon. The electron population in one bunch estimated to be 0.6×10^{10} e$^-$. The result indicates that the SCL phenomenon does not also appear in the beam with peak current of ≥ 1 A and time scale of ~3 ns.

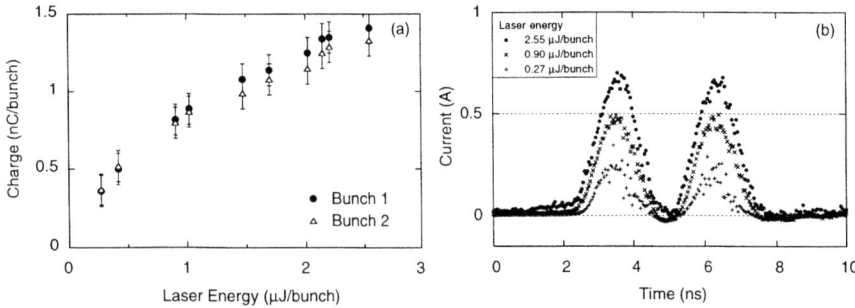

FIGURE 6. Charge saturation curves (a) and temporal beam profiles (b) measured for GaAs-GaAsP strained-layer superlattice using sub-nanosecond double-bunch laser.

SUMMARY AND REMAINING SUBJECTS

Photocathode : the strained-layer superlattice could satisfy the requirements for polarization ($\geq 80\%$) and QE ($\geq 0.3\%$). They are expected to be improved further by the systematic research of the GaAs-GaAsP strained-layer superlattice [9].

Multi-bunch generation : no SCL phenomena in the beam with peak current of ≥ 1 A and both time scales of nanosecond and sub-nanosecond have been observed for the superlattice photocathodes with heavily p-doped surface. It means that the charges are limited only by the space-charge-limit of the gun. Therefore, the charge requirements for JLC (2×10^{10} e$^-$/micro-bunch and 1.4×10^{12} e$^-$/macro-pulse) should be fulfilled by raising a bias voltage of the gun. The achieved performance and JLC requirements are summarized in Table 1.

TABLE 1. Summary of achieved performance and JLC requirements.

	Electrons / bunch	Bunch Width	Bunch Separation	Bunch Number	Total Electrons / pulse	Intensity Jitters
Experiment (Nanosec.)	1.3×10^{11}	12 ns	25 ns	4	0.5×10^{12}	$\geq 10\%$
Experiment (Sub-nanosec.)	0.6×10^{10}	1.4 ns	2.8 ns	2	1.2×10^{10}	$\geq 10\%$
Linear Collider	2×10^{10}	≤ 1 ns	2.8 ns	70	1.4×10^{12}	$\leq 1\%$

New gun system : for the present gun the operational voltage was limited below 70 kV, because the accumulated cesium atoms deposited on the cathode electrode during the many NEA activation processes increased the dark current. In order to avoid this problem, the new polarized DC-gun system which posses a load-lock system has been constructed and is being tested [10].

New laser system : the remaining important subject is to accomplish the polarized beam with real multi-bunch structure. The good stability, especially extremely low fluctuation of bunch-by-bunch intensity of the charge ($\leq 1\%$) is required. They are determined by the specifications of the laser light source. Since no multi-bunch laser system that meets the specifications of JLC is available at this moment, the special design and the construction technique are needed for this issue.

REFERENCES

1. *JLC Design Study*, KEK-Report 97-01 (1997)
2. M. Woods et al., *J. Appl. Phys.* **73** 8531-8535 (1993)
3. A. Herrera-Gómez et al., *J. Appl. Phys.* **79** 7318-7323 (1996)
4. K. Togawa et al., *Nucl. Instr. Meth.* **A414** 431-445 (1998)
5. These sample numbers are defined in T. Nakanishi et al., *AIP Conf. Proc.* **421** 300-310 (1998)
6. R. Alley et al., *Nucl. Instr. Meth.* **A365** 1-27 (1995), T. Maruyama et al., this workshop
7. O. Watanebe et al., this workshop
8. K. Togawa et al., *Nucl. Instr. Meth.* **A455** 118-122 (2000)
9. T. Nishitani et al., this workshop
10. K. Wada et al., this workshop, M. Yamamoto et al., this workshop

Present status of experimental S-band GaAs-photogun driven by the Solid State GaAs Pulse Laser[1]

N.S. Dikansky*, R.G. Gromov*, E.S. Konstantinov*,
P.V. Logatchov*, A.V. Alexandrov[†]

*Budker Institute of Nuclear Physics, Novosibirsk, Russia.
[†]Los-Alamos National Laboratory, Los-Alamos, USA

Abstract. This paper concerns the present status of experimental RF-photogun at BINP (Novosibirsk). The bulk GaAs-photocathode to investigate some none well-known phenomenon on the photocathode surface in high RF-field is used. An interesting question concerned with S-band photogun is to achieve bright beam with high level of polarization. For this goal two solid state GaAs lasers (SSL) at 820 nm wavelength with 2-4 W power and pulse duration on 2 μs and 200 ps were designed and produced.

INTRODUCTION

The majority of semiconducting RF-photogun technology have been developed for the last decade. Appeared in 1984 year [2] this idea received a number of research directions for high current beam improvement including high current, polarization level etc. One of them is investigation of cathode materials to have as high as possible polarization level of electrons. Nowadays, there is a set of good candidates for photocathode production including cathodes for polarized beams. The most attractive is GaAs and its family. The remarkable features of it is combining possibility to emit electrons having high level of polarization, high Quantum Efficiency, very short response time.

The intrinsic ability of GaAs is width of E_{th} – energy level between valence zone and conductive zone. It equals to $1.45 eV$. This gap corresponds to wavelength $820 nm$. To achieve as high as possible level of polarization one need to use infrared laser, because deviation of photon energy from $1.45 eV$ decrease the polarization level, or to pick-up E_{th} up to about $2 eV$ by doping Al or In to use a laser having working wavelength in the visible spectrum region. In our case we have choose the first direction and have developed SSL having photons energy precisely correspond to E_{th}.

---

[1] Work supported by Russian Fondation for Basic Research under contract No 98-02-17887.

INSTALLATION SETUP

Vacuum chamber and magnetic system

The first design of S-band RF-photogun was described on EPAC'96 conference [1]. The photogun divided into two main volumes – activation chamber and accelerating chamber. Activation chamber has build-in cathode manipulator, Cs and O dispensers, anode ring and window for laser spot on the cathode observing. Accelerating chamber has RF-cavity and RF-power input of waveguide. These chambers are divided by mobility blind to protect cathode during RF-processing or RF-cavity surface from Cs and Oxygen pollution during activation process.

For the time being some improvements on the installation have been done. First, the Faraday cup was moved for more long distance from RF-cavity to improve vacuum condition. An additional pumping vessel on the transport channel was installed to protect photocathode surface from Faraday cup's intense gas release. The magnetic spectrometer able to resolve electron beams of energy up to $1.3 MeV$ was installed.

The vacuum level into photogun after heating up to 300 grad during 70 hours reached $5 \cdot 10^{-11}$ Torr. It is sufficient for the photocathode lifetime up to tens hours.

The laser input window is made from quartz glass. The laser beam is goes through electron transport channel. It helps to simplify installation construction and RF-cavity geometry.

RF-system

The choose of RF-cavity geometry depends on type of RF-power supply, required beam parameters and photocathode features.

The gun is powered by a $60MV$ KIU-12 klystron. This klystron together with its modulator was used as a power supply for VEPP-5 preinjector prototype. After

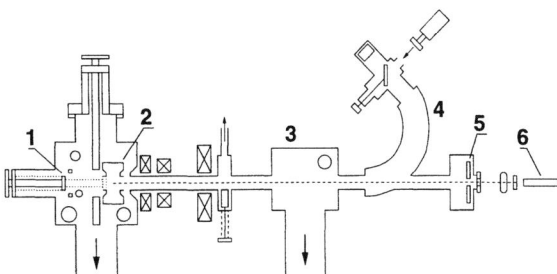

FIGURE 1. Layout of S-band GaAs-Photogun 1 – GaAs Photocathode, 2 – RF cavity, 3 – additional pumping vessel, 4 – spectrometer, 5 – Faraday cup and NEG pumping, 6 – laser system

mounting the regular preinjector klystron, KIU-12 is used only for experimental photogun performing. The main frequency of it if $2856 MHz$, peak power applied to the RF-cavity was not exceeded $2MW$ during processing and $0.8MW$ while photocathode is installed into the cavity. This power level corresponds to $65 MV/m$ accelerating electric fields and excite electron energy about $1 MeV$.

RF-cavity shape was optimized to have the shunt impedance maximal with the minimal electric field on the cavity wall relatively to the field on the cathode with the URMEL code.

GaAs-photocathode

The photocathode is mounted on the manipulator operated by hand. The cathode handler includes heating spiral, thermocontrole, heatguide to GaAs-cathode for clearing before activation procedure.

All things directly contacting with the cathode made from Ta because it has the same expansion thermo coefficient like GaAs.

Laser System

The RF photogun laser system is designed to provide a number of experiments on GaAs photocathode with dark current, field emission, secondary emission and to solve the problems of the laser beam transfer to the cathode.

In the Novosibirsk solid state lasers based on GaAsInP produce the light on $820 nm$ wavelength with $2-4W$ power have been made. The wavelength chosen in order to have the maximum QE both positive and negative electron affinity regimes of the cathode.

The two sets of SSL developed. The first one is to produce the light pulses with $2-4\mu s$ duration. The second laser set is developed to produce the light pulses with $100-200 ps$ duration on the same wavelength and power.

The both lasers are equipped with pulse power supplies and synchronization systems allowing to keep the laser pulse linked with the RF phase. The problem of the correct phasing is not important very much for the $2-4\mu s$ laser and is crucial for the $100-200 ps$ one. For the last case the easily obtained 100 ps jitter is to be dumped for a few times with the thermostabilization of the synchronization block.

The both lasers are also equipped with the optics allowing to produce the circularity polarized light in order to have polarized electrons. The further activities concerning the development of the RF photogun laser system suggests the use of the SSL arrays with 10-100 lasers with the focusing optics to grow the emitted electrons current.

OBTAINED EXPERIMENTAL RESULTS

After high vacuum level receiving the activation experiments, QE and life time measurements have been performed. During first experiments the vacuum condition in the photogun was about $2 \cdot 10{-10} Torr$.

There was obtained the photocathode activated for $QE = 0.5\%$ can performs into strength RF accelerating electric field up to $60 MeV/m$ without irreversible damages. In this case the photocathode lifetime was about 3 hour. After last modernization the vacuum level improved up to $5 \cdot 10^{-11} Torr$. In this case the life time growed up to 10 hours for the some level of QE.

The dark current depends on quantum efficiency of the cathode and for NEA GaAs photocathode exceeds by $2-3$ orders of magnitude dark current of PEA photocathode in a field of the same strength. The possible mechanisms of dark current generation are considered in [3].

FUTURE PLANES

In the nearest future we are planning a set of experiments with new SSL. The $2-4\mu s$ duration laser should provide the experiments with bunched trains. It concerns the measurements of the dark current and the life time of the cathode. The second $100-200 ps$ laser is to be used in the field emission, secondary emission and in experiments on effective lossless acceleration.

Although this experiments not demand polarized beams, the experimental photogun may produce produce longitudinally polarized electrons. After this generation of experiments with SSL we are schedule preparation of S-band photogun to the polarized investigations. For this reconstruction we have already Mott-polarimeter. Spin-manipulator and new magnetic system now are under design and needed in making.

REFERENCES

1. A.V.Alexandrov at all, *A prototype of RF photogun with GaAs photocathode for injector of VEPP-5*, EPAC'96, Sitges, June 1996.
2. G.A.Westenskow, J.M.J.Madey *Microwave Electron Gun* Laser and Particle Beams (1984), Vol.2, Part 2, pp. 223-225.
3. A.V. Aleksandrov, N.S. Dikansky, R.G. Gromov, E.S. Konstantinov, P.V. Logachev *Experimental Study on GaAs Photocathode Performance in RF Gun* Proceedings of the EPAC'2000, Conference, Vienna, Austria, 26-30 July 2000.

Fabrication of Photocathode Test-stand

G. N. Kim[*,+], M. W. Lee[*], D. Son[*], Y. J. Park[+], S. J. Park[+], M. H. Cho[+], I. S. Ko[+], and W. Namkung[+]

*Center for High Energy Physics, Kyungpook National University, Taegu 702-701, Korea
+Physics Department, Pohang University of Science and Technology, Pohang 790-784, Korea

Abstract. We have designed and fabricated a photocathode test-stand in order to develop the polarized electron source in Korea. The test-stand consists of a photocathode gun, a light source, a cylindrical condenser and a mini-Mott chamber. We report the status of the test-stand and the future plan for the development of a polarized electron source.

INTRODUCTION

The 2-GeV electron linac was constructed as a full energy injector to the Pohang Light Source (PLS). The prime mission for providing beams to the storage ring requires a few minutes for each injection, and it will be happen once or twice in a day. Therefore, we planned to use electron beams of various energies to promote other branches of basic and applied sciences. The electron linac of the Pohang Accelerator Laboratory (PAL) accelerates up to 2.5 GeV with a pulse width of 3-μs and a repetition rate of 120 Hz with little modification, which corresponds to the duty-factor of 3.6×10^{-4}[1]. The poor duty-factor of the present electron linac makes nuclear experiments difficult to perform. In order to increase the low duty-factor we need to construct a stretcher ring. The stretcher ring is a very cost-effective approach for upgrading electron linac for nearly continuous wave operation. Meanwhile, the other way to utilize the poor duty-factor electron beams is to employ the polarized electron source. When spin dependence is included, new terms in the cross section arise. Thus, polarized electrons open a new dimension in the field of nuclear and high-energy physics. In recently, the polarized electron source becomes to play the very active role in both high energy and nuclear experiments.

In the first stage, we designed and constructed a test-stand to develop the polarized electron source. In this report, we describe the design and fabrication of the test-stand and the future plan for the development of a polarized electron source.

DESIGN AND FABRICATION OF THE TEST-STAND

A photocathode test-stand to develop the polarized electron source has been designed and its layout is shown in Fig. 1.

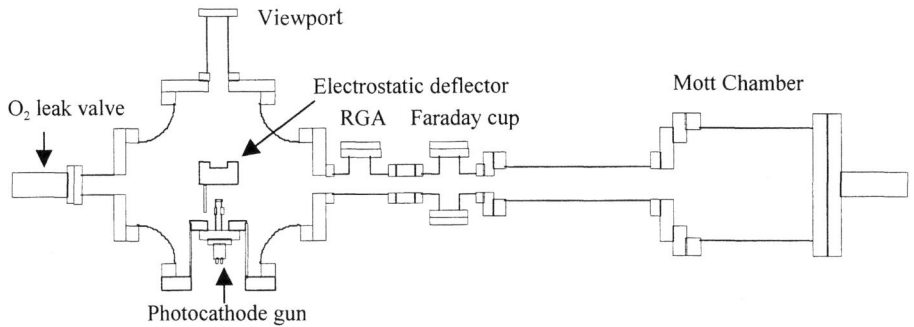

FIGURE 1. Schematic layout of the test-stand.

The test-stand consists of four subsystems: the photocathode gun, a light source, an electrostatic deflector, and a Mott polarimeter. We will employ a Zn-doped GaAs as a photocathode in the test-stand. We used a He-Ne laser and a QTH lamp for the light source. The spin direction should be transverse (i.e., the direction of the electron spin is perpendicular to that of the electron momentum) in the Mott polarimeter. The transverse spin can be obtained by the 90-degree bending of the electron in a static electric field. The electrostatic deflector serves to provide this bending electric field, and additionally a transverse focusing in the direction that lies in the plane of the bending. The schematic diargram of the Mott polarimeter[2] is show in Fig. 2.

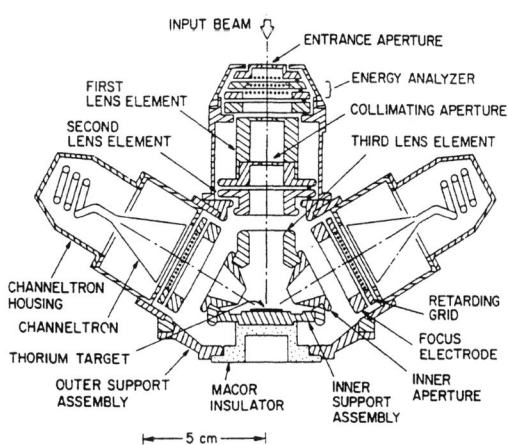

FIGURE 2. Schematic diagram of the Mott polarimeter.

Two 25-mm diameter channeltrons manufactured from Dr. Sjuts Optotechnik were used in the Mott polarimeter. The channeltron housings and outer support assembly are operated at ground potential. We mounted the Mott polarimeter to a 10-inch diameter conflat flange (CFF) with various electrical feedthroughs. All necessary terminals at the polarimeter were connected to the feedthroughs by use of clean copper wires. The Mott polarimeter and 10-inch CFF were mounted on 8-inch diameter vacuum chamber, Mott chamber. In order to provide appropriate high voltages to the channeltrons, a high-voltage resistor circuit was prepared and installed near the polarimeter chamber. Required high voltage power supplies were also prepared.

In order to test the photocathode gun, we measured the quantum efficiency (QE) using the experimental setup as shown in Fig. 3. As a photocathode, we used a Zn-doped GaAs of 1.2cm x 0.5 cm, which was given by T. Nakanishi in Nagoya University. The pressure of the gun chamber after bakeout for about 100 hours at 150-200°C is achieved about 10^{-10} Torr. The cathode is heated to about 600°C for 1.5 hours. After heat treatment the GaAs cathode is left to cool down. The activation is performed by passing a current 4.5A through a Cs dispenser. While maintaining Cs deposition, oxygen is let in through a leak valve and the O_2 leak rate is adjusted so as to obtain a maximum emission current. The GaAs cathode is illuminated during the activation by a He-Ne laser (632.8 nm). The maximum emission current obtained from the GaAs with the He-Ne laser operated at approximately 0.9 mW was 3 µA, which correspond to QE=0.68 %. We measured QE varying the wavelength from 400 nm to 800 nm using the quartz-tungsten-halogen (QTH) lamp. The QE was varied from 4% to 0.5% according to the wavelength.

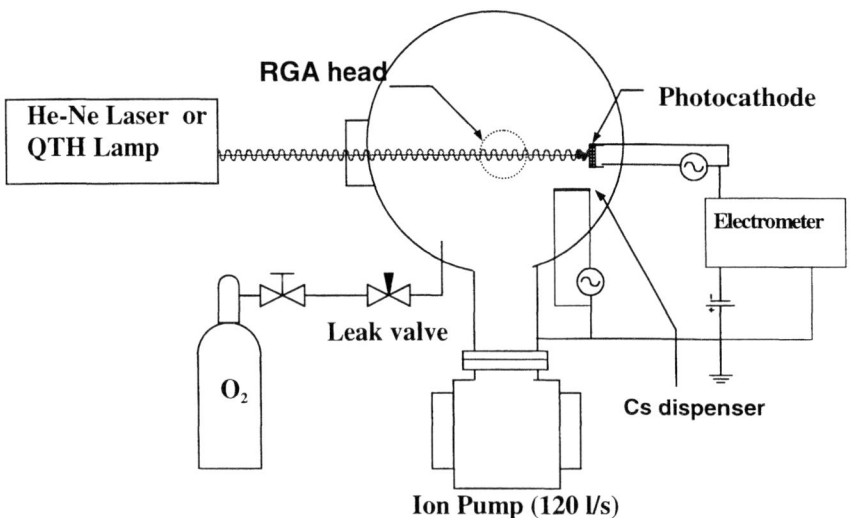

FIGURE 3. Experimental setup to measure the quantum efficiency

SUMMARY AND FUTURE PLAN FOR THE POLARIZED ELECTRON SOURCE

The 2-GeV electron linac at PAL is the only facility for the experimental nuclear and high-energy physics in Korea. Since the low duty-factor of the linac makes difficult to perform the meaningful experiments, the development of a polarized electron source is one way to utilize the 2-GeV electron linac. In addition to this, the experience from the development of a polarized electron source is of crucial importance to future collaboration work with the international accelerator group for the very high-energy linear collider.

In this report, we briefly described the status of the photocathode test-stand. There are many subjects to be solved in the photocathode test-stand, i.e., UHV system, activation method and a polarimeter. Until now, we just stepped in the first stage of the polarized electron source development.

The photocathode gun of the PEGGY system [3] was moved from the Stanford Linear Accelerator Center (SLAC) early in 1998. Recently, the PEGGY system was no more used in any experiment at SLAC. However, the photocathode gun of the PEGGY system is still useful for the polarized electron source development. The photocathode gun is very similar to the ordinary PLS electron gun, with the conventional thermionic cathode replaced with a GaAs cathode. The entire system of the PEGY consists of a photocathode gun, a laser system, a spin manipulation system including the beam transport line and the beam quality monitors, and a Mott polarimeter. In order to complete the polarized electron source, we need to design and construct other subsystem based on the operation experience from the test stand.

After completing the polarized electron source in the future, we want to accelerate the polarized electron beams in the 100-MeV test-linac built in April of 1998 at PAL. For a future e^+/e^- linear collider such as NLC and JLC, polarized electron beams will again play a powerful role in its physics programs. Therefore, the experience from the development of a polarized electron source is important for accelerator community in Korea to involve the future high-energy e^+/e^- linear collider projects.

REFERENCES

1. G. N. Kim et al., "Activities and Plans for Nuclear Physics Experiments with PLS 2-GeV Electron Linac," in Frontier 96, edited by H. Toki, T. Kishimoto and M. Fujiwara, Proceedings of XV RCNP Osaka International Symposium, Osaka, 1996, pp123-129.
2. G. C. Burnett, T. J. Monroe, and F. B. Dunning, Rev. Sci. Instrum. **65**, 1893-1896 (1994).
3. M. J. Alguard, et al., Nucl. Instr. and Meth. **163**, 29 (1979).

A pulsed electron source for atomic collision experiments

C. D. Schröter[*][†], A. Dorn[*], J. Deipenwisch[*], C. Höhr[*], R. Moshammer[*][†] and J. Ullrich[*][†]

[*] *Fakultät für Physik, Universität Freiburg,*
79104 Freiburg, Germany
[†] *Max-Planck-Institut für Kernphysik,*
69029 Heidelberg, Germany

Abstract. A pulsed photoelectron gun has been constructed to satisfy the beam specifications required by "reaction microscope" studies of atomic collisions [1,3].

Short pulses of electrons (< 500 ps) are produced using GaAs/AlGaAs heterostructure crystals. The laser diode for the gun operates at 650 nm and can generate up to 60 μA average current. Quantum efficiencies of 3 % and a photocathode lifetime of one week have been achieved.

To study the emission characteristics of the photoelectrons a novel imaging spectrometer has been constructed and first energy-distribution curves have been measured at moderate resolution. Computer simulations show, that an excellent energy (< 1 meV) and angular resolution is attainable with such a spectrometer.

INTRODUCTION

In the recent past, a many-particle momentum analyzer has been developed for investigating the dynamics of atomic and molecular reactions [1,2]. With a modified version of this novel analyzer, the first kinematically complete experiments for double ionization of helium using a pulsed electron beam have been performed [3,4]. For these studies, requiring high detection efficiency and excellent momentum resolution of all particles emerging from the reaction, the quality of the electron beam is decisive.

Optimum conditions require pulse widths less than 500 ps, a beam diameter of 100 μm at the target and small currents. The charge per pulse has to be less than 1 fC, at a maximum pulse-repetition rate of about 3 MHz.

To fulfill these requirements an electron source based on laser-pulsed electron emission from a GaAs crystal with negative electron affinity (NEA) has been built.

EXPERIMENTAL SET-UP

To ensure a long lifetime of the cathodes, UHV technology and extremely careful preparation of the crystal surface is mandatory. For this purpose a multi-chamber preparation system consisting of four UHV chambers has been built. The experimental set-up is shown in figure 1.

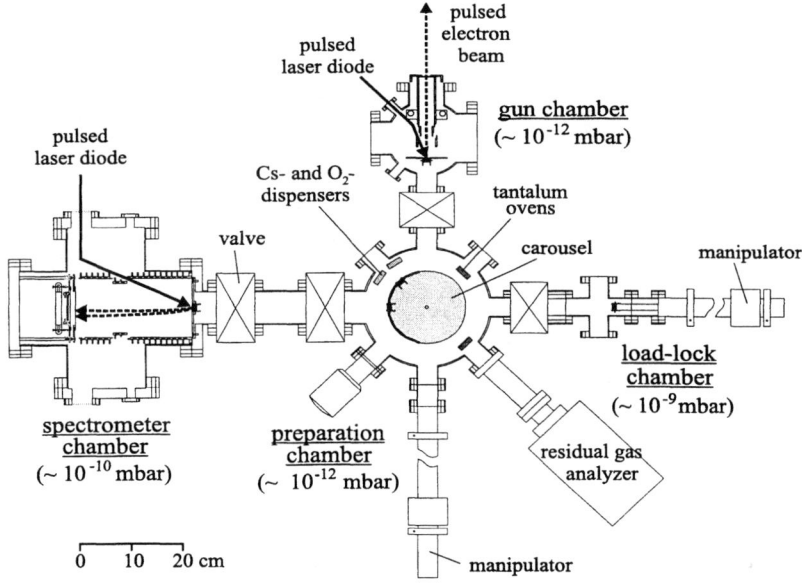

FIGURE 1. Experimental set-up.

The surface of the GaAs crystal has to be prepared in such a way that it obtains negative electron affinity. This is done by activation, i.e. a deposition of cesium atoms and oxygen on the atomically clean surface.

Clean surfaces are prepared by an oxide-free technique. The GaAs crystals are chemically cleaned under argon atmosphere in a solution of HCl in isopropyl alcohol and transferred without exposure to air in a transfer vessel to the load-lock chamber and from there to the preparation chamber. The crystals are mounted on a carousel, which can be turned into appropriate positions for the final heat-cleaning procedure (at $\sim 550\,°C$) and for the activation process.

The preparation chamber is evacuated by an ion-getter pump. For effective reduction of residual gas components volume-getter stripes are mounted inside the chamber. After baking the stainless-steel chamber at a temperature of $300\,°C$ for the duration of one week, a pressure of 2×10^{-12} mbar was reached. When required, the chamber can be pumped additionally by a titanium-sublimation pump. The composition of the residual gas is controlled by a residual-gas analyzer.

After preparation the photocathodes can be transferred to the gun chamber or to the spectrometer chamber, which are both separated from the preparation chamber by gate valves.

PULSED ELECTRON GUN

In the gun chamber the activated cathode is used for the production of a short-pulsed, low-emittance, cold electron beam. The gun is designed for electron energies ranging from about 100 eV up to 5 keV.

The laser diode for this gun operates at 650 nm and generates up to 60 μA average current. Using doped p-type reflection mode GaAs/AlGaAs heterostructure crystals (6×10^{18} Zn/cm^3), quantum efficiencies of 3 % and a photocathode lifetime of one week have been achieved. The active GaAs layer of this heterostructure has a thickness of only 0.9 μm, therefore the long tail of the generated electron pulse extends out to less than 200 ps [5]. Because the laser diode has a pulse width of 100 to 300 ps (dependent on the laser power), an electron-pulse width of less than 500 ps is expected.

Experiments with transversely polarized electrons are planned in the near future. For this purpose a 90° electrostatic deflector has already been constructed. Using strained GaAs photocathodes together with circularly polarized light from a laser head working at a wavelength of 840 nm, the source should provide a beam of spin polarized electrons with a high degree of polarization (up to 80 %).

EMISSION CHARACTERISTICS OF PHOTOELECTRONS

To investigate the photoelectron-escape mechanism from a NEA-photocathode a spectrometer is required, that allows the simultaneous measurement of the energy and angular distributions of very low energy electrons. The set-up of our UHV compatible spectrometer is shown in figure 2.

Here the photoelectrons are imaged by a homogeneous electric field onto a position-sensitive multi-channel plate (MCP) detector. The time-of-flight (TOF) of the electrons as well as the position on the detector is measured. From the TOF the longitudinal momentum of the electrons can be determined, from the position on the MCP detector and the TOF the transversal momentum of the electrons can be extracted. This is equal to the determination of the electrons energy and emission angle.

The spectrometer is in operation and preliminary energy-distribution curves have been measured. The first tests have been performed under non-ideal vacuum-pressure conditions ($\sim 10^{-10}$ mbar). Furthermore, due to the use of different materials for the photocathode mounting contact-potential differences arose. Therefore, the pilot-experiments have been performed at high extraction fields, hence the energy-distribution curves extracted so far have lower resolution than ultimately achievable.

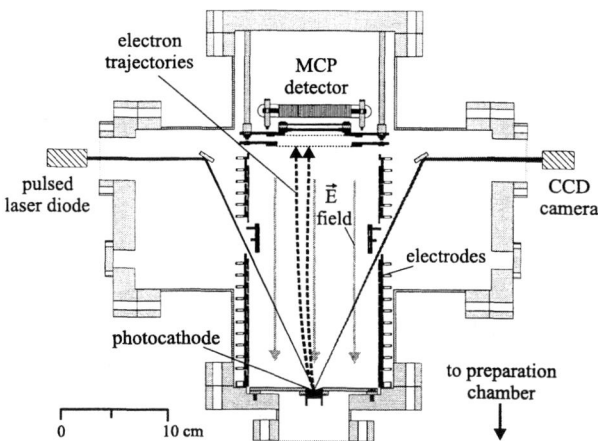

FIGURE 2. UHV chamber containing spectrometer for photoelectron detection.

Computer simulations show, that with this spectrometer an excellent energy resolution in the transverse direction can be achieved ($<1\,\mathrm{meV}$), even at high extraction fields. In the longitudinal direction high resolution can be reached only for the smallest extraction fields. For further improvement the electric-field strength must be decreased. This leads to a very limited transversal energy acceptance, i.e. acceptance for the emission angle of the photoelectrons, but this dilemma can be solved by overlapping a homogeneous magnetic field to the detection system [1,2].

This work was partially supported by the Deutsche Forschungsgemeinschaft within the Leibniz-program and the SFB 276. The heterostructure material was kindly put at our disposal by A. S. Terekhov. We are grateful to the MPI group of D. Schwalm and A. Wolf as well as to A. S. Terekhov for detailed information on the basic conception of the photocathode preparation set-up. We would like to thank D. Orlov and S. Pastuszka for fruitful discussions.

REFERENCES

1. Moshammer R., Unverzagt M., Schmitt W., Ullrich J., Schmidt-Böcking H., *Nucl. Instr. and Meth.* B **108**, 425 (1996).
2. Ullrich J., Moshammer R., Dörner R., Jagutzki O., Mergel V., Schmidt-Böcking H., and Spielberger L., *J. Phys.* B **30**, 2917 (1997).
3. Dorn A., Moshammer R., Schröter C. D., Zouros T. J. M., Schmitt W., Kollmus H., Mann R., and Ullrich J., *Phys. Rev. Lett.* **82**, 2496 (1999).
4. Dorn A., Kheifets A., Schröter C. D., Najjari B., Höhr C., Moshammer R., and Ullrich J., *Phys. Rev. Lett.* submitted (2000).
5. Hartmann P., Bermuth J., v. Harrach D., Hoffmann J., Köbis S., Reichert E., Aulenbacher K., Schuler J., and Steigerwald M., *J. Appl. Phys.* **86**, 2245 (1999).

Surface photovoltage effect on clean and negative electron-affinity surfaces of GaAs and its superlattice

Senku Tanaka [1], Sam D Moré [2], Tomohiro Nishitani [3], Kazutoshi Takahashi [2], Tsutomu Nakanishi [3] and Masao Kamada [1,2]

[1] *Department of Structural Molecular Science, The Graduate University for Advanced Studies, Okazaki 444-8585, Japan*
[2] *UVSOR Facility, Institute for Molecular Science, Okazaki 444-8585, Japan*
[3] *Department of Physics, Nagoya University, Nagoya 464-8602, Japan*

Abstract. Photoelectron spectroscopy with the combination of synchrotron radiation and laser has been used for exploring the surface photovoltage effect of GaAs photocathodes. It has been observed that the laser induced surface photovoltage causes core-level shifts of photoelectron spectra without any spectral change. The surface photovoltage effects for core-levels and its temporal profiles on clean and negative electron-affinity surfaces of GaAs (100) and GaAs/GaAsP superlattice are reported.

INTRODUCTION

The negative electron-affinity (NEA) surface of GaAs(100) and its superlattice is known as a useful emitter for spin-polarized electrons. In studies of GaAs photocathodes, some groups have reported the charge-limit effect for high-intensity laser-pulse irradiation[1-3]. It has been suggested that the surface photovoltage (SPV) effect plays an important role for the charge-limit effect. The purpose of the present work is therefore to investigate the SPV effect and its temporal profile.

Figure 1(a) shows the schematic diagram of the SPV effect for p-type semiconductor. The absorbed photons induce the formation of free carriers, resulting in a change of the surface potential. This SPV effect causes the core-level shift. Thus, core-level photoelectron spectroscopy is a good way to investigate the SPV effect directly. The combination of SR and laser is a powerful tool to observe the temporal profiles of the SPV effect.

EXPERIMENTAL RESULTS AND DISCUSSIONS

The experiments were performed on beam line BL5A of the UVSOR Facility, Institute for Molecular Science, Okazaki, where an SGM-TRAIN type monochromator and a hemispherical electron-analyzer have been installed. The base pressure of the analyzing chamber was less than 1×10^{-8} Pa. A Zn doped (1×10^{19} atoms/cm^3) p-type

GaAs (100) wafer and a GaAs/GaAsP strained superlattice were used in the present experiments. The laser output from the regenerative amplifier with 10 kHz was introduced in the chamber through an optical fiber and was focused to a 5-mm spot concentric with the synchrotron radiation spot.

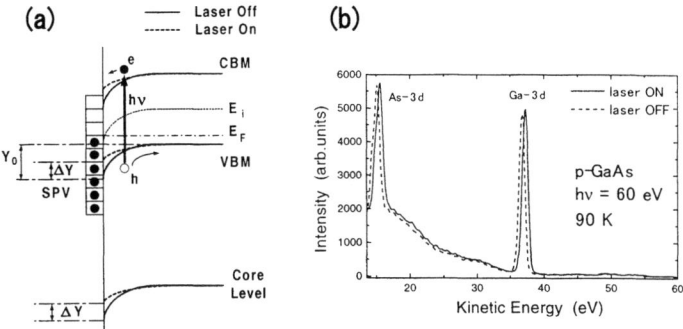

FIGURE 1. (a) Schematic diagram of band bending and the SPV effect (ΔY). (b) Photoelectron spectra with and without laser for clean p-GaAs(100) at 90 K (solid and dashed curves, respectively). Under laser illumination, each core-level is shifted to the higher kinetic-energy side due to the SPV effect.

Figure 1(b) shows photoelectron spectra with and without laser for clean p-GaAs (100) at 90 K. It has been observed that the Ga-3d and As-3d core-levels are shifted to higher kinetic-energy side due to the SPV effect. We have performed the detailed measurements of the SPV effect for the clean and NEA of the bulk and the superlattice samples. From these results, it was found that the SPV value of clean surface is larger than that of NEA surface, and the SPV value of the bulk sample is larger than that of superlattice. This can lead us to conclude that the SPV values are affected by both of the surface condition and the bulk structure.

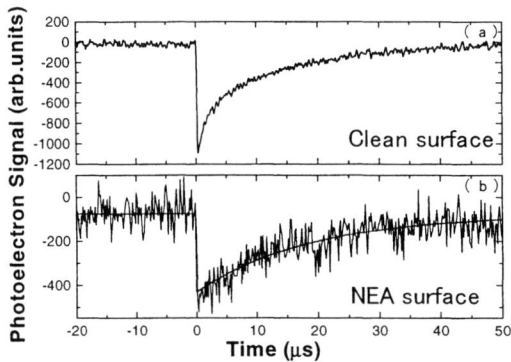

FIGURE 2. The time-dependence of the Ga-3d photoelectron intensity at a fixed kinetic energy. (a) clean GaAs surface. (b) NEA surface. Solid line is a guide for eyes.

Figure 2 shows the time-dependence of the Ga-3d core-level photoelectrons for the clean and the NEA surface. This temporal profiles of the photoelectron signal were obtained by using the time-to-amplitude converter system combined with a hemispherical electron-analyzer. The measured photoelectron energy was fixed at the lower kinetic-energy side of the core-level peak. At the moment samples were illuminated by the laser, photoelectron counts decreased, with a subsequent gradual recovering. It is obvious that there are two time-constants in the time-responses of the SPV effect on the clean surface, but the NEA surface has only one time-constant. Similar behavior has been also observed on the superlattice. These results indicate that there are two processes (a fast recovering process and a slow one) to recover from the non-equilibrium to the equilibrium conditions for clean surfaces. On the other hand, only a slow recovering process is dominant for the NEA surfaces.

CONCLUSIONS

The SPV effects on a GaAs and its superlattice have been studied by the photoelectron spectroscopy with the combination of SR and laser. The SPV values for clean GaAs are larger than those of their respective NEA surface, and also the SPV effect for bulk GaAs is stronger than that of the superlattice. It is suggested that the SPV effect has two origins, bulk and surface. The temporal profiles of photoelectron signal have been also investigated. Two recovering processes were observed on clean surfaces, while only one process was found on NEA surface. This means that the dynamical process of the recovering from non-equilibrium to equilibrium has also two origins.

ACKNOWLEDGMENTS

This work was partially supported by a Grand-in-Aid for Scientific Research from the Ministry of Science, Education, Sports and Culture, Japan.

REFERENCES

1. Woods, M., Clendenin, J., Frisch, J., Kulikov, A., Saez, P., Schultz, D., Turner, J., Witte, K., and Zolotorev, M., *J. Appl. Phys.* **73**, 8531-8535 (1993).
2. Herrera-Gómez, A., Vergara, G., and Spicer, W.E., *J. Appl. Phys.* **79**, 7318-7323 (1996).
3. Togawa, K., Nakanishi, T., Baba, T., Furuta, F., Horinaka, H., Ida, T., Kurihara, Y., Matsumoto, H., Matsuyama, T., Mizuta, M., Okumi, S., Omori, S., Suzuki, C., Takeuchi, Y., Wada, K., Wada, K., Yoshioka, M., *Nucl. Instr. and Meth .in Phys. Res. A.* **414**, 431-445 (1998).

Structure and magnetism of Fe thin films grown on Rh(001) studied by spin-resolved photoelectron spectroscopy

K. Hayashi[1], M. Sawada[1], A. Harasawa[1], A. Kimura[2] and A. Kakizaki[3]

[1] *Institute for Solid State Physics, University of Tokyo, Chiba 277-8581, Japan*
[2] *Department of Physical Sciences, Hiroshima University, Hiroshima 739-8526, Japan*
[3] *Institute of Materials structure Science, KEK, Ibaraki 305-0801, Japan*

Bulk Fe is known to be a bcc structure at room temperature and ferromagnetic below 920 K. On the other hand Fe films grown on non-magnetic substrates show a variety of structures and magnetism depending on the degree of the lattice constant (a_0) difference between substrate and a bulk bcc Fe (a_0=2.87Å). On Au(001) (a_0=4.07Å) and Ag(001), Fe films grow in bcc structure and are ferromagnetic due to the small lattice mismatch to the Fe(110). On Cu(001) and Co(001), Fe films thinner than 5 ML reveal a face centred tetragonal (fct) structure and ferromagnetism, while in 6-10 ML region Fe films show fcc structure and only the topmost few layers were found to be ferromagnetic. The origin of this complicated magnetic behaviour has been considered due to the lattice mismatch at the interface and investigated by a first principle calculation of the total energy and magnetic moments [1]. In this report, we present the structural and electronic properties of the Fe films epitaxially grown on a Rh(001) surface. The atomic distance of Rh(001) is an intermediate value between Cu(001) and Au(001), and the structure and magnetism of Fe films on Rh(001) are still controversial [2, 3]. We have measured the angle- (ARPES) and spin-resolved photoemission (SARPES) spectra to investigate the valence band structure of the Fe films and present Fe $2p_{3/2}$ photoelectron diffraction (XPD) patterns to study the geometric structure of the Fe films.

The ARPES and XPD experiments were carried out at the beamline 18A of the Photon Factory equipped with a standard ARPES measurement apparatus and a sample preparation system, to which RHEED system was installed for the present study. The SARPES spectra were measured using a He discharge lamp and a SARPES spectrometer consisting of a hemispherical electron energy analyzer and a compact retarding type Mott detector. The XPD patterns were observed using a MgKα (1253.6 eV) x-ray source and rotating the analyzer around the sample in mirror planes with 1° angular step.

Figure 1(a) shows valence band spectra of 4 ML film grown at 350 K. The angle of incident light was 45° off-normal and spectra were measured in normal emission mode. The experimental energy band structure obtained by assuming the free electron final state for the photoelectrons and the inner potential to be 10 eV is shown in Fig. 1(b) together with a theoretical band calculation along the Γ-X direction in the Brilouin zone (B.z.) for fcc Fe [4]. The good agreement between the experimental results and the theoretical band calculation implies that the Fe film in low coverage region resembles fcc structure or probably fct. We also measured the ARPES spectra of the Fe film in the high coverage region (8 ML) and compared with a theoretical band calculation, and found that the observed valence band structure of 8 ML film resembles of those along the Γ-N direction of the B.z. of bcc Fe. From the results of ARPES, we could conclude that the electronic structure and magnetic properties of the Fe film epitaxially grown on Rh(001) varies depending on the film thickness, which may be originated from the thickness dependence of geometric structure of Fe films. Since we did not observe thickness dependence in 1x1 RHEED patterns below 12 ML, the interlayer spacing of the Fe film would cause the change of the structure of the films from fcc(001) in the low coverage region to bcc(110) in high coverage region.

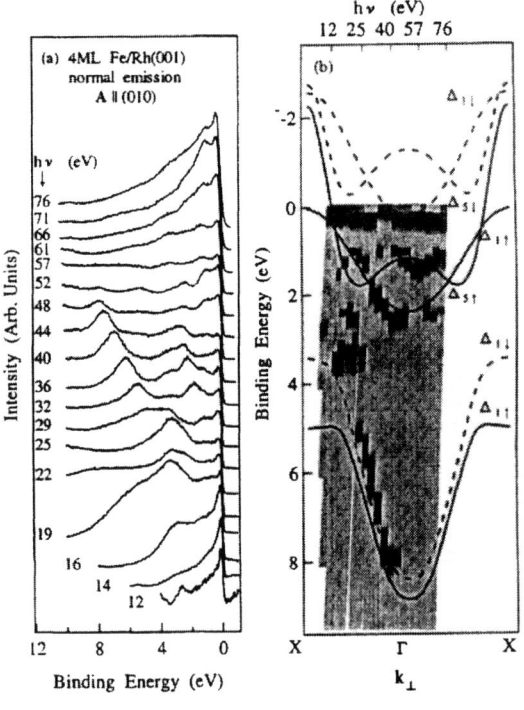

FIGURE 1.

Figure 2 shows the Fe $2p_{3/2}$ XPD patterns observed in the Rh(010) plane. In the figure, each pattern consists of a few peaks. It should be remarked in the figure that the peak around 50° and 30° tend to shift their positions toward smaller angles as the film thickness increases, which corresponds to the increase of the interlayer spacing of the Fe film. We evaluated the interlayer spacing of Fe films by assuming the initial growth stage to be 1.53±0.05Å and it increases with film thickness and reaches to a constant value of 1.66±0.06Å at 6 ML.

The result shows that the Fe films grow in compressed structure along the direction perpendicular to the surface and the degree of compression is released as the film thickness increases. According to the first principle calculation of the total energy and the magnetic moment of Fe films [3, 5], it has been predicted that ferromagnetism is stable in fcc Fe for $v>11.57\text{Å}^3$, while in bcc Fe $v>11.7\text{Å}^3$. Based on the RHEED and XPD observations, we have evaluated the atomic volume of Fe for each film thickness and realized that Fe film on Rh(001) substrate reveals ferromagnetism in fcc structure with thickness over 4 ML and bcc ferromagnetism over 6 ML.

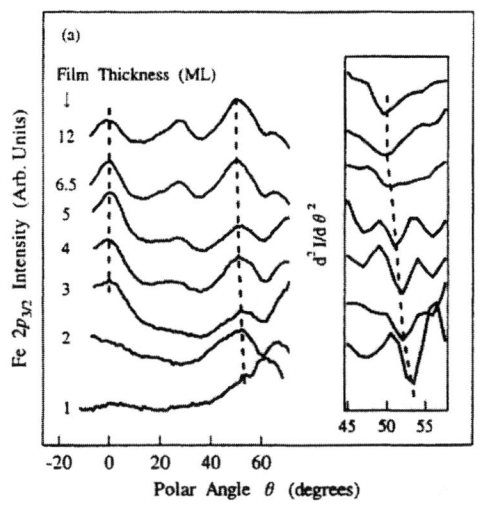

FIGURE 2.

REFERENCES

1. V. L. Moruzzi, P. M. Marcus and P. C. Pattnaik, Phys. Rev. B **37**, 8003 (1988).
2. C. Egawa, Y. Tezuka, S. Oki and Y. Murata, Surf. Sci. **283**, 388 (1993).
3. A. M. Begley, S. K. Kim, F. Jona and P. M. Marcus, Phys. Rev. B **48**, 1786 (1993).
4. M. Podgorny, J. Magn. Magn. Mater. **78**, 352 (1989).
5. V. L. Moruzzi, P. M. Marcus K. Schwartz and P. Mohn, Phys. Rev. B **34**, 1783 (1989).

Electronic structure and magnetic anisotropy of Co/Au(111) : a spin-resolved photoelectron spectroscopy study

M. Sawada[1], K. Hayashi[1] and A. Kakizaki[2]

[1] *Institute for Solid State Physics, University of Tokyo, Chiba 277-8581, Japann*
[2] *Institute of Materials Structure Science, KEK, Ibaraki 305-0801, Japan*

The interest in epitaxial grown Co films and Co/Au multilayers are due to their peculiar magnetic properties such as oscillatory behavior of interlayer magnetic coupling and perpendicular magnetic anisotropy (PMA), etc. [1]. In Co/Au(111) system, a large lattice mismatch (ca. 14 %) between Co and Au(111) causes the characteristic growth mode of Co films, which were observed in detail by ion scattering, STM, LEED and AES experiments. The results suggest the close connection between the structure and electronic and hence magnetic properties of Co films at the interface. However, only a few works were devoted to directly observe the electronic structures of Co films on Au and their interfaces, so far. In this report, we present spin- and angle-resolved photoemission (SARPES) spectra of Co/Au(111) system in order to investigate the thickness dependence of the electronic and magnetic properties of Co films.

The SARPES measurements were carried out at the undulator beamline BL-19A of the Photon Factory using spin- and angle-resolved photoelectron spectrometer consisting of a hemispherical electron energy analyzer and a compact retarding type Mott detector [2]. In SARPES measurements, samples were magnetized remanently by applying a magnetic field along the direction parallel or perpendicular to the sample surface. We have adopted s-polarized incident light and collected photoelectrons emitted normal to the sample surface.

Figure 1 shows SARPES spectrum and the spin polarization of Co/Au(111) in low coverage region together with the spectrum of clean Au(111). They were measured at room temperature with applying the magnetic field perpendicular to the surface. The remarkable point in the figure is that the shoulder at the Au 5d bands is diminished after Co deposition (Fig. 1(b)) and shows a considerably large positive spin polarization perpendicular to the surface. Since the Au 5d states are almost fully occupied at the initial state, the observation indicates that the Co deposition alters the surface resonant states originated from upper Au 5d states and causes the charge transfer from the Au 5d occupied states to the Co 3d empty states at the Au/Co interface. This leads the occupancy difference between the majority and minority spin electrons and hence the positive spin polarization in the upper 5d states.

Figure 2 shows the thickness dependence of SARPES spectra. The reorientation of the magentization from PMA to in-plane magnetization ocurrs at about 6ML. The specific range of the Co film thickness with PMA is comparable with the previous reports [3, 4] within the limitation of errors of the measured thickness originated from the characteristic growth mode of Co on Au(111) substrate. In Fig. 2, two

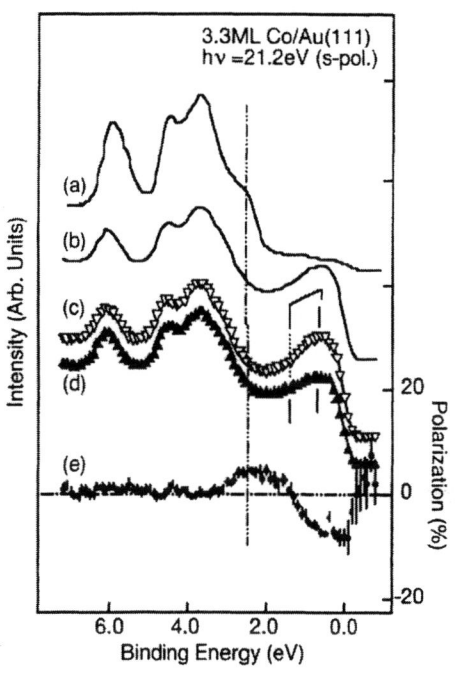

FIGURE 1.

spectral features are observed in each majority spin spectrum. The feature at lower binding energy corresponds to e_g states and the one at higher binding energy to t_{2g} states. The binding energies of both features show thickness dependence; the binding energy of upper band peak slightly increases from 0.37 to 0.46 eV and the binding energy of the lower band changes from 0.69 to 1.92 eV as the film thickness increases from 2.1 to 12 ML. In the minority spin spectra, the observed peaks correspond to the lower 3d band. Its binding energy increases from 0.23 to 0.54 eV with increase of the Co thickness. For the Co film with thickness over 7 ML, an additional feature appears near the Fermi level and its binding energy and spectral intensity increase with the film thickness. The above observation suggest that the strain induced in the Co at the interface plays a significant role in the phenomenon and that the reorientation of the magnetization direction could be understood as a results of the restoration of the strain [5]. In Fig. 2, the intensity and the width of Co 3d states increase with increase of the film thickness. Since the probing depth of the SARPES in the present work is about 3 ML or less, the observed thickness dependence implies that the density of the Co valence states in the topmost few layers vary with the film thickness, the origin of which is possibly due to the relaxation of the lattice constant of the Co film. Since the STM and XAFS [6] studies showed that the structure of Co on Au(111) was hcp and the in-plane lattice constant did not vary with Co film thickness, it is reasonable to regard that the relaxation of the lattice distance of Co occurs in the direction perpendicular to the surface and the restoration to the lattice constant of bulk Co causes the thickness dependence of the electronic structure. In the momentum space, the relaxation causes the restoration, i.e. the shrink of the reciprocal lattice along the Γ-L direction in the B.z. and hence, leads to the increase of the density of states observed by SARPES with a certain energy and momentum resolution.

The additional feature corresponds to the upper Λ_3 band and mainly consists of out-of-plane Co $3d_{yz}$ and $3d_{zx}$ states. In Fig. 2, the binding energy of this band increases with increase of the film thickness. This implies the increase of the occupancy of this band and the increase of the in-plane orbital moment. Hence, the relaxation of the Co atomic distance perpendicular to the surface causes the increasing contribution of $3d_{yz}$ and $3d_{zx}$ states with the film thickness to the Co valence states, which leads to the reorientation of the PMA occurred at the Co/Au(111) interface.

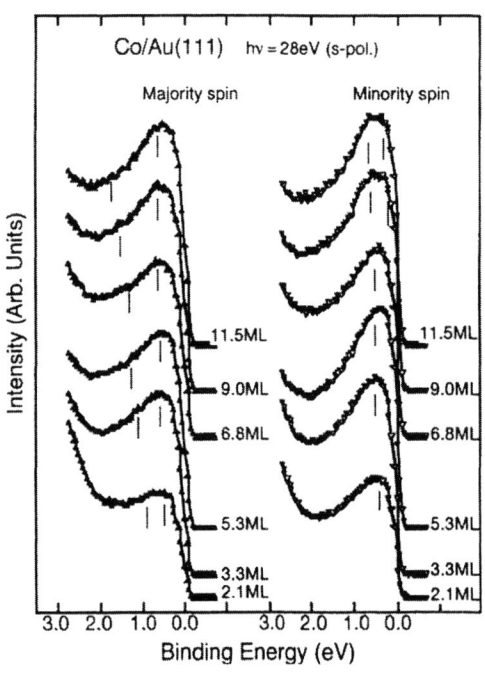

FIGURE 2.

REFERENCES

1. see for example, *Ultrathin Magnetic Structures*, eds. B. Heinrich and J. A. C. Bland, (Springer-Verlag, Berlin, 1994).
2. S. Qiao, et al., Rev. Sci. Instrum. **68**, 4390 (1997).
3. R. Allenspach, M. Stampanoni and A. Bischof, Phys. Rev. Lett. **65**, 3344 (1990).
4. T. Duden and E. Bauer, Surf. Rev. Lett. **5**, 1213 (1998), and references therein.
5. C. Uiberacker, et al., Phys. Rev. Lett. **82**, 1289 (1999).
6. C. Chapper, et al., J. Mag. Mag. Mater. **93**, 319 (1991).

Reduction of field emission current from Stainless steel and copper surface

C. Suzuki, T. Nakanishi, S. Okumi, , T. Gotoh, K. Togawa, F. Furuta,
K. Wada, T. Nishitani, M. Yamamoto, J. Watanabe, S. Kurahashi,
H. Matsumoto[a], M. Yoshioka[a], K. Asano[b], and H. Kobayakawa[c].

Nagoya University, Department of Physics, Nagoya 464-8602, Japan
a) KEK High Energy Accelerator Research Organization, Tsukuba 305-0801, Japan
b) Akita National College of Technology, Electrical Engineering Department, Akita 011-8511, Japan
c) Nagoya University, Department of Materials Processing Engineering, Nagoya 464-8603, Japan

Abstract. The field emission dark currents from stainless steel and copper electrodes were measured under DC high field gradient condition. The stainless steel electrode was made from special material called Clean-Z and was polished by electro-chemical buffing. The electrodes achieved 34MV/m with the dark current of 90pA at the gap separation of 1 mm. The copper electrodes were treated by four different types of surface cleaning procedures. The best results were obtained by using the electrode rinsed with ultra-pure water after diamond turning. A field gradient of 47MV/m was achieved with dark current at the level of 1nA, and the microscopic field enhancement factor was estimated to be a very low value of 56. The dark current from this electrode was dependent only on the field gradient at the cathode and not affected by the total voltage applied to the gap. This fact suggests that the surface fabricated by diamond turning method with ultra-pure water rinsing creates few secondary ions by the electron bombardment.

INTRODUCTION

The reduction of dark current is very important for development of a "polarzied DC-gun" and the feasibility study of a "polarized RF-gun". In the DC-gun case NEA (negative electron affinity) surface of a GaAs type semiconductor is usually used to extract polarized electrons from the crystal into vacuum. The performance of NEA surface is extremely sensitive to contamination, and the NEA state is easily degraded by back bombardment of ionized atoms and molecules, which are produced by electron beam itself and by dark currents. A vacuum level of 10^{-11} torr and a dark current level below 10 nA must be kept for the stable operation of polarized DC-guns which have field gradients of a few MV/m [1]. The technical difficulties are much greater for the polarized RF-gun than for the polarized DC-gun because the extraction field at the cathode surface is about two orders of magnitude higher (~100 MV/m).

At first, the reduction method of dark current for the stainless steel electrode of polarized DC-gun is described, and next, the method for the copper surface of RF-cavity is reported.

TEST APPARATUS

A test apparatus was built to investigate the relationship between the metal surface conditions and the amount of dark current. It was designed to supply a high DC field gradient (~100 MV/m) for measurement of field emission dark current from the sample electrode under ultra high vacuum (~10^{-11} torr) conditions. A schematic view of this test stand and the geometrical shapes of the cathode and anode are shown in Fig. 1. The sample electrode can be replaced to compare the performance of various kinds of electrodes. The field gradients can be changed by control both of the gap separation of the electrodes (0.5~20 mm) and of the bias voltage (~150 kV). The field emission current emitted from the cathode was collected at the anode and measured by a pico-ampere meter [2].

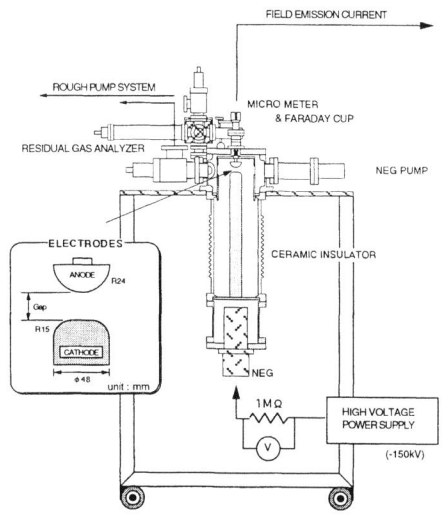

FIGURE 1. A test apparatus for high gradient test.

EXPERIMENT

The SUS electrode was made from new type SUS material called Clean Z [3], which contains much fewer impurities such as C, Si, P, S etc. than normal SUS 316L. The surface of electrode was finished to mirror-like surface by electro-chemical buffing method. The surface rinsing was done in the class-1 clean room with hot ultra-pure water (70°C, 18MΩ*cm). The dark current was measured in UHV of <3×10^{-11}Torr. Fig. 2 shows the results. The field gradient of 34MV/m was achieved with the dark current level of 90pA at the gap separation of 1mm. The field enhancement factor β was calculated to be 42 by FN-plot of the data. By applying these reduction methods to the operational polarized DC-gun, long cathode lifetime of over 50 hours was achieved with successful reduction of the dark current level of 10nA at the bias voltage of 70kV [2].

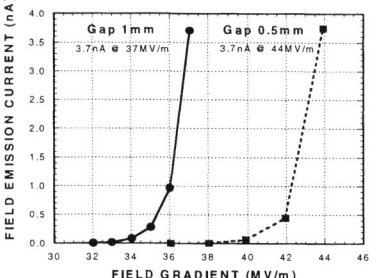

FIGURE 2. The dark current measurements from SUS electrodes.

All the copper electrodes tested in this experiment were machined from a class-1 OFHC (Oxygen Free High Conductivity) copper block, which has the purity of 99.996% and was forged by HIP method (hot isostatic pressing). The final machining to mirror-like finish was done by the diamond turning method in a clean room with an air filter of 0.3 μm. The ethanol was used as the lubricant instead of machine oils. The average surface roughness (Ra) was measured to be as good as 0.05 μm. Four different types of combination of surface treatment and rinsing method were employed for the test electrodes. Specifications of four electrodes are shown in Table 1. The results obtained by four measurement at the gap separation of 1.0mm is summarized in Fig. 3 (left), where the dark currents are plotted as a function of the maximum field gradient at the cathode surface treated by EP, EP+OUR, no treatment, and no treatment+OUR, respectively. We compare the magnitude of the surface field gradient for which total dark current exceeds the level of 1nA, and the highest value of 47MV/m was achieved by the electrode without EP treatment. This electrode gave also the lowest β-value of 56. As far as we know, this value is the lowest one under high DC field gradient condition and similar to the best value obtained by the KEK experiment for RF-cavity [3]. Next, The dark current for the different gap separations of 0.5 mm, 1.0 mm, and 3.0 mm were measured, the results are shown in Fig. 3 (right). The electrode without EP treatment was not affected by the total voltage. Almost the same behavior was observed for gap separations of 0.5 mm and 1.0 mm. This means that the amount of field emission current was dependent only on the cathode field gradient. For this electrode, no vacuum increase was observed during the high voltage measurement. Therefore it is suggested that very few secondary particles was produced even when the anode was bombarded by the dark current.

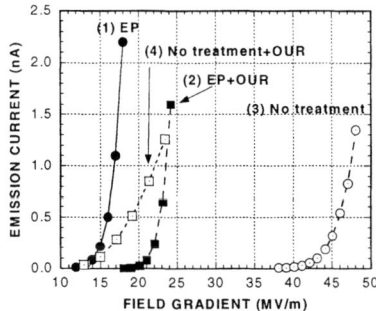 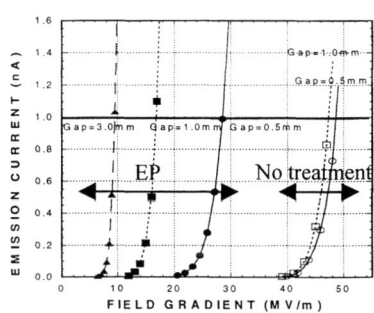

FIGURE 3. The dark current from four different electrodes (left) and the dark measurements at different gap length.(right)

TABLE 1. Surface treatment procedures for copper electrodes.

Surface Treatment	Diamond Turning	EP [1]	OUR [2]	UWR [3]
1 EP	○	○		○
2 EP + OUR	○	○	○	○
3 None	○			○
4 None + OUR	○		○	○

1) Electro polishing (EP): The 5μm was removed by this method.
2) Ozonized ultra-pure water rinsing (OUR): The concentration rate of ozone was 3ppm, 30minutes.
3) Ultra-pure water rinsing (UWR): The resistance of water is 18MΩ*cm.

REFERENCES

1. K. Togawa et al., *Nucl. Instr. and Meth.* A 414 (1998) 431.
2. C. Suzuki et al., "Fabrication of Ultra-clean copper surface to Minimize Field Emission Dark Currents" accepted to *Nucl. Instr. and Meth A*
3. H. Matsumoto, Proc. Int. Linac Conference 96 (LC96), Geneva, 1996, pp. 626.

Development of 200 keV polarized electron gun

K. Wada, M. Yamamoto, T. Nakanishi, S. Okumi, T. Gotoh, K. Togawa,
C. Suzuki, F. Furuta, T. Nishitani, J. Watanabe, S. Kurahashi,
M. Miyamoto, H. Matsumoto [a], Y. Takeuchi [a] and M. Yoshioka [a]

Nagoya University, Department of Physics, Nagoya 464-8602, Japan
a) KEK High Energy Accelerator Research Organization, Tsukuba 305-0801, Japan

Abstract. A 200 keV polarized electron gun system with a load lock mechanism has been developed to produce the high-intensity and low-emittance beam required by Japan Linear Collider. The construction of this system has been completed and the performance tests are in progress. Up to now, DC high voltage of 150 kV could be successfully applied to the accelerating electrodes with an extremely low dark current of < 0.1 nA.

INTRODUCTION

A high field gradient gun is indispensable not only for an electron injector of e^+- e^- linear collider but also for that of free electron laser system. The higher field gradient makes it possible to produce the beam that satisfies two conflicting characteristics, high-intensity and low-emittance, because the current density limited by space charge effect becomes higher and the beam divergence caused by space charge force is suppressed. Japan Linear Collider (JLC) requires such a beam with multi-bunch structure. At the source, 72 micro-bunches with a bunch separation time of 2.8 ns must be produced at 150 Hz and each micro-bunch must have 2×10^{10} e^- in ~ 700 ps bunch width, which corresponds to about 3 A peak current [1]. The required normalized emittance at the exit of the gun is < 10 π mm mrad (r.m.s.).

A polarized electron beam extracted from GaAs-type photocathodes with Negative Electron Affinity (NEA) surface is expected to play an important role in JLC. We have demonstrated that the superlattice photocathodes can produce the high-intensity and multi-bunch beam without surface-charge-limit phenomenon [2][3]. However, the gun bias voltage was limited below 70 kV due to the dark currents caused by cesium atoms accumulated on the electrodes during the repetitive NEA activation process and thus the peak current was limited to be 1.6 A by space charge effect. The dark currents stimulate gas desorption from the electrodes. They will be ionized by high intense beam and these positive ions will significantly degrade the NEA surface.

In order to solve this problem, a new DC high voltage (HV) gun system with a load-lock mechanism has been constructed. The schematic view of this apparatus is shown in Figure 1. It consists of three vacuum isolated chambers: a photocathode loading chamber, a photocathode activation chamber and a HV gun chamber. The details of the photocathode preparation system were reported by M. Yamamoto in this workshop [4]. The design value of bias voltage was 200 kV at maximum.

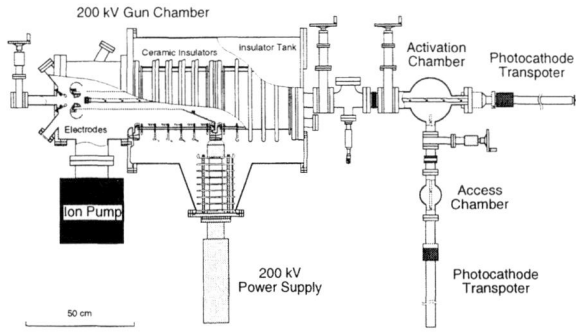

Figure 1. Schematic view of the 200 kV polarized electron gun system.

HIGH VOLTAGE INSULATION

In order to operate the photocathode preparation system at grand level, HV is supplied to the center of two ceramic insulator-columns and the cathode electrode support tube is connected to there. Each ceramic insulator is divided into five segments to avoid HV breakdown, where five 500 MΩ divider resistors are used for HV distribution. A HV power supply can supply the maximum voltage of 250 kV, which is used for aging the electrode surface to assure the stable operation at 200 kV.

These HV components are enclosed by the insulation gas tank. Dry nitrogen is flowed into the tank to remove the humid air that causes the leakage currents along the ceramic surface, and then it is pressurized to suppress corona discharge. The necessary pressure at 200 kV is estimated to be about 3.0 atm by extrapolation of the data taken at the lower voltage.

ACCELERATING ELECTRODES

To extract high current from the photocathode, the cathode-anode gap distance and the photocathode diameter were chosen to be 35 mm and 18 mm φ. The space-charge-limited current at 200 kV in the case of full laser illumination was estimated to be about 30 A using the simulation code of EGUN. This value is considered to be enough to satisfy the charge requirement of JLC.

To reduce the dark current of accelerating electrodes, the curvature of the cathode electrode was enlarged (~ 30 mm) so that the field gradient become smaller. The maximum field gradient on the cathode electrode surface was estimated to be 7.8 MV/m at 200 kV using the simulation code of POISSON.

Fabrication of the electrodes is so important to reduce the dark current. Non-metallic impurities on the electrode surface would become the emitting sites of the dark current. To preserve the NEA surface, the dark current must be kept below about 10 nA together with the UHV of ~ 10^{-11} torr. A basic research on this issue has been made by a test-apparatus build at KEK [1]. To reduce the non-metallic impurities, super-clean stainless steel made by the melting method was used as an electrode material. Surface polishing by electrochemical buffing method and the subsequent rinsing with ultra-pure water were also employed to remove the contaminations. As a result, the dark current from the cathode area of ~ 7 mm^2 could be reduced to ~ 90 pA under a high field gradient of 34 MV/m. These techniques have been applied to fabricate the accelerating electrodes.

ULTRA-HIGH VACUUM

The gun chamber has two main pumps: 360 l/s ion pump and 850 l/s NEG pump. During the baking process, the temperature was ramped up or down at the rate of ± 4 ℃/hour to avoid the mechanical breaking of HV insulators due to the difference of thermal expansions between metal and ceramics. After baking at 200 ℃ for 100 hours, the total pressure at room temperature decreased to 4.8×10^{-11} torr which was monitored by an extractor gage.

HIGH VOLTAGE TEST

In order to measure the field emission dark current (I_f) of the accelerating electrodes with good accuracy, this current must be distinguished from the discharge current of other HV elements. For this purpose, both of the HV power supply and insulation tank are electrically isolated from ground. The discharge current (I_d) generated at the corona rings is collected in the insulation tank and then goes to ground through a current meter. The loop current that flows the divider resistors returns to the HV power supply without passing through ground. I_f and I_d, which are leave from this loop, go to ground and then come back to the HV power supply through another current meter, because the HV power supply gets back this total lost current (I_t) from ground to keep the HV constant. I_f can be estimated simply as $I_f = I_t - I_d$.

At the first high voltage test, a dummy photocathode of aluminum was installed and the gas pressure of the insulation tank was increased to 2.2 atm that can suppress corona discharge up to 160 kV. The dark current could be kept at a level of < 0.1 nA up to 138 kV, where a sudden HV breakdown occurred. Then the dark current was stepped up to 140 nA even at 65 kV. To improve this situation, the HV processing method was employed under the vacuum condition of 1×10^{-6} torr by introducing 99.9999 % pure nitrogen [5]. As a result, the dark current could be reduced below 0.1 nA at 150 kV.

SUMMARY

The first HV test of the new polarized DC-gun has been performed. Up to now, 150 kV has been successfully supplied to the accelerating electrode with an extremely low dark current level of < 0.1 nA. The basic research to operate the gun at 200 kV is continued.

REFERENCES

1. T. Nakanishi et al., "Polarized Electron Source" in *JLC Design Study*, KEK Report 97-01, 1997, pp. 36-48.
2. K. Togawa et al., *Nucl. Inst. and Meth. in Phys. Res.* **A 414**, 431-445 (1998)
3. K. Togawa et al., *Nucl. Inst. and Meth. in Phys. Res.* **A 455**, 118-122 (2000)
4. M. Yamamoto et al., "Atomic hydrogen cleaning for GaAs surface with load lock system" in this *Workshop*
5. R. Alley et al., *Nucl. Inst. and Meth. in Phys. Res.* **A 365**, 1-27 (1995)

Test of Cesium Telluride Photocathode as A Feasibility Study on Polarized RF-gun

F.Furuta, H.Sugiyama[1], T.Nakanishi, S.Okumi, K.Togowa, C.Suzuki,
K.Wada, M.Yamaoto, T.Nishitani, J.Watanabe, S.Kurahashi,
M.Miyamoto, M.Kuwahara, R.Mizuno, T.Hirose, K.Kimura[1]
H.Kobayakawa[1], Y.Takashima[1], M.Yoshioka[2], and H.Matsumoto[2]

Department of Physics, Nagoya Univercity,Furo-cho,Chikusa-ku,Naagoya,464-8602,Japan
(1) Faculty of Engineering, Nagoya University,Furo-cho,Chikusa-ku,Nagoya,464-8603,Japan
(2) High Energy Accelerator Research Organization(KEK),Tsukuba,305-0801,Japan

Abstract. Our collaboration group have continued to develop the polarized DC-gun system for future electron-positron linear colliders [1]. In parallel, the feasibility study to develop the polarized RF-gun is planned. As the first step, the test of Cs_2Te photocathode was performed.

INTRODUCTION

The future linear collider, like Japan linear collider (JLC) requires a high intensity polarized beam with a multi bunch structure to obtain high luminosity of $\sim 10^{34}$ cm^{-1} s^{-1} [2]. A concept of polarized RF-gun was proposed about 10 years ago, and is still attractive for this purpose [3]. By an RF-gun the field gradient can be much higher (~100MV/m) than that of a conventional DC-gun. Such higher RF-field makes it possible to increase the space-charge-limited current density, together with low emittance and short bunch at the source. The low emittance beam will greatly relax the required operating condition of the damping ring and eliminate the need of bunching system. However, for GaAs-type polarized electron source (PES) there is a serious question on survivability of NEA (Negative Electron Affinity) surface in an RF-gun condition. The dark current generated by field emission from RF cavity will be increased and NEA surface will be destroyed. In fact, it was shown at Novosibirsk that the quantum efficiency (QE) of NEA GaAs photocathode dropped suddenly when several tens of RF-pulse was applied [4]. To the contrary, the Cs_2Te photocathode can keep

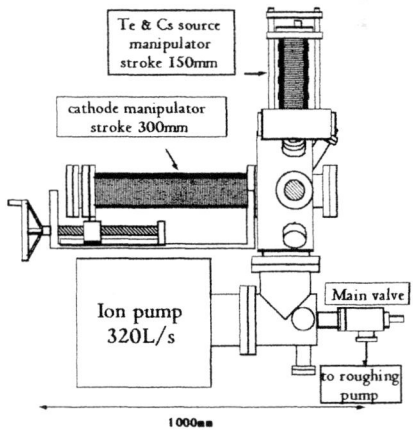

Figure 1 Schematic of the Cs_2Te photocathode fabrication chamber

the high QE even in the RF-gun as demonstrated at CERN and Los Alamos [5],[6], although it can not produce the polarized electron beam. In addition Cs_2Te has another advantage as an excellent photoemitter. The QE of ~10% is obtained from the very thin Cs_2Te layer of 10 nm thickness when it is irradiated by light with ~250nm wavelength. Therefore it seems important to investigate the physical mechanism which brings the high QE and the long lifetime to the Cs_2Te photocathode. Especially the mechanism to give the long lifetime should give us a hint to find the new photoemitter material instead of NEA-GaAs.

EXPERIMENTAL SET-UP

Schematic of the Cs_2Te fabrication chamber is shown in Figure 1. In this chamber, Cs_2Te was fabricated on the Mo substrate. First, ~10 nm thickness of Te was deposited on the Mo substrate. This thickness was monitored by a crystal oscillation type deposition monitor (resolution 1 Å). Then Cs was evaporated and photoemission current was measured by irradiating the cathode ultraviolet light. As the light source, we use a Xe lump with a monochromator. The Cs deposition was stopped when photocurrent reached the maximum. These processes were done under UHV conditions (base pressure $\sim 2 \times 10^{-10}$ Torr) at room temperature. The Cs_2Te on the Mo substrate could be swept out by the 600°C heating of about 1 hour and we could repeat the tests with different thickness of Te.

RESULTS and DISCUSSIONS

Figure 2 and 3 show the QE of Cs_2Te photocathode against wavelength of light and thickness of Te, respectively. 10nm thickness of Te is enough to give the maximum QE (~10%) at 245nm. Figure 4 shows the data of lifetime measurement by continuous extraction of photocurrent from Cs_2Te photocathode. Although the extracted current was only about 100nA for this experiment, the QE was constant over 100 hours.

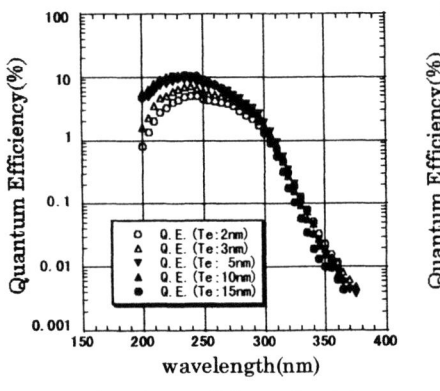

Figure 2 QE vs light wavelength for Cs_2Te photocathode

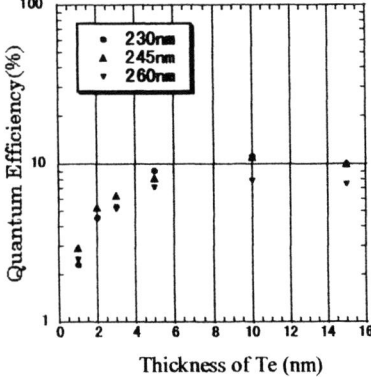

Figure 3 QE dependence on the Te layer thickness

The detailed study of photoemission from Cs_2Te was already done by R.A.Powell et al, and the processes are explained by three steps [7]. First, electrons are optically exited from valence band into conduction band, then migrate to the solid surface, and escape across the surface potential barrier into vacuum. Figure 5 shows a picture of the band-gap and the state-density structure of Cs_2Te drawn by K.Frottmann for interpretation of Powell's data [8]. The wavelength giving the maximum QE corresponds to the transition between both of high density states in valence band and conduction band. Thus the photoabsorption length is very short (~10nm) as shown in Figure 3.

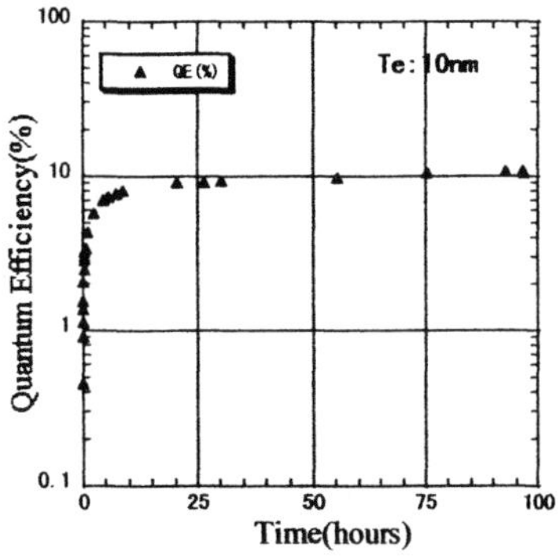

Figure 4 Lifetime measurement of Cs_2Te photocathode

The Cs_2Te has the positive electron affinity (PEA), but it is very small (~0.2eV). In addition the energy level where electrons were optically exited is higher than this, so it is easy for electrons to escape into vacuum. Both of this fact and the short absorption length are the reason of high QE. It is also expected that PEA surface is strong against the degradation of surface under RF conditions.

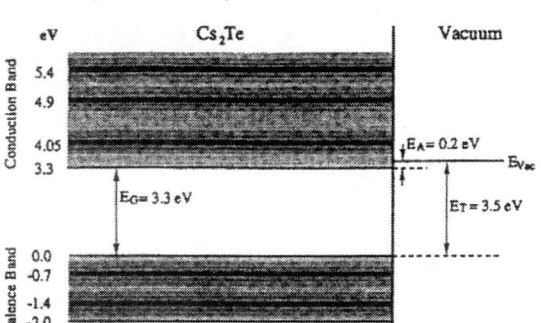

Figure 5 A picture of the band-gap and the state-density structure of Cs_2Te photocathode

CONCLUSIONS

We had succeeded to fabricate Cs_2Te photocathode which had high QE (~10%). The long lifetime over 100 hours was also demonstrated under low current extraction. We could understand that small PEA surface is very useful for achieving the long lifetime. In order to make a feasibility study of polarized RF-gun, we probably need to find the small PEA semiconductor with the GaAs like band-structure.

REFERENCES

1. K.Togawa et al., *Nucl. Instr. and Meth.* A455 (2000) 109.
2. JLC group, JLC Design Study, KEK Report 97-1, 1997.
3. J.Clendenin et al., SLAC-PUB-6172 May 1993.
4. A.V.Aleksandrov et al., EPAC98 Contributions to the Proceedings, p.1450.
5. G.Suberlucq et al., CERN-PS-98-036-LP, 1998.
6. Steven H. Kong et al, LA-UR-94-2851, 1994.
7. R.A.Powell et al., *Phys. Rev.* B8, No8 (1973) 3987.
8. K.Flottmann "Note on the thermal emittance of electrons emitted by Cesium Telluride photo cathodes" TESRA group Internal Report.

Atomic Hydrogen Cleaning of GaAs Photocathode with a Load-Lock System

M.Yamamoto, K.Wada, T.Nakanishi, S.Okumi, K.Togawa, C.Suzuki,
F.Furuta, T.Nishitani, J.Watanabe, S.Kurahashi, M.Miyamoto

Department of Physics, Nagoya University, Nagoya 464-8602, Japan

Abstract. We are constructing a new polarized electron source for Japan Linear Collider. It is designed to operate the gun at 200kV. The "load-lock" mechanism is employed to avoid the dark current due to the Cs accumulation on the electrodes and to exchange the activated NEA photocathode quickly. We have developed superlattice photocathode which has advantages of high spin polarization, high quantum efficiency and high resistance against surface charge limit phenomenon. However, it seems difficult to clean the surface of such a thin layer photocathode by the normal etching procedure without destruction of its delicate structure if it has no As cap-layer. Atomic hydrogen is expected to clean the surface of superlattice effectively. We have introduced these techniques to the new source design.

INTRODUCTION

We have been developing the polarized electron source to generate the multi-bunch beam with high electron spin polarization ($\geq$85%) and high peak current ($\geq$3A for 0.7ns bunch width) for Japan Linear Collider. We already built a 70kV test gun and succeeded in extraction of the sub-nanosecond multi-bunch electron beam with peak current of 1.6A (space-charge-limit) from GaAs-GaAsP superlattice photochathode.

It is well-known that dark current decrease the lifetime of NEA (negative electron affinity) photocathode and should be kept below ~10nA. Repetitive depositions of Cs atoms on cathode electrode cause increase of the dark current, and this is the reason why the bias voltage of the test gun is limited below 70kV. The use of another chamber for NEA activation, which is first employed at SLAC, seems to be a good solution, especially in our case to supply the bias voltage up to 200kV to cathode.

High polarization can be achieved using the GaAs-GaAsP superalttice photocathode fabricate by the MOCVD apparatus. However, it is difficult for this method to cover the surface with arsenic layer, which is effective to protect the surface against air exposure. Therefore, before NEA activation, the crystal was annealed at near growth temperature (550~600°C) for several hours, although such high temperature seems quite risky for the delicate superlattice structure. The oxidized GaAs surface consists of arsenic oxide and Ga_2O_3-like oxide. It is suggested that the removal of Ga_2O_3 is possible through formation of more unstable state of Ga_2O. It is reported that atomic hydrogen (H·) can deoxidize Ga_2O_3 and change it to Ga_2O at relatively low temperature (<200°C). This process is expressed as follows,

$$Ga_2O_3 + 4H \cdot \rightarrow Ga_2O\uparrow + 2H_2O\uparrow$$

Ga_2O can removed at 350°C~400°C. In this respect, the clean surface should be created by atomic hydrogen without destruction of the highly doped surface and the superlattice structure.

PREPARATION CHAMBERS WITH LOAD-LOCK SYSTEM

The 200kV polarized electron source is composed of a gun chamber and two preparation chambers (activation chamber and access chamber). The preparation chambers are connected to the gun chamber at the ground level even when the bias high voltage is applied to the cathode electrode. The details of the gun system are shown in another paper in this workshop.

A schematic view of preparation chambers is shown is Figure 1. The access chamber was opened when the photocathode is installed into the system. The photocathode is held on a "pack" made of molybdenum. The access chamber was baked for one day, then pumped to a vacuum level of $\sim 10^{-9}$ torr by a 100 l/s ion pump. After the baking, the photocathode surface was then cleaned by atomic hydrogen which is generated by RF with ~100MHz frequency at a pyrex H_2 dissociation chamber connected with the access chamber. The pressure of hydrogen gas was maintained to be ~20 mtorr.

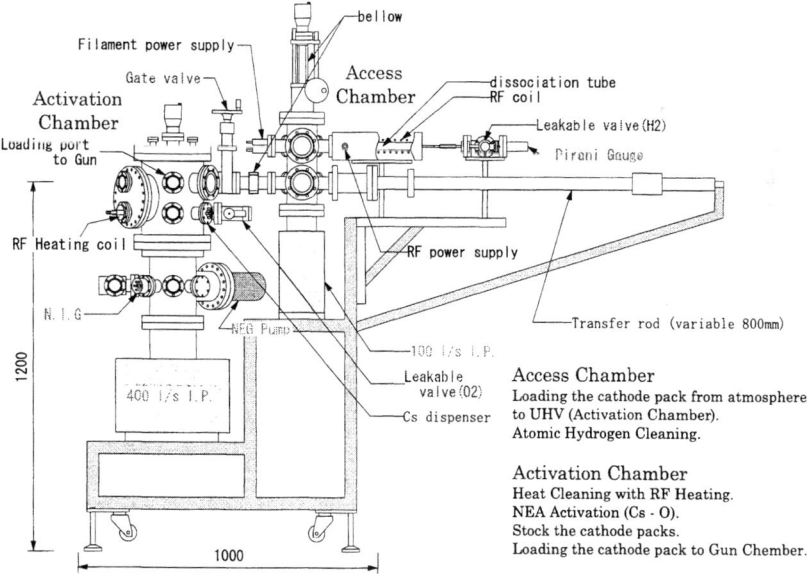

FIGURE 1. Schematic view of Activation and Access chamber.

In the Activation chamber, the NEA surface of the photocathodes is prepared. The vacuum pressure of $\sim 10^{-11}$ torr was obtained using a 400 l/s ion pump and a 450 l/s NEG pump. The photocatode puck is finally heat-cleaned at 400°C for 30 minutes, by induction heating method. Cs is provided from a cesium dispenser and oxygen is introduced through a leakable valve. After activation, the photocathode pack is loaded to the cathode electrode of the gun chamber without breaking the ultra-high-vacuum.

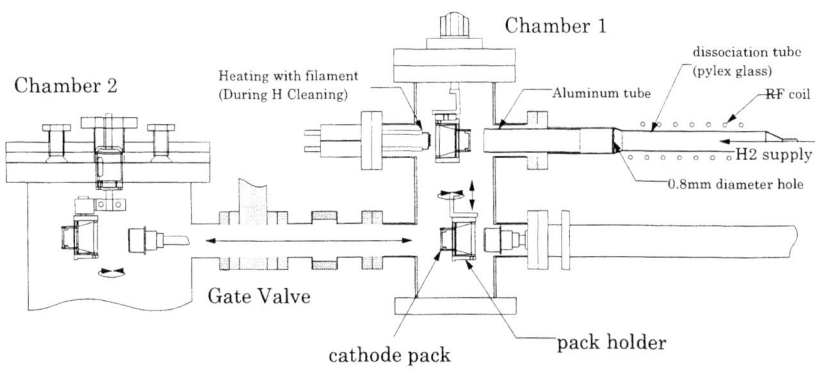

FIGURE 2. Process of photocathode installation from atmosphere to NEA activation chamber.

We have started to investigate the adequate parameters of atomic hydrogen cleaning in this system. In the future, we consider to make tests to produce atomic hydrogen by means of hot tungsten filament or tube instead of RF dissociation. Those methods would be advantageous because energetic ions, which might give damage to the surface of superlattice structure, are not generated.

ACKNOWLEDGMENTS

We wish to thank Drs. C.Sinclair and M.Poelker of TJNAF for giving a sample of pyrex dissociation chamber and a lot of information about hydrogen cleaning.

REFERENCES

1. JLC design study, KEK-Report 97-1, 1997.
2. K.Togawa et al., *Nucl. Instr. and Meth.* **A** 414 (1998) 431.
3. M.Yamada and Y.Ide, *Jpn. J. Appl. Phys.*, **33** (1994) L671
4. Y.Ide and M.Yamada, *J.Vac.Sci.Technol.*, **A12** (1994) 1858
5. T.Nakanisi et al, Proc. Low Energy Polarized Electron Workshop, St. Petersburg, Russia, 1998, p.118
6. T.Nakanisi et al., Proc. 7th Int. Workshop on Polarized Gas Targets and Polarized Beams, Urbana-Champaign, USA, AIP Conf. Proc. 421, 1997, p.300.
7. T.Nakanisi et al., Proc. 12th Int. Symp. On High-Energy-Spin Physics, Amsterdam, Netherlands, World Scientific, 1996, p.712.

Development of spin polarized electron photocathodes : GaAs-GaAsP superlattice and GaAs-AlGaAs superlattice with DBR

T. Nishitani [a], O. Watanabe [b], T. Nakanishi [a], S. Okumi [a], K. Togawa [a],
C. Suzuki [a], F. Furuta [a], K. Wada [a], M. Yamamoto [a], J. Watanabe [a],
S. Kurahashi [a], M. Miyamoto [a], H. Kobayakawa [b], Y. Takeda [b], T. Saka [c],
K. Kato [d], A. K. Bakarov [e], A. S. Jaroshevich [e], H. E. Scheibler [e],
A. I. Toropov [e], A. S. Terekhov [e]

(a) Department of Physics, Nagoya University 464-8602, Japan
(b) Faculty of Engineering, Nagoya University 464-8603, Japan
(c) Daido Institute of Technology, Nagoya 457-8531, Japan
(d) Daido Steel Co. Ltd., Nagoya 457-8531, Japan
(e) Institute of Semiconductor Physics, 630090 Novosibrisk, Russia

Abstract. We have tested two kinds of spin-polarized electron photocathodes, the GaAs-GaAsP strained layer superlattice and the GaAs-AlGaAs superlattice with distributed Bragg reflector. The experimental results of these photocathodes are briefly reported.

1. INTRODUCTION

We have been conducting the developments of superlattice photocathodes for spin-polarized electron source. Up to now, the material combinations of GaAs-AlGaAs, InGaAs-GaAs and InGaAs-AlGaAs have been investigated systematically. It was clarified that the strained-layer superlattice with a wide band-gap energy (E_{th}) can achieve high electron spin polarization (ESP≥80%) and high quantum efficiency (QE≥0.3%) simultaneously [1]. It has been also demonstrated that the high-intensity multi-bunch beam can be produced from the superlattice without the surface charge limit (SCL) phenomenon [2].

The GaAs-GaAsP strained-layer superlattice has an advantage to have the large energy splitting between heavy- and light-holes (δs >50meV) due to the large valence band offset, and thus the high ESP can be expected. High QE is also anticipated because the larger amount of negative electron affinity should be obtained by relativity larger E_{th} (~1.6eV).

In order to enhance QE, we had introduced a distributed Bragg reflector (DBR) into the thin strained GaAs layer [3]. Incident laser light is confined in a Fabry-Perot cavity which consists of the GaAs surface and the DBR, and thus the optical absorption rate is enhanced by a factor of ~10. By this technique, QE of ~1.0 % and ESP of ~85% were achieved. Recently, Novosibrisk group has applied this technique to the thin GaAs-AlGaAs superlattice layer.

The ESP and QE of these superlattice samples have been also measured by the cathode test system at Nagoya University.

2. GaAs-GaAsP STRAINED-LAYER SUPERLATTICE

The GaAs-GaAsP superlattice samples were fabricated by MOCVD apparatus at Nagoya Univ. Fig.1 shows the structure of test samples (SLSP#9 and SLSP#16). The thickness of the well and barrier layers were selected to be 3nm for SLSP#9 and 4nm for SLSP#16 to obtain large δs. Only the GaAs well layers are strained (strain~1.3%), since the superlattice layer is grown on the GaAsP buffer. The phosphorus fraction was chosen to be 0.36 to achieve the wide Eth. The δs of 84meV for SLSP#9 and 95meV for SLSP#16 and Eth of 1.60eV for SLSP#9 and 1.62eV for SLSP#16 were estimated by the calculation using Kronig-Penny-Bastard model. In order to overcome the

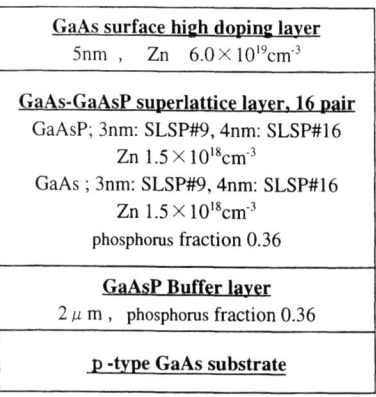

Figure.1 Structure of GaAs-GaAsP superlattices

SCL phenomenon, the GaAs surface layer was heavily doped (Zn: 6.0×10^{19} cm^{-3}), while, in the interior of superlattice it was chosen to be low (Zn: 1.5×10^{18} cm^{-3}) to avoid the spin-flip depolarization.

Fig.2 shows the ESP and QE spectra of these samples.

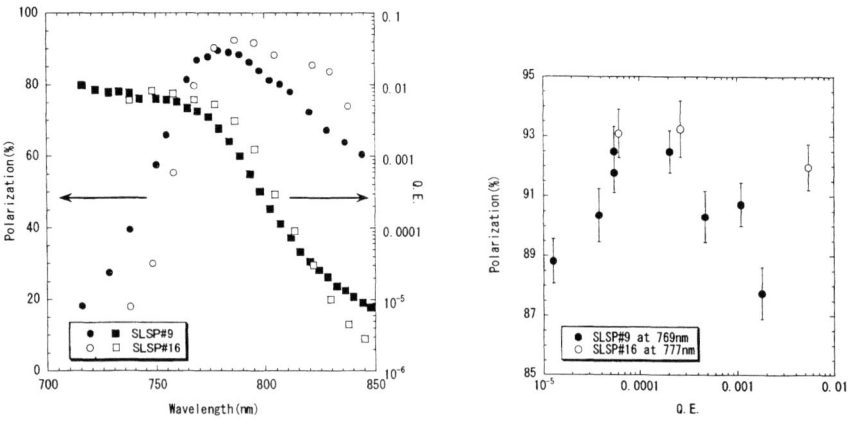

Figure.2 ESP and QE spectra of GaAs-GaAsP superlattice, and the relation between ESP and QE

The high ESP (≥85%) and high QE (≥0.4%) were achieved by both samples. The relation between ESP and QE indicates that ESP tends to be maximum at lower QE rather than the best QE. The production of sub-nanosecond multi-bunch beam has been achieved without the SCL phenomenon with SLSP#9 photocathode[4].

3. GaAs-AlGaAs SUPERLATTICE WITH DBR

The GaAs-AlGaAs superlattice with DBR was fabricated by a MBE apparatus of Novosibrisk group. Fig.3 shows the sample structure. The parameters of this superlattice layer are same as those of the superlattice developed by Nagoya group (SL#7; ESP= ~68%, QE= ~0.5%), except the thickness of superlattice layer (22nm for Novosibrisk and 90nm for SL#7) [1]. The DBR consists of 18 pairs of the alternative AlAs and AlGaAs layers. Fig.4 shows the ESP and QE spectra.

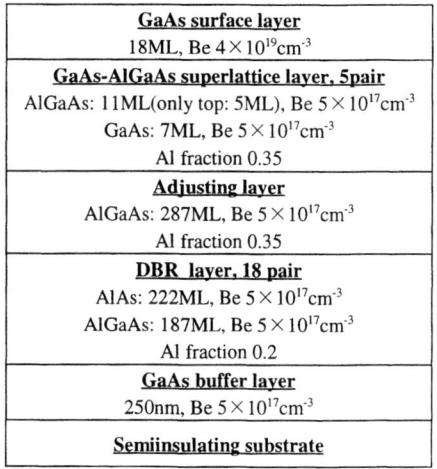

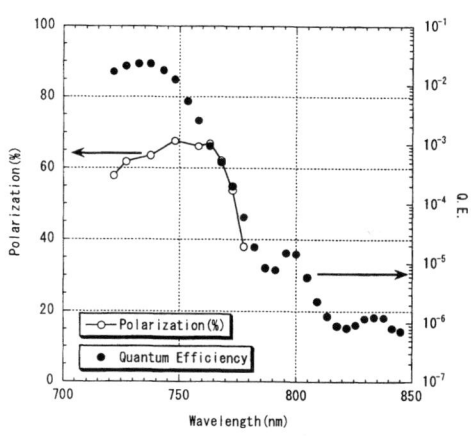

Figure.3 Structure of GaAs-AlGaAs superlattice with DBR

Figure.4 ESP and QE spectra of GaAs-AlGaAs superlattice with DBR

The QE enhancement was clearly observed at 738nm, 796nm and 834nm. The QE of 2.3% was obtained at 738nm which is near the ESP maximum. The best QE of this sample achieved by Novosibrisk group was ~8% at ~740nm. The maximum ESP (68%) is consistent with the value by SL#7.

4. CONCLUSION

We have tested the GaAs-GaAsP strained-layer superlattice and the GaAs-AlGaAs superlattice with DBR. Since the technique of DBR can be applied for every thin layer photocathode, the GaAs-GaAsP superlattice with DBR is expected to give the higher ESP than 80%.

5. REFERENCES

1. T. Nakanishi et al., *AIP. Conf. Proc.* **421** 300-310 (1998)
2. K. Togawa et al., *Nucl. Instr. and Meth. in Phys. Res.* **A 414** 431-445 (1998)
3. T. Saka et al., *Jpn. J. Appl. Phys.* **32** L1837-L1840 (1993)
4. K. Togawa et al., this workshop

Fabrication of GaAs/GaAsP superlattice photocathode

O. WATANABE, T. NISHITANI[a], K. TOGAWA[a], Y. TAKASHIMA[b],
T. NAKANISHI[a], Y. TAKEDA[c], and H. KOBAYAKAWA.

Nagoya University, Department of Materials Processing Engineering, Nagoya 464-8603, Japan
a) Nagoya University, Department of Physics, Nagoya 464-8602, Japan
b)Institute for Molecular Science, Okazaki 444-8585, Japan
c) Nagoya University, Department of Materials Science & Engineering, Nagoya 464-8603, Japan

Abstract. Several samples of GaAs/GaAsP superlattice were fabricated using a method of the Metalorganic Chemical Vapor Deposition (MOCVD) growth for the purpose of investigating various properties as a photocathode of spin-polarized electron sources. The MOCVD growth is easy to control in comparison with the Molecular Beam Epitaxial (MBE) growth in the fabrication procedure of the GaAs/GaAsP superlattices. We used Tertiarybutylarsine (TBAs) and Tertiarybutylphophine (TBP) as V-group sources to make the samples. It is for this reason that the toxicity of TBAs and TBP is lower than that of arsine (AsH_3) and phosphine (PH_3) which are commonly used, and the pyrolysis temperature for TBAs and TBP is lower than that for AsH_3 and PH_3 [1]. A large spin-polarization exceeding 90% was observed using the sample made in this method. We also obtained large quantum efficiencies of approximately 0.4% in the wavelength range from 760nm to 780nm.

INTRODUCTION

Spin-polarized electrons play important roles as a new probe in the fields of the materials science and the high-energy physics. At present, most of polarized-electron sources are based on the photo-emission process on the GaAs-type semiconductors. In order to produce electrons with high degree of polarization and high-quantum efficiency (QE), we generally use the GaAs-type semiconductors as photocathodes, which have structures of strained GaAs-layer, or of superlattice. Polarization of 70-80% are able to obtain under those systems which have relatively high-QE of 0.1-0.5%, and have reasonably long lifetime as the polarized-electron gun. We fabricated ten samples of GaAs/GaAsP superlattice using MOCVD growth method under several conditions of Zn-doping and various layer thicknesses. Values of polarization and QE's of extracted electrons were measured for the purpose of critical performance tests as the photocathode. We confirmed that high-polarization of approximately 90% and large QE's are obtainable under the superlattices made with the MOCVD method.

MOCVD GROWTH STSTEM

GaAs/GaAsP superlattice was made under the MOCVD growth process. Figure1 shows the gas-flow system of the growth system, which is able to precisely adjust the pressure at the reactor, in which the crystal grows. During the procedure of crystal growth, the reactor pressure was kept 76Torr within a fluctuation of 0.2%. The gas-flow system consists of two main sections, source-section and acceleration-section, as shown in Fig.1.

The acceleration-section serves as preserving proper gas-flow-rate at the reactor. The acceleration-section has two paths, the reactor-side line and the exhaust-side line. The carrier gas (H_2) in each line passes through a vent-run system, and flows into the reactor or the exhaust unit.

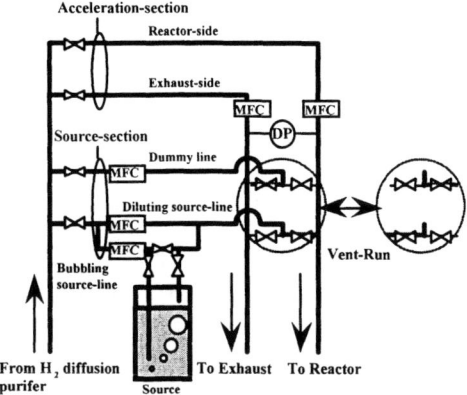

FIGURE 1. Gas-flow system of Growth System.

The source-line supplies gases containing sources to the reactor. This section has three lines; the babbling-source line, the diluting-source line, and the dummy line.

The vent-run system is a joint section of the source-lines and the acceleration-lines. This system helps joining in a proper rate of the source-gases and acceleration gas, and minimizes an initial fluctuation of the gas-flow due to the bulb switching. This precision control of the reactor pressure assures creation of the superlattices having abrupt interfaces.

EXPERIMENT

Several samples of GaAs/GaAs$_{0.64}$P$_{0.36}$ superlattice were fabricated using the MOCVD growth system. Growth temperature was 650℃, and V/III ratios were 10 for GaAs and 15 for GaAsP. One of the samples has a structure of GaAsP buffer layer of 2μm built on a p-GaAs substrate. GaAs layers (3nm) and GaAsP layers (3nm) were piled up as forming a superlattice (16pairs) on the buffer layer. GaAs of 5nm was piled on the superlattice as a surface layer. The amounts of Zn-doping were $1.5 \times 10^{18}/cm^3$ in the superlattice layers and $6.0 \times 10^{19}/cm^3$ in the surface layer. Photoluminescence (PL), polarization,

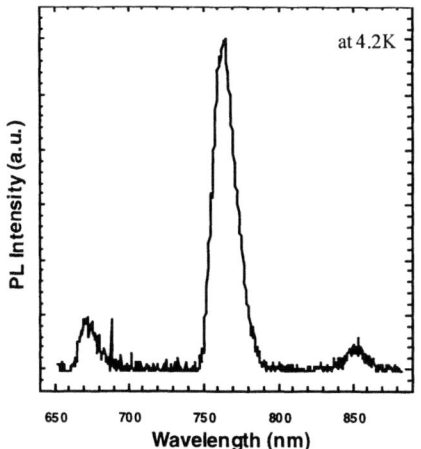

FIGURE 2. PL spectrum of GaAs/GaAsP superlattice.

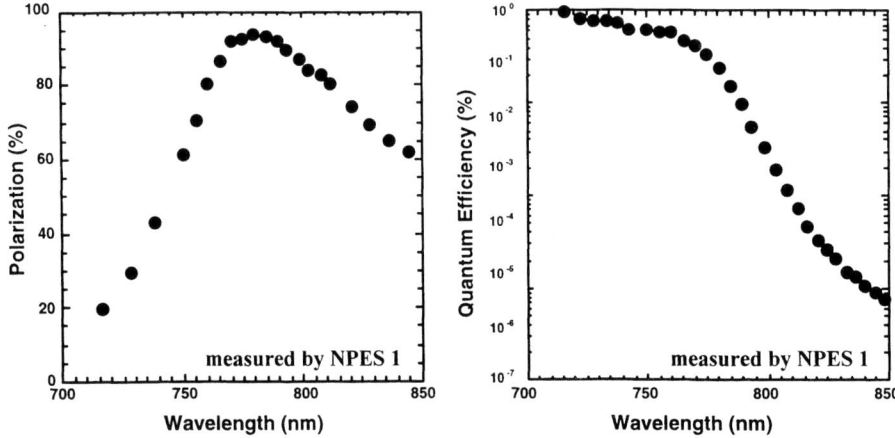

FIGURE 3. Polarization of GaAs/GaAsP superlattice.   **FIGURE 4.** Q.E. of GaAs/GaAsP superlattice

QE were measured on this sample. We observed a large luminous intensity in the wavelength range around 770nm (the estimated energy gap between the heavy hole state and the conduction band of the GaAs wells) in PL as shown in Fig.2. It is for this reason that the mixed crystalization due to the diffusion of Zn was small in our samples. Moreover, though the crosshatchings and the crackings were observed in the buffer layer, the abruption of layer interfaces was not lost. Polarization exceeding 90% was observed using the sample made in MOCVD method. We also observed large QE's of approximately 0.4% in the wavelength range from 760nm to 780nm. These data show that the effect of the negative electron affinity on the surface appeared substantially, because large amounts of Zn were doped on the surface layers in these samples.

SUMMARY

We fabricated ten samples of GaAs/GaAsP superlattice for the spin-polarized electron source using the MOCVD growth system. The rate of V-group mixed crystal (As:P) was 0.64:0.36. Spin-polarizations of photo-emitted electrons and quantum efficiencies were measured using the polarized-electron-source (PES) system at Nagoya University. Values of spin-polarization exceeded 90%, and quantum efficiencies were approximately 0.4% in the wavelength range between 760nm and 780nm. Present experiment shows that the GaAs/GaAsP superlattice made with the MOCVD growth is very promising as photocathode of spin-polarised electron sources.

REFERENCES

1. G. B. STRINGFELLOW, ., *Orgaanometallic Vapor-Phase Epitaxy* 2nd Edition (1999).

APPENDICES

SPIN2000 Schedule

Monday, October 16

8:30–9:00	Registration
	Opening
9:00–9:05	Opening Address
	Ejiri, H. (RCNP/JASRI), Chairperson of SPIN2000
9:05–9:15	Welcome Address
	Yamaguchi, Y (Univ. of Tokyo), Former president of IUPAP
9:15–9:20	Spin Physics 2000
	Prescott, C.Y. (SLAC), Chairperson of International Committee
	Plenary Session
	Chair: Fidecaro, G. (CERN), Secretary: Toki, H. (RCNP)
9:20–10:10	Spin Physics: Progress and Prospects
	Jaffe, R.L. (MIT)
10:10–10:50	HERMES at the Turn of the Millennium
	Aschenauer, E.C. (DESY-Zeuthen)
10:50–11:20	Coffee Break
	Plenary Session
	Chair: Soffer, J. (CPT-CNRS), Secretary: Toki, H. (RCNP)
11:20–12:00	Highlights from the SMC and COMPASS Experiments
	Bradamante, F. (CERN/PPE)
12:00–12:40	The Muon g-2 Experiment at Brookhaven
	Bunce, G. (BNL)
14:00–17:40	**Parallel Sessions**

Tuesday, October 17

8:30–9:00	Cont. Breakfast
	Plenary Session
	Chair: Vigdor, S.E. (Indiana Univ.), Secretary: Yosoi, M. (RCNP)
9:00–9:40	Weak Nucleon Form Factors
	Souder, P. (Syracuse Univ.)
9:40–10:20	Polarization in Photo and Electro Disintegration of the Deuteron
	Gilman, R. (Rutgers Univ.)
10:20–11:00	Nucleon Electromagnetic Form Factors
	Gao, H. (MIT)
11:00–11:20	Coffee Break
	Plenary Session
	Chair: van Oers, W.T.H. (Univ. of Manitoba), Secretary: Yosoi, M. (RCNP)
11:20–12:00	The Beta-Neutrino Correlation Using a Neutral Atom Trap
	Vetter, P. (LBL)
12:00–12:40	Fundamental Symmetry and Polarized Muon
	Kuno, Y. (KEK)
14:00–16:00	**Parallel Sessions**
16:00–18:00	**Poster Sessions at Icho Kaikan**

Wednesday, October 18

8:30–9:00	Cont. Breakfast
	Plenary Session
	Chair: Masaike, A. (Fukui Univ. of Tech.), Secretary: Sakaguchi, A. (Osaka Univ.)
9:00–9:40	Nuclear Moment Studies with Polarized Radioactive Nuclear Beams at RIKEN
	Asahi, K. (Tokyo Inst. of Tech)
9:40–10:20	High Energy Hadron Spin Observables
	Kubo, K.-I. (TMU/TMCAE)
10:20–11:00	Spin Physics in Hypernuclei and Hyperon-Nucleon Interactions
	Kishimoto, T. (Osaka Univ.)
11:00–11:20	Coffee Break
	Plenary Session
	Chair: Courant, E.D. (BNL), Secretary: Sakaguchi, A. (Osaka Univ.)
11:20–12:00	Nucleon Spin Structure Functions
	Vogelsang, W. (RIKEN BNL Center)
12:00–12:40	Spin Physics at RHIC
	Saito, N (RIKEN/RIKEN BNL Center)
14:00–18:00	**Parallel Sessions**

Thursday, October 19

8:30–9:00	Cont. Breakfast
	Plenary Session*
	Chair: Krisch, A. (Univ. of Michigan), Secretary: Itahashi, T. (RCNP)
9:00–9:40	Electron-Positron Spin Dynamics between 40 and 100 GeV at LEP
	Assmann, R. (CERN)
9:40–10:20	Beam Polarization Aspects of Muon-Muon Collider
	Shatunov, Yu.M. (BINP)
10:20–11:00	High Intensity Polarized Ion Source
	Zelenski, A.N. (TRIUMF)
11:00–11:20	Coffee Break
	Plenary Session
	Chair: Horikawa, N. (Nagoya Univ.), Secretary: Itahashi, T. (RCNP)
11:20–12:00	Polarized Photon Beam Experiments at SPring-8
	Nakano, T. (RCNP)
12:00–12:40	Polarized Photon Beam Experiments at GRAAL
	Hourany, E. (Orsay)
14:00–18:00	**Excursion to the National Museum of Ethnology**
18:00–20:00	**Banquet at Senri Hankyu Hotel**

* The memorial session in honor of Dr. L.G. Ratner.

Friday, October 20

8:30–9:00	Cont. Breakfast

Plenary Session
Chair: Hatanaka, K. (RCNP), Secretary: Fujita, Y. (Osaka Univ.)

9:00–9:40	Three-Nucleon Spin Observables: Signatures for Three-Nucleon Force Effects
	Witala, H. (Krakow U.)
9:40–10:20	Spin-Isospin Responses in Nuclei via the Polarization Measurements
	Sakai, H. (Univ. of Tokyo)
10:20–11:00	Nuclear Medium Effects Studied by Nucleon Quasifree Scattering
	Noro, T. (RCNP)
11:00–11:20	Coffee Break

Plenary Session
Chair: Solovianov, V.L. (IHEP), Secretary: Fujita, Y. (Osaka Univ.)

11:20–12:00	Effective Field Theory for Non-Relativistic Bound States
	Manohar, A. (UCSD)
12:00–12:40	Effective Theory for Heavy Quarkonium Production
	Lee, J. (DESY)
14:00–16:00	**Parallel Sessions**
16:00–18:00	**Poster Sessions at Icho Kaikan**

Saturday, October 21

8:30–9:00	Cont. Breakfast

Plenary Session
Chair: Lowenstein, D.I. (BNL), Secretary: Shimizu, H. (RCNP)

9:00–9:40	Recent Developments in the Field of Polarized Solid State Target Materials
	Goertz, S. (Bochum)
9:40–10:20	Workshop Report on Spin Polarized Electron Source and Polarimeter
	Nakanishi, T. (Nagoya Univ.)
10:20–11:00	Polarized Sources and Targets - Erlangen 1999
	Steffens, E. (Erlangen)
11:00–11:20	Coffee Break

Plenary Session
Chair: Haeberli, W. (Univ. of Wisconsin), Secretary: Shimizu, H. (RCNP)

11:20–12:00	Prospects and Uses of a Polarized ep Collider at HERA and at eRHIC
	Hughes, V. (Yale)
12:00–12:40	Symposium Highlights
	Roser, T. (BNL)

Adjourn

Parallel Sessions

Monday, October 16, 14:00–15:40

Session 3, Hall 2 Chair: Toki, H (RCNP)

14:00–14:25 Measurement of the Spin Structure Function g1 of the Proton and Deuteron
Stoesslein, U. (Univ. of Colorado)

14:25–14:50 Measurement of Polarized Quark Distributions in the Nucleon at Hermes
Lindemann, T. (DESY)

14:50–15:15 Polarized Structure Functions of the Deuteron
Kumano, S. (Saga Univ.)

15:15–15:40 Extraction of g_1 in the Resonance Region using the First Polarized Target Data from CLAS
Fatemi, R. (Univ. of Virginia)

Session 5, Hall 3 Chair: Morii, T (Kobe Univ.)

14:00–14:25 Spin Effects in Diffractive Hadron Photoproduction
Goloskokov, S. (JINR)

14:25–14:50 Spin-Density Matrix Elements for Diffractive Vector Meson Production
Kinney, E. (Univ. of Colorado)

14:50–15:15 Nucleon Resonances in Polarized ω Photoproduction
Titov, A. (JINR)

15:15–15:40 Polarization Phenomena in Vector Meson Production on Nucleons near Threshold
Gokalp, A. (Middle East Tech. Univ.)

Session 6, Main Hall Chair: Bacher, A. (IUCF)

14:00–14:25 Precise Determination of the Spin-transfer Coefficient $K_{NN'}$ for $\vec{n}p$ Elastic Scattering at 187 MeV
Wissink, S.W. (Indiana Univ.)

14:25–14:50 Neutron Densities in ^{120}Sn Observed by Polarized Proton Scattering
Sakaguchi, H. (Kyoto Univ.)

14:50–15:15 Spin-flip Probability for the ^{26}Mg(^{3}He,t)^{26}Al*(1^+;1.058 MeV) Reaction at 177 MeV
Sakoda, S. (Univ. of Tokyo)

15:15–15:40 Study of Isospin Structure of 1^+ spin states in ^{58}Ni and ^{58}Cu by the Comparisons of ^{58}Ni(p,p') and ^{58}Ni$(^3$He,t$)$ Reactions
Fujita, H. (Osaka Univ.)

Session 7, Hall 1 Chair: Derbenev, Y.S. (Univ. of Michigan)

14:00–14:20 Polarized Proton Beam Development at COSY with EDDA as a Fast Internal Polarimeter
Hinterberger, F. (ISKP, Univ. of Bonn)

14:20–14:40 Spin-flipping with an Rf-dipole and a Full Siberian Snake
Lin, A.M.T. (Univ. of Michigan)

14:40–15:00 Crossing a Coupling Spin Resonance With an RF Dipole
Bai, M. (BNL)

15:00–15:20 Beam Polarization Distributions for RHIC
Lehrach, A. (BNL)

15:20–15:40 The Application of the Amplitude Dependent Spin Tune for the Study of High Order Spin–orbit Resonances in Storage Rings
Barber, D.P. (DESY)

Monday, October 16, 16:00–18:00

Session 3, Hall 2 Chair: Imai, K. (Kyoto Univ.)
16:00–16:25 Preliminary Results for the Spin Structure Function g_2 from SLAC E-155X
Prepost, R. (Univ. of Wisconsin)
16:25–16:50 Generalized GDH Sum and Spin Structure of ^{3}He and the Neutron
Averett, T. (Coll. of William and Mary)
16:50–17:15 Spin-Azimuthal Asymmetries in Semi-Inclusive Pion Production
Di Nezza, P. (INFN)
17:15–17:40 Estimation of the Proton Transversity from Azimuthal Asymmetries in DIS
Efremov, A.V. (JINR)

Session 4, Hall 3 Chair: Sivers, D. (Portland Physics Inst.)
15:55–16:20 Measurement of the Λ^0 Polarization in ν_μ Charged Current
Naumov, D.V. (JINR)
16:20–16:45 Hyperon Polarization in Inclusive Hadronic Production
Kanazawa, Y. (Niigata Univ.)
16:45–17:10 Polarization of Hyperons in Photon induced Reaction at High Energy
Suzuki, K. (Univ. of Tokyo)
17:10–17:35 Longitudinal and Transverse Lambda Polarization at Hermes
Bernreuther, S. (Tokyo Inst. of Tech.)
17:35–18:00 Spin Transfer in High Energy Fragmentation Processes
Liang, Z.-T. (Shandong Univ.)

Session 6, Main Hall Chair: Tanifuji, M. (Hosei Univ.)
16:00–16:25 Momentum Transfer Dependence of Spin Isospin Modes in Quasi-Elastic $(\vec{p},\vec{n})$ Reactions
Wakasa, T. (RCNP)
16:25–16:50 Relativistic Calculations of Quasielastic Proton-nucleus Spin Observables based on a Complete Lorentz Invariant Representation of the NN Scattering Matrix
Van Der Ventel, B. (Univ. of Stellenbosch)
16:50–17:15 Calculation of the Complete Set of Spin Transfer Coefficients including One- and Two-step Processes in (p,nx) Reaction at 346 MeV
Ogata, K. (Kyushu Univ.)
17:15–17:40 Relativistic Plane Wave Model for Complete Sets of $(p,2p)$ Spin Transfer Observables
Wyngaardt, S. (Univ. of Stellenbosch)

Session 7, Hall 1 Chair: Makdisi, Y. (BNL)
16:00–16:25 The Polarized Electron Beam at ELSA
Hoffmann, M. (Bonn)
16:25–16:50 First Electron Beam Polarization Measurements with the Compton Polarimeter using High Finesse Fabry-Pérot Cavity at JLab
Pussieux, T. (CEA Saclay)
16:50–17:15 Instrumentation for the Polarization Transfer Experiment in Proton Inelastic Scattering at 0°
Yosoi, M. (RCNP)
17:15–17:40 Deuteron Polarimeter DPOL and Calibration of the System
Kato, H. (Univ. of Tokyo)

Tuesday, October 17, 14:00–16:05

Session 2, Hall 2 Chair: Kuno, Y. (KEK)
14:00–14:25 Results from the HyperCP Experiment at Fermilab
Dukes, E.C. (Univ. of Virginia)

14:25–14:50	Symmetry Tests in Polarized Z0 Decays to b bbar g Maruyama, T. (SLAC)
14:50–15:15	Polarized Muon Decay: Measurement of the Polarization Vector of the Decay Positrons as a Test of Time Reversal Invariance Fetscher, W. (IPP/ETH Zurich)
15:15–15:40	Measuring the Michel Parameter ξ'' in Polarized Muon Decay Morelle, X. (PSI)
15:40–16:05	How Could CP-Invariance and Physics Beyond SM Be Tested In Polarized Proton Collisions at RHIC? Rykov, V.L. (Wayne State Univ.)

Session 5, Hall 3 Chair: Shibata, T.-A. (Tokyo Inst. of Tech.)

14:00–14:25	Hard Exclusive Meson Production at HERMES Ryckbosch, D. (Univ. of Gent)
14:25–14:50	Spin Structure Function of the Virtual Photon and Polarized Parton Distributions Sasaki, K. (Yokohama Nat'l Univ)
14:50–15:15	Polarization Observables in Wide Angle Compton Scattering Wojtsekhowski, B. (TJNAF)
15:15–15:40	Fragmentation of Transversely Polarized and Unpolarized Quarks Anselmino, M. (Univ. of Torino)

Session 6, Main Hall Chair: Noro, T. (RCNP)

14:00–14:25	Spin-dependent Effective Interaction Studied by the ^{12}C,^{28}Si$(\vec{p},\vec{p}')$ Reactions at Zero Degrees Tamii A. (Univ. of Tokyo)
14:25–14:50	Conventional and Non-conventional Medium Effects in (p,p') Reactions Stephenson, E. (Indiana Univ.)
14:50–15:15	Study of Spin-dipole Resonances in ^{16}O by measuring ^{16}O$(\vec{p},\vec{p}')$ Reaction at Extremely Forward Angles Kawabata, T. (Kyoto Univ.)
15:15–15:40	Measurement of Single and Double Spin-Flip Probabilities in Inelastic Deuteron Scattering on ^{12}C and ^{28}Si at 270 MeV Satou Y. (RIKEN)
15:40–16:05	DWIA Calculations for Inelastic Scattering of Deuterons at E_d=400 MeV Suzuki, T. (Tokyo Metro. Univ.)

Session 7, Hall 1 Chair: Courant, E.D. (BNL)

14:00–14:25	Radiative Polarization in the BATES South Hall Ring Shatunov, Yu.M. (BINP)
14:25–14:50	Polarized electrons at HERA: Experience and Expectations after the Luminosity Upgrade Gianfelice, E. (DESY)
14:50–15:15	Concepts for Stern-Gerlach Polarization of Antiproton Beam in a Strage Ring Derbenev, Y.S. (Univ. of Michigan)
15:15–15:40	Stern-Gerlach Interaction in Fermion Beams Conte, M. (Univ. of Genova/INFN)
15:40–16:05	Physics with EPIC, a high luminosity electron-polarized ion collider Kinney, E. (Univ. of Colorado)

Wednesday, October 18, 14:00–15:40

Session 3, Hall 2 Chair: Tannenbaum, M.J. (BNL)

14:00–14:25	Spin physics by HERMES using hadron identification with the Ring Imaging Cerenkov Counter Sakemi, Y. (Tokyo Inst. of Tech.)

14:25–14:50	Deeply Virtual Compton Scattering at HERMES Amarian, M. (INFN)
14:50–15:15	Polarized Gluon Distributions from High-p_T Pair Hadron Productions in Polarized Deep Inelastic Scattering Yamanishi, T. (Fukui Univ. of Tech.)
15:15–15:40	Commissioning of RHIC p-Carbon CNI Polarimeter Huang, H. (BNL)

Session 5, Hall 3 Chair: Souder, P. (Syracuse Univ.)

14:00–14:25	Precise Measurement of the Spin-Dependent Transverse Asymmetry in Quasi-elastic $^3\vec{\mathrm{He}}(\vec{e},e')$ and the Neutron Magnetic Form Factor Hansen, J.-O. (TJNAF)
14:25–14:50	Single π^0 Electroproduction in $\Delta(1232)$ from CLAS at Jefferson Lab. Joo, K. (TJNAF)
14:50–15:15	Results and Status of the Inelastic ed-Scattering Experiments at the Internal Polarized Deuterium Targets of the VEPP-3 Stibunov, V.N. (NPI)
15:15–15:40	Measurement of Spin Correlation Parameters in the Δ Region for the $^1\mathrm{H}(\vec{e},e')$ Reaction van Buuren, L.D. (NIKHEF/VU Amsterdam)

Session 6, Main Hall Chair: Stephenson, E. (Indiana Univ.)

14:00–14:25	Nuclear Spectroscopy by means of $(\vec{p},\alpha)$ Reactions on Magic and Near Magic Nuclei: the $^{122}\mathrm{Sn}(\vec{p},\alpha)^{119}\mathrm{In}$ Reaction at 26 MeV Guazzoni, P. (Univ. Degli Studi)
14:25–14:50	Measurement of Analysing Power for $pp \rightarrow pp\pi^0$ Reaction at the Beam Energy 392 MeV Maeda, Yoshikazu (RCNP)
14:50–15:15	Energy Dependence of $^{12,13}\mathrm{C}(\vec{p},\pi^-)^{13,14}\mathrm{O}_{g.s.}$ Reactions in Δ_{1232} Resonance Region Kamiya, J. (RCNP)
15:15–15:40	Effective Charge Anomaly in Neutron-rich Nuclei revealed from Spin-polarizd RI Beam Experiments Ogawa, H. (Tokyo Inst. Tech.)

Session 7, Hall 1 Chair: Plis, Yu.A. (JINR)

14:00–14:25	Measurement of the Analyzing Power for Proton-Carbon Elastic Scattering in the CNI Region with a 22 GeV/c Polarized Proton Beam Tojo, J. (Kyoto Univ.)
14:25–14:50	Deuteron Beam Polarimetry at JINR Accelerator Facility Zolin, L.S. (JINR)
14:50–15:15	Absolute Calibration of the Deuteron Beam Polarization at Intermediate Energies via the $^{12}\mathrm{C}(\vec{d},\alpha)^{10}\mathrm{B}^*(2^+)$ Reaction Suda, K. (Saitama Univ.)
15:15–15:40	A Spin Polarizer for Radioactive Beams by Taking Advantage of Polarized Electron Transfer Reaction Shimoda, T. (Osaka Univ.)

Wednesday, October 18, 15:55–18:00

Session 1, Hall 3 Chair: Kishimoto, T. (Osaka Univ.)

15:55–16:20	Constraints of a Parity-Conserving/Time-Reversal-Non-conserving Interaction van Oers, W.T.H. (Univ. of Manitoba)

16:20–16:45	Neutrinos by Means of Double Beta Decay by ELEGANT V and Spin-Isospin Responses
	Kudomi, N. (RCNP)
16:45–17:10	Search for Spin Coupled Dark Matter with the Large Volume NaI(Tl) Scintillators
	Yoshida, S. (RCNP)
17:10–17:35	Hyperon Beta-Decay Analysis and the Recent KTeV Data
	Ratcliffe, P.G. (Univ. dell'Insubria)
17:35–18:00	Novel Spin Maser Mechanism Studied for High-Precision Measurement of Neutron Electric Dipole Moment
	Yoshimi, A. (RIKEN)

Session 3, Hall 2 Chair: En'yo, H. (Kyoto Univ.)

15:55–16:20	The Shape and Experimental Tests of the Scale-Invariant Polarized Gluon Asymmetry
	Ramsey, G. (Loyola Univ.)
16:20–16:45	Measurement of the Gluon Polarization in the Proton at PHENIX
	Goto, Y. (RIKEN BNL Center)
16:45–17:10	Extracting $\Delta G(x)$ from the $\vec{p}\vec{p} \rightarrow \gamma + \text{jet} + X$ Reaction with the STAR detector at RHIC
	Wissink, S. (Indiana Univ.)
17:10–17:35	$W^{\pm}$ Producton in Polarized Proton Collisions with the STAR Detector at RHIC
	Vigdor, S.E. (Indiana Univ.)
17:35–18:00	Transversity Measurement at RHIC
	Grosse Perdekamp, M. (RIKEN BNL Center)

Session 6, Main Hall Chair: Sakai, H. (Univ. of Tokyo)

15:55–16:20	Role of Deuteron Internal Variables in the $^3\text{He}(d,p)^4\text{He}$ Reaction
	Uesaka, T. (Saitama Univ.)
16:20–16:45	Tensor Analysing Power A_{yy} in Deuteron Breakup on Hydrogen and Nuclei at Large Transverse Momenta of Proton
	Ladygin, V.P. (JINR)
16:45–17:10	Polarization Transfer for the $^2\text{H}(d,p)^3\text{H}$ Reaction at $\theta = 0°$ at a Very Low Energy
	Katabuchi, T. (Univ. of Tsukuba)
17:10–17:35	Calculation of Low Energy $^2\text{H}(d,p)^3\text{H}$ Reaction by the Four-body Faddeev-Yakubovsky Equation
	Uzu, E. (Science Univ. of Tokyo)
17:35–18:00	Measurement of Analyzing Power T_{20} in Elastic Electron-Deuteron Scattering in the Momentum Transfer Range of 0.3 - 0.8 $(GeV/c)^2$
	Toporkov, D.K. (Budker INP)

Session 8, Hall 1 Chair: Iwata, T. (Nagoya Univ.)

15:55–16:20	NMR System with Non-resonant Cable Arrangment
	Crabb, D. (Univ. of Virginia)
16:20–16:45	A High Intensity Stern-Gerlach Polarized Hydrogen Source Using ECR Ionization and Charge Exchange in Cs Vapour for the Munich MP-Tandem Laboratory
	Hertenberger, R. (Muenchen)
16:45–17:10	The Polarized Internal Gas Target for ANKE at COSY
	Seyfarth, H. (Jülich)
17:10–17:35	Development of Polarized Negative Hydrogen Ion Source with Resonant Charge-exchange Plasma Ionizer
	Belov, A.S. (INR, Moscow)
17:35–18:00	Development of polarized ^{3}He ion source - From OPPIS to Spin-exchange
	Tanaka, M. (Kobe Tokiwa Coll.)

Friday, October 20, 14:00–16:05

Session 3, Hall 2 Chair: Kumano, S. (Saga Univ.)

14:00–14:25 Positivity Constraints in Spin Physics
Soffer, J. (CPT-CNRS)

14:25–14:50 Twist-2 Polarized Fragmentation Function in the Open Charm Production in DIS
Arestov, Yu (IHEP)

14:50–15:15 Polarized Gluon Distribution Function of Nucleon in Diffractive Leptoproduction of Charmonium
Hayashigaki, A. (Univ. of Tokyo)

15:15–15:40 Determination of polarized parton distribution functions
Miyama, M. (Tokyo Metro. Univ.)

15:40–16:05 Solving the Nucleon Spin Puzzle based on the Chiral Quark Soliton Model
Wakamatsu, M. (Osaka Univ.)

Session 4, Hall 3 Chair: Shapiro, G. (Univ. of California)

14:00–14:25 Top physics at the LHC with the CMS experiment: Determination of the top Quark Mass, Spin Correlations and Anomalous interactions
Sonnenschein, L. (RWTH Aachen)

14:25–14:50 Total and Differential Cross Sections and Polarization Effects in Proton-Proton Elastic Scattering at RHIC
Guryn, W. (BNL)

14:50–15:15 New Approaches to the pp Total Cross Section Measurements at Polarized Colliders
Nurushev, S. (IHEP)

15:15–15:40 Measurement of Analyzing Powers and Spin Correlation Coefficients for Elastic pp Scattering
Bauer, F. (Univ. of Hamburg)

15:40–16:05 A_N for Inclusive $\pi^{\pm}$ at 21.6 GeV from ^{12}C and H_2
Spinka, H. (ANL)

Session 6, Main Hall Chair: Fujiwara, M. (RCNP)

14:00–14:25 Difference between $nd\ A_y$ and $pd\ A_y$ below 16 MeV
Sagara, K. (Kyushu Univ.)

14:25–14:50 Measurement of Cross Sections and Analyzing Powers for $d-p$ Elastic Scattering at Intermediate Energies and Three-Nucleon Force Effects
Sekiguchi, K. (Univ. of Tokyo)

14:50–15:15 Three-Body Force Effects in Proton-Deuteron Elastic Scattering
Kinney, E. (Univ. of Colorado)

15:15–15:40 Measurement of Differential Cross Sections and Vector Analyzing Powers for $\vec{n}+d$ Reaction at 250 MeV
Maeda, Yukie (Univ. of Tokyo)

15:40–16:05 Forward Scattering Amplitudes and Contributions of Three Nucleon Forces in nd Elastic Scattering
Ishikawa, S. (Hosei Univ.)

Session 8, Hall 1 Chair: Crabb, D. (Univ. of Virginia)

14:00–14:25 The new LEGS Highly Polarized Frozen-Spin Solid HD Target Facility
Wei, X. (BNL)

14:25–14:50 The HERMES Internal Polarized Deuterium Target
Lenisa, P. (Ferrara/INFN)

14:50–15:15 Michigan Ultra-Cold Polarized Atomic Hydrogen Jet Target
Luppov, V.G. (Univ. of Michigan)

15:15–15:40 Development of a Polarized Proton Target in a Low Magnetic Field at High Temperature
Wakui, T. (RIKEN)

15:40–16:05 Polarized Nuclei in Plastic Scintillators: a New Class of Polarized Targets
van den Brandt, B. (PSI)

Poster Sessions
(Icho Kaikan)

Tuesday, October 17, 16:00–18:00

1-1 To the physical nature of spin and unified theory of matter
Chashihin, J. (Moscow Insti. for Phys. & Eng.)

1-2 Isospin dependences of the double beta decay
Takahisa, K. (RCNP)

1-3 Helicity Dependence in Photodisintegration of the Deuteron
Wojtsekhowski, B. (TJNAF)

1-4 Spin-Statics Transmutation in Quantum Field Theory of Dyons
Marchetti, P.A. (Univ. of Padova)

1-5 A new method of UCN production using a spatially alternating magnetic field with spin flips
Sakai, K. (Tokyo Inst. of Tech.)

1-6 Search for the Charge Symmetry Breaking Reaction $dd \to {}^4\text{He}\pi^0$
Bacher, A.D. (IUCF)

2-1 Nuclear Responses for Solar-ν, Supernova-ν and $\beta\beta - \nu$
Ejiri, H. (RCNP)

2-2 Search for T-Violation in $K^+ \to \pi^0 \mu^+ \nu$ Decay
Asano, Y. (Univ. of Tsukuba)

2-3 Experimental Sensirivity of Contact interaction at RHIC
Murata, J. (RIKEN)

6-1 A search for Non-conventional Medium Effects in (p,p') Reactions
Stephenson, E. (IUCF)

6-2 Isoscalar, Isovector, Spin and Orbital Contributions in M1 Transitions
Fujita, Y. (Osaka Univ.)

6-3 J^π Decomposition of a Bump at $E_x \simeq 7$ MeV in ^{12}N
Tamii, A. (Univ. of Tokyo)

6-4 Polarization Transfer Invariants at 0° in Nuclear Reactions
Suzuki, T. (Tokyo Metro. Univ.)

6-5 Systematic study of spin-isospin excitations in neutron rich light nuclei via the (d,^{2}He) reaction at 270 MeV
Ohnishi, T. (RIKEN)

6-6 A complete set of total cross sections for imaginary parts of nd forward scattering amplitudes
Tanifuji, M. (Hosei Univ.)

6-7 Implications for the πNN coupling constant from spin transfer measurements in pp elastic scattering at 200 MeV
Wissink, S.W. (Indiana Univ.)

6-8 Isoscalar Giant Monopole Resonance and M1 State of ^{58}Ni via Proton Inelastic Scattering at and near Zero Degrees
Ishikawa, T. (Kyoto Univ.)

6-9 Isospin Identification for A=25 Mirror Nuclei by High Resolution (p,p') and $(^3\text{He},t)$ Experiments
Shimbara, Y. (Osaka Univ.)

6-10 Incident Energy Dependence of the Polarization Observables in Deuteron Elastic Scattering at $E_d = 50 \sim 700$ MeV
Iseri, Y. (Chiba-Keizai Coll.)

6-11 Effect of Δ-Isobar Excitation on Spin-Dependent Observables of Elastic Nucleon-Deuteron Scattering
Gojuki, S. (Science Univ. of Tokyo)

6-12 CDCC analysis of Vector and Tensor Analyzing Powers in ^{208}Pb(d,d) Elastic Scattering at $Ed = 8$ MeV
Aoki, Y. (Univ. of Tsukuba)

6-13 Measurement of Spin Flip Probabilities in Highly Excited Continuum of ^{12}C
Kato, H. (Univ. of Tokyo)

6-14 Tensor analyzing powers of ^{3}He$(d,p)^4$He reactions around 430 keV resonances
Kameyama, H. (Chiba-Keizai Coll.)

6-15 Measurement of H$(d,^3$He$)\gamma$ reaction using large acceptance spectrograph
Yagita T. (Kyushu Univ.)

6-16 Evidence for the existence of supersymmetry in atomic nuclei
Hertenberger, R. (Ludwig-Maximilians)

6-17 Spin effects at fragmentation of polarized deuterons into pions
Zolin, L.S. (JINR)

7-1 Quasi-periodicity of spin motion in strage rings – a new look at spin tune
Barber, D.P. (DESY)

7-2 High Resolution Measurement by Dispersion Matching between New WS Beam Line and Grand-Raiden Spectrometer at RCNP
Wakasa, T. (RCNP)

7-3 A Vector and Tensor Polarimeter for Intermediate Energy Deuterons
Satou, Y. (RIKEN)

7-4 RCNP (n,p) Facility
Yako, K. (Univ. of Tokyo)

7-5 Application of Internal Gas Target for Beam Polarization Measurement in the Electron Strage Ring
Toporkov, D.K. (Budker INP)

7-6 Orientation of Radioactive Nuclei
Finger, M. (Charles Univ.)

7-7 Suppressing Intrinsic Spin Harmonics in the AGS
A. Lehrach (BNL)

7-8 Spin Flipping in RHIC
Bai, M. (BNL)

7-9 Pair-Polarimeter for High Energy Photons
Wojtsekhowski, B. (TJNAF)

7-10 Feasibility of Exact Measurement of the Low Energy Atom Polarization in the Atomic Decay of Neutrons
Plis, Yu.A. (JINR)

7-11 Proposal of Experiments on Spin Polarized Nuclear Fusion
Plis, Yu.A. (JINR)

7-12 Sychrotron-sideband snake depolarizing resonances
Kageya, T. (Univ. of Michigan)

7-13 Beam-line Polarimeter for Intermediate-energy Deuteron
Uesaka, T. (Saitama Univ.)

7-14 Development of deuteron beam polarimeter at RCNP
Yagita, T. (Kyushu Univ.)

7-15 CNI effects and high energy proton beam polarimetry
Selugin, O.V. (JINR)

7-16 Spin Tracking with "Real" Siberian Snakes in RHIC
Xiao, M. (Fermilab)

7-17 Measurements of Analyzing power of pC-scattering and proton polarization in quasi-elastic scattering on carbon
Svirida, D. (ITEP)

Friday, October 20, 16:00–18:00

3-1 Prompt Photon Reconstruction for Gluon Polarization Measurements with the PHENIX Detector
Bazilevsky, A. (RIKEN BNL Center)

3-2 Spin Content of Proton in Quark-Soliton Model
Musulmanbekov, G. (JINR)

3-3 The Q^2 Dependence of the Generalized Gerasimov-Drell-Hearn Integral for the Proton
Seitz, B. (Univ. of Alberta)

3-4 Measurement of the Gluon Polarization with the PHENIX Muon Arms
Sato, H. (Kyoto Univ.)

3-5 cancelled

3-6 A new approach to parton-density evolution and applications to small-x data analysis of $g_1(x)$
Goshtasbpour, M. (Shahid Beheshti Univ.)

3-7 Measurements of $\bar{u}$ and $\bar{d}$ polarization in the proton at RHIC-PHENIX with polarized proton beams
Kurita, K. (RIKEN BNL Center)

3-8 Spin Physics with a Charged Particle Trigger at PHENIX
Barish, K. (UC Riverside)

3-9 The Measurements of Transversal Hardedness in 3π Diffractive Production
Tkatchev, L.G. (JINR)

4-1 Chiral-Odd Contribution to Single Transverse-Spin Asymmetry in Hadronic Pion Production
Kanazawa, Y. (Niigata Univ.)

4-2 Structure of spin-flip amplitude in the diffraction range
Selugin, O.V. (PNPI)

4-3 Λ polarization in unpolarized hadron collisions
Murgia, F. (INFN, Cagliari)

4-4 Uncertainties of partial wave analyses and experimental data on spin rotaion parameters in the elastic πp-scattering
Svirida, D. (ITEP)

4-5 Study of Heavy Meson Production in NN Collisions with Polarized Beam and Target at COSY
Rathmann, F

4-6 The Observation of Stable Dibaryons
Aslanyan, P.Z. (JINR/Yerevan Univ.)

5-1 Semi-inclusive Λ_C^+ electroproduction inpolarized reactions and polarized gluon distributions
Oyama, S. (Kobe Univ)

5-2 Electron Scattering at TESLA with Polarized and Unpolarized Nucleon Targets
E.C. Aschenauer (ESY-Zeuthen)

5-3 Semi-inclusive $\Lambda_c^+(\bar{\Lambda}_c^+)$ leptoproduction and polarized s-quark distribution
Sudoh, K. (Kobe Univ.)

8-1 Status of the HERMES Atomic Beam Source and Possible Improvements
Steffens, E. (Erlangen)

8-2 How to improve performance of electron pumping polarized ^{3}He ion source - Study on relaxation mechanism of optically pumped Rb vapor
Yonehara, K. (RCNP/Univ. of Michigan)

8-3 A Polarized Beam Facility at ISAC
Levy, P. (TRIUMF)

8-4 Polarization at the NUCLEON
Pilipenko, Yu.K. (JINR)

8-5 Proposal for polarized ^{3}He ion source based on spin-exchange collision
     Arimoto, Y. (JASRI)
8-6 Production of thick CD_2 targets for measurements of the n+d reaction at 250 MeV
     Maeda, Yukie (Univ. of Tokyo)
8-7 Calibration of ^{3}He polarization using electron-spin-resonance method
     Nishikawa, J. (Saitama Univ.)
8-8 The Bochum Polarized Target
     Reicherz, G. (Bochum)
8-9 Polarized Deuteron Target System for Low Energy D(d,p)T Measurement
     Daito, I. (Nagoya Univ.)

PES2000 Program

Thursday, October 12

9:00–9:15	**Opening Address**
	C. Prescott (SLAC)

Session 1) NEA Surface Physics
Chair: Y. Mamaev (St. Petersburg Tech. Univ.)

9:15–9:45	Atomic and Electronic Engineering of p-GaAs-(Cs,O)-vacuum Interface
	A. Terekhov (Institute of Semiconductor Physics)
9:45–10:15	STM, STS and Workfunction Study of Cs/ p-GaAs (110)
	T. Mizoguchi (Gakushu-in Univ.)
10:15–10:35	Longitudinal and Transverse Energy Distributions of Electrons Emitted from GaAs (Cs, O)
	D. Orlov (Max Planck Institute)
10:35–10:55	Cesiumoxide - GaAs Interface and Activation Layer Thickness in NEA Surface Formation - Do We Have to Expand the Dipole Model?
	S. More (IMS-UVSOR)
10:55–11:10	Coffee Break

Session 2) Photocathode Spin Physics
Chair: A. Terekhov (Institute of Semiconductor Physics)

11:10–11:40	Temperature Dependence of Electron Spin Dynamics in Strained Heterostructures
	Y. Mamaev (St. Petersburg Tech. Univ.)
11:40–12:00	Time-resolved Intensity and Polarization Measurements in the Photoemission from Highly Polarized Superlattice Photocathodes
	J. Schuler (Mainz Univ.)
12:00–12:20	Photo-luminescence Study of Superlattice Photocathode
	T. Matsuyama (Osaka Pref. Univ.)
12:20–14:00	Lunch

Session 3) Polarimeter
Chair: T. Maruyama (SLAC)

14:00–14:30	Transmission Sources and Spin Filter Detectors
	G. Lampel (Ecole Polytechnique)
14:30–15:00	5 MeV Mott Polarimeter at Jefferson Lab.
	M. Steigerwald (TJNAF)
15:00–15:15	Coffee Break
15:15–17:15	**Poster Session**
18:30–21:00	**Welcome Party**

Friday, October 13

Session 4) Spin Polarized Beam for High Energy Physics
Chair: D. Shultz (SLAC)

9:00–9:30 Polarized Source Performance and Development at Jefferson Lab
M. Poelker (TJNAF)

9:30–10:00 New Results from the Mainz Polarized Electron Facilities
K. Aulenbacher (Mainz Univ.)

10:00–10:30 New Results from the MIT-Bates Polarized Source and the Test Beam Setup
M. Farkhondeh (MIT-Bates)

10:30–10:50 The 50 kV Inverted Source of Polarized Electrons at ELSA
W. Hillert (Bonn Univ.)

10:50–11:05 Coffee Break

Session 5) Applied Spin Physics
Chair: G. Lampel (Ecole Polytechnique)

11:05–11:35 Spin Polarized Low Energy Electron Microscopy
E. Bauer (Arizona State Univ.)

11:35–11:55 Atomic Force Microscopy of Antiferromagnetic NiO(100) Surfaces and Exchange Force Microscope
K. Sueoka (Hokkaido Univ.)

11:55–12:15 Spin Polarized Inverse Photoemission Spectroscopy at Hiroshima University
A. Kimura (Hiroshima Univ.)

12:15–14:00 Lunch

Session 6) New Spin Technology
Chair: A. Kakizaki (KEK)

14:00–14:20 Application of Micro-Fabricated GaAs Tip to Spin-Polarized Scanning Tunneling Microscope
K. Yamaguchi (Univ. of Electro-Communication)

14:20–14:40 Electron Emission from Na/Fe(100) Surfaces by Deexcitation of Spin-Polarized Helium Metastable Atoms
Y. Yamauchi (National Research Institute for Metals)

Session 7) Development of Polarized Electron Source
Chair: K. Aulenbacher (Mainz Univ.)

14:40–15:10 Investigations of the Charge Limit Phenomenon in GaAs Photocathodes
T. Maruyama (SLAC)

15:10–15:40 Status of Polarized Beams at SLAC
D. Schultz (SLAC)

15:40–15:55 Coffee Break

15:55–16:25 Polarized Electron Source for Japan Linear Collider
K. Togawa (Nagoya Univ.)

16:25–16:45 Present Status of Experimental S-band GaAs-photogun Driven by the Solid State GaAs Pulse Laser
R. Gromov (BINP)

16:45–17:05 Fabrication of Photocathode Test-Stand
G. Kim (Kyungpook National Univ.)

17:05–17:15 **Summary**

Poster Session

A Pulsed Photo-electron Source for Kinematically Complete Collision Experiments
C. D. Shroeter (Freiburg Univ.)
Deporalization Study of ELSA
M. Hoffmann (Bonn Univ.)
Construction of Spin Polarized Inverse Photoemission System
A. Morihara (Hiroshima Univ.)
Electron Gun Design for Spin Polarized Inverse Photoemission Spectroscopy
S. Hasui (Hiroshima Univ.)
Spin Polarized Photoemission at Hiroshima Synchrotron Radiation Center
H. Narita (Hiroshima Univ.)
Surface Photo-voltage Effect on Clean and NEA Surfaces of GaAs and its Superlattice
S. Tanaka (IMS-UVSOR)
Fe Epitaxial Thin Film and Spin-polarized STM with Semiconductor Probe
K.Sueoka (Hokkaido Univ.)
Structure and Magnetism of Fe Thin Films on Rh(001) Studied by Spin-resolved Photoelectron Spectroscopy
A. Kakizaki (KEK)
Electronic Structure and Magnetic Anisotropy of Co/Au(111): a Spin-resolved Photoelectron Spectroscopy Study
A. Kakizaki (KEK)
Reduction of Field Emission Current from Stainless Steel and Copper Surface
C. Suzuki (Nagoya Univ.)
Development of 200 keV Polarized Electron Gun
K. Wada (Nagoya Univ.)
Feasibility Study on Polarized RF-gun; Test of Cesium Telluride Photocathode
F. Furuta (Nagoya Univ.)
Atomic Hydrogen Cleaning for GaAs Surface with Load Lock System
M. Yamamoto (Nagoya Univ.)
Development of Spin Polarized Electron Photocathodes; GaAs-GaAsP Superlattice and Superlattice with DBR
T. Nishitani (Nagoya Univ.)
Fabrication of GaAs-GaAsP Superlattice Photocathode
O. Watanabe (Nagoya Univ.)

SPIN2000 Participants List

J.K. Ahn
RCNP
Osaka University
10-1 Mihogaoka, Ibaraki
Osaka 567-0047
Japan
jkahn@rcnp.osaka-u.ac.jp

M. Amarian
INFN
Sezione di Roma, Gruppo Sanita'
viale Regina Elena
299-00161 Roma
Italy
amarian@hermes.desy.de

M. Anselmino
Univ. of Torino
Dept. of Theoretical Physics
Via Giuria 1, Torino 10125
Italy
anselmino@to.infn.it

Y. Aoki
Univ. of Tsukuba
Tandem Accelerator Center
Tsukuba, Ibaraki 305-8577
Japan
yaoki@tac.tsukuba.ac.jp

Yu. Arestov
IHEP
142281 Protvino, Moscow Region
Russia
arestov@rampex.ihep.su

Y. Arimoto
JASRI
Kouto 1-1-1, Mikazuki, Sayo
Hyogo 679-5198
Japan
arimoto@spring8.or.jp

K. Asahi
Tokyo Inst. of Tech
2-12-1 Ookayama, Meguro-ku
Tokyo 152-8551
Japan
asahi@yap.nucl.ap.titech.ac.jp

Y. Asano
Univ. of Tsukuba
Dept. of Physics
Tsukuba, Ibaraki 305-8577
Japan
asanoy@mail.kek.jp

E.C. Aschenauer
DESY Zeuthen
DESY-HERMES Building 1e/407
Notkestrasse 85, Hamburg, D-22603
Germany
elke@hermes.desy.de

P. Aslanyan
JINR/Yerevan Univ.
Dubna, Moscow Region 141980
Russia
aslanyan@sunhe.jinr.ru

R. Assmann
CERN
SL-Division, Geneva 23, 1211
Switzerland
ralph.assmann@cern.ch

K. Aulenbacher
Univ. of Mainz
Institute for Nuclear Physics
D-55099-Mainz
Germany
aulenbac@kph.uni-mainz.de

T. Averett
Coll. of William and Mary
Dept. of Physics, Williamsburg
VA 23187-8795
USA
averett@physics.wm.edu

Andrew D. Bacher
IUCF
2401 Milo B. Sampson Lane
Bloomington, IN 47408
USA
bacher@iucf.indiana.edu

M. Bai
BNL
Bldg 911B, C-A Department
Upton, NY 11973-5000
USA
mbai@bnl.gov

D.P. Barber
DESY
Notkestrasse 85, Hamburg, D-22607
Germany
mpybar@mail.desy.de

K. Barish
UC Riverside
Department of Physics
CA 92521
USA
kenneth.barish@ucr.edu

F. Bauer
Univ. of Hamburg
I. Institut f. Experimentalphysik
Luruper Chaussee 149
22761 Hamburg
Germany
bauer@kaa.desy.de

A.S. Belov
INR, Academy of Sciences of Russia
60th October Anniversary Prospect, 7a
117312 Moscow
Russia
belov@al20.inr.troitsk.ru

G.P. Berg
RCNP/IUCF
Osaka University
10-1 Mihogaoka, Ibaraki
Osaka 567-0047
Japan
gpberg@rcnp.osaka-u.ac.jp

S. Bernreuther
Tokyo Inst. of Tech.
2-12-1 Ookayama, Meguro-ku
Tokyo 152-8551
Japan
bernreut@hermes.desy.de

Daniel Boer
RIKEN BNL Research Center
BNL, Bldg. 510A
Upton, NY 11973-5000
USA
dboer@bnl.gov

F. Bradamante
University of Trieste
Dipartimento di Fisica
via A. Valerio 2, Trieste
I-34127
Italy
franco.bradamante@ts.infn.it

G. Bunce
BNL
Bldg. 510A
Upton, NY 11973-5000
USA
bunce@bnl.gov

V.V. Burov
RCNP/JINR
Osaka University
10-1 Mihogaoka, Ibaraki
Osaka 567-0047
Japan
burov@rcnp.osaka-u.ac.jp

W.C. Chang
Academia Sinica
Institute of Physics
Taipei, 11529
Taiwan
changwc@phys.sinica.edu.tw

S. Choe
Hiroshima Univ.
Dept. of Physics
Higashi-Hiroshima 739-8526
Japan
schoe@hirohe.hepl.hiroshima-u.ac.jp

M. Conte
Univ. of Genova/INFN
Via Dodecaneso 33, Genova, 16146
Italy
Mario.Conte@ge.infn.it

E.D. Courant
BNL
Bldg. 911B
Upton, NY 11973-5000
USA
courant@bnl.gov

D. Crabb
Univ. of Virginia
Physics Department
382 McCormick Rd.
Charlottesville
VA 22903
USA

I. Daito
Nagoya Univ.
Dept. of Physics, Furo-chou, Chikusa-ku
Nagoya 464-8602
Japan
daito@kiso.phys.nagoya-u.ac.jp

Y.S. Derbenev
Univ. of Michigan
Physics Dept., 500 East University
Ann Arbor MI 48109-1120
USA
derbenev@umich.edu

P. Di Nezza
INFN
INFN-LNF, via E. Fermi 40 Frascati
Rome I-00044
Italy
Pasquale.DiNezza@lnf.infn.it

E.C. Dukes
Univ. of Virginia
High Energy Physics Lab.
382 McCormick Rd.
Charlottesville, VA 22904-4714
USA
dukes@uvahea.phys.Virginia.EDU

A.V. Efremov
Bogolyubov Lab. Theor. Physics, JINR
Dubna
Moscow Region 141980
Russia
efremov@thsun1.jinr.ru

H. Ejiri
RCNP
Osaka University
10-1 Mihogaoka, Ibaraki
Osaka 567-0047
Japan
ejiri@rcnp.osaka-u.ac.jp

H. En'yo
Kyoto Univ.
Dept. of Phys., Kitashirakawa-Oiwake
Kyoto 606-8502
Japan
enyo@pn.scphys.kyoto-u.ac.jp

R. Fatemi
Univ. of Virginia
115-102 Wood Duck Place
Charlottesville,VA 22902
USA
rdh2x@virginia.edu

J. Felix
Univ. of Guanajuato
Instituto de Fisica
Lomas del bosque 103
frac. Lomas del campestre
Leon Guanajuato 37150
Mexico

W. Fetscher
IPP/ETH Zurich
Institute for Particle Physics
c/o Paul Scherrer Institut, WLGA E25
CH-5232 Villigen PSI
Switzerland
wulf.fetscher@psi.ch

G. Fidecaro
CERN
CH-1211 Geneva 23
Switzerland
fideg@mail.cern.ch

M. Finger
Charles University
Faculty of Mathematics and Physics
V Holesovickach 2, Praha 8
Czech Republic
finger@mbox.troja.mff.cuni.cz

H. Fujimura
RCNP
Osaka University
10-1 Mihogaoka, Ibaraki
Osaka 567-0047
Japan
fujimura@rcnp.osaka-u.ac.jp

H. Fujita
Osaka Univ.
Dept. of Phys., 1-1 Machikaneyama
Toyonaka, Osaka 560-0043
Japan
hfujita@lns.sci.osaka-u.ac.jp

Y. Fujita
Osaka Univ.
Dept. of Phys., 1-1 Machikaneyama
Toyonaka, Osaka 560-0043
Japan
fujita@rcnp.osaka-u.ac.jp

M. Fujiwara
RCNP
Osaka University
10-1 Mihogaoka, Ibaraki
Osaka 567-0047
Japan
fujiwara@rcnp.osaka-u.ac.jp

H. Gao
MIT
26-413, 77 Massachusetts Ave.
Cambridge, MA 02139
USA
haiyan@mit.edu

E. Gianfelice
DESY
Notkestrasse 85, Hamburg, D-22607
Germany
mpyeli@mail.desy.de

R. Gilman
Rutgers Univ.
Jefferson Lab.
Newport News, VA 23606
USA
gilman@jlab.org

S. Goertz
Ruhr University Bochum
Institute for Experimantal Physics I
Bldg. NB Room 2/30, Universitaetsstr. 150
D-44780 Bochum
Germany
goertz@ep1.ruhr-uni-bochum.de

S. Gojuki
Science Univ. of Tokyo
Dept. of Physics, 2641 Yamazaki
Noda, 278-8510
Japan
j6299704@ed.noda.sut.ac.jp

A. Gokalp
Middle East Tech. Univ.
Physics Dept., 06531 Ankara
Turkey
agokalp@metu.edu.tr

S. Goloskokov
Bogolyubov Lab. Theor. Physics, JINR
Dubna
Moscow Region 141980
Russia
goloskkv@thsun1.jinr.ru

M. Goshtasbpour
Shahid Behesgti Univ.
Dept. of Physics, Evin, Tehran 19834
Iran
goshtasb@Alborz.sbu.ac.ir

Y. Goto
RIKEN BNL Research Center
BNL, Bldg. 510A
Upton, NY 11973-5000
USA
goto@bnl.gov

M. Gowin
Univ. of Bonn
Physics Institute
Nussallee 12, D-53115 Bonn
Germany
gowin@physik.uni-bonn.de

M. Greenfield
ICU
10-2, Osawa 3-chome, Mitaka
Tokyo 181-8585
Japan
green@icu.ac.jp

M. Grosse Perdekamp
RIKEN BNL Research Center
BNL, Bldg. 510A
Upton, NY 11973-5000
USA
matthias@bnl.gov

P. Guazzoni
Universita' degli Studi
Dipartimento di Fisica
via Celoria 16
I20133 Milano
Italy
PAOLO.GUAZZONI@MI.INFN.IT

W. Guryn
BNL
Bldg. 510C
Upton, NY 11973-5000
USA
guryn@bnl.gov

W. Haeberli
Univ. of Wisconsin
Dept. of Physics
1150 University Ave.
Madison, WI 53706
USA
whaeberli@uwnuc0.physics.wisc.edu

J.-O. Hansen
TJNAF
12000 Jefferson Ave.
Newport News, VA 23606
USA
ole@jlab.org

K. Hara
RCNP
Osaka University
10-1 Mihogaoka, Ibaraki
Osaka 567-0047
Japan
hara@rcnp.osaka-u.ac.jp

T. Hasegawa
Miyazaki Univ.
Faculty of Engineering
1-1
Gakuen-Kibanadai-Nishi, Miyazaki 889-21
Japan
hasegawa@phys.miyazaki-u.ac.jp

K. Hatanaka
RCNP
Osaka University
10-1 Mihogaoka, Ibaraki
Osaka 567-0047
Japan
hatanaka@rcnp.osaka-u.ac.jp

M. Hatano
Univ. of Tokyo
Department of Phys.
Hongo 7-3-1, Bunkyo
Tokyo 113-0033
Japan
hatano@nucl.phys.s.u-tokyo.ac.jp

A. Hayashigaki
Univ. of Tokyo
Department of Phys.
Hongo 7-3-1, Bunkyo
Tokyo 113-0033
Japan
arata@nt.phys.s.u-tokyo.ac.jp

R. Hertenberger
Sektion Physik LMU Muenchen
Am Coulombwall 1
D-85748 Garching
Germany
ralf.hertenberger@physik.uni-muenchen.de

W. Hillert
Univ. of Bonn
Physics Institute
Nussallee 12, D-53115 Bonn
Germany
hillert@physik.uni-bonn.de

F. Hinterberger
ISKP, Univ. of Bonn
Nussallee 14-16
D-53115 Bonn
Germany
fh@iskp.uni-bonn.de

M. Hirai
Saga Univ.
1 Honjyoumati, Saga 840-8502
Japan
98td25@edu.cc.saga-u.ac.jp

Y. Hirayama
Osaka Univ.
Dept. of Phys., 1-16 Machikaneyama
Toyonaka, Osaka 560-0041
Japan
kazu@adam.phys.wani.osaka-u.ac.jp

D. Hirooka
RCNP
Osaka University
10-1 Mihogaoka, Ibaraki
Osaka 567-0047
Japan
hirooka@rcnp.osaka-u.ac.jp

M. Hoffmann
Univ. of Bonn
Physics Institute
Nussallee 12, D-53115 Bonn
Germany
mhoffman@physik.uni-bonn.de

N. Horikawa
Nagoya Univ.
CIRSE
Furo-chou, Chikusa-ku
Nagoya 464-8603
Japan
horikawa@kiso.phys.nagoya-u.ac.jp

T. Hotta
RCNP
Osaka University
10-1 Mihogaoka, Ibaraki
Osaka 567-0047
Japan
hotta@rcnp.osaka-u.ac.jp

E. Hourany
IN2P3, Institut de Physique Nucléaire
91406 Orsay Cedex
France
hourany@ipno.in2p3.fr

H. Huang
BNL
C-A Dept., Bldg. 911B
Upton, NY 11973-5000
USA
huanghai@bnl.gov

V. Hughes
Yale University
Dept. of Physics
New Haven, CT 06520-8121
USA
hughes@hepmail.physics.yale.edu

K. Imai
Kyoto Univ.
Dept. of Phys., Kitashirakawa-Oiwake
Kyoto 606-8502
Japan
imai@nh.scphys.kyoto-u.ac.jp

Y. Iseri
Chiba-Keizai Coll.
Todoroki-cho 4-3-30, Inage-ku
Chiba 263-0021
Japan
iseri@chiba-kc.ac.jp

T. Ishida
Kyushu Univ.
Dept. of Physics
6-10-1 Hakozaki, Higashi-ku
Fukuoka 812-8581
Japan
ishida@kutl.kyushu-u.ac.jp

M. Ishihara
RIKEN
2-1, Hirosawa, Wako
Saitama 351-0198
Japan
ishihara@rikaxp.riken.go.jp

S. Ishikawa
Hosei Univ.
Department of Physics
Fujimi 2-17-1, Chiyoda
Tokyo 102-8160
Japan
ishikawa@i.hosei.ac.jp

T. Ishikawa
Kyoto Univ.
Dept. of Phys., Kitashirakawa-Oiwake
Kyoto 606-8502
Japan
takatugu@ne.scphys.kyoto-u.ac.jp

Takahisa Itahashi
RCNP
Osaka University
10-1 Mihogaoka, Ibaraki
Osaka 567-0047
Japan
itahasi@rcnp.osaka-u.ac.jp

Tetsuro Itahashi
Osaka Univ.
Dept. of Phys., 1-1 Machikaneyama
Toyonaka, Osaka 560-0043
Japan
itahashi@km.phys.sci.osaka-u.ac.jp

M. Itoh
Kyoto Univ.
Dept. of Phys., Kitashirakawa-Oiwake
Kyoto 606-8502
Japan
itoh@ne.scphys.kyoto-u.ac.jp

T. Iwata
Nagoya Univ.
Dept. of Physics, Furo-chou, Chikusa-ku
Nagoya 464-8602
Japan
iwata@kiso.phys.nagoya-u.ac.jp

H. Izumi
Osaka Univ.
Dept. of Phys., 1-16 Machikaneyama
Toyonaka, Osaka 560-0041
Japan
izumi@phys.wani.osaka-u.ac.jp

R.L. Jaffe
MIT
6-311
77 Massachusetts Ave.
Cambridge, MA 02139
USA
jaffe@mit.edu

K. Joo
TJNAF
12000 Jefferson Ave. 12H
Newport News, VA 23606
USA
kjoo@jlab.org

T. Kageya
Univ. of Michigan
Spin Physics Center, C.S.S. Bldg.1239
Kipke Drive Rorm 2341
Ann Arber MI 48109-1010
USA
kageya@umich.edu

H. Kameyama
Chiba-Keizai Coll.
4-3-30 Todoroki, Inage-ku
Chiba 263-0021
Japan
kameyama@chiba-kc.ac.jp

J. Kamiya
RCNP
Osaka University
10-1 Mihogaoka, Ibaraki
Osaka 567-0047
Japan
kamiya@rcnp.osaka-u.ac.jp

Y. Kanazawa
Niigata Univ.
Dept. of Phys., 2 no chou 8050, Ikarashi
Niigata 950-2181
Japan
yasu@nt.sc.niigata-u.ac.jp

T. Katabuchi
Univ. of Tsukuba
Tandem Accelerator Center
Tsukuba, Ibaraki 305-8577
Japan
buchi@tac.tsukuba.ac.jp

A. Katcharava
RCNP/JINR
Osaka University
10-1 Mihogaoka, Ibaraki
Osaka 567-0047
Japan
andro@rcnp.osaka-u.ac.jp

H. Kato
Univ. of Tokyo
Department of Phys.
Hongo 7-3-1, Bunkyo
Tokyo 113-0033
Japan
hkato@nucl.phys.s.u-tokyo.ac.jp

A. Katsuki
Osaka Univ.
Dept. of Phys., 1-1 Machikaneyama
Toyonaka, Osaka 560-0043
Japan
katsuki@km.phys.sci.osaka-u.ac.jp

T. Kawabata
Kyoto Univ.
Dept. of Phys., Kitashirakawa-Oiwake
Kyoto 606-8502
Japan
kawabata@ne.scphys.kyoto-u.ac.jp

G. Kim
Kyungpook National University
Center for High Energy Physics
1370 Sangyek-dong, Puk-ku
Taegu 702-701
Korea
gnkim@postech.ac.kr

E. Kinney
Univ. of Colorado
CB 390 Boulder, CO 80309-0390
USA
edward.kinney@colorado.edu

T. Kishimoto
Osaka Univ.
Dept. of Phys., 1-1 Machikaneyama
Toyonaka, Osaka 560-0043
Japan
kisimoto@km.phys.sci.osaka-u.ac.jp

Y. Kitamura
RCNP
Osaka University
10-1 Mihogaoka, Ibaraki
Osaka 567-0047
Japan
yasuyuki@rcnp.osaka-u.ac.jp

A. Kobushkin
RCNP
Osaka University
10-1 Mihogaoka, Ibaraki
Osaka 567-0047
Japan
akob@rcnp.osaka-u.ac.jp

Y. Koike
Niigata Univ.
Dept. of Physics, Ikarashi
Niigata 950-2181
Japan
koike@nt.sc.niigata-u.ac.jp

H. Kolster
MIT
77 Massachusetts Ave.
Cambridge, MA 02139
USA
hauke@pierre.mit.edu

K. Kondo
Nagoya Univ.
Dept. of Physics, Furo-chou, Chikusa-ku
Nagoya 464-8602
Japan
kao@kiso.phys.nagoya-u.ac.jp

M. Kondo
Wakasa ERC
64-52-1 Nagaya, Tsuruga
Fukui 914-0192
Japan
kondo@werc.or.jp

A.D. Krisch
Univ. of Michigan
Randall Lab of Physics
Ann Arbor, MI 48109-1120
USA
krisch@umich.edu

K.-I. Kubo
TMCAE/TMU
8-52-1 Minami-senju, Arakawa
Tokyo 116-0003
Japan
kubo@comp.metro-u.ac.jp

N. Kudomi
RCNP
Osaka University
10-1 Mihogaoka, Ibaraki
Osaka 567-0047
Japan
kudomi@rcnp.osaka-u.ac.jp

S. Kumano
Saga Univ.
Department of Physics
Honjo-1, Saga 840-8502
Japan
kumanos@cc.saga-u.ac.jp

Y. Kuno
Osaka Univ.
Dept. of Phys., 1-1 Machikaneyama
Toyonaka, Osaka 560-0043
Japan
kuno@phys.sci.osaka-u.ac.jp

K. Kurita
RIKEN BNL Research Center
BNL, Bldg. 510A
Upton, NY 11973-5000
USA
kurita@bnl.gov

V.P. Ladygin
JINR
Dubna, Moscow Region 141980
Russia
ladygin@sunhe.jinr.ru

J. Lee
DESY
Theory Group
Notkestr. 85
D-22603 Hamburg
Germany
jungil@mail.desy.de

T. Lee
MIT
26-402 77 Mass. Ave.
Cambridge, MA 02139
USA
tong@mitlns.mit.edu

A. Lehrach
Forschungszentrum Jülich
Institut für Kernphysik
Postfach 1913
D-52425 Jülich
Germany
a.lehrach@fz-juelich.de

P. Lenisa
INFN - SEZ. DI FERRARA
Via Paradiso, 12
44100 FERRARA
Italy
lenisa@hermes.desy.de

C.D.P. Levy
TRIUMF
4004 Wesbrook Mall
Vancouver, BC V6T 2A3
Canada
levy@triumf.ca

Z.-T. Liang
Shandong Univ.
Dept. of Physics, Jinan
Shandong 250100
China
liang@sdu.edu.cn

A.M.T. Lin
Univ. of Michigan
Physics Dept., 500 East University
Ann Arbor MI 48109-1120
USA
alilin@umich.edu

T. Lindemann
DESY
DESY/HERMES, Building 1e/403
Notkestrasse 85
Hamburg, D-22607
Germany
thore.lindemann@desy.de

D.I. Lowenstein
BNL
C-A Dept.
Bldg. 911B
Upton, NY 11973-5000
USA
Lowenstein@bnl.gov

V.G. Luppov
Univ. of Michigan
Physics Dept., 500 East University
Ann Arbor MI 48109-1120
USA
vluppov@umich.edu

Yoshikazu Maeda
RCNP
Osaka University
10-1 Mihogaoka, Ibaraki
Osaka 567-0047
Japan
ymaeda@rcnp.osaka-u.ac.jp

Yukie Maeda
Univ. of Tokyo
Department of Phys.
Hongo 7-3-1, Bunkyo
Tokyo 113-0033
Japan
yukie@nucl.phys.s.u-tokyo.ac.jp

Y. Makdisi
BNL
Bldg. 911B
Upton, NY 11973-5000
USA
makdisi@bnl.gov

H. Makii
RCNP
Osaka University
10-1 Mihogaoka, Ibaraki
Osaka 567-0047
Japan
makii@rcnp.osaka-u.ac.jp

S. Makino
Wakayama Medical Coll.
Kimiidera 811-1, Wakayama 641-0012
Japan
makino@wakayama-med.ac.jp

G. Mallot
CERN
EP Division
CH-1211 Geneva
Switzerland
gerhard.mallot@cern.ch

A. Manohar
UCSD
Physics Department 0319
9500 Gilman Drive
La Jolla, CA 92093
USA
manohar@ucsd.edu

P.A. Marchetti
Univ. of Padova
Dipartimento di Fisica
Via Marzolo 8 Padova
35131
Italy
pieralberto.marchetti@pd.infn.it

A. Martin
Univ. of Trieste
Dipartimento di Fisica
via A. Valerio 2, Trieste
I-34127
Italy
anna.martin@ts.infn.it

T. Maruyama
SLAC
2575 Sand Hill Rd., Menlo Park
CA 94025
USA
tvm@slac.stanford.edu

A. Masaike
Nara Sangyo University
3-12-1 Tatsuno-kita, Sango-cho, Ikoma-gun
Nara, 636-8503
Japan
masaike@mvb.biglobe.ne.jp

T. Matsumura
RCNP
Osaka University
10-1 Mihogaoka, Ibaraki
Osaka 567-0047
Japan
toru@rcnp.osaka-u.ac.jp

T. Mibe
RCNP
Osaka University
10-1 Mihogaoka, Ibaraki
Osaka 567-0047
Japan
mibe@rcnp.osaka-u.ac.jp

S. Minami
Osaka Univ.
Dept. of Phys., 1-1 Machikaneyama
Toyonaka, Osaka 560-0043
Japan
minami@km.phys.sci.osaka-u.ac.jp

Y. Miyachi
Nagoya Univ.
Dept. of Physics, Furo-chou, Chikusa-ku
Nagoya 464-8602
Japan
miyachi@kiso.phys.nagoya-u.ac.jp

M. Miyake
Osaka Univ.
Dept. of Phys., 1-1 Machikaneyama
Toyonaka, Osaka 560-0043
Japan
miyake@km.phys.sci.osaka-u.ac.jp

M. Miyama
Tokyo Metro. Univ.
Dept. of Phys., 1-1 Minami-Ohsawa, Hachioji
Tokyo 192-0397
Japan
miyama@comp.metro-u.ac.jp

X. Morelle
Paul Scherrer Institute
OLGA 017, CH-5232 Villigen PSI
Switzerland
xavier.morelle@psi.ch

Y. Mori
KEK
1-1 Oho, Tsukuba, Ibaraki 305-0801
Japan
yoshiharu.mori@kek.jp

T. Morii
Kobe Univ.
Faculty of Human Development
3-11 Tsurukabuto, Nada
Kobe 657-8501
Japan
morii@kobe-u.ac.jp

M. Morita
RCNP
Osaka University
10-1 Mihogaoka, Ibaraki
Osaka 567-0047
Japan
mmorita@rcnp.osaka-u.ac.jp

N. Muramatsu
RCNP
Osaka University
10-1 Mihogaoka, Ibaraki
Osaka 567-0047
Japan
mura@rcnp.osaka-u.ac.jp

J. Murata
RIKEN
Hirosawa 2-1, Wako
Saitama 351-0198
Japan
jiro@bnl.gov

F. Murgia
INFN, Sezione di Cagliari
C.P. 170, Monserrato (CA)
I-09042
Italy
francesco.murgia@ca.infn.it

R. Muto
Kyoto Univ.
Dept. of Phys., Kitashirakawa-Oiwake
Kyoto 606-8502
Japan
muto@nh.scphys.kyoto-u.ac.jp

Y. Nagai
RCNP
Osaka University
10-1 Mihogaoka, Ibaraki
Osaka 567-0047
Japan
nagai@rcnp.osaka-u.ac.jp

T. Nagao
Osaka Univ.
Dept. of Phys., 1-1 Machikaneyama
Toyonaka, Osaka 560-0043
Japan
nagao@km.phys.sci.osaka-u.ac.jp

Y. Nagashima
Osaka Univ.
Dept. of Phys., 1-1 Machikaneyama
Toyonaka, Osaka 560-0043
Japan
naga@hep.sci.osaka-u.ac.jp

M. Nakamura
Kyoto Univ.
Dept. of Phys., Kitashirakawa-Oiwake
Kyoto 606-8502
Japan
nakamura@nh.scphys.kyoto-u.ac.jp

T. Nakanishi
Nagoya Univ.
Dept. of Physics, Furo-cho Chikusa-ku
Nagoya, 464-8602
Japan
nakanisi@spin.phys.nagoya-u.ac.jp

T. Nakano
RCNP
Osaka University
10-1 Mihogaoka, Ibaraki
Osaka 567-0047
Japan
nakano@rcnp.osaka-u.ac.jp

D.V. Naumov
Laboratory of Nuclear Problems, JINR
Dubna
Moscow Region 141980
Russia
naumov@thsun1.jinr.ru

S. Ninomiya
RCNP
Osaka University
10-1 Mihogaoka, Ibaraki
Osaka 567-0047
Japan
ninomiya@rcnp.osaka-u.ac.jp

T. Noro
RCNP
Osaka University
10-1 Mihogaoka, Ibaraki
Osaka 567-0047
Japan
noro@rcnp.osaka-u.ac.jp

S. Nurushev
IHEP
142281 Protvino, Moscow Region
Russia
nurushev@mx.ihep.su

E. Obayashi
RCNP
Osaka University
10-1 Mihogaoka, Ibaraki
Osaka 567-0047
Japan
obayashi@rcnp.osaka-u.ac.jp

K. Ogata
Kyushu Univ.
Dept. of Physics
6-10-1 Hakozaki, Higashi-ku
Fukuoka 812-8581
Japan
kazu2scp@mbox.nc.kyushu-u.ac.jp

A. Ogawa
Penn State
c/o Cora Feliciano, Bldg. 510A
BNL, Upton
NY 11973-5000
USA
akio@bnl.gov

H. Ogawa
Tokyo Inst. of Tech.
2-12-1 Ookayama, Meguro-ku
Tokyo 152-8551
Japan
hogawa@rarfaxp.riken.go.jp

Y. Ohashi
LEPS
JASRI, Kouto 1-1-1, Mikazuki, Sayo
Hyogo 679-5198
Japan
ohashi@spring8.or.jp

T. Ohnishi
RIKEN
2-1, Hirosawa, Wako
Saitama 351-0198
Japan
oonishi@rarfaxp.riken.go.jp

H. Okamura
Saitama Univ.
Dept. of Physics, Shimo-okubo 255, Urawa
Saitama 338-8570
Japan
okamura@phy.saitama-u.ac.jp

M. Okamura
RIKEN
2-1, Hirosawa, Wako
Saitama 351-0198
Japan
mokamura@postman.riken.go.jp

S. Okumi
Nagoya Univ.
Dept. of Physics, Furo-cho, Chikusa-ku
Nagoya 464-8602
Japan
okumi@spin.phys.nagoya-u.ac.jp

N. Okumura
Univ. of Tsukuba
Tandem Accelerator Center
Tsukuba, Ibaraki 305-8577
Japan
nori@tac.tsukuba.ac.jp

D. Orlov
MPI für Kernphysik
Saupfercheckweg 1
69117 Heidelberg
Germany
Dmitry.Orlov@mpi-hd.mpg.de

S. Oryu
Science Univ. of Tokyo
Dept. of Physics, 2641 Yamazaki, Noda, 278-8510
Japan
oryu@ph.noda.sut.ac.jp

S. Oyama
Kobe Univ.
Faculty of Human Development
3-11 Tsurukabuto, Nada
Kobe 657-8501
Japan
satoshi@radix.h.kobe-u.ac.jp

Yu. Pilipenko
JINR
Dubna, Moscow Region 141980
Russia
pilipen@sunhe.jinr.ru

Yu.A. Plis
JINR
Dubna, Moscow Region 141980
Russia
plis@nusun.jinr.ru

B. Poelker
TJNAF
12000 Jefferson Ave.
Newport News, VA 23606
USA
poelker@jlab.org

R. Prepost
Univ. of Wisconsin
Dept. of Physics, 1150 University Av.
Madison, WI 53706
USA
prepost@hep.physics.wisc.edu

C. Prescott
SLAC
MS78
P. O. Box 4349
Stanford, CA 94309
USA
prescott@slac.stanford.edu

T. Pussieux
CEA Saclay
DAPNIA/SPhN, F-91191 Gif sur Yvette
France
pussieux@cea.fr

G. Ramsey
Loyola Univ.
Physics Dept.
6525 N. Sheridan
Chicago, IL 60626
USA
gpr@hep.anl.gov

P.G. Ratcliffe
Univ. dell'Insubria
Dip. di Scienze, via Lucini 3
22100 Como
Italy
pgr@fis.unico.it

G. Reicherz
Ruhr-Universitat Bochum
Universitatsstr. 150, D-44780 Bochum
Germany
reicherz@ep1.ruhr-uni-bochum.de

T. Roser
BNL
Bldg. 911B
Upton, NY 11973-5000
USA
roser@bnl.gov

D. Ryckbosch
Univ. of Gent
Proeftuinstaat 86, B-9000 Gent
Belgium
Dirk.Ryckbosch@rug.ac.be

V.L. Rykov
Wayne State Univ.
Dept. of Physics and Astronomy
666 W. Hancock, Detroit
MI 48202
USA
rykov@physics.wayne.edu

K. Sagara
Kyushu Univ.
Dept. of Physics
6-10-1 Hakozaki, Higashi-ku
Fukuoka 812-8581
Japan
sagara@kutl.kyushu-u.ac.jp

N. Saito
RIKEN BNL Research Center
BNL, Bldg. 510A
Upton, NY 11973-5000
USA
saito@bnl.gov

T. Saito
RCNP
Osaka University
10-1 Mihogaoka, Ibaraki
Osaka 567-0047
Japan
saito@rcnp.osaka-u.ac.jp

A. Sakaguchi
Osaka Univ.
Dept. of Phys., 1-1 Machikaneyama
Toyonaka, Osaka 560-0043
Japan
sakaguch@km.phys.sci.osaka-u.ac.jp

H. Sakaguchi
Kyoto Univ.
Dept. of Phys., Kitashirakawa-Oiwake
Kyoto 606-8502
Japan
sakaguchi@ne.scphys.kyoto-u.ac.jp

Hideyuki Sakai
Univ. of Tokyo
Department of Phys.
Hongo 7-3-1, Bunkyo
Tokyo 113-0033
Japan
sakai@phys.s.u-tokyo.ac.jp

Hitoshi Sakai
Osaka Univ.
Dept. of Phys., 1-1 Machikaneyama
Toyonaka, Osaka 560-0043
Japan
sakai@km.phys.sci.osaka-u.ac.jp

K. Sakai
Tokyo Inst. of Tech.
2-12-1 Ookayama, Meguro-ku
Tokyo 152-8551
Japan
sakai@yap.nucl.ap.titech.ac.jp

Y. Sakemi
Tokyo Inst. of Tech.
2-12-1 Ookayama, Meguro-ku
Tokyo 152-8551
Japan
sakemi@nucl.phys.titech.ac.jp

S. Sakoda
Univ. of Tokyo
Department of Phys.
Hongo 7-3-1, Bunkyo
Tokyo 113-0033
Japan
sakoda@nucl.phys.s.u-tokyo.ac.jp

E. Sano
RCNP
Osaka University
10-1 Mihogaoka, Ibaraki
Osaka 567-0047
Japan
esano@rcnp.osaka-u.ac.jp

K. Sasaki
Yokohama Nat'l Univ.
Dept. of Physics, 79-5 Tokiwadai, Hodogaya-ku
Yokohama 240-8501
Japan
sasaki@phys.ynu.ac.jp

H. Sato
Kyoto Univ.
BNL, Bldg. 510A
Upton, NY 11973-5000
USA
satohiro@bnl.gov

Y. Satou
RIKEN
2-1, Hirosawa, Wako
Saitama 351-0198
Japan
ysatou@rikaxp.riken.go.jp

P. Schmelzbach
Paul Scherrer Institute
Accelerator Division, WBGA/C19
CH-5232 Villigen PSI
Switzerland
Pierre.Schmelzbach@psi.ch

M. Segawa
RCNP
Osaka University
10-1 Mihogaoka, Ibaraki
Osaka 567-0047
Japan
segawa@rcnp.osaka-u.ac.jp

B. Seitz
Univ. of Alberta
Dept. of Physics
Edmonton, Alberta T6G 2J1
Canada
bjoern.seitz@desy.de

K. Sekiguchi
Univ. of Tokyo
Department of Phys.
Hongo 7-3-1, Bunkyo
Tokyo 113-0033
Japan
kimiko@nucl.phys.s.u-tokyo.ac.jp

H. Seyfarth
Forschungszentrum Jülich
Institut für Kernphysik
D-52425 Jülich
Germany
h.seyfarth@fz-juelich.de

G. Shapiro
Univ. of California
Physics Dept.
Berkeley, CA 94720-7300
USA
g_shapiro@lbl.gov

Yu.M. Shatunov
Budker INP
11 Lavrentyev prospect
Novosibirsk 630090
Russia
shatunov@inp.nsk.su

T.-A. Shibata
Tokyo Inst. of Tech.
Department of Physics
2-12-1 Ookayama
Meguro-ku Tokyo 152-8551
Japan
shibata@nucl.phys.titech.ac.jp

T. Shima
RCNP
Osaka University
10-1 Mihogaoka, Ibaraki
Osaka 567-0047
Japan
shima@rcnp.osaka-u.ac.jp

Y. Shimbara
Osaka Univ.
Dept. of Phys., 1-1 Machikaneyama
Toyonaka, Osaka 560-0043
Japan
shimbara@lns.sci.osaka-u.ac.jp

H. Shimizu
RCNP
Osaka University
10-1 Mihogaoka, Ibaraki
Osaka 567-0047
Japan
hshimizu@rcnp.osaka-u.ac.jp

Y. Shimizu
RCNP
Osaka University
10-1 Mihogaoka, Ibaraki
Osaka 567-0047
Japan
yshimizu@rcnp.osaka-u.ac.jp

T. Shimoda
Osaka Univ.
Dept. of Phys., 1-16 Machikaneyama
Toyonaka, Osaka 560-0041
Japan
shimoda@rcnp.osaka-u.ac.jp

D. Sivers
Portland Physics Inst.
4730 SW Macadam Portland
OR 97201
USA
densivers@sivers.com

J. Soffer
Centre de Physique Theorique, CNRS
Luminy Case 907
13288 Marseille Cedex 09
France
soffer@cpt.univ-mrs.fr

V.L. Solovianov
IHEP
142281 Protvino, Moscow Region
Russia
solovianov@mx.ihep.su

L. Sonnenschein
RWTH Aachen
III Phys. Inst.
D-52056 Aachen
Switzerland
Lars.Sonnenschein@cern.ch

P. A. Souder
Syracuse Univ.
201 Physics Building
Syracuse, NY 13244-1130
USA
souder@phy.syr.edu

T. Specuner
University of Erlangen-Nuernberg
Physikalisches Institut
Erwin-Rommel Str. 1
91058 Erlangen
Germany
thorsten.speckner@physik.uni-erlangen.de

H. Spinka
ANL
Bldg. 362-HEP, 9700 S.
Cass Avenue Argonne, IL 60439
USA
hms@anl.gov

E. Steffens
Univ. of Erlangen
Phys. Inst., Erwin-Rommel-Str. 1
Erlangen D-91058
Germany
steffens@physik.uni-erlangen.de

M. Steigerwald
Michael Steigerwald
Jochgasse 45
73434 Aalen
Germany
m.steigerwald@zeiss.de

E. Stephenson
Indiana Univ.
IUCF, 2401 Milo B. Sampson Lane
Bloomington, IN 47408
USA
stephenson@iucf.indiana.edu

V.N. Stibunov
Nuclear Physics Institute
at Tomsk Polytechnical Univ.
Lenin Avenue 2 A
Tomsk, 634050
Russia
stib@npi.tpu.ru

U. Stoesslein
University of Colorado
Nuclear Physics Lab
Boulder, CO 80309-0446
USA
uta.stoesslein@ifh.de

K. Suda
Saitama Univ.
Dept. of Physics, Shimo-okubo 255, Urawa
Saitama 338-8570
Japan
suda@ne.phy.saitama-u.ac.jp

Katsuhiko Suzuki
Univ. of Tokyo
Department of Phys.
Hongo 7-3-1, Bunkyo
Tokyo 113-0033
Japan
ksuzuki@nt.phys.s.u-tokyo.ac.jp

T. Suzuki
Tokyo Metro. Univ.
Dept. of Physics, 1-1 Minami-Ohsawa, Hachioji
Tokyo 192-0397
Japan
suzukit@phys.metro-u.ac.jp

D. Svirida
ITEP
B. Cheremushkinskaya, 25
Moscow 117259
Russia
Dmitry.Svirida@itep.ru

Y. Tagishi
Univ. of Tsukuba
Tandem Accelerator Center
Tsukuba, Ibaraki 305-8577
Japan
tagishi@tac.tsukuba.ac.jp

N. Takahashi
Osaka Univ./Osaka Gakuin U.
Kishibe-Minami 2-36-1, Suita
Osaka 564-8511
Japan
ntakahas@utc.osaka-gu.ac.jp

K. Takahisa
RCNP
Osaka University
10-1 Mihogaoka, Ibaraki
Osaka 567-0047
Japan
takahisa@rcnp.osaka-u.ac.jp

H. Takeda
Kyoto Univ.
Dept. of Phys., Kitashirakawa-Oiwake
Kyoto 606-8502
Japan
takeda@nh.scphys.kyoto-u.ac.jp

A. Tamii
Univ. of Tokyo
Department of Phys.
Hongo 7-3-1, Bunkyo
Tokyo 113-0033
Japan
tamii@phys.s.u-tokyo.ac.jp

M. Tanaka
Kobe Tokiwa Coll.
Ohtani-cho 2-6-2, Nagata-ku, Kobe 653-0838
Japan
tanaka@rcnp.osaka-u.ac.jp

M. Tanifuji
Hosei Univ.
Department of Physics
Fujimi 2-17-1, Chiyoda
Tokyo 102-8160
Japan

M.J. Tannenbaum
BNL
Physics, Bldg. 510C
Upton, NY 11973-5000
USA
mjt@bnl.gov

A. Titov
Bogolyubov Lab. Theor. Physics, JINR
Dubna
Moscow Region 141980
Russia
atitov@thsun1.jinr.ru

L. Tkatchev
JINR
Dubna, Moscow Region 141980
Russia
tkatchev@vxjinr.jinr.ru

H. Togawa
RCNP
Osaka University
10-1 Mihogaoka, Ibaraki
Osaka 567-0047
Japan
togawa@rcnp.osaka-u.ac.jp

J. Tojo
Kyoto Univ.
Dept. of Phys., Kitashirakawa-Oiwake
Kyoto 606-8502
Japan
tojo@bnl.gov

H. Toki
RCNP
Osaka University
10-1 Mihogaoka, Ibaraki
Osaka 567-0047
Japan
toki@rcnp.osaka-u.ac.jp

K. Tokushuku
KEK
1-1 Oho, Tsukuba, Ibaraki 305-0801
Japan
katsuo.tokushuku@kek.jp

A. Tomyo
RCNP
Osaka University
10-1 Mihogaoka, Ibaraki
Osaka 567-0047
Japan
tomyo@rcnp.osaka-u.ac.jp

D.K. Toporkov
Budker INP
11 Lavrentyev prospect
Novosibirsk 630090
Russia
D.K.Toporkov@inp.nsk.su

M. Uchida
Kyoto Univ.
Dept. of Phys., Kitashirakawa-Oiwake
Kyoto 606-8502
Japan
uchida@nh.scphys.kyoto-u.ac.jp

T. Uematsu
FIHS, Kyoto Univ.
Yoshida Nihonmatsu-cho
Kyoto 606-8501
Japan
uematsu@phys.h.kyoto-u.ac.jp

T. Uesaka
Saitama Univ.
Dept. of Physics, Shimo-okubo 255, Urawa
Saitama 338-8570
Japan
uesaka@phy.saitama-u.ac.jp

E. Uzu
Science Univ. of Tokyo
Dept. of Physics
Fac. of Science and Technology
2641 Yamazaki, Noda
Chiba 278-8510
Japan

B. Van Der Ventel
Univ. of Stellenbosch
Dept. of Physics, Private Bag X1 Matieland
7602 Stellenbosch
South Africa
ventel@physics.sun.ac.za

P. Vetter
LBL
MS 88-205
Berkley, CA 94720
USA
pvetter@lbl.gov

S.E. Vigdor
Indiana Univ.
IUCF, 2401 Milo B. Sampson Lane
Bloomington, IN 47408
USA
vigdor@iucf.indiana.edu

W. Vogelsang
RIKEN BNL Research Center
BNL, Bldg. 510A
Upton, NY 11973-5000
USA
wvogelsang@bnl.gov

E. Voutier
ISN
53 avenue des Martyrs, Grenoble
38026 cedex
France
Voutier@isn.in2p3.fr

A. Wakai
Nagoya Univ.
CIRSE
Furo-chou, Chikusa-ku
Nagoya 464-8603
Japan
wakaia@kiso.phys.nagoya-u.ac.jp

M. Wakamatsu
Osaka Univ.
Dept. of Phys., 1-1 Machikaneyama
Toyonaka, Osaka 560-0043
Japan
wakamatu@rcnp.osaka-u.ac.jp

T. Wakasa
RCNP
Osaka University
10-1 Mihogaoka, Ibaraki
Osaka 567-0047
Japan
wakasa@rcnp.osaka-u.ac.jp

T. Wakui
RIKEN
2-1, Hirosawa, Wako
Saitama 351-0198
Japan
wakui@rarfaxp.riken.go.jp

Xiangdong Wei
BNL
1-233 Bldg. 510A
Upton, NY 11973-5000
USA
xwei@bnl.gov

U. Weigel
MPI für Kernphysik
Saupfercheckweg 1
69117 Heidelberg
Germany
Udo.Weigel@mpi-hd.mpg.de

T. Wise
Univ. of Wisconsin
Dept. of Physics, 1150 University Av.
Madison, WI 53706
USA
wise@uwnuc0.physics.wisc.edu

S.W. Wissink
Indiana Univ.
IUCF, 2401 Milo B. Sampson Lane
Bloomington, IN 47408
USA
wissink@iucf.indiana.edu

H. Witala
Jagellonian Univ.
Inst. of Phys., ul. Reymonta 4
PL-30059 Cracow
Poland
witala@if.uj.edu.pl

B. Wojtsekhowski
TJNAF
12000 Jefferson Ave.
Newport News, VA 23606
USA
bogdanw@jlab.org

S. Wyngaardt
Univ. of Stellenbosch
Dept. of Physics, 7600 Stellenbosch
South Africa
smw@land.sun.ac.za

M. Xiao
Fermilab
MS220, P.O.Box 500
Batavia, IL 60510
USA
meiqin@fnal.gov

T. Yagita
Kyushu Univ.
Dept. of Physics
6-10-1 Hakozaki, Higashi-ku
Fukuoka 812-8581
Japan
yagita@kutl.kyushu-u.ac.jp

K. Yako
Univ. of Tokyo
Department of Phys.
Hongo 7-3-1, Bunkyo
Tokyo 113-0033
Japan
yakou@nucl.phys.s.u-tokyo.ac.jp

S. Yamada
KEK
1-1 Oho, Tsukuba, Ibaraki 305-0801
Japan
sakue.yamada@kek.jp

M. Yamaguchi
Univ. of Tsukuba
Tandem Accelerator Center
Tsukuba, Ibaraki 305-8577
Japan
honey@tac.tsukuba.ac.jp

Y. Yamaguchi
Univ. of Tokyo
2-22-11 Hikawadai, Higashi-kurume
Tokyo 203-0004
Japan

T. Yamanishi
Fukui Univ. of Tech.
3-6-1, Gakuen, Fukui 910-8505
Japan
yamanisi@ccmails.fukui-ut.ac.jp

Y. Yasuda
Kyoto Univ.
Dept. of Phys., Kitashirakawa-Oiwake
Kyoto 606-8502
Japan
yuusuke@nh.scphys.kyoto-u.ac.jp

Y. Yasui
Radiation Lab., RIKEN
2-1, Hirosawa, Wako
Saitama 351-0198
Japan
yasui@taro02.riken.go.jp

D. Yokoyama
Osaka Univ.
Dept. of Phys., 1-1 Machikaneyama
Toyonaka, Osaka 560-0043
Japan
yokoyama@km.phys.sci.osaka-u.ac.jp

K. Yonehara
Univ. of Michigan
CSSB, 1239 Kipke Dr.
Ann Arbor, MI 48109-2036
USA
yonehara@umich.edu

H. Yoshida
RCNP
Osaka University
10-1 Mihogaoka, Ibaraki
Osaka 567-0047
Japan
hidetomo@rcnp.osaka-u.ac.jp

S. Yoshida
RCNP
Osaka University
10-1 Mihogaoka, Ibaraki
Osaka 567-0047
Japan
sei@rcnp.osaka-u.ac.jp

A. Yoshimi
RIKEN
2-1, Hirosawa, Wako
Saitama 351-0198
Japan
yoshimi@rarfaxp.riken.go.jp

M. Yoshimura
RCNP
Osaka University
10-1 Mihogaoka, Ibaraki
Osaka 567-0047
Japan
yoshimur@rcnp.osaka-u.ac.jp

M. Yosoi
RCNP
Osaka University
10-1 Mihogaoka, Ibaraki
Osaka 567-0047
Japan
yosoi@rcnp.osaka-u.ac.jp

G. Zeitler
University of Erlangen-Nuernberg
Physikalisches Institut
Erwin-Rommel Str. 1
91058 Erlangen
Germany
guenter.zeitler@physik.uni-erlangen.de

A.N. Zelenski
BNL
Bldg. 930
Upton, NY 11973-5000
Canada
zelenski@triumf.ca

L. Zetta
Universita' di Milano
Dipartimento Di Fisica
Via Celoria 16, Milano
I-20133
Italy
LUISA.ZETTA@MI.INFN.IT

L.S. Zolin
JINR
Dubna, Moscow Region 141980
Russia
zolin@moonhe.jinr.ru

L.D. van Buuren
NIKHEF/VU Amsterdam
Kruislaan 409 Amsterdam
1098 SJ Amsterdam
Holland
buuren@nikhef.nl

Willem T.H. van Oers
Univ. of Manitoba
Dept. of Physics and Astronomy
Winnipeg, MB, Q3T 2N2
Canada
vanoers@physics.umanitoba.ca

B. van den Brandt
Paul Scherrer Institute
WLGA-D23
CH-5232 Villigen PSI
Switzerland
Ben.vandenBrandt@psi.ch

W. von Drachenfels
Univ. of Bonn
Physics Institute
Nussallee 12, D-53115 Bonn
Germany
drachen@physik.uni-bonn.de

PES2000 Workshop Participants List

<Country>
Name / Institute / E-mail Address

<France>
T. Pussieux / CEA Saclay / pussieux@cea.fr
G. Lampel / Ecole Polyechnique / georges.lampel@polytechnique.fr

<Germany>
W. Drachenfels / Bonn Univ. / drachen@physik.uni-bonn.de
M. Gowin / Bonn Univ. / gowin@physik.uni-bonn.de
W. Hillert / Bonn Univ. / hillert@physik.uni-bonn.de
M. Hoffman / Bonn Univ. / m.hoffmann@uni-bonn.de
C. Schroeter / Freiburg Univ. / schroecd@uni-freiburg.de
K. Aulenbacher / Mainz Univ. / aulenbac@kph.uni-mainz.de
J. Schuler / Mainz Univ. / jschuler@mail.uni-mainz.de
D. Orlov / Max Planck Institute / Dmitry.Orlov@mpi-hd.mpg.de
U. Weigel / Max Planck Institute / Udo.Weigel@mpi-hd.mpg.de

<Korea>
G. Kim / Kyungpook National Univ. / gnkim@postech.ac.kr

<Russia>
R. Gromov / BINP / gromov@inp.nsk.su
A. Terekhov / Institute of Semiconductor Physics / terek@thermo.isp.nsc.ru
Y. Mamaev / St. Petersburg Tech. Univ. / mamaev@spes.stu.neva.ru

<U.S.A>
E. Bauer /Arizona state Univ. / ernst.bauer@asu.edu
D. Schultz / SLAC / DCS@SLAC.Stanford.edu
T. Maruyama / SLAC / tvm@SLAC.Stanford.EDU
M. Poelker / TJNAF / poelker@jlab.org
M. Steigerwald / TJNAF / steigerw@jlab.org

<Japan>
T. Saka / Daido Inst. Tech. / saka@daido-it.ac.jp
J. Fujii / Gakushu-in Univ. / 940143@gakushuin.ac.jp
T. Mizoguchi / Gakushu-in Univ. / tadashi.mizoguchi@gakushuin.ac.jp
O. Morimoto / Graduate Univ. for Advanced Studies / morihon@post.kek.jp
S. Hasui / Hiroshima Univ. / hasui@hiroshima-u.ac.jp
A. Kimura / Hiroshima Univ. / akimura@hisor.material.sci.hiroshima-u.ac.jp
E. Kotani / Hiroshima Univ. / m1279007@hiroshima-u.ac.jp
A. Morihara / Hiroshima Univ. / atsushi@hisor.material.sci.hiroshima-u.ac.jp
H. Namatame / Hiroshima Univ. / namatame@hisor.material.sci.hiroshima-u.ac.jp
H. Narita / Hiroshima Univ. / narimac@hiroshima-u.ac.jp
Q. Shan / Hiroshima Univ. / qiao@hisor.material.sci.hiroshima-u.ac.jp
K. Shimada / Hiroshima Univ. / kshimada@hisor.material.sci.hiroshima-u.ac.jp
K. Sueoka / Hokkaido Univ. / sueoka@nano.hokudai.ac.jp
M. Kamada / IMS-UVSOR / kamada@ims.ac.jp
S. More / IMS-UVSOR / more@ims.ac.jp

K. Takahashi / IMS-UVSOR / ktakahashi@ims.ac.jp
S. Tanaka / IMS-UVSOR / senku@ims.ac.jp
Y. Yamazaki / Japan Nuclear CycleDevelop Institute / yamazaki@oec.jnc.go.jp
A. Kakizaki / KEK / akito.kakizaki@kek.jp
M. Yoshioka / KEK / masakazu.yoshioka@kek.jp
Y. Yamauchi / National Research Institute for Metals / yamauchi@nrim.go.jp
H. Horinaka / Osaka pref. Univ. / horinaka@pe.osakafu-u.ac.jp
T. Matsuyama / Osaka pref. Univ. / matsu@pe.osakafu-u.ac.jp
K. Wada / Osaka pref. Univ. / wada@pe.osakafu-u.ac.jp
K. Yamaguchi / Univ. of Electro-Communications / kyama@ee.uec.ac.jp
K. Hayashi / Univ. of Tokyo / hayashik@sakura.issp.u-tokyo.ac.jp
F. Furuta / Nagoya Univ. / furuta@spin.phys.nagoya-u.ac.jp
H. Kobayakawa / Nagoya Univ. / kobayaka@numse.nagoya-u.ac.jp
T. Nakanishi / Nagoya Univ. / nakanisi@spin.phys.nagoya-u.ac.jp
T. Nishitani / Nagoya Univ. / nisitani@spin.phys.nagoya-u.ac.jp
S. Okumi / Nagoya Univ. / okumi@spin.phys.nagoya-u.ac.jp
C. Suzuki / Nagoya Univ. / chihiro@spin.phys.nagoya-u.ac.jp
K. Togawa / Nagoya Univ. / togawa@spin.phys.nagoya-u.ac.jp
K. Wada / Nagoya Univ. / wada@spin.phys.nagoya-u.ac.jp
O. Watanabe / Nagoya Univ. / watanabe@numse.nagoya-u.ac.jp
M. Yamamoto / Nagoya Univ. / yamamoto@spin.phys.nagoya-u.ac.jp

AUTHOR INDEX

A

Adachi, T., 884, 886
Adderley, P., 943
Afanasief, S. V., 689, 889
Ajaka, J., 198
Akimune, H., 604, 639, 649, 765
Akiyoshi, H., 709, 888, 894
Alarcon, R., 591
Alekseev, I., 790, 795
Alexandrov, A. V., 988
Allgower, C., 534
Amarian, M., 428
Ambrajei, A. N., 920
Andreev, V. E., 901
Anferov, V. A., 736, 893
Angelov, V., 895
Anghinolfi, M., 198
Anselmino, M., 571, 880
Aoi, N., 109, 679
Aoki, Y., 694
Arenhövel, H., 704
Arestov, Y. I., 468
Arimoto, Y., 841
Arkhipov, V. V., 689, 889
Asahi, K., 109, 353, 679, 877
Asai, M., 811
Asano, K., 1009
Asano, Y., 881
Aschenauer, E. C., 24, 883
Aslanyan, P. Z., 882
Assafiri, Y., 198
Assmann, R., 169
Aulenbacher, K., 926, 949
Averett, T., 412
Azhgirey, L. S., 689

B

Baba, T., 926, 930, 982
Babacan, H., 551
Babakan, T., 551
Bacher, A. D., 614, 886
Badier, J., 169
Bai, M., 534, 741, 790
Bakarov, A. K., 1021

Bakin, V. V., 901
Barber, D. P., 751, 780, 889
Bartalini, O., 198
Bassalleck, B., 790
Battaglieri, M., 198
Baturine, V., 534
Bauer, E., 965
Bauer, T. S., 591
Bauer, F., 529
Belikov, N. I., 534
Bellini, V., 198
Belostotsky, S. L., 704
Belov, A. S., 835, 895
Berg, G. P. A., 614, 884, 886
Berglund, M., 780
Bernreuther, S., 504
Blasi, N., 609
Blinov, B. B., 736, 856, 893
Blondel, A., 169
Bocquet, J. P., 198
Bodek, K., 366
Boer, D., 571, 880
Boersma, D., 591
Bogdanov, A. A., 524
Böge, M., 169
Bondarev, V. K., 689, 889
Borisov, N. S., 856
Borzounov, Y. T., 689
Bradamante, F., 34
Brown, B. A., 884
Brüggemann, B., 830
Budzanowski, A., 366
Bulten, H. J., 591
Bültmann, S., 819
Bunce, G., 46, 534, 790, 795
Bunyatova, E. I., 866

C

Calvat, P., 198
Camera, F., 609
Cameron, P., 785
Capogni, M., 198
Castoldi, M., 198
Cho, M. H., 992
Choi, S., 599

Chu, C. M., 736, 893
Clark, J., 943
Clendenin, J. E., 976
Conte, M., 785
Corvisiero, P., 198
Courant, E. D., 323
Court, G., 819
Cowley, A. A., 634
Crabb, D. G., 819
Crozon, M., 169

D

Daito, I., 639, 765, 889, 896
D'Alesio, U., 571, 880
D'Angelo, D., 198
Danneberg, N., 366, 371
Day, A., 943
Day, D. B., 819
Dehning, B., 169
de Huu, M. A., 609
Deipenwisch, J., 996
de Jager, C. W., 704
De Kock, P. R., 624
Derbenev, Y. S., 736, 893
Derevschikov, A. A., 534
Deshpande, A., 790, 795
Deutsch, J., 371
de Vries, H., 591, 704
Didelez, J. P., 198
Dikansky, N. S., 988
Di Nezza, P., 417
DiSalvo, R., 198
Dmitriev, V. F., 704
Dorn, A., 996
Doskow, J., 790
Doumoto, E., 811
Doushita, H., 896
Doushita, N., 889
Drachenfels, W. v., 756
Dupák, J., 891
Düren, M., 882
Duval, M. A., 198
Dyug, M. V., 586, 704

E

Efremov, A. V., 422, 880
Egger, J., 371
Eilerts, S., 790

Eisermann, Y., 664, 825, 888
Ejiri, H., 0, 338, 343, 881, 884
Ellingh, F., 609
Ellinghaus, F., 883
Ellison, J., 889
Emmerich, R., 830
Engels, R., 830
En'yo, H., 534
Ershov, V. P., 895
Esin, S. K., 835
Euteneuer, H., 949

F

Farkhondeh, M., 955
Fatemi, R., 402
Ferro, M., 785
Ferro-Luzzi, M., 591
Fetscher, W., 366, 371
Fichen, L., 198
Fields, D. E., 790, 795
Filipov, G., 689
Fimushkin, V. V., 856, 895
Finger, Jr., M., 891
Finger, M., 891
Foroughi, F., 371
Foster, C. C., 614, 886
Franklin, W. A., 599
Freedman, S. J., 89
Frekers, D., 609
Frommberger, F., 756
Fujii, J., 908
Fujikawa, B. K., 89
Fujimura, H., 604, 649, 765, 884, 886
Fujita, H., 614, 649, 884, 886
Fujita, T., 709
Fujita, Y., 614, 639, 649, 765, 884, 886
Fujiwara, M., 639, 649, 765
Fukuda, N., 109, 679
Fukui, S., 889
Fukusaka, S., 654, 884, 885, 890
Furuta, F., 982, 1009, 1012, 1015, 1018, 1021
Fushimi, K., 338, 343

G

Gai, G. I., 895
Gao, H., 79
Garwin, E. L., 976

Gaulard, C., 198
Gemme, G., 785
Gervino, G., 198
Ghazikhanian, V., 534
Ghio, F., 198
Gianfelice, E., 780
Gilman, R., 69
Girolami, B., 198
Gladycheva, S. E., 856
Glenn, J. W., 891
Glöckle, W., 208
Goertz, S., 261, 896
Gokalp, A., 551
Golak, J., 208
Goloskokov, S. V., 541
Golovanov, L. B., 689
Goshtasbpour, M., 879
Goto, A., 109, 679, 877
Goto, Y., 442, 534, 790
Gotoh, T., 1009, 1012
Govaerts, J., 371
Gowin, M., 756, 961
Grames, J., 943
Graw, G., 664, 825, 888
Greenfield, M. B., 609, 719, 890, 895
Grishin, V. N., 736, 856
Gröger, J., 888
Gromov, R. G., 988
Grosse Perdekamp, M., 457
Grote, H., 169
Gu, J. N., 664
Guazzoni, P., 664
Guidal, M., 198
Guinault, E., 198
Günther, C., 888
Guryn, W., 519

H

Hagemann, M., 609
Hannen, V. M., 609
Hansen, J.-O., 576
Hansknecht, J., 943
Hara, K., 614, 649, 884, 886
Harada, K., 614, 886
Harakeh, M. N., 609
Harasawa, A., 1003
Harmsen, J., 261, 896
Harrach, D. v., 926, 949

Hartmann, P., 943
Hasegawa, S., 896
Hatanaka, K., 614, 619, 639, 649, 674, 719, 765, 884, 884, 886, 888, 890, 894, 895
Hatano, M., 609, 619, 684, 714, 719, 770, 806, 861, 890, 895
Hautle, P., 866
Hayashi, K., 338, 343, 1003, 1006
Hayashi, N., 534
Hayashigaki, A., 472
Heckmann, J., 261, 896
Heinemann, K., 889
Helbing, K., 756
Hertenberger, R., 664, 825, 888
Heyse, J., 609
Higinbotham, D. W., 591
Hilbes, C., 366, 371
Hillert, W., 756, 961
Hillhouse, G. C., 624, 634
Hino, M., 877
Hinterberger, F., 731
Hirabayashi, Y., 659
Hirai, M., 477
Hirayama, Y., 811
Hirooka, D., 649, 674, 719, 888, 890, 894, 895
Hirose, T., 1015
Hoffmann, M., 756
Hoffstätter, G. H., 751
Höhr, C., 996
Holt, R. J., 704
Honda, T., 679
Hoppe, M., 912
Horie, K., 811
Horikawa, N., 889, 896
Horikawa, S., 896
Horinaka, H., 926, 930, 982
Hosono, K., 639, 649, 765
Houlden, M., 819
Hourany, E., 198
Huang, H., 534, 790, 795
Hughes, V. W., 296, 790
Husmann, D., 756

I

Iami, K., 534
Ichihara, T., 534, 654, 890

Ichimura, M., 619
Igo, G., 534
Ihara, F., 639
Ihloff, E., 955
Iizuka, T., 694
Imai, H., 109
Imai, I., 795
Imai, K., 790
Imai, N., 679
Inomata, T., 639, 765
Iseri, Y., 724, 887
Ishida, S., 654, 890
Ishida, T., 669, 888, 894
Ishihara, M., 109, 679, 790
Ishikawa, S., 724
Ishikawa, T., 604, 639, 649, 765
Isupov, A. Y., 689, 889
Itoh, K., 684, 885, 893
Itoh, K. S., 654, 770, 890
Itoh, M., 604, 639, 649, 765
Ivanov, V. I., 689
Ivanshin, Y. I., 880
Iwata, T., 889, 896
Izumi, H., 811

J

Jacobs, K., 736
Jacobs, W. W., 599
Jaffe, R. L., 3
Janata, A., 891
Jänecke, J., 614, 886
Jansen, P., 882
Jarczyk, L., 366
Jaroshevich, A. S., 901, 1021
Jaskola, M., 664
Jennewein, P., 949
Jolie, J., 888
Joo, K., 581

K

Kachaev, I., 880
Kageya, T., 736, 856, 893
Kaiser, K.-H., 949
Kakizaki, A., 1003, 1006
Kamada, H., 208
Kamada, M., 916, 1000

Kameda, D., 877
Kameyama, H., 887
Kamiya, J., 619, 649, 674, 719, 884, 886, 888, 890, 894, 895
Kanavets, V., 790, 795
Kanazawa, Y., 494
Kantsyrev, D. Y., 736, 856
Kartamyshev, A. A., 689
Kashirin, V. A., 689, 889
Kasprzyk, T., 534
Katabuchi, T., 694
Kato, H., 619, 684, 714, 719, 770, 806, 890, 895
Kato, K., 1021
Kato, T., 982
Katori, K., 614, 884, 886
Kawabata, M., 639, 765, 811
Kawabata, T., 604, 614, 639, 649, 765, 884, 886
Kawahigashi, K., 619
Kawai, M., 629
Kazimi, R., 943
Keil, J., 756
Keith, C., 819
Khrenov, A. N., 689, 889
Kilvington, I., 198
Kim, G. N., 992
Kimura, A., 1003
Kimura, K., 1015
Kinney, E. R., 704
Kirby, R. E., 976
Kirch, K., 366, 371
Kishimoto, T., 128, 338, 343
Kistryn, S., 366
Klehr, F., 882
Kleines, H., 830
Klement, J., 366
Kleppner, D., 856
Klous, S., 591
Knowles, P., 371
Ko, I. S., 992
Kobayakawa, H., 982, 1009, 1015, 1021, 1024
Kobayashi, H., 477
Kobayashi, Y., 109, 679
Koch, N., 894
Köhler, K., 366, 371
Kohno, M., 629
Koike, Y., 494
Kolesnikov, V. I., 689

Kolster, H., 591
Komori, M., 338
Kondo, K., 889, 896
Kondo, M., 888, 894
Kondo, Y., 534
Konstantinov, E. S., 988
Konter, J. A., 866
Koptev, V., 830
Korostelev, M., 775
Koutchouk, J. P., 169
Kouznetsov, V., 198
Kovalenko, A. D., 895
Kozela, A., 366, 371
Kracíková, T. I., 891
Kravtsov, P., 830
Kreidel, H. J., 949
Krisch, A. D., 736, 856, 893
Krueger, K., 534
Krüsemann, B., 609
Kubo, K.-I., 119, 499
Kubo, T., 109, 679
Kudo, K., 694
Kudomi, N., 338, 343
Kumano, S., 397
Kume, K., 338, 343
Kunne, R., 198
Kuno, Y., 99
Kurahashi, M., 972
Kurahashi, S., 1009, 1012, 1015, 1018, 1021
Kuramoto, H., 338, 343
Kurihara, Y., 982
Kurita, K., 790, 795
Kuroś-Żolnierczuk, J., 208
Kutuzova, L. V., 895
Kuwahara, M., 1015
Kuznezov, V. A., 689
Kwiatkowski, K., 790

L

Ladygin, V. P., 689, 889
Ladygina, N. B., 689
Lang, J., 366, 371, 591
Lapik, A., 198
Lazarenko, B. A., 586, 704
Lebedev, N. A., 891
Lee, J., 251
Lee, M. W., 992

Lee, S., 893
Lee, S. Y., 534
Lee, T.-S. H., 546
Lehrach, A., 746, 891
Lemaître, S., 830
Lenisa, P., 851
Levi Sandri, P., 198
Lewis, B., 790
Ley, J., 830
Liang, Z-t., 509
Lin, A. M. T., 736
Lincoln, F., 846
Lindemann, T., 392
Litvinenko, A. G., 689, 889
Litvinov, A. N., 901
Liu, C-x., 509
Liu, Y. W., 371
Llácer, G. L., 366
Lleres, A., 198
Logatchov, P. V., 988
Loginov, A. Y., 586
Lorentz, B., 830
Lorenz, S., 830
Lorenzon, W., 893
Lowry, M., 846
Lozowski, B., 790, 795
Luccio, A. U., 746, 785
Luppov, V. G., 856

M

MacKay, W. W., 746, 785, 795
Maeda, Y., 609, 619, 674, 684, 714, 719, 770, 806, 890, 895
Maeda, Y., 669, 674
Makdisi, Y., 534, 790, 795
Malakhov, A. I., 889, 895
Mamaev, Y. A., 920
Mango, S., 866
Mano, J., 634
Manohar, A. V., 238
Markiewicz, M., 366
Martin, S., 882
Maruyama, T., 361, 976
Masuno, K., 694
Matsumoto, H., 982, 1009, 1012, 1015
Matsuoka, K., 338, 343
Matsuyama, T., 930, 982
Matulenko, Y. A., 534

Medve, R., 371
Meier, A., 261, 896
Metz, A., 825, 888
Meyer, H. O., 790, 882
Meyer, W., 261, 896
Michailov, V. A., 895
Michel, T., 756
Mikirtytchiants, M., 830
Minami, S., 888, 894
Mishnev, S. I., 586, 704, 892
Miyachi, Y., 896
Miyama, M., 477
Miyamoto, M., 1012, 1015, 1018, 1021
Miyoshi, H., 109, 679, 877
Mizoguchi, T., 908
Mizuno, R., 1015
Mizutori, S., 884, 886
Moré, S. D., 916, 1000
Morelle, X., 366, 371
Mori, K., 896
Moricciani, D., 198
Morii, T., 433
Morozov, B. V., 790
Morozov, V. S., 736, 856
Moshammer, R., 996
Mul, F. A., 591
Mulai, M., 930
Mulhollan, G. A., 976
Murata, J., 878
Murgia, F., 571, 880
Murray, J. R., 736, 856
Mysnik, A. I., 856

N

Nagakura, M., 353, 679
Nakada, Y., 534
Nakajima, N., 499
Nakamura, H., 709
Nakamura, M., 534, 639, 649, 765, 790, 811
Nakanishi, T., 274, 916, 926, 930, 982, 1000, 1009, 1012, 1015, 1018, 1021, 1024
Nakano, T., 189
Nakaoka, Y., 619
Nakashima, T., 709
Namkung, W., 992
Nass, A., 894

Naumann, J., 756
Naumov, D. V., 489
Naviliat, O., 371
Nedorezov, V., 198
Neff, B., 961
Nekipelov, M., 830
Nelyubin, V. V., 704, 830
Nemchonok, I. B., 866
Netchaeva, L. P., 835
Neyens, G., 109, 679
Nicoletti, L., 198
Niizeki, T., 654, 885, 890
Nikitin, S. A., 892
Nikolenko, D. M., 586, 591, 704, 892
Ninane, A., 371
Nishikawa, J., 684, 714, 770, 806
Nishimori, K. S., 654
Nishimori, N., 709, 890
Nishitani, T., 916, 982, 1000, 1009, 1012, 1015, 1018, 1021, 1024
Nogach, L. V., 534
Nogga, A., 208
Nomachi, M., 669
Nonaka, T., 654, 890
Noro, T., 228, 604, 614, 639, 649, 674, 765, 884, 886, 888, 894
Norum, B. E., 591
Nurushev, S. B., 524, 534

O

Obayashi, E., 604, 639, 649, 674, 765, 888, 894
Ogata, K., 629
Ogawa, A., 379, 457, 534
Ogawa, H., 109, 353, 679, 877
Oh, Y., 546
Ohnishi, T., 609, 619, 654, 684, 714, 770, 806, 884, 885, 890, 893
Ohnuma, H., 885
Ohsumi, H., 338, 343
Okamura, H., 534, 619, 654, 674, 684, 714, 719, 770, 806, 884, 885, 890, 890, 893, 895
Okamura, M., 534
Okumi, S., 926, 930, 982, 1009, 1012, 1015, 1018, 1021
Okumura, K., 877
Omori, T., 982

Orlov, D. A., 912
Oryu, S., 699
Osipov, A. V., 586, 704
Otsu, H., 619, 654, 890

P

Pakhnevich, A. A., 901
Palazzi, M., 785
Park, S. J., 992
Park, Y. J., 992
Parodi, R., 785
Passchier, I., 591
Pavlinov, A. I., 534
Penev, V. N., 889, 895
Penttilä, S., 819
Penzo, A., 524
Peterson, T., 599
Pilipenko, Y. K., 800, 889, 895
Placidi, M., 169
Plis, Y. A., 892
Poelker, M., 943
Ponomarev, V. Y., 664
Poolman, H. R., 591
Popov, V. G., 704
Potterveld, D. H., 704
Prepost, R., 407, 976
Prescott, C. Y., 0, 899, 976
Prieels, R., 371
Prok, Y., 819
Protopopov, I. Y., 892
Przewoski, B., 790
Pussieux, T., 761
Pusterla, M., 785

R

Rachek, I. A., 586, 591, 704, 892
Radtke, E., 261, 896
Raithel, M., 894
Rakers, S., 609
Ramsey, G. P., 437
Ranjbar, V., 891
Rapaport, J., 719, 895
Ratcliffe, P. G., 348, 879
Rathmann, F., 830, 882
Raymond, R. S., 856
Rebreyend, D., 198

Reichert, E., 926, 949
Reicherz, G., 261, 896
Renard, F., 198
Reznikov, S. G., 689, 889
Richter, A., 884
Rikhvitskiy, V. S., 882
Rinckel, T., 599, 893
Rinkel, T., 790
Ripani, M., 198
Rith, K., 882
Roberts, D. A., 614, 886
Rochansky, A. V., 920
Roethgen, J., 926
Rose, T., 795
Roser, T., 312, 534, 741, 746, 790, 891
Rotter, M., 891
Rudnev, N., 198
Rukoyatkin, P. A., 689, 889
Rusanov, I., 889
Rusek, A., 790
Rutt, P., 943
Rutzo, M. F., 524
Ryckbosch, D., 556
Rykov, V. L., 379

S

Sadykov, R. S., 586, 704
Sagara, K., 709, 888, 894
Saito, N., 154, 379, 534, 790, 795
Saito, T., 684, 719, 890, 895
Saitoh, T., 846
Saka, T., 982, 1021
Sakaguchi, H., 604, 614, 639, 649, 765
Sakai, H., 218, 534, 609, 619, 654, 674, 684, 714, 719, 770, 806, 861, 884, 885, 890, 890, 893, 895
Sakai, K., 109, 353, 679, 877
Sakamoto, N., 654, 684, 714, 770, 806, 885, 890, 893
Sakemi, Y., 423
Sakoda, S., 609, 619, 684, 714, 719, 770, 806, 884, 890, 893, 895
Sammarruca, F., 644, 883
Sandorfi, A. M., 846
Sanzone, M., 198
Sarkadi, J., 830
Sasaki, K., 561
Sato, H., 893

Sato, H. D., 534, 879
Satou, Y., 654, 684, 714, 770, 806, 885, 890, 893
Sawada, M., 1003, 1006
Schaerf, C., 198
Scheibler, H. E., 901, 1021
Schieck, H. P., 830
Schiemenz, P., 825
Schmidt, R., 169
Schmidt-Ott, W.-D., 109, 679
Schröter, C. D., 996
Schuler, J., 926, 949
Schwalm, D., 912
Schwandt, P., 736, 893
Schweizer, T., 366
Scielzo, N. D., 89
Segawa, M., 669
Seitz, B., 878
Sekiguchi, K., 619, 654, 684, 714, 719, 770, 806, 884, 885, 890, 890, 893, 895
Selyugin, O. V., 524
Semenov, A. Y., 689
Semenova, I. A., 689
Seyfarth, H., 830, 882
Shatunov, Y. M., 775, 892
Shestakov, Y. V., 586, 704
Shigematsu, T., 811
Shigenaga, K., 709
Shimakura, N., 841
Shimbara, Y., 614, 649, 669, 884, 886
Shimizu, S., 811
Shimoda, T., 811
Shinada, T., 614, 884, 886
Sidorov, A. A., 586
Simani, M. C., 591
Sinclair, C., 943
Sivers, D., 893
Sivers, D. W., 736
Six, E., 591
Skibiński, R., 208
Skrinsky, A. N., 892
Slepnev, V. M., 800, 895
Slunečka, M., 891
Smith, B., 790
Smyrski, J., 366
Soffer, J., 461
Solovianov, V. L., 736
Son, D., 992
Sonnemann, F., 169
Sonnenschein, L., 514

Souder, P. A., 59
Sourkont, K., 893
Sowinski, J., 599
Speckner, T., 756
Sperisen, F., 893
Spinka, H., 534, 795
Sromicki, J., 366
Staudt, G., 664
Steffens, E., 286, 830, 882, 894
Steigerwald, M., 935, 943
Stepanov, A. D., 895
Stephan, E., 366
Stephenson, E. J., 614, 644, 883, 886
Stewart, I. W., 238
Stibunov, V. N., 586, 892
Stoletov, G. D., 689
Stösslein, U., 387
Strikhanov, M. N., 524
Ströher, H., 830, 882
Strzałkowski, A., 366
Subashiev, A. V., 920, 976
Suda, K., 619, 684, 714, 719, 770, 806, 890, 893, 895
Suga, T., 109, 679, 877
Sugaya, Y., 669
Sugiyama, H., 982, 1015
Suzuki, C., 982, 1009, 1012, 1015, 1018, 1021
Suzuki, K., 472, 499
Suzuki, T., 353, 679
Suzuki, T., 659, 885
Suzuki, T., 972
Svirida, D., 790, 795
Syphers, M., 534, 790
Szczerba, D., 591

T

Tagishi, Y., 694
Taiuti, M., 198
Takabayashi, N., 896
Takahashi, K., 1000
Takahashi, N., 811
Takahisa, K., 338, 343, 888, 894
Takashima, Y., 982, 1015, 1024
Takeda, H., 604, 639, 649, 765
Takeda, Y., 982, 1021, 1024
Taketani, A., 790
Takeuchi, Y., 982, 1012

Taki, T., 604, 614, 639, 649, 765
Tamii, A., 604, 609, 619, 639, 649, 674, 684, 714, 719, 765, 770, 806, 861, 884, 890, 893, 895
Tamura, K., 669, 674
Tanaka, M., 841
Tanaka, S., 916, 1000
Tanaka, S.-i., 916
Tanifuji, M., 659, 699, 724, 887, 887
Tecker, F., 169
Terekhov, A. S., 901, 912, 1021
Tereshchenko, O. E., 901
Teughels, S., 109, 679
Thomas, T. L., 790
Tioukine, V., 949
Titov, A. I., 546
Tkatchev, L. G., 880
Togawa, K., 926, 930, 982, 1009, 1012, 1015, 1018, 1021, 1024
Tojo, J., 790, 795
Tojyo, T., 896
Toki, H., 499
Toporkov, D. K., 586, 704, 892
Toropov, A. I., 1021
Toyokawa, H., 639, 649, 765
Trentalange, S., 534
Tsentalovich, E., 955
Tsujimoto, Y., 343
Tsukahara, N., 604, 649
Tsuruta, K., 709
Tumaikin, G. M., 892
Turbabin, A. V., 835
Turinge, A., 198
Tzvinev, A. P., 689

U

Uchida, M., 604, 649, 765
Uchigashima, N., 684, 719, 890, 895
Uematsu, T., 561
Ueno, H., 109, 614, 649, 679, 884, 886
Uesaka, T., 609, 654, 684, 714, 770, 806, 861, 885, 890, 893
Ullrich, J., 996
Umehara, S., 338, 343
Underwood, D., 534, 790, 795
Utsuro, M., 877
Uzu, E., 699

V

Vadeev, V. P., 895
Valevich, A. I., 895
van Buuren, L. D., 591
van den Berg, A. M., 609
van den Brand, J. F., 591
van den Brandt, B., 866
van der Ventel, B. I. S., 624, 634
van Hove, P., 371
van Oers, W. T. H., 331, 887
Vasil'ev, G. A., 835
Vasiliev, A. N., 524, 534
Vassiliev, A., 830
Vetter, P. A., 89
Vigdor, S. E., 452
Vikhrov, V. V., 704
Vitturi, A., 664
Vogelsang, W., 138
Vogt, M., 751
Volkov, V. I., 895
von Neumann-Cosel, P., 884
von Przewoski, B., 736, 893

W

Wada, K., 930, 982, 982, 1012, 1015, 1018, 1021
Wakai, A., 889, 896
Wakamatsu, M., 482
Wakasa, T., 534, 619, 649, 654, 674, 684, 714, 719, 770, 884, 888, 890, 890, 893, 894, 895
Wakui, T., 684, 861
Wang, K., 591
Warr, N., 888
Watanabe, J., 1009, 1012, 1015, 1018, 1021
Watanabe, O., 982, 1021, 1024
Watanabe, Y., 629
Watanabe, Y. X., 109, 679
Wei, X., 846
Weigel, U., 912
Weili, S., 629
Weiss, A., 884
Wenninger, J., 169
Whitten, C., 534
Wiessner, M., 949
Winkler, K., 949

Wissink, S. W., 447, 599, 886
Witała, H., 208
Wojtsekhowski, B. B., 566, 877
Wolanski, M., 599
Wolf, A., 912
Wolfe, D., 790
Wong, V. K., 736, 893
Wörtche, H. J., 609
Wyngaardt, S. M., 634

Y

Yagi, M., 811
Yagita, T., 619, 669, 709, 888, 894
Yako, K., 609, 619, 654, 674, 684, 714, 719, 770, 806, 884, 890, 890, 893, 895
Yakou, K., 885
Yamada, T., 908
Yamagata, T., 841
Yamaguchi, Y., 0
Yamamoto, K., 790
Yamamoto, M., 982, 1009, 1012, 1015, 1018, 1021
Yamamoto, S., 684
Yamanishi, T., 433
Yamasaki, K., 649
Yamauchi, Y., 972
Yang, H., 599
Yashin, Y. P., 920
Yasuda, K., 669
Yasuda, Y., 649
Yemelyanenko, V. N., 882
Yilmaz, O., 551
Yogo, K., 109, 353, 679, 877
Yokosawa, A., 534
Yokoyama, G., 654, 890
Yoneda, K., 109, 679

Yonehara, K., 736, 841
Yoshida, A., 109, 679
Yoshida, H. P., 604, 639, 649, 669, 674, 765, 888, 894
Yoshida, S., 338, 343
Yoshifuku, M., 884, 886
Yoshimi, A., 109, 353, 679, 877
Yoshimura, M., 604, 639, 765, 888, 894
Yoshioka, M., 982, 1009, 1012, 1015
Yosoi, M., 604, 614, 639, 649, 765, 884, 886
Youssof, S., 893
Yu-Bing, D., 433
Yudin, N. P., 689
Yushkevich, Y. V., 891

Z

Zaitsev, A., 880
Zegers, R. G. T., 609
Zeitler, G., 756
Zejma, J., 366
Zelenski, A. N., 179
Zetta, L., 664
Zevakov, S. A., 586, 704
Zhao, Q., 198
Zhilich, E. N., 892
Zhmyrov, V. N., 689
Zhou, Z.-L., 591
Zhu, L., 790
Zolin, L. S., 689, 800, 889, 895
Zucchiatti, A., 198
Zulkarneev, R. Y., 880
Zwart, G. T., 736
Zwart, T., 955
Zwoll, K., 830